Bioinformatics and Functional Genomics

BIOINFORMATICS AND FUNCTIONAL GENOMICS

Second Edition

Jonathan Pevsner

Department of Neurology, Kennedy Krieger Institute

and

Department of Neuroscience and Division of Health Sciences Informatics, The Johns Hopkins School of Medicine, Baltimore, Maryland

WILEY-BLACKWELL

A JOHN WILEY & SONS, INC., PUBLICATION

Wiley-Blackwell is an imprint of John Wiley & Sons, formed by the merger of Wiley's global Scientific, Technical, and Medical business with Blackwell Publishing.

Published by John Wiley & Sons, Inc., Hoboken, New Jersey
Published simultaneously in Canada

For general information on our other products and services or for technical support, please contact our Customer Care Department within the United States at (800) 762–2974, outside the United States at (317) 572–3993 or fax (317) 572–4002.

Wiley also publishes its books in variety of electronic formats. Some content that appears in print may not be available in electronic format. For more information about Wiley products, visit our web site at www.wiley.com.

Cover illustration includes detail from Leonardo da Vinci (1452–1519), dated c.1506–1507, courtesy of the Schlossmuseum (Weimar).

ISBN: 978-0-470-08585-1

Library of Congress Cataloging-in-Publication Data is available.

Printed in the United States of America

10 9 8 7 6 5 4 3

For Barbara, Ava and Lillian with all my love.

Contents in Brief

Contents

PART II   GENOMEWIDE ANALYSIS OF RNA AND PROTEIN

Preface to the Second Edition

The Neurobehavioral Unit of the Kennedy Krieger Institute has 16 hospital beds. Most of the patients are children who have been diagnosed with autism, and most engage in self-injurious behavior. They engage in self-biting, self-hitting, head-banging, and other destructive behaviors. In most cases, we do not understand the genetic contributions to such behaviors, limiting the available strategies for treatment. In my research, I am motivated to understand molecular changes that underlie childhood brain diseases. The field of bioinformatics provides tools we can use to understand disease processes through the analysis of molecular sequence data. More broadly, bioinformatics facilitates our understanding of the basic aspects of biology including development, metabolism, adaptation to the environment, genetics (e.g., the basis of individual differences), and evolution.

Since the publication of the first edition of this textbook in 2003, the fields of bioinformatics and genomics have grown explosively. In the preface to the first edition (2003) I noted that tens of billions of base pairs (gigabases) of DNA had been deposited in GenBank. Now in 2009 we are reaching tens of trillions (terabases) of DNA, presenting us with unprecedented challenges in how to store, analyze, and interpret sequence data. In this second edition I have made numerous changes to the content and organization of the book. All of the chapters are rewritten, and about 90% of the figures and tables are updated. There are two new chapters, one on functional genomics and one on the eukaryotic chromosome. I now focus on the globins as examples throughout the book. Globins have a special place in the history of biology, as they were among the first proteins to be identified (in the 1830s) and sequenced (in the 1950s and 1960s). The first protein to have its structure solved by X-ray crystallography was myoglobin (Chapter 11); molecular phylogeny was applied to the globins in the 1960s (Chapter 7); and the globin gene loci were among the first to be sequenced (in the 1980s; see Chapter 16).

The fields of bioinformatics and genomics are far too broad to be understood by one person. Thus many textbooks are written by multiple authors, each of whom brings a deeper knowledge of the subject matter. I hope that this book at least offers the benefit of a single author's vision of how to present the material. This is essentially two textbooks: one on bioinformatics (parts I and II) and one on genomics (part III). I feel that presenting bioinformatics on its own would be incomplete without further applying those approaches to sequence analysis of genomes across the tree of life. Similarly I feel that it is not possible to approach genomics without first treating the bioinformatics tools that are essential engines of that field.

As with the previous edition a companion website is available which provides up-to-date web links referred to in the book and PowerPoint slides arranged by chapter (www.bioinfbook.org). A resource site for instructors is also available giving detailed solutions to problems (www.wiley.com/go/pevsnerbioinformatics).

In preparing each edition of this book I read many papers and reviewed several thousand websites. I sincerely apologize to those authors, researchers and others whose work I did not cite. It is a great pleasure to acknowledge my colleagues who have helped in the preparation of this book. Some read chapters including Jef Boeke (Chapter 12), Rafael Irizarry (Chapter 9), Stuart Ray (Chapter 7), Ingo Ruczinski (Chapter 11), and Sarah Wheelan (Chapters 3 and 5–7). I thank many students and faculty at Johns Hopkins and elsewhere who have provided critical feedback, including those who have lectured in bioinformatics and genomics courses (Judith Bender, Jef Boeke, Egbert Hoiczyk, Ingo Ruczinski, Alan Scott, David Sullivan, David Valle, and Sarah Wheelan). Many others engaged in helpful discussions including Charles D. Cohen, Bob Cole, Donald Coppock, Laurence Frelin, Hugh Gelch, Gary W. Goldstein, Marjan Gucek, Ada Hamosh, Nathaniel Miller, Akhilesh Pandey, Elisha Roberson, Kirby D. Smith, Jason Ting, and N. Varg. I thank my wife Barbara for her support and love as I prepared this book.

Preface to the First Edition

ORIGINS OF THIS BOOK

This book emerged from lecture notes I prepared several years ago for an introductory bioinformatics and genomics course at the Johns Hopkins School of Medicine. The first class consisted of about 70 graduate students and several hundred auditors, including postdoctoral fellows, technicians, undergraduates, and faculty. Those who attended the course came from a broad variety of fields—students of genetics, neuroscience, immunology or cell biology, clinicians interested in particular diseases, statisticians and computer scientists, virologists and microbiologists. They had a common interest in wanting to understand how they could apply the tools of computer science to solve biological problems. This is the domain of bioinformatics, which I define most simply as the interface of computer science and molecular biology. This emerging field relies on the use of computer algorithms and computer databases to study proteins, genes, and genomes. Functional genomics is the study of gene function using genome-wide experimental and computational approaches.

COMPARISON

At its essence, the field of bioinformatics is about comparisons. In the first third of the book we learn how to extract DNA or protein sequences from the databases, and then to compare them to each other in a pairwise fashion or by searching an entire database. For the student who has a gene of particular interest, a natural question is to ask "what other genes (or proteins) are related to mine?"

In the middle third of the book, we move from DNA to RNA (gene expression) and to proteins. We again are engaged in a series of comparisons. We compare gene expression in two cell lines with or without drug treatment, or a wildtype mouse heart versus a knockout mouse heart, or a frog at different stages of development. These comparisons extend to the world of proteins, where we apply the tools of proteomics to complex biological samples under assorted physiological conditions. The alignment of multiple, related DNA or protein sequences is another form of comparison. These relationships can be visualized in a phylogenetic tree.

The last third of the book spans the tree of life, and this provides another level of comparison. Which forms of human immunodeficiency virus threaten us, and how can we compare the various HIV subtypes to learn how we might develop a vaccine? How are a mosquito and a fruitfly related? What genes do vertebrates such as fish and humans share in common, and which genes are unique to various phylogenetic lineages?

I believe that these various kinds of comparisons are what distinguish the newly emerging fields of bioinformatics and genomics from traditional biology. Biology has always concerned comparisons; in this book I quote 19th century biologists such as Richard Owen, Ernst Haeckel, and Charles Darwin who engaged in comparative studies at the organismal level. The problems we are trying to solve have not changed substantially. We still seek a more complete understanding of the unifying concepts of biology, such as the organization of life from its constituent parts (e.g., genes and proteins), the behavior of complex biological systems, and the continuity of life through evolution. What *has* changed is how we pursue this more complete understanding. This book describes databases filled with raw information on genes and gene products and the tools that are useful to analyze these data.

THE CHALLENGE OF HUMAN DISEASE

My training is as a molecular biologist and neuroscientist. My laboratory studies the molecular basis of childhood brain disorders such as Down syndrome, autism, and lead poisoning. We are located at the Kennedy Krieger Institute, a hospital for children for developmental disorders. (You can learn more about this Institute at http://www.kennedykrieger.org.) Each year over 10,000 patients visit the Institute. The hospital includes clinics for children with a variety of conditions including language disorders, eating disorders, autism, mental retardation, spina bifida, and traumatic brain injury. Some have very common disorders, such as Down syndrome (affecting about 1:700 live births) and mental retardation. Others have rare disorders, such as Rett syndrome or adrenoleukodystrophy.

We are at a time when the number of base pairs of DNA deposited in the world's public repositories has reached tens of billions, as described in Chapter 2. We have obtained the first sequence of the human genome, and since 1995 hundreds of genomes have been sequenced. Throughout the book, you can follow the progress of science as we learn how to sequence DNA, and study its RNA and protein products. At times the pace of progress seems dazzling.

Yet at the same time we understand so little about human disease. For thousands of diseases, a defect in a single gene causes a pathological effect. Even as we discover the genes that are defective in diseases such as cystic fibrosis, muscular dystrophy, adrenoleukodystrophy, and Rett syndrome, the path to finding an effective treatment or cure is obscure. But single gene disorders are not nearly as common as complex diseases such as autism, depression, and mental retardation that are likely due to mutations in multiple genes. And all genetic disease is not nearly as common as infectious disease. We know little about why one strain of virus infects only humans, while another closely related species infects only chimpanzees. We do not understand why one bacterial strain may be pathogenic, while another is harmless. We have not learned how to develop an effective vaccine against any eukaryotic pathogen, from protozoa (such as *Plasmodium falciparum* that causes malaria) to parasitic nematodes.

The prospects for making progress in these areas are very encouraging specifically because of the recent development of new bioinformatics tools. We are only now beginning to position ourselves to understand the genetic basis of both disease-causing agents and the hosts that are susceptible. Our hope is that the information so rapidly accumulating in new bioinformatics databases can be translated through research into insights into human disease and biology in general.

NOTE TO READERS

This book describes over 1,000 websites related to bioinformatics and functional genomics. All of these sites evolve over time (and some become extinct). In an effort to keep the web links up-to-date, a companion website (http://www.bioinfbook.org) maintains essentially all of the website links, organized by chapter of the book. We try our best to maintain this site over time. We use a program to automatically scan all the links each month, and then we update them as necessary.

An additional site is available to instructors, including detailed solutions to problems (see http://www.wiley.com).

ACKNOWLEDGMENTS

Writing this book has been a wonderful learning experience. It is a pleasure to thank the many people who have contributed. In particular, the intellectual environment at the Kennedy Krieger Institute and the Johns Hopkins School of Medicine has been extraordinarily rich. These chapters were developed from lectures in an introductory bioinformatics course. The Johns Hopkins faculty who lectured during its first three years were Jef Boeke (yeast functional genomics), Aravinda Chakravarti (human disease), Neil Clarke (protein structure), Kyle Cunningham (yeast), Garry Cutting (human disease), Rachel Green (RNA), Stuart Ray (molecular phylogeny), and Roger Reeves (the human genome). I have benefited greatly from their insights into these areas.

I gratefully acknowledge the many reviewers of this book, including a group of anonymous reviewers who offered extremely constructive and detailed suggestions. Those who read the book include Russ Altman, Christopher Aston, David P. Leader, and Harold Lehmann (various chapters), Conover Talbot (Chapters 2 and 18), Edie Sears (Chapter 3), Tom Downey (Chapter 7), Jef Boeke (Chapter 8 and various other chapters), Michelle Nihei and Daniel Yuan (Chapter 8), Mario Amzel and Ingo Ruczinski (Chapter 9), Stuart Ray (Chapter 11), Marie Hardwick (Chapter 13), Yukari Manabe (Chapter 14), Kyle Cunningham and Forrest Spencer (Chapter 15), and Roger Reeves (Chapter 16). Kirby D. Smith read Chapter 18 and provided insights into most of the other chapters as well. Each of these colleagues offered a great deal of time and effort to help improve the content, and each served as a mentor. Of the many students who read the chapters I mention Rong Mao, Ok-Hee Jeon, and Vinoy Prasad. I particularly thank Mayra Garcia and Larry Frelin who provided invaluable assistance throughout the writing process. I am grateful to my editor at John Wiley & Sons, Luna Han, for her encouragement.

I also acknowledge Gary W. Goldstein, President of the Kennedy Krieger Institute, and Solomon H. Snyder, my chairman in the Department of Neuroscience at Johns Hopkins. Both provided encouragement, and allowed me the opportunity to write this book while maintaining an academic laboratory.

On a personal note, I thank my family for all their love and support, as well as N. Varg, Kimberly Reed, and Charles Cohen. Most of all, I thank my fiancée Barbara Reed for her patience, faith, and love.

Foreword

Ask 10 investigators in human genetics what resources they need most and it is highly likely that computational skills and tools will be at the top of the list. Genomics, with its reliance on microarrays, genotyping, high throughput sequencing and the like, is intensely data-rich and for this reason is impossible to disentangle from bioinformatics. This text, with its clear descriptions, practical examples and focus on the overlaps and interdependence of these two fields, is thus an essential resource for students and practitioners alike.

Interestingly, bioinformatics and genomics are both relatively recent disciplines. Each emerged in the course of the Human Genome Project (HGP) that was conceived in the mid-1980s and began officially on October 1, 1990. As the HGP matured from its initial focus on gene maps in model organisms to the massive efforts to produce a reference human whole genome sequence, there was an increasing need for computational biology tools to store, analyze and disseminate large amounts of sequence data. For this reason, genomics increasingly relied on bioinformatics and, in turn, the field of bioinformatics flourished. Today, no serious student of genomics can imagine life without bioinformatics. This interdependence continues to grow by leaps and bounds as the questions and activities of investigators in genomics become bolder and more expansive; consider, for example, whole genome association studies (GWAS), the ENCODE project, the challenge of copy number variants, the 1000 Genomes project, epigenomics, and the looming growth of personal genome sequences and their analysis.

This textbook provides a clear and timely introduction to both bioinformatics and genomics. It is organized so that each chapter can correspond to a lecture for a course on bioinformatics or genomics and, indeed, we have used it this way for our students. Also, for readers not taking courses, the book provides essential background material. For computer scientists and biologists alike the book offers explanations of available methods and the kinds of problems for which they can be used. The sections on bioinformatics in the first part of the book describe many of the basic tools that are used to analyze and compare DNA and protein sequences. The tone is inviting as the reader is guided to learn to use different software by example. Multiple approaches for solving particular problems, such as sequence alignment and molecular phylogeny, are presented. The middle part of the book introduces functional genomics. Here again the focus is on helping the reader to learn how to do analyses (such as microarray data analysis or protein structure prediction) in a practical way. A companion website provides many data sets, so the student can get experience in performing analyses. Chapter 12 provides a roadmap to the very complicated topic of functional genomics, spanning a range of techniques and model organisms used to study gene function. The last third of the book provides a survey of the tree of life from a genomics perspective. There is an attempt to be comprehensive, and at the same time, to present the material in an interesting way, highlighting the fascinating features that make each genome unique.

Far from being a dry account of the facts of genomics and bioinformatics, the book offers many features that highlight the vitality of this field. There are discussions throughout about how to critically evaluate the performance of different software. For example, there are 'competitions' in which different research groups perform computational analyses on data sets that have been validated with some 'gold standard', allowing false positive and false negative error rates to be determined. These competitions are described in areas such as microarray data analysis (Chapter 9), mass spectrometry (Chapter 10), protein structure prediction (Chapter 11), or gene prediction (Chapter 16). The book also includes descriptions of important movements in the fields of bioinformatics and genomics, ranging from the RefSeq project for organizing sequences to the ENCODE and HapMap projects. Similarly, there is a rich description of the historical context for different aspects of bioinformatics and genomics, such as Garrod's views on disease (Chapter 20); Ohno's classic 1970 book on genome duplication (Chapter 17); and, the earliest attempts to create alignments and phylogenetic trees of the globins.

Where will the fields of bioinformatics and genomics go in the next five to 10 years? The opportunities are vast and any prediction will certainly be incomplete, but it is certain that the rapid technological advances in sequencing will provide an unprecedented view of human genetic variation and how this relates to phenotype. In the area of human disease studies, genome-wide association studies can be expected to lead to the identification of hundreds of genes underlying complex disorders. Finally, our understanding of evolution and its relevance to medicine will expand dramatically. Dr Pevsner's valuable book will help the student or researcher access the tools and learn the principles that will enable this exciting research.

David Valle, M.D.
Henry J. Knott Professor and Director McKusick-Nathans Institute of Genetic Medicine,
Johns Hopkins University School of Medicine

Part I

Analyzing DNA, RNA, and Protein Sequences in Databases

account of this very identity of composition. Hence the opinion is not unworthy of a closer investigation, that gelatine, when taken in the dissolved state, is again converted, in the body, into cellular tissue, membrane and cartilage; that it may serve for the reproduction of such parts of these tissues as have been wasted, and for their growth.

And when the powers of nutrition in the whole body are affected by a change of the health, then, even should the power of forming blood remain the same, the organic force by which the constituents of the blood are transformed into cellular tissue and membranes must necessarily be enfeebled by sickness. In the sick man, the intensity of the vital force, its power to produce metamorphoses, must be diminished as well in the stomach as in all other parts of the body.

In this condition, the uniform experience of practical physicians shows that gelatinous matters in a dissolved state exercise a most decided influence on the state of the health. Given in a form adapted for assimilation, they serve to husband the vital force, just as may be done, in the case of the stomach, by due preparation of the food in general. Brittleness in the bones of graminivorous animals is clearly owing to a weakness in those parts of the organism whose function it is to convert the constituents of the blood into cellular tissue and membrane; and if we can trust to the reports of physicians who have resided in the East, the Turkish women, in their diet of rice, and in the frequent use of enemata of strong soup, have united the conditions necessary for the formation both of cellular tissue and of fat.

PART II.

THE METAMORPHOSIS OF TISSUES.

1. The absolute identity of composition in the chief constituents of blood and the nitrogenized compounds in vegetable food would, some years ago, have furnished a plausible reason for denying the accuracy of the chemical analysis leading to such a result. At that period, experiment had not as yet demonstrated the existence of numerous compounds, both containing nitrogen and devoid of that element, which with the greatest diversity in external characters, yet possess the very same composition in 100 parts; nay, many of which even contain the same absolute amount of equivalents of each element. Such examples are now very frequent, and are known by the names of *isomeric* and *polymeric* compounds.

2. Cyanuric acid, for example, is a nitrogenized compound which crystallizes in beautiful transparent octahedrons, easily soluble in water and in acids, and very permanent. Cyamelide is a second body, absolutely insoluble in water and acids, white and opaque like porcelain or magnesia. Hydrated cyanic acid is a third compound, which is a liquid more volatile than pure acetic acid, which blisters the skin, and cannot be brought in contact with water without being instantaneously resolved into new products. These three substances not only yield, on analysis, absolutely the same relative weights of the same elements, but they may be converted and reconverted into one another, even in hermetically closed vessels —that is, without the aid of any foreign matter. (See Appendix, 21.) Again, among those substances which contain no nitrogen, we have aldehyde, a combustible liquid miscible with water, which boils at the temperature of the hand, attracts oxygen from the atmosphere with avidity, and is thereby

changed into acetic acid. This compound cannot be preserved, even in close vessels; for after some hours or days, its consistence, its volatility, and its power of absorbing oxygen, all are changed. It deposits long, hard, needle-shaped crystals, which at 212° are not volatilized, and the supernatant liquid is no longer aldehyde. It now boils at 140°, cannot be mixed with water, and when cooled to a moderate degree crystallizes in a form like ice. Nevertheless, analysis has proved, that these three bodies, so different in their characters, are identical in composition. (21.)

3. A similar group of three occurs in the case of albumen, fibrine, and caseine. They differ in external character, but contain exactly the same proportions of organic elements. ◄

When animal albumen, fibrine, and caseine are dissolved in a moderately strong solution of caustic potash, and the solution is exposed for some time to a high temperature, these substances are decomposed. The addition of acetic acid to the solution causes, in all three, the separation of a gelatinous translucent precipitate, which has exactly the same characters and composition, from whichever of the three substances above mentioned it has been obtained.

Mulder, to whom we owe the discovery of this compound, found, by exact and careful analysis, that it contains the same organic elements, and exactly in the same proportion, as the animal matters from which it is prepared; insomuch, that if we deduct from the analysis of albumen, fibrine, and caseine, the ashes they yield when incinerated, as well as the sulphur and phosphorus they contain, and then calculate the remainder for 100 parts, we obtain the same result as

---

The study of bioinformatics includes the analysis of proteins. In the first half of the nineteenth century the Dutch researcher Gerardus Johannes Mulder (1802–1880), advised by the Swedish chemist Jöns Jacob Berzelius (1779–1848), studied the "albuminous" substances or proteins fibrin, albumin from blood, albumin from egg (ovalbumin), and the coloring matter of blood (hemoglobin). Mulder and others extracted and purified these proteins and believed that they all shared the same elemental composition ($C_{400}H_{260}N_{100}O_{120}$), with varying amounts of phosphorus and sulfur. Justus Liebig (1803–1873) believed that the composition of protein was $C_{48}H_{36}N_6O_{14}$. This page, from Liebig's Animal Chemistry, or Organic Chemistry in its Applications to Physiology and Pathology (1847, p. 36), discusses albumin, fibrin, and casein (see arrowhead).

1

Introduction

Bioinformatics represents a new field at the interface of the twentieth-century revolutions in molecular biology and computers. A focus of this new discipline is the use of computer databases and computer algorithms to analyze proteins, genes, and the complete collections of deoxyribonucleic acid (DNA) that comprises an organism (the genome). A major challenge in biology is to make sense of the enormous quantities of sequence data and structural data that are generated by genome-sequencing projects, proteomics, and other large-scale molecular biology efforts. The tools of bioinformatics include computer programs that help to reveal fundamental mechanisms underlying biological problems related to the structure and function of macromolecules, biochemical pathways, disease processes, and evolution.

According to a National Institutes of Health (NIH) definition, bioinformatics is "research, development, or application of computational tools and approaches for expanding the use of biological, medical, behavioral or health data, including those to acquire, store, organize, analyze, or visualize such data." The related discipline of computational biology is "the development and application of data-analytical and theoretical methods, mathematical modeling and computational simulation techniques to the study of biological, behavioral, and social systems."

While the discipline of bioinformatics focuses on the analysis of molecular sequences, genomics and functional genomics are two closely related disciplines. The goal of genomics is to determine and analyze the complete DNA sequence of an organism, that is, its genome. The DNA encodes genes, which can be expressed as ribonucleic acid (RNA) transcripts and then in many cases further translated into

The NIH Bioinformatics Definition Committee findings are reported at ► http://www.bisti.nih.gov/CompuBioDef.pdf. For additional definitions of bioinformatics and functional genomics, see Boguski (1994), Luscombe et al. (2001), Ideker et al. (2001), and Goodman (2002).

Bioinformatics and Functional Genomics, Second Edition. By Jonathan Pevsner
Copyright © 2009 John Wiley & Sons, Inc.

protein. Functional genomics describes the use of genomewide assays in the study of gene and protein function.

The aim of this book is to explain both the theory and practice of bioinformatics and genomics. The book is especially designed to help the biology student use computer programs and databases to solve biological problems related to proteins, genes, and genomes. Bioinformatics is an integrative discipline, and our focus on individual proteins and genes is part of a larger effort to understand broad issues in biology, such as the relationship of structure to function, development, and disease. For the computer scientist, this book explains the motivations for creating and using algorithms and databases.

ORGANIZATION OF THE BOOK

There are three main sections of the book. The first part (Chapters 2 to 7) explains how to access biological sequence data, particularly DNA and protein sequences (Chapter 2). Once sequences are obtained, we show how to compare two sequences (pairwise alignment; Chapter 3) and how to compare multiple sequences (primarily by the Basic Local Alignment Search Tool [BLAST]; Chapters 4 and 5). We introduce multiple sequence alignment (Chapter 6) and show how multiply aligned sequences can be visualized in phylogenetic trees (Chapter 7). Chapter 7 thus introduces the subject of molecular evolution.

The second part of the book describes functional genomics approaches to RNA and protein and the determination of gene function (Chapters 8 to 12). The central dogma of biology states that DNA is transcribed into RNA then translated into protein. We will examine bioinformatic approaches to RNA, including both noncoding and coding RNAs. We then describe the technology of DNA microarrays and examine microarray data analysis (Chapter 9). From RNA we turn to consider proteins from the perspective of protein families, and the analysis of individual proteins (Chapter 10) and protein structure (Chapter 11). We conclude the middle part of the book with an overview of the rapidly developing field of functional genomics (Chapter 12).

Since 1995, the genomes have been sequenced for several thousand viruses, prokaryotes (bacteria and archaea), and eukaryotes, such as fungi, animals, and plants. The third section of the book covers genome analysis (Chapters 13 to 20). Chapter 13 provides an overview of the study of completed genomes and then descriptions of how the tools of bioinformatics can elucidate the tree of life. We describe bioinformatics resources for the study of viruses (Chapter 14) and bacteria and archaea (Chapter 15; these are two of the three main branches of life). Next we examine the eukaryotic chromosome (Chapter 16) and explore the genomes of a variety of eukaryotes, including fungi (Chapter 17), organisms from parasites to primates (Chapter 18), and then the human genome (Chapter 19). Finally, we explore bioinformatic approaches to human disease (Chapter 20).

BIOINFORMATICS: THE BIG PICTURE

We can summarize the fields of bioinformatics and genomics with three perspectives. The first perspective on bioinformatics is the cell (Fig. 1.1). The central dogma of molecular biology is that DNA is transcribed into RNA and translated into protein. The focus of molecular biology has been on individual genes, messenger RNA

Central dogma of molecular biology

DNA $\longrightarrow$ RNA $\longrightarrow$ protein $\longrightarrow$ cellular phenotype

Central dogma of genomics

genome $\longrightarrow$ transcriptome $\longrightarrow$ proteome $\longrightarrow$ cellular phenotype

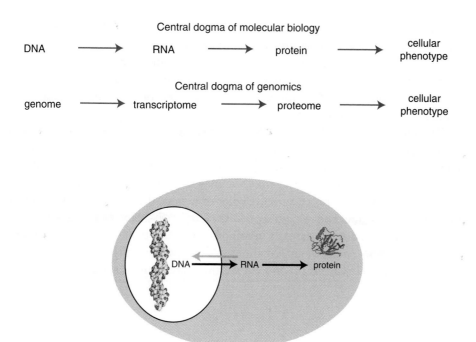

DNA $\longrightarrow$ RNA $\longrightarrow$ protein

FIGURE 1.1. *The first perspective of the field of bioinformatics is the cell. Bioinformatics has emerged as a discipline as biology has become transformed by the emergence of molecular sequence data. Databases such as the European Molecular Biology Laboratory (EMBL), GenBank, and the DNA Database of Japan (DDBJ) serve as repositories for hundreds of billions of nucleotides of DNA sequence data (see Chapter 2). Corresponding databases of expressed genes (RNA) and protein have been established. A main focus of the field of bioinformatics is to study molecular sequence data to gain insight into a broad range of biological problems.*

(mRNA) transcripts as well as noncoding RNAs, and proteins. A focus of the field of bioinformatics is the complete collection of DNA (the genome), RNA (the transcriptome), and protein sequences (the proteome) that have been amassed (Henikoff, 2002). These millions of molecular sequences present both great opportunities and great challenges. A bioinformatics approach to molecular sequence data involves the application of computer algorithms and computer databases to molecular and

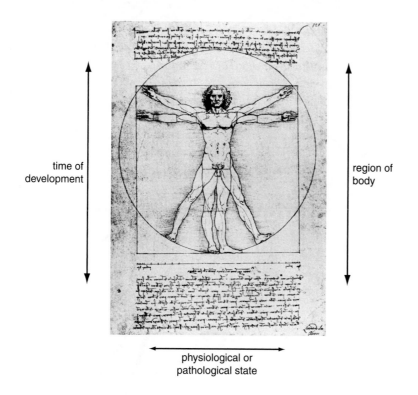

time of development

region of body

physiological or pathological state

FIGURE 1.2. *The second perspective of bioinformatics is the organism. Broadening our view from the level of the cell to the organism, we can consider the individual's genome (collection of genes), including the genes that are expressed as RNA transcripts and the protein products. Thus, for an individual organism bioinformatics tools can be applied to describe changes through developmental time, changes across body regions, and changes in a variety of physiological or pathological states.*

cellular biology. Such an approach is sometimes referred to as functional genomics. This typifies the essential nature of bioinformatics: biological questions can be approached from levels ranging from single genes and proteins to cellular pathways and networks or even whole genomic responses (Ideker et al., 2001). Our goals are to understand how to study both individual genes and proteins and collections of thousands of genes or proteins.

From the cell we can focus on individual organisms, which represents a second perspective of the field of bioinformatics (Fig. 1.2). Each organism changes across different stages of development and (for multicellular organisms) across different regions of the body. For example, while we may sometimes think of genes as static entities that specify features such as eye color or height, they are in fact dynamically regulated across time and region and in response to physiological state. Gene expression varies in disease states or in response to a variety of signals, both intrinsic and environmental. Many bioinformatics tools are available to study the broad biological questions relevant to the individual: there are many databases of expressed

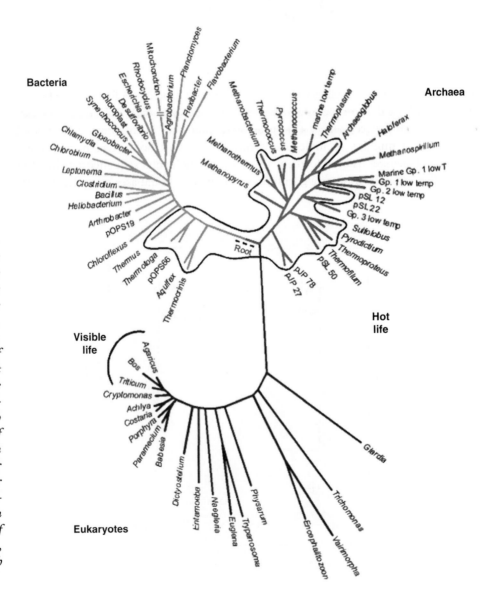

FIGURE 1.3. The third perspective of the field of bioinformatics is represented by the tree of life. The scope of bioinformatics includes all of life on Earth, including the three major branches of bacteria, archaea, and eukaryotes. Viruses, which exist on the borderline of the definition of life, are not depicted here. For all species, the collection and analysis of molecular sequence data allow us to describe the complete collection of DNA that comprises each organism (the genome). We can further learn the variations that occur between species and among members of a species, and we can deduce the evolutionary history of life on Earth. (After Barns et al., 1996 and Pace, 1997.) Used with permission.

genes and proteins derived from different tissues and conditions. One of the most powerful applications of functional genomics is the use of DNA microarrays to measure the expression of thousands of genes in biological samples.

At the largest scale is the tree of life (Fig. 1.3) (Chapter 13). There are many millions of species alive today, and they can be grouped into the three major branches *←Third perspective* of bacteria, archaea (single-celled microbes that tend to live in extreme environments), and eukaryotes. Molecular sequence databases currently hold DNA sequences from over 150,000 different organisms. The complete genome sequences of thousands of organisms are now available, including organellar and viral genomes. One of the main lessons we are learning is the fundamental unity of life at the molecular level. We are also coming to appreciate the power of comparative genomics, in which genomes are compared. Through DNA sequence analysis we are learning how chromosomes evolve and are sculpted through processes such as chromosomal duplications, deletions, and rearrangements, as well as through whole genome duplications (Chapters 16 to 18).

Figure 1.4 presents the contents of this book in the context of these three perspectives of bioinformatics.

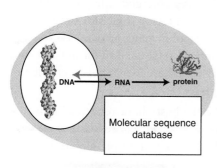

Part 1: Analyzing DNA, RNA, and protein sequences

Chapter 1: Introduction
Chapter 2: How to obtain sequences
Chapter 3: How to compare two sequences
Chapters 4 and 5: How to compare a sequence
      to all other sequences in databases
Chapter 6: How to multiply align sequences
Chapter 7: How to view multiply aligned sequences
      as phylogenetic trees

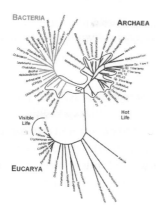

Part 2: Genome-wide analysis of RNA and protein

Chapter 8: Bioinformatics approaches to RNA
Chapter 9: Microarray data analysis
Chapter 10: Protein analysis and protein families
Chapter 11: Protein structure
Chapter 12: Functional genomics

Part 3: Genome analysis

Chapter 13: The tree of life
Chapter 14: Viruses
Chapter 15: Prokaryotes
Chapter 16: The eukaryotic chromosome
Chapter 17: The fungi
Chapter 18: Eukaryotes from parasites to plants to primates
Chapter 19: The human genome
Chapter 20: Human disease

FIGURE 1.4. *Overview of the chapters in this book.*

A CONSISTENT EXAMPLE: HEMOGLOBIN

Throughout this book, we will focus on the globin gene family to provide a consistent example of bioinformatics and genomics concepts. The globin family is one of the best characterized in biology.

- Historically, hemoglobin was one of the first proteins to be studied, having been described in the 1830s and 1840s by Mulder, Liebig, and others.
- Myoglobin, a globin that binds oxygen in the muscle tissue, was the first protein to have its structure solved by x-ray crystallography (Chapter 11).
- Hemoglobin, a tetramer of four globin subunits (principally $\alpha_2\beta_2$ in adults), is the main oxygen carrier in blood of vertebrates. Its structure was also one of the earliest to be described. The comparison of myoglobin, alpha globin, and beta globin protein sequences represents one of the earliest applications of multiple sequence alignment (Chapter 6), and led to the development of amino acid substitution matrices used to score protein relatedness (Chapter 3).
- In the 1980s as DNA sequencing technology emerged, the globin loci on human chromosomes 16 (for α globin) and 11 (for β globin) were among the first to be sequenced and analyzed. The globin genes are exquisitely regulated across time (switching from embryonic to fetal to adult forms) and with tissue-specific gene expression. We will discuss these loci in the description of the control of gene expression (Chapter 16).
- While hemoglobin and myoglobin remain the best-characterized globins, the family of homologous proteins extends to two separate classes of plant globins, invertebrate hemoglobins (some of which contain multiple globin domains within one protein molecule), bacterial homodimeric hemoglobins (consisting of two globin subunits), and flavohemoglobins that occur in bacteria, archaea, and fungi. Thus the globin family is useful as we survey the tree of life (Chapters 13 to 18).

Another protein we will use as an example is retinol-binding protein (RBP4), a small, abundant secreted protein that binds retinol (vitamin A) in blood (Newcomer and Ong, 2000). Retinol, obtained from carrots in the form of vitamin A, is very hydrophobic. RBP4 helps transport this ligand to the eye where it is used for vision. We will study RBP4 in detail because it has a number of interesting features:

- There are many proteins that are homologous to RBP4 in a variety of species, including human, mouse, and fish ("orthologs"). We will use these as examples of how to align proteins, perform database searches, and study phylogeny.
- There are other human proteins that are closely related to RBP4 ("paralogs"). Altogether the family that includes RBP4 is called the lipocalins, a diverse group of small ligand-binding proteins that tend to be secreted into extracellular spaces (Akerstrom et al., 2000; Flower et al., 2000). Other lipocalins have fascinating functions such as apoliprotein D (which binds cholesterol), a pregnancy-associated lipocalin, aphrodisin (an "aphrodisiac" in hamsters), and an odorant-binding protein in mucus.

- There are bacterial lipocalins, which could have a role in antibiotic resistance (Bishop, 2000). We will explore how bacterial lipocalins could be ancient genes that entered eukaryotic genomes by a process called lateral gene transfer.

- Because the lipocalins are small, abundant, and soluble proteins, their biochemical properties have been characterized in detail. The three-dimensional protein structure has been solved for several of them by x-ray crystallography (Chapter 11).

- Some lipocalins have been implicated in human disease.

ORGANIZATION OF THE CHAPTERS

The chapters of this book are intended to provide both the theory of bioinformatics subjects as well as a practical guide to using computer databases and algorithms. Web resources are provided throughout each chapter. Chapters end with brief sections called Perspective and Pitfalls. The perspective feature describes the rate of growth of the subject matter in each chapter. For example, a perspective on Chapter 2 (access to sequence information) is that the amount of DNA sequence data deposited in GenBank is undergoing an explosive rate of growth. In contrast, an area such as pairwise sequence alignment, which is fundamental to the entire field of bioinformatics (Chapter 3), was firmly established in the 1970s and 1980s. But even for fundamental operations such as multiple sequence alignment (Chapter 6) and molecular phylogeny (Chapter 7) dozens of novel, ever-improving approaches are introduced at a rapid rate. For example, hidden Markov models and Bayesian approaches are being applied to a wide range of bioinformatics problems.

The pitfalls section of each chapter describes some common difficulties encountered by biologists using bioinformatics tools. Some errors might seem trivial, such as searching a DNA database with a protein sequence. Other pitfalls are more subtle, such as artifacts caused by multiple sequence alignment programs depending upon the type of paramters that are selected. Indeed, while the field of bioinformatics depends substantially on analyzing sequence data, it is important to recognize that there are many categories of errors associated with data generation, collection, storage, and analysis. We address the problems of false positive and false negative results in a variety of searches and analyses.

Each chapter offers multiple-choice quizzes, which test your understanding of the chapter materials. There are also problems that require you to apply the concepts presented in each chapter. These problems may form the basis of a computer laboratory for a bioinformatics or genomics course.

The references at the end of each chapter are accompanied by an annotated list of recommended articles. This suggested reading section includes classic papers that show how the principles described in each chapter were discovered. Particularly helpful review articles and research papers are highlighted.

A TEXTBOOK FOR COURSES ON BIOINFORMATICS AND GENOMICS

This is a textbook for two separate courses: one is an introduction to bioinformatics (and uses Chapters 1 to 12 [Parts 1 and 2]), and one is an introduction to genomics (and uses Chapters 13 to 20 [Part 3]). In a sense, the discipline of bioinformatics serves biology, facilitating ways of posing and then answering questions about proteins, genes, and genomes. The third part of this book surveys the tree of life from the perspective of genes and genomes. Progress in this field could not occur at its current pace without the bioinformatics tools described in the first parts of the book.

Often, students have a particular research area of interest, such as a gene, a physiological process, a disease, or a genome. It is hoped that in the process of studying globins and other specific proteins and genes throughout this book, students can also simultaneously apply the principles of bioinformatics to their own research questions.

Web material for this book is available at ▶ http//www.wiley. com/go/pevsnerbioinformatics.

In teaching courses on bioinformatics and genomics at Johns Hopkins, it has been helpful to complement lectures with computer labs. These labs and many other resources are posted on the website for this book (▶ http://www.bioinfbook. org). That site contains many relevant URLs, organized by chapter. Each chapter makes references to web documents posted on the site. For example, if you see a figure of a phylogenetic tree or a sequence alignment, you can easily retrieve the raw data and make the figure yourself.

Another feature of the Johns Hopkins bioinformatics course is that each student is required to discover a novel gene by the last day of the course. The student must begin with any protein sequence of interest and perform database searches to identify genomic DNA that encodes a protein no one has described before. This problem is described in detail in Chapter 5 (and summarized in web document 5.15 at ▶ http://www.bioinfbook.org/chapter5). The student thus chooses the name of the gene and its corresponding protein and describes information about the organism and evidence that the gene has not been described before. Then, the student creates a multiple sequence alignment of the new protein (or gene) and creates a phylogenetic tree showing its relation to other known sequences.

Each year, some beginning students are slightly apprehensive about accomplishing this exercise, but in the end all of them succeed. A benefit of this exercise is that it requires a student to actively use the principles of bioinformatics. Most students choose a gene (or protein) relevant to their own research area, while others find new lipocalins or globins.

For a genomics course, students select a genome of interest and describe five aspects in depth (described at the start of Chapter 13): (1) What are the basic feature of the genome, such as its size, number of chromosomes, and other features? (2) A comparative genomic analysis is performed to study the relation of the species to its neighbors. (3) The student describes biological principles that are learned through genome analysis. (4) The human disease relevance is described. (5) Bioinformatics aspects are described, such as key databases or algorithms used for genome analysis.

Teaching bioinformatics and genomics is notable for the diversity of students learning this new discipline. Each chapter provides background on the subject matter. For more advanced students, key research papers are cited at the end of each chapter. These papers are technical, and reading them along with the chapters will provide a deeper understanding of the material. The suggested reading section also includes review articles.

KEY BIOINFORMATICS WEBSITES

The field of bioinformatics relies heavily on the Internet as a place to access sequence data, to access software that is useful to analyze molecular data, and as a place to integrate different kinds of resources and information relevant to biology. We will describe a variety of websites. Initially, we will focus on the three main publicly accessible databases that serve as repositories for DNA and protein data. In Chapter 2 we begin with the National Center for Biotechnology Information (NCBI), which hosts GenBank. The NCBI website offers a variety of other bioinformatics-related tools. We will gradually introduce the European Bioinformatics Institute (EBI) web server, which hosts a complementary DNA database (EMBL, the European Molecular Biology Laboratory database). We will also introduce the DNA Database of Japan (DDBJ). The research teams at GenBank, EMBL, and DDBJ share sequence data on a daily basis. Throughout this book we will highlight the key genome browser hosted by the University of California, Santa Cruz (UCSC). A general theme of the discipline of bioinformatics is that many databases are closely interconnected. Throughout the chapters of this book we will introduce over 1,000 additional websites that are relevant to bioinformatics.

SUGGESTED READING

Overviews of the field of bioinformatics have been written by Mark Gerstein and colleagues (Luscombe et al., 2001), Claverie et al. 2001, and Yu et al. 2004. Kaminski 2000 also introduces bioinformatics, with practical suggestions of websites to visit. Russ Altman 1998 discusses the relevance of bioinformatics to medicine, while David Searls 2000 introduces bioinformatics tools for the study of genomes. An approach to learning about the current state of bioinformatics education is to read about the perspectives of the programs at Yale (Gerstein et al., 2007), Stanford (Altman and Klein, 2007), and in Australia (Cattley, 2004).

REFERENCES

Akerstrom, B., Flower, D. R., and Salier, J. P. Lipocalins: Unity in diversity. *Biochim. Biophys. Acta* **1482**, 1–8 (2000).

Altman, R. B. Bioinformatics in support of molecular medicine. *Proc. AMIA Symp.*, 53–61 (1998).

Altman, R. B., and Klein, T. E. Biomedical informatics training at Stanford in the 21st century. *J. Biomed. Inform.* **40**, 55–58 (2007).

Barns, S. M., Delwiche, C. F., Palmer, J. D., and Pace, N. R. Perspectives on archaeal diversity, thermophily and monophyly from environmental rRNA sequences. *Proc. Natl. Acad. Sci. USA* **93**, 9188–9193 (1996).

Bishop, R. E. The bacterial lipocalins. *Biochim. Biophys. Acta* **1482**, 73–83 (2000).

Boguski, M. S. Bioinformatics. *Curr. Opin. Genet. Dev.* **4**, 383–388 (1994).

Cattley, S. A review of bioinformatics degrees in Australia. *Brief. Bioinform.* **5**, 350–354 (2004).

Claverie, J. M., Abergel, C., Audic, S., and Ogata, H. Recent advances in computational genomics. *Pharmacogenomics* **2**, 361–372 (2001).

Flower, D. R., North, A. C., and Sansom, C. E. The lipocalin protein family: Structural and sequence overview. *Biochim. Biophys. Acta* **1482**, 9–24 (2000).

Gerstein, M., Greenbaum, D., Cheung, K., and Miller, P. L. An interdepartmental Ph.D. program in computational biology and bioinformatics: The Yale perspective. *J. Biomed Inform.* **40**, 73–79 (2007).

Goodman, N. Biological data becomes computer literate: New advances in bioinformatics. *Curr. Opin. Biotechnol.* **13**, 68–71 (2002).

Henikoff, S. Beyond the central dogma. *Bioinformatics* **18**, 223–225 (2002).

Ideker, T., Galitski, T., and Hood, L. A new approach to decoding life: Systems biology. *Annu. Rev. Genomics Hum. Genet.* **2**, 343–372 (2001).

Kaminski, N. Bioinformatics. A user's perspective. *Am. J. Respir. Cell Mol. Biol.* **23**, 705–711 (2000).

Liebig, J. *Animal Chemistry, or Organic Chemistry in its Applications to Physiology and Pathology.* James M. Campbell, Philadelphia, 1847.

Luscombe, N. M., Greenbaum, D., and Gerstein, M. What is bioinformatics? A proposed definition and overview of the field. *Methods Inf. Med.* **40**, 346–358 (2001).

Newcomer, M. E., and Ong, D. E. Plasma retinol binding protein: Structure and function of the prototypic lipocalin. *Biochim. Biophys. Acta* **1482**, 57–64 (2000).

Pace, N. R. A molecular view of microbial diversity and the biosphere. *Science* **276**, 734–740 (1997).

Searls, D. B. Bioinformatics tools for whole genomes. *Annu. Rev. Genomics Hum. Genet.* **1**, 251–279 (2000).

Yu, U., Lee, S. H., Kim, Y. J., and Kim, S. Bioinformatics in the post-genome era. *J. Biochem. Mol. Biol.* **37**, 75–82 (2004).

admodum fecernantur, exponam. Res eft parvi laboris. Fa-
rina fumitur ex optimo tritico, modice trita, ne cribrum
furfures fubeant; oportet enim ab his effe quam expurga-
tiffimam, ut omnis mitturæ tollatur fufpicio. Tum aquæ pu-
riffimæ permifcetur, ac fubigitur. Quod reliquum eft ope-
ris, lotura abfolvit. Aqua enim partes omnes, quafcumque
poteft folvere, fecum avehit; alias intactas relinquit.

Porro hæ, quas aqua relinquit, contrectatæ manibus,
preffæque fub aqua reliqua, paullatim in maffam coguntur
mollem, & fupra, quam credi poteft, tenacem: egregium glu-
tinis genus, & ad opificia multa aptiffimum; in quo illud
notatu dignum eft, quod aquæ permifceri fe amplius non
finit. Illæ aliæ, quas aqua fecum avehit, aliquandiu inna-
tant, & aquam lacteam reddunt; poft paullatim deferuntur
ad fundum, & fubfidunt; nec admodum inter fe cohærent;
fed quafi pulvis vel leviffimo concuffu furfum redeunt. Ni-
hil his affinius eft amylo; vel potius ipfæ veriffimum funt
amylum. Atque hæc fcilicet duo funt illa partium genera,
quæ fibi Beccarius propofuit ad chymicum opus faciendum,
quæque ut fuis nominibus diftingueret, glutinofum alterum
appellare folebat, alterum amylaceum.

Tanta eft autem horum generum diverfitas, ut fi utrum-
que vel digeftione, vel deftillatione refolvas, & principia,
unde conftant, chymicorum more, elicias, non ex una ac
fimplici, fed ex duabus longiffimeque inter fe diverfis rebus
prodiiffe videantur; cum enim amylacea pars fuum præ fe
genus ferat, eaque principia oftendat, quæ a vegetabili natura duci folent; glutinofa originem quafi detrectat fuam,
ac fe per omnia fic præbet, quafi effet ab animante quo-
piam profecta. Quod ut melius intelligatur, generatim pri-
mum fcire convenit, quam diffimiliter vegetabilia atque ani-
mantia in digeftionibus deftillationibufque fe præftent.

In digeftionibus, quas lenis & diuturnus calor facit,
animantium partes numquam ad veram abfolutamque fer-
mentationem perducuntur; fed putrefiunt teterrime femper.
Vegetabilia quafi fua fponte fermentantur, neque putrefcunt,
nifi ars adiuvet; eaque inter fermentandum manifefta acoris
indicia præbent, quæ nulla funt in animalibus, dum putrefcunt. Fermentatione autem confecta, vinofum aut acetofum
liquorem vegetabilia largiuntur; animalia, fi putrefiant, uri-
nofum.

Q 2

Chapter 2 introduces ways to access molecular data, including information about DNA and proteins. One of the first scientists to study proteins was Iacopo Bartolomeo Beccari (1682–1776), an Italian philosopher and physician who discovered protein as a component of vegetables. This image is from page 123 of the Bologna Commentaries, published in 1745 and written by a secretary on the basis of a 1728 lecture by Beccari. Beccari separated gluten (plant proteins) from wheaten flour. The passage beginning Res est parvi laboris *("it is a thing of little labor"; see solid arrowhead) is translated as follows (Beach, 1961, p. 362):*

"It is a thing of little labor. Flour is taken of the best wheat, moderately ground, the bran not passing though the sieve, for it is necessary that this be fully purged away, so that all traces of a mixture have been removed. Then it is mixed with pure water and kneaded. What is left by this procedure, washing clarifies. Water carries off with itself all it is able to dissolve, the rest remains untouched. After this, what the water leaves is worked with the hands, and pressed upon in the water that has stayed. Slowly it is drawn together in a doughy mass, and beyond what is possible to be believed, tenacious, a remarkable sort of glue, and suited to many uses; and what is especially worthy of note, it cannot any longer be mixed with water. The other particles, which water carries away with itself, for some time float and render the water milky; but after a while they are carried to the bottom and sink; nor in any way do they adhere to each other; but like powder they return upward on the lightest contact. Nothing is more like this than starch, or rather this truly is starch. And these are manifestly the two sorts of bodies which Beccari displayed through having done the work of a chemist and he distinguished them by their names, one being appropriately called glutinous (see open arrowhead) and the other amylaceous."

In addition to purifying gluten, Beccari identified it as an "animal substance" in contrast to starch, a "vegetable substance," based on differences on how they decomposed with heat or distillation. A century later Jons Jakob Berzelius proposed the word protein, and he also posited that plants form "animal materials" that are eaten by herbivorous animals.

2

Access to Sequence Data and Literature Information

INTRODUCTION TO BIOLOGICAL DATABASES

All living organisms are characterized by the capacity to reproduce and evolve. The genome of an organism is defined as the collection of DNA within that organism, including the set of genes that encode proteins. In 1995 the complete genome of a free-living organism was sequenced for the first time, the bacterium *Haemophilus influenzae* (Fleischmann et al., 1995; Chapters 13 and 15). In the few years since then the genomes of thousands of organisms have been completely sequenced, ushering in a new era of biological data acquisition and information accessibility. Publicly available databanks now contain billions of nucleotides of DNA sequence data collected from over 260,000 different organisms (Kulikova et al., 2007). The goal of this chapter is to introduce the databases that store these data and strategies to extract information from them.

Three publicly accessible databases store large amounts of nucleotide and protein sequence data: GenBank at the National Center for Biotechnology Information (NCBI) of the National Institutes of Health (NIH) in Bethesda (Benson et al., 2009), the DNA Database of Japan (DDBJ) at the National Institute of

Bioinformatics and Functional Genomics, Second Edition. By Jonathan Pevsner
Copyright © 2009 John Wiley & Sons, Inc.

GenBank is at ▶ http://www. ncbi.nlm.nih.gov/Genbank; DDBJ is at ▶ http://www.ddbj. nig.ac.jp/; and EMBL/EBI is at ▶ http://www.ebi.ac.uk/. You can visit the INSDC at ▶http://www. insdc.org/. By November 2008 the total number of sequenced bases had passed 97 billion.

Genetics in Mishima (Miyazaki et al., 2004), and the European Molecular Biology Laboratory (EMBL) Nucleotide Sequence Database at the European Bioinformatics Institute (EBI) in Hinxton, England (Kulikova et al., 2007). These three databases share their sequence data daily. They are coordinated by the International Nucleotide Sequence Database Collaboration (INSDC), which announced in August 2005 that the total amount of sequenced DNA had reached 100 billion base pairs.

In addition to GenBank, DDBJ, and EBI, there are other categories of bioinformatics databases that contain DNA and/or protein sequence data:

- Whole-genome shotgun (WGS) sequences and the Short Read Archive (Chapter 13 and discussed below) are not formally part of GenBank, but contain even more DNA sequences.

- Databases such as Ensembl, NCBI, and the genome browser at the University of California, Santa Cruz (UCSC) provide annotation of the human genome and other genomes (see below).

- Some contain nucleotide and/or protein sequence data that are relevant to a particular gene or protein (such as kinases). Other databases are specific to particular chromosomes or organelles (Chapters 16 to 18).

Pfam (▶ http://www.sanger.ac. uk/Software/Pfam/) and other related databases are described in Chapters 6 (multiple sequence alignment) and 10 (protein families).

- A variety of databases include information on sequences sharing common properties that have been grouped together. For example, the Protein Family (Pfam) database consists of several thousand families of homologous proteins.

- Hundreds of databases contain sequence information related to genes that are mutated in human disease. These databases are described in Chapter 20.

- Many specialized databases focus on particular organisms (such as yeast); examples are listed in the section on genomes (Chapters 13 to 20).

- There are databases devoted to particular types of nucleic acids or proteins or properties of these macromolecules. Examples are databases of gene expression (see Chapters 8 and 9), databases of transfer RNA (tRNA) molecules, databases of tissue-specific protein expression (see Chapter 10), or databases of gene regulatory regions such as 3'-untranslated regions (see Chapter 16).

Some bioinformatics databases do not contain nucleotide or protein sequence data as their main function. Instead, they contain information that may link to individual genes or proteins.

- Literature databases contain bibliographic references relevant to biological research and in some cases contain links to full-length articles. We will describe two of these databases, PubMed and the Sequence Retrieval System (SRS), in this chapter.

- Structure databases contain information on the structure of proteins and other macromolecules. These databases are described in Chapter 10 (on proteins) and Chapter 11 (on protein structure).

GENBANK: DATABASE OF MOST KNOWN NUCLEOTIDE AND PROTEIN SEQUENCES

While the sequence information underlying DDBJ, EBI, and GenBank is equivalent, we begin our discussion with GenBank. GenBank is a database consisting of most known public DNA and protein sequences (Benson et al., 2009). In addition to storing these sequences, GenBank contains bibliographic and biological annotation. Data from GenBank are available free of charge from the National Center for Biotechnology Information (NCBI) in the National Library of Medicine at the NIH (Wheeler et al., 2007).

Amount of Sequence Data

GenBank currently contains about 100 billion nucleotides from 100 million sequences (release 168). The growth of GenBank in terms of both nucleotides of DNA and number of sequences from 1982 to 2008 is summarized in Fig. 2.1*a*. Over the period 1982 to the present, the number of bases in GenBank has doubled approximately every 18 months.

The WGS division consists of sequences generated by high throughput sequencing efforts. Since 2002, WGS sequences have been available at NCBI, but they are not considered part of the GenBank releases. As indicated in Fig. 2.1, the number of base pairs of DNA included among WGS sequences (136 billion base pairs in release 168, October 2008) is larger than the size of GenBank.

While the amount of sequence data in GenBank has risen rapidly, the arrival of next-generation sequencing technology, described in Chapter 13, is instantly leading

Between December 2007 and December 2008, over 15 billion base pairs (bp) of DNA were added to GenBank, an average of 42 million bp per day. In comparison, the first eukaryotic genome to be completed (*Saccharomyces cerevisiae;* Chapter 17) is about 13 million bp in size.

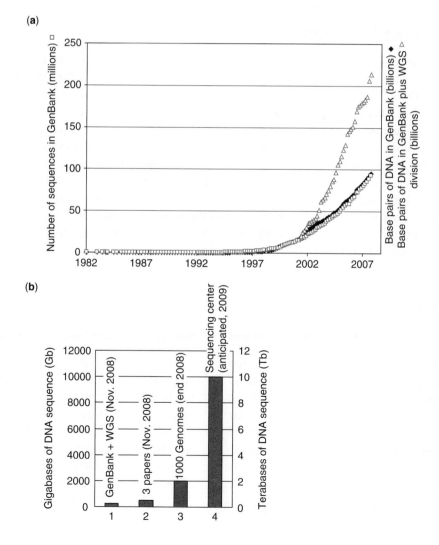

(a)

(b)

FIGURE 2.1. (a) Growth of GenBank from release 3 (1982) to release 168 (October 2008). Data were plotted from the GenBank release notes at ▶ *http://www.ncbi.nlm.nih.gov/Genbank/. Additional DNA sequences from the whole genome shotgun sequencing projects, begun in 2002, are shown. (b) The amount of sequenced DNA is vastly increasing. Bar 1 indicates the amount of DNA in GenBank plus WGS as shown in panel (a). Bar 2 indicates the amount of DNA sequence (492 gigabases) reported in three research articles published in a single issue of Nature (Bentley et al., 2008; Wang et al., 2008; Ley et al., 2008). Bar 3 indicates the amount of DNA sequence (2 terabases) expected to be generated by the end of 2008 as part of the 1000 Genomes Project (Chapter 13); 1 terabase was reported by the Wellcome Trust Sanger Institute in a six-month period in 2008. Bar 4 indicates the amount of sequence data (10 terabases) it is anticipated will be generated in 2009 alone by a typical major genome sequencing center.*

to a vast new influx of DNA sequence data (Fig. 2.1*b*). Next-generation sequencing involves the generation of massive amounts of sequence data, such as 1 billion bases (1 Gb) in a single experiment that is completed in a matter of days. In a single issue of the journal *Nature* in November 2008 Bentley et al. described the sequencing of an individual of Nigerian ancestry, Wang et al. reported the DNA sequence of an Asian individual, and Ley et al. analyzed the genome sequence of a tumor sample. Together, these three papers involved the generation and analysis of 492 gigabases (Gb) of DNA sequence. By the end of 2008 the 1000 Genomes Project generated several terabases of data. For major sequencing centers (such as those at the Wellcome Trust Sanger Institute, Beijing Genomics Institute Shenzhen, the Broad Institute of MIT and Harvard, Washington University School of Medicine's Genome Sequencing Center, and Baylor College of Medicine's Human Genome Sequencing Center) it is estimated that each will generate approximately 10 terabases in the year 2009. According to a Wellcome Trust Sanger Institute press release in 2008, that center now produces as much sequence data every 2 minutes as was generated in the first five years at GenBank. Thus the amount of DNA sequence generated by next-generation sequencing technologies has already dwarfed the amount of sequence in GenBank. Such data are available through the Trace Archive at NCBI and the Ensembl Trace Server at EBI, including the Short Read Archive that was initiated in 2007.

Organisms in GenBank

Over 260,000 different species are represented in GenBank, with over 1000 new species added per month (Benson et al., 2009). The number of organisms represented in GenBank is shown in Table 2.1. We will define the bacteria, archaea, and eukaryotes in detail in Chapters 13 to 18. Briefly, eukaryotes have a nucleus and are often multicellular, whereas bacteria do not have a nucleus. Archaea are single-celled organisms, distinct from eukaryotes and bacteria, which constitute a third major branch of life. Viruses, which contain nucleic acids (DNA or RNA) but can only replicate in a host cell, exist at the borderline of the definition of living organisms.

We have seen so far that GenBank is very large and growing rapidly. From Table 2.1 we see that the organisms in GenBank consist mostly of eukaryotes. Of the microbes, there are currently over 25 times more bacteria than archaea represented in GenBank.

You can download all of the sequence data in GenBank at the website ► ftp://ftp.ncbi.nih.gov/genbank. For release 158.0 in February 2007, the total size of these files is about 250 gigabytes (250×10^9 bytes). By comparison, all the words in the United States Library of Congress add up to 20 terabytes (20×10^{12} bytes; 20 trillion bytes). And the particle accelerator used by physicists at CERN near Geneva (► http://public.web.cern.ch/Public/) collects petabytes of data each year (10^{15} bytes; 1 quadrillion bytes).

TABLE 2-1 Taxa Represented in GenBank				
Ranks:	Higher Taxa	Genus	Species	Lower Taxa	Total
Archaea	89	106	502	105	802
Bacteria	996	1,857	13,973	4,973	21,799
Eukaryota	15,205	45,066	167,764	13,200	241,235
Fungi	1,096	3,307	18,699	1,058	24,160
Metazoa	11,113	27,222	73,062	6,643	118,040
Viridiplantae	1,849	12,557	69,729	4,869	89,004
Viruses	445	294	5,054	33,909	39,702
All taxa	16,756	47,331	191,956	52,217	308,260

Source: From ► http://www.ncbi.nlm.nih.gov/Taxonomy/txstat.cgi (November 2008).

TABLE 2-2 Twenty Most Sequenced Organisms in GenBank

Entries	Bases	Species	Common Name
11,550,460	13,148,670,755	*Homo sapiens*	Human
7,255,650	8,361,230,436	*Mus musculus*	Mouse
1,757,685	6,060,823,765	*Rattus norvegicus*	Rat
2,086,880	5,235,078,866	*Bos taurus*	Cow
3,181,318	4,600,009,751	*Zea mays*	Corn
2,489,204	3,551,438,061	*Sus scrofa*	Pig
1,591,342	2,978,804,803	*Danio rerio*	Zebrafish
1,205,529	1,533,859,717	*Oryza sativa*	Rice
228,091	1,352,737,662	*Strongylocentrotus purpuratus*	Purple sea urchin
1,673,038	1,142,531,302	*Nicotiana tabacum*	Tobacco
1,413,112	1,088,892,839	*Xenopus (Silurana)*	Western clawed frog
212,967	996,533,885	*Pan troglodytes*	Chimpanzee
780,860	913,586,921	*Drosophila melanogaster*	Fruit fly
2,211,104	912,500,625	*Arabidopsis thaliana*	Thale cress
650,374	905,797,007	*Vitis vinifera*	Wine grape
804,246	871,336,795	*Gallus gallus*	Chicken
77,069	803,847,320	*Macaca mulatta*	Rhesus macaque
1,215,319	748,031,972	*Ciona intestinalis*	Sea squirt
1,224,224	744,373,069	*Canis lupus*	Dog
1,725,913	680,988,452	*Glycine max*	Soybean

Source: From ► ftp://ftp.nebinith.gov/genbank/gbrel.txt (GenBank release 168.0, October 2008).

The number of entries and bases of DNA/RNA for the 20 most sequenced organisms in GenBank is provided in Table 2.2 (excluding chloroplast and mitochondrial sequences). This list includes some of the most common model organisms that are studied in biology. Notably, the scientific community is studying a series of mammals (e.g., human, mouse, cow), other vertebrates (chicken, frog), and plants (corn, rice, bread wheat, wine grape). Different species are useful for a variety of different studies. Bacteria, archaea, and viruses are absent from the list in Table 2.2 because they have relatively small genomes.

We will discuss how genomes of various organisms are selected for complete sequencing in Chapter 13.

To help organize the available information, each sequence name in a GenBank record is followed by its data file division and primary accession number. (Accession numbers are defined below.) The following codes are used to designate the data file divisions:

The International Human Genome Sequencing Consortium adopted the Bermuda Principles in 1996, calling for the rapid release of raw genomic sequence data. You can read about recent versions of these principles at ► http://www.genome.gov/10506376.

1. PRI: primate sequences

2. ROD: rodent sequences

3. MAM: other mammalian sequences

4. VRT: other vertebrate sequences

5. INV: invertebrate sequences

6. PLN: plant, fungal, and algal sequences

7. BCT: bacterial sequences

8. VRL: viral sequences

9. PHG: bacteriophage sequences

The terms STS, GSS, EST, and HTGS are defined bellow.

10. SYN: synthetic sequences

11. UNA: unannotated sequences

12. EST: EST sequences (expressed sequence tags)

13. PAT: patent sequences

14. STS: STS sequences (sequence-tagged sites)

15. GSS: GSS sequences (genome survey sequences)

16. HTG: HTGS sequences (high throughput genomic sequences)

17. HTC: HTC sequences (high throughput cDNA sequences)

18. ENV: environmental sampling sequences

Beta globin is sometimes called hemoglobin-beta. In general, a gene does not always have the same name as the corresponding protein. Indeed there is no such thing as a "hemoglobin gene" because globin genes encode globin proteins, and the combination of these globins with heme forms the various types of hemoglobin. Often, multiple investigators study the same gene or protein and assign different names. The human genome organization (HUGO) Gene Nomenclature Committee (HGNC) has the critical task of assigning official names to genes and proteins. See ▶ http://www.gene.ucl.ac.uk/nomenclature/.

Types of Data in GenBank

There is an enormous number of molecular sequences in GenBank. We will next look at some of the basic kinds of data present in GenBank. Afterward, we will address strategies to extract the data you want from GenBank.

We start with an example. We want to find out the sequence of human beta globin. A fundamental distinction is that DNA, RNA-based, and protein sequences are stored in discrete databases. Furthermore, within each database, sequence data are represented in a variety of forms. For example, beta globin may be described at the DNA level (e.g. as a gene), at the RNA level (as a messenger RNA [mRNA] transcript), and at the protein level (see Fig. 2.2). Because RNA is relatively unstable, it is typically converted to complementary DNA (cDNA), and a variety

FIGURE 2.2. Types of sequence data in GenBank and other databases using human beta globin as an example. Note that "globin" may refer to a gene or other DNA feature, an RNA transcript (or its corresponding complementary DNA), or a protein. There are specialized databases corresponding to each of these three levels. See text for abbreviations. There are many other databases (not listed) that are not part of GenBank and NCBI; note that SwissProt, PDB, and PIR are protein databases that are independent of GenBank. The raw nucleotide sequence data in GenBank, DDBJ, and EBI are equivalent.

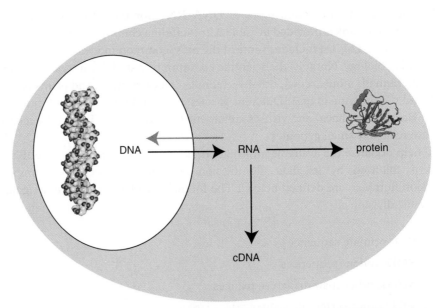

GenBank DNA databases containing beta globin data
non-redundant (nr)
dbGSS
dbHTGS
dbSTS

GenBank DNA databases, derived from RNA, containing beta globin data
Entrez Gene
dbEST
UniGene
Gene Expression Omnibus

Protein databases containing beta globin data
Entrez Protein
non-redundant (nr)
UniProt
Protein Data Bank
SCOP
CATH

of databases contain cDNA sequences corresponding to RNA transcripts. Thus for our example of beta globin, the various forms of sequence data include the following.

Genomic DNA Databases

- Beta globin is part of a chromosome. In the case of human RBP we will see that its gene is situated on chromosome 11 (Chapter 16, on the eukaryotic chromosome).

- Beta globin may be a part of a large fragment of DNA such as a cosmid, bacterial artificial chromosome (BAC), or yeast artificial chromosome (YAC) that may contain several genes. A BAC is a large segment of DNA (typically about 200,000 base pairs [bp], or 200 kilobases [kb]) that is cloned into bacteria. Similarly, YACs are used to clone large amounts of DNA into yeast. BACs and YACs are useful vectors with which to sequence large portions of genomes.

- Beta globin is present in databases as a gene. The gene is the functional unit of heredity (further defined in Chapter 16), and it is a DNA sequence that typically consists of regulatory regions, protein-coding exons, and introns. Often, human genes are 10 to 100 kb in size.

- Beta globin is present as a sequence-tagged site (STS)—that is, as a small fragment of DNA (typically 500 bp long) that is used to link genetic and physical maps and which is part of a database of sequence-tagged sites (dbSTS).

Human chromosome 11, which is a mid-sized chromosome, contains about 1800 genes and is about 134,000 kilobases (kb) in length.

cDNA Databases Corresponding to Expressed Genes

Beta globin is represented in databases as an expressed sequence tag (EST), that is, a cDNA sequence derived from a particular cDNA library. If one obtains a tissue such as liver, purifies RNA, then converts the RNA to the more stable form of cDNA, some of the cDNA clones contained in that cDNA are likely to encode beta globin.

In GenBank, the convention is to use the four DNA nucleotides when referring to DNA derived from RNA.

Expressed Sequence Tags (ESTs)

The database of expressed sequence tags (dbEST) is a division of GenBank that contains sequence data and other information on "single-pass" cDNA sequences from a number of organisms (Boguski et al., 1993). An EST is a partial DNA sequence of a cDNA clone. All cDNA clones, and thus all ESTs, are derived from some specific RNA source such as human brain or rat liver. The RNA is converted into a more stable form, cDNA, which may then be packaged into a cDNA library (refer to Fig. 2.2). ESTs are typically randomly selected cDNA clones that are sequenced on one strand (and thus may have a relatively high sequencing error rate). ESTs are often 300 to 800 bp in length. The earliest efforts to sequence ESTs resulted in the identification of many hundreds of genes that were novel at the time (Adams et al., 1991).

In November, 2008 GenBank had over 58,000,000 ESTs. We discuss ESTs further in Chapter 8.

Currently, GenBank divides ESTs into three major categories: human, mouse, and other. Table 2.3 shows the 10 organisms from which the greatest number of ESTs has been sequenced. Assuming that there are 22,000 human genes (see Chapter 19) and given that there are about 8.1 million human ESTs, there is currently an average of over 300 ESTs corresponding to each human gene.

TABLE 2-3 Top Ten Organisms for Which ESTs Have Been Sequenced

Organisms	Common Name	Number of ESTs
Homo sapiens	Human	8,138,094
Mus musculus + domesticus	Mouse	4,850,602
Zea mays	Maize	2,002,585
Arabidopsis thaliana	Thale cress	1,526,133
Bos taurus	Cattle	1,517,139
Sus scrofa	Pig	1,476,546
Danio rerio	Zebrafish	1,379,829
Glycine max	Soybean	1,351,356
Xenopus (Silurana) tropicalis	Western clawed frog	1,271,375
Oryza sativa	Rice	1,220,908

Many thousand of cDNA libraries have been generated from a variety of organism, and the total number of public entries is currently over 58 million.
Source: ▶ http://www.nebi.nlm.nin.gov/dbEST/dbEST_summary.html (dbEST release 022307, November 2008).

ESTs and UniGene

To find the entry for beta globin, go to ▶ http://www.ncbi.nlm.nih.gov, select All Databases then click UniGene, select human, then enter beta globin or HBB. The UniGene accession number is Hs.523443; note that Hs refers to *Homo sapiens*. To see the DNA sequence of a typical EST, click on an EST accession number from the UniGene page (e.g., AA970968.1), then follow the link to the GenBank entry in Entrez Nucleotide.

We are using beta globin as a specific example. If you want to type "globin" as a query, you will simply get more results from any database—in UniGene, you will find over 100 entries corresponding to a variety of globin genes in various species.

The UniGene project has become extremely important in the effort to identify protein-coding genes in newly sequenced genomes. We discuss this in Chapters 13 and 16.

The goal of the UniGene (unique gene) project is to create gene-oriented clusters by automatically partitioning ESTs into nonredundant sets. Ultimately there should be one UniGene cluster assigned to each gene of an organism. There may be as few as one EST in a cluster, reflecting a gene that is rarely expressed, to tens of thousands of ESTs, associated with a highly expressed gene. We discuss UniGene clusters further in Chapter 8 (on gene expression). There are over 100 organisms currently represented in UniGene, 71 of which are listed in Table 2.4.

For human beta globin, there is only a single UniGene entry. This entry currently has 2400 human ESTs that match the beta globin gene. This large number of ESTs reflects how abundantly the beta globin gene has been expressed in cDNA libraries that have been sequenced. A UniGene cluster is a database entry for a gene containing a group of corresponding ESTs (Fig. 2.3).

There are now thought to be approximately 22,000 human genes (see Chapter 19). One might expect an equal number of UniGene clusters. However, in practice, there are more UniGene clusters than there are genes—currently, there are about 120,000 human UniGene clusters. This discrepancy could occur for three reasons. (1) Clusters of ESTs could correspond to distinct regions of one gene. In that case there would be two (or more) UniGene entries corresponding to a single gene (see Fig. 2.3). Two UniGene clusters may properly cluster into one, and the number of UniGene clusters may collapse over time. (2) In the past several years it has become appreciated that much of the genome is transcribed at low levels (see Chapter 8). Currently, 40,000 human UniGene clusters consist of a single EST, and over 76,000 UniGene clusters consist of just one to four ESTs. These could reflect authentic genes that have not yet been appreciated by other means of gene identification. Alternatively they may represent rare transcription events of unknown biological relevance. (3) Some DNA may be transcribed during the creation of a cDNA library without corresponding to an authentic transcript. Thus it is a cloning artifact. We discuss the criteria for defining a eukaryotic gene in Chapter 16. Alternative splicing (Chapter 8) may introduce apparently new clusters of genes because the spliced exon is not homologous to the rest of the sequence.

TABLE 2-4   Seventy-One Organisms Represented in UniGene

Group	No.	Species
Chordata: Mammalia	12	*Bos taurus* (cattle), *Canis familiaris* (dog), *Equus caballus* (horse), *Homo sapiens* (human), *Macaca fascicularis* (crab-eating macaque), *Macaca mulatta* (rhesus monkey), *Mus musculus* (mouse), *Oryctolagus cuniculus* (rabbit), *Ovis aries* (sheep), *Rattus norvegicus* (Norway rat), *Sus scrofa* (pig), *Trichosurus vulpecula* (silver-gray brushtail possum)
Chordata: Actinopterygii	8	*Danio rerio* (zebrafish), *Fundulus heteroclitus* (killifish), *Gasterosteus aculeatus* (three spined stickleback), *Oncorhynchus mykiss* (rainbow trout), *Oryzias latipes* (Japanese medaka), *Pimephales promelas* (fathead minnow), *Salmo salar* (Atlantic salmon), *Takifugu rubripes* (pufferfish)
Chordata: Amphibia	2	*Xenopus laevis* (African clawed frog), *Xenopus tropicalis* (western clawed frog)
Chordata: Ascidiacea	3	*Ciona intestinalis*, *Ciona savignyi*, *Molgula tectiformis*
Chordata: Aves	2	*Gallus gallus* (chicken), *Taeniopygia guttata* (zebra finch)
Chordata: Cephalochordata	1	*Branchiostoma floridae* (Florida lancelet)
Chordata: Hyperoartia	1	*Petromyzon marinus* (sea lamprey)
Echinodermata: Echinoidea	1	*Strongylocentrotus purpuratus* (purple sea urchin)
Arthropoda: Insecta	6	*Aedes aegypti* (yellow fever mosquito), *Anopheles gambiae* (African malaria mosquito), *Apis mellifera* (honey bee), *Bombyx mori* (domestic silkworm), *Drosophila melanogaster* (fruit fly), *Tribolium castaneum* (red flour beetle)
Nematoda: Chromadorea	1	*Caenorhabditis elegans* (nematode)
Platyhelminthes: Trematoda	2	*Schistosoma japonicum*, *Schistosoma mansoni*
Cnidaria: Hydrozoa	1	*Hydra magnipapillata*
Streptophyta: Bryopsida	1	*Physcomitrella patens*
Streptophyta: Coniferopsida	3	*Picea glauca* (white spruce), *Picea sitchensis* (Sitka spruce), *Pinus taeda* (loblolly pine)
Streptophyta: Eudicotyledons	18	*Aquilegia formosa* × *Aquilegia pubescens*, *Arabidopsis thaliana* (thale cress), *Brassica napus* (rape), *Citrus sinensis* (Valencia orange), *Glycine max* (soybean), *Gossypium hirsutum* (upland cotton), *Gossypium raimondii*, *Helianthus annuus* (sunflower), *Lactuca sativa* (garden lettuce), *Lotus japonicus*, *Malus* × *domestica* (apple), *Medicago truncatula* (barrel medic), *Nicotiana tabacum* (tobacco), *Populus tremula* × *Populus tremuloides*, *Populus trichocarpa* (western balsam poplar), *Solanum lycopersicum* (tomato), *Solanum tuberosum* (potato), *Vitis vinifera* (wine grape)
Streptophyta: Liliopsida	6	*Hordeum vulgare* (barley), *Oryza sativa* (rice), *Saccharum officinarum* (sugarcane), *Sorghum bicolor* (sorghum), *Triticum aestivum* (wheat), *Zea mays* (maize)
Chlorophyta: Chlorophyceae	1	*Chlamydomonas reinhardtii*
Dictyosteliida: Dictyostelium	1	*Dictyostelium discoideum* (slime mold)
Apicomplexa: Coccidia	1	*Toxoplasma gondii*

Source: UniGene ▶ http://www.ncbi.nlm.nih.gov/entrez/query.fcgi?db=unigene (November 2008).

FIGURE 2.3. Schematic description of UniGene clusters. Expressed sequence tags (ESTs) are mapped to a particular gene and to each other. The number of ESTs that constitute a UniGene cluster ranges from 1 to tens of thousands; on average there are 300 human ESTs per cluster. Sometimes, as shown in the diagram, separate UniGene clusters correspond to distinct regions of a gene. Eventually, as genome sequencing increases our ability to define and annotate full-length genes, these two UniGene clusters would be collapsed into one single cluster. Ultimately, the number of UniGene clusters should equal the number of genes in the genome.

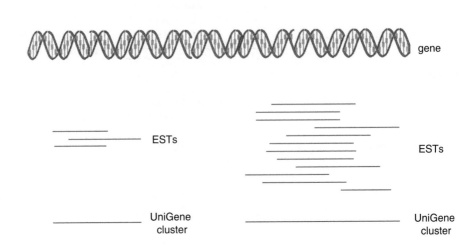

As of November 2008 there are 1.3 million STSs, derived from 300 organisms.

Sequence-Tagged Sites (STSs)

The dbSTS is an NCBI site containing STSs, which are short genomic landmark sequences for which both DNA sequence data and mapping data are available (Olson et al., 1989). STSs have been obtained from several hundred organisms, including primates and rodents (Table 2.5). A typical STS is approximately the size of an EST. Because they are sometimes polymorphic, containing short sequence repeats (Chapter 16), STSs can be useful for mapping studies.

Genome Survey Sequences (GSSs)

There are currently 24 million GSS entries from over 800 organisms (November 2008). The top four organisms (Table 2.6) account for about a third of all entries. This database is accessed via ▶ http://www.ncbi.nlm.nih.gov/projects/dbGSS/.

The GSS division of GenBank is similar to the EST division, except that its sequences are genomic in origin, rather than cDNA (mRNA). The GSS division contains the following types of data (see Chapters 13 and 16):

- Random "single-pass read" genome survey sequences
- Cosmid/BAC/YAC end sequences
- Exon-trapped genomic sequences
- The *Alu* polymerase chain reaction (PCR) sequences

TABLE 2-5   Organisms from Which STSs Have Been Obtained

Organism	Approximate Number of STSs
Homo sapiens	324,000
Pan troglodytes	161,000
Macaca mulatta	72,000
Mus musculus	56,000
Rattus norvegicus	50,000

These are the organisms with the most UniSTS entries.
Source: ▶ http://www.ncbi.nlm.nih.gov/genome/sts/unists_stats.html (November 2008).

TABLE 2-6 Selected Organisms from Which GSSs Have Been Obtained. For a discussion of Metagenomes see Chapter 13

Organism	Approximate Number of Sequences
Marine metagenome	2,643,000
Zea mays + subsp. *mays* (maize)	2,091,000
Mus musculus + *domesticus* (mouse)	1,864,000
Nicotiana tabacum (tobacco)	1,421,000
Homo sapiens (human)	1,214,000
Canis lupus familiaris (dog)	854,000

Source: ► http://www.ncbi.nlm.nih.gov/dbGSS/dbGSS_summary.html (November 2008).

All searches of the Entrez Nucleotide database provide results that are divided into three sections: GSS, ESTs, and "CoreNucleotide" (that is, the remaining nucleotide sequences). Recent holdings of the GSS database are listed in Table 2.6.

High Throughput Genomic Sequence (HTGS)

The HTGS division was created to make "unfinished" genomic sequence data rapidly available to the scientific community. It was done in a coordinated effort between the three international nucleotide sequence databases: DDBJ, EMBL, and GenBank. The HTGS division contains unfinished DNA sequences generated by the high throughput sequencing centers.

The HTGS home page is ► http://www.ncbi.nlm.nih.gov/HTGS/ and its sequences can be searched via BLAST (see Chapters 4 and 5).

Protein Databases

The name beta globin may refer to the DNA, the RNA, or the protein. As a protein, beta globin is present in databases such as the nonredundant (nr) database of GenBank (Benson et al., 2009), the SwissProt database (Boeckmann et al., 2003), UniProt (UniProt Consortium 2007), and the Protein Data Bank (Kouranov et al., 2006).

We have described some of the basic kinds of sequence data in GenBank. We will next turn our attention to Entrez and the other programs in NCBI and elsewhere, which allow you to access GenBank, EMBL, and DDBJ data and related literature information. In particular, we will introduce the NCBI website, one of the main web-based resources in the field of bioinformatics.

NATIONAL CENTER FOR BIOTECHNOLOGY INFORMATION

Introduction to NCBI: Home Page

The NCBI creates public databases, conducts research in computational biology, develops software tools for analyzing genome data, and disseminates biomedical information (Wheeler et al., 2007). The NCBI home page is shown in Fig. 2.4. Across the top bar of the website, there are seven categories: PubMed, Entrez, BLAST, OMIM, Books, Taxonomy, and Structure.

PubMed

PubMed is the search service from the National Library of Medicine (NLM) that provides access to over 18 million citations in MEDLINE (Medical Literature,

Extremely useful tutorials are available for Entrez, PubMed, and other NCBI resources at ► http://www.ncbi.nlm.nih.gov/Education/. You can also access this from the education link on the NCBI home page (► http://www.ncbi.nlm.nih.gov).

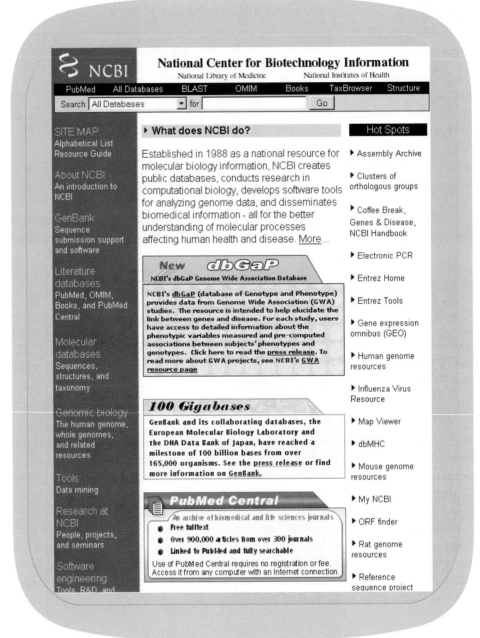

FIGURE 2.4. *The main page of the National Center for Biotechnology Information (NCBI) website (▶ http://www.ncbi.nlm.nih. gov). Across the top bar, sections include PubMed, Entrez and Books (described in this chapter), BLAST (Chapters 3–5), Taxonomy (Chapters 13–19), Structure (Chapter 11), and Online Mendelian Inheritance in Man (OMIM, Chapter 20). Note that the left sidebar includes tutorials within the Education section.*

Analysis, and Retrieval System Online) and other related databases, with links to participating online journals.

Entrez

Entrez integrates the scientific literature, DNA and protein sequence databases, three-dimensional protein structure data, population study data sets, and assemblies of complete genomes into a tightly coupled system. PubMed is the literature component of Entrez.

BLAST

BLAST (Basic Local Alignment Search Tool) is NCBI's sequence similarity search tool designed to support analysis of nucleotide and protein databases (Altschul et al., 1990, 1997). BLAST is a set of similarity search programs designed to explore all of the available sequence databases regardless of whether the query is protein or DNA. We explore BLAST in Chapters 3 to 5.

OMIM

Online Mendelian Inheritance in Man (OMIM) is a catalog of human genes and genetic disorders. It was created by Victor McKusick and his colleagues and developed for the World Wide Web by NCBI (Hamosh et al., 2005). The database contains detailed reference information. It also contains links to PubMed articles and sequence information. We describe OMIM in Chapter 20 (on human disease).

Books

NCBI offers several dozen books online. These books are searchable, and are linked to PubMed.

Taxonomy

The NCBI taxonomy website includes a taxonomy browser for the major divisions of living organisms (archaea, bacteria, eukaryota, and viruses). The site features taxonomy information such as genetic codes and taxonomy resources and additional information such as molecular data on extinct organisms and recent changes to classification schemes. We will visit this site in Chapters 7 (on evolution) and 13 to 18 (on genomes and the tree of life).

Structure

The NCBI structure site maintains the Molecular Modelling Database (MMDB), a database of macromolecular three-dimensional structures, as well as tools for their visualization and comparative analysis. MMDB contains experimentally determined biopolymer structures obtained from the Protein Data Bank (PDB). Structure resources at NCBI include PDBeast (a taxonomy site within MMDB), Cn3D (a three-dimensional structure viewer), and a vector alignment search tool (VAST) which allows comparison of structures. (See Chapter 11, on protein structure.)

The Protein Data Bank (▶ http://www.rcsb.org/pdb/) is the single worldwide repository for the processing and distribution of biological macromolecular structure data. We explore the PDB in Chapter 11.

THE EUROPEAN BIOINFORMATICS INSTITUTE (EBI)

The EBI website is comparable to NCBI in its scope and mission, and it represents a complementary, independent resource. EBI features six core molecular databases (Brooksbank et al., 2003), as follows. (1) EMBL-Bank is the repository of DNA and RNA sequences that is complementary to GenBank and DDBJ (Kulikova et al., 2007). (2) SWISS-PROT and (3) TrEMBL are two protein databases that are described further below. (4) MSD is a protein structure database (see Chapter 11). (5) Ensembl is one of the three main genome browsers (described below). (6) ArrayExpress is one of the two main worldwide repositories for gene expression

You can access EBI at ▶EBI at http://www.ebi.ac.uk/.

data, along with the Gene Expression Omnibus at NCBI; both are described in Chapter 8.

Throughout this book we will focus on both the NCBI and EBI websites. In many cases those sites begin with similar raw data and then provide distinct ways of organizing, analyzing, and displaying data across a broad range of bioinformatics applications. When you work on a problem, such as studying the structure or function of a particular gene, it is often helpful to explore the wealth of resources on both these sites. For example, each offers expert functional annotation of particular sequences and expert curation of the database. The NCBI and EBI websites increasingly offer an integration of their database resources so that one can link to information between the two sites with reasonable effort.

ACCESS TO INFORMATION: ACCESSION NUMBERS TO LABEL AND IDENTIFY SEQUENCES

When you have a problem you are studying that involves any gene or protein, it is likely that you will need to find information about some database entries. You may begin your research problem with information obtained from the literature or you may have the name of a specific sequence of interest. Perhaps you have raw amino acid and/or nucleotide sequence data; we will explore how to analyze these (e.g. Chapters 3 to 5). The problem we will address now is how to extract information about your gene or protein of interest from databases.

An essential feature of DNA and protein sequence records is that they are tagged with accession numbers. An accession number is a string of about 4 to 12 numbers and/or alphabetic characters that are associated with a molecular sequence record. An accession number may also label other entries, such as protein structures or the results of a gene expression experiment (Chapters 8 and 9). Accession numbers from molecules in different databases have characteristic formats (Box 2.1). These formats vary because each database employs its own system. As you explore databases from which you extract DNA and protein data, try to become familiar with the different formats for accession numbers. Some of the various databases (Fig. 2.2) employ accession numbers that tell you whether the entry contains nucleotide or protein data.

For a typical molecule such as beta globin there are thousands of accession numbers (Fig. 2.5). Many of these correspond to ESTs and other fragments of DNA that match beta globin. How can you assess the quality of sequence or protein data? Some sequences are full-length, while others are partial. Some reflect naturally occurring variants such as single nucleotide polymorphisms (SNPs; Chapter 16) or alternatively spliced transcripts (Chapter 8). Many of the sequence entries contain errors, particularly in the ends of EST reads. When we compare beta globin sequences derived from mRNA and from genomic DNA, we may expect them to match perfectly (or nearly so), but as we will see, discrepancies routinely occur.

In addition to accession numbers, NCBI also assigns unique sequence identification numbers that apply to the individual sequences within a record. GI numbers are assigned consecutively to each sequence that is processed. For example, the human beta globin DNA sequence associated with the accession number NM_000518.4 has a gene identifier GI:28302128. The suffix .4 on the accession number refers to a version number; NM_000518.3 has a different gene identifier, GI: 13788565.

DNA is usually sequenced on both strands. However, ESTs are often sequenced on one strand only, and thus they have a high error rate. We will discuss sequencing error rates in Chapter 13.

BOX 2-1
Types of Accession Numbers

Type of Record	Sample Accession Format
GenBank/EMBL/DDBJ nucleotide sequence records	One letter followed by five digits, e.g., X02775
	Two letters followed by six digits, e.g., AF025334
GenPept sequence records (which contain the amino acid translations from GenBank/EMBL/DDBJ records that have a coding region feature annotated on them)	Three letters and five digits, e.g., AAA12345
Protein sequence records from SwissProt and PIR	Usually one letter and five digits, e.g., P12345. SwissProt numbers may also be a mixture of numbers and letters.
Protein sequence records from the Protein Research Foundation	A series of digits (often six or seven) followed by a letter, e.g., 1901178A
RefSeq nucleotide sequence records	Two letters, an underscore bar, and six or more digits, e.g., mRNA records (NM_*): NM_006744; genomic DNA contigs (NT_*): NT_008769
RefSeq protein sequence records	Two letters (NP), an underscore bar, and six or more digits, e.g., NP_006735
Protein structure records	PDB accessions generally contain one digit followed by three letters, e.g., 1TUP. They may contain other mixtures of numbers and letters (or numbers only). MMDB ID numbers generally contain four digits, e.g., 3973.

The Reference Sequence (RefSeq) Project

One of the most important recent developments in the management of molecular sequences is RefSeq. The goal of RefSeq is to provide the best representative sequence for each normal (i.e., nonmutated) transcript produced by a gene and for each normal protein product (Pruitt et al., 2009; Maglott et al., 2000). There may be hundreds of GenBank accession numbers corresponding to a gene, since GenBank is an archival database that is often highly redundant. However, there will be only one RefSeq entry corresponding to a given gene or gene product, or several RefSeq entries if there are splice variants or distinct loci.

To see and compare the three myoglobin RefSeq entries at the DNA and the protein levels, visit ► http://www.bioinfbook.org/ chapter2 and select webdocument 2.1.

Consider human myoglobin as an example. There are three RefSeq entries (NM_005368, NM_203377, and NM_203378), each corresponding to a distinct splice variant. Each splice variant involves the transcription of different exons from a single gene locus. In this example, all three transcripts happen to encode an identical protein having the same amino acid sequence. Because the source of the transcript varies distinctly, each identical protein sequence is assigned its own protein accession number (NP_005359, NP_976311, and NP_976312, respectively).

Allelic variants, such as single base mutations in a gene, are not assigned different RefSeq accession numbers. However, OMIM and dbSNP (Chapters 16 and 20) do catalog allelic variants.

FIGURE 2.5. There are thousands of accession numbers corresponding to many genes and proteins. A search with the query "beta globin" from the main page of NCBI shows the results across the databases of the Entrez search engine. There are over 1000 each of core nucleotide sequences, expressed sequence tags (ESTs), and proteins. The RefSeq project is particularly important in trying to provide the best representative sequence of each normal (nonmutated) transcript produced by a gene and of each distinct, normal protein sequence.

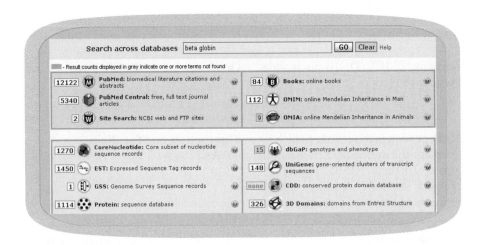

A GenBank or RefSeq accession number refers to the most recent version of a given sequence. For example NM_000558.3 is currently a RefSeq identifier for human alpha globin. The suffix ".3" is the version number. By default, if you do not specify a version number then the most recent version is provided. Try doing an Entrez nucleotide search for NM_000558.1 and you can learn about the revision history of that accession number. In Chapter 3 we will learn how to compare two sequences; you can blast NM_000558.1 against NM_000558.3 to see the differences, or view the results in web document 2.2 at ▶ http://www.bioinfbook.org/chapter2.

RefSeq entries are curated by the staff at NCBI, and are nearly nonredundant. However, there can be two proteins encoded by distinct genes sharing 100% amino acid identity. Each is assigned its own unique RefSeq identifier. For example, the alpha-1 globin and alpha-2 globin genes in human are physically separate genes that encode proteins with identical sequences. The encoded alpha-1 globin and alpha-2 globin proteins are assigned the RefSeq identifiers NP_000549 and NP_000508.

Refseq entries have different status levels (predicted, provisional, and reviewed), but in each case the RefSeq entry is intended to unify the sequence records. You can recognize a RefSeq accession by its format, such as NP_000509 (P stands for beta globin protein) or NM_006744 (for beta globin mRNA). A variety of RefSeq identifiers are shown in Table 2.7, and examples of beta globin identifiers are given in Table 2.8.

TABLE 2-7 Formats of Accession Numbers for RefSeq Entries

Molecule	Accession Format	Genome
Complete genome	NC_123456	Complete genomic molecules, including genomes, chromosomes, organelles, and plasmids
Genomic DNA	NW_123456 NW_123456789	Intermediate genomic assemblies
Genomic DNA	NZ_ABCD12345678	Collection of whole genome shotgun sequence data
Genomic DNA	NT_123456	Intermediate genomic assemblies (BAC and/or WGS sequence data)
mRNA	NM_123456 or NM_123456789	Transcript products; mature mRNA protein-coding transcripts
Protein	NP_123456 or NM_123456789	Protein products (primarily full-length)
RNA	NR_123456	Noncoding transcripts (e.g. structural RNAs, transcribed pseudogenes)

There are currently 21 different RefSeq accession formats. The methods include expert manual curation, automated curation, or a combination. Abbreviations: BAC, bacterial artificial chromosome; WGS, whole genome shotgun (see Chapter 13).

Source: Adapted from ▶ http://www.ncbi.nlm.nih.gov/RefSeq/key.html#accessions (March 2007).

TABLE 2-8	RefSeq Accession Numbers Corresponding to Human Beta Globin		
Category	Accession	Size	Description
DNA	NC_000011	134,452,384 bp	Genomic contig
DNA	NM_000518.4	626 bp	DNA corresponding to mRNA
DNA	NG_000007.3	81,706 bp	Genomic reference
DNA	NW_925006.1	1,606 bp	Alternate assembly
Protein	NP_000509.1	147 amino acids	Protein

The Consensus Coding Sequence (CCDS) Project

The Consensus Coding Sequence (CCDS) project was established to identify a core set of protein coding sequences that provide a basis for a standard set of gene annotations. The CCDS project is a collaboration between four groups (EBI, NCBI, the Wellcome Trust Sanger Institute, and the University of California, Santa Cruz [UCSC]). Currently, the CCDS project has been applied to the human and mouse genomes, and thus its scope is considerably more limited than RefSeq.

You can learn about the CCDS project at ▶ http://www.ncbi.nlm.nih.gov/projects/CCDS/.

ACCESS TO INFORMATION VIA ENTREZ GENE AT NCBI

How can one navigate through the bewildering number of protein and DNA sequences in the various databases? An emerging feature is that the various databases are increasingly interconnected, providing a variety of convenient links to each other and to algorithms that are useful for DNA, RNA, and protein analysis. Entrez Gene (formerly LocusLink) is particularly useful as a major portal. It is a curated database containing descriptive information about genetic loci (Maglott et al., 2007). You can obtain information on official nomenclature, aliases, sequence accessions, phenotypes, EC numbers, OMIM numbers, UniGene clusters, HomoloGene (a database that reports eukaryotic orthologs), map locations, and related websites.

Entrez Gene is accessed from the main NCBI web page (by clicking All Databases). Currently (November 2008), Entrez Gene encompasses about 5,700 taxa and 4.6 million genes. We will explore many of the resources within Entrez Gene in later chapters.

To illustrate the use of Entrez Gene we will search for human myoglobin. The result of entering an Entrez Gene search is shown in Fig. 2.6. Note that in performing this search, it can be convenient to restrict the search to a particular organism of interest. (This can be done using the "limits" tab on the Entrez Gene page.) The "Links" button (Fig. 2.6, top right) provides access to various other database entries on myoglobin. Clicking on the main link to the human myoglobin entry results in the following information (Fig. 2.7):

- At the top right, there is a table of contents for the Entrez Gene myoglobin entry. Below it are further links to myoglobin entries in NCBI databases (e.g. protein and nucleotide databases and PubMed), as well as external databases (e.g. Ensembl and UCSC; see below and Chapter 16).

- Entrez Gene provides the official symbol and name for human myoglobin, MB.

- A schematic overview of the gene structure is provided, hyperlinked to the Map Viewer (see below).

- There is a brief description of the function of MB, defining it as a carrier protein of the globin family.

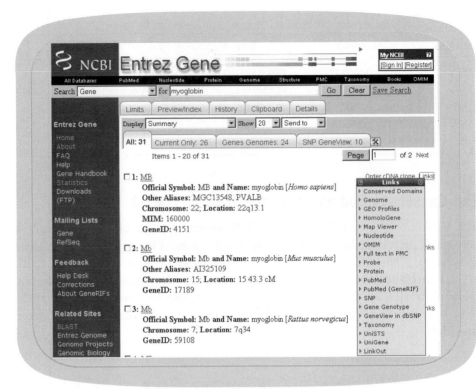

FIGURE 2.6. *Result of a search for "myoglobin" in Entrez Gene. Information is provided for a variety of organisms, including Homo sapiens, Mus musculus, and Rattus norvegicus. The links button (top right) provides access to information on myoglobin from a variety of other databases.*

FIGURE 2.7. *Portion of the Entrez Gene entry for human myoglobin. Information is provided on the gene structure, chromosomal location, as well as a summary of the protein's function. RefSeq accession numbers are also provided (not shown); you can access them by clicking "Reference sequences" in the table of contents (top right). The menu (right sidebar) provides extensive links to additional databases, including PubMed, OMIM, UniGene, a variation database (dbSNP), HomoloGene (with information on homologs), a gene ontology database, and Ensembl viewers at EBI. We will describe these resources in later chapters.*

- The Reference Sequence (RefSeq) accession numbers are provided: NM_005368 for the DNA sequence encoding the longest myoglobin transcript and NP_005359 for the protein entry. GenBank accession numbers corresponding to myoglobin (both nucleotide and protein) are also provided.

Figure 2.8 shows the standard, default form of a typical Entrez Protein record (for myoglobin). It is simple to obtain a variety of formats by changing the Entrez display options. By using the Display pulldown menu (Fig. 2.8a) one can obtain

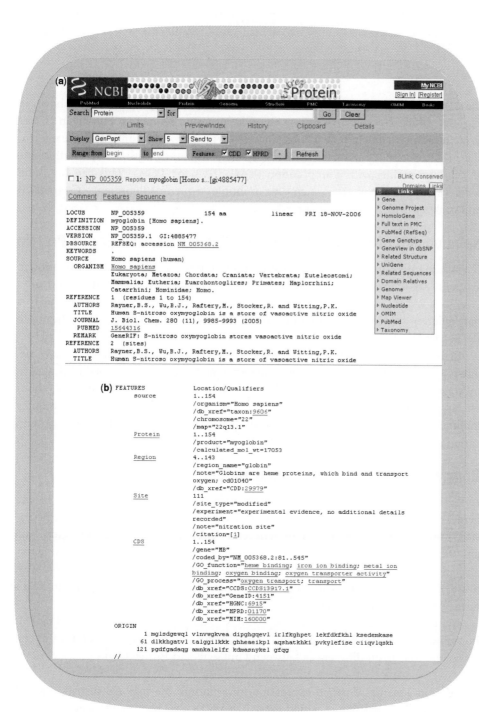

FIGURE 2.8. *Display of an Entrez Protein record for human myoglobin. This is a typical entry for any protein. (a) Top portion of the record. Key information includes the length of the protein (154 amino acids), the division (PRI, or primate), the accession number (NP_005359), the organism (H. sapiens), literature references, comments on the function of globins, and many links to other databases (right side). At the top of the page, the display option allows you to obtain this record in a variety of formats, such as FASTA (Figure 2.9). (b) Bottom portion of the record. This includes features such as the coding sequence (CDS). The amino acid sequence is provided at the bottom in the single letter amino acid code.*

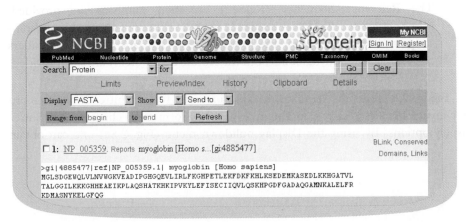

FIGURE 2.9. The protein entry for human myogobin can be displayed in the FASTA format. This is easily accomplished by adjusting the "Display" pull-down menu from an Entrez protein record. The FASTA format is used in a variety of software programs that we will use in later chapters.

FASTA is both an alignment program (described in Chapter 3) and a commonly used sequence format (further described in Chapter 4).

the commonly used FASTA format for protein (or DNA) sequences, as shown in Fig. 2.9. Note also that by clicking the CDS (coding sequence) link of an Entrez Protein or Entrez Nucleotide record (shown in Fig. 2.8b), you can obtain the nucleotides that encode a particular protein, typically beginning with a start methionine (ATG) and ending with a stop codon (TAG, TAA, or TGA). This can be useful for a variety of applications including multiple sequence alignment (Chapter 6) and molecular phylogeny (Chapter 7).

Relationship of Entrez Gene, Entrez Nucleotide, and Entrez Protein

If you are interested in obtaining information about a particular DNA or protein sequence, it is reasonable to visit Entrez Nucleotide or Entrez Protein and do a search. A variety of search strategies are available, such as limiting the output to a particular organism or taxonomic group of interest, or limiting the output to RefSeq entries.

There are also many advantages to beginning your search through Entrez Gene. There, you can identify the official gene name, and you can be assured of the chromosomal location of the gene (thus providing unambigous information about which particular gene you are studying). Furthermore, each Entrez Gene entry includes a section of reference sequences that provides all the DNA and protein variants that are assigned RefSeq accession numbers.

Comparison of Entrez Gene and UniGene

As described above, the UniGene project assigns one cluster of sequences to one gene. For example, for *RBP4* there is one UniGene entry with the UniGene accession number Hs.50223. This UniGene entry includes a list of all the GenBank entries, including ESTs, that correspond to the *RBP4* gene. The UniGene entry also includes mapping information, homologies, and expression information (i.e., a list of the tissues from which cDNA libraries were generated that contain ESTs corresponding to the RBP gene).

Entrez Gene now has about 40,000 human gene entries (as of November 2008).

UniGene and Entrez Gene have features in common, such as links to OMIM, homologs, and mapping information. They both show RefSeq accession numbers. There are four main differences between UniGene and Entrez Gene:

1. UniGene has detailed expression information; the regional distributions of cDNA libraries from which particular ESTs have been sequenced are listed.

2. UniGene lists ESTs corresponding to a gene, allowing one to study them in detail.

3. Entrez Gene may provide a more stable description of a particular gene; as described above, UniGene entries may be collapsed as genome-sequencing efforts proceed.

4. Entrez Gene has fewer entries than UniGene, but these entries are better curated.

Entrez Gene and HomoloGene

The HomoloGene database provides groups of annotated proteins from a set of completely sequenced eukaryotic genomes. Proteins are compared (by blastp; see Chapter 4), placed in groups of homologs, and then the protein alignments are matched to the corresponding DNA sequences. This allows distance metrics to be calculated such as Ka/Ks, the ratio of nonsynonymous to synonymous mutations (see Chapter 7). You can find a HomoloGene entry for a gene/protein of interest by following a link on the Entrez Gene page.

HomoloGene is available by clicking All Databases from the NCBI home page, or at ▶ http://www.ncbi.nlm.nih.gov/entrez/query.fcgi?db=homologene. Release 53 (March 2007) has over 170,000 groups. We will define homologs in Chapter 3.

A search of HomoloGene with the term hemoglobin results in dozens of matches for myoglobin, alpha globin, and beta globin. By clicking on the beta globin group one gains access to a list of proteins with RefSeq accession numbers from human, chimpanzee, dog, mouse, and chicken. The pairwise alignment scores (see Chapter 3) are summarized and linked to, and the sequences can be displayed as a multiple sequence alignment (Chapter 6), or in the FASTA format.

ACCESS TO INFORMATION: PROTEIN DATABASES

In many cases you are interested in obtaining protein sequences. The Entrez Protein database at NCBI consists of translated coding regions from GenBank as well as sequences from external databases (the Protein Information Resource [PIR], SWISS-PROT, Protein Research Foundation [PRF], and the Protein Data Bank [PDB]). The EBI also provides information on proteins via these major databases. We will next explore ways to obtain protein data through UniProt, an authoritative and comprehensive protein database.

EBI offers access to over a dozen different protein databases, listed at ▶ http://www.ebi.ac.uk/Databases/protein.html.

UniProt

The Universal Protein Resource (UniProt) is the most comprehensive, centralized protein sequence catalog (UniProt Consortium, 2009). Formed as a collaborative effort in 2002, it consists of a combination of three key databases. (1) Swiss-Prot is considered the best-annotated protein database, with descriptions of protein structure and function added by expert curators. (2) The translated EMBL (TrEMBL) Nucleotide Sequence Database Library provides automated (rather than manual) annotations of proteins not in Swiss-Prot. It was created because of the vast number of protein sequences that have become available through genome sequencing projects. (3) PIR maintains the Protein Sequence Database, another protein database curated by experts.

The European Bioinformatics Institute (EBI) in Hinxton and the Swiss Institute of Bioinformatics (SIB) in Geneva created Swiss-Prot and TrEMBL. PIR is a division of the National Biomedical Research Foundation (▶ http://pir.georgetown.edu/) in Washington, D.C. PIR was founded by Margaret Dayhoff, whose work is described in Chapter 3. The UniProt web site is ▶ http://www.uniprot.org. It contains over 7 million entries (release 14.4, November 2008).

UniProt is organized in three database layers. (1) The UniProt Knowledgebase (UniProtKB) is the central database that is divided into the manually annotated UniProtKB/Swiss-Prot and the computationally annotated UniProtKB/TrEMBL.

To access UniProt from EBI, visit ▶ http://www.ebi.ac.uk/uniprot/. To access UniProt from ExPASy, visit ▶ http://www.expasy.org/sprot/.

(2) The UniProt Reference Clusters (UniRef) offer nonredundant reference clusters based on UniProtKB. UniRef clusters are available with members sharing at least 50%, 90%, or 100% identity. (3) The UniProt Archive, UniParc, consists of a stable, nonredundant archive of protein sequences from a wide variety of sources (including model organism databases, patent offices, RefSeq, and Ensembl).

You can access UniProt directly from its website, or from EBI or ExPASy.

The Sequence Retrieval System at ExPASy

One of the most useful resources available to obtain protein sequences and associated data is provided by ExPASy, the Expert Protein Analysis System. The ExPASy server is a major resource for proteomics-related analysis tools, software, and databases. In addition to providing access to the UniProt database, ExPASy serves as a portal for the Sequence Retrieval System (SRS). The query page has four rectangular boxes (Fig. 2.10). Each has an associated pull-down menu, and as a default condition each says "AllText." In the first box, type "retinol-binding." (Note that queries should consist of one word.) In the second box, type "human," change the corresponding pull-down menu to "organism," then click "do query." You see 10 entries listed. Click the link in which we are interested (SWISS_PROT: RETB_HUMAN P02753).

ExPASy is a proteomics server of the Swiss Institute of Bioinformatics (▶ http://www.expasy.ch/), another portal from which the Sequence Retrieval System (SRS) is accessed. From ▶ http://www.expasy.ch/srs5/, click "Start a new SRS session," then click "continue." SRS was created by Lion Biosciences, and a list of several dozen publicly available SRS servers is at ▶ http://downloads.lionbio.co.uk/publicsrs.html.

An output consists of a SwissProt record. This provides very useful, well-organized information, including alternative names and accession numbers; literature links; functional data and information about cellular localization; links to GenBank and other database records for both the RBP protein and gene; and links to many databases such as OMIM, InterPro, Pfam, Prints, GeneCards, PROSITE, and two-dimensional protein gel databases. We will describe these resources later (Chapters 6 and 10). The record includes features; note that by clicking on any of the linked features, you can see the protein sequence with that feature highlighted in color. While we have mentioned several key ways to acquire sequence data, there are dozens of other useful servers. As an example, the Protein Information Resource (PIR) provides access to sequences (Wu et al., 2002). PIR is especially useful for its efforts to annotate functional information on proteins.

FIGURE 2.10. Format of a query at the Sequence Retrieval System (SRS) of the Expert Protein Analysis System (ExPASy) (▶ http://www.expasy.ch/srs5/). This website provides one of the most useful resources for protein analysis. You can also access the SRS through other sites such as the European Bioinformatics Institute (▶ http://srs6.ebi.ac.uk/).

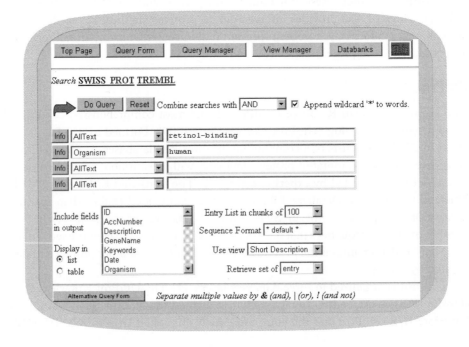

ACCESS TO INFORMATION: THE THREE MAIN GENOME BROWSERS

Genome browsers are databases with a graphical interface that presents a representation of sequence information and other data as a function of position across the chromosomes. We will focus on viral, prokaryotic, and eukaryotic chromosomes in Chapters 14 to 19. Genome browsers have emerged as an essential tool for organizing information about genomes. We will now briefly introduce the three principal genome browsers and describe how they may be used to acquire information about a gene or protein of interest.

Genomes are analyzed over time in assemblies (see Chapter 13). The main human genome browsers share the same underlying assemblies, and differ in the ways they annotate and present information. NCBI Build 36 (November, 2005) is an example of a human assembly.

The Map Viewer at NCBI

The NCBI Map Viewer includes chromosomal maps (both physical maps and genetic maps; see Chapter 16) for a variety of organisms, including metazoans (animals), fungi, and plants. Map Viewer allows text-based queries (e.g., "beta globin") or sequence-based queries (e.g., BLAST; see Chapter 4). For each genome, four levels of detail are available: (1) the home page of an organism; (2) the genome view, showing ideograms (representations of the chromosomes); (3) the map view, allowing you to view regions at various levels of resolution; and (4) the sequence view, displaying sequence data as well as annotation of interest such as the location of genes.

The Map Viewer is accessed from the main page of NCBI or via ► http://www.ncbi.nlm.nih.gov/mapview/. Records in Entrez Gene, Entrez Nucleotide, and Entrez Protein also provide direct links to the Map Viewer.

The University of California, Santa Cruz (UCSC) Genome Browser

The UCSC browser currently supports the analysis of three dozen vertebrate and invertebrate genomes, and it is perhaps the most widely used genome browser for human and other prominent organisms such as mouse. The Genome Browser provides graphical views of chromosomal locations at various levels of resolution (from several base pairs up to hundreds of millions of base pairs spanning an entire chromosome). Each chromosomal view is accompanied by horizontally oriented annotation tracks. There are hundreds of available tracks in categories such as mapping and sequencing, phenotype and disease associations, genes, expression, comparative genomics, and genomic variation. These annotation tracks offer the Genome Browser tremendous depth and flexibility. The Genome Browser has a complementary, interconnected Table Browser that provides tabular output of information.

The UCSC genome browser is available from the UCSC bioinformatics site at ► http://genome.ucsc.edu. You can see examples of it in Figs. 5.17, 5.20, 6.10, 8.8, 12.8, 16.4, and 9.20.

As an example of how to use the browser, go the UCSC bioinformatics site, click Genome Browser, set the clade (group) to Vertebrate, the genome to human, the assembly to March 2006 (or any other build date), and under "position or search term" type beta globin (Fig. 2.11a). Click submit and you will see a list of known genes and a RefSeq gene entry for beta globin on chromosome 11 (Fig. 2.11b). By following this RefSeq link you will view the beta globin gene (spanning about 1600 base pairs) on chromosome 11, and can perform detailed analyses of the beta globin gene (including neighboring regulatory elements), the messenger RNA (see Chapter 8), and the protein (Fig. 2.11c).

The Ensembl Genome Browser

The Ensembl project offers a series of comprehensive websites for a variety of eukaryotic organisms (Hubbard et al., 2007). The project's goals are to automatically analyze and annotate genome data (see Chapter 13) and to present genomic data via its

FIGURE 2.11. *Using the UCSC Genome Browser. (a) One can select from dozens of organisms (mostly vertebrates) and assemblies, then enter a query such as "beta globin" (shown here) or an accession number or chromosomal position. (b) By clicking submit, a list of known genes as well as RefSeq genes is displayed. (c) Following the link to the RefSeq gene for beta globin, a browser window is opened showing 1606 base pairs on human chromosome 11. A series of horizontal tracks is displayed including a list of RefSeq genes and Ensembl gene predictions; exons are displayed as thick bars, and arrows indicate the direction of transcription (from right to left, toward the telomere or end of the short arm of chromosome 11). See* ▶ *http:// genome.ucsc.edu.*

Ensembl (▶ http://www.ensembl. org) is supported by EMBL and the EBI (▶ http://www.ebi.ac.uk/) in cooperation with the Wellcome Trust Sanger Institute (WTSI; ▶ http://www.sanger.ac.uk/). Ensembl focuses on vertebrate genomes, although its genome browser format is being adopted for the analysis of many additional eukaryotic genomes.

web browser. Ensembl is in some ways comparable in scope to the UCSC Genome Browser, although the two offer distinct resources.

We can begin to explore Ensembl from its home page by selecting *Homo sapiens* and doing a text search for "hbb," the gene symbol for beta globin. This yields a link to the beta globin protein and gene; we will return to the Ensembl resource in later chapters. This entry contains a large number of features relevant to HBB, including identifiers, the DNA sequence, and convenient links to many other database resources.

EXAMPLES OF HOW TO ACCESS SEQUENCE DATA

We will next explore two practical problems in accessing data: the human immunodeficiency virus-1 (HIV-1) pol protein, and human histones. Each presents distinct challenges.

HIV *pol*

We explore bioinformatics approaches to HIV-1 in detail in Chapter 14 on viruses.

As of November 2008 there are about 250,000 entries in Entrez Nucleotide for the query "hiv-1."

Consider reverse transcriptase, the RNA-dependent DNA polymerase of HIV-1 (Frankel and Young, 1998). The gene-encoding reverse transcriptase is called *pol* (for polymerase). How do you obtain its DNA and protein sequence?

From the home page of NCBI enter "hiv-1" (do not use quotation marks; the use of capital letters is optional). All Entrez databases are searched. Under the Nucleotide category, there are several hundred thousand entries. Click Nucleotide to see these entries. Over 800 entries have RefSeq identifiers; while this narrows the search considerably, there are still too many matches to easily find HIV-1 pol. One reason for the large number of entries in Entrez Nucleotide is that the HIV-1 genome has been resequenced thousands of times in efforts to identify variants. Another reason for the many hits is that entries for a variety of organisms, including mouse and human, refer to HIV-1 and thus are listed in the output. Performing a search with the query "hiv-1 pol" further reduces the number of matches, but there are still several thousand.

A useful alternate strategy is to limit the search to the organism you are interested in. Begin the search again from the home page of NCBI by clicking "Taxonomy Browser" (along the top bar), and entering Hiv-1. Next follow the link to the taxonomy page specific to HIV-1 (Fig. 2.12). Here you will find the taxonomy identifier for HIV-1; each organism or group in GenBank (e.g., kingdom, phylum, order, genus, species) is assigned a unique identifier. Also, there is an extremely useful table of links to Entrez records. By clicking on the link to Entrez Nucleotide (Fig. 2.11, right side), you will find all the records of sequences from HIV-1, but no records from any other organisms. There is now only one RefSeq entry (NC_001802). This entry refers to the 9181 bases that constitute HIV-1, encoding just nine genes including gag-pol. Given the thousands of HIV-1 pol variants that exist, this example highlights the usefulness of the RefSeq project, allowing the research community to have a common reference sequence to explore.

As an alternative strategy, from the Entrez table on the HIV-1 taxonomy page one can link to the single Entrez Genome record for HIV-1, and find a table of the nine genes (and nine proteins) encoded by the genome. Each of these nine Entrez Genome records contains detailed information on the genes; in the case of gag-pol, there are seven separate RefSeq entries, including one for the gag-pol precursor (NP_057849, 1435 amino acids in length) and one for the mature HIV-1 pol protein (NP_789740, 995 amino acids).

Note that other NCBI databases are not appropriate for finding the sequence of a viral reverse transcriptase: UniGene does not incorporate viral records, while OMIM is limited to human entries. UniGene and OMIM, however, do have links to genes that are related to HIV, such as eukaryotic reverse transcriptases.

We will see that BLAST searches (Chapter 4) can be limited by any Entrez query; you can enter the taxonomy identifier into a BLAST search to restrict the output to any organism or taxonomic group of interest.

From the Entrez Genome or other Entrez pages, try exploring the various options under the Display pull-down menu. For example, for the Entrez Genome entry for NC_001802 you can display a convenient protein table; from Entrez Nucleotide or Entrez Protein you can select Graph to obtain a schematic view of the HIV-1 genome and the genes and proteins it encodes.

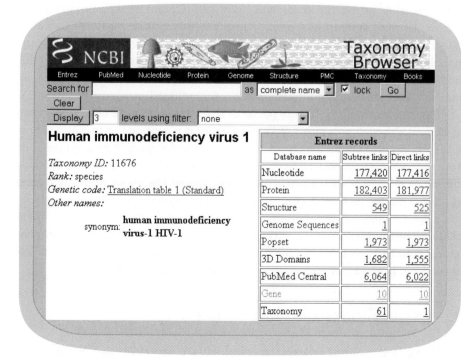

FIGURE 2.12. The entry for human immunodeficiency virus 1 (HIV-1) at the NCBI Taxonomy Browser displays information about the genus and species as well as a variety of links to Entrez records. By following these links, one can obtain a list of proteins, genes, DNA sequences, structures, or other data types that are restricted to this organism. This can be a useful strategy to find a protein or gene from a particular organism (e.g., a species or subspecies of interest), excluding data from all other species. By following the Entrez Genome Sequences link, one can access a list of nine known HIV-1 protein-coding genes.

In a separate approach, one can obtain the HIV-1 reverse transcriptase sequence from SRS. Select the SwissProt database to search. In the four available dialog boxes, set one row to "organism" and "HIV-1," then set another row to "AllText" and "reverse." Upon clicking "Do query," a list of several dozen entries is returned; many of these are identified as fragments and may be ignored. One entry is SWISS_PROT:POL_ HV1A2 (SWISS-PROT accession P03369), a protein of 1437 amino acids. Following the SwissProt link, one finds the "NiceProt" for this database entry. This information includes entry and modification dates, names of this protein and synonyms, references (with PubMed links), comments (including a brief functional description), cross-references to over a dozen other useful databases, a keyword listing, features such as predicted secondary structure, and finally, the amino acid sequence in the single-letter amino acid code and the predicted molecular weight of the protein. For this case, the gene encodes a protein as an unprocessed precursor that is further cleaved to generate many smaller proteins, including matrix protein p17, capsid protein p24, nucleocapsid protein p7, a viral protease, a reverse transcriptase/ribonuclease H multifunctional protein, and an integrase. These features are clearly described in the UniProtKB/Swiss-Prot entry for P03369.

Histones

By clicking the Details tab on an Entrez Protein search, you can see that the command is interpreted as "txid9606[Organism:exp] AND histone[All Fields]". The Boolean operator AND is included between search terms by default.

The Histone Sequence Database is available at ▶ http://research. nhgri.nih.gov/histones/ (Sullivan et al., 2002). It was created by David Landsman, Andy Baxevanis, and colleagues at the National Human Genome Research Institute.

You can find links to a large collection of specialized databases at ▶ http://www.expasy.org/links. html, the Life Science Directory at the ExPASy (Expert Protein Analysis System) proteomics server of the Swiss Institute of Bioinformatics (SIB).

The NLM website is ▶ http:// www.nlm.nih.gov/.

The biological complexity of proteins can be astonishing, and accessing information about some proteins can be extraordinarily challenging. Histones are among the most familiar proteins by name. They are small proteins (12 to 20 kilodaltons) that are localized to the nucleus where they interact with DNA. There are five major histone subtypes as well as additional variant forms; the major forms serve as core histones (the H2A, H2B, H3, and H4 families) which $\sim$147 base pairs of DNA wrap around, and linker histones (the H1 family). Suppose you want to inspect a typical human histone for the purpose of understanding the properties of a representative gene and its corresponding protein. A challenge is that there are currently 80,000 histone entries in Entrez Protein (November 2008). Restricting the output to human histone proteins (using the command "txid9606[Organism:exp] histone") there are currently 5000 human histone proteins, of which 1200 have RefSeq accession numbers. Some of these are histone deacetylases and histone acetyltransferases; by expanding the query to "txid9606[Organism:exp] AND histone[All Fields] NOT deacetylase NOT acetyltransferase" there are 800 proteins with RefSeq accession numbers. There are many additional strategies for limiting Entrez searches (Box 2.2).

How can the search be further pursued? (1) You may select a histone at random and study it although you may not know whether it is representative. (2) There are specialized, expert-curated databases available online for many genes, proteins, diseases, and other molecular features of interest. The Histone Sequence Database (Sullivan et al., 2002) shows that the human genome has about 86 histone genes, including a cluster of 68 adjacent genes on chromosome 6p. This information is useful to understand the scope of the family. (3) There are databases of protein families, including Pfam and InterPro. We will introduce these in Chapters 6 (multiple sequence alignment) and 10 (proteomics). Such databases offer succinct descriptions of protein and gene families and can orient you toward identifying representative members.

ACCESS TO BIOMEDICAL LITERATURE

The NLM is the world's largest medical library. In 1971 the NLM created MEDLINE (Medical Literature, Analysis, and Retrieval System Online), a

BOX 2-2
Tips for Using Entrez Databases

The Boolean operators AND, OR, and NOT must be capitalized. By default, AND is assumed to connect two terms; subject terms are automatically combined.

You can perform a search of a specific phrase by adding quotation marks. This may potentially restrict the output, so it is a good idea to repeat a search with and without quotation marks.

Boolean operators are processed from left to right. If you add parentheses, the enclosed terms will be processed as a unit rather than sequentially. A search of Entrez Gene with the query "globin AND promoter OR enhancer" yields 4800 results; however, by adding parentheses, the query "globin AND (promoter or enhancer)" yields just 70 results.

If you are interested in obtaining results from a particular organism (or from any taxonomic group such as the primates or viruses), try beginning with TaxBrowser to select the organism first. See Fig. 2–11 for a detailed explanation. Adding the search term human[ORGN] will restrict the output to human. Alternatively, you can use the taxonomy identifier for human, 9606, as follows: txid9606[Organism:exp]

A variety of limiters can be added. In Entrez Protein, the search 500000:999999[Molecular weight] will return proteins having a molecular weight from 500,000 to 1 million daltons. If you would like to see proteins between 10,000 and 50,000 daltons that I have worked on, enter 010000:050000[Molecular weight] pevsner j (or, equivalently, 010000[MOLWT]: 050000[MOLWT] AND pevsner j[Author]).

By truncating a query with an asterisk, you can search for all records that begin with a particular text string. For example, a search of Entrez Nucleotide with the query "globin" returns 5800 results; querying with "glob*" returns 8.2 million results. These include entries with the species *Chaetomium globosum* or the word global.

Keep in mind that any Entrez query can be applied to a BLAST search to restrict its output (Chapter 4).

---

bibliographic database. MEDLINE currently contains over 18 million references to journal articles in the life sciences with citations from over 4300 biomedical journals in 70 countries. Free access to MEDLINE is provided on the World Wide Web through PubMed (▶ http://www.ncbi.nlm.nih.gov/PubMed/), which is developed by NCBI. While MEDLINE and PubMed both provide bibliographic citations, PubMed also contains links to online full-text journal articles. PubMed also provides access and links to the integrated molecular biology databases maintained by NCBI. These databases contain DNA and protein sequences, genome-mapping data, and three-dimensional protein structures.

PubMed Central and Movement toward Free Journal Access

The biomedical research community has steadily increased access to literature information. Groups such as the Association of Research Libraries (ARL) monitor the migration of publications to an electronic form. Thousands of journals are currently available online. Increasingly, online versions of articles include supplementary material such as molecular data (e.g., the sequence of complete

MEDLINE is also accessible through the SRS at the European Bioinformatics Institute via ▶ http://srs.ebi.ac.uk/. A PubMed tutorial is offered at ▶ http://www.nlm.nih.gov/bsd/ pubmed_tutorial/m1001.html. The growth of MEDLINE is described at ▶ http://www.nlm. nih.gov/bsd/medline_growth. html. Despite the multinational contributions to MEDLINE, the percentage of articles written in English has risen from 59% at its inception in 1966 to 92% in the year 2008 (▶ http://www.nlm. nih.gov/bsd/medline_lang_distr. html).

The National Library of Medicine also offers access to PubMed through NLM Gateway (▶ http://gateway.nlm.nih.gov). This comprehensive service includes access to a variety of NLM databases not offered through PubMed, such as meeting abstracts and a medical encyclopedia.

The ARL website is ▶ http://www.arl.org/index.shtml.

genomes, or gene expression data) or videotapes illustrating an article. PubMed Central provides a central repository for biological literature (Roberts, 2001). All these articles have been peer reviewed and published simultaneously in another journal. As of 2008, publications resulting from research funded by the NIH, Wellcome Trust, and Medical Research Council must be made freely available in PubMed Central.

Example of PubMed Search: RBP

A search of PubMed for information about "RBP" yields 1700 entries. Box 2.3 describes the basics of using Boolean operators in PubMed. There are many additional ways to limit this search. Press "limits" and try applying features such as restricting the output to articles that are freely available through PubMed Central.

BOX 2-3
Venn Diagrams of Boolean Operators AND, OR, and NOT for Hypothetical Search Terms 1 and 2

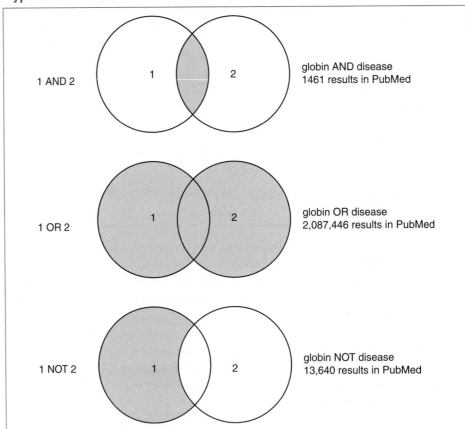

1 AND 2
globin AND disease
1461 results in PubMed

1 OR 2
globin OR disease
2,087,446 results in PubMed

1 NOT 2
globin NOT disease
13,640 results in PubMed

The AND command restricts the search to entries that are both present in a query. The OR command allows either one or both of the terms to be present. The NOT command excludes query results. The shaded areas represent search queries that are retrieved. Examples are provided for the queries "lipocalin" or "retinol-binding protein" in PubMed. The Boolean operators affect the searches as indicated.

The Medical Subject Headings (MeSH) browser provides a convenient way to focus or expand a search. MeSH is a controlled vocabulary thesaurus containing 25,000 descriptors (headings). From PubMed, click "MeSH Database" on the left sidebar and enter "retinol-binding protein." The result suggests a series of possibly related topics. By adding MeSH terms, a search can be focused and structured according to the specific information you seek. Lewitter (1998) and Fielding and Powell (2002) discuss strategies for effective MEDLINE searches, such as avoiding inconsistencies in MeSH terminology and finding a balance between sensitivity (i.e., finding relevant articles) and specificity (i.e., excluding irrelevant citations). For example, for a subject that is not well indexed, it is helpful to combine a text keyword with a MeSH term. It can also be helpful to use truncations; for example, the search "therap*" introduces a wildcard that will retrieve variations such as therapy, therapist, and therapeutic. Figure 2.13 provides an example of sensitivity and specificity in a PubMed search for articles on hemoglobin.

The MeSH website at NLM is ► http://www.nlm.nih.gov/mesh/meshhome.html; you can also access MeSH via the NCBI website including its PubMed page.

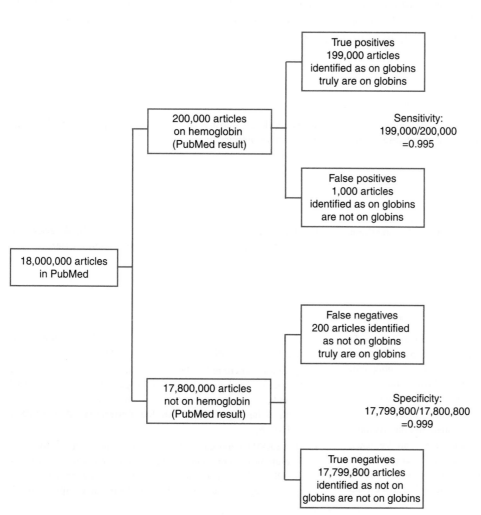

FIGURE 2.13. Sensitivity and specificity in a database search. We will describe sensitivity and specificity in Chapter 3 (see Fig. 3.27) but can begin thinking about those concepts in terms of a hypothetical search of PubMed for hemoglobin. Each search of a database yields results that are reported (positives) or not (negatives). According to some "gold standard" or objective measure of the truth, these results may be true positives (e.g., a search for globins does return literature citations on globins) or false positives (e.g., a search for glob* returns information about the species C. globosum but those citations are irrelevant to globins). The sensitivity is defined as the proportion of true positives relative to true plus false positives. There also will be many negative results (lower portion of figure). These may include true negatives (e.g., articles that do not describe globins and are not included in the search results) and false negatives (e.g., articles that do discuss globins but are not part of the search results; this might occur if the title and abstract do not mention globins but the body of the article does). Specificity may be defined as the proportion of true negative results divided by the sum of true negative and false positive results.

PERSPECTIVE

Bioinformatics is a young, emerging field whose defining feature is the accumulation of biological information in databases. The three major DNA databases—GenBank, EMBL, and DDBJ—are adding several million new sequences each year as well as billions of nucleotides. Beginning in 2008, terabases (thousands of gigabases) of DNA sequence are arriving.

In this chapter, we described ways to find information on the DNA and/or protein sequence of globins, RBP4, and the HIV *pol* gene. In addition to the three major databases, a variety of additional resources are available on the web. Increasingly, there is no single correct way to find information—many approaches are possible. Moreover, resources such as those described in this chapter—NCBI, ExPASy, EBI/EMBL, and Ensembl—are closely interrelated, providing links between the databases.

PITFALLS

There are many pitfalls associated with the acquisition of both sequence and literature information. In any search, the most important first step is to define your goal: for example, decide whether you want protein or DNA sequence data. A common difficulty that is encountered in database searches is receiving too much information; this problem can be addressed by learning how to generate specific searches with appropriate limits.

WEB RESOURCES

You can visit the website for this book (▶ http://www.bioinfbook.org) to find many of the URLs, organized by chapter. The Wiley-Blackwell website for this book is http://www.wiley.com/go/pevsnerbioinformatics.

DISCUSSION QUESTIONS

[2-1] What categories of errors occur in databases? How are these errors assessed?

[2-2] How is quality control maintained in GenBank, given that thousands of individual investigators submit data?

PROBLEMS

[2-1] In this chapter we explored histones as an example of a protein that can be challenging to study because it is part of a large gene family. Another challenging example is ubiquitin. How many ubiquitins are there in the human genome, and what is the sequence of a prototypical (that is, representative) ubiquitin?

[2-2] How many human proteins are bigger than 300,000 daltons? *Hints:* Try to first limit your search to human by using TaxBrowser. Then follow the link to Entrez Protein, where all the results will be limited to human. Enter a command in the format xxxxxx:yyyyyy[molwt] to restrict the output to a certain number of daltons; for example, 002000:010000[molwt] will select proteins of molecular weight 2,000 to 10,000.

[2-3] You are interested in learning about genes involved in breast cancer. Which genes have been implicated? What are the DNA and protein accession numbers for several of these genes? Try all of these approaches: PubMed, Entrez, OMIM, and SRS at ExPASy.

[2-4] An ATP (adenosine triphosphate) binding cassette (ABC) is an example of a common protein domain that is found in many so-called ABC transporter proteins. However, you are not familiar with this motif and would like to learn more. Approximately

how many human proteins have ABC domains? Approximately how many bacterial proteins have ABC domains? Which of the resources you used in problem 2.3 is most useful in providing you a clear definition of an ABC motif? (We will discuss additional resources to solve this problem in Chapter 10.)

[2-5] Find the accession number of a lipocalin protein (e.g., retinol-binding protein, lactoglobulin, any bacterial lipocalin, glycodelin, or odorant-binding protein). First, use Entrez, then UniGene, then OMIM. Which approach is most effective? What is the function of this protein?

[2-6] Three prominent tools for *text*-based searching of molecular information are:

- the National Center for Biotechnology Information's PubMed, Entrez, and OMIM tools (► http://www.ncbi.nlm.nih.gov),

- the European Bioinformatics Institute (EBI) Sequence Retrieval System (SRS) (► http://srs.ebi.ac.uk) or its related SRS site (► http://www.expasy.ch/srs5/), and

- DBGET, the GenomeNet tool of Kyoto University, and the University of Tokyo (► http://www.genome.ad.jp/dbget/dbget2.html) literature database LitDB.

You are interested in learning more about West Nile virus. What happens when you use that query to search each of these three resources?

[2-7] You would like to know what articles about viruses have been published in the journal *BMC Bioinformatics*. Do this search using PubMed.

SELF-TEST QUIZ

[2-1] Which of the following is a RefSeq accession number corresponding to an mRNA?

(a) J01536

(b) NM_15392

(c) NP_52280

(d) AAB134506

[2-2] Approximately how many human clusters are currently in UniGene?

(a) About 8,000

(b) About 25,000

(c) About 100,000

(d) About 300,000

[2-3] You have a favorite gene, and you want to determine in what tissues it is expressed. Which one of the following resources is likely the most direct route to this information?

(a) UniGene

(b) Entrez

(c) PubMed

(d) PCR

[2-4] Is it possible for a single gene to have more than one UniGene cluster?

(a) Yes

(b) No

[2-5] Which of the following databases is derived from mRNA information?

(a) dbEST

(b) PBD

(c) OMIM

(d) HTGS

[2-6] Which of the following databases can be used to access text information about human diseases?

(a) EST

(b) PBD

(c) OMIM

(d) HTGS

[2-7] What is the difference between RefSeq and GenBank?

(a) RefSeq includes publicly available DNA sequences submitted from individual laboratories and sequencing projects.

(b) GenBank provides nonredundant curated data.

(c) GenBank sequences are derived from RefSeq.

(d) RefSeq sequences are derived from GenBank and provide nonredundant curated data.

[2-8] If you want literature information, what is the best website to visit?

(a) OMIM

(b) Entrez

(c) PubMed

(d) PROSITE

[2-9] Compare the use of Entrez and ExPASy to retrieve information about a protein sequence.

(a) Entrez is likely to yield a more comprehensive search because GenBank has more data than EMBL.

(b) The search results are likely to be identical because the underlying raw data from GenBank and EMBL are the same.

(c) The search results are likely to be comparable, but the SwissProt record from ExPASy will offer a different output format with distinct kinds of information.

SUGGESTED READING

Bioinformatics databases are evolving extremely rapidly. Each January, the first issue of the journal *Nucleic Acids Research* includes nearly 100 brief articles on databases. These include descriptions of NCBI (Wheeler et al., 2007), GenBank (Benson et al., 2009), and EMBL (Cochrane et al., 2008).

REFERENCES

Adams, M. D., et al. Complementary DNA sequencing: Expressed sequence tags and human genome project. *Science* **252**, 1651–1656 (1991).

Altschul, S. F., Gish, W., Miller, W., Myers, E. W., and Lipman, D. J. Basic local alignment search tool. *J. Mol. Biol.* **215**, 403–410 (1990).

Altschul, S. F., et al. Gapped BLAST and PSI-BLAST: A new generation of protein database search programs. *Nucleic Acids Res.* **25**, 3389–3402 (1997).

Beach, E. F. Beccari of Bologna. The discoverer of vegetable protein. *J. Hist. Med.* **16**, 354–373 (1961).

Bentley, D. R., et al. Accurate whole human genome sequencing using reversible terminator chemistry. *Nature* **456**, 53–39 (2008).

Benson, D. A., Karsch-Mizrachi, I., Lipman, D. J., Ostell, J., and Sayers, E. W. GenBank. *Nucl. Acids Res.* **37**, D26–D31 (2009).

Boeckmann, B., Bairoch, A., Apweiler, R., Blatter, M. C., Estreicher, A., Gasteiger, E., Martin, M. J., Michoud, K., O'Donovan, C., Phan, I., Pilbout, S., and Schneider, M. The SWISS-PROT protein knowledgebase and its supplement TrEMBL in 2003. *Nucleic Acids Res.* **31**, 365–370 (2003).

Boguski, M. S., Lowe, T. M., and Tolstoshev, C. M. dbEST—database for "expressed sequence tags." *Nat. Genet.* **4**, 332–333 (1993).

Brooksbank, C., Camon, E., Harris, M. A., Magrane, M., Martin, M. J., Mulder, N., O'Donovan, C., Parkinson, H., Tuli, M. A., Apweiler, R., Birney, E., Brazma, A., Henrick, K., Lopez, R., Stoesser, G., Stoehr, P., and Cameron, G. The European Bioinformatics Institute's data resources. *Nucleic Acids Res.* **31**, 43–50 (2003).

Cochrane, G., et al. Priorities for nucleotide trace, sequence and annotation data capture at the Ensembl Trace Archive and the EMBL Nucleotide Sequence Database. *Nucleic Acids Res.* **36**, D5–D12 (2008).

Fielding, A. M., and Powell, A. Using Medline to achieve an evidence-based approach to diagnostic clinical biochemistry. *Ann. Clin. Biochem.* **39**, 345–350 (2002).

Fleischmann, R. D., et al. Whole-genome random sequencing and assembly of *Haemophilus influenzae* Rd. *Science* **269**, 496–512 (1995).

Frankel, A. D., and Young, J. A. HIV-1: Fifteen proteins and an RNA. *Annu. Rev. Biochem.* **67**, 1–25 (1998).

Hamosh, A., Scott, A. F., Amberger, J. S., Bocchini, C. A., and McKusick, V. A. Online Mendelian Inheritance in Man (OMIM), a knowledgebase of human genes and genetic disorders. *Nucleic Acids Res.* **33**, D514–D517 (2005).

Hubbard, T. J., et al. Ensembl 2007. *Nucleic Acids Res.* **35**, D610–D617 (2007).

Kouranov, A., Xie, L., de la Cruz, J., Chen, L., Westbrook, J., Bourne, P. E., and Berman, H. M. The RCSB PDB information portal for structural genomics. *Nucleic Acids Res.* **34**, D302–D305 (2006).

Kulikova, T., et al. EMBL Nucleotide Sequence Database in 2006. *Nucleic Acids Res.* **35**, D16–D20 (2007).

Lewitter, F. Text-based database searching. *Bioinformatics: A Trends Guide* **19**, 3–5 (1998).

Ley, T. J., et al. DNA sequencing of a cytogenetically normal acute myeloid leukaemia genome. *Nature* **456**, 66–72 (2008).

Maglott, D. R., Katz, K. S., Sicotte, H., and Pruitt, K. D. NCBI's LocusLink and RefSeq. *Nucleic Acids Res.*, **28**, 126–128 (2000).

Maglott, D., Ostell, J., Pruitt, K. D., and Tatusova, T. Entrez Gene: Gene-centered information at NCBI. *Nucleic Acids Res.* **35**, D26–D31 (2007).

Miyazaki, S., Sugawara, H., Ikeo, K., Gojobori, T., and Tateno, Y. DDBJ in the stream of various biological data. *Nucleic Acids Res.* **32**, D31–D34 (2004).

Olson, M., Hood, L., Cantor, C., and Botstein, D. A common language for physical mapping of the human genome. *Science* **245**, 1434–1435 (1989).

Pruitt, K. D., Tatusova, T., Klimke, W., and Maglott, D. R. NCBI Reference Sequences: current status, policy and new initiatives. *Nucl. Acids Res.* **37**, D32–D36 (2009).

Roberts, R. J. PubMed Central: The GenBank of the published literature. *Proc. Natl. Acad. Sci. USA* **98**, 381–382 (2001).

Sullivan, S., Sink, D. W., Trout, K. L., Makalowska, I., Taylor, P. M., Baxevanis, A. D., and Landsman, D. The Histone Database. *Nucleic Acids Res.* **30**, 341–342 (2002).

UniProt Consortium. The Universal Protein Resource (UniProt). *Nucleic Acids Res.* **35** (Database issue), D193–D197 (2007).

UniProt Consortium. The Universal Protein Resource (UniProt) 2009. *Nucl. Acids Res.* **37**, D169–D174 (2009).

Wang, J., et al. The diploid genome sequence of an Asian individual. *Nature* **456**, 60–65 (2008).

Wheeler, D. L., et al. Database resources of the National Center for Biotechnology Information. *Nucleic Acids Res.* **35** (Database issue), D5–D12 (2007).

Wu, C. H., et al. The Protein Information Resource: An integrated public resource of functional annotation of proteins. *Nucleic Acids Res.* **30**, 35–37 (2002).

Adrenocorticotropin (ACTH)

The complete amino acid sequences are known for corticotropins isolated from the anterior pituitary glands of three different species, pig, beef, and sheep. The structure of sheep ACTH was discussed in the last chapter, and the sequences shown in Table 9 include only those areas of the three molecules where differences are to be found. Although some difference between the content of amide nitrogen groups has been reported for the three species, these are not included in the figure since it has not been possible to rule out, with certainty, the possibility that these variations are due, in part, to the rigors of the isolation and purification techniques employed.

TABLE 9
Variations in Amino Acid Sequences Among Different Preparations of ACTH

Preparation	Species	Residue No.								
		25	26	27	28	29	30	31	32	33
β-Corticotropin	sheep } beef[a]	Ala.	Gly.	Glu.	Asp.	Asp.	Glu	Ala.	Ser.	Glu.NH$_2$
Corticotropin A	pig	Asp.	Gly.	Ala.	Glu.	Asp.	Glu	Leu.	Ala.	Glu

[a] Identity with sheep hormone not absolutely certain but very probable as judged from the nearly complete sequence analysis by J. S. Dixon and C. H. Li (personal communication to the author).

Two points are of particular interest in regard to the sequences shown. First, the corticotropins of sheep and beef are identical and differ from that of the pig. This finding is consonant with the closer phylogenetic relationship of sheep and cows to each other than of either to pigs. Second, chemical differences are found only in that portion of the ACTH molecule which has been shown to be unessential for hormonal activity. Genetic mutations leading to such differences might, therefore, not be expected to impose significant disadvantages in terms of survival, and these genes could become established in the gene pools of the species.

Melanotropin (MSH)

Melanotropin, like the other hormones considered in this chapter, is a typically chordate polypeptide. Indeed, the demonstration of melanocyte-stimulating activity in extracts of tunicates constitutes an

Pairwise alignment involves matching two protein or DNA sequences. The first proteins that were sequenced include insulin (by Frederick Sanger and colleagues; see fig. 7.1) and globins. This figure is from The Molecular Basis of Evolution by the Nobel laureate Christian Anfinsen (1959, p. 153). It shows the results of a pairwise alignment of a portion of adrenocorticotropic hormone (ACTH) from sheep or cow (top) with that of pig (below). Such alignments, performed manually, led to the realization that amino acid sequences of proteins reflect the phylogenetic relatedness of different species. Furthermore, pairwise alignments reveal the portions of a protein that may be important for its biological function. Used with permission.

3

Pairwise Sequence Alignment

INTRODUCTION

One of the most basic questions about a gene or protein is whether it is related to any other gene or protein. Relatedness of two proteins at the sequence level suggests that they are homologous. Relatedness also suggests that they may have common functions. By analyzing many DNA and protein sequences, it is possible to identify domains or motifs that are shared among a group of molecules. These analyses of the relatedness of proteins and genes are accomplished by aligning sequences. As we complete the sequencing of many organisms' genomes, the task of finding out how proteins are related within an organism and between organisms becomes increasingly fundamental to our understanding of life.

Two genes (or proteins) are homologous if they have evolved from a common ancestor.

In this chapter we will introduce pairwise sequence alignment. We will adopt an evolutionary perspective in our description of how amino acids (or nucleotides) in two sequences can be aligned and compared. We will then describe algorithms and programs for pairwise alignment.

Protein Alignment: Often More Informative Than DNA Alignment

Given the choice of aligning a DNA sequence or the sequence of the protein it encodes, it is often more informative to compare protein sequences. There are several reasons for this. Many changes in a DNA sequence (particularly at

Bioinformatics and Functional Genomics, Second Edition. By Jonathan Pevsner

the third position of a codon) do not change the amino acid that is specified. Furthermore, many amino acids share related biophysical properties (e.g., lysine and arginine are both basic amino acids). The important relationships between related (but mismatched) amino acids in an alignment can be accounted for using scoring systems (described in this chapter). DNA sequences are less informative in this regard. Protein sequence comparisons can identify homologous sequences from organisms that last shared a common ancestor over 1 billion years ago (BYA) (e.g., glutathione transferases) (Pearson, 1996). In contrast, DNA sequence comparisons typically allow lookback times of up to about 600 million years ago (MYA).

When a nucleotide coding sequence is analyzed, it is often preferable to study its translated protein. In Chapter 4 (on BLAST searching), we will see that we can move easily between the worlds of DNA and protein. For example, the tblastn tool from the National Center for Biotechnology Information (NCBI) BLAST website allows one to search with a protein sequence for related proteins derived from a DNA database (see Chapter 4). This query option is accomplished by translating each DNA sequence into all of the six proteins that it potentially encodes.

Nevertheless, in many cases it is appropriate to compare nucleotide sequences. This comparison can be important in confirming the identity of a DNA sequence in a database search, in searching for polymorphisms, in analyzing the identity of a cloned cDNA fragment, or in many other applications.

Definitions: Homology, Similarity, Identity

Let us consider the globin family of proteins. We will begin with human myoglobin (accession number NP_005359) and beta globin (accession number NP_000509) as two proteins that are distantly but significantly related. The accession numbers are obtained from Entrez Gene (Chapter 2). Myoglobin and the hemoglobin chains (alpha, beta, and other) are thought to have diverged some 600 million years ago, near the time the vertebrate and insect lineages diverged.

Some researchers use the term *analogous* to refer to proteins that are not homologous, but share some similarity by chance. Such proteins are presumed to have not descended from a common ancestor.

Two sequences are *homologous* if they share a common evolutionary ancestry. There are no degrees of homology; sequences are either homologous or not (Reeck et al., 1987; Tautz, 1998). Homologous proteins almost always share a significantly related three-dimensional structure. Myoglobin and beta globin have very similar structures as determined by x-ray crystallography (Fig. 3.1). When two sequences are homologous, their amino acid or nucleotide sequences usually share significant identity. Thus, while homology is a qualitative inference (sequences are homologous or not), identity and similarity are quantities that describe the relatedness of sequences. Notably, two molecules may be homologous without sharing statistically significant amino acid (or nucleotide) identity. For example, in the globin family, all the members are homologous, but some have sequences that have diverged so greatly that they share no recognizable sequence identity (e.g., human beta globin and human neuroglobin share only 22% amino acid identity). Perutz, Kendrew and others demonstrated that individual globin chains share the same overall shape as myoglobin (see Ingram, 1963), even though the myoglobin and alpha globin proteins share only about 26% amino acid identity. In general, three-dimensional structures diverge much more slowly than amino acid sequence identity between two proteins

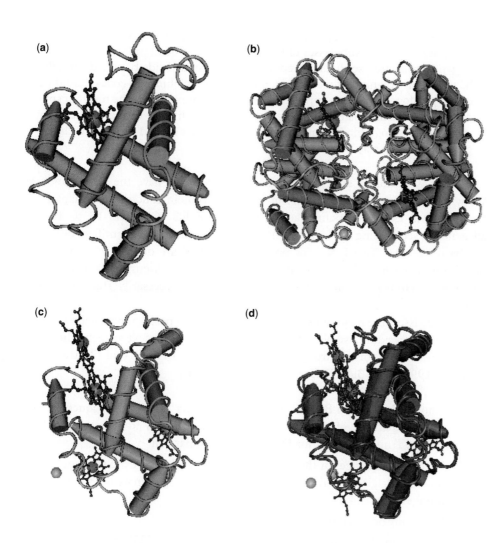

FIGURE 3.1. *Three-dimensional structures of (a) myoglobin (accession 2MM1), (b) the tetrameric hemoglobin protein (2H35), (c) the beta globin subunit of hemoglobin, and (d) myoglobin and beta globin superimposed. The images were generated with the program Cn3D (see Chapter 11). These proteins are homologous (descended from a common ancestor), and they share very similar three-dimensional structures. However, pairwise alignment of these proteins' amino acid sequences reveals that the proteins share very limited amino acid identity.*

(Chothia and Lesk, 1986). Recognizing this type of homology is an especially challenging bioinformatics problem.

Proteins that are homologous may be orthologous or paralogous. *Orthologs* are homologous sequences in different species that arose from a common ancestral gene during speciation. Figure 3.2 shows a tree of myoglobin orthologs. There is a human myoglobin gene and a rat gene. Humans and rodents diverged about 80 MYA (see Chapter 18), at which time a single ancestral myoglobin gene diverged by speciation. Orthologs are presumed to have similar biological functions; in this example, human and rat myoglobins both transport oxygen in muscle cells. *Paralogs* are homologous sequences that arose by a mechanism such as gene duplication. For example, human alpha-1 globin (NP_000549) is paralogous to alpha-2 globin (NP_000508); indeed, these two proteins share 100% amino acid identity. Human alpha-1 globin and beta globin are also paralogs (as are all the proteins shown in Fig. 3.3). All of the globins have distinct properties, including regional distribution in the body, developmental timing of gene expression, and abundance. They are all thought to have distinct but related functions as oxygen carrier proteins.

You can see the protein sequences used to generate Figs. 3.2 and 3.3 in web documents 3.1 and 3.2 at ► http://www.bioinfbook.org/chapter3.

In general when we consider other paralogous families they are presumed to share common functions. Consider the lipocalins: all are about 20 kilodalton proteins that have a hydrophobic binding pocket that is thought to be used to transport a hydrophobic ligand. Members include retinol binding protein (a retinol transporter), apolipoprotein D (a cholesterol transporter), and odorant-binding protein (an odorant transporter secreted from a nasal gland).

We thus define homologous genes within the same organism as paralogous. But consider further the case of globins. Human α-globin and β-globin are paralogs, as are mouse α-globin and mouse β-globin. Human α-globin and mouse α-globin are orthologs. What is the relation of human α-globin to mouse β-globin? These could be considered paralogs, because α-globin and β-globin originate from a gene duplication event rather than from a speciation event. However, they are not paralogs because they do not occur in the same species. It may thus be most appropriate to simply call them "homologs," reflecting their descent from a common ancestor. Fitch (1970, p. 113) notes that phylogenies require the study of orthologs (see also Chapter 7).

Richard Owen (1804–1892) was one of the first biologists to use the term homology. He defined homology as "the same organ in different animals under every variety of form and function" (Owen, 1843, p. 379). Charles Darwin (1809–1882) also discussed homology in the sixth edition of *The Origin of Species by means of Natural Selection or, The Preservation of Favoured Races in the Struggle for Life* (1872). He wrote: "That relation between parts which results from their development from corresponding embryonic parts, either in different animals, as in the case of the arm of man, the foreleg of a quadruped, and the wing of a bird; or in the same individual, as in the case of the fore and hind legs in quadrupeds, and the segments or rings and their appendages of which the body of a worm, a centipede, &c., is composed. The latter is called serial homology. The parts which stand in such a relation to each other are said to be homologous, and one such part or organ is called the homologue of the other. In different plants the parts of the flower are homologous, and in general these parts are regarded as homologous with leaves."

Walter M. Fitch (1970, p. 113) defined these terms. He wrote: there should be two subclasses of homology. Where the homology is the result of gene duplication so that both copies have descended side by side during the history of an organism (for example, α and β hemoglobin) the genes should be called paralogous (para = in parallel). Where the homology is the result of speciation so that the history of the gene reflects the history of the species (for example α hemoglobin in man and mouse) the genes should be called orthologous (ortho = exact).

Notably, orthologs and paralogs do not necessarily have the same function. We will provide various definitions of gene and protein function in Chapter 10. Later we will explore genomes across the tree of life (Chapters 13 to 19). In all genome sequencing projects, orthologs and paralogs are identified based on database searches. Two DNA (or protein) sequences are defined as homologous based on achieving significant alignment scores, as discussed below and in Chapter 4. However, homologous proteins do not necessarily share the same function.

We can assess the relatedness of any two proteins by performing a *pairwise alignment*. In this procedure, we place the two sequences directly next to each other. One practical way to do this is through the NCBI pairwise BLAST tool (Tatusova and Madden, 1999) (Fig. 3.4). Perform the following steps:

1. Choose the protein BLAST program and select "BLAST 2 sequences" for our comparison of two proteins. An alternative is to select blastn (for "BLAST nucleotides") for DNA–DNA comparison.

2. Enter the sequences or their accession numbers. Here we use the sequence of human beta globin in the fasta format, and for myoglobin we use the accession number (Fig. 3.4).

3. Select any optional parameters.
   - You can choose from five scoring matrices: BLOSUM62, BLOSUM45, BLOSUM80, PAM70, and PAM30. Select PAM250.
   - You can change the gap creation penalty and gap extension penalty.
   - For blastn searches you can change reward and penalty values.
   - There are other parameters you can change, such as word size, expect value, filtering, and dropoff values. We will discuss these more in Chapter 4.

4. Click "BLAST." The output includes a pairwise alignment using the single-letter amino acid code (Fig. 3.5a).

Note that the fasta format uses the single-letter amino acid code; those abbreviations are shown in Box 3.1.

It is extremely difficult to align proteins by visual inspection. Also, if we allow gaps in the alignment to account for deletions or insertions in the two sequences, the number of possible alignments rises exponentially. Clearly, we will need a computer algorithm to perform an alignment (see Box 3.2). In the pairwise alignments shown in Fig. 3.5a, beta globin is on top (on the line labeled query) and myoglobin is below (on the subject line). An intermediate row indicates the presence of *identical* amino acids in the alignment. For example, notice that near the beginning of the alignment the residues WGKV are identical between the two proteins. We can count the total number of identical residues; in this case, the two proteins share

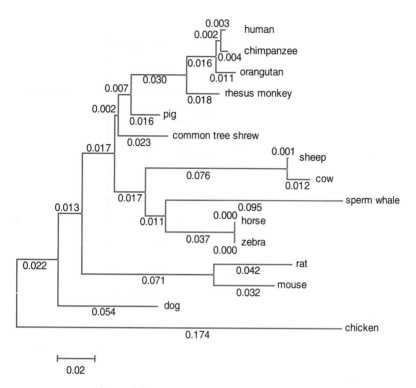

FIGURE 3.2. *A group of myoglobin orthologs, visualized by multiply aligning the sequences (Chapter 6) then creating a phylogenetic tree by neighbor-joining (Chapter 7). The accession numbers and species names are as follows: human, NP_005359 (Homo sapiens); chimpanzee, XP_001156591 (Pan troglodytes); orangutan, P02148 (Pongo pygmaeus); rhesus monkey, XP_001082347 (Macaca mulatta); pig, NP_999401 (Sus scrofa); common tree shrew, P02165 (Tupaia glis); horse, P68082 (Equus caballus); zebra, P68083 (Equus burchellii); dog, XP_850735 (Canis familiaris); sperm whale, P02185 (Physeter catodon); sheep, P02190 (Ovis aries); rat, NP_067599 (Rattus norvegicus); mouse, NP_038621 (Mus musculus); cow, NP_776306 (Bos taurus); chicken_XP_416292 (Gallus gallus). The sequences are shown in web document 3.1 (▶ http://www.bioinfbook.org/chapter3). In this tree, sequences that are more closely related to each other are grouped closer together. Note that as entire genomes continue to be sequenced (Chapters 13 to 19), the number of known orthologs will grow rapidly for most families of orthologous proteins.*

25% identity (37 of 145 aligned residues). Identity is the extent to which two amino acid (or nucleotide) sequences are invariant. Note that this particular alignment is called *local* because only a subset of the two proteins is aligned: the first and last few amino acid residues of each protein are not displayed. A global pairwise alignment includes all residues of both sequences.

Another aspect of this pairwise alignment is that some of the aligned residues are similar but not identical; they are related to each other because they share similar biochemical properties. *Similar* pairs of residues are structurally or functionally related. For example, on the first row of the alignment we can find threonine and serine (T and S connected by a + sign in Fig. 3.5a); nearby we can see a leucine and a valine residue that are aligned. These are *conservative substitutions*. Amino acids with similar properties include the basic amino acids (K, R, H), acidic amino acids (D, E), hydroxylated amino acids (S, T), and hydrophobic amino acids (W, F, Y, L, I, V, M, A). Later in this chapter we will see how scores are assigned to aligned amino acid residues.

You can access the pairwise BLAST program at the NCBI blast site, ▶ http://www.ncbi.nlm.nih.gov/BLAST/. We discuss various options for using the Basic Local Alignment Search Tool (BLAST) in Chapter 4. We discuss global and local alignments below.

FIGURE 3.3. *Paralogous human globins: Each of these proteins is human, and each is a member of the globin family. This unrooted tree was generated using the neighbor-joining algorithm in MEGA (see Chapter 7). The proteins and their RefSeq accession numbers (also shown in web document 3.2) are delta globin (NP_000510), G-gamma globin (NP_000175), beta globin (NP_000509), A-gamma globin (NP_000550), epsilon globin (NP_005321), zeta globin (NP_005323), alpha-1 globin (NP_000549), alpha-2 globin (NP_000508), theta-1 globin (NP_005322), hemoglobin mu chain (NP_001003938), cytoglobin (NP_599030), myoglobin (NP_005359), and neuroglobin (NP_067080). A Poisson correction model was used (see Chapter 7).*

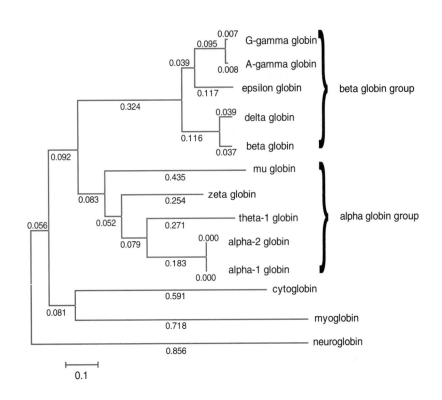

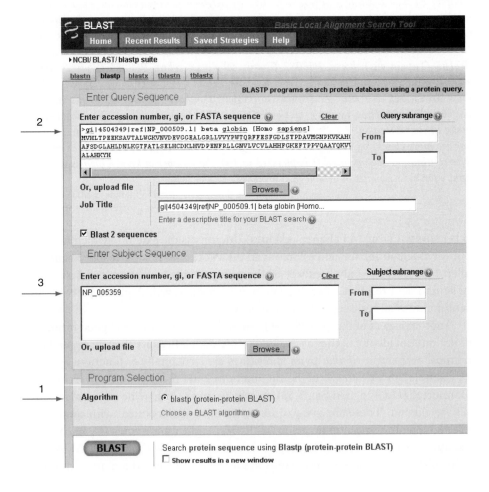

FIGURE 3.4. *The BLAST program at the NCBI website allows the comparison of two DNA or protein sequences. Here the program is set to blastp for the comparison of two proteins (arrow 1). Human beta globin (NP_000509) is input in the fasta format (arrow 2), while human myoglobin (NP_005359) is input as an accession number (arrow 3).*

(a)

```
   Score = 43.9 bits (102),   Expect = 1e-09, Method: Composition-based stats.
   Identities = 37/145 (25%), Positives = 57/145 (39%), Gaps = 2/145 (1%)

                          ▼                        ▼
Query    4   LTPEEKSAVTALWGKVNVD--EVGGEALGRLLVVYPWTQRFFESFGDLSTPDAVMGNPKV   61
             L+  E   V  +WGKV  D     G E L  RL    +P T   F+  F   L +D +  +  +
Sbjct    3   LSDGEWQLVLNVWGKVEADIPGHGQEVLIRLFKGHPETLEKFDKFKHLKSEDEMKASEDL   62

Query   62   KAHGKKVLGAFSDGLAHLDNLKGTFATLSELHCDKLHVDPENFRLLGNVLVCVLAHHFGK  121
             K HG  VL A    L      + +     L++ H  K  +   +      ++ VL
Sbjct   63   KKHGATVLTALGGILKKKGHHEAEIKPLAQSHATKHKIPVKYLEFISECIIQVLQSKHPG  122

Query  122   EFTPPVQAAYQKVVAGVANALAHKY  146
             +F    Q  A  K +      +A Y
Sbjct  123   DFGADAQGAMNKALELFRKDMASNY  147
```

(b)

```
   Score = 18.1 bits (35),   Expect = 0.015, Method: Composition-based stats.
   Identities = 11/24 (45%), Positives = 12/24 (50%), Gaps = 2/24 (8%)

Query   12   VTALWGKVNVD--EVGGEALGRLL   33
             V  +WGKV  D     G E L RL
Sbjct   11   VLNVWGKVEADIPGHGQEVLIRLF   34
```

match	4	11	5	6		6	5	4	5	sum of matches: +60
			6	4					4	
mismatch	-1	1		0		-2	-2	-4	0	sum of mismatches: -13
	-2			0			-3		0	
gap open				-11						sum of gap penalties: -12
gap extend				-1						

total raw score: 60 - 13 - 12 = 35

FIGURE 3.5. *Pairwise alignment of human beta globin (the "query") and myoglobin (the "subject"). Panel (a) shows the alignment from the search shown in Fig. 3.4. Note that this alignment is local (i.e., the entire lengths of each protein are not compared), and there are many positions of identity between the two sequences (indicated with amino acids intervening between the query and subject lines). The alignment contains an internal gap (indicated by two dashes). Panel (b) illustrates how raw scores are calculated, using the result of a separate search with just amino acids 10–34 of HBB (corresponding to the region between the arrowheads in panel a). The raw score is 35; this represents the sum of the match scores (from a BLOSUM62 matrix in this case), the mismatch scores, the gap opening penalty (set to −11 for this search), and the gap extension penalty (set to −1).*

In the pairwise alignment of a segment of HBB and myoglobin, you can see that each pair of residues is assigned a score that is relatively high for matches, and often negative for mismatches.

The *percent similarity* of two protein sequences is the sum of both identical and similar matches. In Fig. 3.5a, there are 57 aligned amino acid residues that are similar. In general, it is more useful to consider the identity shared by two protein sequences, rather than the similarity, because the similarity measure may be based on a variety of definitions of how related (similar) two amino acid residues are to each other.

In summary, pairwise alignment is the process of lining up two sequences to achieve maximal levels of identity (and maximal levels of conservation in the case of amino acid alignments). The purpose of a pairwise alignment is to assess the degree of similarity and the possibility of homology between two molecules. We may say that two proteins share, for example, 25% amino acid identity and 39% similarity. If the amount of sequence identity is sufficient, then the two sequences are probably homologous. It is never correct to say that two proteins share a certain percent homology, because they are either homologous or not. Similarly, it is not appropriate to describe two sequences as "highly homologous"; instead one can say that they share a high degree of similiarity. We will discuss the statistical significance of sequence alignments below, including the use of expect values to assess whether an alignment of two sequences is likely to have occurred by chance (Chapter 4).

Two proteins could have similar structures due to convergent evolution. Molecular evolutionary studies are essential (based on sequence analyses) to assess this possibility.

Box 3.1
Structures and One- and Three-Letter Abbreviations of Twenty Common Amino Acids

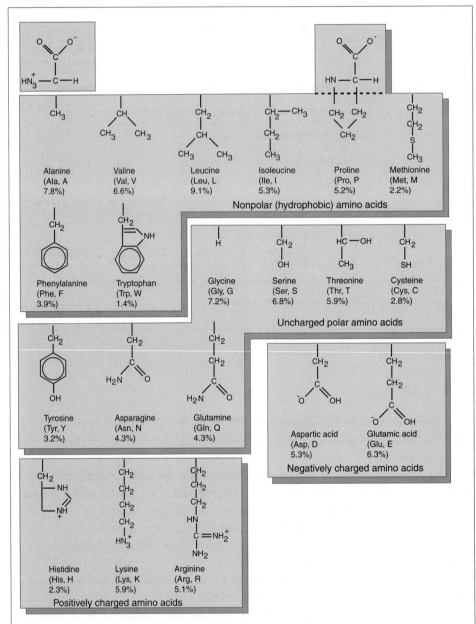

It is very helpful to memorize these abbreviations and to become familiar with the physical properties of the amino acids. The percentages refer to the relative abundance of each amino acid in proteins.

Such analyses provide evidence to assess the hypothesis that two proteins are homologous. Ultimately the strongest evidence to determine whether two proteins are homologous comes from structural studies in combination with evolutionary analyses.

Box 3.2
Algorithms and Programs

An *algorithm* is a procedure that is structured in a computer program (Sedgewick, 1988). For example, there are many algorithms used for pairwise alignment. A computer *program* is a set of instructions that uses an algorithm (or multiple algorithms) to solve a task. For example, the BLAST program (Chapters 3 to 5) uses a set of algorithms to perform sequence alignments. Other programs that we introduce in Chapter 7 use algorithms to generate phylogenetic trees.

Computer programs are essential to solve a variety of bioinformatics problems because millions of operations may need to be performed. The algorithm used by a program provides the means by which the operations of the program are automated. Throughout this book, note how many hundreds of programs have been developed using many hundreds of different algorithms. Each program and algorithm is designed to solve a specific task. An algorithm that is useful to compare one protein sequence to another may not work in a comparison of one sequence to a database of 10 million protein sequences.

Why is it that an algorithm that is useful for comparing two sequences cannot be used to compare millions of sequences? Some problems are so inherently complex that an exhaustive analysis would require a computer with enormous memory or the problem would take an unacceptably long time to complete. A *heuristic algorithm* is one that makes approximations of the best solution without exhaustively considering every possible outcome. The 13 proteins in Fig. 3.2 can be arranged in a tree over a billion distinct ways (see Chapter 7)—and finding the optimal tree is a problem that a heuristic algorithm can solve in a second.

Gaps

Pairwise alignment is useful as a way to identify mutations that have occurred during evolution and have caused divergence of the sequences of the two proteins we are studying. The most common mutations are *substitutions, insertions*, and *deletions*. In protein sequences, substitutions occur when a mutation results in the codon for one amino acid being changed into that for another. This results in the alignment of two nonidentical amino acids, such as serine and threonine. Insertions and deletions occur when residues are added or removed and are typically represented by dashes that are added to one or the other sequence. Insertions or deletions (even those just one character long) are referred to as *gaps* in the alignment.

In our alignment of human beta globin and myoglobin there is one gap (Fig. 3.5a), between the D and E residues of the query). Gaps can occur at the ends of the proteins or in the middle. Note that one of the effects of adding gaps is to make the overall length of each alignment exactly the same. The addition of gaps can help to create an alignment that models evolutionary changes that have occurred. In a typical scoring scheme there are two gap penalties: one for creating a gap (−11 in the example of Fig. 3.5b) and one for each additional residue that a gap extends (−1 in Fig. 3.5b).

Pairwise Alignment, Homology, and Evolution of Life

If two proteins are homologous, they share a common ancestor. Generally, we observe the sequence of proteins (and genes) from organisms that are extant. We

It is possible to infer the sequence of the common ancestor (see Chapter 7).

can compare myoglobins from species such as human, horse, and chicken, and see that the sequences are homologous (Fig. 3.2). This implies that an ancestral organism had a myoglobin gene and lived sometime before the divergences of the lineages that gave rise to human and chicken (over 300 MYA; see Chapter 18). Descendants of that ancestral organism include many vertebrate species. The study of homologous protein (or DNA) sequences by pairwise alignment involves an investigation of the evolutionary history of that protein (or gene).

For a brief overview of the time scale of life on Earth, see Fig. 3.6 (refer to Chapter 13 for a more detailed discussion). The divergence of different species is established through the use of many sources of data, especially the fossil record. Fossils of prokaryotes have been discovered in rocks 3.5 billion years old or even older (Schopf, 2002). Fossils of methane-producing archaea, representative of a second domain of life, are found in rocks over 3 billion years old. The other main domain of life, the eukaryotes, emerged soon after. In the case of globins, in addition to the vertebrate proteins represented in Fig. 3.2, there are plant globins that must have shared a common ancestor with the metazoan (animal) globins some 1.5 billion years ago. There are also many bacterial and archaeal globins suggesting that the globin family arose earlier than two billion years ago.

As we examine a variety of homologous protein sequences, we can observe a wide range of conservation between family members. Some are very ancient and well conserved, such as the enzyme glyceraldehyde-3-phosphate dehydrogenase (GAPDH). A multiple sequence alignment, which is essentially a series of pairwise alignments between a group of proteins, reveals that GAPDH orthologs are extraordinarily well conserved (Fig. 3.7). Such highly conserved proteins may have any degree of representation across the tree of life, from being present in most known species to only a select few.

Databases such as Pfam (Chapter 6) and COGS (Chapter 15) summarize the phylogenetic distribution of gene/protein families across the tree of life.

The GAPDH sequences used to generate Fig. 3.7 and the kappa casein sequences used to generate fig. 3.8 are shown in web documents 3.3 and 3.4 at ▶ http://www.bioinfbook.org/chapter3.

Orthologous kappa caseins from various species provide an example of a less well-conserved family (Fig. 3.8). Some columns of residues in this alignment are perfectly conserved among the selected species, but most are not, and many gaps needed to be introduced. Several positions at which four or even five different residues occur in an aligned column are indicated.

We can see from the preceding examples that pairwise sequence alignment between any two proteins can exhibit widely varying amounts of conservation. We will next examine how the information in such alignments can be used to decide how to quantitate the relatedness of any two proteins.

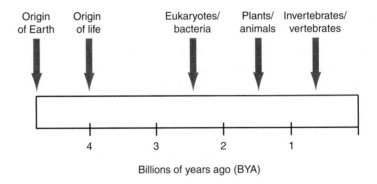

FIGURE 3.6. Overview of the history of life on Earth. See Chapter 13 for details. Gene/protein sequences are analyzed in the context of evolution: Which organisms have orthologous genes? When did these organisms evolve? How related are human and bacterial globins?

```
                          ▼▼▼▼▼           ▼▼            ▼  ▼▼  ▼   ▼▼
NP 002037.2      164   IHDNFGIVEGLMTTVHAITATQKTVDGPSGKLWRDGRGALQNII   207
XP 001162057.1   164   IHDNFGIVEGLMTTVHAITATQKTVDGPSGKLWRDGRGALQNII   207
NP 001003142.1   162   IHDHFGIVEGLMTTVHAITATQKTVDGPSGKMWRDGRGAAQNII   205
XP 893121.1      168   IHDNFGIMEGLMTTVHAITATQKTVDGPSGKLWRDGRGAAQNII   211
XP 576394.1      162   IHDNFGIVEGLMTTVHAITATQKTVDGPSGKLWRDGRGAAQNII   205
NP 058704.1      162   IHDNFGIVEGLMTTVHAITATQKTVDGPSGKLWRDGRGAAQNII   205
XP 001070653.1   162   IHDNFGIVEGLMTTVHAITATQKTVDGPSGKLWRDGRGAAQNII   205
XP 001062726.1   162   IHDNFGIVEGLMTTVHAITATQKTVDGPSGKLWRDGRGAAQNII   205
NP 989636.1      162   IHDNFGIVEGLMTTVHAITATQKTVDGPSGKLWRDGRGAAQNII   205
NP 525091.1      161   INDNFEIVEGLMTTVHAITATQKTVDGPSGKLWRDGRGAAQNII   204
XP 318655.2      161   INDNFGILEGLMTTVHATTATQKTVDGPSGKLWRDGRGAAQNII   204
NP 508535.1      170   INDNFGIIEGLMTTVHAVTATQKTVDGPSGKLWRDGRGAGQNII   213
NP 595236.1      164   INDTFGIEEGLMTTVHATTATQKTVDGPSKKDWRGGRGASANII   207
NP 011708.1      162   INDAFGIEEGLMTTVHSLTATQKTVDGPSHKDWRGGRTASGNII   205
XP 456022.1      161   INDEFGIDEALMTTVHSITATQKTVDGPSHKDWRGGRTASGNII   204
NP 001060897.1   166   IHDNFGIIEGLMTTVHAITATQKTVDGPSSKDWRGGRAASFNII   209
```

FIGURE 3.7. *Multiple sequence alignment of a portion of the glyceraldehyde 3-phosphate dehydrogenase (GAPDH) protein from thirteen organisms:* Homo sapiens *(human),* Pan troglodytes *(chimpanzee),* Canis lupus *(dog),* Mus musculus *(mouse),* Rattus norvegicus *(rat; four variants),* Gallus gallus *(chicken),* Drosophila melanogaster *(fruit fly),* Anopheles gambiae *(mosquito),* Caenorhabditis elegans *(worm),* Schizosaccharomyces pombe *(fission yeast),* Saccharomyces cerevisiae *(baker's yeast),* Kluyveromyces lactis *(a fungus), and* Oryza sativa *(rice). Columns in the alignment having even a single amino acid change are indicated with arrowheads. The accession numbers are given in the figure. The alignment was created by searching HomoloGene at NCBI with the term gapdh. The full alignment is given in Web Document 3.3 at* ► *http://www. bioinfbook.org/chapter3.*

```
                ▼                  ▼             ▼             ▼
                              ▼             ▼▼           ▼
mouse   AIPNPSFLAMPTNENQDNTAIPTIDPITPIVST--PVPTM------ESIVNTVANPEAST
rabbit  S--HPFFMAILPNKMQDKAVTPTTNTIAAVEPT--PIPTT------EPVVSTEVIAEASP
sheep   PHPHLSFMAIPPKKDQDKTEIPAINTIASAEPTVHSTPTT------EAVVNAVDNPEASS
cattle  PHPHLSFMAIPPKKNQDKTEIPTINTIASGEPT--STPTT------EAVESTVATLEDSP
pig     PRPHASFIAIPPKKNQDKTAIPAINSIATVEPT--IVPATEPIVNAEPIVNAVVTPEASS
human   PNLHPSFIAIPPKKIQDKIIIPTINTIATVEPT--PAPAT------EPTVDSVVTPEAFS
horse   PCPHPSFIAIPPKKLQEITVIPKINTIATVEPT--PIPTP------EPTVNNAVIPDASS
         . :  *:*: .:: *:    *   :.*:. .* *:      *.  .   : .
```

FIGURE 3.8. *Multiple sequence alignment of seven kappa caseins, representing a protein family that is relatively poorly conserved. Only a portion of the entire alignment is shown. Note that just eight columns of residues are perfectly conserved (indicated with asterisks), and gaps of varying length form part of the alignment. In several columns, there are four different aligned amino acids (arrowheads); in two instances there are five different residues (double arrowheads). The sequences were aligned with MUSCLE 3.6 (see Chapter 6) and were human (NP_005203), equine (Equus caballus; NP_001075353), pig (Sus scrofa NP_001004026), ovine (Ovis aries NP_001009378), rabbit (Oryctolagus cuniculus P33618), bovine (Bos taurus NP_776719), and mouse (Mus musculus NP_031812). The full alignment is available as web document 3.3 at* ► *http://www.bioinfbook.org/chapter3.*

SCORING MATRICES

When two proteins are aligned, what scores should they be assigned? For the alignment of beta globin and myoglobin in Fig. 3.5a there were specific scores for matches and mismatches; how were they derived? Margaret Dayhoff (1978) provided a model of the rules by which evolutionary change occurs in proteins. We will now examine the Dayhoff model, which provides the basis of a quantitative scoring system for pairwise alignments. This system accounts for scores between any proteins, whether they are closely or distantly related. We will then describe the BLOSUM matrices of

The Dayhoff (1978) reference is to the *Atlas of Protein Sequence and Structure*, a book with 25 chapters (and various coauthors) describing protein families. The 1966 version of the *Atlas* described the sequences of just several dozen proteins (cytochromes c, other respiratory proteins, globins, some enzymes such as lysozyme and ribonucleases, virus coat proteins, peptide hormones, kinins, and fibrinopeptides). The 1978 edition included about 800 protein sequences.

Dayhoff et al. (1972) focused on proteins sharing 85% or more identity. Thus, they could construct their alignments with a high degree of confidence. Later in this chapter, we will see how the Needleman and Wunsch algorithm (described in 1970) permits the optimal alignment of protein sequences.

Steven Henikoff and Jorja G. Henikoff (1992). Next, we will discuss the two main kinds of pairwise sequence algorithms, global and local. Many database searching methods such as BLAST (Chapters 4 and 5) depend in some form on the evolutionary insights of the Dayhoff model.

Dayhoff Model: Accepted Point Mutations

Dayhoff and colleagues considered the problem of how to assign scores to aligned amino acid residues. Their approach was to catalog thousands of proteins and compare the sequences of closely related proteins in many families. They considered the question of which specific amino acid substitutions are observed to occur when two homologous protein sequences are aligned. They defined an *accepted point mutation* as a replacement of one amino acid in a protein by another residue that has been accepted by natural selection. Accepted point mutation is abbreviated *PAM* (which is easier to pronounce than APM). An amino acid change that is accepted by natural selection occurs when (1) a gene undergoes a DNA mutation such that it encodes a different amino acid and (2) the entire species adopts that change as the predominant form of the protein.

Which point mutations are accepted in protein evolution? Intuitively, conservative replacements such as serine for threonine would be most readily accepted. In order to determine all possible changes, Dayhoff and colleagues examined 1572 changes in 71 groups of closely related proteins (Box 3.3). Thus, their definition of "accepted" mutations was based on empirically observed amino acid substitutions. Their approach involved a phylogenetic analysis: rather than comparing two amino acid residues directly, they compared them to the inferred common ancestor of those sequences (Fig. 3.9 and Box 3.4).

For the PAM1 matrix, the proteins have undergone 1% change (that is, 1 accepted point mutation per 100 amino acid residues). The results are shown in Fig. 3.10, which describes the frequency with which any amino acid pairs i, j are aligned. Inspection of this table reveals which substitutions are unlikely to occur (for example, cysteine and tryptophan have noticeably few substitutions), while others such as asparagine and serine tolerate replacements quite commonly. Today, we could generate a table like this with vastly more data (refer to Fig. 2.1 and the explosive growth of GenBank). Several groups have produced updated versions of the PAM matrices (Gonnet et al., 1992; Jones et al., 1992). Nonetheless the findings from 1978 are essentially correct.

The main goal of Dayhoff's approach was to define a set of scores for the comparison of aligned amino acid residues. By comparing two aligned proteins, one can then tabulate an overall score, taking into account identities as well as mismatches, and also applying appropriate penalties for gaps. A scoring matrix defines scores for the interchange of residues i and j. It is given by the probability $q_{i,j}$ of aligning original amino acid residue j with replacement residue i relative to the likelihood of observing residues i by chance (p_i). The scoring matrix further incorporates a logarithm to generate log-odds scores. For the Dayhoff matrices, this takes the following form:

$$s_{i,j} = 10 \times \log\left(\frac{q_{i,j}}{p_i}\right) \tag{3.1}$$

Here the score $s_{i,j}$ refers to the score for aligning any two residues (including an amino acid with itself) along the length of a pairwise alignment. The probability $q_{i,j}$ is the

Box 3.3
Dayhoff's Protein Superfamilies

Dayhoff (1978, p. 3) and colleagues studied 34 protein "superfamilies" grouped into 71 phylogenetic trees. These proteins ranged from some that are very well conserved (e.g., histones and glutamate dehydrogenase; see Fig. 3.7) to others that have a high rate of mutation acceptance (e.g., immunoglobulin [Ig] chains and kappa casein; see Fig. 3.8). Protein families were aligned (compare Fig. 3.7); then they counted how often any one amino acid in the alignment was replaced by another. Here is a partial list of the proteins they studied, including the rates of mutation acceptance. For a more detailed list, see Table 11.1. There is a range of almost 400-fold between the families that evolve fastest and slowest, but within a given family the rate of evolution (measured in PAMs per unit time) varies only two- to threefold between species. Used with permission.

Protein	PAMs per 100 million years
Immunoglobulin (Ig) kappa chain C region	37
Kappa casein	33
Epidermal growth factor	26
Serum albumin	19
Hemoglobin alpha chain	12
Myoglobin	8.9
Nerve growth factor	8.5
Trypsin	5.9
Insulin	4.4
Cytochrome c	2.2
Glutamate dehydrogenase	0.9
Histone H3	0.14
Histone H4	0.10

observed frequency of substitution for each pair of amino acids. The values for q_{ij} are called the "target frequencies," and they are estimated in reference to a particular amount of evolutionary change. For example, in a comparison of human beta globin versus the closely related chimpanzee beta globin, the likelihood of any particular residue matching another in a pairwise alignment is extremely high, while in a comparison of human beta globin and a bacterial globin the likelihood of a match is low. If in a particular comparison of closely related proteins an aligned serine were to change to a threonine 5% of the time, then that target frequency $q_{S,T}$ would be 0.05. If in a different comparison of differently related proteins serine were to change to threonine more often, say 40% of the time, then that target frequency $q_{S,T}$ would be 0.4.

Equation 3.1 describes an odds ratio (Box 3.5). For the numerator, Dayhoff et al. (1972) considered an entire spectrum of models for evolutionary change in determining target frequencies. We begin with the PAM1 matrix, which describes substitutions that occur in very closely related proteins. For the denominator of Equation 3.1, $p_i p_j$, is the probability of amino acid residues i and j occurring by chance. We will

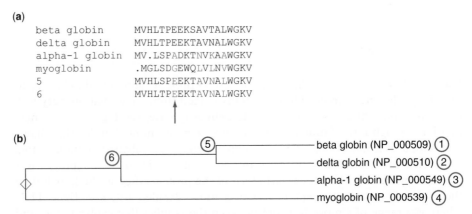

(a)

```
beta globin      MVHLTPEEKSAVTALWGKV
delta globin     MVHLTPEEKTAVNALWGKV
alpha-1 globin   MV.LSPADKTNVKAAWGKV
myoglobin        .MGLSDGEWQLVLNVWGKV
5                MVHLSPEEKTAVNALWGKV
6                MVHLTPEEKTAVNALWGKV
```

(b)

beta globin (NP_000509) ①
delta globin (NP_000510) ②
alpha-1 globin (NP_000549) ③
myoglobin (NP_000539) ④

FIGURE 3.9. *Dayhoff's approach to determining amino acid substitutions. Panel (a) shows a partial multiple sequence alignment of human alpha-1 globin, beta globin, delta globin, and myoglobin. Four columns in which alpha-1 globin and myoglobin have different amino acid residues are indicated in red. For example, A is aligned with G (arrow). Panel (b) shows a phylogenetic tree that shows the four extant sequences (labeled 1 to 4), as well as two internal nodes that represent the ancestral sequences (labeled 5 and 6). The inferred ancestral sequences were identified by maximum parsimony analysis using the software PAUP (Chapter 7), and are displayed in panel (a). From this analysis it is apparent that at each of the columns labeled in red, there was not a direct interchange of two amino acids between alpha-1 globin and myoglobin. Instead, an ancestral residue diverged. For example, the arrow in panel (a) indicates an ancestral glutamate that evolved to become alanine or glycine, but it would not be correct to suggest that alanine had been converted directly to glycine.*

Box 3.4
A Phylogenetic Approach to Aligning Amino Acids

Dayhoff and colleagues did not compare the probability of one residue mutating directly into another. Instead, they constructed phylogenetic trees using parsimony analysis (see Chapter 7). Then, they described the probability that two aligned residues derived from a common ancestral residue. With this approach, they could minimize the confounding effects of multiple substitutions occurring in an aligned pair of residues. As an example, consider an alignment of the four human proteins alpha-1 globin, beta globin, delta globin, and myoglobin. A direct comparison of alpha-1 globin to myoglobin would suggest several amino acid replacements, such as ala ↔ gly, asn ↔ leu, lys ↔ leu, and ala ↔ val (Fig. 3.9a, residues highlighted in red). However, a phylogenetic analysis of these four proteins results in the estimation of internal nodes that represent ancestral sequences. In Fig. 3.9b the external nodes (corresponding to the four existing proteins) are labeled, as are internal nodes 5 and 6, which correspond to inferred ancestral sequences. In one of the four cases that are highlighted in Fig. 3.9a, the ancestral sequences suggest that a glu residue changed to ala and gly in alpha-1 globin and myoglobin, but ala and gly never directly interchanged (Fig. 3.9a, arrow). Thus, the Dayhoff approach was more accurate by taking an evolutionary perspective.

In a further effort to avoid the complicating factor of multiple substitutions occurring in alignments of protein families, Dayhoff et al. also focused on using multiple sequence alignments of closely related proteins. Thus, for example, their analysis of globins considered the alpha globins and beta globins separately.

	A Ala	R Arg	N Asn	D Asp	C Cys	Q Gln	E Glu	G Gly	H His	I Ile	L Leu	K Lys	M Met	F Phe	P Pro	S Ser	T Thr	W Trp	Y Tyr	V Val
A																				
R	30																			
N	109	17																		
D	154	0	532																	
C	33	10	0	0																
Q	93	120	50	76	0															
E	266	0	94	831	0	422														
G	579	10	156	162	10	30	112													
H	21	103	226	43	10	243	23	10												
I	66	30	36	13	17	8	35	0	3											
L	95	17	37	0	0	75	15	17	40	253										
K	57	477	322	85	0	147	104	60	23	43	39									
M	29	17	0	0	0	20	7	7	0	57	207	90								
F	20	7	7	0	0	0	0	17	20	90	167	0	17							
P	345	67	27	10	10	93	40	49	50	7	43	43	4	7						
S	772	137	432	98	117	47	86	450	26	20	32	168	20	40	269					
T	590	20	169	57	10	37	31	50	14	129	52	200	28	10	73	696				
W	0	27	3	0	0	0	0	0	3	0	13	0	0	10	0	17	0			
Y	20	3	36	0	30	0	10	0	40	13	23	10	0	260	0	22	23	6		
V	365	20	13	17	33	27	37	97	30	661	303	17	77	10	50	43	186	0	17	
	A Ala	R Arg	N Asn	D Asp	C Cys	Q Gln	E Glu	G Gly	H His	I Ile	L Leu	K Lys	M Met	F Phe	P Pro	S Ser	T Thr	W Trp	Y Tyr	V Val

FIGURE 3.10. *Numbers of accepted point mutations, multiplied by 10, in 1572 cases of amino acid substitutions from closely related protein sequences. This figure is modified from Dayhoff (1978, p. 346). Amino acids are presented alphabetically according to the three-letter code. Notice that some substitutions are very commonly accepted (such as V and I or S and T). Other amino acids, such as C and W, are rarely substituted by any other residue. Used with permission.*

Box 3.5
Statistical Concept: The Odds Ratio

Dayhoff et al. (1972) developed their scoring matrix by using odds ratios. The mutation probability matrix has elements M_{ij} that give the probability that amino acid j changes to amino acid i in a given evolutionary interval. The normalized frequency f_i gives the probability that amino acid i will occur at that given amino acid position by chance. The relatedness odds matrix in Equation 3.1 may also be expressed as follows:

$$R_{ij} = \frac{M_{ij}}{f_i}$$

Here, R_{ij} is the relatedness odds ratio. Equation 3.1 may also be represented:

$$\text{Probability of an authentic alignment} = \frac{p(\text{aligned} \mid \text{authentic})}{p(\text{aligned} \mid \text{random})}$$

The right side of this equation can be read, "the probability of an alignment given that it is authentic (i.e. the substitution of amino acid j with amino acid i) divided by the probability that the alignment occurs given that it happened by chance. An *odds ratio* can be any positive ratio. The *probability* that an event will occur is the fraction of times it is expected to be observed over many trials; probabilities have values ranging from 0 to 1. Odds and probability are closely related concepts. A probability of 0 corresponds to an odds of 0; a probability of 0.5 corresponds to an odds of 1.0; a probability of 0.75 corresponds to odds of 75:25 or 3. Odds and probabilities may be converted as follows:

$$odds = \frac{probability}{1 - probability} \quad and \quad probability = \frac{odds}{1 + odds}$$

next explain how they calculated these values, resulting in the creation of an entire series of scoring matrices.

Dayhoff et al. calculated the relative mutabilities of the amino acids (Table 3.1). This simply describes how often each amino acid is likely to change over a short evolutionary period. (We note that the evolutionary period in question is short because this analysis involves protein sequences that are closely related to each other.) To calculate relative mutability, they divided the number of times each amino acid was observed to mutate by the overall frequency of occurrence of that amino acid. Table 3.2 shows the frequency with which each amino acid is found.

Why are some amino acids more mutable than others? The less mutable residues probably have important structural or functional roles in proteins, such that the consequence of replacing them with any other residue could be harmful to the organism. (We will see in Chapter 20 that many human diseases, from cystic fibrosis to the autism-related Rett syndrome to hemoglobinopathies, can be caused by a single amino acid substitution in a protein.) Conversely, the most mutable amino acids—asparagine, serine, aspartic acid, and glutamic acid—have functions in proteins that are easily assumed by other residues. The most common substitutions seen in Fig. 3.10 are glutamic acid for aspartic acid (both are acidic), serine for

You can look up a recent estimate of the frequency of occurrence of each amino acid at the SwissProt website ► http://www.expasy.ch/sprot/relnotes/relstat.html. From the UniProtKB/Swiss-Prot protein knowledgebase (release 51.7), the amino acid composition of all proteins is shown in web document 3.5 (► http://www.bioinfbook.org/chapter3).

TABLE 3-1	Relative Mutabilities of Amino Acids		
Asn	134	His	66
Ser	120	Arg	65
Asp	106	Lys	56
Glu	102	Pro	56
Ala	100	Gly	49
Thr	97	Tyr	41
Ile	96	Phe	41
Met	94	Leu	40
Gln	93	Cys	20
Val	74	Trp	18

The value of alanine is arbitrarily set to 100.
Source: From Dayhoff (1978). Used with permission.

TABLE 3-2	Normalized Frequencies of Amino Acid		
Gly	0.089	Arg	0.041
Ala	0.087	Asn	0.040
Leu	0.085	Phe	0.040
Lys	0.081	Gln	0.038
Ser	0.070	Ile	0.037
Val	0.065	His	0.034
Thr	0.058	Cys	0.033
Pro	0.051	Tyr	0.030
Glu	0.050	Met	0.015
Asp	0.047	Trp	0.010

These values sum to 1. If the 20 amino acids were equally represented in proteins, these values would all be 0.05 (i.e., 5%); instead, amino acids vary in their frequency of occurrence
Source: From Dayhoff (1978). Used with permission.

alanine, serine for threonine (both are hydroxylated), and isoleucine for valine (both are hydrophobic and of a similar size).

The substitutions that occur in proteins can also be understood with reference to the genetic code (Box 3.6). Observe how common amino acid substitutions tend to require only a single nucleotide change. For example, aspartic acid is encoded by GAU or GAC, and changing the third position to either A or G causes the codon to encode a glutamic acid. Also note that four of the five least mutable amino acids (tryptophan, cysteine, phenylalanine, and tyrosine) are specified by only one or two codons. A mutation of any of the three bases of the W codon is guaranteed to change that amino acid. The low mutability of this amino acid suggests that substitutions are not tolerated by natural selection. Of the eight least mutable amino acids (Table 3.1), only one (leucine) is specified by six codons, and only two (glycine and proline) are specified by four codons. The others are specified by one or two codons.

PAM1 Matrix

Dayhoff and colleagues next used the data on accepted mutations (Fig. 3.10) and the probabilities of occurrence of each amino acid to generate a *mutation probability*

Box 3.6
The Standard Genetic Code

Second nucleotide

		T	C	A	G	
First nucleotide	**T**	TTT Phe 171 TTC Phe 203 TTA Leu 73 TTG Leu 125	TCT Ser 147 TCC Ser 172 TCA Ser 118 TCG Ser 45	TAT Tyr 124 TAC Tyr 158 TAA Ter 0 TAG Ter 0	TGT Cys 99 TGC Cys 119 TGA Ter 0 TGG Trp 122	T C A G
	C	CTT Leu 127 CTC Leu 187 CTA Leu 69 CTG Leu 392	CCT Pro 175 CCC Pro 197 CCA Pro 170 CCG Pro 69	CAT His 104 CAC His 147 CAA Gln 121 CAG Gln 343	CGT Arg 47 CGC Arg 107 CGA Arg 63 CGG Arg 115	T C A G
	A	ATT Ile 165 ATC Ile 218 ATA Ile 71 ATG Met 221	ACT Thr 131 ACC Thr 192 ACA Thr 150 ACG Thr 63	AAT Asn 174 AAC Asn 199 AAA Lys 248 AAG Lys 331	AGT Ser 121 AGC Ser 191 AGA Arg 113 AGG Arg 110	T C A G
	G	GTT Val 111 GTC Val 146 GTA Val 72 GTG Val 288	GCT Ala 185 GCC Ala 282 GCA Ala 160 GCG Ala 74	GAT Asp 230 GAC Asp 262 GAA Glu 301 GAG Glu 404	GGT Gly 112 GGC Gly 230 GGA Gly 168 GGG Gly 160	T C A G

Third nucleotide

In this table, the 64 possible codons are depicted along with the frequency of codon utilization and the three-letter code of the amino acid that is specified. There are four bases (A, C, G, U) and three bases per codon, so there are $4^3 = 64$ codons.

Several features of the genetic code should be noted. Amino acids may be specified by one codon (M, W), two codons (C, D, E, F, H, K, N, Q, Y), three codons (I), four codons (A, G, P, T, V), or six codons (L, R, S). UGA is rarely read as a selenocysteine (abbreviated sec, and the assigned single-letter abbreviations is U).

For each block of four codons that are grouped together, one is often used dramatically less frequently. For example, for F, L, I, M, and V (i.e., codons with a U in the middle, occupying the first column of the genetic code), adenine is used relatively infrequently in the third-codon position. For codons with a cytosine in the middle position, guanine is strongly underrepresented in the third position.

Also note that in many cases mutations cause a conservative change (or no change at all) in the amino acid. Consider threonine (ACX). Any mutation in the third position causes no change in the specified amino acid, because of "wobble." If the first nucleotide of any threonine codon is mutated from A to U, the conservative replacement to a serine occurs. If the second nucleotide C is mutated to a G, a serine replacement occurs. Similar patterns of conservative substitution can be seen along the entire first column of the genetic code, where all of the residues are hydrophobic, and for the charged residues D, E and K, R as well.

Codon usage varies between organisms and between genes within organisms. Note also that while this is the standard genetic code, some organisms use alternate genetic codes. A group of two dozen alternate genetic codes are listed at the NCBI Taxonomy website, ▶ http://www.ncbi.nlm.nih.gov/Taxonomy/taxonomyhome.html/. As an example of a nonstandard code, vertebrate mitochondrial genomes use AGA and AGG to specify termination (rather than arg in the standard code), ATA to specify met (rather than ile), and TGA to specify trp (rather then termination).

Source: Adapted from the International Human Genome Sequencing Consortium (2001), fig. 34. Used with permission.

matrix M (Fig. 3.11). Each element of the matrix M_{ij} shows the probability that an original amino acid *j* (see the columns) will be replaced by another amino acid *i* (see the rows) over a defined evolutionary interval. In the case of Fig. 3.11 the interval is one PAM, which is defined as the unit of evolutionary divergence in which 1% of the amino acids have been changed between the two protein sequences. Note that the evolutionary interval of this PAM matrix is defined in terms of percent amino acid divergence and not in units of years. One percent divergence of protein sequence may occur over vastly different time frames for protein families that undergo substitutions at different rates.

Examination of Fig. 3.11 reveals several important features. The highest scores are distributed in a diagonal from top left to bottom right. The values in each column sum to 100%. The value 98.67 at the top left indicates that when the original sequence consists of an alanine there is a 98.67% chance that the replacement amino acid will also be an alanine over an evolutionary distance of one PAM. There is a 0.28% chance that it will be changed to serine. The most mutable amino acid (from Table 3.1), asparagine, has only a 98.22% chance of remaining unchanged; the least mutable amino acid, tryptophan, has a 99.76% chance of remaining the same.

For each original amino acid, it is easy to observe the amino acids that are most likely to replace it if a change should occur. These data are very relevant to pairwise sequence alignment because they will form the basis of a scoring system (described below) in which reasonable amino acid substitutions in an alignment are rewarded while unlikely substitutions are penalized. These concepts are also relevant to database searching algorithms such as BLAST (Chapters 4 and 5) which depend on rules to score the relatedness of molecular sequences.

Almost all molecular sequence data are obtained from extant organisms. We can infer ancestral sequences, as described in Box 3.4 and Chapter 7. But in general, for an aligned pair of residues *i, j* we do not know which one mutated into the other. Dayhoff and colleagues used the assumption that accepted amino acid mutations are undirected, that is, they are equally likely in either direction. In the PAM1 matrix, the close relationship of the proteins makes it unlikely that the ancestral residue is entirely different than both of the observed, aligned residues.

PAM250 and Other PAM Matrices

The PAM1 matrix was based on the alignment of closely related protein sequences, all of which were at least 85% identical within a protein family. We are often interested in exploring the relationships of proteins that share far less than 85% amino acid identity. We can accomplish this by constructing probability matrices for proteins that share any degree of amino acid identity. Consider closely related proteins, such as the GAPDH proteins shown in Fig. 3.7. A mutation from one residue to another

	A Ala	R Arg	N Asn	D Asp	C Cys	Q Gln	E Glu	G Gly	H His	I Ile	L Leu	K Lys	M Met	F Phe	P Pro	S Ser	T Thr	W Trp	Y Tyr	V Val
A	98.67	0.02	0.09	0.10	0.03	0.08	0.17	0.21	0.02	0.06	0.04	0.02	0.06	0.02	0.22	0.35	0.32	0.00	0.02	0.18
R	0.01	99.13	0.01	0.00	0.01	0.10	0.00	0.00	0.10	0.03	0.01	0.19	0.04	0.01	0.04	0.06	0.01	0.08	0.00	0.01
N	0.04	0.01	98.22	0.36	0.00	0.04	0.06	0.06	0.21	0.03	0.01	0.13	0.00	0.01	0.02	0.20	0.09	0.01	0.04	0.01
D	0.06	0.00	0.42	98.59	0.00	0.06	0.53	0.06	0.04	0.01	0.00	0.03	0.00	0.00	0.01	0.05	0.03	0.00	0.00	0.01
C	0.01	0.01	0.00	0.00	99.73	0.00	0.00	0.00	0.01	0.01	0.00	0.00	0.00	0.00	0.01	0.05	0.01	0.00	0.03	0.02
Q	0.03	0.09	0.04	0.05	0.00	98.76	0.27	0.01	0.23	0.01	0.03	0.06	0.04	0.00	0.06	0.02	0.02	0.00	0.00	0.01
E	0.10	0.00	0.07	0.56	0.00	0.35	98.65	0.04	0.02	0.03	0.01	0.04	0.01	0.00	0.03	0.04	0.02	0.00	0.01	0.02
G	0.21	0.01	0.12	0.11	0.01	0.03	0.07	99.35	0.01	0.00	0.01	0.02	0.01	0.01	0.03	0.21	0.03	0.00	0.00	0.05
H	0.01	0.08	0.18	0.03	0.01	0.20	0.01	0.00	99.12	0.00	0.01	0.01	0.00	0.02	0.03	0.01	0.01	0.01	0.04	0.01
I	0.02	0.02	0.03	0.01	0.02	0.01	0.02	0.00	0.00	98.72	0.09	0.02	0.21	0.07	0.00	0.01	0.07	0.00	0.01	0.33
L	0.03	0.01	0.03	0.00	0.00	0.06	0.01	0.01	0.04	0.22	99.47	0.02	0.45	0.13	0.03	0.01	0.03	0.04	0.02	0.15
K	0.02	0.37	0.25	0.06	0.00	0.12	0.07	0.02	0.02	0.04	0.01	99.26	0.20	0.00	0.03	0.08	0.11	0.00	0.01	0.01
M	0.01	0.01	0.00	0.00	0.00	0.02	0.00	0.00	0.00	0.05	0.08	0.04	98.74	0.01	0.00	0.01	0.02	0.00	0.00	0.04
F	0.01	0.01	0.01	0.00	0.00	0.00	0.00	0.01	0.02	0.08	0.06	0.00	0.04	99.46	0.00	0.02	0.01	0.03	0.28	0.00
P	0.13	0.05	0.02	0.01	0.01	0.08	0.03	0.02	0.05	0.01	0.02	0.02	0.01	0.01	99.26	0.12	0.04	0.00	0.00	0.02
S	0.28	0.11	0.34	0.07	0.11	0.04	0.06	0.16	0.02	0.02	0.01	0.07	0.04	0.03	0.17	98.40	0.38	0.05	0.02	0.02
T	0.22	0.02	0.13	0.04	0.01	0.03	0.02	0.02	0.01	0.11	0.02	0.08	0.06	0.01	0.05	0.32	98.71	0.00	0.02	0.09
W	0.00	0.02	0.00	0.00	0.00	0.00	0.00	0.00	0.00	0.00	0.00	0.00	0.00	0.01	0.00	0.01	0.00	99.76	0.01	0.00
Y	0.01	0.00	0.03	0.00	0.03	0.00	0.01	0.00	0.04	0.01	0.01	0.00	0.00	0.21	0.00	0.01	0.01	0.02	99.45	0.01
V	0.13	0.02	0.01	0.01	0.03	0.02	0.02	0.03	0.03	0.57	0.11	0.01	0.17	0.01	0.03	0.02	0.10	0.00	0.02	99.01

FIGURE 3.11. The PAM1 mutation probability matrix. From Dayhoff (1978, p. 348, fig. 82). The original amino acid j is arranged in columns (across the top), while the replacement amino acid i is arranged in rows. Used with permission.

is a relatively rare event, and a scoring system used to align two such closely related proteins should reflect this. In the PAM1 mutation probability matrix (Fig. 3.11) some substitutions such as tryptophan to threonine are so rare that they are never observed in the data set. But next consider two distantly related proteins, such as the kappa caseins shown in Fig. 3.8. Here, substitutions are likely to be very common. PAM matrices such as PAM100 or PAM250 were generated to reflect the kinds of amino acid substitutions that occur in distantly related proteins.

How are PAM matrices other than PAM1 derived? Dayhoff et al. multiplied the PAM1 matrix by itself, up to hundreds of times, to obtain other PAM matrices (see Box 3.7). Thus they extrapolated from the PAM1 matrix.

To make sense of what different PAM matrices mean, consider the extreme cases. When PAM equals zero, the matrix is a unit diagonal (Fig. 3.12), because no amino acids have changed. PAM can be extremely large (e.g., PAM greater than 2000, or the matrix can even be multiplied against itself an infinite number of times). In the resulting PAM∞ matrix there is an equal likelihood of any amino acid being present and all the values consist of rows of probabilities that approximate the background probability for the frequency of occurrence of each amino acid (Fig. 3.12, lower panel). We described these background frequencies in Table 3.2.

The PAM250 matrix is of particular interest (Fig. 3.13). It is produced when the PAM1 matrix is multiplied against itself 250 times, and it is one of the common matrices used for BLAST searches of databases (Chapter 4). This matrix applies

Box 3.7
Matrix Multiplication

A matrix is an orderly array of numbers. An example of a matrix with rows i and columns j is:

$$\begin{bmatrix} 1 & 2 & 4 \\ 2 & 0 & -3 \\ 4 & -3 & 6 \end{bmatrix}$$

In a symmetric matrix, such as the one above, $a_{ij} = a_{ji}$. This means that all the corresponding nondiagonal elements are equal. Matrices may be added, subtracted, or manipulated in a variety of ways. Two matrices can be multiplied together provided that the number of columns in the first matrix M_1 equals the number of rows in the second matrix M_2. Following is an example of how to multiply M_1 by M_2.

Successively multiply each row of M_1 by each column of M_2:

$$M_1 = \begin{bmatrix} 3 & 4 \\ 0 & 2 \end{bmatrix} \quad M_2 = \begin{bmatrix} 5 & -2 \\ 2 & 1 \end{bmatrix}$$

$$M_{12} = \begin{bmatrix} (3)(5) + (4)(2) & (3)(-2) + (4)(1) \\ (0)(5) + (2)(2) & (0)(-2) + (2)(1) \end{bmatrix} = \begin{bmatrix} 23 & -2 \\ 4 & 2 \end{bmatrix}$$

If you want to try matrix multiplication yourself, enter the PAM1 mutation probability matrix of Fig. 3.11 into a program such as MATLAB® (Mathworks), divide each value by 10,000, and multiply the matrix times itself 250 times. You will get the PAM250 matrix of Fig. 3.13.

FIGURE 3.12. Portion of the matrices for a zero PAM value (PAM0; upper panel) or for an infinite PAM∞ value (lower panel). At PAM∞ (i.e., if the PAM1 matrix is multiplied against itself an infinite number of times), all the entries in each row converge on the normalized frequency of the replacement amino acid (see Table 3.2). A PAM2000 matrix has similar values that tend to converge on these same limits. In a PAM2000 matrix, the proteins being compared are at an extreme of unrelatedness. In constrast, at PAM0, no mutations are tolerated, and the residues of the proteins are perfectly conserved.

original amino acid

PAM0	A	R	N	D	C	Q	E	G
A	100	0	0	0	0	0	0	0
R	0	100	0	0	0	0	0	0
N	0	0	100	0	0	0	0	0
D	0	0	0	100	0	0	0	0
C	0	0	0	0	100	0	0	0
Q	0	0	0	0	0	100	0	0
E	0	0	0	0	0	0	100	0
G	0	0	0	0	0	0	0	100

(replacement amino acid)

original amino acid

PAM∞	A	R	N	D	C	Q	E	G
A	8.7	8.7	8.7	8.7	8.7	8.7	8.7	8.7
R	4.1	4.1	4.1	4.1	4.1	4.1	4.1	4.1
N	4.0	4.0	4.0	4.0	4.0	4.0	4.0	4.0
D	4.7	4.7	4.7	4.7	4.7	4.7	4.7	4.7
C	3.3	3.3	3.3	3.3	3.3	3.3	3.3	3.3
Q	3.8	3.8	3.8	3.8	3.8	3.8	3.8	3.8
E	5.0	5.0	5.0	5.0	5.0	5.0	5.0	5.0
G	8.9	8.9	8.9	8.9	8.9	8.9	8.9	8.9

(replacement amino acid)

to an evolutionary distance where proteins share about 20% amino acid identity. Compare this matrix to the PAM1 matrix (Fig. 3.11) and note that much of the information content is lost. The diagonal from top left to bottom right tends to contain higher values than elsewhere in the matrix, but not in the dramatic fashion of the PAM1 matrix. As an example of how to read the PAM250 matrix, if the original amino acid is an alanine, there is just a 13% chance that the second sequence will also have an alanine. In fact, there is a nearly equal probability (12%) that the alanine will have been replaced by a glycine. For the least mutable amino acids, tryptophan and cysteine, there is more than a 50% probability that those residues will remain unchanged at this evolutionary distance.

FIGURE 3.13. The PAM250 mutation probability matrix. From Dayhoff (1978, p. 350, fig. 83). At this evolutionary distance, only one in five amino acid residues remains unchanged from an original amino acid sequence (columns) to a replacement amino acid (rows). Note that the scale has changed relative to Fig. 3.11, and the columns sum to 100. Used with permission.

	A	R	N	D	C	Q	E	G	H	I	L	K	M	F	P	S	T	W	Y	V
A	13	6	9	9	5	8	9	12	6	8	6	7	7	4	11	11	11	2	4	9
R	3	17	4	3	2	5	3	2	6	3	2	9	4	1	4	4	3	7	2	2
N	4	4	6	7	2	5	6	4	6	3	2	5	3	2	4	5	4	2	3	3
D	5	4	8	11	1	7	10	5	6	3	2	5	3	1	4	5	5	1	2	3
C	2	1	1	1	52	1	1	2	2	2	1	1	1	1	2	3	2	1	4	2
Q	3	5	5	6	1	10	7	3	7	2	3	5	3	1	4	3	3	1	2	3
E	5	4	7	11	1	9	12	5	6	3	2	5	3	1	4	5	5	1	2	3
G	12	5	10	10	4	7	9	27	5	5	4	6	5	3	8	11	9	2	3	7
H	2	5	5	4	2	7	4	2	15	2	2	3	2	2	3	3	2	2	3	2
I	3	2	2	2	2	2	2	2	2	10	6	2	6	5	2	3	4	1	3	9
L	6	4	4	3	2	6	4	3	5	15	34	4	20	13	5	4	6	6	7	13
K	6	18	10	8	2	10	8	5	8	5	4	24	9	2	6	8	8	4	3	5
M	1	1	1	1	0	1	1	1	1	2	3	2	6	2	1	1	1	1	1	2
F	2	1	2	1	1	1	1	1	3	5	6	1	4	32	1	2	2	4	20	3
P	7	5	5	4	3	5	4	5	5	3	3	4	3	2	20	6	5	1	2	4
S	9	6	8	7	7	6	7	9	6	5	4	7	5	3	9	10	9	4	4	6
T	8	5	6	6	4	5	5	6	4	6	4	6	5	3	6	8	11	2	3	6
W	0	2	0	0	0	0	0	0	1	0	1	0	0	1	0	1	0	55	1	0
Y	1	1	2	1	3	1	1	1	3	2	2	1	2	15	1	2	2	3	31	2
V	7	4	4	4	4	4	4	5	4	15	10	4	10	5	5	5	7	2	4	17

From a Mutation Probability Matrix to a Log-Odds Scoring Matrix

Our goal in studying PAM matrices is to derive a scoring system so that we can assess the relatedness of two sequences. When we perform BLAST searches (Chapters 4 and 5) or pairwise alignments, we employ a scoring matrix, but it is not in the form we have described so far. The PAM250 mutation probability matrix (Fig. 3.13) is useful because it describes the frequency of amino acid replacements between distantly related proteins. We next need to convert the elements of a PAM mutation probability matrix into a scoring matrix, also called a log-odds matrix or relatedness odds matrix.

The cells in a log-odds matrix consist of scores as defined in Equation 3.1 above. The target frequencies q_{ij} are derived from a mutation probability matrix, such as those shown in Figs. 3.11 (for PAM1) and 3.13 (for PAM250). These values consist of positive numbers that sum to 1. The background frequencies $p_i p_j$ reflect the independent probabilities of each amino acid i, j occurring in this position. Its values were given in Table 3.2.

For this scoring system Dayhoff and colleagues took 10 times the base 10 logarithm of the odds ratio (Equation 3.1). Using the logarithm here is helpful because it allows us to sum the scores of the aligned residues when we perform an overall alignment of two sequences. (If we did not take the logarithm, we would need to multiply the ratios at all the aligned positions, and this is computationally more cumbersome.)

A log-odds matrix for PAM250 is shown in Fig. 3.14. The values have been rounded off to the nearest integer. Try using Equation 3.1 to make sure you understand how the mutation probability matrix (Fig. 3.13) is converted into the log-odds scoring matrix (Fig. 3.14). As an example, to determine the score assigned to two aligned tryptophan residues, the PAM250 mutation probability matrix value is 0.55 (Fig. 3.13), and the normalized frequency of tryptophan is 0.010 (Table 3.2). Thus,

Note that this scoring matrix is symmetric, in contrast to the mutation probability matrix in Fig. 3.13. In a comparison of two sequences it does not matter which is given first. In problem [3-6] of this chapter we will calculate the likelihood of changing cys to glu, then of changing glu to cys.

$$S_{(tryptophan, tryptophan)} = 10 \times \log_{10}\left(\frac{0.55}{0.01}\right) = +17.4 \qquad (3.2)$$

A	2																		
R	-2	6																	
N	0	0	2																
D	0	-1	2	4															
C	-2	-4	-4	-5	12														
Q	0	1	1	2	-5	4													
E	0	-1	1	3	-5	2	4												
G	1	-3	0	1	-3	-1	0	5											
H	-1	2	2	1	-3	3	1	-2	6										
I	-1	-2	-2	-2	-2	-2	-2	-3	-2	5									
L	-2	-3	-3	-4	-6	-2	-3	-4	-2	-2	6								
K	-1	3	1	0	-5	1	0	-2	0	-2	-3	5							
M	-1	0	-2	-3	-5	-1	-2	-3	-2	2	4	0	6						
F	-3	-4	-3	-6	-4	-5	-5	-5	-2	1	2	-5	0	9					
P	1	0	0	-1	-3	0	-1	0	0	-2	-3	-1	-2	-5	6				
S	1	0	1	0	0	-1	0	1	-1	-1	-3	0	-2	-3	1	2			
T	1	-1	0	0	-2	-1	0	0	-2	0	-2	0	-1	-3	0	1	3		
W	-6	2	-4	-7	-8	-5	-7	-7	-3	-5	-2	-3	-4	0	-6	-2	-5	17	
Y	-3	-4	-2	-4	0	-4	-4	-5	0	-1	-1	-4	-2	7	-5	-3	-3	0	10
V	0	-2	-2	-2	-2	-2	-2	-1	-2	4	2	-2	2	-1	-1	-1	0	-6	-2	4
	A	R	N	D	C	Q	E	G	H	I	L	K	M	F	P	S	T	W	Y	V

FIGURE 3.14. *Log-odds matrix for PAM250. High PAM values (e.g., PAM250) are useful for aligning very divergent sequences. A variety of algorithms for pairwise alignment, multiple sequence alignment, and database searching (e.g., BLAST) allow you to select an assortment of PAM matrices such as PAM250, PAM70, and PAM30.*

We state that a score of +17 for tryptophan indicates that the correspondence of two tryptophans in an alignment of homologous proteins is 50 times more likely than a chance alignment of two tryptophan residues. How do we derive the number 50? From Equation 3.1, let $S_{i,j} = +17$ and let the probability of replacement $q_{ij}/p_i = x$. Then $+17 = 10 \log_{10} x$, $+1.7 = \log_{10} x$, and $10^{1.7} = x = 50$.

This value is rounded off to 17 in the PAM250 log-odds matrix (Fig. 3.14). What do the scores in the PAM250 matrix signify? A score of -10 indicates that the correspondence of two amino acids in an alignment that accurately represents homology (evolutionary descent) is one-tenth as frequent as the chance alignment of these amino acids. This assumes that each was randomly selected from the background amino acid frequency distribution. A score of zero is neutral. A score of $+17$ for tryptophan indicates that this correspondence is 50 times more frequent than the chance alignment of this residue in a pairwise alignment. A score of $+2$ indicates that the amino acid replacement occurs 1.6 times as frequently as expected by chance. The highest values in this particular matrix are for tryptophan (17 for an identity) and cysteine (12), while the most severe penalties are associated with substitutions for those two residues. When two sequences are aligned and a score is given, that score is simply the sum of the scores for all the aligned residues across the alignment.

It is easy to see how different PAM matrices score amino acid substitutions by comparing the PAM250 matrix (Fig. 3.14) with a PAM10 matrix (Fig. 3.15). In the PAM10 matrix, identical amino acid residue pairs tend to produce a higher score than in the PAM250 matrix; for example, a match of alanine to alanine scores 7 versus 2, respectively. The penalties for mismatches are greater in the PAM10 matrix; for example, a mutation of aspartate to arginine scores -17 (PAM10) versus -1 (PAM250). PAM10 even has negative scores for substitutions (such as glutamate to asparagine, -5) that are scored positively in the PAM250 matrix ($+1$).

Practical Usefulness of PAM Matrices in Pairwise Alignment

We can demonstrate the usefulness of PAM matrices by performing a series of global pairwise alignments of both closely related proteins and distantly related proteins. For the closely related proteins we will use human beta globin (NP_000509) and beta globin from the chimpanzee *Pan troglodytes* (XP_508242); these proteins share 100% amino acid identity. The bit scores proceed in a fairly linear, decreasing fashion from about 590 bits using the PAM10 matrix to 200 bits using the PAM250 matrix and 100 bits using the PAM500 matrix (Fig. 3.16, black line). In this pairwise alignment there are no mismatches or gaps, and the high bit scores associated with low PAM matrices (such as PAM10) are accounted for by the lower relative entropy (defined below). The PAM10 matrix is thus appropriate for comparisons of closely related proteins. Next consider pairwise alignments of two relatively divergent proteins, human beta globin and alpha globin (NP_000549) (Fig. 3.16, red line). The PAM70 matrix yields the highest score. Lower PAM matrices (e.g., PAM10 to PAM60) produce lower bit scores because the sequences share only 42% amino acid identity, and mismatches are assigned large negative scores. We conclude that different scoring matrices vary in their sensitivity to protein sequences (or DNA sequences) of varying relatedness. When you compare two sequences you may need to repeat the search using several different scoring matrices. Alignment programs cannot be preset to choose the right matrix for each pair of sequences. Instead they begin with the most broadly useful scoring matrix such as BLOSUM62, which we describe next.

Important Alternative to PAM: BLOSUM Scoring Matrices

In addition to the PAM matrices, another very common set of scoring matrices is the blocks substitution matrix (BLOSUM) series. Henikoff and Henikoff (1992, 1996)

	A	R	N	D	C	Q	E	G	H	I	L	K	M	F	P	S	T	W	Y	V
A	7																			
R	-10	9																		
N	-7	-9	9																	
D	-6	-17	-1	8																
C	-10	-11	-17	-21	10															
Q	-7	-4	-7	-6	-20	9														
E	-5	-15	-5	0	-20	-1	8													
G	-4	-13	-6	-6	-13	-10	-7	7												
H	-11	-4	-2	-7	-10	-2	-9	-13	10											
I	-8	-8	-8	-11	-9	-11	-8	-17	-13	9										
L	-9	-12	-10	-19	-21	-8	-13	-14	-9	-4	7									
K	-10	-2	-4	-8	-20	-6	-7	-10	-10	-9	-11	7								
M	-8	-7	-15	-17	-20	-7	-10	-12	-17	-3	-2	-4	12							
F	-12	-12	-12	-21	-19	-19	-20	-12	-9	-5	-5	-20	-7	9						
P	-4	-7	-9	-12	-11	-6	-9	-10	-7	-12	-10	-10	-11	-13	8					
S	-3	-6	-2	-7	-6	-8	-7	-4	-9	-10	-12	-7	-8	-9	-4	7				
T	-3	-10	-5	-8	-11	-9	-9	-10	-11	-5	-10	-6	-7	-12	-7	-2	8			
W	-20	-5	-11	-21	-22	-19	-23	-21	-10	-20	-9	-18	-19	-7	-20	-8	-19	13		
Y	-11	-14	-7	-17	-7	-18	-11	-20	-6	-9	-10	-12	-17	-1	-20	-10	-9	-8	10	
V	-5	-11	-12	-11	-9	-10	-10	-9	-9	-1	-5	-13	-4	-12	-9	-10	-6	-22	-10	8

FIGURE 3.15. Log-odds matrix for PAM10. Low PAM values such as this are useful for aligning very closely related sequences. Compare this with the PAM250 matrix (Fig. 3.14) and note that there are larger positive scores for identical matches in this PAM10 matrix and larger penalties for mismatches.

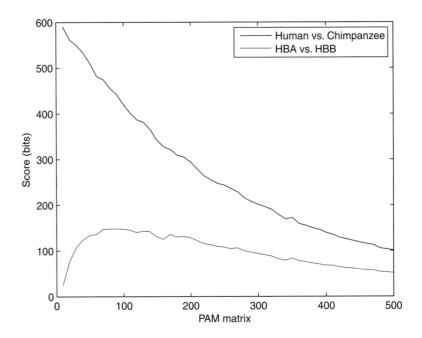

FIGURE 3.16. *Global pairwise alignment scores using a series of PAM matrices. Two closely related globins (human and chimpanzee beta globin; black line) were aligned using a series of PAM matrices (x axis) and the bit scores were measured (y axis). For two distantly related globins (human alpha versus beta globin; red line) the bit scores are smaller for low PAM matrices (such as PAM1 to PAM20) because mismatches are severely penalized.*

used the BLOCKS database, which consisted of over 500 groups of local multiple alignments (blocks) of distantly related protein sequences. Thus the Henikoffs focused on conserved regions (blocks) of proteins that are distantly related to each other. The BLOSUM scoring scheme employs a log-odds ratio using the base 2 logarithm:

$$s_{ij} = 2 \times \log_2 \left(\frac{q_{ij}}{p_{ij}} \right) \tag{3.3}$$

Equation 3.3 resembles Equation 3.1 in its format. Karlin and Altschul (1990) and Altschul (1991) have shown that substitution matrices can be described in general in a log-odds form as follows:

$$s_{ij} = \left(\frac{1}{\lambda} \right) \ln \left(\frac{q_{ij}}{p_i p_j} \right) \tag{3.4}$$

The PAM matrix is given as 10 times the log base 10 of the odds ratio. The BLOSUM matrix is given as 2 times the log base 2 of the odds ratio. Thus, BLOSUM scores are not quite as large as they would be if given on the same scale as PAM scores. Practically, this difference in scales is not important because alignment scores are typically converted from raw scores to normalized bit scores (Chapter 4).

Here s_{ij} refers to the score of amino acid i aligning with j. q_{ij} are the positive target frequencies; these sum to 1. λ is a positive parameter that provides a scale for the matrix. We will again encounter λ when we describe the basic statistical measure of a BLAST result (Chapter 4, Equation 4.5).

The BLOSUM62 matrix is the default scoring matrix for the BLAST protein search programs at NCBI. It merges all proteins in an alignment that have 62% amino acid identity or greater into one sequence. If a block of aligned globin orthologs includes several that have 62%, 80%, and 95% amino acid identity, these would all be weighted (grouped) as one sequence. Substitution frequencies for the BLOSUM62 matrix are weighted more heavily by blocks of protein sequences having less than 62% identity. (Thus, this matrix is useful for scoring proteins that share less than 62% identity.) The BLOSUM62 matrix is shown in Fig. 3.17.

Henikoff and Henikoff (1992) tested the ability of a series of BLOSUM and PAM matrices to detect proteins in BLAST searches of databases. They found that

	A	R	N	D	C	Q	E	G	H	I	L	K	M	F	P	S	T	W	Y	V
A	4																			
R	-1	5																		
N	-2	0	6																	
D	-2	-2	1	6																
C	0	-3	-3	-3	9															
Q	-1	1	0	0	-3	5														
E	-1	0	0	2	-4	2	5													
G	0	-2	0	-1	-3	-2	-2	6												
H	-2	0	1	-1	-3	0	0	-2	8											
I	-1	-3	-3	-3	-1	-3	-3	-4	-3	4										
L	-1	-2	-3	-4	-1	-2	-3	-4	-3	2	4									
K	-1	2	0	-1	-1	1	1	-2	-1	-3	-2	5								
M	-1	-2	-2	-3	-1	0	-2	-3	-2	1	2	-1	5							
F	-2	-3	-3	-3	-2	-3	-3	-3	-1	0	0	-3	0	6						
P	-1	-2	-2	-1	-3	-1	-1	-2	-2	-3	-3	-1	-2	-4	7					
S	1	-1	1	0	-1	0	0	0	-1	-2	-2	0	-1	-2	-1	4				
T	0	-1	0	-1	-1	-1	-1	-2	-2	-1	-1	-1	-1	-2	-1	1	5			
W	-3	-3	-4	-4	-2	-2	-3	-2	-2	-3	-2	-3	-1	1	-4	-3	-2	11		
Y	-2	-2	-2	-3	-2	-1	-2	-3	2	-1	-1	-2	-1	3	-3	-2	-2	2	7	
V	0	-3	-3	-3	-1	-2	-2	-3	-3	3	1	-2	1	-1	-2	-2	0	-3	-1	4

FIGURE 3.17. The BLOSUM62 scoring matrix of Henikoff and Henikoff (1992). This matrix merges all proteins in an alignment that have 62% amino acid identity or greater into one sequence. BLOSUM62 performs better than alternative BLOSUM matrices or a variety of PAM matrices at detecting distant relationships between proteins. It is thus the default scoring matrix for most database search programs such as BLAST (Chapter 4). Used with permission.

BLOSUM62 performed slightly better than BLOSUM60 or BLOSUM70 and dramatically better than PAM matrices at identifying various proteins. Their matrices were especially useful for identifying weakly scoring alignments. BLOSUM50 and BLOSUM90 are other commonly used scoring matrices in BLAST searches. (For an alignment of two proteins sharing about 50% identity, try using the BLOSUM50 matrix. The fasta family of sequence comparison programs use BLOSUM50 as a default.)

The relationships of the PAM and BLOSUM matrices are outlined in Fig. 3.18. To summarize, BLOSUM and PAM matrices both use log-odds values in their scoring systems. In each case, when you perform a pairwise sequence alignment (or when you search a query sequence against a database), you specify the exact matrix to use based on the suspected degree of identity between the query and its matches. PAM matrices are based on data from the alignment of closely related protein families, and they involve the assumption that substitution probabilities for highly related proteins (e.g., PAM40) can be extrapolated to probabilities for distantly related proteins (e.g., PAM250). In contrast, the BLOSUM matrices are based on empirical observations of more distantly related protein alignments. Note that a PAM30 matrix, which is available as an option on standard blastp searches at NCBI (Chapter 4), may be

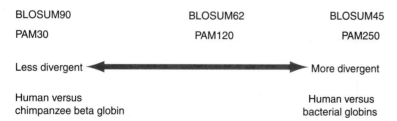

FIGURE 3.18. *Summary of PAM and BLOSUM matrices. High-value BLOSUM matrices and low-value PAM matrices are best suited to study well-conserved proteins such as mouse and rat globins. BLOSUM matrices with low numbers (e.g., BLOSUM45) or high PAM numbers are best suited to detect distantly related proteins. Remember that in a BLOSUM45 matrix all members of a protein family with greater than 45% amino acid identity are grouped together, allowing the matrix to focus on proteins with less than 45% identity.*

useful for identifying significant conservation between two closely related proteins. However, a BLOSUM matrix with a high value (such as the BLOSUM80 matrix that is available at the NCBI blastp site) is not necessarily suitable for scoring closely related sequences. This is because the BLOSUM80 matrix is adapted to regions of sequences that share up to 80% identity, but beyond that limited region two proteins may share dramatically less amino acid identity (Pearson and Wood, 2001).

Pairwise Alignment and Limits of Detection: The "Twilight Zone"

A hit is a change in an amino acid residue that occurs by mutation. We discuss mutations (including multiple hits at a nucleotide position) in Chapter 7 (see Fig. 7.11). We discuss mutations associated with human disease in Chapter 20.

When we compare two protein sequences, how many mutations can occur between them before their differences make them unrecognizable? When we compared glyceraldehyde-3-phosphate dehydrogenase proteins, it was easy to see their relationship (Fig. 3.7). However, when we compared human beta globin and myoglobin, the relationship was much less obvious (Fig. 3.5). Intuitively, at some point two homologous proteins are too divergent for their alignment to be recognized as significant.

The plot in Fig. 3.19 reaches an asymptote below about 15% amino acid identity. This asymptote would reach about 5% (or the average background frequency of the amino acids) if no gaps were allowed in the comparison between the proteins.

The best way to determine the detection limits of pairwise alignments is through statistical tests that assess the likelihood of finding a match by chance. These are described below, and in Chapter 4. In particular we will focus on the expect (E) value. It can also be helpful to compare the percent identity (and percent divergence) of two sequences versus their evolutionary distance. Consider two protein sequences, each 100 amino acids in length, in which various numbers of mutations are introduced. A plot of the two diverging sequences has the form of a negative exponential (Fig. 3.19) (Doolittle, 1987; Dayhoff, 1978). If the two sequences have 100% amino acid identity, they have zero changes per 100 residues. If they share 50% amino acid identity, they have sustained an average of 80 changes per 100 residues. One might have expected 50 changes per 100 residues in the case of two proteins that share 50% amino acid identity. However, any position can be subject to multiple hits. Thus, percent identity is not an exact indicator of the number of mutations that have occurred across a protein sequence. When a protein sustains about 250 hits

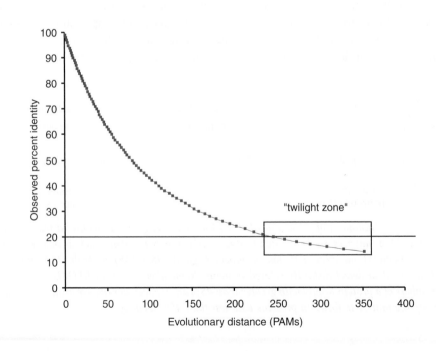

FIGURE 3.19. Two randomly diverging protein sequences change in a negatively exponential fashion. This plot shows the observed number of amino acid identities per 100 residues of two sequences (y axis) versus the number of changes that must have occurred (the evolutionary distance in PAM units). The twilight zone (Doolittle, 1987) refers to the evolutionary distance corresponding to about 20% identity between two proteins. Proteins with this degree of amino acid sequence identity may be homologous, but such homology is difficult to detect. This figure was constructed using MATLAB® software with data from Dayhoff (1978) (see Table 3.3).

TABLE 3-3 Relationship between Observed Number of Amino Acid Differences per 100 Residues of Two Aligned Protein Sequences and Evolutionary Difference[a]

Observed Differences in 100 Residues	Evolutionary Distance in PAMs
1	1.0
5	5.1
10	10.7
15	16.6
20	23.1
25	30.2
30	38.0
35	47
40	56
45	67
50	80
55	94
60	112
65	133
70	159
75	195
80	246

[a]The number of changes that must have occurred, in PAM units.
Source: Adapted from Dayhoff (1978, p. 375). Used with permission.

per 100 amino acids, it may have about 20% identity with the original protein, and it can still be recognizable as significantly related. If a protein sustains 360 changes per 100 residues, it evolves to a point at which the two proteins share about 15% amino acid identity and are no longer recognizable as significantly related in a direct, pairwise comparison.

The PAM250 matrix assumes the occurrence of 250 point mutations per 100 amino acids. As shown in Fig. 3.19, this corresponds to the "twilight zone." At this level of divergence, it is usually difficult to assess whether the two proteins are homologous. Other techniques, including multiple sequence alignment (Chapter 6) and structural predictions (Chapter 11), are often very useful to assess homology in these cases. PAM matrices are available from PAM1 to PAM250 or higher, and a specific number of observed amino acid differences per 100 residues is associated with each PAM matrix (Table 3.3 and Fig. 3.19). Consider the case of the human alpha globin compared to myoglobin. These proteins are approximately 150 amino acid residues in length, and they may have undergone over 300 amino acid substitutions since their divergence (Dayhoff et al., 1972, p. 19). If there were 345 changes (corresponding to 230 changes per 100 amino acids), then an additional 100 changes would result in only 10 more observable changes (Dayhoff et al., 1972; Table 3.3).

There are about $2^{2n}/\sqrt{\pi n}$ possible global alignments between two sequences of length n (Durbin et al., 2000; Ewens and Grant, 2001). For two sequences of length 1000, there are about 10^{600} possible alignments. For two proteins of length 200 amino acid residues, the number of possible alignments is over 6×10^{58}.

ALIGNMENT ALGORITHMS: GLOBAL AND LOCAL

Our discussion so far has focused on matrices that allow us to score an alignment between two proteins. This involves the generation of scores for identical matches, mismatches, and gaps. We also need an appropriate algorithm to perform the alignment. When two proteins are aligned, there is an enormous number of possible alignments.

There are two main types of alignment: global and local. We will explore these approaches next. A *global alignment* such as one produced by the method of Needleman and Wunsch (1970) contains the entire sequence of each protein or DNA sequence. A *local alignment* such as the method of Smith and Waterman (1981) focuses on the regions of greatest similarity between two sequences. We saw a local alignment of human beta globin and myoglobin in Fig. 3.5. For many purposes, a local alignment is preferred, because only a portion of two proteins aligns. (We will study the modular nature of proteins in Chapter 10.) Most database search algorithms, such as BLAST (Chapter 4), use local alignments.

Each of these methods is guaranteed to find one or more optimal solutions to the alignment of two protein or DNA sequences. We will then describe two rapid-search algorithms, BLAST and FASTA. BLAST represents a simplified form of local alignment that is popular because the algorithm is very fast and easily accessible.

Global Sequence Alignment: Algorithm of Needleman and Wunsch

One of the first and most important algorithms for aligning two protein sequences was described by Saul Needleman and Christian Wunsch (1970), with subsequent modifications by Sellers (1974), Gotoh (1982), and others. This algorithm is important because it produces an optimal alignment of two protein or DNA sequences, even allowing the introduction of gaps. The result is optimal, but nonetheless not all possible alignments need to be evaluated. The Needleman–Wunsch (sometimes called Needleman–Wunsch–Sellers) algorithm is an example of dynamic programming in which the optimal alignment is identified by reducing the problem to a series of smaller alignments on a residue-by-residue basis. An exhaustive pairwise comparison would be too computationally expensive to perform.

We can describe the Needleman–Wunsch approach to global sequence alignment in three steps: (1) setting up a matrix, (2) scoring the matrix, and (3) identifying the optimal alignment. We will illustrate this process using two globin sequences.

Step 1: Setting Up a Matrix

First, we compare two sequences in a two-dimensional matrix (Fig. 3.20 and following figures). The first sequence, of length m, is listed vertically along the y axis, with its amino acid residues corresponding to rows. The second sequence, of length n, is listed horizontally along the x axis so that its amino acid residues correspond to the columns.

In our two-dimensional matrix, a perfect alignment between two identical sequences is represented by a diagonal line extending from the top left to the bottom right (Fig. 3.20a). Any mismatch between two sequences is still represented on this diagonal path (Fig. 3.20b). In the example of Fig. 3.20b, the mismatch of L and M residues might be assigned a score lower than the perfect match of L and L shown in Fig. 3.20a. Gaps are represented in this matrix using horizontal or vertical paths, as shown in Fig. 3.20c,d. Any gap in the second sequence is represented as a vertical line (Fig. 3.20c), while any gap in the first sequence is drawn as a horizontal line (Fig. 3.20d). These gaps can be of any length, and gaps can be internal or terminal. Sellers (1974) introduced a modification to allow linear gap penalties, while Gotoh (1982) allowed affine gap penalties in which there is a large penalty for introducing a gap and a small penalty for each position that the gap is extended (see Chapter 4).

This algorithm is also sometimes called the Needleman–Wunsch–Sellers algorithm. Sellers (1974) provided a related alignment algorithm (one that focuses on minimizing differences, rather than on maximizing similarities). Smith et al. (1981) showed that the Needleman–Wunsch and Sellers approaches are mathematically equivalent.

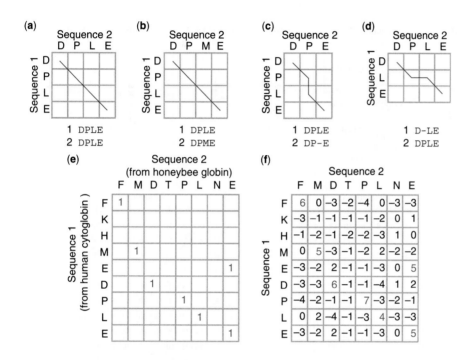

FIGURE 3.20. Pairwise alignment of two amino acid sequences using a dynamic programming algorithm of Needleman and Wunsch (1970) for global alignment. (a) Two identical sequences can be assigned a diagonal path through the matrix. (b) A mismatch in one sequence still results in a diagonal path, but the score of a mismatch may be lower than that of a perfect match. (c) A deletion in sequence 2 (or an insertion in sequence 1) results in the insertion of a gap position and a resulting vertical path in the optimal alignment. (d) A gap in the first sequence is represented by a horizontal path through the matrix. (e) A portion of the sequences of human cytoglobin (NP_599030) and a honeybee globin (NP_001071291) are used to demonstrate global alignment. An identity matrix uses a simple scoring system of +1 at each position in which the sequences share an identical amino acid residue, and 0 (not shown) in all other cells. (f) A BLOSUM62 scoring matrix is applied to provide scores for the same two sequences as in (e).

We will use a specific example of globally aligning a portion of human cytoglobin (sequence 1 in Figs. 3.20e,f and 3.21) and a honeybee globin (sequence 2).

Step 2: Scoring the Matrix

The Needleman–Wunsch approach (from 1970) begins with the creation of an identity matrix (also called a unitary matrix). To do this, we simply place a value of +1 in each cell of the matrix where the two proteins share an identical amino acid residue (Fig. 3.20e). The identity matrix uses the simplest scheme for assigning scores, but we could apply any scoring matrix such as Blosum62 (as shown in Fig. 3.20f).

Next, we set up a scoring matrix (Fig. 3.21), distinct from the identity matrix. For our example, we employ a simple scoring system in which each amino acid identity gains a score of +1, each mismatch scores −2, and each gap position scores −2 (Fig. 3.21b). Our goal in finding an optimal alignment is to determine the path through the matrix that maximizes the score. This usually entails finding a path through as many positions of identity as possible while introducing as few gaps as

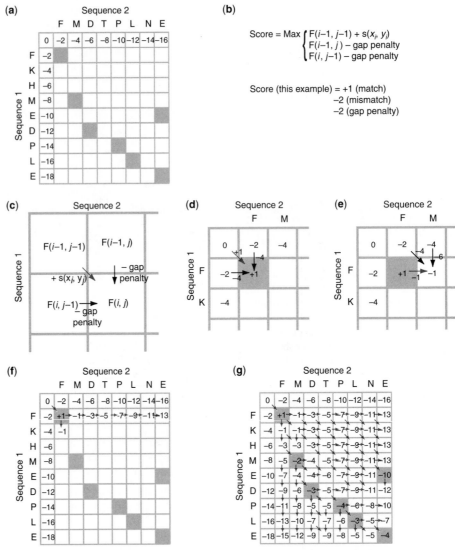

FIGURE 3.21. *Pairwise alignment of two amino acid sequences using the dynamic programming algorithm of Needleman and Wunsch (1970) for global alignment. (a) For sequences of length m and n we form a matrix of dimensions m+1 by n +1 and add gap penalties in the first row and column. Each gap position receives a score of −2. The cells having identity are shaded gray. (b) The scoring system in this example is +1 for a match, −2 for a mismatch, and −2 for a gap penalty. In each cell, the score is assigned using the recursive algorithm that identifies the highest score from three calculations. (c) In each cell F(i, j) we calculate the scores derived from following a path from the upper left cell (we add the score of that cell + F(i, j)), the score of the cell to the left (including a gap penalty), and the cell directly above (again including a gap penalty). (d) To calculate the score in the cell of the second row and column, we take the maximum of the three scores +1, −4, −4. This best score (+1) follows the path of the red arrow, and we maintain the information of the best path resulting in each cell's score in order to later reconstruct the pair wise alignment. (e) To calculate the score in the second row, third column we again take the maximum of the three scores −4, −1, −4. The best score follows from the left cell (red arrow). (f) We proceed to fill in scores across the first row of the matrix. (g) The completed matrix includes the overall score of the optimal alignment (−4; see cell at bottom right, corresponding to the carboxy termini of each protein). Red arrows indicate the path(s) by which each cell's highest score was obtained.*

possible. As shown in Fig. 3.20a–e, there are four possible occurrences at each position i, j (i.e., in each cell in the matrix):

1. Two residues may be perfectly matched (i.e., identical).
2. They may be mismatched.
3. A gap may be introduced from the first sequence.
4. A gap may be introduced from the second sequence.

The Needleman–Wunsch–Sellers algorithm provides a score, corresponding to each of these possible outcomes, for each position of the aligned sequences. The algorithm also specifies a set of rules describing how we can move through the matrix. We define two sequences of length m and n and create a scoring matrix of dimensions $m+1$ and $n+1$ (Fig. 3.21a). The upper left cell is assigned a score of 0. Subsequent cells down the first column and across the first row correspond to terminal gap assignments. On the first row, the cell with a score of -2 corresponds to the insertion of one gap position in the final alignment; a score of -4 indicates two terminal gap positions.

Let us next introduce a nomenclature for the cells of the matrix. Define i corresponding to rows (in sequence 1) and j corresponding to columns (sequence 2). Consider any given cell at position (i, j) for which we want to assign a score (Fig. 3.21c, lower right). The score is the maximal value of three scores derived from the three adjacent cells that are (1) positioned diagonally up to the left, (2) directly to the left, or (3) directly above.

(1) The cell diagonally up to the left, at position $F(i-1, j-1)$, corresponds to an alignment having either a match or mismatch. We take the score in that cell $i-1, j-1$ and add to it the score $s(x_i, y_i)$ in the lower right quadrant (Fig. 3.21c). This score may take a negative value if the residues are mismatched. To calculate the score in row 2, column 2 in Fig. 3.21d, we add 0 (the score in the upper left cell) plus 1 (the score assigned to a match between F and F). To calculate the score in row 2, column 3 in Fig. 3.21e, we add -2 (the score in the upper left quadant) to -2 (the score assigned to a mismatch between F and M) for a total of -4.

(2) The cell directly to the left of i, j (that is, $i, j-1$) has some score (see Fig. 3.21c). As shown in Fig. 3.21b, we take this score and subtract the gap penalty. Choosing this path (described below) corresponds to the insertion of a gap position in sequence 1. To calculate the score in row 2, column 2 (Fig. 3.21d), we sum -2 (the score in row 2, column 1) and -2 (the gap penalty) for a score of -4. In Fig. 3.21e, we sum $+1$ (the score to the left of row 2, column 3) and -2 (the gap penalty) for a score of -1.

(3) The cell directly above i, j is at position $i-1, j$. To move from $i-1, j$ to i, j requires inserting a gap in sequence 2. We calculate scores as described for a gap insertion in sequence 1. To score the cell in row 2, column 2 of Fig. 3.21d, we sum -2 and the gap penalty -2 for a total of -4. To score the cell in row 2, column 3, we sum -4 and the gap penalty of -2 for a total of -6 (Fig. 3.21e).

We can add scores to the matrix moving systematically across the rows (Fig. 3.21f). As we do this, we keep track of which of the three possible paths lead to the optimal score in each cell. To help show which cells have $+1$ scores from the identity matrix of Fig. 3.20e, we have shaded those same cells gray in Fig. 3.21. When the process of filling the scoring matrix is complete (Fig. 3.21g), the score in the lower right hand cell is the overall score of the alignment.

Step 3: Identifying the Optimal Alignment

After the matrix is filled, we know the overall score of the alignment but we do not know that optimal alignment itself. That is determined by a trace-back procedure.

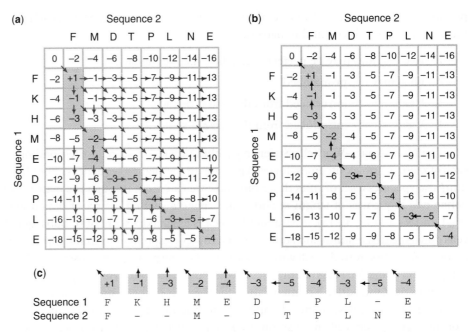

FIGURE 3.22. *Global pairwise alignment of two amino acid sequences using a dynamic programming algorithm: scoring the matrix and using the trace-back procedure to obtain the alignments. (a) The alignment of Fig. 3.22g is shown. The cells highlighted represent the source of the optimal scores. (b) In an equivalent representation, we use arrows to point back to the source of each cell's optimal score. (c) This traceback allows us to determine the sequence of the optimal alignment. Vertical or horizontal arrows correspond to positions of gap insertions, while diagonal lines correspond to exact matches (or mismatches). Note that the final score (−4) equals the sum of matches (6×1=6), mismatches (none in this example), and gaps (5 × −2=−10).*

The needle program for global pairwise alignment is part of the EMBOSS package available online at the European Bioinformatics Institute (▶ http://www.ebi.ac.uk/emboss/align/). It is further described at the EMBOSS website under applications (▶ http://emboss.sourceforge.net/). The *E. coli* and *S. cerevisiae* proteins are available in the fasta format, as well as globally and locally aligned in web document 3.6 (▶ http://www.bioinfbook.org/chapter3).

For this, we begin with the cell at the lower right of the matrix (carboxy termini of the proteins or 3 end of the nucleic acid sequences). In our example, this has a score of −4 and represents an alignment of two glutamate residues. For this and every cell we can determine from which of the three adjacent cells the best score was derived. This procedure is outlined in Fig. 3.22a in which red arrows indicate the paths from which the best scores were obtained for each cell. We thus define a path (see red-shaded cells) that will correspond to the actual alignment. Sometimes, two or even all three paths will give the same score at that cell; in such cases all equivalent traceback pointers are kept. In Fig. 3.22b, we further show just the arrows indicating which cell each best score was derived from. This is a different way of defining the optimal path of the pairwise alignment. We build that alignment, including gaps in either sequence, proceeding from the carboxy to the amino terminus. The final alignment (Fig. 3.22c) is guaranteed to be optimal, given this scoring system. There may be multiple alignments that share an optimal score, although this rarely occurs when scoring matrices such as BLOSUM62 are employed.

A variety of programs implement global alignment algorithms (see Web Resources at the end of this chapter). An example is the Needle program from EMBOSS (Box 3.8). Two bacterial globin family sequences are entered: one from *Streptomyces avermitilis* MA-4680 (NP_824492, 260 amino acids), and another from *Mycobacterium tuberculosis* CDC1551 (NP_337032, 134 amino acids). Penalties are selected for gap creation and extension, and each sequence is pasted into an input box in the fasta format. The resulting global alignment includes descriptions of the percent identity and similarity shared by the two proteins, the length of the alignment, and the number of gaps introduced (Fig. 3.23a).

Box 3.8
EMBOSS

EMBOSS (European Molecular Biology Open Software Suite) is a collection of freely available programs for DNA, RNA, or protein sequence analysis (Rice et al., 2000). There are over 200 available programs in three dozen categories. The home page of EMBOSS (► http://emboss.sourceforge.net/) describes the various packages. A variety of web servers offer EMBOSS, including the following.

	http://bips.u-strasbg.fr/EMBOSS/
	http://bioinfo.hku.hk/EMBOSS/
	http://sbcr.bii.a-star.edu.sg/emboss/
Virginia Bioinformatics Institute	http://phytophthora.vbi.vt.edu/ EMBOSS/
Weizmann Institute	http://inn.weizmann.ac.il/EMBOSS/
Strasbourg Bioinformatics Platform	http://bips.u-strasbg.fr/EMBOSS/

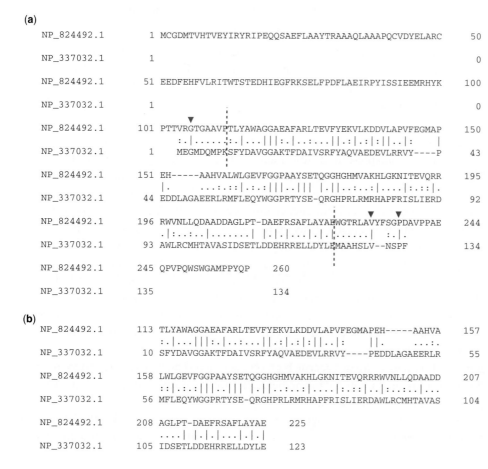

(a)

```
NP_824492.1     1 MCGDMTVHTVEYIRYRIPEQQSAEFLAAYTRAAAQLAAAPQCVDYELARC    50

NP_337032.1     1                                                         0


NP_824492.1    51 EEDFEHFVLRITWTSTEDHIEGFRKSELFPDFLAEIRPYISSIEEMRHYK   100

NP_337032.1     1                                                         0


NP_824492.1   101 PTTVRGTGAAVPTLYAWAGGAEAFARLTEVFYEKVLKDDVLAPVFEGMAP   150
                  ::|......:::.|...|||::|.....::.||.:|.:|:||..|:   |
NP_337032.1     1    MEGMDQMPKSFYDAVGGAKTFDAIVSRFYAQVAEDEVLRRVY----P    43


NP_824492.1   151 EH-----AAHVALWLGEVFGGPAAYSETQGGHGHMVAKHLGKNITEVQRR   195
                  |.     ...:.::||.:::|||.|||| |.||..:.::|....|:.::|.
NP_337032.1    44 EDDLAGAEERLRMFLEQYWGGPRTYSE-QRGHPRLRMRHAPFRISLIERD    92


NP_824492.1   196 RWVNLLQDAADDAGLPT-DAEFRSAFLAYAEWGTRLAVYFSGPDAVPPAE   244
                  .|:....:.||....| |.|.|..|.|.|      | :.|.
NP_337032.1    93 AWLRCMHTAVASIDSETLDDEHRRELLDYLEMAAHSLV--NSPF        134


NP_824492.1   245 QPVPQWSWGAMPPYQP       260

NP_337032.1   135                134
```

(b)

```
NP_824492.1   113 TLYAWAGGAEAFARLTEVFYEKVLKDDVLAPVFEGMAPEH-----AAHVA   157
                  ::|...|||:.|...::..||.:|.:|:||..|:   ||.    ...:.
NP_337032.1    10 SFYDAVGGAKTFDAIVSRFYAQVAEDEVLRRVY----PEDDLAGAEERLR    55


NP_824492.1   158 LWLGEVFGGPAAYSETQGGHGHMVAKHLGKNITEVQRRWVNLLQDAADD   207
                  ::|.:::|||..|||.||.|..:.::|:::|.|:::|..|:.....|.
NP_337032.1    56 MFLEQYWGGPRTYSE-QRGHPRLRMRHAPFRISLIERDAWLRCMHTAVAS   104


NP_824492.1   208 AGLPT-DAEFRSAFLAYAE       225
                  ....| |.|.|...|.|.|
NP_337032.1   105 IDSETLDDEHRRELLDYLE       123
```

FIGURE 3.23. (a) Global pairwise alignment of bacterial proteins containing globin domains from Streptomyces avermitilis MA-4680 (NP_824492) and Mycobacterium tuberculosis CDC1551 (NP_337032). The scoring matrix was BLOSUM62. The aligned proteins share 14.7% identity (39/266 aligned residues), 22.6% similarity (60.266), and 51.9% gaps (138/266). (b) A local pairwise alignment of these two sequences lacks the unpaired amino- and carboxy-terminal extensions and shows 30% identity (35/115 aligned residues). The alignment in (b) corresponds to the region within the dotted vertical lines of (a). The arrowheads in (a) indicate aligned residues that not seen in the local alignment. Thus, in performing local alignments (as is done in BLAST, Chapter 4) some authentically aligned regions may be missed.

Local Sequence Alignment: Smith and Waterman Algorithm

The local alignment algorithm of Smith and Waterman (1981) is the most rigorous method by which subsets of two protein or DNA sequences can be aligned. Local alignment is extremely useful in a variety of applications, such as database searching where we may wish to align domains of proteins (but not the entire sequences). A local sequence alignment algorithm resembles that for global alignment in that two proteins are arranged in a matrix and an optimal path along a diagonal is sought. However, there is no penalty for starting the alignment at some internal position, and the alignment does not necessarily extend to the ends of the two sequences.

For the Smith–Waterman algorithm a matrix is again constructed with an extra row along the top and an extra column on the left side. Thus for sequences of lengths m and n, the matrix has dimensions $m + 1$ and $n + 1$. The rules for defining the value in each position of the matrix differ slightly from those used in the Needleman–Wunsch algorithm. The score in each cell is selected as the maximum of the preceding diagonal or the score obtained from the introduction of a gap. However, the score cannot be negative: a rule introduced by the Smith–Waterman algorithm is that if all other score options produce a negative value, then a zero must be inserted in the cell. Thus, the score $s(i, j)$ is given as the maximum of four possible values (Fig. 3.24):

1. The score from the cell at position $i - 1, j - 1$; that is, the score diagonally up to the left. To this score, add the new score at position $s[i, j]$, which consists of either a match or a mismatch.

2. $s(i, j - 1)$ (i.e., the score one cell to the left) minus a gap penalty.

3. $s(i - 1, j)$ (i.e., the score immediately above the new cell) minus a gap penalty.

4. The number zero. This condition assures that there are no negative values in the matrix. In contrast negative numbers commonly occur in global alignments because of gap or mismatch penalties (note the log-odds matrices in this chapter).

An example of the use of a local alignment algorithm to align two nucleic acid sequences, adapted from Smith and Waterman (1981), is shown in Fig. 3.24. The topmost row and the leftmost column of the matrix are filled with zeros. The maximal alignment can begin and end anywhere in the matrix (within reason; the linear order of the two amino acid sequences cannot be violated). The procedure is to identify the highest value in the matrix (this value is 3.3 in Fig. 3.24a). This represents the end (3′ end for nucleic acids, or carboxy-terminal portion proteins) of the alignment. This position is not necessarily at the lower right corner, as it must be for a global alignment. The trace-back procedure begins with this highest value position and proceeds diagonally up to the left stopping when a cell is reached with a value of zero. This defines the start of the alignment, and it is not necessarily at the extreme top left of the matrix.

A requirement of the Smith–Waterman algorithm is that the expected score for a random match is negative. This condition ensures that alignments between very long unrelated sequences do not accrue high scores. Such sequences could otherwise produce spurious alignments having higher scores than the authentic match between two proteins over a shorter region.

An example of a local alignment of two proteins using the Smith–Waterman algorithm is shown in Fig. 3.23b. Compare this with the global alignment of Fig.

(a)

Sequence 1

		C	A	G	C	C	U	C	G	C	U	U	A	G
	0.0	0.0	0.0	0.0	0.0	0.0	0.0	0.0	0.0	0.0	0.0	0.0	0.0	0.0
A	0.0	0.0	1.0	0.0	0.0	0.0	0.0	0.0	0.0	0.0	0.0	0.0	1.0	0.0
A	0.0	0.0	1.0	0.7	0.0	0.0	0.0	0.0	0.0	0.0	0.0	0.0	1.0	0.7
U	0.0	0.0	0.0	0.7	0.3	0.0	1.0	0.0	0.0	0.0	1.0	1.0	0.0	0.7
G	0.0	0.0	0.0	1.0	0.3	0.0	0.0	0.7	1.0	0.0	0.0	0.7	0.7	1.0
C	0.0	1.0	0.0	0.0	2.0	1.3	0.3	1.0	0.3	2.0	0.7	0.3	0.3	0.3
C	0.0	1.0	0.7	0.0	1.0	3.0	1.7	1.3	1.0	1.3	1.7	0.3	0.0	0.0
A	0.0	0.0	2.0	0.7	0.3	1.7	2.7	1.3	1.0	0.7	1.0	1.3	1.3	0.0
U	0.0	0.0	0.7	1.7	0.3	1.3	2.7	2.3	1.0	0.7	1.7	2.0	1.0	1.0
U	0.0	0.0	0.3	0.3	1.3	1.0	2.3	2.3	2.0	0.7	1.7	2.7	1.7	1.0
G	0.0	0.0	0.0	1.3	0.0	1.0	1.0	2.0	3.3	2.0	1.7	1.3	2.3	2.7
A	0.0	0.0	1.0	0.0	1.0	0.3	0.7	0.7	2.0	3.0	1.7	1.3	2.3	2.0
C	0.0	1.0	0.0	0.7	1.0	2.0	0.7	1.7	1.7	3.0	2.7	1.3	1.0	2.0
G	0.0	0.0	0.7	1.0	0.3	0.7	1.7	0.3	2.7	1.7	2.7	2.3	1.0	2.0
G	0.0	0.0	0.0	1.7	0.7	0.3	0.3	1.3	1.3	2.3	1.3	2.3	2.0	2.0

(left axis label: Sequence 2)

(b)

```
sequence 1    GCC-UCG
sequence 2    GCCAUUG
```

(c)

```
sequence 1    CAGCC-UCGCUUAG
sequence 2    AAUGCCAUUGACGG
```

FIGURE 3.24. *Local sequence alignment method of Smith and Waterman (1981). (a) In this example, the matrix is formed from two RNA sequences (CAGCCUCGCUUAG and AAUGCCAUUGACGG). While this is not an identity matrix (such as that shown in Fig. 3.20e), positions of nucleotide identity are shaded gray. The scoring system here is +1 for a match, minus one-third for a mismatch, and a gap penalty of the difference between a match and a mismatch (−1.3 for a gap of length one). The matrix is scored according to the rules outlined on the bottom of page 82. The highest value in the matrix (3.3) corresponds to the beginning of the optimal local alignment, and the aligned residues (shaded red) extend up and to the left until a value of zero is reached. (b) The local alignment derived from this matrix is shown. Note that this alignment includes identities, a mismatch, and a gap. (c) A global alignment of the two sequences is shown for comparison to the local alignment. Note that it encompasses the entirety of both sequences, and that the local alignment is not a subsequence of the global alignment. Used with permission.*

3.23a and note that the aligned region is shorter for the local alignment, while the percent identity and similarity are higher. Note also that the local alignment ignores several identically matching residues (Fig. 3.23a, arrowheads). Since database searches such as BLAST (Chapter 4) rely on local alignments, there may be conserved regions that are not reported as aligned, depending on the particular search parameters you choose.

Rapid, Heuristic Versions of Smith–Waterman: FASTA and BLAST

While the Smith–Waterman algorithm is guaranteed to find the optimal alignment(s) between two sequences, it suffers from the fact that it is relatively slow. For pairwise alignment, speed is usually not a problem. But when a pairwise alignment algorithm is applied to the problem of comparing one sequence (a "query") to an entire database, the speed of the algorithm becomes a significant issue and may vary by orders of magnitude.

The modified alignment algorithms introduced by Gotoh (1982) and Myers and Miller (1988) require only $O(nm)$ time and occupy $O(n)$ in space. Instead of committing the entire matrix to memory, the algorithms ignore scores below a threshold in order to focus on the maximum scores that are achieved during the search.

Most algorithms have a parameter N that refers to the number of data items to be processed (see Sedgewick, 1988). This parameter can greatly affect the time required for the algorithm to perform a task. If the running time is proportional to N, then doubling N doubles the running time. If the running time is quadratic (N^2), then for $N = 1000$, the running time is one million. For both the Needleman–Wunsch and the Smith–Waterman algorithms, both the computer space and the time required to align two sequences is proportional to at least the length of the two query sequences multiplied against each other, $m \times n$. For the search of a database of size N, this is $m \times N$.

Another useful descriptor is O-notation (called "big-Oh notation") which allows one to approximate the upper bounds on the running time of an algorithm. The Needleman–Wunsch algorithm requires $O(mn)$ steps, while the Smith–Waterman algorithm requires $O(m^2n)$ steps. Subsequently, Gotoh (1982) and Myers and Miller (1988) improved the algorithms so they require less time and space.

FASTA stands for FAST-All, referring to its ability to perform a fast alignment of all sequences (i.e., proteins or nucleotides).

Two popular local alignment algorithms have been developed that provide rapid alternatives to Smith–Waterman: FASTA (Pearson and Lipman, 1988) and BLAST (Basic Local Alignment Search Tool) (Altschul et al., 1990). Each of these algorithms requires less time to perform an alignment. The time saving occurs because FASTA and BLAST restrict the search by scanning a database for likely matches before performing more rigorous alignments. These are heuristic algorithms (Box 3.2) that sacrifice some sensitivity in exchange for speed; in contrast to Smith–Waterman, they are not guaranteed to find optimal alignments.

The FASTA search algorithm introduced by Pearson and Lipman (1988) proceeds in four steps.

The parameter *ktup* refers to multiples such as duplicate, triplicate, or quadruplicate (for $k = 2$, $k = 3$, $k = 4$). The *ktup* values are usually 3 to 6 for nucleotide sequences and 1 to 2 for amino acid sequences. A small *ktup* value yields a more sensitive search but requires more time to complete. William Pearson of the University of Virginia provides FASTA online. Visit ▶ http://fasta.bioch. virginia.edu/fasta_www2/fasta_ list2.shtml. Another place to try FASTA is at the European Bioinformatics Institute website, ▶ http://www.ebi.ac.uk/fasta33/.

1. A lookup table is generated consisting of short stretches of amino acids or nucleotides from a database. The size of these stretches is determined from the *ktup* parameter. If *ktup* = 3 for a protein search, then the query sequence is examined in blocks of three amino acids against matches of three amino acids found in the lookup table. The FASTA program identifies the 10 highest scoring segments that align for a given *ktup*.

2. These 10 aligned regions are rescored, allowing for conservative replacements, using a scoring matrix such as PAM250.

3. High-scoring regions are joined together if they are part of the same proteins.

4. FASTA then performs a global (Needleman–Wunsch) or local (Smith–Waterman) alignment on the highest scoring sequences, thus optimizing the alignments of the query sequence with the best database matches. Thus, dynamic programming is applied to the database search in a limited fashion, allowing FASTA to return its results very rapidly because it evaluates only a portion of the potential alignments.

BLAST was introduced as a local alignment search tool that identifies alignments between a query sequence and a database without the introduction of gaps (Altschul et al., 1990). The version of BLAST that is available today allows gaps in the alignment. We gave an example of an alignment of two proteins (Figs. 3.4 and 3.5) and we introduce BLAST in more detail in Chapter 4, where we describe its heuristic algorithm.

Pairwise Alignment with Dot Plots

In addition to displaying a pairwise alignment, the output of pairwise BLAST at NCBI includes a dot plot (or dot matrix), which is a graphical method for comparing two sequences. One protein or nucleic acid sequence is placed along the x axis, and the other is placed along the y axis. Positions of identity are scored with a dot. A region of identity between two sequences results in the formation of a diagonal

Dotlet is a web-based diagonal plot tool available from the Swiss Institute of Bioinformatics (▶ http://www.isrec.isb-sib.ch/java/dotlet/Dotlet.html). It was written by Marco Pagni and Thomas Junier. The website provides examples of the use of Dotlet to visualize repeated domains, conserved domains, exons and introns, terminators, frameshifts, and low-complexity regions.

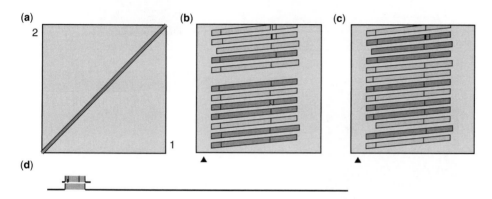

```
Score = 40.4 bits (93),  Expect = 0.070
Identities = 25/134 (18%),  Positives = 59/134 (44%),  Gaps = 4/134 (2%)

Query    25  KAVQAMWARLYANCE---DVGVAILVRFFVNFPSAKQYFSQFKHMEDPLEMERSPQLRKH   81
             +A+++ WA L A  +   + G   ++       P+ ++ F +F   +    ++   + K
Sbjct   129  QALRSSWATLTAGADGRNNFGNNFVLWLLNTIPNIRERFEKFNAHQSDEALKNDNEFVKQ  188

Query    82  ACRVMGALNTVVENLHDPDKVSSVLALVGKAH-ALKHKVEPVYFKILSGVILEVVAEEFA  140
             ++G L + ++NL +P ++ + +  +    H  ++  +   YF+ L    I + VA
Sbjct   189  VKLIVGGLQSFIDNLENPGQLQATIERLASVHLKMRPTIGLEYFRPLQENIAQYVASALG  248

Query   141  SDFPPETQRAWAKL   154
             +AW +L
Sbjct   249  VGADDAAPKAWERL   262
```

FIGURE 3.25. Dot plots in the output of the NCBI BLAST pairwise alignment algorithm permit visualization of matching domains in pairwise protein alignments. The program is used as described in Fig. 3.4. (a) For a comparison of human cytoglobin (NP_599030, length 190 amino acids) with itself, the output includes a dot plot shown with sequences 1 and 2 (both cytoglobin) on the x and y axes, and the data points showing amino acid identities appear as a diagonal line. (b) For a comparison of cytoglobin with a globin from the snail Biomphalaria glabrata (accession CAJ44466, length 2,148 amino acids), the cytoglobin sequence (x axis) matches 12 times with internal globin repeats in the snail protein. (c) Changing the scoring matrix to PAM250 enables all 13 globin repeats of the snail protein to be aligned with cytoglobin. (d) A pairwise alignment of the sequences shows that the snail globin repeats align with residues 25–154 of cytoglobin. This is reflected in the dot plots, where the portion on the x axis corresponding to cytoglobin residues 1–24 [see arrowheads, panels (b) and (c)] does not align to the snail sequence. A broader perspective on the pairwise alignment is also shown by the output (panel d) showing the cytoglobin (top) aligned to a portion of the large snail protein (see aligned rectangles).

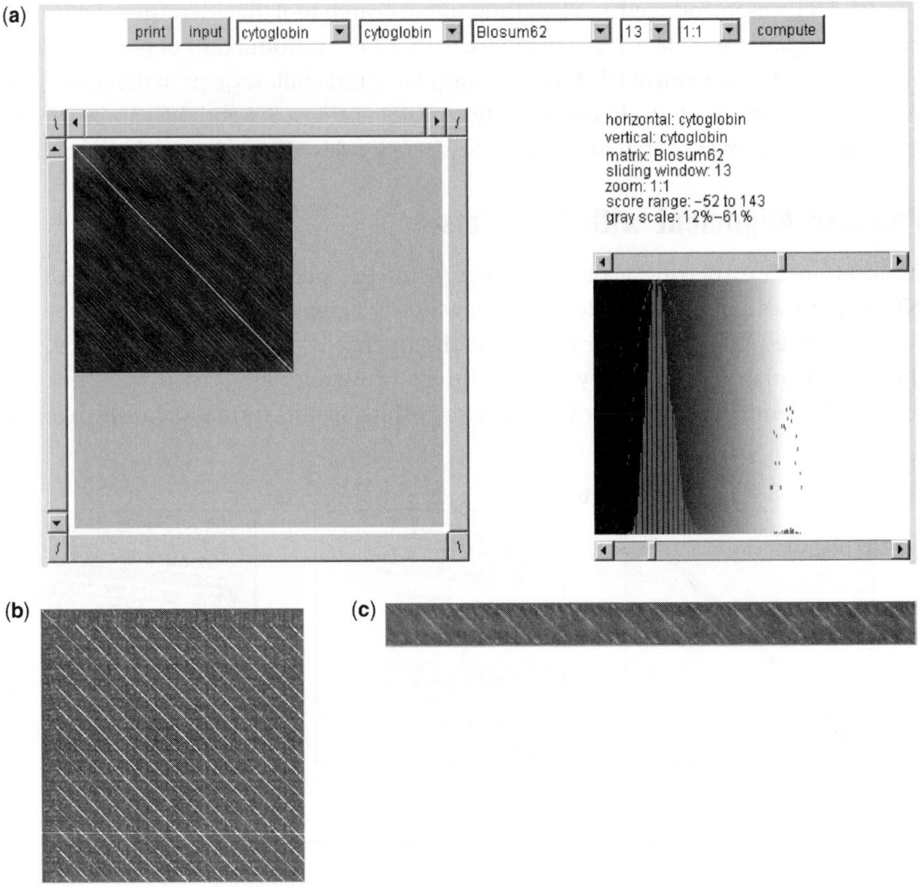

FIGURE 3.26. *Pairwise alignment with the Dotlet program. (a) Comparison of cytoglobin with itself provides a result comparable to that shown in Fig. 3.25 using BLAST 2 Sequences. Dotlet includes user-controlled zoom and window features allowing the background intensity to be adjusted. This allows one to maximize views such as (b) globin from the snail Biomphalaria glabrata (CAJ44466) aligned with itself, or (c) globin from the snail Biomphalaria glabrata (CAJ44466) aligned with human cytoglobin.*

We will encounter dot plots in Chapter 15 when we compare bacterial genome sequences to each other. We will also see a dot plot in Chapter 17 (on fungi). Protein sequences from *Saccharomyces cerevisiae* chromosomes were systematically BLAST searched against each other. The resulting dot plot showed many diagonal lines indicating homologous regions. This provided evidence that, surprisingly, the entire yeast genome duplicated over 100 million years ago.

line. This is illustrated for an alignment of human cytoglobin with itself as part of the BLAST output (Fig. 3.25a). We also illustrate a dot plot using the web-based Dotlet program of Junier and Pagni (2000) (Fig. 3.26a). That program features an adjustable sliding window size, a zoom feature, a variety of scoring matrices, and a histogram window to adjust the pixel intensities (Fig. 3.26a, right side) in order to manually optimize the signal to noise ratio.

We can further illustrate the usefulness of dot plots by examining an unusual hemoglobin protein of 2,148 amino acids from the snail *Biomphalaria glabrata*. It consists of 13 globin repeats (Lieb et al., 2006). When we compare it to human cytoglobin (190 amino acids) with a default BLOSUM62 matrix, the BLAST output shows cytoglobin (x axis) matching the snail protein 12 times (y axis) (Fig. 3.25b); one repeat is missed. By changing the scoring matrix to BLOSUM45 we can now see all 13 snail hemoglobin repeats (Fig. 3.25c). The gap at the start of the dot plot (Figs. 3.25b and c, arrowheads) is evident in the pairwise alignment of that region (Fig. 3.25d): the first 128 amino acids of the snail protein are unrelated and thus not aligned with cytoglobin. Using Dotlet, all 13 globin repeats are evident in a comparison of the snail protein with itself (Fig. 3.26b) or with cytoglobin (Fig. 3.26c).

THE STATISTICAL SIGNIFICANCE OF PAIRWISE ALIGNMENTS

How can we decide whether the alignment of two sequences is statistically significant? We address this question for local alignments, and then for global alignments.

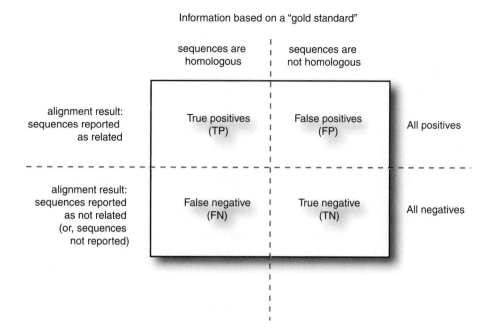

Information based on a "gold standard"

	sequences are homologous	sequences are not homologous	
alignment result: sequences reported as related	True positives (TP)	False positives (FP)	All positives
alignment result: sequences reported as not related (or, sequences not reported)	False negative (FN)	True negative (TN)	All negatives

FIGURE 3.27. Sequence alignments, whether pairwise (this chapter) or from a database search (Chapter 4), can be classified as true or false and positives or negatives. Statistical analyses of alignments provide the main way that you can evaluate whether an alignment represents a true positive, that is, an alignment of homologous sequences. Ideally, an alignment algorithm can maximize both sensitivity and specificity.

Consider two proteins that share limited amino acid identity (e.g., 20% to 25%). Alignment algorithms report the score of a pairwise alignment or the score of the best alignments of a query sequence against an entire database of sequences (Chapter 4). We need statistical tests to decide whether the matches are true positives (i.e., whether the two aligned proteins are genuinely homologous) or whether they are false positives (i.e., whether they have been aligned by the algorithm by chance) (Fig. 3.27). For the alignments that are not reported by an algorithm, for instance because the score falls below some threshold, we would like to evaluate whether the sequences are true negatives (i.e., genuinely unrelated) or whether they are false negatives, that is, homologous sequences that receive a score suggesting that they are not homologous.

A main goal of alignment algorithms is thus to maximize the sensitivity and specificity of sequence alignments (Fig. 3.27). Sensitivity is the number of true positives divided by the sum of true positive and false negative results. This is a measure of the ability of an algorithm to correctly identify genuinely related sequences. Specificity is the number of true negative results divided by the sum of true negative and false positive results. This describes the sequence alignments that are not homologous.

Statistical Significance of Global Alignments

When we align two proteins, such as human beta globin and myoglobin, we obtain a score. We can use hypothesis testing to assess whether that score is likely to have occurred by chance. To do this, we first state a null hypothesis (H_0) that the two sequences are not related. According to this hypothesis, the score S of beta globin and myoglobin represents a chance occurrence. We then state an alternative hypothesis (H_1) that they are indeed related. We choose a significance value α, often set to 0.05, as a threshold for defining statistical significance. One approach to determining whether our score occurred by chance is to compare it to the scores of beta globin or myoglobin relative to a large number of other proteins (or DNA sequences) known to not be homologous. Another approach is to compare the query to a set of randomly generated sequences. A third approach is to randomly scramble the sequence of one of the two query proteins (e.g., myoglobin) and obtain a score relative to beta globin;

by repeating this process 100 times, we can obtain the sample mean ($\bar{x}$) and sample standard deviation (s) of the scores for the randomly shuffled myoglobin relative to beta globin. We can express the authentic score in terms of how many standard deviations above the mean it is. A Z score (Box 3.9) is calculated as:

$$Z = \frac{x - \mu}{\sigma} \tag{3.5}$$

PRSS, written by William Pearson, is available online at ▶ http://fasta.bioch.virginia.edu/fasta/prss.htm. For an example of PRSS output for a comparison of human beta globin and myoglobin, see web document 3.7 at ▶ http://www.bioinfbook.org/chapter3.

where x is the score of two aligned sequences, μ is the mean score of many sequence comparisons using a scrambled sequence, and σ is the standard deviation of those measurements obtained with random sequences. We can do the shuffle test using a computer algorithm such as PRSS. This calculates the score of a global pairwise alignment, and also performs comparisons of one protein to a randomized (jumbled) version of the other.

If the scores are normally distributed, then the Z statistic can be converted to a probability value. If $Z = 3$, then we can refer to a table in a standard statistics resource to see that 99.73% of the population (i.e., of the scores) are within three standard deviations of the mean, and the fraction of scores that are greater than three standard deviations beyond the mean is only 0.13%. We can expect to see this particular score by

Box 3.9
Statistical Concepts: *Z* Scores

The familiar bell-shaped curve is a Gaussian distribution or normal distribution. The x axis corresponds to some measured values, such as the alignment score of beta globin versus 100 randomly shuffled versions of myoglobin. The y axis corresponds to the probability density (when considering measurements of an exhaustive set of shuffled myoglobins) or to the number of trials (when considering a number of shuffled myoglobins). The mean value is obtained simply by adding all the scores and dividing by the number of pairwise alignments; it is apparent at the center of a Gaussian distribution. For a set of data points $x_1, x_2, x_3, \ldots, x_n$ the mean $\bar{x}$ is the sum divided by n, or:

$$\bar{x} = \frac{\sum_{i=1}^{n} x_i}{n}$$

The sample variance s^2 describes the spread of the data points from the mean. It is related to the squares of the distances of the data points from the mean, and it is given by:

$$s^2 = \frac{1}{n-1} \sum_{i=1}^{n} (x_i - \bar{x})^2$$

The sample standard deviation s is the square root of the variance, and thus its units match those of the data points. It is given by:

$$s = \sqrt{\frac{\sum_{i=1}^{N} (Y_i - m)^2}{N - 1}}$$

Here s is the sample standard deviation (rather than the population standard deviation, σ); note that s^2 is the sample variance. Population variance refers to the average of the square of the deviations of each value from the mean, while the sample variance includes an adjustment from number of measurements N. m is the sample mean (rather than the population mean, μ).

Z scores (also called standardized scores) describe the distance from the mean per standard deviation:

$$Z_i = \frac{x_i - \bar{x}}{s}$$

If you compare beta globin to myoglobin, you can get a score (such as 43.9 as shown in Fig. 3.5a) based on some scoring system. Randomly scramble the sequence of myoglobin 1000 times (maintaining the length and composition of the myoglobin), and measure the 1000 scores of beta globin to these scrambled sequences. You can obtain a mean and standard deviation of the comparison to shuffled sequences.

Many books and articles introduce statistical concepts, including Motulsky (1995) and Cumming et al. (2007).

chance about 1 time in 750 (i.e., 0.13% of the time). The problem in adopting this approach is that if the distribution of scores deviates from a Gaussian distribution, the estimated significance level will be wrong. For global (but not local) pairwise alignments, the distribution is generally not Gaussian, and hence there is not a strong statistical basis for assigning significance values to pairwise alignments. What can we conclude from a Z score? If 100 alignments of shuffled proteins all have a score less than the authentic score of RBP4 and β-lactoglobulin, this indicates that the probability (p) value is less than 0.01 that this occurred by chance. (Thus we can reject the null hypothesis that the two protein sequences are not significantly related.) However, because of the concerns about the applicability of the Z score to sequence scores, conclusions about statistical significance should be made with caution.

Another consideration involves the problem of multiple comparisons. If we compare a query such as beta globin to one million proteins in a database, we have a million opportunities to find a high-scoring match between the query and some database entry. In such cases it is appropriate to adjust the significance level α, that is, the probability at which the null hypothesis is rejected, to a more stringent level. One approach, called a Bonferroni correction, is to divide α (nominally $p < 0.05$) by the number of trials (10^6) to set a new threshold for defining statistical significance at a level of $0.05/10^6$, or 5×10^{-8}. The equivalent of a Bonferroni correction is applied to the probability value calculation of BLAST statistics (see Chapter 4), and we will also encounter multiple comparison corrections in microarray data analysis (see Chapter 9).

Statistical Significance of Local Alignments

Most database search programs such as BLAST (Chapter 4) depend on local alignments. Additionally, many pairwise alignment programs compare two sequences using local alignment. For local pairwise alignments, the best approach to defining statistical significance is to estimate an expect value (E value), which is closely related to a probability value (p value). In contrast to the situation with global alignment, for local alignment there is a thorough understanding of the distribution of scores. An E value describes the number of matches having a particular score (or better) that are expected to occur by chance. For example, if a pairwise alignment of a beta globin and a myoglobin has some score with an associated E value of 10^{-3}, one can expect to obtain that particular score (or better) one time in one thousand by chance. This is the approach taken by the BLAST family of programs; we discuss E values in detail in Chapter 4.

Percent Identity and Relative Entropy

The accession numbers of rat and bovine odorant-binding proteins are NP_620258 and P07435; the human protein closest to rat has accession EAW50553. The alignments of these proteins are shown in web document 3.8 at ▶ http://www.bioinfbook.org/chapter3.

One approach to deciding whether two sequences are significantly related from an evolutionary point of view is to consider their percent identity. It is very useful to consider the percent identity that two proteins share in order to get a sense of their degree of relatedness. As an example, a global pairwise alignment of odorant-binding protein from rat and cow share only 30% identity, although both are functionally able to bind odorants with similar affinities (Pevsner et al., 1985). The rat protein shares just 26% identity to its closest human ortholog. From a statistical perspective the inspection of percent identities has limited usefulness in the "twilight zone" because it does not provide a rigorous set of rules for inferring homology, and it is associated with false positive or false negative results. A high degree of identity over a short region might sometimes not be evolutionarily significant, and conversely a low percent identity could reflect homology. Percent amino acid identity alone is not sufficient to demonstrate (nor to rule out) homology.

Still, it may be useful to consider percent identity. Some researchers have suggested that if two proteins share 25% or more amino acid identity over a span of 150 or more amino acids, they are probably significantly related (Brenner et al., 1998). If we consider an alignment of just 70 amino acids, it is popular to consider the two sequences "significantly related" if they share 25% amino acid identity. However, Brenner et al. (1998) have shown that this may be erroneous, in part because the enormous size of today's molecular sequence databases increases the likelihood that such alignments occur by chance. For an alignment of 70 amino acid residues, 40% amino acid identity is a reasonable threshold to estimate that two proteins are homologous (Brenner et al., 1998). If two proteins share about

Box 3.10
Relative Entropy

Altschul (1991) estimated that about 30 bits of information are required to distinguish an authentic alignment from a chance alignment of two proteins of average size (given that one protein is searched against a database of a particular size). For each substitution matrix with its unique target frequencies q_{ij} and background distributions $p_i p_j$, it is possible to derive the relative entropy H as follows (Altschul, 1991):

$$H = \sum_{i,j} q_{i,j} s_{i,j} = \sum_{i,j} q_{i,j} \log_2 \frac{q_{ij}}{p_i p_j}$$

H corresponds to the information content of the target and background distributions associated with a particular scoring matrix. The units of H are nats. As shown in Fig. 3.28, for higher H values, it is easier to distinguish the target from background frequencies. This analysis is consistent with the analysis of the diagonals for the PAM1 and PAM250 mutation probability matrices (Figs. 3.11 and 3.13) in which there is far less signal apparent in the PAM250 matrix.

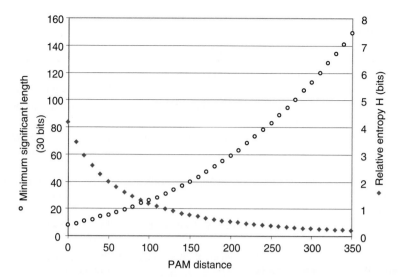

FIGURE 3.28. *Relative entropy (H) as a function of PAM distance. For PAM matrices with low value (e.g., PAM10), the relative entropy in bits is high and the minimum length required to detect a signifantly aligned pair of sequences is short (e.g., about 10 amino acids). Thus, using a PAM10 matrix, two very closely related proteins can be detected as homologous even if only a relatively short region of amino acid residues is compared. For PAM250 and other PAM matrices with high values, the relative entropy (or information content in the sequence) is low, and it is necessary to have a longer region of amino acids (e.g., 80 residues) aligned in order to detect significant relationships between two proteins. Adapted from Altschul (1991).*

20%–25% identity over a reasonably long stretch (e.g., 70 to 100 amino acid residues), they are in the "twilight zone" (Fig. 3.19), and it is more difficult to be sure. Two proteins that are completely unrelated often share about 10% to 20% identity when aligned. This is especially true because the insertion of gaps can greatly improve the alignment of any two sequences.

Altschul (1991) evaluated alignment scores from an information theory perspective. Target frequencies vary as a function of evolutionary distance. Recall that an alignment of alanine with threonine is assigned a different score in a PAM10 matrix (−3; see Fig. 3.15) than in a PAM250 matrix (+1; see Fig. 3.14). The relative entropy (H) of the target and background distributions measures the information that is available per aligned amino acid position that, on average, distinguishes a true alignment from a chance alignment (Box 3.10). For a PAM10 matrix, the value of H is 3.43 bits. Assuming that 30 bits of information are sufficient to distinguish a true rather than a chance alignment in a database search, an alignment of at least 9 residues is needed using a PAM10 matrix (Fig. 3.28). But for a PAM250 matrix, the relative entropy is 0.36 and an alignment of at least 83 residues is needed to distinguish authentic alignments.

We will see in Chapter 6 that multiple sequence alignments can offer far more sensitivity than pairwise sequence alignment. We will also see in Chapter 5 that scoring matrices ("profiles") can be customized to a sequence alignment, greatly increasing the sensitivity of a search.

PERSPECTIVE

The pairwise alignment of DNA or protein sequences is one of the most fundamental operations of bioinformatics. Pairwise alignment allows one to determine the

TABLE 3-4 Global Pairwise Alignment Algorithms

Resource	Description	URL
ALIGN	At the GENESTREAM server, France	► http://www2.igh.cnrs.fr/bin/align-guess.cgi
GAP	From the Genetics Computer Group (GCG)	► http://www.gcg.com
Needle	From the Institut Pasteur; implements Needleman–Wunsch global alignment	► http://bioweb.pasteur.fr/docs/EMBOSS/needle.html
Pairwise alignment (various)	From the University of Virginia (Bill Pearson)	► http://alpha10.bioch.virginia.edu/fasta/
Pairwise	Two Sequence Alignment Tool (global and local options)	► http://informagen.com/Applets/Pairwise/
Pairwise Sequence Alignment	From the Baylor College of Medicine; various tools	► http://searchlauncher.bcm.tmc.edu/
Stretcher	From the Institut Pasteur; global alignment	► http://bioweb.pasteur.fr/docs/EMBOSS/stretcher.html
Vector NTI Suite 7	From Informax	► http://www.informaxinc.com

Abbreviations: EMBOSS, The European Molecular Biology Open Software Suite (► http://www.uk.embnet.org/Software/EMBOSS/); ISREC, Swiss Institute for Experimental Cancer Research (► http://www.isrec.isb-sib.ch/).

TABLE 3-5 Local Pairwise Alignment Algorithms

Resource	Description	URL
BestFit	From the Genetics Computer Group (GCG)	► http://www.gcg.com
BLAST	At NCBI	► http://www.ncbi.nlm.nih.gov/BLAST/
est2genome	EMBOSS program from the Institut Pasteur; aligns expressed sequence tags to genomic DNA	► http://bioweb.pasteur.fr/docs/EMBOSS/est2genome.html
LALIGN	Finds multiple matching subsegments in two sequences	► http://www.ch.embnet.org/software/LALIGN_form.html
Pairwise	Two Sequence Alignment Tool (global and local options)	► http://informagen.com/Applets/Pairwise/
Pairwise Sequence Alignment	From the Baylor College of Medicine; various tools	► http://searchlauncher.bcm.tmc.edu/
PRSS	From the University of Virginia (Bill Pearson)	► http://fasta.bioch.virginia.edu/fasta/prss.htm
SIM	Alignment tool for protein sequences from ExPASy	► http://www.expasy.ch/tools/sim-prot.html
SIM	SIM, gap at the Department of Computer Science, Michigan Tech	► http://genome.cs.mtu.edu/align.html
SSEARCH	At the Protein Information Resource	► http://pir.georgetown.edu/pirwww/

Abbreviations: EMBOSS, The European Molecular Biology Open Software Suite (► http://www.uk.embnet.org/Software/EMBOSS/); ISREC, Swiss Institute for Experimental Cancer Research (► http://www.isrec.isb-sib.ch/).

Substituent residue
(Percentage of total residue sites at which the substitution occurs)

Sequence (original amino acid)	A	R	N	D	C	Q	E	G	H	I	L	K	M	F	P	S	T	W	Y	V
A	■			28			31	33							31					
R		■							50			58			25					
N	33		■	47				33				33				33	33			
D	44	22		■		47	34	22				28			25					
C	(66)				■															
Q				56		■	30		40			70								
E	50			44			■		38			41	24							
G	51		33				30	■				27			36					
H			26						■	26	30				22	22				
I	39									■	58									46
L	21								23		23	28	■							30
K	23	21		28		31	23			21		■			21					
M	22								22	89	22		■							45
F								22		61				■						
P	50			43		57	43					21			■					
S	49			24		24	36					24				■	40			
T	32					28	24					24				52	■			
W	(40)								(40)		(60)							■		
Y								(33)			(50)								■	
V	36								21	43	21									■

FIGURE 3.29. Substitution frequencies of globins (adapted from Zuckerkandl and Pauling, 1965, p. 118). Amino acids are presented alphabetically according to the three letter abbreviations. The rows correspond to an original amino acid in an alignment of several dozen hemoglobin and myoglobin protein sequences from human, other primates, horse, cattle, pig, lamprey, and carp. Numbers represent the percentages of residue sites at which a given substitution occurs. For example, a glycine substitution was observed to occur in 33% of all the alanine sites. Substitutions that were never observed to occur are indicated by squares colored red. Rarely occurring substitutions (percentages <20%) are indicated by empty white squares (numerical values are not given). "Very conservative" substitutions (percentages ≥40%) are in boxes shaded gray. For example, in 89% of the sites containing a methionine, leucine was also observed to be present. Identities are indicated by black solid squares. Values in parentheses indicate a very small available sample size, suggesting that conclusions about those data should be made cautiously. Used with permission.

relationship between any two sequences, and the degree of relatedness that is observed helps one to form a hypothesis about whether they are homologous (descended from a common evolutionary ancestor). Almost all of the topics in the rest of this book are heavily dependent on sequence alignment. In Chapter 4, we introduce the searching of large DNA and/or protein databases with a query sequence. Database searching typically involves an extremely large series of local pairwise alignments, with results returned as a rank order beginning with most related sequences.

The algorithms used to perform pairwise alignment were developed in the 1970s, beginning with the global alignment procedure of Needleman and Wunsch (1970). Dayhoff (1978) introduced PAM scoring matrices that permit the comparison and evaluation of distantly related molecular sequences. Scoring matrices are an integral part of all pairwise (or multiple) sequence alignments, and the choice of a scoring matrix can strongly influence the outcome of a comparison. By the 1980s, local alignment algorithms were introduced (see especially the work of Sellers [1974], Smith and Waterman [1981], and Smith et al. [1981]). Practically, pairwise alignment is performed today with a limited group of software packages, most of which are freely available on the World Wide Web.

The sensitivity and specificity of the available pairwise sequence alignment algorithms continue to be assessed. Recent areas in which pairwise alignment has been further developed include methods of masking low-complexity sequences (discussed in Chapter 4) and theoretical models for penalizing gaps in alignments.

PITFALLS

The optional parameters that accompany a pairwise alignment algorithm can greatly influence the results. A comparison of human RBP4 and bovine β-lactoglobulin using BLAST results in no match detected if the default parameters are used.

Any two sequences can be aligned, even if they are unrelated. In some cases, two proteins that share even greater than 30% amino acid identity over a stretch of 100 amino acids are not homologous (evolutionarily related). It is always important to assess the biological significance of a sequence alignment. This may involve searching for evidence for a common cellular function, a common overall structure, or if possible a similar three-dimensional structure.

WEB RESOURCES

Pairwise sequence alignment can be performed using software packages that implement global or local alignment algorithms. In all cases, two protein or two nucleic acid sequences are directly compared.

Many websites offer freely available pairwise local alignment algorithms based on global alignment (Table 3.4) or local alignment (Table 3.5). These sites include the NCBI's BLAST, the Baylor College of Medicine (BCM) launcher, the SIM program at ExPASy, and SSEARCH at the Protein Information Resource (PIR) at Georgetown University.

DISCUSSION QUESTIONS

[3-1] If you want to compare any two proteins, is there any one "correct" scoring matrix to choose? Is there any way to know which scoring matrix is best to try?

[3-2] Many protein (or DNA) sequences have separate domains. (We discuss domains in Chapter 10.) Consider a protein that has one domain that evolves rapidly and a second domain that evolves slowly. In doing a pairwise alignment with another protein (or DNA) sequence, would you use two separate alignments with scoring matrices such as PAM40 and PAM250 or would you select one "intermediate" matrix? Why?

[3-3] Years before Margaret Dayhoff and colleagues published a protein atlas with scoring matrices, Emile Zuckerkandl and Linus Pauling (1965) produced a scoring matrix for several dozen available globin sequences (Fig. 3.29). The rows (y axis) of this Figure show the original globin amino acid, and the columns show substitutions that were observed to occur. Numerical values are entered in cells for which the substitutions occur in at least 20% of the sites. Note that for cells shaded red, these amino acid substitutions were never observed, while for cells shaded gray the amino acid substitutions were defined as very conservative.

How do the data in this matrix compare to those described by Dayhoff and colleagues? Which substitutions occur most rarely, and which most frequently? How would you go about filling in this table today?

Joshua Lederberg helped Zuckerkandl and Pauling (1965) make this matrix. They used an IBM 7090 computer, one of the first commercial computers based on transistor technology. The computer cost about $3 million. Its memory consisted of 32,768 binary words or about 131,000 bytes. (To read about Lederberg's Nobel Prize from 1958, see ▶ http://nobelprize.org/nobel_prizes/medicine/laureates/1958/.)

PROBLEMS/COMPUTER LAB

[3-1] Viral reverse transcriptases, such as the *pol* gene product encoded by HIV-1, have human homologs. The GenBank accession number for HIV-1 reverse transcriptase is NP_057849. (Use Entrez to confirm this is the correct accession number.) A search of Entrez reveals many human viral-related gene products, including a retrovirus-related Pol polyprotein of 874 amino acid residues (P10266). Perform a pairwise alignment using the blastp program.

The default conditions for this search include the use of the BLOSUM62 scoring matrix. The expect value is about 1×10^{-67}, indicating that the proteins are closely related even though they share only 28% identity over a span of 761 amino acids. Repeat the analysis using the BLOSUM62, BLOSUM50, and BLOSUM90 scoring matrices. What is the effect of changing the search parameters?

[3-2] Next perform pairwise alignments of the proteins described in problem 3-1 using the PAM30, PAM70, and PAM250 matrices. What are the expect values? What span of amino acid residues is aligned? Are the search results using different PAM matrices similar or different to the results of using different BLOSUM matrices?

[3-3] Perform a local pairwise alignment between RBP and β-lactoglobulin using BLAST 2 Sequences (or any of the programs listed in Table 3.5). Repeat the alignment using lower gap penalties. What is the result?

[3-4] Compare modern human mitochondrial DNA to extinct Neanderthal DNA. First obtain the nucleotide sequence of a mitochondrially encoded gene, cytochrome oxidase. (Begin by searching the taxonomy division of the NCBI website,
► http://www.ncbi.nlm.nih.gov/Taxonomy/taxonomyhome.html/, and select "extinct organisms" to find Neanderthal DNA.) Next, perform pairwise alignments and record the percent nucleotide identities.

[3-5] We have seen that some gene families change slowly (e.g., GAPDH in Fig. 3.8) while others change rapidly (e.g., see Box 3.3). How can you determine whether the cytochrome oxidase gene you studied in problem 3.4 changes relatively rapidly or slowly? Try using pairwise blast p.

[3-6] Calculate the PAM250 log-odds scoring matrix score for the alignment of a cysteine to a glutamate. Then, calculate the score for changing a glutamate to a cysteine. Use Equation 3.1, the PAM250 mutation probability matrix (Figure 3.13), and the table of normalized amino acid frequencies (Table 3.2).

[3-7] A pairwise alignment of human alpha and beta globin is described in Fig. 3.16. Try this pairwise alignment at the NCBI protein BLAST site, using the available matrices (PAM30, PAM70, BLOSUM45, BLOSUM62, BLOSUM80). Which gives the highest bit score?

[3-8] Aphrodisin and odorant-binding protein are both examples of lipocalins. First obtain the accession numbers for rodent forms of these proteins, and then perform a pairwise sequence alignment. (Use pairwise BLAST.) Record the percent amino acid identity, the percent similarities, the expect values, and bit scores. Which metric is most useful in helping you evaluate their relatedness?

[3-9] The coelacanth *Latimeria chalumnae* is a lobe-finned fish that has been called a "living fossil." Long thought to be extinct for at least 90 million years, several specimens have now been discovered lurking in the ocean. Surprisingly, some phylogenetic analyses of mitochondrial DNA sequences indicate that the coelacanth is more closely related to humans than to herrings (Lewin, 2001) (see also Chapter 18). Find the accession numbers for some mitochrondrial DNA from human, herring, and coelacanth, and then perform pairwise alignments to decide if you agree. *Hint:* Use PubMed to find the genus and species name of an organism; herring is *Clupea harengus*. Next use this species name in a search of Entrez nucleotides, such as "*Clupea harengus* mitochondrion."

[3-10] The PAM1 matrix (Fig. 3.11) is nonreciprocal: the probability of changing an amino acid such as alanine to arginine is not equal to the probability of changing an arginine to an alanine. Why?

[3-11] Is a hippopotamus more closely related to a pig or to a whale? To answer this question, first find the protein sequence of hemoglobin from each of these three organisms. Next, perform pairwise sequence alignments and record the percent amino acid identities.

SELF-TEST QUIZ

[3-1] Match the following amino acids with their single-letter codes:

Asparagine	Q
Glutamine	W
Tryptophan	Y
Tyrosine	N
Phenylalanine	F

[3-2] Orthologs are defined as:

(a) Homologous sequences in different species that share an ancestral gene

(b) Homologous sequences that share little amino acid identity but share great structural similarity

(c) Homologous sequences in the same species that arose through gene duplication

(d) Homologous sequences in the same species which have similar and often redundant functions

[3-3] Which of the following amino acids is least mutable according to the PAM scoring matrix?

(a) Alanine

(b) Glutamine

(c) Methionine

(d) Cysteine

[3-4] The PAM250 matrix is defined as having an evolutionary divergence in which what percentage of amino acids between two homologous sequences have changed over time?

(a) 1%

(b) 20%

(c) 80%

(d) 250%

[3-5] Which of the following sentences best describes the difference between a global alignment and a local alignment between two sequences?

(a) Global alignment is usually used for DNA sequences, while local alignment is usually used for protein sequences.

(b) Global alignment has gaps, while local alignment does not have gaps.

(c) Global alignment finds the global maximum, while local alignment finds the local maximum.

(d) Global alignment aligns the whole sequence, while local alignment finds the best subsequence that aligns.

[3-6] You have two distantly related proteins. Which BLOSUM or PAM matrix is best to use to compare them?

(a) BLOSUM45 or PAM250

(b) BLOSUM45 or PAM10

(c) BLOSUM80 or PAM250

(d) BLOSUM80 or PAM10

[3-7] How does the BLOSUM scoring matrix differ most notably from the PAM scoring matrix?

(a) It is best used for aligning very closely related proteins.

(b) It is based on global multiple alignments from closely related proteins.

(c) It is based on local multiple alignments from distantly related proteins.

(d) It combines local and global alignment information.

[3-8] True or false: Two proteins that share 30% amino acid identity are 30% homologous.

[3-9] A global alignment algorithm (such as the Needleman–Wunsch algorithm) is guaranteed to find an optimal alignment. Such an algorithm:

(a) Puts the two proteins being compared into a matrix and finds the optimal score by exhaustively searching every possible combination of alignments

(b) Puts the two proteins being compared into a matrix and finds the optimal score by iterative recursions

(c) Puts the two proteins being compared into a matrix and finds the optimal alignment by finding optimal subpaths that define the best alignment(s)

(d) Can be used for proteins but not for DNA sequences %

[3-10] In a database search or in a pairwise alignment, sensitivity is defined as:

(a) The ability of a search algorithm to find true positives (i.e., homologous sequences) and to avoid false positives (i.e., unrelated sequences having high similarity scores)

(b) The ability of a search algorithm to find true positives (i.e., homologous sequences) and to avoid false positives (i.e., homologous sequences that are not reported)

(c) The ability of a search algorithm to find true positives (i.e., homologous sequences) and to avoid false negatives (i.e., unrelated sequences having high similarity scores)

(d) The ability of a search algorithm to find true positives (i.e., homologous sequences) and to avoid false negatives (i.e., homologous sequences that are not reported)

SUGGESTED READING

We introduced this chapter with the concept of homology, an often misused term. A one-page article by Reeck et al. (1987) provides authoritative, standard definitions of the terms homology and similarity. Another discussion of homology in relation to phylogeny is provided by Tautz (1998).

All of the papers describing sequence alignment consider the divergence of two homologous sequences in the context of a model of molecular evolution. Russell F. Doolittle (1981) has written a clear, thoughtful overview of sequence alignment. William Pearson (1996) has reviewed sequence alignment. He provides descriptions of the statistics of similarity scores, sensitivity and selectivity, and search programs such as Smith–Waterman and FASTA. Other reviews of pairwise alignment include short articles by Altschul (1998) and Brenner (1998) in *Bioinformatics: A Trends Guide*.

For studies of pairwise sequence alignment algorithms, an important historical starting point is the 1978 book by Margaret O. Dayhoff and colleagues (Dayhoff, 1978). Most of this book consists of an atlas of protein sequences with accompanying phylogenetic reconstructions. Chapter 22 introduces the concept of accepted point mutations, while Chapter 23 describes various PAM matrices. By the early 1990s, when far more protein sequence data were available, Steven and Jorja Henikoff (1992) described the BLOSUM matrices. This article provides an excellent technical introduction to the use of scoring matrices, usefully contrasting the performance of PAM and BLOSUM matrices. Later (in Chapters 4 and 5) we will use these matrices extensively in database searching.

The algorithms originally describing global alignment are presented technically by Needleman and Wunsch (1970) and later local alignment algorithms were introduced by Smith and

Waterman (1981) and Smith, Waterman, and Fitch (1981). The problem of both sensitivity (the ability to identify distantly related sequences) and selectivity (the avoidance of unrelated sequences) of pairwise alignments was addressed by Pearson and Lipman in a 1988 paper introducing the FASTA program.

More recent articles address technical aspects of sequence-scoring statistics. Marco Pagni and C. Victor Jongeneel (2001) of the Swiss Institute of Bioinformatics provide an excellent overview. This includes a discussion of BLAST scoring statistics that is relevant to Chapters 4 and 5.

Finally, Steven Brenner, Cyrus Chothia, and Tim Hubbard (1998) have compared several pairwise sequence methods. This article is highly recommended as a way to learn how different algorithms can be assessed (we will see similar approaches for multiple sequence alignment in Chapter 10, for example). Reading this paper can help to show why statistical scores are more effective than other search parameters, such as raw scores or percent identity in interpreting pairwise alignment results.

REFERENCES

Altschul, S. F. Amino acid substitution matrices from an information theoretic perspective. *J. Mol. Biol.* **219**, 555–565 (1991).

Altschul, S. F. Fundamentals of database searching. *Bioinformatics: A Trends Guide* **1998**, 7–9 (1998).

Altschul, S. F., Gish, W., Miller, W., Myers, E. W., and Lipman, D. J. Basic Local Alignment Search Tool. *J. Mol. Biol.* **215**, 403–410 (1990).

Anfinsen, C. *The Molecular Basis of Evolution.* Wiley, New York, 1959.

Brenner, S. E. Practical database searching. *Bioinformatics: A Trends Guide* **1998**, 9–12 (1998).

Brenner, S. E., Chothia, C., and Hubbard, T. J. Assessing sequence comparison methods with reliable structurally identified distant evolutionary relationships. *Proc. Natl. Acad. Sci. USA* **95**, 6073–6078 (1998).

Chothia, C., and Lesk, A. M. The relation between the divergence of sequence and structure in proteins. *EMBO J.* **5**, 823–826 (1986).

Cumming, G., Fidler, F., and Vaux, D. L. Error bars in experimental biology. *J. Cell Biol.* **177**, 7–11 (2007).

Dayhoff, M. O. (ed.). *Atlas of Protein Sequence and Structure.* National Biomedical Research Foundation, Silver Spring, MD, 1978.

Dayhoff, M. O., Hunt, L. T., McLaughlin, P. J., and Jones, D. D. (1972) Gene duplications in evolution: The globins. In Dayhoff, M. O. (ed.), *Atlas of Protein Sequence and Structure 1972*, Vol. 5. National Biomedical Research Foundation, Washington, D.C., 1972.

Doolittle, R. F. Similar amino acid sequences: Chance or common ancestry? *Science* **214**, 149–159 (1981).

Doolittle, R. F. *Of URFS AND ORFS: A Primer on How to Analyze Derived Amino Acid Sequences.* University of Science Books, Mill Valley, CA, 1987.

Durbin, R., Eddy, S., Krogh, A., and Mitchison, G. *Biological Sequence Analysis.* Cambridge University Press, Cambridge, 2000.

Ewens, W. J. and Grant, G. R. *Statistical Methods in Bioinformatics: An Introduction.* Springer-Verlag, New York, 2001.

Fitch, W. M. Distinguishing homologous from analogous proteins. *Syst. Zool.* **19**, 99–113 (1970).

Gonnet, G. H., Cohen, M. A., and Benner, S. A. Exhaustive matching of the entire protein sequence database. *Science* **256**, 1443–1445 (1992).

Gotoh, O. An improved algorithm for matching biological sequences. *J. Mol. Biol.* **162**, 705–708 (1982).

Henikoff, S., and Henikoff, J. G. Amino acid substitution matrices from protein blocks. *Proc. Natl. Acad. Sci. USA* **89**, 10915–10919 (1992).

Henikoff, J. G., and Henikoff, S. Blocks database and its applications. *Methods Enzymol.* **266**, 88–105 (1996).

Ingram, V. M. *The Hemoglobins in Genetics and Evolution.* Columbia U. Press, New York, 1963, 18–28.

International Human Genome Sequencing Consortium. Initial sequencing and analysis of the human genome. *Nature* **409**, 860–921 (2001).

Jones, D. T., Taylor, W. R., and Thornton, J. M. The rapid generation of mutation data matrices from protein sequences. *Comput. Appl. Biosci.* **8**, 275–282 (1992).

Junier, T., and Pagni, M. Dotlet: diagonal plots in a web browser. *Bioinformatics* **16**, 178–179 (2000).

Karlin, S., and Altschul, S. F. Methods for assessing the statistical significance of molecular sequence features by using general scoring schemes. *Proc. Natl. Acad. Sci. USA* **87**, 2264–2268 (1990).

Lewin, R. A. Why rename things? *Nature* **410**, 637 (2001).

Lieb, B., Dimitrova, K., Kang, H. S., Braun, S., Gebauer, W., Martin, A., Hanelt, B., Saenz, S. A., Adema, C. M., and Markl, J. Red blood with blue-blood ancestry: intriguing structure of a snail hemoglobin. *Proc. Natl. Acad. Sci. USA* **103**, 12011–12016 (2006).

Motulsky, H. *Intuitive Biostatistics.* Oxford University Press, New York, 1995.

Myers, E. W., and Miller, W. Optimal alignments in linear space. *Comput. Appl. Biosci.* **4**, 11–17 (1988).

Needleman, S. B., and Wunsch, C. D. A general method applicable to the search for similarities in the amino acid sequence of two proteins. *J. Mol. Biol.* **48**, 443–453 (1970).

Owen, R. *Lectures on the Comparative Anatomy and Physiology of the Invertebrate Animals.* Longman, Brown, Green, and Longmans, London, 1843.

Pagni, M., and Jongeneel, C. V. Making sense of score statistics for sequence alignments. *Brief Bioinform.* **2**, 51–67 (2001).

Pearson, W. R., Effective protein sequence comparison. *Methods Enzymol.* **266**, 227–258 (1996).

Pearson, W. R., and Lipman, D. J. Improved tools for biological sequence comparison. *Proc. Natl. Acad. Sci. USA* **85**, 2444–2448 (1988).

Pearson, W. R., and Wood, T. C. Statistical significance in biological sequence comparison. In D. J. Balding, M. Bishop, and C. Cannings (eds.), *Handbook of Statistical Genetics*, 39–65. Wiley, London, 2001.

Pevsner, J., Trifiletti, R. R., Strittmatter, S. M., and Snyder, S. H. Isolation and characterization of an olfactory receptor protein for odorant pyrazines. *Proc. Natl. Acad. Sci. USA* **82**, 3050–3054 (1985).

Reeck, G. R., et al. "Homology" in proteins and nucleic acids: A terminology muddle and a way out of it. *Cell* **50**, 667 (1987).

Rice, P., Longden, I., and Bleasby, A. EMBOSS: The European Molecular Biology Open Software Suite. *Trends Genet.* **16**, 276–277 (2000).

Schopf, J. W. When did life begin? In Schopf, J. W. (ed.), *Life's Origin: The Beginnings of Biological Evolution.* University of California Press, Berkeley, 2002.

Sedgewick, R. *Algorithms.* 2nd edition. Addison-Wesley, Boston, 1988.

Sellers, P. H. On the theory and computation of evolutionary distances. *SIAM J. Appl. Math.* **26**, 787–793 (1974).

Smith, T. F., and Waterman, M. S. Identification of common molecular subsequences. *J. Mol. Biol.* **147**, 195–197 (1981).

Smith, T. F., Waterman, M. S., and Fitch, W. M. Comparative biosequence metrics. *J. Mol. Evol.* **18**, 38–46 (1981).

Tatusova, T. A., and Madden, T. L. BLAST 2 Sequences, a new tool for comparing protein and nucleotide sequences. *FEMS Microbiol. Lett.* **174**, 247–250 (1999).

Tautz, D. Evolutionary biology. Debatable homologies. *Nature* **395**, 17, 19 (1998).

Zuckerkandl, E., and Pauling, L. Evolutionary divergence and convergence in proteins. In V. Bryson and H. J. Vogel (eds.), *Evolving Genes and Proteins*, 97–166. Academic Press, New York, 1965.

of the variate, the first double exponential distribution has larger (smaller) densities than the normal one. The opposite is true for the second double exponential distribution.

Table 5.2.7. Selected Probabilities for Normal and
Largest Values

Value	Reduced Variate		Probabilities		Return Periods	
	Largest	Normal	Largest	Normal	Largest	Normal
$\bar{x} - \sigma$	−.70533	−1	.13206	.15866	7.57	6.30
$\bar{x} + \sigma$	1.85977	1	.85581	.84134	6.93	6.30
$\bar{x} \pm \sigma$	—	—	.72375	.68268	—	—
$\bar{x} - 2\sigma$	−1.98788	−2	.00068	.02275	1480.	43.96
$\bar{x} + 2\sigma$	3.14232	2	.95773	.97725	23.7	43.96
$\bar{x} \pm 2\sigma$	—	—	.95705	.95450	—	—
$\bar{x} - 3\sigma$	−3.27043	−3	$3.7 \cdot 10^{-12}$	.00135	$.27 \cdot 10^{11}$	741
$\bar{x} + 3\sigma$	4.42486	3	.98810	.99865	84.01	741
$\bar{x} \pm 3\sigma$	—	—	.98810	.99730	—	—

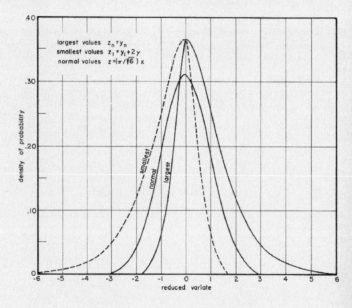

largest values $z_n = y_n$
smallest values $z_1 = y_1 + 2\gamma$
normal values $z = (\pi/\sqrt{6})\, x$

Graph 5.2.7(1). **Extreme and Normal Distributions**

In Graph 5.2.7(2) the probabilities of the largest and the smallest values and the normal probabilities for the same mean and standard deviation

Chapter 4 describes the principal database search tool, BLAST. While BLAST was first described by Altschul et al. in 1990, the statistical interpretation of the scores you get in a BLAST search are based on mathematical models developed by the 1950s. In many instances, the distribution of values in a population assumes a normal (Gaussian) distribution, as shown in this figure (see curve labeled "normal"). However, for a wide variety of natural phenomena the distribution of extreme values is not normal. Such is the case for database searches in which you search with a protein or DNA sequence of interest (the query) against a large database, as will be described in this chapter. The maximum scores fit an extreme value distribution (EVD) rather than a normal distribution. In 1958 Emil Gumbel described the statistical basis of the EVD in his book Statistics of Extremes. *This figure (Gumbel, 1958, p. 180) shows the EVD. Note that for the curve marked "largest" the tail is skewed to the right. Also, as shown in the table, for a normal distribution, values that are up to three standard deviations above the mean occupy 99.865% of the area under the curve, while for the EVD values up to three standard deviations occupy only 98.810%. In other words, the EVD is characterized by a larger area under the curve at the extreme right portion of the plot. We will see how this analysis is applied to BLAST search results to let you assess whether a query sequence is significantly related to a match in the database. Used with permission.*

4

Basic Local Alignment Search Tool (BLAST)

INTRODUCTION

Basic Local Alignment Search Tool (BLAST) is the main tool of the National Center for Biotechnology Information (NCBI) for comparing a protein or DNA sequence to other sequences in various databases (Altschul et al., 1990, 1997). BLAST searching is one of the fundamental ways of learning about a protein or gene: the search reveals what related sequences are present in the same organism and other organisms. The NCBI website includes several excellent resources for learning about BLAST.

In Chapter 3, we described how to perform a pairwise sequence alignment between two protein or nucleotide sequences. BLAST searching allows the user to select one sequence (termed the *query*) and perform pairwise sequence alignments between the query and an entire database (termed the *target*). Typically, this means that millions of alignments are analyzed in a BLAST search, and only the most closely related matches are returned. The Needleman–Wunsch (1970) global alignment algorithm is not used for database searches because we are usually more interested in identifying locally matching regions such as protein domains. The Smith–Waterman (1981) local alignment algorithm finds optimal pairwise alignments, but we cannot generally use it

NCBI resources include a tutorial and a course that can be accessed through the main BLAST page (▶ http://www.ncbi.nlm.nih.gov/BLAST/).

Bioinformatics and Functional Genomics, Second Edition. By Jonathan Pevsner
Copyright © 2009 John Wiley & Sons, Inc.

for database searches because it is too computationally intensive. BLAST offers a local alignment strategy having both speed and sensitivity, as described in this chapter. It also offers convenient accessibility on the World Wide Web.

BLAST is a family of programs that allows you to input a query sequence and compare it to DNA or protein sequences in a database. A DNA sequence can be converted into six potential proteins (see below), and the BLAST algorithms include strategies to compare protein sequences to dynamically translated DNA databases or vice versa. The programs produce high-scoring segment pairs (HSPs) that represent local alignments between your query and database sequences. BLAST searching has a wide variety of uses. These include:

You can go directly to the BLAST site via ► http://www.ncbi.nlm.nih.gov/BLAST/. Or go the main page of NCBI (► http://www.ncbi.nlm.nih.gov), then select BLAST from the toolbar.

- *Determining what orthologs and paralogs are known for a particular protein or nucleic acid sequence.* Besides alpha and beta globin and myoglobin, what other globins are known? When a new bacterial genome is sequenced and several thousand proteins are identified, how many of these proteins are paralogous? How many of the predicted genes have no significantly related matches in GenBank?

- *Determining what proteins or genes are present in a particular organism.* Are there any globins in plants? Are there any reverse transcriptase genes (such as HIV-1 *pol* gene) in fish? In some cases searching for remote homlogs requires the use of specialized BLAST-like approaches; we describe some of these in Chapter 5, including strategies to align entire genomes.

- *Determining the identity of a DNA or protein sequence.* For example, you may perform a subtractive hybridization experiment or a microarray experiment and learn that a particular DNA sequence is dramatically regulated under the experimental conditions that you are using. This DNA sequence may be searched against a protein database to learn what proteins are most related to the protein encoded by your DNA sequence.

- *Discovering new genes.* A BLAST search of genomic DNA may reveal that the DNA encodes a protein that has not been described before. In Chapter 5, we show how BLAST searching can be used to find novel, previously uncharacterized genes.

- *Determining what variants have been described for a particular gene or protein.* For example, many viruses are extremely mutable; what HIV-1 *pol* variants are known?

- *Investigating expressed sequence tags that may exhibit alternative splicing.* There is an EST database that can be explored by BLAST searching. Indeed, there are dozens of specialized databases that can be searched. For example, specialized databases consist of sequences from a specific organism, a tissue type, a chromosome, a type of DNA (such as untranslated regions), or a functional class of nucleic acids or proteins.

- *Exploring amino acid residues that are important in the function and/or structure of a protein.* The results of a BLAST search can be multiply aligned (Chapter 6) to reveal conserved residues such as cysteines that are likely to have important biological roles.

There are four components to performing any BLAST search:

1. Selecting a sequence of interest and pasting, typing, or uploading it into the BLAST input box.

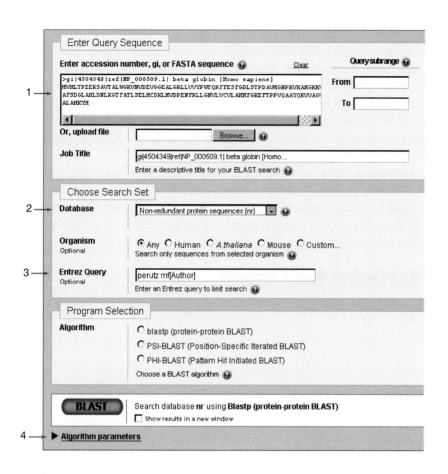

FIGURE 4.1. *Main page for a blastp search at NCBI. The sequence can be input as an accession number, GI identifier, or fasta-formatted sequence as shown here (arrow 1). The database must be selected (arrow 2). The search can be restricted to a particular organism or taxonomic group, and Entrez queries can be used to further focus the search (arrow 3). We discuss the blastp algorithm in this chapter, and PSI-BLAST and PHI-BLAST in Chapter 5. Many of the search parameters can be modified (arrow 4).*

2. Selecting a BLAST program (most commonly blastp, blastn, blastx, tblastx, tblastn).

3. Selecting a database to search. A common choice is the nonredundant (nr) database, but there are many other databases.

4. Selecting optional parameters, both for the search and for the format of the output. These options include choosing a substitution matrix, filtering of low-complexity sequences, and restricting the search to a particular set of organisms.

As we describe the steps of BLAST searching, we can begin with a specific example. Select the link for protein BLAST and enter the accession number of human beta globin (NP_000509), then click the "BLAST" button (Fig. 4.1). The result lists the proteins that are most closely related to beta globin. We will now describe the practical aspects of BLAST searching in detail.

As of January 2009, you can search a database of over 7 million protein sequences (and about 2.6 billion amino acid residues) within several seconds.

BLAST SEARCH STEPS

Step 1: Specifying Sequence of Interest

A BLAST search begins with the selection of a DNA or protein sequence. There are two main forms of data input: (1) cutting and pasting DNA or protein sequence (e.g., in the FASTA format), and (2) using an accession number (e.g., a RefSeq or

The FASTA format is further described at ▶ http://www.ncbi.nlm.nih.gov/BLAST/fasta.html. Do not confuse the FASTA format with the FASTA program, which we described briefly in Chapter 3.

GenBank Identification [GI] number). A sequence in FASTA format begins with a single-line description followed by lines of sequence data. The description line is distinguished from the sequence data by a greater than ("$>$") symbol in the first column. It is recommended that all lines of text be shorter than 80 characters in length. An example of a sequence in FASTA format was shown in Fig. 2.9.

For BLAST searches, your query can be in uppercase or lowercase, with or without intervening spaces or numbers. If the query is DNA, BLAST algorithms will search both strands.

It is often convenient to input the accession number to a BLAST search. Note that the BLAST programs can recognize and ignore numbers that appear in the midst of the letters of your input sequence. The BLAST search also allows you to select a subset of an entire query sequence, such as a region or domain of interest.

Step 2: Selecting BLAST Program

The NCBI BLAST family of programs includes five main programs, as summarized in Fig. 4.2.

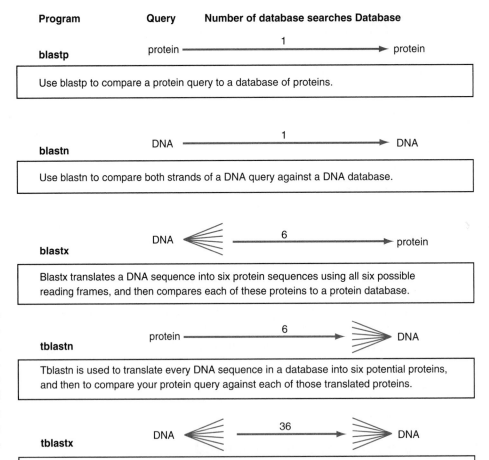

Program	Query	Number of database searches	Database
blastp	protein	1	protein

Use blastp to compare a protein query to a database of proteins.

| blastn | DNA | 1 | DNA |

Use blastn to compare both strands of a DNA query against a DNA database.

| blastx | DNA | 6 | protein |

Blastx translates a DNA sequence into six protein sequences using all six possible reading frames, and then compares each of these proteins to a protein database.

| tblastn | protein | 6 | DNA |

Tblastn is used to translate every DNA sequence in a database into six potential proteins, and then to compare your protein query against each of those translated proteins.

| tblastx | DNA | 36 | DNA |

Tblastx is the most computational intensive BLAST algorithm. It translates DNA from both a query and a database into six potential proteins, then performs 36 protein-protein database searches.

FIGURE 4.2. *Overview of the five main BLAST algorithms. Note that the suffix p refers to protein (as in blastp), n refers to nucleotide, and x refers to a DNA query that is dynamically translated into six protein sequences. The prefix t refers to "translating," in which a DNA database is dynamically translated into six proteins. We will discuss the use of these BLAST algorithms later in this chapter.*

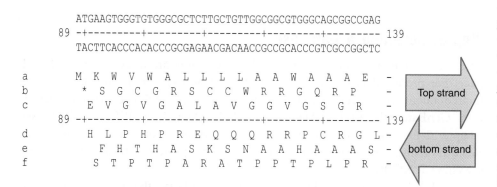

ATGAAGTGGGTGTGGGCGCTCTTGCTGTTGGCGGCGTGGGCAGCGGCCGAG
89 -+---------+---------+---------+---------+--------- 139
TACTTCACCCACACCCGCGAGAACGACAACCGCCGCACCCGTCGCCGGCTC

```
a      M K W V W A L L L L A A W A A A E  -
b      * S G C G R S C C W R R G Q R P    -    Top strand
c      E V G V G A L A V G G V G S G R    -
89 -+---------+---------+---------+---------+--------- 139
d      H L P H P R E Q Q Q R R P C R G L  -
e      F H T H A S K S N A A H A A A S    -    bottom strand
f      S T P T P A R A T P P T P L P R    -
```

FIGURE 4.3. DNA can potentially encode six different proteins. The two strands of DNA sequence of human retinol-binding protein (accession NM_006744) are shown. From the top strand, three potential proteins are encoded (frames a, b, c). The protein encoded in frame a is authentic RBP. The first codon of potential protein b, TGA, encodes a stop codon (asterisk). On the bottom strand, three additional proteins are potentially encoded (frames d, e, f). For example, the first amino acid of protein d is leucine, encoded by CTC.

1. The first program is *blastp*. This program compares an amino acid query sequence against a protein sequence database. Note that for this type of search there are optional parameters (see below) that are specifically relevant to protein searches, such as the choice of various PAM and BLOSUM scoring matrices.

2. The second program is *blastn*. This is used to compare a nucleotide query sequence against a nucleotide sequence database.

Three additional BLAST algorithms rely on the fundamental relationship of DNA to protein. Any DNA sequence can be transcribed and translated into six potential reading frames (three on the top strand and three on the bottom strand; Fig. 4.3). For BLAST searching, the query DNA sequence may be translated into potential proteins, an entire DNA database may be translated, or both. In all three cases, these algorithms perform protein–protein alignments.

3. The program *blastx* compares a nucleotide query sequence translated in all reading frames against a protein sequence database. If you have a DNA sequence and you want to know what protein (if any) it encodes, you can perform a blastx search. This automatically translates the DNA into six potential proteins (see Figs. 4.2 and 4.3). The blastx program then compares each of the six translated protein sequences to all the members of a protein database (Gish and States, 1993).

4. The program *tblastn* compares a protein query sequence against a nucleotide sequence database dynamically translated in all reading frames. One might use this program to ask whether a DNA database encodes a protein that matches your protein query of interest. Does a query with beta globin yield any matches in a database of genomic DNA from the genome-sequencing project of a particular organism?

5. The program *tblastx* compares the six-frame translations of a nucleotide query sequence against the six-frame translations of a nucleotide sequence database. The tblastx program is computationally very intensive. Consider a

UniGene uses blastx to compare each nucleotide sequence in its database to all known proteins from eight organisms (*Homo sapiens*, *Mus musculus*, *Rattus norvegicus*, *Drosophila melanogaster*, *Caenorhabditis elegans*, *Saccharomyces cerevisiae*, *Escherichia coli*, and *Arabidopsis thaliana*). The E value cutoff (discussed below) is 10^{-6}.

We discuss expressed sequence tags (ESTs) in Chapter 8. Tblastx can help you identify frameshifts in ESTs, since all reading frames are compared.

situation in which you have a DNA sequence with no obvious database matches, and you want to know if it encodes a protein with distant, statistically significant database matches in a database of expressed sequence tags. A blastx search would be more sensitive than blastn and thus useful to reveal genes that encode proteins homologous to your query.

Step 3: Selecting a Database

The databases that are available for BLAST searching are listed on each BLAST page. For protein database searches (blastp and blastx), the default option is the nonredundant (nr) database. This consists of the combined protein records from GenBank, the Protein Data Bank (PDB), SwissProt, PIR, and PRF (see Chapter 2 for descriptions of these resources). Another option is to search only Refeq proteins. Table 4.1 summarizes the available protein databases for BLAST searching at NCBI.

For DNA database searches (blastn, tblastn, tblastx) the default option is to search the human (or mouse) genomic plus transcript database. Other commonly used options include the nucleotide nr database or the EST database. Nr includes nucleotide sequences from GenBank, EMBL, DDBJ, and PDB. However, the nr database does not have records from the EST, STS, GSS, or high-throughput genomic sequence (HTGS) databases.

The nr databases are derived by merging several main protein or DNA databases. These databases often contain identical sequences. Generally only one of these sequences is retained by the nr database, along with multiple accession numbers. (Even if two sequences in the nr database appear to be identical, they usually have at least some subtle difference.) The nr databases are often the preferred sites for searching the majority of available sequences.

A summary of all the nucleotide sequence databases that can be searched by standard BLAST searching at NCBI is provided in Table 4.2.

Step 4a: Selecting Optional Search Parameters

We will initially focus our attention on a standard protein–protein BLAST search. In addition to deciding on which sequence to input and which database to search, there are many optional parameters that you can adjust (see Figs. 4.1 and 4.4).

TABLE 4-1 Protein Sequence Databases That Can Be Searched by BLAST Searching at NCBI

Database	Description
nr	Nonredundant GenBank coding sequences + RefSeq proteins + PDB + SwissProt + PIR + PRF
Month	Sequence data released in the previous 30 days
Swissprot	Most recent release from SwissProt
RefSeq	RefSeq protein sequences from NCBI's Reference Sequence Project
Pdb	Protein data bank at Brookhaven (▶ http://www.rcsb.org/pdb/)
pat	Proteins from the Patent division of GenPept.
env_nr	Protein sequences from environmental samples

Source: Modified from ▶ http://www.ncbi.nlm.nih.gov/blast/.

TABLE 4-2	Nucleotide Sequence Databases That Can Be Searched Using BLAST at NCBI
Database	Description
nr	All GenBank + RefSeq Nucleotides + EMBL + DDBJ + PDB sequences (but no EST, STS, GSS, or phase 0, 1, or 2 HTGS sequences); no longer "nonredundant"
chromosome	A database with complete genomes and chromosomes from the NCBI Reference Sequence project.
dbsts	Database of GenBank + EMBL + DDBJ sequences from STS divisions
env_nt	Nucleotide sequences from environmental samples, including those from Sargasso Sea and Mine Drainage projects
est	Database of GenBank + EMBL + DDBJ sequences from EST divisions
est_human	Human subset of EST
est_mouse	Mouse subset of EST
est_others	Nonhuman, nonmouse subset of EST
gss	Genome survey sequence, includes single-pass genomic data, exon-trapped sequences, and *Alu* PCR sequences
htgs	Unfinished high-throughput genomic sequences
month	All new or revised GenBank + EMBL + DDBJ + PDB sequences released in the last 30 days
pat	Nucleotides from the Patent division of GenBank
pdb	Sequences derived from the three-dimensional structure from Brookhaven Protein Data Bank (▶ http://www.rcsb.org/pdb/)
refseq_rna	RNA entries from the NCBI Reference Sequence project
refseq_genomic	Genomic entries from NCBI's Reference Sequence project
wgs	A database for whole genome shotgun sequence entries

Source: Modified from documentation in ▶ http://www.ncbi.nlm.nih.gov/blast/.

1. Query.
In addition to a choice of formats (accession number, gi identifier, or fasta format), you can select a range of amino acid or nucleotide residues to search.

2. Limit by Entrez Query.
Any NCBI BLAST search can be limited using any terms that are used in an Entrez search. Enter the term "perutz mf[Author]" and perform a blastp search using beta globin as a query (Fig. 4.1, arrow 3). Instead of obtaining hundreds of hits, the matches are to entries that refer to Nobel laureate Max Perutz. BLAST searches can also be restricted by organism. Some popular groups are Archaea, Metazoa (multicellular animals), Bacteria, Vertebrata, Eukaryota, Mammalia, Embryophyta (higher plants), Rodentia, Fungi, and Primates. BLAST searches can be restricted to any genus and species or other taxonomic grouping. We will illustrate some effects of applying optional features of blastp by using human insulin (NP_000198) as a query, and restricting the output to the worm *C. elegans* (type "elegans" into the Entrez Query box to choose *C. elegans* from a pull-down menu, or enter txid6239).

3. Short Queries.
If you select this option, the expect value and word size are automatically adjusted. We discuss these concepts below.

4. Expect Threshold.
The expect value E is the number of different alignments with scores equal to or greater than some score S that are expected to occur in a

If you want to restrict your blast search to a particular organism (or group of organisms), use the box labeled "organism" and type at least part of the name to access a dynamic pull-down menu. You can also access a specific taxonomy identifier. To do this, try beginning at the home page of NCBI and selecting Taxonomy Browser from the top bar (or visit ▶ http://www.ncbi.nlm.nih.gov/Taxonomy/taxonomyhome.html/). Select from the list of commonly studied organisms, or perform a query in the taxonomy page. You can thus find the appropriate taxonomy identifier (txid) for any organism. Examples include txid10090 for mouse, txid9606 for human, and txid33090 for Viridiplantae (the plant kingdom).

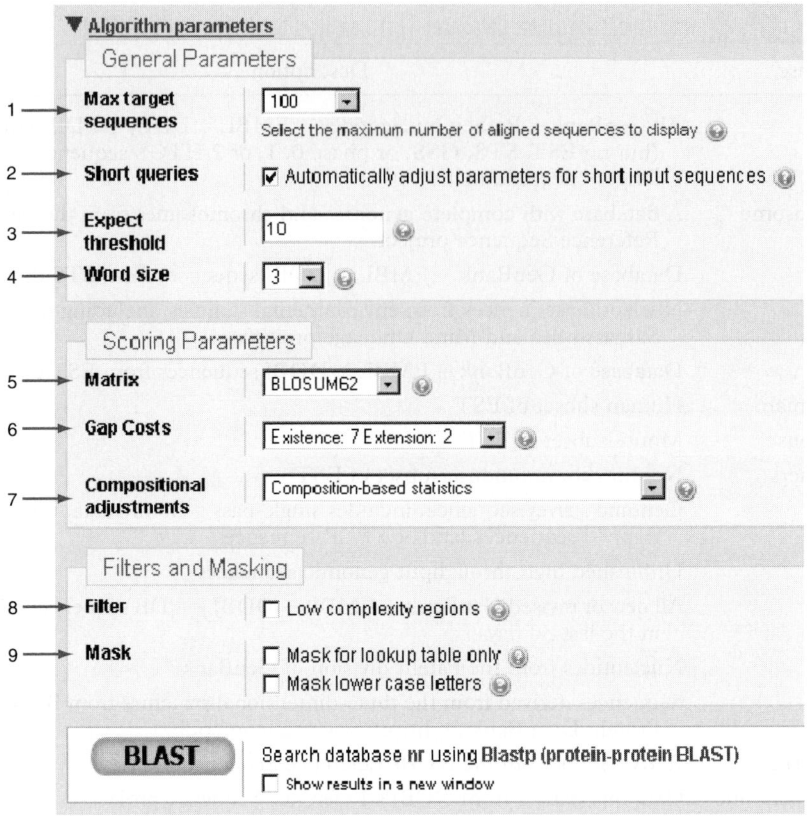

FIGURE 4.4. Optional blastp parameters. Numbered arrows refer to discussion in the text.

database search by chance. Look at the best match in Fig. 4.5a (a match between human insulin and a worm ortholog). The score is 32.7 bits, and the E value is 0.034. This indicates that based on the particular search parameters used (including the size of the database and the choice of the scoring matrix), a score of 32.7 bits or better is expected to occur by chance 3.4 in 100 times (i.e., about 1 in 30 times). A reasonable general guideline is that database matches having E values of ≤ 0.05 are statistically significant.

The default setting for the expect value is 10 for blastn, blastp, blastx, and tblastn. At this E value, 10 hits with scores equal to or better than the alignment score S are expected to occur by chance. (This assumes that you search the database using a random query with similar length to your actual query.) By changing the expect option to a lower number (such as 0.01), fewer database hits are returned; fewer chance matches are reported. Increasing E returns more hits. Consider a very short protein or nucleotide query (e.g., 10 amino acids). There is no opportunity for that query to accumulate a large score, and since the score is inversely related to the expect value (see Equation 4.5 below), the E value cannot be very small. Indeed an E value of 50 or 100 might occur for a database match of considerable biological interest. Hence when you select the optional parameter "short queries" in blastp, the E value is set to 20,000, or $E = 1,000$ in blastn. We will describe the E value in more detail in a discussion of BLAST search statistics later in this chapter, including a comparison of searches with varying E values.

The expect value is sometimes also referred to as the expectation value. We will discuss practical examples of interpreting E values later in this chapter. Note that E values of higher than 0.05 may be biologically relevant, homologous matches, as discussed below.

5. Word Size.   For protein searches, a window size of 3 (default) or 2 may be set. When a query is used to search a database, the BLAST algorithm first divides the

(a) Default

```
>□ref|NP_501926.1| U G INSulin related family member (ins-1) [Caenorhabditis elegans]
Length=109

  Score = 32.7 bits (73), Expect = 0.034, Method: Composition-based stats.
  Identities = 30/101 (29%), Positives = 41/101 (40%), Gaps = 14/101 (13%)

Query  10  LLALLALWGPDPAAAFVNQHLCGSHLVEALYLVCGERGFFYTPKTRREAEDLQVGQVELG  69
           LA+L L  P P+ A +   LCGS L   L  VC +        +R A+
Sbjct  16  FLAILLLSSPTPSDASI--RLCGSRLTTTLLAVCRNQLCTGLTAFKRSADQSY-------  66

Query  70  GGPGAGSLQPLALEGSLQKRG-IVEQCCTSICSLYQLENYC  109
              A + + L        QKRG I +CC   CS   L+ +C
Sbjct  67  ----APTTRDLFHIHHQQKRGGIATECCEKRCSFAYLKTFC  103
```
                                                                    1

(b) No compositional adjustment

```
>□ref|NP_501926.1| U G INSulin related family member (ins-1) [Caenorhabditis elegans]
Length=109

  Score = 34.7 bits (78), Expect = 0.009
  Identities = 30/100 (30%), Positives = 41/100 (41%), Gaps = 14/100 (14%)

Query  11  LALLALWGPDPAAAFVNQHLCGSHLVEALYLVCGERGFFYTPKTRREAEDLQVGQVELGG  70
           LA+L L  P P+ A +   LCGS L   L  VC +        +R A+
Sbjct  17  LAILLLSSPTPSDASIR--LCGSRLTTTLLAVCRNQLCTGLTAFKRSADQSY--------  66

Query  71  GPGAGSLQPLALEGSLQKRG-IVEQCCTSICSLYQLENYC  109
             A + + L        QKRG I +CC   CS   L+ +C
Sbjct  67  ---APTTRDLFHIHHQQKRGGIATECCEKRCSFAYLKTFC  103
```

(c) Conditional compositional score matrix adjustment

```
>□ref|NP_501926.1| U G INSulin related family member (ins-1) [Caenorhabditis elegans]
Length=109

  Score = 33.5 bits (75), Expect = 0.020, Method: Composition matrix adjust.
  Identities = 27/100 (27%), Positives = 39/100 (39%), Gaps = 12/100 (12%)

Query  10  LLALLALWGPDPAAAFVNQHLCGSHLVEALYLVCGERGFFYTPKTRREAEDLQVGQVELG  69
           LA+L L  P P+ A +   LCGS L   L  VC +        +R A+
Sbjct  16  FLAILLLSSPTPSDASIR--LCGSRLTTTLLAVCRNQLCTGLTAFKRSADQ--------S  65

Query  70  GGPGAGSLQPLALEGSLQKRGIVEQCCTSICSLYQLENYC  109
           P   L    +  ++ GI +CC   CS   L+ +C
Sbjct  66  YAPTTRDL--FHIHHQQKRGGIATECCEKRCSFAYLKTFC  103
```
                                                                    2

(d) Filter low-complexity region

```
>□ref|NP_501926.1| U G INSulin related family member (ins-1) [Caenorhabditis elegans]
Length=109

  Score = 25.4 bits (54), Expect = 6.3, Method: Composition-based stats.
  Identities = 11/24 (45%), Positives = 14/24 (58%), Gaps = 1/24 (4%)

Query  87  QKRG-IVEQCCTSICSLYQLENYC  109
           QKRG I +CC   CS   L+ +C
Sbjct  80  QKRGGIATECCEKRCSFAYLKTFC  103
```

(e) Mask for lookup table only

```
>□ref|NP_501926.1| U G INSulin related family member (ins-1) [Caenorhabditis elegans]
Length=109

  Score = 32.7 bits (73), Expect = 0.034, Method: Composition-based stats.
  Identities = 30/101 (29%), Positives = 41/101 (40%), Gaps = 14/101 (13%)

Query  10  llallalwgpdpaaaFVNQHLCGSHLVEALYLVCGERGFFYTPKTRREAEDLQVGQVELG  69
           LA+L L  P P+ A +   LCGS L   L  VC +        +R A+
Sbjct  16  FLAILLLSSPTPSDASI--RLCGSRLTTTLLAVCRNQLCTGLTAFKRSADQSY-------  66

Query  70  GGPGAGSLQPLALEGSLQKRG-IVEQCCTSICSLYQLENYC  109
              A + + L        QKRG I +CC   CS   L+ +C
Sbjct  67  ----APTTRDLFHIHHQQKRGGIATECCEKRCSFAYLKTFC  103
```

FIGURE 4.5. Pairwise alignments from blastp searches illustrating the effects of changing compositional matrices and filtering options. Human insulin (NP_000198) was used as a query in a blastp search restricted to RefSeq proteins in the nematode Caenorhabditis elegans. (a) Default settings show a match to a worm protein with a score of 32.7 bits and an E value of 0.034. Results are shown using (b) no compositional adjustments and (c) conditional compositional score matrix adjustment. (d) The effect of filtering low complexity regions is shown, as well as (e) masking for lookup table only, in which the query is masked while producing seeds to scan the database, but the extensions of database hits are not masked. Applying these various adjustments can have dramatic effects on the expect value, score, percent identity, gap length and gap placement (see arrows 1 and 2).

query into a series of smaller sequences (words) of a particular length (word size), as will be described below. For blastp, a larger word size yields a more accurate search. For any word size, matches made to each word are then extended to produce the BLAST output. In practice, the word size can remain at 3 and should be reduced to 2 only when your query is a very short peptide (i.e., a short string of amino acids). Changing the size from 3 to 2 has no effect on the alignment (or the scores) of human insulin with its nematode homolog.

For nucleotide searches, the default word size is 11 and can be raised (word size 15) or reduced (word size 7). Lowering the word size yields a more accurate but slower search. Raising the word size is applied in MegaBLAST and discontiguous MegaBLAST (see Chapter 5), two alternate programs at NCBI that perform nucleotide searches. For MegaBLAST the default word size is 28, and can be set as high as 64. This is useful for speed when searching with long queries (e.g., many thousands of nucleotides) for nearly exact matches in a database. Very long word sizes match relatively infrequently, encouraging a much faster search.

6. Matrix.

Five amino acid substitution matrices are available for blastp protein–protein searches: PAM30, PAM70, BLOSUM45, BLOSUM62 (default), and BLOSUM80. Some alternative BLAST servers (discussed in Chapter 5, on advanced BLAST searching) offer many more choices for substitution matrices such as PAM250. It is usually advisable to try a BLAST search using several different scoring matrices. For example, as described in Chapter 3, PAM40 and PAM250 matrices (Fig. 3.16) have entirely distinct properties as scoring matrices for sequences sharing varying degrees of similarity. For very short queries (e.g., 15 or fewer amino acid residues), a PAM30 matrix is recommended.

For blastn, the default scoring system is +2 for a match and -3 for a mismatch. A variety of other scoring schemes are available, including the default +1, −1 for Megablast (Chapter 5). For each scoring system, the BLAST family offers appropriate gap opening and extension penalties.

7. Gap Penalties.

A gap is a space introduced into an alignment to compensate for insertions and deletions in one sequence relative to another (Chapter 3). Since a single mutational event may cause the insertion or deletion of more than one residue, the presence of a gap is frequently ascribed more significance than the length of the gap. Hence, the gap introduction is penalized heavily, whereas a lesser penalty is ascribed to each subsequent residue in the gap. To prevent the accumulation of too many gaps in an alignment, introduction of a gap causes the deduction of a fixed amount (the gap score) from the alignment score. Extension of the gap to encompass additional nucleotides or amino acid is also penalized in the scoring of an alignment.

Gap scores are typically calculated as the sum of G, the gap-opening penalty, and L, the gap extension penalty. For a gap of length n, the gap cost would be $G + Ln$. The choice of gap costs is typically 10 to 15 for G and 1 to 2 for L. These are called affine gap penalties, in which the penalty for introducing a gap is far greater than the penalty for extending one.

MegaBLAST (Chapter 5) uses non-affine gap penalties, that is, there is no cost for opening a gap. We will further discuss the problem of gaps in multiple sequence alignments in Chapter 6.

8. Composition-Based Statistics.

This option, which is selected as default, generally improves the calculation of the E value statistic (see below). Some proteins (whether queries or database matches) have nonstandard compositions such as having hydrophobic or cysteine-rich regions. For some organisms, the entire genome has a very high guanine plus cytosine (GC) or adenine plus thymine (AT) content. For example, the entire genome of the malaria parasite *Plasmodium falciparum* is 80.6% AT, biasing its proteins towards having amino acids encoded by AT-rich codons. A standard matrix such as BLOSUM62 is not appropriate for the comparison of two proteins with nonstandard composition, and the target frequencies q_{ij} (see Equation 3.4) need to be adjusted in the context of new background frequencies $p_i p_j$ (Yu et al., 2003; Yu and Altschul, 2005). In performing a blastp search, a default option is to use composition-based statistics. This implements a slightly different scoring system for each database sequence in which all scores are scaled by an analytically determined constant (Schäffer et al., 2001). It is applicable to any BLAST protein search, including the position-specific scoring matrix of PSI-BLAST (Chapter 5).

For examples of proteins that are highly hydrophobic, very cysteine-rich, or from *P. falciparum*, see web documents 4.1, 4.2, and 4.3 at ▶ http://www.bioinfbook.org/chapter4. We discuss *P. falciparum* in Chapter 18; it is responsible for over 2 million deaths a year.

Compositional adjustments generally increase the accuracy of BLAST searches considerably (Schäffer et al., 2001; Altschul et al., 2005). The improvement can be quantitated using receiver operating characteristic (ROC) curves that plot the number of true positives (based on an independent criterion such as expert manual curation) versus false positives (Gribskov and Robinson, 1996). In addition to using a composition-based statistics, a "compositional score matrix adjustment" can be applied to blastp searches. This can reduce false positive search results in specialized circumstances such as subjects matching queries of very different lengths (Altschul et al., 2005). In that case the longer sequence may have a substantially different composition than the shorter.

In the particular example of our insulin search, removing compositional adjustments actually increases the bit score slightly and reduces the E value more than threefold from 0.034 to 0.009 (Fig. 4.5b). Invoking a conditional compositional score matrix adjustment alters the E value to 0.02 (Fig. 4.5c). The magnitude of these effects depends on the composition of the particular query you choose, and for some searches it is helpful to try a series of compositional adjustments.

9. Filtering and Masking.

Filtering masks portions of the query sequence that have low complexity (or highly biased compositions) (Wootton and Federhen, 1996). Low-complexity sequences are defined as having commonly found stretches of amino acids (or nucleotides) with limited information content. Examples are dinucleotide repeats (e.g., the repeating nucleotides CACACACA...), *Alu* sequences, or regions of a protein that are extremely rich in one or two amino acids. Stretches of hydrophobic amino acid residues that form a transmembrane domain are very common, and a database search with such sequences results in many database matches that are statistically significant but biologically irrelevant. Other motifs that are masked by filtering include acidic-, basic-, and proline-rich regions.

The blastp and blastn programs offer several main options. Note that filtering is applied to the query sequence, and not to the entire database. (1) Filter low-complexity regions. For protein sequence queries, the SEG program is used; for nucleic acid sequences, the DUST program is employed. (2) Filter repeats (for blastn only). This is useful to avoid matching a query with *Alu* repeats or other repetitive DNA to spurious database entries. (3) Mask for lookup table only. This option masks the matching of words above threshold to database hits (discussed below). This avoids matches to low-complexity sequences or repeats. Then, BLAST extensions occur without masking (so hits can be extended even if they contain low complexity sequence). (4) Mask lower case. This allows you to enter a query in the fasta format using upper case characters for the search but filtering those residues you choose to filter by entering them in lower case.

Adjusting the filtering option can have dramatic effects on BLAST search results. When human insulin is searched against worm proteins, filtering low-complexity regions results in an alignment of only 24 residues (in contrast to 101 aligned residues in the default search) (Fig. 4.5d). Using the "mask for lookup table only" option results in displaying a portion of the hydrophobic leader sequence on human insulin in lower-case characters (Fig. 4.5e). Those hydrophobic residues can potentially match thousands of database entries and in most cases would not reflect homology. Thus, as in this case, using filters can be helpful to avoid producing spurious database matches. But in some cases, as shown in Fig. 4.5d, authentic matches may be missed.

The NCBI BLAST site offers a variety of other programs and options for searching, such as the use of position-specific scoring matrices (PSSMs), pattern

We explore repetitive DNA sequences in Chapter 16. Web document 4.4 (at ▶ http://www.bioinfbook.org/chapter4) offers over a dozen spectacular examples of repetitive DNA and protein sequences.

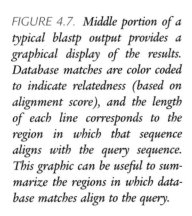

FIGURE 4.6. Top portion of a BLAST output describes the search that was performed (BLASTP 2.2.16 in this case; arrow 1), the database (arrow 2; the RefSeq database has about 3 million sequences and 1 billion amino acid residues), a link to a taxonomy report that organizes the search results by species (arrow 3), and the query (human beta globin) (arrow 4). There is an option to display conserved domains (arrow 5).

hit-initiated (PHI) blast, and searches of the Conserved Domain Database (CDD). These are described in Chapter 5.

Step 4b: Selecting Formatting Parameters

There are many options for formatting the output of a BLAST search. These are illustrated by performing a protein–protein blastp search with human beta globin (NP_000509) as a query and restricting the search to RefSeq proteins from the mouse (*Mus musculus*). The results of the search occur in several main parts. In the top (Fig. 4.6), details of the search are provided, including the type of BLAST search (Fig. 4.6, arrow 1), a description of the query and the database that were searched (arrows 2 and 4), and a taxonomy link to the results organized by species (arrow 3). There is also an option to display conserved domains (arrow 5).

The middle portion of a typical BLAST output provides a list of database sequences that match the query sequence (Fig. 4.7). A graphical overview provides a color-coded summary, with the length of the query sequence represented across the x axis. Each bar drawn below the map represents a database protein (or nucleic

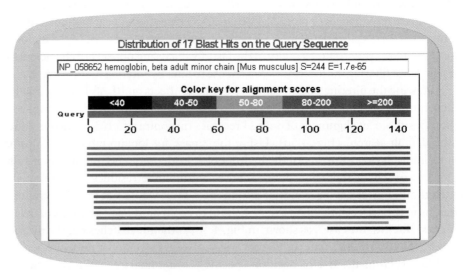

FIGURE 4.7. Middle portion of a typical blastp output provides a graphical display of the results. Database matches are color coded to indicate relatedness (based on alignment score), and the length of each line corresponds to the region in which that sequence aligns with the query sequence. This graphic can be useful to summarize the regions in which database matches align to the query.

Distance tree of results ^{NEW}

Sequences producing significant alignments:	Score (Bits)	E Value	
ref|NP_058652.1| hemoglobin, beta adult minor chain [Mus musculu	244	2e-65	U G
ref|NP_032246.2| hemoglobin, beta adult major chain [Mus musculu	228	2e-60	U G
ref|XP_978992.1| PREDICTED: similar to Hemoglobin epsilon-Y2 ...	226	3e-60	G
ref|NP_032247.1| hemoglobin Y, beta-like embryonic chain [Mus mu	223	4e-59	U G
ref|NP_032245.1| hemoglobin Z, beta-like embryonic chain [Mus mu	223	6e-59	U G
ref|XP_998314.1| PREDICTED: similar to Hemoglobin beta-H1 sub...	203	4e-53	G
ref|XP_978924.1| PREDICTED: similar to Hemoglobin epsilon-Y2 ...	187	2e-48	G
ref|XP_912634.1| PREDICTED: similar to Hemoglobin beta-2 subu...	161	2e-40	G
ref|XP_488069.1| PREDICTED: similar to Hemoglobin beta-2 subu...	154	3e-38	U G
ref|NP_032244.1| hemoglobin alpha 1 chain [Mus musculus]	105	1e-23	U G
ref|XP_994669.1| PREDICTED: similar to Hemoglobin alpha subun...	101	3e-22	G
ref|XP_356935.3| PREDICTED: similar to Hemoglobin alpha subun...	100	4e-22	U G
ref|NP_034535.1| hemoglobin X, alpha-like embryonic chain in ...	94.0	4e-20	U G
ref|NP_001029153.1| similar to hemoglobin, theta 1 [Mus musculus	88.2	2e-18	U G
ref|NP_778165.1| hemoglobin, theta 1 [Mus musculus]	73.9	5e-14	U G
ref|XP_978150.1| PREDICTED: similar to hemoglobin, beta adult...	41.6	2e-04	G
ref|NP_795942.2| 5'-nucleotidase, cytosolic II-like 1 protein [M	28.9	1.5	U G

FIGURE 4.8. A typical blastp output includes a list of database sequences that match the query. Links are provided to that database entry (e.g., an Entrez Protein entry) and to the pairwise alignment to the query. The bit score and E value for each alignment are also provided. Note that the best matches at the top of the list have large bit scores and small E values. To the right, links are given to UniGene (U) and Entrez Gene (G).

acid) sequence that matches the query sequence. The position of each bar relative to the linear map of the query allows the user to see instantly the extent to which the database matches align with a single or multiple regions of the query. The most similar hits are shown at the top in red. Hatched areas (when present) correspond to the nonsimilar sequence between two or more distinct regions of similarity found within the same database entry.

The alignments are next described by a list of one-line summaries called "descriptions" (Fig. 4.8). The description lines are sorted by increasing E value; thus, the most signficant alignments (lowest E values) are at the top. The description consists of the following columns:

1. Identifier for the database sequence. Some of the common sequence identifiers are:

   RefSeq (e.g., ref|NP_006735.1|),

   Protein database (e.g., pdb|1RBP|),

   Protein Information Resource (pir|A27786),

   Swiss-Prot database (sp|P27485|),

   GenBank (e.g., gb|AAF69622.1|),

   European Molecular Biology Lab (EMBL) (e.g., emb|CAB64947.1|), and

   Database of Japan (DBJ)(dbj|BAA13453.1|).

2. Brief description of the sequence.

3. The bit score of the highest scoring match found for each database sequence (bit scores will be defined below).

4. The expect value E. The identifer is linked to the full GenBank entry. Clicking on the score in a given description line will take the user to the corresponding sequence alignment. The alignment can also be reached by scrolling down the output page.

5. Links to Entrez Gene (G) and UniGene (U).

FIGURE 4.9. The lower part of a blastp search (or other BLAST family search) consists of a series of pairwise sequence alignments such as those shown in Fig. 4.5. Using the reformat option, the results can be displayed as a multiple sequence alignment as shown here for a group of murine globins. Other output format options are available, allowing the user to inspect regions of similarity as well as divergent regions within protein families.

```
☐ Query        61  VKAHGKKVLGAFSDGLAHLDNLKGTFATLSELHCDKLHVDPENFRLLGNVLVCVLAHHFG  120
☐ NP_058652    61  VKAHGKKVITAFNEGLKNLDNLKGTFASLSELHCDKLHVDPENFRLLGNAIVIVLGHHLG  120
☐ NP_032246    61  VKAHGKKVITAFNDGLNHLDSLKGTFASLSELHCDKLHVDPENFRLLGNMIVIVLGHHLG  120
☐ XP_978992    61  VKAHGKKVLTAFGESIKNLDNLKSALAKLSELHCDKLHVDPENFKLLGNVLVIVLASHFG  120
☐ NP_032247    61  VKAHGKKVLTAFGESIKNLDNLKSALAKLSELHCDKLHVDPENFKLLGNVLVIVLASHFG  120
☐ NP_032245    61  IRAHGKKVLTSLGLGVKNMDNLKETFAHLSELHCDKLHVDPENFKLLGNMLVIVLSTHFA  120
☐ XP_998314    61  IRAHGKKVLTSLGLGVKNMDNLKETFAHLSELHCDKLHVDPENFKLLGNMLVIVLSTHFA  120
☐ XP_978924    41  VKAHGKKVLTAFGESIKNLDNLKSALAKLSELHCDKLHVDPENFKLLGNVLVIVLASHFG  100
☐ XP_912634    60  MKALGKKMIESFSEGLQPLDNLNYTFSSLELHHDKLHMDPENFKLLGSMIVIVLSPHFG   119
☐ XP_488069    61  LKALGKKMIESFSEDLQSLDNLHYTFASLSELHHDKLHMDPENFKLLGSMIVIVMSPHFG   120
☐ NP_032244    56  VKGHGKKVADALASAAGHLDDLPGALSALSDLHAHKLRVDPVNFKLLSHCLLVTLASHHP   115
☐ XP_994669    55  VQAHGKKVMDVLTDAVNHVNDLPDALSTLSDLHAHKLCVDPANFKLLSHCLLVTLAIHHS   114
☐ XP_356935    55  VQAHGKKVMDVLTDAVNHVNDLPDALSTLSDLHAHKLCVDPANFKLLSHCLLVTLAIHHS   114
☐ NP_034535    56  LRAHGFKIMTAVGDAVKSIDNLSSALTKLSELHAYILRVDPVNFKLLSHCLLVTMAARFP   115
☐ NP_001029153 56  VKAHQKVADALTLATQHLDDLPASLSALSDLHAHKLCVDPANFQFFSHCLLVTLARHYP   115
☐ NP_778165    56  VKAHQKVADALTLATQHLDDLPASLSALSDLHAHKLCVDPANFQFFSCCLLVTLARHYP   115
☐ XP_978150     1                                                        MIVIVLGHHLG  11
```

The lower portion of a BLAST search output consists of a series of pairwise sequence alignments, such as the ones in Fig. 4.5. Here, one can inspect the pairwise match between the query (input sequence) and the subject (i.e., the particular database match that is aligned to the query). Four scoring measures are provided: the bit score, the expect score, the percent identity, and the positives (percent similarity).

Without reperforming an entire BLAST search, the output can be reformatted to provide a range of different output options. The number of descriptions and of alignments can be modified. There are several options for visualizing the aligned sequences as a multiple sequence alignment (Fig. 4.9). This is an especially useful way to identify specific amino acid residues that are conserved (or divergent) within a protein or DNA family. For nucleotide searches (e.g., blastn), by selecting the CDS (coding sequence) feature, the pairwise alignments also show the positions of the corresponding protein, when that information is available. For example, a search of human beta globin DNA (NM_000518) against human RefSeq nucleotide sequences includes a match to delta globin (NM_000519). That alignment includes information about the corresponding proteins (Fig. 4.10).

```
> ☐ ref|NM_000519.3|  U E G M  Homo sapiens hemoglobin, delta (HBD), mRNA
Length=774


Score =  706 bits (356),  Expect = 0.0
Identities = 449/480 (93%),  Gaps = 0/480 (0%)
Strand=Plus/Plus

CDS:beta globin [Hom  1                                              M  V  H  L  T  P
Query                 8  CTTCTGACACAACTGTGTTCACTAGCAACCTCAAACAGACACCATGGTGCATCTGACTCC  67
                         |||||||||| ||| |||||||||||||||||||||||||||||||||||||||||||||
Sbjct               153  CTTCTGACATAACAGTGTTCACTAGCAACCTCAAACAGACACCATGGTGCATCTGACTCC  212
CDS:delta globin [Ho  1                                              M  V  H  L  T  P

CDS:beta globin [Hom  7     E  E  K  S  A  V  T  A  L  W  G  K  V  N  V  D  E  V  G  G
Query                68  TGAGGAGAAGTCTGCCGTTACTGCCCTGTGGGGCAAGGTGAACGTGGATGAAGTTGGTGG  127
                         ||||||||||| || |||| ||| || ||||||||||||||||||||||| ||||||||
Sbjct               213  TGAGGAGAAGACTGCTGTCAATGCCCTGTGGGGCAAAGTGAACGTGGATGCAGTTGGTGG  272
CDS:delta globin [Ho  7     E  E  K  T  A  V  N  A  L  W  G  K  V  N  V  D  A  V  G  G

CDS:beta globin [Hom 27     E  A  L  G  R  L  L  V  V  Y  P  W  T  Q  R  F  F  E  S  F
Query               128  TGAGGCCCTGGGCAGGCTGCTGGTGGTCTACCCTTGGACCCAGAGGTTCTTTGAGTCCTT  187
                         ||||||||||||| | |||||||||||||||||||||||||||||||||||||||||||
Sbjct               273  TGAGGCCCTGGGCAGGATTACTGGTGGTCTACCCTTGGACCCAGAGGTTCTTTGAGTCCTT  332
CDS:delta globin [Ho 27     E  A  L  G  R  L  L  V  V  Y  P  W  T  Q  R  F  F  E  S  F

CDS:beta globin [Hom 47     G  D  L  S  T  P  D  A  V  M  G  N  P  K  V  K  A  H  G  K
Query               188  TGGGGATCTGTCCACTCCTGATGCTGTTATGGGCAACCCTAAGGTGAAGGCTCATGGCAA  247
                         |||||||||| ||||||||||||||||||||||||||||||||||||||||||||||||
Sbjct               333  TGGGGATCTGTCCTCTCCTGATGCTGTTATGGGCAACCCTAAGGTGAAGGCTCATGGCAA  392
CDS:delta globin [Ho 47     G  D  L  S  S  P  D  A  V  M  G  N  P  K  V  K  A  H  G  K
```

FIGURE 4.10. For blastn searches, the coding sequence (CDS) option in the reformat page allows the amino acid sequence of the coding regions of the query and the subject (i.e., the database match) to be displayed. Here, human beta globin DNA (NM_000518) was used as a query, and a match to the closely related delta globin is shown.

BLAST ALGORITHM USES LOCAL ALIGNMENT SEARCH STRATEGY

The BLAST search identifies the matches in a database to an input query sequence. Global similarity algorithms optimize the overall alignment of two sequences. These algorithms are best suited for finding matches consisting of long stretches of low similarity. In contrast, local similarity algorithms such as BLAST identify relatively short alignments. Local alignment is a useful approach to database searching because many query sequences have domains, active sites, or other motifs that have local but not global regions of similarity to other proteins. Also, databases typically have fragments of DNA and protein sequences that can be locally aligned to a query.

BLAST Algorithm Parts: List, Scan, Extend

The BLAST search algorithm finds a match between a query and a database sequence and then extends the match in either direction (Altschul et al., 1990, 1997). The search results consist of both highly related sequences from the database as well as marginally related sequences, along with a scoring scheme to describe the degree of relatedness between the query and each database hit. The blastp algorithm can be described in three phases (Fig. 4.11):

1. BLAST compiles a preliminary list of pairwise alignments, called word pairs.
2. The algorithm scans a database for word pairs that meet some threshold score T.
3. BLAST extends the word pairs to find those that surpass a cutoff score S, at which point those hits will be reported to the user. Scores are calculated from scoring matrices (such as BLOSUM62) along with gap penalties.

In the first phase, the blastp algorithm compiles a list of "words" of a fixed length w that are derived from the query sequence. A threshold value T is established for the score of aligned words. Those words either at or above the threshold are collected and used to identify database matches; those words below threshold are not further pursued. For protein searches the word size typically has a default value of 3. Since there are 20 amino acids, there are $20^3 = 8000$ possible words. The word size parameter can be modified by the BLAST user, as described above (see option 3). The threshold score T can be lowered to identify more initial pairwise alignments. This will increase the time required to perform the search and may increase the sensitivity. You can modify the threshold and dozens of other parameters using the command-line program netblast (Box 4.1).

For blastn, the first phase is slightly different. Threshold scores are not used in association with words. Instead, the algorithm demands exact word matches. The default word size is 11 (and can be adjusted by the user to values of 7 or 11). Lowering the word length effectively accomplishes the same thing as lowering the threshold score. Specifying a smaller word size induces a slower, more accurate search.

In the second phase, after compiling a list of word pairs at or above threshold T, the BLAST algorithm scans a database for hits. This requires BLAST to search an index of the database to find entries that correspond to words on the compiled list. In the original implementation of BLAST, one hit was sufficient. In the current versions of BLAST, the algorithm seeks two separate word pairs (i.e., two

In the BLAST papers by Steven Altschul, David Lipman, and colleagues, the threshold parameter is denoted T (Altschul et al., (1990), 1997). In the BLAST program (e.g., in netblast), the threshold parameter is called f.

FIGURE 4.11. Schematic of the original BLAST algorithm. In the first phase a query sequence (such as human beta globin) is analyzed with a given word size (e.g., w = 3), and a list of words is compiled having a threshold score (e.g., T = 11). Several possible words derived from the query sequence are listed in the figure (from VTA to NVD); in an actual BLAST search there are 8000 words compiled for w = 3. For a given word, such as the portion of the query sequence consisting of LWG, a list of words is compiled with scores greater than or equal to some threshold T (e.g., 11). In this example, 15 words are shown along with their scores from a BLOSUM62 matrix; ten of these are above the threshold, and five are below. In phase 2, a database is scanned to find entries that match the compiled word list. In phase 3, the database hits are extended in both directions to obtain a high-scoring segment pair (HSP). If the HSP score exceeds a particular cutoff score S, it is reported in the BLAST output. Note that in this particular example the word pair that triggers the extension step is not an exact match (see boxed residues LWG aligned to AWG). The main idea of the threshold T is to also allow both exact and related but non-exact word hits to trigger an extension.

Phase 1: compile a list of words (w = 3) above threshold T

• Query sequence: human beta globin NP_000509 (includes ... `VTALWGKVNVD`...)

• words derived from query sequence (HBB):

`VTA TAL ALW LWG WGK GKV KVN VNV NVD`

• generate a list of words matching query (both above and below T). Consider `LWG` in the query and the scores (derived from a BLOSUM62 matrix) for various words:

LWG	4+11+6=21
IWG	2+11+6=19
MWG	2+11+6=19
VWG	1+11+6=18
examples of FWG	0+11+6=17
words above AWG	0+11+6=17
threshold 11 LWS	4+11+0=15
LWN	4+11+0=15
LWA	4+11+0=15
LYG	4+ 2+6=12
threshold ——— LFG	4+ 1+6=11
examples of FWS	0+11+0=11
words below AWS	-1+11+0=10
threshold 11 CWS	-1+11+0=10
IWC	2+11-3=10

Phase 2:

select all the words above threshold T

scan the database for entries that match the compiled list

Phase 3: extend the hits in either direction. Stop when the score drops.

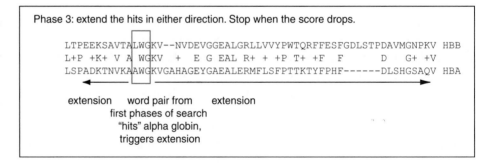

extension   word pair from   extension
first phases of search
"hits" alpha globin,
triggers extension

For the parameter *A*, the default value is 0 (for blastn and mega-blast) and 40 (for other programs such as blastp).

nonoverlapping hits) within a certain distance *A* from each other. It then generates an ungapped extension of these hits (Altschul et al., 1997). The two-hit approach greatly speeds up the time required to do a BLAST search. Compared to the one-hit approach, the two-hit method generates on average about three times as many hits, but the algorithm then needs to perform only one-seventh as many extensions (Altschul et al., 1997).

In the third phase, BLAST extends hits to find alignments called high-scoring segment pairs (HSPs). For sufficiently high-scoring alignments, a gapped extension is triggered. The extension process is terminated when a score falls below a cutoff.

In summary, the main strategy of the BLAST algorithm is to compare a protein or DNA query sequence to each database entry and to form pairwise alignments

Box 4.1
Netblast

Most BLAST users rely on the convenient web-based server at the NCBI website. An alternative is to use Netblast, a stand-alone client that you download directly from the NCBI BLAST website. It is available for Windows, Macintosh, Linux, or other platforms. The installation takes only a matter of seconds, and the download includes extensive documentation and examples. Netblast operates from the command line only. First, save a text file containing your query of interest in the fasta format. We provide several examples in the problems at the end of this chapter, along with web documents that provide sample text files. The second step is to invoke a command line editor and execute a blast search of interest. There are four required parameters: the input, the database to be searched, the BLAST program, and the output file. A typical Netblast search has a syntax as follows:

>blastcl3 -i hbb.txt -p blastp -d nr -o hbb1.txt

Here blastcl3 (for blast client 3) is the program; the four required parameters are -i (for the input or query file), -p (for the program), -d (the database), and -o (the output file; here it is called hbb1.txt). The results are returned as a text file in the folder in which you installed Netblast. You can modify 40 different optional parameters of a Netblast search. Examples include:

-p tblastn (changes the program to tblastn)

-f 16 (changes the threshold of a blastp search)

-d "nr refseq_protein" (searches both the nr and the refseq_protein databases)

-e 2e-5 (changes the expect value cutoff to 2×10^{-5})

Other options control gap penalties, the output format, the filters used for masking sequences, the scoring matrix, and dozens of additional parameters.

A potential disadvantage of Netblast is that it uses only a command line, and does not have a graphical user interface (GUI). However, this limitation is offset by the great flexibility it offers in specifying the parameters of a BLAST search.

Another strength of Netblast is that it allows batch queries. Simply create a text file with multiple queries in the fasta format; the output file includes a series of BLAST results. We provide an example in Problem 4.4.

Netblast is available at ► http://www.ncbi.nlm.nih.gov/BLAST/download.shtml with documentation at ► ftp://ftp.ncbi.nlm.nih.gov/blast/documents/netblast.html.

(HSPs). As a heuristic algorithm, BLAST is designed to offer both speed and sensitivity. When the threshold parameter is raised, the speed of the search is increased, but fewer hits are registered, and so distantly related database matches may be missed. When the threshold parameter is lowered, the search proceeds far more slowly, but many more word hits are evaluated, and thus sensitivity is increased.

We can demonstrate the effect of different threshold levels on a blastp search by changing the f parameter from its default value (11) to a range of other values. The results are dramatic (Fig. 4.12). With the default threshold value of 11, there are about 47 million hits to the database and 1.8 million extensions. When the threshold is lowered to just 3, there are about 1.9 billion hits to the database and 582 million extensions. This occurs because many additional words have scores above T. With the threshold raised to 15 or higher, there are only about 6 million hits and 50,000

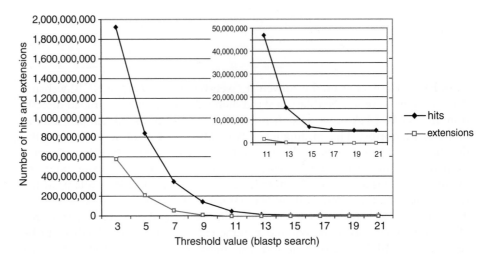

FIGURE 4.12. The effect of varying the threshold (x axis) on the number of database hits (black line) and extensions (red line). A series of blastp searches were performed using human beta globin as a query.

extensions. The final results of the search are not dramatically different with the default value compared to the lowered or raised threshold values, as the number of gapped HSPs is comparable. With the high threshold some matches were missed, although the reported matches are more likely to be true positives; with the lower threshold values there were somewhat more successful extensions. This supports the conclusion that a lower threshold parameter yields a more accurate search, though a slower one. This trade-off between sensitivity and speed is central to the BLAST algorithm. Practically, for most users of BLAST the default threshold parameters are always appropriate.

BLAST Algorithm: Local Alignment Search Statistics and *E* Value

We care about the statistical significance of a BLAST search because we want some quantitative measure of whether the alignments represent significant matches or whether they would be expected to occur by chance alone. For local alignments (including BLAST searches), rigorous statistical tests have been developed (Altschul and Gish, 1996; Altschul et al., 1990, 1994, 1997; Pagni and Jongeneel, 2001).

We have described how local, ungapped alignments between two protein sequences are analyzed as HSPs. Using a substitution matrix, specific probabilities are assigned for each aligned pair of residues, and a score is obtained for the overall alignment. For the comparison of a query sequence to a database of random sequences of uniform length, the scores can be plotted and shown to have the shape of an extreme value distribution (see Fig. 4.13, where it is compared to the normal distribution). The normal, or Gaussian, distribution forms the familiar, symmetric bell-shaped curve. The extreme value distribution is skewed to the right, with a tail that decays in x (rather than x^2, which describes the decay of the normal distribution). The properties of this distribution are central to our understanding of BLAST statistics because they allow us to evaluate the likelihood that the highest scores from a search (i.e., the values at the right-hand tail of the distribution) occurred by chance.

We will next examine the extreme value distribution so that we can derive a formula (Equation 4.5 below) that describes the likelihood that a particular BLAST score occurs by chance.

Because of the rapid tailing of the normal distribution in x^2, if we tried to use the normal distribution to describe the significance of a BLAST search result (for example by estimating how many standard deviations above the mean a search result occurs) we would tend to overestimate the significance of the alignment.

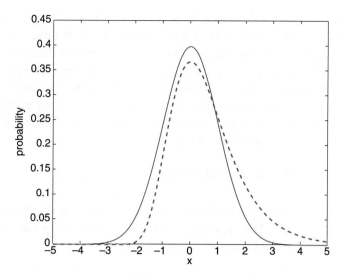

FIGURE 4.13. *Normal distribution (solid line) is compared to the extreme value distribution (dotted line). Comparing a query sequence to a set of uniform-length random sequences usually generates scores that fit an extreme value distribution (rather than a normal distribution). The area under each curve is 1. For the normal distribution, the mean (μ) is centered at zero, and the probability Z of obtaining some score x is given in terms of units of standard deviation (σ) from x to the mean: Z = (x − μ)/σ. In contrast to the normal distribution, the extreme value distribution is asymmetric, with a skew to the right. It is fit to the equation f(x) = (e^{−x})(e^{−e−x}). The shape of the extreme value distribution is determined by the characteristic value u and the decay constant λ (u = 0, λ = 1).*

The shape of the extreme value distribution shown in Fig. 4.13 is described by two parameters: the characteristic value u and the decay constant λ. The extreme value distribution is sometimes called the Gumbel distribution, after the person who described it in 1958. The application of the extreme value distribution to BLAST searching has been reviewed by Altschul and colleagues (1994, 1996) and others (Pagni and Jongeneel, 2001). For two random sequences m and n, the cumulative distribution function of scores S is described by the formula

The characteristic value u relates to the maximum of the distribution, although it is not the mean μ (mu).

$$P(S < x) = \exp\left(-e^{-\lambda(x-u)}\right) \qquad (4.1)$$

To use this equation, we need to know (or estimate) the values of the parameters u and λ. For ungapped alignments, the parameter u is dependent on the lengths of the sequences being compared and is defined as

$$u = \frac{\ln Kmn}{\lambda} \qquad (4.2)$$

In Equation 4.2, m and n refer to the lengths of the sequences being compared and K is a constant. Combining Equations 4.1 and 4.2, the probability of observing a score equal to or greater than x by chance is given by the formula

$$P(S \geq x) = 1 - \exp\left(-Kmne^{-\lambda x}\right) \qquad (4.3)$$

Our goal is to understand the likelihood that a BLAST search of an entire database produces a result by chance alone. The number of ungapped alignments with a score of at least x is described by the parameter $Kmne^{-\lambda x}$. In the context of a database search, m and n refer to the length (in residues) of the query sequence and the length of the entire database, respectively. The product $m \cdot n$ defines the size of the search space. The search space represents all the sites at which a query sequence can be aligned to any sequence in the database. Because the ends of a sequence are not as likely to participate in an average-sized alignment, the BLAST algorithm calculates the effective search space in which the average length of an alignment L is subtracted from m and n (Altschul and Gish, 1996):

We will see how BLAST uses this definition of search space in Fig. 4.14.

$$\text{Effective Search Space} = (m - L) \cdot (n - L) \qquad (4.4)$$

We now arrive at the main mathematical description of the significance of scores from a BLAST search. The expected number of HSPs having some score S (or better) by chance alone is described using the equation

$$E = Kmne^{-\lambda S} \qquad (4.5)$$

Equation 4.5 is described online in the document "The Statistics of Sequence Similarity Scores" available in the help section of the NCBI BLAST site.

Here, E refers to the expect value, which is the number of different alignments with scores equivalent to or better than S that are expected to occur by chance in a database search. This provides an estimate of the number of false positive results from a BLAST search. From Equation 4.5 we see that the E value depends on the score and λ, which is a parameter that scales the scoring system. Also, E depends on the length of the query sequence and the length of the database. The parameter K is a scaling factor for the search space. The parameters K and λ were described by Karlin and Altschul (1990) and so are often called Karlin–Altschul statistics.

Note several important properties of Equation 4.5:

- The value of E decreases exponentially with increasing S. The score S reflects the similarity of each pairwise comparison and is based in part on the scoring matrix selected. Higher S values correspond to better alignments; we saw in Fig. 4.8 that BLAST results are ranked by score. Thus, a high score on a BLAST search corresponds to a low E value. As E approaches zero, the probability that the alignment occurred by chance approaches zero. We will relate the E value to probability (p) values below.

- The expected score for aligning a random pair of amino acids must be negative. Otherwise, very long alignments of two sequences could accumulate large positive scores and appear to be significantly related when they are not.

- The size of the database that is searched—as well as the size of the query— influences the likelihood that particular alignments will occur by chance. Consider a BLAST result with an E value of 1. This value indicates that in a database of this particular size one match with a similar score is expected to occur by chance. If the database were twice as big, there would be twice the likelihood of finding a score equal to or greater than S by chance.

- The theory underlying Equation 4.5 was developed for ungapped alignments. For these, BLAST calculates values for λ, K, and H (entropy; see Fig. 3.28). Equation 4.5 can be successfully applied to gapped local alignments as well (such as the results of a BLAST search). However, for gapped alignments λ, K and H cannot be calculated analytically, but instead they are estimated by simulation and looked up in a table of precomputed values.

Making Sense of Raw Scores with Bit Scores

A typical BLAST output reports both E values and scores. There are two kinds of scores: raw and bit scores. Raw scores are calculated from the substitution matrix and gap penalty parameters that are chosen. The bit score S' is calculated from the raw score by normalizing with the statistical variables that define a given scoring system. Therefore, bit scores from different alignments, even those employing different scoring matrices in separate BLAST searches, can be compared. A raw score from a BLAST search must be normalized to parameters such as the size of the database being queried. The raw score is related to the bit score by the equation

$$S' = \frac{\lambda S - \ln K}{\ln 2} \qquad (4.6)$$

where S' is the bit score, which has a standard set of units. The E value corresponding to a given bit score is given by

$$E = mn \times 2^{-S'} \qquad (4.7)$$

Why are bit scores useful? First, raw scores are unitless and have little meaning alone. Bit scores account for the scoring system that was used and describe the information content inherent in a pairwise alignment. Thus, they allow scores to be compared between different database searches, even if different scoring matrices are employed. Second, bit scores can tell you the E value if you know the size of the search space, $m \cdot n$. (The BLAST algorithms use the effective search space size, described above.)

Bit scores are displayed in a column to the right in the BLAST output window, next to corresponding E values for each database match.

BLAST Algorithm: Relation between E and p Values

The p value is the probability of a chance alignment occurring with the score in question or better. It is calculated by relating the observed alignment score S to the expected distribution of HSP scores from comparisons of random sequences of the same length and composition as the query to the database. The most highly significant p values are those close to zero. The p and E values are different ways of representing the significance of the alignment. The probability of finding an HSP with a given E value is

$$p = 1 - e^{-E} \qquad (4.8)$$

Table 4.3 lists several p values corresponding to E values. While BLAST reports E values rather than p values, the two measures are nearly identical, especially for very small values associated with strong database matches. An advantage of using E values is that it is easier to think about E values of 5 versus 10 rather than 0.99326205 versus 0.99995460.

Some BLAST servers (such as those at the European Molecular Biology Laboratory; see Chapter 5) use p values in the output.

A p value below 0.05 is traditionally used to define statistical significance (i.e., to reject the null hypothesis that your query sequence is not related to any database sequence). If the null hypothesis is true, then 5% of all random alignments will result in an apparently significant score. Thus, an E value of 0.05 or less may be considered significant.

It is also possible to approach E values with conservative corrections. We discussed probability (p) values in Chapter 3, and we will return to the topic in Chapter 9 when we discuss microarray data analysis. The significance level α is typically set to 0.05, such that a p value of 0.05 suggests that some observation

TABLE 4-3 Relationship of E to p Values in BLAST Using Equation 4.8

E	p
10	0.99995460
5	0.99326205
2	0.86466472
1	0.63212056
0.1	0.09516258
0.05	0.04877058
0.001	0.00099950
0.0001	0.0001000

Small E values (0.05 or less) correspond closely to the P values.

See ► http://www.ncbi.nlm.nih. gov/Education/BLASTinfo/ rules.html for a description of significant scores.

(e.g., the score of a protein query to a match in a database) is likely to have occurred by chance 1 time in 20. The null hypothesis is that your query is not homologous to the database match, and the alternate hypothesis is that they are homologous. If the p value is sufficiently small (e.g., <0.05), we can reject the null hypothesis. When you search a database that has one million proteins, there are many opportunities for your query to find matches. Five percent of 1 million proteins is 50,000 proteins, and we might expect to obtain that many matches (with $\alpha = 0.05$) by chance. A related issue arises in microarray data analysis when we compare two conditions (e.g., normal versus diseased sample) and measure the RNA transcript levels of 20,000 genes: 1,000 transcripts (i.e., 5%) may be differentially expressed by chance.

This situation involves multiple comparisons: you are not hypothesizing that your query will match *one* particular database entry, you are interested in knowing if it matches *any* entries. A solution is to correct for multiple comparisons by adjusting the α level. A very conservative way to do this (called the Bonferroni correction) is to divide α by the number of measurements (e.g., divide α by the size of the database). In the case of BLAST searches this is done automatically, as shown in Equation 4.5, because the E value is multiplied by the effective search space.

Beyond this multiple comparison correction inherent in BLAST, some researchers consider it appropriate to adjust the significance level α for search results from 0.05 to some even lower value. In analyses of completed microbial genomes, BLAST or FASTA search E values were reported as significant if they were below 10^{-4} (Ferretti et al., 2001) or below 10^{-5} (Chambaud et al., 2001; Tettelin et al., 2001; Ermolaeva et al., 2001). In the public consortium analysis of the human genome, Smith–Waterman alignments were reported with an E value threshold of

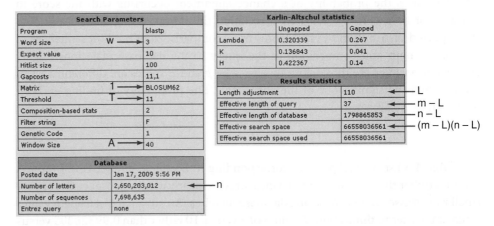

FIGURE 4.14. *BLAST search statistics. The NCBI search summary (using human beta globin NP_000509 as a query) includes search parameters (such as word size, arrow labeled w; scoring matrix, arrow 1; threshold value, arrow T; window size indicating the length that separates two independent hits to trigger an extension, arrow A). Database features include the number of sequences in the database and the total database size (arrow n). Karlin–Altschul statistics are provided and can be used following equation 4.5 to relate scores to expect values. Result statistics include the length adjustment of the query (here beta globin is 147 amino acids and the length adjustment is 110; arrow L), the effective length of the query (m − L = 147 − 110 = 37 in this example), the effective length of the database, and the effective search space (obtained by multiplying the effective length of the query by the effective length of the database). For a related figure, see Pagni and Jongeneel (2001).*

10^{-3}, and tblastn searches used a threshold of 10^{-6} (International Human Genome Sequencing Consortium, 2001). You can choose how conservatively to interpret BLAST results.

Parameters of a BLAST Search

Each of the various BLAST algorithms provides a summary of the search statistics. An example based on a blastp search using beta globin (NP_000509) as a query is shown in Fig. 4.14. In the problems at the end of this chapter, we will try varying several parameters in a BLAST search. The summary statistics provide a way to evaluate how the parameters modified the search results.

BLAST SEARCH STRATEGIES

General Concepts

BLAST searching is a tool to explore databases of protein and DNA sequence. We have introduced the procedure. It is essential that you define the question you want to answer, the DNA or protein sequence you want to input, the database you want to search, and the algorithm you want to use. We will now address some basic principles regarding strategies for BLAST searching (Altschul et al., 1994). We will illustrate these issues with globin, lipocalin, and HIV-1 *pol* searches. Key issues include how to evaluate the statistical significance of BLAST search results and how to modify the optional parameters of the BLAST programs when your search yields too little or too much information. An overview of the kinds of searches that can be performed with RBP4 DNA (NM_006744) or protein (NP_006735) sequence is presented in Fig 4.15.

Principles of BLAST Searching

How to Evaluate Significance of Your Results

When you perform a BLAST search, which database matches are authentic? To answer this question, we first define a true positive as a database match that is homologous to the query sequence (descended from a common ancestor). Homology is inferred based on sequence similarity, with support from statistical evaluation of the search results. A consistent finding of several research groups is that the error rate of database search algorithms is reduced by using statistical scores such as expect values rather than relying on percentage identity (or percent similarity) (Brenner, 1998; Park et al., 1998; Gotoh, 1996). Thus, we focus on inspection of E values.

The problem of assigning homology between genes or proteins is not solved by sequence analysis alone: it is also necessary to apply biological criteria to support the inference of homology. One can supplement BLAST results with evaluations of protein structure and function. The sequences of genuinely related proteins can diverge greatly, even while these proteins retain a related three-dimensional structure. Thus, we expect that database searches (and pairwise protein alignments) will result in a number of false negative matches. Many members of the lipocalin family, such as RBP4 and odorant-binding protein (OBP), share very limited sequence identity, although their three-dimensional structures are closely related and their functions as carriers of hydrophobic ligands are thought to be the same.

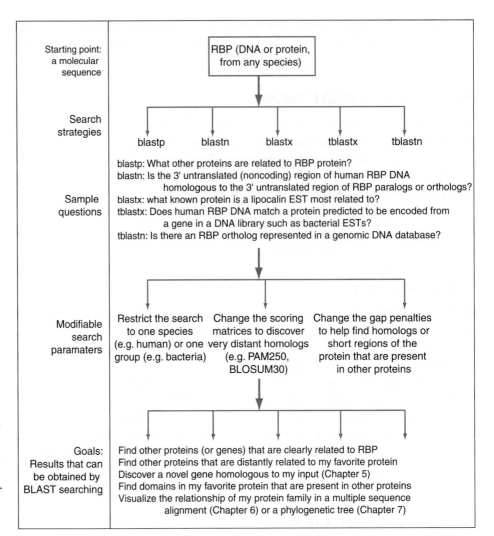

FIGURE 4.15. *Overview of BLAST searching strategies. There are many hundreds of questions that can be addressed with BLAST searching, from characterizing the genome of an organism to evaluating the sequence variation in a single gene.*

We will see in Chapter 5 how the PSI-BLAST program at NCBI can generate a score of over 100 bits and an E value of about 10^{-21} for the same match of RBP4 to complement component 8 gamma. In general, PSI-BLAST is preferable to blastp for many typical database searches.

Consider a blastp search of the nr database restricted to human entries using human RBP4 as a query. There are 29 entries in this case. The first way to evaluate the results is to inspect the E value list (Fig. 4.16). The entries with the lowest E values (beginning with $7e10^{-116}$) are all named RBP. This redundancy occurs because the alignments involve closely related versions of RBP. In some cases, the alignment may involve identical regions of the RBP database sequences that differ elsewhere in the sequence outside the aligned region. (For example, a fragment of a protein sequence may be deposited in the database.) Inspecting the alignment of the 14th entry (E value 2×10^{-13}) shows that it is a very short fragment of 36 amino acids that is aligned to the full-length RBP (Fig. 4.17). The next database match is apolipoprotein D. The RBP fragment and apolipoprotein D have similar expect values, but in this case they have very different percent identities to RBP (94% versus 31%; Fig. 4.17). In analyzing any BLAST search results, it is important to carefully inspect the alignments as well as the scores.

Further down the list (Fig. 4.16) we see a database match of complement component 8 gamma (NP_000597). The alignment has a high, nonsignificant E value of 0.27 and a low score of 33.9 bits, and the protein shares only 25% amino acid identity with RBP over a span of 114 amino acid residues—including three

```
                                                   Score    E
Sequences producing significant alignments:       (Bits)  Value

gb|AAH20633.1|  Retinol binding protein 4, plasma [Homo sapiens]    414    7e-116  [UG]
ref|NP_006735.2|  retinol-binding protein 4, plasma precursor ...   398    4e-111  [UG]
emb|CAH72329.1|  retinol binding protein 4, plasma [Homo sapiens]   384    1e-106
gb|EAW50068.1|  retinol binding protein 4, plasma, isoform CRA_b    382    3e-106
emb|CAA24959.1|  unnamed protein product [Homo sapiens]             380    2e-105  [UG]
pdb|1JYD|A  Chain A, Crystal Structure Of Recombinant Human Se...   379    2e-105  [S]
pdb|1RBP|  Chain , Retinol Binding Protein >pdb|1BRP| Chain...       379    2e-105  [S]
pdb|1JYJ|A  Chain A, Crystal Structure Of A Double Variant (W6...   374    7e-104  [S]
pdb|1QAB|E  Chain E, The Structure Of Human Retinol Binding Pr...   369    3e-102  [S]
gb|AAF69622.1|AF119917_30  PRO2222 [Homo sapiens]                   330    1e-90
emb|CAA26553.1|  RBP [Homo sapiens]                                 204    1e-52   [G]
emb|CAB46489.1|  unnamed protein product [Homo sapiens]             152    5e-37   [G]
gb|AAC02945.1|  mutant retinol binding protein [Homo sapiens]       90.9   2e-18   [G]
gb|AAC02946.1|  mutant retinol binding protein [Homo sapiens]       74.3   2e-13   [G]
gb|AAB32200.1|  apolipoprotein D, apoD [human, plasma, Peptide, 2   57.4   2e-08
ref|NP_001638.1|  apolipoprotein D precursor [Homo sapiens] >s...   57.4   3e-08   [UG]
gb|AAB35919.1|  apolipoprotein D; apoD [Homo sapiens]               43.9   3e-04   [UG]
prf||0801163A  complex-forming glycoprotein HC                      38.1   0.017
gb|EAW88163.1|  progestagen-associated endometrial protein (pl...   37.4   0.027
pdb|1IW2|A  Chain A, X-Ray Structure Of Human Complement Prote...   36.2   0.066   [S]
gb|EAW88165.1|  progestagen-associated endometrial protein (pl...   36.2   0.067
ref|NP_001018059.1|  glycodelin precursor [Homo sapiens] >ref|...   36.2   0.067   [UG]
emb|CAB43305.1|  hypothetical protein [Homo sapiens]                36.2   0.067
ref|NP_001624.1|  alpha-1-microglobulin/bikunin precursor [Hom...   35.8   0.074   [UG]
gb|AAI13627.1|  Complement component 8, gamma polypeptide [Homo s   35.0   0.13    [UG]
ref|NP_000597.1|  complement component 8, gamma polypeptide [H...   33.9   0.27    [UG]
gb|AAA60147.1|  placental protein 14                                32.7   0.67    [UG]
emb|CAI12748.1|  complement component 8, gamma polypeptide [Homo    31.6   1.6
sp|P07360|C08G_HUMAN  Complement component C8 gamma chain prec...    31.2   2.0     [G]
```

FIGURE 4.16. *Results of a blastp nr search using human RBP as a query, restricting the output to human proteins. Note that there are 21 hits, and inspection of the E values suggests that in addition to RBP itself, several authentic paralogs may have been identified by this search. Are complement component 8 and a progestagen-associated protein likely to be homologous to RBP?*

gaps in the alignment (Fig. 4.18). One might conclude that these two proteins are not homologous. But in this case they are. In deciding whether two proteins (or DNA sequences) are homologous, one can ask several questions:

- Is the expect value significant? In this particular case it is not, because the proteins are distantly related. Search techniques such as PSI-BLAST or HMMer (based on hidden Markov models), introduced in Chapter 5, can typically assign higher scores and lower *E* values to distant matches.

```
>  gb|AAC02946.1|  [G] mutant retinol binding protein [Homo sapiens]
Length=36

 Score = 74.3 bits (181),  Expect = 2e-13, Method: Composition-based stats.
 Identities = 34/36 (94%),  Positives = 35/36 (97%),  Gaps = 0/36 (0%)

Query  84   NWDVCADMVGTFTDTEDPAKFKMKYWGVASFLQKGN   119
            NWDVCADMV TFTDTEDPAKFKMKYWGVASFLQKG+
Sbjct  1    NWDVCADMVDTFTDTEDPAKFKMKYWGVASFLQKGS   36
```

```
>  gb|AAB32200.1|  apolipoprotein D, apoD [human, plasma, Peptide, 246 aa]
Length=246

 Score = 57.4 bits (137),  Expect = 2e-08, Method: Composition-based stats.
 Identities = 47/151 (31%),  Positives = 78/151 (51%),  Gaps = 39/151 (25%)

Query  29   VKENFDKARFSGTWYAMAKKDPEGLFLQDNIVAEFSVDETGQMSATAKGRVRLLNNWDVC   88
            V+ENFD  ++ G WY + +K P      I A +S+ E G     ++++LN ++
Sbjct  13   VQENFDVNKYLGRWYEI-EKIPTTFENGRCIQANYSLMENG-------KIKVLNQ-ELR   62

Query  89   ADMVGTFTDTE---------DPAKFKMKY-WGVASFLQKGNDDHWIVDTDYDTYAVQYSC  138
            AD  GT   E          +PAK ++K+ W + S       +WI+ TDY+ YA+ YSC
Sbjct  63   AD--GTVNQIEGEATPVNLTEPAKLEVKFSWFMPS------APYWILATDYENYALVYSC  114

Query  139  ----RLLNLDGTCADSYSFVFSRDPNGLPPE   165
                +L ++D    ++++ +R+PN LPPE
Sbjct  115  TCIIQLFHVD------FAWILARNPN-LPPE   138
```

FIGURE 4.17. *Two pairwise alignments returned from the human RBP4 search (see Fig. 4.16, halfway down the list). An RBP4 fragment of just 36 amino acids yields a similar score and expect value as the longer match between RBP4 and apolipoprotein D. This result highlights the need to inspect each pairwise alignment from a BLAST search.*

FIGURE 4.18. *Alignment of human RBP4 (query) with progestagen-associated endometrial protein. The bit score is relatively low, the expect value (0.27) is not significant, and in the local alignment the two proteins share only 25% amino acid identity over 114 amino acids. Nonetheless, these proteins are homologous. Their homology can be confirmed because (1) the two proteins are approximately the same size; (2) they share a lipocalin signature, including a GXW motif; (3) they can be multiply aligned; (4) they are both soluble, hydrophilic, abundant proteins that probably share similar functions as carrier proteins; and (6) they are very likely to share a similar three-dimensional structure (see Chapter 11).*

```
>┌ref|NP_000597.1| U G  complement component 8, gamma polypeptide [Homo sapiens]
  emb|CAA29773.1| U G  unnamed protein product [Homo sapiens]
  gb|AAA51863.1| U G  complement component C8-gamma precursor
  gb|AAA51888.1| U G  complement protein C8 gamma subunit precursor
  gb|AAA18482.1| G  complement C8 gamma subunit precursor
  gb|AAI13625.1| U G  Complement component 8, gamma polypeptide [Homo sapiens]
  gb|EAW88316.1| G  complement component 8, gamma polypeptide [Homo sapiens]
Length=202

 Score = 33.9 bits (76),  Expect = 0.27, Method: Composition-based stats.
 Identities = 29/114 (25%), Positives = 50/114 (43%), Gaps = 8/114 (7%)

Query  24  VSSFRVKENFDKARFSGTWYAMAKKDPEGLFLQDNIVAEFSVDETG-QMSATAKGRVRLL  82
           +S+ + K NFD +F+GTW +A        + AE +   Q +A A   R L
Sbjct  33  ISTIQPKANFDAQQFAGTWLLVAVGSACRFLQEQGHRAEATTLHVAPQGTAMAVSTFRKL  92

Query  83  NNWDVCADMVGTFTDTEDPAKFKMKYWGVASFLQKGNDDHWIVDTDYDTYAVQY  136
           +    +C + + DT  +F ++ G     +G   + +TDY ++AV Y
Sbjct  93  DG--ICWQVRQLYGDTGVLGRFLLQARGA-----RGAVHVVVAETDYQSFAVLY  139
```

The accession number for the x-ray structure of human complement protein C8γ is 1IW2, while for RBP4 an accession is 1RBP. We discuss Protein Data Bank accession numbers (such as these) in Chapter 11.

- Are the two proteins approximately the same size? It is not at all required that homologous proteins have similar sizes, and it is possible for two proteins to share only a limited domain in common. Indeed, local alignments search tools such as BLAST are specialized to find limited regions of overlap. However, it is also important to develop a biological intuition about the likelihood that two proteins are homologous. A 1000 amino acid protein with transmembrane domains is relatively unlikely to be homologous to RBP, and the vast majority of lipocalins are approximately 200 amino acids in length (20 to 25 kilodaltons).

- Do the proteins share a common motif or signature? In this case, both RBP4 and complement component 8 gamma have a glycine-X-tryptophan (GXW) signature that is characteristic of the lipocalin superfamily.

- Are the proteins part of a reasonable multiple sequence alignment? We will see in Chapter 6 that this is the case.

- Do the proteins share a similar biological function? Like all lipocalins, both proteins are small, hydrophilic, abundant, secreted molecules.

- Do the proteins share a similar three-dimensional structure? Although there is great diversity in lipocalin sequences, they share a remarkably well-conserved structure. This structure, a cuplike calyx, allows them to transport hydrophobic ligands across an aqueous compartment (see Chapter 11).

- Is the genomic context informative? The human complement component gamma gene has a similar number and length of exons as other lipocalins (Kaufman and Sodetz, 1984). It is mapped to chromosome 9q34.3, immediately adjacent to another lipocalin gene (LCN12) in the vicinity of 10 other lipocalin genes on 9q34. This information suggests that the blastp match is biologically significant, even if the *E* value is not statistically significant.

- If a BLAST search results in a marginal match to another protein, perform a new BLAST search using that distantly related protein as a query. A blastp nr search using complement component 8 gamma as a query results in the identification of several proteins (complex-forming glycoprotein HC and α-1-microglobulin/bikunin) that are also detected by RBP4 (Fig. 4.19). This finding increases our confidence that RBP4 and complement component 8 gamma are in fact homologous members of a protein superfamily. If the blastp search using complement component 8 gamma had shown that protein to be part of another characterized family, this would have greatly lessened our confidence that it is authentically related to RBP4.

We will define motifs and signatures in Chapter 10 and trees in Chapter 7.

Historically, early database searches yielded results that were entirely unexpected. In 1984, the β-adrenergic receptor was found to be homologous to rhodopsin (Dixon et al., 1986). This was surprising because of the apparent differences between these receptors in terms of function and localization: rhodopsin is a retina-specific receptor for light, and the adrenergic receptors were known to bind epinephrine (adrenalin) and norepinephrine, stimulating a signal transduction cascade that results in cyclic adenosine monophosphate (AMP) production. Alignment of the protein sequences revealed that they share similar structural features (seven predicted transmembrane domains). It is now appreciated that rhodopsin and the β-adrenergic receptor are prototypic members of a superfamily of receptors that bind ligands, initiating a second messenger cascade. Another surprising finding was that some viral genes that are involved in transforming mammalian cells are actually derived from

Go to Entrez Gene and enter "rhodopsin," restricting the organism to human. There are about 700 entries, mostly consisting of members of this family of receptors thought to have seven transmembrane spans. We will see how to explore protein families in Chapter 10.

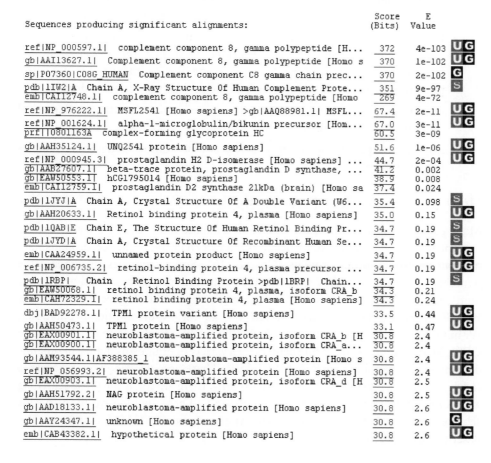

FIGURE 4.19. *Results of a blastp nr search (restricted to human proteins) using progestagen-associated endometrial protein as a query (NP_002562). This protein is related to several proteins (complex-forming glycoprotein HC, α-1-microglobulin/bikunin) that also appear in the output of an RBP4 blastp search. This overlap supports the hypothesis that progestagen-associated endometrial protein and RBP4 are indeed homologous. It is often important to perform reciprocal BLAST searches separately using two possibly related sequences as queries.*

the host species. The human epidermal growth factor receptor was sequenced and found to be homologous to an avian retroviral oncogene, *v-erb-B* (Downward et al., 1984). There are many more examples of database searches that revealed unexpected relationships. In many other cases, the reported relationships represented false positive results. The false positive error rate will yield occasional matches that are not authentic, and comparison of the three-dimensional structures of the potential homologs can be used as a criterion for deciding whether two proteins are in fact homologous.

How to Handle Too Many Results

A common situation that is encountered in BLAST searching is that too many results are returned. There are many strategies available to limit the number of results, but to make the appropriate choices, you must focus on the question you are trying to answer.

- Select a "refseq" database and all the hits that are returned will have RefSeq accession numbers. This will often eliminate redundant database matches.
- Limit the database returns by organism, when applicable. One convenient approach is to select the taxonomy identifier (txid) of interest through the Taxonomy Browser. This may eliminate extraneous information. If you use the options feature of the BLAST server to limit a search by organism, the same size search is performed; in contrast, if you choose an organism-specific database, this may increase the speed of the search. (We present some organism-specific BLAST servers in Chapter 5.) You can use the Boolean operator NOT to ignore matches from an organism or group of interest.
- Use just a portion of the query sequence, when appropriate. A search of a multidomain protein can be performed with just the isolated domain sequence. If you are studying HIV-1 pol, you may be interested in the entire protein or in a specific portion such as the reverse transcriptase domain.
- Adjust the scoring matrix to make it more appropriate to the degree of similarity between your query and the database matches.
- Adjust the expect value; lowering E reduces the number of database matches that are returned.

How to Handle Too Few Results

Many genes and proteins have no significant database matches or have very few. As new microbial genomes are sequenced, as many as half the predicted proteins have no matches to any other proteins (Chapter 15). Some strategies to increase the number of database matches from BLAST searching are obvious: remove Entrez limits, raise the expect values, and try scoring matrices with higher PAM or lower BLOSUM values. One can also search a large variety of additional databases. Within the NCBI website, one can search all available databases (e.g., HTGS and GSS). Many genome-sequencing centers for a variety of organisms maintain separate databases that can be searched by BLAST. These are described in Chapter 5 (advanced BLAST searching). Most importantly, there are many database searching algorithms that are more sensitive than BLAST. These include position-specific scoring matrices (PSSMs) and hidden Markov models (HMMs) and are also described in Chapter 5.

BLAST Searching With Multidomain Protein: HIV-1 pol

The gag-pol protein of HIV-1 (NP_057849) has 1435 amino acid residues and includes separate protease, reverse transcriptase, and integrase domains. Thus, it is an example of a multidomain protein. Figure 4.20 previews the kinds of searches we can perform with a viral protein such as this one.

What happens upon blastp searching the nonredundant protein database with this protein? We input the RefSeq accession number (NP_057849) and click submit. The program reports that putative conserved domains have been detected, and a schematic of the protein indicates the location of each domain (Fig. 4.21). Clicking on any of these domains links to the NCBI conserved domain database as well as to the Pfam and SMART databases (see Chapters 6 and 10). Continuing with the BLAST search, we see that there are many hits, all with extremely low expect values, and all correspond to HIV matches from various isolates (Fig. 4.22). Reformatting the output to "query-anchored with letters for identities" is one way to view the dramatic conservation of these viral proteins (Fig. 4.23).

These highly conserved HIV-1 variants of gag-pol obscure our ability to evaluate non-HIV-1 matches. We can repeat the blastp search, setting the database to RefSeq proteins. Now gag-pol orthologs are evident across a variety of virus species. Clicking Taxonomy Reports from the main blastp search result page shows that surprisingly

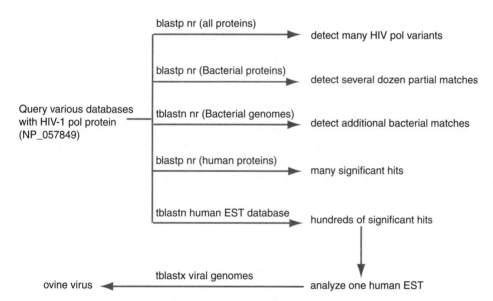

FIGURE 4.20. Overview of BLAST searches beginning with HIV-1 pol protein. A series of BLAST searches can often be performed to pursue questions about a particular gene, protein, or organism. The number of database matches returned by a BLAST search can vary from none to thousands and depends entirely on the nature of the query, the database, and the search parameters.

FIGURE 4.21. *A blastp search with viral* pol *(NP_057849) shows conserved domains in the protein. These blocks are clickable and link to the Conserved Domain Database at NCBI (Chapter 6). The links are to protein domains (Gag_p17, Gag_p24) and abbreviations include rvp, retroviral aspartyl protease; rvt, reverse transcriptase (RNA-dependent DNA polymerase); rnaseH, ribonuclease H; rve, integrase core domain.*

```
                                                                          Score    E
                         Sequences producing significant alignments:      (Bits)  Value

ref|NP_057849.4|  Gag-Pol [Human immunodeficiency virus 1] >sp...         2884    0.0    G
sp|P03366|POL_HV1B1  Gag-Pol polyprotein (Pr160Gag-Pol) [Conta...         2856    0.0
sp|P03367|POL_HV1BR  Gag-Pol polyprotein (Pr160Gag-Pol) [Conta...         2852    0.0
sp|P04587|POL_HV1B5  Gag-Pol polyprotein (Pr160Gag-Pol) [Conta...         2845    0.0
gb|AAD03191.1|  gag-pol fusion polyprotein [Human immunodefici...         2818    0.0
sp|P35963|POL_HV1Y2  Gag-Pol polyprotein (Pr160Gag-Pol) [Conta...         2805    0.0
sp|P12497|POL_HV1N5  Gag-Pol polyprotein (Pr160Gag-Pol) [Conta...         2804    0.0
sp|P20875|POL_HV1JR  Gag-Pol polyprotein (Pr160Gag-Pol) [Conta...         2800    0.0
gb|AAD03200.1|  gag-pol fusion polyprotein [Human immunodefici...         2794    0.0
sp|P20892|POL_HV1OY  Gag-Pol polyprotein (Pr160Gag-Pol) [Conta...         2793    0.0
dbj|BAB85751.1|  Gag-pol fusion polyprotein [Human immunodeficien        2789    0.0
sp|Q73368|POL_HV1B9  Gag-Pol polyprotein (Pr160Gag-Pol) [Conta...         2785    0.0
sp|P03369|POL_HV1A2  Gag-Pol polyprotein (Pr160Gag-Pol) [Conta...         2775    0.0
gb|AAG30116.1|AF286365_2  gag-pol fusion polyprotein [Human im...         2771    0.0
sp|P05959|POL_HV1RH  Gag-Pol polyprotein (Pr160Gag-Pol) [Conta...         2767    0.0
gb|AAD03217.1|  gag-pol fusion polyprotein [Human immunodefici...         2758    0.0
sp|P05961|POL_HV1MN  Gag-Pol polyprotein (Pr160Gag-Pol) [Conta...         2753    0.0
gb|AAD03225.1|  gag-pol fusion polyprotein [Human immunodefici...         2749    0.0
gb|AAD03326.1|  gag-pol fusion polyprotein [Human immunodefici...         2739    0.0
dbj|BAC77511.1|  Gag-Pol fusion protein [Human immunodeficiency v        2722    0.0
dbj|BAC77486.1|  Gag-Pol fusion polyprotein [Human immunodeficien        2722    0.0
sp|P18802|POL_HV1ND  Gag-Pol polyprotein (Pr160Gag-Pol) [Conta...         2717    0.0
dbj|BAC77477.1|  Gag-Pol fusion polyprotein [Human immunodeficien        2714    0.0
gb|AAD03241.1|  gag-pol fusion polyprotein [Human immunodefici...         2713    0.0
gb|AAD03233.1|  gag-pol fusion polyprotein [Human immunodefici...         2709    0.0
gb|AAN73492.1|AF484483_1  gag-pol fusion polyprotein [Human immun        2706    0.0
```

FIGURE 4.22. *A blastp nr search with HIV-1 pol results in a very large number of database hits that all appear to be variants of HIV-1. Note that all the E values shown are zero. This result obscures any possible hits that are not from HIV-1.*

there are even some homologs in the chimpanzee *Pan troglodytes*, the chicken *Gallus gallus*, and the opossum *Monodelphis domestica* (Fig. 4.24).

To learn more about the distribution of pol proteins throughout the tree of life, we may further ask what bacterial proteins are related to the viral HIV-1 pol polyprotein. Repeat the blastp search with NP_057849 as the query, but limit the search to "Bacteria" (txid2[Organism]). Here, the graphical overview of BLAST search results is extremely helpful to show that two domains of viral pol have the majority of matches to known bacterial sequences, corresponding to amino acids 1000–1100 and 1200–1300 of pol (Fig. 4.25). Comparison of this output to the domain architecture of HIV-1 pol (Fig. 4.21) suggests that the two viral protein domains with matches to bacterial proteins are RNAse H and an integrase. Indeed, the bacterial matches aligned to viral pol include ribonuclease H and integrases (Fig. 4.26). Inspection of the pairwise alignments indicates that the viral

```
Query     181   PQDLNTMLNTVGGHQAAMQMLKETINEEAAEWDRVHPVHAGPIAPGQMREPRGSDIAGTT   240
P03366    181   PQDLNTMLNTVGGHQAAMQMLKETINEEAAEWDRVHPVHAGPIAPGQMREPRGSDIAGTT   240
P03367    181   PQDLNTMLNTVGGHQAAMQMLKETINEEAAEWDRVHPVHAGPIAPGQMREPRGSDIAGTT   240
P04587    181   PQDLNTMLNTVGGHQAAMQMLKETINEEAAEWDRVHPVHAGPIAPGQMREPRGSDIAGTT   240
AAD03191  181   PQDLNTMLNTVGGHQAAMQMLKETINEEAAEWDRLHPVHAGPIAPGQMREPRGSDIAGTT   240
P35963    181   PQDLNTMLNTVGGHQAAMQMLKETINEEAAEWDRLHPVHAGPIAPGQMREPRGSDIAGTT   240
P12497    181   PQDLNTMLNTVGGHQAAMQMLKETINEEAAEWDRLHPVHAGPIAPGQMREPRGSDIAGTT   240
P20875    181   PQDLNTMLNTVGGHQAAMQMLKETINEEAAEWDRLHPVHAGPIAPGQMREPRGSDIAGTT   240
AAD03200  182   PQDLNTMLNTVGGHQAAMQMLKETINEEAAEWDRVHPVHAGPIAPGQMREPRGSDIAGTT   241
P20892    181   PQDLNTMLNTVGGHQAAMQMLKETINEEAAEWDRLHPVHAGPIAPGQMREPRGSDIAGTT   240
BAB85751  181   PQDLNTMLNTVGGHQAAMQMLKETINEEAAEWDRLHPVHAGPIAPGQMREPRGSDIAGTT   240
Q73368    181   PQDLNTMLNTVGGHQAAMQMLKETINEEAAEWDRLHPVQAGPVAPGQMREPRGSDIAGTT   240
P03369    183   PQDLNTMLNTVGGHQAAMQMLKETINEEAAEWDRVHPVHAGPIAPGQMREPRGSDIAGTT   242
AAG30116  181   PQDLNTMLNTVGGHQAAMQMLKETINEEAAEWDRLHPVQAGPVAPGQMREPRGSDIAGTT   240
P05959    181   PQDLNTMLNTVGGHQAAMQMLKETINEEAAEWDRLHPVHAGPIAPGQMREPRGSDIAGTT   240
AAD03217  181   PQDLNTMLNTVGGHQAAMQMLKETINEEAGEWDRLHPAQAGPVAPGQMREPRGSDIAGTT   240
P05961    184   PQDLNTMLNTVGGHQAAMQMLKETINEEAAEWDRLHPVHAGPITPGQMREPRGSDIAGTT   243
AAD03225  181   PQDLNTMLNTVGGHQAAMQMLKETINEEAAEWDRLHPVHAGPIAPGQMREPRGSDIAGTT   240
AAD03326  181   PQDLNTMLNTVGGHQAAMQMLKETINEEAAEWDRLHPVHAGPIAPGQMREPRGSDIAGTT   240
BAC77511  184   PQDLNTMLNTVGGHQAAMQMLKETINEEAAEWDRLHPVHAGPVAPGQMREPRGSDIAGTT   243
BAC77486  184   PQDLNTMLNTVGGHQAAMQMLKETINEEAAEWDRLHPVHAGPAAPGQMREPRGSDIAGTT   243
P18802    178   PQDLNTMLNTVGGHQAAMQMLKETINDEAAEWDRLHPVHAGPVAPGQMREPRGSDIAGTT   237
BAC77477  180   PQDLNTMLNTVGGHQAAMQMLKETINEEAAEWDRLHPVHAGPVAPGQMREPRGSDIAGTT   239
AAD03241  186   PQDLNTMLNTVGGHQAAMQMLKETINEEAAEWDRLHPVHAGPVAPGQMRDPRGSDIAGTT   245
AAD03233  181   PQDLNTMLNTVGGHQAAMQMLKETINEEAAEWDRLHPVQAGPVAPGQMREPRGSDIAGTT   240
AAN73492  179   PQDLNTMLNTVGGHQAAMQMLKETINEEAAEWDRLHPVHAGPPAPGQMREPRGSDIAGTT   238
```

FIGURE 4.23. *Portion of the output of a blastp search using the HIV-1 pol protein as a query (NP_057849). The flat query-anchored output format reveals substituted amino acid residues as well as those that remain invariant.*

Lineage Report

```
root
. Orthoretrovirinae              [viruses]
. . Lentivirus             [viruses]
. . . Primate lentivirus group [viruses]
. . . . Human immunodeficiency virus 1 ---------  2884 11 hits [viruses]   Gag-Pol [Human immunodeficiency virus 1]
. . . . Simian immunodeficiency virus ..........  1651  2 hits [viruses]   Gag-Pol [Simian immunodeficiency virus]
. . . . Human immunodeficiency virus 2 .........  1646  2 hits [viruses]   gag-pol fusion polyprotein [Human immunodeficiency virus 2]
. . . . Simian immunodeficiency virus SIV-and 2.  1349  2 hits [viruses]   pol protein [Simian immunodeficiency virus SIV-and 2]
. . . . Simian-Human immunodeficiency virus ....  1175  2 hits [viruses]   pol [Simian-Human immunodeficiency virus]
. . . Jembrana disease virus --------------------   489  3 hits [viruses]   gag-pol precursor [Jembrana disease virus]
. . . Feline immunodeficiency virus ............   489  3 hits [viruses]   pol polyprotein [Feline immunodeficiency virus]
. . . Equine infectious anemia virus ...........   421  3 hits [viruses]   pol polyprotein [Equine infectious anemia virus]
. . . Caprine arthritis-encephalitis virus .....   413  3 hits [viruses]   pol protein [Caprine arthritis-encephalitis virus]
. . . Ovine lentivirus .........................   402  3 hits [viruses]   pol protein [Ovine lentivirus]
. . . Visna/Maedi virus ........................   389  3 hits [viruses]   pol polyprotein [Visna/Maedi virus]
. . . Bovine immunodeficiency virus ............   369  3 hits [viruses]   reverse transcriptase [Bovine immunodeficiency virus]
. . Ovine enzootic nasal tumour virus ---------   252  1 hit  [viruses]   gag-pro-pol fusion [Ovine enzootic nasal tumour virus]
. . Enzootic nasal tumour virus of goats ......   246  1 hit  [viruses]   gag-pro-pol fusion [Enzootic nasal tumour virus of goats]
. . Jaagsiekte sheep retrovirus ...............   206  1 hit  [viruses]   pol protein [Jaagsiekte sheep retrovirus]
. . Rous sarcoma virus ........................   199  3 hits [viruses]   Pr180 polyprotein precursor [Rous sarcoma virus]
. . Bovine leukemia virus .....................   182  1 hit  [viruses]   RT-IN [Bovine leukemia virus] >gi|4523901|ref|NP_777384.2|
. . Simian T-lymphotropic virus 2 .............   180  4 hits [viruses]   gag-pro-pol polyprotein [Simian T-lymphotropic virus 2]
. . Human T-lymphotropic virus 1 ..............   175  4 hits [viruses]   Pr gag-pro-pol [Human T-lymphotropic virus 1]
. . Avian leukosis virus - RSA ................   159  2 hits [viruses]   polymerase [Avian leukosis virus - RSA]
. . Mouse mammary tumor virus .................   158  5 hits [viruses]   p30DU-p13PR-RT-IN [Mouse mammary tumor virus]
. . Mason-Pfizer monkey virus .................   155  3 hits [viruses]   RT-IN [Mason-Pfizer monkey virus] >gi|42766920|ref|NP_95456
. . Simian T-lymphotropic virus 3 .............   141  2 hits [viruses]   polymerase [Simian T-lymphotropic virus 3]
. . Human T-lymphotropic virus 2 ..............   127  2 hits [viruses]   pol polyprotein [Human T-lymphotropic virus 2]
. . Woolly monkey sarcoma virus ...............   119  2 hits [viruses]   pol protein [Woolly monkey sarcoma virus]
. . Reticuloendotheliosis virus ...............   102  2 hits [viruses]   protease/polymerase [Reticuloendotheliosis virus]
. . Rauscher murine leukemia virus ............    96  2 hits [viruses]   Pol [Rauscher murine leukemia virus]
. . Friend murine leukemia virus ..............    96  2 hits [viruses]   RNA-dependent DNA polymerase [Friend murine leukemia virus]
. Bos taurus (cow) ---------------------------   275  3 hits [mammals]   PREDICTED: similar to pol polyprotein [Bos taurus]
. Monodelphis domestica .......................   264 50 hits [marsupials] PREDICTED: similar to pol polyprotein [Monodelphis domestic
. Gallus gallus (bantam) ......................   152  2 hits [birds]     PREDICTED: similar to pol protein [Gallus gallus]
. Pan troglodytes .............................   112  3 hits [primates]  PREDICTED: similar to truncated polyprotein [Pan troglodyte
```

FIGURE 4.24. The taxonomy report for a blastp search shows an overview of which species have proteins matching the HIV-1 query. These include cow, opossum, chicken, and chimpanzee.

and bacterial proteins are homologous, sharing up to 30% amino acid identity over spans of up to 166 amino acids.

Let us now turn our attention to human proteins that may be homologous to HIV-1 pol. The search is identical to our search of bacteria, except that we restrict the organism to *Homo sapiens*. Interestingly, there are many human matches (Fig. 4.27a). These human proteins have been annotated as zinc finger proteins

Some researchers have suggested that neuropsychiatric diseases such as schizophrenia are associated with elevated levels of endogenous retroviral gene expression (Karlsson et al., 2001).

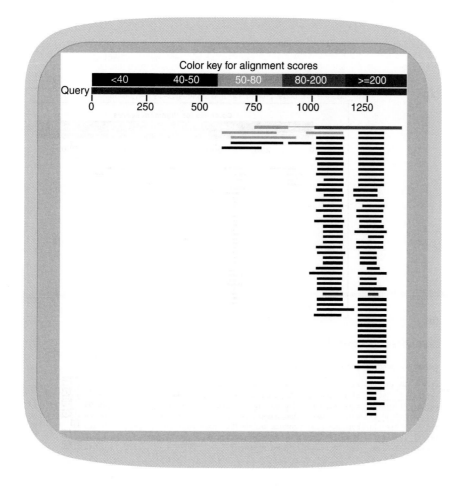

FIGURE 4.25. Result of a blastp search with HIV-1 pol as a query, restricting the output to bacteria. The graphical output of the BLAST search allows easy identification of the domains within HIV-1 that have bacterial matches.

```
                                                              Score    E
Sequences producing significant alignments:                  (Bits)   Value

ref|ZP_01298044.1|  hypothetical protein CburD_01002090 [Coxie...  88.6    1e-15
ref|ZP_00374187.1|  pol protein [Wolbachia endosymbiont of Drosop  75.5    1e-11
ref|ZP_01298039.1|  hypothetical protein CburD_01002094 [Coxie...  62.8    6e-08
ref|YP_445237.1|    ribonuclease H [Salinibacter ruber DSM 13855]  54.3    2e-05  G
ref|NP_840540.1|    Integrase, catalytic core [Nitrosomonas europae 48.5    0.001  G
ref|NP_841129.1|    Integrase, catalytic core [Nitrosomonas euro... 48.5    0.001  G
ref|NP_841960.1|    Integrase, catalytic core [Nitrosomonas euro... 48.5    0.001  G
ref|NP_842525.1|    Integrase, catalytic core [Nitrosomonas europae 48.5    0.001  G
ref|NP_840427.1|    Integrase, catalytic core [Nitrosomonas euro... 48.5    0.001  G
ref|NP_842038.1|    Integrase, catalytic core [Nitrosomonas europae 48.5    0.001  G
ref|NP_840377.1|    Integrase, catalytic core [Nitrosomonas europae 48.5    0.001  G
ref|NP_840793.1|    Integrase, catalytic core [Nitrosomonas europae 48.5    0.001  G
ref|NP_841866.1|    Integrase, catalytic core [Nitrosomonas europae 48.5    0.001  G
ref|NP_840338.1|    Integrase, catalytic core [Nitrosomonas europae 48.1    0.002  G
ref|NP_842406.1|    Integrase, catalytic core [Nitrosomonas europae 48.1    0.002  G
ref|ZP_01217503.1|  hypothetical protein PCNPT3_00010 [Psychromon  47.4    0.002
ref|YP_045841.1|    bifunctional protein [Includes: ribonuclease...  46.6    0.005  G
ref|YP_001113791.1| ribonuclease H [Desulfotomaculum reducens MI   44.3    0.021
ref|YP_555940.1|    Putative integrase [Burkholderia xenovorans LB4 44.3    0.022  G
ref|YP_555298.1|    Putative integrase [Burkholderia xenovorans LB4 43.9    0.030  G
ref|YP_169613.1|    Ribonuclease H [Francisella tularensis subsp... 43.5    0.034  G
ref|YP_319519.1|    integrase, catalytic region [Nitrobacter winogr 43.5    0.037  G
ref|ZP_01530841.1|  ribonuclease H [Roseiflexus castenholzii DSM   43.5    0.037
ref|YP_949974.1|    ISAau1, transposase orfB [Arthrobacter aures... 43.5    0.038  G
ref|YP_513576.1|    Ribonuclease H [Francisella tularensis subsp... 43.5    0.044  G
ref|YP_754778.1|    ribonuclease H [Syntrophomonas wolfei subsp.... 43.1    0.052  G
ref|YP_932276.1|    RnhA1 protein [Azoarcus sp. BH72]              43.1    0.055  G
```

FIGURE 4.26. *Bacterial proteins that are identified in a blastp search with HIV-1 pol include integrases and ribonuclease H proteins.*

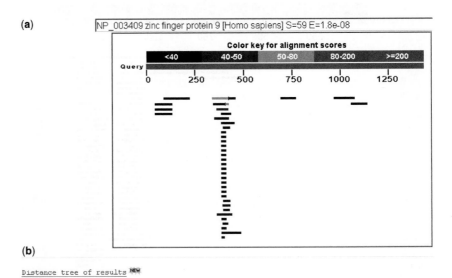

FIGURE 4.27. *(a) Graphical output of a blastp search using HIV-1 pol protein to search for matches against human proteins. (b) Note that some human hits have very low expect values.*

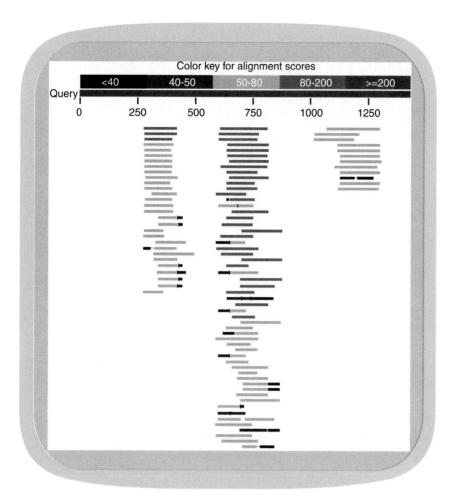

FIGURE 4.28. *Are human transcripts expressed that encode proteins homologous to HIV-1 pol protein? The results of a tblastn search with viral pol protein against a human EST database are shown. Many human genes are actively transcribed to generate transcripts predicted to make proteins homologous to HIV-1 pol.*

(Fig. 4.27). Are these human genes expressed? If so, they should produce RNA transcripts that may be characterized as ESTs from cDNA libraries. Perform a search of human ESTs with the viral pol protein; it is necessary to use the translating BLAST website with the tblastn algorithm, and the database must be set to EST (Fig. 4.28). There are hundreds of human transcripts, actively transcribed, that are predicted to encode proteins homologous to viral pol (Fig. 4.29). In Chapter 8, we will see how to evaluate these human ESTs to determine where in the body they are expressed and when during development they are expressed.

Could the human ESTs that are homologous to HIV-1 *pol* be even more closely related to other viral *pol* genes? To answer this question, select a human EST that we found is related to HIV-1 *pol* (from Fig. 4.29; we will choose accession AI636743). Perform the computationally intensive tblastx search using this EST's accession as an input and restrict the organism of the search to viruses. At the present time, this search results in the identification of many endogenous human retroviral sequences (i.e., DNA sequences that are part of the human genome) that are mistakenly left in the viruses division of GenBank. The virus that is most closely related to this particular human EST is a virus that afflicts sheep (accession AF105220). We initially performed a BLAST search with an HIV query and have used a further series of BLAST searches to gain insight into the biology of HIV-1 *pol*.

(a)

```
                                                                      Score    E
        Sequences producing significant alignments:                  (Bits) Value

        emb|BX509809.1|  DKFZp686K0981_r1 686 (synonym: hlcc3) Homo sa...  122   4e-26
        dbj|DA092449.1|  DA092449 BRACE3 Homo sapiens cDNA clone BRACE...  120   2e-25
        gb|CX166029.1|   HESC2_35_F01.g1_A035 NIH_MGC_258 Homo sapiens ...  119   4e-25
        gb|AI636743.1|   ts89e05.x1 NCI_CGAP_GC6 Homo sapiens cDNA clon...  114   2e-23
        gb|BE467547.1|   hz72c07.x1 NCI_CGAP_Lu24 Homo sapiens cDNA clo...  114   4e-22
        dbj|DA170606.1|  DA170606 BRAMY2 Homo sapiens cDNA clone BRAMY...   109   5e-22
        emb|BX481216.1|  DKFZp686B20225_r1 686 (synonym: hlcc3) Homo s...   105   7e-21
        gb|BF905435.1|   IL3-MT0267-261200-410-C08 MT0267 Homo sapiens ...  103   2e-20
        gb|AI680833.1|   tx35c11.x1 NCI_CGAP_Lu24 Homo sapiens cDNA clo...  103   3e-20
        gb|AI633818.1|   tt28h04.x1 NCI_CGAP_GC6 Homo sapiens cDNA clon...  103   4e-20
        gb|BE466420.1|   hz21e05.x1 NCI_CGAP_GC6 Homo sapiens cDNA clon...  100   2e-19
        dbj|DB152087.1|  DB152087 THYMU3 Homo sapiens cDNA clone THYMU...   97.4  2e-18
        gb|BG674607.1|   602620646F1 NCI_CGAP_Skn3 Homo sapiens cDNA cl...   95.1  1e-17
        gb|BF062180.1|   7k73e04.x1 NCI_CGAP_GC6 Homo sapiens cDNA clon...   92.4  2e-17
        gb|BU428879.1|   UI-HF-BNO-aed-g-12-0-UI.r1 NIH_MGC_50 Homo sap...   92.4  5e-17
        gb|BE550633.1|   7a30g04.x1 NCI_CGAP_GC6 Homo sapiens cDNA clon...   92.0  8e-17
        gb|AA584179.1|   nolla02.s1 NCI_CGAP_Phel Homo sapiens cDNA clo...   90.5  2e-16
        gb|BF590446.1|   naa38g04.x1 NCI_CGAP_Kid11 Homo sapiens cDNA c...   89.7  4e-16
        emb|BX955997.1|  DKFZp781F1975_r1 781 (synonym: hlcc4) Homo sa...    88.6  9e-16
        gb|AA740931.1|   oc89d02.s1 NCI_CGAP_GCB1 Homo sapiens cDNA clo...   88.6  9e-16
```

(b)

```
        >emb|BX509809.1|  DKFZp686K0981_r1 686 (synonym: hlcc3) Homo sapiens cDNA clone
        DKFZp686K0981 5', mRNA sequence.
        Length=736

         Score =  122 bits (307), Expect = 4e-26
         Identities = 75/213 (35%), Positives = 115/213 (53%), Gaps = 14/213 (6%)
         Frame = -3

        Query  611  WPLTEEKIKALVEICTEMEKEGKISKIGPENPYNTPVFAIKKKDSTKWRKLVDFRELNKR  670
                    WPL++EK++AL ++ TE  ++G  I  I   +P N+PVF IKKK S KWR L D R +N
        Sbjct  641  WPLSKEKLEALEDLVTEQLEKGHIVPIF--SP*NSPVFVIKKK-SGKWRMLTDLRAINSV  471

        Query  671  TQDFWEVQLGIPHPAGLKKKKSVTVLDVGDAYFSVPLDEDFRKYTAFTIPSINNETPGIR  730
                    Q    +Q G+P PA ++ +   + V+D+ D +F++PL E    AFTI ++NN  P
        Sbjct  470  IQPMGTLQPGLPSPAMIPRNWPLIVIDLKDCFFTIPLAEQDCEWFAFTILAVNNLQPAKH  291

        Query  731  YQYNVLPQGWKGSPAIFQSSMTKILEPFRKQNPDIVIYQYMDDLYVGSD-----LEIGQH  785
                    + + VLPQG  SP I Q+ + + +EP K+   I  YMDD+  +    L+   H
        Sbjct  290  FHWKVLPQGMLNSPTICQTYIGQAIEPTLKKFSQCYIIHYMDDILCAAPTREILLQCYDH  111

        Query  786  RTKIEELRQHLLRWGLTTPDKKHQKEPPFLWMG  818
                    L+ +   GL    K Q   P+ ++G
        Sbjct  110  ------LQNSISHTGLIIAPDKIQTTTPYSYLG  30
```

FIGURE 4.29. (a) Result of a tblastn search showing human ESTs matching HIV-1 pol. Many ESTs encode proteins having high scoring matches to the query (low E values). (b) The highest-scoring human match shares 35% amino acid identity.

PERSPECTIVE

BLAST searching has emerged as an indispensable tool to analyze the relation of a DNA or protein sequence to millions of sequences in public databases. All database search tools confront the issues of sensitivity (i.e., the ability to minimize false negative results), selectivity (i.e., the ability to minimize false positive results), and time. As the size of the public databases has grown exponentially in recent years, the BLAST tools have evolved to provide a rapid, reliable way to screen the databases. For protein searches we have focused on blastp. However, for most biologists performing even routine searches with a protein query, the PSI-BLAST program described in Chapter 5 is strongly preferred. This is because of its more optimally constructed scoring matrices.

PITFALLS

There are several common pitfalls to avoid in BLAST searching. The most common error among novice BLAST users is to search protein or DNA sequences against the wrong database. It is also important to understand the basic BLAST algorithms. These concepts are summarized in Fig. 4.3.

An important issue in BLAST searching is deciding whether an alignment is significant. Each potential BLAST match should be compared to the query sequence to evaluate whether it is reasonable from both a statistical and a biological point of view. It is more likely that two proteins are homologous if they share similar domain architecture (i.e., motifs or domains; Chapter 10).

WEB RESOURCES

The main website for BLAST searching is that of the National Center for Biotechnology Information (► http://www.ncbi.nlm.nih.gov/blast/). Within this site are links to the main programs (blastn, blastp, blastx, tblastn, and tblastx). There are other specialized BLAST programs at NCBI, which are discussed in Chapter 5. As an alternative approach, one may query other databases that employ position-specific scoring matrices (rather than fixed matrices such as BLOSUM62). These database search tools are also discussed in Chapter 5.

An important web resource is the set of BLAST tutorials, courses, and references available at the NCBI BLAST site.

DISCUSSION QUESTIONS

[4-1] Why doesn't anyone offer "Basic Global Alignment Search Tool" (BGAST) to complement BLAST? Would BGAST be a useful tool? What computational difficulties might there be in setting it up?

[4-2] Should you consider a significant expect value to be 1, 0.05, or 10^{-5}? Does this depend on the particular search you are doing?

[4-3] Why is it that database programs such as BLAST must make a trade-off between sensitivity and selectivity? How does the blastp algorithm address this issue? Refer to Altschul et al. (1990).

COMPUTER LAB/PROBLEMS

[4-1] Perform a blastp search at NCBI using the following query of just 12 amino acids: PNLHGLFGRKTG. By default, the parameters are adjusted for short queries. Inspect the search summary of the output. What is the E value cutoff? What is the word size? What is the scoring matrix? How do these settings compare to the default parameters?

[4-2] Protein searches are usually more informative than DNA searches. Do a blastp search using RBP4 (NP_006735), restricting the output to Arthropoda (insects). Next, do a blastn search using the RBP4 nucleotide sequence (NM_006744; select only the nucleotides corresponding to the coding region of the DNA). Which search is more informative? How many databases matches have an E value less than 1.0 in each search?

[4-3] The NCBI BLAST site offers the netblast program. Download this and locally install it (on a Windows, Apple, or Linux platform). This process takes only a matter of seconds, and the download includes extensive documentation and examples. Netblast operates from the command line only. First, save a text file (called hbb.txt) containing human beta globin protein in the fasta format. This is available as web document 4.5 at ► http://www.bioinfbook.org/chapter4. Next, invoke a command line editor and execute a search of the nonredundant (nr) database as follows: ">blastcl3 -i hbb.txt -p blastp -d nr -o hbb1.txt" (do not use the quotation marks, and the >

sign indicates a command editor prompt). blastcl3 is the program; the four required parameters are -i (for the input or query file), -p (for the program), -d (the database), and -o (the output file; here it is called hbb1.txt). The results are returned as a text file in the folder in which you installed netblast. Note the E value of the best match, and also note the effective length of the search space (shown in the search summary of the output). Repeat this search adding the command -Y 40000000. This changes the effective length of the search space to 40,000,000. What is the E value of the best match, using this search space?

[4-4] Netblast is useful to do batch queries. Create a text file named 3proteins.txt having three protein sequences: human beta globin, bovine odorant-binding protein, and cytochrome b from the malaria parasite *Plasmodium falciparum*. (These are available at web document 4.6.) Enter the search "blastcl3 -i 3proteins.txt -p blastp -d refseq_protein -o 3proteins_out.txt"; this performs blastp against the RefSeq protein database. The output file includes the results of three separate blastp searches.

[4-5] The largest gene family in humans is said to be the olfactory receptor family. Do a BLAST search to evaluate how large the family is. *Hint:* As one strategy, first go to Entrez Gene and enter "olfactory receptor" limiting the organism to *Homo sapiens*. There are over 2600 entries, but this does not tell you

whether they are related to each other. Select one accession number and perform a blastp search restricting the organism to human.

[4-6] For the search you just performed in problem 4.5, what happens if you use a scoring matrix that is more suited to finding distantly related proteins?

[4-7] Is the pol protein of HIV-1 more closely related to the pol protein of HIV-2 or to the pol protein of simian immunodeficiency virus (SIV)? Use the blastp program to decide. *Hint*: try the Entrez

command "NOT hiv-1[organism]" to focus the search away from HIV-1 matches.

[4-8] "The Iceman" is a man who lived 5300 years ago and whose body was recovered from the Italian Alps in 1991. Some fungal material was recovered from his clothing and sequenced. To what modern species is the fungal DNA most related?

[4-9] You perform a BLAST search and a result has an E value of about 1×10^{-4}. What does this E value mean? What are some parameters on which an E value depends?

SELF-TEST QUIZ

[4-1] You have a reasonably short, typical, double-stranded DNA sequence. Basically, how many proteins can it *potentially* encode?

(a) 1

(b) 2

(c) 3

(d) 6

[4-2] You have a DNA sequence. You want to know which protein in the main protein database ("nr," the nonredundant database) is most similar to some protein encoded by your DNA. Which program should you use?

(a) blastn

(b) blastp

(c) blastx

(d) tblastn

(e) tblastx

[4-3] Which output from a BLAST search provides an estimate of the number of false positives from a BLAST search?

(a) E value

(b) Bit score

(c) Percent identity

(d) Percent positives

[4-4] Match up the following BLAST search programs with their correct descriptions:

blastp_(a) Nucleotide query against a nucleotide sequence database

blastn_(b) Protein query against a translated nucleotide sequence database

blastx_(c) Translated nucleotide query against a protein database

tblastn_(d) Protein query against a protein database

tblastx_(e) Translated nucleotide query against a translated nucleotide database

[4-5] Changing which of the following BLAST parameters would tend to yield fewer search results?

(a) Turning off the low-complexity filter

(b) Changing the expect value from 1 to 10

(c) Raising the threshold value

(d) Changing the scoring matrix from PAM30 to PAM70

[4-6] You can limit a BLAST search using any Entrez term. For example, you can limit the results to those containing a researcher's name.

(a) True

(b) False

[4-7] An extreme value distribution:

(a) Describes the distribution of scores from a query against a database

(b) Has a larger total area than a normal distribution

(c) Is symmetric

(d) Has a shape that is described by two constants: μ (mu, the mean) and λ (a decay constant)

[4-8] As the E value of a BLAST search becomes smaller:

(a) The value K also becomes smaller

(b) The score tends to be larger

(c) The probability P tends to be larger

(d) The extreme value distribution becomes less skewed

[4-9] The BLAST algorithm compiles a list of "words" typically of three amino acids (for a protein search). Words at or above a threshold value T are defined as:

(a) "Hits" and are used to scan a database for exact matches that may then be extended

(b) Hits and are used to scan a database for exact or partial matches that may then be extended

(c) Hits and are aligned to each other

(d) Hits and are reported as raw scores

[4-10] Normalized BLAST scores (also called bit scores):

(a) Are unitless

(b) Are not related to the scoring matrix that is used

(c) Can be compared between different BLAST searches, even if different scoring matrices are used

(d) Can be compared between different BLAST searches, but only if the same scoring matrices are used

SUGGESTED READING

BLAST searching was introduced in a classic paper by Stephen Altschul, David Lipman, and colleagues (1990). This paper describes the theoretical basis for BLAST searching and describes basic issues of BLAST performance, including sensitivity (accuracy) and speed. Fundamental modifications to the original BLAST algorithm were later introduced, including the introduction of gapped BLAST (Altschul et al., 1997). This paper includes a discussion of specialized position-specific scoring matrices that we will consider in Chapter 5.

William Pearson 1996 provides an excellent description of database searching in an article entitled "Effective Protein Sequence Comparison." Altschul et al. (1994) provide a highly recommended article, "Issues in Searching Molecular Sequence Databases." Marco Pagni and C. Victor Jongeneel (2001) of the Swiss Institute of Bioinformatics provide a technical overview of sequence alignment statistics. This article includes sections on the extreme value distribution, the use of random sequences, local alignment with and without gaps, and BLAST statistics. Another excellent review of alignment statistics was written by Stephen Altschul and Warren Gish (1996).

REFERENCES

Altschul, S. F., Boguski, M. S., Gish, W., and Wootton, J. C. Issues in searching molecular sequence databases. *Nat. Genet.* **6**, 119–129 (1994).

Altschul, S. F., and Gish, W. Local alignment statistics. *Methods Enzymol.* **266**, 460–480 (1996).

Altschul, S. F., Gish, W., Miller, W., Myers, E. W., and Lipman, D. J. Basic Local Alignment Search Tool. *J. Mol. Biol.* **215**, 403–410 (1990).

Altschul, S. F., et al. Gapped BLAST and PSI-BLAST: A new generation of protein database search programs. *Nucleic Acids Res.* **25**, 3389–3402 (1997).

Altschul, S. F., Wootton, J. C., Gertz, E. M., Agarwala, R., Morgulis, A., Schaffer, A. A., and Yu, Y.-K. Protein database searches using compositionally adjusted substitution matrices. *FEBS J.* **272**, 5101–5109 (2005).

Brenner, S. E. Practical database searching. *Bioinformatics: A Trends Guide* **1998**, 9–12 (1998).

Chambaud, I., Heilig, R., Ferris, S., Barbe, V., Samson, D., Galisson, F., Moszer, I., Dybvig, K., Wroblewski, H., Viari, A., Rocha, E. P., and Blanchard, A. The complete genome sequence of the murine respiratory pathogen *Mycoplasma pulmonis*. *Nucleic Acids Res.* **29**, 2145–2153 (2001).

Dixon, R. A., et al. Cloning of the gene and cDNA for mammalian beta-adrenergic receptor and homology with rhodopsin. *Nature* **321**, 75–79 (1986).

Downward, J., et al. Close similarity of epidermal growth factor receptor and v-*erb*-B oncogene protein sequences. *Nature* **307**, 521–527 (1984).

Ermolaeva, M. D., White, O., and Salzberg, S. L. Prediction of operons in microbial genomes. *Nucleic Acids Res.* **29**, 1216–1221 (2001).

Ferretti, J. J., et al. Complete genome sequence of an M1 strain of *Streptococcus pyogenes*. *Proc. Natl. Acad. Sci. USA* **98**, 4658–4663 (2001).

Gish, W., and States, D. J. Identification of protein coding regions by database similarity search. *Nat. Genet.* **3**, 266–272 (1993).

Gotoh, O. Significant improvement in accuracy of multiple protein sequence alignments by iterative refinement as assessed by reference to structural alignments. *J. Mol. Biol.* **264**, 823–838 (1996).

Gribskov, M., and Robinson, N. L. Use of receiver operating characteristic (ROC) analysis to evaluate sequence matching. *Comput. Chem.* **20**, 25–33 (1996).

Gumbel, E. J. *Statistics of Extremes.* Columbia University Press, New York, 1958.

International Human Genome Sequencing Consortium. Initial sequencing and analysis of the human genome. *Nature* **409**, 860–921 (2001).

Karlin, S., and Altschul, S. F. Methods for assessing the statistical significance of molecular sequence features by using general scoring schemes. *Proc. Natl. Acad. Sci. USA* **87**, 2264–2268 (1990).

Karlsson, H., et al. Retroviral RNA identified in the cerebrospinal fluids and brains of individuals with schizophrenia. *Proc. Natl. Acad. Sci. USA* **98**, 4634–4639 (2001).

Kaufman, K. M., and Sodetz, J. M. Genomic structure of the human complement protein C8γ: homology to the lipocalin gene family. *Biochemistry* **33**, 5162–5166 (1994).

Needleman, S. B., and Wunsch, C. D. A general method applicable to the search for similarities in the amino acid sequence of two proteins. *J. Mol. Biol.* **48**, 443–453 (1970).

Pagni, M., and Jongeneel, C. V. Making sense of score statistics for sequence alignments. *Brief. Bioinform.* **2**, 51–67 (2001).

Park, J., et al. Sequence comparisons using multiple sequences detect three times as many remote homologues as pairwise methods. *J. Mol. Biol.* **284**, 1201–1210 (1998).

Pearson, W. R. Effective protein sequence comparison. *Methods Enzymol.* **266**, 227–258 (1996).

Schäffer, A. A., Aravind, L., Madden, T. L., Shavirin, S., Spouge, J. L., Wolf, Y. I., Koonin, E.V., and Altschul, S. F. Improving the accuracy of PSI-BLAST protein database searches with composition-based statistics and other refinements. *Nucleic Acids Res.* **29**, 2994–3005 (2001).

Smith, T. F., and Waterman, M. S. Identification of common molecular subsequences. *J. Mol. Biol.* **147**, 195–197 (1981).

Tettelin, H., et al. Complete genome sequence of a virulent isolate of *Streptococcus pneumoniae*. *Science* **293**, 498–506 (2001).

Wootton, J. C., and Federhen, S. Analysis of compositionally biased regions in sequence databases. *Methods Enzymol.* **266**, 554–571 (1996).

Yu, Y.-K., and Altschul, S. F. The construction of amino acid substitution matrices for the comparison of proteins with nonstandard compositions. *Bioinformatics* **21**, 902–911 (2005).

Yu, Y.-K., Wootton, J. C., and Altschul, S. F. The compositional adjustment of amino acid substitution matrices. *Proc. Natl. Acad. Sci. USA* **100**, 15688–15693 (2003).

NOTE 8.   *Ultimate composition of fibrin from ox-blood.* (Mulder.)

					Atoms.	Calculated.
Carbon	.	.	.	54·56	400	54·90
Hydrogen	.	.	.	6·90	310	6·95
Nitrogen	.	.	.	15·72	50	15·89
Oxygen	.	.	.	22·13	120	21·55
Phosphorus	.	.	.	0·33	1	0·35
Sulphur	.	.	.	0·36	1	0·36

Hence, in its composition, it is identical with the albumen of eggs.

NOTE 9.   *Ultimate composition of casein from cows' milk.*
(Mulder.)

					Atoms.	Calculated.
Carbon	.	.	.	54·96	400	55·10
Hydrogen	.	.	.	7·15	310	6·97
Nitrogen	.	.	.	15·80	50	15·95
Oxygen	.	.	.	21·73	120	21·62
Sulphur	.	.	.	0·36	1	0·36

NOTE 10.   *Ultimate composition of crystallin from the eye.*
(Mulder.)

Carbon	.	.	55·39	
Hydrogen	.	.	6·94	
Nitrogen	.	.	16·51	hence it closely resembles casein.
Oxygen	.	.	20·91	
Sulphur	.	.	0·25	

NOTE 11.   *Ultimate composition of globulin.*

The analysis referred to in the text was published by Mulder in the 'Bulletin' for 1839. In his recent work on the 'Chemistry of Animal and Vegetable Physiology,' he states that, although a protein-compound, its real composition is not yet known.

NOTE 12.   *Ultimate composition of pepsin.* (Vogel.)

Carbon	.	.	.	57·718
Hydrogen	.	.	.	5·666
Nitrogen	.	.	.	21·088
Oxygen	.	.	.	16·064

Mulder and other biochemists in the 1830s and 1840s hypothesized that all proteins had the same composition of carbon, hydrogen, nitrogen, oxygen, phosphorus, and sulfur. Simon (1846) summarized the composition of known proteins, including fibrin, casein, crystalline, and pepsin. He noted that the composition of globulin (i.e., hemoglobin) was not yet known.

5

Advanced Database Searching

INTRODUCTION

In Chapters 3 and 4 we introduced pairwise alignments and BLAST searching. BLAST searching allows one to search a database to find what proteins or genes are present. BLAST searches can be very versatile, and in this chapter we will cover several advanced database searching techniques.

Let us introduce two problems for which the five main NCBI BLAST programs are not sufficient. (1) We know that myoglobin is homologous to alpha globin and beta globin; all are vertebrate members of a globin superfamily. We have seen in Fig. 3.1 that myoglobin shares a very similar three-dimensional structure with alpha and beta globin. However, if you use beta globin (NP_000509) as a query and perform a blastp search (restricting the output to human and setting the database to nr [nonredundant] or RefSeq), myoglobin does not appear in the results. Fortunately there are programs such as PSI-BLAST and HMMER that can easily find such homologous but distantly related proteins. (2) Suppose we want to compare very long query sequences (e.g., 20,000 base pairs or more) against a database. We might also want to perform a pairwise alignment between two long sequences, such as human chromosome 20 (62 million base pairs long) versus mouse chromosome 2. We will need an algorithm that is faster than blastn, and we need to explore both global and local strategies. For this problem we can expect some regions of the alignment to have regions of high conservation, but other regions will have diverged substantially. Finding solutions to such searching and alignment problems becomes

Using human myoglobin (NP_005359) as a query in a blastp result against human RefSeq proteins, beta globin does not appear.

Bioinformatics and Functional Genomics, Second Edition. By Jonathan Pevsner

more critical with the very recent availability of thousands of completed genome sequences.

We begin with a brief overview of the kinds of specialized BLAST resources that are available to help solve many kinds of research questions. Next, we will introduce PSI-BLAST and hidden Markov models as tools to find distantly related proteins. We will then consider BLAST-like tools for the alignment of genomic DNA. We will also address the problem of how to discover novel genes using BLAST searches.

SPECIALIZED BLAST SITES

So far, we have used two BLAST resources, both from the NCBI website: pairwise BLAST (Chapter 3) and the five standard BLAST programs (Chapter 4). Other, related programs are available, including organism-specific BLAST sites, BLAST sites that allow searches of specific molecules, and specialized database search algorithms.

Organism-Specific BLAST Sites

You can access the Map Viewer from the home page of NCBI or ▶ http://www.ncbi.nlm.nih.gov/mapview/.

We have seen that for standard BLAST searches at the NCBI website the output can be restricted to a particular organism. BLAST searches focused on dozens of prominent organisms can also be performed through the NCBI Map Viewer site.

Web document 5.1 at ▶ http://www.bioinfbook.org/chapter5 lists several dozen organism-specific BLAST servers.

There are many other databases that consist of molecular sequence data from a specific organism, and many of these offer organism-specific BLAST servers. In some cases the data include unfinished sequences that have not yet been deposited in GenBank. If you have a protein or DNA sequence with no apparent matches in standard NCBI BLAST searches, then searching these specialized databases can provide a more exhaustive search. Also, as described below, some of these databases present unique output formats and/or search algorithms.

Ensembl BLAST

The Wellcome Trust Sanger Institute website is ▶ http://www.sanger.ac.uk/. The EBI is at ▶ http://www.ebi.ac.uk/. Ensembl's human BLAST server is at ▶ http://www.ensembl.org/Homo_sapiens/blastview, and Ensembl BLAST servers for mouse and other organisms can also be found through ▶ http://www.ensembl.org/.

Project Ensembl is a joint effort of the Wellcome Trust Sanger Institute (WTSI) and the European Bioinformatics Institute (EBI). The Ensembl website provides a comprehensive resource for studying the human genome and other genomes (see Chapters 16, 18, and 19). The Ensembl BLAST server allows the user to search the Ensembl database. As an example, paste in the FASTA-formatted amino acid sequence of human beta globin (accession NP_000509) and perform a tblastn search. The output also consists of a graphical output showing the location of the database matches by chromosome (Fig. 5.1). This conveniently shows the chromosomal location of the best hits, including chromosome 11 for beta globin. An alignment summary is provided (Fig. 5.2) with an emphasis on genomic loci. One can see reasonably high-scoring matches to chromosome 16, corresponding to alpha globin. The output links include pairwise alignments between the query and each match, and a link to the ContigView. That is the genome browser consisting of assorted graphics and dozens of fields of information (e.g., an ideogram of the chromosome band, a view of neighboring genes, protein and DNA database links, polymorphisms, mouse homologies, and expression data).

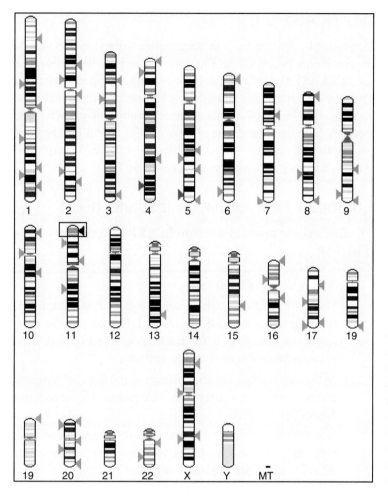

FIGURE 5.1. *Output of a BLAST search of the Ensembl database using human beta globin as a query. The results are presented in a graphical format by chromosome, showing the best match to the short arm of chromosome 11 (boxed). Weaker matches to paralogs on other chromosomes are also evident.*

Wellcome Trust Sanger Institute

The WTSI has a major role in genome sequencing. Its website offers BLAST searches specific for over 100 organisms. There are also BLAST servers for the Vertebrate Genome Annotation (VEGA) project that focuses on high quality, manual annotation of selected vertebrate genomes (currently human, mouse, dog, pig, and zebrafish) (Ashurst et al., 2005; Loveland, 2005).

The WSTI blast resources are available at ► http://www.sanger.ac.uk/DataSearch/blast.shtml. The VEGA homepage is ► http://vega.sanger.ac.uk/.

Specialized BLAST-Related Algorithms

We have focused on the standard BLAST algorithms at NCBI, but many other algorithms are available.

Links	Query			Chromosome				Chromosome				Stats				
	Start	End	Ori	Name	Start	End	Ori	Name	Start	End	Ori	Score	E-val	P-val	%ID	Length
[A] [S] [G] [C]	31	106	+	Chr:11	5204380	5204607	-	Chr:11	5204380	5204607	-	652	1.2e-100	1.2e-100	98.68	76
[A] [S] [G] [C]	31	124	+	Chr:11	5211731	5212021	-	Chr:11	5211731	5212021	-	646	4.4e-70	4.4e-70	81.63	98
[A] [S] [G] [C]	31	110	+	Chr:11	5232080	5232322	-	Chr:11	5232080	5232322	-	532	7.4e-89	7.4e-89	75.31	81
[A] [S] [G] [C]	13	121	+	Chr:11	5247182	5247556	-	Chr:11	5247182	5247556	-	529	2.4e-43	2.4e-43	56.25	128
[A] [S] [G] [C]	31	110	+	Chr:11	5227156	5227398	-	Chr:11	5227156	5227398	-	527	2.3e-88	2.3e-88	75.31	81
[A] [S] [G] [C]	32	104	+	Chr:11	5220915	5221133	-	Chr:11	5220915	5221133	-	436	1.9e-79	1.9e-79	72.97	74
[A] [S] [G] [C]	101	147	+	Chr:11	5203407	5203538	-	Chr:11	5203407	5203538	-	360	1.2e-100	1.2e-100	91.49	47
[A] [S] [G] [C]	65	147	+	Chr:11	5210773	5210994	-	Chr:11	5210773	5210994	-	323	1.4e-38	1.4e-38	55.95	84
[A] [S] [G] [C]	1	45	+	Chr:11	5204699	5204827	-	Chr:11	5204699	5204827	-	272	1.8e-45	1.8e-45	80.00	45
[A] [S] [G] [C]	105	147	+	Chr:11	5246278	5246406	-	Chr:11	5246278	5246406	-	266	2.3e-29	2.3e-29	74.42	43
[A] [S] [G] [C]	65	147	+	Chr:11	5231086	5231304	-	Chr:11	5231086	5231304	-	263	4.7e-29	4.7e-29	50.59	85
[A] [S] [G] [C]	31	143	+	Chr:16	166926	167237	+	Chr:16	166926	167237	+	260	1.2e-19	1.2e-19	35.54	121
[A] [S] [G] [C]	31	143	+	Chr:16	163122	163433	+	Chr:16	163122	163433	+	256	3.1e-19	3.1e-19	35.59	118

FIGURE 5.2. *The Ensembl BLAST server (available at ► http://www.ensembl.org/) provides an output summary with scores, E values, and links to pairwise alignments (labeled A), the query sequence (S), the genome (matching) sequence (G), and an Ensembl ContigView (C).*

WU BLAST 2.0

You can learn more about WU BLAST 2.0 at ▶ http://blast.wustl.edu. This includes a comparison of its function to the NCBI BLAST tools.

Developed by Warren Gish at Washington University, WU BLAST 2.0 is related to the traditional NCBI BLAST algorithms, as both were developed from the original NCBI BLAST algorithms that did not permit gapped alignments. WU BLAST 2.0 may provide faster speed and increased sensitivity, and it includes a variety of options, such as a full Smith–Waterman alignment on some pairwise alignments of database matches. The command line version of WU BLAST 2.0 offers dozens of options, comparable in scope to the client blastall at NCBI (Chapter 4). WU BLAST 2.0 runs on a variety of computer servers (Table 5.1).

European Bioinformatics Institute (EBI)

The EBI website provides access to BLAST and other related database search tools (Fig. 5.3):

You can also access MPsearch through the EBI Sequence and Retrieval System (SRS) site at ▶ http://srs.ebi.ac.uk/. Look under tools for similiarity search tools.

- The WU BLAST 2.0 tools
- FASTA (fAST-All), which we introduced in Chapter 3, is, like BLAST, a heuristic algorithm for searching DNA or protein databases. The default word size (ktup) is 6 for nucleotide searches and 2 for protein searches. A larger ktup is faster but less sensitive.
- MPsearch is a fast implementation of the Smith–Waterman algorithm. While the run time is relatively slow, this provides a more sensitive algorithm than BLAST or FASTA.

ScanPS was written by Geoffrey J. Barton of the University of Dundee.

- ScanPS (Scan Protein Sequence) also implements the Smith–Waterman algorithm. It also includes iterative profile searching (similar to PSI-BLAST, which we introduce below).
- PSI-BLAST and PHI-BLAST (introduced below)

The alternative splicing database (Stamm et al., 2006) is available at ▶ http://www.ebi.ac.uk/asd/.

- BLAST and/or FASTA searches of specialized databases are offered, such as the Alternative Splicing Database (ASD), a ligand gated ion channel database, and a single nucleotide polymporphism database.

Specialized NCBI BLAST Sites

The main BLAST site at NCBI offers access to specialized searches of immunoglobulins, vectors, single nucleotide polymorphisms (SNPs; see Chapter 16), or the trace archives of raw genomic sequence (see Chapter 13). For example, IgBLAST reports the three germline V genes, two D and two J genes that show the closest match to the query sequence.

TABLE 5-1 Examples of Servers Running WU-BLAST 2.0

Site	URL
Institut Pasteur	http://bioweb.pasteur.fr/seqanal/interfaces/wublast2.html
European Bioinformatics Institute (EBI)	http://www.ebi.ac.uk/Tools/similarity.html
Saccharomyces Genome Database (SGD)	http://seq.yeastgenome.org/cgi-bin/blast-sgd.pl

General DNA and Protein Searches

Tool	Description
Blast2-WU Protein ⓘ	Washington University (WU) blast2 for protein databases. (blast 2.0 with gaps)
Blast2-WU Nucleotide ⓘ	Washington University (WU) blast2 for nucleotide databases. (blast 2.0 with gaps)
Blast2-NCBI Protein ⓘ	NCBI blast2 (blastall) program for protein databases.
Blast2-NCBI Nucleotide ⓘ	NCBI blast2 (blastall) program for nucleotide databases.
Blast2-NCBI EVEC ⓘ	European blast2 Vector Searches. Check your sequences for vector contamination.
PSI-Blast	Position specific iterative **Blast** (**PSI-Blast**) refers to a feature of Blast 2.0 in which a profile is automatically constructed from the first set of Blast alignments.
PHI-Blast	Pattern Hit Initiated **Blast** (**PHI-Blast**) treats two occurrence of the same pattern within the query sequence as two independent sequences.
Fasta Nucleotide ⓘ	Sequence similarity searching against nucleotide databases using Fasta.
Fasta Protein ⓘ	Sequence similarity searching against protein databases using Fasta.
Fasta-Proteome server ⓘ	Completed Proteomes Fasta server.
Fasta-Genome server ⓘ	Completed Genomes Fasta server.
Fasta-WGS server ⓘ	Whole genome shotgun (WGS) Fasta server.

Rigorous Protein Searches

Tool	Description
MPsrch ⓘ	Anedabio, formerly Edinburgh Biocomputing Systems' very fast implementation of the true Smith and Waterman algorithm.
Scanps2.3 ⓘ	NEW! Version 2.3 of Scanps Fast implementation of the true Smith & Waterman algorithm for protein database searches.

FIGURE 5.3. *The European Bioinformatics Institute (EBI;* ► *http://www.ebi.ac.uk/) offers a variety of programs for searching DNA and protein sequences, including BLAST and more rigorous (but slower) Smith–Waterman implementations.*

FINDING DISTANTLY RELATED PROTEINS: POSITION-SPECIFIC ITERATED BLAST (PSI-BLAST)

Many homologous proteins share only limited sequence identity. Such proteins may adopt the same three-dimensional structures (based on methods such as x-ray crystallography), but in pairwise alignments they may have no apparent similarity. We have seen that scoring matrices are sensitive to protein matches at various evolutionary distances. For example, we compared the PAM250 to the PAM10 log-odds matrices (Figs. 3.14 and 3.15) and saw that the PAM250 matrix provides a superior scoring system for the detection of distantly related proteins. In performing a database search with BLAST, we can adjust the scoring matrix to try to detect distantly related proteins. Even so, many proteins in a database are too distantly related to a query to be detected using a standard blastp search. In many other cases protein matches are detected but are so distant that the inference of homology is unclear. We saw that a blastp search using RBP4 as a query returned a statistically marginal match to complement component 8 gamma (Fig. 4.19). We would like an algorithm that can assign statistical significance to distantly related proteins that are true positives, while minimizing the numbers of both false positive results (e.g., reporting two proteins as related when they are not) and false negative results (e.g., failing to report that two proteins are significantly related, as was done by blastp).

You can access PSI-BLAST at ▶ http://www.ncbi.nlm.nih.gov/ BLAST and at other servers such as the Pasteur Institute (▶ http:// bioweb.pasteur.fr/seqanal/ interfaces/psiblast.html) and CMBI, Netherlands (▶ http:// www.cmbi.kun.nl/bioinf/tools/ psiblast.shtml).

We have seen a multiple sequence alignment from a BLAST output in Fig. 4.12, and we will examine this topic in Chapter 6.

PSSM is sometimes pronounced "possum."

Position-specific iterated BLAST (abbreviated PSI-BLAST or ψ-BLAST) is a specialized kind of BLAST search that is often more sensitive than a regular BLAST search (Zhang et al., 1998; Altschul et al., 1997; Schaffer et al., 2001). The purpose of using PSI-BLAST is to look deeper into the database to find distantly related proteins that match your protein of interest. In many cases, when a complete genome is sequenced and the predicted proteins are analyzed to search for homologs, PSI-BLAST is the algorithm of choice.

PSI-BLAST is performed in five steps:

1. A normal blastp search uses a scoring matrix (such as BLOSUM62, the default scoring matrix) to perform pairwise alignments of your query sequence (such as RBP) against the database. PSI-BLAST also begins with a protein query that is searched against a database at the NCBI website.

2. PSI-BLAST constructs a multiple sequence alignment from an initial blastp-like search using composition-based statistics (Schaffer et al., 2001). It then creates a specialized, individualized search matrix (also called a profile) based on that multiple alignment.

3. This position-specific scoring matrix (PSSM) is then used as a query to search the database again. (Your original query is not used.)

4. PSI-BLAST estimates the statistical significance of the database matches, essentially using the parameters we described for gapped alignments.

5. The search process is continued iteratively, typically about five times. At each step a new profile is used as the query. You must decide how many iterations to perform; simply press the button labeled "Run PSI-BLAST Iteration." You can stop the search process at any point—whenever few new results are returned or when the program reports convergence because no new results are found.

You can adjust the inclusion threshold. Try E values of 0.5 or 0.00005 to see what happens to your search results. If you set the E value too low, you will only see very closely related homologs. If you set E too high, you will probably find false positive matches.

We can illustrate the dramatic results of the PSI-BLAST process as follows. Go to the protein blast page at NCBI, enter the protein accession number of human RBP4 (NP_006735), and select the PSI-BLAST option and the RefSeq database. Using the default parameters, there are about 100 hits (as of October 2007; your results will likely vary) (Table 5.2). About half of these have significant E value lower than the inclusion threshold (set as a default at $E = 0.005$), and by inspection these are all called RBP (from various species) or apolipoprotein D (another lipocalin). There are also dozens of database matches that are worse than the inclusion

TABLE 5-2 PSI-BLAST Produces Dramatically More Hits With Significant E Values Than blastp

Iteration	Hits with $E < 0.005$	Hits with $E > 0.005$
1	34	61
2	314	79
3	416	57
4	432	50
5	432	50

Human retinol-binding protein 4 (RBP4; NP_006735) was used as a query in a PSI-BLAST search of the RefSeq database.

threshold: These do not have significant *E* values. Some of these distantly related matches (such as insecticyanins) are authentic lipocalins, based on criteria such as having similar three-dimensional structures and related biological functions as carrier proteins. Other proteins on this list are viral and appear to be true negatives.

Through this initial step, the PSI-BLAST search is performed in a manner nearly identical to a standard blastp search, using some amino acid substitution matrix such as BLOSUM62. However, the program creates a multiple sequence alignment from the initial database matches. By analyzing this alignment, the PSI-BLAST program then creates a PSSM. The original query sequence serves as a template for this profile.

Consider a portion of the BLAST output so far viewed as a multiple sequence alignment (Fig. 5.4). In one column, the amino acid residues R, I, and K are found. Substitutions of R and K are quite common in general, but it is rare to substitute I for either of these basic residues. The key idea of PSI-BLAST is that the aligned residues at each position of the multiple sequence alignment provide information about the unique profile of accepted mutations in the query and its nearest database matches. That information forms the basis of a matrix that can be used to search the database with more sensitivity than a standard BLOSUM or PAM matrix can provide.

For a query of length *L*, PSI-BLAST generates a PSSM of dimension $L \times 20$. The rows of each matrix have a length *L* equal to the query sequence. Redundant sequences (having at least 94% amino acid identity in a pairwise alignment of any two sequences in the matrix) are eliminated. This ensures that a group of very closely related sequences will not overly bias the construction of the PSSM. The same gap scores are applied as in blastp, rather than implementing position-specific gap scores. A unique scoring matrix (profile) is derived from the multiple sequence alignment (Box 5.1). For each iteration of PSI-BLAST, a separate scoring matrix is created.

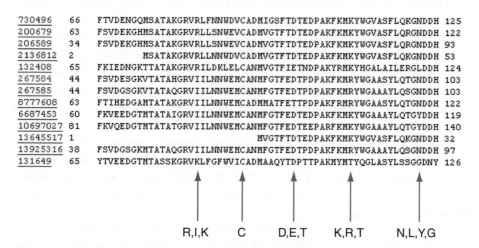

FIGURE 5.4. PSI-BLAST search begins with a standard blastp-like search. The output is used to generate a profile or PSSM. A PAM or BLOSUM matrix describes the likelihood that one amino acid will be substituted for another, based on a statistical analysis of thousands of proteins. The PSSM is created specifically for the protein query of the PSI-BLAST search. The figure shows a portion of the initial PSI-BLAST output (shown using a query anchored without identities alignment view). The arrows point to examples of columns of amino acids in the alignment and the actual amino acid residues that are tolerated in each position. Some of the positions are invariant (such as C), while other columns show aligned residues (such as R, I, K) that tolerate amino acid substitutions that may be unique to this particular group of proteins.

Box 5.1
PSI-BLAST Target Frequencies

Scores are derived for each specific column position in the form $\log(q_i/p_i)$, where q_i is the estimated probability for residue i to be found in that column position and p_i is the background probability for that residue (Altschul et al., 1997). The key problem is to estimate the target frequencies q_i. This is accomplished in two steps using a method of pseudocounts (Tatusov et al., 1994). First, pseudocount frequencies g_i are obtained for each column position as follows:

$$g_i = \sum_j \frac{f_j}{p_j} q_{ij}$$

where f_j are the observed frequencies, p_j are the background frequencies, and q_{ij} are the target frequencies implicit in the substitution matrix (as described in Equation 3.1). Next, the target frequencies q_i (corresponding to the likelihood of observing residue i in the position of a column) are given by:

$$q_i = \frac{\alpha f_i + \beta g_i}{\alpha + \beta}$$

In this equation, α and β are relative weights assigned to the observed frequencies f_i and the pseudocount residue frequencies g_i. Having estimated the target frequencies, it is now possible to assign a score for a given aligned column as $(\ln[q_i/p_i]/\lambda)$.

A portion of a PSSM derived from a PSI-BLAST search using RBP4 as a query is shown in Fig. 5.5. The columns have the 20 common amino acids. Look at the scores given to alanine (at positions 6, 11, 12, 14–16, and 42). In some cases, alanine scores a +4 (this is also the score from a BLOSUM62 matrix, as shown in Fig. 3.17). However, alanine occasionally scores a +2, +3, or +5 at various positions in the query sequence, indicating that this residue is more or less favored at these positions. The scoring matrix is thus customized to a query. Next consider the tryptophan at position 40 that is part of the nearly invariant GXW pattern present in hundreds of lipocalins. This W (as well as several other tryptophans) scores a +12 in position 40, but a different W (at position 13) scores only a +7. These examples illustrate one of the main advantages of PSI-BLAST: the PSSM reflects a more customized estimate of the probabilities with which amino acid substitutions occur at various positions.

The unique profile that PSI-BLAST identifies is next used to perform an iterative search. Press the button "run PSI-BLAST iteration 2." The search is repeated using the customized profile, and new proteins are often added to the alignment. This is seen in the second iteration as the number of database hits better than the threshold rises from 34 to 314 (Table 5.2). In subsequent iterations, the number of database hits better than the threshold rose to over 400. By inspection, all of these are authentic members of the lipocalin family. One can halt the search once such a plateau is reached, or continue the iteration process until the program reports that convergence has been reached. This indicates that no more database matches are found, and the PSI-BLAST search is ended.

		A	R	N	D	C	Q	E	G	H	I	L	K	M	F	P	S	T	W	Y	V
1	M	-1	-2	-2	-3	-2	-1	-2	-3	-2	1	2	-2	6	0	-3	-2	-1	-2	-1	1
2	K	-1	1	0	1	-4	2	4	-2	0	-3	-3	3	-2	-4	-1	0	-1	-3	-2	-3
3	W	-3	-3	-4	-5	-3	-2	-3	-3	-3	-3	-2	-3	-2	1	-4	-3	-3	[12]	2	-3
4	V	0	-3	-3	-4	-1	-3	-3	-4	-4	3	1	-3	1	-1	-3	-2	0	-3	-1	4
5	W	-3	-3	-4	-5	-3	-2	-3	-3	-3	-3	-3	-3	-2	1	-4	-3	-3	[12]	2	-3
6	A	[5]	-2	-2	-2	-1	-1	-1	0	-2	-2	-2	-1	-1	-3	-1	1	0	-3	-2	0
7	L	-2	-2	-4	-4	-1	-2	-3	-4	-3	2	4	-3	2	0	-3	-3	-1	-2	-1	1
8	L	-1	-3	-3	-4	-1	-3	-3	-4	-3	2	2	-3	1	3	-3	-2	-1	-2	0	3
9	L	-1	-3	-4	-4	-1	-2	-3	-4	-3	2	4	-3	2	0	-3	-3	-1	-2	-1	2
10	L	-2	-2	-4	-4	-1	-2	-3	-4	-3	2	4	-3	2	0	-3	-3	-1	-2	-1	1
11	A	[5]	-2	-2	-2	-1	-1	-1	0	-2	-2	-2	-1	-1	-3	-1	1	0	-3	-2	0
12	A	[5]	-2	-2	-2	-1	-1	-1	0	-2	-2	-2	-1	-1	-3	-1	1	0	-3	-2	0
13	W	-2	-3	-4	-4	-2	-2	-3	-4	-3	1	4	-3	2	1	-3	-3	-2	[7]	0	0
14	A	[3]	-2	-1	-2	-1	-1	-2	4	-2	-2	-2	-1	-2	-3	-1	1	-1	-3	-3	-1
15	A	[2]	-1	0	-1	-2	2	0	2	-1	-3	-3	0	-2	-3	-1	3	0	-3	-2	-2
16	A	[4]	-2	-1	-2	-1	-1	-1	3	-2	-2	-2	-1	-1	-3	-1	1	0	-3	-2	-1
...																					
37	S	2	-1	0	-1	-1	0	0	0	-1	-2	-3	0	-2	-3	-1	4	1	-3	-2	-2
38	G	0	-3	-1	-2	-3	-2	-2	6	-2	-4	-4	-2	-3	-4	-2	0	-2	-3	-3	-4
39	T	0	-1	0	-1	-1	-1	-1	-2	-2	-1	-1	-1	-1	-2	-1	1	5	-3	-2	0
40	W	-3	-3	-4	-5	-3	-2	-3	-3	-3	-3	-2	-3	-2	1	-4	-3	-3	[12]	2	-3
41	Y	-2	-2	-2	-3	-3	-2	-2	-3	2	-2	-1	-2	-1	3	-3	-2	-2	2	7	-1
42	A	[4]	-2	-2	-2	-1	-1	-1	0	-2	-2	-2	-1	-1	-3	-1	1	0	-3	-2	0
...																					

FIGURE 5.5. Portion of a PSSM from a PSI-BLAST search using RBP4 (NP_006735) as a query. The 199 amino acid residues of the query are represented in rows; the 20 amino acids are in columns. Note that for a given residue such as alanine the score can vary (compare A14, A15, and A16, which receive scores of 3, 2, and 4). The tryptophan in position 40 is invariant in several hundred lipocalins. Compare the score of W40, W3, or W5 (each receives +12) with W13 (+7); in the W3, W5, and W40 positions a match is rewarded more highly, and the penalties for mismatches are substantially greater. A PSSM such as this one allows PSI-BLAST to perform with far greater sensitivity than standard blastp searches.

What did this search achieve? After a series of position-specific iterations, hundreds of additional database matches were identified. Many distantly related proteins are now shown in the alignment. We can understand how the sensitivity of the search increased by examining the pairwise alignment of the query (RBP4) with a match, human apolipoprotein D (Fig. 5.6). In the first PSI-BLAST iteration, the bit score was 57.4, the expect value was 3e-07 (i.e., 3×10^{-7}), and there were 47 identities and 39 gaps across an alignment of 151 residues (Fig. 5.6a). After the second iteration, the score rose to 175 bits, the E value dropped (to 10^{-42}), the length of the alignment increased (to 163 residues), and the number of gaps decreased. In the second iteration, larger portions of the amino- and carboxy-terminials of the two proteins were included in the alignment. We previously discussed a questionable match between retinol binding protein 4 (RBP4) and complement component 8 γ (Fig. 4.19). The E value was 0.27 and the score was 33.9 bits. Here in the third PSI-BLAST iteration the E value for this pairwise alignment is 2×10^{-21} (Fig. 5.6c). The E value was dramatically lower as a result of using a scoring matrix specially constructed for this family of proteins.

We can visualize the PSI-BLAST process by imagining each lipocalin in the database as a point in space (Fig. 5.7). An initial search with RBP4 detects other RBP homologs as well as several apolipoprotein D proteins. The PSSM of PSI-BLAST allows the detection of other lipocalins related to apolipoprotein D. Odorant-binding proteins are not detected by a blastp search using RBP4 as a query, but they are found by PSI-BLAST.

As another example of the usefulness of PSI-BLAST, consider a search using RBP4 as a query, with the output restricted to bacteria. Currently (January 2009), there is no match better than threshold after the first iteration. By the second iteration there are two matches better than threshold, and by the third iteration there are over 300 sequences.

You can see the results of nine iterations for the pairwise alignment of RBP4 to apolipoprotein D in web document 5.2 at ▶ http://www.bioinfbook.org/chapter5.

The accession number for human RBP4 is NP_006735. To restrict the search to bacteria, use the command Bacteria (taxid:2).

(a) Iteration 1

```
>ref|NP_001638.1| apolipoprotein D precursor [Homo sapiens]
Length=189

 Score = 57.4 bits (137),  Expect = 3e-07, Method: Composition-based stats.
 Identities = 47/151 (31%), Positives = 78/151 (51%), Gaps = 39/151 (25%)

Query  29   VKENFDKARFSGTWYAMAKKDPEGLFLQDNIVAEFSVDETGQMSATAKGRVRLLNNWDVC  88
            V+ENFD  ++ G WY + +K P      I A +S+ E G        ++++LN  ++
Sbjct  33   VQENFDVNKYLGRWYEI-EKIPTTFENGRCIQANYSLMENG-------KIKVLNQ-ELR  82

Query  89   ADMVGTFTDTE---------DPAKFKMKY-WGVASFLQKGNDDHWIVDTDYDTYAVQYSC  138
            AD  GT   E         +PAK ++K+ W + S      +WI+ TDY+ YA+ YSC
Sbjct  83   AD--GTVNQIEGEATPVNLTEPAKLEVKFSWFMPS------APYWILATDYENYALVYSC  134

Query  139  ----RLLNLDGTCADSYSFVFSRDPNGLPPE  165
             +L ++D    ++++ +R+PN LPPE
Sbjct  135  TCIIQLFHVD------FAWILARNPN-LPPE  158
```

(b) Iteration 2

```
>ref|NP_001638.1| apolipoprotein D precursor [Homo sapiens]
Length=189

 Score =  175 bits (443),  Expect = 1e-42, Method: Composition-based stats.
 Identities = 45/163 (27%), Positives = 77/163 (47%), Gaps = 31/163 (19%)

Query  14   GSGRAERDCRVSSFRVKENFDKARFSGTWYAMAKKDPEGLFLQDNIVAEFSVDETGQMSA  73
            G+A  +   + V+ENFD  ++ G WY + +K P      I A +S+ E G++
Sbjct  18   AEGQAFHLGKCPNPPVQENFDVNKYLGRWYEI-EKIPTTFENGRCIQANYSLMENGKIKV  76

Query  74   TAK-----GRVRLLNNWDVCADMVGTFTDTEDPAKFKMKY-WGVASFLQKGNDDHWIVDT  127
            +        G V +    T  + +PAK ++K+ W + S      +WI+ T
Sbjct  77   LNQELRADGTVNQIEG-------EATPVNLTEPAKLEVKFSWFMPS------APYWILAT  123

Query  128  DYDTYAVQYSCR----LLNLDGTCADSYSFVFSRDPNGLPPEA  166
            DY+ YA+ YSC    L ++D    ++++ +R+PN LPPE
Sbjct  124  DYENYALVYSCTCIIQLFHVD------FAWILARNPN-LPPET  159
```

(c) Iteration 3

```
>ref|NP_000597.1| complement component 8, gamma polypeptide [Homo sapiens]
Length=202

 Score =  104 bits (260),  Expect = 2e-21, Method: Composition-based stats.
 Identities = 40/186 (21%), Positives = 74/186 (39%), Gaps = 29/186 (15%)

Query  24   VSSFRVKENFDKARFSGTWYAMAKKDPEGLFLQDNIVAEFSVDETG-QMSATAKGRVRLL  82
            +S+ + K NFD +F+GTW  +A      +     AE +     Q +A A    R L
Sbjct  33   ISTIQPKANFDAQQFAGTWLLVAVGSACRFLQEQGHRAEATTLHVAPQGTAMAVSTFRKL  92

Query  83   NNWDVCADMVGTFTDTEDPAKFKMKYWGVASFLQKGNDDHWIVDTDYDTYAVQY------  136
            +  +C +   DT   +F ++   +G     +  +TDY ++AV Y
Sbjct  93   DG--ICWQVRQLYGDTGVLGRFLLQARGA-----RGAVHVVVAETDYQSFAVLYLERAGQ  145

Query  137  -SCRLLNLDGTCADSYSFVFSRDPNGLPPEAQKIVRQRQEELCLARQYRLIVHNGYCDGR  195
             S +L    +DS  F +      EA  ++++     +Y     G+C+
Sbjct  146  LSVKLYARSLPVSDSVLSGFEQRVQ----EA---HLTEDQIFYFPKY------GFCEAA  191

Query  196  SERNLL  201
            + ++L
Sbjct  192  DQFHVL  197
```

FIGURE 5.6. *PSI-BLAST search detects distantly related proteins using progressive iterations with a PSSM. (a) A search with RBP4 as a query (NP_006735) detects the lipocalin apolipoprotein D (NP_001638) in the first iteration. (b) As the search progresses to the second iteration, the length of the alignment increases, the bit score becomes higher, the expect value decreases, and the number of gaps in the alignment decreases. (c) By the third iteration, the match to human complement component 8 gamma achieves a significant E value (2e-21), while previously (Fig. 4.19) in a standard blastp search it had been 0.27.*

The number of iterations that a PSI-BLAST search performs relates to the number of hits (sequences) in the database that running the program reports. After each PSI-BLAST iteration, the results that are returned describe which sequences match the input PSSM.

Assessing Performance of PSI-BLAST

There are several ways to assess the performance of PSI-BLAST. When a query is searched against a large database such as SwissProt, the PSSMs can be searched against versions of the database that either are shuffled or have the order of each sequence reversed. When this is done, the PSI-BLAST expect values are not significant (Altschul et al., 1997).

In another approach, several groups have compared the relationships detected using PSI-BLAST to those detected by the rigorous structural analysis of homologous proteins that share limited amino acid identity. Park and colleagues (1998) used the structural classification of proteins (SCOP) database. They found

In a related approach, Schaffer et al. (2001) plotted the number of PSI-BLAST false positives versus true positives to generate a sensitivity curve. They used this plot to assess the accuracy of PSI-BLAST using a variety of adjustments to the parameters.

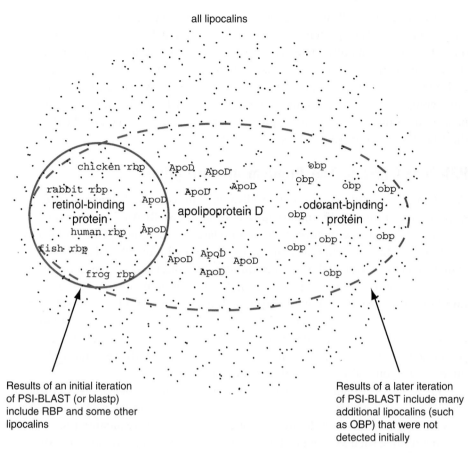

all lipocalins

Results of an initial iteration
of PSI-BLAST (or blastp)
include RBP and some other
lipocalins

Results of a later iteration
of PSI-BLAST include many
additional lipocalins (such
as OBP) that were not
detected initially

FIGURE 5.7. PSI-BLAST algorithm increases the sensitivity of a database search by detecting homologous matches with relatively low sequence identity. In this figure, each dot represents a single protein, some of which are labeled RBP (retinol-binding protein), ApoD (apolipoprotein D), or OBP (odorant-binding protein). All these proteins are homologous by virtue of their membership in the lipocalin family. A standard blastp search with RBP returns matches that are relatively close to RBP in sequence identity, and the result (represented by the circle at left) may include additional matches to lipocalins such as ApoD. However, many other lipocalins such as OBP are not detected. The fundamental limitation in standard BLAST search sensitivity is the reliance on standard PAM and BLOSUM scoring matrices. In a PSI-BLAST search, a PSSM generates a scoring system that is specific to the group of matches detected using the initial query sequence (e.g., RBP). While the initial iteration of a PSI-BLAST search results in an identical number of database matches as a standard BLAST search, subsequent PSI-BLAST iterations (represented by the dashed oval) using a customized matrix extend the results to allow the detection of more distantly related homologs.

that PSI-BLAST searches of this database were more accurate using an E value of 0.0005 (the default inclusion threshold for E is 0.005). They estimated the false-positive rate for PSI-BLAST matches by assessing how many false predictions were made out of 432,680 possible matches. At a low rate of false positives (1 in 100,000), PSI-BLAST detected 27% of homologous matches in the database; at a higher rate of false positives (1 in 1000), PSI-BLAST detected 44% of the homologous matches. This performance is comparable to that observed for SAM-T98, an implementation of a hidden Markov model procedure (see below), and PSI-BLAST was far more sensitive than the standard gapped BLAST or FASTA algorithms.

SCOP (Chapter 11) is available at
► http://scop.mrc-lmb.cam.ac.uk/scop/. It was developed by Cyrus Chothia and colleagues. Park et al. (1998) used the PDBD40-J database, which contains proteins of known structure with ≤40% amino acid identity.

Friedberg et al. (2000) used the fold classification based on structure–structure alignment of proteins (FSSP) and Distant Aligned Protein Sequences (DAPS) (see Chapter 11). Their study included several lipocalins: bovine RBP identified bovine odorant-binding protein; bovine RBP detected both mouse major urinary protein and a bilin-binding protein.

This definition of corruption is adapted from Schaffer et al. (2001).

SEG was described by Wootton and Federhen (1996). An example of its output is shown at ▶ http://www.ncbi.nlm.nih.gov/Education/BLASTinfo/Seg.html.

Friedberg and colleagues (2000) assessed the accuracy of PSI-BLAST alignments. They selected proteins from two structure databases. Fifty-two sequences out of 123 successfully detected their known structural matches using PSI-BLAST, even though the aligned pairs shared less than 25% amino acid identity and none of the pairs could be detected by the Smith–Waterman algorithm. They then compared the alignments generated by PSI-BLAST with the alignments measured using x-ray crystallography and found that, on average, about 44% of the residues were correctly aligned.

PSI-BLAST Errors: The Problem of Corruption

PSI-BLAST is useful to detect statistically weak but biologically meaningful relationships between proteins. The main source of error in PSI-BLAST searches is the spurious amplification of sequences that are unrelated to the query. This problem most often arises when the query (or the profile generated after PSI-BLAST iterations) contains regions with highly biased amino acid composition. Once the program finds even one new protein hit having an E value even slightly above the inclusion threshold, that new hit will be incorporated into the next profile and will reappear in the next PSI-BLAST iteration. If the hit is to a protein that is not homologous to the original query sequence, then the PSSM has been corrupted. We can define corruption as occurring when, after five iterations of PSI-BLAST, the PSSM produces at least one false positive alignment of $E < 10^{-4}$.

There are three main approaches to stopping corruption of PSI-BLAST queries. (1) You can apply a filtering algorithm that removes biased amino acid regions. These "low entropy" regions include stretches of amino acids that are highly basic, acidic, or rich in a residue such as proline. The NCBI PSI-BLAST site employs the SEG program for this purpose, applying the filtering algorithm to database sequences that are detected by the query. (2) You can adjust the expect level from its default value (e.g., $E = 0.005$) to a lower value (e.g., $E = 0.0001$). This may suppress the appearance of false positives, although it could also interfere with the detection of true positives. (3) You can visually inspect each PSI-BLAST iteration. Each protein listed in the PSI-BLAST output has a checkbox; you can select and remove suspicious ones. As an example, your query protein may have a generic coiled-coil domain, and this may cause other proteins sharing this motif (such as myosin) to score better than the inclusion threshold even though they are not homologous.

If a protein has several motifs, such as both a kinase domain and a C2 domain, PSI-BLAST may find database matches related to both. The results must be interpreted carefully. One should not conclude that the kinase domain is related to the C2 domain. For PSI-BLAST searches with multidomain proteins, it may be helpful to search using just one region of interest, such as the reverse transcriptase domain of HIV-1 *pol* protein. In problem 5.1 at the end of this chapter, we consider a hybrid protein as a query having both a lipocalin and a C2 domain. In problem 5.4 we explore fungal globins that are typically multidomain proteins as large as 1000 residues in length. PSI-BLAST searches using a vertebrate globin as a query successfully identify many fungal globin-containing proteins, but the PSSM does not extend into the non-globin regions of these fungal proteins, and so corruption does not occur.

Reverse Position-Specific BLAST

Reverse position-specific BLAST (RPS-BLAST) is used to search a single protein query against a large database of predefined PSSMs. The purpose is to identify

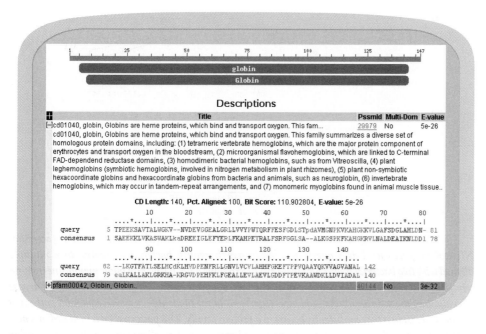

FIGURE 5.8. *Reverse position-specific BLAST is used to search a query (here human beta globin) against a collection of predefined position-specific scoring matrices. The result includes an E value, a pairwise alignment of the query to the consensus PSSM, and a description of the family of proteins in the PSSM. This BLAST tool is available at NCBI.*

conserved protein domains in the query. RPS-BLAST searches are implemented in the Conserved Domain Database at NCBI (Marchler-Bauer et al., 2009). A typical result, based on using human beta globin as a query, shows the globin family (Fig. 5.8). Annotations are by CDD and by the protein family database PFAM, which we describe in Chapter 6.

CDD is available at ▶ http://www. ncbi.nlm.nih.gov/Structure/cdd/ cdd.shtml or through the main BLAST page (▶ http://www.ncbi. nlm.nih.gov/BLAST/). Currently (April 2007) there are over 12,500 PSSMs in the CDD database. We will discuss protein domains in Chapter 10.

PHI-BLAST is launched from the NCBI blastp web page.

Pattern-Hit Initiated BLAST (PHI-BLAST)

Often a protein you are interested in contains a pattern, or "signature," of amino acid residues that help define that protein as part of a family. For example, a signature might be an active site of an enzyme, a string of amino acid residues that define a structural or functional domain of a protein family, or even a characteristic signature of unknown function (such as the three amino acids GXW that is almost always present in the lipocalin family, where X refers to any residue). Pattern-hit initiated BLAST (PHI-BLAST) is a specialized BLAST program that allows you to search with a query and to find database matches that both match a pattern and are significantly related to the query (Zhang et al., 1998). PHI-BLAST may be preferable to simply searching a database with a short query corresponding to a pattern, because such a search could result in the detection of many random matches or proteins that are unrelated to your query protein.

Consider a blastp search of bacterial sequences using human RBP4 as a query (NP_006735), restricted to the refseq database. The result (as of April 2007) is that there are two database matches having small *E* values (0.004 and 0.059). We know that there are many bacterial lipocalins distantly related to human RBP4; one way to confirm this is to perform an Entrez protein query with the words "bacteria lipocalin." Select the two best-scoring bacterial lipocalin protein sequences and align them with human RBP4 (Fig. 5.9a; we describe how to make multiple sequence alignments in Chapter 6). This alignment shows us which amino acid residues are actually shared between RBP4 and the two bacterial proteins. Focusing on the GXW motif that is shared between almost all lipocalins, we can

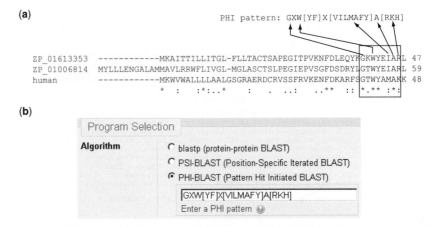

FIGURE 5.9. *Choosing a pattern for a PHI-BLAST search. (a) Human RBP4 (accession NP_006735) was used as a query in a blastp search against bacterial sequences, then multiply aligned with two bacterial lipocalins (these are Alteromonadales bacterium outer membrane lipoprotein [ZP_01613353] and Prochlorococcus marinus lipoprotein Blc [ZP_01006814]). The purpose of evaluating these three protein sequences together is to try to identify a short, sequential pattern of amino acid residues that consistently occurs in a protein family. This pattern then is included in a new PHI-BLAST search to increase its sensitivity and specificity. The alignment was performed using ClustalW (Chapter 6), and a portion of the alignment is shown. By inspection, the invariant GXW motif that is typical of lipocalins is evident (red box). A PHI pattern can be selected that includes these residues and several more. As an example, we select the pattern GXW[YF]X[VILMAFY]A[RKH], in which, following GXW, the next position contains either Y or its closely related residue F, then X denoting any residue, then a set of hydrophobic residues, then an alanine, and finally a basic residue (R, K, or H). The user can select any pattern by trial and error. (b) A PHI-BLAST search is selected from the NCBI protein blast page, and the PHI pattern is entered. The database will then be searched, with a requirement that all database matches include the selected pattern.*

The syntax for a PHI-BLAST pattern is derived from the Prosite dictionary (Chapter 10) and is described at ► http://www.ncbi.nlm.nih.gov/blast/html/PHIsyntax.html. We provide a detailed example of using PHI-BLAST in web document 5.3 at ► http://www.bioinfbook.org/chapter5.

try to define a pattern (or signature) of amino acids that is shared by RBP4, the two bacterial lipocalins, and possibly many other bacterial lipocalins. The purpose of defining a signature is to customize a PSI-BLAST algorithm to search for proteins containing that signature.

How is the signature or pattern defined? We do not expect the signature to be exactly identical between all bacterial lipocalins, and so we want to include freedom for ambiguity. We can define any pattern we want; as an example we will examine the multiple alignment in Fig. 5.9a and create the pattern GXW[YF]X[VILMAFY] A[RKH]. Note that the pattern you choose must not occur too commonly; the algorithm only allows patterns that are expected to occur less frequently than one time for every 5000 database residues. In general, it is acceptable to choose any pattern with four completely specified residues or three residues with average background frequencies of $\leq 5.8\%$ (Zhang et al., 1998).

The BLAST search output is restricted to a subset of the database consisting of proteins that contain that specified pattern. By inspection of the pattern we have chosen, each database match must have a G; the X allows any residue to come next; the W specifies that the third amino acid residue of the pattern must be a W. Next, we write [YF] to specify that the next amino acid must be either a Y or an F; we choose this because it is very common for tyrosines to be substituted with phenylalanine (see Chapter 3). In the next position, we select X to accommodate any residue. We select [VILMAFY] to correspond to the residues in the alignment (I, M) as well as additional hydrophobic residues that we add based on the intuition that any

(a)

Sequences with pattern at position 40 and E-value BETTER than threshold

			Score	E
Sequences producing significant alignments:			(Bits)	Value
NEW ☑	ref|ZP_01613353.1|	outer membrane lipoprotein (lipocalin) [Al...	24.7	1e-05
NEW ☑	ref|YP_212549.1|	hypothetical protein BF2935 [Bacteroides fragil	22.5	5e-05 G
NEW ☑	ref|NP_813404.1|	putative sugar nucleotide epimerase [Bactero...	22.1	7e-05 G
NEW ☑	ref|YP_100376.1|	putative sugar nucleotide epimerase [Bactero...	21.4	1e-04 G
NEW ☑	ref|YP_677044.1|	outer membrane lipoprotein (lipocalin) [Cyto...	20.3	3e-04 G
NEW ☑	ref|ZP_01006814.1|	lipoprotein Blc [Prochlorococcus marinus str.	19.2	5e-04
NEW ☑	ref|YP_001101638.1|	outer membrane lipoprotein (lipocalin) [H...	18.8	7e-04 G
NEW ☑	ref|ZP_01244727.1|	conserved hypothetical protein [Flavobacte...	18.8	7e-04
NEW ☑	ref|YP_341541.1|	outer membrane lipoprotein (lipocalin) [Pseu...	18.8	7e-04 G
NEW ☑	ref|ZP_01065301.1|	lipoprotein Blc [Vibrio sp. MED222]	18.1	0.001
NEW ☑	ref|ZP_01578373.1|	Lipocalin-like [Delftia acidovorans SPH-1]	17.3	0.002
NEW ☑	ref|ZP_01474068.1|	hypothetical protein VEx2w_02003361 [Vibrio s	16.6	0.003
NEW ☑	ref|ZP_00592335.1|	Lipocalin-related protein and Bos/Can/Equ ...	16.3	0.004

Run PSI-Blast iteration 2

(b)

Sequences with E-value BETTER than threshold

			Score	E
Sequences producing significant alignments:			(Bits)	Value
NEW ☑	gi|119470685|ref|ZP_01613353.1|	outer membrane lipoprotein (l...	42.7	0.005

Run PSI-Blast iteration 2

(c)

```
> ref|ZP_01613353.1|  outer membrane lipoprotein (lipocalin) [Alteromonadales bacterium
TW-7]
Length=173

 Score = 24.7 bits (72),  Expect = 1e-05
 Identities = 21/80 (26%), Positives = 40/80 (50%), Gaps = 1/80 (1%)

Pattern              ********
Query   31   ENFDKARFSGTWYAMAKKDPEGLFLQDNIVAEFSVDETGQMSATAKGRVRLLNNWDVCAD  90
             +NFD ++ G WY +A+ D     + + A ++V++ G +   KG +     WD  A+
Sbjct   30   KNFDLEQYKGKWYEIARLDHSFEEGMEQVTATYTVNDDGTVKVLNKGFITKEQKWDE-AE  88

Query   91   MVGTFTDTEDPAKFKMKYWG  110
             + F + D   FK+ ++G
Sbjct   89   GLAKFVEGTDTGHFKVSFFG  108
```

FIGURE 5.10. Results of a PHI-BLAST search of bacterial RefSeq proteins using human RBP4 as a query. (a) The output includes about a dozen proteins having both an E value better than the default threshold and a successful match to the PHI pattern. (b) The same search performed using PSI-BLAST (but without a PHI pattern) yielded just one hit better than the threshold. (c) The pairwise alignments in the output include a row of asterisks showing where the selected PHI pattern occurs in both the query (here human RBP4) and a database match (here a bacterial lipocalin).

hydrophobic residue might occur in this position. We conclude the PHI pattern by specifying an invariant alanine, then a basic residue (arginine, lysine, or histidine).

We next use PHI-BLAST and enter the "PHI pattern" GXW[YF]X[VILMAFY] A[RKH] (Fig. 5.9b). The result of this search is about a dozen database matches consisting of bacterial lipocalins having scores better than threshold (Fig. 5.10a). This contrasts with the results in the absence of a PHI-BLAST pattern, where only one match is found (Fig. 5.10b). The pairwise alignment output of the PHI-BLAST search has the identical format to the PSI-BLAST output, except that information about where both the query and each database sequence match the PHI pattern is shown by a series of asterisks (Fig. 5.10c). The ensuing PSI-BLAST iteration, which no longer uses the PHI pattern but instead uses a search-specific PSSM, will successfully identify a large family of bacterial lipocalins.

The PHI-BLAST algorithm employs a statistical analysis based on identifying alignment A_0 spanned by the input pattern and regions A_1 and A_2 to either side of the pattern, which are scored by gapped extensions. Scores S_0, S_1, and S_2 corresponding to these regions are calculated, and PHI-BLAST scores are ranked by the score $S' = S_1 + S_2$ (ignoring S_0). The alignment statistics are closely related to those used for blastp searches (Zhang et al., 1998).

PROFILE SEARCHES: HIDDEN MARKOV MODELS

PSI-BLAST employs scoring matrices that are customized because of their position-specific nature in a manner that is dependent on the particular input sequence(s). Hence PSI-BLAST is more sensitive at detecting significantly related aligned residues than PAM or BLOSUM matrices. PSSMs are examples of profiles, a concept introduced by Gribskov and others (Gribskov et al., 1987, 1990). Profile hidden Markov models (HMMs) are even more versatile than PSSMs to generate a position-specific scoring system useful for the detection of remote sequence similarities (Krogh et al., 1994; Eddy, 1998; Baldi et al., 1994). HMMs have been widely used in a variety of signal detection problems ranging from speech detection to sonar. Within the field of bioinformatics they have been used for applications as diverse as sequence alignment, prediction of protein structure, prediction of transmembrane domains in proteins, analysis of chromosomal copy number changes, and gene-finding algorithms.

The main strength of a profile HMM is that it is a probabilistic model. This means that it assesses the likelihood of matches, mismatches, insertions and deletions (i.e., gaps) at a given position of an alignment. By developing a statistical model that is based on known sequences, we can use a profile HMM to describe the likelihood that a particular sequence (even one that was previously unknown) matches the model. In contrast, PSI-BLAST does not specify a full probabilistic model. A profile HMM can convert a multiple sequence alignment into a position-specific scoring system. A common application of profile HMMs is the query of a single protein sequence of interest against a database of profile HMMs. Another application is to use a profile HMM as the query in a database search. PFAM and SMART (Chapters 6 and 10) are examples of prominent databases that are based on HMMs.

A Markov chain is a data structure that consists of a computational model with a start state, a finite, discrete set of possible states, and transition functions that describe how to move from one state to the next. This type of computational model is also called a finite state machine. A basic feature of Markov chains is that the process occupies one state at any given unit of time, and remains in that state or moves to another allowable state.

In the case of a hidden Markov model (HMM), we cannot observe the states directly. However, we do have observations from which we can infer the hidden states. A common introduction to the use of HMMs is the scenario of predicting the weather in a distant city. The hidden states have discrete values, such as 1 for sunny and 2 for rainy across a series of days. Assume you have no access to that information, but your goal is to infer the weather over time. The observed state might be the information you get from a friend in that city who tells you whether his dog has gone outside. This represents the observed output that depends on the hidden states. In the case of molecular sequences, the observed states are the positions of amino acids (or nucleotides) in a multiple sequence alignment. The hidden states are the match states, insert states, and delete states. Together, such states define a model for the sequence of that protein family.

An HMM thus consists of a series of defined states. Consider the five amino acid residues taken from an alignment of five globin proteins (Fig. 5.11a). An HMM can be calculated by estimating the probability of occurrence of each amino acid in the five positions (Fig. 5.11b). In this sense, the HMM approach resembles the position-specific scoring matrix calculation of PSI-BLAST. From the HMM probabilities, a score can be derived for the occurrence of any specific pattern of a related query, such as HARTV (Fig. 5.11c). The HMM is a model that can be described in terms of "states" at each position of a sequence (Fig. 5.11d).

A profile HMM is constructed from an initial multiple sequence alignment to define a set of probabilities. The structure of a profile HMM is shown in Fig. 5.12a (Krogh et al., 1994). Along the bottom row is a series of main states

Markov chains were introduced by Andrei Andreyevich Markov (1856–1922), a Russian mathematician. HMMs were introduced into the field of bioinformatics by Anders Krogh, David Haussler, and colleagues (Krogh et al., 1994), Gary Churchill (1989) and others.

(a)

```
1D8U      HAMSV
1OJ6A     HIRKV
2hhbB     HGKKV
1FSL      HAEKL
2MM1      HGATV
```

(b)

Probability	position				
	1	2	3	4	5
p(H)	1.0				
p(A)		0.4			
p(I)		0.2			
p(G)		0.4			
p(M)			0.2		
p(R)			0.2		
p(K)			0.2		
p(E)			0.2		
p(A)			0.2		
p(S)				0.2	
p(K)				0.6	
p(T)				0.2	
p(V)					0.8
p(L)					0.2

(c)

p(HARTV) = (1.0)(0.4)(0.2)(0.2)(0.8) = 0.0128
Log odds score = ln(1.0) + ln(0.4) + ln(0.2) + ln(0.2) + ln(0.8) = −4.4

(d)

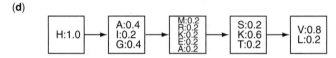

FIGURE 5.11. Hidden Markov models describe alignments based on the probability of amino acids occurring in an aligned column. This is conceptually related to the position-specific scoring matrix used by PSI-BLAST. (a) An alignment of five globins is shown (see Fig. 6.3, open arrowhead). The five proteins are a nonsymbiotic plant hemoglobin from rice (Oryza sativa) (1D8U), human neuroglobin (1OJ6A), human beta globin (2hhbB), leghemoglobin from the soybean Glycine max (1FSL), and human myoglobin (2MM1). (b) The probability of each residue occurring in each aligned column of residues is calculated. (c) From these probabilities, a score is derived for any query such as HARTV. Note that the actual score will also account for gaps and other parameters. Also note that this is a position-specific scoring scheme; for example, there is a different probability of the amino acid residue lysine occurring in position 3 versus 4. (d) The probabilities associated with each position of the alignment can be displayed in boxes representing states.

(from "begin" to *m*1 to *m*5, then "end"). These states might correspond to residues of an amino acid sequence such as HAEKL. The second row consists of insert states (Fig. 5.12, diamond-shaped objects labeled *i*1 to *i*5). These states model variable regions in the alignment, allowing sequences to be inserted as necessary. The third row, at the top, consists of circles called delete states. These correspond to gaps: they provide a path to skip a column (or columns) in the multiple sequence alignment. The emissions lead to the observed sequences in the alignment.

The sequence of an HMM is defined by a series of states that are influenced by two main parameters: the transition probability and the emission probability. The transition probability describes the path followed along the hidden state sequence of the Markov chain (Fig. 5.12a, solid arrows). Each state also has a "symbol emission" probability distribution for matching a particular amino acid residue. The

FIGURE 5.12. *The structure of a hidden Markov model. (a) The HMM consists of a series of states associated with probabilities (adapted from Krogh et al., 1994). The "main states" are shown in boxes along the bottom (from begin to end, with m1 to m5 in between). These main states model the columns of a multiple sequence alignment, and the probability distribution is the frequency of amino acids (see Fig. 5.11d). The "insert states" are in diamond-shaped objects and represent insertions. For example, in a multiple sequence alignment some of the proteins might have an inserted region of amino acids, and these would be modeled by insert states. The "deletion states" (d1 to d5) represent gaps in the alignment. (b) Pair hidden Markov model (Pair-HMM) for the alignment of sequences X and Y having residues x_i and y_j. State M corresponds to the alignment between two amino acids; this state emits two letters. State I corresponds to a position in which a residue x_i is aligned to a gap, while state J corresponds to an alignment of y_j to a gap. The logarithm of the emission probability function $P(x_i, y_j)$ at state M corresponds to a substitution scoring matrix. The transition penalties δ and ε define the transition probabilities.*

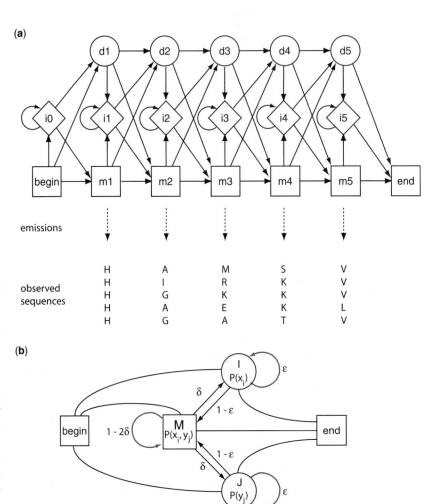

HMMER is available at ▶ http://hmmer.janelia.org/. It was written by Sean Eddy. The program is designed to run on UNIX or MacOS platforms. We will discuss how to create multiple sequence alignments (used as an input to HMMER) in Chapter 6. You can obtain the two sets of vertebrate and bacterial/fungal/vertebrate globin sequences, as web documents 5.4 and 5.5 at ▶ http://www.bioinfbook.org/chapter5. The multiple sequence alignments that we use as input to HMMER are in web documents 5.6 and 5.7.

symbol sequence of an HMM is an observed sequence that resembles a consensus for the multiple sequence alignment. Note that profile HMMs, unlike PSI-BLAST, include probabilities associated with insertions and deletions. The HMM is called a "hidden" model because it consists of both observed symbols (such as the amino acid residues in a sequence modeled by the HMM) and a hidden state sequence which is inferred probabilistically from the observed sequence.

HMMs can also be applied to pairwise alignments (Fig. 5.12b). In addition to beginning and end states, there is a match state M with emission probability x_i, y_j for emitting an aligned pair of residues i, j. State I has the emission probability p_i for emitting a symbol i aligned to a gap; state J corresponds to residue j aligned to a gap. The gaps may be extended with probability ε. The alignment is modeled through a process of choosing sequential states from beginning to end according to the highest transition probabilities, with aligned residues added according to the emission probabilities.

```
Mac:/globins plab$ hmmbuild 01.hmm vertebrate_globins.fasta
hmmbuild - build a hidden Markov model from an alignment
HMMER 2.3.2 (Oct 2003)
Copyright (C) 1992-2003 HHMI/Washington University School of Medicine
Freely distributed under the GNU General Public License (GPL)
- - - - - - - - - - - - - - - - - - - - - - - - - - - - - - - - - - - -
Number of sequences: 10
Number of columns:   149

Determining effective sequence number    ... done. [2]
Weighting sequences heuristically         ... done.
Constructing model architecture           ... done.
Converting counts to probabilities        ... done.
Setting model name, etc.                  ... done. [vertebrate_abglobins]

Constructed a profile HMM (length 148)
Average score:    356.46 bits
Minimum score:    339.74 bits
Maximum score:    367.43 bits
Std. deviation:    11.40 bits

Finalizing model configuration            ... done.
Saving model to file                      ... done.
```

FIGURE 5.13. The HMMER program can be used to create a profile HMM using a multiple sequence alignment as input. The program was obtained from ▶ http:// hmmer.wustl.edu and downloaded on a MacO/S X machine. Ten vertebrate globin proteins were multiply aligned with T-Coffee (see Chapter 6) and entered into HMMER. A profile HMM was built with the hmmbuild program (then in a separate step it is calibrated against a database of 5000 sequences to provide statistical evaluations of the HMM). The average score of the profile HMM was 356 bits; bit scores vary according to the sequences you input.

(a)

```
Scores for complete sequences (score includes all domains):
Sequence                          Description              Score   E-value   N
--------                          -----------              -----   -------  ---
gi|4504349|ref|NP_000509.1|       beta globin [Homo sa     361.9    4e-105   1
gi|4504351|ref|NP_000510.1|       delta globin [Homo s     349.1  2.7e-101   1
gi|4504347|ref|NP_000549.1|       alpha 1 globin [Homo     336.8   1.3e-97   1
gi|4504345|ref|NP_000508.1|       alpha 2 globin [Homo     336.8   1.3e-97   1
gi|6715607|ref|NP_000175.1|       G-gamma globin [Homo     331.3   6.3e-96   1
gi|28302131|ref|NP_000550.2|      A-gamma globin [Homo     328.8   3.5e-95   1
gi|4885393|ref|NP_005321.1|       epsilon globin [Homo     328.8   3.7e-95   1
gi|4885397|ref|NP_005323.1|       zeta globin [Homo sa     247.3   1.2e-70   1
gi|4885395|ref|NP_005322.1|       theta 1 globin [Homo     233.6   1.7e-66   1
gi|51510893|ref|NP_001003938.1|   hemoglobin mu chain      184.5     1e-51   1
gi|19549331|ref|NP_599030.1|      cytoglobin [Homo sap      87.2   1.9e-22   1
gi|4885477|ref|NP_005359.1|       myoglobin [Homo sapi      55.4     7e-13   1
gi|44955888|ref|NP_976312.1|      myoglobin [Homo sapi      55.4     7e-13   1
gi|44955885|ref|NP_976311.1|      myoglobin [Homo sapi      55.4     7e-13   1
gi|10864065|ref|NP_067080.1|      neuroglobin [Homo sa      -0.8    0.0004   1
gi|48675813|ref|NP_038461.2|      transportin 2 (impor     -40.5       1.5   1
```

(b)

```
gi|4885477|ref|NP_005359.1|: domain 1 of 1, 1 to 148: score 55.4, E = 7e-13
                *->vvLsaeeKsnvkglWgKvggnvdEvGaEALeRllvvYPwTkryFpsf
                   + Ls+ e ++v ++WgKv ++    G E L Rl+   P T   F+ f
    gi|4885477    1  MGLSDGEWQLVLNVWGKVEADIPGHGQEVLIRLFKGHPETLEKFDKF 47


                GDLSsadAimGsaqVKaHGKKVldalaealkhlDdlkgtlakLSdLHadK
                   L s d +  s   K HG  Vl+al+  lk       + L + Ha K
    gi|4885477   48  KHLKSEDEMKASEDLKKHGATVLTALGGILKKKGHHEAEIKPLAQSHATK 97


                LrVDPvNFkLLgnvLlvvLAsHfpkdfTPavqAaldKflasVatvLahkY
                   ++     + +++++ vL s +p+df ++q a +K l   + +a+ Y
    gi|4885477   98  HKIPVKYLEFISECIIQVLQSKHPGDFGADAQGAMNKALELFRKDMASNY 147


                r<-*
                   +
    gi|4885477  148  K      148
```

FIGURE 5.14. (a) The output of a HMMER search against all human RefSeq proteins includes a variety of globins. Results vary when the same database is searched with different HMMs. (b) The alignment of the HMM to myoglobin has a significant E value.

Profile HMMs are important because they provide a powerful way to search databases for distantly related homologs. Thus HMM methods complement standard BLAST searching. Profile HMMs can define a protein or gene family, and databases of profile HMMs are searchable. Practically, HMMs can be created using the HMMER program. You can build a profile HMM with the hmmbuild program, which reads a multiple sequence alignment as input (Fig. 5.13). We illustrate the use of HMMER by selecting a group of vertebrate alpha and beta globin proteins and aligning them to use as input. Separately, we use a group of highly divergent globins, including human, fungal, and bacterial proteins. By default, for each model that is built, the resulting profile HMM is global with respect to the HMM and local with respect to the sequences it matches in a database search. The HMM model does not invoke Needleman–Wunsch (global) and Smith–Waterman (local) algorithms separately, but rather uses a model that has the properties of both (and has sometimes been called "glocal"). You can adjust the sensitivity of a HMMer search by building an HMM that is, for example, local with respect to both the sequence and the HMM, thus focusing on local alignments rather than on complete domains.

Next, the hmmcalibrate program matches a set of 5000 random sequences to the profile HMM, fits the scores to an extreme value distribution (EVD; Chapter 4), and calculates the EVD parameters that are necessary to estimate the statistical significance of database matches. The profile HMM can then be used to search a database using the hmmsearch program.

When the profile HMM was built from a multiple sequence alignment of vertebrate alpha and beta globins and used to search the human RefSeq database, there were many database matches (Fig. 5.14a), including myoglobin, that we

The full results of the HMMER searches for (1) vertebrate, (2) bacterial plus fungal, and (3) bacterial plus fungal plus vertebrate globins are shown in web documents 5.8, 5.9, and 5.10 at ▶ http://www.bioinfbook.org/ chapter5. The HMM match to human myoglobin had a higher score and lower *E* value in search (3) than in (1). HMMer searches are run locally. This search was run against all human RefSeq proteins. You can download NCBI databases such as RefSeq by visiting the file transfer protocol (FTP) site from the home page of NCBI or going directly to ▶ http://www. ncbi.nlm.nih.gov/ftp/. Place the downloaded database into the same directory as your input sequences for HMMER.

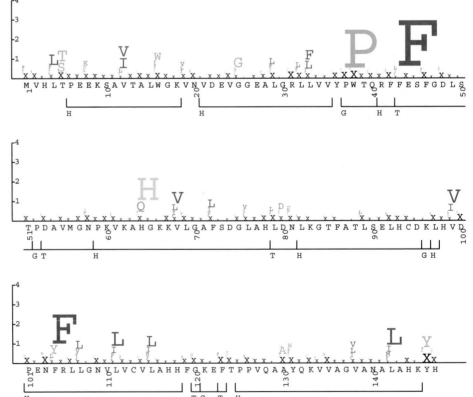

FIGURE 5.15. *The SAM program generates linear HMMs for sequence alignment. Human beta globin protein sequence (NP_000509) was submitted to the SAM web server. The output files included the graphic of the HMM model structure created the makelogo program. The y axis corresponds to the negative entropy in bits of each match state. The relative frequencies of the amino acids at each position are displayed along the x axis. The overall height of each amino acid residue reflects the degree of conservation at that position. Highly conserved residues include a histidine at position 64 that participates in coordination to the heme group.*

could not detect with blastp (Fig. 5.14b). In contrast, when an alignment of bacterial and fungal globins was used to generate a profile HMM, the output consisted of one result with a nonsignficant expect value. Combining several human globins with the bacterial and fungal globins in a multiple sequence alignment resulted in the creation of an HMM that readily detected human globins. Thus, the profile HMM is a model that is sensitive to the choice of sequences that are used as input for the multiple sequence alignment.

As an alternative to HMMER, you can use the Sequence Alignment and Modeling Software System (SAM). SAM includes a suite of tools that can train a group of sequences and create a new model using a linear HMM to represent the family. The target2k program allows you to input a single protein sequence to create a model. We can submit human beta globin protein sequence as a query to the SAM web server. The model structure can be viewed using makelogo (Fig. 5.15) or drawmodel. Makelogo displays the negative entropy (in bits) of each match state and the relative frequencies of the amino acids at each position. This model is built with a target entropy weighting of 0.5 bits per column. An alternative SAM output format shows an HMM architecture such as those in Fig. 5.12.

BLAST-LIKE ALIGNMENT TOOLS TO SEARCH GENOMIC DNA RAPIDLY

As genomic DNA databases grow in size (Chapters 13 to 19), it becomes increasingly common to search them using protein queries or DNA sequences corresponding to expressed transcripts as queries. This is a specialized problem:

1. The genomic DNA includes both exons (regions corresponding to the coding sequence) and introns (intervening, noncoding regions of genes). Ideally, an alignment tool should find the exons in genomic DNA.

2. Genomic DNA often has sequencing errors that should be taken into account.

3. We may want to compare genomic DNA between closely related organisms, such as mouse and rat, or distantly related organisms (e.g., fish and tomato). In any comparison, genomic changes may have occurred, such as deletions, duplications, inversions, or translocations. Algorithms should solve problems such as the alignment of 10 million base pairs containing a 1 million base pair inversion.

4. Algorithms are needed to find small differences between DNA sequences, such as single-nucleotide polymorphisms (SNPs; Chapter 16).

Several BLAST-like algorithms have been written to address these needs. The algorithms are available in programs that are useful for pairwise alignments and/or searches of entire databases with a query. We will illustrate several of these programs using a query of 50,000 base pairs from human chromosome 11p (the short arm of chromosome 11). This region contains five globin genes (HBE1, HBD, HBB, HBG2, and HBG1 corresponding to ε, δ, β, γ2, and γ1 globins) and a beta globin pseudogene (HBBP1). A convenient way to view this region of 50,000 base pairs is to visit the UCSC Genome Browser.

SAM was developd by Richard Hughey, Kevin Karplus, Anders Krogh, and others at the University of California, Santa Cruz (UCSC). A SAM web server is available at ▶ http://www.soe.ucsc.edu/research/compbio/SAM_T06/T06-query.html. Documentation is available at ▶ http://www.soe.ucsc.edu/research/compbio/sam.html. We will discuss exons and introns in Chapters 8 (on gene expression), and 16 (on the eukaryotic chromosome).

The UCSC Genome Browser is available at ▶ http://genome.ucsc.edu. Set the genome to human (May 2004 build), and enter chr11:5,200,001-5,250,000 to specify the genomic position. For an explanation of how to view and obtain this in the UCSC Genome Brower, see web document 5.11 at ▶ http://www.bioinfbook.org/chapter5. For some software such as BLAT (see below), the query cannot be longer than 25,000 base pairs; files with both 50 kilobase (kb) and 25 kb queries are available in web documents 5.12 and 5.13.

Benchmarking to Assess Genomic Alignment Performance

Throughout this book we will describe benchmark datasets that allow one to assess the specificity and sensitivity of a method. We discussed this for PSI-BLAST above. For the multiple sequence alignment of proteins (Chapter 6) several databases contain information on the trusted members of homologous protein families based on their three-dimensional crystal structures as rigorously determined by x-ray crystallography. And in finding genes in genomic DNA (Chapter 16) we will describe the EGASP project that provides a "gold standard" for assessing gene prediction software. In the case of tools for the alignment of genomic DNA, often including large regions of noncoding DNA, there are not experimentally derived databases of correct alignments of large genomic regions. Nonetheless in each case sensitivity and specificity are evaluated. For example, Schwartz et al. (2003) compared human to mouse genomic DNA using BLASTZ, finding that about 39% of the human and mouse genomes could be aligned. Then they reversed the mouse sequence (without complementing it), obtaining a mouse test set with the same size and compositional complexity as the real mouse sequences; only 0.164% of the human sequence now aligned to this reversed set.

The Pollard et al., data are available at ▶ http://rana.lbl.gov/AlignmentBenchmarking/data.html. ROSE is available at ▶ http://bibiserv.techfak.uni-bielefeld.de/rose/.

A benchmark dataset for noncoding genomic DNA can be created using a strategy of computational simulations rather than using experimentally obtained standards. Pollard et al. (2004a) examined noncoding DNA in the fruitfly *Drosophila melanogaster* (a well-characterized genome that lacks many of the ancestral repeats and lineage-specific transposition events found in vertebrates), assembling a group of 10 kilobase fragments. They used the ROSE software package (Stoye et al., 1998) to create a set of simulated sequences having a variety of insertions, deletions, point substitutions, and interspersed blocks of constrained sequences, that is, variations in evolutionary rate estimated across a range of species divergence times. They then tested the ability of eight pairwise genomic alignments tools, including BLASTZ (Pollard et al., 2004b). They concluded that global alignment tools (such as Lagan) have the highest overall sensitivity, while local alignment tools (such as BLASTZ) more accurately align variable regions.

PatternHunter

Other implementations of PatternHunter use slightly different models such as 111010010100110111. PatternHunter software is available commercially at ▶ http://www.bioinformaticssolutions.com.

Blastn uses a short seed, typically consisting of a word size of 11 consecutive nucleotides. Exact matches are identified in a DNA database and then extended into longer alignments (Chapter 4). PatternHunter (Ma et al., 2002) achieves improvements in both speed and sensitivity by creatively using nonconsecutive letters as seeds. If we denote 1 for a match and 0 for a mismatch, the blastn word (w = 11) has a form 11111111111 (Fig. 5.16a). No mismatches are tolerated. For PatternHunter, the pattern of its seed is 110100110010101111 (Fig. 5.16b). There are still 11 matches, but they are distributed over a range of 18 nucleotide positions. If a query aligns with a database entry having the sequence a mismatch corresponding to a 0 position, which is ignored, an extension can still occur. The reason for the improved sensitivity of a nonconsecutive mismatch becomes clear if we consider a particular region of length 64 nucleotides having 70% identity, as described by Ma et al. (2002). For blastn the probability of having at least one hit is 0.30, while for the nonconsecutive seed model the probability is 0.466. This is illustrated in Fig. 5.16c, which shows greater sensitivity for a given of amount similarity. Within some region of 64 nucleotides, the consecutive seed model is disrupted for a mismatch across a group of adjacent seeds which all share a group of 1s. For the nonconsecutive seed model, the seed

(a)

```
11111111111
ATGGTGCATCT (example of a seed)(extended)
ATTGTGCATCT (example of a mismatch)(not extended)
```

(b)

```
110100110010101111
ATGGTGCATCTGACTCCT (example of a seed)(extended)
ATTGTGCATCTGACTCCT (example of an acceptable match)(extended)
```

(c)

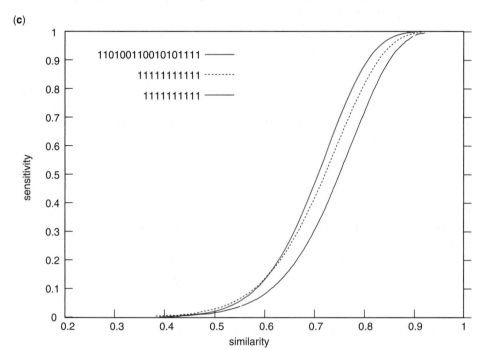

FIGURE 5.16. *Nonconsecutive seeds in PatternHunter improve its sensitivity in database searches. (a) In a typical blastn search with a word size of 11, the matching nucleotides occur consecutively and may be represented with a series of 1s. An example of a seed from a database query is shown; if the database target has a single nucleotide substitution, there is no perfect match and an extension does not occur. (b) The Ma et al. (2005) approach uses nonconsecutive letters as seeds. The values 1 correspond to matches, while the 0 positions are ignored. For some nucleotide mismatches, as shown here, the seed nonetheless matches successfully and extension occurs. (c) A plot of similarity versus sensitivity for the consecutive model with 10 letters (black line), 11 letters (black dotted line), or the spaced model having 11 matches (red line). The sensitivity is higher over a range of similarities for the nonconsecutive seed approach. Adapted from Ma et al. (2005). Used with permission.*

matches occur at different positions, helping to increase sensitivity. This occurs because fewer bases are shared between neighboring seed matches, making the matches more independent than for a consecutive seed strategy.

The innovative approach to seed models introduced by Ma et al. (2002) has been adopted by other homology search algorithms, including BLASTZ and Megablast, discussed below.

BLASTZ

BLASTZ was developed to align human and mouse genomic DNA sequences based on modifications of the gapped BLAST program (Schwartz et al., 2003). It is useful for comparing long genomic sequences from a variety of organisms. Like gapped BLAST, it searches for short near-exact matches, extends them without allowing gaps, and then performs further extensions using dynamic programming. BLASTZ occurs as follows (Schwartz et al., 2003):

1. Lineage-specific interspersed repeats (further described in Chapter 16) are removed from both sequences. Also, to improve its execution speed, when one region of the human genome aligns to multiple regions of the mouse genome, that human segment is dynamically masked. This

Transitions are substitutions between the purines (the nucleotides A↔G) and between the pyrimidines (C ↔T). Transversions are substitutions between purines and pyrimidines (A↔C, A↔T, G↔C, or G↔T). Transitions occur more commonly than transversions (see Chapter 7).

You can obtain BLASTZ at Webb Miller's web site at Pennsylvania State University, ► http://www.bx.psu.edu/miller_lab/. The laj interactive viewer (used to create Fig. 5.18) is also available at that site. For a text file containing 20,000 base pairs of *Macaca mulatta* chromosome 14 sequence that you can use to align to the human beta globin region, see web document 5.14 at ► http://www.bioinfbook.org/chapter5. That region is chr14:68,158,001-68,177,377 at the UCSC Genome Browser (► http://genome.ucsc.edu, January 2006 Rhesus build).

was helpful in processing regions of the mouse genome that have a large number of highly related genes (e.g., zinc finger genes or olfactory receptor genes).

2. Matches are identified using a word size of 12 (either identically matching or allowing one transition), and are extended without allowing gaps. When the score exceeds some threshold, extensions are repeated with gaps allowed. Following the innovation introduced in PatternHunter, BLASTZ uses a seed of 12 matches in 19 consecutive positions having the string 1110100110010101111.

3. Step (2) is repeated for regions adjacent to successful alignments using a lower (more sensitive) word size, such as 7.

BLASTZ was used to align the mouse genome (2.5 billion bases or gigabases [Gb]) with the human genome (2.8 Gb of sequence) (Schwartz et al., 2003). To accomplish this, the human genome was divided into ~3000 segments of about 1 megabase (1 Mb) each, while the mouse genome was divided into ~100 segments of 30 Mb. BLASTZ alignments between a variety of species are represented as tracks on the UCSC Genome Browser. For the 50,000 base pairs containing the human HBB region, an example of the features one can view are as follows (Fig. 5.17): (1) the chromosome band (11p15.4); (2) the genes in the region (HBB, HBD, HBG1, HBE1); (3) a vertebrate conservation track, showing in particular high-scoring regions of conservation across multiple species at the location of the globin genes (and in some noncoding regions having regulatory functions); (4) mouse chains, showing the regions aligned well by BLASTZ; and (5) mouse nets, showing a summary of the best-scoring chains. By clicking on the chains or nets you can access the pairwise sequence alignments. In this example, there are 380 distinct blocks of aligned genomic DNA sequence, separated by regions that could not be reliably aligned.

BLASTZ has been employed for various projects, including an analysis of 13 million base pairs of DNA from the extinct woolly mammoth (*Mammuthus primigenius*) to the modern African elephant (*Loxodonta africana*) (Poinar et al., 2006) and an analysis of transcription units on human chromosome 22 (Lipovich and King, 2006). The BLASTZ program is available for local use. An alignment of 25,000 nucleotides of human chromosome 11, including a portion of the beta globin locus, to the corresponding region of the rhesus monkey (*Macaca mulatta*) is shown in Fig. 5.18.

MegaBLAST and Discontiguous MegaBLAST

MegaBLAST is an NCBI program optimized for the rapid alignment of very large DNA queries (Zhang et al., 2000). The program offers a default word size of 28 (and can accommodate a word size as large as 64), in contrast to the default word size of 11 for blastn. This greatly increases the speed of MegaBLAST, since the word size corresponds to the minimal length of an exact match required to initiate an extension. Blastn, with its smaller word size, is more sensitive but slower. For MegaBLAST you can also specify the percent identity threshold to be reported (e.g., only alignments sharing values such as 99%, 90%, or 80% identity) as well as the corresponding match and mismatch scores. For example, for sequences sharing 95% to 99% identity, a match score of $+1$ and mismatch of -3 is applied; for alignments sharing 85% or 90% identity the mismatch score is instead set to -2.

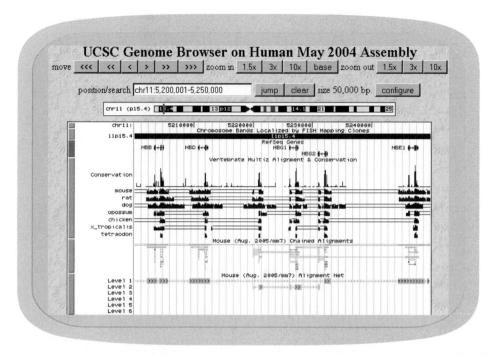

FIGURE 5.17. Precomputed alignments of genomic sequence, aligned by BLASTZ, can be visualized using the UCSC Genome Browser. The genome browser is set to the May 2004 assembly of the human genome, and 50,000 base pairs on chromosome 11p are displayed. The tracks include the following: (1) base pair positions; (2) the chromosome band (11p15.4); (3) RefSeq genes in this region (there are five); (4) vertebrate multiz alignment and conservation (precomputed BLASTZ results showing an overall conservation score as well as alignments from human to mouse, rat, dog, opossum, chicken, frog, and fish); (5) mouse chained alignments, showing BLASTZ alignment results; and (6) Mouse Alignment Net showing a summary of the highest-scoring alignments between mouse and human genomic DNA using BLASTZ. Note that the UCSC Genome Browser annotation tracks can be interactively added or removed, and information can be displayed in a more or less compressed form. Here it is evident that the most highly conserved segments in this 50 kilobase region correspond to the five globin genes, while intergenic regions tend to be less well conserved.

Non-affine gapping parameters are used: the gap opening penalty is 0 (thus causing MegaBLAST to have alignments with more gaps but with the benefit of enhanced speed), and the gap extension penalty is based on the selected match and mismatch scores.

Discontiguous MegaBLAST is a related algorithm at NCBI that is designed to align more distantly related genomic sequences. It employs a "discontiguous word" strategy of Ma et al. (2002) described for PatternHunter. It is useful for comparing relatively divergent sequences (e.g., from different organisms).

We can demonstrate the use of MegaBLAST selecting as a query 50,000 base pairs of DNA from the short arm of human chromosome 11, and selecting an orangutan (*Pongo pygmaeus*) nonredundant nucleotide collection (abbreviated nr/nt). This query region contains five globin genes (HBE1, HBD, HBB, HBG2, HBG1) and a beta globin pseudogene (HBBP1). Using the default settings of MegaBLAST (word size 28, match score $+1$, mismatch score -2, and gap opening and extension penalties zero), we find matches ranging from about 80% to 97% nucleotide identity to the human genomic DNA query (Fig. 5.19).

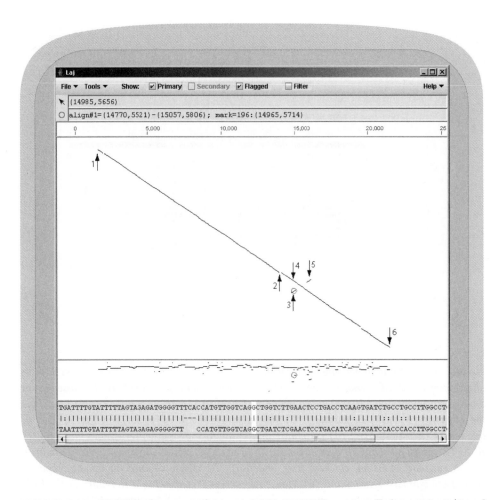

FIGURE 5.18. *BLASTZ alignment of genomic DNA. BLASTZ was installed on a MacO/S and used to align 23,000 base pairs of genomic DNA from the human beta globin locus on chromosome 11 with 20,000 base pairs of genomic DNA from rhesus monkey (Macaca mulatta) chromosome 14. The alignment was visualized using Laj software. The alignment proceeds along the top strand of the human sequence (x axis) and the bottom strand of the monkey sequence (y axis) along a diagonal (beginning at arrow 1 and proceeding to arrow 6). Gaps are evident (e.g., arrow 2), as well as two inverted repeats (arrows 3 and 5). The cursor is placed on one inverted repeat (see circle at arrow 3); the graphic directly below confirms that the selected region is duplicated. The bottom-most panel shows the nucleotide alignment in the region indicated by arrow 3. It proceeds from nucleotides 14,806 to 15,027 along the human sequence (i.e., along the top strand), and from nucleotides 5601 to 5821 of the monkey sequence (i.e., also along the top strand). Most of the monkey sequence aligns to human along the bottom strand (from nucleotide 46 at arrow 6 to nucleotide 19,956 at arrow 1), showing that the repeats at arrows 3 and 5 are indeed inverted.*

BLAT

BLAT is designed to perform extremely rapid genomic DNA searches (Kent, 2002). Like SSAHA, the BLAT algorithm is in some ways a mirror image of BLAST. BLAST parses a query sequence into words and then searches a database with words above a threshold score. Two proximal hits are extended. BLAT parses an entire genomic DNA database into an index of words. These words consist of all nonoverlapping 11-mers in the genome (excluding repetitive DNA sequences). BLAT then searches a query using words from the database.

(a)

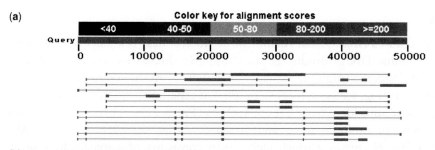

(b)

Sequences producing significant alignments:
(Click headers to sort columns)

Accession	Description	Max score	Total score	Query coverage	E value	Max ident
M92296.1	Pongo pygmaeus gamma-1 and gamma-2 globin genes, co	1.805e+04	2.046e+04	26%	0.0	95%
M18038.1	Orangutan (P.pygmaeus) beta- and eta-globin pseudogene:	1.095e+04	1.156e+04	15%	0.0	94%
X05035.1	Orangutan epsilon-globin gene with Alu repeats in flanking	6547	8198	10%	0.0	96%
M18796.1	Orangutan beta- and delta-globin gene intergenic region wi	5171	5889	7%	0.0	96%
M21825.1	Orangutan delta globin gene, complete cds	3616	4516	5%	0.0	97%
M16209.1	Orangutan gamma-2-fetal globin gene, complete cds	2950	6424	9%	0.0	94%
M16208.1	Orangutan gamma-1-fetal globin gene, complete cds	2935	6667	9%	0.0	94%

FIGURE 5.19. Megablast is an NCBI tool specialized for rapidly searching long DNA queries against genomic DNA databases.

BLAT offers a variety of additional features (Kent, 2002):

- While BLAST triggers an extension with two hits, BLAT triggers extensions on multiple strong hits.

- BLAT is designed to find matches between queries that share 95% nucleotide identity or more. While it is in some ways similar to the Megablast, Sim4, and SSAHA programs, it is orders of magnitude faster.

- BLAT searches for intron–exon boundaries, essentially building a model of a gene structure. It only uses each nucleotide derived from an mRNA query once (as is appropriate from a biological perspective), rather than searching only for highest scoring segment pairs.

An example of a BLAT search using human beta globin protein as a query is shown in Fig. 5.20. Human genomic DNA is translated in six frames, and the best match is to the HBB gene on chromosome 11 that encodes the HBB protein. By adjusting the coordinates on the genome browser to display 50,000 base pairs in the beta globin locus region, we can see that the BLAT search resulted in matches to genes encoding other globin proteins as well.

BLAT is accessible on the web at ▶ http://genome.ucsc.edu. This is one of the main human genome browsers, introduced in Chapter 2. We will explore it in Chapter 16.

(a)

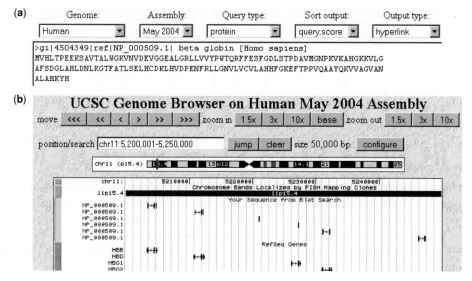

FIGURE 5.20. The BLAST-Like Tool (BLAT) at the UCSC Genome Bioinformatics website. (a) DNA or protein sequence can be pasted or uploaded from a text file. The output settings include a hyperlink option to access the Genome Browser view. (b) The Browser view includes a custom track ("Your Sequence from BLAT Search") which shows a series of matches to five globin genes (HBB, HBD, HBG1, HBG2, HBE1) in a 50,000 base pair segment of human chromosome 11. The genomic location is indicated as well as the cytoband.

LAGAN

LAGAN (Limited Area Global Alignment of Nucleotides) is a pairwise alignment tool for genomic DNA (Brudno et al., 2003). We discuss its companion Multi-LAGAN in Chapter 6 (multiple sequence alignment). LAGAN creates a global pairwise alignment (Fig. 5.21a) in three steps. First, it generates local alignment between two sequences, thus identifying a set of anchors (Fig. 5.21b). This strategy permits the matching of multiple short inexact words rather than long, exact words. Second, LAGAN creates a rough global map consisting of a maximally scoring ordered subset of the alignments (anchors) (Fig. 5.21c). Third, it computes a final global alignment, restricting the operation to the limited area defined by the rough map (Fig. 5.21d). This focused search strategy avoids the inefficiency of performing a global alignment with the Needleman–Wunsch algorithm on the two input sequences.

SSAHA

SSAHA is available at the Ensembl web server (▶ http://www.ensembl.org). The SSAHA home page is ▶ http://www.sanger.ac.uk/Software/analysis/SSAHA/. A hash table contains data (e.g., a list of words having a length of 14 nucleotides in a DNA database) and associated information (e.g., the positions in genomic DNA of each of those words).

Sequence Search and Alignment by Hashing Algorithm, abbreviated SSAHA, is designed to search large DNA databases very rapidly (Ning et al., 2001). The SSAHA converts a DNA database into a hash table, which can then be searched quickly for matches.

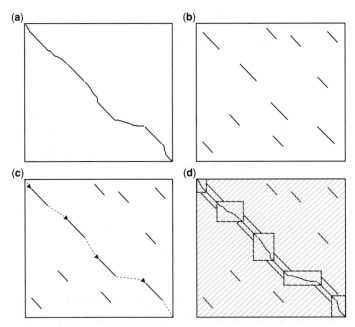

FIGURE 5.21. *The LAGAN algorithm for pairwise alignment of genomic DNA sequences. (a) LAGAN uses a combined local/global strategy to produce a global alignment of two sequences. The x and y axes correspond to the physical position (e.g., chromosomal coordinates) of two DNA queries. (b) A local alignment search strategy identifies conserved regions (solid downward-slanting lines). Note that an inversion in one of the sequences would be represented by a line having a positive slope. (c) Locally aligned segments are joined in chains. Anchors, or maximally scoring ordered subsets of locally aligned regions, are identified and joined to create a rough global map. (d) LAGAN computes an optimal alignment within the boxed areas, ignoring the hatched regions. Adapted from Brudno et al. (2003). Used with permission.*

SIM4

Sim4 uses a BLAST-like algorithm to determine high-scoring segment pairs (HSPs; Chapter 4) and to extend them in both directions (Florea et al., 1998). A dynamic programming algorithm identifies a chain of HSPs that could represent a gene. For example, the program searches for potential splice signals at exon–intron boundaries.

You can access sim4 through a server at ► http://pbil.univ-lyon1. fr/sim4.php. Alternatively, you can download the program from a website of the sim4 authors (including Webb Miller at Pennsylvania State University) at ► http://globin.cse.psu.edu/.

USING BLAST FOR GENE DISCOVERY

A common problem in biology is finding a new gene. Traditionally, genes and proteins were identified using the techniques of molecular biology and biochemistry. Complementary DNAs were cloned from libraries, or proteins were purified then sequenced based on some biochemical criteria such as enzymatic activity. Such experimental biology approaches will always remain essential. Bioinformatics approaches can also be useful to provide evidence for the existence of new genes. For our purposes a "new" gene refers to the discovery of some DNA sequence in a database that is not annotated (described). You may want to find new genes for many reasons:

- You want to study a globin or lipocalin that no one has characterized before, perhaps in a specific organism of interest such as a plant or archaeon.

- You are interested in the lipocalins, and you see that one has been described in the tears of hamsters. Could there be a new, undiscovered gene that encodes a lipocalin protein expressed in human tears? (At present, there is one!)

- You want to know if bacteria have globins or lipocalins. If so, this might give you insight into the evolution of these families of carrier proteins.

- You study diseases in which sugars are not processed properly, and as part of this research, you study sugar transport in cell lines from some organism. You know that glucose transporters have been characterized by biochemical assays (e.g., sugar uptake). You also know that there is a family of glucose transporter genes (and proteins) that have been deposited in GenBank. You cloned all the known transporters, expressed them in cells, and found that none of the recombinant proteins transports your sugar. You hypothesize that there must be at least one more transporter that has not yet been described. Is there a way to search the database to find genes encoding novel transporters?

- You are studying the HIV pol protein, in particular its reverse transcriptase domain. You would like to identify an example of this domain in a eukaryotic protein. However, rather than studying a known eukaryotic protein with this motif, you would like to study a novel one that has never been characterized.

A general strategy to solve any of these problems is presented in Fig. 5.22. I have called this the "find-a-gene" project and have used it as a teaching exercise since the year 2000. All 500 students who attempted it completed it successfully. Each student summarizes the results in a word document. The steps are as follows.

(1) Choose a protein you are interested in. Include the species and the accession number. As an example, we will select human beta globin (NP_000509).

(2) Perform a tblastn search against a DNA database consisting of genomic DNA or ESTs. The BLAST server can be at NCBI or elsewhere. Include the output of that BLAST search in your document.

The "find-a-gene project" is summarized at web document 5.15. The beta globin "find-a-gene project" described here is available as web document 5.16 (► http://www.bioinfbook.org/ chapter5).

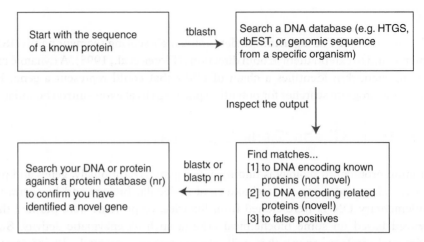

FIGURE 5.22. *How to discover a novel gene by BLAST searching. Begin with the sequence of a known protein such as human beta globin. Perform a tblastn search of a DNA database. It is unlikely that there are many "novel" genes in the well-characterized genomes of organisms such as human, yeast, or E. coli. Thus, it may be helpful to search databases of organisms that are poorly characterized or not fully annotated. The tblastn search may result in two types of significant matches: (1) matches of your query to known proteins that are already annotated and (2) homologous proteins that have not yet been annotated ("novel" genes and corresponding novel proteins). (3) The DNA sequence corresponding to the putative novel gene may be searched using the blastx algorithm against the nonredundant (nr) database. This may confirm that the DNA does indeed encode a protein that has no perfect match to any described protein.*

If appropriate, change the font to Courier size 10 so that the results are displayed neatly. You can also screen capture a BLAST output. It is not necessary to print out all of the blast results if there are many pages.

On the BLAST results, clearly indicate a match that represents a protein sequence, encoded from some DNA sequence, that is homologous to your query protein. It is important to be able to inspect the pairwise alignment you have selected, including the E value and score. In general, this step is the most difficult for students because it requires you to have a "feel" for how to interpret BLAST results. You need to distinguish between a perfect match to your query (i.e., a sequence that is not "novel"), a near match (something that might be "novel," depending on the results of step [4] below), and a nonhomologous result.

Perform a tblastn search of expressed sequence tags (ESTs) restricted to plants ("Viridiplantae (taxid:33090)"). We introduce ESTs in Chapter 8; they are short fragments of DNA (typically up to 800 base pairs) corresponding to genes that have been expressed in a particular organism in some region and at some time of development. For example, libraries of ESTs are available from human fetal liver or adult mouse brain. By restricting the output to plant ESTs, we find many matches with significant E values. One, shown in Fig. 5.23, is EST DT731130, an 878 base pair clone from *Aquilegia formosa* × *Aquilegia pubescens* (a hybrid eudicot from the genus known as columbine). This plant EST encodes a protein that shares 22% amino acid identity with human beta globin, with an E value of 0.20. We will next pursue the possibility that this plant protein is "novel" in the sense that it has never been annotated as a globin.

(3) Gather information about this "novel" protein. At a minimum, identify the protein sequence of the "novel" protein as displayed in the BLAST results from step (2). In some cases, you will be able to do further BLAST searches to obtain even more sequence of your novel gene.

Here, propose a name of the novel protein (e.g., "columbine globin"), and the species from which it derives. It is very unlikely (but still definitely possible) that

(a) >gb|DT731130.1| EST1164980 Aquilegia cDNA library Aquilegia formosa x Aquilegia
pubescens cDNA clone C01T402, mRNA sequence.
Length=878

```
     Score = 38.1 bits (87),  Expect = 0.20
     Identities = 33/146 (22%), Positives = 63/146 (43%), Gaps = 11/146 (7%)
     Frame = +2

Query  5    TPEEKSAVTALWG--KVNVDEVGGEALGRLLVVYPWTQRFFESFGDLSTPDAVMGNPKVK  62
            T ++++ V   W   K N+ E+ +    +L + P + F   D T +   NPK+K
Sbjct  95   TEQQEALVKESWEIMKQNIPELSLQFFTTILEIAPAAKGLFSFLKD--TDEVPQNNPKLK  268

Query  63   AHGKKVLGAFSDGLAHL------DNLKGTFATLSELHCDKLHVDPENFRLLGNVLVCVLA  116
            AH  KV    +    L       D +T  L +H  K  +DP +F  ++    L+ +
Sbjct  269  AHAVKVFKMTCEAAVQLREKGAVDLPESTLKYLGAVHVKKGVIDP-HFEVVKEALLRTIK  445

Query  117  HHFGKEFTPPVQAAYQKVVAGVANAL  142
             G++++ +  A+ +    +A A+
Sbjct  446  DGVGEKWSEELCGAWSEAYDQLATAI  523
```

(b) >ref|NP_187663.1| **UG** AHB2 (NON-SYMBIOTIC HAEMOGLOBIN 2) [Arabidopsis thaliana]
sp|O24521|HBL2_ARATH **G** Non-symbiotic hemoglobin 2 (Hb2) (ARAth GLB2)
gb|AAG51381.1|AC011560_13 **G** class 2 non-symbiotic hemoglobin; 69592-70841 [Arabidopsis thaliana]
gb|AAB82770.1| **G** class 2 non-symbiotic hemoglobin [Arabidopsis thaliana]
gb|AAF76353.1| **G** class 2 non-symbiotic hemoglobin [Arabidopsis thaliana]
gb|AAM65188.1| **UG** Non-symbiotic hemoglobin Hb2 [Arabidopsis thaliana]
gb|ABI49456.1| **UG** At3g10520 [Arabidopsis thaliana]
Length=158

```
     Score =  243 bits (619),  Expect = 1e-62
     Identities = 116/154 (75%), Positives = 141/154 (91%), Gaps = 0/154 (0%)
     Frame = +2

Query  83   EMVFTEQQEALVKESWEIMKQNIPELSLQFFTTILEIAPAAKGLFSFLKDTDEVPQNNPK  262
            E+ FTE+QEALVKESWEI+KQ+IP+ SL FF+ ILEIAPAAKGLFSFL+D+DEVP NNPK
Sbjct  3    EIGFTEKQEALVKESWEILKQDIPKYSLHFFSQILEIAPAAKGLFSFLRDSDEVPHNNPK  62

Query  263  LKAHAVKVFKMTCEAAVQLREKGAVDLPESTLKYLGAVHVKKGVIDPHFEVVKEALLRTI  442
            LKAHAVKVFKMTCE A+QLRE+G V + ++TL+YLG+H+K GVIDPHFEVVKEALLRT+
Sbjct  63   LKAHAVKVFKMTCETAIQLREEGKVVVADTTLQYLGSIHLKSGVIDPHFEVVKEALLRTL  122

Query  443  KDGVGEKWSEELCGAWSEAYDQLATAIKTEMKEE  544
            K+G+GEK++EE+ GAWS+AYD LA AIKTEMK+E
Sbjct  123  KEGLGEKYNEEVEGAWSQAYDHLALAIKTEMKQE  156
```

FIGURE 5.23. The find-a-gene project was performed using human beta globin (NP_000509) as a query and searching a database of expressed sequence tags (ESTs) restricted to plants. (a) The matches included one to an EST from Aquilegia formosa × Aquilegia pubescens (a hybrid eudicot) (GenBank accession DT731130). (b) Using this accession as a query, a blastx nr search revealed matches to known beta globins. The best match, shown here, was to nonsymbiotic hemoglobin 2 from the thale cress Arabidopsis thaliana. However, since there was not a match to an Aquilegia globin, this suggests that the find-a-gene project resulted in the identification of a DNA sequence that encodes a previously undescribed plant globin. One can then proceed to characterize this novel globin in terms of its full-length sequence, homologs, evolution, structure, and function.

you will find a novel gene from an organism such as *S. cerevisiae*, human or mouse, because those genomes have already been thoroughly annotated. It is more likely that you will discover a new gene in a genome that is currently being sequenced, such as bacteria or primates or protozoa.

(4) Demonstrate that this gene, and its corresponding protein, are novel. For the purposes of this project, "novel" is defined as follows. Use the DNA sequence of the EST and perform a blastx query against the nonredundant (nr) database. The best match is not to an *Aquilegia* protein, and thus this DNA sequence does indeed encode a novel globin. As an alternative strategy, take the encoded *Aquilegia* protein sequence (step [3]), and use it as a query in a blastp search of the nonredundant (nr) database at NCBI.

- If there is a match with 100% amino acid identity to a protein in the database, from the same species, then your protein is NOT novel (even if the match is to a protein with a name such as "unknown"). Someone has already found and annotated this sequence, and assigned it an accession number.

- If the best match is to a protein with less than 100% identity to your query, then it is likely that your protein is novel, and you have succeeded.

- If there is a match with 100% identity, but to a different species than the one you started with, then you have succeeded in finding a novel gene.

- If there are no database matches to the original query from step (1), this indicates that you have found a DNA/protein that is not homologous to the original query. You should start over.

There are several further steps for this project, involving themes we will cover in later chapters. (5) Generate a multiple sequence alignment with your novel protein, your original query protein, and a group of other members of this family. A typical number of proteins to use in a multiple sequence alignment is a minimum of 5 or 10 and a reasonable maximum is 30. We will describe multiple sequence alignment in Chapter 6. (6) Create a phylogenetic tree, using a method such as neighbor-joining, maximum parsimony, maximum likelihood, or Bayesian inference (see Chapter 7). Bootstrapping and tree rooting are optional. Use any program such as MEGA, PAUP, Phylip, or MrBayes. (7) Predict the secondary and tertiary structure of your novel protein (see Chapter 11), and compare it to that of a known structure. (8) Determine whether this gene is under positive or negative evolutionary selection (see Chapter 7). (9) Discuss the significance of your novel gene. What have you learned about this gene/protein family?

The main benefits of the find-a-gene project as a teaching tool are that (1) it requires you to know when and how to use the main family of BLAST programs (e.g., tblastn, blastx); (2) it allows you to become familiar with a variety of searchable databases (e.g., EST, genomic DNA, and nonredundant); and (3) it requires you to interpret different kinds of BLAST output. For many initial tblastn searches with a

(a) >gb|AAAA02036297.1| D Oryza sativa (indica cultivar-group) Ctg036297, whole genome
shotgun sequence
Length=20073

```
    Score = 35.4 bits (80),  Expect = 0.58
    Identities = 20/57 (35%), Positives = 32/57 (56%), Gaps = 1/57 (1%)
    Frame = +1

Query  22    IPGHGQEVLIRLFKGHPETLEKFDKFKHLKSEDEMKASED-LKKHGATVLTALGGIL  77
             +P H  ++L+R   G P TLE F  F H+ + +  SE  L+++G +V  A G +L
Sbjct  9040  VPHHFLQILLRHIGGRPLTLETFSLFCHGRESNRLIRSEK*LERNGKSVRWARGEVL  9210
```

(b) >sp|P02193|MYG_DIDMA Myoglobin
Length=154

```
    Score = 36.2 bits (82),  Expect = 0.60, Method: Composition-based stats.
    Identities = 20/57 (35%), Positives = 32/57 (56%), Gaps = 1/57 (1%)

Query  1     VPHHFLQILLRHIGGRPLTLETFSLFCHGRESNRLIRSEKXLERNGKSVRWARGEVL  57
             +P H  ++L+R   G P TLE F  F H+ + +  SE  L+++G +V  A G +L
Sbjct  22    IPGHGQEVLIRLFKGHPETLEKFDKFKHLKSEDEMKASED-LKKHGATVLTALGNIL  77
```

>sp|P02161|MYG_ZALCA Myoglobin
Length=154

```
    Score = 35.8 bits (81),  Expect = 0.78, Method: Composition-based stats.
    Identities = 21/54 (38%), Positives = 32/54 (59%), Gaps = 1/54 (1%)

Query  4     HFLQILLRHIGGRPLTLETFSLFCHGRESNRLIRSEKXLERNGKSVRWARGEVL  57
             H  ++L+R   G P TLE F  F H+ + + RSE  L+++GK+V  A G +L
Sbjct  25    HGQEVLIRLFKGHPETLEKFDKFKHLKSEDEMKRSED-LKKHGKTVLTALGGIL  77
```

FIGURE 5.24. *The find-a-gene project was performed beginning with human myoglobin (NP_005359) as a query, performing a tblastn search against the database of whole genome shotgun sequences (WGS) and restricting the search to plants. (a) One of the pairwise alignments is to a rice clone (accession AAAA02036297) of about 20,000 base pairs. It encodes a protein that matches human myoglobin. Although the E value (0.58) is unconvincing, this still could represent a rice myoglobin. (b) Performing a blastp search against the nonredundant database using the putative rice myoglobin as query, matches to many known myoglobins are evident, including the two best matches shown here to proteins from Didelphis virginiana (North American opossum) and Zalophus californianus (California sealion). However, rice myoglobin is not present in the nr database, suggesting that we have identified a novel myoglobin in rice. Indeed, a blastp search using human myoglobin as a query results in no significant matches in plants, although there is a match to a rice protein (NP_001056053) having a hemoglobin domain. That rice protein has no significant similarity to the protein predicted to be encoded by the 20,000 base pair WGS clone.*

protein query of interest, it is easy to find "novel" genes; for some cases it is not easy to find new genes, perhaps because relevant homologs do not exist, or because the appropriate database is not searched. One can begin again with a different protein query.

In the example above we used human beta globin as a query of an EST database restricted to plants. As a related example, search human myoglobin against whole genome shotgun (WGS) reads of genomic DNA. Restrict the output to the Viridiplantae (plants). One of the output results is to a rice (*Oryza sativa*) clone of over 20,000 base pairs (Fig. 5.24a). This clone encodes a protein fragment of 56 residues that shares 35% identity with human myoglobin. In our earlier example in step (4) we used blastx of the nonredundant database to determine whether this match corresponds to a novel globin. In the case of this rice WGS clone, a blastx search would yield hundreds of irrelevant matches to proteins encoded by various parts of the 20,000 base pairs of DNA. A better strategy is to select the predicted protein sequence and perform a blastp search against the nr database. The best matches are to myoglobins from the opossum (*Didelphis virginiana*) and California sea lion (*Zalophus californianus*), but not to a previously reported rice globin (Fig. 5.24b). Thus, we have indeed identified a novel myoglobin from rice. This example typifies how the find-a-gene project may depend on choosing blastp, blastx, or other programs to evaluate potential homologs.

We will discuss WGS in Chapter 13. To restrict a BLAST search to plants, enter "Viridiplantae (taxid:33090)" in the organism box; by entering the first two letters ("vi") in the box, a pulldown menu allows you to select this taxonomy identifier.

PERSPECTIVE

While BLAST searching has emerged as a fundamental tool for studying proteins and genes (Chapter 4), many specialized BLAST applications have also been developed. These applications include variant algorithms (such as the PSSM of PSI-BLAST and the hidden Markov models of HMMER and SAM) and specialized databases (such as a variety of organism-specific databases). PSI-BLAST has been used extensively to characterize proteins encoded by complete genomes (Chapters 14 to 19). PSI-BLAST can only be successfully applied to cases in which a blastp search results in at least some statistically significant result.

The exponential rise in DNA sequence data (Fig. 2.1) presents us with massive amounts of information about genes and proteins. BLAST searching is a fundamental tool for searching these databases. A BLAST search is often more definitive than a literature search for answering questions about protein or gene families across the tree of life. In this chapter, we described several ways to use alternative BLAST databases and alternative BLAST algorithms to perform database searches. These tools will continue to be fundamentally important to biology for many years to come, especially as the pace of genomic sequencing continues to accelerate. Currently, a blastp search using human beta globin as a query fails to identify human myoglobin as a significant match. In contrast, using PSI-BLAST or HMMer myoglobin is easily detected. This highlights the need for position-specific scoring matrices as well as databases built upon HMMs. We will highlight one such database, Pfam, in Chapters 6 and 10.

PITFALLS

As with any bioinformatics problem, it is essential to define the purpose of a database search. What are you trying to accomplish? Once you have decided this, you can select the appropriate database and search algorithm.

For PSI-BLAST, the biggest problem is obtaining false positives. Once a spurious sequence has been detected that is better than some expect value cutoff, it will be included in the PSSM for the next iteration. This iteration will almost certainly find the spurious sequence again and will probably expand the number of database matches. To avoid this problem:

- Inspect the results for apparently spurious database matches. If you see them, remove such spurious matches by deselecting them.
- Adjust the expect value as appropriate.
- Perform "reverse" searches in which you evaluate a potentially spurious PSI-BLAST result by using that sequence as a query in a BLAST search.
- Further evaluate a marginal database match by performing pairwise sequence alignment as described in Chapter 3.

HMMer is likely to be more sensitive than PSI-BLAST. However, it is slower, it is not web-based (although queries can be sent to the Pfam web server and other servers that employ databases of HMMs), and it requires that the user perform iterations to calibrate an HMM. As with PSI-BLAST, corruption is a potential problem.

For PHI-BLAST, the most common problem encountered is that new users do not have a feel for the rules involved in creating a PHI-BLAST pattern. The best approach is to practice using a variety of signatures.

Web Resources

We have described different kinds of BLAST and related search tools, including organism-specific databases for BLAST searching, BLAST sites that focus on specialized molecules, and alternative algorithms for database searching, including PSI-BLAST, HMMER, MEGABLAST, BLAT. Links to these resources are provided at ► http://www.bioinfbook.org/chapter5.

Discussion Questions

[5-1] BLAT is an extremely fast, accurate program. Why will it not replace BLAST or at least become as commonly used as BLAST? Is it applicable to protein searches?

[5-2] In the original implementation of PSI-BLAST, the algorithm performed a multiple sequence alignment and deleted all but one copy of aligned sequence segments having $\geq 98\%$ identity (Altschul et al., 1997). In a recent modification, the program now purges segments having $\geq 94\%$ identity. What do you think would happen if this percentage were adjusted to $\geq 75\%$? How could you test this idea in practice?

Problems/Computer Lab

[5-1] Create an artificial protein sequence consisting of human RBP4 followed by the C2 domain of human protein kinase Cα. Enter this combined sequence into a PSI-BLAST search. The result is shown in Fig. 5.21. In general, are multiple domains always detected by the PSI-BLAST program? Do any naturally occurring proteins have both lipocalin and C2 domains?

[5-2] The malaria parasite *Plasmodium vivax* has a multigene family called *vir* that is specific to that organism (del Portillo et al., 2001). There are 600 to 1000 copies of these genes, and they may have a role in causing chronic infection through antigenic variation. Select *vir1* and perform a blastp search of the nonredundant database. Then perform a PSI-BLAST search with the same entry.

(a) In an initial search, approximately how many proteins have an E value less than 0.002, and how many have a score greater than 0.002?

(b) What is the score of the best new sequence that is added between the first iteration and the second iteration of PSI-BLAST?

[5-3] Are there globins in fungi? Perform a PSI-BLAST search using human beta globin (NP_000509) as a query, restricting the output to sequences from fungi (taxid:4751) in the nr database. What is the approximate range of lengths of fungal proteins having globin domains? What non-globin domains are often present in fungal globins? Does the presence of these unrelated domains lead to corruption? Why or why not? In the first iteration there are several hits (with the *E* values below the 0.005 threshold). After several more iterations there are many dozens of hits, including flavohemoproteins, that include a globin domain. These fungal proteins have globin domains that are more related to bacterial than vertebrate orthologs. Most of the fungal flavohemoproteins are quite long (over 400 amino acids and sometimes about 1000 amino acids long), having multiple domains. However, only the globin domain is used for the continued PSI-BLAST iterations.

[5-4] We previously performed a series of BLAST searches using HIV-1 pol as a query (NP_057849). Perform a blastp search using this query. Look at the taxonomy report to see which viruses match this query. Next, repeat the search using several iterations of PSI-BLAST. Compare this taxonomy report to that of the blastp search. What do you observe? Are there any nonviral sequences detected in the PSI-BLAST search? Did you expect to find any?

[5-5] Explore PHI-BLAST using human RBP4 (NP_006735) as a query, restricting the output to bacteria and the RefSeq database. Use the PHI pattern GXW[YF]X[VILMAFY]A[RKH]. Perform this search, and save the results. Then repeat the search using the PHI pattern GXW[YF][EA][IVLM]. How do the results differ? Select one protein that appears as a bacterial protein in a pairwise alignment with the human RBP4 query; what are the *E* values, and why do they differ?

SELF-TEST QUIZ

[5-1] A PSI-BLAST search is most useful when you want to do the following:

(a) Find the rat ortholog of a human protein

(b) Extend a database search to find additional proteins

(c) Extend a database search to find additional DNA sequences

(d) Use a pattern or signature to extend a protein search

[5-2] Which of the following BLAST programs uses a signature of amino acids to find proteins within a family?

(a) PSI-BLAST

(b) PHI-BLAST

(c) MS BLAST

(d) WormBLAST

[5-3] In a position-specific scoring matrix, the column headings can have the 20 amino acids, and the rows can represent the residues of a query sequence. Within the matrix, the score for any given amino acid residue is assigned based on:

(a) A PAM or BLOSUM matrix

(b) Its frequency of occurrence in a multiple sequence alignment

(c) Its background frequency of occurrence

(d) The score of its neighboring amino acids

[5-4] As part of a PSI-BLAST search, a score is assigned to an alignment between a query sequence and a database match over some length (such as 50 amino acid residues). It is possible for this pairwise alignment to receive a higher or lower score over successive PSI-BLAST iterations, even though there is no change in which amino acid residues are aligned.

(a) True

(b) False

[5-5] A position-specific scoring matrix is said to be "corrupted" when it incorporates a spurious sequence (i.e., a false positive result).

Which of the following choices is the best way to reduce corruption?

(a) Lower the *E* value

(b) Remove filtering

(c) Use a shorter query

(d) Run fewer iterations

[5-6] What is the main advantage of employing reverse position-specific BLAST?

(a) Reversing a query and/or a set of database sequences provides a set of null alignments from which the statistical significance of a PSI-BLAST search can be estimated.

(b) This method precomputes a large collection of position-specific scoring matrices, allowing a query to be rapidly assigned to a protein family.

(c) This method allows critical conserved residues in the query sequence to be identified.

(d) This method facilitates the comparison of multiple position-specific scoring matrices.

[5-7] What capability does a profile hidden Markov model (HMM) offer that PSI-BLAST does not offer for protein queries?

(a) A profile HMM can model the likelihood of insertions and deletions in aligned residues.

(b) A profile HMM can identify distantly related homologs that are not identified by standard blast searching.

(c) A profile HMM can estimate the probability of achieving particular scores for aligned residues across the length of a multiple sequence alignment.

(d) A profile HMM can model both protein relationships that are either conserved or distant.

[5-8] Why do algorithms used to align genomic DNA (such as PatternHunter, BLASTZ, and Megablast) use nonconsecutive letters as seeds, rather than the conventional word size (such

as a word size of 11 exactly matching nucleotides) employed by blastn?

(a) This novel strategy improves specificity because longer seed lengths (such as those typically found in genomic DNA) are matched more efficiently.

(b) This novel strategy improves sensitivity by not tolerating any mismatches within the seed region.

(c) This novel strategy relies on a longer word size to match query sequences far more rapidly than is possible for shorter seeds.

(d) This novel strategy improves both speed and sensitivity by tolerating mismatches within a seed region.

[5-9] How does BLAT differ from BLAST?

(a) BLAT includes both global and local alignment.

(b) BLAT employs a database that is parsed into a set of words that are matched to the DNA query.

(c) BLAT only identifies genomic regions that exactly match a query sequence.

(d) BLAT cannot accept a protein sequence as a query.

SUGGESTED READING

In this chapter we introduced a variety of BLAST servers and BLAST-related software. In most cases the websites contain documentation online. PSI-BLAST was introduced in an excellent paper by Altschul and colleagues (1997) (see also Suggested Readings for Chapter 4). Further modifications of PSI-BLAST are introduced by Schaffer et al. (2001). The PHI-BLAST algorithm is described by Zhang et al. (1998).

REFERENCES

Altschul, S. F., et al. Gapped BLAST and PSI-BLAST: A new generation of protein database search programs. *Nucleic Acids Res.* **25**, 3389–3402 (1997).

Ashurst, J. L., Chen, C. K., Gilbert, J. G., Jekosch, K., Keenan, S., Meidl, P., Searle, S. M., Stalker, J., Storey, R., Trevanion, S., Wilming, L., and Hubbard, T. The Vertebrate Genome Annotation (VEGA) database. *Nucleic Acids Res.* **33**, D459–D465 (2005).

Baldi, P., Chauvin, Y., Hunkapiller, T., and McClure, M. A. Hidden Markov models of biological primary sequence information. *Proc. Natl. Acad. Sci. USA* **91**, 1059–1063 (1994).

Brudno, M., Do, C. B., Cooper, G. M., Kim, M. F., Davydov, E., NISC Comparative Sequencing Program, Green, E.D., Sidow, A., and Batzoglou, S. LAGAN and Multi-LAGAN: Efficient tools for large-scale multiple alignment of genomic DNA. *Genome Res.* **13**, 721–731 (2003).

Churchill, G. A. Stochastic models for heterogeneous DNA sequences. *Bull. Math. Biol.* **51**, 79–94 (1989).

del Portillo, H. A., et al. A superfamily of variant genes encoded in the subtelomeric region of *Plasmodium vivax*. *Nature* **410**, 839–842 (2001).

Eddy, S. R. Profile hidden Markov models. *Bioinformatics* **14**, 755–763 (1998).

Florea, L., Hartzell, G., Zhang, Z., Rubin, G. M., and Miller, W. A computer program for aligning a cDNA sequence with a genomic DNA sequence. *Genome Res.* **8**, 967–974 (1998).

Friedberg, I., Kaplan, T., and Margalit, H. Evaluation of PSI-BLAST alignment accuracy in comparison to structural alignments. *Protein Sci.* **9**, 2278–2284 (2000).

Gribskov, M., Lüthy, R., and Eisenberg, D. Profile analysis. *Methods Enzymol.* **183**, 146–159 (1990).

Gribskov, M., McLachlan, A. D., and Eisenberg, D. Profile analysis: detection of distantly related proteins. *Proc. Natl. Acad. Sci. USA* **84**, 4355–4358 (1987).

Kent, W. J. BLAT—the BLAST-like alignment tool. *Genome Res.* **12**, 656–664 (2002).

Krogh, A., Brown, M., Mian, I. S., Sjölander, K., and Haussler, D. Hidden Markov models in computational biology. Applications to protein modeling. *J. Mol. Biol.* **235**, 1501–1531 (1994).

Lipovich, L., and King, M. C. Abundant novel transcriptional units and unconventional gene pairs on human chromosome 22. *Genome Res.* **16**, 45–54 (2006).

Loveland, J. VEGA, the genome browser with a difference. *Brief. Bioinform.* **6**, 189–193 (2005).

Ma, B., Tromp, J., and Li, M. PatternHunter: Faster and more sensitive homology search. *Bioinformatics* **18**, 440–445 (2002).

Marchler-Bauer, A., Anderson, J. B., Chitsaz, F., Derbyshire, M. K., DeWeese-Scott, C., Fong, J. H., Geer, L. Y., Geer, R. C., Gonzales, N. R., Gwadz, M., et al. CDD: specific functional annotation with the Conserved Domain Database. *Nucl. Acids Res.* **37**, D205–D210 (2009).

Ning, Z., Cox, A. J., and Mullikin, J. C. SSAHA: A fast search method for large DNA databases. *Genome Res.* **11**, 1725–1729 (2001).

Park, J., et al. Sequence comparisons using multiple sequences detect three times as many remote homologues as pairwise methods. *J. Mol. Biol.* **284**, 1201–1210 (1998).

Pearson, W. R., and Lipman, D. J. Improved tools for biological sequence comparison. *Proc. Natl. Acad. Sci. USA* **85**, 2444–2448 (1988).

Poinar, H. N., Schwarz, C., Qi, J., Shapiro, B., Macphee, R. D., Buigues, B., Tikhonov, A., Huson, D. H., Tomsho, L. P., Auch, A., Rampp, M., Miller, W., and Schuster, S. C. Metagenomics to paleogenomics: Large-scale sequencing of mammoth DNA. *Science* **311**, 392–394 (2006).

Pollard, D. A., Bergman, C. M., Stoye, J., Celniker, S. E., and Eisen, M. B. Benchmarking tools for the alignment of functional noncoding DNA. *BMC Bioinformatics* **5**, 1–17 (2004a).

Pollard, D. A., Bergman, C. M., Stoye, J., Celniker, S. E., and Eisen, M. B. Correction: Benchmarking tools for the alignment of functional noncoding DNA. *BMC Bioinformatics* **5**, 73 (2004b).

Schaffer, A. A., et al. Improving the accuracy of PSI-BLAST protein database searches with composition-based statistics and other refinements. *Nucleic Acids Res.* **29**, 2994–3005 (2001).

Schwartz, S., Kent, W. J., Smit, A., Zhang, Z., Baertsch, R., Hardison, R. C., Haussler, D., and Miller, W. Human-mouse alignments with BLASTZ. *Genome Res.* **13**, 103–107 (2003).

Simon, J. F. *Animal Chemistry with References to the Physiology and Pathology of Man*. Sydenham Society, London, 1846.

Stamm, S., Riethoven, J.-J.M., Le Texier, V., Gopalakrishnan, C., Kumanduri, V., Tang, Y., Barbosa-Morais, N.L., and Thanaraj, T. A. ASD: A bioinformatics resource on alternative splicing. *Nucleic Acids Res.* **34**, D46–D55 (2006).

Stoye, J., Evers, D., and Meyer, F. ROSE: Generating sequence families. *Bioinformatics* **14**, 157–163 (1998).

Tatusov, R. L., Altschul, S. F., and Koonin, E. V. Detection of conserved segments in proteins: Iterative scanning of sequence databases with alignment blocks. *Proc. Natl. Acad. Sci. USA* **91**, 12091–12095 (1994).

Wootton, J. C., and Federhen, S. Analysis of compositionally biased regions in sequence databases. *Methods Enzymol.* **266**, 554–571 (1996).

Zhang, Z., et al. Protein sequence similarity searches using patterns as seeds. *Nucleic Acids Res.* **26**, 3986–3990 (1998).

Zhang, Z., Schwartz, S., Wagner, L., and Miller, W. A greedy algorithm for aligning DNA sequences. *J. Comput. Biol.* **7**, 203–214 (2000).

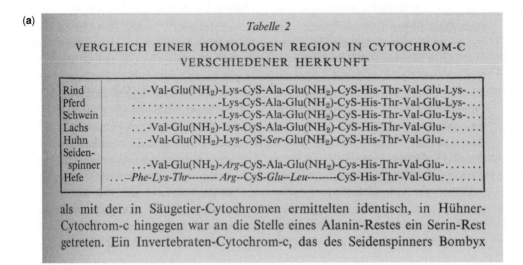

(a)

Tabelle 2

VERGLEICH EINER HOMOLOGEN REGION IN CYTOCHROM-C VERSCHIEDENER HERKUNFT

Rind	...-Val-Glu(NH$_2$)-Lys-CyS-Ala-Glu(NH$_2$)-CyS-His-Thr-Val-Glu-Lys-...
Pferd	................-Lys-CyS-Ala-Glu(NH$_2$)-CyS-His-Thr-Val-Glu-Lys-...
Schwein	................-Lys-CyS-Ala-Glu(NH$_2$)-CyS-His-Thr-Val-Glu-Lys-...
Lachs	...-Val-Glu(NH$_2$)-Lys-CyS-Ala-Glu(NH$_2$)-CyS-His-Thr-Val-Glu- ......
Huhn	...-Val-Glu(NH$_2$)-Lys-CyS-*Ser*-Glu(NH$_2$)-CyS-His-Thr-Val-Glu-.......
Seiden-spinner	...-Val-Glu(NH$_2$)-*Arg*-CyS-Ala-Glu(NH$_2$)-Cys-His-Thr-Val-Glu-.......
Hefe	...-*Phe-Lys-Thr*------- *Arg*--CyS-*Glu*--*Leu*-------CyS-His-Thr-Val-Glu-.......

als mit der in Säugetier-Cytochromen ermittelten identisch, in Hühner-Cytochrom-c hingegen war an die Stelle eines Alanin-Restes ein Serin-Rest getreten. Ein Invertebraten-Cytochrom-c, das des Seidenspinners Bombyx

(b)

Tabelle 3

ARTUNTERSCHIEDE IN INSULIN (NACH HARRIS, SANGER UND NAUGHTON, 1956)

Rind	...-CyS-*Ala-Ser-Val*-CyS-...
Schwein	...-CyS-*Thr-Ser-Ileu*-CyS-...
Schaf	...-CyS-*Ala-Gly-Val*-CyS-...
Pferd	...-CyS-*Thr-Gly-Ileu*-CyS-...
Wal	...-CyS-*Thr-Ser-Ileu*-CyS-...

As the linear amino acid sequences of proteins were determined in the 1950s and 1960s, it became of obvious interest to try to align them. (a) Hans Tuppy (1958, p. 71) described the alignment of cytochromes c from Rind (beef), Pferd (horse), Schwein (pig), Lachs (salmon), Huhn (chicken), Seiden-spinner (silkworm), and Hefe (yeast). This alignment showed that even though gaps had to be introduced, protein sequences from organisms as distantly related as mammals and yeast could still be aligned. (b) Tuppy (1958, p. 73) also described an alignment of insulin amino acid sequences from beef, pig, Schaf (sheep), horse, and Wal (whale). In this case, he noted the lack of conservation of several amino acid residues in a region between two cysteine residues. For more details on the alignment of insulins, see Fig. 7.1. Used with permission.

6

Multiple Sequence Alignment

INTRODUCTION

When we consider a protein (or gene), one of the most fundamental questions is what other proteins are related. Biological sequences often occur in families. These families may consist of related genes within an organism (paralogs), sequences within a population (e.g., polymorphic variants), or genes in other species (orthologs). Sequences diverge from each other for reasons such as duplication within a genome or speciation leading to the existence of orthologs. We have studied pairwise comparisons of two protein (or DNA) sequences (Chapter 3), and we have also seen multiple related sequences in the form of profiles or as the output of a BLAST or other database search (Chapters 4 and 5). We will also explore multiple sequence alignments in the context of molecular phylogeny (Chapter 7), protein domains (Chapter 10), and protein structure (Chapter 11).

In this chapter, we consider the general problem of multiple sequence alignment from three perspectives. First, we describe five approaches to making multiple sequence alignments from a group of homologous sequences of interest. Second, we discuss multiple alignment of genomic DNA. This is typically a comparative genomics problem of aligning large chromosomal regions from different species. Third, we explore databases of multiply aligned sequences, such as Pfam, the protein family database. While multiple sequence alignment is commonly performed for both protein and DNA sequences, most databases consist of protein families only. Nucleotides corresponding to coding regions are typically less well conserved than

Bioinformatics and Functional Genomics, Second Edition. By Jonathan Pevsner
Copyright © 2009 John Wiley & Sons, Inc.

proteins because of the degeneracy of the genetic code. Thus they can be harder to align with with high confidence.

Multiple sequence alignments are of great interest because homologous sequences often retain similar structures and functions. Pairwise alignments may suffice to create links between structure and function. Multiple sequence alignments are very powerful because two sequences that may not align well to each other can be aligned via their relationship to a third sequence, thereby integrating information in a way not possible using only pairwise alignments. We can thus define members of a gene or protein family, and identify conserved regions. If we know a feature of one of the proteins (e.g., RBP4 transports a hydrophobic ligand), then when we identify homologous proteins, we can predict that they may have similar function. The overwhelming majority of proteins have been identified through the sequencing of genomic DNA or complementary DNA (cDNA; Chapter 8). Thus, the function of most proteins is assigned on the basis of homology to other known proteins rather than on the basis of results from biochemical or cell biological (functional) assays.

Definition of Multiple Sequence Alignment

Domains or motifs that characterize a protein family are defined by the existence of a multiple sequence alignment of a group of homologous sequences. A multiple sequence alignment is a collection of three or more protein (or nucleic acid) sequences that are partially or completely aligned. Homologous residues are aligned in columns across the length of the sequences. These aligned residues are homologous in an evolutionary sense: they are presumably derived from a common ancestor. The residues in each column are also presumed to be homologous in a structural sense: aligned residues tend to occupy corresponding positions in the three-dimensional structure of each aligned protein.

Multiple sequence alignments are easy to generate, even by eye, for a group of very closely related protein (or DNA) sequences. We have seen an alignment of closely related sequences (Fig. 3.7, GAPDH). As soon as the sequences exhibit some divergence, the problem of multiple alignment becomes extraordinarily difficult to solve. In particular, the number and location of gaps is difficult to assess. We saw an example of this with kappa caseins (Fig. 3.8), and in this chapter we will examine a challenging region of five distantly related globins. Practically, you must (1) choose homologous sequences to align, (2) choose software that implements an appropriate objective scoring function (i.e., a metric such as maximizing the total score of a series of pairwise alignments), and (3) choose appropriate parameters such as gap opening and gap extension penalties.

There is not necessarily one "correct" alignment of a protein family. This is because while protein structures tend to evolve over time, protein sequences generally evolve even more rapidly than structures. Looking at the sequences of human beta globin and myoglobin, we saw that they share only 25% amino acid identity (Fig. 3.5), but the three-dimensional structures are nearly identical (Fig. 3.1). In creating a multiple sequence alignment, it may be impossible to identify the amino acid residues that should be aligned with each other as defined by the three-dimensional structures of the proteins in the family. We often do not have high-resolution structural data available, and we rely on sequence data to generate the alignment. Similarly, we often do not have functional data to identify domains (such as the specific amino acids that form the catalytic site of an enzyme), so again we rely on sequence data. It is possible to compare the results of multiple sequence alignments that are generated solely from sequence data and to then examine known structures for those proteins. For a given pair of divergent but significantly related protein sequences (e.g., for two proteins sharing 30% amino acid identity), Chothia and Lesk (1986) found that about 50% of the individual amino acid residues are superposable in the two structures.

Aligned columns of amino acid residues characterize a multiple sequence alignment. This alignment may be determined because of features of the amino acids such as the following:

- There are highly conserved residues such as cysteines that are involved in forming disulfide bridges.

- There are conserved motifs such as a transmembrane domain or an immunoglobulin domain. We will encounter examples of protein domains and motifs (such as the PROSITE dictionary) in Chapter 10.

- There are conserved features of the secondary structure of the proteins, such as residues that contribute to α helices, β sheets, or transitional domains.

- There are regions that show consistent patterns of insertions or deletions.

Typical Uses and Practical Strategies of Multiple Sequence Alignment

When and why are multiple sequence alignments used?

- If a protein (or gene) you are studying is related to a larger group of proteins, this group membership can often provide insight into the likely function, structure, and evolution of that protein.

- Most protein families have distantly related members. Multiple sequence alignment is a far more sensitive method than pairwise alignment to detect homologs (Park et al., 1998). Profiles (such as those described for PSI-BLAST and hidden Markov models in Chapter 5) depend on accurate multiple sequence alignments.

- When one examines the output of any database search (such as a BLAST search), a multiple sequence alignment format can be extremely useful to reveal conserved residues or motifs in the output.

- If one is studying cDNA clones, it is common practice to sequence them. Multiple sequence alignment can show whether there are any variants or discrepancies in the sequences. Alignments of genomic DNA containing single nucleotide polymorphisms (SNPs; Chapter 16) are of interest, for example, in the identification of nonsynonymous SNPs.

- Analysis of population data can provide insight into many biological questions involving evolution, structure, and function. The PopSet portion of Entrez (described below) contains nucleotide (and protein) population data sets that are viewed as multiple alignments.

- When the complete genome of any organism is sequenced, a major portion of the analysis consists of defining the protein families to which all the gene products belong. Database searches effectively perform multiple sequence alignments, comparing each novel protein (or gene) to the families of all other known genes.

- We will see in Chapter 7 how phylogeny algorithms begin with multiple sequence alignments as the raw data with which to generate trees. The most critical part of making a tree is to produce an optimal multiple sequence alignment.

- The regulatory regions of many genes contain consensus sequences for transcription factor-binding sites and other conserved elements. Many such regions are identified based on conserved noncoding sequences that are detected using multiple sequence alignment.

Benchmarking: Assessment of Multiple Sequence Alignment Algorithms

We will describe five different approaches to creating multiple sequence alignments. How can we assess the accuracy and performance properties of the various algorithms? The performance depends on factors including the number of sequences being aligned, their similarity, and the number and position of insertions or deletions (McClure et al., 1994).

A convincing way to assess whether a multiple sequence alignment program produces a "correct" alignment is to compare the result with the alignment of known three-dimensional structures as established by x-ray crystallography (Chapter 11). Several databases have been constructed to serve as benchmark data sets. These are reference sets in which alignments are created from proteins having known structures. Thus, one can study proteins that are by definition structurally homologous. This allows an assessment of how successfully assorted multiple sequence alignment algorithms are able to detect distant relationships among proteins. For proteins sharing about 40% amino acid identity or more, most multiple sequence alignment programs produce closely similar results. For more distantly related proteins, the programs can produce markedly different alignments, and benchmarks are useful to compare accuracy.

The performance of a multiple sequence alignment algorithm relative to a benchmark data set is measured by some objective scoring function. One commonly used metric is the sum-of-pairs score (Box 6.1). This involves counting the number of

Box 6.1
Evaluating Multiple Sequence Alignments

Thompson et al. (1999) described two main ways to assess multiple sequence alignments. The first is the sum-of-pairs scores (SPS). This score increases as a program succeeds in aligning sequences relative to the BAliBASE or other reference alignment. The SPS assumes statistical independence of the columns. For an alignment of N sequences in M columns, the ith column is designated A_{i1}, $A_{i2}, \ldots, A_{iN}$. For each pair of residues A_{ij} and A_{ik}, a score of 1 is assigned ($p_{ijk} = 1$) if they are also aligned in the reference, and a score of 0 is assigned if they are not aligned ($p_{ijk} = 0$). Then for the entire ith column, the score S_i is given by:

$$S_i = \sum_{j=1, j \neq k}^{N} \sum_{k=1}^{N} p_{ijk}$$

For the entire multiple sequence alignment, the SPS is given by:

$$\text{SPS} = \frac{\sum\limits_{i=1}^{M} S_i}{\sum\limits_{i=1}^{Mr} S_{ri}}$$

Here S_{ri} is the score S_i for the ith column in the reference alignment, and Mr corresponds to the number of columns in the reference alignment.

A second approach is to create a column score (CS). For the ith column, $C_i = 1$ if all the residues in the column are aligned in the reference, and $C_i = 0$ if not.

$$\text{CS} = \sum_{i=1}^{M} \frac{C_i}{M}$$

Sum-of-pairs scores and column scores have been used to assess the performance of multiple sequence alignment algorithms. Gotoh (1995) and others further described weighted sum-of-pairs scores that correct for biased contributions of sequences caused by divergent members of a group being aligned. Lassmann and Sonnhammer (2005) note that a column score becomes zero if even a single sequence is misaligned; thus it may be too stringent.

pairs of aligned residues that occur in the target and reference alignment, divided by the total number of pairs of residues in the reference.

Benchmark data sets may contain separate categories of multiple sequence alignments, such as those having proteins of varying length, varying divergence, insertions or deletions (indels) of various lengths, and varying motifs (such as internal repeats). Investigators routinely employ benchmark data sets to assess the performance of alignment algorithms (e.g., Morgenstern et al., 1996; McClure et al., 1994; Thompson et al., 1999; Gotoh, 1996; Briffeuil et al., 1998). Blackshields et al. (2006) compared the properties of six benchmark datasets (Table 6.1).

Another approach to benchmarking is to use a program such as ROSE (Stoye et al., 1998) that simulates the evolution of sequences. We introduced ROSE in

You can examine typical benchmark entries for the globins and the lipocalins from the HOMSTRAD database (Mizuguchi et al., 1998) in Web documents 6.1 and 6.2 at ▶ http://www.bioinfobook.org/chapter6. HOMSTRAD (the homologous structure alignment database) contains aligned three-dimensional structures of homologous proteins from over 1000 families. Later in this chapter, studying the T-Coffee suite of programs, we will introduce a new approach to benchmarking that is based on structural data but does not employ a benchmark database.

TABLE 6-1 Benchmark Data Sets to Assess Multiple Sequence Alignment Accuracy

Database	Reference	URL
BAliBASE	Thompson et al. (2005)	http://www-bio3d-igbmc.u-strasbg.fr/balibase/
HOMSTRAD	Mizuguchi et al. (1998)	http://www-cryst.bioc.cam.ac.uk/~homstrad/
IRMBASE	Subramanian et al. (2005)	http://dialign-t.gobics.de/main
OxBench	Raghava et al. (2003)	http://www.compbio.dundee.ac.uk/Software/Oxbench/oxbench.htm
Prefab	Edgar (2004b)	http://www.drive5.com/muscle/prefab.htm
SABmark	Van Walle et al. (2005)	http://bioinformatics.vub.ac.be/databases/content.html

ROSE software is available at ► http://bibiserv.techfak.uni-bielefeld.de/rose/.

Chapter 5 as a benchmark for analyzing genomic alignment software. It has also been used to assess multiple sequence alignment software such as Kalign (Lassmann and Sonnhammer, 2005) and MUSCLE (Edgar, 2004a).

FIVE MAIN APPROACHES TO MULTIPLE SEQUENCE ALIGNMENT

There are many approaches to multiple sequence alignment; in the past decade many dozens of programs have been introduced. We may consider five algorithmic approaches: (1) exact methods, (2) progressive alignment (e.g., ClustalW), (3) iterative approaches (e.g., PRALINE, IterAlign, MUSCLE), (4) consistency-based methods (e.g., MAFFT, ProbCons), and (5) structure-based methods that include information about one or more known three-dimensional protein structures to facilitate creation of a multiple sequence alignment (e.g., Expresso). The programs we will describe in categories (3) to (5) are often overlapping; for example, all rely on progressive alignment and some combine iterative and structure-based approaches. All the programs offer trade-offs in speed and accuracy. MUSCLE and MAFFT are fastest, and are thus most useful for aligning large numbers of sequences. ProbCons and T-Coffee, although slower, are more accurate in many applications.

We will explore sets of distantly and closely related globin sequences in the FASTA format. These are available as web documents 6.3 and 6.4 at ► http://www.bioinfbook.org/chapter6. There are many ways that you can easily obtain a group of sequences in the FASTA format. Examples include HomoloGene at NCBI (for eukaryotic proteins), or you can select any subset of the results of a BLAST search and view the sequences in Entrez Protein (or Entrez Nucleotide) in the FASTA format.

We will explore how one set of globin sequences can be aligned differently using various programs, and we will try to assess which alignments are most accurate. A related question is the consequence of a misalignment. Potentially, the conservation of critical residues (such as active site amino acids of an enzyme, the heme-binding residues of a globin, or conserved residues that cause disease when mutated) may be missed. Phylogenetic inference (Chapter 7) may be compromised because all molecular phylogeny algorithms depend on a multiple sequence alignment as input. Protein structure prediction (Chapter 11) is severely compromised by faulty multiple sequence alignment, which is often a first step in homology-based modeling.

The programs we will explore can be used by web interfaces, although local installation of the programs typically allows you access to a more complete package of options. All the web interfaces allow you to paste in a set of DNA, RNA, or protein sequences in the FASTA format, or to upload a text file containing these sequences.

Exact Approaches to Multiple Sequence Alignment

Dynamic programming as described by Needleman and Wunsch (1970) for pairwise alignment is guaranteed to identify the optimal global alignment(s). Exact methods for multiple sequence alignment employ dynamic programming, although the matrix is multidimensional rather than two-dimensional. The goal is to maximize the summed alignment score of each pair of sequences. Exact methods generate optimal alignments but are not feasible in time or space for more than a few sequences. For N sequences, the computational time that is required is $O(2^N L^N)$ where N is the number of sequences and L is the average sequence length. An exact multiple sequence alignment of more than four or five average sized proteins would consume prohibitively too much time. Nonexact methods, which we will discuss next, are computationally feasible. For example, ClustalW has time complexity $O(N^4 + L^2)$ and MUSCLE has time complexity $O(N^4 + NL^2)$. Although they are faster, these heuristic approaches are not guaranteed to produce optimal alignments.

Progressive Sequence Alignment

The most commonly used algorithms that produce multiple alignments are derived from the progressive alignment method. This was proposed by Fitch and Yasunobu (1975) and described by Hogeweg and Hesper (1984) who applied it to the alignment of 5S ribosomal RNA sequences. The method was popularized by Feng and Doolittle (1987, 1990). It is called "progressive" because the strategy entails calculating pairwise sequence alignment scores between all the proteins (or nucleic acid sequences) being aligned, then beginning the alignment with the two closest sequences and progressively adding more sequences to the alignment. A benefit of this approach is that it permits the rapid alignment of even hundreds of sequences. A major limitation is that the final alignment depends on the order in which sequences are joined. Thus, it is not guaranteed to provide the most accurate alignments.

Perhaps the most popular web-based program for performing progressive multiple sequence alignment is ClustalW (Thompson et al., 1994). There are many ways to access the program (Box 6.2). The ClustalW algorithm proceeds in three stages. We can illustrate the procedure by aligning five distantly related globins, selected from Entrez and pasted into a text document in the FASTA format (Fig. 6.1). The results are shown in Figs. 6.2 and 6.3. Later we will also align five closely related globins (Figs. 6.4 and 6.5). In this particular example we select proteins for which the corresponding three-dimensional structure has been solved by x-ray crystallography. This will help us to interpret the accuracy of the alignment from a structural perspective as well as an evolutionary perspective.

1. In stage 1, the global alignment approach of Needleman and Wunsch (1970; Chapter 3) is used to create pairwise alignments of every protein that is to be included in a multiple sequence alignment (Fig. 6.2, stage 1). As shown in the figure, for an alignment of five sequences, 10 pairwise alignment scores are generated.

Algorithms that perform pairwise alignments generate raw similarity scores. Note that for the default setting of ClustalW the scores are simply the percent identities. Many progressive sequence alignment algorithms including ClustalW use a distance matrix rather than a similarity matrix to describe the relatedness of the proteins. The conversion of similarity scores for each pair of sequences to distance scores is outlined in Box 6.3. The purpose of generating distance measures is to generate a guide tree (stage 2, below) to construct the alignment.

Note that while most database searches such as BLAST rely on local alignment strategies, many multiple sequence alignments focus on global alignments, or a combination of global and local strategies.

For N sequences that are multiply aligned, the number of pairwise alignments that must be calculated for the initial matrix equals $\frac{1}{2}(N-1)(N)$. For five proteins, 10 pairwise alignments are made. For a multiple sequence alignment of 500 proteins, $(499)(500)/2 = 12,250$ pairwise alignments are made; this is why the speed of an algorithm can be a concern. ClustalW is slow relative to other approaches such as MUSCLE, described below, but for most typical applications its speed is quite reasonable.

To confirm that the ClustalW scores are percent identities, perform pairwise alignments between any two of the sequences in Fig. 6.2 or 6.4 using BLAST at NCBI (Chapter 3).

Box 6.2
Using ClustalW

ClustalW is accessed online at many servers, including ▶ http://www.ebi.ac.uk/clustalw/, where it is hosted by the European Bioinformatics Institute.

Another way to access ClustalW is through the EMBOSS program emma. A variety of EMBOSS servers hosting emma are available, including ▶ http://phytophthora.vbi.vt.edu/EMBOSS/, ▶ http://bioportal.cgb.indiana.edu/cgi-bin/emboss/emma and ▶ http://embossgui.sourceforge.net/demo/emma.html.

ClustalX is a downloadable stand-alone program related to ClustalW (Thompson et al., 1997). ClustalX offers a graphical user interface for editing multiple sequence alignments. You can obtain ClustalX at ▶ http://bips.u-strasbg.fr/fr/Documentation/ClustalX/. An introductory tutorial for using ClustalX in conjunction with phylogeny software has been written by Hall (2001).

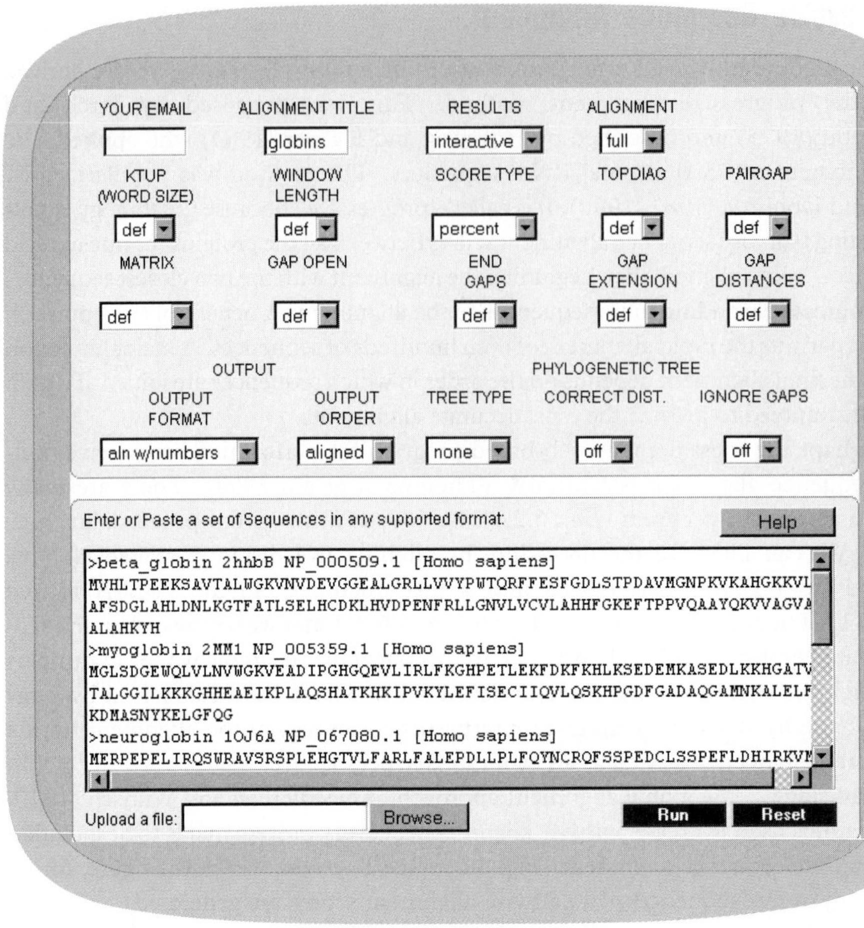

FIGURE 6.1. *Multiple sequence alignment of five distantly related globins using the ClustalW server at EBI (▶ http://www.ebi.ac.uk/clustalw/). Five distantly related globin proteins were pasted in using the FASTA format from Entrez (NCBI).*

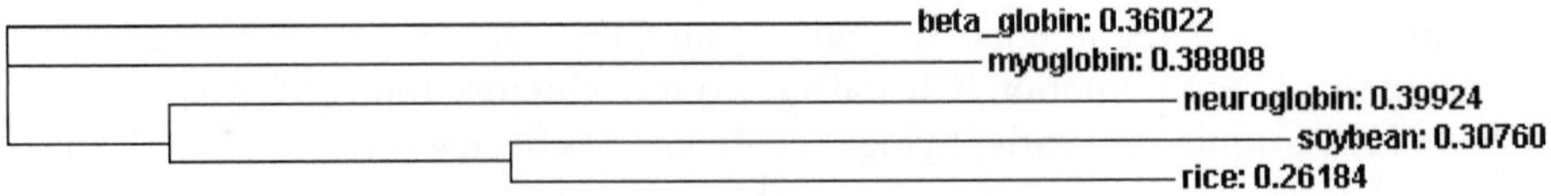

FIGURE 6.2. *Progressive alignment method of Feng and Doolittle (1987) used by many multiple alignment programs such as ClustalW. In stage 1, a series of pairwise alignments is generated for five distantly related globins (see Fig. 6.1). Note that the best score is for an alignment of two plant globins (score = 43; arrow 1). In stage 2, a guide tree is calculated describing the relationships of the five sequences based on their pairwise alignment scores. A graphical representation of the guide tree is shown using the JalView tool at the ClustalW web server. Branch lengths (rounded off) reflect distances between sequences and are indicated on the tree; compare to Fig. 6.4.*

Stage 1: generate a series of pairwise alignments

SeqA	Name	Len(aa)	SeqB	Name	Len(aa)	Score
1	beta_globin	147	2	myoglobin	154	25
1	beta_globin	147	3	neuroglobin	151	15
1	beta_globin	147	4	soybean	144	13
1	beta_globin	147	5	rice	166	21
2	myoglobin	154	3	neuroglobin	151	16
2	myoglobin	154	4	soybean	144	8
2	myoglobin	154	5	rice	166	12
3	neuroglobin	151	4	soybean	144	17
3	neuroglobin	151	5	rice	166	18
4	soybean	144	5	rice	166	43

← 1

Stage 2: create a guide tree, calculated from a distance matrix

```
(
beta_globin:0.36022,
myoglobin:0.38808,
(
neuroglobin:0.39924,
(
soybean:0.30760,
rice:0.26184)
:0.13652)
:0.06560);
```

beta_globin: 0.36022
myoglobin: 0.38808
neuroglobin: 0.39924
soybean: 0.30760
rice: 0.26184

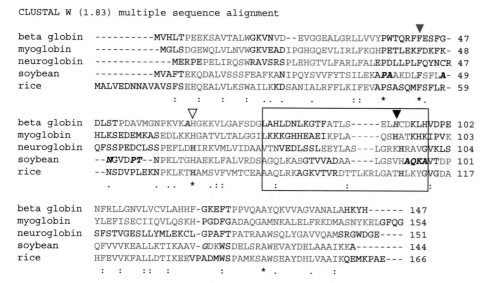

```
CLUSTAL W (1.83) multiple sequence alignment
                                                            ▼
beta globin    ----------MVHLTPEEKSAVTALWGKVNVD--EVGGEALGRLLVVYPWTQRFFESFG-  47
myoglobin      ----------MGLSDGEWQLVLNVWGKVEADIPGHGQEVLIRLFKGHPETLEKFDKFK-  48
neuroglobin    ------------MERPEPELIRQSWRAVSRSPLEHGTVLFARLFALEPDLLPLFQYNCR  47
soybean        ----------MVAFTEKQDALVSSSFEAFKANIPQYSVVFYTSILEKAPAAKDLFSFLA-  49
rice           MALVEDNNAVAVSFSEEQEALVLKSWAILKKDSANIALRFFLKIFEVAPSASQMFSFLR-  59
                       :    :     :    :  ..  .        ::   *    *.

                          ▽                           ▼
beta globin    DLSTPDAVMGNPKVKAHGKKVLGAFSDGLAHLDNLKGTFATLS-----ELHCDKLHVDPE 102
myoglobin      HLKSEDEMKASEDLKKHGATVLTALGGILKKKGHHEAEIKPLA-----QSHATKHKIPVK 103
neuroglobin    QFSSPEDCLSSPEFLDHIRKVMLVIDAAVTNVEDLSSLEEYLAS---LGRKHRAVGVKLS 104
soybean        --NGVDPT--NPKLTGHAEKLFALVRDSAGQLKASGTVVADAA----LGSVHAQKAVTDP 101
rice           --NSDVPLEKNPKLKTHAMSVFVMTCEAAAQLRKAGKVTVRDTTLKRLGATHLKYGVGDA 117
                .       .  ..  *   .::       :          :          .
```

```
beta globin    NFRLLGNVLVCVLAHHF-GKEFTPPVQAAYQKVVAGVANALAHKYH------ 147
myoglobin      YLEFISECIIQVLQSKH-PGDFGADAQGAMNKALELFRKDMASNYKELGFQG 154
neuroglobin    SFSTVGESLLYMLEKCL-GPAFTPATRAAWSQLYGAVVQAMSRGWDGE---- 151
soybean        QFVVVKEALLKTIKAAV-GDKWSDELSRAWEVAYDELAAAIKKA-------- 144
rice           HFEVVKFALLDTIKEEVPADMWSPAMKSAWSEAYDHLVAAIKQEMKPAE--- 166
                :   :   ::    :        *  .    .  .
```

FIGURE 6.3. *Multiple sequence alignment of five distantly related globins. The output is from ClustalW using the progressive alignment algorithm of Feng and Doolittle (1987). In stage 3, a multiple sequence alignment is created by performing progressive sequence alignments. First, the two closest sequences are aligned (soybean and rice globins). Next, further sequences are added in an order based on their position in the guide tree. An asterisk indicates positions in which the amino acid residue is 100% conserved in a column; a colon indicates conservative substitutions; a dot indicates less conservative substitutions. The proteins are human beta globin (accession NP_000509; Protein Data Bank identifier 2hhb), human myoglobin (NP_005359; 2MM1), human neuroglobin (NP_067080; 1OJ6A), leghemoglobin (from the soybean Glycine max; 1FSL), and nonsymbiotic plant hemoglobin (from rice; 1D8U). Regions of alpha helices (defined in Chapter 11) based on x-ray crystallography are indicated in red letters. Three highly conserved residues are indicated by arrowheads: phe44 of myoglobin (red arrowhead), his65 (open arrowhead); and his93 (black arrowhead). These two histidines are important in coordinating protein binding to the heme group. A box surrounds the second histidine including five amino acids downstream (to the carboxy-terminal) and 17 amino acids upstream (to the end of an alpha helical region). We will discuss the alignment within this box for ClustalW in comparison to other alignment programs (Fig. 6.6).*

In our example, note that the best pairwise global alignment score is for rice versus soybean hemoglobin (Fig. 6.2, arrow 1). For a group of closely related beta globins, all have high scores (Fig. 6.4), even for sequences from avian and mammalian species that diverged over 300 million years ago.

2. In the second stage, a guide tree is calculated from the distance (or similarity) matrix. There are two principal ways to construct a guide tree: the unweighted pair group method of arithmetic averages (UPGMA) and the neighbor-joining method. We will define these algorithms in Chapter 7. The two main features of a tree are its topology (branching order) and branch lengths (which can be drawn so that they are proportional to evolutionary distance). Thus, the tree reflects the relatedness of all the proteins to be multiply aligned.

In ClustalW, the tree is described with a written syntax called the Newick format, as well as with a graphical output (Figs. 6.2 and 6.4, stage 2). The chicken sequence has the lowest score relative to the human, chimpanzee, dog, and mouse beta globins, and this is reflected in its position in the guide tree (Fig. 6.4, stages 1 and 2). A tree can also be displayed graphically at the ClustalW site by using the JalView option.

Stage 1: generate a series of pairwise alignments

SeqA	Name	Len(aa)	SeqB	Name	Len(aa)	Score
1	human_NP_000509	147	2	Pan_troglodytes_XP_508242	147	100
1	human_NP_000509	147	3	Canis_familiaris_XP_537902	147	89
1	human_NP_000509	147	4	Mus_musculus_NP_058652	147	80
1	human_NP_000509	147	5	Gallus_gallus_XP_444648	147	69
2	Pan_troglodytes_XP_508242	147	3	Canis_familiaris_XP_537902	147	89
2	Pan_troglodytes_XP_508242	147	4	Mus_musculus_NP_058652	147	80
2	Pan_troglodytes_XP_508242	147	5	Gallus_gallus_XP_444648	147	69
3	Canis_familiaris_XP_537902	147	4	Mus_musculus_NP_058652	147	78
3	Canis_familiaris_XP_537902	147	5	Gallus_gallus_XP_444648	147	71
4	Mus_musculus_NP_058652	147	5	Gallus_gallus_XP_444648	147	66

FIGURE 6.4. Example of a multiple sequence alignment of closely related globin proteins using the progressive sequence alignment method of Feng and Doolittle (1987) as implemented by ClustalW. Compare these scores to those for distantly related proteins (Fig. 6.2), and note that the pairwise alignment scores are consistently higher and the distances (reflected in branch lengths on the guide tree) are much shorter.

Stage 2: create a guide tree, calculated from a distance matrix

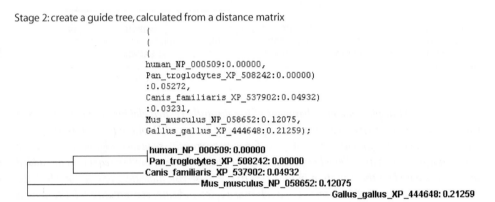

```
(
 (
  (
  human_NP_000509:0.00000,
  Pan_troglodytes_XP_508242:0.00000)
  :0.05272,
  Canis_familiaris_XP_537902:0.04932)
  :0.03231,
  Mus_musculus_NP_058652:0.12075,
  Gallus_gallus_XP_444648:0.21259);
```

human_NP_000509: 0.00000
Pan_troglodytes_XP_508242: 0.00000
Canis_familiaris_XP_537902: 0.04932
Mus_musculus_NP_058652: 0.12075
Gallus_gallus_XP_444648: 0.21259

Guide trees are usually not considered true phylogenetic trees, but instead are templates used in the third stage of ClustalW to define the order in which sequences are added to a multiple alignment. A guide tree is estimated from a distance matrix based on the percent identities between sequences you are aligning. In constrast, a phylogenetic tree almost always includes a model to account for multiple substitutions that commonly occur at the position of aligned amino acids (or nucleotides), as discussed in Chapter 7.

3. In stage 3, the multiple sequence alignment is created in a series of steps based on the order presented in the guide tree. The algorithm first selects the two most closely related sequences from the guide tree and creates a pairwise alignment. These two sequences appear at the terminal nodes of the tree, that is, the locations of extant sequences. For example, rice globin and soybean globin are aligned. The next sequence is either added to the pairwise alignment (to generate an aligned group of three sequences, sometimes called a profile) or used in another pairwise alignment. At some point, profiles are aligned with profiles. The alignment continues progressively until the root of the tree is reached, and all sequences have been aligned. At this point a full multiple sequence alignment is obtained (Figs. 6.3 and 6.5, stage 3).

In the alignment of five distantly related globins, we can note that a highly conserved phenylalanine is aligned (Fig. 6.3, red arrowhead) as is a histidine that coordinates heme binding in most globins (open arrowhead). However, an even more highly conserved histidine (black arrowhead) is aligned in beta globin and myoglobin, but is placed in a separate column for neuroglobin and two plant globins. This represents a misalignment, and we will explore how other programs treat this region. For a group of closely related globins, the level of conservation is so high that there are no gaps and thus no ambiguities about how to perform the alignment (Fig. 6.5).

```
CLUSTAL W (1.83) multiple sequence alignment


human_NP_000509          MVHLTPEEKSAVTALWGKVNVDEVGGEALGRLLVVYPWTQRFFESFGDLS 50
Pan_troglodytes_XP_508242 MVHLTPEEKSAVTALWGKVNVDEVGGEALGRLLVVYPWTQRFFESFGDLS 50
Canis_familiaris_XP_537902 MVHLTAEEKSLVSGLWGKVNVDEVGGEALGRLLIVYPWTQRFFDSFGDLS 50
Mus_musculus_NP_058652   MVHLTDAEKSAVSCLWAKVNPDEVGGEALGRLLVVYPWTQRYFDSFGDLS 50
Gallus_gallus_XP_444648  MVHWTAEEKQLITGLWGKVNVAECGAEALARLLIVYPWTQRFFASFGNLS 50
                         *** *  **. :: **.*** * *.***.***:*******:* ***:**


human_NP_000509          TPDAVMGNPKVKAHGKKVLGAFSDGLAHLDNLKGTFATLSELHCDKLHVD 100
Pan_troglodytes_XP_508242 TPDAVMGNPKVKAHGKKVLGAFSDGLAHLDNLKGTFATLSELHCDKLHVD 100
Canis_familiaris_XP_537902 TPDAVMGNPKVKAHGKKVLGAFSDGLKNLDNLKGTFAKLSELHCDKLHVD 100
Mus_musculus_NP_058652   SASAIMGNPKVKAHGKKVITAFNEGLKNLDNLKGTFASLSELHCDKLHVD 100
Gallus_gallus_XP_444648  SPTAILGNPMVRAHGKKVLTSFGDAVKNLDNIKNTFSQLSELHCDKLHVD 100
                         :. *::.*. *;******: :*.:.: :***:*.**: ************


human_NP_000509          PENFRLLGNVLVCVLAHHFGKEFTPPVQAAYQKVVAGVANALAHKYH 147
Pan_troglodytes_XP_508242 PENFRLLGNVLVCVLAHHFGKEFTPPVQAAYQKVVAGVANALAHKYH 147
Canis_familiaris_XP_537902 PENFKLLGNVLVCVLAHHFGKEFTPQVQAAYQKVVAGVANALAHKYH 147
Mus_musculus_NP_058652   PENFRLLGNAIVIVLGHHLGKDFTPAAQAAFQKVVAGVATALAHKYH 147
Gallus_gallus_XP_444648  PENFRLLGDILIIVLAAHFSKDFTPECQAAWQKLVRVVAHALARKYH 147
                         ****:***: :: **. *:.*:*** ***:**:* ** ***:***
```

FIGURE 6.5. Multiple sequence of five closely related beta globin orthologs (see Fig. 6.4). The output is a screen capture from ClustalW using the progressive alignment algorithm of Feng and Doolittle. The arrowheads (red, open, and black) correspond to the human beta globin phe44, his72, and his104 residues, respectively. These are highly conserved among the globin superfamily.

Box 6.3
Similarity versus Distance Measures

Trees that represent protein or nucleic acid sequences usually display the differences between various sequences. One way to measure distances is to count the number of mismatches in a pairwise alignment. Another method, employed by the Feng and Doolittle progressive alignment algorithm, is to convert similarity scores to distance scores. Similarity scores are calculated from a series of pairwise alignments among all the proteins being multiply aligned. The similarity scores S between two sequences (i, j) are converted to distance scores D using the equation

$$D = -\ln S_{eff}$$

where

$$S_{eff} = \frac{S_{real(ij)} - S_{rand(ij)}}{S_{iden(ij)} - S_{rand(ij)}} \times 100$$

Here, $S_{real(ij)}$ describes the observed similarity score for two aligned sequences i and j, $S_{iden(ij)}$ is the average of the two scores for the two sequences compared to themselves (if score i compared to i receives a score of 20 and score j compared to j receives a score of 10, then $S_{iden(ij)} = 15$); $S_{rand(ij)}$ is the mean alignment score derived from many (e.g., 1000) random shufflings of the sequences; and S_{eff} is a normalized score. If sequences i, j have no similarity, then $S_{eff} = 0$ and the distance is infinite. If sequences i, j are identical, then $S_{eff} = 1$ and the distance is 0.

The Feng–Doolittle approach includes the rule "once a gap, always a gap." The most closely related pair of sequences is aligned first. As further sequences are added to the alignment, there are many ways that gaps could be included. The rationale for the "once a gap, always a gap" rule is that the two most closely related sequences that are initially aligned should be weighted most heavily in assigning gaps. ClustalW dynamically assigns position-specific gap penalties that increase the likelihood of having a new gap occur in the same position as a preexisting gap. That serves to give the overall alignment a block-like structure that often appears efficient in terms of minimizing the number of gap positions.

Should an insertion be penalized the same amount as a deletion? No, according to Loytynoja and Goldman (2005): a single deletion event is typically penalized once where it occurs, but a single insertion event that occurs once inappropriately results in multiple penalties to all the other sequences. The result of these high penalties is that many multiple sequence alignments are unrealistically aligned with too few gaps. Loytynoja and Goldman (2005) introduced a pair hidden Markov model approach that distinguishes insertions from deletions. They showed that their method creates gaps that are consistent with phylogeny, even though the alignments appear less compact than with ClustalW. Their approach applies to the alignment of protein, RNA, or DNA sequences, but it may be especially useful for the alignment of genomic DNA. There, overfitting may occur with traditional progressive alignment, for example when one sequence has long insertions. The approach of Loytynoja and Goldman (2005), reviewed in Higgins et al. (2005), provides multiple sequence alignments that have more gaps but are likely to be more accurate, based on criteria such as correct alignment of exons.

ClustalW implements a series of additional features to optimize the alignment (Thompson et al., 1994). The distance of each protein (or DNA) sequence from the root of the guide tree is calculated, and those sequences that are most closely related are downweighted by a multiplicative factor. This adjustment assures that if an alignment includes a group of very closely related sequences as well as another group of divergent sequences, the closely related ones will not overly dominate the final multiple sequence alignment. Other adjustments include the use of a series of scoring matrices that are applied to pairwise alignments of proteins depending on their similarity, and compensation for differences in sequence length.

Many other algorithms use variants of progressive alignment. For example, Kalign employs a string-matching algorithm to achieve speeds ten times faster than ClustalW (Lassmann and Sonnhammer, 2005). Kalign aligns 100 protein sequences of length 500 residues in less than a second.

The website ► http://msa.cgb.ki. se includes Kalign for alignment, Kalignvu as a viewer, and Mumsa to assess the quality of a multiple sequence alignment (Lassmann and Sonnhammer, 2006). Kalign is also offered through the European Bioinformatics Institute (► http://www.ebi.ac.uk/kalign/).

Iterative Approaches

Iterative methods compute a suboptimal solution using a progressive alignment strategy, and then modify the alignment using dynamic programming or other methods until a solution converges. Thus, they create an initial alignment and then modify it to try to improve it. Progressive alignment methods have the inherent limitation that once an error occurs in the alignment process it cannot be corrected, and iterative approaches can overcome this limitation. In standard dynamic programming the branching order of the guide tree may be suboptimal, or the scoring parameters may cause gaps to be misplaced. Iterative refinement can search for more optimal solutions stochastically (seeking higher maximal scores according to some metric such as the sum-of-pairs scores; Box 6.1) or by systematically extracting and realigning sequences from an initial profile that is generated. Examples of programs employing iterative approaches are MAFFT (Multiple Alignment using Fast Fourier Transform) (Katoh et al., 2005), Iteralign (Karlin and Brocchieri, 1998), Praline (Profile ALIgNmEnt) (Heringa, 1999; Simossis and Heringa, 2005), and MUSCLE (MUltiple Sequence Comparison by Log-Expectation) (Edgar, 2004a, 2004b).

MAFFT offers a suite of tools with choices of more speed or accuracy. The fastest version involves progressive alignment using matching 6-tuples (strings of six residues) to calculate pairwise distances. This approach is called k-mer counting. A k-mer (also called a k-tuple or word) is a contiguous subsequence of length k. k-mer counting is extremely fast because it requires no alignment. The initial distance matrix can optionally be recalculated once all pairwise alignments are calculated, yielding a more reliable progressive alignment. In the iterative refinement step, a weighted sum-of-pairs score is calculated and optimized. MAFFT allows options including global or local pairwise alignment.

MAFFT and PRALINE can both incorporate information from homologous sequences that are analyzed in addition to those you submit for multiple sequence alignment. These sequences are used to improve the multiple sequence alignment; in the case of MAFFT, the extra sequences are then removed. PRALINE performs a PSI-BLAST search (Chapter 5) on the query protein sequences and then performs progressive alignment using the PSI-BLAST profiles. PRALINE also permits the incorporation of predicted secondary structure information.

Since its introduction in 2004, the MUSCLE program of Robert Edgar (2004a, 2004b) has become popular because of its accuracy and its exceptional speed, especially for multiple sequence alignments involving large numbers of sequences. For example, 1000 protein sequences of average length 282 residues were aligned in 21 seconds on a desktop computer (Edgar, 2004a). MUSCLE operates in a series of three stages. First, a draft progressive alignment is generated. To achieve this, the algorithm calculates the similarity between each pair of sequences using either the fractional identity (calculated from a global alignment of each pair of sequences), or k-mer counting. Based on the similarities, MUSCLE calculates a triangular distance matrix, then constructs a rooted tree using UPGMA or neighbor-joining (see Chapter 7). Sequences are added progressively to the multiple sequence alignment following the branching order of the tree. In the second stage, MUSCLE improves the tree and builds a new progressive alignment (or a new set of alignments). The similarity of each pair of sequences is assessed using the fractional identity, and a tree is constructed using a Kimura distance matrix (discussed in Chapter 7). In a comparison of two sequences there is some likelihood that multiple amino acid (or nucleotide) substitutions occurred at any given position, and the Kimura distance matrix provides a model for such changes. As each tree is constructed it is compared to the tree from stage 1, and the process results in an improved progressive alignment. In stage 3 the guide tree is iteratively refined by systematically partitioning the tree to obtain subsets; an edge (branch) of the tree is deleted to create a bipartition. Next, MUSCLE extracts a pair of profiles (multiple sequence alignments), and realigns them (performing profile-profile alignment; see Box 6.4). The algorithm accepts or rejects the newly generated alignment based on whether the sum-of-pairs score increases. All edges of the tree are systematically visited and deleted to create bipartitions. This iterative refinement step is rapid and had been shown earlier to increase the accuracy of the multiple sequence alignment (Hirosawa et al., 1995).

The alignments of five distantly related globins using PRALINE (Fig. 6.6a) and MUSCLE (Fig. 6.6b) show a somewhat different result than we saw with ClustalW (Fig. 6.3). In the boxed region there are only 10 total gaps with PRALINE and 4 with MUSCLE, compared with 17 using ClustalW. This reflects a more compact overall alignment. Both these programs still fail to align the highly conserved histidine (Fig. 6.6a and b, black arrowhead).

MAFFT is available at the EBI website, ► http://www.ebi.ac.uk/mafft/, or with more options from its project home page, ► http://align.bmr.kyushu-u.ac.jp/mafft/software/. PRALINE can be accessed from ► http://zeus.cs.vu.nl/programs/pralinewww/.

The idea of a triangular distance matrix in stage 1 is that the distance measure between sequences (A,B) equals the distance of (A,C) plus (B,C). This is a good approximation for closely related sequences, but the accuracy is further increased using the Kimura distance correction in stage 2.

MUSCLE can be downloaded or accessed via web servers at ► http://www.drive5.com/muscle/ or at the European Bioinformatics website, ► http://www.ebi.ac.uk/muscle/.

Box 6.4
Profile-Profile Alignment with the MUSCLE Algorithm

The name MUSCLE (multiple sequence comparison by log expectation) includes the phrase "log expectation." Like ClustalW, MUSCLE measures the distance between sequences (Edgar, 2004a, 2004b). In its third stage, MUSCLE iteratively refines a multiple sequence alignment by deleting the edge of the guide tree to form a bipartition, then extracting a pair of profiles and realigning them. It does this using several scoring functions to optimally align pairs of columns. For amino acid types i and j, p_i is the background probability of i, p_{ij} is the joint probability of i and j being aligned, S_{ij} is the score from a substitution matrix, f_i^x is the observed frequency of i in column x of the first profile, f_G^x is the observed frequency of gaps in column x, and α_i^x is the estimated probability of observing residue i in position x in the family based on the observed frequencies f. (Note that $S_{ij} = \log(p_{ij}/p_ip_j)$ as discussed in Chapter 3.) MUSCLE, ClustalW, and MAFFT use a profile sum-of-pairs (PSP) scoring function:

$$PSP^{xy} = \sum_i \sum_j f_i^x f_j^y S_{ij}$$

PSP is a sequence-weighted sum of substitution matrix scores for each pair of letters (one from each column that is being aligned in a pairwise fashion). The PSP function maximizes the sum-of-pairs objective score. MUSCLE applies two PAM matrices for its PSP function. MUSCLE also employs a novel log-expectation (LE) score that is defined as follows:

$$LE^{xy} = (1 - f_G^x)(1 - f_G^y) \log \sum_i \sum_j f_i^x f_j^y \frac{p_{ij}}{p_ip_j}$$

The factor $(1 - f_G)$ is the occupancy of a column. This promotes the alignment of columns that are highly occupied (i.e., that have fewer gaps) while downweighting column pairs with many gaps. Edgar (2004a) reported that this significantly improved the accuracy of the alignment.

Consistency-Based Approaches

In progressive alignments using the Feng–Doolittle approach, pairwise alignment scores are generated and used to build a tree. Consistency-based methods adopt a different approach by using information about the multiple sequence alignment as it is being generated to guide the pairwise alignments. We will discuss two consistency-based multiple sequence alignment programs: ProbCons (Do et al., 2005) and T-Coffee (Notredame et al., 2000). The MAFFT program also includes an iterative refinement approach with consistency-based scores (Katoh et al., 2005).

The idea of consistency is that for sequences x, y, and z, if residue x_i aligns with z_k and z_k aligns with y_j, then x_i should align with y_j. Consistency-based techniques score pairwise alignments in the context of information about multiple sequences, for example, adjusting the score of x_i to y_j based on the knowledge that z_k aligns to both x_i and to y_j. This approach is distinctive because it incorporates evidence from multiple sequences to guide the creation of a pairwise alignment (Do et al., 2005). Using the notation given in a review by Wallace and colleagues (2005), the likelihood that residue i from sequence x and residue j from sequence y are aligned, given the sequences of x and y, is given by:

$$P(x_i \sim y_j | x, y) \qquad (6.1)$$

This is the posterior probability, and it is calculated for each pair of amino acids. The consistency transformation further incorporates data from additional residues to improve the estimate of two residues aligning (that is, given information about how x and y each align with z):

$$P(x_i \sim y_j | x, y, z) \approx \sum_k P(x_i \sim z_k | x, z) P(y_i \sim z_k | y, z) \qquad (6.2)$$

The consistency-based approach often generates final multiple sequence alignments that are more accurate than those achieved by progressive alignments, based on benchmarking studies.

The ProbCons algorithm has five steps. First, the algorithm calculates the posterior probability matrices for each pair of sequences. This involves a pair hidden Markov model as described in Fig. 5.12. This HMM has three states: M (corresponding to two aligned positions of sequences x and y), I_x (a residue in sequence x that is aligned to a gap), and I_y (a residue in y that is aligned to a gap). There is an initial probability of starting in a particular state, a transition probability from the initial state to the next residue, and an emission probability for the next residue to be aligned. Second, the expected accuracy of each pairwise alignment is computed. The expected accuracy is the number of correctly aligned pairs of residues divided by the length of the shorter sequence. The alignment is performed according to the Needleman–Wunsch dynamic programming method, but instead of using a PAM or BLOSUM scoring matrix, scores are assigned based on the posterior probability terms for the corresponding residues and gap penalties are set to zero. Third, the quality scores for each pairwise alignment are reestimated by applying a "probabilistic consistency transformation." This step applies information about conserved residues that were identified through all the pairwise alignments, resulting in the use of more accurate substitution scores. Fourth, an expected accuracy guide tree is constructed using hierarchical clustering (similar to the approach adopted by ClustalW). The guide tree is based on similarities (rather than distances). Fifth, the sequences are progressively aligned (as in ClustalW) by following the order specified by the guide tree. Further iterative refinements may be applied. Do et al. (2005) reported that ProbCons outperformed six other multiple sequence alignment programs, including ClustalW, DIALIGN, T-Coffee, MAFFT, MUSCLE, and Align-m, based on testing on the BAliBASE, PREFAB, and SABmark benchmark databases.

ProbCons is available at ▶ http://probcons.stanford.edu/.

T-Coffee is an acronym for tree based consistency objective function for alignment evaluation. T-Coffee first computes a library consisting of pairwise alignments. By default these include all possible pairwise global alignments of the input sequences (using the Needleman–Wunsch algorithm), and the ten highest-scoring local alignments. Every pair of aligned residues is assigned a weight. These weights are recalculated to generate an "extended library" that serves as a position-specific substitution matrix. The program then computes a multiple sequence alignment by progressive alignment, creating a distance matrix, calculating a neighbor-joining

T-Coffee was developed by Cédric Notredame, Desmond Higgins, Jaap Heringa, and colleagues. It is available at ▶ http://www.tcoffee.org. It is also mirrored at the European Bioinformatics Institute (▶ http://www.ebi.ac.uk/t-coffee/), the Swiss Institute of Bioinformatics, and the Centre National de la Recherche Scientifique (Paris).

(a) Praline multiple sequence alignment

```
                                                                    ▼
beta globin    ..........MVHLTPEEKSAVTALWGKV..NVDEVGGEALGRLLVVYPWTQRFFES.FG
myoglobin      ...........MGLSDGEWQLVLNVWGKVEADIPGHGQEVLIRLFKGHPETLEKFDK.FK
neuroglobin    ............MERPEPELIRQSWRAVSRSPLEHGTVLFARLFALEPDLLPLFQYNCR
soybean        ..........MVAFTEKQDALVSSSFEAFKANIPQYSVVFYTSILEKAPAAKDLFS..FL
rice           MALVEDNNAVAVSFSEEQEALVLKSWAILKKDSANIALRFFLKIFEVAPSASQMFS..FL
Consistency    0000000000142654382579345734633643436244536864 33*35344*50063

                                   ▽                    ┌──────────▼──────┐
beta globin    DLSTPDAVMGNPKVKAHGKKVLGAFSDG│LAHLDNLKGTFATLSEL..HCDKLH│....VDP
myoglobin      HLKSEDEMKASEDLKKHGATVLTALGGI│LKKKGHHEAEIKPLAQS..HATKHK│....IPV
neuroglobin    QFSSPEDCLSSPEFLDHIRKVMLVIDAA│VTNVEDLSSLEEYLASLGRKHRAVG│....VKL
soybean        A.NGVDP..TNPKLTGHAEKLFALVRDS│AGQL.KASGTVVADAA....LGSVHAQKAVTD│
rice           R.NSDVPLEKNPKLKTHAMSVFVMTCEA│AAQL.RKAGKVTVRDTTLKRLGATHLKYGVGD│
Consistency    3166354224776653*436863542445445133563433354200333544│0000922
```

(b) MUSCLE (3.6) multiple sequence alignment

```
                                                                   ▼
beta globin    -----------MVHLTPEEKSAVTALWGKVNVD--EVGGEALGRLLVVYPWTQRFFES-FG
myoglobin      -----------MGLSDGEWQLVLNVWGKVEADIPGHGQEVLIRLFKGHPETLEKFDK-FK
neuroglobin    ------------MERPEPELIRQSWRAVSRSPLEHGTVLFARLFALEPDLLPLFQYNCR
soybean        ----------MVAFTEKQDALVSSSFEAFKANIPQYSVVFYTSILEKAPAAKDLFSF-LA
rice           MALVEDNNAVAVSFSEEQEALVLKSWAILKKDSANIALRFFLKIFEVAPSASQMFSF-LR
                       :    :   ..   .     :        ::   *    *.

                                  ▽              ┌────────▼─────────┐
beta globin    DLSTPDAVMGNPKVKAHGKKVLGAF---SDG│LAHLDNLKGTFATLSELHCDKLH│--VDPE
myoglobin      HLKSEDEMKASEDLKKHGATVLTAL---GGI│LKKKGHHEAEIKPLAQSHATKHK│--IPVK
neuroglobin    QFSSPEDCLSSPEFLDHIRKVMLVI---DAA│VTNVEDLSSLEEYLASLGRKHRA│VGVKLS
soybean        NGVDP----TNPKLTGHAEKLFALVRDSAGQ│LKASGTVVAD----AALGSVHAQKA│VTDP
rice           NSDVP--LEKNPKLKTHAMSVFVMTCEAAAQ│LRKAGKVTVRDTTLKRLGATHLKY│GVGDA
                 .  ..  *  .::      │  :        :         : │  :

beta globin    NFRLLGNVLVCVLAHHFGKE-FTPPVQAAYQKVVAGVANALAHKYH------
myoglobin      YLEFISECIIQVLQSKHPGD-FGADAQGAMNKALELFRKDMASNYKELGFQG
neuroglobin    SFSTVGESLLYMLEKCLGPA-FTPATRAAWSQLYGAVVQAMSRGWDGE----
soybean        QFVVVKEALLKTIKAAVGDK-WSDELSRAWEVAYDELAAAIKKA--------
rice           HFEVVKFALLDTIKEEVPADMWSPAMKSAWSEAYDHLVAAIKQEMKPAE---
               : : :  ::   :        :       *.   . .  :
```

FIGURE 6.6. Multiple sequence alignment of five distantly related globins using four different programs. The alignments were performed with (a) PRALINE, (b) MUSCLE, (c) ProbCons, and (d) T-Coffee. The proteins used to make the alignments and the symbols used to illustrate the figure are the same as those described in Fig. 6.3. Note that the programs differ in their abilities to align corresponding regions of alpha helical secondary structure (red lettering); in their alignment of a highly conserved histidine residue (black arrowhead); and in the number and placement of gaps (see boxed regions).

You can see an output of the five distantly related globins using M-Coffee in web document 6.5.

PipeAlign is available at ▶ http://bips.u-strasbg.fr/PipeAlign/.

guide tree, and using dynamic programming and the substitution matrix derived from the extended library.

T-Coffee includes a suite of related alignment and evaluation tools. M-Coffee (Meta-Coffee) combines the output of as many as 15 different multiple sequence alignment methods (Wallace et al., 2006; Moretti et al., 2007). These include T-Coffee, ClustalW, MAFFT, MUSCLE, and ProbCons. M-Coffee employs a consistency-based approach to estimate a consensus alignment that is more accurate than any of the individual methods. By adding structural information (discussed next), even further accuracy is achieved.

Structure-Based Methods

Tertiary structures evolve more slowly than primary sequences. Thus, for example, human beta globin and myoglobin share limited sequence identity (in the "twilight zone") yet share structures that are clearly related. It is possible to improve the accuracy of multiple sequence alignments by including information about the three-dimensional structure of one or more members of the group of proteins being

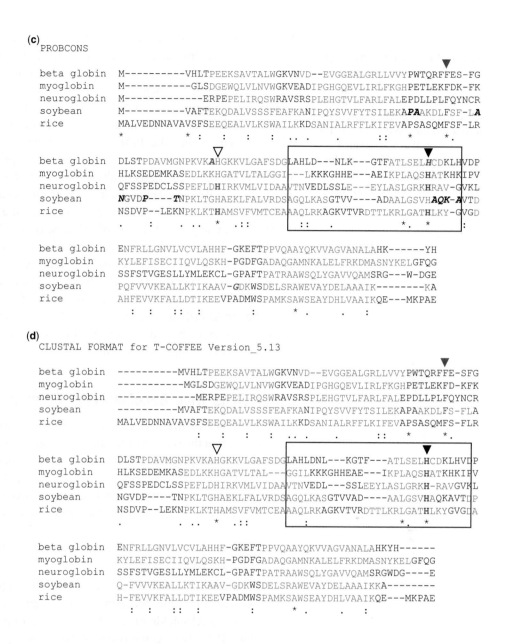

(c)
PROBCONS

```
beta globin   M----------VHLTPEEKSAVTALWGKVNVD--EVGGEALGRLLVVYPWTQRFFES-FG
myoglobin     M-----------GLSDGEWQLVLNVWGKVEADIPGHGQEVLIRLFKGHPETLEKFDK-FK
neuroglobin   M------------ERPEPELIRQSWRAVSRSPLEHGTVLFARLFALEPDLLPLFQYNCR
soybean       M----------VAFTEKQDALVSSSFEAFKANIPQYSVVFYTSILEKAPAAKDLFSF-LA
rice          MALVEDNNAVAVSFSEEQEALVLKSWAILKKDSANIALRFFLKIFEVAPSASQMFSF-LR
               *           *   :   :   :  ..  .     ::   *    *.
```

```
beta globin   DLSTPDAVMGNPKVKAHGKKVLGAFSDGLAHLD---NLK---GTFATLSELHCDKLHVDP
myoglobin     HLKSEDEMKASEDLKKHGATVLTALGGI---LKKKGHHE---AEIKPLAQSHATKHKIPV
neuroglobin   QFSSPEDCLSSPEFLDHIRKVMLVIDAAVTNVEDLSSLE---EYLASLGRKHRAV-GVKL
soybean       NGVDP----TNPKLTGHAEKLFALVRDSAGQLKASGTVV----ADAALGSVHAQK-AVTD
rice          NSDVP--LEKNPKLKTHAMSVFVMTCEAAQLRKAGKVTVRDTTLKRLGATHLKY-GVGD
               .   :   . ..  *  .::        ::   .        *. *
```

```
beta globin   ENFRLLGNVLVCVLAHHF-GKEFTPPVQAAYQKVVAGVANALAHK------YH
myoglobin     KYLEFISECIIQVLQSKH-PGDFGADAQGAMNKALELFRKDMASNYKELGFQG
neuroglobin   SSFSTVGESLLYMLEKCL-GPAFTPATRAAWSQLYGAVVQAMSRG---W-DGE
soybean       PQFVVVKEALLKTIKAAV-GDKWSDELSRAWEVAYDELAAAIK-------KA
rice          AHFEVVKFALLDTIKEEVPADMWSPAMKSAWSEAYDHLVAAIKQE---MKPAE
               :  :   ::  :      *  .   :
```

(d)
CLUSTAL FORMAT for T-COFFEE Version_5.13

```
beta globin   ----------MVHLTPEEKSAVTALWGKVNVD--EVGGEALGRLLVVYPWTQRFFE-SFG
myoglobin     ----------MGLSDGEWQLVLNVWGKVEADIPGHGQEVLIRLFKGHPETLEKFD-KFK
neuroglobin   ------------MERPEPELIRQSWRAVSRSPLEHGTVLFARLFALEPDLLPLFQYNCR
soybean       ----------MVAFTEKQDALVSSSFEAFKANIPQYSVVFYTSILEKAPAAKDLFS-FLA
rice          MALVEDNNAVAVSFSEEQEALVLKSWAILKKDSANIALRFFLKIFEVAPSASQMFS-FLR
               ::    :      :   :  ...   .     ::   *    *.
```

```
beta globin   DLSTPDAVMGNPKVKAHGKKVLGAFSDGLAHLDNL---KGTF---ATLSELHCDKLHVDP
myoglobin     HLKSEDEMKASEDLKKHGATVLTAL---GGILKKKGHHEAE---IKPLAQSHATKHKIPV
neuroglobin   QFSSPEDCLSSPEFLDHIRKVMLVIDAAVTNVEDL---SSLEEYLASLGRKH-RAVGVKL
soybean       NGVDP----TNPKLTGHAEKLFALVRDSAGQLKASGTVVAD----AALGSVHAQKAVTDP
rice          NSDVP--LEKNPKLKTHAMSVFVMTCEAAQLRKAGKVTVRDTTLKRLGATHLKYGVGDA
               .     . ..  *  .::       :          *. *
```

```
beta globin   ENFRLLGNVLVCVLAHHF-GKEFTPPVQAAYQKVVAGVANALAHKYH------
myoglobin     KYLEFISECIIQVLQSKH-PGDFGADAQGAMNKALELFRKDMASNYKELGFQG
neuroglobin   SSFSTVGESLLYMLEKCL-GPAFTPATRAAWSQLYGAVVQAMSRGWDG----E
soybean       Q-FVVVKEALLKTIKAAV-GDKWSDELSRAWEVAYDELAAAIKKA--------
rice          H-FEVVKFALLDTIKEEVPADMWSPAMKSAWSEAYDHLVAAIKQE---MKPAE
               :  :   ::  :      :     *  .    . :
```

FIGURE 6.6. (Continued)

aligned. Programs that enable you to incorporate structural information include PRALINE (Simossis and Heringa, 2005), the T-Coffee module Expresso (Armougom et al., 2006b), and PipeAlign (Plewniak et al., 2003).

When you use the Expresso program at the T-Coffee website, you submit a series of sequences (typically in the fasta format). Each sequence is automatically searched by BLAST against the Protein Data Bank (PDB) database, and matches (sharing >60% amino acid identity) are used to provide a template to guide the creation of the multiple sequence alignment.

We described BLAST in Chapter 4, and we will describe PDB in Chapter 11.

Structural information can also be used to assess the accuracy of a multiple sequence alignment after it has been made. This is done in benchmarking studies (described above) for protein families having known structures. In another approach you can incorporate structural information and assess the quality of a protein multiple sequence alignment that you make at the iRMSD-APDB ("Analyze alignments with Protein Data Bank") server of the T-Coffee package (O'sullivan et al., 2003;

The iRMSD-APDB server is part of the T-Coffee suite of tools (► http://www.tcoffee.org). Examples of five divergent and five closely related globin sequences, formatted for input to the APDB server, as well as the detailed output, are available in web documents 6.6 and 6.7 at ► http://www.bioinfbook.org/chapter6.

Armougom et al., 2006c). It is necessary to obtain the accession numbers corresponding to the Protein Data Bank (PDB) file having the known structures of at least two of the proteins you are aligning. As an example, we can obtain the PDB accession numbers for each of the five distantly related globins described above by performing a blastp search at NCBI, restricting the output to PDB. Next, perform a multiple sequence alignment using T-Coffee or any other program. Finally, input this alignment (using the PDB accession number in place of the name) to the APDB server at the T-Coffee website. The output provides an analysis of the quality of the alignment on the basis of all pairwise comparisons of those sequences having structures as well as an average quality assessment for each protein. The main approach to assessing how well two structures align is to measure the root mean square deviation (RMSD) (see Chapter 11). The RMSD is a measure of how closely the alpha carbons of two aligned amino residues are positioned. Notredame and colleagues introduced iRMSD as an intra molecular RMSD measure (Armougom et al., 2006a).

For the case of five divergent globins analyzed with the iRMSD-APDB server, 79% of the pairwise columns could be evaluated, 51% of the columns were aligned correctly (according to APDB), and the average iRMSD over all the evaluated columns was 1.07 Ångstroms. This analysis did not depend on a reference alignment, but instead involved a calculation of the superposition of the structures in the alignment.

Conclusions from Benchmarking Studies

We have discussed some of the programs for making multiple sequence alignments, and we have seen that they can produce differing results for a set of distantly related globins. Nonetheless most programs produce reasonably consistent alignments, especially for relatively closely related protein or DNA sequences. Comparative studies of multiple sequence alignment algorithms have been performed based on tests against benchmark databases. Some of the general conclusions include the following.

- Adding more homologs to a multiple sequence alignment improves its accuracy (Katoh et al., 2005).

- As the group of sequences being multiply aligned begins to share less amino acid identity, the accuracy of the alignments decreases (Briffeuil et al., 1998; Blackshields et al., 2006). For groups of sequences that share less than 25% identity, the problem becomes especially severe. Thompson et al. (1999) found that the best programs available at the time (PRRP, ClustalX, and SAGA) aligned about 60% to 70% of the amino acid residues for groups of proteins with <25% identity. For multiple sequence alignments of proteins sharing more identity (20% up to 40%), they found that on average 80% of the residues were aligned properly (Thompson et al., 1999).

- For highly divergent DNA sequences, programs that use local alignment (such as DiAlign and LAGAN) perform better than those using global alignment (such as ClustalW) (Kumar and Filipski, 2007).

- Orphan sequences are proteins that are highly divergent members of a family. If we examined a multiple sequence alignment of retinol-binding protein (RBP) from 10 species, then added the distantly related odorant-binding protein (OBP) to that multiple sequence alignment, OBP would be considered an orphan. Orphans might be expected to disrupt the organization of a multiple sequence alignment, and yet they do not. Global alignment algorithms outperform local alignment methods for the introduction of orphans to an alignment (Thompson et al., 1999).

- Separate multiple sequence alignments can be combined, such as a group of closely related myoglobins and a group of closely related neuroglobins. Iterative algorithms performed this task better than progressive alignment methods (Thompson et al., 1999). However, many programs have difficulty in accurately producing a single alignment from a subset of alignments.

- Often, some proteins in a family contain large extensions at the amino- and/or carboxy-terminals. Overall, local alignment programs dramatically outperformed global alignment programs at this task. For most multiple sequence alignment applications, global alignments are superior.

DATABASES OF MULTIPLE SEQUENCE ALIGNMENTS

We have discussed different methods for creating multiple sequence alignments. We will next examine databases of precomputed multiple sequence alignments, many of which are available. These may be searched using text (i.e., a keyword search) or using any query sequence. The query may be an already known sequence (such as myoglobin or RBP) or any novel protein (such as the raw sequence of a new lipocalin or globin you have identified). In some databases, the query sequence you provide is incorporated into the multiple sequence alignment of a particular precomputed protein family.

Pfam: Protein Family Database of Profile HMMs

Pfam is one of the most comprehensive databases of protein families (Bateman et al., 2004; Finn et al., 2006). It is a compilation of both multiple sequence alignments and profile HMMs of protein families. The database can be searched using text (keywords or protein names) or by entering sequence data. Its combination of HMM-based approach and expert curation makes Pfam one of the most trusted and widely used resources for protein families.

Pfam consists of two databases. Pfam-A is a manually curated collection of protein families in the form of multiple sequence alignments and profile HMMs. HMMER software (Chapter 5) is used to perform searches. For each family, Pfam provides four features: annotation, a seed alignment, a profile HMM, and a full alignment. The full alignment can be quite large; currently the top 20 Pfam families each contain over 20,000 sequences in their full alignment. The seed alignments contain a smaller number of representative family members. Sequences in Pfam-A are grouped in families, assigned stable accession numbers (such as PF00042 for globins) and expertly curated. Additional protein sequences are automatically aligned and deposited in Pfam-B where they are not annotated or assigned permanent accession numbers. Pfam-B serves as a useful supplement that makes the database more comprehensive. For all Pfam families, the underlying HMM is accessible from the main output page.

We can see the main features of Pfam in a search for globins using the Wellcome Trust Sanger Institute site. There are three main ways to access the database: by browsing for families, by entering a protein sequence search (with a protein accession number or sequence), and by entering a text search. From the front page, select a text-based search and enter "globin." The results summary includes links to the Pfam entry and to related databases (InterPro, described below; the Protein Data

Pfam is maintained by a consortium of researchers, including Alex Bateman, Ewan Birney, Lorenzo Cerrutti, Richard Durbin, Sean Eddy, and Erik Sonnhammer, and others. Five sites host Pfam: ► http://www.sanger.ac.uk/Software/Pfam/ (U.K.), ► http://pfam.janelia.org/ (U.S.), ► http://pfam.cgb.ki.se/ (Sweden), ► http://pfam.jouy.inra.fr/ (France), and ► http://pfam.ccbb.re.kr/index.shtml (South Korea). Version 23.0 (July 2008) has 10,340 protein families. Pfam is based on sequences in Swiss-Prot and SP-TrEMBL (Chapter 2). Currently (May 2007), 74% of the proteins in those databases have at least one domain that matches to a Pfam family.

Mycobacterium leprae is a bacterium that causes leprosy. Its globin has accession number NP_301903.

You can also search Pfam with a DNA query. Go to ▶ http://www.sanger.ac.uk/Software/Pfam/dnasearch.shtml.

Bank, introduced in Chapter 11; and clans). Each protein in Pfam can have membership in exactly one family. Some proteins, such as sperm whale myoglobin and a globin from the leprosy-causing bacterium *Mycobacterium leprae*, belong to distinct families (globins and bacterial-like globins, respectively). Those two families are distantly related and are defined as members of a larger clan.

The output includes an overview of the globin family, including description of the structure of a typical member, a Pfam accession number, clan membership, and a description of the globin family from the InterPro database (discussed below) (Fig. 6.7a). The Pfam entry further includes access to the alignment, domain organization, species distribution, and a phylogenetic tree (Fig. 6.7b). The alignment can be viewed for the seed set, consisting of a core group of representative members of the family, or the full set, consisting of all known family members. The alignment can be retrieved in a variety of formats, including gapped alignments (useful for viewing aligned regions of the family) or ungapped alignments (useful as input into other multiple sequence alignment programs such as those discussed

FIGURE 6.7. The Pfam database is a comprehensive resource for studying protein families. (a) A typical entry is shown for globins, including a representative three-dimensional structure, a list of related Pfam families (e.g., the bacteria-like globins), and a description of the protein (from InterPro). (b) The output options include viewing the seed alignment (with a core of 76 representative globin sequences) or the full alignment (with 2,039 globins in this particular case). Various format options are provided (see Figs. 6.8 and 6.9). Other output options include the domain architecture, species distribution, and a phylogenetic tree. HMMER-derived hidden Markov models corresponding to the globin family can also be viewed.

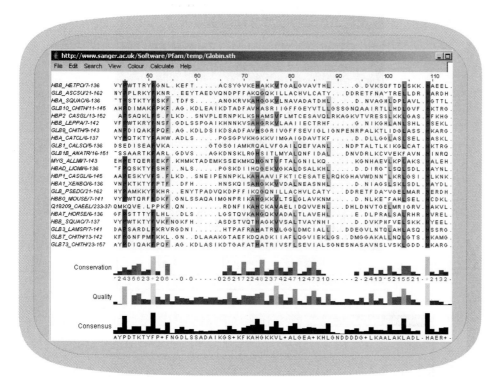

FIGURE 6.8. A Pfam alignment can be retrieved in the JalView Java viewer format. The Pfam JalView applet displays a multiple sequence alignment of any Pfam protein family. The aligned residues can be viewed with a variety of color schemes. The relationships of the proteins within the family can be explored using a variety of algorithms, including principal components analysis and phylogenetic trees (Fig. 6.9).

earlier in this chapter). One of the versatile output formats is JalView. After selecting this option, press the JalView button. A Java applet allows the multiple sequence alignment to be viewed, analyzed, and saved in a variety of ways (Fig. 6.8). The applet will display a principal components analysis (PCA) on the aligned family (Fig. 6.9a). We describe PCA, a technique to reduce highly dimensional data into two- (or three-) dimensional space, in Chapter 9 (Fig. 9.16). Here, each protein in a multiple sequence alignment is represented as a point in space based on a distance metric, and outliers are easily identified. Similar information can be represented with a phylogenetic tree (Fig. 6.9b; see Chapter 7) using the Java applet.

Smart

The Simple Modular Architecture Research Tool (SMART) is a database of protein families implicated in cellular signaling, extracellular domains, and chromatin function (Schultz et al., 1998; Ponting et al., 1999; Letunic et al., 2006). Like Pfam, SMART employs profile HMMs using HMMER software. SMART can be used in normal mode (providing searches against Swiss-Prot, SP-TrEMBL, and stable Ensembl proteomes) or in genomic mode (providing searches against proteomes of completely sequenced metazoan organisms from Ensembl or other organisms from Swiss-Prot, including eukaryotes, bacteria, and archaea).

Also like Pfam, the SMART database is searchable by sequence or by keyword, or by browsing the available domains. Domains identified in a SMART search are extensively annotated with information on functional class, tertiary structure, and taxonomy.

Conserved Domain Database

The Conserved Domain Database (CDD) is an NCBI tool that allows sequence-based or text-based queries of Pfam and SMART. CDD uses reverse position-specific BLAST

SMART (► http://smart.embl-heidelberg.de/) currently (May 2007) has 726 profile HMMs (domains). It was developed by Peer Bork and colleagues.

CDD is available at ► http://www.ncbi.nlm.nih.gov/Structure/cdd/cdd.shtml or through the main BLAST page (► http://www.ncbi.nlm.nih.gov/BLAST/). CDD can also be searched by entering a protein query sequence into the Domain Architecture Retrieval Tool (DART) at NCBI. DART is available at ► http://www.ncbi.nlm.nih.gov/Structure/lexington/html/overview.html.

(a)

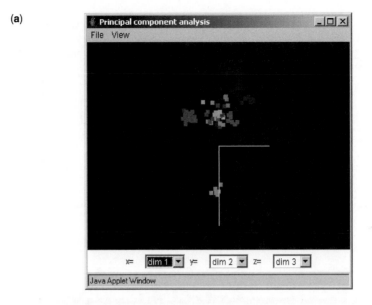

(b)

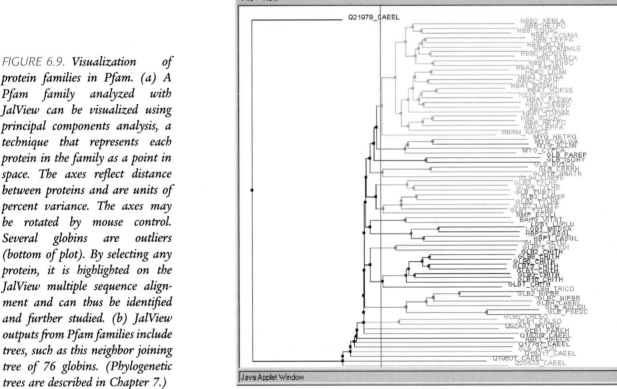

FIGURE 6.9. *Visualization of protein families in Pfam. (a) A Pfam family analyzed with JalView can be visualized using principal components analysis, a technique that represents each protein in the family as a point in space. The axes reflect distance between proteins and are units of percent variance. The axes may be rotated by mouse control. Several globins are outliers (bottom of plot). By selecting any protein, it is highlighted on the JalView multiple sequence alignment and can thus be identified and further studied. (b) JalView outputs from Pfam families include trees, such as this neighbor joining tree of 76 globins. (Phylogenetic trees are described in Chapter 7.)*

(RPS-BLAST) by comparing a query sequence to a set of many position-specific scoring matrices (PSSMs). RPS-BLAST is related to PSI-BLAST (Chapter 5) but is distinct because it searches against profiles generated from preselected alignments. The main purpose of CDD (and RPS-BLAST) is to identify conserved domains in the query sequence. We provided an example in Chapter 5 (Fig. 5.8).

Prints

The PRINTS database consists of protein "fingerprints" that define families in the SwissProt/TrEMBL databases (Attwood et al., 2003). A hyperlink to PRINTS outputs is the Colour Interactive Editor for Multiple Alignments (CINEMA) editor (Parry-Smith et al., 1998). This is a Java applet application that is integrated with software for analysis of the alignments.

Release 36.0 of PRINTS (2007) has 1800 database entries covering a total of 10,931 individual motifs. It is available at ▶ http://www.bioinf.manchester.ac.uk/dbbrowser/PRINTS/.

Integrated Multiple Sequence Alignment Resources: InterPro and iProClass

A main theme of multiple sequence alignment databases is that while each employs a unique algorithm and search format, they are well integrated with each other. Another important idea is that individual databases such as Pfam and PROSITE have evolved specific approaches to the problem of protein classification and analysis. Some databases employ HMMs; some focus on protein domains, while others assess smaller motifs. Integrated resources allow you to explore the features of a protein using several related algorithms in parallel.

InterPro is available at ▶ http://www.ebi.ac.uk/interpro/. Release 18.0 (September 2008) contains over 16,500 entries representing about 11,000 families and 5,000 domains.

At least two comprehensive resources have been developed to integrate most of the major alignment databases. The InterPro database provides an integration of PROSITE, PRINTS, ProDom, Pfam, and TIGRFAMs with cross-references to BLOCKS (Table 6.2) (Mulder et al., 2007). The project is coordinated by eight centers, including EBI and the Wellcome Trust Sanger Institute (Apweiler et al., 2001).

The iProClass organizes 200,000 nonredundant Protein Information Resource (PIR) and SwissProt proteins in 28,000 superfamilies, 2600 domains, 1300 motifs, and 280 posttranslational modification sites (Wu et al., 2004). iProClass has links to 30 other databases. Resources such as iProClass and InterPro can be useful to identify conflicts between a variety of databases and to define the size of protein families.

You can access iProClass at ▶ http://pir.georgetown.edu/iproclass/.

TABLE 6-2	Databases on Which InterPro (Release 15.0) Is Based
Database	Contents (Entries)
PANTHER 6.1	30,128
Pfam 21.0	8,957
PIRSF 2.68	1,748
PRINTS 38.0	1,900
ProDom 2005.1	1,522
PROSITE 20.0	2,006
SMART 5.0	706
TIGRFAMs 6.0	2,946
GENE3D 3.0.0	2,147
SUPERFAMILY 1.69	1,538
UniProtKB/Swiss-Prot 52.0	261,513
UniProtKB/TrEMBL 35.0	3,987,044
InterPro 15.0	14,764
GO Classification	23,937

Source: From ▶ http://www.ebi.ac.uk/interpro/release_notes.html, May 2007.

PopSet

You can access PopSet via ► http://www.ncbi.nlm.nih.gov/Entrez/.

We turn now to a specialized resource for multiple sequence alignments. PopSet (Population Data Study Sets) is a collection of aligned protein or DNA sequences within the NCBI Entrez site. These sequences are derived from studies of closely related sequences from population, phylogenetic, or mutation studies. From the home page of NCBI, enter the search "globin" and follow the link to the PopSet results. One of them is entitled "The molecular basis of high-altitude adaptation in deer mice." This entry includes a multiple sequence alignment of DNA encoding alpha globins in deer mice (*Peromyscus maniculatus*). In a subset of those mice that have adapted to a high-altitude environment, the alpha globin genes have undergone mutations that confer a mixture of oxygen-binding affinities (Storz et al., 2007). This may provide a selective advantage, as globins with such properties have previously been found only in other high-dwelling animals such as yaks and birds. PopSet provides a convenient repository for experimentally derived multiple sequence alignments.

Multiple Sequence Alignment Database Curation: Manual versus Automated

Some databases are curated manually. This requires expert annotation; Sean Eddy and colleagues have curated Pfam, while Amos Bairoch and colleagues have

TABLE 6-3 Multiple Sequence Alignment Programs Available on the World Wide Web		
Program	Description	URL
AMAS (Analyse Multiply Aligned Sequences)	At the European Bioinformatics Institute; used to analyze premade MSAs	► http://barton.ebi.ac.uk/servers/amas_server.html
CINEMA	Colour INteractive Editor for Multiple Alignments	► http://utopia.cs.manchester.ac.uk/
ClustalW	At the European Bioinformatics Institute and other sites	► http://www.ebi.ac.uk/clustalw/
ClustalX	Download by FTP	► http://bips.u-strasbg.fr/fr/Documentation/ClustalX/
DIALIGN	Especially useful for local MSA; from the University of Bielefeld, Germany	► http://bibiserv.techfak.uni-bielefeld.de/dialign/
Match-Box Web Server 1.3	From the University of Namur, Belgium	► http://www.fundp.ac.be/sciences/biologie/bms/matchbox_submit.shtml
MultAlin	From INRA (► http://www.inra.fr/), Toulouse	► http://bioinfo.genopole-toulouse.prd.fr/multalin/multalin.html
Multiple Sequence Alignment version 2.0	At the GeneStream server of the Institut de Génétique Humaine	► http://xylian.igh.cnrs.fr/msa/msa.html
Musca	From the IBM Bioinformatics Group	► http://cbcsrv.watson.ibm.com/Tmsa.html
PileUp	Commercial package available through the SeqWeb or UNIX versions of the Genetics Computer Group (GCG)	► http://www.accelrys.com/products/gcg/

Note: Additional algorithms are listed at ExPASy (► http://www.expasy.org/tools/#align).
Abbreviation: MSA, multiple sequence alignment.

curated PROSITE. BLOCKS and PRINTS are also manually annotated. Expert annotation is obviously difficult but has the great advantage of allowing judgments to be made on the protein family members. Programs such as DOMO and ProDom use automated annotation. Errors in the alignment or the addition of unrelated sequences can be problematic, as discussed for PSI-BLAST (Chapter 5). However, automated annotation is valuable for exhaustive analyses of large data sets such as the thousands of predicted protein sequences derived from genome-sequencing projects.

Many dozens of multiple sequence alignment programs are available on the internet, including several that are listed in Table 6.3.

MULTIPLE SEQUENCE ALIGNMENTS OF GENOMIC REGIONS

Complete genomes are being sequenced at a rapid pace, with thousands of projects now completed or in progress. These are described in Part III of this book (Chapters 13 to 20). A basic problem is the alignment of entire genomes, or parts of genomes. In some cases closely related species are compared, such as humans and the chimpanzee *Pan troglodytes* (these diverged 5 to 7 million years ago), or different strains of the yeast *Saccharomyces cerevisiae*. In other cases highly divergent genomes are compared, such as *Homo sapiens* and the monotremes (e.g., the platypus *Ornithorhynchus anatinus*) which diverged about 210 million years ago. We examine the alignment of prokaryotic genomes in Chapter 15 (on bacteria and archaea).

One basic motivation for performing multiple sequence alignments of genomic regions is to identify DNA sequences that are under the influence of positive selection (and thus are changing rapidly in a given lineage) or negative selection (and thus are highly conserved and accumulate mutations slower than the neutral rate). We will introduce the concepts of positive and negative selection in Chapter 7, and we will see in the last third of this book that comparative genome analyses are used to identify highly conserved regions between genomes that are presumed to be functionally important. Practically, multiple sequence alignment of genomic regions typically uses modifications of the progressive alignment strategy we have discussed. The problem differs from that of conventional multiple sequence alignment in several ways.

- We have been considering programs that typically are used for a set of many protein or nucleic acid sequences, ranging up to hundreds or even thousands of sequences that typically have a length of no more than 1000 or 2000 residues. For genomic alignments, we typically have only a few sequences (in unusual cases as many as several dozen) that may have lengths of millions or tens of millions of base pairs. The addition of sequences from multiple species improves the accuracy of multiple sequence alignments of orthologous regions, relative to pairwise alignments or to the use of a limited number of species (Margulies et al., 2006).

- Aligning the genomic DNA of closely related organisms (e.g., those that diverged less than 10 million years ago) is often straightforward, but for more divergent organisms (e.g., human to mouse or human to fish) there are often islands of appreciable conservation (typically consisting of exons and conserved noncoding elements) separated by regions of extremely low conservation. This led to the idea of "anchors" for multiple sequence alignment of genomic regions, discussed below.

- Eukaryotic genomes are riddled with repetitive DNA elements such as DNA transposons and long and short interspersed nuclear elements (LINEs, SINEs) (Chapter 16). Such repeats occur in a lineage-specific fashion and can occupy a substantial portion of a genome. They must be accounted for in multiple sequence alignment.

- Chromosomal loci are subject to dynamic rearrangements such as duplications, deletions, inversions, and translocations. These often involve millions of base pairs. Such genomic changes occur frequently in individuals (serving as a major source of human disease) and as features of a species that become fixed (e.g., human chromosome 2 corresponds to two separate acrocentric chromosomes of the chimpanzee, following a chromosomal fusion event early in the hominoid lineage perhaps 5 to 7 million years ago). In the multiple sequence alignment of genomic regions it is common to find large stretches of apparent deletions or inversions, presenting a challenge for alignment algorithms.

- There are no benchmark data sets for genomic alignments comparable to those described above based on protein structures. However, for each algorithm it is essential to define both the sensitivity (the fraction of all truly orthologous relationships that are detected) and specificity (the fraction of predictions of an orthologous relationship that are correct) (Margulies et al., 2007). Two approaches have been adopted (Blanchette et al., 2004). First, biological sequences with known features such as exons are studied, although this approach does not provide information on how to correctly align poorly conserved regions. Second, simulations have been used, although a challenge is to faithfully model varying evolutionary rates and assorted genomic features such as repetitive elements.

Consider the human beta globin locus on chromosome 11 as an illustration of the usefulness of creating and exploring multiple sequence alignments of genomic DNA. We visited this region in Chapter 5 when we introduced the BLASTZ algorithm for pairwise alignments of genomic DNA. We used the UCSC Genome Browser to visualize the extent of conservation in a region of 50,000 base pairs across multiple species relative to human (Fig. 5.18). This browser allows the user to select a region of interest across many scales (from single nucleotides to whole chromosomes) and across many eukaryotic organisms while displaying a user-selected set of annotation tracks. We can now revisit this region, focusing on a span of 1800 base pairs that includes the beta globin gene (Fig. 6.10a). The peak heights for a conservation track indicate that the coding exons are highly conserved among a group of vertebrates (including mouse, rat, rabbit, dog, frog, and chicken), while much of the intergenic regions tend to be poorly conserved. Some conserved noncoding regions are apparent (e.g., Fig. 6.10a, arrow 2) which could represent conserved regulatory domains. By further zooming in to view just several dozen base pairs, a multiple sequence alignment appears (Fig. 6.10b), in this case including the ATG codon that encodes the start methionine (indicated by asterisks).

The multiple sequence alignments in the UCSC Genome Browser were generated in several steps. First, optimal pairwise alignments were created using blastz (Chapter 5). A species tree was generated for a group of species containing

---

Human chromosome 2, the second largest chromosome, is 243 million base pairs (Mbp) in size. It corresponds to chromosomes 2a and 2b of the chimpanzee *Pan troglodytes*.

To find this genomic region, go to ▶ http://genome.ucsc.edu and select ENCODE (discussed further in Chapter 16). Click on the region for the beta globin gene (on chromosome 11), and select coordinates chr11:5,200,001-5,250,000. Upon further reducing the displayed region to 30,000 base pairs or less, you can then view the nucleotide multiple sequence alignment of over a dozen species. Click on the conservation track (Fig. 6.10a, arrow 2) to download the multiple sequence alignment.

orthologous regions; this is analogous to a guide tree from progressive alignment. The program MULTIZ, a component of the Threaded Blockset Aligner (TBA) program (Blanchette et al., 2004), was used to generate the multiple sequence alignment. MULTIZ implements dynamic programming to align blocks of sequences.

Two other programs useful for the alignment of genomic DNA are MAVID (Bray and Pachter, 2004) and Multi-LAGAN (MLAGAN; Brudno et al., 2003). We discussed the LAGAN program in Chapter 5 (Fig. 5.19). MLAGAN uses progressive alignment (analogous to ClustalW), and makes the reasonable assumption that the phylogenetic tree is known (so that a guide tree does not need to be estimated). The multiple sequence alignment by MLAGAN (1) generates rough global maps (as described for LAGAN), (2) performs progressive multiple sequence alignment using anchors, and (3) performs iterative refinement with the anchors.

The UCSC Genome Browser provides multiple sequence alignments of genomic DNA from dozens of species. It is possible to directly compare the results of the TBA, MLAGAN, and MAVID programs (Fig. 6.11).

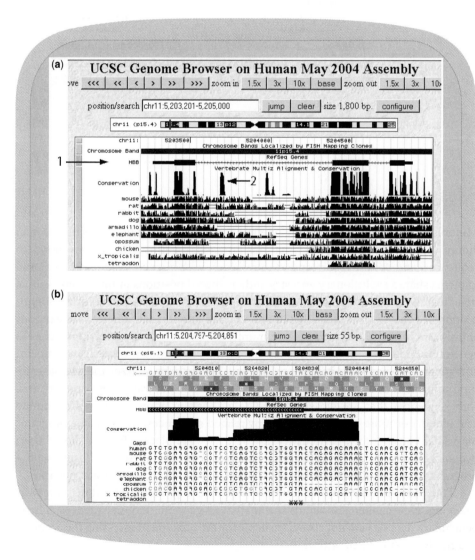

FIGURE 6.10. *Multiple sequence alignment of the human beta globin locus compared to other vertebrate genomic sequences. (a) A view in the UCSC Genome Browser of the beta globin gene is indicated. Exons are represented by blocks (arrow 1) and tend to be highly conserved among a group of vertebrate genomes. Additionally, several regions of high conservation occur in noncoding areas (e.g., arrow 2). (b) A view of 55 base pairs at the beta globin locus. At this magnification (fewer than 30,000 base pairs), the UCSC genome browser displays the nucleotides of genomic DNA in the multiple sequence alignment of a group of vertebrates. The ATG codon (oriented from right to left) is indicated (three asterisks), and the human protein product is shown (amino acids from right to left matching the start of protein NP_000509, MVHLTPEEKS).*

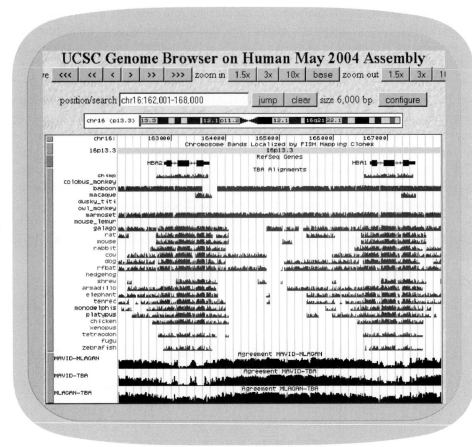

FIGURE 6.11. *Comparison of three programs for multiple alignment of genomic sequences. The UCSC Genome Browser includes dozens of annotation tracks that contain genomic data. Tracks for the TBA, MLAGAN, and MAVID were displayed for the beta globin locus, allowing direct comparisons of results generated with these different methods. By clicking on the graphic output, the underlying multiple sequence alignments can be displayed and downloaded.*

PERSPECTIVE

Multiple sequence alignment is the operation by which all the members of a protein family (or nucleic acid family) may be grouped together. Rows correspond to sequences, and columns correspond to residues, with aligned residues in a column implying shared evolutionary ancestry and/or shared positions in three-dimensional structures. Multiple sequence alignment serves many purposes, including the identification of conserved residues that are functionally important. There is tremendous enthusiasm in the bioinformatics community for the variety of novel approaches to generating accurate multiple sequence alignments, including progressive alignment, and approaches based on iterative refinement, consistency, and/or the use of structural information. A general conclusion is that most programs perform very well with sequences that are closely related (e.g., sharing approximately 40% amino acid identity or more). For more distantly related sequences, the available programs may differ considerably, particularly in where gaps are placed. For the typical user, two suggestions are to try performing multiple sequence alignments using several programs, and also using a variety of alternative parameters such as gap penalties.

Thus, the subdiscipline of multiple sequence alignment algorithms is rapidly changing. New challenges include the analysis of genomic DNA sequences. There, benchmark data sets are not always available for the purpose of assessing the accuracy of a newly developed algorithm.

Databases of multiply aligned protein families such as Pfam and InterPro are rapidly expanding in size and are increasingly important tools. These databases are often accompanied by careful expert annotation. A general trend is that databases offer the integration of many alignment resources.

PITFALLS

A very basic pitfall to avoid is the use of a group of sequences for multiple sequence alignment in which one or more sequences are not homologous to the rest. For multiple sequence alignments with relatively divergent members, it is common for different programs to give dramatically different results. A challenge is that you may not be able to assess which is most accurate based on criteria such as structure or shared evolutionary history. Gaps are particularly hard to place, and the most compact alignment (with the fewest gaps) is not necessarily the one most faithful to the evolutionary history of the sequences you are aligning. As an example of a challenge, ClustalW, MAFFT, and MUSCLE all adopt different approaches to the problem of what gap penalties to assign to terminal gaps (deletions) relative to internal gaps. There may not be a single correct approach, but this is an example of why different programs will produce different alignments.

It is especially important to perform a proper multiple sequence alignment for molecular phylogeny studies. The alignment constitutes the raw data that go into making a tree (see Chapter 7). Kumar and Filipski (2007) have reviewed many of the issues of multiple sequence alignment of DNA. They describe the accuracy of different multiple sequence alignment software in relation to the effects on subsequent phylogenetic analyses, and also as a function of evolutionary distance, guanine plus cytosine content, gap parameters affecting insertions/deletions, and choice of species.

WEB RESOURCES

In this chapter, we described programs for multiple sequence alignments. Links to these and additional programs are summarized in web document 6.8 (▶ http://www.bioinfbook. org/chapter6).

DISCUSSION QUESTIONS

[6-1] Feng and Doolittle introduced the "once a gap, always a gap" rule, saying that the two most closely related sequences that are initially aligned should be weighted most heavily in assigning gaps. Why was it necessary to introduce this rule?

[6-2] What are some of the issues associated with adapting multiple sequence alignment programs to large genomic DNA sequences?

PROBLEMS/COMPUTER LAB

[6-1] Practice using three NCBI resources to obtain groups of sequences in the fasta format that you can use for multiple sequence alignment. Select a keyword such as cytochrome (other suggestions are ferritin, S100, or trypsin). In a first approach, enter this search from the home page of NCBI, and follow the link to HomoloGene. By default, the entries are displayed in the summary format. Using the pull-down menu change the display to Multiple Alignment. This allows you to scroll through a series of multiple sequence alignments. Select one for further study. It is helpful to choose one in which there are some gaps, so that you can evaluate the performance of various software programs (in problem [10-2]). Once you identify a group of proteins of interest, click to view that HomoloGene group, and change the display to FASTA. Copy these sequences and/or save them to a text document. In a second approach, repeat this exercise beginning at the home page of NCBI, but select the link to CDD (the Conserved Domain Database). Here, there are pfam, cdd, smart, and/or COG identifiers. Select an entry with a CDD identifier (such as cd00904 for ferritin). Here, a multiple sequence alignment is shown. Change the format to obtain the desired number of proteins in this family (e.g., up to 5, 10, or 20) in the FASTA format; you may select the most diverse members of this group. In a third approach, perform a blastp search using a query such as ferritin light chain (NP_000137) and inspect the pairwise alignments to the query. Select a group of ten proteins by clicking on the box next to each, and click "Get selected sequences." These ten proteins appear on an Entrez Protein page; change the display option to FASTA and use the pull-down menu option "send to text." The sequences are now available in the FASTA format for further study.

[6-2] Using the FASTA-formatted sequences from problem [6-1], perform multiple sequence alignments using programs available at the European Bioinformatics Institute: ClustalW, MAFFT, Muscle, and T-Coffee. Save and compare each result. How do they differ? How can you assess which is likely to be the most accurate? When applicable, try adjusting the parameters such as the scoring matrices, gap opening and extension penalties, or number of iterations to see the effects on the alignments.

[6-3] We described how ClustalW applies a correction factor to downweight the influence of closely related proteins. Test the performance of ClustalW: take the globins in web documents 6.3 and/or 6.4 and align. Then repeat the alignment with the additional input one divergent sequence repeated varying number of times. For example, in the closely related group of beta globins, add five copies of the chicken sequence to see its influence on the alignment.

[6-4] Use the T-Coffee programs to evaluate the effect of structural information on your alignments. Follow these steps.

- Obtain a group of five distantly related lipocalins from web document 6.9 (▶ http://www.bioinfbook.org/chapter6). These include rat odorant-binding protein and human retinol-binding protein.
- Align the sequences using T-Coffee (▶ http://www.tcoffee.org/), or use another program.
- Evaluate the alignment with the iRMSD program (▶ http://www.tcoffee.org/). Include the information on two known lipocalin structures. Note the score.
- Align the same sequences again using Expresso (▶ http://www.tcoffee.org/) to incorporate structural information. Note the score. Did it improve? Do the alignments differ?

[6-5] X-linked adrenoleukodystrophy (X-ALD) is the most common inherited disease affecting peroxisomes (a subcellular organelle involved in lipid metabolism and other metabolic functions). The disease is caused by mutations in the ABCD1 gene on chromosome Xq28 encoding ALD protein (ALDP). In humans, there are thought to be four ALDP-related proteins on peroxisomes: ALDP (NP_000024; 745 amino acid residues), ALDR (NP_005155, 740 residues), PMP70 (NP_002849, 659 residues), and PMP70R (NP_005041, 606 residues). Two yeast ALDP-like proteins have also been identified, Pxa1p (NP_015178) and Pxa2p (NP_012733). These proteins are all part of a much larger family of ATP-binding cassette (ABC) transporters, including the cystic fibrosis transmembrane regulator (CFTR) and multidrug-resistant proteins (MDR).

Create a multiple sequence alignment of the human, mouse, and yeast ALDP family of proteins. Identify the presumed nucleotide binding site, GPNGCGKS. Is this motif perfectly conserved?

SELF-TEST QUIZ

[6-1] Benchmarking refers to

(a) Making a set of multiple sequence alignments (MSAs) from closely related proteins that form a trusted alignment

(b) Making a set of MSAs from proteins that have had their tertiary structure determined, thus allowing the MSA to be validated based on structural criteria

(c) Making a set of MSAs with an algorithm that are subsequently employed to refine tertiary structure predictions

(d) Making a set of MSAs from proteins that are known, based on structural criteria, to be members of distinct protein families

[6-2] Why doesn't ClustalW (a program that employs the Feng and Doolittle progressive sequence alignment algorithm) report expect values?

(a) ClustalW *does* report expect values.

(b) ClustalW uses global alignments for which E value statistics are not available.

(c) ClustalW uses local alignments for which E value statistics are not available.

(d) E value statistics are not relevant to multiple sequence alignments.

[6-3] The "once a gap, always a gap" rule for the Feng–Doolittle method:

(a) Assures that gaps will not be filled in inappropriately with inserted sequences

(b) Assures that sequences that diverged early in evolution will be given priority in establishing the order in which a multiple sequence alignment is constructed

(c) Assures that gaps occurring between sequences that are most closely related in a multiple sequence alignment will be preserved

(d) Assures that gaps occurring between sequences that are distantly related will be maintained in the multiple sequence alignment

[6-4] How can multiple sequence alignment programs improve performance?

(a) By performing PSI-BLAST

(b) By incorporating data on secondary structure

(c) By incorporating data on three-dimensional structures

(d) all of the above

[6-5] What is a main strength of consistency-based approaches (such as ProbCons)?

(a) They include information based on position-specific scoring matrices.

(b) They include information based on three-dimensional protein structures, typically obtained from x-ray crystallography studies.

(c) They perform profile-profile alignments and are extremely fast algorithms.

(d) They include information based on multiple sequence alignments to guide the determination of pairwise alignments.

[6-6] If you perform a multiple sequence alignment of a group of proteins and include a distantly related protein (a divergent member called an "orphan"):

(a) The orphan is typically aligned with the group of proteins.

(b) The orphan is typically not aligned with the group of proteins.

[6-7] The main difference between Pfam-A and Pfam-B is that:

(a) Pfam-A is manually curated while Pfam-B is automatically curated.

(b) Pfam-A uses hidden Markov models while Pfam-B does not.

(c) Pfam-A provides full-length protein alignments while Pfam-B aligns protein fragments.

(d) Pfam-A incorporates data from SMART and PORSITE while Pfam-B does not.

[6-8] If you perform a multiple sequence alignment of a group of proteins and include a distantly related protein (a divergent member called an "orphan"):

(a) The orphan is typically aligned with the group of proteins.

(b) The orphan is typically not aligned with the group of proteins.

[6-9] What is a feature of algorithms that align large tracts of genomic DNA, in contrast to programs such as ClustalW that align smaller blocks of DNA or protein?

(a) They are generally unable to align DNA from organisms that are highly divergent, such as those that speciated several hundred million years ago.

(b) They generally use progressive alignment and so are fundamentally similar.

(c) They often employ anchors that help to align regions of conservation that are interspersed with less conserved regions (such as those arising in noncoding regions, deleted regions, or inverted regions).

(d) They are specialized to accept very long inputs.

SUGGESTED READING

Da-Fei Feng and Russell F. Doolittle's (1987) progressive alignment approach to multiple sequence alignment is an important paper. This work stresses the relationship between multiple sequence alignment and the evolutionary relationships of proteins. It is thus relevant to our treatment of phylogeny in Chapter 7. Doolittle (2000) also wrote a personal account of his interest in sequence analysis, phylogeny, and bioinformatics, including mention of the historical context in which he developed his alignment algorithm.

A few research groups have systematically compared multiple sequence alignment algorithms, and reading any of these papers provides a deeper insight into the strengths and weaknesses of the algorithms (Blackshields et al., 2006; Briffeuil et al., 1998; Park et al., 1998). A study by Julie Thompson and colleagues (1999) is particularly informative.

For alignment of genomic DNA, excellent sources are Margulies et al. (2006, 2007) and Kumar and Filipski (2007).

REFERENCES

Apweiler, R., et al. The InterPro database, an integrated documentation resource for protein families, domains and functional sites. *Nucleic Acids Res.* **29**, 37–40 (2001).

Armougom, F., Moretti, S., Keduas, V., and Notredame, C. The iRMSD: A local measure of sequence alignment accuracy using structural information. *Bioinformatics* **22**, e35–e39 (2006a).

Armougom, F., Moretti, S., Poirot, O., Audic, S., Dumas, P., Schaeli, B., Keduas, V., and Notredame, C. Expresso: Automatic incorporation of structural information in multiple sequence alignments using 3D-Coffee. *Nucleic Acids Res.* **34**(Web Server issue), W604–W608 (2006b).

Armougom, F., Poirot, O., Moretti, S., Higgins, D. G., Bucher, P., Keduas, V., and Notredame, C. APDB: A web server to evaluate the accuracy of sequence alignments using structural information. *Bioinformatics* **22**, 2439–2440 (2006c).

Attwood, T. K., Bradley, P., Flower, D. R., Gaulton, A., Maudling, N., Mitchell, A. L., Moulton, G., Nordle, A., Paine, K., Taylor, P., Uddin, A., and Zygouri, C. PRINTS and its automatic supplement, prePRINTS. *Nucleic Acids Res.* **31**, 400–402 (2003).

Bateman, A., Coin, L., Durbin, R., Finn, R. D., Hollich, V., Griffiths-Jones, S., Khanna, A., Marshall, M., Moxon, S., Sonnhammer, E. L., Studholme, D. J., Yeats, C., and Eddy, S. R. The Pfam protein families database. *Nucleic Acids Res.* **32**(Database issue), D138–D141 (2004).

Blackshields, G., Wallace, I. M., Larkin, M., and Higgins, D. G. Analysis and comparison of benchmarks for multiple sequence alignment. *In Silico Biol.* **6**, 321–339 (2006).

Blanchette, M., Kent, W. J., Riemer, C., Elnitski, L., Smit, A. F., Roskin, K. M., Baertsch, R., Rosenbloom, K., Clawson, H., Green, E. D., Haussler, D., and Miller, W. Aligning multiple genomic sequences with the threaded blockset aligner. *Genome Res.* **14**, 708–715 (2004).

Bray, N., and Pachter, L. MAVID: Constrained ancestral alignment of multiple sequences. *Genome Res.* **14**, 693–699 (2004).

Briffeuil, P., et al. Comparative analysis of seven multiple protein sequence alignment servers: Clues to enhance reliability of predictions. *Bioinformatics* **14**, 357–366 (1998).

Brudno, M., Do, C. B., Cooper, G. M., Kim, M. F., Davydov, E., NISC Comparative Sequencing Program, Green, E.D., Sidow, A., and Batzoglou, S. LAGAN and Multi-LAGAN: Efficient tools for large-scale multiple alignment of genomic DNA. *Genome Res.* **13**, 721–731 (2003).

Chenna, R., Sugawara, H., Koike, T., Lopez, R., Gibson, T. J., Higgins, D. G., and Thompson, J. D. Multiple sequence alignment with the Clustal series of programs. *Nucleic Acids Res.* **31**, 3497–3500 (2003).

Chothia, C., and Lesk, A. M. The relation between the divergence of sequence and structure in proteins. *EMBO J.* **5**, 823–826 (1986).

Do, C. B., Mahabhashyam, M. S., Brudno, M., and Batzoglou, S. ProbCons: Probabilistic consistency-based multiple sequence alignment. *Genome Res.* **15**, 330–340 (2005).

Doolittle, R. F. On the trail of protein sequences. *Bioinformatics* **16**, 24–33 (2000).

Edgar, R. C. MUSCLE: A multiple sequence alignment method with reduced time and space complexity. *BMC Bioinformatics* **5**, 113 (2004a).

Edgar, R. C. MUSCLE: Multiple sequence alignment with high accuracy and high throughput. *Nucleic Acids Res.* **32**, 1792–1797 (2004b).

Feng, D. F., and Doolittle, R. F. Progressive sequence alignment as a prerequisite to correct phylogenetic trees. *J. Mol. Evol.* **25**, 351–360 (1987).

Feng, D. F., and Doolittle, R. F. Progressive alignment and phylogenetic tree construction of protein sequences. *Methods Enzymol.* **183**, 375–387 (1990).

Finn, R. D., et al. Pfam: Clans, web tools and services. *Nucleic Acids Res.* **34**(Database issue), D247–D251 (2006).

Fitch, W. M., and Yasunobu, K. T. Phylogenies from amino acid sequences aligned with gaps: The problem of gap weighting. *J. Mol. Evol.* **5**, 1–24 (1975).

Gotoh, O. A weighting system and algorithm for aligning many phylogenetically related sequences. *Comput. Appl. Biosci.* **11**, 543–551 (1995).

Gotoh, O. Significant improvement in accuracy of multiple protein sequence alignments by iterative refinement as assessed by reference to structural alignments. *J. Mol. Biol.* **264**, 823–838 (1996).

Hall, B. G. *Phylogenetic Trees Made Easy. A How-To for Molecular Biologists.* Sinauer Associates, Sunderland, MA, 2001.

Heringa, J. Two strategies for sequence comparison: Profile-preprocessed and secondary structure-induced multiple alignment. *Comput Chem.* **23**, 341–364 (1999).

Higgins, D. G., Blackshields, G., and Wallace, I. M. Mind the gaps: Progress in progressive alignment. *Proc. Natl. Acad. Sci. USA.* **102**, 10411–10412 (2005).

Hirosawa, M., Totoki, Y., Hoshida, M., and Ishikawa, M. Comprehensive study on iterative algorithms of multiple sequence alignment. *Comput. Appl. Biosci.* **11**, 13–18 (1995).

Karlin, S., and Brocchieri, L. Heat shock protein 70 family: Multiple sequence comparisons, function, and evolution. *J. Mol. Evol.* **47**, 565–577 (1998).

Katoh, K., Kuma, K., Toh, H., and Miyata, T. MAFFT version 5: Improvement in accuracy of multiple sequence alignment. *Nucleic Acids Res.* **33**, 511–518 (2005).

Krogh, A. An introduction to hidden Markov models for biological sequences. In S. L. Salzberg, D. B. Searls, and S. Kasif (eds.), *Computational Methods in Molecular Biology*, Chapter 4. Elsevier, New York, 1998.

Kumar, S., and Filipski, A. Multiple sequence alignment: in pursuit of homologous DNA positions. *Genome Res.* 17, 127–135 (2007).

Lassmann, T., and Sonnhammer, E. L. Kalign: An accurate and fast multiple sequence alignment algorithm. *BMC Bioinformatics* 6, 298 (2005).

Lassmann, T., and Sonnhammer, E. L. Kalign, Kalignvu and Mumsa: Web servers for multiple sequence alignment. *Nucleic Acids Res.* 34(Web Server issue), W596–W599 (2006).

Letunic, I., Copley, R. R., Pils, B., Pinkert, S., Schultz, J., and Bork, P. SMART 5: Domains in the context of genomes and networks. *Nucleic Acids Res.* 34(Database issue), D257–D260 (2006).

Loytynoja, A., and Goldman, N. An algorithm for progressive multiple alignment of sequences with insertions. *Proc Natl Acad Sci USA* 102, 10557–10562 (2005).

Margulies, E. H., Chen, C. W., and Green, E. D. Differences between pair-wise and multi-sequence alignment methods affect vertebrate genome comparisons. *Trends Genet.* 22, 187–193 (2006).

Margulies, E. H., et al. Analyses of deep mammalian sequence alignments and constraint predictions for 1% of the human genome. *Genome Res.* 17, 760–774 (2007).

McClure, M. A., Vasi, T. K., and Fitch, W. M. Comparative analysis of multiple protein-sequence alignment methods. *Mol. Biol. Evol.* 11, 571–592 (1994).

Mizuguchi, K., Deane, C. M., Blundell, T. L., and Overington, J. P. HOMSTRAD: A database of protein structure alignments for homologous families. *Protein Sci.* 7, 2469–2471 (1998).

Moretti, S., Armougom, F., Wallace, I. M., Higgins, D. G., Jongeneel, C. V., and Notredame, C. The M-Coffee web server: A meta-method for computing multiple sequence alignments by combining alternative alignment methods. *Nucleic Acids Res.* 35, W645–W648 (2007).

Morgenstern, B., Dress, A., and Werner, T. Multiple DNA and protein sequence alignment based on segment-to-segment comparison. *Proc. Natl. Acad. Sci. USA* 93, 12098–12103 (1996).

Mulder, N. J., et al. New developments in the InterPro database. *Nucleic Acids Res.* 35(Database issue), D224–D228 (2007).

Needleman, S. B., and Wunsch, C. D. A general method applicable to the search for similarities in the amino acid sequence of two proteins. *J. Mol. Biol.* 48, 443–453 (1970).

Notredame, C., Higgins, D. G., and Heringa, J. T-Coffee: A novel method for fast and accurate multiple sequence alignment. *J Mol Biol.* 302, 205–217 (2000).

O'sullivan, O., Zehnder, M., Higgins, D., Bucher, P., Grosdidier, A., and Notredame, C. APDB: A novel measure for benchmarking sequence alignment methods without reference alignments. *Bioinformatics* 19 **Suppl 1**, i215–i221 (2003).

Park, J., et al. Sequence comparisons using multiple sequences detect three times as many remote homologues as pairwise methods. *J. Mol. Biol.* 284, 1201–1210 (1998).

Parry-Smith, D. J., Payne, A. W., Michie, A. D., and Attwood, T. K. CINEMA—a novel colour INteractive editor for multiple alignments. *Gene* 221, GC57–GC63 (1998).

Plewniak, F., et al. PipeAlign: A new toolkit for protein family analysis. *Nucleic Acids Res.* 31, 3829–3832 (2003).

Ponting, C. P., Schultz, J., Milpetz, F., and Bork, P. SMART: Identification and annotation of domains from signalling and extracellular protein sequences. *Nucleic Acids Res.* 27, 229–232 (1999).

Raghava, G. P., Searle, S. M., Audley, P. C., Barber, J. D., and Barton, G. J. OXBench: A benchmark for evaluation of protein multiple sequence alignment accuracy. *BMC Bioinformatics* 4, 47 (2003).

Schultz, J., Milpetz, F., Bork, P., and Ponting, C. P. SMART, a simple modular architecture research tool: Identification of signaling domains. *Proc. Natl. Acad. Sci. USA* 95, 5857–5864 (1998).

Simossis, V. A., and Heringa, J. PRALINE: A multiple sequence alignment toolbox that integrates homology-extended and secondary structure information. *Nucleic Acids Res.* 33 (Web Server issue), W289–W294 (2005).

Storz, J. F., Sabatino, S. J., Hoffmann, F. G., Gering, E. J., Moriyama, H., Ferrand, N., Monteiro, B., and Nachman, M. W. The molecular basis of high-altitude adaptation in deer mice. *PLoS Genet.* 3, 448–459 (2007).

Stoye, J., Evers, D., and Meyer, F. ROSE: Generating sequence families. *Bioinformatics* 14, 157–163 (1998).

Subramanian, A. R., Weyer-Menkhoff, J., Kaufmann, M., Morgenstern, B. DIALIGN-T: an improved algorithm for segment-based multiple sequence alignment. *BMC Bioinformatics* 6, 66 (2005).

Thompson, J. D., Higgins, D. G., and Gibson, T. J. CLUSTALW: Improving the sensitivity of progressive multiple sequence alignment through sequence weighting, position-specific gap penalties and weight matrix choice. *Nucleic Acids Res.* 22, 4673–4680 (1994).

Thompson, J. D., Gibson, T. J., Plewniak, F., Jeanmougin, F., and Higgins, D. G. The CLUSTAL_X windows interface: Flexible strategies for multiple sequence alignment aided by quality analysis tools. *Nucleic Acids Res.* 25, 4876–4882 (1997).

Thompson, J. D., Koehl, P., Ripp, R., and Poch, O. BAliBASE 3.0: Latest developments of the multiple sequence alignment benchmark. *Proteins* 61, 127–136 (2005).

Thompson, J. D., Plewniak, F., and Poch, O. A comprehensive comparison of multiple sequence alignment programs. *Nucleic Acids Res.* **27**, 2682–2690 (1999).

Tuppy, H. Über die Artspezifität der Proteinstruktur. In A. Neuberger (ed.), *Symposium on Protein Structure*, 66–76. Wiley, New York, 1958.

Van Walle, I., Lasters, I., and Wyns, L. SABmark – a benchmark for sequence alignment that covers the entire known fold space. *Bioinformatics* **21**, 1267–1268 (2005).

Wallace, I. M., Blackshields, G., and Higgins, D. G. Multiple sequence alignments. *Curr. Opin. Struct. Biol.* **15**, 261–266 (2005).

Wallace, I. M., O'sullivan, O., Higgins, D. G., and Notredame, C. M-Coffee: combining multiple sequence alignment methods with T-Coffee. *Nucleic Acids Res.* **34**, 1692–1699 (2006).

Wu, C. H., Huang, H., Nikolskaya, A., Hu, Z., and Barker, W. C. The iProClass integrated database for protein functional analysis. *Comput. Biol. Chem.* **28**, 87–96 (2004).

(a)

(b)

In the following pages the results of precipitin tests with haemato-sera are given in the zoological order of the antisera, the tests made by other observers being summarized in each case, the results of my tests following.

The number of tests, made by me with 30 antisera produced, is given in the following table, the total number of tests being 16,000.

Antiserum for	No. of tests therewith	Antiserum for	No. of tests therewith
Man	835	Ox	790
Chimpanzee	47	Sheep	701
Ourang	81	Horse	790
Cercopithecus	733	Donkey	94
Hedgehog	383	Zebra	94
Cat	785	Whale	94
Hyaena	378	Wallaby	691
Dog	777	Fowl	792
Seal	358	Ostrich	649
Pig	818	Fowl-egg	789
Llama	363	Emu-egg	650
Mexican Deer	749	Turtle	666
Reindeer	69	Alligator	468
Hog Deer	699	Frog	551
Antelope	686	Lobster	450
	7751		8249

	7751
	8249
Total number of tests	16,000

(c)

Class **MAMMALIA**
1. *Order* **PRIMATES**
ANTISERA FOR......

(Data table of precipitin test results for the suborder ANTHROPOIDEA, Fam. Hominidae, with columns for Mammalia: Man, Chimpanzee, Ourang, Monkey, Hedgehog, Cat, Hyaena, Dog, Seal, Pig, Llama, Hog-deer, Mexican-deer, Antelope, Sheep, Ox, Horse, Wallaby.)

For the first half of the twentieth century, the main phylogenetic analyses based on molecular data were the remarkable precipitin tests pioneered by George Nuttall and colleagues. Antisera were incubated with serum samples from a variety of species, and the time required for a precipitation reaction was recorded, as well as the strength of the reaction. *(a) Sample test tubes in which the reactions were conducted (Nuttall, (1904), plate I) (b) Excerpt from Nuttall ((1904), p. 160) describing the 16,000 tests he performed. (c) Portion of the 92-page data summary of Nuttall ((1904), pp. 222–223). The 900 rows (of which 11 are shown here) represent blood samples that were tested, and the columns correspond to antisera obtained from 30 organisms (of which 18 are shown here). The values represent the time (in minutes) required for a reaction. The symbols indicate the degree of reaction (+ being greatest, and − indicating no reaction). The letter D indicates the presence of deposits in the test tube. Nuttall used these data to infer the phylogenetic relationships of assorted mammals, birds, reptiles, amphibians, and crustaceans. In the 1950s and 1960s, amino acid sequence comparisons largely replaced immunological tests for phylogenetic analysis.*

7

Molecular Phylogeny and Evolution

INTRODUCTION TO MOLECULAR EVOLUTION

Evolution is the theory that groups of organisms change over time so that descendants differ structurally and functionally from their ancestors. Evolution may also be defined as the biological process by which organisms inherit morphological and physiological features that define a species. In 1859 Charles Darwin published his landmark book, *On the Origin of Species by Means of Natural Selection, or the Preservation of Favoured Races in the Struggle for Life*. "As many more individuals of each species are born than can possibly survive; and as, consequently, there is a frequently recurring struggle for existence, it follows that any being, if it vary however slightly in any manner profitable to itself, under the complex and sometimes varying conditions of life, will have a better chance of surviving, and thus be naturally selected. From the strong principle of inheritance, any selected variety will tend to propagate its new and modified form."

Evolution is a process of change. Heredity is generally conservative—offspring resemble their parents—and yet the structure and function of bodies change over the course of generations. There are three main mechanisms by which changes may occur (Simpson, 1952):

- Conditions of growth affect development. Environmental factors such as accidents and disease-causing infections are not hereditary in nature

We will explore the tree of life in Chapter 13. You can read *The Origin of Species* by Charles Darwin on-line at ▶ http://www.literature.org/authors/darwin-charles/the-origin-of-species/.

Bioinformatics and Functional Genomics, Second Edition. By Jonathan Pevsner
Copyright © 2009 John Wiley & Sons, Inc.

The word phylogeny is derived from the Greek *phylon* ("race, class") and *geneia* ("origin"). Ernst Haeckel, whose tree of life is shown on the frontis to Chapter 13, coined the terms *phylogeny*, *phylum*, and *ecology*. He also wrote that "ontogenesis is a brief and rapid recapitulation of phylogenesis, determined by the physiological functions of heredity (generation) and adaptation (maintenance)" (Haeckel, 1990, p. 81). See also ▶ http://www.ucmp.berkeley.edu/history/haeckel.html.

(although an individual's response to disease or environmental stimuli is genetically controlled to some extent, as discussed in Chapter 20).

• The mechanism of sexual reproduction assures change from one generation to the next. DNA sequences including genes are "shuffled" in a unique combination when an offspring inherits chromosomes from two parents.

• Mutation with selection as well as genetic drift can produce changes in genes and more generally in chromosomes.

At the molecular level, evolution is a process of mutation with selection. Molecular evolution is the study of changes in genes and proteins throughout different branches of the tree of life. This discipline also uses data from present-day organisms to reconstruct the evolutionary history of species.

Phylogeny is the inference of evolutionary relationships. Traditionally, phylogeny was assessed by comparing morphological features between organisms from a variety of species (Mayr, 1982). However, molecular sequence data can also be used for phylogenetic analysis. The evolutionary relationships that are inferred, which are usually depicted in the form of a tree, can provide hypotheses of past biological events.

Goals of Molecular Phylogeny

All life forms share a common origin and are part of the tree of life. More than 99% of all species that ever lived are extinct (Wilson, 1992). Of the extant species, closely related organisms are descended from more recent common ancestors than distantly related organisms. In principle, there may be one single tree of life that accurately describes the evolution of species. One object of phylogeny is to deduce the correct trees for all species of life. Historically, phylogenetic analyses were based on easily observable features, such as the presence or absence of wings or a spinal cord. More recently, phylogenetic analyses also rely on molecular sequence data that define families of genes and proteins. Another object of phylogeny is to infer or estimate the time of divergence between organisms since the time they last shared a common ancestor.

While the tree of life provides an appealing metaphor, evolution is not predicated on there necessarily being a single tree. Instead, evolution is based on a process of mutation and selection. We will see in Chapter 15 that genes can be laterally transferred between species, complicating the ways organisms can acquire genes and traits. In many situations the tree of life has been described as a densely interconnected bush (or reticulated tree) rather than a simple tree with well-defined branches (e.g., Doolittle, 1999).

A *true tree* depicts the actual, historical events that occurred in evolution. It is essentially impossible to generate a true tree. Instead, we generate *inferred trees*, which depict a hypothesized version of the historical events. Such trees describe a series of evolutionary events that are inferred from the available data, based on some model.

Viruses are generally not considered to be part of the tree of life (see Chapter 14), although phylogenetic trees have been studied for all subgroups of viruses.

The tree of life has three major branches: bacteria, archaea, and eukaryotes. We explore the global tree in Chapter 13. In this chapter we address the topic of phylogenetic trees that are used to assess the relationships of homologous proteins (or homologous nucleic acid sequences) in a family. Any group of homologous proteins (or nucleic acid sequences) can be depicted in a phylogenetic tree.

In Chapter 3, we defined two proteins as homologous if they share a common ancestor. You may perform a BLAST search and observe several proteins with high scores (low expect values) and simply view these database matches as related proteins that possibly have related function. However, it is also useful to view orthologs and paralogs in an evolutionary context. We have applied a variety of approaches to study the relations of proteins: pairwise alignment using Dayhoff's scoring matrices (Chapter 3), BLAST searching (Chapters 4 and 5), and multiple sequence alignment (Chapter 6), and we will address the identification of related protein folds (Chapters 10 and 11). All these approaches rely on evolutionary models to account for the observed similarities and differences between molecular sequences:

In 1973 Theodosius Dobzhansky wrote an article entitled "Nothing in biology makes sense except in the light of evolution."

- Dayhoff et al. (1978, p. 345) introduce scoring matrices in explicit evolutionary terms: "An accepted point mutation in a protein is a replacement of one amino acid by another, accepted by natural selection. It is the result of two distinct processes: the first is the occurrence of a mutation in the portion of the gene template producing one amino acid of a protein; the second is the acceptance of the mutation by the species as the new predominant form. To be accepted, the new amino acid usually must function in a way similar to the old one: chemical and physical similarities are found between the amino acids that are observed to interchange frequently." Dayhoff et al. compare observed amino acid sequences from two proteins not with each other but with their inferred ancestor obtained from phylogenetic trees.

- Feng and Doolittle (1987, p. 351) used the Needleman and Wunsch pairwise alignment progressively "to achieve the multiple alignment of a set of protein sequences and to construct an evolutionary tree depicting their relationship. The sequences are assumed a priori to share a common ancestor, and the trees are constructed from different matrices derived directly from the multiple alignment. The thrust of the method involves putting more trust in the comparison of recently diverged sequences than in those evolved in the distant past."

- In our description of protein families, we provided the example of the Pfam JalView tool that allows distance information from the multiple sequence alignment of any Pfam family to be depicted as a tree (Fig. 6.9).

In this chapter, we use multiple sequence alignments of protein (or DNA or RNA) to generate phylogenetic trees. These trees provide a visualization of the evolutionary history of molecular sequences.

Historical Background

Historically, the globins have been among the protein families most important to our understanding of biochemistry and molecular evolution, from the identification of hemoglobin in the 1830s and myoglobin in the 1860s to their crystallization in the nineteenth century for the purpose of comparative studies across species (Box 7.1). Globins were among the first proteins to be sequenced and to be analyzed using x-ray crystallography (Chapter 11). Following earlier work by Ingram (1961) and others to determine globin protein sequences, Eck and Dayhoff (1966) used parsimony analysis (defined below) to generate trees of the globin family. We provided phylogenetic trees to introduce the concepts of paralogs (various human globins in Fig. 3.2) and orthologs (myoglobins in various species; Fig. 3.3). Figure 7.1 shows a phylogenetic analysis of 13 globin proteins from various species, redrawn from Dayhoff (1972). We will return to these 13 sequences for phylogenetic analyses in

Thirteen protein sequences corresponding to the proteins in Fig. 7.1 are provided in web document 7.1 at ▶ http://www.bioinfbook.org/chapter7. We will use these sequences as examples throughout this chapter. A similar phylogenetic tree was reported by Zuckerkandl and Pauling (1965).

this chapter. Figure 7.2 (also from Dayhoff, 1972) further provides a timeline of the events in which globin genes duplicated (e.g., an ancestral globin gene duplicated to form the lineages leading to modern alpha globin and beta globin), and also a timeline for speciation events (e.g., the modern fish and humans shared a common vertebrate ancestor ~400 million years ago). These studies focused on two aspects of phylogenetic trees. First, trees can depict the relatedness of particular protein subfamilies, such as the alpha globins, beta globins, and myoglobins. Second, trees can depict the relatedness of species, providing inferences about the evolutionary history of life forms as well as the history of genes and gene products. We expand on the relation of gene trees and species trees below.

Box 7.1
A Brief History of the Globins

The globins have a leading position in the history of protein biochemistry. There was early interest in identifying the coloring matter of blood, with studies by Antoine-Laurent de Lavoisier (in 1777) and with the discovery of iron in blood by Antoine François Fourcroy (1759–1809). Hemoglobin was among the earliest proteins to be identified: it was named haematosine by Louis-René Lecanu (1800–1871) in 1838, and renamed "globulin" by Jöns Jacob Berzelius (1779–1848), who also coined the term protein. By 1864, when Hoppe-Seyler again renamed it haemoglobin, detailed spectroscopic properties of the hemoglobins were published, and soon after, Charles MacMunn (1884) discovered myoglobin. Crystals of purified hemoglobin were first prepared in the 1840s, and by 1909 Reichert and Brown produced an extensive catalog of hemoglobin crystals and used them as a basis for phylogeny. Once globin proteins were sequenced, they were used as a basis for molecular phylogeny, for example by Ingram (1961) and Dayhoff and colleagues (1966, 1972).

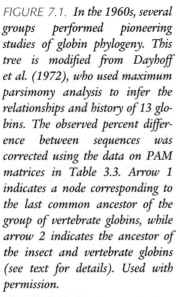

FIGURE 7.1. In the 1960s, several groups performed pioneering studies of globin phylogeny. This tree is modified from Dayhoff et al. (1972), who used maximum parsimony analysis to infer the relationships and history of 13 globins. The observed percent difference between sequences was corrected using the data on PAM matrices in Table 3.3. Arrow 1 indicates a node corresponding to the last common ancestor of the group of vertebrate globins, while arrow 2 indicates the ancestor of the insect and vertebrate globins (see text for details). Used with permission.

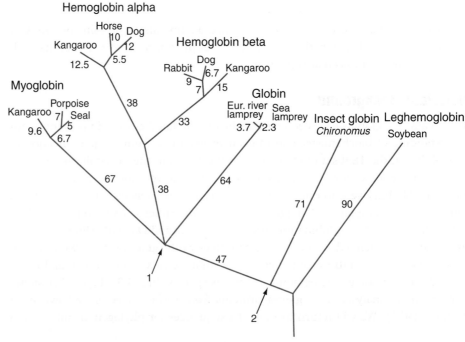

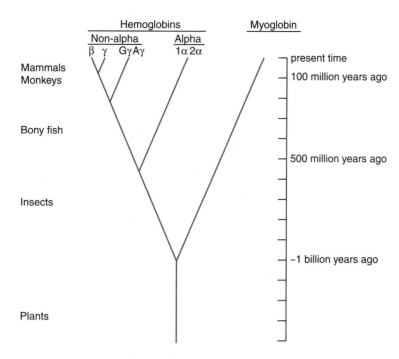

FIGURE 7.2. Dayhoff et al. (1972) summarized the relationship of the globin subfamilies in the context of evolutionary time. The dates of speciation events were inferred from fossil-based studies. Used with permission.

Tremendous progress was also made in our understanding of molecular evolution through the study of insulin beginning in the 1950s. Insulin is a small protein secreted by pancreatic islet cells that stimulates glucose uptake on binding to an insulin receptor on muscle and liver cells. In 1953 Frederick Sanger and colleagues determined the primary amino acid sequence of insulin, the first time this feat had been accomplished for any protein. The mature, biologically active protein consists of two subunits, the A chain and B chain, that are covalently attached through intermolecular disulfide bridges. More recently, the structure of the human preproinsulin molecule was shown to consist of a signal peptide, the B chain, an intervening sequence called the C peptide, and the A chain (Fig. 7.3a). The C peptide is flanked by dibasic residues (arg-arg or lys-arg; see Fig. 7.3a and b) at which proteolytic cleavage occurs.

Sanger and others sequenced insulin proteins from five species (cow, sheep, pig, horse, and whale). It became clear immediately that the A chain and B chain residues are highly conserved. Furthermore, amino acid differences were restricted to three residues within a disulfide "loop" region of the A chain (Fig. 7.3b, shaded red). This suggested that amino acid substitutions occur nonrandomly, some changes affecting biological activity dramatically and other changes having negligible effects (Anfinsen, 1959). The differences within the disulfide loop are termed "neutral" changes (Jukes and Cantor, 1969, p. 86; Kimura, 1968). Later, when the biologically active A and B chain sequences were compared to the functionally less important C peptide, even more dramatic differences were seen. Kimura (1983) reported that the C peptide evolves at a rate of 2.4×10^{-9} per amino acid site per year, sixfold faster than the rate for the A and B chains (0.4×10^{-9} per amino acid site per year). At the nucleotide level, the rate of evolution is similarly about sixfold faster for the DNA region encoding the C peptide (Li, 1997).

As insulin was sequenced from additional species, a surprising finding emerged. Insulin from guinea pig and a closely related species of the family Caviidae (the coypu) appeared to evolve seven times faster than insulin from other species. As shown in the alignment of Fig. 7.3c, the guinea pig insulin sequence differs from

Frederick Sanger won the Nobel Prize in Chemistry (1958) "for his work on the structure of proteins, especially that of insulin" (▶ http://nobelprize.org/nobel_prizes/chemistry/laureates/1958/). In 1980, he shared the Nobel Prize in Chemistry (with Paul Berg and Walter Gilbert) for his "contributions concerning the determination of base sequences in nucleic acids."

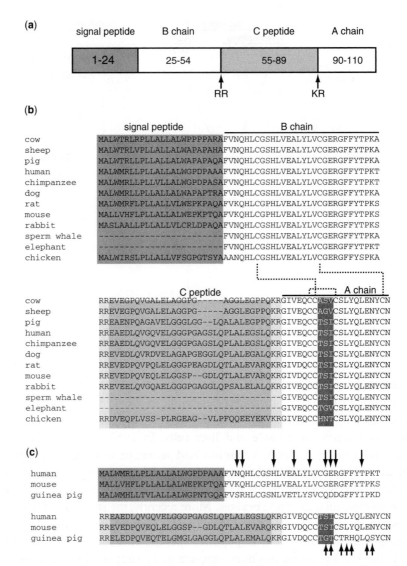

FIGURE 7.3. *Since the 1950s, studies of insulin have facilitated our understanding of molecular evolution. (a) The human insulin molecule consists of a signal peptide (required for intracellular transport; amino acid residues 1–24), the B chain, the C peptide, and the A chain. Dibasic residues (RR, KR) flank the C peptide and are the sites at which proteases cleave the protein. The A chain and B chain are then covalently linked through disulfide bridges, forming mature insulin. (b) Multiple sequence alignment of insulin from 12 species. Amino acid substitutions occur in nonrandom patterns. Note that within the A chain of insulin the amino acid residues are almost perfectly conserved between different species, except for three divergent columns of amino acids (A chain, colored region in a "disulfide loop"). However, the rate of nucleotide substitution is about sixfold higher in the region encoding the intervening C peptide than in the region encoding the B and A chains (Kimura, 1983), and gaps in the multiple sequence alignment are evident here. Disulfide bridges between cysteine residues are indicated by dashed lines. The accession numbers are NP_000198 (human), P30410 (chimpanzee), NP_062002 (rat), P01321 (dog), NP_032412 (mouse), P01311 (rabbit), P01315 (pig), P01332 (chicken), NP_776351 (cow), P01318 (sheep), INEL (elephant), and INWHP (sperm whale). (c) Guinea pig insulin (Cavia porcellus, accession P01329) evolves about sevenfold faster than insulin from other species. Human, mouse, and guinea pig insulins are aligned. Arrows indicate 16 amino acid positions at which the guinea pig sequence varies from that of human and/or mouse.*

Arginine vasporessin   `CYFQNCPRG`
Oxytocin   `CYIQNCPLG`

FIGURE 7.4. Human oxytocin (NP_000906, residues 20–28) and arginine vasopressin (NP_000481, residues 20–28) differ at only two amino acid positions, yet they have vastly different biological functions. The comparison of these peptide sequences in the 1960s led to the appreciation of the importance of primary amino acid sequences in determining protein function.

human and mouse insulin at 16 different amino acid positions within the A and B chains. The explanation for this phenomenon (Jukes, 1979) is that guinea pig and coypu insulin do not bind two zinc ions, whereas insulin from all the other species does. There is presumably a strong functional constraint on most insulin molecules to maintain amino acid residues that are able to complex zinc. Guinea pig and coypu insulin have less selective constraint.

In the early 1950s, other laboratories sequenced vasopressin and oxytocin and found that peptides differing by only two amino acid residues have vastly different biological function (Fig. 7.4). And in 1960 Max Perutz and John Kendrew solved the structures of hemoglobin and myoglobin. These proteins, both of which serve as oxygen carriers, are homologous and share related structures (see Fig. 3.1). Thus, it became clear by the 1960s that there are significant structural and functional consequences to variation in primary amino acid sequence.

Perutz and Kendrew won the 1962 Nobel Prize in Chemistry "for their studies of the structures of globular proteins." You can read about their accomplishments at ► http://www.nobel.se/chemistry/laureates/1962/.

Molecular Clock Hypothesis

In the 1960s, primary amino acid sequence data were accumulated for abundant, soluble proteins such as hemoglobins, cytochrome c, and fibrinopeptides in a variety of species. Some proteins, such as cytochrome c from many organisms, were found to evolve very slowly, while other protein families accumulated many substitutions. Emil Zuckerkandl and Linus Pauling (1962) as well as Emanuel Margoliash (1963) proposed the concept of a molecular clock. This hypothesis states that for every given gene (or protein), the rate of molecular evolution is approximately constant. In a pioneering study, Zuckerkandl and Pauling observed the number of amino acid differences between human globins, including beta and delta (about 6 differences), beta and gamma (~36 differences), beta and alpha (~78 differences), and alpha and gamma (~83 differences). They could also compare human to gorilla (both alpha and beta globins), observing either 2 or 1 differences respectively, and they knew from fossil evidence that humans and gorillas diverged from a common ancestor about 11 million years ago. Using this divergence time as a calibration point, they estimated that gene duplications of the common ancestor to beta and delta occurred 44 million years ago (MYA); beta and gamma derived from a common ancestor 260 MYA; alpha and beta 565 MYA; and alpha and gamma 600 MYA.

A related study demonstrating the existence of a molecular clock was performed by Richard Dickerson in 1971 (Fig. 7.5). He analyzed three proteins for which a large amount of sequence data were available: cytochrome c, hemoglobins, and fibrinopeptides. For each, he plotted the relationship between the number of amino acid differences for a protein in two organisms versus the divergence time (in millions of years, MY) for the organisms. These divergence times were estimated from paleontology.

When estimating the number of amino acid (or nucleic acid) differences between a group of sequences, one needs a model to explain the process by which

For alignments of these globin proteins and a summary (including the correct number of differences) of the Zuckerkandl and Pauling (1962) data, see web document 7.2 at ► http://www.bioinfbook.org/chapter7.

FIGURE 7.5. *A comparison of the number of amino acid changes that occur between proteins (y axis) versus the time since the species diverged (x axis) reveals that individual proteins evolve at distinct rates. Some proteins, such as cytochrome c from a variety of organisms, evolve very slowly; others such as hemoglobin evolve at an intermediate rate; and proteins such as fibrinopeptides undergo substitutions rapidly. This behavior is described by the molecular clock hypothesis, proposed by Pauling, Margoliash, and others in the 1960s. The time of divergence of various organisms (arrows) is estimated primarily from fossil evidence. Abbreviation: MY, millions of years in the past. Adapted from Dickerson (1971); some data points and the standard deviation measurements are omitted. Used with permission.*

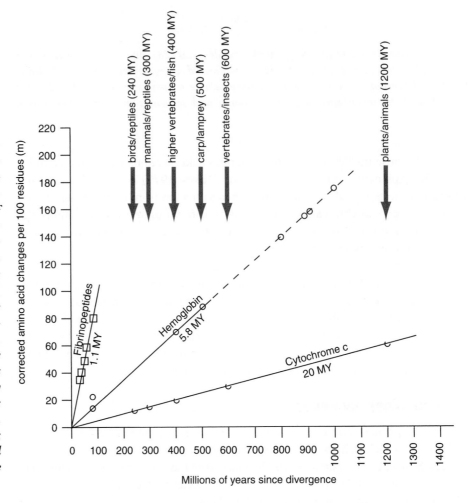

The correction formula of Equation 7.1 was written incorrectly in the original Margoliash and Smith (1965) article, but was used correctly by Dickerson (1971) and is further discussed by Fitch and Ayala (1994).

substitutions occur; we will address this subject later in this chapter. We have already encountered the idea that more mutational events occur than can be observed directly when we examined PAM matrices (Chapter 3). There, we saw that two proteins of length 100 that share 50% amino acid identity have sustained an average of 80 changes (Fig. 3.19). Notably Zuckerkandl and Pauling (1962) had assumed for the purpose of their analyses that the number of observed differences reflects the number of substitutions that have actually occurred. However, they acknowledged that the situation is more complicated because multiple substitutions may occur at any given site: "Thus the number of effective mutational events that have actually occurred since the α- and β-chains have evolved from their common ancestor may be significantly greater than is presently apparent" (Zuckerkandl and Pauling, 1962, p. 204). Margoliash and Smith (1965, p. 233) as well as Zuckerkandl and Pauling (1965, p. 150) proposed a correction for the relationship between observed changes and actual changes. This correction was employed by Dickerson (1971) (Fig. 7.5). The y axis of this plot consists of the corrected number of amino acid changes per 100 residues, m. The value of m is calculated

$$\frac{m}{100} = -\ln\left(1 - \frac{n}{100}\right) \tag{7.1}$$

This equation can be restated as

$$\frac{n}{100} = 1 - e^{-(m/100)} \tag{7.2}$$

where m is the total number of amino acid changes that have occurred in a 100 amino acid segment of a protein and n is the observed number of amino acid changes per 100 residues. This correction adjusts for amino acid changes that occur but are not directly observed, such as two or more amino acid changes occurring in the same position (see Fig. 7.16 and the discussion below).

The results of this plot (Fig. 7.5) allow several conclusions (Dickerson, 1971):

- For each protein, the data lie on a straight line. This suggests that the rate of change of amino acid sequence has remained constant for each protein.

- The average rates of change are distinctly different for each protein family. For example, fibrinopeptides evolve with a much higher rate of substitution. The time (in millions of years) for a 1% change in amino acid sequence to occur between two divergent lines of evolution is 20.0 MY for cytochrome c, 5.8 MY for hemoglobin, and 1.1 MY for fibrinopeptides.

- The observed variations in rate of change between protein families reflect functional constraints imposed by natural selection.

TABLE 7-1 Rates of Amino Acid Substitutions per Amino Acid Site per 10^9 Years $(\lambda \times 10^9)$ in Various Proteins

Protein	Rate	Protein	Rate
Fibrinopeptides	9.0	Thyrotropin beta chain	0.74
Growth hormone	3.7	Parathyrin	0.73
Immunoglobulin (Ig) kappa chain C region	3.7	Parvalbumin	0.70
Kappa casein	3.3	Trypsin	0.59
Ig gamma chain C region	3.1	Melanotropin beta	0.56
Lutropin beta chain	3.0	Alpha crystallin A chain	0.50
Ig lambda chain C region	2.7	Endorphin	0.48
Lactalbumin	2.7	Cytochrome b_5	0.45
Epidermal growth factor	2.6	Insulin (except guinea pig and coypu)	0.44
Somatotropin	2.5	Calcitonin	0.43
Pancreatic ribonuclease	2.1	Neurophysin 2	0.36
Serum albumin	1.9	Plastocyanin	0.35
Phospholipase A2	1.9	Lactate dehydrogenase	0.34
Prolactin	1.7	Adenylate kinase	0.32
Carbonic anhydrase C	1.6	Cytochrome c	0.22
Hemoglobin alpha chain	1.2	Troponin C, skeletal muscle	0.15
Hemoglobin beta chain	1.2	Alpha crystallin B chain	0.15
Gastrin	0.98	Glucagon	0.12
Lysozyme	0.98	Glutamate dehydrogenase	0.09
Myoglobin	0.89	Histone H2B	0.09
Amyloid AA	0.87	Histone H2A	0.05
Nerve growth factor	0.85	Histone H3	0.014
Acid proteases	0.84	Ubiquitin	0.010
Myelin basic protein	0.74	Histone H4	0.010

Dayhoff (1978) expressed these rates as accepted point mutations (PAMs) per 100 amino acid residues that are estimated to have occurred in 100 million years of evolution (compare Box 3.3). Thus, the rate of mutation acceptance for serum albumin is 19 PAMs per 100 million years.
Source: Dayhoff (1978, p. 3) as adapted by Nei (1987, p. 50). Used with permission.

The rate of amino acid substitution is measured by the number of substitutions per amino acid site per year, λ. Some values for λ are given in Table 7.1. Note that some proteins such as histones and ubiquitin undergo substitutions extraordinarily slowly. For reference, Table 7.2 lists the 20 most conserved proteins present in the *Homo sapiens*, *Caenorhabditis elegans* (nematode), and *Saccharomyces cerevisiae* (yeast) proteomes as determined by reciprocal BLAST searches.

Note that we say that histones undergo substitution very slowly, but we do not say that they *mutate* very slowly. Mutation is the biochemical process that results in a change in sequence. For example, a polymerase copies DNA (or RNA) with a particular mutation rate. Substitution is the observed change in nucleic acid or protein sequences (e.g., between various histones). The observed substitutions that are fixed in a population occur at a rate that reflects both mutation and selection, the process by which characters are selected for (or against) in evolution. If the rate of mutation of the DNA or RNA polymerases among an organism's genes is relatively constant, then variation in substitution rates among those genes may be due primarily to positive or negative selection. In the language of Susumu Ohno (1970), some substitutions are *forbidden* because they are deleterious to the organism and are selected against. For example, substitutions in histones are almost always not tolerated, that is, they are lethal.

TABLE 7-2   Twenty Most Conserved Proteins Present in *C. Elegans* (worm), *H. Sapiens*, and *S. Cerevisiae* (Yeast)

Protein	Pairwise Percent Identity		
	Worm/ Human	Worm/ Yeast	Yeast/ Human
1. H4 histone	99	91	92
2. H3.3 histone	99	89	90
3. Actin B	98	88	89
4. Ubiquitin	98	95	96
5. Calmodulin	96	59	58
6. Tubulin 2	94	75	76
7. Ubiquitin-conjugating enzyme UBC4	93	80	80
8. Clathrin coat associated protein	93	48	48
9. Tubulin	93	73	74
10. ADP ribosylation factor 1	93	77	77
11. Dynein light chain 1	92	51	50
12. GTP-binding nuclear protein RAN	89	82	81
13. Ser/Thr protein phosphatase PP1γ	89	84	85
14. Ser/Thr protein phosphatase PP2β	89	74	76
15. Ubiquitin-conjugating enzyme UBE2 N	88	67	70
16. Histone H2A.Z	88	69	69
17. Histone H2A.2	87	79	76
18. DIM1P homolog	86	61	65
19. G25K GTP-binding protein	86	76	80
20. 40S ribosomal protein S15A	86	76	77

Note: The data are adapted from Peer Bork and colleagues (Copley et al., 1999), who performed reciprocal BLAST searches against the completed proteomes of these three organisms, as available at the time. As we will see in Chapter 16, genome and proteome annotation change substantially over time. Used with permission.

A significant implication of the molecular clock hypothesis is that if protein sequences evolve at constant rates, then they can be used to estimate the time that the sequences diverged. In this way phylogenetic relationships can be established between organisms. This is analogous to the dating of geological specimens using radioactive decay. An example of how the molecular clock may be used is given in Box 7.2.

Box 7.2
Rate of Nucleotide Substitution r and Time of Divergence T

The rate of nucleotide substitution r is the number of nucleotide substitutions that occur per site per year. Similar calculations can be made for the rate of amino acid substitutions. These rates vary considerably and it is of interest to characterize whether a region evolves slowly or rapidly. The rate is given by

$$r = \frac{K}{2T} \qquad (1)$$

Here T is the time of divergence of two extant sequences from a common ancestor. $2T$ is given in the equation to reflect the time of divergence from a common ancestor on two separate lineages, as depicted in Fig. 7.16a. T can sometimes be established based on fossil (paleontological) data. As an example, the lineages leading to modern humans and rodents diverged about 80 million years ago. K is the number of substitutions per site. The alpha globins from rat and human differ by 0.093 nonsynonymous substitutions per site (Graur and Li, 2000); as discussed below, nonsynonymous changes are DNA substitutions in coding regions that result in a change in the amino acid that is specified. Given values for K and T we can estimate r:

$$r = \frac{0.093 \; substitutions/site}{2(8 \times 10^7 \; years)} \qquad (2)$$

Thus we calculate that the alpha chain of hemoglobin undergoes 0.58×10^{-9} nonsynonymous nucleotide substitutions per site per year. We can also use Equation (1) to estimate the time of divergence of two sequences, given values for r and K; T is given by $K/2r$.

Source: Graur and Li (2000).

The molecular clock hypothesis does not apply to all proteins, and a variety of exceptions and caveats have been noted:

- The rate of molecular evolution varies among different organisms. For example, some viral sequences tend to change extremely rapidly compared to other life forms.
- The clock varies among different genes (see Table 7.1) and across different parts of an individual gene (see, e.g., Fig. 7.3 and see the discussion on the gamma parameter below). The main force guiding the molecular clock is selection. Rodents tend to have a faster molecular clock than primates; this may be because their generation times are shorter and they have high metabolic rates.

We will discuss the duplication of an entire genome, followed by subsequent, rapid mutation and gene loss, in Chapters 16 (the eukaryotic chromosome), 17 (on the yeast *S. cerevisiae*) and 18 (eukaryotic chromosomes).

- The clock is only applicable when a gene in question retains its function over evolutionary time. Genes may become nonfunctional (e.g., pseudogenes) leading to rapid changes in nucleotide (and amino acid) sequence. The rate of evolution sometimes accelerates after gene duplication occurs. For example, after gene duplication generated alpha and beta globins, high rates of amino acid substitution occurred that presumably altered the function of the gene, allowing some globin proteins to be expressed at highly specific developmental stages.

Despite these issues, the molecular clock hypothesis has proven useful and valid in many cases to which it is applied. Fitch and Ayala (1994) described a reasonably accurate molecular clock for Cu,Zn superoxide dismutase from a group of 67 protein sequences. However, obtaining correct inferences from the clock required tuning a variety of parameters.

As one practical approach to testing whether a molecular sequence has clocklike behavior, we can use the relative rate test of Tajima (1993) (Box 7.3). For sequences A, B, and C of the same protein or DNA/RNA from three species, let A and B be from two species from which we wish to compare the relative rates of evolution. Let C be a sequence from an outgroup, and let O be the common ancestor of A and B (Fig. 7.6a). Tajima's test determines whether there is accelerated evolution in lineage A or B, in which case we reject the null hypothesis that A and B exhibit equal evolutionary rates. Given the observed number of substitutions in sequence pairs AB, AC, and BC we can infer distances OA, OB, and OC and thus test the null hypothesis that the relative rates OA and OB are the same (Fig. 7.6b, c, and d).

Box 7.3
Tajima's Relative Rate Test

Tajima (1993) introduced a test for whether DNA or protein sequences in two lineages (such as human and chimpanzee) are undergoing evolution at equal rates. This is a test of the molecular clock: the null hypothesis is that there is an equal rate, and if we reject the null hypothesis at the 0.05 level then one of the lineages is evolving significantly faster or slower. For three protein or DNA sequences A, B, C, let A and B be from two species we wish to compare and C is from an outgroup. For example, we can compare human and chimpanzee mitochondrial DNA using orangutan mitochondrial DNA as an outgroup. The relationships of A, B, and the outgroup C are shown in the form of a tree in Fig. 7.6a. The observed number of sites n_{ijk} have the nucleotides i, j, k, respectively. The expectation of n_{ijk} must equal that of n_{jik}; that is,

$$E(n_{ijk}) = E(n_{jik})$$

If this equality occurs, the rate is constant per year; if it does not hold, the rate is not constant. We can measure the number of sites m_1 in which residues in sequence A differ from those in B and C; similarly m_2 corresponds to sites in B that are different than A and C. Given that C is an outgroup, the expectation of m_1 must equal the expectation of m_2, that is,

$$E(m_1) = E(m_2)$$

This equality is tested with a chi-square analysis:

$$\chi^2 = \frac{(m_1 - m_2)^2}{m_1 + m_2}$$

This test results in a p value. If $p < 0.05$, the molecular evolutionary clock hypothesis is rejected at the 5% level, regardless of the substitution model. Tajima's relative rate test is implemented in Molecular Evolutionary Genetics Analysis (MEGA) software (Tamura et al., 2007). For the mitochondrial sequences analyzed in Fig. 7.6 there were 31 unique sequence differences in A (human) and 49 unique differences in B (chimpanzee) so the χ^2 test statistic was 4.05. This was obtained from:

$$\chi^2 = \frac{(31 - 49)^2}{31 + 49}$$

This corresponds to a $P = 0.04$ with one degree of freedom, suggesting that we may reject the null hypothesis of equal rates between lineages. In using Tajima's test it is important to select an outgroup that is an appropriate evolutionary distance from the two organisms you are comparing. For example the bonobo or pygmy chimpanzee (*Pan paniscus*) may be too closely related to human and chimpanzee, as all three species diverged about 5 to 7 million years ago; it is a problem for an intended outgroup to have the properties of an ingroup. At the other extreme, rat or mouse are too divergent as they diverged from the primate lineage about 80 million years ago. Suitable choices may include primates such as orangutan or gorilla; one wants to select the closest true outgroup that is available.

Tajima's relative rate test is implemented in MEGA software program (Tamura et al., 2007; Kumar et al., 2008). We will use MEGA for phylogenetic analyses later in this chapter. We provide a specific, detailed example of using the test in Problems/Computer Lab 7-1, including an explanation of how to enter the sequences into MEGA, align them, and perform Tajima's test.

MEGA is available from ► http://www.megasoftware.net/.

Positive and Negative Selection

Darwin's theory of evolution suggests that, at the phenotypic level, traits in a population that enhance survival are selected for (positive selection), while traits that reduce fitness are selected against (negative selection). For example, among a group of giraffes millions of years in the past, those giraffes that had longer necks were able to reach higher foliage and were more reproductively successful than their shorter-necked group members, that is, there was positive selection for height.

At the molecular level, a conventional evolutionary point of view is that positive and negative selection also operate on DNA sequences. A gene encoding an enzyme may duplicate (see Chapters 16 and 17), and then subsequent nucleotide changes may allow one of the duplicated genes to encode an enzyme with a novel function that is advantageous and hence selected for. This process of positive selection is thought to have occurred on two occasions in the evolution of lysozyme, an enzyme that breaks down bacterial peptidoglycan linkages and thus serves as an antimicrobial protein in sources such as milk, saliva, and tears. About 25 million years ago the lysozyme gene duplicated and assumed a novel digestive function in stomach in the ancestor of goats, cows, and deer. The emergence of this novel function

FIGURE 7.6. A relative rate test to determine if two sequences follow the molecular clock hypothesis of approximately constant rates of amino acid or nucleotide substitution over evolutionary time. (a) Tajima (1993) proposed a relative rate test to determine whether protein or nucleic acid sequences from two organisms (A and B) have evolved at a similar relative rate. A and B share a common ancestor (O), and the sequence of an outgroup (C) is known. By measuring the substitution rates AB, AC, and BC it is possible to infer the rates OA and OB and to perform a chi square (χ^2) test to determine whether these rates are comparable (the null hypothesis) or whether one lineage has evolved at a relative accelerated or decelerated rate, thus violating the behavior of a molecular clock. Details of this test are presented in Tajima (1993) and Nei and Kumar (2000, pp. 193–195). (b) Tajima's test is implemented in MEGA software. The pull-down menu for phylogenetic analysis is shown. (c) The test in MEGA allows the user to specify groups A, B, and C (outgroup). In this example mitochondrial DNA sequences from human and chimpanzee are compared using orangutan DNA as an outgroup. (d) The output consists of a table listing the number of substitutions and an associated p value from a χ^2 test. In this example the p value is <0.05 suggesting that the null hypothesis can be rejected and the human and chimpanzee sequences do not exhibit molecular clock-like behavior. This specific example is presented in detail in problem 7-1 at the end of this chapter.

(a)

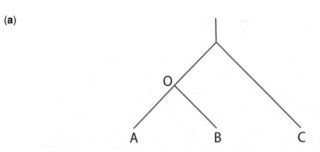

(b)

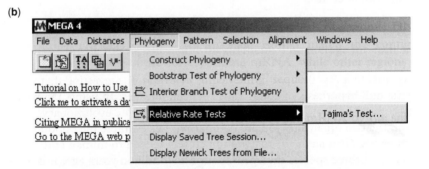

(c)

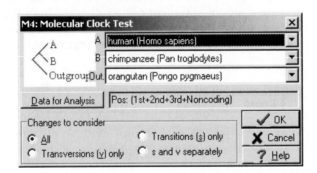

(d)

Table. Results from the Tajima test for 3 Sequences [1].

Configuration	Count
Identical sites in all three sequences (m_{iii})	712
Divergent sites in all three sequences (m_{ijk})	3
Unique differences in Sequence A (m_{iij})	31
Unique differences in Sequence B (m_{iji})	49
Unique differences in Sequence C (m_{ijj})	100

Note: The equality of evolutionary rate between *human (Homo sapiens)* and *chimpanzee (Pan troglodytes)* is tested using *orangutan (Pongo pygmaeus)* as an outgroup in Tajima' relative rate test in MEGA4 [1, 2]. The χ^2 test statistic was 4.05 ($P = 0.04417$ with 1 degree[s] of freedom). *P*-value less than 0.05 is often used to reject the null hypothesis of equal rates between lineages.

```
human            M   V   H   L   T   P   E   E   K   S   A   V
chimpanzee       M   V   H   L   T   P   E   E   K   S   A   V
                                     ▼
mouse            M   V   H   L   T   D   A   E   K   S   A   V
                                         ▼
dog              M   V   H   L   T   A   E   E   K   S   L   V
```

```
human        5' AACAGACACC ATG GTG CAT CTG ACT CCT GAG GAG AAG TCT GCC GTT 3'
chimpanzee   5' AACAGACACC ATG GTG CAC CTG ACT CCT GAG GAG AAG TCT GCC GTT 3'
mouse        5' AACAGACATC ATG GTG CAC CTG ACT GAT GCT GAG AAG TCT GCT GTC 3'
dog          5' AACAGACACC ATG GTG CAT CTG ACT GCT GAA GAG AAG AGT CTT GTC 3'
codon                 ↑     1   2   3   4   5   6   7   8   9  10  11  12
```

FIGURE 7.7. *Phylogenetic trees can be constructed using DNA, RNA, or protein sequence data. Often, the DNA sequence is more informative than protein in phylogenetic analysis. As an example, the sequences of beta globin from three species are aligned at the 5' end of the DNA (with the corresponding amino-terminals of the proteins). In the 5'- and 3'-untranslated regions, where no protein is encoded, there is typically less selective pressure to maintain particular nucleotide residues. (Some regulatory elements may be highly conserved.) Here, just one nucleotide position varies (arrow). Within the protein-coding region, there are variant amino acid residues at amino acid positions 6, 7, and 11 (see black arrowheads). These variants may be informative in performing phylogeny. However, there is an even greater number of informative nucleotide changes, restricting our attention to the coding region. There are seven synonymous nucleotide changes (nucleotides labeled red; see codons 3, 7, 10–12) that do not cause a different amino acid to be specified. There are also seven nonsynonymous changes that do cause an amino acid change (red arrowheads). For one of these (codon 6 of the dog sequence), a single nucleotide change of C→G, relative to the primate sequences, accounts for the amino acid change. For three other nonsynonymous codons, two nucleotides are changed relative to the primate sequences. The beta globin sequences are from human (GenBank accession NM_000518), chimpanzee (Pan troglodytes; XM_508242), mouse (Mus musculus; NM_016956), and dog (Canis lupus familiaris; XM_537902).*

occurred independently in leaf-eating monkeys such as the langur some 15 million years ago (Jollès et al., 1990). In each of these instances the rate of amino acid replacement increased due to positive selection as the lysozyme assumed a novel function. Other examples of positive selection include the primate ribonuclease genes (Zhang and Gu, 1998) and the MEDEA genes in plants (Spillane et al., 2007).

There are several ways to assess whether selection has occurred in sequence data. One approach relies on the fact that the portion of DNA that codes for a protein can have both synonymous and nonsynonymous substitutions. For a nucleotide change in a given codon, a synonymous substitution does not result in a change in the amino acid that is specified. For example, consider an alignment of human, chimpanzee, mouse, and dog beta globin at their 5' ends (amino-terminals of the proteins) (Fig. 7.7). In the third codon the nucleotides CAT in the human and dog sequences encode a histidine. Changing the third position to yield CAC in the chimpanzee and mouse sequences does not alter the amino acid that is encoded. Other synonymous changes are evident (Fig. 7.7, red-colored nucleotides). A nonsynonymous substitution does change the amino acid that is specified. For example, human and chimpanzee beta globin have a CCT codon that specifies a proline, but the corresponding canine sequence has a single substitution resulting in a codon (GCT) that specifies an alanine (Fig. 7.11, codon 6).

Comparison of the rates of nonsynonymous substitution per nonsynonymous site ($\hat{d}_N$) versus synonymous substitution per synonymous site ($\hat{d}_S$) may reveal evidence of positive or negative selection. If $\hat{d}_S$ is greater than $\hat{d}_N$, this suggests that

Refer to the genetic code in Box 3.6.

the DNA sequence is under negative or purifying selection. Negative selection is selection that limits change in a corresponding amino acid sequence; this occurs when some aspect of the structure and/or function of a protein is critical and cannot tolerate substitutions. When $\hat{d}_N$ is greater than $\hat{d}_S$, this suggests that positive selection occurs. An example of positive selection is with a duplicated gene that is under pressure to evolve new functions.

A variety of computer programs assess the ratio of synonymous to nonsynonymous substitutions. One is Synonymous Non-synonymous Analysis Program (SNAP), which requires as its input codon-aligned nucleotide sequences (Korber, 2000). Datamonkey is a suite of tools including robust maximum likelihood approaches to determining positive or negative selection (Pond and Frost, 2005). Another program is MEGA (Tamura et al., 2007). Upon entering a group of nucleotide coding sequences, MEGA employs the Nei and Gojobori (1986) method to test the null hypothesis that the sequences are under either positive, negative, or neutral selection.

There is considerable interest in measuring positive or negative selection on a genome-wide basis. Many approaches have been adopted (Sabeti et al., 2006; Nielsen, 2005). For example, Bustamante et al. (2005) studied the DNA sequence of 11,000 genes in 39 individuals and reported rapid amino acid evolution at 9% of the informative loci. For many of the genomes that have recently been sequenced (e.g., human, chimpanzee, dog, chicken, rat), a description of those genes that are under positive selection is a basic part of the genome analysis (see Chapter 18).

Positive and negative can also be studied on a highly compressed time scale in viruses. In 1978, 500 women were inadvertently infected with hepatitis C virus (HCV). Stuart Ray and colleagues (2005) sequenced a 5.2 kilobase portion of the HCV genome from the original inoculum and from 22 women about 20 years after the infection. They showed loci with both positive and negative selection reflecting the evolution of the virus to optimize its fitness in each host. For example, amino acid substitutions in known epitopes diverged from the consensus sequence in individuals having the human leukocyte antigen (HLA) allele for that epitope, indicating a mechanism of immune selection. In another study Cox et al. (2005) studied sequence variation of HCV both before, during, and after HCV infection. They showed that amino acid substitutions reflect escape from T cell recognition; in those individuals with persistent infection, there were selection pressures on epitopes that resulted in nonsynonymous changes. The Ray et al. (2005) and Cox et al. (2005) results provide examples of the usefulness of longitudinal studies in phylogeny, and they reveal mechanisms through which positive and natural selection shape the fitness of viruses.

Neutral Theory of Molecular Evolution

There is a tremendous amount of DNA polymorphism in all species that is difficult to account for by conventional natural selection. We will examine this throughout the tree of life in the last third of this book. In Chapter 16, we examine single nucleotide polymorphisms (SNPs), an extremely common form of polymorphism that does not appear to be under selection in most instances. Similarly, many chromosomal copy number variants occur in apparently normal individuals (Chapter 16). These involve multiple regions of up to millions of base pairs of DNA that are deleted or duplicated, and the majority of copy number variants appear to be sporadic, benign, and not under positive or negative selective pressure.

SNAP is available at the HIV database website (▶ http://www.hiv.lanl.gov) in the tools menu. Web document 7.3 introduces 12 globin DNA coding sequences (11 myoglobin orthologs plus one cytoglobin sequence as an outgroup); see ▶ http://www.bioinfbook.org/chapter7. That file includes multiple sequence alignments of those sequences. We will use these sequences as examples later in this chapter. Web document 7.4 provides an example of how to use four of those globin coding sequences to test for selection using SNAP software, while web document 7.5 shows an example of tests for selection in MEGA software. Datamonkey is available at ▶ http://www.datamonkey.org/.

In the decades up to the 1960s the prevailing model of molecular evolution was that most changes in genes are selected for or against in a Darwinian sense. Motoo Kimura (1968, 1983) proposed a different model to explain evolution at the DNA level. Kimura (1968) noted that the rate of amino acid substitution averages approximately one change per 28×10^6 years for proteins of 100 residues. He further estimated that the corresponding rate of nucleotide substitution must be extremely high (one base pair of DNA replaced in the genome of a population every 2 years on average).

Kimura's conclusion was that most observed DNA substitutions must be neutral or nearly neutral, and that the main cause of evolutionary change (or variability) at the molecular level is random drift of mutant alleles. Most nonsynonymous mutations are deleterious, and thus are not observed as substitutions in the population. Under this model, called the neutral theory of evolution, positive Darwinian selection plays an extremely limited role. Indeed, the existence of a molecular clock makes sense in the context of the neutral hypothesis because most amino acid substitutions are neutral. (Thus, substitutions are tolerated by natural selection to change in a manner that has clock-like properties. If substitutions occurred primarily in the context of positive or negative selection, it is unlikely that they could account for clock-like evolution.) In the decades since his 1983 publication, the neutral theory continues to be tested in a variety of organisms. We will explore some of these studies when we consider the eukaryotic chromosome in Chapter 16.

Kimura (1968) based his calculations on substitution rates measured within the families of alpha and beta globin, cytochrome c, and triosephosphate dehydrogenase proteins.

MOLECULAR PHYLOGENY: PROPERTIES OF TREES

Molecular phylogeny is the study of the evolutionary relationships among organisms or among molecules using the techniques of molecular biology. Many other techniques are used to study evolution, including morphology, anatomy, paleontology, and physiology. We will focus on phylogenetic trees using molecular sequence data. We begin with an explanation of the nomenclature used to describe trees. There are two main kinds of information inherent in any phylogenetic tree: the topology and the branch lengths. It is necessary to introduce a variety of terms that are used to characterize trees.

Let us first define the main parts of a tree and the main types of trees. A phylogenetic tree is a graph composed of branches and nodes (Fig. 7.8a). Only one branch (also called an edge) connects any two nodes. The nodes represent the taxonomic units (taxa or taxons); the node (from the Latin for "knot") is the intersection or terminating point of two or more branches. For us, taxa will typically be protein sequences. An operational taxonomic unit (OTU) is an extant taxon present at an external node, or leaf; the OTUs are the available nucleic acid or protein sequences that we are analyzing in a tree.

Consider the two trees in Figs. 7.8b and c. Each tree consists of five OTUs (labeled A, B, C, D, and E). These five OTUs define five external nodes. In addition, there are internal nodes at positions F, G, H, and I. Each internal node represents an inferred ancestor of the OTUs. Imagine that the tree is of five globins, and A and B correspond to human and rat beta globin. The internal node that connects to A and B represents an ancestral sequence that existed in an organism that predated the divergence of primates and rodents some 80 MYA.

The topology of a tree defines the relationships of the proteins (or other objects) that are represented in the tree. For example, the topology shows the common ancestor of two homologous protein sequences. The branch lengths sometimes (but not always) reflect the degree of relatedness of the objects in the tree.

In our discussion of trees, it is assumed that the raw data may consist of DNA, RNA, or protein sequence data. These data are presented as a multiple sequence alignment.

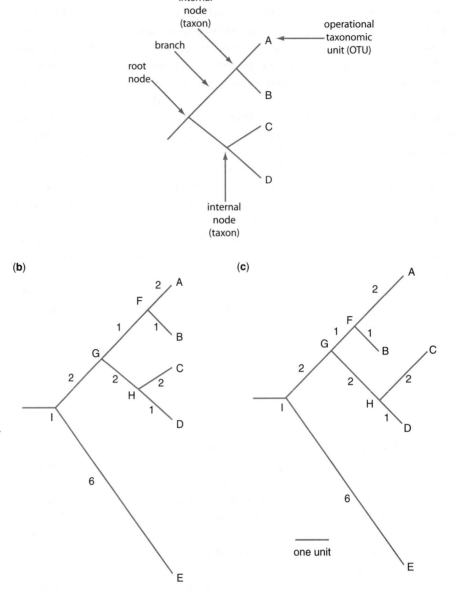

FIGURE 7.8. (a) Phylogenetic trees contain nodes and branches. A node may be external, internal, or at the root of a tree (the root is defined below). A branch connects two nodes. The nodes represent taxa or taxonomic units; the taxa that provide observable features, such as existing protein sequences or morphological features, are called operational taxonomic units (OTUs). Phylogenetic trees may be (b) unscaled or (c) scaled. In an unscaled tree, the branch lengths are not proportional to the number of amino acid (or nucleotide) changes. For example, note that branches FA (two units) and FB (one unit) have the same apparent length. Here, operational taxonomic units (ABCDE) are neatly aligned in a column at the tips of the tree. In the scaled tree in (b), the branch lengths are proportional to the number of substitutions. With this topology it is much easier to visualize the relatedness of proteins (or genes) in the tree. The x axis represents distance and/or time (in units such as millions of years).

Branches define the topology of the tree, that is, the relationships among the taxa in terms of ancestry. In some trees, the branch length represents the number of amino acid changes that have occurred in that branch. Branches of a tree are also called edges. In the trees of Fig. 7.8, the branches leading to each of the OTUs are called external branches (or peripheral branches). The branches leading to F, G, H, and I are called internal branches.

In the example of Fig. 7.8b, the branches are unscaled. This implies that they are not proportional to the number of changes. This form of presenting a tree (called a cladogram) has the advantage of aligning the OTUs neatly in a vertical column. This may be especially useful if the tree has many dozens of OTUs. Also, it is possible to infer a time scale on this tree if we assume particular dates of divergence.

In the tree of Fig. 7.8c, the same raw data are used to generate the tree, but the branch lengths are now scaled. Thus, the branch lengths are proportional to the

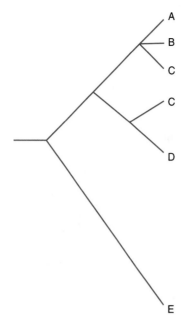

FIGURE 7.9. A phylogenetic tree is said to be multifurcating or polytomous if it has a node with three or more branches (see the node leading to taxa ABC). It is common to make a multifurcating tree when the available data do not provide enough information to define a tree with only bifurcating nodes.

number of amino acid (or nucleotide) changes that occurred between the sequences. This format (called a phylogram) has the helpful feature of conveying a clear visual idea of the relatedness of different proteins within the tree. If one assumes a constant molecular clock, then time and distance are proportional.

An internal node is bifurcating if it has only two immediate descendant lineages (branches). Bifurcating trees are also called binary or dichotomous; any branch that divides splits into two daughter branches. A tree is multifurcating if it has a node with more than two immediate descendants (Fig. 7.9, ABC).

A clade is a group of all the taxa that have been derived from a common ancestor plus the common ancestor itself. A clade is also called a monophyletic group. In our context, a clade is a set of proteins that form a group within a tree. In the example of either tree in Figs. 7.8b and c, C, D, and H form a clade, but B is not a member of this clade. A larger clade is defined by C, D, H, A, B, F, and G. The OTU labeled E is not a member of this larger clade. The taxonomic group ABF that shares a common ancestor (G) with another taxonomic group (CDH) is paraphyletic.

A multifurcation is also called a polytomy. Multifurcating trees are by definition nonbinary. For an example of a multifurcating tree, see Rokas et al. (2005). They reported that many metazoan (animal) phyla are unresolved, reflecting a temporal compression due to the rapid radiation of many animal groups. Philippe and colleagues (Baurain et al., 2007) suggested that such multifurcations occur in a phylogenetic tree because of insufficient sampling.

Tree Roots

A phylogenetic tree has a root representing a most recent common ancestor of all the sequences. Often this root is not known today, and some tree-making algorithms do not provide conjectures about placement of a root. The alternative to a rooted tree is an unrooted tree. An unrooted tree specifies the relationships among the OTUs. However, it does not define the evolutionary path completely or make assumptions about common ancestors. Figure 7.10 shows a binary tree with five OTUs that is either unrooted (Fig. 7.10a) or rooted (Fig. 7.10b). The OTUs (extant taxons, leaves) are numbered 1 to 5. Some OTUs can be swapped (exchanged) without altering the topology of the tree, such as 4 and 5 in either tree. A rule is that OTUs or clades that share an immediate ancestor node can be rotated on that node. But others cannot be swapped, such as 1 and 2. Note that in the unrooted tree the direction of time is undetermined.

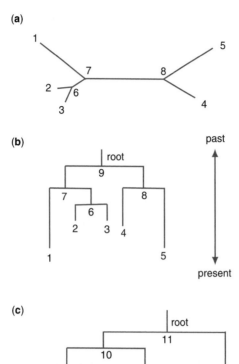

FIGURE 7.10. A phylogenetic tree may be (a) unrooted or (b) rooted. The same raw data are used to generate each type of tree in (a) and (b). The placement of a root implies a hypothesis about the common ancestor of all members of the tree. When this information is not known, an unrooted tree may be more appropriate. Rooting of a tree may be accomplished in two ways. In midpoint rooting the longest branch (here, the branch connecting nodes 7 and 8 in [b]) may be used to define the most likely place to add a root. (c) A single taxon from a phylogenetically distant organism is added to the data set (taxon 6) and used to define an outgroup in a new tree. (In [c] we label the six OTUs 1 to 6, and then label the internal nodes 7 to 11.) An invertebrate lipocalin is an example of a taxon that may form an outgroup relative to a series of mammalian lipocalins.

In web document 7.3 we include human cytoglobin as an outgroup for 12 closely related myoglobin DNA sequences.

Some phylogeny projects involve the generation of trees for thousands of taxa. See the Deep Green plant project (▶ http://ucjeps.berkeley.edu/bryolab/GPphylo/). The Ribosomal Database (▶ http://rdp.cme.msu.edu/) includes an analysis of over 300,000 aligned sequences. For typical analyses, you may analyze several dozen taxa. If you want to make a phylogenetic tree with the globins that are currently in Pfam (version 21.0), you could use the 76 proteins available in the seed alignment or all 2039 proteins available in the full alignment.

The principal way to root a tree is to specify an outgroup. In Fig. 7.10c, imagine that sequences 1 to 5 are mammalian myoglobin orthologs and that the sequence of a homologous bacterial or invertebrate protein (OTU 6) is obtained. This invertebrate sequence is clearly derived from a common ancestor that predates the appearance of all the other OTUs. Thus, it can be used to define the location of the root.

You can select an OTU in order to place a root by identifying the most closely related outgroup. A second way to place a root is through midpoint rooting. Here, the longest branch is determined (such as the branch between internal nodes 7 and 8 in Fig. 7.10a). This longest branch is presumed to be the most reasonable site for a root.

Enumerating Trees and Selecting Search Strategies

The number of possible trees to describe the relationships of a dozen protein sequences is staggeringly large. It is important to know the number of possible trees for any tree you are making. There is only one "true" tree representing the evolutionary path by which molecular sequences (or even species) evolved. The number of potential trees is useful in deciding which tree-making algorithms to apply.

The number of possible rooted and unrooted trees is described in Box 7.4. For two OTUs, there is only one tree possible. For three taxa, it is possible to construct either one unrooted tree or three different rooted trees (Fig. 7.11). For four taxa, the number of possible trees rises to 3 unrooted trees or 15 rooted trees (Fig. 7.12).

Box 7.4
Number of Rooted and Unrooted Trees

The number of bifurcating unrooted trees (N_U) for n OTUs ($n \geq 3$) is given by Cavalli-Sforza and Edwards (1967):

$$N_U = \frac{(2n-5)!}{2^{n-3}(n-3)!}$$

The number of bifurcating rooted trees (N_R) for n OTUs ($n \geq 2$) is

$$N_R = \frac{(2n-3)!}{2^{n-2}(n-2)!}$$

For example, for four OTUs, N_R equals $(8-3)!/(2^2)(2)! = 5!/8 = 15$. The number of possible rooted and unrooted trees (up to 50 OTUs) is as follows. The values were calculated using MATLAB® software (MathWorks).

No. of OTUs	No. of Rooted Trees	No. of Unrooted Trees
2	1	1
3	3	1
4	15	3
5	105	15
6	945	105
7	10,395	945
8	135,135	10,395
9	2,027,025	135,135
10	34,489,707	2,027,025
15	213,458,046,676,875	8×10^{12}
20	8×10^{21}	2×10^{20}
50	2.8×10^{76}	3×10^{74}

To give a sense of the immense number of possible trees corresponding to just a few dozen taxa, there are on the order of 10^{79} protons in the universe.

An exhaustive search examines all possible trees and selects the one with the most optimal features, such as the shortest overall sum of the branch lengths. An important practical limit is reached at around 12 sequences, for which there are over 6.5×10^8 possible unrooted trees and 1.3×10^{10} rooted trees. For about 12 texa (or fewer) it is possible for a standard desktop computer to perform exhaustive searches for which all possible trees are evaluated. For example, PAUP software (introduced below) sets an upper limit of 12 taxa to perform an exhaustive search.

The branch-and-bound method provides an exact algorithm for identifying the optimal tree (or trees) without performing an exhaustive search (Penny et al., 1982; reviewed in Felsenstein, 2004). In one variant of this approach three taxa are

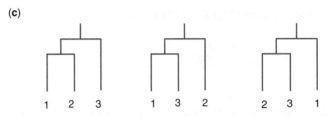

FIGURE 7.11. (a) For three operational taxonomic units (such as three aligned protein sequences 1–3), there is one possible unrooted tree. (b) Any of these edges may be used to select a root (see arrows), from which (c) three corresponding rooted trees are possible.

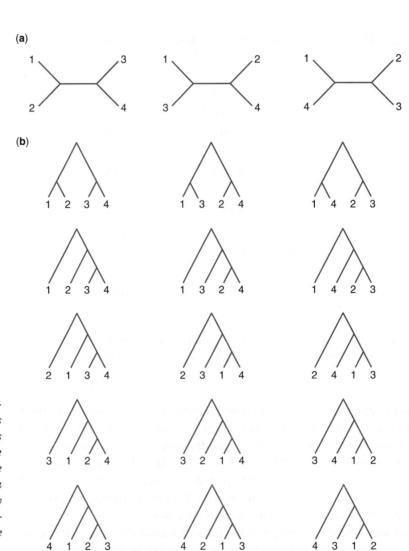

FIGURE 7.12. For four operational taxonomic units (such as four aligned protein sequences 1–4), there are (a) 3 possible unrooted trees and (b) 15 possible rooted trees. Only one of these is a true tree in which the topology accurately describes the evolutionary process by which these sequences evolved.

used to make a tree; only one unrooted tree is possible. A fourth taxon is added, creating three possible trees (as shown in Fig. 7.12a). Upon addition of a fifth taxon there are three times five (i.e., 15) possible trees. By considering the tree in each group having the shortest branch lengths, it is possible to efficiently identify candidates for the optimal tree(s). This allows a strategy of not performing exhaustive searches for trees (or subtrees) having a worse score than the potential optimal tree. The name of this method refers to a boundary that is reached once the search process has identified a subtree with a suboptimal score.

For more than a dozen sequences it is generally necessary to use a heuristic algorithm to identify an optimal tree (or trees). A heuristic algorithm explores a subset of all possible trees, discarding vast numbers of trees that have a topology that is unlikely to be useful. In this way it is possible to create phylogenetic trees having hundreds or even thousands of protein (or DNA) sequences. As an example of how a heuristic algorithm works, consider a data set in which the algorithm seeks a tree with the shortest total branch lengths (i.e., the most parsimonious tree). This search occurs without evaluating all possible trees, but instead by performing a series of rearrangements of the topology. Once a tree with a particular score is obtained, the algorithm can establish that score as an upper limit and discard all trees for which rearrangements are unlikely to yield a shorter tree.

A variety of heuristic approaches are available. Stepwise addition involves the addition of taxa (as described for branch-and-bound) with subsequent branch swapping on the shortest tree(s). The choice of which three taxa are joined initially may be determined arbitrarily (e.g., by the order in which the sequences are input), randomly,

By analogy to the branch-and-bound approach, the Needleman–Wunsch method identifies the optimal subpaths in a pairwise alignment without exhaustively evaluating all possible subpaths (Chapter 3).

(a)

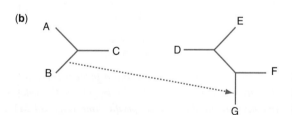

(b)

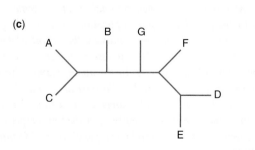

(c)

FIGURE 7.13. Branch swapping using the tree bisection reconnection (TBR) approach. After a tree is made it is bisected along a branch to form two subtrees. These are reconnected by joining one branch from each subtree. All possible bisections are evaluated, as well as all possible reconnection patterns. The goal is to identify the most optimal tree(s). Adapted from the PAUP user's manual. Used with permission.

or based on which three taxa are most closely related. Another heuristic algorithm is branch swapping. In the "tree bisection and reconnection" version, a tree is bisected along a branch, generating two subtrees. These are reconnected by systematically joining all possible pairs of branches with one branch originating from each subtree (Fig. 7.13). Heuristic algorithms have an inherent trade-off between search time and confidence in the search result. One can assume that they provide an approximation of the "best" tree.

TYPE OF TREES

Species Trees versus Gene/Protein Trees

Species evolve and proteins (and genes) evolve. The analysis of protein evolution can be complicated by the time that two species diverged. Speciation, the process by which two new species are created from a single ancestral species, occurs when the species become reproductively isolated (Fig. 7.14). In a species tree, an internal node represents a speciation event. For example, for a species tree containing human and mouse taxa connected by a node, that node corresponds to the last common ancestor of humans and mice, a creature that lived some 80 million years ago. In a gene tree (or protein tree), an internal node represents the divergence of an

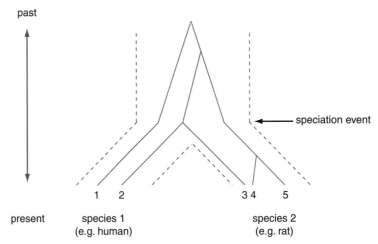

FIGURE 7.14. *A species tree and a protein (or gene) tree can have a complex relationship. A speciation event, such as the divergence of the lineage that generated modern humans and rodents, can be dated to a specific time (e.g., 80 MYA). When speciation occurs, the species become reproductively isolated from one another. This event is represented by dotted lines (see horizontal arrow). Phylogenetic analysis of a specific group of homologous proteins is complicated by the fact that a gene duplication could have preceded or followed the speciation event. In essentially all phylogenetic analyses, the extant proteins (OTUs) are sequences from organisms that are alive today. It is necessary to reconstruct the history of the protein family as well as the history of each species. In the above example, there are two human paralogs and three rat paralogs. Proteins 1 and 5 diverged at a time that greatly predates the divergence of the two species. Proteins 2 and 3 diverged at a time that matches the date of species divergence. Proteins 4 and 5 diverged recently, after the time of species divergence. It is possible to reconstruct both species trees and protein (or gene) trees. Modified from Nei (1987) and Graur and Li (2000).*

ancestral gene into two new genes (or proteins) with distinct sequences. Phylogeny software such as PAUP can reconstruct ancestral DNA or protein sequences that are present at an inferred node. An example is shown for a group of myoglobin sequences (Fig. 7.15). For a tree containing rat and mouse myoglobin sequences, the node connecting those two taxa represents the sequence of an ancestral rodent that existed at the time of the rat-mouse speciation (perhaps 20 million years ago). In almost all cases this ancestral sequence is not known but is inferred. Reconstructions of ancestral states are subject to a variety of artifacts, especially when rates of evolution are rapid in some branches of the tree (Cunningham et al., 1998).

The interpretation of a phylogenetic tree should be in terms of historical events (Baum et al., 2005). Consider the tree of globins shown in Fig. 7.1. Is a globin from lamprey (a fish) more closely related to insect globin than to horse alpha globin? No, it is not: lamprey globin and horse alpha globin are members of a clade that share a

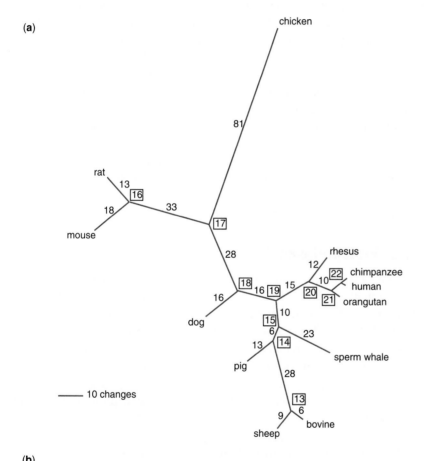

(a)

(b)

| Taxon/ | 1111111111222222222233333333334444444444555555555566666666667777777777 |
Node	1234567890123456789012345678901234567890123456789012345678901234567890123456789012345678
13	atggggctcagcgacggggaatggcagttggtgctgaatgcctggggggaaggtggaggctgatgtcgcaggccatggg
14	atggggctcagcgacggggaatggcagttggtgctgaacgtctggggggaaggtggaggctgatgtcgcaggccatggg
15	atggggctcagcgacggggaatggcagttggtgctgaacgtctggggggaaggtggaggctgatgtcgcaggccatggg
16	atggggctcagtgatggggagtggcagctggtgctgaacatctggggggaaggtggaggccgaccttgctggccatgga
17	atggggctcagcgacggggaatggcagctggtgctgaacatctggggggaaagtggaggccgaccttgctggccatggg
18	atggggctcagcgacggggaatggcagttggtgctgaacatctggggggaaggtggaggctgacctggccatggg
19	atggggctcagcgacggggaatggcagttggtgctgaacgtctggggggaaggtggaggctgacctcgcaggccatggg
20	atggggctcagcgacggggaatggcagttggtgctgaacgtctggggggaaggtggaggctgacatcccaagccacggg
21	atggggctcagcgacggggaatggcagttggtgctgaacgtctggggggaaggtggaggctgacatcccaagccacggg
22	atggggctcagcgacggggaatggcagttggtgctgaacgtctggggggaaggtggaggctgacatcccaggccatggg

FIGURE 7.15. Reconstruction of ancestral sequences. (a) 12 myoglobin DNA coding sequences were aligned and a phylogenetic tree was constructed in PAUP using the maximum parsimony criterion. An unrooted phylogram is shown in which values on the branches correspond to the number of nucleotide changes (scale bar = 10 changes). Several values were removed for clarity. In addition to the 12 terminal nodes for the OTUs, there are 10 internal nodes (assigned numbers 13 to 22 in boxes). (b) PAUP software generates the inferred ancestral sequence at each node. A portion of the output is shown for nucleotides 1 to 78 of nodes 13 to 22.

common ancestor (see the node at arrow 1 having a multifurcation), and that ancestor is the descendant of the last common ancestor of insect globin and lamprey globin (Fig. 7.1, arrow 2). Interpreting trees in phylogeny contrasts with the analysis of trees in other areas of biology such as microarray data analysis (Chapter 9). There the nodes connecting samples or genes do not have a historical meaning.

In a genetically polymorphic population, gene duplication events may occur before or after speciation. A protein (or gene) tree differs from a species tree in two ways (Graur and Li, 2000): (1) The divergence of two genes from two species may have predated the speciation event. This may cause overestimation of branch lengths in a phylogenetic analysis. (2) The topology of the gene tree may differ from that of the species tree. In particular, it may be difficult to reconstruct a species tree from a gene tree. A molecular clock may be applied to a gene tree in order to date the time of gene divergence, but it cannot be assumed that this is also the time that speciation occurred.

Reconstructing a phylogenetic tree based upon a single protein (or gene) can thus give complicated results. For this reason, many researchers construct trees from a variety of distinct protein (or gene) families in order to assess the relationships of different species. Another strategy that has been adopted is to generate concatenated protein (or DNA) sequences. For example, Baldauf et al. (2000) used four concatenated protein sequences to create a comprehensive phylogenetic tree of eukaryotes (see Fig. 18.1). Such a strategy produces a tree that is weighted by the average protein length, and the choice of which sequences are included will impact the outcome.

In looking at phylogenetic trees, it is important to be aware of the type of data that were used to generate the tree. It is also important to inspect the scale bar (if present) which describes whether the units are number of substitutions per site, number of substitutions per branch, elapsed time, or some other measure.

DNA, RNA, or Protein-Based Trees

Synapomorphy is defined as a character state that is shared by several taxa. Homoplasy is defined as a character state that arises independently (e.g., through convergent substitutions or back substitutions) but is not derived from a common ancestor (i.e., is not homologous). See Graur and Li (2000).

When you generate a phylogenetic tree using molecular sequence data, you can use DNA, RNA, or protein sequences. In one common scenario, you may want to evaluate the relationship of a group of molecules such as globins. The choice of whether to study protein or DNA depends in part on the question you are asking. In some cases, protein studies are preferable; you may prefer to study a multiple sequence alignment of proteins, or the lower rate of substitutions in protein relative to DNA may make protein studies more appropriate for comparisons across widely divergent species. In many other cases, studying DNA is more informative than protein. There are several reasons for this.

- DNA allows the study of synonymous and nonsynonymous mutation rates, as dicussed above (Fig. 7.7).
- Substitutions in DNA include those that are directly observed in an alignment, such as single-nucleotide substitutions, sequential substitutions, and coincidental substitutions (depicted in Fig. 7.16). By analyzing two sequences with reference to an ancestral sequence (Fig. 7.16a and b), it is possible to infer a great deal of information about mutations that do not appear in a direct comparison of two (or more) sequences. These mutational processes include parallel substitutions, convergent substitutions, and back substitutions (Fig. 7.16c).

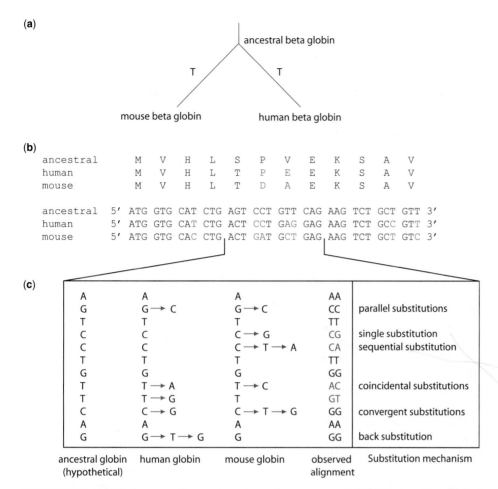

FIGURE 7.16. *Multiple types of mutations occur in sequences. (a) There is a hypothetical, ancestral globin sequence from which human and murine beta globin diverged in the past at time T when these organisms last shared a common ancestor. We can infer the nucleotide and amino acid sequences of the ancestor. (b) Consider a portion of the coding sequence of human and murine beta globin (the data are from Fig. 7.7). There are two observed mismatches at the amino acid level, and seven observed mismatches at the nucleotide level. Many more than seven mutations may have occurred in this region. Hypothetical ancestral protein and DNA sequences are shown, selected for the purpose of illustration. (c) Comparison of 12 nucleotides of the hypothetical ancestral sequence with the observed human and murine sequences illustrates several mutational mechanisms. Single-nucleotide substitution, sequential substitution, and coincidental substitution all could account for observed mutations (red-colored nucleotides). Parallel, convergent, and back substitutions all could occur without producing an observed mismatch. In this example, four mutations are observed (nucleotides colored red) while 13 mutations actually occurred. [(a, c) Adapted from Graur and Li (2000).] Used with permission.*

- Noncoding regions (such as the 5′- and 3′-untranslated regions of genes, or introns; see Fig. 7.7) may be analyzed using molecular phylogeny. For some portions of noncoding DNA, there is little evolutionary pressure to conserve the nucleotide sequence, and these regions may vary greatly. That is, the nucleotide substitution rate equals the neutral mutation rate. In other cases there is tremendous nucleotide conservation, perhaps because of the presence of a regulatory element such as a transcription factor binding motif.

We describe several ribosomal RNA databases in Chapter 8. These serve as excellent sources of sequences for phylogenetic analyses.

- Pseudogenes have been studied using molecular phylogeny, for example to estimate the neutral rate of evolution. By definition, pseudogenes do not encode functional proteins (see Chapter 16). Similarly, inactive DNA transposons and other repetitive DNA elements have been analyzed as "molecular fossils" to explore speciation events and the evolution of chromosomes.

- The rate of transitions and transversions can be evaluated (Box 7.5). In a comparison of mitochondrial DNA among a group of primate species (human, chimpanzee, and gorilla), 92% of the differences were transitions (Brown et al., 1982). Transitions commonly occur far more frequently than transversions in nuclear DNA as well, and this is reflected in various models of nucleotide substitution (see below).

Box 7.5
Transitions and transversions

A transition is a nucleotide substitution between two purines (A to G or G to A) or between two pyrimidines (C to T or T to C). A transversion is the substitution between a purine and a pyrimidine (e.g., A to C, C to A, G to T; there are eight possible transversions). The International Union of Pure and Applied Chemistry (IUPAC; ► http://www.iupac.org) defines many symbols commonly used in science. The abbreviations of the four nucleotides are adenine (A), cytosine (C), guanine (G), and thymine (T). Additional abbreviations are for an unspecified or unknown nucleotide (N), an unspecified purine nucleotide (R), and an unspecified pyrimidine nucleotide (Y). You can assess the rate of transitions and transversions using the MEGA package. Open a protein-coding DNA alignment file in MEGA. Visit the Sequence Data Editor, and under the Statistics pull-down menu choose Nucleotide Pair Frequencies (Directional). The output tabulates the number of identical pairs of nucleotides, the transitional and transversional pairs, and their ratio. Alternatively, use the Pattern pull-down menu, and choose Computer Transition/Transversion Bias.

We will show how the entire genome of a fungus duplicated (Chapter 17). The evidence for this consisted of blastp searches of all *Saccharomyces cerevisiae* proteins against each other, resulting in the detection of conserved blocks of sequence from various chromosomes (see Fig. 15.9). Here, blastn searches would not have been sensitive enough to reveal the homology between different chromosomes.

While the analysis of DNA can offer many advantages, it is sometimes preferable to study proteins for phylogenetic analysis. The evolutionary distance between two organisms may be so great that any DNA sequences are saturated. That is, at many sites all the possible nucleotide changes may occur (even multiple times), so that phylogenetic signal is lost. Proteins have 20 states (amino acids) instead of only four states for DNA, so there is a stronger phylogenetic signal. We saw that blastp searches of human globins against plants were more sensitive than blastn searches (Chapter 4). For closely related sequences, such as mouse versus rat beta globin, DNA-based phylogeny can be more appropriate than protein studies, because of the advantages of DNA discussed above.

Whether nucleotides or amino acids are selected for phylogenetic analysis, the effects of character changes can be defined. An unordered character is a nucleotide or amino acid that changes to another character in one step. An ordered character is one that must pass through one or more intermediate states before it changes to a different character. Partially ordered characters have a variable or indeterminate number of states between the starting value and the ending value. Nucleotides are unordered characters: any one nucleotide can change to any other in one step

(a)

	A	C	T	G
A	0	1	1	1
C	1	0	1	1
T	1	1	0	1
G	1	1	1	0

(b)

	A	C	D	E	F	G	H	I	K	L	M	N	P	Q	R	S	T	V	W	Y
A	0	2	1	1	2	1	2	2	2	2	2	2	1	2	2	1	1	1	2	2
C		0	2	3	1	1	2	2	3	2	3	2	2	3	1	1	2	2	1	1
D			0	1	2	1	1	2	2	3	1	2	2	2	2	2	1	3	1	
E				0	3	1	2	2	1	2	2	2	2	1	2	2	2	1	2	2
F					0	2	2	1	3	1	2	2	3	2	1	2	1	2	1	
G						0	2	2	2	2	2	2	2	1	1	2	1	1	2	
H							0	2	2	1	3	1	1	1	1	2	2	2	3	1
I								0	1	1	1	2	2	1	1	1	1	3	2	
K									0	2	1	1	2	1	1	2	1	2	2	2
L										0	1	2	1	1	1	1	2	1	1	2
M											0	2	2	1	2	1	1	2	2	3
N												0	2	2	1	1	2	3	1	
P													0	1	1	1	2	2	2	
Q														0	1	2	2	2	2	
R															0	1	1	2	1	2
S																0	1	2	1	1
T																	0	2	2	2
V																		0	2	2
W																			0	2
Y																				0

FIGURE 7.17. Step matrices for (a) nucleotides or (b) amino acids describe the number of steps required to change from one character to another. For the amino acids, between one and three nucleotide mutations are required to change any one residue to another. Adapted from Graur and Li (2000). Used with permission.

(Fig. 7.17a). Amino acids are partially ordered. If you inspect the genetic code, you will see that some amino acids can change to a different amino acid in a single step of one nucleotide substitution, while other amino acid changes require two or even three nucleotide mutations (Fig. 7.17b).

FIVE STAGES OF PHYLOGENETIC ANALYSIS

Molecular phylogenetic analyses can be divided into five stages: (1) selection of sequences for analysis, (2) multiple sequence alignment of homologous protein or nucleic acid sequences, (3) specification of a statistical model of nucleotide or amino acid evolution, (4) tree building, and (5) tree evaluation. The remainder of this chapter discusses these stages.

Stage 1: Sequence Acquisition

We have discussed some issues regarding the choice of DNA, RNA, or protein sequences for molecular phylogeny. You can acquire the sequences from many sources, including the following.

- HomoloGene at NCBI includes thousands of eurkaryotic protein families. HomoloGene entries can be viewed as sequences in the fasta format (or as a multiple sequence alignment).

You can access the HIV Sequence Database at ▶ http://www.hiv.lanl.gov/, and the HCV database at ▶ http://hcv.lanl.gov/.

- Results from the BLAST family of proteins can be selected, viewed in Entrez Protein or Entrez Nucleotide, and formatted in the fasta format.
- Sequences from a large variety of databases can be output in the fasta format (or as multiple sequence alignments). For RNA, these databases include Rfam and the Ribosomal Database (Chapter 8). For proteins, these databases include Pfam and InterPro (Chapter 6). For viruses, examples include reference databases for human immunodeficiency virus and hepatitis C virus.

Stage 2: Multiple Sequence Alignment

Web document 7.6 (at ▶ http://www.bioinfbook.org/chapter7) includes 13 quasi-randomly selected protein sequences. If you import these into MEGA you can align them using ClustalW and generate a tree. Can you distinguish that tree from one generated using a group of homologous proteins?

Multiple sequence alignment (Chapter 6) is a critical step of phylogenetic analysis. In many cases, the alignment of nucleotide or amino acid residues in a column implies that they share a common ancestor. If you misalign a group of sequences, you will still be able to produce a tree. However, it is not likely that the tree will be biologically meaningful. And if you create a multiple alignment of sequences and include a non-homologous sequence, it may still be incorporated into the phylogenetic tree.

In preparing a multiple sequence alignment for phylogenetic analysis, there are several important considerations in creating and editing the alignment. Let us introduce these ideas by referring to a specific example of 13 globins. We presented a phylogenetic tree of these proteins in Fig. 7.1. The multiple sequence alignment from which this tree was generated is shown in Fig. 7.18. There are several notable features:

1. Carefully inspect the alignment to be sure that all sequences are homologous. It is sometimes possible to identify a sequence that is so distantly related that it is not homologous. You can further test this possibility by performing pairwise alignments (is the expect value significant?), BLAST searches, or checking whether the proteins are members of a Pfam family. If a sequence is not apparently homologous to the others, it should be removed from the multiple sequence alignment.

2. Some multiple sequence alignment programs may treat distantly related sequences by aligning them outside the block of other sequences. If necessary, lower the gap creation and/or gap extension penalties to accommodate the distantly related homolog(s) into the multiple sequence alignment. As discussed in Chapter 6, include methods that incorporate structural information into the alignment of proteins when possible. In some cases, a group of proteins share a domain (defined in Chapter 10) but are unrelated outside the domain; you can restrict your analyses to the just region of the homologous domain using software such as PAUP or MEGA. These programs allow you to select any specific residues for inclusion or exclusion in the phylogenetic analysis.

3. The complete sequence is not known for many sequences. In general, the multiple sequence alignment data used for phylogenetic analyses should be restricted to portions of the proteins (or nucleic acids) that are available for all the taxa being studied. There are both terminal and internal gaps in this alignment (Fig. 7.18, arrowheads). A gap could represent an insertion in some of the sequences or a deletion in the others. Most phylogeny algorithms are not equipped to evaluate insertions or deletions (also called indels). Many

```
                    ▼▼▼▼▼▼▼▼▼▼▼▼ ▼     ○            ▼▼ ▼▼▼▼        ○      ○○◇     ◇
myoglobin_kanga     -------------MGLSDGEWQLVLNIWGKVETDEGGHGKDVLIRLFKGHPETLEKFDKF
myoglobin_harbo     ------------MGLSEGEWQLVLNVWGKVEADLAGHGQDVLIRLFKGHPETLEKFDKF
myoglobin_gray_     ------------MGLSDGEWHLVLNVWGKVETDLAGHGQEVLIRLFKSHPETLEKFDKF
alpha_globin_ho     ------------MV-LSAADKTNVKAAWSKVGGHAGEYGAEALERMFLGFPTTKTYFPHF
alpha_globin_ka     -------------V-LSAADKGHVKAIWGKVGGHAGEYAAEGLERTFHSFPTTKTYFPHF
alpha_globin_do     -------------V-LSPADKTNIKSTWDKIGGHAGDYGGEALDRTFQSFPTTKTYFPHF
beta_globin_dog     ------------MVHLTAEEKSLVSGLWGKV--NVDEVGGEALGRLLIVYPWTQRFFDSF
beta_globin_rab     ------------MVHLSSEEKSAVTALWGKV--NVEEVGGEALGRLLVVYPWTQRFFESF
beta_globin_kan     ------------VHLTAEEKNAITSLWGKV--AIEQTGGEALGRLLIVYPWTSRFFDHF
globin_riverlam     -PIVDS----GSPAVLSAAEKTKIRSAWAPVYSNYETSGVDILVKFFTSTPAAQEFFPKF
globin_sealampr     MPIVDT----GSVAPLSAAEKTKIRSAWAPVYSTYETSGVDILVKFFTSTPAAQEFFPKF
globin_soybean      ------------VAFTEKQDALVSSSFEAFKANIPQYSVVFYTSILEKAPAAKDLFSFL
globin_insect       MKFLILALCFAAASALSADQISTVQASFDKVKGD----PVGILYAVFKADPSIMAKFTQF
                            ::    :     :    .                         :   *      *   :

                    ▼ ▼   ▼▼▼▼▼▼▼○    ◇            ▼        ○ ▼▼ ▼▼▼◇ ○ ○       ▼
myoglobin_kanga     KHLKSEDEMKASEDLKKHGITVLTALGNILKKKGHHEAELKPLAQS---HATKHKIPVQF
myoglobin_harbo     KHLKTEAEMKASEDLKKHGNTVLTALGGILKKKGHHDAELKPLAQS---HATKHKIPIKY
myoglobin_gray_     KHLKSEDDMRRSEDLRKHGNTVLTALGGILKKKGHHDAELKPLAQS---HATKHKIPIKY
alpha_globin_ho     -DLSHGSA-----QVKAHGKKVGDALTLAVGHLDDLPGALSNLSDL---HAHKLRVDPVN
alpha_globin_ka     -DLSHGSA-----QIQAHGKKIADALGQAVEHIDDLPGTLSKLSDL---HAHKLRVDPVN
alpha_globin_do     -DLSPGSA-----QVKAHGKKVADALTTAVAHLDDLPGALSALSDL---HAYKLRVDPVN
beta_globin_dog     GDLSTPDAVMSNAKVKAHGKKVLNSFSDGLKNLDNLKGTFAKLSEL---HCDKLHVDPEN
beta_globin_rab     GDLSSANAVMNNPKVKAHGKKVLAAFSEGLSHLDNLKGTFAKLSEL---HCDKLHVDPEN
beta_globin_kan     GDLSNAKAVMANPKVLAHGAKVLVAFGDAIKNLDNLKGTFAKLSEL---HCDKLHVDPEN
globin_riverlam     KGMTSADELKKSADVRWHAERIINAVNDAVASMDDTEKMSMK--DLSGKHAKSFQVDPQY
globin_sealampr     KGLTTADQLKKSADVRWHAERIINAVNDAVASMDDTEKMSMKLRDLSGKHAKSFQVDPQY
globin_soybean      ANPTDG----VNPKLTGHAEKLFALVRDSAGQL-KASGTVVADAALGSVHAQKAVTNPEF
globin_insect       AG-KDLESIKGTAPFEIHANRIVGFFSKIIGELPNIEADVNTFVAS---HKPRGVTHDQ-
                              .   *.   .    .     .        .              *

                    ▼▼▼  ▼○○ ○        ▼▼▼▼▼▼▼▼▼   ○        ○        ▼▼▼▼▼▼▼▼▼
myoglobin_kanga     LEFISDAIIQVIQSKHAGNFGADAQAAMKKALELFRHDMAAKYKEFGFQG
myoglobin_harbo     LEFISEAIIHVLHSRHPAEFGADAQGAMNKALELFRKDIATKYKELGFHG
myoglobin_gray_     LEFISEAIIHVLHSKHPAEFGADAQAAMKKALELFRNDIAAKYKELGFHG
alpha_globin_ho     FKLLSHCLLSTLAVHLPNDFTPAVHASLDKFLSSVSTVLTSKYR------
alpha_globin_ka     FKLLSHCLLVTFAAHLGDAFTPEVHASLDKFLAAVSTVLTSKYR------
alpha_globin_do     FKLLSHCLLVTLACHHPTEFTPAVHASLDKFFAAVSTVLTSKYR------
beta_globin_dog     FKLLGNVLVCVLAHHFGKEFTPQVQAAYQKVVAGVANALAHKYH------
beta_globin_rab     FRLLGNVLVIVLSHHFGKEFTPQVQAAYQKVVAGVANALAHKYH------
beta_globin_kan     FKLLGNIIVICLAEHFGKEFTIDTQVAWQKLVAGVANALAHKYH------
globin_riverlam     FKVL-AVIADTVAAG---------DAGFEKLSMCIILMLRSAY-------
globin_sealampr     FKVLAAVIADTVAAG---------DAGFEKLMSMICILLRSAY-------
globin_soybean      --VVKEALLKTIKAAVGDKWSDELSRAWEVAYDELAAAIKAK--------
globin_insect       ---LNNFRAGFVSYMKAHTDFAGAEAAWGATLDTFFGMIFSKM-------
                            :       .      .   .        .  .    :
```

FIGURE 7.18. We will introduce tree-making approaches with a multiple sequence alignment of 13 globins, made using MAFFT at EBI (FFT-NS-1 v5.861). The sequences correspond to those in Fig. 7.1. There are three myoglobins (red kangaroo Macropus rufus, P02194; harbor porpoise Phocoena phocoena, P68278; gray seal Halichoerus grypus, P68081); three alpha globins (horse Equus caballus, P01958; eastern gray kangaroo Macropus giganteus, P01975; dog Canis lupus familiaris, P60529); three beta globins (dog Canis lupus familiaris, XP_537902; rabbit Oryctolagus cuniculus, NP_001075729; eastern gray kangaroo Macropus giganteus, P02106); two fish globins (European river lamprey Lampetra fluviatilis, 690951A; sea lamprey Petromyzon marinus, P02208); an insect globin (midge larva Chironomus thummi thummi, P02229); and a plant leghemoglobin (soybean Glycine max 711674A). Gaps in the alignment (solid arrowheads) are not easily interpretable by phylogenetic algorithms and could represent either insertions or deletions. Four positions are 100% conserved (open diamonds). Amino acids in many other positions distinguish the groups of myoglobins, alpha globins, beta globins, and other globins (examples are shown in columns with open circles; in some cases the groups are perfectly distinguishable in an aligned column). A phylogenetic tree provides a visualization of these relationships (Fig. 3.2 and this chapter).

TABLE 7-3	ReadSeq Servers Available on the Internet
Source	URL
Baylor College of Medicine	▶ http://searchlauncher.bcm.tmc.edu/seq-util/readseq.html
Center for Information Technology, National Institutes of Health	▶ http://bimas.dcrt.nih.gov/molbio/readseq/
Pasteur Institute	▶ http://bioweb.pasteur.fr/seqanal/interfaces/readseq-simple.html
European Bioinformatics Institute	▶ http://www.ebi.ac.uk/readseq/

Many other servers are also available.

experts recommend that any column of a multiple sequence alignment that includes a gap in any position should be deleted, and software programs typically delete columns with incomplete data as a default option.

4. In this example, note that the sequences include three myoglobins, three alpha globins, three beta globins, and four other globins. Intuitively, we expect these globin sequences to be distinguished in a phylogenetic tree, and this is the case (Figs. 7.1 and 7.2). Indeed, we can see such differences by inspecting the multiple sequence alignment. There are positions in which the amino acid in a particular position differs between the myoglobins, alpha globins, and beta globins (Fig. 7.18, columns with open circles and red lettering). Other positions are highly conserved among all these proteins (columns indicated with diamonds), as expected for a family of proteins having closely related structures. The phylogenetic tree (Fig. 7.1) visualizes these various relationships. Any time you inspect a multiple sequence alignment and a tree, you are looking at related information from different perspectives.

First released in 1993, ReadSeq was written by Don Gilbert and is in the public domain.

A variety of tree-building programs accept a multiple sequence alignment as input. ReadSeq is a convenient web-based program that translates multiple sequence alignments into formats compatible with most commonly used phylogeny packages. Several ReadSeq servers are listed in Table 7.3.

Stage 3: Models of DNA and Amino Acid Substitution

Phylogenetic analyses rely on models of DNA or amino acid substitution. These models may be implicit or explicit. For distance-based methods, statistical models are employed to estimate the number of DNA or amino acid changes that occurred in a series of pairwise comparisons of sequences. For maximum likelihood and Bayesian approaches, statistical models are applied to individual characters (residues) in order to assess the most likely topology as well as other features such as substitution rates along individual branches. For maximum parsimony, the criterion for finding the best tree is based on the shortest branch lengths, and while individual characters are also evaluated, many of these statistical models are not applicable.

The simplest approach to defining the relatedness of a group of nucleotide (or amino acid) sequences is to align pairs of sequences and count the number of differences. The degree of divergence is sometimes called the Hamming distance.

For an alignment of length N with n sites at which there are differences, the degree of divergence d is defined as

$$d = \frac{n}{N} \times 100 \qquad (7.3)$$

Earlier in this chapter we discussed an example of this type of calculation by Zuckerkandl and Pauling (1962), who counted the number of amino acid differences between human beta globin and delta, gamma, and alpha globin. The Hamming distance is simple to calculate, but it ignores a large amount of information about the evolutionary relationships among the sequences. The main reason is that character *differences* are not the same as *distances*: the differences between two sequences are easy to measure, but the genetic distance involves many mutations that cannot be observed directly. As shown in Fig. 7.16, there are many kinds of mutations that occur but are not detected in an estimate of divergence based on counting differences. We also discussed a correction implemented by Dickerson (1971) that was proposed by Margoliash and Smith (1965) and by Zuckerkandl and Pauling (1965); see Equations 7.1 and 7.2. In MEGA software this is referred to as the Poisson correction (see Nei and Kumar, 2000, p. 20). The Poisson correction for distance d assumes equal substitution rates across sites and equal amino acid frequencies. It uses the following formula to correct for multiple substitutions at a single site:

$$d = -\ln(1 - p) \qquad (7.4)$$

where d is the distance, and p is the proportion of residues that differ. We make the following assumptions (Uzzell and Corbin, 1971). First, the probability of observing a change is small but nearly identical across the genome. This probability is proportional to the length of the time interval $\lambda \Delta t$ for some constant λ. The probability of observing no changes is thus $1 - \lambda \Delta t$. Second, we assume the number of nucleotide or amino acid changes is constant over the time interval t. When a mutation does occur, this does not alter the probability of another mutation occurring at this same position. Third, we assume that changes occur independently. Equation 7.4 is derived from the Poisson distribution, which describes the random occurrence of events when that probability of occurrence is small. The Poisson distribution is used to model a variety of phenomena, such as the decay of radioactivity over time. It is given by the formula:

$$P(X) = \frac{e^{-\mu} \mu^X}{X!} \qquad (7.5)$$

where $P(X)$ is the probability of X occurrences per unit of time, μ represents the population mean number of changes over time, and e is ~ 2.71828 (Zar, 1999).

Let us consider a practical example of how different substitution models affect the distances that are measured in a set of 13 globin proteins. We enter the proteins into MEGA and select the Distances pull-down menu to compute pairwise distances between the 13 proteins. We can view the number of amino acid differences per sequence (Fig. 7.19a), highlighting several pairwise comparisons that are relatively closely or distantly related. Next we estimate the differences based on the Hamming distance (Equation 7.3; called the p-distance in MEGA) (Fig. 7.19b). When we next use the Poisson correction, the distance values are comparable (relative to the Hamming distance) for closely related sequences such as globins from two lampreys (Fig. 7.19c, dashed red boxes). However, the estimated evolutionary

(a) Number of differences

	1	2	3	4	5	6	7	8	9	10	11	12
1. mbkangaroo P02194 Macropus rufus [red...												
2. mbharbor porpoise P68278 Phocoena pho...	19											
3. mbgray seal P68081 Halichoerus grypus	16	12										
4. alphahorse P01958 Equus caballus	84	84	84									
5. alphakangaroo P01975 Macropus gigante...	85	87	84	24								
6. alphadog P60529 Canis lupus familiari...	88	88	86	22	27							
7. betadog XP 537902 Canis lupus familia...	80	79	78	66	69	67						
8. betarabbit NP 001075729 Oryctolagus c...	80	81	78	64	67	65	16					
9. betakangaroo P02106 Macropus giganteu...	83	82	80	68	69	66	25	28				
10. globinlamprey 690951A Lampetra fluvia...	88	92	88	77	77	76	83	83	81			
11. globinsealamprey P02208 Petromyzon ma...	89	91	89	76	77	76	83	85	81	8		
12. globinsoybean 711674A Glycine max [so...	98	97	97	93	93	93	87	90	90	93	94	
13. globininsect P02229 Chironomus thummi...	87	88	86	92	93	97	92	90	94	88	89	91

(b) p-distance

	1	2	3	4	5	6	7	8	9	10	11	12
1. mbkangaroo P02194 Macropus rufus [red...												
2. mbharbor porpoise P68278 Phocoena pho...	0.17											
3. mbgray seal P68081 Halichoerus grypus	0.14	0.11										
4. alphahorse P01958 Equus caballus	0.74	0.74	0.74									
5. alphakangaroo P01975 Macropus gigante...	0.75	0.77	0.74	0.21								
6. alphadog P60529 Canis lupus familiari...	0.78	0.78	0.76	0.19	0.24							
7. betadog XP 537902 Canis lupus familia...	0.71	0.70	0.69	0.58	0.61	0.59						
8. betarabbit NP 001075729 Oryctolagus c...	0.71	0.72	0.69	0.57	0.59	0.58	0.14					
9. betakangaroo P02106 Macropus giganteu...	0.73	0.73	0.71	0.60	0.61	0.58	0.22	0.25				
10. globinlamprey 690951A Lampetra fluvia...	0.78	0.81	0.78	0.68	0.68	0.67	0.73	0.73	0.72			
11. globinsealamprey P02208 Petromyzon ma...	0.79	0.81	0.79	0.67	0.68	0.67	0.73	0.75	0.72	0.07		
12. globinsoybean 711674A Glycine max [so...	0.87	0.86	0.86	0.82	0.82	0.82	0.77	0.80	0.80	0.82	0.83	
13. globininsect P02229 Chironomus thummi...	0.77	0.78	0.76	0.81	0.82	0.86	0.81	0.80	0.83	0.78	0.79	0.81

(c) Poisson correction

	1	2	3	4	5	6	7	8	9	10	11	12
1. mbkangaroo P02194 Macropus rufus [red...												
2. mbharbor porpoise P68278 Phocoena pho...	0.18											
3. mbgray seal P68081 Halichoerus grypus	0.15	0.11										
4. alphahorse P01958 Equus caballus	1.36	1.36	1.36									
5. alphakangaroo P01975 Macropus gigante...	1.40	1.47	1.36	0.24								
6. alphadog P60529 Canis lupus familiari...	1.51	1.51	1.43	0.22	0.27							
7. betadog XP 537902 Canis lupus familia...	1.23	1.20	1.17	0.88	0.94	0.90						
8. betarabbit NP 001075729 Oryctolagus c...	1.23	1.26	1.17	0.84	0.90	0.86	0.15					
9. betakangaroo P02106 Macropus giganteu...	1.33	1.29	1.23	0.92	0.94	0.88	0.25	0.28				
10. globinlamprey 690951A Lampetra fluvia...	1.51	1.68	1.51	1.14	1.14	1.12	1.33	1.33	1.26			
11. globinsealamprey P02208 Petromyzon ma...	1.55	1.64	1.55	1.12	1.14	1.12	1.33	1.40	1.26	0.07		
12. globinsoybean 711674A Glycine max [so...	2.02	1.95	1.95	1.73	1.73	1.73	1.47	1.59	1.59	1.73	1.78	
13. globininsect P02229 Chironomus thummi...	1.47	1.51	1.43	1.68	1.73	1.95	1.68	1.59	1.78	1.51	1.55	1.64

FIGURE 7.19. *Estimating the evolutionary divergence between sequences. The MEGA software package includes a menu for choosing models of nucleotide or amino acid substitution. Similar options are available in other software packages such as PAUP and PHYLIP. (a) The number of amino acid differences per sequence is displayed below the diagonal, based on pairwise analyses of 13 globins (see Fig. 7.18 legend for their accession numbers). Two closely related globins (with few differences) are highlighted in dashed red boxes, while two divergent globins (with many differences) are highlighted in solid red boxes. Standard error estimates are displayed above the diagonal and were obtained by using analytic formulas. (b) Evolutionary divergence was estimated using the p-distance option to calculate the number of amino acid differences per site. Note that each cell (below the diagonal) represents the number of observed differences divided by the total number of positions in the dataset (113 in this case, with all columns containing gaps eliminated from the final data matrix). For example, the value of 0.87 for a comparison of taxa 1 (myoglobin from kangaroo) and 12 (soybean globin), shown in a red box, is obtained by dividing 98 by 113. (c) Evolutionary divergence was estimated using the Poisson correction. Note that this introduces a substantial increase in the estimated distance for the more divergent sequences. Such larger estimates are likely to be more realistic than simple Hamming distances, and will lead to the creation of trees with different branch lengths and topologies.*

divergence for distantly related sequences is dramatically different using the Poisson correction (Fig. 7.19d, solid red boxes). The consequence of these differences is that entirely different phylogentic trees may be constructed depending on the particular model you choose. We can use this data set of globin proteins to construct a neighbor-joining tree (defined below) using either the *p*-distance (Fig. 7.20a) or the Poisson correction (Fig. 7.20b). Note that the topologies of the two trees differ in this example (inspect soybean and insect globin), and the branch lengths differ.

FIGURE 7.20. *The effect of differing models of amino acid substitution on phylogenetic trees. Phylogenetic trees of 13 globin proteins were made using the neighbor-joining method, which uses the distance information that is presented in Fig. 7.19. The trees were made using (a) the p-distance or (b) the Poisson correction. Branch lengths are in the units of evolutionary distances used to infer each tree. The sum of the branch lengths was 2.81 in (a) and 4.93 in (b). Trees were created using MEGA software. Bootstrapping was performed using 500 bootstrap replicates to identify the percent of instances (indicated in red) in which bootstrap trees support each clade in the inferred tree. For example, in panel (b) in 100% of the bootstrap trials, horse, dog, and kangaroo alpha globin were supported as being in a clade. However, the clade containing horse and dog alpha globin proteins was supported in only 52% of the bootstrap replicates. This means that in 48% of the bootstrap trees, kangaroo alpha globin joined that group of proteins, and we can infer that there is not strong support for a distinct, closely related horse/dog group that shared an ancestor with the kangaroo protein. In general the bootstrap can provide a measure of how well supported an inferred tree topology is upon repeated samplings of the data set.*

(a) Neighbor-joining tree with p-distance correction

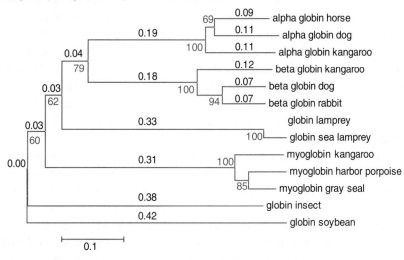

(b) Neighbor-joining tree with Poisson correction

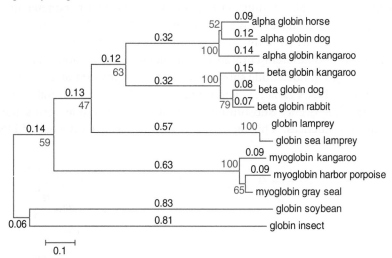

For the optimal tree using the *p*-distance correction, the sum of the branch lengths is 2.81, while for the tree made with the Poisson correction the sum of the branch lengths is 4.93. Such differences can have large effects on the interpretation of a phylogenetic tree, and this example shows how it is important to choose an appropriate model.

In order to model substitutions that occur in DNA sequences, Jukes and Cantor (1969, p. 100) proposed another fundamentally useful corrective formula:

$$D = -\frac{3}{4}\ln\left(1 - \frac{4}{3}p\right) \tag{7.6}$$

As an example of how to use Equation 7.6, consider an alignment where 3 nucleotides out of 60 aligned residues differ. The normalized Hamming distance is $\frac{3}{60} = 0.05$. The Jukes–Cantor correction $D = -\frac{3}{4} \ln [1 - (4 \times 0.05/3)] = 0.052$. In this case, applying the correction causes only a small effect. When $\frac{30}{60}$ nucleotides differ, the Jukes–Cantor correction is $-\frac{3}{4} \ln (1 - [4 \times 0.5/3]) = 0.82$, a far more substantial adjustment.

The Jukes–Cantor one-parameter model describes the probability that each nucleotide will mutate to another (Fig. 7.21a). It makes the simplifying assumption that each residue is equally likely to change to any of the other three residues and that the four bases are present in equal frequencies. Thus, this model assumes that the rate of transitions equals the rate of transversions. The corrections are minimal for very closely related sequences but can be substantial for more distantly related sequences. Beyond about 70% differences, the corrected distances are difficult to estimate. This approaches the percent differences found in randomly aligned sequences.

Dozens of models have been developed that are more sophisticated than Jukes–Cantor. Usually, the transition rate is greater than the transversion rate; for eukaryotic nuclear DNA it is typically twofold higher. The Kimura (1980) two-parameter model adjusts the transition and transversion ratios by giving more weight to transversions to account for their likelihood of causing nonsynonymous changes in protein-coding regions (Fig. 7.21b). In any region of DNA (including noncoding sequence), the transition/transversion ratio corrects for the biophysical threshold for creating a purine-purine or pyrimidine-pyrimidine pair in the double helix. For example, Tamura (1992) extended the two-parameter model to adjust for the guanine and cytosine (GC) content of the DNA sequences (Fig. 7.21c). We will see in Part III of this book that the GC content varies greatly among different organisms and different chromosomal regions within an organism's genome.

Changes in nucleotide substitution at a given position of an alignment represent one kind of DNA variation, and we have been discussing several ways to correct for changes that occur. Substitution rates are often variable across the length of a group of sequences. This represents a second distinct category of DNA variation,

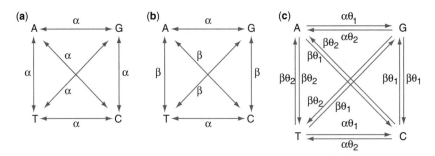

FIGURE 7.21. *Models of nucleotide substitution. (a) The Jukes–Cantor model of evolution corrects for superimposed changes in an alignment. The model assumes that each nucleotide residue is equally likely to change to any of the other three residues and that the four bases are present in equal proportions. The rate of transitions (α) equals the rate of transversions (β). (b) In the Kimura two-parameter model, $\alpha \neq \beta$. Typically, transversions are given more weight. (c) Tamura's model, which accounts for variations in GC content. This is an example of a more complex model of nucleotide substitution. Note that there are distinct parameters for nucleotide substitutions, and that many of these parameters are directional (e.g. the rate of changing from nucleotides T to C differs from the rate for C to T).*

and we can also model these changes. Some sites (columns of aligned residues) are invariant, while others do undergo substitutions.

- Because of the degeneracy of the genetic code, the third position of a codon almost always has a higher substitution rate than the first and second codon position.

- Some regions of a protein have conserved domains. We saw an example of this with the insulin orthologs in Fig. 7.3. Viruses or immunoglobulins often display hypervariable regions of mutation.

- Noncoding RNAs (Chapter 8) often have functional constraints such as stem and loop structures that include highly conserved positions with low substitution rates.

A gamma (Γ) model accounts for unequal substitution rates across variable sites (Box 7.6). The gamma family of distributions can be plotted with the substitution rate (x axis) versus the frequency (y axis) (Fig. 7.22). This shape of the distribution varies as determined by the gamma shape parameter (α). Zhang and Gu (1998) measured α for protein sequences from 51 vertebrate nuclear genes and 13 mammalian mitochondrial genes. They reported a range of values from 0.17 to 3.45 (median value 0.71) for the 51 nuclear genes. There was a negative correlation between the extent of among-site rate variation and the mean substitution rate. Genes with a high level of rate variation among sites (large α) have a low mean substitution rate and thus are slowly evolving. Rapidly evolving proteins have a low level of rate variation among sites.

Box 7.6
The Gamma Distribution

In mathematics, the gamma distribution (Γ) is commonly used to model continuous variables that have skewed distributions. The gamma distribution has been used to model the among-site rate variation of proteins. Given a substitution rate r at a site, the Γ distribution has the following probability density function (Zhang and Gu, 1998):

$$g(r) = \frac{(\alpha/\mu)^{\alpha}}{\Gamma(\alpha)} r^{\alpha-1} e^{-(\alpha/\mu)r}$$

The two parameters in this equation are the mean rate $\mu = E(r)$ and the shape parameter α. Here $E(r)$ is the mean substitution rate (or the expectation of r). Small values of α correspond to a high degree of rate variation among sites. In a study by Zhang and Gu (1998), genes with a high value of α included the C-kit proto-oncogene ($\alpha = 3.45$) and alpha globin ($\alpha = 1.93$) while genes with a low α value included histone H2A.X ($\alpha = 0.19$) and $\beta2$ thyroid hormone receptor ($\alpha = 0.21$). In the R programming language, you can invoke the gamma distribution with the commands `prompt> x = seq(0,10,length = 101)` `Prompt> plot(x,dgamma(x,shape = 2),type = "l")` `Prompt> lines(x,dgamma(x,shape = 0.25))` You can also display the gamma distribution using Microsoft Excel with the formula gammadist.

FIGURE 7.22. *The gamma distribution describes the substitution rate (x axis; from low to high) with a frequency distribution (y axis) that is dependent on shape parameter* α. *For small values of* α *(e.g.,* $\alpha = 0.25$), *most of the nucleotides undergo substitutions at slow rates, and thus most of the observed variation is attributed to relatively few nucleotide sites that evolve rapidly. For large values of* α *(e.g.,* $\alpha = 5$) *few nucleotide sites undergo very fast or very slow evolution, and there is minimal among-site rate variation. For intermediate values of* α *(e.g.,* $\alpha = 2$) *some nucleotides evolve with high substitution rates. This figure was generated in the R programming language as described in Box 7.6.*

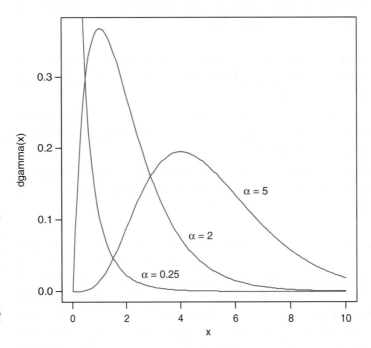

ModelTest was developed by David Posada and colleagues, and is available from ▶ http://darwin.uvigo.es/software/modeltest.html. The site includes a ModelTest server. An example of an output file from ModelTest, showing the results of analyzing 56 substitution models from 11 myoglobin coding sequences, is shown in web document 7.7 at ▶ http://www.bioinfbook.org/chapter7. The Hepatitis C Virus (HCV) sequence database at the Los Alamos National Laboratories (▶ http://hcv.lanl.gov/) offers Findmodel, a web-based implementation of ModelTest that accepts DNA sequences as input. It displays over two dozen models in the format of Fig. 7.21 (▶ http://hcv.lanl.gov/content/hcv-db/findmodel/matrix/all.html).

When we create a phylogenetic tree using 13 globin protein sequences using PAUP or MEGA software, we can specify that there is a uniform rate of variation among sites (thus not invoking the gamma distribution), or we can set the shape parameter of α to any positive value. For a group of globin proteins, there are dramatic differences in the branch lengths and the topologies of trees created using the same neighbor-joining method and the Poisson correction with varying gamma distributions and shape parameters $\alpha = 0.25$, $\alpha = 1$, or $\alpha = 5$ (Fig. 7.23a to c).

There are many choices for nucleotide or amino acid substitution models in programs such as PAUP, Phylip, and MEGA (Fig. 7.24). In addition to and independent of the substitution model, there are many choices for the shape parameter α of the gamma distribution. Several groups have developed strategies to estimate the appropriate models to apply to a data set for phylogenetic analysis. For example, the ModelTest program implements a log likelihood ratio test to compare models (Posada and Crandall, 1998; Posada, 2006). The log likelihood ratio test is a statistical test of the goodness-of-fit between two models. After a DNA data set is executed in PAUP software, ModelTest systematically tests up to 56 models of variation. The likelihood scores of a null model (L_0) and an alternative model (L_1) are calculated for comparisons of a relatively simple model and a relatively complex model. A likelihood ratio test statistic is obtained:

$$\delta = -2 \log \Lambda \tag{7.7}$$

where

$$\Lambda = \frac{\max [L_0(NullModel|Data)]}{\max [L_1(AlternativeModel|Data)]} \tag{7.8}$$

Λ is the Greek letter corresponding to L.

(a) Neighbor-joining tree with Poisson correction and gamma distribution shape parameter α = 0.25

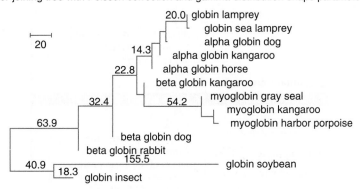

(b) Neighbor-joining tree with Poisson correction and gamma distribution shape parameter α = 1

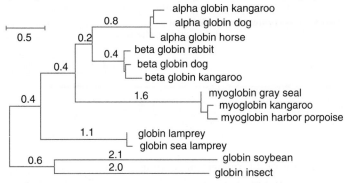

(c) Neighbor-joining tree with Poisson correction and gamma distribution shape parameter α = 5

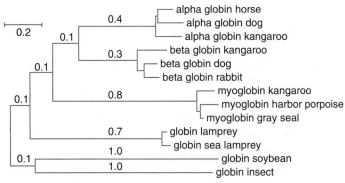

FIGURE 7.23. Effect of changing the α parameter of the Γ distribution on phylogenetic trees. A data set consisting of 13 globin proteins (see fig. 7.1) was aligned and trees were generated in MEGA software using the neighbor-joining technique, the Poisson correction, and α parameters of (a) 0.25, (b) 1, or (c) 5. Note the dramatic effects on the estimated branch lengths. Also note that the topologies differ within the alpha globin, beta globin, and myoglobin clades. The scale bars are in units of number of substitutions.

This test statistic follows a χ^2 distribution, and so given the number of degrees of freedom (equal to the number of additional parameters in the more complex model), a probability value is obtained. As an alternative to log likelihood ratio tests, ModelTest also uses the Akaike information criterion (AIC) (Posada and Buckley, 2004). This measures the best fitting model as that having the smallest AIC value:

$$AIC = -2\ln L + 2N \tag{7.9}$$

where L is the maximum likelihood for a model using N independently adjusted parameters for that model. In this way, good maxium likelihood scores are rewarded, while using too many parameters is penalized.

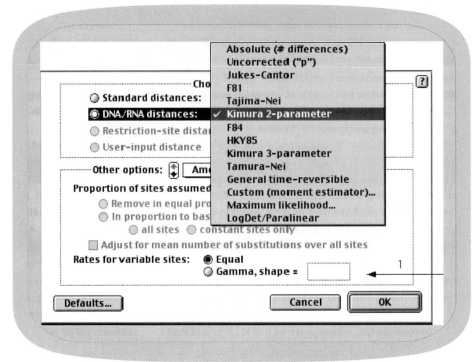

FIGURE 7.24. Models of nucleotide and amino acid substitution. Software packages for phylogenetic analysis such as PAUP (shown here in the Macintosh version) include many choices for models of nucleotide and amino acid substitution. In addition, the shape parameter α of the gamma distribution can be specified (arrow 1).

For the example of 11 myoglobin coding sequences, ModelTest selected the Kimura (1980) two-parameter model with a Γ distribution shape parameter of $\alpha = 0.45$. You can perform a similar analysis using the web-based Findmodel tool of the HCV database.

Stage 4: Tree-Building Methods

There are many ways to build a phylogenetic tree, reviewed in books (Nei, 1987; Graur and Li, 2000; Li, 1997; Maddison and Maddison, 2000; Durbin et al., 1998; Baxevanis and Ouellette, 2001; Clote and Backofen, 2000; Hall, 2001; Felsenstein, 2004) and articles (Bos and Posada, 2005; Felsenstein, 1996, 1988; Hein, 1990; Nei, 1996; Thornton and DeSalle, 2000). We will consider four principal methods of making trees: distance-based methods, maximum parsimony, maximum likelihood, and Bayesian inference. Distance-based methods begin by analyzing pairwise alignments of the sequences and using those distances to infer the relatedness between all the taxa. Maximum parsimony is a character-based method in which columns of residues are analyzed in a multiple sequence alignment to identify the tree with the shortest overall branch lengths that can account for the observed charcter differences. Maximum likelihood and Bayesian inference are model-based statistical approaches in which the best tree is inferred that can account for the observed data.

Molecular phylogeny captures and visualizes the sequence variation that occurs in homologous DNA, RNA, or protein molecules. As we learn how to make trees we will also use some of the most popular software tools for phylogeny. All are extremely versatile and offer a broad range of approaches to making trees.

- PAUP (Phylogenetic Analysis Using Parsimony) was developed by David Swofford.

Phylip is available from ▶ http://evolution.genetics.washington.edu/phylip/general.html. MEGA4 can be downloaded from ▶ http://www.megasoftware.net/. The TREE-PUZZLE site is ▶ http://www.tree-puzzle.de/. MrBayes is available from ▶ http://mrbayes.csit.fsu.edu/download.php. Joseph Felsenstein offers a web page with about 200 phylogeny software links (▶ http://evolution.genetics.washington.edu/phylip/software.html).

- MEGA (Molecular Genetic Evolutionary Analysis) was written by Sudhir Kumar, Koichiro Tamura, and Masatoshi Nei. Many of its concepts are explained in an excellent textbook by Nei and Kumar (2000), *Molecular Evolution and Phylogenetics.*
- PHYLIP (the PHYLogeny Inference Package) was developed by Joseph Felsenstein. It is perhaps the most widely used phylogeny program, and together with PAUP is the most-used software for published phylogenetic trees. Felsenstein has written an outstanding book, *Inferring Phylogenies* (2004).
- TREE-PUZZLE was developed by Korbinian Strimmer, Arndt von Haeseler, and Heiko Schmidt. It implements a maximum likelihood method, which is a model-based approach to phylogeny.
- MrBayes was developed by John Huelsenbeck and Fredrik Ronquist. It implements Bayesian estimation of phylogeny, another model-based approach. MrBayes estimates a quantity called the posterior probability distribution, which is the probability of a tree conditioned on the observed data.

All of these programs are useful. PAUP has a particularly user-friendly interface for the Macintosh platform, although of the programs discussed in this chapter it is the only one that is a commercial package. The others are freely available by download. MEGA is particularly inviting for the PC platform. PHYLIP is perhaps the most popular program, and is command-line driven without an accessible graphical user interface.

PHYLOGENETIC METHODS

Distance

Distance-based methods begin the construction of a tree by calculating the pairwise distances between molecular sequences (Felsenstein, 1984; Desper and Gascuel, 2006). A matrix of pairwise scores for all the aligned proteins (or nucleic acid sequences) is used to generate a tree. The goal is to find a tree in which the branch lengths correspond as closely as possible to the observed distances. The main distance-based methods include the unweighted pair group method with arithmetic mean (UPGMA) and neighbor joining (NJ). Distance-based methods of phylogeny are computationally fast, and thus they are particularly useful for analyses of a larger number of sequences (e.g., >50 or 100).

These methods use some distance metric, such as the number of amino acid changes between the sequences, or a distance score (see Box 6.3). A distance metric is distinguished by three properties: (1) the distance from a point to itself must be zero, that is, $D(x, x) = 0$; (2) the distance from point x to y must equal the distance from y to x, that is, $D(x, y) = D(y, x)$; and (3) the triangle inequality must apply in that $D(x, y) \leq D(x, z) + D(z, y)$. While similarities are also useful, distances (which differ from differences when they obey the above properties) offer appealing properties for describing the relationships between objects (Sneath and Sokal, 1973).

The observed distances between any two sequences i, j can be denoted d_{ij}. The sum of the branch lengths of the tree from taxa i and j can be denoted d'_{ij}. Ideally, these two distance measures are the same, but phenomena such as the occurrence of multiple substitutions at a single position typically cause d_{ij} and d'_{ij} to differ. The goodness of fit of the distances based on the observed data and the branch lengths can be estimated as follows (see Felsenstein, 1984):

$$\sum_i \sum_j w_{ij} \left(d_{ij} - d'_{ij} \right)^2 \tag{7.10}$$

The Clusters of Orthologous Groups (COG) database (Table 10.9 and Chapter 15) relies on distance-based assignments of gene relatedness.

The goal is to minimize this value; it is zero when the branch lengths of a tree match the distance matrix exactly. For Cavalli-Sforza and Edwards (1967) $w_{ij} = 1$ while for Fitch and Margoliash (1967) $w_{ij} = 1/d^2_{ij}$.

We can inspect the multiple sequence alignment in Fig. 7.18 as well as the tree in Fig. 7.1 to think about the essence of distance-based molecular phylogeny. In this approach, one can calculate the percent amino acid similarity between each pair of proteins in the multiple sequence alignment. Some pairs, such as dog and rabbit beta globins, are very closely related and will be placed close together in the tree. Others, such as insect globin and soybean globin, are more distant than the other sequences and will be placed farther away on the tree. In a sense, we can look at the sequences in Fig. 7.18 horizontally, calculating distance measurements between the entire sequences. This approach discards a large amount of information about the characters (i.e., the aligned columns of residues), instead summarizing information about the overall relatedness of sequences. In constrast, character information is evaluated in maximum parsimony, maximum likelihood, and Bayesian approaches. All strategies for inferring phylogenies must make some simplifying assumptions, but nonetheless the simpler approaches of distance-based methods very often produce phylogenetic trees that closely resemble those derived by character-based methods.

The UPGMA Distance-Based Method

We introduce UPGMA here because the tree-building process is relatively intuitive and UPGMA trees are broadly used in the field of bioinformatics. However, the algorithm most phylogeny experts employ to build distance-based trees is neighbor-joining (described below). We can make a distance-based tree in PAUP by selecting the distance criterion from the analysis menu. A dialog box allows you to choose either the UPGMA or neighbor-joining algorithm. MEGA4 similarly offers a pull-down menu for these options. UPGMA clusters sequences based on a distance matrix. As the clusters grow, a tree is assembled. A tree of 13 globins using UPGMA is shown in Fig. 7.25. As we would expect, the alpha globins, beta globins, lamprey globins, and myoglobins are clustered in distinct clades. The two most closely related protein (lamprey globins) are clustered most closely together.

We described the use of a distance matrix to create a guide tree in Chapter 6.

The UPGMA algorithm was introduced by Sokal and Michener (1958) and works as follows. Consider five sequences whose distances can be represented as points in a plane (Fig. 7.26a). We also represent them in a distance matrix. Some protein sequences, such as 1 and 2, are closely similar, while others (such as 1 and 3) are far less related. UPGMA clusters the sequences as follows (adapted from Sneath and Sokal, 1973, p. 230):

1. We begin with a distance matrix. We identify the least dissimilar groups (i.e. the two OTUs i and j that are most closely related). All OTUs are given equal weights. If there are several equidistant minimal pairs, one is picked randomly. In Fig. 7.26a we see that OTUs 1 and 2 have the smallest distance.

2. Combine i and j to form a new group ij. In our example, groups 1 and 2 have the smallest distance (0.1) and are combined to form cluster (1, 2)

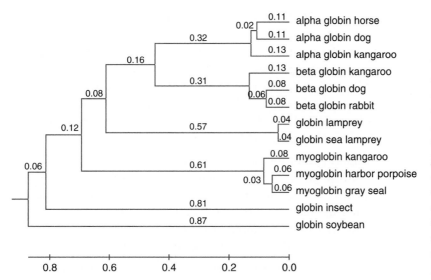

FIGURE 7.25. UPGMA phylogenetic tree of 13 globins. Branch lengths are proportional to evolutionary distances which were calculated using the Poisson correction method. The units are the number of amino acid substitutions per site. Note that UPGMA trees are rooted; the root is placed at the left of the tree. The sum of the branch lengths was 4.93. All positions containing gaps were eliminated from the analysis. The tree was generated using MEGA4 software.

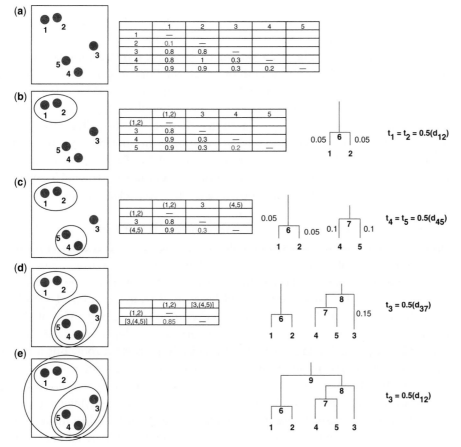

FIGURE 7.26. Explanation of the UPGMA method. This is a simple, fast algorithm for making trees. It is based on clustering sequences. (a) Each sequence is assigned to its own cluster. A distance matrix, based on some metric, quantitates the distance between each object. The circles in the figure represent these sequences. (b) The taxa with the closest distance (sequences 1 and 2) are identified and connected. This allows us to name an internal node [right, node 6, in (b)]. The distance matrix is reconstructed counting taxa 1 and 2 as a group. We can also identify the next closest sequences (4 and 5; their distance is shaded red in the table). (c) These next closest sequences (in the table 4 and 5) are combined into a cluster, and the matrix is again redrawn. In the tree (right side) taxa 4 and 5 are now connected by a new node, 7. We can further identify the next smallest distance (value 0.3, shaded red) corresponding to the union of taxon 3 to cluster (4,5). (d) The newly formed group (cluster 4,5 joined with sequence 3) is represented on the emerging tree with new node 8. Finally, (e) all sequences are connected in a rooted tree.

(see Fig. 7.26b). This results in the formation of a new, clustered distance matrix having one fewer row and column than the initial matrix. Dissimilarities that are not involved in the formation of the new cluster remain unchanged; for example, in the distance matrix of Fig. 7.25b taxa 3 and 4 still maintain a distance of 0.3. The values for the clustered taxa (1,2) reflect the average of OTUs 1 and 2 to each of the other OTUs. The distance of OTU 1 to OTU 4 was initially 0.8, of OTU 2 to OTU 4 was 1.0, and then the distance of OTU (1,2) to OTU 4 becomes 0.9.

3. Connect i and j through a new node on the nascent tree. This node corresponds to group ij. The branches connecting i to ij and j to ij each have a length $D_{ij}/2$. In our example, OTUs 1 and 2 are connected through node 6, and the distance between OTU1 and node 6 is 0.05 (Fig. 7.25b, right side). We label the internal node 6 because we reserve the numbers 1 to 5 on the x axis as the terminal nodes of the tree.

4. Identify the next smallest dissimilarity (between OTUs 4 and 5 in Fig. 7.25b), and combine those taxa to generate a second clustered dissimilarity matrix (Fig. 7.25c). In this step it is possible that two OTUs will be joined (if they share the least dissimilarity), or a single OTU (denoted i) will be joined with a cluster (denoted jk), or two clusters will be joined (ij, kl). The dissimilarity of a single OTU i with a cluster jk is computed simply by taking the average dissimilarity of ij and ik. In this process a new distance matrix is formed, and the tree continues to be constructed. In Fig. 7.25c the smallest distance in the matrix is 0.3 corresponding to the relation of OTU 3 to the combined OTU 4,5. These are joined in Fig. 7.25d in the graphic representation, in the distance matrix, and in the tree.

5. Continue until there are only two remaining groups, and join these.

The tree shown in Fig. 7.25 was made by the UPGMA approach using the sequences of 13 globin proteins, and for which the Poisson-corrected distances are shown in Fig. 7.19c. We demonstrate how to perform UPGMA calculations on this data set in a series of 12 tables available on the supplementary website. Compare Figs. 7.19c and 7.25 and note that the two closest OTUs of the distance matrix (globin from lamprey and sea lamprey) have the shortest branch lengths on the UPGMA tree. The second closest group (myoblobin from harbor porpoise and gray seal) has the next shortest branch lengths. That group of two OTUs collectively has a short branch length to kangaroo myoglobin. These relationships are visualized in the phylogenetic tree.

Visit ▶ http://www.bioinfbook. org/chapter7/for a detailed UPGMA analysis.

A critical assumption of the UPGMA approach is that the rate of nucleotide or amino acid substitution is constant for all the branches in the tree, that is, the molecular clock applies to all evolutionary lineages. If this assumption is true, branch lengths can be used to estimate the dates of divergence, and the sequence-based tree mimics a species tree. An UPGMA tree is rooted because of its assumption of a molecular clock. If it is violated and there are unequal substitution rates along different branches of the tree, the method can produce an incorrect tree. Note that other methods (including neighbor-joining) do not automatically produce a root, but a root can be placed by choosing an outgroup or by applying midpoint rooting.

The UPGMA method is a commonly used distance method in a variety of applications including microarray data analysis (see Chapter 9). In phylogenetic analyses using molecular sequence data its simplifying assumptions tend to make it significantly less accurate than other distance-based methods such as neighbor-joining.

Making Trees by Distance-Based Methods: Neighbor Joining

The neighbor-joining method is used for building trees by distance methods (Saitou and Nei, 1987). It produces both a topology and branch lengths. We begin by defining a neighbor as a pair of OTUs connected through a single interior node X in an unrooted, bifurcating tree. In the tree of globins shown in Fig. 7.1, porpoise and seal myoglobins are neighbors, while kangaroo myoglobin is not a neighbor because it is separated from those two proteins by two nodes. In general, the number of neighbor pairs in a tree depends on the particular topology. For a bifurcating tree with N OTUs, $N - 2$ pairs of neighbors can potentially occur. The neighbor-joining method first generates a full tree with all the OTUs in a starlike structure with no hierarchical structure (Fig. 7.27a). All $N(N - 1)/2$ pairwise comparisons are made to identify the two most closely related sequences. These OTUs give the smallest sum of branch lengths (see taxa 1 and 2 in Fig. 7.27b). OTUs 1 and 2 are now treated as a single OTU, and the method identifies the next pair of OTUs that gives the smallest sum of branch lengths. This could be two OTUs such as 4 and 6, or a single OTU such as 4 paired with the newly formed clade that includes OTUs 1 and 2. The tree has $N - 3$ interior branches, and the neighbor-joining method continues to successively identify nearest neighbors until all $N - 3$ branches are identified.

The process of starting with a star-like tree and finding and joining neighbors is continued until the topology of the tree is completed. The neighbor-joining, algorithm minimizes the sum of branch lengths at each stage of clustering OTUs (see Box 7.7); although the final tree is not necessarily the one with the shortest overall branch lengths. Thus, its results may differ from minimum evolution strategies or maximum parsimony (discussed below). Neighbor joining produces an unrooted tree topology (because it does not assume a constant rate of evolution), unless an outgroup is specified or midpoint rooting is applied.

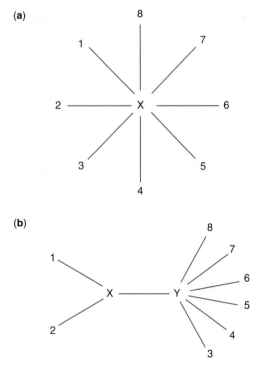

FIGURE 7.27. The NJ method is a distance-based algorithm. (a) The OTUs are first clustered in a starlike tree. "Neighbors" are defined as OTUs that are connected by a single, interior node in an unrooted, bifurcating tree. (b) The two closest OTUs are identified, such as OTUs 1 and 2. These neighbors are connected to the other OTUs via the internal branch XY. The OTUs that are selected as neighbors in (b) are chosen as the ones that yield the smallest sum of branch lengths. This process is repeated until the entire tree is generated. Adapted from Saitou and Nei (1987). Used with permission.

Box 7.7
Branch Lengths in a Neighbor-Joining Tree

Saitou and Nei (1987) defined the sum of the branch lengths as follows. Let D_{ij} equal the distance between OTUs i and j, and let L_{ab} equal the branch lengths between nodes a and b. The sum of the branch lengths S for the tree in Fig. 7.27a is

$$S = \sum_{i=1}^{N} L_{iX} = \frac{1}{N-1} \sum_{i<j} D_{ij}$$

This result follows from the fact that in computing the total distance each branch is counted $N-1$ times. For the tree in Fig. 7.27b the branch length between nodes X and Y (given by L_{XY}) is:

$$L_{XY} = \frac{1}{2(N-2)} \left[\sum_{k=3}^{N} (D_{1k} + D_{2k}) - (N-2)(L_{1X} + L_{2X}) - 2 \sum_{i=3}^{N} L_{iY} \right]$$

In this equation, the first term in the brackets is the sum of all distances that include L_{XY}, and the other terms exclude irrelevant branch lengths. Saitou and Nei (1987) provide further detailed analyses of the total branch lengths of the tree.

We have shown several examples of neighbor-joining trees for 13 globins (Figs. 7.20 and 7.23). This algorithm is especially useful when studying large numbers of taxa. There are many recent examples of its use in the literature, such as studies of the 1918 influenza virus (Taubenberger et al., 2005). There are many alternative distance-based approaches, some of which have been systematically compared (Hollich et al., 2005; Desper and Gascuel, 2006).

Phylogenetic Inference: Maximum Parsimony

The main idea behind maximum parsimony is that the best tree is that with the shortest branch lengths possible (Czelusniak et al., 1990). Parsimony-based phylogeny based on morphological characters was described by Hennig (1966), and Eck and Dayhoff (1966) used a parsimony-based approach to generating phylogenetic trees such as that in Fig. 7.1. According to maximum parsimony theory, having fewer changes to account for the way a group of sequences evolved is preferable to more complicated explanations of molecular evolution. Thus we seek the most parsimonius explanations for the observed data. The assumption of phylogenetic systematics is that genes exist in a nested hierarchy of relatedness, and this is reflected in a hierarchical distribution of shared characters in the sequences. The most parsimonious tree is supposed to best describe the relationships of proteins (or genes) that are derived from common ancestors.

The steps are as follows:

The word parsimony (from the Latin parcere, "to spare") refers to simplicity of assumptions in a logical formulation.

- Identify informative sites. If a site is constant (e.g., Fig. 7.18, diamonds), then it is not informative (see below). MEGA software includes an option to view parsimony-informative sites (Fig. 7.28a, arrow). Noninformative sites include constant sites (Fig. 7.28a, closed arrowheads) and positions in which there are

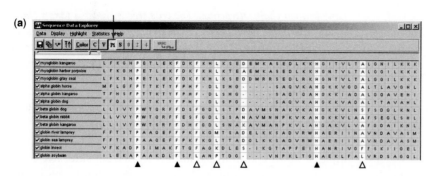

(a)

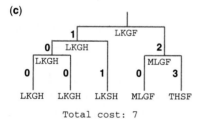

(b)

```
kangaroo     LKGH
porpoise     LKGH
gray seal    LKSH
horse α      MLGF
kangaroo α   THSF
```

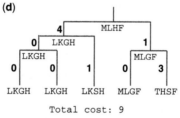

(c) Total cost: 7   **(d)** Total cost: 9

FIGURE 7.28. Principle of maximum parsimony. (a) Considering the columns of aligned residues, many are informative for parsimony analysis. However, columns having entirely conserved residues (filled arrowheads) are not informative, nor are columns in which there are at least two different residues that occur at least two times (open arrowheads). This alignment of 13 globin proteins was viewed in MEGA4 software, with the option to display parsimony-informative characters selected (see arrow); other options include viewing conserved or variable positions. (b) Example of four amino acid residues from five different species (taken from the top left of panel [a]). Maximum parsimony identifies the simplest (most parsimonious) evolutionary path by which those sequences might have evolved from ancestral sequences. (c, d) Two trees showing possible ancestral sequences. The tree in (c) requires seven changes from its common ancestor, while the tree in (d) requires nine changes. Thus, maximum parsimony would select the tree in (c).

not at least two states (e.g. two different amino acid residues) with at least two taxa having each state (such noninformative sites are indicated in Fig. 7.28a, open arrowheads).

- Construct trees. Every tree is assigned a cost, and the tree with the lowest cost is sought. When a reasonable number of taxa are evaluated, such as about a dozen or fewer, all possible trees are evaluated and the one with the shortest branch length is chosen. When necessary, a heuristic search is performed to reduce the complexity of the search by ignoring large families of trees that are unlikely to contain the shortest tree.

- Count the number of changes and select the shortest tree (or trees).

Parsimony analysis assumes that characters are independent of each other. The length L of a full tree is computed as the sum of the lengths l_j of the individual characters:

$$L = \sum_{j=1}^{C} w_j l_j \tag{7.11}$$

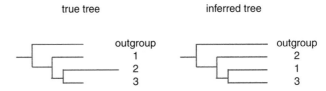

FIGURE 7.29. *Long branch chain attraction. The true tree includes a taxon (labeled 2) that evolves more quickly than the other taxa. It shares a common ancestor with taxon 3. However, in the inferred tree taxon 2 is placed separately from the other taxa because it is attracted by the long branch of the outgroup. Adapted from Philippe and Laurent (1998).*

where C is the total number of characters, and the weight w_j assigned to each character is typically 1. A different weight might be assigned if, for example, nucleotide transversions are more penalized than transitions.

As an example of how maximum parsimony works, consider five aligned amino acid sequences (Fig. 7.28b, taken from the upper left of Fig. 7.28a). Two possible trees describe these sequences (Fig. 7.28c and d); each tree has hypothetical sequences assigned to ancestral nodes. One of the trees (Fig. 7.28c) requires fewer changes to explain how the observed sequences evolved from a hypothetical common ancestor. In this example, each site is treated independently.

In PAUP, you can set the tree-making criterion to parsimony. It is preferable to perform an exhaustive search of all possible trees to find the one with the shortest total branch lengths. In practice, this is not possible for more than 12 taxa, so it is often necessary to perform a heuristic search. Both heuristic and exhaustive searches often result in the identification of several trees having the same minimal value for total branch length of the tree. Trees can be visualized as a phylogram or a cladogram.

An artifact called long-branch attraction sometimes occurs in phylogenetic inference, and parsimony approaches may be particularly susceptible. In a phylogenetic reconstruction of protein or DNA sequences, a branch length indicates the number of substitutions that occur between two taxa. Parsimony algorithms assume that all taxa evolve at the same rate and that all characters contribute the same amount of information. Long-branch attraction is a phenomenon in which rapidly evolving taxa are placed together on a tree, not because they are closely related, but artifactually because they both have many mutations. Consider the true tree in Fig. 7.29, in which taxon 2 represents a DNA or protein that changes rapidly relative to taxa 1 and 3. The outgroup is (by definition) more distantly related than taxa 1, 2, and 3 are to each other. A maximum parsimony algorithm may generate an inferred tree (Fig. 7.29) in which taxon 2 is "attracted" toward another long branch (the outgroup) because these two taxa have a large number of substitutions. Anytime two long branches are present, they may be attracted.

Model-Based Phylogenetic Inference: Maximum Likelihood

Maximum likelihood is an approach that is designed to determine the tree topology and branch lengths that have the greatest likelihood of producing the observed data set. A likelihood is calculated for each residue in an alignment, including some model of the nucleotide or amino acid substitution process. It is among the most computationally intensive but most flexible methods available (Felsenstein, 1981). Maximum parsimony methods sometimes fail when there are large amounts of evolutionary change in different branches of a tree. Maximum likelihood, in contrast, provides a statistical model for evolutionary change that varies across branches. Thus, for example, maximum likelihood can be used to estimate positive and negative selection across individuals branches of a tree. The relative merits of maximum parsimony and maximum likelihood continue to be explored. For example, Kolaczkowski and Thornton (2004) reported that when sequences evolve in a heterogeneous fashion over time maximum parsimony can outperform maximum likelihood.

A computationally tractable maximum likelihood method is implemented in the Tree-Puzzle program (Strimmer and von Haeseler, 1996; Schmidt et al., 2002). The program allows you to specify various models of nucleotide or amino acid substitution and rate heterogeneity (e.g., the Γ distribution). There are three steps. First, Tree-Puzzle reduces the problem of tree reconstruction to a series of quartets of sequences. For quartet A,B,C,D there are three possible topologies (Fig. 7.12a). In the maximum likelihood step the program reconstructs all quartet trees. For N sequences there are

$$\binom{N}{4}$$

possible quartets; for example, for 12 myoglobin DNA sequences there are

$$\binom{12}{4}$$

or 495 possible quartets. The three quartet topologies are weighted by their posterior probabilities.

$$\binom{n}{k}$$

is a binomial coefficient that is read as "n choose k." It describes the number of combinations, that is, how many ways there are to choose k things out of n possible choices. Given the factorial functions $n!$ and $k!$ we can write the binomial coefficient

$$\binom{n}{k} = \frac{n!}{k! \cdot (n-k)!}$$

For

$$\binom{12}{4}$$

this corresponds to

$$\frac{12!}{4!(8)!} \quad \text{or} \quad \frac{12 \cdot 11 \cdot 10 \cdot 9}{4 \cdot 3 \cdot 2 \cdot 1}$$

which is 495.

In the second step, called the quartet puzzling step, a large group of intermediate trees is obtained. The program begins with one quartet tree. Since that tree has four sequences, $N - 4$ sequences remain. These are added systematically to the branches that are most likely based on the quartet results from the first step. Puzzling allows estimates of the support to each internal branch of the tree that is constructed; such estimates are not available for distance- or parsimony-based trees. In the third step, the program generates a majority consensus tree. The branch lengths and maximum likelihood value are estimated. An example of a consensus tree is shown in Fig. 7.30a.

The Tree-Puzzle program also allows an option called likelihood mapping which describes the support of an internal branch as well as a way to visualize the

A file showing how to format 13 globin proteins for input into the Tree-Puzzle program is provided in web document 7.8 at ▶ http://www.bioinfbook.org/chapter7. Web document 7.9 shows the Tree-Puzzle output file.

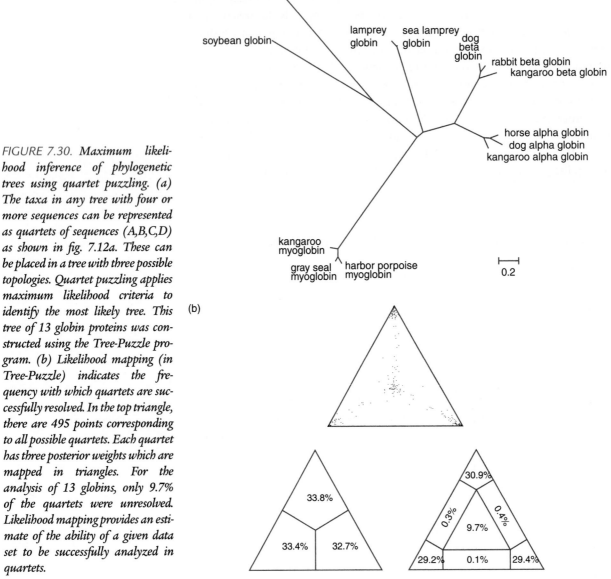

FIGURE 7.30. *Maximum likelihood inference of phylogenetic trees using quartet puzzling. (a) The taxa in any tree with four or more sequences can be represented as quartets of sequences (A,B,C,D) as shown in fig. 7.12a. These can be placed in a tree with three possible topologies. Quartet puzzling applies maximum likelihood criteria to identify the most likely tree. This tree of 13 globin proteins was constructed using the Tree-Puzzle program. (b) Likelihood mapping (in Tree-Puzzle) indicates the frequency with which quartets are successfully resolved. In the top triangle, there are 495 points corresponding to all possible quartets. Each quartet has three posterior weights which are mapped in triangles. For the analysis of 13 globins, only 9.7% of the quartets were unresolved. Likelihood mapping provides an estimate of the ability of a given data set to be successfully analyzed in quartets.*

The Tree-Puzzle program of Korbinian Strimmer and colleagues is available at ▶ http://www.tree-puzzle.de/. You can also perform maximum likelihood (and quartet puzzling) using DNAML (Phylip) and PAUP.

phylogenetic content of a multiple sequence alignment (Strimmer and von Haeseler 1996, 1997). The quartet topology weights sum to 1, and likelihood mapping plots them on a triangular surface. In this plot, each dot corresponds to a quartet that is positioned spatially according to its three posterior weights (Fig. 7.30b). For 13 globin protein sequences, 9.7% of the quartets were unresolved (as indicated in the center of the triangle). An additional 0.3% + 0.4% + 0.1% of the quartets were partially resolved. For 12 myoglobin DNA coding sequences, only 3% of the quartets were unresolved (not shown). Likelihood mapping summarizes the strength (or conversely the ambiguity) inherent in a data set for which you perform tree puzzling.

Tree Inference: Bayesian Methods

Bayesian inference is a statistical approach to modeling uncertainty in complex models. Conventionally we calculate the probability of observing some data (such as the result of a coin toss) given some probability model. This probability is denoted

P(data|model), that is, the probability of the data given the model. (This is also read as "the probability of the data conditional upon the model.") Bayesian inference instead seeks the probability of a tree conditional on the data (that is, based on the observations such as a given multiple sequence alignment). This assumes the form P(model|data), P(hypothesis|data), or in our case P(tree|data). According to Bayes's theorem (Huelsenbeck et al., 2001),

$$\Pr[Tree \,|\, Data] = \frac{\Pr[Data \,|\, Tree] \times \Pr[Tree]}{\Pr[Data]} \qquad (7.12)$$

Bayesian estimation of phylogeny is focused on a quantity called the posterior probability distribution of trees, Pr[Tree | Data]. (This is read as "the probability of observing a tree given the data.") For a given tree, the posterior probability is the probability that the tree is correct, and our goal is to identify the tree with the maximum probability. On the right side of Equation 7.12, the denominator Pr[Data] is a normalizing constant over all possible trees. The numerator consists of the prior probability of a phylogeny (Pr[Tree]) and the likelihood Pr[Data | Tree]. These terms represent a distinctive feature of Bayesian inference of phylogeny: the user specifies a prior probability distribution of trees (although it is allowable for all possible trees to be given equal weight).

Practically, we can apply a Bayesian inference approach using the MrBayes software program. There are four steps. First, read in a Nexus data file. This can be accomplished by performing a multiple sequence alignment of interest, then converting it into the Nexus format with a tool such as ReadSeq. We will use an example of 13 globin protein-coding DNA sequences.

Second, specify the evolutionary model. This includes options for data that are DNA (whether coding or not), ribosomal DNA (for the analysis of paired stem regions; see Chapter 8), and protein. Before performing the analysis, one specifies a prior probability distribution for the parameters of the likelihood model. There are six types of parameters that are set as the priors for the model in the case of the analysis of nucleotide sequences: (1) the topology of the trees (e.g. some nodes can be constrained to always be present), (2) the branch lengths, (3) the stationary frequencies of the four nucleotides, (4) the six nucleotide substitution rates (for $A \leftrightarrow C, A \leftrightarrow G, A \leftrightarrow T, C \leftrightarrow G, C \leftrightarrow T$, and $G \leftrightarrow T$), (5) the proportion of invariant sites, and (6) the shape parameter of the gamma distribution of rate variation. For protein sequences, both fixed-rate and variable-rate models are offered. Your decisions on how to specify these parameters may be subjective. This can be considered either a strength of the Bayesian approach (because your judgment may help you to select optimal parameters) or a weakness (because there is a subjective element to the procedure). All priors do not have to be informative; one can select conservative settings.

Third, run the analysis. This is invoked with the mcmc (Monte Carlo Markov Chain) command. The posterior probability of the possible phylogenetic trees is ideally calculated as a summation over all possible trees, and for each tree, all combinations of branch lengths and substitution model parameters are evaluated. In practice this probability cannot be determined analytically, but it can be approximated using MCMC. This is done by drawing many samples from the posterior distribution (Huelsenbeck et al., 2002). MrBayes runs two simultaneous, independent analyses beginning with distinct, randomly initiated trees. This helps to assure that your analysis includes a good sampling from the posterior probability distribution.

MrBayes is available from
► http://mrbayes.csit.fsu.edu/.
Web document 7.10 shows how to format 13 globin proteins for input into MrBayes, and web document 7.11 shows the output. See ► http://www.bioinfbook.org/chapter7.

Eventually the two runs should reach convergence. An MCMC analysis is performed in three steps: first, a Markov chain is started with a tree that may be randomly chosen. Second, a new tree is proposed. Third, the new tree is accepted with some probability. Typically tens to hundreds of thousands of MCMC iterations are performed. The proportion of time that the Markov chain visits a particular tree is an approximation of the posterior probability of that tree. Some authors have cautioned that MCMC algorithms can give misleading results, especially when data have conflicting phylogenetic signals (Mossel and Vigoda, 2005).

Fourth, summarize the samples. MrBayes provides a variety of summary statistics including a phylogram, branch lengths (in units of the number of expected substitutions per site), and clade credibility values. An example for 13 globin proteins is shown in Fig. 7.31. The summary statistics for a Bayesian analysis are provided. They include a list of all trees sorted by their probabilities; this is used to create a "credible" list of trees (Fig. 7.31a). A consensus tree showing branch lengths and support values for interior nodes is generated (Fig. 7.31b).

Bayesian inference of phylogeny resembles maximum likelihood because each method seeks to identify a quantity called the likelihood which is proportional to observing the data conditional on a tree. The methods differ in that Bayesian inference includes the specification of prior information and uses MCMC to estimate the posterior probability distribution. Although they were introduced relatively recently, Bayesian approaches to phylogeny are becoming increasingly commonplace.

Stage 5: Evaluating Trees

After you have constructed a phylogenetic tree, how can you assess its accuracy? The main criteria by which accuracy may be assessed are consistency, efficiency, and robustness (Hillis, 1995; Hillis and Huelsenbeck, 1992). One may study the accuracy of a tree-building approach or the accuracy of a particular tree. The most common approach is bootstrap analysis (Felsenstein, 1985; Hillis and Bull, 1993). Bootstrapping is not a technique to assess the accuracy of a tree. Instead, it describes the robustness of the tree topology. Given a particular branching order, how consistently does a tree-building algorithm find that branching order using a randomly permuted version of the original data set? Bootstrapping allows an inference of the variability in an unknown distribution from which the data were drawn (Felsenstein, 1985).

Nonparametric bootstrapping is performed as follows. A multiple sequence alignment is used as the input data to generate a tree using some tree-building method. The program then makes an artificial data set of the same size as the original data set by randomly picking columns from the multiple sequence alignment. This is usually performed with replacement, meaning that any individual column may appear multiple times (or not at all). A tree is generated from the randomized data set. A large number of bootstrap replicates are then generated; typically, between 100 and 1000 new trees are made by this process. The bootstrap trees are compared to the original, inferred tree(s). The information you get from bootstrapping is the frequency with which each clade in the original tree is observed.

An example of the bootstrap procedure using MEGA4 is shown in Fig. 7.20. The percentage of times that a given clade is supported in the original tree is provided based on how often the bootstraps supported the original tree topology. Bootstrap values above 70% are sometimes considered to provide support for the clade designations. Hillis and Bull (1993) have estimated that such values provide statistical significance at the $p < 0.05$ level. This approach measures the effect of random

Accuracy refers to the degree to which a tree approximates the true tree. We will define and discuss precision and accuracy in Chapter 9 in the context of microarray data analysis.

Parametric bootstrapping refers to repeated random sampling without replacement from the original sample. It is not used as often as nonparametric bootstrapping.

(a) Phylogram

```
/--- mbkangaro (1)
|
|-- mbharbor_ (2)
|
|- mbgray_se (3)
|
|                                    /--- alphahors (4)
|                                  /+
|                                  |\---- alphadog (6)
|                       /---------+
+                       |          \--- alphakang (5)
|                       |                /-- betadog (7)
|              /-----+   |              /-+
|              |      |   |              | \-- betarabbi (8)
|              |      |   \--------+
|              |      |             \----- betakanga (9)
\-----------------+   |
                  |   |                /- globinlam (10)
                  |   /--------------+
                  |   |               \-- globinsea (11)
                  \-----+                /---------------------- globinsoy (12)
                        \-----------+
                                    \-------------------- globinins (13)

|--------------| 0.500 expected changes per site
```

(b) Radial tree with clade credibility values

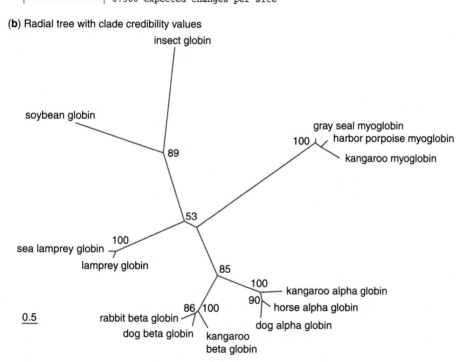

FIGURE 7.31. Bayesian inference of phylogeny for 13 globin proteins using MrBayes software (version 3.1.2). The input sequences were aligned using MAFFT at EBI (see Chapter 6). The amino acid model (using default settings) was Poisson, with 20 states corresponding to the amino acids and equal rates of substitution. Prior parameters included equal, fixed frequencies for the states, an equal probability for all topologies, and unconstrained branch lengths. Monte Carlo Markov Chain estimation of the posterior distribution was achieved using a run of 600,000 trials from which over 11,000 trees were evaluated. Clade credibility values give strong support for the separation of groups containing myoglobins, alpha globins, beta globins, and lamprey globins (all these clades had 100% support). (a) The phylogram output shows clades containing various globin subtypes. (b) Tree files can be exported and viewed using TreeView software. Here a radial tree is shown with clade credibility values added. The scale bar is 0.5 expected changes per amino acid site. The procedure for performing this analysis is described in web document 7.10 at ▶ http://www.bioinfbook.org/chapter7.

weighting of characters in the original data matrix, giving insight into how strongly the phylogenetic signal that produces a tree is distributed through the multiple sequence alignment. In Fig. 7.20a and b, the clade containing three alpha globins has 100% bootstrap support, indicating that in all 500 bootstrap replicates that clade maintained its integrity (with none of the three alpha globins assigned to a different clade, and no non-alpha globin joining that clade). However, the clade containing horse and dog alpha globin received only 52% bootstrap support (Fig. 7.20b), suggesting that about half the time kangaroo alpha globin was in a clade with the dog or horse orthologs. This example shows how viewing the bootstrap percentages can be useful to estimate the robustness of each clade in a tree. Note that bootstrapping supports a model in which alpha globins, beta globins, myoglobins, and lamprey globins each are assigned to a unique clade.

Maximum likelihood approaches report the tree with the greatest likelihood, and they also report the likelihood for internal branches. For Bayesian inference of phylogeny, the result is typically the most probable tree (called a maxiumum a posteriori probability estimate). The results are often summarized using a majority rule consensus tree in which the values represent the posterior probability that each clade is true. The confidence estimates may sometimes be too liberal (Suzuki et al., 2002). For example, Mar et al. (2005) found that Bayesian posterior probabilities reached 100% at bootstrap percentages of 80%.

PERSPECTIVE

Molecular phylogeny is a fundamental tool for understanding the evolution and relationships of protein (and nucleic acid) sequences. The main output of this analysis is a phylogenetic tree, which is a graphical representation of a multiple sequence alignment. The recent rapid growth of DNA and protein sequence data, along with the visual impact of phylogenetic trees, has made phylogeny increasingly important and widely applied. We will show examples of trees in Chapters 13 to 20 as we explore genomes across the tree of life.

The field of molecular phylogeny includes conceptually distinct approaches, including those outlined in this chapter (distance, maximum parsimony, maximum likelihood, and Bayesian methods). For each of these approaches software tools continue to evolve. Thus it is reasonable for you to obtain a multiple sequence alignment and perform phylogenetic analyses with all four tree-making approaches and with a variety of substitution models. The relative merits of these maximum-parsimomy versus model-based approaches continue to be debated (e.g., see Kolaczkowski and Thornton, 2004; Steel, 2005).

PITFALLS

The quality of a phylogenetic tree based on molecular sequence data depends on the quality of the sequence data and the multiple sequence alignment. It is also necessary to choose the appropriate models of nucleotide substitution for the phylogeny. There is an active debate within the field concerning the importance of selecting models without too few or too many parameters. Furthermore, the choice of tree-making approaches (from distance to maximum parsimony, maximum likelihood, and Bayesian frameworks) may produce an optimal tree having different topologies and branch lengths. In contrast to multiple sequence alignments of proteins having known structures, there are few "gold standard" benchmark data sets that allow objective definitions of the true trees.

In practice, for many published phylogenetic trees the underlying multiple sequence alignments are not available and it is challenging to assess the quality of published trees. A group of 28 phylogeny experts has begun to define reporting standards for phylogenetic analysis (called "Minimum Information about a Phylogenetic Analysis" or MIAPA) (Leebens-Mack et al., 2006). Such standards may someday require those who report trees to include the underlying data as well as descriptions of the models used to construct trees.

Finally, your understanding of the output of the phylogenetic analysis is critical. Each of the methods used to reconstruct phylogenetic trees involves many assumptions and suffers from potential weaknesses. It is also important to learn how to interpret trees as graphs that reflect the historical relationships of taxa; in a tree of protein sequences, for example, the nodes correspond to inferred ancestral sequences.

WEB RESOURCES

The best starting point for phylogeny resources on the World Wide Web is the site of Joseph Felsenstein (▶ http://evolution.genetics. washington.edu/phylip/software.html). About 200 links are listed, organized by categories such as phylogenetic methods, computer platforms, and assorted types of data. All of the major software tools listed in this chapter have websites that we have listed, and most of these sites include detailed documentation and examples that further illustrate both the practical use of the software and the conceptual issues addressed by the authors' particular approach to phylogeny.

The HIV Sequence Database at the Los Alamos National Laboratory (discussed in Chapter 14, on viruses) offers a brief online guide to making and interpreting phylogenetic trees (▶ http://www.hiv.lanl.gov/content/hiv-db/TREE_TUTORIAL/ Tree-tutorial.html). This site includes links to PAUP, Phylip, and other tree-making programs. The National Center for Biotechnology Information also offers an online primer (*Systematics and Molecular Phylogenetics*) that introduces molecular trees (▶ http://www.ncbi.nlm.nih.gov/About/primer/phylo.html).

Please note that web links described in this chapter are available online at ▶ http://www.bioinfbook.org/chapter7.

DISCUSSION QUESTIONS

[7-1] Consider a multiple sequence alignment containing a grossly incorrect region. What is the likely consequence of using this alignment to infer a phylogenetic tree using a distance-based or character-based method?

[7-2] Are there gene (or protein) families for which you expect distance-based tree-building methods to give substantially different results than character-based methods?

[7-3] How would you test whether a particular human gene (or protein) is under positive selection? What species would you select for comparison to the human sequence?

PROBLEMS/COMPUTER LAB

[7-1] Determine whether human and chimpanzee mitochondrial DNA sequences have equal evolutionary rates between lineages. To do this, use Tajima's relative rate test as implemented in MEGA.

(1) Obtain MEGA software.

(2) Obtain mitochondrial DNA sequences from human, chimpanzee, bonobo, orangutan, gorilla, and gibbon from web document 7.12 at ▶ http://www.bioinfbook.org/chapter7.

[7-2] Perform phylogenetic analyses using PAUP software.

Alignment Input

(1) Go to Conserved Domain Database (CCD) at NCBI (▶ http://www.ncbi.nlm.nih.gov/Structure/cdd/ cdd.shtml).

(2) On the homepage for CDD you will find the option to "Find CDs in Entrez" by keyword; type in "lipocalins."

The search result will include "pfam00061." Click on the family. As an alternative, do this exercise beginning with a query for "globins" or another family of interest.

(3) The new window will present you with the brief introduction of the lipocalin protein family, including a representative multiple sequence alignment.

(4) Select the mFasta format then click "Reformat."

(5) Copy the FASTA alignment into a simple text file. Change names of the sequences from *gi number* into a name having up to nine characters in the first row of the alignment. It is helpful to change the filenames as follows: convert "gi|809398" to "btrbp" (*Bos taurus* retinol binding protein). PAUP does not usually tolerate "/" and numbers. If your data set includes a consensus sequence, delete the text up to the beginning of the protein sequence and rename it with a nine-character name (such as "consensus").

Convert to Readable Format (via ReadSeq)

(6) Open a new window in your browser and go to a ReadSeq page such as ▶ http://searchlauncher.bcm.tmc.edu/seq-util/readseq.html. You can find this URL easily by entering the query "readseq" into a search engine.

(7) Paste your FASTA sequences into the provided window. Choose the PAUP/NEXUS output format. Double click on "Perform Conversion."

(8) Take the output alignment, and copy it into your computer's clipboard. Alternatively copy it into simple text or a notepad file. Start copying with the #NEXUS and end with the symbols "end;".

Import into PAUP

(9) Open the PAUP program.

(10) Under the File heading, choose "New."

(11) Paste in your alignment.

(12) Under the File menu, choose "Execute."

(13) If you get some error message you may have to try several alternatives such as

(i) Look for the line "format datatype = protein interleave missing = -;" and change the hyphen (-) in this line to a period (·)

(ii) Under the Edit menu choose "Find." Replace all instances of the symbol / with the underscore (-).

(14) When the Execute command is successful, you will be notified that "Data matrix has 10 taxa (i.e., 10 proteins) and 163 characters (amino acids)."

Tree Analysis Based on Maximum Parsimony: (a) Heuristic Search

(15) Under the Analysis heading, select "Parsimony."

(16) Under the Analysis menu, choose "Heuristic search." Click "Search" and use the default setting to find the tree.

(17) Once the search finishes, the small window with "Heuristic Search Status" will have a clickable close icon. The program describes the number of tree rearrangements that were tried and the scores for the best tree(s).

(18) View the tree. Go under the Trees menu and choose "Print Tree." You will have the option to see an unrooted phylogram or cladogram among several options. In phylogram displays, the branch lengths are proportional to amino acid changes, and the tree is accompanied by a scale bar. On the other hand, branch lengths are not proportional to amino acid changes in the cladogram. A cladogram portrays evolutionary relationships within species and populations. In addition, if you have more than one tree found to have the same best score, you will have the option to view the particular tree individually or to display all trees at once.

(19) How many amino acid changes occur in the shortest branch and longest branch in your tree? Which OTUs (taxa) are connected by these branches?

(20) If you have more than one tree, you can choose a consensus tree. Under the Trees menu, select "Compute consensus."

Evaluation of Trees

(i) The principle of the random tree test is to compare the score of the found tree to the score distribution of X randomly generated trees starting with your alignment.

(21) Evaluate the tree by frequency distribution of lengths of 100, 1000, and 10,000 random trees.

(22) Go under the Analysis icon and use the "Evaluate random tree." option. Change the number of random trees and see how the mean and standard deviation change.

(23) Is the score of your tree found by a heuristic search significantly better than the score distribution from randomly generated trees?

(24) Perform the heuristic search for the most parsimonious tree again and check the score. How good is your score this time? Since we are working with the protein alignment of 10 taxa (OTUs), we can perform the exhaustive search for the tree with maximum parsimony. Go under the Analysis icon and choose "Exhaustive search." What score did you get and how does it compare to the score(s) obtained using the heuristic search strategy?

(ii) The bootstrap test is another type of resampling test. The principle is to randomly sample the individual columns of aligned amino acid sequence data from the original alignment. The newly generated data sets will maintain the identical size of the original alignment. The bootstrap describes the percent of instances in which a particular clade designation is supported.

(25) Perform the bootstrap test. Go under the Analysis icon and choose "Bootstrap/Jacknife." Then change the number of sampling. Try 1000 and 10,000 replicate samplings with replacement.

(26) After each test you can view the tree → Trees → Print Bootstrap Consensus → Preview. (The plot type can be changed to "unrooted.")

(27) Analyze the tree with 10,000 replicate samplings. Based on your bootstrap values, how many strongly supported clades (bootstrap value >70%) are present in your tree and what taxa do they comprise?

(28) Can you determine if members of the particular clade are paralogs or orthologs? What kind of information do you need to make this inference?

Change Input Files

(29) Go back into the CDD database and retrieve the lipocalin "pfam00061" family (points 1 to 3 above).

(30) Instead of choosing "10 most diverse sequences" as is set by default (see point 4), choose "top listed sequences" in the

option of sequences in the output alignment. (Alternatively, e.g., your can increase the input of your sequences up to 25.)

(31) Repeat the above exercise and observe the differences.

(32) You can also customize the input of the sequences. Choose the option "Selected sequences" and check the chosen sequences listed below this menu (see point 4).

(33) *Analysis of trees based on distance method.* This method is based on comparing the number of pairwise differences in sequences and using the computed distances between the sequences to construct a tree. Unfortunately, some of these mutations (especially if you are constructing a DNA rather than a protein tree) can become overlooked if a mutation occurs following another mutation back to the original character.

Under the Analysis icon choose the "Distance" option. Then perform the heuristic search for the tree. View the tree and see if the taxa (OTUs) separated as they did in a tree based on maximum parsimony. In addition, perform an evaluation of your tree with randomness testing and bootstrap as you did with the tree above.

(34) *Analysis of trees based on maximum likelihood method.* This is a character-based tree-building method (as is maximum parsimony). In this case, the trees are evaluated based on the likelihood of producing the observed data. The PAUP program will let you to construct the tree only with aligned DNA sequences.

One way to obtain aligned DNA sequences is to retrieve the alignment from the PopSet database. (Go to the main NCBI site and choose Entrez → click PopSet → browse the Popset for your favorite alignment). Paste the PHYLIP formatted alignment from PopSet into ReadSeq and proceed as above.

A tree based on the maximum-likelihood method from protein alignment can be created with program Puzzle (► http://www.tree-puzzle.de).

[7-3] Perform phylogenetic analyses using MEGA software.

[7-4] Perform Bayesian inference of phylogeny using MrBayes software. A detailed analysis for 13 globin proteins is provided in web documents 7.10 and 7.11. A detailed analysis for DNA coding sequences from a group of myoglobins (and cytoglobin as an outgroup) is provided in web document 7.3.

SELF-TEST QUIZ

[7-1] According to the molecular clock hypothesis:

(a) All proteins evolve at the same, constant rate.

(b) All proteins evolve at a rate that matches the fossil record.

(c) For every given protein, the rate of molecular evolution gradually slows down like a clock that runs down.

(d) For every given protein, the rate of molecular evolution is approximately constant in all evolutionary lineages.

[7-2] The two main features of any phylogenetic tree are:

(a) The clades and the nodes

(b) The topology and the branch lengths

(c) The clades and the root

(d) The alignment and the bootstrap

[7-3] Which one of the following is a character-based phylogenetic algorithm?

(a) Neighbor joining

(b) Kimura

(c) Maximum likelihood

(d) PAUP

[7-4] Two basic ways to make a phylogenetic tree are distance based and character based. A fundamental difference between them is:

(a) Distance-based methods essentially summarize relatedness across the length of protein or DNA sequences while character-based methods do not.

(b) Distance-based methods are only used for DNA data while character-based methods are used for DNA or protein data.

(c) Distance-based methods use parsimony while character-based methods do not.

(d) Distance-based methods have branches that are proportional to time while character-based methods do not.

[7-5] An example of an operational taxonomic unit (OTU) is:

(a) Multiple sequence alignment

(b) Protein sequence

(c) Clade

(d) Node

[7-6] For a given pair of OTUs, which of the following is true?

(a) The corrected genetic distance is greater than or equal to the proportion of substitutions.

(b) The proportion of substitutions is greater than or equal to the corrected genetic distance.

[7-7] Transitions are almost always weighted more heavily than transversions.

(a) True

(b) False

[7-8] One of the most common errors in making and analyzing a phylogenetic tree is:

(a) Using a bad multiple sequence alignment as input

(b) Trying to infer the evolutionary relationships of genes (or proteins) in the tree

(c) Trying to infer the age at which genes (or proteins) diverged from each other

(d) Assuming that clades are monophyletic

[7-9] ClustalX can be used to generate neighbor-joining trees with or without bootstrap values.

(a) True

(b) False

[7-10] You have 200 viral DNA sequences of 500 residues each, and you want to know if there are any pairs that are identical (or nearly identical). Which of the following is the most efficient method to use?

(a) BLAST

(b) Maximum-likelihod phylogenetic analysis

(c) Neighbor-joining phylogenetic analysis

(d) Popset

SUGGESTED READING

An excellent overview of evolution from a history-of-science perspective is provided by Mayr (1982). There are many superb textbooks on molecular evolution, including those by Felsenstein (2004), Graur and Li (2000), and Li (1997). A book by Maddison and Maddison (2000) describing the MacClade software package provides an extensive, clear introduction to phylogeny. Hall's book (2001) provides another excellent, practical introduction.

A thorough, detailed overview of phylogenetic inference is provided in a chapter by David Swofford and colleagues (Swofford et al., 1996). A highly recommended overview of phylogenetics is by Thornton and DeSalle (2000). They describe phylogenetic principles (e.g., character-based and distance-based approaches), the assessment of orthology and paralogy, and the use of phylogeny in comparative genomics. Moritz and Hillis (1996) provide an excellent introduction to molecular systematics.

For Bayesian inference of phylogeny, excellent articles are from Ronquist (2004) and Huelsenbeck et al. (2002).

REFERENCES

Anfinsen, C. B. *The Molecular Basis of Evolution*. Wiley, New York, 1959.

Baldauf, S. L., Roger, A. J., Wenk-Siefert, I., and Doolittle, W. F. A kingdom-level phylogeny of eukaryotes based on combined protein data. *Science* **290**, 972–977 (2000).

Baum, D. A., Smith, S. D., and Donovan, S. S. Evolution. The tree-thinking challenge. *Science* **310**, 979–980 (2005).

Baurain, D., Brinkmann, H., and Philippe, H. Lack of resolution in the animal phylogeny: Closely spaced cladogeneses or undetected systematic errors? *Mol. Biol. Evol.* **24**, 6–9 (2007).

Baxevanis, A. D., and Ouellette, B. F. *Bioinformatics*, 2nd ed. Wiley-Interscience, New York, 2001.

Bos, D. H., and Posada, D. Using models of nucleotide evolution to build phylogenetic trees. *Dev. Comp. Immunol.* **29**, 211–227 (2005).

Brown, W. M., Prager, E. M., Wang, A., and Wilson, A. C. Mitochondrial DNA sequences of primates: Tempo and mode of evolution. *J. Mol. Evol.* **18**, 225–239 (1982).

Bustamante, C. D., Fledel-Alon, A., Williamson, S., Nielsen, R., Hubisz, M. T., Glanowski, S., Tanenbaum, D. M., White, T. J., Sninsky, J. J., Hernandez, R. D., Civello, D., Adams, M. D., Cargill, M., and Clark, A. G. Natural selection on protein-coding genes in the human genome. *Nature* **437**, 1153–1157 (2005).

Cavalli-Sforza, L. L., and Edwards, A. W. F. Phylogenetic analysis: Models and estimation procedures. *Am. J. Hum. Genet.* **19**, 233–257 (1967).

Clote, P., and Backofen, R. *Computational Molecular Biology. An Introduction*. Wiley, New York, 2000.

Copley, R. R., Schultz, J., Ponting, C. P., and Bork, P. Protein families in multicellular organisms. *Curr. Opin. Struct. Biol.* **9**, 408–415 (1999).

Cox, A. L., Mosbruger, T., Mao, Q., Liu, Z., Wang, X. H., Yang, H. C., Sidney, J., Sette, A., Pardoll, D., Thomas, D. L., and Ray, S. C. Cellular immune selection with hepatitis C virus persistence in humans. *J. Exp. Med.* **201**, 1741–1752 (2005).

Cunningham, C. W., Omland, K. E., and Oakley, T. H. Reconstructing ancestral character states: A critical reappraisal. *Tree* **13**, 361–366 (1998).

Czelusniak, J., Goodman, M., Moncrief, N. D., and Kehoe, S. M. Maximum parsimony approach to construction of evolutionary trees from aligned homologous sequences. *Methods Enzymol.* **183**, 601–615 (1990).

Darwin, Charles. *The Origin of Species by Means of Natural Selection*. John Murray, London, 1859.

Dayhoff, M. O. *Atlas of Protein Sequence and Structure 1972*. National Biomedical Research Foundation, Silver Spring, MD, 1972.

Dayhoff, M. O. *Atlas of Protein Sequence and Structure*. National Biomedical Research Foundation, Silver Spring, MD, 1978.

Dayhoff, M. O., Schwartz, R. M., and Orcutt, B. C. A model of evolutionary change in proteins. In M. O. Dayhoff, *Atlas of Protein Sequence and Structure 1972*. National Biomedical Research Foundation, Silver Spring, MD, 1972, pp. 345–352.

Desper, R., and Gascuel, O. Getting a tree fast: Neighbor Joining, FastME, and distance-based methods. *Curr. Protoc. Bioinformatics* Chapter 6, Unit 6.3 (2006).

Dickerson, R. E. Sequence and structure homologies in bacterial and mammalian-type cytochromes. *J. Mol. Biol.* **57**, 1–15 (1971).

Dobzhansky, T. Nothing in biology makes sense except in the light of evolution. *Am. Biol. Teacher* **35**, 125–129 (1973).

Doolittle, W. F. Phylogenetic classification and the universal tree. *Science* **284**, 2124–2129 (1999).

Durbin, R., Eddy, S., Krogh, A., and Mitchison, G. *Biological Sequence Analysis*. Cambridge University Press, Cambridge, 1998.

Eck, R. V. and Dayhoff, M. O. *Atlas of Protein Sequence and Structure*. National Biomedical Research Foundation, Silver Spring, MD, 1966.

Felsenstein, J. Evolutionary trees from DNA sequences: A maximum likelihood approach. *J. Mol. Evol.* **17**, 368–376 (1981).

Felsenstein, J. Distance methods for inferring phylogenies: A justification. *Evolution* **38**, 16–24 (1984).

Felsenstein, J. Confidence limits on phylogenies: An approach using the bootstrap. *Evolution* **39**, 783–791 (1985).

Felsenstein, J. Phylogenies from molecular sequences: Inference and reliability. *Annu. Rev. Genet.* **22**, 521–565 (1988).

Felsenstein, J. Inferring phylogenies from protein sequences by parsimony, distance, and likelihood methods. *Methods Enzymol.* **266**, 418–427 (1996).

Felsenstein, J. *Inferring Phylogenies*. Sinauer Associates, Sunderland, MA, 2004.

Fitch, W. M., and Ayala, F. J. The superoxide dismutase molecular clock revisited. *Proc. Natl. Acad. Sci. USA.* **91**, 6802–6807 (1994).

Fitch, W. M., and Margoliash, E. Construction of phylogenetic trees. *Science* **155**, 279–284 (1967).

Feng, D. F., and Doolittle, R. F. Progressive sequence alignment as a prerequisite to correct phylogenetic trees. *J. Mol. Evol.* **25**, 351–360 (1987).

Graur, D., and Li, W.-H. *Fundamentals of Molecular Evolution*, 2nd ed. Sinauer Associates, Sunderland, MA, 2000.

Haeckel, E. *The Riddle of the Universe*. Harper and Brothers, New York, 1900.

Hall, B. G. *Phylogenetic Trees Made Easy. A How-To for Molecular Biologists*. Sinauer Associates, Sunderland, MA, 2001.

Hein, J. Unified approach to alignment and phylogenies. *Methods Enzymol.* **183**, 626–645 (1990).

Hennig, W. *Phylogenetic systematics*. University of Illinois Press, Urbana, 1966.

Hillis, D. M. Approaches for assessing phylogenetic accuracy. *Syst. Biol.* **44**, 3–16 (1995).

Hillis, D. M., and Bull, J. J. An empirical test of bootstrapping as a method for assessing confidence in phylogenetic analysis. *Systematic Biol.* **42**, 182–192 (1993).

Hillis, D. M., and Huelsenbeck, J. P. Signal, noise, and reliability in molecular phylogenetic analyses. *J. Hered.* **83**, 189–195 (1992).

Hollich, V., Milchert, L., Arvestad, L., and Sonnhammer, E. L. Assessment of protein distance measures and tree-building methods for phylogenetic tree reconstruction. *Mol. Biol. Evol.* **22**, 2257–6422 (2005).

Huelsenbeck, J. P., Ronquist, F., Nielsen, R., and Bollback, J. P. Bayesian inference of phylogeny and its impact on evolutionary biology. *Science* **294**, 2310–2314 (2001).

Huelsenbeck, J. P., Larget, B., Miller, R. E., and Ronquist, F. Potential applications and pitfalls of Bayesian inference of phylogeny. *Syst. Biol.* **51**, 673–688 (2002).

Ingram, V. M. Gene evolution and the haemoglobins. *Nature* **189**, 704–708 (1961).

Jollès, J., Prager, E. M., Alnemri, E. S., Jollès, P., Ibrahimi, I. M., and Wilson, A. C. Amino acid sequences of stomach and non-stomach lysozymes of ruminants. *J. Mol. Evol.* **30**, 370–382 (1990).

Jukes, T. H. Dr. Best, insulin, and molecular evolution. *Can. J. Biochem.* **57**, 455–458 (1979).

Jukes, T. H., and Cantor, C. Evolution of protein molecules. In H. N. Munro and J. B. Allison, (eds.), *Mammalian Protein Metabolism*. 21–132, Academic Press, New York, 1969.

Kimura, M. Evolutionary rate at the molecular level. *Nature* **217**, 624–626 (1968).

Kimura, M. A simple method for estimating evolutionary rates of base substitutions through comparative studies of nucleotide sequences. *J. Mol. Evol.* **16**, 111–120 (1980).

Kimura, M. *The Neutral Theory of Molecular Evolution*. Cambridge University Press, Cambridge, 1983.

Kolaczkowski, B., and Thornton, J. W. Performance of maximum parsimony and likelihood phylogenetics when evolution is heterogeneous. *Nature* **431**, 980–984 (2004).

Korber, B. HIV signature and sequence variation analysis. In A. G. Rodrigo and G. H. Learn (eds.), *Computational Analysis of HIV Molecular Sequences*, Kluwer Academic Publishers, Dordrecht, Netherlands, 2000, pp. 55–72.

Kumar, S., Dudley, J., Nei, M., and Tamura, K. MEGA: A biologist-centric software for evolutionary analysis of DNA and protein sequences. *Briefings in Bioinformatics* **9**, 299–306 (2008).

Leebens-Mack, J., Vision, T., Brenner, E., Bowers, J. E., Cannon, S., Clement, M. J., Cunningham, C. W., dePamphilis, C., deSalle, R., Doyle, J. J., Eisen, J. A., Gu. X., Harshman, J., Jansen, R. K., Kellogg, E. A., Koonin, E. V., Mishler, B. D., Philippe, H., Pires, J. C., Qiu, Y. L., Rhee, S. Y., Sjolander, K., Soltis, D. E., Soltis, P. S., Stevenson, D. W., Wall, K., Warnow, T., and Zmasek, C. Taking the first steps towards a standard for reporting on phylogenies: Minimum Information About a Phylogenetic Analysis (MIAPA). *OMICS* **10**, 231–237 (2006).

Li, W.-H. *Molecular Evolution*. Sinauer Associates, Sunderland, MA, 1997.

Maddison, D., and Maddison, W. *MacClade 4: Analysis of Phylogeny and Character Evolution*. Sinauer Associates, Sunderland, MA, 2000.

MacMunn, C. A. On myohaematin, an intrinsic muscle pigment of vertebrates and invertebrates, on histohaematin, and on the spectrum of the supra-renal bodies. *Proc. Physiol. Soc.* **5**, 267–298 (1886).

Mar, J. C., Harlow, T. J., and Ragan, M. A. Bayesian and maximum likelihood phylogenetic analyses of protein sequence data under relative branch-length differences and model violation. *BMC Evol. Biol.* **5**, 1–20 (2005).

Margoliash, E. Primary structure and evolution of cytochrome *c*. *Proc. Natl. Acad. Sci. USA* **50**, 672–679 (1963).

Margoliash, E. and Smith, E. L. Structural and functional aspects of cytochrome *c* in relation to evolution. In V. Bryson and H. J. Vogel (eds.), *Evolving Genes and Proteins*, Academic Press, New York, 1965, pp. 221–242.

Mayr, E. *The Growth of Biological Thought. Diversity, Evolution, and Inheritance*. Belknap Harvard, Cambridge, MA, 1982.

Moritz, C., and Hillis, D. M. Molecular systematics: Context and controversies. In D. M. Hillis, C. Moritz, and B. K. Mable (eds.), *Molecular Systematics*, 1–13, Sinauer Associates, Sunderland, MA, 1996.

Mossel, E., and Vigoda, E. Phylogenetic MCMC algorithms are misleading on mixtures of trees. *Science* **309**, 2207–2209 (2005).

Nei, M. *Molecular Evolutionary Genetics*. Columbia University Press, New York, 1987.

Nei, M. Phylogenetic analysis in molecular evolutionary genetics. *Annu. Rev. Genet.* **30**, 371–403 (1996).

Nei, M. and Gojobori, T. Simple methods for estimating the numbers of synonymous and nonsynonymous nucleotide substitutions. *Mol. Biol. Evol.* **3**, 418–426 (1986).

Nei, M. and Kumar, S. *Molecular Evolution and Phylogenetics*. Oxford University Press, New York, 2000.

Nielsen, R. Molecular signatures of natural selection. *Annu. Rev. Genet.* **39**, 197–218 (2005).

Nuttall, G.H.F. *Blood Immunity and Blood Relationship*. Cambridge University Press, Cambridge, 1904.

Ohno, S. *Evolution by Gene Duplication*. Springer-Verlag, New York, 1970.

Penny, D., Foulds, L. R., and Hendy, M. D. Testing the theory of evolution by comparing phylogenetic trees constructed from five different protein sequences. *Nature* **297**, 197–200 (1982).

Philippe, H., and Laurent, J. How good are deep phylogenetic trees? *Curr. Opin. Genet. Dev.* **8**, 616–623 (1998).

Pond, S. L., and Frost, S. D. Datamonkey: Rapid detection of selective pressure on individual sites of codon alignments. *Bioinformatics* **21**, 2531–2533 (2005).

Posada, D. ModelTest Server: A web-based tool for the statistical selection of models of nucleotide substitution online. *Nucleic Acids Res.* **34**(Web Server issue), W700–W703 (2006).

Posada, D., and Buckley, T. R. Model selection and model averaging in phylogenetics: Advantages of Akaike information criterion and Bayesian approaches over likelihood ratio tests. *Syst. Biol.* **53**, 793–808 (2004). PMID: 15545256

Posada, D., and Crandall, K. A. MODELTEST: Testing the model of DNA substitution. *Bioinformatics* **14**, 817–818 (1998).

Ray, S. C., Fanning, L., Wang, X. H., Netski, D. M., Kenny-Walsh, E., and Thomas, D. L. Divergent and convergent evolution after a common-source outbreak of hepatitis C virus. *J. Exp. Med.* **201**, 1753–1759 (2005).

Rokas, A., Kruger, D., and Carroll, S. B. Animal evolution and the molecular signature of radiations compressed in time. *Science* **310**, 1933–1938 (2005).

Ronquist, F. Bayesian inference of character evolution. *Trends Ecol. Evol.* **19**, 475–481 (2004).

Sabeti, P. C., Schaffner, S. F., Fry, B., Lohmueller, J., Varilly, P., Shamovsky, O., Palma, A., Mikkelsen, T. S., Altshuler, D., and Lander, E. S. Positive natural selection in the human lineage. *Science* **312**, 1614–1620 (2006).

Saitou, N., and Nei, M. The neighbor-joining method: A new method for reconstructing phylogenetic trees. *Mol. Biol. Evol.* **4**, 406–425 (1987).

Schmidt, H. A., Strimmer, K., Vingron, M., and von Haeseler, A. TREE-PUZZLE: Maximum likelihood phylogenetic analysis using quartets and parallel computing. *Bioinformatics* **18**, 502–504 (2002).

Simpson, G. G. *The Meaning of Evolution: A Study of the History of Life and of Its Significance for Man*. Yale University Press, New Haven, 1952.

Sneath, P. H. A. and Sokal, R. R. *Numerical Taxonomy*. W.H. Freeman & Co., San Francisco, 1973.

Sokal, R. R. and Michener, C. D. A statistical method for evaluating systematic relationships. *Univ. Kansas Science Bull.* **38**, 1409–1437 (1958).

Spillane, C., Schmid, K. J., Laoueille-Duprat, S., Pien, S., Escobar-Restrepo, J. M., Baroux, C., Gagliardini, V., Page, D. R., Wolfe, K. H., and Grossniklaus, U. Positive darwinian selection at the imprinted MEDEA locus in plants. *Nature* **448**, 349–352 (2007).

Steel, M. Should phylogenetic models be trying to "fit an elephant"? *Trends Genet.* **21**, 307–309 (2005).

Strimmer, K., and von Haeseler, A. Quartet puzzling: A quartet maximun likelihood method for reconstructing tree topologies. *Mol. Biol. Evol.* **13**, 964–969 (1996).

Strimmer, K., and von Haeseler, A. Likelihood-mapping: a simple method to visualize phylogenetic content of a sequence alignment. *Proc. Natl. Acad. Sci. USA.* **94**, 6815–6819 (1997).

Suzuki, Y., Glazko, G. V., and Nei, M. Overcredibility of molecular phylogenies obtained by Bayesian phylogenetics. *Proc. Natl. Acad. Sci. USA* **99**, 16138–16143 (2002).

Swofford, D. L., Olsen, G. J., Waddell, P. J., and Hillis, D. M. Molecular systematics: Context and controversies. In

D. M. Hillis, C. Moritz, and B. K. Mable (eds.), *Molecular Systematics*, 2nd ed. Sinauer Associates, Sunderland, MA, 1996.

Tajima, F. Simple methods for testing the molecular evolutionary clock hypothesis. *Genetics* **135**, 599–607 (1993).

Tamura, K. Estimation of the number of nucleotide substitutions when there are strong transition-transversion and $G + C$-content biases. *Mol. Biol. Evol.* **9**, 678–687 (1992).

Tamura, K., Dudley, J., Nei, M., and Kumar, S. MEGA4: Molecular Evolutionary Genetics Analysis (MEGA) software version 4.0. *Mol. Biol. Evol.* **10**, 1093 (2007).

Taubenberger, J. K., Reid, A. H., Lourens, R. M., Wang, R., Jin, G., and Fanning, T. G. Characterization of the 1918 influenza virus polymerase genes. *Nature* **437**, 889–893 (2005).

Thornton, J. W., and DeSalle, R. Gene family evolution and homology: Genomics meets phylogenetics. *Annu. Rev. Genomics Hum. Genet.* **1**, 41–73 (2000).

Uzzell, T., and Corbin, K. W. Fitting discrete probability distributions to evolutionary events. *Science* **172**, 1089–1096 (1971).

Wilson, E. O. *The Diversity of Life*. W.W. Norton, New York, 1992.

Zar, J. H. *Biostatistical Analysis*, 4th edition. Prentice Hall, Upper Saddle River, NJ, 1999.

Zhang, J., and Gu, X. Correlation between the substitution rate and rate variation among sites in protein evolution. *Genetics* **149**, 1615–1625 (1998).

Zuckerkandl, E., and Pauling, L. Molecular disease, evolution, and genic heterogeneity. In M. Kasha and B. Pullman (eds.), *Horizons in Biochemistry*, Albert Szent-Gyorgyi Dedicatory Volume. Academic Press, New York, 1962.

Zuckerkandl, E., and Pauling, L. Evolutionary divergence and convergence in proteins. In V. Bryson and H. J. Vogel (eds.), *Evolving Genes and Proteins*, Academic Press, New York, 1965, pp. 97–166.

Part II

Genomewide Analysis of RNA and Protein

In the first third of this book, we described how to find sequences (and other information) in databases, how to align DNA or protein sequences in a pairwise fashion or in a multiple sequence alignment, and how to perform evolutionary studies, including the visualization of aligned sequences through molecular phylogeny. In this middle third of the book we follow the flow of central dogma of molecular biology by moving from DNA to RNA (Chapters 8 and 9) to protein (Chapters 10 and 11). We then discuss functional genomics (Chapter 12), which is the genomewide study of the function of genes and gene products.

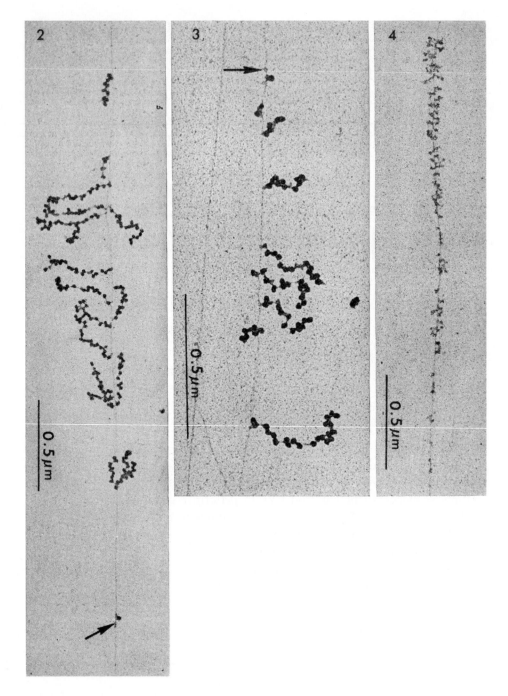

Miller and colleagues (1970, p. 394) visualized gene expression. They showed Escherichia coli *chromosomal DNA (oriented vertically as a thin strand in each figure) in the process of transcription and translation. As mRNA is transcribed from the genomic DNA and extends off to the side, polyribosomes (dark objects) appear like beads on a string, translating the mRNA to protein. Used with permission.*

8

Bioinformatic Approaches to Ribonucleic Acid (RNA)

INTRODUCTION TO RNA

The word "gene" was introduced by Johannsen in 1909 to describe the entity that determines how characteristics of an organism are inherited. Classic studies by Beadle and Tatum (1941) in the fungus *Neurospora* showed that genes direct the synthesis of enzymes in a 1:1 ratio. As early as 1944, Oswald T. Avery demonstrated that deoxyribonucleic acid (DNA) is the genetic material. Avery et al. (1944) showed that DNA from bacterial strains with high pathogenicity could transform strains with low to high pathogenicity. Further experiments involving bacterial transformation, performed by Avery, McLeod, McCarthy, Hotchkiss, and Hershey confirmed that DNA is the genetic material.

James Watson and Francis Crick proposed the double helical nature of DNA in 1953 (Fig. 8.1). Soon after, Crick 1958 could formulate the central dogma of molecular biology that DNA is transcribed into RNA then translated into protein. Crick wrote (1958, p. 153) that the central dogma "states that once 'information' has passed into protein *it cannot get out again*. In more detail, the transfer of

You can learn about some of the original discoveries concerning nucleic acids by reading about their Nobel Prize awards. Albrecht Kossel was awarded the Nobel Prize in 1910 for characterizing nucleic acids (▶ http://nobelprize.org/nobel_prizes/medicine/laureates/1910/). Beadle and Tatum were awarded Nobel Prizes in 1958 for their one gene-one enzyme hypothesis (see ▶ http://nobelprize.org/nobel_prizes/medicine/laureates/1958/). Severo Ochoa and Arthur Kornberg shared a 1959 Nobel Prize "for their discovery of the mechanisms in the biological synthesis of ribonucleic acid and deoxyribonucleic acid" (▶ http://nobelprize.org/nobel_prizes/medicine/laureates/1959/). Although Oswald Avery was the first to show that DNA is the genetic material, he did not receive a Nobel Prize.

Bioinformatics and Functional Genomics, Second Edition. By Jonathan Pevsner
Copyright © 2009 John Wiley & Sons, Inc.

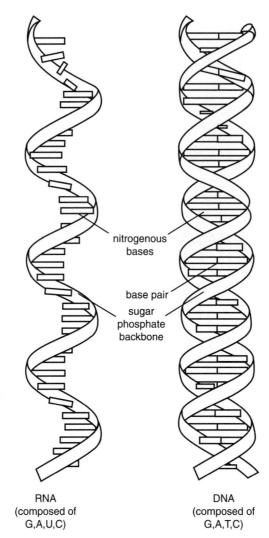

nitrogenous bases

base pair

sugar phosphate backbone

RNA
(composed of
G,A,U,C)

DNA
(composed of
G,A,T,C)

FIGURE 8.1.   *Deoxyribonucleic acid (DNA) and ribonucleic acid (RNA). While DNA usually adopts a double helical conformation, RNA tends to be single stranded. A notable exception is the double-stranded base pairing of many noncoding RNAs to form stem-loop structures, described in this chapter. The image is adapted from the RNA entry of the National Human Genome Research Institute (NHGRI) Talking Glossary (▶ http://www.genome.gov/glossary.cfm).*

Francis Crick, James Watson, and Maurice Wilkins shared the 1962 Nobel Prize in Physiology or Medicine "for their discoveries concerning the molecular structure of nucleic acids and its significance for information transfer in living material." See ▶ http://nobelprize.org/nobel_prizes/medicine/laureates/1962/.

The 1968 Nobel Prize in Physiology or Medicine was awarded to Robert Holley, Har Khorana, and Marshall Nirenberg "for their interpretation of the genetic code and its function in protein synthesis." Visit ▶ http://nobelprize.org/nobel_prizes/medicine/laureates/1968/.

information from nucleic acid to nucleic acid, or from nucleic acid to protein may be possible, but transfer from protein to protein, or from protein to nucleic acid is impossible. Information means here the *precise* determination of sequence, either of bases in the nucleic acid or of amino acid residues in the protein." In this article Crick further postulated the existence of an adaptor molecule to convert the information from codons in RNA to amino acids in proteins; transfer RNA (tRNA) was indeed later identified.

During the 1960s the genetic code was solved (e.g., Nirenberg, 1965), showing the relationship between messenger RNA codons and the amino acids that are specified. This completed a detailed model for the flow of genetic information from DNA to RNA to protein. However, even by the 1950s, this model was called into question by the nature of RNA. Why did hybridization experiments suggest that only a minute fraction of RNA was complementary to DNA of genes? RNA could be purified from DNA and proteins, and then shown to separate into discrete bands on density gradients having sedimentation coefficients of 23S, 16S, and 4S. The 23S and 16S species were found to localize to ribosomes and constituted about 85% of all RNA in bacteria. tRNA was found to constitute about 15% of all RNA. Thus, surprisingly, mRNA was found to represent only a small percentage of total RNA (about 1% to 4%).

DNA consists of the four nucleotides adenine, guanine, cytosine, and thymidine (A, G, C, T). It can be transcribed into ribonucleic acid (RNA), consisting of the nucleotides A, G, C, and U (uracil) (Fig. 8.2). RNA has a backbone consisting of the five-carbon sugar ribose with a purine or pyrimidine base attached to each sugar residue. A phosphate group links the nucleoside (i.e., the sugar with base) to form a nucleotide.

The process of transcription of DNA results in the formation of RNA molecules in two broad classes. The first class is coding RNA, formed when DNA is transcribed into messenger RNA (mRNA). This mRNA is subsequently translated into protein on the surface of a ribosome in a process mediated by transfer RNA (tRNA) and

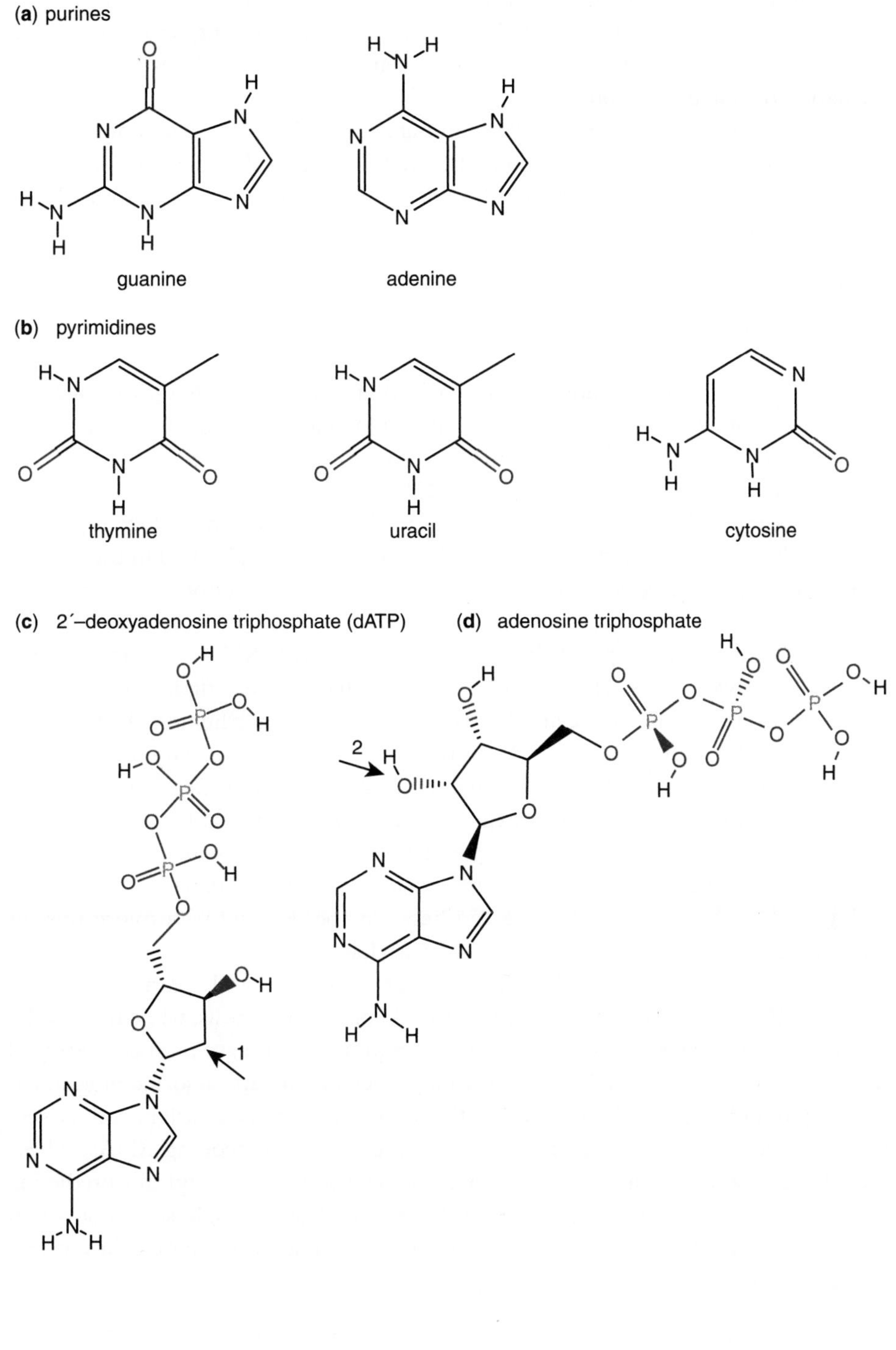

The purines include adenine and guanine, and the pyrimidines include cytosine, thymine and uracil. To view their structures at the NCBI website, enter their names into a search of Entrez, and view the results in the PubChem database.

(a) purines

guanine   adenine

(b) pyrimidines

thymine   uracil   cytosine

(c) 2′–deoxyadenosine triphosphate (dATP)   **(d)** adenosine triphosphate

FIGURE 8.2. *The nucleotide bases. (a) The purine bases are guanine and adenine. (b) The pyrimidine bases are thymine, uracil (which substitutes for thymine in RNA), and cytosine. (c) 2′-deoxyadenosine triphosphate (dATP) and (d) adenosine triphosphate (ATP). The nitrogenous bases (a, b) are attached to ribose sugars and triphosphate groups. In the case of DNA, the ribose lacks an oxygen side group (arrow 1) that is present in RNA (arrow 2).*

Sidney Altman and Thomas Cech shared the 1989 Nobel Prize in Chemistry "for their discovery of the catalytic properties of RNA." See ▶ http://nobelprize.org/nobel_prizes/chemistry/laureates/1989/. Altman characterized RNA enzymes (ribozymes) in the bacterium *Escherichia coli*, while Cech studied ribozymes in *Tetrahymena thermophila*. An example of a human gene encoding a noncoding RNA with enzymatic activity is RNA component of mitochondrial RNA processing endoribonuclease (RMRP, accession NR_003051, assigned to chromosome 9p21-p12).

ribosomal RNA (rRNA) as well as by proteins. A second class is noncoding RNA in which RNA products that are transcribed from DNA function without being further translated into protein. We will next discuss noncoding and coding RNA from a bioinformatics perspective. There is considerable excitement about many recent advances in our understanding of all classes of RNAs, as we begin to recognize their diverse functional properties. By the 1980s the extraordinary versatility of RNA began to be appreciated when, in addition to the three major RNAs (rRNA, tRNA, mRNA), RNAs with catalytic properties were discovered. Previously, nucleic acids had been considered molecules underlying heredity while proteins functioned as enzymes or other modulators of cellular processes (see Chapter 10). The discovery of ribozymes is consistent with a model of the early evolution of life on earth in which RNA was the first genetic material, prior to the emergence of DNA. Another implication is that RNA has many potential functional roles in the cell beyond serving as an intermediary between DNA and protein. For example, rRNA catalyzes peptide bond formation during translation.

Throughout this chapter we will use human chromosome 21 to demonstrate the nature of various RNAs. This is among the smallest human chromosomes (about 47 million base pairs) and one of the five human chromosomes having ribosomal DNA clusters that produce rRNA.

NONCODING RNA

The RefSeq accession for the human XIST is NR_001564, spanning 19,271 base pairs. The accession for murine "antisense Igf2r RNA" (AIR) on chromosome 17 is NR_002853 (3699 base pairs).

The major classes of noncoding RNAs are tRNA and rRNA, which together account for approximately 95% of all RNAs. Other noncoding RNAs, discussed below, include small nuclear RNA (snRNA), small nucleolar RNA (snoRNA), microRNA, and short interfering RNA (siRNA). Beyond tRNA and rRNA, relatively few noncoding RNAs have had their functions defined. A prominent example of a functionally characterized noncoding RNA is *Xist*. This gene, located in the X inactivation center of the X chromosome, encodes an RNA transcript (called X (inactive)-specific transcript or *Xist*) that functions in X chromosome inactivation. While males have one copy of the X chromosome (with XY sex chromosomes), females have two copies, of which one is inactivated in every diploid cell. Xist is expressed from the inactive X and binds to its chromatin facilitating chromosome inactivation (Borsani et al., 1991). Another functional noncoding RNA is *Air* which functions at the *Igf2R* locus (Sleutels et al., 2002). Some genes that are present in two copies are imprinted, that is, expressed selectively from an allele of one parent. In mouse, noncoding Air RNA is required to suppress expression of three genes (*Igf2r/Slc22a2/Slc22a3*) from the paternal chromosome. It is notable that many noncoding RNAs are very poorly conserved between species, and we explore this for *XIST* and *Air* in problem 8-1 at the end of this chapter.

In addition to Rfam and MirBase there are many other excellent noncoding RNA databases such as RNAdb (Pang et al., 2005) at ▶ http://research.imb.uq.edu.au/rnadb/.

Considering all the noncoding RNAs, the abundant and well-characterized ones (tRNA, rRNA, and mRNA) have central roles in protein synthesis, while the smaller and more poorly characterized ones have been proposed to have a broad variety of functions in the regulation of gene expression, development, and assorted physiological and pathophysiological processes. In the following sections we will introduce several prominent databases that collect information about noncoding RNAs. These include Rfam and MirBase. We will also introduce two main methods for predicting RNA structures: a comparative method that is based on multiple sequence alignments of RNAs, and the thermodynamic approach that seeks the minimum free energy of a structure.

Noncoding RNAs in the Rfam Database

We introduced the Pfam database for protein families in Chapter 6 as an important bioinformatics resource. The Rfam database serves a comparable role in characterizing RNA families (Griffiths-Jones et al., 2005). Rfam includes RNA alignments, consensus secondary structures, and covariance models (discussed below). Each Rfam family has a covariance model that is a statistical model of that family's sequence and structure.

The contents of Rfam permit a survey of all currently known noncoding RNAs (Fig. 8.3). These include several well-characterized families that span all three domains of life: tRNAs, rRNAs, SRP RNA (responsible for protein export), and RNaseP (necessary for tRNA maturation). Table 8-1 lists the most abundant RNA families in Rfam for all species.

We can further survey typical noncoding RNAs by viewing an Rfam summary of those present on the long arm of human chromosome 21 (Fig. 8.4). This has 19 distinct families in 35 regions. These include a tRNA gene, an rRNA gene, small nuclear genes involved in splicing, small nucleolar genes, and microRNAs. We will next examine these various noncoding RNA types.

Transfer RNA

Transfer RNA molecules carry a specific amino acid and match it to its corresponding codon on an mRNA during protein synthesis. tRNAs occur in 20 amino acid acceptor groups corresponding to the 20 amino acids specified in the genetic code. tRNA forms a structure consisting of about 70 to 90 nucleotides folded into a characteristic

You can access the Rfam database at ▶ http://www.sanger.ac.uk/Software/Rfam/ or ▶ http://rfam.janelia.org/. Release 8.0 (February 2007) has 574 models and over 13,400 candidate noncoding RNA genes.

Chromosome 21p (the short arm of chromosome 21) is about 12 million base pairs in length and contains rDNA clusters (described below) and a total of eight RefSeq genes. Chromosome 21q (the long arm) extends for about 35 million base pairs and has 322 RefSeq genes.

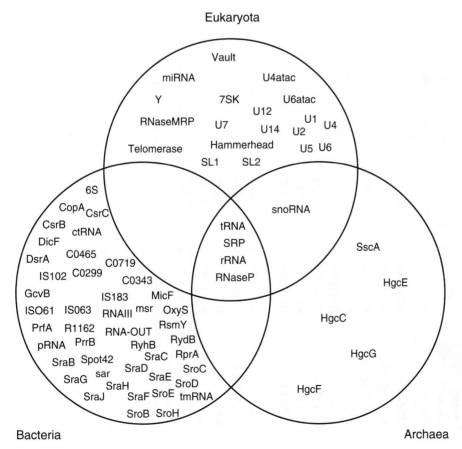

FIGURE 8.3. The Rfam family includes alignments and descriptions of RNA families from the three domains of life. Adapted from Griffiths-Jones et al. (2005). Used with permission.

TABLE 8-1 Twelve RFAM Entries with the Largest Number of Members

Name	Accession	No. full	Av.len (full)	Id	Type	Description
tRNA	RF00005	84,974	72	44	Gene; tRNA	tRNA
5_8S_rRNA	RF00002	84,358	153	78	Gene; rRNA	5.8S ribosomal RNA
Intron_gpI	RF00028	20,085	397	53	Intron	Group I catalytic intron
5S_rRNA	RF00001	12,117	116	61	Gene; rRNA	5S ribosomal RNA
Intron_gpII	RF00029	8189	77	54	Intron	Group II catalytic intron
HIV_GSL3	RF00376	6923	81	87	Cis-reg	HIV gag stem loop 3 (GSL3)
UnaL2	RF00436	5255	54	78	Cis-reg	UnaL2 LINE 3' element
HIV_FE	RF00480	4437	51	90	Cis-reg; frameshift_element	HIV Ribosomal frameshift signal
RRE	RF00036	3972	311	87	Cis-reg	HIV Rev response element
IRES_HCV	RF00061	3632	206	90	Cis-reg; IRES	Hepatitis C virus internal ribosome entry site
U6	RF00026	2599	105	77	Gene; snRNA; splicing	U6 spliceosomal RNA
TAR	RF00250	2004	56	91	Cis-reg	Trans-activation response element (TAR)

Abbreviations: No. full, number of members of the Rfam family (for the full data set rather than the seed alignment of representative members); Av. len, average length; Id, the average percent identity of the full alignments.
Source: RFAM (September, 2007), ▶ http://www.sanger.ac.uk/Software/Rfam/.

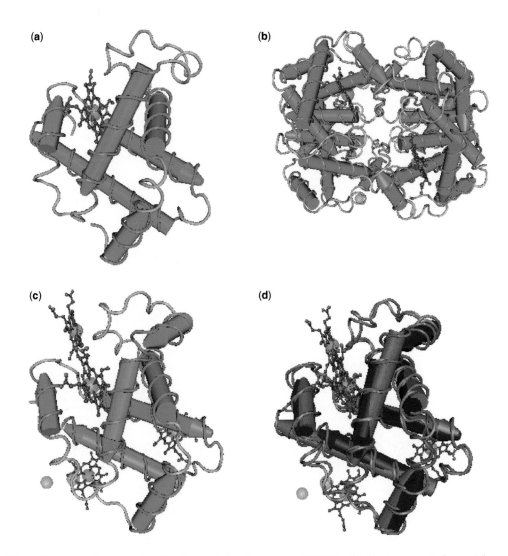

FIGURE 3.1. *Three-dimensional structures of (a) myoglobin (accession 2MM1), (b) the tetrameric hemoglobin protein (2H35), (c) the beta globin subunit of hemoglobin, and (d) myoglobin and beta globin superimposed. The images were generated with the program Cn3D (see Chapter 11). These proteins are homologous (descended from a common ancestor), and they share very similar three-dimensional structures. However, pairwise alignment of these proteins' amino acid sequences reveals that the proteins share very limited amino acid identity.*

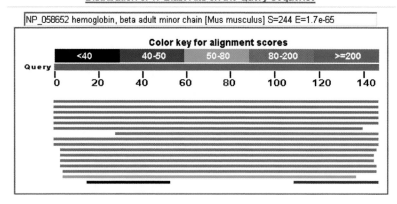

FIGURE 4.7. *Middle portion of a typical blastp output provides a graphical display of the results. Database matches are color coded to indicate relatedness (based on alignment score), and the length of each line corresponds to the region in which that sequence aligns with the query sequence. This graphic can be useful to summarize the regions in which database matches align to the query.*

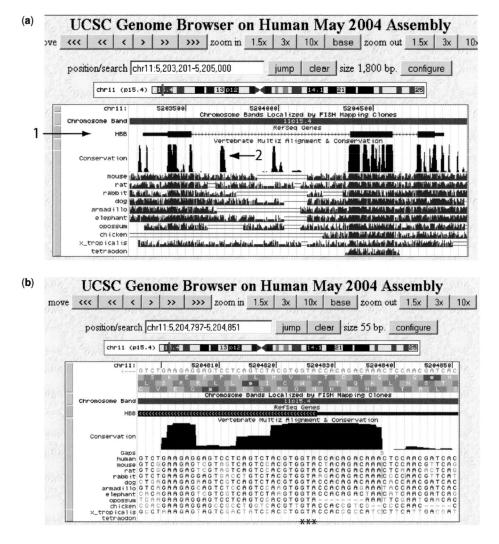

FIGURE 6.10. *Multiple sequence alignment of the human beta globin locus compared to other vertebrate genomic sequences. (a) A view in the UCSC Genome Browser of the beta globin gene is indicated. Exons are represented by blocks (arrow 1) and tend to be highly conserved among a group of vertebrate genomes. Additionally, several regions of high conservation occur in noncoding areas (e.g., arrow 2). (b) A view of 55 base pairs at the beta globin locus. At this magnification (fewer than 30,000 base pairs), the UCSC genome browser displays the nucleotides of genomic DNA in the multiple sequence alignment of a group of vertebrates. The ATG codon (oriented from right to left) is indicated (three asterisks), and the human protein product is shown (amino acids from right to left matching the start of protein NP_000509, MVHLTPEEKS).*

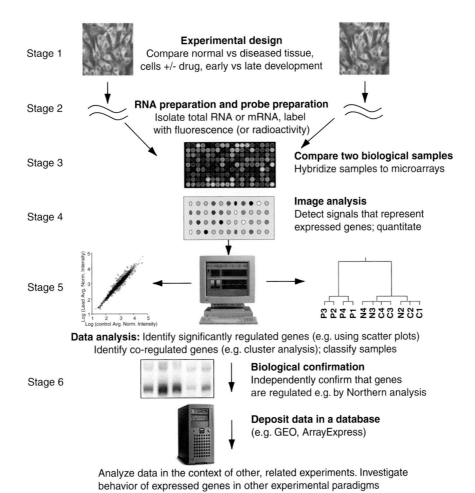

Stage 1

Experimental design
Compare normal vs diseased tissue, cells +/- drug, early vs late development

Stage 2

RNA preparation and probe preparation
Isolate total RNA or mRNA, label with fluorescence (or radioactivity)

Stage 3

Compare two biological samples
Hybridize samples to microarrays

Stage 4

Image analysis
Detect signals that represent expressed genes; quantitate

Stage 5

Log (Lead Avg. Norm. Intensity)
Log (control Avg. Norm. Intensity)

P3 P2 P4 P1 N4 N3 C4 C3 N2 C2 C1

Data analysis: Identify significantly regulated genes (e.g. using scatter plots)
Identify co-regulated genes (e.g. cluster analysis); classify samples

Stage 6

Biological confirmation
Independently confirm that genes are regulated e.g. by Northern analysis

Deposit data in a database
(e.g. GEO, ArrayExpress)

Analyze data in the context of other, related experiments. Investigate behavior of expressed genes in other experimental paradigms

FIGURE 8.17. Overview of the process of generating high throughput gene expression data using microarrays. In stage 1, biological samples are selected for a comparison of gene expression. In stage 2, RNA is isolated and labeled, often with fluorescent dyes. These samples are hybridized to microarrays, which are solid supports containing complementary DNA or oligonucleotides corresponding to known genes or ESTs. In stage 4, image analysis is performed to evaluate signal intensities. In stage 5, the expression data are analyzed to identify differentially regulated genes (e.g., using ANOVA [Chapter 9] and scatter plots; stage 5, at left) or clustering of genes and/or samples (right). Based on these findings, independent confirmation of microarray-based findings is performed (stage 6). The microarray data are deposited in a database so that large-scale analyses can be performed.

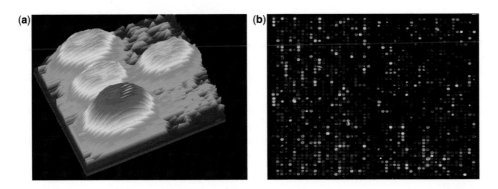

FIGURE 8.21. Microarray images. (a) A nitrocellulose filter is probed with [^{32}P]cDNA derived from the hippocampus of a postmortem brain of an individual with Down syndrome. There are 5000 cDNAs spotted on the array. The pattern in which genes are represented on any array is randomized. (b) Six of the signals are visualized using NIH Image software. Image analysis software must define the properties of each signal, including the likelihood that an intense signal (lower left) will "bleed" onto a weak signal (lower right). (c) A microarray from NEN Perkin-Elmer (representing 2400 genes) was probed with the same Rett syndrome and control brain samples used in Fig. 8.20. This technology employs cDNA samples that are fluorescently labeled in a competitive hybridization.

(a) Primary structure

```
MVHLTPEEKSAVTALWGKVNVDEVGGEALGRLLVVYPWTQRFFESFGDLSTPDAVMGNPKVKAHGKKVLGAFSD
GLAHLDNLKGTFATLSELHCDKLHVDPENFRLLGNVLVCVLAHHFGKEFTPPVQAAYQKVVAGVANALAHKYH
```

(b) Secondary structure

```
                  10        20        30        40        50        60        70
                   |         |         |         |         |         |         |
UNK_257900  MVHLTPEEKSAVTALWGKVNVDEVGGEALGRLLVVYPWTQRFFESFGDLSTPDAVMGNPKVKAHGKKVLG
DSC         cccchhhhhhhhhhhhhcccccchhhhhhhhhhhhccccchhhhhhhhcccccccccccccchhhhhhhhhhhh
MLRC        ccccchhhhhhhhhhhcccccccccchhhhhhheeecccchhhhhcccccccccccccccccccccchhhhh
PHD         cccchhhhhhhhhhhhccchhhcchhhhhhhheeeccchhhhhhhhhcccchhhhhecchhhhhhhhhhhhhh
Sec.Cons.   cccchhhhhhhhhhhhhccccchcchhhhhhhheeecccchhhhhhhhhcccccccccccccchhhhhhhhhhhh
```

```
                  80        90        100       110       120       130       140
                   |         |         |         |         |         |         |
UNK_257900  AFSDGLAHLDNLKGTFATLSELHCDKLHVDPENFRLLGNVLVCVLAHHFGKEFTPPVQAAYQKVVAGVAN
DSC         hhhhhhhhhhhhhhhhhhhhhhhhhcccccchhhhhhhhhhhhhhhhhhhcccccchhhhhhhhhhhhhhhh
MLRC        hhhhhhhhhhhhhhhhhhhhhhhhccccccccchhhhhhhhhhhhhhhhhhcccccchhhhhhhhhhhhhhhh
PHD         hhhhhhhhhhhhhhhhhhhhhhhhhhhccccchhhhhhhhhhhhhhhhhhhhcccccchhhhhhhhhhhhhhhh
Sec.Cons.   hhhhhhhhhhhhhhhhhhhhhhhhhhhcccccchhhhhhhhhhhhhhhhhhhcccccchhhhhhhhhhhhhhhh
```

```
UNK_257900  ALAHKYH
DSC         hhhhccc
MLRC        hhhhccc
PHD         hhhhhcc
Sec.Cons.   hhhhccc
```

(c) Tertiary structure   **(d)** Quaternary structure

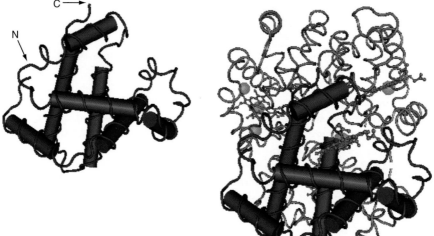

FIGURE 11.1. *A hierarchy of protein structure. (a) The primary structure of a protein refers to the linear polypeptide chain of amino acids. Here, human beta globin is shown (NP_000539). (b) The secondary structure includes elements such as alpha helices and beta sheets. Here, beta globin protein sequence was input to the POLE server for secondary structure (▶ http://pbil.univ-lyon1.fr/) where three prediction algorithms were run and a consensus was produced. Abbreviations: h, alpha helix; c, random coil; e, extended strand. (c) The tertiary structure is the three-dimensional structure of the protein chain. Alpha helices are represented as thickened cylinders. Arrows labeled N and C point to the amino- and carboxy-terminals, respectively. (d) The quarternary structure includes the interactions of the protein with other subunits and heteroatoms. Here, the four subunits of hemoglobin are shown (with an α2β2 composition and one beta globin chain highlighted) as well as four noncovalently attached heme groups. Panels (c) and (d) were produced using Cn3D software from NCBI.*

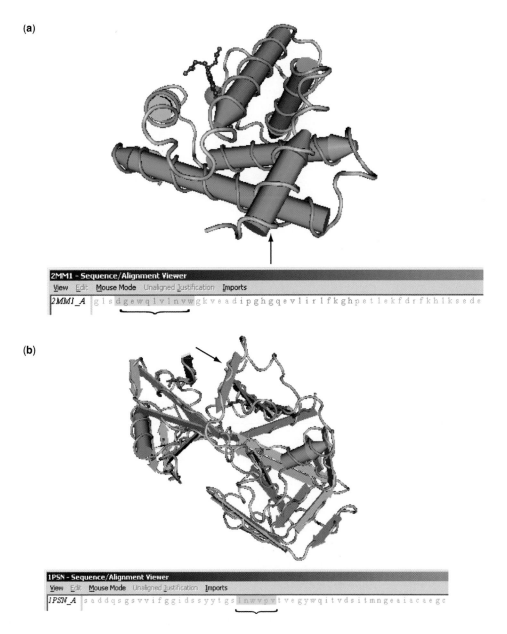

(a)

(b)

FIGURE 11.3. *Examples of secondary structure. (a) Myoglobin (Protein Data Bank ID 2MM1) is composed of large regions of α helices, shown as strands wrapped around barrel-shaped objects. By entering the accession 2MM1 into NCBI's structure site, one can view this three-dimensional structure using Cn3D software. The accompanying sequence viewer shows the primary amino acid sequence. By clicking on a colored region (bracket) corresponding to an alpha helix, that structure is highlighted in the structure viewer (arrow). (b) Human pepsin (PDB 1PSN) is an example of a protein primarily composed as β strands, drawn as large arrows. Selecting a region of the primary amino acid sequence (bracket) results in a highlighting of the corresponding β strand.*

(a)

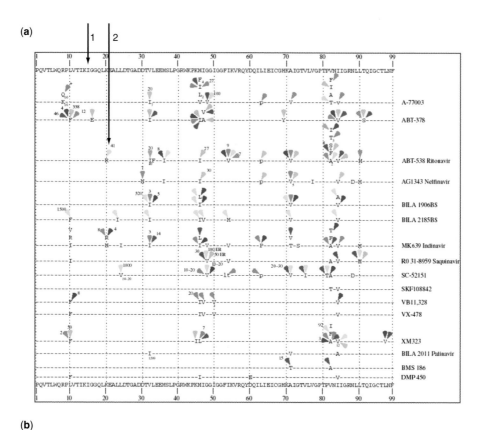

(b)

Gene: **Protease**		DrugClass: Protease Inhibitor
Compound: **ABT-538**		Synonyms: ABT-538, Ritonovir, Norvir
AAMutation: **K 20 R**		Codon mutation: AAG -> AGG
In Vitro: N In Vivo: Y		
FoldResist:		CrossResist:
Comment: K20R/M36I/I54V/V82A: 41-fold		

A. Molla, M. Korneyeva, Q. Gao, S. Vasavanonda, P. J. Schipper, H. M. Mo, M. Markowitz, T. Chernyavskiy, P. Niu, N. Lyons, A. Hsu, G. R. Granneman, D. D. Ho, C. A. Boucher, J. M. Leonard, D. W. Norbeck, D. J. Kempf
Ordered accumulation of mutations in HIV protease confers resistance to ritonavir.
Nat Med **2**: 760-6 (1996) Medline link: 96266327

FIGURE 14.17. *(a) The LANL website offers a map of HIV-1 protease mutations versus drugs. Each row represents a drug (labeled at right). The wild-type (strain HXB2) HIV-1 protease sequence is listed at top and bottom (arrow 1). Dashes indicate wild-type amino acid positions, while mutations that confer resistance to the drug are indicated. An example of a K-to-R (lysine-to-arginine) mutation is indicated (arrow 2). The small number (41) indicates the "fold resistance" of that particular mutation. Mutations that have a colored shape pointing to them are also part of a synergistic combination of mutations. (b) By clicking on the position of a mutation (arrow 2), the map links to a detailed report of the effects of that mutation.*

FIGURE 18.8. *Whole genome duplication in the ciliate* Paramecium tetraurelia *is inferred by analysis of protein paralogs. The outer circle displays all chromosome-sized scaffolds from the genome sequencing project. Lines link pairs of genes with a "best reciprocal hit" match. The three interior circles show the reconstructed ancestral sequences obtained by combining the paired sequences from each previous step. The inner circles are progressively smaller and reflect fewer conserved genes with a smaller average similarity. From Aury et al. (2006). Used with permission.*

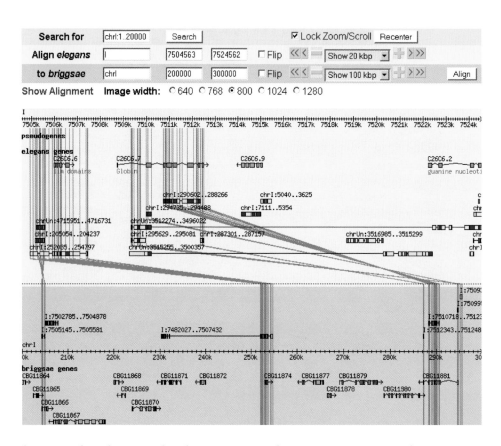

FIGURE 18.18. *Alignment of* C. elegans *and* C. briggsae *conserved syntenic regions using the synteny viewer at WormBase* (▶ *http://www.wormbase.org*). *Regions of chromosome I are aligned from* C. elegans *(above) and* C. briggsae *(below).*

Rfam family	Type	Count	View
U6	Gene;snRNA;splicing;	8	☐
Y	Gene;	6	☐
5S_rRNA	Gene;rRNA;	3	☐
SRP_euk_arch	Gene;	2	☐
U1	Gene;snRNA;splicing;	2	☐
SNORA51	Gene;snRNA;guide;HACA-box;	1	☐
SNORA32	Gene;snRNA;guide;HACA-box;	1	☐
SNORA42	Gene;snRNA;guide;HACA-box;	1	☐
snoMBI-28	Gene;snRNA;guide;HACA-box;	1	☐
sno26	Gene;snRNA;guide;CD-box;	1	☐
snoZ18	Gene;snRNA;guide;CD-box;	1	☐
SNORA70	Gene;snRNA;guide;HACA-box;	1	☐
mir-10	Gene;miRNA;	1	☐
lin-4	Gene;miRNA;	1	☐
SECIS	Cis-reg;	1	☐
let-7	Gene;miRNA;	1	☐
U4	Gene;snRNA;splicing;	1	☐
U3	Gene;snRNA;guide;CD-box;	1	☐
tRNA	Gene;tRNA;	1	☐

FIGURE 8.4. Summary of noncoding RNA families in the Rfam database that are assigned to the long arm of human chromosome 21. This table was accessed by browsing genomes from the Rfam website (► http://www.sanger.ac.uk/Software/Rfam/index.shtml, September 2007).

cloverleaf. Key features of this structure include a D loop, an anticodon loop which is responsible for recognizing messenger RNA codons, a T loop and a 3′ end to which aminoacyl tRNA synthetases attach the appropriate amino acid specific for each tRNA.

We will demonstrate a computational approach to identifying tRNAs using the tRNAscan-SE program (Lowe and Eddy, 1997). As an input, we will use a tRNA known to be assigned to human chromosome 21. The output includes the anticodon counts (Fig. 8.5a), listing the anticodons that have been identified corresponding to the 20 amino acids as well as stop codons and the modified amino acid selenocysteine. In this example, the isotype is GCC indicating that this is a glycine tRNA (in the genetic code glycine is encoded by GGG, GGA, GGT, or GGC; the GCC anticodon matches the GGC codon). Other information in the output shows the predicted tRNA secondary structure in a bracket notation (Fig. 8.5b) as well as a model of its structure (Fig. 8.5c).

tRNAscan-SE produces just one false positive per 15 billion nucleotides of random DNA sequence. It achieves high sensitivity and specificity by combining the output of three separate methods of tRNA identification (Lowe and Eddy, 1997). There are three

The tRNAscan-SE server is available at ► http://lowelab.ucsc.edu/tRNAscan-SE/, ► http://selab.janelia.org/tRNAscan-SE/, or ► http://bioweb.pasteur.fr/seqanal/interfaces/trnascan-simple.html. You can also visit Todd Lowe's site to download tRNAscan-SE and run it locally. The human chromosome 21 tRNA is given in web document 8.1 at ► http://www.bioinfbook.org/chapter8. (This 71 base pair sequence also matches nucleotides 84511 to 84581 of clone AP001670.1.) In Chapter 3 we introduced Dotlet for pairwise alignments. Try using it (► http://myhits.isb-sib.ch/cgi-bin/dotlet)

(a)   Isotype / Anticodon Counts:

Ala	: 0	AGC:	GGC:	CGC:	TGC:			
Gly	: 1	ACC:	GCC: 1	CCC:	TCC:			
Pro	: 0	AGG:	GGG:	CGG:	TGG:			
Thr	: 0	AGT:	GGT:	CGT:	TGT:			
Val	: 0	AAC:	GAC:	CAC:	TAC:			
Ser	: 0	AGA:	GGA:	CGA:	TGA:	ACT:		GCT:
Arg	: 0	ACG:	GCG:	CCG:	TCG:	CCT:		TCT:
Leu	: 0	AAG:	GAG:	CAG:	TAG:	CAA:		TAA:
Phe	: 0	AAA:	GAA:					
Asn	: 0	ATT:	GTT:					
Lys	: 0			CTT:	TTT:			
Asp	: 0	ATC:	GTC:					
Glu	: 0			CTC:	TTC:			
His	: 0	ATG:	GTG:					
Gln	: 0			CTG:	TTG:			
Ile	: 0	AAT:	GAT:		TAT:			
Met	: 0			CAT:				
Tyr	: 0	ATA:	GTA:					
Supres:	0			CTA:	TTA:			
Cys	: 0	ACA:	GCA:					
Trp	: 0			CCA:				
SelCys:	0				TCA:			

(b)   Your-seq.trna1 (1-71)   Length: 71 bp
Type: Gly   Anticodon: GCC at 33-35 (33-35) Score: 71.03
```
            *   |   *   |   *   |   *   |   *   |   *   |   *   |
Seq: GCATGGGTGGTTCAGTGGTAGAATTCTCGCCTGCCACGCGGGAGGCCCGGGTTCGATTCCCGGCCCATGCA
Str: >>>>>>>..>>>>.......<<<<.>>>>>.......<<<<<....>>>>>.......<<<<<<<<<<<<.
```

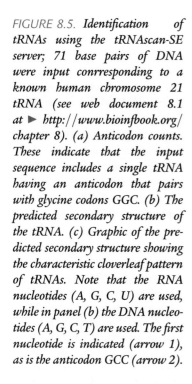

FIGURE 8.5. *Identification of tRNAs using the tRNAscan-SE server; 71 base pairs of DNA were input conrresponding to a known human chromosome 21 tRNA (see web document 8.1 at* ▶ *http://www.bioinfbook.org/ chapter 8). (a) Anticodon counts. These indicate that the input sequence includes a single tRNA having an anticodon that pairs with glycine codons GGC. (b) The predicted secondary structure of the tRNA. (c) Graphic of the predicted secondary structure showing the characteristic cloverleaf pattern of tRNAs. Note that the RNA nucleotides (A, G, C, U) are used, while in panel (b) the DNA nucleotides (A, G, C, T) are used. The first nucleotide is indicated (arrow 1), as is the anticodon GCC (arrow 2).*

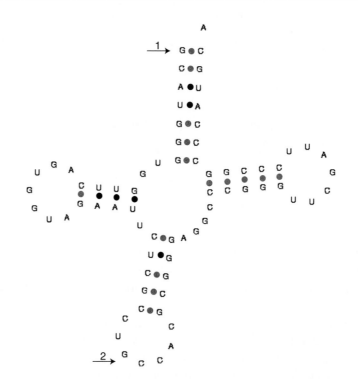

with the human tRNA as a query against itself, employing a small window size to find the internally matching stem-loop structures.

stages. First, it runs two programs that find tRNAs in DNA (or RNA) sequences. One program identifies conserved intragenic promoter sequences found in prototypic tRNAs, and also requires base pairings that occur in tRNA stem-loop "cloverleaf" structures (Fichant and Burks, 1991). The other program searches for signals that occur in eukaryotic RNA polymerase III promoters and terminators (Pavesi et al., 1994). The results of these two programs are merged. In the second stage, tRNAscan-SE analyzes the sequences using a covariance model or stochastic context-free grammar (SCFG) (Eddy and Durbin, 1994). A covariance model or SCFG is a probabilistic model of RNA secondary structure and sequence consensus, allowing insertions, deletions, and mismatches (Box 8.1). The covariance model includes a training step based on over 1000 previously characterized tRNAs. In the third stage tRNAscan-SE performs a secondary structure prediction and identifies the anticodon of the tRNA. tRNAs with introns and tRNA pseudogenes are further identified.

The approach adopted by tRNAscan-SE involves the alignment of multiple RNA sequences in order to infer a common structure of each family based on the two inter-related properties of primary sequence and secondary structure. Such an approach is motivated by the fact that noncoding RNAs may diverge over time in a way that preserves each molecule's base-paired structure while conserving only a limited amount of sequence similarity between homologous RNAs.

A distinct approach to determining RNA structures is to estimate the minimum free energy of folding. This thermodynamic approach was pioneered by Zuker and Stiegler 1981. It is implemented in a variety of programs, including the Vienna RNA package (Hofacker, 2003) which incorporates several folding algorithms. A sample output using the Vienna RNA webserver, using a chromosome 21 tRNA sequence as input, is shown in Fig. 8.6.

The Vienna RNA package is available at ▶ http://www.tbi.univie.ac.at/RNA/.

BOX 8.1
Stochastic Context-Free Grammars, or Covariance Models

Hidden Markov models (HMMs) are probabilistic models that are useful in many areas of bioinformatics to identify features in sequences such as conserved residues that define a particular protein family (Chapter 5), or nucleotide residues that constitute the structure of a gene (Chapter 16). Stochastic context-free grammars (SCFG; Sakakibara et al., 1994) or covariance models (Eddy and Durbin, 1994) constitute another class of probabilistic models that account for long-range correlations along a sequence that occur because of base pairing of noncoding RNA sequences. Such base pairing is required to form appropriate secondary structure such as a stem. Eddy and Durbin 1994 introduced a covariance model in which an RNA sequence is described as an ordered tree in which there are states M (including match states, insert states, and delete states), symbol emission probabilities (these are assigned to specific bases according to the 16 possible pairwise nucleotide combinations or the four unpaired nucleotides), and state transition probabilities (scores assigned to changing states such as entering an insert state). They found that the information content in the secondary structure of tRNA molecules is comparable to that of the primary sequences.

SCFGs are comparable to covariance models. The input of a SCFG is a multiple sequence alignment of noncoding RNAs (such as tRNAs) (Sakakibara et al., 1994). The SCFG models how to derive the observed sequences based on a set of "production rules." Production rules and their associated probabilities define a grammar. The advantages of a SCFG are that its parameters are derived from known RNA sequences and structures, and its probabilistic framework yields confidence estimates on its predictions. SCFGs (like HMMs) originate in the field of language processing (speech recognition).

The Rfam covariance models generated by software called Infernal do not provide expect (E) values, but they do offer bit scores. These are derived from log-odds ratios of the probability that a sequence matches a covariance model divided by the probability that the sequence was generated by a random model (Griffith-Jones, 2005).

(a)   GCAUGGGUGGUUCAGUGGUAGAAUUCUCGCCUGCCACGCGGGAGGCCCGGGUUCGAUUCCCGGCCCAUGCA
(((((((......(((((((.........)))))))).((((((...((....)).))))))))))))).

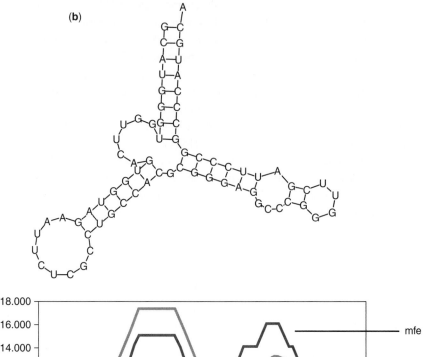

(b)

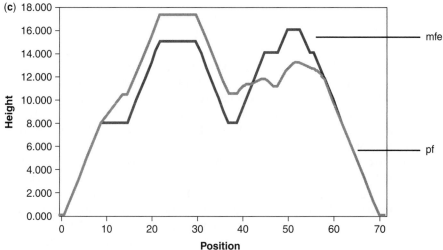

(c)

FIGURE 8.6. RNA structure pre-diction based on the minimum free energy of folding. A chromo-some 21 sequence known to encode a tRNA (see Fig. 8.5 and web document 8.1) was analyzed using the Vienna RNA web server. (a) Optimal predicted struc-ture of an RNA using bracket nota-tion. Unpaired nucleotides are represented as dots, while base paired nucleotides are represented by a pair of matching parentheses. The minimum free energy was −35.96 kcal/mol. (b) Predicted structure of the RNA including stems (double-stranded regions with base pairing) and loops (single-stranded regions). (c) Plot of the minimum free energy (mfe) and a positional entropy measure (pf) (y axis) versus the nucleotide position of the input DNA sequence (x axis).

The Genomic tRNA Database (G~tRNA~db) from the labora-tory of Todd Lowe is available at ▶ http://lowelab.ucsc.edu/ GtRNAdb/ and contains tRNA identifications of many genomes made using tRNAscan-SE. Web document 8.1 at ▶ http://www. bioinfbook.org/chapter8 includes all human tRNAs from the Lowe database. TFAM is available at ▶ http://www.lcb.uu.se/~dave/ TFAM and is especially useful for classifying tRNAs having unusual modifications. Another very useful resource is the tRNA data-base of Mathias Sprinzl and

In sequencing complete genomes it is of interest to identify all the tRNA genes. In the human genome, there are over 600 tRNA genes, making this among the largest gene families. The reason for so many genes is the necessity for large amounts of tRNAs to enable protein synthesis to occur in all cells throughout life. Two major resources are the Genomic tRNA Database and TFAM (Tåquist et al., 2007). A sum-mary of the number of tRNA genes in selected organisms is presented in Table 8-2.

Ribosomal RNA

Ribosomal RNA molecules form structural and functional components of ribo-somes, the subcellular units responsible for protein synthesis. rRNA constitutes approximately 80% to 85% of the total RNA in a cell. In eukaryotes, synthesis of rRNA occurs in the nucleolus, a specialized structure within the nucleus. Purified ribosomes include particles that migrate at characteristic sedimentation coefficients upon centrifugation through a gradient (Table 8-3). In bacteria these include the

| TABLE 8-2 | Summary of the Number of tRNA Genes in Selected Organisms |

Organism	Common Name	# tRNAs Decoding the 20 Amino Acids	# Predicted Pseudogenes	Other	Total
Homo sapiens	Human	448	171	3	622
Pan troglodytes	Chimpanzee	451	103	15	569
Mus musculus	Mouse	431	25,606	3	26,040
Canis familiaris	Dog (canFam1)	889	0	8	897
Drosophila melanogaster	Fruit fly	289	4	2	295
Saccharomyces cerevisiae	Baker's yeast	273	0	2	275
Arabidopsis thaliana	Plant	630	8	1	639
Plasmodium falciparum	Malaria parasite	42	1	1	44
Methanococcus jannaschii	Archaeon	36	0	1	37
Escherichia coli K12	Bacterium	86	1	1	88
Mycobacterium leprae	Bacterium	45	0	0	45

The "other" category refers to selenocysteine tRNAs (TCA), suppressor tRNAs (CTA,TTA) or tRNAs with undetermined or unknown isotypes. Additionally, some organisms have tRNAs with introns (e.g., human, 32; *P. falciparum*, 1; *Arabidopsis*, 83).

Source: G~tRNA~db at ▶ http://lowelab.ucsc.edu/GtRNAdb/, September 2007.

Konstantin Vassilenkoat ▶ http://www.uni-bayreuth.de/departments/biochemie/trna/.

70S ribonucleoprotein particle that is composed of 30S and 50S subunits, further containing three major rRNA forms (16S, 23S, and 5S). In eukaryotes the 80S ribonucleoprotein particle consists of a 45S ribosomal RNA subunit that is further processed to generate 18S, 28S, and 5.8S subunits.

rRNA derives from a multicopy ribosomal DNA (rDNA) gene family. In humans these families are localized to the p arms (i.e., short arms) of the five acrocentric chromosomes (13, 14, 15, 21, and 22) (Henderson et al., 1972). The rDNA loci consist of a repeat unit, about 43 kilobases in length, of which 13 kilobases are transcribed and the remainder are nontranscribed spacers (Fig. 8.7). The rDNA genes are identified as RNR1 (mitochondrially encoded 12S RNA), RNR2 (mitochondrially encoded 16S RNA), RNR3, RNR4, and RNR5. In the human genome, there are typically ~400 copies of the rDNA repeat. These loci share a high degree of sequence conservation in a process of homogenization that involves both concerted evolution through recombination and gene conversion.

Ribosomal DNA genes have a complex repetitive structure, tremendous conservation across loci on different chromosomes, and enormous variability in the size of the loci between individuals. Thus, they are not currently incorporated into the reference human genome at NCBI, UCSC, or Ensembl. To identify human rRNA RefSeq sequences from GenBank, take the following steps. (1) From the home page of NCBI, select TaxBrowser; click *Homo sapiens* then choose Nucleotides and follow the link for Core Nucleotides. This is equivalent to beginning in Entrez Nucleotide and restricting the search to human using the command "txid9606[Organism:exp]." (2) Currently (September 2007) there are over 12 million entries. Click "limits" and under the molecule option select rRNA. (3) There are now three RefSeq entries corresponding to 5.8S rRNA (NR_003285; 156 base

For an example of a human genomic DNA sequence that you can use as an input to search Rfam for rRNA families, see web document 8.2 at ▶ http://www.bioinfbook.org/chapter8.

We discuss the structure of the chromosome in Chapter 16, including explanations of mechanisms for conserving sequence identity across chromosomal loci, such as concerted evolution and gene conversion. The five acrocentric chromosomes have a centromere positioned near an end of the chromosome rather than in the center.

RefSeq accession numbers having the format NR_123456 consist of noncoding transcripts, including structural RNAs and transcribed pseudogenes. The three human RefSeq entries for rRNA are given in web document 8.3 at ▶ http://www.bioinfbook.org/chapter8.

TABLE 8-3   Major Forms of rRNA in Bacteria and Eukaryotes

	Bacteria			Eukaryotes		
Ribonucleoprotein	70S			80S		
MW	2.6×10^6			4.3×10^6		
% RNA	~60			~50		
Ribosomal subunits	30S	50S		40S	60S	
MW	0.7×10^6	1.8×10^6			2.7×10^6	
Function	Binding mRNA	Peptide bond formation		Binding mRNA	Peptide bond formation	
RNAs from subunits	16S	23S	5S	18S	28S	5.8S
Accession numbers	M25588	M25458	M24300	NR_003286	NR_003287	NR_003285
Rfam accession	–	–	RF00001	–	–	RF00002
Base pairs	1504	542	120	1871		156
MW of RNA	0.55×10^6	1.07×10^6	0.04×10^6	0.7×10^6	1.8×10^6	0.04×10^6

Abbreviations: S, sedimentation coefficient; MW, molecular weight.
Accession numbers are provided for *E. coli* and human rRNAs.
Source: Adapted from Dayhoff (1972, p. D-252).

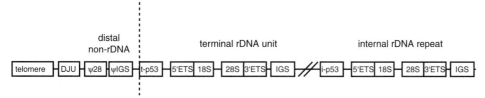

FIGURE 8.7. *Structure of a eukaryotic ribosomal DNA repeat unit. A region of an acrocentric chromosome is depicted from the telomere (left side, denoted an end of the chromosome) to a distal non-rDNA region (containing sequences DJU and two pseudogene regions), then a distal junction (vertical dotted line). To the right (3′ end) of this distal junction a terminal rDNA unit is shown; this unit is repeated internally many times, with each unit sharing identical or nearly identical DNA sequence. This region is found in GenBank accession U67616 (8,353 base pairs including a variety of repetitive DNA elements and 28S rDNA pseudogenes) and U13369 (42,999 base pairs including transcribed spacers, DNA encoding 18S, 5.8S, and 28S rRNA, and various repetitive DNA elements). Abbreviation: IGS, intergenic spacer (also called nontranscribed spacer). Adapted from Gonzalez and Sylvester (2001).*

pairs), 28S rRNA (NR_003287; 5035 base pairs), and 18S rRNA (NR_003286; 1871 base pairs). For each, the chromosomal assignment is to the acrocentric p-arms.

rRNA sequences are particularly important for phylogenetic analyses across life forms (including the three domains of bacteria, archaea, and eukaryotes). They are uniquely useful because they are closely conserved enough to permit trusted multiple sequence alignments, while they are specific enough to each species that they permit accurate classification. Furthermore, rRNA can be sequenced from environmental samples such as soil or water, in which vast numbers of species exist but cannot be cultured (see Chapter 15). Also rRNA genes are generally not subject to lateral gene transfer (discussed in Chapter 15). That is a form of inheritance in which genes are transmitted horizontally across species rather than being inherited through generations within a species, and it can confound phylogenetic analyses.

The Ribosome Data Project is online at ▶ http://rdp.cme.msu.edu/index.jsp. Release 9.53 (August 2007) contains over 400,000 16S rRNA sequences.

Currently there are over 700,000 16S rRNA sequences in GenBank from about 100,000 different species (see Schloss and Handelsman, 2004). There are several major databases of rRNA sequences, including the Ribosome Database Project (RDB) (Cole et al., 2007). RDP is approaching half a million aligned and annotated rRNA sequences, one third from cultivated bacterial strains and two thirds from environmental samples. Alignment is performed against a bacterial rRNA alignment model using a stochastic context free grammar (Box 8.1) as described by Sakakibara et al. (1994).

The ARB project is another major resource for RNA studies (Ludwig et al., 2004). It is a UNIX-based program with a graphical interface that provides software tools to analyze large rRNA databases (including those imported from the RDP). The related SILVA database includes small subunit (16S, 18S) and large submuit (23S, 28S) rRNA from bacteria, archaea, and eukaryotes. Sequences are downloadable from a browser in the fasta or other formats.

You can access the ARB project at ▶ http://www.arb-home.de/. It was developed by Wolfgang Ludwig and colleagues at the Technical University, Munich. ARB refers to arbor (Latin for tree) while silva is Latin for forest. The SILVA website (including a browser) is ▶ http://silva.mpi-bremen.de/.

RNAmmer is a hidden Markov model approach to identifying rRNA genes, particularly in newly sequenced genomes (Lagesen et al., 2007). It is useful for searching with large amounts of DNA (e.g., up to 20 million nucleotides) to identify the genomic loci of rRNA genes.

You can access RNAmmer at ▶ http://www.cbs.dtu.dk/services/RNAmmer/.

Small Nuclear RNA

Small nuclear RNA (snRNA) is localized to the nucleus and consists of a family of RNAs that are responsible for functions such as RNA splicing (in which introns

TABLE 8-4   Examples of Human Noncoding Spliceosomal RNAs

Name	Accession	Chromosome	Length (Base Pairs)
U2	NR_002716	17 q12–q21	186
U4	NR_002760	12	144
U4B2	NR_003925	12	144
U5F	NR_002753	1p34.1	116
U6	NR_002752	10p13	45

are removed from genomic DNA to generate mature mRNA transcripts) and the maintenance of telomeres (chromosome ends). snRNAs associate with proteins to form small nuclear ribonucleoproteins (SNRNPs).

The spliceosome is a nuclear complex that includes hundreds of proteins and the five snRNAs U1, U2, U4, U5, and U6 (Valadkhan, 2005). Properties of several of these snRNAs are given in Table 8-4. In humans there are about 100 copies of the U4 gene (Bark et al., 1986; Rfam family RF00015) and there are about 1200 copies of the U6 snRNA, including many pseudogenes (nonfunctional genes) (Rfam family RF00026). Pseudogenes of protein-coding genes are relatively straightforward to detect for because one can recognize the interruption of an open reading frame (see below and Chapter 16). The identification of nonfunctional, noncoding RNAs presents a far greater challenge because there are no such landmarkers as open reading frames, and functional noncoding RNAs are routinely found to have divergent sequences.

Small Nucleolar RNA

In eukaryotes, ribosome biogenesis occurs in the nucleolus. This process is facilitated by small nucleolar RNAs (snoRNAs), a group of noncoding RNAs that process and modify rRNA and small nuclear spliceosomal RNAs. The two main classes of snoRNAs are C/D box RNAs, which methylate rRNA on a $2'$-O-ribose position, and H/ACA box RNAs, which convert uridine to pseudouridine in rRNA. Table 8-5 presents several online databases that list snoRNAs.

Computational approaches have facilitated the discovery of snoRNAs. For example, after the genome of the yeast *Saccharomyces cerevisiae* was completely sequenced (see Chapter 17), snoRNAs remained challenging to identify. Lowe and Eddy 1999 used a covariance model to identify 22 snoRNAs whose function in methylating rRNA they subsequently confirmed.

As another example of a strategy of combining computational and experimental approaches, Omer et al. (2000) considered the problem that snoRNAs were known to

TABLE 8-5   Small Nucleolar RNA (snoRNA) Resources

Database	Focus	URL
Plant snoRNA database		http://bioinf.scri.sari.ac.uk/cgi-bin/ plant_snorna/home
Yeast snoRNA database		http://people.biochem.umass.edu/ fournierlab/snornadb/main.php
SnoRNABase	Human H/ACA and C/D box snoRNAs	http://www-snorna.biotoul.fr/
SnoRNA database	Yeast, archaeal, *Arabidopsis*	http://lowelab.ucsc.edu/ snoRNAdb/

occur in eukaryotes, but had not been identified in prokaryotes (bacteria or archaea). They cloned 18 snoRNAs from the archaeon *Sulfolobus acidocaldarius* (based on coimmunoprecipitation experiments using antisera against nucleolar proteins fibrillarin and Nop56). They then trained a probabilistic model and identified several hundred additional sno-like RNAs from archaea.

MicroRNA

MicroRNAs (miRNAs) are noncoding RNA molecules of approximately 22 nucleotides that have been identified in animals and plants. Since their discovery in the 1990s they have gained tremendous interest because of their potential functional roles in regulating gene expression. The earliest members of this family to be identified were the *lin-4* and *let-7* gene products of the worm *Caenorhabditis elegans* (Pasquinelli and Ruvkun, 2002). Those genes were identified through positional cloning in a forward genetics strategy: a worm mutant having a defective cell lineage was identified, and a mutation in the lin-4 RNA was shown to account for the phenotype (Lee et al., 1993). Subsequently many other miRNA candidates have been identified by complementary DNA (cDNA) cloning of size-selected RNA samples. The major function of microRNAs appears to be the downregulation of protein function by inhibiting the translation of protein from mRNA or by promoting the degradation of mRNA.

We can examine a typical microRNA by visiting miRBase, a repository of miRNA data (Griffiths-Jones, 2004; Griffiths-Jones et al., 2006). One can browse by organism and find a group of microRNAs assigned to human chromosome 21. Currently, these include five microRNAs: hsa-let-7c, hsa-mir-99a, hsa-mir-125b-2, hsa-mir-155, and hsa-mir-802. The entry for let-7c includes the predicted stem-loop structure, the genomic coordinates on chromosome 21, a description of neighboring microRNAs (e.g., hsa-mir-99a is less than 10 kilobases away), and database links (e.g., to the European Molecular Biology Laboratory, Rfam, and the Human Genome Organization official nomenclature).

MiRBase also provides links to predicted targets of each microRNA. These targets are RNA transcripts that are potentially regulated by a given microRNA, and predictions are linked from three databases: MiRanda, TargetScan, and Pictar (Krek et al., 2005). These predictions serve as useful guides to potential targets, but most have not been experimentally validated. At present, about 60 microRNA families, conserved in vertebrates, are proposed to regulate at least 30% of all human protein-coding genes (Rajewsky, 2006).

It can be challenging to distinguish an authentic microRNA from other classes of noncoding (or coding) RNA. Ambros et al. (2003) proposed a series of definitions of microRNAs based on two criteria regarding their expression:

1. microRNAs consist of an RNA transcript of about 22 nucleotides based on hybridization of the transcript to size-fractionated RNA. Typically, this is accomplished by a Northern blot in which total RNA is purified from a sample such as a cell line, electrophoresed on an agarose gel, transferred to a membrane, and probed with a radioactively labeled form of the candidate miRNA. This experiment shows the size of the RNA, its abundance, and whether the probe hybridizes to multiple RNA species in a sample.

2. The ~22 nucleotide candidate should be present in a library of cDNAs that is prepared from size-fractionated RNA.

miRBase is available at ► http://microrna.sanger.ac.uk/sequences/. Release 10.0 (August 2007) includes 5071 entries, including 533 from human. The five chromosome 21 microRNAs are available in web document 8.4 at ► http://www.bioinfbook.org/chapter8. For target predictions, MiRanda is part of MiRBase and makes 1480 target predictions for let-7c. TargetScan, available at ► http://www.targetscan.org, makes 691 target predictions (of which 691 are well conserved and 83 are poorly conserved). Pictar (► http://pictar.bio.nyu.edu/) makes 602 target predictions. Other software for predictions includes DIANA (► http://diana.pcbi.upenn.edu/cgi-bin/micro_t.cgi/) and RNAHybrid (► http://bibiserv.techfak.uni-bielefeld.de/rnahybrid/).

We describe cDNA libraries later in this chapter.

Ambros et al. (2003) proposed three additional criteria concerning miRNA biogenesis:

3. The miRNA should have a precursor structure (typically 60 to 80 nucleotides in animals) that potentially folds into a stem (or hairpin) with the ~22 nucleotide mature miRNA located in one arm of the hairpin. Such a structure is predicted by RNA-folding programs such as mfold (Mathews et al., 1999).

4. Both the ~22 nucleotide miRNA sequence and its predicted fold-back precursor secondary structure must be phylogenetically conserved.

A RefSeq accession for the human Dicer protein is NP_085124.

5. Dicer is a protein that functions as a ribonuclease and is involved in processing small noncoding RNAs. There should be increased precursor accumulation in organisms having reduced Dicer function.

Ideally, a putative miRNA meets all five of these criteria, although in practice a subset (such as 1 and 4) may be sufficient.

Short Interfering RNA

Andrew Fire and Craig Mello were awarded the 2006 Nobel Prize in Physiology or Medicine "for their discovery of RNA interference—gene silencing by double-stranded RNA." See ▶ http://nobelprize.org/nobel_prizes/medicine/laureates/2006/.

In 1998 Andrew Fire, Craig Mello, and colleagues reported that double-stranded RNA introduced into the nematode *Caenorhabditis elegans* can suppress the activity of a gene (Fire et al., 1988). This process is called RNA interference (RNAi). They found that gene silencing occurred when they injected annealed, double-stranded RNA, but not either sense or antisense RNA alone. The silencing was specific to each target gene they studied (such as *unc-22*), and depended on the injection of double-stranded RNA corresponding to exons rather than introns or promoter sequences. Messenger RNA that is targeted by RNAi is degraded prior to translation, with double-stranded RNA targeting homologous mRNAs in a catalytic manner. This process depends on an RNA-inducing silencing complex (RISC) that includes an endonuclease (to cleave mRNA) and the nuclease Dicer that converts large double stranded RNA precursors to short interfering RNA.

It is now recognized that RNA interference has many functional implications for eukaryotic cells. RNAi can protect plant and animal cells against infection by single-stranded RNA viruses. RNAi further protects cells from the harmful action of endogenous transposons. These are mobile genetic elements that comprise portions of the human and other genomes. The RNAi mechanism also offers an experimental approach to systematically inhibit the function of genes in mammalian systems; we will consider this approach in Chapter 12 (Functional Genomics).

Noncoding RNAs in the UCSC Genome and Table Browser

As the human genome and other vertebrate genomes continue to be sequenced and analyzed in increasing depth, the UCSC Genome Browser has emerged as an essential tool for visualizing genomic data (Hinrichs et al., 2006). For noncoding RNAs we can view human chromosome 21 and display a series of user-selected annotation tracks. The following tracks are visible at the resolution of the entire chromosome 21 (46 million base pairs; Fig. 8.8a) and a zoomed in region of 1000 base pairs from nucleotides 16,833,201 to 16,834,200 (Fig. 8.8b):

- Evofold (Pedersen et al., 2006) shows RNA secondary structure predictions based on phylogenetic stochastic context-free grammars.

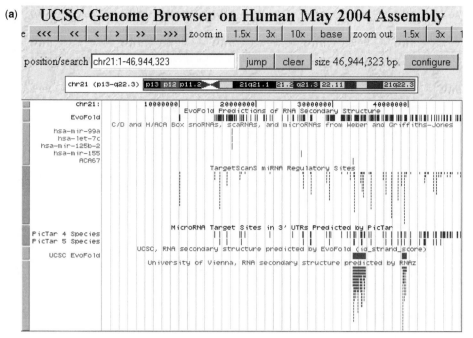

(a)

(b)

FIGURE 8.8. *Viewing the genomic landscape of noncoding RNAs on human chromosome 21. To recreate this display, visit ▶ http:// genome.ucsc.edu and select Genome Browser. Set the clade to vertebrate, the genome to human, the assembly to May 2004 (different assemblies have varying annotation tracks available), the position to chr21, and click submit. (a) All of chromosome 21 is displayed (about 47 million base pairs). You can specify which annotation tracks to select using a series of pull-down menus; under the Genes and Gene Prediction Tracks category select EvoFold (full display option) and sno/ miRNA (full). (b) A region of 1000 base pairs at coordinates chr21:16,833,201-16,834,200 is shown. This includes two miRNAs, hsa-mir99a and hsa-let7c. Arrows indicate the sense orientation.*

- The sno/miRNA track displays information on four types of noncoding RNAs: (1) microRNAs from the miRNA registry (Griffiths-Jones, 2004; Weber, 2005); (2) C/D box, (3) H/ACA box snoRNAs, and (4) Cajal body-specific RNAs (scaRNAs) from the snoRNA-LBME-DB (Lestrade and Weber, 2006). In this particular genome assembly four miRNAs are evident (hsa-let-7c, hsa-mir99, hsa-mir-125b-2, and hsa-mir-155). Also one snoRNA is present, ACA67. Clicking on it within the Genome Browser leads to a page of information confirming that it is an H/ACA Box snoRNA of 136 base pairs on chromosome 21q22.11.

- The TargetScanS miRNA Regulatory Sites track shows putative miRNA binding sites in the 3′ untranslated region of RefSeq genes. These sites are predicted by the TargetScanS program (Lewis et al., 2005).

- The PicTar track also displays putative miRNA target sites. Note that there is partial overlap with the TargetScanS predictions: there are 230 TargetScanS predictions on chromosome 21; 172 of these overlap picTarMiRNA4Way predictions, and 78 overlap picTarMiRNA5Way predictions. You can determine these overlaps using the Table Browser (see below).

- Evofold and RNAz predictions are made in the ENCODE regions that currently span 1% of the human genome (Washietl et al., 2007; Gruber et al., 2007). We describe ENCODE below and in Chapter 16.

The miRNA Registry at the Wellcome Trust Sanger Institute is available at ▶ http://microrna. sanger.ac.uk/sequences/. The snoRNABase is online at ▶ http://www-snorna.biotoul.fr/.

FIGURE 8.9. Analyzing noncoding RNAs using the UCSC Table Browser. (a) From a view in the UCSC Genome Browser there is a link to the Table Browser allowing a tabular output of features of interest. For example, setting the track to EvoFold (arrow 1) you can click summary/statistics (arrow 4) to obtain a count of the number of EvoFold items in any region of interest (arrow 2) such as the genome, the chromosome, or the region just viewed in the Genome Browser. Additional options include features to filter the results, or to intersect two tables (arrow 3). (b) Sample output of the summary for EvoFold items on chromosome 21. The UCSC site can be used in the opposite direction: starting with an appropriately formatted table of data, custom tracks can be generated in the Genome Browser.

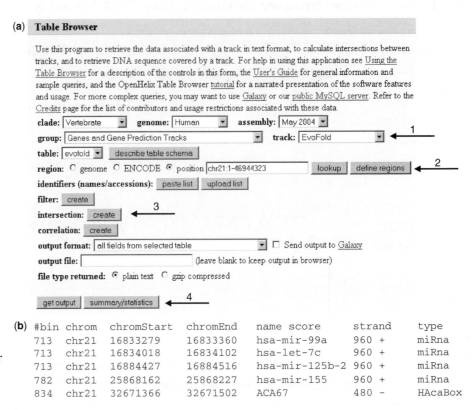

The UCSC Table Browser is complementary to the Genome Browser. Suppose we want to know the exact number of EvoFold entries that occur on chromosome 21. From the full view of chromosome 21 click the "Tables" link on the top bar. Choose the table of interest (e.g., EvoFold, Fig. 8.9a, arrow 1) and click "summary/statistics" (arrow 4) to see that there are 329 EvoFold items. For sno/miRNAs there are five items; by clicking "get output" you can obtain their genomic coordinate positions (Fig. 8.9b). How many RefSeq genes overlap with these Evofold regions on chromosome 21? To answer this, simply click the "intersection" button (Fig. 8.9a, arrow 3) and from the Genes and Gene Prediction Tracks group select RefSeq genes; the current answer is 140.

INTRODUCTION TO MESSENGER RNA

Gene expression occurs when DNA is transcribed into RNA. Each eukaryotic cell contains a nucleus with some 2,000 to 60,000 protein-coding genes, depending on the organism. However, at any given time the cell expresses only a subset of those genes as mRNA transcripts. The set of genes expressed by a genome is sometimes called the transcriptome. A conventional view that emerged since the "one gene, one enzyme" hypothesis of Beadle and Tatum, and continued through the establishment of the central dogma of molecular biology, is that genes correspond to discrete loci and are transcribed to mRNA in order to make a protein product. We now appreciate that the situation is vastly more complex because of the existence of noncoding RNAs, the interruption of genes by introns, the existence of alternative splicing to generate different mRNA transcripts that often produce distinct protein products, and the pervasive transcription of most nucleotide bases in the genome. We discuss these topics below. Furthermore, while humans, chimpanzees, and mice all have an extremely closely related set of about 20,000 to 25,000 genes per genome, what distinguishes the phenotypic expression of each species may depend on the intricacies of the regulation of gene expression. Gene expression is typically regulated in several basic ways:

For the range of gene content in eukaryotic genomes see Chapters 16 to 18.

- By region (e.g., brain vs. kidney)
- In development (e.g., fetal vs. adult tissue)
- In dynamic response to environmental signals (e.g., immediate–early response genes that are activated by a drug)
- In disease states
- By gene activity (e.g., mutant vs. wild-type bacterium)

In addition to viewing gene expression as a dynamically regulated process, we can also view proteins and metabolites as regulated dynamically in every cell. See Chapter 10.

The comparison of gene expression profiles has been used to address a variety of biological questions in an assortment of organisms. For viruses and bacteria, studies have focused both on viral and bacterial gene expression and also on the host response to pathogenic invasion. Among eukaryotes, gene expression studies and in particular microarrays have been employed to address fundamental questions such as the identification of genes activated during the cell cycle or throughout development. In multicellular animals cell-specific gene expression has been investigated, and the effect of disease on gene expression has been studied in rodents and primates, including humans. In recent years, gene expression profiling has become especially important in the annotation of genomic DNA sequences. When the genome of an organism is sequenced, one of the most fundamental issues is to determine which genes it encodes (Chapters 13 to 19). Large-scale sequencing of expressed genes, such as those isolated from cDNA libraries (described in this chapter), is invaluable in helping to identify gene sequences in genomic DNA.

In recent decades, gene expression has been studied using a variety of techniques such as Northern blotting, the polymerase chain reaction with reverse transcription (RT-PCR), and the RNase protection assay. Each of these approaches is used to study one transcript at a time. In Northern blotting, RNA is isolated, electrophoresed on an agarose gel, and probed with a radioactive cDNA derived from an individual gene. RT-PCR employs specific oligonucleotide primers to exponentially amplify specific transcripts as cDNA products. RNase protection is used to quantitate the amount of an RNA transcript in a sample based on the ability of a specific in vitro transcribed cDNA to protect an endogenous transcript from degradation with a ribonuclease. Gene expression may be compared in several experimental conditions (such as normal vs. diseased tissue, cell lines with or without drug treatment). The signals are typically fluorescent or radioactive and may be quantitated. Signals are also normalized to a number of housekeeping genes or other controls that are expected to remain unchanged in their expression levels.

The enzyme reverse transcriptase, often present in retroviruses, is an RNA-dependent DNA polymerase (i.e., it converts RNA to DNA).

In contrast to these approaches, several high throughput techniques have emerged that allow a broad survey of gene expression. A global approach to gene expression offers two important advantages over the study of the expression of individual genes:

- A broad survey may identify individual genes that are expressed in a dramatic fashion in some biological state. For example, global comparisons of gene expression in assorted human tissues can reveal which individual transcripts are expressed in a region-specific manner. Milner and Sutcliffe 1983 performed 191 Northern blots, although this experimental approach is not normally employed to measure the expression of so many genes in one study. They found that 30% of the genes were expressed in brain but not in liver or kidney. This type of approach can now be repeated on a larger scale and far more easily with high throughput techniques that are described in this chapter.

- High throughput analyses of gene expression can reveal patterns or signatures of gene expression that occur in biological samples. This may include the coordinate expression of genes whose protein products are functionally related. We examine tools for the analysis of gene expression data (such as clustering trees) in Chapter 9.

Several high throughput approaches to gene expression are displayed in Fig. 8.10. In each case, total RNA or mRNA is isolated from two (or more) biological samples that are compared. The RNA is typically converted to cDNA using reverse transcriptase. Complementary DNA is inherently less susceptible to proteolytic or chemical degradation than RNA, and cDNA can readily be cloned, propagated, and sequenced. In this chapter we explore three computer-based approaches to the analysis of gene expression: the comparison of cDNA libraries in UniGene, the comparison of serial analysis of gene expression (SAGE) libraries, and the most popular approach to gene expression studies, DNA microarrays. Another very recent approach, depicted in Fig. 8.10, is high throughput sequencing of cDNA. Once RNA is isolated, converted to cDNA, and cloned into a library, recently developed technologies such as Solexa (Chapter 13) can be used to sequence large amounts of cDNA (e.g., 1 billion base pairs) rapidly at a relatively modest cost.

There are other technologies available for the measurement of gene expression, such as differential display and subtractive hybridization. While these approaches have been technically successful for many gene expression problems, they differ from the techniques shown in Fig. 8.10 because they generally do not involve high throughput or the establishment of electronic databases. These earlier techniques used to measure gene expression have been reviewed by Sagerstrom et al. (1997), Vietor and Huber (1997), and Carulli et al. (1998).

Although databases of gene expression have been established, it is important to contrast them with DNA databases. A DNA database such as GenBank contains information about the sequence of DNA fragments, ranging in size from small clones to entire chromosomes or entire genomes. The error rate involved in genomic DNA sequencing can be measured (see Chapter 13), and independent laboratories can further confirm the quality of DNA sequence data. In general, DNA sequence does not change for an individual organism across time or in different body regions. In contrast, gene expression is context dependent. A database of gene expression contains some quantitative measurement of the expression level of a specified gene. If two laboratories attempt to describe the expression level of beta globin from a cell line, the measurement may vary based on many variables, such as the source of the cell line (e.g., liver or kidney), the cell-culturing conditions (e.g., cells grown to subconfluent or confluent levels), the cellular environment (e.g., choice of growth media), the age of the cells, the type of RNA that is studied (total RNA vs. mRNA, each with varying amounts of contaminating biomaterials), the measurement technique, and the approach to statistical analysis. Thus, while it has been

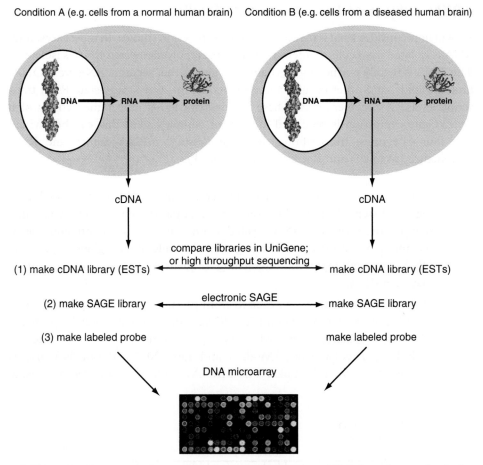

FIGURE 8.10. *Gene expression can be measured with a variety of high throughput technologies. In most cases, two biological samples are compared, such as a cell line with or without drug treatment, cells with or without viral infection, or aged versus neonatal rat brain. RNA can be converted to cDNA allowing broader surveys of transcription in a cell. In this chapter and the next we will examine several approaches to gene expression. (1) cDNA libraries can be constructed, generating expressed sequence tags (ESTs). These can be electronically compared in UniGene. (2) Serial analysis of gene expression (SAGE) is another technology in which the abundance of transcripts can be compared. This can also be studied electronically. (3) Complex cDNA mixtures can be labeled with a fluorescent molecule and hybridized on DNA microarrays, which contain cDNA or oligonucleotide fragments corresponding to thousands of genes. (4) High throughput sequencing of cDNA libraries represents a recent, powerful approach to comparing transcripts in two samples. For example, Solexa (Illumina, Inc.) sequencing permits the assessment of genome-wide expression profiles by sequencing millions of cDNAs per sample.*

possible to create a project such as RefSeq to identify high-quality representative DNA sequences of genes, any similar attempt to describe a standard expression profile for genes must account for many variables related to the context in which transcription occurs.

In an effort to provide a reference set of RNA transcripts that can serve as a "gold standard," the External RNA Controls Consortium has been established. This project includes the goals of providing access to clones, protocols, and bioinformatics tools (Baker et al., 2005).

You can read about the progress of the External RNA Controls Consortium at ▶ http://www.nist.gov.

mRNA: Subject of Gene Expression Studies

Richard J. Roberts and Phillip A. Sharp received the 1993 Nobel Prize in Physiology or Medicine for their discovery of "split genes." See ▶ http://www.nobel.se/medicine/laureates/1993/.

A molecule in *Drosophila* provides an extraordinary example of alternative splicing. The Down syndrome cell adhesion molecule (DSCAM) gene product potentially exists in more than 38,000 distinct isoforms (Schmucker et al., 2000; Celotto and Graveley, 2001). The gene contains 95 alternative exons that are organized into clusters. Functionally, multiple DSCAM proteins may confer specificity to neuronal connections in *Drosophila*.

We will now consider what is measured in gene expression studies. In most cases, total RNA is isolated from cells of interest. (Sometimes, polyadenylated RNA is isolated.) This RNA is readily purified using chaotropic agents that separate RNA from DNA, protein, lipids, and other cellular components. In this way steady-state RNA transcript levels can be measured. These steady-state levels reflect the activity of a gene. Gene expression is regulated in a set of complex steps that can be divided into four categories: transcription, RNA processing, mRNA export, and RNA surveillance (Maniatis and Reed, 2002) (Fig. 8.11).

1. *Transcription.* Genomic DNA is transcribed into RNA in a set of highly regulated steps. In the 1970s, sequence analysis of genomic DNA revealed that portions of the DNA (called exons) match the contiguous open reading frame of the corresponding mRNA, while other regions of genomic DNA (introns) represent intervening sequences that are not present in mature mRNA.

2. *RNA processing.* Introns are excised from pre-mRNA by the spliceosome, a complex of five stable small nuclear RNAs (snRNAs) and over 70 proteins. Alternative splicing occurs when the spliceosome selectively includes or excludes particular exons (Modrek and Lee, 2002). Pre-mRNA also is capped at the 5′ end. (Eukaryotic mRNAs contain an inverted guanosine

FIGURE 8.11. RNA processing of eukaryotic genes. Genomic DNA contains exons (corresponding to the mature mRNA) and introns (intervening sequences). After DNA is transcribed, pre-mRNA is capped at the 5′ end, and splicing removes the introns. A polyadenylation signal (most commonly AAUAAA) is recognized, the RNA is cleaved by an endonuclease about 10 to 35 nucleotides downstream, and a polyA polymerase adds a polyA tail (typically 100 to 300 residues in length). Polyadenylated mRNA is exported to the cytoplasm where it is translated on ribosomes into protein. An RNA surveillance system involving nonsense-mediated decay degrades aberrant mRNAs; a dashed line indicates that RNA surveillance machinery can also degrade pre-mRNAs.

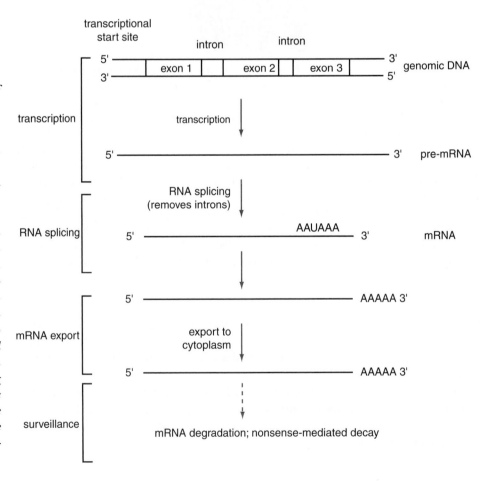

called a cap.) Mature mRNA has the unique property among nucleic acids of having a long string of adenine residues attached to its 3' end. This tract is typically preceded by the polyadenylation signal AAUAAA or AUUAAA, located 10 to 35 nucleotides upstream. Polyadenylation of mRNA is extremely convenient from an experimental point of view, because an oligonucleotide (consisting of a string of thymidine residues attached to a solid support [oligo(dT) resin]) can be used to rapidly isolate mRNA to a high degree of purity. In some cases gene expression studies employ total RNA, while many others employ mRNA.

3. *RNA export.* After splicing occurs, RNA is exported from the nucleus to the cytoplasm where translation occurs. Note that the phrase "gene expression profiling" is commonly used to describe the measurement of steady-state cytoplasmic RNA transcript levels, but may not be precisely correct. "RNA transcript level profiling" is what is performed, and the actual expression of genes is an activity that is not directly measured.

4. *RNA surveillance.* An extensive RNA surveillance process allows eukaryotic cells to scan pre-mRNA and mRNA molecules for nonsense mutations (inappropriate stop codons) or frame-shift mutations (Maquat, 2002). This nonsense-mediated decay mechanism is important in the maintenance of functional mRNA molecules. Additional mechanisms control the half-life of mRNAs, targeting them for degradation and thus regulating their availability.

Let us consider human alpha-2 globin mRNA as an example of a transcript. The function of the globin genes has been characterized in detail. The two alpha globin genes, HBA1 and HBA2, encode proteins sharing 100% identical amino acid sequence. However, the HBA2 mRNA transcript and protein are expressed at levels about threefold higher than the mRNA and protein products of the HBA1 gene (Liebhaber et al., 1986). We can view the HBA2 gene using the UCSC Genome Browser. There are three exons, as shown in Fig. 8.12a. The exons are interrupted by introns; to view this, try performing a blastn search of the RefSeq DNA sequence for HBA2 against the corresponding region of genomic DNA (Fig. 8.12b). Matches to the exons are evident as pairwise alignments, but the introns (absent from the mature mRNA and thus not part of the NM_000517 entry) do not match the genomic reference. By zooming in on the first exon of HBA2, we can see that it is transcribed along the top strand (from left to right beginning at the short arm of chromosome 16) (Fig. 8.12c). The RefSeq track shows the portion of the first exon that is at the 5'-untranslated end (left side), then the coding portion of the exon is displayed with a thickened bar (Fig. 8.12c). Here the third or bottom reading frame begins with a methionine and continues to correspond to the protein sequence of HBA2.

The HBA2 gene locus includes portions corresponding to the coding region as well as 5'- and 3'-untranslated regions (UTRs). These UTRs typically contain regulatory signals such as a ribosome binding site near the start methionine and a polyadenylation signal (often AATAAA) in the 3' UTR. In the case of alpha-2 globin, the 3' UTR contains three cytosine-rich (C-rich) segments that are critical for maintaining the stability of the mRNA (Waggoner and Liebhaber, 2003). Specific RNA-binding proteins interact with the 3' UTR, which adopts a stem-loop structure. Mutations that disrupt this region can lead to destabilization of alpha globin mRNA, causing a form of the disease α-thalassemia (Chapter 20).

Some alignments of RNA-derived sequences and the corresponding genomic DNA, such as those analyzed in Fig. 8.12b, have mismatches. These discrepancies reflect polymorphisms or errors associated with either the sequencing of genomic DNA or cDNA. One way to decide which sequence has an error is to look for consistency: if multiple, independently derived genomic DNA clones or expressed sequence tags have the identical nucleotide sequence in a region of interest, you can be more confident that sequence is correct. See Chapter 13 for a further discussion.

The RefSeq entry for HBA2 and a list of genomic features are presented in web document 8.5 at ▶ http://www.bioinfbook.org/chapter8.

FIGURE 8.12. The HBA2 mRNA in the context of the corresponding genomic DNA. (a) The HBA2 gene region of human chromosome 16 is displayed using the UCSC Genome Browser. The ideogram (chromosomal diagram) shows that the region zoomed in on is at the telomeric region of the p arm of chromosome 16. A window size of 1500 base pairs is displayed. The RefSeq Genes track shows the three exons of HBA2. (b) To compare the mRNA sequence of HBA2 to its corresponding genomic DNA sequence, blastn was performed using the blast 2 sequences program at NCBI (Chapter 3). The sequences were NM_000517 and a genomic contig, RefSeq accession NT_037887.4, nucleotides 162875-163708 from chromosome 16 that spans the HBA2 gene locus. (c) A detailed view of the first exon of HBA2, including the beginning of the protein coding sequence (the start methionine is on the bottom of the three reading frames).

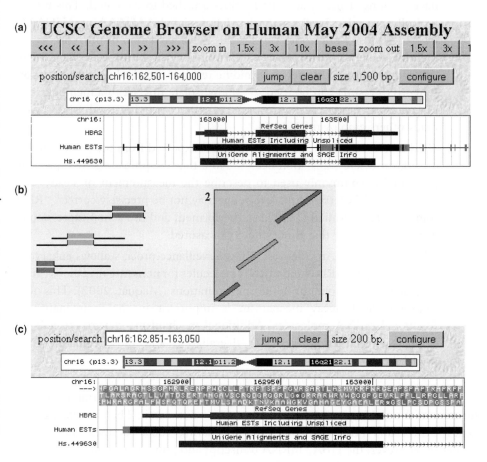

A summary of the number of ESTs in GenBank is available at ► http://www.ncbi.nlm.nih.gov/dbEST/dbEST_summary.html. UniGene is accessed via ► http://www.ncbi.nlm.nih.gov/UniGene/.

Analysis of Gene Expression in cDNA Libraries

How can we study the majority of mRNA molecules that are expressed from a tissue sample or other biological system of interest? It is technically straightforward to isolate RNA and/or mRNA from a small tissue sample in reasonably large quantities (e.g., hundreds of micrograms of total RNA). However, RNA is both unstable and highly complex, typically containing thousands of distinct transcripts. One way to solve this problem is to generate a cDNA library (Fig. 8.13). In brief, RNA is converted to double-stranded cDNA, cloned into a vector, and propagated in a bacterial cell line. A vector such as a plasmid has the properties of small size, rapid growth, and the ability to contain a single cDNA insert derived from the starting tissue sample or other biological source. Thousands of cDNA libraries are available commercially; each is derived from a particular organism, cell type, developmental stage, and physiological condition. The clones in a cDNA library may be plated onto Petri dishes. The cDNA inserts, called expressed sequence tags (ESTs), may then be sequenced.

Millions of ESTs have been sequenced, usually as a *single-pass read* of approximately 500 bp from the 3' end and/or the 5' end of the cDNA clone. Adams et al. (1991, 1993) pioneered the approach of sequencing thousands of ESTs to identify genes expressed in a particular tissue. (This is called a *shotgun single-pass* approach.) These studies revealed which genes are expressed at the highest relative levels (such as

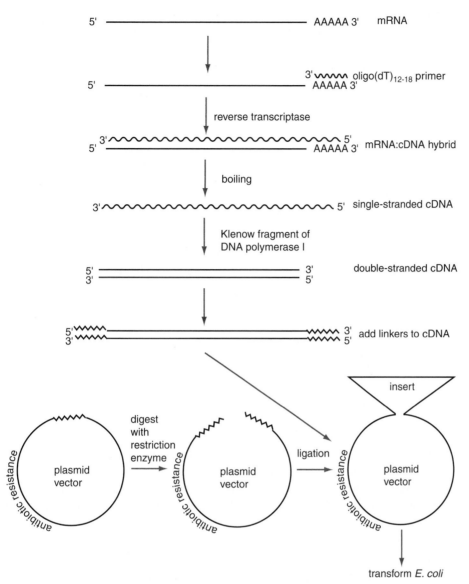

5' ——————————————— AAAAA 3' mRNA

3' ∿∿∿∿ oligo(dT)₁₂₋₁₈ primer

$3'$ ∿∿∿∿ oligo(dT)$_{12-18}$ primer
5' ——————————————— AAAAA 3'

reverse transcriptase

5' 3' ∿∿∿∿∿∿∿∿∿∿∿∿∿∿∿∿∿∿ 5' mRNA:cDNA hybrid
5' ——————————————— AAAAA 3'

boiling

3' ∿∿∿∿∿∿∿∿∿∿∿∿∿∿∿∿∿∿∿ 5' single-stranded cDNA

Klenow fragment of
DNA polymerase I

5' ——————————————— 3' double-stranded cDNA
3' ——————————————— 5'

5' ∿∿∿————————————∿∿∿ 3' add linkers to cDNA
3' ∿∿∿————————————∿∿∿ 5'

insert

plasmid vector — digest with restriction enzyme → plasmid vector — ligation → plasmid vector

antibiotic resistance

transform E. coli

FIGURE 8.13. *Construction of a cDNA library. mRNA is hybridized to an oligo(dT) primer at its 3'-polyadenylated tail, and an mRNA:cDNA hybrid is generated by reverse transcription. After boiling to denature the RNA, cDNA is made double stranded with a DNA polymerase. Linkers (e.g., nucleotides recognized by a restriction endonuclease) are added to cDNA so that after appropriate digestion of both the cDNA (also termed the insert) and a plasmid or bacteriophage (also called the vector), the two can be ligated. Escherichia coli bacteria are then transformed and selected for antibiotic resistance. In this way, a cDNA library is formed, containing up to thousands of unique cDNA inserts derived from the starting mRNA population.*

β-actin and myelin basic protein in human brain), and they also described the regional variation in gene expression across different brain regions.

We previously described UniGene, a system for partitioning ESTs into a nonredundant set of clusters (see Fig. 2.3). In principle, each unique gene ("unigene") is assigned a single UniGene entry. UniGene encompasses both well-characterized genes and those inferred by the existence of ESTs; all ESTs corresponding to a gene are assigned to that particular UniGene accession number. UniGene clusters containing ESTs that are similar to a known gene are categorized as "highly similar" to that gene (defined as >90% identity in the aligned region), "moderately similar" (70% to 90% identity), or "weakly similar" (<70% identity).

The number of human UniGene clusters provides one estimate for the number of human genes, although UniGene clusters are retired when two clusters can be joined into one. Each cluster has some number of sequences associated with it, from one (*singletons*) to almost 50,000 (Table 8-6). Of the 124,104 clusters in Table 8-6, one-third are singletons, suggesting that these may be genes expressed

TABLE 8-6 Histogram of Cluster Sizes for Human Entries in UniGene (Build 204, *Homo sapiens*)

Cluster size	Number of clusters
1	40,649
2	19,301
3–4	18,369
5–8	13,447
9–16	8,059
17–32	5,011
33–64	3,926
65–128	4,485
129–256	3,953
257–512	3,997
513–1024	1,886
1,025–2,048	719
2,049–4,096	223
4,097–8,192	55
8,193–16,384	16
16,385–32,768	7
32769-65536	1

Source: ► http://www.ncbi.nlm.nih.gov/unigene/, September 2007.

so rarely that they have been observed only one time. These singletons may represent portions of the genome that are transcribed without representing functional genes (see the discussion below of pervasive transacription of the genome). Indeed, the presence of ~125,000 UniGene clusters is inconsistent with the estimate of some 20,000 to 25,000 protein-coding genes in the human genome (Chapter 19).

The presence of thousands of UniGene entries with small cluster sizes suggests that some genes are expressed only rarely. Other genes (such as actin and tubulin) are expressed at very high levels. Even some EST clusters that do not correspond to known, annotated genes are highly represented. The largest cluster sizes represented in UniGene are described in Table 8-7 for humans and in Table 8-8 for nonhuman organisms.

DNA hybridization studies from the 1970s suggest that a typical mammalian cell expresses about 300,000 mRNA transcripts, expressed from between 10,000 and 30,000 distinct genes (Hastie and Bishop, 1976). Bishop and colleagues 1974 grouped mRNA into three classes based on relative abundance: (1) genes that are expressed at highly abundant levels (accounting for 10% of the overall transcripts), (2) medium-abundance genes (45% of the mRNA), and (3) low-abundance genes (45% of the mRNA). These three classes correspond to cluster sizes in UniGene (Table 8-6), although the cluster sizes are not formally labeled.

Since all ESTs are derived from a specific region of the body at a particular time of development and a particular physiological state, there is inherently a large amount of information associated with the analysis of many ESTs (Schuler, 1997). There are several approaches to extracting information from UniGene. First, if we want to know where in the body a particular gene (such as RBP or beta globin) is expressed, we can survey UniGene. The number of ESTs associated with that gene reflects the

TABLE 8-7 Ten Largest Cluster Sizes in UniGene for Human Entries

UniGene Identifier	Cluster Size	Gene Symbol	Gene Name
Hs.644639	47,048	EEF1A1	Eukaryotic translation elongation factor 1 alpha 1
Hs.586423	29,028	EEF1A1	Eukaryotic translation elongation factor 1 alpha 1
Hs.520640	25,622	ACTB	Actin, beta
Hs.551713	21,908	MBP	Myelin basic protein
Hs.520348	19,251	UBC	Ubiquitin C
Hs.524390	18,880	TUBA1B	Tubulin, alpha 1b
Hs.514581	17,219	ACTG1	Actin, gamma 1
Hs.544577	17,004	GAPDH	Glyceraldehyde-3-phosphate dehydrogenase
Hs.418167	16,306	ALB	Albumin
Hs.696053	15,512	HSPA8	Heat shock 70 kDa protein 8

Source: UniGene, September 2007.

TABLE 8-8 Ten Largest Cluster Sizes in UniGene for Nonhuman Entries

UniGene Identifier	Species	Cluster Size	Gene Name
Str.4908	*Xenopus tropicalis* (frog)	39,694	Transcribed locus, weakly similar to XP_001103729.1 similar to Transcription factor 19 (Transcription factor SC1) [*Macaca mulatta*]
Ta.447	*Triticum aestivum* (wheat)	32,606	Ribulose-bisphosphate carboxylase small unit
Bfl.12870	*Branchiostoma floridae* (Florida lancelet)	19,456	Transcribed locus, weakly similar to NP_001038525.1 protein LOC564619 [*Danio rerio*]
At.46639	*Arabidopsis thaliana* (thale cress)	15,482	Ribulose bisphosphate carboxylase small chain 1A / RuBisCO small subunit 1A (RBCS-1A) (ATS1A)
Str.64706	*Xenopus tropicalis* (frog)	13,008	Eukaryotic translation elongation factor 1 alpha 1
Dr.31797	*Danio rerio* (zebrafish)	12,653	Elongation factor 1-alpha
Ssa.709	*Salmo salar* (Atlantic salmon)	12,519	Transcribed locus, weakly similar to XP_001111161.1 similar to mitogen-activated protein kinase kinase kinase 6 isoform 3 [*Macaca mulatta*]
Dr.75552	*Danio rerio* (zebrafish)	12,420	Actin, alpha 1, skeletal muscle
Mm.441437	*Mus musculus* (mouse)	12,413	CDNA clone IMAGE:40049146
Bmo.418	*Bombyx mori* (domestic silkworm)	11,838	Clone 1-15 mRNA sequence

Source: UniGene, September 2007.

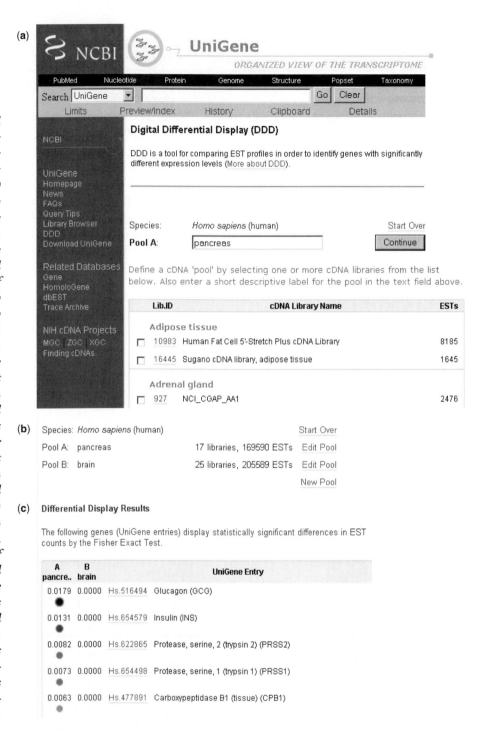

FIGURE 8.14. *Digital differential display (DDD) is used to compare the content of expressed sequence tags (ESTs) in cDNA libraries from UniGene. Over 1000 libraries have been generated by isolating RNA from a tissue source (such as pancreas in [a]), synthesizing cDNA, packaging the cDNA in a cDNA library, and sequencing up to thousands of cDNA clones (ESTs) from each library. (b) The clones in each library (or in pools of libraries) may be compared using DDD. This site is accessed from the NCBI UniGene site; on the left sidebar click* Homo sapiens, *then select "Library digital differential display." At this site, click boxes corresponding to any library (or set of libraries) then press "Accept changes". You will then be given the opportunity to select a second library (or second set of libraries) for comparison. (c) Result of an electronic comparison of cDNA libraries using the DDD tool of UniGene. The results are displayed as a list of genes (with UniGene accession numbers) for transcripts that are preferentially expressed in one or the other pool of libraries. Here, a variety of pancreas-specific transcripts are displayed (e.g., glucagon, insulin). Other transcripts (not shown) are more highly represented in brain-derived libraries.*

abundance of the transcript (but see the discussion below), and the tissue source of the libraries from which ESTs are derived reflects the regional distribution of ESTs. This approach is sometimes called an *electronic Northern blot*.

Another approach to extracting information from UniGene is to electronically subtract cDNA libraries. Electronic cDNA library subtraction in UniGene also allows the cDNA sequences in two populations to be compared using the digital differential display (DDD) tool. You can access this tool from the main UniGene page by selecting DDD and then an organism (e.g., *H. sapiens*). Select two pools,

BOX 8.2
Fisher's Exact Test

Fisher's exact test is used to test the null hypothesis that the number of sequences for any given gene in the two pools (e.g., insulin in pancreas versus brain) is the same in either pool (Table 8-9).

The p value for a Fisher's exact test is given by

$$p = \frac{N_A! N_B! c! C!}{(N_A + N_B)! g1_B! (N_A - g1_A)! (N_B - g1_B)!}$$

The null hypothesis (that gene 1 is not differentially regulated between brain and muscle) is rejected when the probability value p is less than $0.05/G$, where 0.05 is the nominal threshold for declaring significance and G is the number of UniGene clusters analyzed (thus, G is a conservative Bonferroni correction; see Chapter 9).

While the NCBI website employs Fisher's exact test, other statistical approaches to cDNA library comparison have been described. In particular, Stekel et al. (2000) developed a log-likelihood procedure to assess the probability that gene expression differences observed in a comparison of two or even multiple cDNA libraries are due to genuine transcriptional differences and not sampling errors.

such as a series of cDNA libraries from pancreas (Fig. 8.14a) and brain. This search reveals many transcripts selectively associated with pancreas (e.g., glucagon, insulin) (Fig. 8.14b) and others selectively associated with brain. The output shows a dot for each gene whose intensity corresponds to the expression level of that gene. A probability value is associated with each transcript using a Fisher's exact test (Box 8.2).

The comparison of gene expression profiles using databases of libraries may be considered a tool that rapidly provides candidate genes for further analysis. In our pancreas versus brain library comparison, some of the regulated genes identified with digital differential display correspond to "hypothetical proteins" that have not been functionally characterized. These could be studied in further detail. Schmitt and colleagues 1999 used a similar approach to identify 139 transcripts that are selectively upregulated in breast cancer tissue. Such transcripts could provide markers for the early detection of breast cancer or they could reflect changes in tumor tissue that offer targets for therapeutic intervention. Many other studies have employed EST sequencing and/or electronic analyses of sequenced cDNA libraries (e.g., Carulli et al., 1998).

Experimentally, the differential display technique allows two RNA (or corresponding cDNA) sources to be subtracted from each other. One population is labeled selectively (e.g., by ligating an oligonucleotide sequence to the ends of one population of cDNAs), the two populations are hybridized to form duplexes between clones shared in common in the two populations, and cDNA clones that are strongly overrepresented in one of the two original populations are selectively amplified. These clones are then sequenced.

TABLE 8-9 Fisher's 2 × 2 Exact Test Used to Test Null Hypothesis that a Given Gene (Gene 1) Is Not Differentially Regulated in Two Pools

	Gene 1	All Other Genes	Total
Pool A (e.g., brain)	Number of sequences assigned to gene 1 ($g1_A$)	Number of sequences in this pool NOT gene 1 ($N_A - g1_A$)	N_A
Pool B (e.g., pancreas)	Number of sequences assigned to gene 1 ($g1_B$)	Number of sequences in this pool NOT gene 1 ($N_B - g1_B$)	N_B
Total	$c = g1_A + g1_B$	$C = (N_A - g1_A) + (N_B - g1_B)$	

Source: Adapted from Claverie (1999).

Pitfalls in Interpreting Expression Data from cDNA Libraries

The contents of cDNA libraries in UniGene and elsewhere must be analyzed with caution for several reasons:

- Investigators choose which libraries to construct, and there is likely to be bias toward familiar tissues (such as brain and liver) and bias away from more unusual tissues. The rat nose contains over two dozen secretory glands, almost all of which are of unknown function, but for most of these glands cDNA libraries have never been constructed.

- The depth to which a library is sequenced affects its ability to represent the contents of the original cell or tissue. A cDNA library is expected to contain a frequency of clones that faithfully reflects the abundance of transcripts in the source material. By sequencing only 500 clones, it is unlikely that many low-abundance transcripts will be represented when the contents of the entire library are analyzed. In practice, cDNA libraries are sequenced to varying depths.

- Another source of bias is in library construction. Many libraries are normalized, a process in which abundant transcripts become relatively underrepresented while rare transcripts are represented more frequently. The goal in normalizing a library is to minimize the redundant sequencing of highly expressed genes and to thus discover rare transcripts (Bonaldo et al., 1996). It would be inappropriate to compare normalized and nonnormalized libraries directly using a tool such as UniGene's differential display.

- ESTs are often sequenced on one strand only, rather than thoroughly sequencing both top and bottom strands. Thus, there is a substantially higher error rate than is found in finished sequence. (We discuss sequencing error rates in Chapter 13.)

- Chimeric sequences can contaminate cDNA libraries. For example, two unrelated inserts are occasionally cloned into a vector during library construction.

In UniGene, click *Homo sapiens*, then "library browser," to see the range of clones that are sequenced in typical libraries. Currently (September 2007) there are almost 2000 human cDNA libraries in UniGene having at least 1000 sequences, and many thousands of smaller libraries.

Full-Length cDNA Projects

While UniGene is an example of a database that incorporates information on ESTs and protein-coding genes, it is also of interest to catalog, characterize, and make available collections of cDNAs. There are two main forms of cDNAs: those having full-length protein coding sequences (typically including some portions of the 5'- and 3'-untranslated regions), and expression clones in which the protein coding portion of the cDNA is cloned into a vector that permits protein expression in the appropriate cell type (Temple et al., 2006). There are many important resources for obtaining cloned, high quality, full-length cDNAs. We will next introduce five of the many available cDNA resources.

1. The Functional Annotation of the Mouse (FANTOM) project provides functional annotation of the mammalian transcriptome (Maeda et al., 2006). Currently, over 100,000 full-length mouse cDNAs have been annotated, including both coding and non-protein-coding transcripts. These have been mapped to genomic loci using BLAT, blastn, and other search tools. The annotation categories included artifacts (such as contaminants from other

species or chimeric mRNAs) and coding sequences (complete, 5′- or 3′-truncated, 5′- or 3′-untranslated regions only, immature, with or without insertion/deletion errors, stop codons, coding for selenoproteins, or mitochondrial transcripts). Upon analyzing transcription start and stop sites, the 5′ and 3′ boundaries of over 180,000 transcripts were identified (Carninci et al., 2005). This study led to the identification of over 5000 previously unidentified mouse proteins. Another astonishing conclusion of the FANTOM project is that antisense transcription, in which clustered cDNA sequences on one strand at least partially match to the opposite strand, occurs for 72% of all genome-mapped transcriptional units (Katayama et al., 2005).

You can access the FANTOM project at ▶ http://fantom.gsc.riken.go.jp/.

2. The H-Invitational Database provides an integrative annotation of human genes, including gene structures, alternative splicing isoforms, coding as well as noncoding RNAs, single nucleotide polymorphisms (Chapter 16), and comparative results with the mouse (Imanishi et al., 2004): 21,037 human gene candidates were analyzed corresponding to 41,118 full-length cDNAs. Information from this database is available as an optional annotation track in the UCSC Genome Browser.

The H-invitational database is available at ▶ http://www.h-invitational.jp/. Hosted by the Japan Biological Information Research Center (JBIRC), this site features a highly informative genome browser.

3. The Gene Index Project (formerly the Institute for Genomic Research [TIGR] Gene Indices) is a collection of ESTs organized into several dozen species-specific databases (Lee et al., 2005). The approach taken by this project is to focus on the analysis of EST sequences to assemble unique genes called tentative consensus sequences. This emphasis on clustering and assembly results in a collection of consensus sequences corresponding to genes. TIGR Gene Indices are then useful for a variety of purposes not as readily available with UniGene. (While UniGene does not assemble ESTs into a single cluster, NCBI does provide a list of the longest EST sequence from each cluster.) The Eukaryotic Gene Orthologs (EGO) database consists of orthologous genes identified by pairwise alignments of tentative consensus sequences, allowing the comparison of homologous genes across dozens of organisms.

The Gene Index Project at the Dana Farber Cancer Institute is available at ▶ http://compbio.dfci.harvard.edu/tgi/ along with EGO and related bioinformatics tools. It was developed by John Quackenbush and colleagues. Currently (September 2007) the Gene Index Project includes indices from animals (34 species), plants (34 species), protists (15 species), and fungi (9 species).

4. The Mammalian Gene Collection (MGC) is an NIH project that originally aimed to gather full-length cDNA clones for all human and mouse genes, but subsequently has expanded to include rat, cow, frog, and zebrafish (Gerhard et al., 2004; Baross et al., 2004). Its site can be searched by BLAST. MGC clones are distributed through the Integrated Molecular Analysis of Genomes and Their Expression (IMAGE) consortium.

The Mammalian Gene Collection (MGC) website is ▶ http://mgc.nci.nih.gov/. It currently includes over 26,000 human clones (corresponding to about 16,000 nonredundant genes). The IMAGE consortium website (▶ http://image.llnl.gov/) can be queried for clones from a number of species.

5. Another important cDNA resource is the Kazusa mammalian cDNA set, called "KIAA" genes (Nagase et al., 2006). This project focuses on characterizing full-length cDNAs that encode particularly large genes. Clones are described and distributed through the HUGE database (Kikuno et al., 2004).

The HUGE database is at ▶ http://www.kazusa.or.jp/huge/.

Serial Analysis of Gene Expression (SAGE)

Serial analysis of gene expression allows the quantitative measurement of gene expression by measuring large numbers of RNA transcripts from tissues of interest. Short tags of 9 to 14 bp of DNA are isolated from the 3′ end of transcripts, sequenced, and assigned to genes. Major benefits of SAGE experiments are that (unlike microarrays) there is no need for prior knowledge of which mRNA transcripts to study, and (unlike Northern blots or PCR) it is a high throughput technology. Also, many useful variants of SAGE have been developed (see Wang, 2007). Perhaps the major limitation is that construction and analysis of SAGE libraries remains a relatively specialized skill. Also in a typical SAGE experiment as many as half the SAGE tags do not match to known transcripts or genes. For low abundance transcripts the reproducibility is poor. Wang 2007 provides a discussion of these issues, including a comparison of SAGE and microarray technologies.

The SAGE site at NCBI is at ► http://www.ncbi.nlm.nih.gov/ SAGE/.

The procedure for producing SAGE tags is outlined in Fig. 8.15 (Velculescu et al., 1995). RNA is isolated from a source of interest and converted to cDNA with a biotinylated oligo(dT) primer. A restriction enzyme (the "anchoring enzyme") is used to

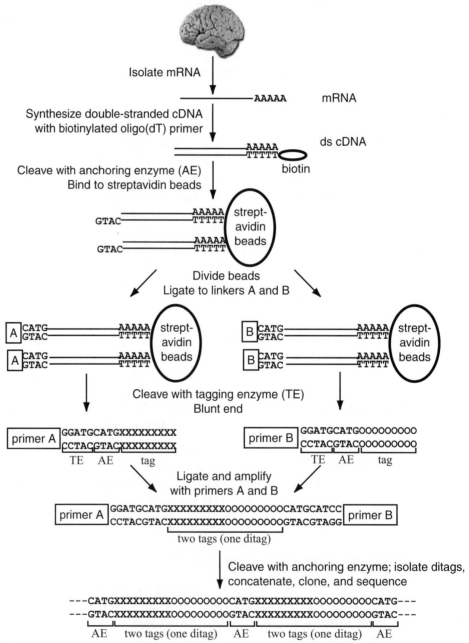

FIGURE 8.15. *Description of serial analysis of gene expression (SAGE). Messenger RNA is isolated from a source (such as brain), and double-stranded cDNA is synthesized using a biotinylated oligo(dT) primer. The cDNA is cleaved with a four-cutter restriction endonuclease ("anchoring enzyme," AE) that cleaves most transcripts in a cell. The 3' portion of each transcript is immobilized on streptavidin beads (large ovals), and linkers (A or B) are added containing restriction endonuclease recognition site ("tagging enzyme," TE). Cleavage of the ligated clones with the tagging enzyme releases the linker with the cDNA tags (Xs at left, Os at right). The ligated tag pairs are concatenated, cloned, and sequenced. Each tag represents a fragment of 9 bp of a transcript. Modified from Velculescu et al. (1995). Used with permission.*

(a)

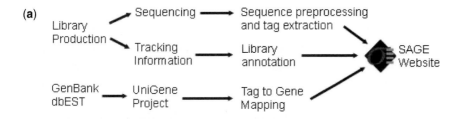

(b) █ **SAGE Data Analysis**

For this query, there are **67532** unique SAGE tags of which the 100 most likely different by greater than 2 fold are shown. For each of these tags, the probability that there is greater than a **2 fold difference in expression levels** between Groups A and B is given.

To download the entire list, see the bottom of this page.

Group A: brain (total tags: 189431)

> SAGE_Duke_H392 : Brain, Duke glioblastoma multiforme cell line derived from a 65 yo female. Cells harvested at passage 32 *(total tags: 57529)*
> SAGE_Duke_GBM_H1110 : Brain, Duke glioblastoma multiforme primary tumor derived from a 51 yo male *(total tags: 70061)*
> SAGE_pooled_GBM : Brain, 5 pooled Duke glioblastoma multiforme primary tumors *(total tags: 61841)*

Group B: lung (total tags: 88901)

> SAGE_normal_lung : Lung, normal *(total tags: 88901)*

(c)

Color = **RED** if expression of tag in **Group A > Group B**		Color = **GREEN** if expression of tag in **Group B > Group A**

#	SAGE tag	UniGene id	Gene description	A:B	Grp A (CoV)	Grp B (CoV)	A:B > 2x
1	CGCAGCGGGT	Hs.322854	pronapsin A	ᴀB	0 (0%)	269 (0%)	101%
2	CTTTGAGTCC	Hs.2240	uteroglobin	ᴀB	0 (0%)	226 (0%)	100%
3	GGCAAGAAAA	Hs.111611	ribosomal protein L27	ᴀB	0 (0%)	25 (0%)	100%
4	TGCCAGGTCT	Hs.177582	surfactant, pulmonary-associated protein A2	ᴀB	0 (0%)	50 (0%)	100%
5	GAGTTAAAAA	Hs.180255	major histocompatibility complex, class II, DR beta 1	ᴀB	28 (116%)	152 (0%)	100%
		Hs.318720	major histocompatibility complex, class II, DR beta 4				
6	ATCAAGAATC	Hs.14623	interferon, gamma-inducible protein 30	ᴀB	23 (105%)	129 (0%)	100%
7	AAGATAGCTC	Hs.1074	surfactant, pulmonary-associated protein C	ᴀB	0 (0%)	74 (0%)	100%
8	TTGGTCCTCT	Hs.108124	ribosomal protein S4, X-linked	ᴀB	384 (85%)	517 (0%)	100%

FIGURE 8.16. (a) Build process for the NCBI SAGE database includes a data generation portion (top; library production) and a mapping portion (bottom). (b) The NCBI SAGE website library comparison tool (obtained by clicking "Analyze by library" at ▶ http://www.ncbi.nlm.nih.gov/SAGE/) allows two pools of SAGE tags to be evaluated in order to compare gene expression profiles. In this case, boxes were clicked under headings A and B that correspond to human brain and lung SAGE libraries. (c) Result of an electronic comparison of SAGE libraries from brain and lung shows SAGE tags corresponding to transcripts that are present in different abundance in the pools. In this example, group B (lung) includes tags corresponding to surfactant and genes known to be expressed preferentially in lung. Additional tags correspond to genes expressed in brain such as a neurotrophic metallothionein.

digest the total population of transcripts so that only short fragments are isolated, and the tight interaction between biotin and avidin allows the 3′ end of each transcript to be tethered to streptavidin beads. Two populations of linkers are added, allowing the cDNA to be digested with a specialized restriction enzyme that releases the linker with a short fragment of cDNA (the "tag"). Tags are concatenated, cloned, and sequenced. This process results in the description of thousands (or millions) of tags from a biological source.

A variety of SAGE libraries have been constructed. Each tag in a library is likely to correspond to a single gene. For a 9 bp tag, there are 4^9, or 262,144, transcripts that can be distinguished, assuming a random nucleotide distribution at the tag site. In practice, tags are mapped to genes using UniGene. In some cases, a tag may be present on more than one gene. In other cases, a gene may have more than one tag (e.g., there may be alternative splicing of a transcript such that there are multiple tags for that gene). An assumption of SAGE is that the number of tags found in a SAGE library is directly proportional to the number of mRNA molecules in that biological sample.

SAGE has been used to describe the properties of the yeast transcriptome (Velculescu et al., 1997). The expression of 4665 genes was characterized, the majority of which had not been functionally characterized. Consistent with the analysis of UniGene clusters (Table 8-6), these data showed that many transcripts are expressed only rarely. Zhang et al. 1997 used SAGE to profile gene expression in normal and neoplastic gastrointestinal tissue. They estimated that the number of distinct transcripts that were expressed in each cell type ranged from about 14,000 to 20,000, and the expression levels ranged from one copy per cell to 5300 copies per cell. The abundance of each gene was estimated by dividing the observed number of tags for a transcript by the total number of tags obtained.

SAGE libraries can be queried electronically at the NCBI website, allowing the comparison of gene expression in any tissues for which SAGE libraries have been generated (Lash et al., 2000; Lal et al., 1999). The website includes tag data from SAGE libraries and annotation data in which tags are mapped to genes (Fig. 8.16). SAGE libraries can be selected in a manner similar to using digital differential display. The genes that correspond to tags differentially present in lung include surfactant, pronapsin A, and secretoglobin, with hundreds of tags in lung but none in brain. Assorted brain-enriched transcripts are also identified. Examination of surfactant (by clicking its link) shows that the mapping of this particular tag (TGCCAGGTCT) to the surfactant gene (UniGene Hs.177852) appears unambiguous, and 50 tags corresponding to surfactant have been identified selectively in a lung library (Fig. 8.16). In some cases tags map to multiple UniGene clusters, and only one or several clusters may be reported as "reliable" as determined by NCBI using criteria such as the availability of corresponding genomic DNA data.

MICROARRAYS: GENOMEWIDE MEASUREMENT OF GENE EXPRESSION

DNA microarrays have emerged as a powerful technique to measure mRNA transcripts (gene expression). While EST sequencing projects and SAGE allow high throughput analyses of gene expression, it is microarrays that have been used most broadly to assess differences in mRNA abundance in different biological samples. The use of microarrays has increased rapidly since the pioneering work of Patrick Brown and colleagues at Stanford University, Jeffrey Trent and colleagues at the National Institutes of Health, and others (DeRisi et al., 1996).

A microarray is a solid support (such as a glass microscope slide or a nylon membrane) on which DNA of known sequence is deposited in a regular gridlike array. The DNA may take the form of cDNA or oligonucleotides, although other materials (such as genomic DNA clones; Chapter 16) may be deposited as well. Typically,

TABLE 8-10 Major Advantages of Microarray Experiments

Advantage	Comment
Fast	One can obtain data on the RNA levels of over 20,000 transcripts within one week.
Comprehensive	Entire transcriptomes can be represented on a microarray.
Flexible	cDNAs or oligonucleotides corresponding to any gene can be represented on a chip. Dozens of applications have been developed, such as microarrays to measure microRNAs or methylated DNA.

several nanograms of DNA are immobilized on the surface of an array. RNA is extracted from biological sources of interest, such as cell lines with or without drug treatment, tissues from wild-type or mutant organisms, or samples studied across a time course. The RNA (or mRNA) is often converted to cDNA, labeled with fluorescence or radioactivity, and hybridized to the array. During this hybridization, cDNAs derived from RNA molecules in the biological starting material can hybridize selectively to their corresponding nucleic acids on the microarray surface. Following washing of the microarray, image analysis and data analysis are performed to quantitate the signals that are detected. Through this process, microarray technology allows the simultaneous measurement of the expression levels of thousands of genes represented on the array.

The term *functional genomics* refers to the large-scale analysis of the genomewide function of genes, in contrast to the study of individual protein and nucleic acid molecules (see Chapter 12). Microarray-based gene expression experiments form one core of functional genomics. There has been enthusiasm for microarrays in the research community because of their potential to yield large amounts of information (Table 8-10). Notably, the rapid accumulation of molecular sequence data in GenBank has led to the availability of many thousands of clones of unknown function. DNA sequences corresponding to both known genes and poorly characterized ESTs have been deposited on microarrays, potentially allowing their function to be determined. The potential of this technology is great, and fundamental biological insights have already been obtained (see Chapter 9).

It is also important to realize the limitations of microarray technology (Table 8-11). The costs of microarrays have caused many investigators to perform relatively

TABLE 8-11 Major Disadvantages of Microarray Experiments

Disadvantage	Comment
Cost	Many researchers find it prohibitively expensive to perform sufficient replicates and other controls, and thus experiments lack statistical power.
Unknown significance of RNA	The final product of gene expression is protein, not RNA.
Uncertain quality control	There are many artifacts associated with image analysis and data analysis.
Common occurrence of artifacts	Because RNA levels are context-dependent, observed differences may be due to nuisance factors (such as differences attributable to the date of RNA isolation, or differences in RNA due to the handling of RNA by different laboratory scientists) rather than attributable to the underlying biological comparison (e.g., normal versus diseased sample).

FIGURE 8.17. Overview of the process of generating high throughput gene expression data using microarrays. In stage 1, biological samples are selected for a comparison of gene expression. In stage 2, RNA is isolated and labeled, often with fluorescent dyes. These samples are hybridized to microarrays, which are solid supports containing complementary DNA or oligonucleotides corresponding to known genes or ESTs. In stage 4, image analysis is performed to evaluate signal intensities. In stage 5, the expression data are analyzed to identify differentially regulated genes (e.g., using ANOVA [Chapter 9] and scatter plots; stage 5, at left) or clustering of genes and/or samples (right). Based on these findings, independent confirmation of microarray-based findings is performed (stage 6). The microarray data are deposited in a database so that large-scale analyses can be performed.

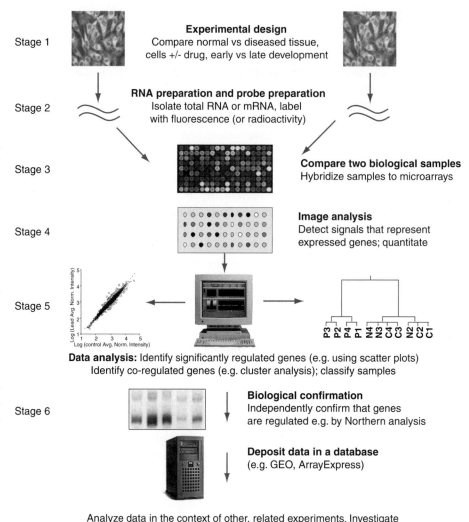

Stage 1
Experimental design
Compare normal vs diseased tissue, cells +/- drug, early vs late development

Stage 2
RNA preparation and probe preparation
Isolate total RNA or mRNA, label with fluorescence (or radioactivity)

Stage 3
Compare two biological samples
Hybridize samples to microarrays

Stage 4
Image analysis
Detect signals that represent expressed genes; quantitate

Stage 5
Data analysis: Identify significantly regulated genes (e.g. using scatter plots)
Identify co-regulated genes (e.g. cluster analysis); classify samples

Stage 6
Biological confirmation
Independently confirm that genes are regulated e.g. by Northern analysis

Deposit data in a database
(e.g. GEO, ArrayExpress)

Analyze data in the context of other, related experiments. Investigate behavior of expressed genes in other experimental paradigms

few control experiments to assess the reliability and validity of their findings. There are many potential quality control issues, such as artifacts associated with microarray manufacture or image acquisition. Perhaps the greatest limitation is that experimental design may be flawed; for example, if untreated samples are processed on day one and treated samples are processed on day two then the design is confounded and one cannot distinguished between observed differences in transcript levels that are due to treatment or date. Many other factors need to be controlled for, such as the operator (i.e., the individuals who isolate RNA and otherwise process the samples) and the batch. A further possible limitation of microarrays is that the ultimate product of gene expression is not mRNA but protein, and those levels are not necessarily correlated (discussed below).

An overview of the procedures used in a microarray experiment is shown in Fig. 8.17, arbitrarily divided into six stages. We will consider each of the stages below.

Stage 1: Experimental Design for Microarrays

In the first stage, total RNA or mRNA is isolated from samples. Notably, experiments have been performed for organisms as diverse as viruses, bacteria, fungi, and humans.

The amount of starting material that is required is typically several hundreds of milligrams (wet weight) or several flasks of cells. For many currently available microarrays, about 1 to 5 μg of total RNA is required. With the amplification of RNA or cDNA products, it is possible to use substantially less starting material. However, the amplified population may not faithfully represent the original RNA population.

The experimental design of a microarray experiment can be considered in three parts (Churchill, 2002). Different sources of variation are associated with each of these three areas. We will discuss experimental design further in Chapter 9.

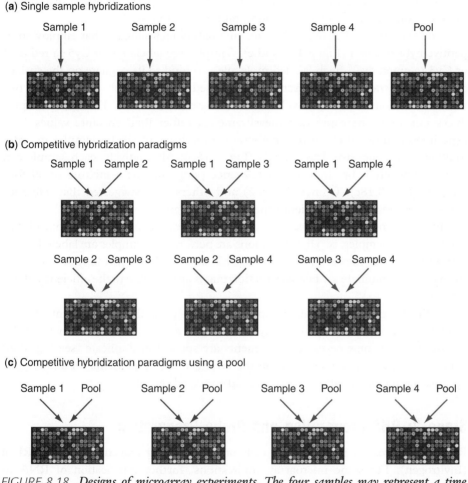

(a) Single sample hybridizations

Sample 1   Sample 2   Sample 3   Sample 4   Pool

(b) Competitive hybridization paradigms

Sample 1 Sample 2   Sample 1 Sample 3   Sample 1 Sample 4

Sample 2 Sample 3   Sample 2 Sample 4   Sample 3 Sample 4

(c) Competitive hybridization paradigms using a pool

Sample 1   Pool   Sample 2   Pool   Sample 3   Pool   Sample 4   Pool

FIGURE 8.18. *Designs of microarray experiments. The four samples may represent a time course, normal versus diseased tissue, or any other paradigm. (a) Design of arrays to which only one sample is hybridized (e.g., nylon filters from Clontech that are probed with radiolabeled cDNA or chips from Affymetrix that are probed with cRNA probes that are visualized with a fluorescent dye). Each sample is hybridized to one chip. The use of a pool of all the samples is not necessary. Additional hybridizations may be performed to increase the number of replicates. (b) For technologies that use competitive hybridization, such as NEN Life Sciences arrays, samples are labeled with Cy5 (red) and Cy3 (green) dyes and competitively hybridized. A set of hybridizations may be performed to compare every combination of samples. The data do not allow intensities from one chip (e.g., sample 1 vs. 2) to be compared across chips. (c) In order to perform data analysis of every sample compared to every other sample, it is also possible to hybridize each sample to a pool consisting of a reference.*

For microarrays from Affymetrix, RNA is converted to cDNA and transcribed to make biotin-labeled complementary RNA (cRNA).

First, the biological samples are selected for comparison, such as a cell line with or without drug treatment. If multiple biological samples are used, these are called "biological replicates." When experimental subjects are selected for treatment, it is appropriate to assign them to groups randomly.

Second, RNA is extracted and labeled (typically as complementary DNA) with radioactivity or fluorescence. When two RNA extractions are obtained from a biological sample and hybridized to microarrays, these are called "technical replicates." For technologies in which one sample is hybridized to one microarray, the transcripts are labeled with radioactivity or with a fluorescent dye. Figure 8.18a shows an example of an experimental design in which four samples are hybridized to arrays. These samples may represent time points in an assay, different pharmacological treatments, or any other paradigm. Optionally, a reference pool consisting of multiple samples from the same treatment condition may be employed.

The experimental design differs for microarray technologies that employ competitive hybridization (Fig. 8.18b and c). Samples are labeled with Cy5 (a red dye) or Cy3 (a green dye). Each labeled molecule can bind to its cognate on the surface of the microarray. If the transcript is expressed at comparable levels in the two samples, the color of the spot will be intermediate (i.e., yellow). This approach produces ratios of gene expression measurements, rather than absolute values. The experimental design may involve pairwise comparisons of all the samples in order to allow comprehensive data analysis (Fig. 8.18b). Alternatively, each sample can be competitively hybridized with a reference pool, such as a mixture of all four samples (Fig. 8.18c). Churchill (2002) and others have suggested that reference pools can represent an inefficient experimental design.

For any two-color experimental paradigm, dye swap is an important control. For two biological samples, two hybridizations are performed. Samples are labeled twice, first with Cy5 and then with Cy3, and used in independent hybridizations. The dye swap helps to eliminate artifactual variation that is attributable to the efficiency of dye labeling.

A third aspect of microarray experimental design is the arrangement of array elements on a slide. Ideally, the array elements are arranged in a randomized order on the slide. In some cases, array elements are spotted in duplicate (see Fig. 8.20 below). Artifacts can occur based on the arrangement of elements on an array, or because a microarray surface is not washed (or dried) evenly.

Stage 2: RNA Preparation and Probe Preparation

RNA can be readily purified from cells or tissues using reagents such as TRIzol (Invitrogen). For some microarray applications, further purification of RNA to mRNA [poly(A)$^+$ RNA] is necessary. In comparing two samples (e.g., cells with or without a drug), it is essential to purify RNA under closely similar conditions. For example, for cells in culture, conditions such as days in culture and percent confluence must be controlled for.

The purity and quality of RNA should also be assessed spectrophotometrically (by measuring $a260/a280$ ratio) and by gel electrophoresis. Fluorescent dyes such as RiboGreen (Molecular Probes) can be used to quantitate yields. Purity of RNA may also be confirmed by Northern analysis or PCR. RNA preparations that are contaminated with genomic DNA, rRNA, mitochondrial DNA, carbohydrates, or other macromolecules may be responsible for impure probes that give high backgrounds or other experimental artifacts.

FIGURE 8.19.   The surface of a typical microarray chip contains oligonucleotides at a density of 0.1 pmol/mm² on a glass slide or one molecule per 39 Å² (from Southern et al., 1999). A typical microarray from Affymetrix contains 20 separate oligonucleotide 25-mers, each corresponding to a single gene. The extent to which an endogenous transcript has been expressed in a sample is assessed by analyzing the fluorescence signal from all 40 oligonucleotides corresponding to that expressed gene. Other microarray platforms, such as arrays from Agilent or Clontech Laboratories, employ oligonucleotides up to 80 nucleotides long. Used with permission.

The RNA is converted to cDNA or to complementary RNA, then labeled with fluorescence (or less commonly with radioactivity) to permit detection.

Stage 3: Hybridization of Labeled Samples to DNA Microarrays

A microarray is a solid support such as a nylon membrane or glass slide to which DNA fragments of known sequence are immobilized. In some cases, the immobilized DNA consists of approximately 5 ng of cDNA (length 100 to 2000 bp) arrayed in rows and columns. In other cases, oligonucleotides rather than cDNAs are immobilized (Lipshutz et al., 1999). This has been accomplished by Affymetrix using a modified process of photolithography (Fodor et al., 1991). An example of this in Fig. 8.19 shows the density of oligonucleotides on the surface of a chip. Depending on the nature of the solid support used to immobilize DNA, the microarray is often called a blot, membrane, chip, or slide. The DNA on a microarray is referred to as "target DNA." In a typical microarray experiment, the gene expression patterns from two samples are compared. RNA from each sample is labeled with fluorescence or radioactivity to generate a "probe."

After RNA is converted into cDNA or cRNA labeled with fluorescence, the efficient labeling of probe must be confirmed. This is followed by hybridization of the probe overnight to the filter or slide and washing of the microarray. The next stage is image analysis.

Stage 4: Image Analysis

After washing, image analysis is performed to obtain a quantitative description of the extent to which each mRNA in the sample is expressed (Duggan et al., 1999). For experiments using radioactive probes (typically using [33P] or [32P] isotopes),

Photolithography is a technique with many applications, including the microelectronics industry, in which substances are deposited on a solid support. For microarray technology, oligonucleotides are synthesized in situ on a silicon surface by combining standard oligonucleotide synthesis protocols with photolabile nucleotides that permit thousands of specific oligonucleotides to be immobilized to a chip surface. Many researchers refer to the DNA on a microarray as the probe and the labeled DNA derived from a biological sample as the target. Thus, there are opposite definitions of probe and target, and the research community has not reached a consensus. We will call the labeled material derived from RNA or mRNA the "probe."

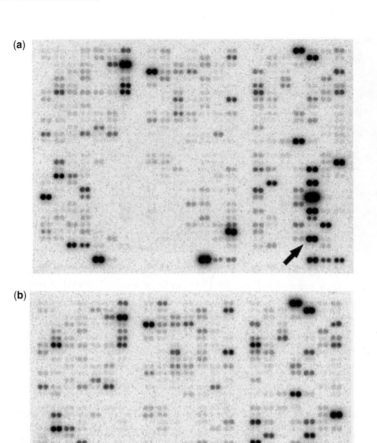

FIGURE 8.20. Example of a microarray experiment using radioactive probes: 588 genes are represented on each array and are spotted in adjacent pairs. Dark dots represent genes expressed at high levels. The filters were hybridized, washed, and exposed to a phosphorimager screen for 6 hours. The output includes a quantitation (in pixel units) of the signals. (a) Clontech Atlas Neurobiology array probed with cDNA derived from the postmortem brain of a girl with Rett syndrome, and (b) the profile from a matched control. The arrows point at an RNA transcript (β-crystallin) that is upregulated in the disease. Note that overall the RNA transcript profiles appear similar in the two brains.

image analysis is performed by quantitative phosphorimaging (Fig. 8.20). Image analysis involves aligning the pixels to a grid and manually adjusting the grid to align the spots. Each spot represents the expression level of an individual transcript. The intensity of a spot is presumed to correlate with the amount of mRNA in the sample. However, many artifacts are possible. The spot may not have a uniform shape. An intense signal may "bleed" to a neighboring spot, artifactually lending it added signal intensity (Fig. 8.21a). Pixel intensities near background may lead to spuriously high ratios. For example, if a control value is 100 units above background and an experimental value is 200 units, this suggests that the experimental condition is upregulated twofold. However, if the pixel values are 50,100 versus 50,200, then no regulation is described.

For fluorescence-based microarrays, the array is excited by a laser and fluorescence intensities are measured (Fig. 8.21b). Data for Cy5 and Cy3 channels may be sequentially obtained and used to obtain gene expression ratios.

Stage 5: Data Analysis

Analysis of microarray data is performed to identify individual genes that have been differentially regulated. It is also used to identify broad patterns of gene expression. In some experiments groups of genes are coregulated, suggesting functional relatedness.

(a)

(b)

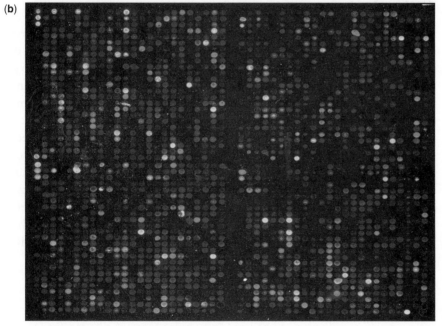

FIGURE 8.21. Microarray images. (a) Six signals are visualized using NIH Image software. Image analysis software must define the properties of each signal, including the likelihood that an intense signal (lower left) will "bleed" onto a weak signal (lower right). (b) A microarray from NEN Perkin-Elmer (representing 2400 genes) was probed with the same Rett syndrome and control brain samples used in Fig. 8.20. This technology employs cDNA samples that are fluorescently labeled in a competitive hybridization.

Samples (rather than genes) may be analyzed and classified into discrete groups. The analysis of microarray data is described in Chapter 9.

In an effort to standardize microarray data analysis, Brazma and colleagues (2001) at 17 different institutions proposed a system for storing and sharing microarray data. Minimum Information About a Microarray Experiment (MIAME) provides a framework for researchers to describe information in six areas: the experimental design, the microarray design, the samples (and how they are prepared), the hybridization procedures, the image analysis, and the controls used for normalization.

The MIAME project is described at the Microarray Gene Expression Database Group website (▶ http://www.mged.org/).

Stage 6: Biological Confirmation

Microarray experiments result in the quantitative measurement of thousands of mRNA transcript values. Data analysis typically reveals that dozens or hundreds of genes are significantly regulated, depending on the particular experimental paradigm and the statistical analysis approach. A list of regulated transcripts may include true positives (those that are authentically regulated) as well as false positives (transcripts reported as significantly regulated even though they were found by chance). It is important to independently confirm the differential regulation of at least some of the most regulated transcripts.

Microarray Databases

ArrayExpress is available at
► http://www.ebi.ac.uk/arrayexpress/, while GEO is at
► http://www.ncbi.nlm.nih.gov/geo/.

Raw as well as processed microarray data are routinely deposited in public repositories upon publication. The two main public repositories are ArrayExpress at the European Bioinformatics Institute and the Gene Expression Omnibus (GEO) at NCBI. Another example of a repository is the Stanford Microarray Database, which offers links to the complete raw and processed data sets from a variety of microarray experiments. We will describe how to acquire data from GEO in Chapter 9.

Further Analyses

Eventually, it is likely that uniform standards will be adopted for all microarray experiments. The greatest variables in these studies are likely to be the quality of the RNA isolated by each investigator and the nature of the microarray that is used to generate data. An ongoing trend in the field of bioinformatics is the unification and cross-referencing of many databases, such as has occurred for databases of molecular sequences and for databases of protein domains. In the arena of gene expression, the lack of acceptable standards may limit the extent to which an integrated view of gene expression is obtained. Nonetheless, it is likely that each gene in each organism will be indexed so that in addition to "stable" data on molecular sequence and chromosomal location, "dynamic" information on the mRNA corresponding to each gene will be cataloged. This information will include the abundance level of each transcript, the temporal and regional locations of gene expression, and other information on the behavior of gene expression in a variety of states. Some initial efforts to integrate information on gene expression are presented in Chapter 9.

INTERPRETATION OF RNA ANALYSES

We began this chapter with a description of noncoding RNA, then described coding (messenger) RNA. We conclude with several issues about the nature and interpretation of RNA.

The Relationship of DNA, mRNA, and Protein Levels

Many human diseases are associated with changes in the number of chromosomes (termed aneuploidy); the most well known of these is Down syndrome, associated with a third copy of chromosome 21. Many diseases are caused by the duplication or deletion of a small chromosome region (e.g., several million base pairs), and copy number changes are also commonly associated with cancers. A variety of evidence suggests that an increase in copy number (i.e., of genomic DNA) is associated with a corresponding increase in mRNA transcript levels. My laboratory (Mao et al., 2003, 2005) and others have shown this for Down syndrome brain and heart, and similar findings have been reported in cancers.

Once mRNA levels are present at elevated or reduced levels, are the corresponding proteins differentially expressed in a similar manner? Perhaps surprisingly, there appears to be only a weak positive correlation between mRNA and protein levels. At present, high throughput protein analyses are technically more difficult to perform (especially protein arrays) than transcriptional profiling studies. We will discuss several high throughput approaches to protein identification and quantitation (e.g. mass spectrometry) in Chapters 10 and 12.

Several groups have reported a weak positive correlation between mRNA levels and levels of the corresponding proteins in the yeast *Saccharomyces cerevisiae* and other systems (Futcher et al., 1999; Greenbaum et al., 2002). Greenbaum et al. 2002 performed a meta-analysis of gene expression and protein abundance data sets and suggested that there is a broad agreement between mRNA and protein levels. Waters et al. 2006 reviewed eight studies and described correlation coefficients that were relatively high when highly abundant proteins were considered (e.g., $r = 0.935$, $r = 0.86$ in two studies) but lower when highly abundant proteins were excluded (e.g., $r = 0.356$, $r = 0.49$, $r = 0.21$, $r = 0.18$).

The correlation coefficient r ranges from $+1$ (perfectly positively correlated) to -1 (negatively correlated), with $r = 0$ indicating that two variables are uncorrelated.

One conclusion from these studies is that it might be appropriate to determine experimentally whether observed changes in RNA correspond to changes in the levels of the corresponding proteins. At present, it is common in the scientific literature for observed changes in RNA transcripts, derived from genes encoding a category of proteins such as those involved in glycolysis, to be cited as evidence that glycolysis has changed in the system being studied. Such a finding represents a hypothesis that can be tested experimentally.

The Pervasive Nature of Transcription

In recent decades, the transcription of DNA to mRNA has been conceptualized in terms of a relatively straightforward model in which protein-coding genes are transcribed into mRNA precursors which are then spliced (to remove introns) and processed (to facilitate export) into mature mRNA. The number of distinct mRNA transcripts was assumed to approximate the number of protein-coding genes, and the exons have been estimated to occupy about 1% of the human genome. More recently, compelling evidence has emerged that this view is overly simple. Instead, it is becoming apparent that the majority of the genomic DNA (comprising the genome) is transcribed.

Strong evidence for pervasive transcription comes from the ENCODE project, in which 30 megabase pairs (that is 30 million base pairs, spanning 1% of the human genome in 44 regions) has been analyzed in depth with over 200 experimental data sets (ENCODE Consortium, 2004, 2007). Transcriptional activity was measured using three technologies. (1) Total RNA or poly(A) RNA was hybridized to tiling arrays. Tiling arrays contain oligonucleotides or PCR products that correspond to positions along each chromosome that are regularly spaced at extremely short intervals such as five base pairs or 30 base pairs. This contrasts with conventional expression arrays that are targeted to previously annotated exons. Genomic tiling arrays do not depend on prior genome annotations of gene structures, and they also offer good sensitivity. (2) Cap-selected RNA was tag sequenced at the 5' or

You can learn more about CAGE at the FANTOM website (▶ http://fantom3.gsc.riken.jp/), including access to CAGE databases.

joint 5′/3′ ends. 5′ cap analysis gene expression (CAGE) is a method of enriching for full-length cDNA by priming the first strand cDNA synthesis with an oligo-dT primer (to capture the 3′ end of a polyadenylated transcript) or a random primer, and "trapping" the cap that commonly occurs at the 5′ end of mRNAs. (3) EST and cDNA sequences were annotated using computational, manual, and experimental approaches. The ENCODE study showed that 14.7% of all the nucleotides in the tiling arrays were transcribed in at least one tissue, with the majority residing outside previously annotated exons.

Other studies convincingly show that much of the genome is transcribed. For example, genomic tiling arrays have demonstrated about ten times more transcriptional activity in human chromosomes 21 and 22 than expected based on current gene annotations (Kapranov et al., 2002).

There are several main explanations for the existence of pervasive transcription (Johnson et al., 2005). (1) There may be many protein-coding genes beyond those described in the RefSeq project. Studies of cDNA projects such as FANTOM (described above) are consistent with such an interpretation. (2) There are likely to be many noncoding RNAs that have yet to be annotated and characterized. Only a subset of these may be evolutionarily conserved (e.g. between human and mouse), and their functions remain to be established. (3) There may be biological "noise" associated with low levels of transcription. Such "noise" could, for example, include retained introns (that are not spliced to form a mature mRNA product). (4) The widespread transcriptional activity could represent an experimental artifact such as genomic contamination of RNA samples. However, several different groups using diverse technological approaches have reached similar conclusions about the extent of transcription, so artifacts are not likely to account for the observations.

We expand on this definition of a gene in Chapter 16 (p. 662).

Studies from the ENCODE project led Gerstein et al. 2007 to propose a novel definition of a gene as "a union of genomic sequences encoding a coherent set of potentially overlapping functional products."

Perspective

Genes in all organisms are expressed in a variety of developmental, environmental, or physiological conditions. The field of functional genomics includes the high throughput study of gene expression. Before the arrival of this new approach, the expression of one gene at a time was typically studied. Functional genomics may reveal the transcriptional program of entire genomes, allowing a global view of cellular function.

Three major shifts have occurred in recent years in our understanding of genes and their expression. First, complementary DNA microarrays and oligonucleotide-based microarrays were introduced in the mid-1990s and have emerged as a powerful and popular tool for the rapid, quantitative analysis of RNA transcript levels in a variety of biological systems. The use of microarrays is likely to increase in the near future as the number of organisms represented on arrays increases, and the experimental applications of microarray technology expand. Second, recent studies, including those of the ENCODE project, have indicated that much of the genome is transcribed, although the biological significance of this is not yet understood. Third, since the 1990s many small noncoding RNAs such as microRNAs have been identified and are beginning to be functionally classified. Together these discoveries and technological advances are leading to a new appreciation of the tremendous structural and functional diversity of RNAs.

PITFALLS

The recent discovery of the pervasive nature of transcription leads to the question of how many mRNA transcripts have functional roles. For small noncoding RNAs we are only beginning to appreciate the range of possible biological functions. The computational challenge of noncoding RNA identification is great, and many more are likely to be identified.

For studies of gene expression with techniques such as EST analysis, SAGE, or microarrays, there are many basic concerns. The mRNA molecules are not measured directly; rather, they are converted to cDNA, and that cDNA is analyzed by sequence analysis or by visualization of fluorescent tags. It is important to assess whether the amount of substance that is actually measured corresponds to the amount of mRNA in the biological sample.

- When RNA (or mRNA) is isolated, is it representative of the entire population of mRNA molecules in the cell?
- If two conditions are being compared, was the RNA isolated under appropriately matched conditions? Any variations in the experimental protocol may lead to artifactual differences (discussed in Chatper 9).
- Has degradation of the RNA occurred in any of the samples?
- For microarrays, there are additional concerns. Most researchers cannot confirm the identity of what is immobilized on the surface of a microarray.

One response to these assorted concerns about microarrays is that with appropriate experimental design one may obtain results with confidence. After microarray data analysis results in the identification of significantly regulated genes (Chapter 9), it is important to perform independent biochemical assays (such as RT-PCR) to validate the findings.

WEB RESOURCES

The RNA World website (▶ http://www.imb-jena.de/RNA.html) organizes many links related to RNA and is an excellent starting point. The main portal for the ENCODE Project is ▶ http://genome.ucsc.edu/ENCODE.

DISCUSSION QUESTIONS

[8-1] There has been an explosion of interest in small noncoding RNAs in plant, animal, and other genomes. Why were these small RNAs not identified and studied in earlier decades?

[8-2] If you have a human cell line and you want to measure gene expression changes induced by a drug treatment, what are some of the advantages and disadvantages of using a subtraction approach versus SAGE versus microarrays? How are your answers different if you want to study gene expression in a less well characterized organism such as a parasite?

[8-3] When you use a microarray, how can you assess what has been deposited on the surface of the array? How do you know the DNA is of the length and composition that the manufacturer of the array specifies?

PROBLEMS

[8-1] We introduced the noncoding RNAs *XIST* and AIR in this chapter. We also discussed how many noncoding RNAs are poorly conserved. Perform a series of blast searches to try to identify human, mouse and other homologs of Xist and AIR. Try searching the RefSeq, nonredundant, or other nucleotide databases.

[8-2] Choose a human rRNA sequence, then perform blastn searches against human genomic DNA databases. How many matches do you find, and to what chromosomes are the rDNA sequences assigned?

[8-3] How many noncoding RNAs are in the vicinity of the human beta globin gene? To assess this, go to the UCSC bioinformatics site (► http://genome.ucsc.edu), select the Genome Browser, set the organism to human and choose a particular genome build, then enter the search term hbb to find that gene on chromosome 11. Then display annotation tracks related to noncoding RNAs, and set the view to 10 million base pairs surrounding the HBB gene.

[8-4] Telomerase is a ribonucleoprotein polymerase that in humans maintains active telomere ends by adding many copies of the repetitive sequence TTAGGG. The enzyme (which is a protein) includes an RNA component that serves as a template for the telomere repeat. To what chromosome is this noncoding RNA gene assigned? As one approach, find the entry in Entrez Nucleotide at NCBI. As another approach, search Rfam with the keyword telomerase.

[8-5] Perform digital differential display:

- Go to UniGene (► http://www.ncbi.nlm.nih.gov/UniGene/).
- Go to *Homo sapiens.*
- Click library differential display.
- Click some brain libraries, then "Accept changes."
- Choose a second pool of libraries to compare.

[8-6] Perform digital SAGE:

- Go to ► http://www.ncbi.nlm.nih.gov/ and click Serial Analysis of Gene Expression.
- Click "Analyze by library."
- Compare two SAGE library collections (e.g., brain and ovary).
- Next, go to Entrez Gen and select a DNA sequence (any will do).
- Copy the DNA to the clipboard, and return to the SAGE page.
- Click "Virtual Northern."
- Paste in your sequence, and submit.

SELF-TEST QUIZ

[8-1] Which are the most abundant RNA types?

(a) rRNA and tRNA

(b) rRNA and mRNA

(c) tRNA and mRNA

(d) mRNA and microRNA

[8-2] MicroRNAs may be distinguished from other RNAs because of the following properties:

(a) They are localized to the nucleolus

(b) Each microRNA is thought to regulate a small number of homologous target messenger RNAs

(c) They are coding RNAs, each of which is thought to regulate the function of a large number of messenger RNAs to which they are homologous

(d) They have a length of about 22 nucleotides, derived from a larger precursor, and regulated messenger RNA function

[8-3] The stages of mRNA processing include all of the following except:

(a) Splicing

(b) Export

(c) Methylation

(d) Surveillance

[8-4] Digital differential display (DDD) is used to compare the content of expressed sequence tags (ESTs) in UniGene's cDNA libraries. ESTs are also represented on microarrays. Which statement best describes ESTs?

(a) Clusters of nonredundant sequences (approximately 500 bp in length)

(b) Stretches of DNA sequence that are repeated many times throughout the genome

(c) Sequences corresponding to expressed genes that are obtained by sequencing complementary DNAs

(d) A "tag" (i.e., a fragment of DNA) derived from complementary DNA (cDNA) that corresponds to a transcript that has not been identified

[8-5] UniGene has cluster sizes from very small (e.g., 1) to very large (e.g., >10,000). What does it mean for there to be a cluster of size 1?

(a) One sequence has been identified that has a very large number of EST transcripts (e.g., over 10,000) associated with it.

(b) One sequence has been identified that corresponds to a gene that has been expressed one time.

(c) One sequence has been identified (presumably it is an EST) that matches one other known sequence (thus allowing it to be identified as a UniGene cluster).

(d) One sequence has been identified (presumably it is an EST) that is thought to correspond to a known gene, but it matches no other known sequences in UniGene (i.e., it does not align to any other ESTs).

[8-6] In analyzing cDNA libraries, a pitfall is that:

(a) The libraries may be derived from different tissues.

(b) The libraries may contain thousands of sequences.

(c) The libraries may have been normalized differently.

(d) The libraries may contain many rarely expressed transcripts.

[8-7] What advantage do oligonucleotide-based microarrays have over cDNA-based arrays?

(a) Two samples can be simultaneously and competitively hybridized to the same chip.

(b) It is easier for the experimenter to verify the identity of each gene that is represented on the array.

(c) It is possible to identify expression of alternatively spliced transcripts.

(d) They are far more sensitive.

[8-8] Most microarrays consist of a solid support on which is immobilized:

(a) DNA

(b) RNA

(c) Genes

(d) Transcripts

[8-9] The purpose of the MIAME project is to provide:

(a) A unified system for the description of microarray manufacture

(b) A unified system for the description of microarray experiments from design to hybridization to image analysis

(c) A unified system for the description of microarray probe preparation including fluorescence- and radioactivity-based approaches

(d) A unified system for microarray databases including standards for data storage, analysis, and presentation

[8-10] The expression of thousands of genes can be measured using cDNA libraries, SAGE, and DNA microarrays. A unique advantage of using DNA microarrays is that:

(a) The expression levels can be described quantitatively.

(b) It is possible to measure the expression levels of thousands of genes in two particular conditions of interest.

(c) It is more practical than the other experimental approaches to compare the expression levels of thousands of genes in two particular conditions of interest.

(d) It can be used to survey the expression levels of essentially all genes in a genome.

SUGGESTED READING

An article by the ENCODE Project Consortium (2007) is important because it introduces a paradigm shift in our understanding of the roles of RNA. This study presents evidence for pervasive transcription of the human genome, and describes assorted roles for noncoding RNAs and the regulation of transcription and chromatin architecture, and assesses regions of the genome under positive or negative selection.

An early report on cDNAs by Mark Adams, J. Craig Venter, and colleagues (1991) described the sequence analysis of over 600 randomly selected human brain cDNAs. Remarkably, 337 of these represented novel genes. They subsequently identified over 6000 more human genes in further screens (Adams et al., 1992, 1993).

In this chapter we described serial analysis of gene expression (SAGE). Victor Velculescu and colleagues 1995 at Johns Hopkins introduced this technology. Figure 8.15 is derived from that paper. Velculescu and colleagues (1997) further used SAGE in an insightful study of gene expression in yeast. In the February 2001 human genome issue in *Science*, Caron et al. describe a human transcriptome map based on SAGE data, showing variations in gene expression based on chromosomal location.

There are many introductions to microarray technology (Duggan et al., 1999; Hegde et al., 2000; Zhang, 1999). The procedure by which oligonucleotides can be immobilized on solid supports is described by Stephen Fodor and colleagues (1991). This approach, adopted by Affymetrix, depends on light-directed, spatially addressable, massively parallel chemical synthesis of oligonucleotides using techniques of photolithography. This paper describes the application of this technology to immobilizing both peptides and oligonucleotides onto surfaces.

REFERENCES

Adams, M. D., et al. Complementary DNA sequencing: Expressed sequence tags and human genome project. *Science* **252**, 1651–1656 (1991).

Adams, M. D., et al. Sequence identification of 2,375 human brain genes. *Nature* **355**, 632–634 (1992).

Adams, M. D., Kerlavage, A. R., Fields, C., and Venter, J. C. 3,400 new expressed sequence tags identify diversity of transcripts in human brain. *Nat. Genet.* **4**, 256–267 (1993).

Ambros, V., et al. A uniform system for microRNA annotation. *RNA* **9**, 277–279 (2003).

Avery, O. T., MacLeod, C. M., and McCarty, M. Studies on the chemical nature of the substance inducing transformation of pneumococcal types. *J. Exp. Med.* **79**, 137–158 (1944).

Baker, S. C., et al. The External RNA Controls Consortium: A progress report. *Nat. Methods* **2**, 731–734 (2005).

Bark, C., Weller, P., Zabielski, J., and Pettersson, U. Genes for human U4 small nuclear RNA. *Gene* **50**, 333–344 (1986).

Baross, A., et al. Systematic recovery and analysis of full-ORF human cDNA clones. *Genome Res.* **14**, 2083–2092 (2004). PMID: 15489330.

Beadle, G. W., and Tatum, E. L. Genetic control of biochemical reactions in *Neurospora*. *Proc. Natl. Acad. Sci. USA* **27**, 499–506 (1941).

Bishop, J. O., Morton, J. G., Rosbash, M., and Richardson, M. Three abundance classes in Hela cell messenger RNA. *Nature* **250**, 199–204 (1974).

Bonaldo, M. F., Lennon, G., and Soares, M. B. Normalization and subtraction: Two approaches to facilitate gene discovery. *Genome Res.* **6**, 791–806 (1996).

Borsani, G., et al. Characterization of a murine gene expressed from the inactive X chromosome. *Nature* **351**, 325–329 (1991).

Brazma, A., et al. Minimum information about a microarray experiment (MIAME): Toward standards for microarray data. *Nat. Genet.* **29**, 365–371 (2001).

Carninci, P., et al. The transcriptional landscape of the mammalian genome. *Science* **309**, 1559–1563 (2005).

Caron, H., et al. The human transcriptome map: Clustering of highly expressed genes in chromosomal domains. *Science* **291**, 1289–1292 (2001).

Carulli, J. P., et al. High throughput analysis of differential gene expression. *J. Cell. Biochem. Suppl.* **31**, 286–296 (1998).

Celotto, A. M., and Graveley, B. R. Alternative splicing of the *Drosophila* Dscam pre-mRNA is both temporally and spatially regulated. *Genetics* **159**, 599–608 (2001).

Churchill, G. A. Fundamentals of experimental design for cDNA microarrays. *Nature Genetics Suppl.* **32**, 490–495 (2002).

Claverie, J. M. Computational methods for the identification of differential and coordinated gene expression. *Hum. Mol. Genet.* **21**, 1821–1832 (1999).

Cole, J. R., Chai, B., Farris, R. J., Wang, Q., Kulam-Syed-Mohideen, A. S., McGarrell, D. M., Bandela, A. M., Cardenas, E., Garrity, G. M., and Tiedje, J. M. The ribosomal database project (RDP-II): Introducing myRDP space and quality controlled public data. *Nucleic Acids Res.* **35**, D169–D172 (2007).

Crick, F. H. On protein synthesis. *Symp. Soc. Exp. Biol.* **12**, 138–163 (1958).

Dayhoff, M. (1972). *Atlas of Protein Sequence and Structure*, Vol. 5. Georgetown University Medical Center, Washington, D.C., 1972.

DeRisi, J., et al. Use of a cDNA microarray to analyse gene expression patterns in human cancer. *Nat. Genet.* **14**, 457–460 (1996).

Duggan, D. J., Bittner, M., Chen, Y., Meltzer, P., and Trent, J. M. Expression profiling using cDNA microarrays. *Nat. Genet.* **21**, 10–14 (1999).

Eddy, S. R., and Durbin, R. RNA sequence analysis using covariance models. *Nucl. Acids Res.*, **22**, 2079–2088 (1994).

ENCODE Project Consortium. The ENCODE (ENCyclopedia Of DNA Elements) Project. *Science* **306**, 636–640 (2004).

ENCODE Project Consortium. Identification and analysis of functional elements in 1% of the human genome by the ENCODE pilot project. *Nature* **447**, 799–816 (2007).

Fichant, G. A., and Burks, C. Identifying potential tRNA genes in genomic DNA sequences. *J. Mol. Biol.* **220**, 659–671 (1991).

Fire, A., Xu, S., Montgomery, M. K., Kostas, S. A., Driver, S. E., and Mello, C. C. Potent and specific genetic interference by double-stranded RNA in *Caenorhabditis elegans*. *Nature* **391**, 806–811 (1998).

Fodor, S. P., et al. Light-directed, spatially addressable parallel chemical synthesis. *Science* **251**, 767–773 (1991).

Futcher, B., Latter, G. I., Monardo, P., McLaughlin, C. S., and Garrels, J. I. A sampling of the yeast proteome. *Mol. Cell. Biol.* **19**, 7357–7368 (1999).

Gerhard, D. S., et al. The status, quality, and expansion of the NIH full-length cDNA project: The Mammalian Gene Collection (MGC). *Genome Res.* **14**, 2121–2127 (2004). PMID: 15489334.

Gerstein, M. B., Bruce, C., Rozowsky, J. S., Zheng, D., Du, J., Korbel, J. O., Emanuelsson, O., Zhang, Z. D., Weissman, S., and Snyder, M. What is a gene, post-ENCODE? History and updated definition. *Genome Res.* **17**, 669–681 (2007).

Gonzalez, I. L., and Sylvester, J. E. Human rDNA: evolutionary patterns within the genes and tandem arrays derived from multiple chromosomes. *Genomics* **73**, 255–263 (2001).

Greenbaum, D., Jansen, R., and Gerstein, M. Analysis of mRNA expression and protein abundance data: An approach for the comparison of the enrichment of features in the cellular population of proteins and transcripts. *Bioinformatics* **18**, 585–596 (2002).

Griffiths-Jones, S. The microRNA Registry. *Nucl. Acids Res.* **32**, D109–D111 (2004).

Griffiths-Jones, S. Annotating non-coding RNAs with Rfam. *Curr. Protocol. Bioinf.* **12**, 5.1–5.12 (2005).

Griffiths-Jones, S., Grocock, R. J., van Dongen, S., Bateman, A., and Enright, A. J. miRBase: microRNA sequences, targets and gene nomenclature. *Nucleic Acids Res.* **34**, D140–D144 (2006).

Griffiths-Jones, S., Moxon, S., Marshall, M., Khanna, A., Eddy, S. R., and Bateman, A. Rfam: Annotating non-coding RNAs in complete genomes. *Nucleic Acids Res.* **33**(Database issue), D121–124 (2005).

Gruber, A. R., Neubock, R., Hofacker, I. L., and Washietl, S. The RNAz web server: Prediction of thermodynamically stable and evolutionarily conserved RNA structures. *Nucleic Acids Res.* **35**(Web Server issue), W335–W338 (2007).

Hastie, N. D., and Bishop, J. O. The expression of three abundance classes of messenger RNA in mouse tissues. *Cell* **9**, 761–774 (1976).

Hegde, P., et al. A concise guide to cDNA microarray analysis. *Biotechniques* **29**, 548–550, 552–554, 556 passim (2000).

Henderson, A. S., Warburton, D., and Atwood, K. C. Location of ribosomal DNA in the human chromosome complement. *Proc. Natl. Acad. Sci. USA* **69**, 3394–3398 (1972).

Hinrichs, A. S., et al. The UCSC Genome Browser Database: Update 2006. *Nucleic Acids Res.* **34**(Database issue), D590–D598 (2006).

Hofacker, I. L. Vienna RNA secondary structure server. *Nucleic Acids Res.* **31**, 3429–3431 (2003).

Imanishi, T., Itoh, T., Suzuki, Y., O'Donovan, C., Fukuchi, S., Koyanagi, K. O., Barrero, R. A., Tamura, T., Yamaguchi-Kabata, Y., Tanino, M., et al. Integrative annotation of 21,037 human genes validated by full-length cDNA clones. *PLoS Biol.* **2**, e162 (2004).

Johnson, J. M., Edwards, S., Shoemaker, D., and Schadt, E. E. Dark matter in the genome: Evidence of widespread transcription detected by microarray tiling experiments. *Trends Genet.* **21**, 93–102 (2005).

Kapranov, P., Cawley, S. E., Drenkow, J., Bekiranov, S., Strausberg, R. L., Fodor, S. P., and Gingeras, T. R. Large-scale transcriptional activity in chromosomes 21 and 22. *Science* **296**, 916–919 (2002).

Katayama, S., et al. Antisense transcription in the mammalian transcriptome. *Science* **309**, 1564–1566 (2005).

Kikuno, R., Nagase, T., Nakayama, M., et al. HUGE: A database for human KIAA proteins, a 2004 update integrating HUGEppi and ROUGE. *Nucleic Acids Res.* **32**, D502–D504, (2004).

Krek, A., Grun, D., Poy, M. N., Wolf, R., Rosenberg, L., Epstein, E. J., MacMenamin, P., da Piedade, I., Gunsalus, K. C., Stoffel, M., and Rajewsky, N. Combinatorial microRNA target predictions. *Nat. Genet.* **37**, 495–500 (2005).

Lagesen, K., Hallin, P., Rodland, E. A., Staerfeldt, H. H., Rognes, T., and Ussery, D. W. RNAmmer: Consistent and rapid annotation of ribosomal RNA genes. *Nucleic Acids Res.* **35**, 3100–3108 (2007).

Lal, A., et al. A public database for gene expression in human cancers. *Cancer Res.* **59**, 5403–5407 (1999).

Lash, A. E., et al. SAGEmap: A public gene expression resource. *Genome Res.* **10**, 1051–1060 (2000).

Lee, R. C., Feinbaum, R. L., and Ambros, V. The *C. elegans* heterochronic gene lin-4 encodes small RNAs with antisense complementarity to lin-14. *Cell* **75**, 843–854 (1993).

Lee, Y., Tsai, J., Sunkara, S., Karamycheva, S., Pertea, G., Sultana, R., Antonescu, V., Chan, A., Cheung, F., and Quackenbush, J. The TIGR Gene Indices: Clustering and assembling EST and known genes and integration with eukaryotic genomes. *Nucleic Acids Res.* **33**(Database issue), D71–D74 (2005).

Lestrade, L., and Weber, M. J. snoRNA-LBME-db, a comprehensive database of human H/ACA and C/D box snoRNAs. *Nucleic Acids Res.* **34**(Database issue), D158–D162 (2006).

Lewis, B. P., Burge, C. B., and Bartel, D. P. Conserved seed pairing, often flanked by adenosines, indicates that thousands of human genes are microRNA targets. *Cell* **120**, 15–20 (2005).

Liebhaber, S. A., Cash, F. E., and Ballas, S. K. Human alpha-globin gene expression. The dominant role of the alpha 2-locus in mRNA and protein synthesis. *J. Biol. Chem.* **261**, 15327–15333 (1986).

Lipshutz, R. J., Fodor, S. P., Gingeras, T. R., and Lockhart, D. J. High density synthetic oligonucleotide arrays. *Nat. Genet.* **21**, 20–24 (1999).

Lowe, T. M., and Eddy, S. R. tRNAscan-SE: A program for improved detection of transfer RNA genes in genomic sequence. *Nucleic Acids Res.* **25**, 955–964 (1997).

Lowe, T. M., and Eddy, S. R. A computational screen for methylation guide snoRNAs in yeast. *Science* **283**, 1168–1171 (1999).

Ludwig, W., et al. ARB: A software environment for sequence data. *Nucleic Acids Res.* **32**, 1363–1371 (2004).

Maeda, N., et al. Transcript annotation in FANTOM3: Mouse gene catalog based on physical cDNAs. *PLoS Genet.* **2**, e62 (2006).

Maniatis, T., and Reed, R. An extensive network of coupling among gene expression machines. *Nature* **416**, 499–506 (2002).

Mao, R., Wang, X., Spitznagel, E. L. Jr., Frelin, L. P., Ting, J. C., Ding, H., Kim, J. W., Ruczinski, I., Downey, T. J., and Pevsner, J. Primary and secondary transcriptional effects in the developing human Down syndrome brain and heart. *Genome Biol.* **6**, R107 (2005).

Mao, R., Zielke, C. L., Zielke, H. R., and Pevsner, J. Global up-regulation of chromosome 21 gene expression in the developing Down syndrome brain. *Genomics* **81**, 457–467 (2003).

Maquat, L. E. Molecular biology. Skiing toward nonstop mRNA decay. *Science* **295**, 2221–2222 (2002).

Mathews, D. H., Sabina, J., Zuker, M., and Turner, D. H. Expanded sequence dependence of thermodynamic parameters improves prediction of RNA secondary structure. *J. Mol. Biol.* **288**, 911–940 (1999).

Miller, O. L., Hamkalo, B. A., and Thomas, C. A. Visualization of bacterial genes in action. *Science* **169**, 392–395, 1970.

Milner, R. J., and Sutcliffe, J. G. Gene expression in rat brain. *Nucleic Acids Res.* **11**, 5497–5520 (1983).

Modrek, B., and Lee, C. A genomic view of alternative splicing. *Nat. Genet.* **30**, 13–19 (2002).

Nagase, T., Koga, H., and Ohara, O. Kazusa mammalian cDNA resources: Towards functional characterization of KIAA gene products. *Brief. Funct. Genomic. Proteomics* **5**, 4–7 (2006). PMID: 16769670.

Nirenberg, M. Protein synthesis and the RNA code. *Harvey Lect.* **59**, 155–185 (1965).

Omer, A. D., Lowe, T. M., Russell, A. G., Ebhardt, H., Eddy, S. R., and Dennis, P. P. Homologs of small nucleolar RNAs in Archaea. *Science* **288**, 517–522 (2000).

Pang, K. C., Stephen, S., Engstrom, P. G., Tajul-Arifin, K., Chen, W., Wahlestedt, C., Lenhard, B., Hayashizaki, Y., and Mattick, J. S. RNAdb: A comprehensive mammalian noncoding RNA database. *Nucl. Acids Res.* **33**(Database issue), D125–D130 (2005).

Pasquinelli, A. E., and Ruvkun, G. Control of developmental timing by microRNAs and their targets. *Annu. Rev. Cell Dev. Biol.* **18**, 495–513 (2002).

Pavesi, A., Conterio, F., Bolchi, A., Dieci, G., and Ottonello, S. Identification of new eukaryotic tRNA genes in genomic DNA databases by a multistep weight matrix analysis of trnascriptional control regions. *Nucl. Acids Res.* **22**, 1247–1256 (1994).

Pedersen, J. S., Bejerano, G., Siepel, A., Rosenbloom, K., Lindblad-Toh, K., Lander, E. S., Kent, J., Miller, W., and Haussler, D. Identification and classification of conserved RNA secondary structures in the human genome. *PLoS Comput. Biol.* **2**, e33 (2006).

Rajewsky, N. MicroRNA target predictions in animals. *Nat Genet.* **38**(Suppl), S8–S13 (2006).

Sagerstrom, C. G., Sun, B. I., and Sive, H. L. Subtractive cloning: Past, present, and future. *Annu. Rev. Biochem.* **66**, 751–783 (1997).

Sakakibara, Y., Brown, M., Hughey, R., Mian, I. S., Sjolander, K., Underwood, R. C., and Haussler, D. Stochastic context-free grammars for tRNA modeling. *Nucleic Acids Res.* **22**, 5112–5120 (1994).

Schmitt, A. O., et al. Exhaustive mining of EST libraries for genes differentially expressed in normal and tumour tissues. *Nucleic Acids Res.* **27**, 4251–4260 (1999).

Schloss, P. D., and Handelsman, J. Status of the microbial census. *Microbiol. Mol. Biol. Rev.* **68**, 686–691 (2004). PMID: 15590780.

Schmucker, D., et al. *Drosophila* Dscam is an axon guidance receptor exhibiting extraordinary molecular diversity. *Cell* **101**, 671–684 (2000).

Schuler, G. D. Pieces of the puzzle: Expressed sequence tags and the catalog of human genes. *J. Mol. Med.* **75**, 694–698 (1997).

Sleutels, F., Zwart, R., and Barlow, D. P. The non-coding Air RNA is required for silencing autosomal imprinted genes. *Nature* **415**, 810–813 (2002).

Southern, E., Mir, K., and Shchepinov, M. Molecular interactions on microarrays. *Nat. Genet.* **21**, 5–9 (1999).

Stekel, D. J., Git, Y., and Falciani, F. The comparison of gene expression from multiple cDNA libraries. *Genome Res.* **10**, 2055–2061 (2000).

Tåquist, H., Cui, Y., and Ardell, D. H. TFAM 1.0: An online tRNA function classifier. *Nucleic Acids Res.* **35**(Web Server issue), W350–W353 (2007).

Temple, G., Lamesch, P., Milstein, S., Hill, D. E., Wagner, L., Moore, T., and Vidal, M. From genome to proteome: Developing expression clone resources for the human genome. *Hum. Mol. Genet.* **15**, R31–R43 (2006). PMID: 16651367.

Valadkhan, S. snRNAs as the catalysts of pre-mRNA splicing. *Curr. Opin. Chem. Biol.* **9**, 603–608 (2005).

Velculescu, V. E., Zhang, L., Vogelstein, B., and Kinzler, K. W. Serial analysis of gene expression. *Science* **270**, 484–487 (1995).

Velculescu, V. E., et al. Characterization of the yeast transcriptome. *Cell* **88**, 243–251 (1997).

Vietor, I., and Huber, L. A. In search of differentially expressed genes and proteins. *Biochim. Biophys. Acta* **1359**, 187–199 (1997).

Waggoner, S. A., and Liebhaber, S. A. Regulation of alpha-globin mRNA stability. *Exp. Biol. Med. (Maywood)* **228**, 387–395 (2003).

Wang, S. M. Understanding SAGE data. *Trends Genet.* **23**, 42–50 (2007).

Washietl, S., et al. Structured RNAs in the ENCODE selected regions of the human genome. *Genome Res.* **17**, 852–864 (2007).

Waters, K. M., Pounds, J. G., and Thrall, B. D. Data merging for integrated microarray and proteomic analysis. *Brief. Funct. Genomic. Proteomic.* **5**, 261–272 (2006).

Watson, J. D., and Crick, F. H. C. A structure for deoxyribose nucleic acid. *Nature* **171**, 737–738 (1953).

Weber, M. J. New human and mouse microRNA genes found by homology search. *FEBS J.* **272**, 59–73 (2005).

Zhang, L., et al. Gene expression profiles in normal and cancer cells. *Science* **276**, 1268–1272 (1997).

Zhang, M. Q. Large-scale gene expression data analysis: A new challenge to computational biologists. *Genome Res.* **9**, 681–688 (1999).

Zuker, M., and Stiegler, P. Optimal computer folding of large RNA sequences using thermodynamics and auxiliary information. *Nucleic Acids Res.* **9**, 133–148 (1981).

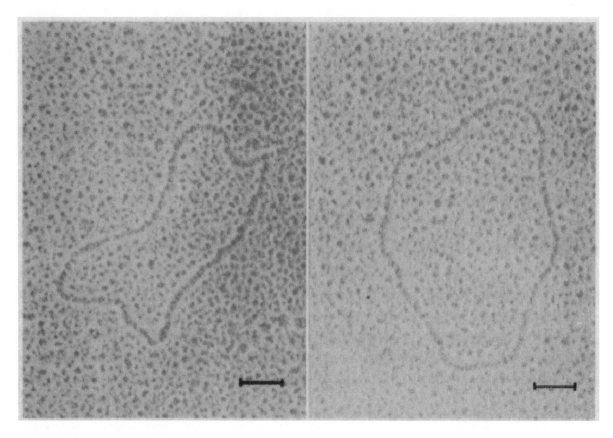

The main idea behind microarrays is that one nucleic acid (DNA) is immobilized on a solid support on a solid surface in a predefined location, and then another nucleic acid (RNA or a derivative such as fluorescently labeled complementary DNA) is hybridized to the surface. Microarrays were first developed in the 1990s by the laboratories of Patrick Brown at Stanford University and Jeffrey Trent, then at the National Institutes of Health (NIH). Beginning in the 1950s, Sol Spiegelman (1914–1983) pioneered the study of RNA hybridization to DNA (see ▶ http://profiles.nlm.nih.gov/PX/). By the early 1960s several groups had immobilized DNA on a solid support then hybridized purified RNA molecules under a variety of conditions. This figure shows electron micrographic images of circular DNA-RNA hybrids by Spiegelman and colleagues (Bassel et al., 1964). The bacteriophage φX174 was shown to transcribe RNA, which bound to DNA in a ribonuclease resistant complex. Studies such as these established the mechanisms by which DNA is transcribed to RNA, and ultimately led to the development of hybridization-based assays, including microarrays. The scale bar is 0.1 μm.

9

Gene Expression: Microarray Data Analysis

INTRODUCTION

DNA microarray experiments have emerged as one of the most popular tools for the large-scale analyisis of gene expression (i.e., mRNA transcript levels). When a microarray experiment is completed and the data arrive, the first question most investigators ask is: Which genes were most dramatically up- or downregulated in my experiment? This can be answered using inferential statistics, a branch of data analysis in which probabilities are assigned to the likelihood that a gene is significantly regulated:

- A spreadsheet listing all the genes represented on the array and all the expression values can be sorted to show the most differentially regulated genes.

- A scatter plot (see below) can help to quickly profile the behavior of the most regulated genes.

- A t-test can be used to describe the probability that a gene is regulated.

Bioinformatics and Functional Genomics, Second Edition. By Jonathan Pevsner
Copyright © 2009 John Wiley & Sons, Inc.

Another fundamental question that may be asked is: What signatures (or patterns or profiles) of gene expression can be found in all the gene expression values obtained in this experiment? This type of question is addressed using descriptive statistics or exploratory analysis. Clustering trees can show the relationships between samples (such as normal vs. diseased cells), between genes, or both. Other tools for the analysis of gene expression include principal components analysis, multidimensional scaling, and self-organizing maps. We will consider all these tools for the analysis of array data.

Microarray experiments typically involve the measurement of the expression levels of many thousands of genes in only a few biological samples. Often, there are few technical replicates (i.e., measuring gene expression with the same starting material on independent arrays), usually because of the relatively high cost of performing microarray experiments. There are also few biological replicates (e.g., measuring gene expression from multiple cell lines, each of which has been given an experimental treatment or a control treatment) relative to the large number of genes represented on the microarray. The challenge to the biologist is to apply appropriate statistical techniques to determine which changes are relevant. There is unlikely to be a single best approach to microarray data analysis, and the tools applied to microarray data analysis are evolving rapidly.

Regardless of the microarray platform that is used, we may begin data analysis by creating a matrix of genes (along rows) and samples (arranged in columns) (Fig. 9.1, top). The values in the matrix consist of intensity measurements that are assumed to be directly proportional to the abundance of mRNA that has been transcribed from each gene. We can consider two-channel and one-channel platforms. For two-channel microarray technologies that rely on competitive hybridization of two samples on an array, the gene expression values are usually presented as ratios (or relative intensities). Typically, these are ratios of intensity values for the Cy3 (green) dye and the Cy5 (red) dye, and the raw data consist of separate intensity values for each channel (dye). In the case of single-channel technologies, a single sample is hybridized to a microarray. This is the case (1) for platforms using oligonucleotides immobilized on a microarray (e.g., the Affymetrix platform), (2) for platforms using beads immobilized on a microarray (e.g., the Illumina platform), or, less commonly, (3) for platforms using radioactivity-labeled cDNA. Absolute values are obtained for two (or more) experimental conditions.

We will describe microarray data analysis in three areas (Fig. 9.1). First, data are "preprocessed." This is essential to allow data sets from two (or more) samples to be compared to each other. Second, inferential statistics are applied. This is also called hypothesis testing, and it allows us to make statements about the likelihood that particular genes are significantly regulated according to statistical criteria. Third, exploratory statistics (also called descriptive statistics) are applied. This set of approaches includes clustering and principal components analysis and is used to inspect the complex data set for biologically meaningful patterns. In some microarray studies classification is applied in order to diagnose physiological states (e.g., cancerous vs. control cells) based on gene expression profiles. It is appropriate to begin data analysis with preprocessing, but inferential statistics and exploratory techniques are commonly applied in parallel during the weeks (or months) that one analyzes a microarray data set.

Excellent reviews of microarray data analysis include Quackenbush (2001, 2006), Sherlock (2001), Dopazo et al. (2001), Brazma and Vilo (2000), Ayroles and Gibson (2006), Olson (2006), Lee and Saeed (2007), as well as a succinct book by Causton et al. (2003).

The Microarray Gene Expression Data (MGED) Society has been formed to establish standards for the analysis, annotation, exchange, and reporting of microarray data. Its website (▶ http://www.mged.org) and publications by the MGED group include important information relevant to standardization of microarray experiments (Brazma, 2001; Ball et al., 2002a, 2002b), the Miniumum Information About a Microarray Experiment (MIAME) standards (Brazma et al., 2001), the design of a microarray gene expression markup language (MAGE-ML) for standardizing the storage and exchange of microarray data (Spellman et al., 2002), and the need for public microarray data repositories (Brazma et al., 2000).

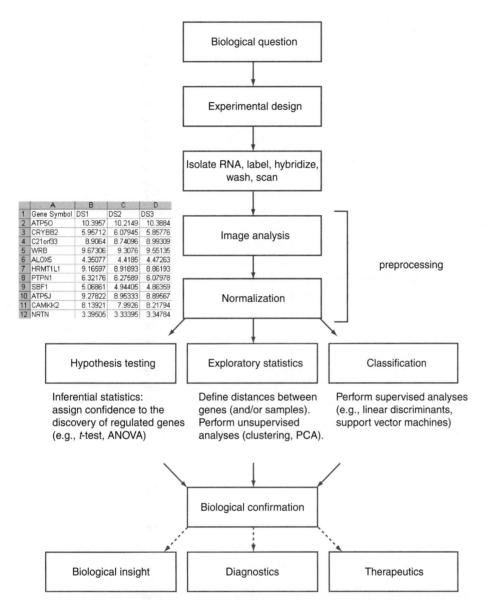

	A	B	C	D
1	Gene Symbol	DS1	DS2	DS3
2	ATP5O	10.3957	10.2149	10.3884
3	CRYBB2	5.95712	6.07945	5.85776
4	C21orf33	8.9064	8.74096	8.99309
5	WRB	9.67306	9.3076	9.55135
6	ALOX5	4.35077	4.4185	4.47263
7	HRMT1L1	9.16597	8.91893	8.86193
8	PTPN1	6.32176	6.27589	6.07978
9	SBF1	5.06861	4.94405	4.86359
10	ATP5J	9.27822	8.95333	8.89567
11	CAMKK2	8.13921	7.9926	8.21794
12	NRTN	3.39505	3.33395	3.34784

FIGURE 9.1. *Overview of microarray data analysis. First, a biological question is formulated and then experimental design is created (preferably with the collaboration of biostatisticians). After RNA is isolated and microarray data are generated, there are three main stages of micro-array data analysis. First, preprocessing is performed in which raw image data are analyzed, normalized, and a matrix of genes and samples is created. For Affymetrix arrays, an additional preprocessing step is summarization in which the expression value of a given gene (mRNA transcript) is estimated based on the results from a series of hybridizations to olignonucleotides corresponding to that gene. Second, hypothesis testing is performed in which t-tests, ANOVA, or other statistical tests are applied to determine which transcripts were significantly up- or down-regulated in the experiment. Third, exploratory (descriptive) statistics may be applied. The simi-larities of the data points are compared with a metric such as a correlation coefficient. This pattern of gene expression may be visualized using unsupervised approaches in which patterns are sought in the representation of genes (or samples). For supervised approaches, samples (or genes) are associated with labels from a preexisting classification (such as normal vs. diseased tissue) and gene expression measurements are used to predict which unknown samples are dis-eased. Finally, after microarray data analysis is performed, biological confirmation experiments may be performed. This may lead to insight about biological processes, or to outcomes relevant to disease such as identifying diagnostic markers or strategies for therapeutic intervention. Adapted in part from Brazma and Vilo (2000).*

Microarray Data Analysis Software and Data Sets

There are three main types of software available for microarray data analysis. These are (1) commercial software packages associated with microarray manufacturers. For example, Affymetrix currently offers MAS 5.0 software for its arrays. (2) Other commercial software packages include BioDiscovery, GeneSifter, MATLAB®, Partek Genomics Suite, and Spotfire, as well as spreadsheet programs such as Microsoft Excel and S-PLUS. Statistics packages include STATA and SAS. (3) There is great enthusiasm in the bioinformatics community for open source software. In this chapter we will introduce the BioConductor project that offers a variety of freely available packages. These are implemented in the freely available R software environment for statistical computing and graphics. Many of the figures in this chapter were generated in R, and we provide step-by-step instructions for installing R and getting started with Bioconductor (Box 9.3 below).

Many dozens of other software resources are available, such as BioArray Software Environment (BASE; Troein et al., 2006), the TM4 suite (Saeed et al., 2006), and MARS (Maurer et al., 2005). Simple analysis tools are also incorporated in the Gene Expression Omnibus (GEO) at the National Center for Biotechnology Information (Barrett and Edgar, 2006; Barrett et al., 2007), as well as ArrayExpress at the European Bioinformatics Institute (Brazma et al., 2006) and CIBEX at the DNA Database of Japan (Ikeo et al., 2003).

We will illustrate data analysis approaches in this chapter using several data sets (all available as web documents). (1) We will examine data from seven trisomy 21 (Down syndrome) brain samples and seven controls from a larger study by Mao et al. (2005); they are available from GEO (Box 9.1) and as a series of web documents introduced below. (2) To illustrate principles of exploratory data analysis we will

For commercial software, BioDiscovery (▶ http://www.biodiscovery.com/) offers products such as ImaGene and GeneDirector. GeneSifter (▶ http://www.genesifter.net/web/) is from VizX Labs. MATLAB® is a product of MathWorks (▶ http://www.mathworks.com/). Partek® Genomics Suite™ is available through Partek, Inc. (▶ http://www.partek.com/). S-PLUS® is from Insightful Corp. (▶ http://www.insightful.com/). SAS is available from ▶ http://www.sas.com, and STATA from ▶ http://www.stata.com/. For freely available software, BASE is at ▶ http://base.thep.lu.se/ and TM4 is available at ▶ http://www.tm4.org/. For the major public repositories, ArrayExpress is at ▶ http://www.ebi.ac.uk/arrayexpress/, CIBEX is at ▶ http://cibex.nig.ac.jp, and GEO is at ▶ http://www.ncbi.nlm.nih.gov/geo/.

BOX 9.1
Obtaining Microarray Data from GEO and ArrayExpress

We provide 14 .cel files of Affymetrix microarray data ($n = 7$ brain samples from individuals with trisomy 21, and $n = 7$ samples from controls) at ▶ http://www.bioinfbook.org. My lab posts various data sets on a website (see microarrays at ▶ http://pevsnerlab.kennedykrieger.org). However, while it can be useful to post data on an individual's website, the use of a centralized repository offers many advantages, including long-term stability, common standards for data entry, and broader accessibility to the research community.

The two main public repositories of microarray data are the Gene Expression Omnibus at the National Center for Biotechnology Information (GEO at NCBI; ▶ http://www.ncbi.nlm.nih.gov/geo/) and ArrayExpress at the European Bioinformatics Institute (EBI; ▶ http://www.ebi.ac.uk/arrayexpress/). Many journals now require that investigators submit raw (and processed) microarray data upon publishing articles. To download a microarray dataset follow these steps.

[1] From the home page of NCBI enter "down syndrome" including the quotation marks; there are currently several dozen GEO DataSets. To further focus the search enter "down syndrome TS13" as a query to find a DataSet including expression data from the brains of individuals with trisomy 21 and trisomy 13.

[2] This leads to the GSE1397 record. From here, you can also download .cel files directly.

analyze a set of just eight RNA transcripts in 14 samples from the Mao et al. (2005) study, highlighting a small set of transcripts that are expressed at very high levels ($n = 4$) or that are differentially regulated transcripts from chromosome 21 genes ($n = 4$). (3) In the web exercises we will explore a very heavily studied data set of expression profiles from 60 cancer cell lines from the National Cancer Institute (NCI60) (Lee et al., 2003). These cell lines have been used to screen over 100,000 compounds, and they have been used to study gene expression as well as genomic DNA with multiple platforms. The data we will examine are in GEO (record GDS1761; see Box 9.1). This provides an example of data from a two-color experiment.

The NCI60 Cancer Microarray Project homepage is at ▶ http://genome-www.stanford.edu/nci60/.

Reproducibility of Microarray Experiments

Microarray experiments can generate large amounts of data, and the question has arisen whether studies across different platforms and/or across different laboratories are reproducible. For example, in the late 1990s when microarrays were first introduced it was widely reported that the cDNA clones deposited on microarrays were often contaminated or represented the wrong gene. More recently Tan et al. (2003) compared gene expression measurements from three commercial platforms (Affymetrix, Agilent, and Amersham) using the same RNA as starting material, and included both biological and technical replicates. They reported that there was only limited overlap in the RNA transcripts identified by the three platforms, with an average Pearson's correlation coefficient r for measurements between the three platforms of only 0.53 (see Box 9.2). Other have raised concerns about microarray data reproducibility and broader issues regarding data analysis (Draghici et al.,

The External RNA Controls Consortium has developed platform-independent RNA controls to facilitate performance assessment of microarrays (Baker et al., 2005).

BOX 9.2
Pearson Correlation Coefficient r

When two variables vary together they are said to correlate. The Pearson correlation coefficient r has values ranging from -1 (a perfect negative correlation) to 0 (no correlation) to 1 (perfect positive correlation). It is possible to state a null hypothesis that two variables are not correlated, and an alternative hypothesis that they are correlated. A probability (p) value can be derived to test the significance of the correlation. The Pearson correlation coefficient is perhaps the most common metric used to define similarity between gene expression data points. It is used by tree-building programs such as Cluster (described below). For any two series of numbers $X = \{X_1, X_2, \ldots, X_N\}$ and $Y = \{Y_1, Y_2, \ldots, Y_N\}$,

$$r = \frac{\sum_{i=1}^{N} \left[\frac{(X_i - \overline{X})}{\sigma_x} \cdot \frac{(Y_i - \overline{Y})}{\sigma_y} \right]}{N - 1}$$

where $\overline{X}$ is the average of the values in X and σ_x is the standard deviation of these values.

For a scatter plot, r describes how well a line fits the values. The Pearson correlation coefficient always has a value between $+1$ (two series are identical), 0 (completely independent sets), and -1 (two sets are perfectly uncorrelated).

The square of the correlation coefficient, r^2, has a value between 0 and 1. It is also smaller than r; $r^2 \leq |r|$. For two variables having a correlation coefficient

$r = 0.9$ (such as two microarray data sets measured in different laboratories using the same RNA starting material), r^2 is 0.81. This means that 81% of the variability in the gene expression measurements in the two data sets can be explained by the correspondence of the results between the two laboratories, while just 19% can be explained by other factors such as error.

Correlation coefficients have been widely misused (Bland and Altman, 1986, 1999). r measures the strength of a relation between two variables, but it does not measure how well those variables agree. (Picture a scatter plot showing the correlation of two measures, as shown in Fig. 9.3; a perfect correlation occurs if the points fall on any straight line, but perfect agreement occurs only if the points fall on a 45° line.) See Bland and Altman (1986, 1999) for additional caveats in interpreting r values.

Source: Motulsky (1995).

2006; Miron and Nadon, 2006; Shields, 2006), with accompanying responses (Quackenbush and Irizarry, 2006).

A far more optimistic assessment was provided by the MicroArray Quality Consortium (MAQC, 2006). This project was established to evaluate the performance of a broad set of microarray platforms and data analysis techniques using identical RNA samples. Twenty microarray products and three technologies were evaluated for 12,000 RNA transcripts expressed in human tumor cell lines or brain. There was substantial agreement between sites and platforms for regulated transcripts, with various measures of concordance ranging from 60% to over 90% and a median rank correlation of 0.87 for comparability across platforms based on a log-ratio measurement. Microarray data were also validated using polymerase chain reaction-based methods, again showing a high correlation (Canales et al., 2006).

> The MAQC project involved over 100 researchers at over 50 institutions. The study included 60 hybridizations on seven different platforms. The MAQC website is ► http://www.fda.gov/nctr/ science/centers/ toxicoinformatics/maqc/.

In another study Irizarry et al. (2005) reported generally good agreement between laboratories using three platforms (Affymetrix, two-color cDNA, and two-color oligonucleotide), although there were prominent differences between platforms and between laboratories. They note three key aspects of comparison studies. (1) An underlying reason that some laboratories report low reproduciblility is that they compare absolute expression measurement of gene expression across platforms. When relative expression is instead studied (comparing between samples within a particular microarray platform), platform-dependent artifacts can be accounted for and removed. (2) The choice of preprocessing approaches has a critical effect on the final results. (3) There is a substantial laboratory effect; for example, if two technicians process samples in an experiment, there may be large differences in expression measurements.

> We discuss further Equation 9.1 in the section on ANOVA below.

These issues of variation in expression data can also be understood by inspection of a statistical model that is commonly used in microarray data analysis:

$$Y_{ijk} = \theta_i + \phi_{ij} + \varepsilon_{ijk} \qquad (9.1)$$

where Y_{ijk} represents a preprocessed probe intensity measurement k (in the $\log_2$ scale) of transcript i measured by platform j; if there are 20,000 transcripts represented on a microarray there will be that many Y_{ijk} values. θ_i is the absolute gene expression value in the $\log_2$ scale, ϕ_{ij} is a platform-specific probe effect, and ε_{ijk} represents a term for measurement error. As noted by Irizarry et al. (2005), a large probe effect ϕ_{ij} (with a large associated variance) tends to inflate the large correlations that have been reported when comparing absolute gene expression measurements within a given platform, while yielding lower correlations between two platforms that differ in their probe effects. A solution is to evaluate relative (rather than absolute) expression within a platform to cancel the ϕ_{ij} terms.

Thus, while many in the community appreciate the demonstrated ability of microarray experiments to generate reproducible results, many factors can strongly influence the results. These factors include appropriate experimental design (e.g., avoiding confounding variables), consistent approach to preparation of RNA through the hybridization steps, appropriate image analysis (in which it is determined which pixels are part of the transcript-associated features), and finally the topics discussed next in this chapter: preprocessing (including global and local background signal correction), identification of differentially expressed transcripts, application of multiple comparison correction, and other analyses such as clustering and classification.

MICROARRAY DATA ANALYSIS: PREPROCESSING

Gene expression changes that are identified could reflect selective, biologically relevant alterations in transcription or they could reflect variations caused by many kinds of experimental artifacts. These artifacts can include the following:

- Different labeling efficiencies of fluorescently (or radioactively) labeled nucleotides.
- Technical artifacts associated with printing tips that deposit cDNAs onto a solid support, such as uneven spotting of DNA onto the array surface.
- Variations in the performance of a fluorescence scanner (used to detect and quantitate fluorescent dyes) or phosphorimager (used for radioactivity-based arrays).
- Variations in the RNA (or mRNA) purity or quantity among the biological samples being studied. For example, there may be heterogeneity in the cell types that are dissected for studies of gene expression in a complex tissue such as the brain.
- Variations in the way the RNA is purified, labeled, and hybridized to the microarray. For example, if gene expression is measured in a cell line at two different time points and the researcher purifies different quantities of RNA, this could lead to experimental artifacts.
- Variations in the way the microarray is washed to remove nonspecific binding.
- Variations in the way the signal is measured.

The main idea of data preprocessing is to remove the systematic bias in the data as completely as possible while preserving the variation in gene expression that occurs because of biologically relevant changes in transcription (Schuchhardt et al., 2000). Some of the key steps in preprocessing are (1) image quantification (referred to in Chapter 8); (2) data exploration, such as scatter plots (discussed next); (3) background adjustment, normalization, and summarization; and (4) quality assessment. Summarization refers to the case of platforms such as Affymetrix for which information about multiple probes is integrated to yield a single measurement for the expression level of one transcript. In Affymetrix expression arrays, 25-mer oligonucleotide probes (that is single-stranded DNA of length 25 bases) are immobilized on an array, and each gene of interest has about a dozen 25-mers that match different portions of the mRNA.

Scatter Plots and MA Plots

One of the most common visualization methods for microarray data is the scatter plot. This shows the comparison of gene expression values for two samples. Most data points typically fall on a 45° line, but transcripts that are up- or downregulated are positioned off the line. The scatter plot displays which transcripts are most dramatically and differentially regulated in the experiment.

We will illustrate scatter plots in more detail using a data set from Mao et al. (2005). The raw intensity values of the mean of seven trisomy 21 (i.e., Down syndrome) brain samples is plotted relative to seven control samples (Fig. 9.2a).

FIGURE 9.2. Scatter plots provide a basic way of analyzing gene expression data from microarray experiments. Data consist of the mean expression values for 22,284 transcripts from trisomy 21 (n = 7) and control (n = 7) brain samples (see web documents 9.1 and 9.2). Plots were made using Microsoft Excel. (a) Plot of raw intensity values for trisomy 21 (Down syndrome) samples (x axis) versus controls (y axis). Each dot represents a transcript. Genes with expression that are upregulated (arrow 1) or downregulated (arrow 2) are indicated. Transcripts expressed at low levels or at background are at the lower left (arrow 3), while transcripts expressed at high levels are at top right (arrow 4). Most data points lie along a 45° line that bisects the data. Note that the scale of the plot is in linear rather than logarithmic units. (b) Transformation of the scale to logarithmic has the effect of distributing the data points more evenly, rather than clumping most values in the lower left corner, as in (a). (c) A plot of the mean logarithmic intensity (x axis) versus the log of the gene expression value ratios (y axis) results in this plot, which is tilted 45° relative to (b). Here, the x axis reflects levels of gene expression, and the y axis reflects up- or down-regulation of gene expression. This is referred to as an MA plot.

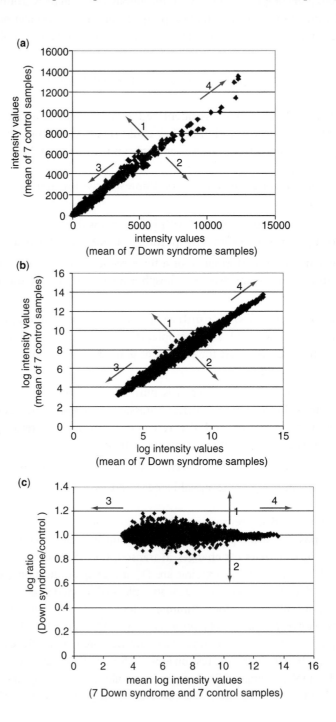

The main feature of this scatter plot (and most such plots of microarray data) is the substantial correlation between the expression values in the two conditions being compared. Transcripts that are upregulated and downregulated fall off the 45° line (arrows 1 and 2). Another feature of this plot is the preponderance of low-intensity values (Fig. 9.2a, lower left). This means that the majority of genes are expressed at only a low level, and relatively few transcripts (arrow 4 at the top right of the scatter plot) are expressed at a high level.

We can $\log_2$ transform the data, producing a plot in which the data points are spread out more evenly (Fig. 9.2b). This creates a more centered distribution in which the properties of the data set are easy to analyze. Also, it is far easier to describe the fold regulation of genes (e.g., twofold upregulated or 1.5-fold downregulated) using a logarithmic scale.

We can review some of the basic values and properties of logarithms in Table 9-1. Consider an example in which expression ratios are measured in a microarray experiment (Table 9-2). Gene expression values are obtained at times $t = 0, 1, 2, 3$. At $t = 1$ the relative value may be unchanged, while at time point $t = 2$ the gene is upregulated twofold and at $t = 3$ the gene is downregulated twofold. The raw ratio values are 1.0, 2.0, and 0.5. Twofold upregulation and twofold downregulation have the same magnitude of change, but in an opposite direction. In raw ratio space, the difference between 1 and 2 is $+1.0$, while the difference between 1 and 0.5 (i.e., time points 1 and 3) is -0.5. In log space (e.g., log base 2 space), the data points are conveniently symmetric about zero. Another key feature of logarithmic transformations is that, in addition to providing symmetry in expression ratios, they stabilize the variance across a wide range of intensity measurements.

A further adjustment is to create an MA plot, which essentially tilts a scatter plot on its side (Fig. 9.2c). The x axis represents the mean of the log intensity values, so that transcripts expressed from low to high levels vary from left to right (arrows 3 and 4). The y axis represents the ratio of the signal intensities in two samples. Here, transcripts that are more upregulated in trisomy 21 have higher y axis values

The plots in Fig. 9.2 were made in Microsoft Excel. Web document 9.1 at ► http://www.bioinfbook. org/chapter9 consists of a spreadsheet with 22,284 raw intensity values from trisomy 21 ($n = 7$) and control ($n = 7$) brain samples. Web document 9.2 includes $\log_2$ intensity values for these same samples and genes, as well as the mean log intensity values of the trisomic ($n = 7$) and control ($n = 7$) groups, and the log ratios. The 14 Affymetrix.cel files containing data from trisomy 21 and control brain samples are posted at ► http://www. bioinfbook.org/chapter9; these .cel files can also be downloaded from the GEO website (accession GSE1397). If you make the plots shown in Fig. 9.2 in S-PLUS (► http://www.insightful.com) or R (► http://www-r-project.org), then you can display all three plots simultaneously. In contrast to Excel, when highlighting a single point or set of data points you can view them in all three plots simultaneously.

To create the plot shown in Fig. 9.2c, you can use the data in columns K and L of the Excel spreadsheet in web document 9.2 (► http://www.bioinfbook.org/ chapter9). To recreate Fig. 9.3, see computer lab problem 9.1 for instructions on how to load .cel files in R and create these plots.

TABLE 9-1 Common Values of Logarithms in Base 2 and Base 10		
Value	Log_{10}	Log_2
1000	3.00	9.97
100	2.00	6.64
50	1.70	5.64
10	1.00	3.32
5	0.70	2.32
2	0.30	1.00
1	0.00	0.00
0.5	-0.30	-1.00
0.2	-0.70	-2.32
0.1	-1.00	-3.32
0.01	-2.00	-6.64
0.001	-3.00	-9.97

Recall that for any positive number b (where b $\neq$ 1), $\log_b y = x$ when $y = b^x$. Thus, $\log_2 8 = 3$ and $2^3 = 8$. Note also that $\log_b b = 1$; $\log_b 1 = 0$; $\log_b xy = \log_b x + \log_b y$; and $\log_b (x/y) = \log_b x - \log_b y$.

TABLE 9-2 Ratios from Microarray Experiments

Time (t)	Behavior of Gene	Raw Ratio Value	Log$_2$ Ratio Value
0	Basal level of expression	1.0	0.0
1	No change	1.0	0.0
2	Twofold upregulation	2.0	1.0
3	Twofold downregulation	0.5	−1.0

Log ratios of gene expression values are often easier to interpret than raw ratios.

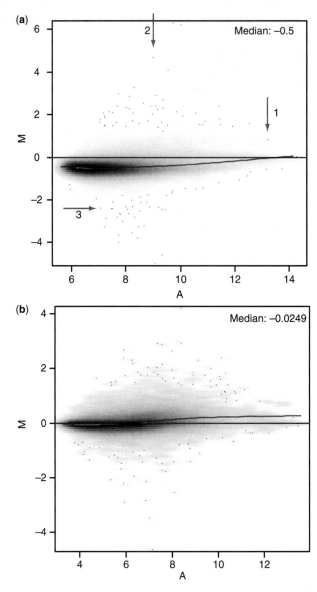

FIGURE 9.3. *MA plots using Bioconductor. (a) Using the R language command* > MAplot (abatch.raw,plot.method="smoothScatter",which=c(14)) *an MA plot of array sample 14 relative to a control was generated. The x axis corresponds to the mean log intensity values. The y axis corresponds to the log ratios of the expression values in sample 14 to a reference. Up-regulated (arrows 1, 2) and down-regulated (arrow 3) transcripts are indicated. (b) Following normalization using RMA, the median value is close to 0, and skewing of the data (evident in (a)) has been improved.*

(arrow 1) while downregulated transcripts have lower y axis values (arrow 2). Such plots help to highlight regulated transcripts, and they also help to visualize aberrant structures in the data such as "smiling" or "frowning" in which there is curvature to the data. This is shown in Fig. 9.3a in which we create an MA plot for a comparison of two data sets in Bioconductor using the R language. The data do not fall along a straight line at the y axis value $M = 0$, but instead are skewed, with a median value of -0.5.

In the R language you can create a variety of plots that complement the scatter plot. Histograms of raw intensity values for 14 .cel files shows that microarray has a distinct profile of intensity values (this further motivates us to next normalize these data) (Fig. 9.4). Boxplots also show the range of raw log intensity values (Fig. 9.5a).

BOX 9.3
The Bioconductor Project and R

R is a freely available computer language and environment for statistical computing and graphics. It is related to the language S which is implemented in the commercial package S-Plus (Insightful Corp.). The R environment allows you to perform calculations (e.g., on matrices), to analyze and graph data, and to implement a series of packages that facilitate data analysis in a wide variety of bioinformatics applications, including microarray data analysis. To get started with R visit its website (▶ http://www.r-project.org) and link to the Comprehensive R Archive Network (CRAN). There you can download R locally onto the Linux, MacOS/X or Windows platforms.

Bioconductor (▶ http:// bioconductor.org) is an open source software project for the analysis of genomic data. While we will describe Bioconductor methods for microarray data analysis in this chapter, it has a wide range of applications, including genomics and proteomics. You can install BioConductor by opening an R session then typing the following:

```
prompt>source("http://bioconductor.org/biocLite.R">
prompt>biocLite()
```

This will invoke biocLite, resulting in the local installation of a group of packages, including the following: affy, affydata, affyPLM, annaffy, annotate, Biobase, Biostrings, DynDoc, gcrma, genefilter, geneplotter, hgu95av2, limma, marray, matchprobes, multtest, ROC, vsn, xtable, affyQCReport. You can further use the command

```
biocLite(c("pkg1", "pkg2"))
```

where pkg1 and pkg2 refer to specific packages you would like to install.

You should create a directory where you will perform your R analyses. Place 14 .cel files in it (seven Down syndrome and seven control samples), obtained from ▶ http://www.bioinfbook.org/chapter9. You will then be able to recreate the figures in this chapter using R.

There are many excellent books and online guides to using R and BioConductor packages, including Gentleman et al. (2005). Sean and Meltzer (2007) describe GEOquery, an R tool that facilitates the import and analysis of data files from GEO.

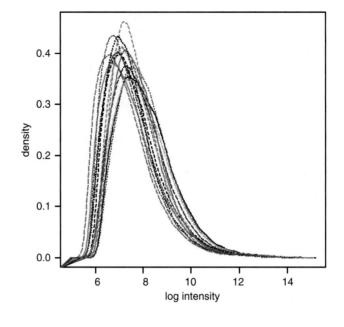

FIGURE 9.4. *Histograms of 14 .cel files. A histogram of unprocessed intensities was created in R using the command* >hist (abatch.raw). *The x axis plots the log intensities, while the y axis plots the density. Differences in intensity values across microarray data sets from different samples highlight the need for normalization steps. Normalization then allows analyses across samples such as the identification of regulated genes in various treatments.*

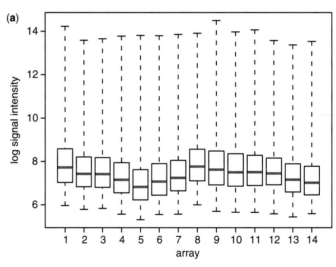

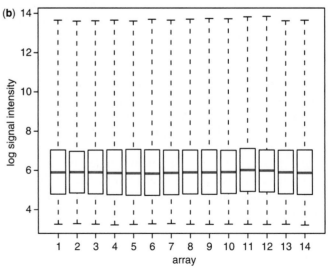

FIGURE 9.5. *Boxplots in R. (a) Analysis of raw data. First, the command* >library(affyPLM) *was used to load the necessary libraries. Then, 14 .cel files were read into an object called* AffyBatch *with the R language command* >abatch.raw <-ReadAffy(). *Next, using the command* >boxplot (abatch.raw) *unprocessed intensities were plotted. The y axis displays the log intensity values, while the x axis corresponds to the 14 data sets (.cel files) that were analyzed. In a boxplot, for each entry the minimum value, lower quartile, median, upper quartile, and maximum values are indicated. (b) Boxplots of data after normalization using RMA.*

Global and Local Normalization

The term "normalization" as applied to microarray data does not refer to the normal (Gaussian) distribution, but instead it refers to the process of correcting two or more data sets prior to comparing their gene expression values.

As an example of why it is necessary to normalize microarray data, the Cy3 and Cy5 dyes are incorporated into cDNA with different efficiencies. Without normalization, it would not be possible to accurately assess the relative expression of samples that are labeled with those dyes; genes that are actually expressed at comparable levels would have a ratio different than one (when considering unlogged data) or zero (for logged data; see below). Normalization is also essential to allow the comparison of gene expression across multiple microarray experiments. Thus, normalization is required for both one- and two-channel microarray experiments.

As a first step, the background intensity signal is measured and subtracted from the signal for each gene (Beissbarth et al., 2000). Empty spots on the array may be used to estimate the background. This background may be constant across the surface of an array or it may vary locally. (We will discuss local background correction below.)

Most investigators apply a global normalization to raw array element intensities so that the average ratio for gene expression is one. The main assumption of microarray data normalization is that the average gene does not change in its expression level in the biological samples being tested. The procedure for global normalization can be applied to two-channel data sets (e.g., Cy3- and Cy5-labeled samples) or one-channel data sets (e.g., Affymetrix chip data). Two-channel data are treated as two individual one-channel data sets such that each element signal intensity is divided by a correction factor specific to the channel from which it was derived. For the two or more data sets being normalized, the intensity for all the gene expression measurements in one channel (Cy5) are multiplied by a constant factor so that the total red and green intensity measurements are equal. As an example, if the mean expression value for samples in the green channel is 10,000 arbitrary units and the mean value for samples in the red channel is 5000, then the expression value for each gene in the red channel would be multiplied by 2. If the data are not log transformed, the mean ratio is then 1. Once the data are log transformed, the mean ratio is zero. Computer lab problem 9-2 at the end of this chapter provides a detailed example of how to perform global normalization of two-channel data from the NCI60 data set.

Other approaches to global normalization are possible. Some investigators normalize all expression values to a set of "housekeeping genes" that are represented on the array. This approach is also used by Affymetrix software. Housekeeping genes might include β-actin and glyceraldehyde-3-phosphate dehydrogenase (GAPDH) and dozens of others. Then each gene expression value in a single array experiment is divided by the mean expression value of these housekeeping genes. A major assumption of this approach is that such genes do not change in their expression values between two conditions. In some cases, this assumption fails. One way to define good candidates for housekeeping genes is to analyze gene expression across a broad range of tissues and conditions. In one project, researchers measured the expression of about 7000 genes in 19 human tissues and deposited the values in a public repository, the Human Gene Expression (HuGE) Index database (Haverty et al., 2002). This database includes a list of 451 housekeeping genes that are commonly expressed across all these tissues.

You can access HuGE at ▶ http://www.biotechnologycenter.org/hio/.

Quantile normalization is an approach that produces the same overall distribution for all the arrays within an experiment (Bolstad et al., 2003). It is a nonparametric method. Parametric tests are applied to data sets that are sampled from a normal (Gaussian) distribution. Common parametric tests include the *t*-test and ANOVA (discussed below). Nonparametric tests do not make assumptions about the population distribution. They rank the outcome variable (here, gene expression measurements) from high to low and analyze the ranks. In quantile normalization, for each array each signal intensity value is assigned to a quantile. We then consider a pooled distribution of each probe across all chips: for each probe, the average intensity is calculated across all the samples. Normalization is performed for each chip by converting an original probe set value to that quantile's value. Figure 9.5b shows a result of using quantile normalization on the trisomy 21 data set. Quantile normalization is incorporated into Robust Multiarray Analysis (RMA) (see below).

Sometimes variance present in gene expression data is not constant across the range of element signal intensities, as shown in Fig. 9.3a. This variation represents an artifact that can be addressed by global and also local normalization processes, which correct bias and variance that are nonuniformly distributed across absolute signal intensity. Many software packages can correct for such variance. One of these, Standardization and Normalization of Microarray Data (SNOMAD), was written by Carlo Colantuoni when he was a graduate student in my laboratory.

Accuracy and Precision

Preprocessing steps are designed to improve accuracy (that is, to lower bias) of gene expression measurements, and to improve precision (that is, to lower the variance). Accuracy (bias) is estimated two ways: by using spike-in samples of known concentrations of RNA, or by diluting known concentrations of RNA. These methods allow an objective assessment of the true positive measurements. The precision (variance) is estimated by using replicate measures of the same sample. We can think of accuracy and precision in terms of a series of arrows hitting a target: accuracy refers to how close the arrows are to the bull's-eye, while precision refers to how reliably the arrows hit the same spot (Fig. 9.6) (Cope et al., 2004). Irizarry et al.

SNOMAD is a web-based tool written in R and available at ▶ http://www.snomad.org; see Colantuoni et al. (2002a, 2002b). Skewing sometimes reflects experimental artifacts, such as the contamination of one RNA source with genomic DNA or rRNA. (Such contaminating nucleic acid could bind to elements on the microarray.) Another source of artifact is the use of unequal amounts of fluorescent probes on the microarray.

A website offering a benchmark for Affymetrix GeneChip expression algorithms is available at ▶ http://affycomp.biostat. jhsph.edu/. You can select algorithms and generate a series of analytic plots at this site, in order to directly compare the performance of dozens of algorithms.

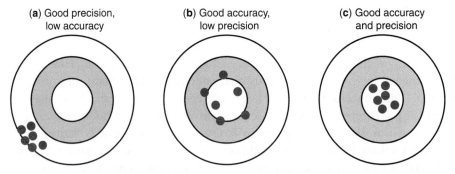

(a) Good precision, low accuracy   **(b)** Good accuracy, low precision   **(c)** Good accuracy and precision

FIGURE 9.6. *Accuracy and precision. (a) Good precision is characterized by reproducible results. It is assessed by repeated measurements of samples (technical replicates). (b) Good accuracy is characterized by measurements that correspond to an independently known result. It can be assessed by measurement of known ("spiked in") concentrations of RNA to an experiment, or by measuring dilutions of known concentrations of RNA. (c) A goal of preprocessing algorithms is to achieve both accuracy and precision.*

(2006) performed a benchmarking study using 31 algorithms for the analysis of Affymetrix probe sets. They concluded that background correction has a large effect on performance, and tends to improve accuracy but worsen precision. The RMA and GCRMA algorithms, introduced below, have consistently performed well in terms of both accuracy and precision and have emerged as leading approaches for the preprocessing of Affymetrix gene expression data.

Robust Multiarray Analysis (RMA)

RMA is a method of performing background subtraction, normalization, and averaging of probe-level feature intensities extracted from .cel files using the Affymetrix

RMA was introduced by Terry Speed, Rafael Irizarry, and colleagues. Early versions of Affymetrix arrays included both perfect match oligonucleotide probes and mismatch probes containing a single base mismatch that are used to estimate background. RMA considers only perfect match values, because mismatch values contribute noise and can have values greater than perfect match probes across as much as one third of a microarray.

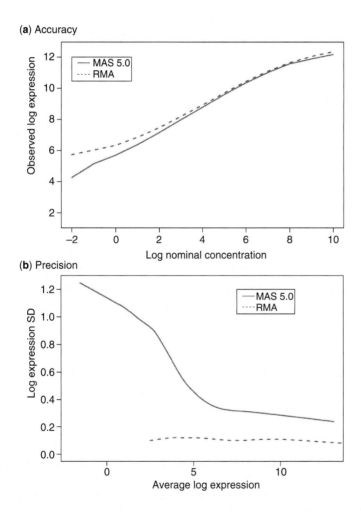

FIGURE 9.7. *Improvements in accuracy and precision using RMA (relative to MAS 5.0 software from Affymetrix). (a) Accuracy is measured by plotting known concentrations of RNA (x axis) versus observed concentrations (y axis). The two methods are comparable. RMA performs slightly worse at low concentrations, a situation that is improved by the GCRMA algorithm. (b) Precision is measured by plotting the average log expression value (x axis) versus the log expression standard deviation (y axis). MAS 5.0 software yields a high standard deviation, particularly for transcripts expressed at low levels, while RMA has a dramatically improved measurement across a broad range of signal intensities.*

platform. It includes steps for background correction, quantile normalization across arrays, a probe-level model fit to each probe set across multiple arrays, and quality assessment. RMA has accuracy comparable to MAS 5.0 software (Affymetrix) (Fig. 9.7a), while its precision is far greater (Fig. 9.7b).

The RMA background correction step includes a convolution model in which the observed signal for each probeset is broken into components of true signal and noise. GCRMA further introduces an adjustment for the presence of nonspecific hybridization that improves accuracy (relative to RMA) while maintaining large gains in precision (relative to other preprocessing techniques). In computer lab problem 9.1 at the end of this chapter, you can use RMA and GCRMA on a trisomy 21 data set, and also invoke a series of quality control plots that demonstrate the effectiveness of these algorithms.

RMA is freely available through the affy package of BioConductor, and it has also been incorporated into a variety of commercial microarray data analysis packages. GCRMA was developed by Zhijin Wu and Rafael Irizarry.

MICROARRAY DATA ANALYSIS: INFERENTIAL STATISTICS

Expression Ratios

The spreadsheet in Fig. 9.8 is available as web document 9.3 at ▶ http://www.bioinfbook.org/ chapter9. The transposed spreadsheet (with 14 samples arranged in rows) is in web document 9.4.

How can you decide which genes are significantly regulated in a microarray experiment? One approach is to calculate the expression ratio in control and experimental cases and to rank order the genes. You might apply an arbitrary cutoff such as a threshold of at least twofold up- or downregulation and define those as genes of interest. Figure 9.8a presents data from a small subset of a Down syndrome experiment with just eight genes and 14 samples; the average value of the Down syndrome samples and the controls is shown (columns C and D) as well as their ratio (column F). In this experiment one might be interested in focusing on the question of whether transcripts derived from chromosome 21 are present at 1.5-fold elevated levels relative to controls, since chromosome 21 is present in a 3:2 ratio.

One problem with a cutoff is that it is usually an arbitrary threshold. In some experiments, no genes (or few) will meet a particular criterion such as twofold; in other experiments, there may be thousands of genes regulated more than twofold in either direction. Also, if the background signal level of a microarray experiment is 50 (in arbitrary units), a gene may be expressed at levels of 150 and 100 in two conditions. After background subtraction, those levels are 100 and 50, and the gene has been regulated twofold. This could have biological significance, but because the absolute values of the expression levels are so close to background, the differences could also represent noise. It is more credible that a gene that is regulated twofold with levels and 4000 versus 2000 units is regulated in a meaningful way.

Expression ratios are important to consider, and they can reveal which genes are most dramatically regulated. But these ratios cannot be converted into probability values to test the hypothesis that particular genes are significantly regulated. Many groups use expression ratios as one of several criteria to apply to microarray data analysis. For example, Iyer et al. (1999) studied the transcriptional response of human fibroblasts to serum and selected genes with expression ratios ≥2.2 for subsequent cluster analysis (described below).

Another possible approach to defining which genes are significantly regulated might be to choose the 5% of genes that have the largest expression ratios. For an experiment with 20,000 gene expression values this would represent 1000 genes. A problem with this approach is that it applies no measure of the extent to which a gene has a different mean expression level in the control and experimental groups.

(a)

| | E11 | ▼ | fx | =TTEST(G11:M11,N11:T11,2,2) | | | | | | | | | | | | | | |

	A	B	C	D	E	F	G	H	I	J	K	L	M	N	O	P	Q	R	S	T
1																				
2	gene	chrom	avg	avg	t-test	ratio			Down syndrome samples								control samples			
3			DS	con		DS/con	DS1	DS2	DS3	DS4	DS5	DS6	DS7	C1	C2	C3	C4	C5	C6	C7
4	RPL41	12	9301	9314	0.97	1.00	9110	9578	8931	9385	9606	8708	9787	8945	10178	10327	8431	9077	9315	8922
5	RPL37A	2	10315	10071	0.70	1.02	9575	10104	9733	11106	10718	10641	10325	8879	9097	8892	13289	10239	9795	10308
6	TUBA6	12	11086	10524	0.28	1.05	10864	10133	10864	12807	10191	11966	10780	9766	9934	10040	11165	9648	11800	11315
7	EEF1A1	6	12329	13530	0.05	0.91	12785	12785	13000	11876	12761	12209	10890	13675	12999	13716	14894	15221	12312	11895
8	ATP5O	21	1437	883	1.2E-06	1.63	1347	1189	1340	1579	1496	1552	1555	793	925	850	992	911	799	914
9	C21orf33	21	484	315	6.59E-06	1.54	480	428	510	498	441	473	555	306	371	362	283	258	319	306
10	WRB	21	769	419	2.67E-05	1.84	816	634	750	705	668	791	1016	427	476	450	492	318	399	368
11	HRMT1L1	21	558	373	7.08E-05	1.50	574	484	465	602	548	535	701	365	365	336	388	368	375	417

(b)

	[,1]	[,2]	[,3]	[,4]	[,5]	[,6]	[,7]	[,8]
DS1	9110	9575	10864	12785	1347	480	816	574
DS2	9578	10104	10133	12785	1189	428	634	484
DS3	8931	9733	10864	13000	1340	510	750	465
DS4	9385	11106	12807	11876	1579	498	705	602
DS5	9606	10718	10191	12761	1496	441	668	548
DS6	8708	10641	11966	12209	1552	473	791	535
DS7	9787	10325	10780	10890	1555	555	1016	701
C1	8945	8879	9766	13675	793	306	427	365
C2	10178	9097	9934	12999	925	371	476	365
C3	10327	8892	10040	13716	850	362	450	336
C4	8431	13289	11165	14894	992	283	492	388
C5	9077	10239	9648	15221	911	258	318	368
C6	9315	9795	11800	12312	799	319	399	375
C7	8922	10308	11315	11895	914	306	368	417

FIGURE 9.8. *Small data set used to demonstrate statistical approaches to microarray data analysis. (a) Intensity values for eight genes (i.e. mRNA transcripts; see rows) were selected. The columns include gene names (column A), chromosomal assignment (column B), and the intensity values for seven Down syndrome samples (DS1 to DS7; columns G to M) and seven control samples (C1 to C7; columns N to T). The first four genes were selected for having high expression levels. The next four genes, all assigned to chromosome 21, were selected because they are differentially regulated: they are present in elevated amounts in Down syndrome samples relative to controls because Down syndrome is associated with three copies of chromosome 21 rather than the normal two copies. The data were imported into Microsoft Excel, and additional columns of information include the average Down syndrome expression value (column C), average control value (D), the p value resulting from a t-test (column E), and the Down syndrome to control ratio (column F, derived from C/D). Note that the cursor is on cell E11 showing the t-test formula in the function box at top. (b) The data matrix is imported into R and transposed, yielding 14 rows (samples) and 8 columns (genes). Problem 9-3 in this chapter describes how to transpose matrices in R.*

It is possible that no genes in an experiment have statistically significantly different gene expression. And yet it will always be possible to rank the genes by expression ratios and to find the group consisting of the most extreme expression ratios. Thus this approach has limited usefulness.

Hypothesis Testing

The goal of inferential statistical analysis of microarray data is to test the hypothesis that some genes are differentially expressed in an experimental comparison of two or more conditions. We formulate the null hypothesis H_0 that there is no difference in signal intensity across the conditions being tested. The alternative hypothesis H_1 is that there are differences in gene expression levels. We define and calculate a test statistic which is a value that characterizes the observed gene expression data. We will accept or reject the null hypothesis based on the results of the test statistic. The probability of rejecting the null hypothesis when it is true is the significance level α, which

The results of a series of *t*-tests on these 14 samples are presented in column G of web document 9.2 (► http://www.bioinfbook.org/ chapter9). Also a *t*-test is implemented in GEO data sets; see problem 9.3 in this chapter.

in science is typically set at $p < 0.05$. Under the null hypothesis, for a set of gene expression intensity values in two conditions, the data are normally distributed with mean 0 and standard deviation σ equal to 1. The standard deviation σ can be estimated using the sample standard deviation s.

The test statistic that you apply to a microarray study depends on the experimental design. Consider the basic paradigm of measuring gene expression in 14 brain samples, seven from trisomy 21 cases (experimental condition, x_1 with observations $x_1, \ldots, x_M$) and seven from apparently normal individuals (control condition, x_2 with observations $x_2, \ldots, x_N$) (Fig. 9.8a). You can calculate the mean and standard deviation for the expression of each gene represented on the microarray. A *t*-test is performed to test the null hypothesis that there is no difference in gene expression levels, considered one gene at a time, between the two populations. Compute the mean expression value for each gene from control (x_1) and experimental (x_2) conditions, estimate the variance, and divide them. The average for each sample (e.g., $\bar{x}_1$) is given by:

$$\bar{x}_1 = \frac{1}{M} \sum_{i=1}^{M} x_i \tag{9.2}$$

The variance (or square of the standard deviation, s^2) for x_1 is given by:

$$s_{x1}^2 = \frac{1}{M-1} \sum_{i=1}^{M} (x_i - \bar{x})^2 \tag{9.3}$$

The *t*-test essentially measures the signal-to-noise ratio in your experiment by dividing the signal (difference between the means) by the noise (variability estimated in the two groups).

$$t\text{-}statistic = \frac{\bar{x}_1 - \bar{x}_2}{\sqrt{\left(\frac{s_{x1}^2}{M} + \frac{s_{x2}^2}{N} \right)}} \tag{9.4}$$

From the *t*-statistic we can calculate a p value. This allows us to either reject or accept the null hypothesis that the control and experimental conditions have equal gene expression values (i.e., the null hypothesis is that there is no differential expression). For a *t*-test that provides a p value of 0.01, this means that one time in 100 the observed difference between the control and experimental groups will occur by chance, and we can safely reject the null hypothesis. Figure 9.8a presents the results of *t*-tests performed on eight genes. Four genes assigned to chromosome 21 have very low p values (about 10^{-5} to 10^{-6}), one has a p value of 0.05, and three have nonsignificant p values (0.97, 0.70, 0.28).

We can think about the usefulness of the *t*-test by considering four genes (i.e., RNA transcripts) (Fig. 9.9) for which we have gene expression measurements from seven samples. Genes 1 and 2 are expressed at low levels, but there is considerable "noise" (variability) in the measurement of gene 2. In a comparison of genes 1 and 2, both the mean values (which may differ) and the variability in the measurement (which shows overlap) are accounted for in a *t*-test. For genes expressed at high levels, the variance may also be small (gene 3) or large (gene 4). If we measure gene 4 in seven controls and also in seven diseased cases (indicated by gene 4* in

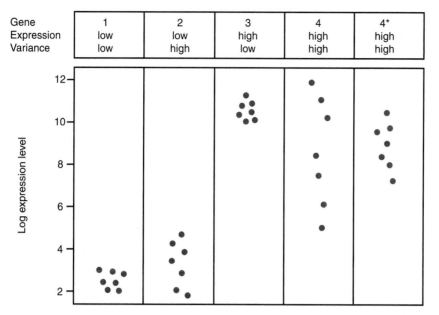

Gene Expression Variance	1 low low	2 low high	3 high low	4 high high	4* high high

FIGURE 9.9. Gene-specific variance is addressed by a t-test. For four hypothetical genes the log expression values are plotted (y axis). Gene 1 has a low absolute expression level and low variance upon repeated measurements in biological replicate samples, while gene 2 has a low expression level and relatively high variance. Genes 3 and 4 are expressed at high levels, with gene 3 having low variance and gene 4 having high variance. Each RNA transcript has a characteristic property of its expression level (although this may vary dramatically across body regions and across developmental stages). When we compare gene 4 in two conditions (indicated by gene 4 and 4*, such as beta globin levels in normal red blood cells and sickle cells), a t-test accounts for the difference in mean between the two measurements, and it also provides an analysis of the variation in expression measurements within each of the two samples.

Fig. 9.8), a relatively large sample size will be necessary to achieve sufficient statistical power to reject the null hypothesis. The power of a statistical test is defined as $1 - \beta$, where β is the probability of concluding there is no significant difference between two means, when in fact the alternative hypothesis is true. (β is the same as the probability of making a type II error.) The larger the sample size, the larger the power. Page et al. (2006) introduced PowerAtlas, a web-based resource to help calculate power estimates for microarray experiments.

An assumption of the *t*-test approach is that gene expression values are normally distributed. If so, the *t*-statistic follows a distribution that allows us to calculate a set of *p* values. (An alternate assumption is that for very large numbers of replicates the *t*-statistic is normally distributed with mean 0 and standard deviation of 1, and again we can compute *p* values. In practice, very large numbers of replicates are rarely available for microarray studies.)

Parametric tests such as the *t*-test assume a normal distribution. In contrast nonparametric tests rank the outcome variables and do not assume a normal distribution. Nonparametric tests, such as the Mann–Whitney and Wilcoxon tests, are less influenced by data points that are extreme outliers. Such tests are not commonly applied to microarray data. Many other approaches have been implemented, such as Bayesian analysis of variance (Ishwaran et al., 2006).

The test that is used depends on the experimental paradigm. Some examples of experimental designs are shown in Fig. 9.10. For a between-subject design (Fig. 9.10a) there are two groups. Golub et al. (1999) measured gene expression in

You can access PowerAtlas at
► http://www.poweratlas.org/.

Many commercial microarray software packages perform hypothesis testing on microarray data, and most include the option to apply conservative corrections. These packages also include a variety of data visualization tools: GeneSight and Genedirector (► http://www.biodiscovery.com), GeneSpring (► http://www.chem.agilent.com/), GeneTraffic (► http://www.iobion.com/), Partek Pro (► http://www.partek.com), SAS (► http://www.sas.com), S-PLUS (► http://www.insightful.com), Spotfire (► http://spotfire.com), and SPSS (► http://www.spss.com).

samples from patients with acute leukemias that occur in two subtypes. In this experimental design it is necessary to control for confounding factors such as differences in age, gender, or weight between individuals in the two groups. For a within-subject design (Fig. 9.10b) a paired *t*-test would be used to test for the differences in mean values between two sets of measurements on paired samples. An example of this is a study by Perou et al. (2000) measuring gene expression in surgical biopsy samples before and after drug treatment of breast tumors. Here the covariates (sometimes called "nuisance variables") such as age and gender are internally controlled.

How can we be sure that the probability value we derive from a test statistic is not just obtained by chance, that is, because of random changes in gene expression? A permutation test can be performed in which the labels associated with each sample (e.g., diseased vs. control) are randomized. The same test statistic is applied to each gene, and *p* values are measured. A large set of permuted tests (e.g., 100 to 1000) is run, and the null hypothesis is rejected if the observed *p* value is smaller than any *p* value from the permutation test.

FIGURE 9.10. Examples of experimental design for microarray experiments involving gene expression profiling. Most such microarray experiments are designed to test the hypothesis that there are significant biological gene expression differences between samples as a function of factors such as tissue type (normal vs. diseased or brain vs. liver), time, or drug treatment. (a) A between-subject design must control for confounding factors such as age, gender, or weight. (b) A within-subject design removes genetic variability and can be used to measure gene expression before and after some treatment. (c) A two-way between-subject design allows the measurement of differences between both treatment and control conditions, and another factor such as gender. (d) A within-subject factorial design might be used to study two treatments over time. (e) In a mixed factorial design there is both a between-subject design (e.g., normal vs. diseased tissue) and a within-subject design (e.g., gene expression measurements over time).

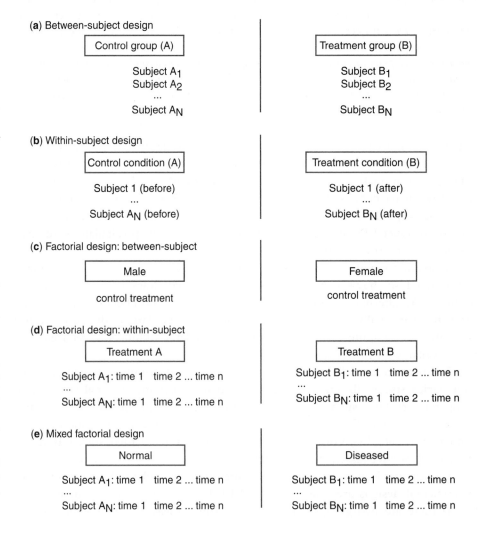

Corrections for Multiple Comparisons

What p value cutoff is appropriate to establish statistical significance? If you measure the expression values for 10,000 transcripts, you can expect to find differences in 5% of them (500 transcripts) purely by chance that are nominally significant at the $p < 0.05$ level. If you hypothesized that one specific gene was significantly regulated, then this α level would be appropriate. However, for 10,000 measurements it is necessary to apply some conservative correction to account for the thousands of repeated, independent measurements you are making. There are two problems we want to avoid. Type I errors (false positive results) involve concluding that a transcript is differentially expressed when it is not; the null hypothesis is true but is inappropriately rejected. Type II errors (false negative results) involve failing to identify a truly regulated transcript; the null hypothesis that is actually false is not rejected as it should be. A p value is defined as the minimum false positive rate at which an observed statistic is categorized as significant.

There are several approaches to accounting for the problem of multiple comparisons. At one extreme, some researchers apply a conservative Bonferroni correction in which the α level for statistical significance is divided by the number of measurements taken (e.g., $p < 0.05/10,000$ is set as the criterion for significance). This correction is considered too severe. A more commonly used approach to the multiple comparisons correction problem is to adjust the false discovery rate (FDR). This is defined as follows:

$$FDR = \frac{\#false\ positives}{\#called\ significant} \qquad (9.5)$$

The FDR represents the rate at which genes identified as significantly regulated are not. For an FDR of 0.05, 5% of the transcripts that are called significant are false positives. For 100 significantly regulated genes and an FDR of 8%, 8 genes out of 100 are expected to represent false positive results. We can contrast the FDR with the false positive rate (FPR) which is:

$$FPR = \frac{\#false\ positives}{\#truly\ null} \qquad (9.6)$$

The FPR measures the rate at which genes that are truly not regulated are called significant.

Significance Analysis of Microarrays (SAM)

Significance analysis of microarrays (SAM) is a modified t-test that finds significantly regulated genes in microarray experiments (Tusher et al., 2001). SAM assigns a score to each gene in a microarray experiment based on its change in gene expression. Statistical significance is assessed using a permutation test in which observed scores are compared to the results of repeated measurements from a shuffled data set.

SAM offers several useful features. The program is convenient to use as a Microsoft Excel plug-in. It accepts microarray data from experiments using a variety of experimental designs such as those outlined in Fig. 9.10. Prior to operating SAM, the user must normalize and scale expression data. (This can be accomplished within Microsoft Excel.) The SAM input data can be in a raw or log-transformed format. Each row of the data matrix contains expression values for one gene, and the columns correspond to samples. SAM uses a modified t-statistic (Equation 9.4) to test the null hypothesis (see Tusher et al., 2001).

If you have just 100 gene expression measurements in a comparison of two groups and there is no difference in gene expression, you would expect to observe $(100)(0.05) = 5$ significantly regulated genes by chance. But assuming that these tests are statistically independent, the probability of obtaining at least one apparently significant result is $1 - 0.95^{100} = 0.994$ (see Olshen and Jain, 2002). It is for this reason that a correction needs to be applied.

You can obtain SAM at ▶ http:// www-stat.stanford.edu/~tibs/ SAM/. Note that effective permutation tests require a large number of permutations (≥ 100) and a reasonably large number of samples (e.g., ≥ 5 in each group). With too few samples, the test is not robust.

A key feature of SAM is its ability to provide information on the false discovery rate (the percent of genes that are expected to be identified by chance). The user can adjust a parameter called delta to adjust the false positive rate: for example, in a typical experiment, for every 100 genes declared significantly regulated according to the test statistic, 10 might be false positives (thus the false discovery rate would be 10%). This false positive rate can be decreased by the user (at the cost of missing true positives) or increased (at the cost of obtaining more false negatives). The SAM algorithm calculates a "q value," which is the lowest false discovery rate at which a gene is described as significantly regulated.

An example of a SAM output is shown in Fig. 9.11a. The genes are ranked according to the test statistic and plotted to show the number of observed genes versus the expected number (Fig. 9.11b). This graph (called a q–q plot) effectively

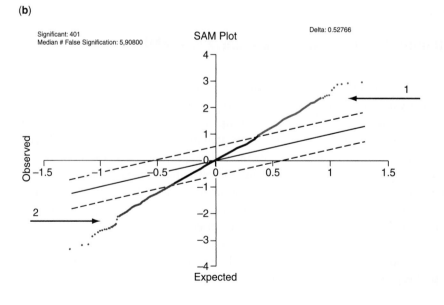

FIGURE 9.11. SAM is a Microsoft Excel plug-in that reports significantly regulated genes using a modified t-statistic. The input to SAM is a matrix of gene expression values and a response variable (e.g., control, experimental). The user selects a parameter delta to determine the cutoff for significance based on the false positive rate. The user can also choose an appropriate fold-change measurement. (a) The output includes a list of significantly regulated genes. The score d is the t-statistic value for each gene; the numerator and denominator in the spreadsheet refer to the difference between the means of the gene expression values being compared and the estimate of the standard deviation of the numerator, respectively. The q value is the false discovery rate. (b) The output includes a plot of expected versus observed expression values; significantly upregulated (arrow 1) and downregulated (arrow 2) genes are color coded as well as listed in the output of (a).

visualizes the outlier genes that are most dramatically regulated. In SAM, a permutation test is used to assess the significance of expressed genes; the test statistic is measured 100 or more times for each gene with the sample labels (e.g., control vs. experimental conditions) randomized.

You can perform similar FDR analyses using the R language or software such as Partek.

From *t*-Test to ANOVA

A variety of test statistics may be applied to microarray data (e.g., Olshen and Jain, 2002). Some of these are listed in Table 9-3. These tests are all used to derive *p* values that help assess the likelihood that particular genes are regulated. For more than two conditions (e.g., analyzing multiple time points or measuring the effects of several drugs on gene expression), the analysis of variance (ANOVA) method is appropriate rather than a *t*-test. The ANOVA identifies differentially expressed genes while accounting for variance that occurs both within groups and between groups (Ayroles and Gibson, 2006; Zolman, 1993). ANOVA is particularly appropriate when a microarray experiment has multiple classes of treatment (e.g., control samples are compared to two different disease states or to five different time points) or multiple factors for each treatment (e.g., gender, age, date of RNA isolation, hybridization batch).

ANOVA is a statistical model called a general linear model such as the one shown in Equation 9.1. A general form of the linear model takes the form:

$$Y = \mu + \beta_1 x_1 + \beta_2 x_2 + \cdots + \beta_j x_j + \varepsilon \tag{9.7}$$

Y is a linear function of X with slope β and intercept μ, and $x_1, x_2, \ldots, x_j$ is a series of independent variables. The terms ϕ and θ in Equation 9.1 are examples of independent variables associated with expression measurement and probe effects. ε is an error term corresponding to residual, unexplained variance. Both fixed and random factors are independent variables accounted for in the linear model. Fixed factors involve treatment effects systematically selected by the experimenter (such as gender or age) that would remain the same if the experiment were replicated. Fixed factors account for the main conditions that an investigator is interested in, such as the change in signal intensity due to a sample coming from Down syndrome rather than control individuals. Random factors provide a model of independent variables that are selected randomly or unsystematically from a population. Examples are

TABLE 9-3   Test Statistics for Microarray Data

Paradigm	Parametric Test	Nonparametric Test
Compare one group to a hypothetical value	One-sample *t*-test	Wilcoxon test
Compare two unpaired groups	Unpaired *t*-test	Mann–Whitney test
Compare two paired groups	Paired *t*-test	Wilcoxon test
Compare three or more unmatched groups	One-way ANOVA	Kruskal–Wallis test
Compare three or more matched groups	Repeated-measures ANOVA	Friedman test

Source: Adapted from Motulsky (1995) and Zolman (1993).

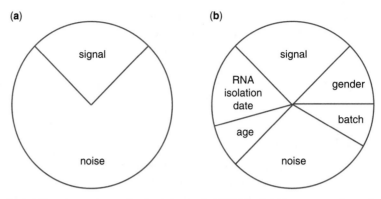

FIGURE 9.12. *Signal-to-noise ratios in t-test and ANOVA. (a) In a t-test, the values from a microarray experiment can be thought of as having components of signal (i.e., intensity measurements that reflect a difference between the means of the two groups being compared) and noise (variations in signal intensity that are not attributable to differences in the means of the two groups). If the RNA from control samples is purified on a Monday, and the RNA from experimental samples is purified on a Tuesday, then there is a perfect confound between date and condition, and some or even all of the observed difference between control and experimental samples could be due to date rather than to treatment. (b) In an ANOVA, fixed and/or random effects can be accounted for. The variable due to factors such as date and gender can be analyzed, as well as the main effect of interest (control versus experimental conditions). By partitioning the data into multiple components, ANOVA improves the signal-to-noise ratio.*

biological replicates, because when we select a group of seven Down syndrome samples we are drawing them in an unbiased manner from the overall population of individuals with Down syndrome. Similarly array effects are random factors because each microarray is randomly selected from the group of all available arrays.

The idea of ANOVA is that differences in gene expression may be due to main effects (e.g., normal versus diseased sample), while other sources of variation (e.g., gender or age) can be identified and accounted for. Analogous to the *t*-statistic of a *t*-test, the *F* statistic of an ANOVA consists of a ratio of signal to noise (Fig. 9.12). However, the ANOVA includes a more detailed estimate of the sources of variation. By partitioning the signal to account for fixed and random effects in the data, the ANOVA boosts the signal-to-noise ratio, often allowing you to more effectively identify regulated transcripts.

MICROARRAY DATA ANALYSIS: DESCRIPTIVE STATISTICS

One of the most fundamental features of microarray experiments is that they generate large amounts of data. There are far more measurements (gene expression values) than samples. How can we evaluate the results of an experiment in which 20,000 gene expression values are obtained in 10 cell lines? Each gene expression value can be conceptualized as a point in 20,000-dimensional space. The human brain is not equipped to visualize highly dimensional space, and so we need to apply mathematical techniques that reduce the dimensionality of the data.

Mathematicians refer to the problems associated with the study of very large numbers of variables as the "curse of dimensionality." In highly dimensional space, the distances between any two points are very large and approximately equal. Descriptive statistics are useful to explore such data. These mathematical approaches typically do not yield statistically significant results because they are not used for hypothesis testing. Rather, they are used to explore the data set and to try to find biologically meaningful patterns. A clustering tree, for example, can show how genes (or samples) form groups. Particular genes can subsequently be used for hypothesis testing.

Several main descriptive techniques are available for the visualization of microarray data, including two that we will focus on, clustering and principal components analysis. Each of these approaches involves the reduction of highly dimensional data to allow conclusions to be reached about the behavior of genes and/or samples in either individual microarray experiments or multiple experiments. In each case, we begin with a matrix of genes (typically arranged in rows) and samples (typically arranged in columns). Appropriate global and/or local normalizations are applied to the data. Then some metric is defined to describe the similarity (or alternatively to describe the distance) between all the data points.

The approaches we will describe first are unsupervised. Here, prior assumptions about the genes and/or samples are not made, and the data are explored to identify groups with similar gene expression behaviors. We will then examine supervised clustering approaches in which the number of clusters are prespecified.

Hierarchical Cluster Analysis of Microarray Data

Clustering is a commonly used tool to find patterns of gene expression in microarray experiments (reviewed in Gollub and Sherlock, 2006). Genes, samples, or both may be clustered in trees. Clustering is the representation of distance measurements between objects. Clusters are commonly represented in scatter plots or in dendrograms, such as those used for phylogenetic analysis (Chapter 7) or for microarray data. The main goal of clustering is to use similarity (or distance) measurements between objects to represent them. Data points within a cluster are more similar, and those in separate clusters are less similar. It is common to use a distance matrix for clustering based on Euclidean distances.

There are several kinds of clustering techniques. The most common form for microarray analysis is hierarchical clustering, in which a sequence of nested partitions is identified resulting in a dendrogram (tree). (We will describe a nonhierarchical clustering technique, k-means clustering, below.) Hierarchical clustering can be performed using agglomerative or divisive approaches (Fig. 9.13). In each case, the result is a tree that depicts the relationships between the objects (genes, samples, or both). In divisive clustering, the algorithm begins at step 1 with all the data in one cluster ($k = 1$). In each subsequent step a cluster is split off, until there are n clusters. In agglomerative clustering, all the objects start apart. Thus, there are n clusters at step 0; each object forms a separate cluster. In each subsequent step two clusters are merged, until only one cluster is left.

Agglomerative and divisive clustering techniques generally produce similar results, although large differences can occur. Let us return to a small portion of the Mao et al. (2003) data set: 14 samples (seven with trisomy 21, seven controls) and just eight genes (four that are expressed at very high levels but are not differentially regulated, and four that are assigned to chromosome 21 and are present at elevated levels in the trisomy 21 samples). We obtain a matrix of 8 genes (rows) by 14 samples (columns) containing raw intensity values (Fig. 9.8a). This matrix may be transposed (Fig. 9.8b).

Agglomerative clustering is sometimes called "bottom up" while divisive clustering is "top down."

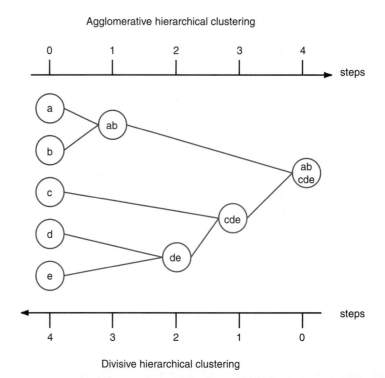

Agglomerative hierarchical clustering

Divisive hierarchical clustering

FIGURE 9.13. *There are two main kinds of hierarchical clustering: agglomerative and divisive. In agglomerative clustering, the data points (genes or samples, represented as the letters a to e) are considered individually (step 0). The two most related data points are joined (circle ab, step 1). The relationship between all the data points is defined by a metric such as Euclidean distance. The next two closest data points are identified (step 2, de). This process continues (steps 3 and 4) until all data points have been combined (agglomerated). The path taken to achieve this structure defines a clustering tree. Divisive hierarchical clustering involves the same process in reverse. The data points are considered as a combined group (step 0, abcde). The most dissimilar object is removed from the cluster. This process is continued until all the objects have been separated. Again, a tree is defined. In practice, agglomerative and divisive clustering strategies often result in similar trees. Adapted from Kaufman and Rousseeuw (1990). Used with permission.*

Using the data matrix in Fig. 9.8b and importing it into S-PLUS software, we can produce three clustering trees (Fig. 9.14). For each tree, the y axis (height) represents dissimilarity. On clustering trees such as these the samples (or genes if the transposed matrix is used) are represented across the x axis so as to be evenly spaced, and the significance of their position depends on the cluster to which they belong. Note that while the overall topologies are similar, several of the samples have distinctly different placements on the tree in agglomerative (Fig. 9.14a) versus divisive clustering (Fig. 9.14b). For example, note that samples C1 and C2 (dashed arrows) cluster close to DS6 and DS7 (solid arrows) in the first tree, but are distantly related to DS6 and DS7 in the second tree.

In general, different exploratory techniques may give subtle or dramatic differences in their representation of the data. Agglomerative techniques tend to give more precision at the bottom of a tree, while divisive techniques offer more precision at the top of a tree and may be better suited for finding relatively few, large clusters. Another feature of a clustering tree is that it may be highly sensitive to the choice of which genes (or samples) to include or exclude. When we remove four

(a) Agglomerative

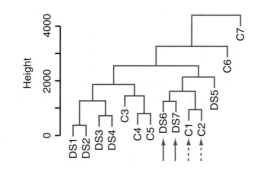

(b) Divisive

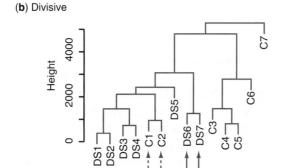

(c) Agglomerative (chromosome 21 transcripts only)

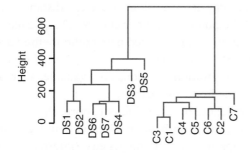

FIGURE 9.14. *(a) Agglomerative hierarchical clustering of microarray data and (b) divisive hierarchical clustering. The data set of 14 samples and eight transcripts (Fig. 9.8b) was clustered using the S-PLUS program. The y axis reflects dissimilarity, and the 14 samples are spaced evenly across the x axis. Note that while most of the groupings are similar between the agglomerative and divisive algorithms, there are notable differences in the placement of samples C1 and C2 (dashed arrows) relative to samples DS6 and DS7 (solid arrows). Discrepancies commonly occur between agglomerative and divisive strategies. (c) The removal of four transcripts derived from non-chromosome 21 genes, leaving only transcripts derived from chromosome 21 genes, now results in a tree that separates Down syndrome from control samples. This illustrates the influence that can be exerted by inclusion or exclusion of selected data points. Some studies identify significantly regulated genes and then perform clustering. Hierarchical clustering is an unsupervised technique, and the preselection of regulated genes in a set of samples guarantees that those samples will be separated in a tree; thus, the tree should not be used as evidence of successful classification.*

non-chromosome 21 genes from the analysis, and include only a set of four chromosome 21-derived transcripts, we can now completely separate the Down syndrome from the control samples (Fig. 9.14c). In some cases the inclusion (or exclusion) of particular data points will reveal fundamental information about the underlying biological phenomena.

Clustering requires two basic operations. One is the creation of a distance matrix (or in some cases a similarity matrix). The two most commonly used metrics used to define the distance between gene expression data points are Euclidean distance (Box 9.4) and the Pearson coefficient of correlation (Box 9.2). Many software packages that perform microarray data analysis allow you to choose between these and other distance measures (such as Manhattan, Canberra, binary, or Minkowski) that describe the relatedness between gene expression values.

The Canberra distance metric is calculated in R by

$$\sum \left(\frac{|x_i - y_i|}{|x_i + y_i|} \right).$$

Terms with zero numerator and denominator are omitted from the sum and treated as if the values were missing.

BOX 9.4
Euclidean Distance

Euclidean distance is defined as the distance d_{12} between two points in three-dimensional space (with coordinates x_1, x_2, x_3 and y_1, y_2, y_3) as follows:

$$d_{12} = \sqrt{(x_1 - y_1)^2 + (x_2 - y_2)^2 + (x_3 - y_3)^2}$$

Euclidean distance thus is the square root of the sum of the squared differences between two features. For n-dimensional expression data, the Euclidean distance is given by

$$d = \sqrt{\sum_{i=1}^{n} (x_i - y_i)^2}$$

Figure 9.15a shows the clustering of eight genes (from 14 samples) in the R statistical package using the hclust command. This procedure is described in detail in problem 9-3 later in this chapter. The distance metric is the default choice of Euclidean distance. Figure 9.15b shows the result in which the distance metric is changed to Canberra, dramatically altering the tree topology.

Given a distance metric, a second operation is the construction of a tree. In addition to selecting agglomerative or divisive approaches as outlined above, we can select a variety of ways of defining clusters. In the hclust package of R the default approach is complete linkage (as is used in Fig. 9.15a and b). The distance between clusters is also commonly defined using the average distance between all the points in one cluster and all the points in another cluster. This is called average-linkage clustering, and it is used in the unweighted pair-group method average (UPGMA). We described the UPGMA procedure in Chapter 7 in the context of phylogenetic trees. However, many additional options are available, such as single linkage (Fig. 9.15c) and Wards's method (Fig. 9.15d).

> Ward's minimum variance method optimizes finding compact, spherical clusters.

What is the significance of these different approaches to making a clustering tree? We can consider the general problem involved in defining a cluster. Objects that are clustered form groups that have homogeneity (internal cohesion) and separation (external isolation) (Sneath and Sokal, 1973; Everitt et al., 2001). The relationships between objects being studied, whether intensity measurements from microarray data or operational taxonomic units (OTUs) in phylogeny, are assessed by similarity or dissimilarity measures. Intuitively, the objects in Fig. 9.16a form two distinct clusters. However, after shifting just two of the data points to create Fig. 9.16b it is not clear whether there are two clusters. Other challenges to identifying the nature of clusters are depicted in Figs. 9.16c and d. Each figure shows two apparent clusters that demonstrate both homogeneity and separation. However, if we identify a central point in each cluster (the centroid) and calculate the distance to the farthest points within a cluster, that distance will also result in overlap with the adjacent cluster.

There are several methods available to calculate the proximity between a single object and a group containing several objects (or to calculate the proximity between two groups). In single linkage clustering, an object that is a candidate to be placed into a cluster has a similarity that is defined as its similarity to the closest member

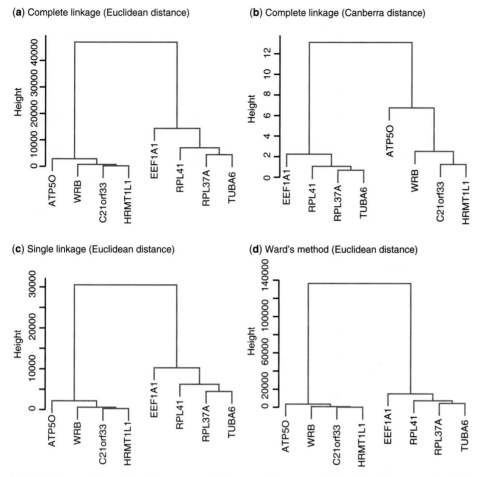

FIGURE 9.15. *Hierarchical clustering of eight genes (in 14 samples) using different methods for creating a distance matrix and for converting the distance matrix as a dendrogram (tree). Data (see Fig. 9.8) were imported into R. A distance matrix was computed using the default distance matrix method (Euclidean distance; panels a, c, and d) or the Canberra method (panel b). The distance matrix was converted into a clustering dendrogram (tree) by performing hierarchical clustering on the set of dissimiliarities from the distance matrix. The methods employed were the default "complete" method (panels a and b), the single linkage method, and Ward's minimum variance method. These four dendrograms provide examples of the many methods available for making distance matrices and converting them into trees; note the differences in the y axis values, and the differences in how the objects (genes) are displayed. All four trees effectively separate the four genes expressed at high levels from those expressed at low levels. You can reproduce these figures by following problem 9-3 at the end of this chapter.*

within that cluster. This method has also been called the minimum method or the nearest neighbor method. It is subject to an artifact called chaining in which "long straggly clusters" form (Sneath and Sokal, 1973, p. 218) as shown in Fig. 9.16f. This can obscure the production of discrete clusters. In complete linkage clustering the most distant OTUs in two groups are joined (Fig. 9.16g); the effect is to tend to form tight, discrete clusters that join other clusters relatively rarely. Many alternative strategies exist (see Sneath and Sokal, 1973). In centroid clustering the central or median object is selected (Fig. 9.16h). These methods often produce different clustering patterns.

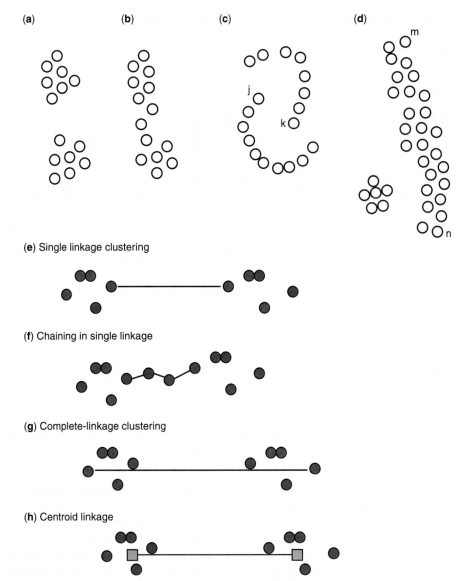

(a)   **(b)**   **(c)**   **(d)**

(e) Single linkage clustering

(f) Chaining in single linkage

(g) Complete-linkage clustering

(h) Centroid linkage

FIGURE 9.16. Examples of the nature of clusters and clustering approaches. (a) Two clusters are intuitively apparent in a group of 14 data points (circles). Good clusters are characterized by internal cohesion and by separation. (b) Two data points are shifted relative to (a), making the assignment of two clusters more questionable. (c) Two clusters are clearly present, by inspection ("c" shapes). However, the separation between each cluster is not robust. For example, point j in the lower cluster may be closer to the center of the upper cluster than point k, even though j is not a member of the upper cluster. (d) Two clusters are again intuitively apparent. The great distance from the long cluster (e.g., points m to n) presents a challenge to finding a rule that distinguishes that cluster from the small one to its left. Such challenges motivate the development of algorithms to define distances between objects and clusters. (e) Single linkage clustering identifies the nearest neighboring objects between clusters. (f) The single linkage approach is sometimes subject to the artifact of chaining in which clusters that might reasonably be expected to remain separate are instead connected and merged. (g) Complete linkage clustering identifies the farthest members of each cluster. This approach tends to generate tight, well separated clusters that exclude objects from clusters. (h) Centroid linkage or average linkage represents a compromise approach to placing objects in clusters. Source: (a)–(d) adapted from Gordon (1980).

Dozens of programs perform cluster analysis. The data in Fig. 9.14 were generated using S-PLUS (Insightful). A popular clustering programs for microarray data is Cluster and its associated tree visualization program, TreeView (Eisen et al., 1998). The input for this software (and other similar programs) is a spreadsheet of expression values for genes and samples (Fig. 9.17). As described above, data can be adjusted by log transformation. Data may also be normalized to set the magnitude (sum of the squares of the values) of a row and/or column vector to 1.0. Data filtering allows genes to be removed, typically because the maximum or minimum values exceed some threshold. The distance metric used by Cluster is the Pearson correlation coefficient r, and the algorithm performs agglomerative hierarchical clustering. Gene expression values are color coded from bright red (most upregulated) to bright green (most downregulated). This allows one to visualize trends or patterns in large data sets.

Two-way clustering of both genes and samples is used to define patterns of genes that are expressed across a variety of samples. A dramatic example is provided by Alizadeh et al. (2000), who defined subtypes of malignant lymphocytes based on gene expression profiling (Fig. 9.18).

Cluster was developed by Michael Eisen and is available through his website, ▶ http://rana.lbl.gov/.

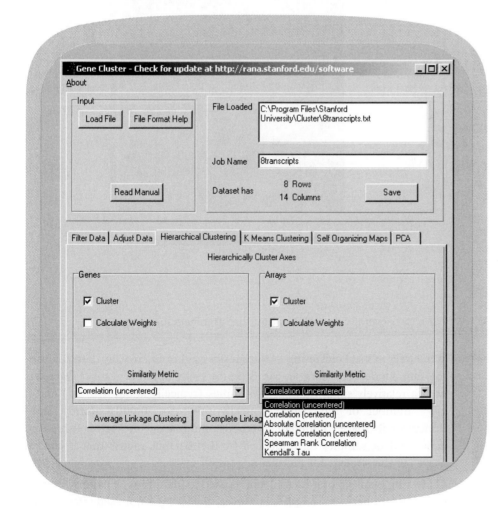

FIGURE 9.17. Two popular programs for the analysis of microarray data are Cluster and Treeview. Data are entered into Cluster as tab-delimited text files. Rows represent genes, and columns represent samples or observations. The program allows a variety of analyses, including hierarchical clustering, k-means clustering, self-organizing maps, and principal components analysis. As shown in this image, options include six different similarity metrics and the choice of average, complete, or single linkage clustering.

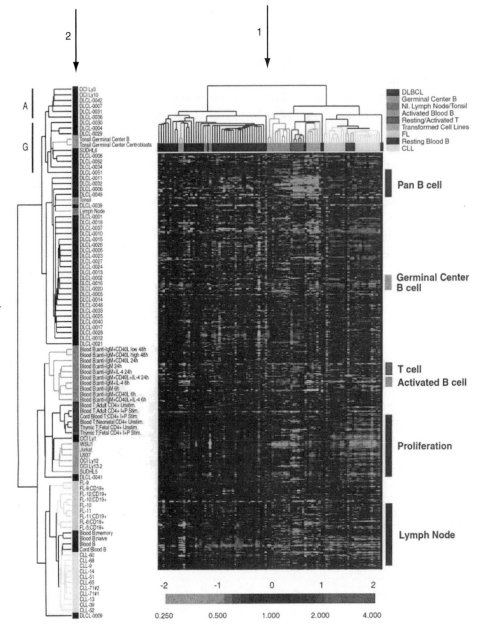

FIGURE 9.18. Example of two-way hierarchical clustering. Alizadeh et al. (2000) made 1.8 million measurements of gene expression in 96 samples of normal and malignant lymphocytes. The cell lines are clustered in columns across the top (arrow 1), and for clarity they are also shown rotated sidewise at left (arrow 2). The genes are arranged in rows. The investigators used a custom-made microarray with 17,856 cDNA clones. This study revealed that tumors from patients with diffuse large B-cell lymphoma can be classified according to their gene expression profiles. Patients with particular tumors have varying severity of phenotype, and such heterogeneity is reflected at the molecular level. These data were generated using Cluster and Treeview. Used with permission.

We can draw several conclusions about hierarchical clustering.

- While hierarchical clustering is commonly used in microarray data analysis, the same underlying data set can produce vastly different results. Data sets with a relatively small number of samples (typically about 4 to 20) and a large number of transcripts (typically 5000 to 30,000) occupy high-dimensional space. Different methods summarize the relationships of genes and/or samples as influenced by the distance metric that is chosen as well as the strategy for producing a tree.

- Clustering is an exploratory tool, and is used to identify associations between genes and/or between samples. However, clustering is not used inferentially.

- Clustering is not a classification method (see below). It is unsupervised in that information about classes (e.g., Down syndrome versus control) is not used to generate the clustering tree.

- The concept of a cluster is not defined well mathematically (see Fig. 9.16). Nonetheless, Thalamuthu et al. (2006) proposed a metric to score the accuracy of clustering algorithms, using both simulated and real data. They found that hierarchical clustering performed poorly relative to other techniques such as k-means clustering and model-based clustering. We will next consider some of these other clustering techniques.

Partitioning Methods for Clustering: *k*-Means Clustering

Sometimes we know into how many clusters our data should fit. For example, we may have treatment conditions we are evaluating, or a set number of time points. An alternative type of unsupervised clustering algorithm is a partitioning method that constructs k clusters (Tavazoie et al., 1999). The steps are as follows: (1) Choose samples and/or genes to be analyzed. (2) Choose a distance metric such as Euclidean. (3) Choose k; data are classified into k groups as specified by the user. Each group must contain at least one object n (e.g., gene expression value), and each object must belong to exactly one group. (In all cases, $k \leq n$.) Two different clusters cannot have any objects in common, and the k groups together constitute the full data set. (4) Perform clustering. (5) Assess cluster fit.

Several types of k-means clustering are performed by programs such as S-PLUS, Partek, and Cluster/TreeView.

How is the value of k selected? If you perform a microarray experiment with two different kinds of diseased samples and one control sample type, you might choose a value for $k = 3$. Also, k may be selected by a computer program that assesses many possible values of k. The output of k-means clustering does not include a dendrogram because the data are partitioned into groups, but without a hierarchical structure.

The k-means clustering algorithm is iterative. It begins by randomly assigning each object (e.g., gene) to a cluster. The center ("centroid") of each cluster is calculated (defined using a distance metric). Other cluster centers are identified by finding the data point farthest from the center(s) already chosen. Each data point is assigned to its nearest cluster. In successive iterations, the objects are reassigned to clusters in a process that minimizes the within-cluster sum of squared distances from the cluster mean. After a large number of iterations, each cluster contains genes with similar expression profiles. Tavazoie et al. (1999) described the use of k-means clustering to discover transcriptional regulatory networks in yeast.

A concern with using k-means clustering is that the cluster structure is not necessarily stable in that it can be sensitive to outliers. Cluster fit has been assessed using a variety of strategies such as the addition of random noise to a data set.

Clustering Strategies: Self-Organizing Maps

The self-organizing map (SOM) algorithm resembles k-means clustering in that it partitions data into a two-dimensional matrix. For SOMs and other structured clustering techniques, you can estimate the number of clusters you expect (e.g., based on the number of experimental conditions) in order to decide on the initial number of clusters to use.

The SOM approach to microarray data analysis has been championed by Todd Golub, Eric Lander, and colleagues from the Whitehead Institute.

Unlike k-means clustering, which is unstructured, SOMs impose a partial structure on the clusters (Tamayo et al., 1999) (Fig. 9.19). Also in contrast to k-means clustering, adjacent partitions in SOMs can influence each other's structure. The principle

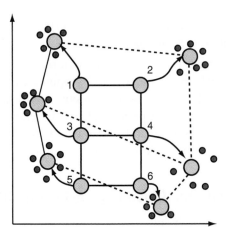

FIGURE 9.19. *Self-organizing maps allow partial structuring to be imposed on clusters. This contrasts with k-means clustering, which imposes a fixed number of clusters. An initial set of nodes (numbered one to six) forms a rectangular grid. During iterations of the self-organizing map algorithm, the nodes migrate to new positions (arrows) to better fit the data. Red dots represent data points. Redrawn from Tamayo et al. (1999). Used with permission.*

of SOMs is as follows. One chooses a number of "nodes" (similar to a value k) and also an initial geometry of nodes such as a 3×2 rectangular grid (indicated by solid lines in the figure connecting the nodes). Clusters are calculated in an iterative process, as in k-means clustering, with additional information from the profiles in adjacent clusters. Nodes migrate to fit the data during successive iterations. The result is a clustering tree with an appearance similar to those produced by hierarchical clustering.

Principal Components Analysis: Visualizing Microarray Data

PCA is also called singular-value decomposition (Alter et al., 2000). It is a linear projection method; this means that the data matrix you start with is "projected" or mapped onto lower dimensional space. Projection methods related to PCA include independent components analysis, factor analysis, multidimensional scaling, and correspondence analysis.

Principal components analysis (PCA) is an exploratory technique used to find patterns in gene expression data from microarray experiments. It is both easy to use and powerful in its ability to represent complex data sets succinctly. PCA is used to reduce the dimensionality of data sets in order to create a two- or three-dimensional plot that reflects the relatedness of the objects that it clusters—that is, the genes and/or samples in your experiment. PCA has been used to analyze expression data in yeast and mammalian systems (Landgrebe et al., 2002; Misra et al., 2002; Alter et al., 2000; Wall et al., 2001; Bouton et al., 2001).

The central idea behind PCA is to transform a number of variables into a smaller number of uncorrelated variables called principal components. The variables that are operated on by PCA may be the expression of many genes (e.g., 20,000 gene expression values), or the results of gene expression across various samples, or even both gene expression values and samples. In a typical microarray experiment, PCA detects and removes redundancies in the data (such as genes whose expression values do not change and thus are not informative about differences in how the samples behave).

We will consider the small matrix of eight gene expression values in 14 samples from Fig. 9.8a. We can convert this matrix into a series of PCA plots of samples (Fig. 9.20a to c) or genes (Fig. 9.20d). The results are comparable to those obtained by hierarchical clustering in terms of how the relationships of genes and/or samples are described. For the visualization of samples, the trisomic cases are clearly distinguished from the control cases (Fig. 9.20a) based on inspection of the first principal component axis (x axis). This separation is even more dramatic when we focus the analysis on the relationships of the 14 samples using only four differentially regulated chromosome 21-derived transcripts (Fig. 9.20b), and the separation vanishes when we analyze the relationships of the samples based on four nonregulated transcripts (Fig. 9.20c).

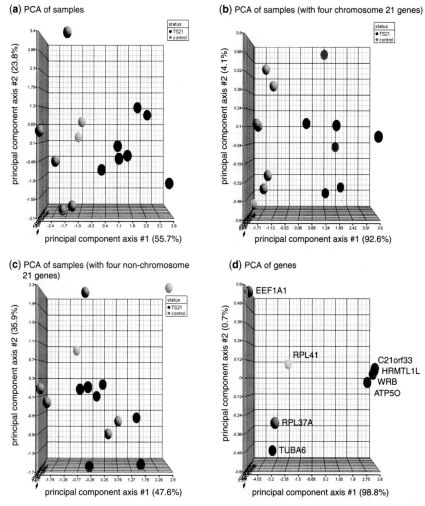

FIGURE 9.20. *Principal components analysis (PCA) reduces the dimensionality of microarray data to visualize the relationship between genes or samples. (a) A PCA plot of 14 samples shows that the trisomy 21 samples (black spheres) can be distinguished from the matched control samples (gray spheres) along the first principal component axis, that is, along the x axis; 55.7% of the variance in the data set is explained along this axis. This is a substantial amount. Along the second principal component axis, 23.8% of the variance is explained, and there is no separation of trisomic and control samples. Note that in PCA space two points that are close together have similar, related properties in the original data matrix. (b) When PCA is performed using data only from four chromosome 21 genes (which are expressed at higher levels in the trisomic samples), the PCA clustering shows a more dramatic separation of controls (gray spheres) and trisomic samples (black spheres). Since the first principal component axis now explains 92.6% of the variance, the separation of control and trisomic samples across the x axis is extremely convincing. (c) When PCA is performed using data from four highly expressed genes (but genes that are not differentially regulated), there is no separation of control and trisomic samples. (d) When the data matrix is transposed (as in Fig. 9.8b), the PCA plot shows the relation of the eight genes based on their expression values across 14 samples. Note that the four highly expressed genes (EEF1A1, RPL41, RPL37A, TUBA6) are extremely well separated along the first principal component axis from the four transcripts assigned to chromosome 21 that are expressed at low levels (C21orf33, HRMTL1, WRB, ATP5O). The separation of the four highly expressed genes on the y axis may occur because of the variability in their expression levels. Note that in PCA the principal component axes may have a straightforward biological interpretation (such as representing gene expression intensity levels or diseased versus normal samples) but the data points are placed on the plot without any supervision or prior assumptions.*

In performing PCA, by default the first principal component accounts for as much as the variability in the data as possible. The second principal component will account for less of the variability than the first. Thus, in Fig. 9.20 the y axis in each plot is associated with a smaller percentage of variance explained than the x axis. The mathematical operations that produce each principal component axis require that they be orthogonal variables; this means that they are uncorrelated to each other (see below). It is typical to display PCA as a two-dimensional plot with the first principal component on the x axis and the second principal component on the y axis. However, a three-dimensional plot is also commonly used (as shown in Fig. 9.20). Additional principal component axes usually account for only a very small amount of variability in the data matrix and are sometimes tabulated.

The starting point for PCA is any matrix of m observations (gene expression values) and n variables (experimental conditions). The goal is to reduce the dimensionality of the data matrix by finding r new variables (where $r < n$). These r variables account for as much of the variance in the original data matrix as possible. The first step of PCA algorithms is to create a new matrix of dimensions $n \times n$. This may be a covariance matrix or a correlation matrix. (In our example in Fig. 9.20a, there is a 14×14 covariance matrix.) The principal components (called eigenvectors) are selected for the biggest variances (called eigenvalues). What this means practically for our example data set is that if a gene's expression values do not vary across the samples, it will not contribute to the formation of the principal components.

How is the first principal component axis related to our raw data? Take the three-dimensional plot of the raw data and redraw the x, y, z coordinate axes so that the origin ("centroid") is at the center of all the data points (Fig. 9.21). Find the line that best fits the data; this corresponds to the first principal component axis. By rotating this axis, it becomes the x axis of the plots in Fig. 9.20. The second principal component axis must also pass through the origin of the graph in Fig. 9.21, and it must be orthogonal to the first axis. In this way, it is uncorrelated. Each axis accounts for successively less of the variability in the data.

This description of PCA is highly simplified. For a description of the vector algebra underlying PCA, see Kuruvilla et al. (2002), Misra et al. (2002), or Landgrebe et al. (2002).

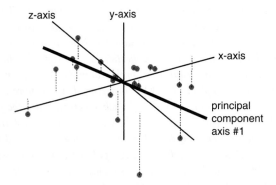

FIGURE 9.21. Principal components analysis. The first principal component axis may be thought of as the best-fit line that traverses the geometric origin of the data set, accounting for most of the variability in the data. The second principal component (not shown) also passes through the origin and is orthogonal to the first component. Cumulatively, all the principal component axis account for 100% of the variance, with each axis accounting for a successively smaller percentage. A large percentage accounted for by the first and/or second principal component axes indicates that the importance of this axis should be given when inspecting the PCA plot.

The final product of PCA is usually a two- or three-dimensional plot consisting of points in space that correspond to either genes or samples. If we use PCA to represent samples, a close distance between any two points implies that their overall pattern of gene expression is similar. Conversely, two points that are separated in a PCA plot have different overall profiles. The initial data set is highly dimensional; for 20,000 gene expression measurements the points could theoretically be described in 20,000-dimensional space. PCA reduces the dimensionality of the data to just two or three dimensions. In reducing the dimensionality, the goal of PCA is to provide as much information as possible about the original data set.

Raychaudhuri and colleagues (2000) used PCA to analyze the full Chu et al. (1998) data set, consisting of 6118 gene expression measurements across seven time points. Their PCA analysis showed that just two principal components accounted for over 90% of the total variability. They further suggested that these components correspond to (1) overall induction of genes and (2) the change in induction level over time. In general, the principal component axes may or may not correspond to variables that have an obvious biological interpretation. This is because the components capture as much information in the data set as possible based strictly on the criterion of variance.

We encountered PCA in an entirely different context, as an application in the protein family database Pfam (Fig. 6.9). There, it is used to describe the relationships between proteins based on a matrix of pairwise sequence alignments. PCA is also used to express the relationships between entire sequenced genomes in the Clusters of Orthologous Genes (COG) database (Chapter 15).

Multidimensional scaling (MDS) is another dimensional reduction technique. MDS plots represent the relationships between objects from a similarity (or dissimilarity) matrix in a manner comparable to principal components analysis, although the axes do not report the percent variance that is captured. The application of MDS to microarray data is discussed by Chen and Meltzer (2005).

The total number of dimensions in PCA can be as large as the sample size in the original data matrix, but most of the information content in PCA is found in the first two or three principal components.

Chen and Meltzer (2005) describe an implementation of multidimensional scaling using MATLAB® software. Visit ► http://research.nhgri.nih.gov/microarray/MDS_Supplement/ for scripts and data sets. Additional software for MDS is available from BRB ArrayTools (► http://linus.nci.nih.gov/BRB-ArrayTools.html), ggobi (► http://www.ggobi.org/), Partek (► http://www.partek.com), R (via the cmdscale function; ► http://www.r-project.org), and S-PLUS (► http://www.insightful.com).

Supervised Data Analysis for Classification of Genes or Samples

The distances and similarities among gene expression values can be described using two types of analysis: supervised or unsupervised. The unsupervised approaches we have described so far are especially useful for finding patterns in large data sets. In supervised analyses, the approach is different because the experimenter assumes some prior knowledge of the genes and/or samples in the experiment. For example, transcriptional profiling has been performed on cell lines or biopsy samples that are either normal or cancerous (e.g., Alizadeh et al., 2000; Shipp et al., 2002; Khan et al., 1998; Perou et al., 1999; West et al., 2001). (In some cases, the cancerous samples are further subdivided into those that are relatively malignant or relatively benign.) Some of these studies apply unsupervised approaches.

The goal of supervised microarray data analysis algorithms is to define a rule that can be used to assign genes (or conditions) into groups. In each case, we begin with gene expression values from known groups (e.g., normal vs. cancerous) and "train" an algorithm to learn a rule. Positive and negative examples are used to train the algorithm. The algorithm is then applied to unknown samples, and its accuracy as a predictor or classifier is assessed. It is critical that the data used for building a classifier are entirely separate from the data used to assess its predictive accuracy.

Some of the most commonly applied supervised data analysis algorithms are k-nearest neighbors, support vector machines, supervised machine learning, neural networks, and linear discriminant analysis. As an example of a supervised approach, Brown et al. (2000) used support vector machines to classify six functional classes of yeast genes: tricarboxylic acid cycle, respiration, cytoplasmic ribosomes, proteasome, histones, and helix–turn–helix proteins. They used a threefold cross-validation method: the data set is divided into thirds (sets 1, 2, and 3). Sets 1 and 2 are used to train the support vector machine; then the algorithm is tested on set 3 as the "unknowns." Next, sets 1 and 3 are used for training and set 2 is tested as the unknowns. Finally, sets 2 and 3 are used for training, and set 1 is tested. They measured the false positive rate and found that support vector machines outperform both unsupervised clustering and alternative supervised clustering approaches.

Dupuy and Simon (2007) described many strategies for properly performing supervised analyses, and also listed many of the common data analysis errors. For example, improperly performing cross-validation leads to overly optimistic prediction accuracy. It is also essential to have an adequate sample size for both the training and the test sets.

FUNCTIONAL ANNOTATION OF MICROARRAY DATA

DRAGON is available at ▶ http://pevsnerlab. kennedykrieger.org. Other websites are for DAVID (▶ http:// david.abcc.ncifcrf.gov/), GenMAPP (▶ http://www. genmapp.org/), Resourcerer (at the Dana Farber Cancer Institute and Harvard University; ▶ http:// compbio.dfci.harvard.edu/tgi/), the Stanford Online Universal Resource for Clones and ESTs (SOURCE; ▶ http://source. stanford.edu), and MILANO (▶ http://milano.md.huji.ac.il).

A major task confronting the user of microarrays is to learn the biological significance of the observed gene expression patterns. Often researchers rely on manual literature searches and expert knowledge to interpret microarray results. Several software tools accept lists of accession numbers (corresponding to genes that are represented on microarrays) and provide annotation. Christopher Bouton, when he was a graduate student in my laboratory, developed the Database Referencing of Array Genes Online (DRAGON) database. This includes a website that allows microarray data to be annotated with data from publicly available databases such as UniGene, Pfam, SwissProt, and KEGG (Bouton and Pevsner, 2000; Bouton et al., 2003). DRAGON offers a suite of visualization tools allowing the user to identify gene expression changes that occur in gene or protein families (Bouton and Pevsner, 2002). The goal of annotation tools such as DRAGON is to provide insight into the biological significance of gene expression findings. We use DRAGON to annotate the chromosomal assignment of genes represented on microarrays. A variety of related annotation tools (many with more features than DRAGON) have been developed, including Database for Annotation, Visualization and Integrated Discovery (DAVID; Dennis et al., 2003), GenMAPP, SOURCE, and Resourcerer (Tsai et al., 2001). Microarray Literature-based Annotation (MILANO; Rubinstein and Simon, 2005) is an example of software that annotates expression data with relevant literature citations. For example, you can input a list of RefSeq entries to retrieve relevant PubMed matches, including the option of guiding the output with secondary search terms.

GSEA software is available from the Broad Institute at ▶ http:// www.broad.mit.edu/gsea/.

An active area of research is the annotation of microarray data based on functional groups such as Gene Ontology categories (we will introduce Gene Ontology in Chapter 10). The premise is that in addition to considering individual transcripts that are significantly regulated, one can identify groups that are functionally related (such as transcripts that encode kinases or that function in mitochondrial biogenesis). Tools to analyze data sets based on annotation groups include GOMiner (Zeeberg et al., 2005) and are reviewed by Osborne et al. (2007). Gene Set Enrichment Analysis (GSEA) represents one increasingly popular approach to identifying regulated sets of genes (Subramanian et al., 2005). That procedure includes over 1000 defined gene sets in which members of each set share common features (such as biological pathways). With all these annotation procedures it is important to keep in mind that the product of mRNAs is protein. Identification of a set of mRNAs encoding proteins in a particular cellular pathway does not mean that the proteins themselves are present in altered levels, nor does it mean that the function of that pathway has been perturbed. Such conclusions can only be drawn from experiments on proteins and pathways performed at the cellular level.

PERSPECTIVE

DNA microarray technology allows the experimenter to rapidly and quantitatively measure the expression levels of thousands of genes in a biological sample. This technology emerged in the late 1990s as a tool to study diverse biological questions. Thousands to millions of data points are generated in microarray experiments. Thus, microarray data analysis employs mathematical tools that have been established in other data-rich branches of science. These tools include cluster analysis, principal components analysis, and other approaches to reduce highly dimensional data to a useful form. The main questions that microarray data analysis seeks to answer are as follows:

- For a comparison of two conditions (e.g., cell lines treated with and without a drug), which genes are dramatically and significantly regulated?
- For comparisons across multiple conditions (e.g., analyzing gene expression in 100 cell lines from normal and diseased individuals), which genes are consistently and significantly regulated?
- Is it possible to cluster data as a function of sample and/or as a function of genes?

A further challenge is to translate the discoveries from microarray experiments into further insight about biological mechanisms.

Dupuy and Simon (2007) reviewed 90 publications in which gene expression profiles were related to cancer outcome. Half of the studies they reviewed in detail had at least one of three flaws: (1) Controls for multiple testing were not properly described or performed. (2) In class discovery a correlation was claimed between clusters and clinical outcomes. However, such correlation is spurious because differentially expressed genes were identified then used to define clusters. (3) Supervised predictions included estimates of accuracy that were biased because of incorrect cross-validation procedures. Dupuy and Simon (2007) offer a useful and practical list of 40 guidelines for the statistical analysis of microarray experiments, spanning topics from data acquisition to identifying differentially regulated genes, class discovery, and class prediction.

Finally, while DNA microarrays have been used to measure gene expression in biological samples, they have also been used in a variety of alternative applications. Microarrays have been used as a tool to detect genomic DNA (e.g., to identify polymorphisms, to obtain DNA sequence, to identify regulatory DNA sequence, to identify deletions and duplications, and to determine the methylation status of DNA; see Chapter 16). Such diverse applications are likely to expand in the near future.

PITFALLS

Errors occur in a variety of stages of microarray experiments:

- Experimental design is a critical but often overlooked stage of a microarray experiment. It is essential to study an adequate number of experimental and control samples. The appropriate number of replicates must also be employed. While there is no consensus on what this number is for every experiment, there must be adequate statistical power and using one to three biological replicates is often insufficient.

- It is difficult to relate intensity values from gene expression experiments to actual copies of mRNA transcripts in a cell. This situation arises because each step of the experiment occurs with some level of efficiency, from total RNA extraction to conversion to a probe labeled with fluorescence and from hybridization efficiency to variability in image analysis. Some groups have introduced universal standards for analysis of a uniform set of RNA molecules, but these have not yet been widely adopted.

- Data analysis requires appropriate attention to global and local background correction. Benchmark studies suggest that while excellent approaches have been developed (such as GCRMA), applying different normalization procedures will lead to different outcomes (such as differing lists of regulated transcripts).

- For exploratory analyses, the choice of distance metric, such as Pearson's correlation coefficient, can have a tremendous influence on outcomes such as clustering of samples.

- Each data analysis approach has advantages and limitations. For example, popular unsupervised methods (such as cluster analysis) sacrifice information about the classes of samples that are studied (such as cell lines derived from patients with different subtypes of cancer). Supervised methods make assumptions about classes that could be false.

- Many experimental artifacts can be revealed through careful data analysis. Skewing of scatter plots may occur because of contamination of the biological sample being studied. Cluster analysis may reveal consistent differences, not between control and experimental conditions, but between samples analyzed as a function of day or operator.

DISCUSSION QUESTIONS

[9-1] A microarray data set can be clustered using multiple approaches, yielding different results. How can you decide which clustering results are "correct" (most biologically relevant)? For microarray data normalization we described the concepts of precision and accuracy; do these apply to clustering as well?

[9-2] What are the best criteria to use to decide if a gene is significantly regulated? If you apply fold change as a criterion, will there be situations in which a fold change is statistically significant but not likely to be significant in a biological sense? If you apply a conservative correction and find that no genes change significantly in their expression levels in a microarray experiment, is this a biologically plausible outcome?

PROBLEMS/COMPUTER LAB

[9-1] Using R and BioConductor:

(a) Install R and Bioconductor packages as described in Box 9.3.

(b) In R, use the command `library(affyPLM)` to load various packages such as affy, Biobase, and gcrma.

(c) Type `abatch.raw <- ReadAffy()` to read all the .cel files in the current directory into an object called AffyBatch. Additional cel definition files (CDF) are automatically downloaded to provide annotation of your data.

(d) Create the plots shown in Figs. 9.3 to 9.5 using the commands `MAplot(abatch.raw)`, `hist(abatch.raw)`, and `boxplot(abatch.raw)`.

(e) Invoke robust multiarray analysis (RMA) by typing `eset.rma <- rma(abatch.raw)`. This function `rma()` returns an expression set `exprSet` that stores expression values. Optionally, use the `gcrma()` command.

(f) Create boxplots and MAplots as described above.

(g) Create additional plots using the commands `Pset <- fitPLM(abatch.raw)` as well as `RLE(Pset)` and `NUSE(Pset)` for various quality control assessments. Information on any of these commands is available in R by typing queries such as `?NUSE`. NUSE is normalized unscaled standard errors. Try the commands `boxplot(eset.rma)`, `Mbox(abatch.raw)`, and `NUSE(Pset)`.

[9-2] In this problem we will obtain ratios as well as raw, two-channel intensity values from the NCI60 data of the National Cancer Institute (NCI).

(a) Data acquisition. The data are available from GEO at NCBI (data set GDS1761) in the form of intensity ratios. These data are also posted as web document 9.5 at ▶ http://www.bioinfbook.org/chapter9 in the form of an Excel document as downloaded in the SOFT format from GEO. In web document 9.6, these same data are presented in an Excel spreadsheet that has been edited to simplify its format. Column A of web document 9.6 contains a list of genes from 1 to 9,706; column B contains names for each of the genes on the microarrays. Most of the remaining columns correspond to the 60 samples that comprise the NCI60 experiment, grouped into samples such as breast cancer and brain cancer. We also include the mean value for the breast cancer set (column L) and the brain set (column S) as well as t-tests comparing these groups (column T). Insert a new column and perform t-tests yourself to compare two groups.

(b) In this problem we will obtain and plot raw, two-channel intensity values from the NCI60 data of the National Cancer Institute (NCI). You can access the NCI60 data in a variety of formats (both raw and processed) at ▶ http://discover.nci.nih.gov/cellminer/ by selecting the raw data option for "RNA: cDNA Array" data. Web document 9.7 presents a Microsoft Excel spreadsheet with the data from one breast cancer sample (BT549) relative to a reference pool. Column A contains names for each of the genes on the microarrays; column B contains IMAGE consortium

identifiers (see Chapter 8); columns C and D list the accession numbers of clones at the 5′ and 3′ ends of each gene. Column E lists the Cy5 channel background intensity for the breast cancer cell line, while F shows the total intensity minus the background intensity; columns G and H show similar information for the Cy3 channel, corresponding to a reference pool of RNA from control cell lines. Column I lists flags (visually identified artifacts, of which there are 19 in this data set out of 9,707 total elements on the microarray). Columns J and K show the Cy5 and Cy3 intensity values after normalization, and column L shows their ratio.

(c) Making scatter plots. Select columns F and H (with headers CH1D and CH2D). Create a scatter plot. What are the mean values in columns F and H? To check, go to the bottom of column F, and in an empty cell use the function command =AVERAGE(F2:F9707). Similarly, calculate the mean of column H. Note that data in columns F and H have not been normalized. After normalization (columns L and M), are the means now the same?

(d) Create logarithmic scatter plots and MA plots. Label four columns (MNOP) as follows: log_CH1_FL; log_CH2_FL; ratio; mean_logs. In column M, create the logarithms (base 2) of CH1_FL; next, create the logarithms of CH2_FL; next add a column of their ratios, and finally a column of the mean log intensity values. Plot the ratios versus the mean log intensities (columns O versus P) in a scatter plot. Web document 9.8 shows the result.

[9-3] Perform t-tests using the NCBI GEO website.

(a) From the home page of NCBI, enter NCI60. Select the "GEO DataSets" link. One of the entries, GDS1761 (NCI60 cancer cell lines), includes an icon of a clustering tree. Click that tree to access the GEO DataSet record (the URL is ▶ http://www.ncbi.nlm.nih.gov/geo/gds/gds_browse.cgi?gds = 1761).

(b) This page includes a table "Find genes differentially expressed between groups." Select a two-tailed t-test at the 0.05 significance level, and choose two groups to compare (e.g. breast tumor in group A versus CNS tumor in group B). Click the button to query A versus B. The result is a set of transcripts (Entrez GEO Profiles) that are differentially regulated according to a t-test ($p < 0.05$).

[9-4] Perform hierarchical clustering using R. Obtain a matrix of genes ($n = 8$) and samples ($n = 14$) from web document 9.3 at ▶ http://www.bioinfbook.org/chapter9. Copy this as a text file into an R working directory. Then use the following commands (# indicates a comment line). You should be able to reproduce the clustering trees shown in Figs. 9.13 and 9.14.

```
dir()
#view the contents of your directory; this should include the file
#myarraydata.txt
z=read.delim("#myarraydata.txt")
#read.delim is a principal way of reading a table of data into R.
#This creates a new file called z
#with 8 rows (genes) and columns including gene name,
#chromosomal locus, and 14 samples.
```

```
z
#view the data matrix z consisting of 8 genes and 14 samples
row.names(z)=z[,1]
clust=hclust(dist(z[,3:16]),method
="complete")
#create a distance matrix using columns 3 to 16; perform hier
#archical clustering using the complete linkage agglomeration
#method
plot(clust)
#generate a plot of the clustering tree, such as a figure shown in
#this chapter
#Note that you can repeat this using a variety of different
#methods (e.g. method="single" or method="median".
#Type ?hclust for more options.
z.back=z[,-c(1,2)]
#create a version of matrix z called z.back in which two columns
#containing the gene names and chromosomal loci are removed.
z.back
#view this matrix
w=t(z.back)
#create a new file called w by transposing z.back.
w
#view matrix w. There are now 4 rows (samples) and 8 columns
#(genes).
clust=hclust(dist(w[,1:8]),method="complete")
plot(clust)
```

```
#perform clustering. The cluster dendrogram now shows 14
#samples (rather than 8 genes).
clust = hclust(dist(z[,3:16],method = "euclidean"),method =
   "complete")
plot(clust)
clust = hclust(dist(z[,3:16],method =
   "manhattan"),method = "complete")
plot(clust)
clust = hclust(dist(z[,3:16],method =
   "minkowski"),method = "complete")
plot(clust)
clust = hclust(dist(z[,3:16],method = "binary"),method =
   "complete")
plot(clust)
clust = hclust(dist(z[,3:16],method = "maximum"),method =
   "complete")
plot(clust)
clust = hclust(dist(z[,3:16],method = "canberra"),method =
   "complete")
plot(clust)
#You can vary the metric by which you create a distance matrix
#(e.g. Euclidean, Manhattan, Minkowski, binary, maximum,
#Canberra) as well as varying the clustering method ("ward",
#"single", "complete", "average", "mcquitty",
#"median" or "centroid").
```

Self-Test Quiz

[9-1] It is necessary to normalize microarray data because:

  (a) Gene expression values are not normally distributed.

  (b) Some experiments use cDNA labeled with fluorescence while others employ cDNA labeled with radioactivity.

  (c) The efficiency of dye incorporation (or radioactivity incorporation) may vary for different samples.

  (d) Housekeeping genes (such as action) may be expressed as varying levels between samples.

[9-2] Microarray data analysis can be performed with scatter plots. The information you get from a scatter plot includes all of the following EXCEPT:

  (a) You can tell whether a gene is expressed at a relatively high level or a low level.

  (b) You can tell whether a gene has been upregulated or downregulated.

  (c) You can tell whether a gene forms a cluster with other genes on the microarray.

  (d) You can tell whether a gene is among the 5% most regulated genes in that experiment.

[9-3] Log ratios of gene expression values are often used rather than raw ratios because:

  (a) Twofold upregulation or twofold downregulation log ratios each have the same absolute value.

  (b) Twofold upregulation or twofold downregulation log ratios each have the same relative value.

  (c) The scale of log ratios is compressed relative to the scale of raw ratios.

  (d) A plot of log ratios compresses the expression values.

[9-4] Inferential statistics can be applied to microarray data sets to perform hypothesis testing:

  (a) In which the probability is assessed that any individual transcript is significantly regulated in a comparison of two samples

  (b) In which the probability is assessed that any individual transcript is significantly regulated in a comparison of two or more samples

  (c) By clustering of array data

  (d) By either supervised or unsupervised analyses

[9-5] Which one of the following statements is FALSE?

  (a) Clustering of microarray data produces a tree that can resemble a phylogenetic tree.

  (b) Clustering of microarray data can be performed on genes and/or samples.

  (c) Clustering of microarray data can be performed with partitioning methods (such as k-means) or hierarchical methods (such as agglomerative or divisive clustering).

  (d) Clustering of microarray data is always performed using principal components analysis.

[9-6] Clustering techniques rely on distance metrics to:

  (a) Describe whether a clustering tree is agglomerative or divisive

(b) Reduce the dimensionality of a highly dimensional data set

(c) Identify the absolute values of gene expression measurements in a matrix of gene expression values versus samples

(d) Define the relatedness of gene expression values from a matrix of gene expression values versus samples

[9-7] A self-organizing map:

(a) Imposes some structure on the formation of clusters

(b) Is unstructured, like k-means clustering

(c) Has neighboring nodes that represent dissimilar clusters

(d) Cannot be represented as a clustering tree

[9-8] Principal components analysis (PCA):

(a) Minimizes entropy to visualize the relationships among genes and proteins

(b) Can be applied to gene expression data from microarrays but not to protein analyses

(c) Can be performed by agglomerative or divisive strategies

(d) Reduces highly dimensional data to show the relationships among genes or among samples

[9-9] The main difference between supervised and unsupervised analyses of microarray data is:

(a) Supervised approaches assign some prior knowledge about function to the genes and/or samples, while unsupervised analyses do not.

(b) Supervised approaches assign a fixed number of clusters, while unsupervised analyses do not.

(c) Supervised approaches cluster genes and/or samples, while unsupervised approaches cluster only genes.

(d) Supervised approaches include algorithms such as support vector machines and decision trees, while unsupervised approaches use clustering algorithms.

Suggested Reading

Miron and Nadon 2006 provide a review of key concepts in microarray data analysis. Other reviews, cited above, are by Quackenbush (2001, 2006), Sherlock (2001), Dopazo et al. (2001), Brazma and Vilo (2000), Ayroles and Gibson (2006), Olson (2006), Lee and Saeed (2007), Raychaudhuri et al. (2001) and Causton et al. (2003).

There are many introductions to R, including an introduction to statistics using R by Verzani (2005) and a book on R and Bioconductor for bioinformatics by Gentleman et al. (2005).

For cluster analysis of microarray data, Gollub and Sherlock (2006) provide an excellent overview. Michael Eisen and colleagues (1998) describe the clustering of 8600 human genes as a function of time. This classic paper includes an excellent description of the metric used to define the relationships of gene expression values and also a discussion of the usefulness of clustering in defining functional relationships among expressed genes.

References

Alizadeh, A. A., et al. Distinct types of diffuse large B-cell lymphoma identified by gene expression profiling. *Nature* **403**, 503–511 (2000).

Alter, O., Brown, P. O., and Botstein, D. Singular value decomposition for genome-wide expression data processing and modeling. *Proc. Natl. Acad. Sci. USA* **97**, 10101–10106 (2000).

Ayroles, J. F., and Gibson, G. Analysis of variance of microarray data. *Methods Enzymol.* **411**, 214–233 (2006).

Baker, S. C., Bauer, S. R., Beyer, R. P., Brenton, J. D., Bromley, B., Burrill, J., Causton, H., Conley, M. P., Elespuru, R., Fero, M., Foy, C., Fuscoe, J., Gao, X., Gerhold, D. L., Gilles, P., Goodsaid, F., Guo, X., Hackett, J., Hockett, R. D., Ikonomi, P., Irizarry, R. A., Kawasaki, E. S., Kaysser-Kranich, T., Kerr, K., Kiser, G., Koch, W. H., Lee, K. Y., Liu, C., Liu, Z. L., Lucas, A., Manohar, C. F., Miyada, G., Modrusan, Z., Parkes, H., Puri, R. K., Reid, L., Ryder, T. B., Salit, M., Samaha, R. R., Scherf, U., Sendera, T. J., Setterquist, R. A., Shi, L., Shippy, R., Soriano, J. V., Wagar, E. A., Warrington, J. A., Williams, M., Wilmer, F., Wilson, M., Wolber, P. K., Wu, X., and Zadro, R. External RNA Controls Consortium. The External RNA Controls Consortium: A progress report. *Nat. Methods* **2**, 731–734 (2005).

Ball, C. A., et al. An open letter to the scientific journals. *Bioinformatics* **18**, 1409 (2002a).

Ball, C. A., et al. Standards for microarray data. *Science* **298**, 539 (2002b).

Barrett, T., and Edgar, R. Mining microarray data at NCBI's Gene Expression Omnibus (GEO)*. *Methods Mol. Biol.* **338**, 175–190 (2006).

Barrett, T., Troup, D. B., Wilhite, S. E., Ledoux, P., Rudnev, D., Evangelista, C., Kim, I. F., Soboleva, A., Tomashevsky, M., and Edgar, R. NCBI GEO: Mining tens of millions of expression profiles—database and tools update. *Nucleic Acids Res.* **35**(Database issue), D760–D765 (2007).

Bassel, A., Hayashi, M., and Spiegelman, S. The enzymatic synthesis of a circular DNA-RNA hybrid. *Proc. Natl. Acad. Sci. USA* **52**, 796–804 (1964).

Beissbarth, T., et al. Processing and quality control of DNA array hybridization data. *Bioinformatics* **16**, 1014–1022 (2000).

Bland, J. M., and Altman, D. G. Statistical methods for assessing agreement between two methods of clinical measurement. *Lancet* **1**, 307–310 (1986).

Bland, J. M., and Altman, D. G. Measuring agreement in method comparison studies. *Stat. Methods Med. Res.* **8**, 135–160 (1999).

Bolstad, B. M., Irizarry, R. A., Astrand, M., and Speed, T. P. A comparison of normalization methods for high density oligonucleotide array data based on variance and bias. *Bioinformatics* **19**, 185–193 (2003).

Bouton, C. M., and Pevsner, J. DRAGON: Database Referencing of Array Genes Online. *Bioinformatics* **16**, 1038–1039 (2000).

Bouton, C. M., and Pevsner, J. DRAGON View: Information Visualization for Annotated Microarray Data. *Bioinformatics* **18**, 323–324 (2002).

Bouton, C. M., Henry, G., Colantuoni, C., and Pevsner, J. DRAGON and DRAGON view: Methods for the annotation, analysis, and visualization of large-scale gene expression data. In G. Parmigiani, E. S. Garrett, R. A. Irizarry, and S. L. Zeger (eds.), *The Analysis of Gene Expression Data: Methods and Software*, 185–209. Springer, New York, 2003.

Bouton, C. M., Hossain, M. A., Frelin, L. P., Laterra, J., and Pevsner, J. Microarray analysis of differential gene expression in lead-exposed astrocytes. *Toxicol. Appl. Pharmacol.* **176**, 34–53 (2001).

Brazma, A. On the importance of standardisation in life sciences. *Bioinformatics* **17**, 113–114 (2001).

Brazma, A., Kapushesky, M., Parkinson, H., Sarkans, U., and Shojatalab, M. Data storage and analysis in ArrayExpress. *Methods Enzymol.* **411**, 370–386 (2006).

Brazma, A., Robinson, A., Cameron, G., and Ashburner, M. One-stop shop for microarray data. *Nature* **403**, 699–700 (2000).

Brazma, A., et al. Minimum information about a microarray experiment (MIAME)—toward standards for microarray data. *Nat. Genet.* **29**, 365–371 (2001).

Brazma, A., and Vilo, J. Gene expression data analysis. *FEBS Lett.* **480**, 17–24 (2000).

Brown, M. P., et al. Knowledge-based analysis of microarray gene expression data by using support vector machines. *Proc. Natl. Acad. Sci. USA* **97**, 262–267 (2000).

Canales, R. D., Luo, Y., Willey, J. C., Austermiller, B., Barbacioru, C. C., Boysen, C., Hunkapiller, K., Jensen, R. V., Knight, C. R., Lee, K. Y., Ma, Y., Maqsodi, B., Papallo, A., Peters, E. H., Poulter, K., Ruppel, P. L., Samaha, R. R., Shi, L., Yang, W., Zhang, L., and Goodsaid, F. M. Evaluation of DNA microarray results with quantitative gene expression platforms. *Nat. Biotechnol.* **24**, 1115–1122 (2006).

Causton, H. C., Quackenbush, J., and Brazma, A. *Microarray Gene Expression Data Analysis: A Beginner's Guide*. Blackwell Publishing, Malden, MA, 2003.

Chen, Y., and Meltzer, P. S. Gene expression analysis via multidimensional scaling. *Curr. Protocols Bioinf.* 7, 7.11.1–7.11.19 (2005).

Chu, S., et al. The transcriptional program of sporulation in budding yeast. *Science* **282**, 699–705 (1998).

Colantuoni, C., Henry, G., Zeger, S., and Pevsner, J. SNOMAD (Standardization and NOrmalization of MicroArray Data): Web-accessible gene expression data analysis. *Bioinformatics* **18**, 1540–1541 (2002a).

Colantuoni, C., Henry, G., Zeger, S., and Pevsner, J. Local mean normalization of microarray element signal intensities across an array surface: Quality control and correction of spatially systematic artifacts. *Biotechniques* **32**, 1316–1320 (2002b).

Cope, L. M., Irizarry, R. A., Jaffee, H. A., Wu, Z., and Speed, T. P. A benchmark for Affymetrix GeneChip expression measures. *Bioinformatics* **20**, 323–331 (2004).

Dennis, G. Jr., Sherman, B. T., Hosack, D. A., Yang, J., Gao, W., Lane, H. C., and Lempicki, R. A. DAVID: Database for Annotation, Visualization, and Integrated Discovery. *Genome Biol.* **4, P3** (2003).

Dopazo, J., Zanders, E., Dragoni, I., Amphlett, G., and Falciani, F. Methods and approaches in the analysis of gene expression data. *J. Immunol. Methods* **250**, 93–112 (2001).

Draghici, S., Khatri, P., Eklund, A. C., and Szallasi, Z. Reliability and reproducibility issues in DNA microarray measurements. *Trends Genet.* **22**, 101–109 (2006).

Dupuy, A., and Simon, R. M. Critical review of published microarray studies for cancer outcome and guidelines on statistical analysis and reporting. *J. Natl. Cancer Inst.* **99**, 147–157 (2007).

Eisen, M. B., Spellman, P. T., Brown, P. O., and Botstein, D. Cluster analysis and display of genome-wide expression patterns. *Proc. Natl. Acad. Sci. USA* **95**, 14863–14868 (1998).

Everitt, B. S., Landau, S., and Leese, M. *Cluster Analysis*, fourth edition. Arnold, London (2001).

Gentleman, R., Carey, V., Huber, W., Irizarry, R., and Sandrine Dudoit, S. (eds.), *Bioinformatics and Computational Biology Solutions Using R and Bioconductor*. Springer, New York, 2005.

Golub, T. R., et al. Molecular classification of cancer: Class discovery and class prediction by gene expression monitoring. *Science* **286**, 531–537 (1999).

Gollub, J., and Sherlock, G. Clustering microarray data. *Methods Enzymol.* **411**, 194–213 (2006).

Gordon, A. D. *Classification*. Chapman and Hall, London, 1980.

Haverty, P. M., et al. HuGEIndex: A database with visualization tools for high-density oligonucleotide array data from normal human tissues. *Nucleic Acids Res.* **30**, 214–217 (2002).

Ikeo, K., Ishi-i, J., Tamura, T., Gojobori, T., and Tateno, Y. CIBEX: Center for information biology gene expression database. *C.R. Biol.* **326**, 1079–1082 (2003).

Irizarry, R. A., Warren, D., Spencer, F., Kim, I. F., Biswal, S., Frank, B. C., Gabrielson, E., Garcia, J. G., Geoghegan, J., Germino, G., Griffin, C., Hilmer, S. C., Hoffman, E.,

Jedlicka, A. E., Kawasaki, E., Martinez-Murillo, F., Morsberger, L., Lee, H., Petersen, D., Quackenbush, J., Scott, A., Wilson, M., Yang, Y., Ye, S. Q., and Yu, W. Multiple-laboratory comparison of microarray platforms. *Nat. Methods* **2**, 345–350 (2005).

Irizarry, R. A., Wu, Z., and Jaffee, H. A. Comparison of Affymetrix GeneChip expression measures. *Bioinformatics* **22**, 789–794 (2006).

Ishwaran, H., Rao, J. S., and Kogalur, U. B. BAMarray: Java software for Bayesian analysis of variance for microarray data. *BMC Bioinformatics* **7**, 59 (2006).

Iyer, V. R., et al. The transcriptional program in the response of human fibroblasts to serum. *Science* **283**, 83–87 (1999).

Kaufman, L., and Rousseeuw, P. J. *FINDING GROUPS IN DATA. An Introduction to Cluster Analysis*, Wiley, New York, 1990.

Khan, J., et al. Gene expression profiling of alveolar rhabdomyosarcoma with cDNA microarrays. *Cancer Res.* **58**, 5009–5013 (1998).

Kuruvilla, F. G., Park, P. J., and Schreiber, S. L. Vector algebra in the analysis of genome-wide expression data. *Genome Biol.* **3**, 11.1–11.11 (2002).

Landgrebe, J., Wurst, W., and Welzl, G. Permutation-validated principal components analysis of microarray data. *Genome Biol.* **3**, 19.1–19.11 (2002).

Lee, J. K., Bussey, K. J., Gwadry, F. G., Reinhold, W., Riddick, G., Pelletier, S. L., Nishizuka, S., Szakacs, G., Annereau, J. P., Shankavaram, U., Lababidi, S., Smith, L. H., Gottesman, M. M., and Weinstein, J. N. Comparing cDNA and oligonucleotide array data: Concordance of gene expression across platforms for the NCI-60 cancer cells. *Genome Biol.* **4**, R82 (2003).

Lee, N. H., and Saeed, A. I. Microarrays: An overview. *Methods Mol. Biol.* **353**, 265–300 (2007).

Mao, R., Zielke, C. L., Zielke, H. R., and Pevsner, J. Global up-regulation of chromosome 21 gene expression in the developing Down syndrome brain. *Genomics* **81**, 457–467 (2003).

Mao, R., Wang, X., Spitznagel, E. L. Jr., Frelin, L. P., Ting, J. C., Ding, H., Kim, J. W., Ruczinski, I., Downey, T. J., and Pevsner, J. Primary and secondary transcriptional effects in the developing human Down syndrome brain and heart. *Genome Biol.* **6**, R107 (2005).

MAQC Consortium, Shi, L., Reid, L. H., Jones, W. D., Shippy, R., Warrington, J. A., Baker, S. C., Collins, P. J., de Longueville, F., Kawasaki, E. S., Lee, K. Y., Luo, Y., Sun, Y. A., Willey, J. C., Setterquist, R. A., Fischer, G. M., Tong, W., Dragan, Y. P., Dix, D. J., Frueh, F. W., Goodsaid, F. M., Herman, D., Jensen, R. V., Johnson, C. D., Lobenhofer, E. K., Puri, R. K., Schrf, U., Thierry-Mieg, J., Wang, C., Wilson, M., Wolber, P. K., Zhang, L., Amur, S., Bao, W., Barbacioru, C. C., Lucas, A. B., Bertholet, V., Boysen, C., Bromley, B., Brown, D., Brunner, A., Canales, R., Cao, X. M., Cebula, T. A., Chen, J. J., Cheng, J., Chu, T. M., Chudin, E., Corson, J., Corton, J. C., Croner, L. J., Davies, C., Davison, T. S., Delenstarr, G., Deng, X., Dorris, D., Eklund, A. C., Fan, X. H., Fang, H., Fulmer-Smentek, S., Fuscoe, J. C., Gallagher, K., Ge, W., Guo, L., Guo, X., Hager, J., Haje, P. K., Han, J., Han, T., Harbottle, H. C., Harris, S. C., Hatchwell, E., Hauser, C. A., Hester, S., Hong, H., Hurban, P., Jackson, S. A., Ji, H., Knight, C. R., Kuo, W. P., LeClerc, J. E., Levy, S., Li, Q. Z., Liu, C., Liu, Y., Lombardi, M. J., Ma, Y., Magnuson, S. R., Maqsodi, B., McDaniel, T., Mei, N., Myklebost, O., Ning, B., Novoradovskaya, N., Orr, M. S., Osborn, T. W., Papallo, A., Patterson, T. A., Perkins, R. G., Peters, E. H., Peterson, R., Philips, K. L., Pine, P. S., Pusztai, L., Qian, F., Ren, H., Rosen, M., Rosenzweig, B. A., Samaha, R. R., Schena, M., Schroth, G. P., Shchegrova, S., Smith, D. D., Staedtler, F., Su, Z., Sun, H., Szallasi, Z., Tezak, Z., Thierry-Mieg, D., Thompson, K. L., Tikhonova, I., Turpaz, Y., Vallanat, B., Van, C., Walker, S. J., Wang, S. J., Wang, Y., Wolfinger, R., Wong, A., Wu, J., Xiao, C., Xie, Q., Xu, J., Yang, W., Zhang, L., Zhong, S., Zong, Y., and Slikker, W. Jr. The MicroArray Quality Control (MAQC) project shows inter- and intraplatform reproducibility of gene expression measurements. *Nat. Biotechnol.* **24**, 1151–1161 (2006).

Maurer, M., Molidor, R., Sturn, A., Hartler, J., Hackl, H., Stocker, G., Prokesch, A., Scheideler, M., and Trajanoski, Z. MARS: Microarray analysis, retrieval, and storage system. *BMC Bioinformatics* **6**, 101 (2005).

Miron, M., and Nadon, R. Inferential literacy for experimental high-throughput biology. *Trends Genet.* **22**, 84–89 (2006).

Misra, J., et al. Interactive exploration of microarray gene expression patterns in a reduced dimensional space. *Genome Res.* **12**, 1112–1120 (2002).

Motulsky, H. *Intuitive Biostatistics*. Oxford University Press, New York, 1995.

Olshen, A. B., and Jain, A. N. Deriving quantitative conclusions from microarray expression data. *Bioinformatics* **18**, 961–970 (2002).

Olson, N. E. The microarray data analysis process: From raw data to biological significance. *NeuroRx.* **3**, 373–383 (2006).

Osborne, J. D., Zhu, L. J., Lin, S. M., and Kibbe, W. A. Interpreting microarray results with gene ontology and MeSH. *Methods Mol. Biol.* **377**, 223–242 (2007).

Page, G. P., Edwards, J. W., Gadbury, G. L., Yelisetti, P., Wang, J., Trivedi, P., and Allison, D. B. The PowerAtlas: A power and sample size atlas for microarray experimental design and research. *BMC Bioinformatics* **7**, 84 (2006).

Perou, C. M., et al. Distinctive gene expression patterns in human mammary epithelial cells and breast cancers. *Proc. Natl. Acad. Sci. USA* **96**, 9212–9217 (1999).

Perou, C. M., et al. Molecular portraits of human breast tumours. *Nature* **406**, 747–752 (2000).

Quackenbush, J. Computational analysis of microarray data. *Nat. Rev. Genet.* **2**, 418–427 (2001).

Quackenbush, J. Computational approaches to analysis of DNA microarray data. *Methods Inf. Med.* **45**, 91–103 (2006).

Quackenbush, J., and Irizarry, RA. Response to Shields: "MIAME, We Have a Problem." *Trends Genet.* **22**, 471–472 (2006).

Raychaudhuri, S., Stuart, J. M., and Altman, R. B. Principal components analysis to summarize microarray experiments: Application to sporulation time series. *Pac. Symp. Biocomput.* 455–466 (2000).

Raychaudhuri, S., Sutphin, P. D., Chang, J. T., and Altman, R. B. Basic microarray analysis: Grouping and feature reduction. *Trends Biotechnol.* **19**, 189–193 (2001).

Rubinstein, R., and Simon, I. MILANO: Custom annotation of microarray results using automatic literature searches. *BMC Bioinformatics* **6**, 12 (2005).

Saeed, A. I., Bhagabati, N. K., Braisted, J. C., Liang, W., Sharov, V., Howe, E. A., Li, J., Thiagarajan, M., White, J. A., and Quackenbush, J. TM4 microarray software suite. *Methods Enzymol.* **411**, 134–193 (2006).

Schuchhardt, J., et al. Normalization strategies for cDNA microarrays. *Nucleic Acids Res.* **28**, E47 (2000).

Sean, D., and Meltzer, P. S. GEOquery: A bridge between the Gene Expression Omnibus (GEO) and BioConductor. *Bioinformatics* **23**, 1846–1847 (2007).

Sherlock, G. Analysis of large-scale gene expression data. *Brief. Bioinform.* **2**, 350–362 (2001).

Shields, R. MIAME, We Have a Problem. *Trends Genet.* **22**, 65–66 (2006).

Shipp, M. A., et al. Diffuse large B-cell lymphoma outcome prediction by gene-expression profiling and supervised machine learning. *Nat. Med.* **8**, 68–74 (2002).

Sneath, P. H. A., and Sokal, R. R. *Numerical Taxonomy.* W.H. Freeman and Co., San Francisco, 1973.

Spellman, P. T., et al. Design and implementation of microarray gene expression markup language (MAGE-ML). *Genome Biol.* **3**, 46.1–46.9 (2002).

Subramanian, A., Tamayo, P., Mootha, V. K., Mukherjee, S., Ebert, B. L., Gillette, M. A., Paulovich, A., Pomeroy, S. L., Golub, T. R., Lander, E. S., and Mesirov, J. P. Gene set enrichment analysis: A knowledge-based approach for interpreting genome-wide expression profiles. *Proc. Natl. Acad. Sci. USA* **102**, 15545–15550 (2005).

Tamayo, P., et al. Interpreting patterns of gene expression with self-organizing maps: Methods and application to hematopoietic differentiation. *Proc. Natl. Acad. Sci. USA* **96**, 2907–2912 (1999).

Tan, P. K., Downey, T. J., Spitznagel, E. L. Jr., Xu, P., Fu, D., Dimitrov, D. S., Lempicki, R. A., Raaka, B. M., and Cam, M. C. Evaluation of gene expression measurements from commercial microarray platforms. *Nucleic Acids Res.* **31**, 5676–5684 (2003).

Tavazoie, S., Hughes, J. D., Campbell, M. J., Cho, R. J., and Church, G. M. Systematic determination of genetic network architecture. *Nat. Genet.* **22**, 281–285 (1999).

Thalamuthu, A., Mukhopadhyay, I., Zheng, X., and Tseng, G. C. Evaluation and comparison of gene clustering methods in microarray analysis. *Bioinformatics* **22**, 2405–2412 (2006).

Tusher, V. G., Tibshirani, R., and Chu, G. Significance analysis of microarray applied to the ionizing radiation response. *Proc. Natl. Acad. Sci. USA* **98**, 5116–5121 (2001).

Troein, C., Vallon-Christersson, J., and Saal, L. H. An introduction to BioArray Software Environment. *Methods Enzymol.* **411**, 99–119 (2006).

Tsai, J., Sultana, R., Lee, Y., Pertea, G., Karamycheva, S., Antonescu, V., Cho, J., Parvizi, B., Cheung, F., and Quackenbush, J. RESOURCERER: A database for annotating and linking microarray resources within and across species. *Genome Biol.* **2**, SOFTWARE0002 (2001).

Verzani, J. *Using R for Introductory Statistics.* Chapman and Hall, New York, 2005.

Wall, M. E., Dyck, P. A., and Brettin, T. S. SVDMAN—singular value decomposition analysis of microarray data. *Bioinformatics* **17**, 566–568 (2001).

West, M., et al. Predicting the clinical status of human breast cancer by using gene expression profiles. *Proc. Natl. Acad. Sci. USA* **98**, 11462–11467 (2001).

Zeeberg, B. R., Qin, H., Narasimhan, S., Sunshine, M., Cao, H., Kane, D. W., Reimers, M., Stephens, R. M., Bryant, D., Burt, S. K., Elnekave, E., Hari, D. M., Wynn, T. A., Cunningham-Rundles, C., Stewart, D. M., Nelson, D., and Weinstein, J. N. High-Throughput GoMiner, an "industrial-strength" integrative gene ontology tool for interpretation of multiple-microarray experiments, with application to studies of Common Variable Immune Deficiency (CVID). *BMC Bioinformatics* **6**, 168. (2005).

Zolman, J. F. *Biostatistics.* Oxford University Press, New York, 1993.

	1 Globin of the Oxyhaemo-globin of Horse's Blood.	2 Serum-albumin of Horse's Blood.	3 Serum-globulin.	4 Egg-white.	5 Egg-albumin crystallised.	6 Albumin of Yolk.	7 Caseinogen (see p. 397 for other products).
Glycocoll	0 [9]	0 [10]	3·52 [95]	...	...	...	0 [45 95]
Alanin	4·19 [9]	2·68 [10]	2·22 [95]	...			0·9 [95]
Leucin	29·04 [9]	20·48 [10]	18·7 [95]	22·6 [40]			10·5 [95]
Phenylalanin	4·24 [9]	3·08 [10]	3·84 [95]	+ [3]			3·2 [95]
α-Pyrrolidin-carboxylic acid	2·34 [9]	1·04 [10]	2·76 [95]	+ [3]	...		3·2 [1 95]
Glutaminic acid	1·73 [9]	1·52 [10]	2·20 [95]	+ [40]	+ [48]		10·7 [95]
Aspartic acid	4·43 [9]	3·12 [10]	2·54 [95]	+ [37]	+ [48]	...	1·2 [95]
Cystin	0·31 [9]	2·53 [43]	1·51 [43]	0·4 [43]	0·29 [43]		0·065
Serin	0·56 [9]	0·6 [10]	...	...	...		0·43 [14]
Oxy-α-Pyrrolidin-carboxylic acid	1·04 [9]	...	...	...	...		0·25 [14]
Tyrosin	1·33 [9]	2·1 [10]	...	0·58 [66]	1·5 [31]		4·5 [36 66]
Lysin	4·28 [9]	...	...	+ [16 75]			5·8 [26]
Histidin	10·96 [9]	...	...	+ [18]	+ [18]	+ [18]	2·6 [26]
Arginin	5·42 [9]	...	...	+ [17]	+ [16 17]	+ [17]	4·84 [26]
Tryptophane	+ [9]	+ [10]	...	+ [80]	...		1·5 [54]
Ammonia	0·93 [50]	1·2 [49]	1·75 [49]	...	1·5 [49]		1·8 [26]
Cystein	...	+ [44]	...	+ [44]	...		0 [44]
Amino-valerianic acid	...	...	...	...	...		1·9 [95]
Glucosamin	...	...	...	...	10–11 [34]		0 [67]
Diamino-trioxydodecanoic acid	...	...	...	...	...		0·75 [92]

While it is obvious to us that most proteins are composed of 20 amino acids, chemists in the late nineteenth century struggled to understand protein composition. At the turn of the century only several dozen proteins were known, including so-called albumins (including serum albumins, lactoglobulins, fibrinogen, myosin, and histones), proteids (e.g., hemoglobin and mucins), and albuminoids (e.g., collagen, keratin, elastin, and amyloid). Of these proteins only a very small group were available in pure form as crystals (e.g., hemoglobin and serum albumin from horse, ovalbumin, and ichthulin [salmon albumin]). Gustav Mann (1906, p. 70–75) described the dissociation products of 51 assorted proteins into their fundamental units. The results are shown for seven proteins (see columns). The rows indicate various compounds found upon dissolving the proteins. Most of these are amino acids; for example, glycocoll is a name formerly given to glycine. This table shows that from the earliest times that proteins could be analyzed, scientists made an effort to understand both the nature of individual protein molecules and the relationships of related proteins from different species.

10

Protein Analysis and Proteomics

INTRODUCTION

A living organism consists primarily of five substances: proteins, nucleic acids, lipids, water, and carbohydrates. Of these essential ingredients, it is the proteins that most define the character of each cell. DNA has often been described as a substance that corresponds to the blueprints of a house, specifying the materials used to build the house. These materials are the proteins, and they perform an astonishing range of biological functions. This includes structural roles (e.g., actin contributes to the cytoskeleton), roles as enzymes (proteins that catalyze biochemical reactions, typically increasing a reaction rate by several orders of magnitude), and roles in transport of materials within and between cells. If DNA is the blueprint of the house, proteins form primary components not just of the walls and floors of the house but also of the plumbing system, the system for generating and transmitting electricity, and the trash removal system.

Proteins are polypeptide polymers consisting of a linear arrangement of amino acids. There is a rich history of attempts to purify proteins and identify their constituent amino acids (Box 10.1). By 1850 a series of proteins had been identified (albumin, hemoglobin, casein, pepsin, fibrin, crystalline) and partially purified. It was not until the 1950s that the complete amino acid sequences of several small proteins were determined. Today, we have access to over 10 million protein sequences.

Bioinformatics and Functional Genomics, Second Edition. By Jonathan Pevsner
Copyright © 2009 John Wiley & Sons, Inc.

BOX 10.1
Brief History of Protein Studies

Protein products have been used for centuries; for example, when Leonardo da Vinci (1452–1519) invented plastics c.1509, his recipe included egg whites (a source of ovalbumin) (Reti, 1952). The word albumin derives from the Latin *albus* (white) and first appeared in English in 1599. In 1720, Beccari became the first person to fractionate proteins, separating the gluten fraction from wheat (see frontis to Chapter 2). The word *protein* was coined in 1828 by the Swedish chemist Jöns Jacob Berzelius (1779–1848); the Greek word *proteios* means "of first rank." Protein purification began in the first half of the nineteenth century upon the discovery of the first proteins: albumin, hemoglobin, casein, pepsin, fibrin, and crystalline (Mulder, 1849).

The modern era of protein purification was ushered in by James Sumner and John Northrup, who purified the first enzymes to homogeneity. The Nobel Prize in Chemistry 1946 was awarded to Sumner "for his discovery that enzymes can be crystallized," and to Northrop and Wendell Stanley "for their preparation of enzymes and virus proteins in a pure form" (see ▶ http://nobelprize.org/nobel_prizes/chemistry/laureates/1946/).

Earlier in this book we learned how to access proteins from databases (Chapter 2), we aligned them and searched them against databases (Chapters 3 to 6), and we visualized multiple sequence alignments as phylogenetic trees (Chapter 7). In this chapter, we discuss techniques to identify proteins (direct sequencing, gel electrophoresis, and mass spectrometry). We then present four perspectives on individual proteins: domains and motifs, physical properties, localization, and function. In Chapter 11, we will consider the structure of proteins. Then in Chapter 12 we will address functional genomics, which is the genome-wide assessment of gene function. Functional genomics encompasses large-scale studies of protein function both in normal conditions and following genetic or environmental perturbations.

Protein Databases

UniProtKB is available at ▶ http://www.uniprot.org; release 12.4 consists of 5,275,429 entries. IPI is at ▶ http://www.ebi.ac.uk/IPI.

Protein sequences were initially obtained directly from purified proteins, but the vast majority of newly identified proteins are predicted from genomic DNA sequence. GenBank (Chapter 2) currently includes ∼100 billion base pairs of sequence, and the separate whole genome shotgun (WGS) division is even larger. The nonredundant database of proteins at NCBI is the largest publicly available source of protein data (Table 4.1). Another major resource is the UniProt database (Chapter 2) which currently includes over 5 million proteins, of which about 300,000 are manually curated as part of the UniProtKB/Swiss-Prot database. The International Protein Index (IPI) at the European Bioinformatics Institute cross references a series of major databases (including RefSeq, UniProt, Ensembl, and the Vertebrate Genome Annotation [VEGA] database) (Kersey et al., 2004). It includes stable identifiers as well as a minimally redundant, maximally complete set of proteins (with one sequence per transcript).

To learn more about the GOS project, visit the homepage of NCBI and enter the term "global ocean sampling" then

In 2007 Craig Venter and colleagues assembled 7.7 million genomic DNA sequence reads as part of the Global Ocean Sampling (GOS) project as well as an earlier Sargasso Sea project (Venter et al., 2004; Yooseph et al., 2007). They used shotgun sequencing (described in Chapter 13) to randomly sample the DNA of microorganisms, including bacteria, archaea, and viruses, in seawater. They predicted the existence of 6.12 million proteins, and thus a single publication doubled the number of known proteins. In Chapters 13 to 15 we will discuss other metagenomics projects in which microorganisms are sequenced from environmental samples. Such projects are intended to explore the relationship between communities of microorganisms and their ecosystems, and will continue to greatly expand the number of known proteins.

Community Standards for Proteomics Research

In all areas of bioinformatics, efforts are underway to standardize the way biological models are formulated and experimental data are generated and described. The Human Proteome Organisation (HUPO) supports a Proteomics Standards Initiative (PSI) with the goals of defining standards for proteomic data representation to facilitate the comparison, exchange, and verification of data (Martens et al., 2007). HUPO-PSI currently has working groups in the areas of gel electrophoresis, mass spectrometry, molecular interactions, protein modifications, proteomics informatics, and sample processing. These groups have proposed a series of guidelines for reporting data, as well as data exchange formats and controlled vocabularies. One example is the minimum information about a proteomics experiment (MIAPE) guidelines (Taylor et al., 2007). MIAPE provides a formal list of items that a researcher should provide when reporting a proteomics experiment, including sufficient information about an experiment to allow others to critically evaluate the conclusions.

follow the link to Genome Projects. The project accession number is AACY000000000. The GOS data are also available at the Community Cyberinfrastructure for Advanced Marine Microbial Ecology Research and Analysis (CAMERA) website (▶ http://camera.calit2.net/). Note that many of the GOS project predicted proteins were not full-length, that is, they were not derived from a DNA segment that included both a start and a stop codon.

The HUPO Protomics Standards Initiative website is ▶ http://www.psidev.info/.

TECHNIQUES TO IDENTIFY PROTEINS

In this section we introduce three fundamental approaches to protein identification: direct protein sequencing, gel electrophoresis, and mass spectrometry.

Direct Protein Sequencing

Per Edman devised the method of systematically degrading proteins, beginning with the amino-terminal residue and proceeding toward the carboxy terminus. Fredrick Sanger was among the first to exploit this technology to determine the primary amino acid sequence of insulin.

The Edman degradation procedure requires purification of a protein to relative homogeneity. This can be achieved by conventional biochemical means such as purification on ion exchange, size exclusion, or other columns, or by electrophoresis. One obtains a portion of the amino acid sequence of a protein by transferring it to a specialized polyvinylidene fluoride (or PVDF) membrane, then performing microsequencing by sequential Edman degradations (Fig. 10.1). About 60% to 85% of the time, the amino terminus of yeast and other eukaryotic proteins is blocked (e.g., acetylated and unavailable for Edman degradations). A standard procedure is to proteolyze (e.g., trypsinize) the protein, purify the proteolytic fragments by reverse-phase high performance liquid chromatography (HPLC), confirm the purity of the fragments, and then perform Edman degradations.

The Edman degradation method has been reviewed by Shively (2000). It remains a fundamental method of protein identification, and is useful to identify sequences of one to 10 picomoles of a protein. It is also well suited to identifying the amino terminus of an intact protein (when unblocked), in contrast to mass

Consider a 10 kilodalton protein; given a molecular weight of $\sim$115 daltons per residue, such a protein consists of about 87 amino acids. To obtain 1 pmol, just 10 nanograms or 10^{-8} g of protein is required.

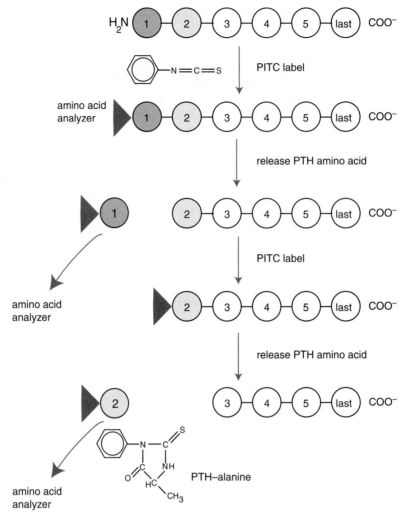

FIGURE 10.1. *Protein sequencing by Edman degradation. The Edman process is illustrated for a protein fragment of six amino acids. The first amino acid reacts through its amino terminus with phenylisothiocyanate (PITC). Under acidic conditions this amino acid residue, derivitized with phenylthiohydantoin (PTH), is cleaved and can be identified in an amino acid analyzer. The peptide now has five amino acid residues, and the cycle is repeated with successive amino-terminal amino acids. The structure of PTH-alanine is shown as an example. The typical result is a readout of 10 to 20 amino acids. The corresponding protein and gene can be evaluated by performing BLAST searches (Chapter 4).*

spectrometry techniques that only analyze peptide fragments. However, the Edman technique has several limitations.

- It is laborious and not amenable to high throughput analyses.
- While it is sensitive, mass spectrometry techniques can be 10 to 100 times more sensitive.
- Direct sequence is not useful for the analysis of posttranslational modifications, in contrast to two-dimensional gel electrophoresis and mass spectrometry.

Gel Electrophoresis

Polyacrylamide gel electrophoresis (PAGE) is a premier tool for the analysis of protein molecular weight. Proteins (like nucleic acids) possess a charge and thus migrate when introduced into an electric field. Proteins are denatured and electrophoresed through a matrix of acrylamide that is inert (so it does not interact with the protein) and porous (so that proteins can move through it). The velocity of a protein as it migrates through an acrylamide gel is inversely proportional to its size, and thus a complex mixture of proteins can be separated in a single experiment. Proteins are almost always electrophoresed through acrylamide under denaturing conditions in the presence of the detergent sodium dodecyl sulfate (SDS), so this technique is commonly abbreviated SDS–PAGE.

O'Farrell (1975) greatly extended the capabilities of this technology by combining it with an initial separation of proteins based on their charge. In the first step, proteins are separated by isoelectric focusing. A gel matrix (or strip) is produced that contains ampholytes spanning a continuous range of pH values, usually between pH 3 and 11. Each protein is zwitterionic (having both positive and negative ions), and when electrophoresed, it migrates to the position at which its total net charge is zero. This is the isoelectric point (abbreviated pI) at which the protein stops migrating. A complex mixture of proteins may thus be separated based on charge, and this corresponds to the first dimension of two-dimensional gel electrophoresis. In the second dimension, proteins are separated by SDS–PAGE.

The technique of two-dimensional gel electrophoresis has matured into an important technology used to analyze proteomes (Dunn, 2000; Görg et al., 2004; Carrette et al., 2006). An example of a two-dimensional gel profile is shown in Fig. 10.2.

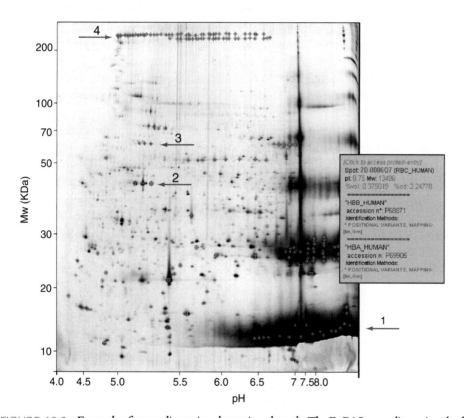

FIGURE 10.2. *Example of a two-dimensional protein gel result. The ExPASy two-dimensional gel resource was searched for beta globin. This profile is of several hundred proteins from human liver. The x axis corresponds to pH; proteins migrate to their isoelectric point (pI) where the net charges are zero. The y axis corresponds to molecular weight; on this particular gel, relatively low molecular weight proteins (10 to 50 kilodaltons) are well resolved, while other gels resolve larger proteins. The highly abundant proteins include alpha and beta globins at molecular weights of about 12 kilodaltons (arrow 1) and 25 kilodaltons. Symbols on the image correspond to identified proteins. By mousing over each identified spot, a dialog box appears with information on the protein (here, HBB with accession P68871) as well as an identifier, a statement of the molecular weight and pI, and a link to further information at ExPASy. Other identified proteins include beta actin (arrow 2), acetylcholinesterase (P22303 in three spots; arrow 3), and a large group of spectrin beta chain variants (arrow 4).*

Several hundred micrograms of protein from human red blood cells were separated by pH (on the x axis) by isoelectric focusing, then by molecular mass (on the y axis) by SDS–PAGE. Thousands of proteins may be visualized with a protein-binding dye such as silver nitrate or Coomassie blue. Note that several proteins are especially abundant, including alpha and beta globin, as well as the structural proteins actin and spectrin. Many proteins have a characteristic pattern of spots that spread along the first dimension. This is a "charge train" that usually represents a series of variants of a protein with differing amounts of charged groups such as phosphates that are covalently attached.

A key property of two-dimensional protein gels is that the individual proteins may be identified by direct protein microsequencing or by sensitive mass spectroscopy techniques (see below) (Farmer and Caprioli, 1998). Applications of two-dimensional SDS–PAGE include a description of hundreds of proteins in human and rat brain (Langen et al., 1999; Fountoulakis et al., 1999) and an analysis of aberrant protein expression profiles in bladder tumors (Østergaard et al., 1997). Grünenfelder et al. (2001) analyzed protein synthesis during the cell cycle of the bacterium *Caulobacter crescentus* and detected about 25% (979) of all the predicted gene products. Many of these were degraded during a single cell cycle.

One of the most important websites for proteomics is the Expert Protein Analysis System (ExPASy) (Fig. 10.3). ExPASy includes the main public database for information on two-dimensional gel electrophoresis (Hoogland et al., 1999; Sanchez et al., 2001). Information is available for gels from a variety of organisms and experimental conditions, including the experiment shown in Fig. 10.2. These profiles may be queried by choosing a two-dimensional gel map by other criteria such as keyword.

There have been many improvements to two-dimensional gel technology. Jonathan Minden and colleagues introduced difference gel electrophoresis (DIGE), a technique in which two (or sometimes three) samples are labeled with amine-reactive, fluorescent dyes (Viswanathan et al., 2006). These samples are mixed, electrophoresed, and then the relative abundance of many proteins is determined based on fluorescence imaging. In some cases, DIGE has been used to detect 0.5 femtomoles of protein (for a 10 kilodalton protein this corresponds to just five picograms).

We may summarize the strengths of two-dimensional gel electrophoresis as follows:

- It offers the ability to describe both isolectric point and molecular mass of intact proteins; this contrasts with mass spectrometry methods that identify molecular mass based on peptide fragments, and that further lose information on pI.

ExPASy is located at ▶ http://www.expasy.ch/. Many of the tools we will explore in this chapter are available at ExPASy, which is part of the Swiss Institute of Bioinformatics.

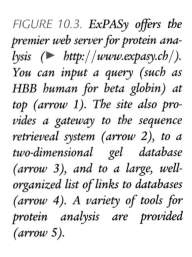

FIGURE 10.3. ExPASy offers the premier web server for protein analysis (▶ http://www.expasy.ch/). You can input a query (such as HBB human for beta globin) at top (arrow 1). The site also provides a gateway to the sequence retrieveal system (arrow 2), to a two-dimensional gel database (arrow 3), and to a large, well-organized list of links to databases (arrow 4). A variety of tools for protein analysis are provided (arrow 5).

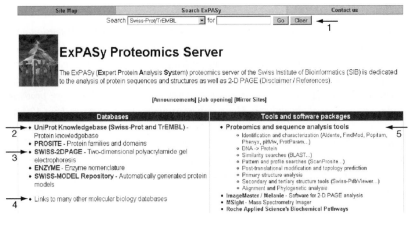

- Several thousand proteins can be resolved and visualized with an appropriate stain

- It is possible to detect and quantitate less than 1 nanogram per spot on the gel. A variety of sensitive stains (dyes) are available to detect proteins.

- Mass spectrometry is commonly used in conjunction with two-dimensional gels for protein identification, as discussed below.

The two-dimensional gel approach has several limitations.

- It is not amenable to high throughput processing of many samples in parallel.

- Sample preparation is a critical step and often requires a great deal of optimization. However, this is true of essentially all proteomics methods.

- Only the most abundant proteins in a sample are usually detected. Hydrophobic proteins, including proteins with transmembrane domains, are underrepresented on two-dimensional gels. Similarly, highly basic or acidic proteins are often excluded.

- It requires considerable expertise to reliably generate consistent results. In comparing two gel profiles, if the polyacrylamide gels vary even slightly in composition or if the samples are electrophoresed under differing conditions, it can be difficult to accurately align the protein spots. An important technical advance in the reproducibility of two-dimensional gel electrophoresis was the introduction of immobilized pH gradients, preformed on dry strips, that replaced an older system of pH gradient formation with ampholytes.

Mass Spectrometry

Mass spectrometry techniques have revolutionized the field of proteomics by allowing proteins to be identified with extraordinary sensitivity (Mann et al., 2001; Cox and Mann, 2007). Mass spectrometry is useful (1) for the identification of proteins (e.g., for identifying protein spots from two-dimensional gels, or from complex mixtures such as extracts of cells, or from other biochemical purification approaches), (2) for the characterization of known proteins (e.g., recombinant proteins), and (3) to characterize posttranslational modifications of proteins. The ability of mass spectrometry to measure the mass of a protein with extremely high accuracy and precision allows it to distinguish even subtle changes in proteins such as the addition of a single phosphate group.

Mass spectrometers analyze charged protein or peptide molecules in the gaseous state. A key step is to transfer proteins into the gas phase and ionize them. This is accomplished using either electrospray or matrix-assisted laser desorption ionization (MALDI). In MALDI-TOF (MALDI with time-of-flight spectroscopy), the analyte molecules (i.e., the material to be analyzed) are dried on a metal substrate, irradiated with a laser, and fragmented (Fig. 10.4). The resulting ions are accelerated in a field that imparts a fixed kinetic energy. The ions traverse a path, are reflected in an ion mirror, and then are detected by a channeltron electron multiplier. The mass-to-charge ratio of an ion determines the time it takes to reach the detector; lighter ions (smaller analytes) have a higher velocity and are detected first. A time-of-flight spectrum is recorded from which the amino acid composition of even one femtomole of peptide can be deduced.

John Fenn and Koichi Tanaka shared half the Nobel Prize in Chemistry 2002 "for their development of soft desorption ionisation methods for mass spectrometric analyses of biological macromolecules." See ▶ http://nobelprize.org/nobel_prizes/chemistry/laureates/2002/.

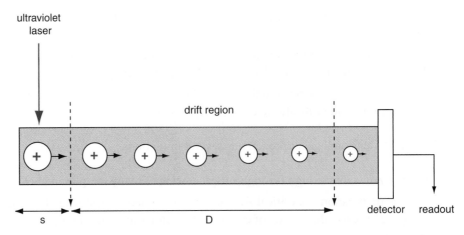

FIGURE 10.4. *Matrix-assisted laser desorption/ionization time-of-flight spectroscopy (MALDI-TOF). Spectroscopy is a technique to measure the mass of protein samples and other macromolecules. A sample is placed in a matrix of material that absorbs ultraviolet light. A laser is fired at the sample in the source region (s), and in the context of the matrix the sample becomes ionized. Some of the protein samples evaporate (i.e., desorption occurs). The ionization occurs in the presence of an electric field that accelerates the ions into a long drift region (D). The acceleration of each protein fragment is proportional to the mass of the ion. A detector records a time-of-flight spectrum that can be analyzed to determine the mass of each fragment. Peptide fragments are then searched against a protein database to determine the identity of the analyte (protein).*

We can consider two common applications of mass spectrometry. (1) Peptide fingerprinting is often used to identify a relatively pure sample such as a spot from a two-dimensional gel (which may contain a few proteins of varying abundance). In one scenario, the protein spot is excised with a razor blade, proteolyzed with trypsin (which cleaves at basic amino acid residues), then subjected to MALDI-TOF (Fig. 10.5a). The application of a laser beam to the sample creates a series of peptide fragments, breaking the peptide bonds of amino acids. One then searches the mass of the observed fragments against a protein database in which all theoretical tryptic peptides are preassembled. The databases that are searched for matches to mass spectrometry spectra typically include RefSeq and dbEST at NCBI, Swiss-Prot, and the Mass Spectrometry protein sequence DataBase (MSDB). The resolution of MALDI-TOF is excellent (about 0.1 to 0.2 daltons), allowing identification of the protein, especially when multiple peptides correspond to the same protein. (2) Liquid chromatography with tandem mass spectrometry (LC-MS/MS) is commonly used to analyze complex mixtures of proteins such as hundreds of proteins, in a purified organelle or thousands of proteins in a cell lysate (Fig. 10.5b). By injecting the sample onto a column (such as a reverse phase column), the complexity of the sample is reduced. A typical column is narrow (with an inner diameter of 75 μm) and has a slow flow rate (such as 300 nanoliters per minute). This affords good resolution of peptide peaks. Peptides from these peaks are fragmented, for example by electrospray, and analyzed twice by mass spectrometry: first to screen the mass of the peptides, and second to obtain fragmentation spectra that can be searched against a database. Typically, one analyzes a series of fractions from the liquid chromatography column.

A key step in mass spectrometry experiments is the identification of proteins by matching of observed mass spectra to the theoretical spectral profiles of peptide

MSDB is available from Imperial College (London) at ▶ http://csc-fserve.hh.med.ic.ac.uk/msdb.html.

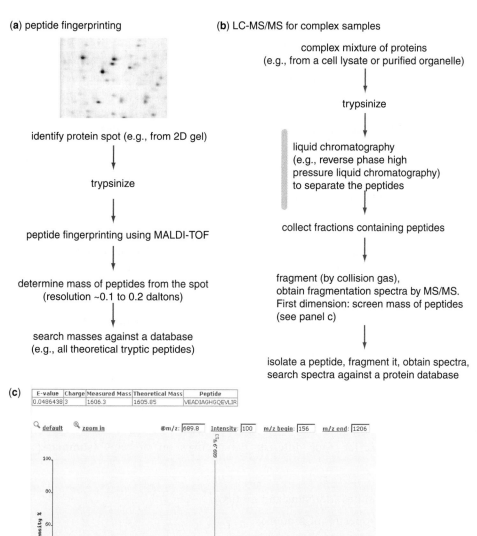

(a) peptide fingerprinting

identify protein spot (e.g., from 2D gel)

↓

trypsinize

↓

peptide fingerprinting using MALDI-TOF

↓

determine mass of peptides from the spot
(resolution ~0.1 to 0.2 daltons)

↓

search masses against a database
(e.g., all theoretical tryptic peptides)

(b) LC-MS/MS for complex samples

complex mixture of proteins
(e.g., from a cell lysate or purified organelle)

↓

trypsinize

↓

liquid chromatography
(e.g., reverse phase high
pressure liquid chromatography)
to separate the peptides

↓

collect fractions containing peptides

↓

fragment (by collision gas),
obtain fragmentation spectra by MS/MS.
First dimension: screen mass of peptides
(see panel c)

↓

isolate a peptide, fragment it, obtain spectra,
search spectra against a protein database

(c)

E-value	Charge	Measured Mass	Theoretical Mass	Peptide
0.0486438	3	1606.3	1605.85	VEADIAGHGQEVLIR

FIGURE 10.5. Mass spectrometry applications. (a) For relatively pure samples, such as a protein band excised from a two-dimensional protein band, peptide fingerprinting is often performed. The sample is proteolyzed with trypsin then MALDI-TOF (or another mass spectrometry technique) is performed to identify the mass of the fragments. These mass values are searched against a protein database. (b) Liquid chromatography-tandem mass spectrometry (LC-MS/MS) is often used to analyze complex protein mixtures such as a cell lysate. A reverse-phase column separates peptides based on hydrophobicity, reducing the complexity of the original sample. Multiple fractions are collected and subjected to two cycles of mass spectrometry, often including electrospray. (c) Sample MS/MS output from an ion trap mass spectrometer. This example shows a fragmentation peptide from equine myoglobin using the Open Mass Spectrometry Search Algorithm (OMSSA) software at NCBI. The figure shows peaks across a range of mass to charge (m/z) ratios. Typically several of these peaks would be collected, fragmented, and the fragmentation spectra are searched against a database to identify the protein(s).

fragments obtained from protein databases (Marcotte, 2007). A variety of software tools are available to do this. An example is the OMSSA tool at the NCBI website. A sample output from a tandem mass spectrometry experiment is shown in Fig. 10.5c. This shows the pattern of peptides following fragmentation of equine myoglobin. Perhaps the most commonly used software is MASCOT® (Perkins et al., 1999). Like other tools it provides a scoring algorithm to evaluate the false positive rate, and an *E* value similar to that used in BLAST (Chapter 4). The main strength of MASCOT® is its integration of three different search methods: peptide mass fingerprinting (in which peptide mass values are obtained), sequence queries (in which peptide mass data are combined with amino acid sequence data and

OMSSA is available at ▶ http://pubchem.ncbi.nlm.nih.gov/omssa/. MASCOT® software is available from Matrix Science (▶ http://www.matrixscience.com/). ProteinPilot is from Applied Biosystems (▶ http://www.appliedbiosystems.com/) and Sequest is from Thermoquest.

The ABRF website is ▶ http://www.abrf.org/. A PubMed search with the term abrf provides access to many studies by this organization in which DNA, RNA, and protein technologies are critically assessed. We will describe the critical assessment of protein structures (CASP) in Chapter 11.

compositional information), and MS/MS data obtained from peptides. Other prominent software includes ProteinPilot and Sequest.

How can we assess the accuracy of protein identification by mass spectrometry? The Association of Biomolecular Resource Facilities (ABRF) has conducted several studies to address this question (Arnott et al., 2002). They prepared five purified proteins at quantities of either 2 picomoles or 200 femtomoles (bovine protein disulfide isomerase [PDI], serum albumin [BSA], and superoxide dismutase; *Escherichia coli* GroEL; and *Schistosoma japonicum* glutathione-*S*-transferase [GST]). They digested the samples with trypsin, mixed them, and sent them "blind" to 41 participating laboratories that performed a total of 55 mass spectrometric analyses. The laboratories tended to use MALDI-TOF or microliquid chromatography with nanospray ionization (μLC-NSI). At the 2 picomole level, 96% (53/55) of the analyses correctly identified PDI, while 80% correctly identified GST. At the 200 femtomole level, 44% identified GroEL, 27% identified BSA, and 11% identified SOD. From one perspective, this is an enormous improvement over earlier mass spectrometry performance; from another perspective, this indicates that it is challenging for many laboratories to detect quantities below one or two picomoles.

There are dozens of important applications of mass spectrometry. We discuss some in this chapter, and also in Chapter 12 when we describe functional genomics as applied to protein–protein interactions.

FOUR PERSPECTIVES ON PROTEINS

We will next describe four different perspectives on proteins (summarized in Fig. 10.6):

1. Protein families (domains and motifs)
2. Physical properties of proteins
3. Protein localization
4. Protein function

The first perspective we will consider is the protein family. We will define terms such as family, domain, and motif. Next, we will consider the physical properties of proteins and how we can assess them. These properties include molecular weight, isoelectric point, and posttranslational modifications (of which several hundred have been described).

The next ways to consider proteins correspond in part to a conceptual framework provided by the Gene Ontology (GO) Consortium. In the past few years 200 billion base pairs of DNA (2×10^{11} bp) have been sequenced, including the complete genomes of thousands of organisms (Chapters 13 to 20). A major challenge to the field of bioinformatics is to identify protein-coding genes (see Chapters 13 and 16). Another great challenge is to annotate them, that is, to provide a description of their nature and function. The GO Consortium (Ashburner et al., 2000) provides a flexible, controlled vocabulary to describe three aspects of proteins: cellular component, biological process, and molecular function. We will provide an overview of the GO project. We will then provide a general description of the protein localization and protein function, corresponding to the GO categories of cellular component and molecular function.

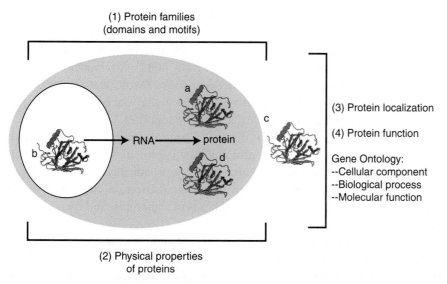

(1) Protein families
(domains and motifs)

(3) Protein localization

(4) Protein function

Gene Ontology:
--Cellular component
--Biological process
--Molecular function

(2) Physical properties
of proteins

FIGURE 10.6. *Overview of proteins. A protein is composed of a series of amino acids specified by a gene. Proteins can be classified by a variety of criteria, including family, localization, physical properties, and function. (1) Protein families are defined by the homology of a protein to other proteins; the proteins may be homologous over a partial region. Databases of protein families and motifs allow hundreds of thousands of proteins to be classified in groups that may be functionally related. (2) Proteins may be described in terms of their physical properties, such as size (molecular weight), shape (e.g., Stokes radius and frictional coefficient), charge (isoelectric point), post-translational modifications (see text), or the existence of isoforms due to proteolytic processing or alternative mRNA splicing. (3) The Gene Ontology (GO) Consortium classifies proteins according to cellular component (i.e., localization), biological process (e.g., transcription or endocytosis), and molecular function (e.g., enzyme or transporter). A protein can belong to multiple categories of any of these groups. The GO system provides a dynamic, controlled vocabulary that can be applied to all eukaryotic proteins. In this figure, a protein is depicted in several possible locations: it may be soluble in the cytosol (label a), in an intracellular organelle such as the nucleus (b), or extracellular as a secreted protein (c). A protein may be bound to membranes on the cell surface (d); membrane localization may be via transmembrane domains or by peripheral attachment. (4) We will also explore the many definitions of protein function.*

PERSPECTIVE 1. PROTEIN DOMAINS AND MOTIFS: MODULAR NATURE OF PROTEINS

Let us begin our discussion of protein domains by considering several types of proteins. In the simplest case, a protein (or gene) has no matches to any other sequences in the available databases. This situation occurs less frequently as increasing numbers of genomes are sequenced, and yet it is still not unusual to find that substantial numbers of predicted proteins have no identifiable homologs (see e.g., Chapters 14, 15, and 18). Even if there are no known homologs, a protein may have features such as a transmembrane domain, potential sites for phosphorylation, or some predicted secondary structure (see below and Chapter 11). Such features may give clues to the structure and/or function of the protein.

For proteins that do have orthologs and/or paralogs, there are regions of significant amino acid identity between at least two proteins (or DNA sequences). Such regions of proteins that share significant structural features and/or sequence identity have a variety of names: signatures, domains, modules, modular elements,

InterPro is accessed at ▶ http://www.ebi.ac.uk/interpro/. It includes 10 member databases: PROSITE (described below),

PRINTS (which uses position-specific scoring matrices), ProDom (which uses automatic sequence clustering), and seven databases that use hidden Markov models (Gene3D, Panther, Pfam, PIRSF, SMART, SUPERFAMILY, TIGRFAMs). InterPro further links to over 20 additional resources, including UniProt.

folds, motifs, patterns, or repeats. These terms have varied definitions, but all refer to the idea that there are closely related amino acid sequences shared by multiple proteins (Bork and Gibson, 1996). Such regions may be considered in terms of protein structure and/or function (Copley et al., 2002). We will primarily adopt the definitions provided by the InterPro Consortium (Mulder et al., 2002, 2007). InterPro is an integrated documentation resource that encompasses a group of databases of protein families, domains, and functional sites.

A *signature* is a broad term that denotes a protein category, such as a domain or family or motif. When you consider a single protein sequence in isolation, there is only a limited amount of information you can infer about its structure or function. However, when you align related sequences, a consensus sequence may be identified. There are two principal kinds of signatures, and each is identified with its own methodology.

A *domain* is a region of a protein that can adopt a particular three-dimensional structure (Doolittle, 1995). Domains are also called modules (Henikoff et al., 1997; Sonnhammer and Kahn, 1994). The term *fold* is commonly used in the context of three-dimensional structure (Jones, 2001). Together, a group of proteins that share a domain is called a family. Many protein domains are further classified based on the subcellular localization of the domain (e.g., intracellular domains of proteins occur in the cytoplasm; extracellular domains are oriented outside the cell) or in terms of the structure of the domain (e.g., zinc finger domains bind the divalent cation zinc).

There are many databases of protein families, such as Pfam and SMART, that we explored in Chapter 6. The definitions of the terms *family, domain, repeat,* and related terms in the InterPro and SMART databases are given in Tables 10.1 and 10.2.

Motifs (or fingerprints) are short, conserved regions of proteins (discussed below). A motif typically consists of a pattern of amino acids that characterizes a protein family (Bork and Gibson, 1996). The size of a defined motif is often 10 to

TABLE 10-1   Definitions from InterPro Database of Protein Families and Related Terms

Term	Definition
Family	An InterPro family is a group of evolutionarily related proteins that share one or more domains/repeats in common. An InterPro entry of "type = family" may contain a signature for a small conserved region that is representative of the family and therefore need not necessarily cover the whole protein.
Domain	A domain is defined as an independent structural unit which can be found alone or in conjunction with other domains or repeats. Domains are evolutionarily related. Even though the structure of a domain is not always known it is still possible to define the boundaries in many cases from sequence alone. Therefore, sequence criteria can be used to define domain boundaries.
Repeat	An InterPro repeat is a region that is not expected to fold into a globular domain on its own. For example, six to eight copies of the *WD40* repeat are needed to form a single globular domain. There also many other short repeat motifs that probably do not form a globular fold that have "type = repeat."
Posttranslational modification	A posttranslational modification includes, for example, an N-glycosylation site. The sequence motif is defined by the molecular recognition of this region in a cell. This may group together proteins that need not be evolutionarily related.

Source: Adapted from ▶ http://www.ebi.ac.uk/interpro/user_manual.html.

TABLE 10-2   Definitions of Protein Domains and Motifs from SMART Database

Term	Definition
Domain	Conserved structural entities with distinctive secondary structure content and a hydrophobic core. In small disulfide-rich and Zn^{2+}-binding or Ca^{2+}-binding domains, the hydrophobic core may be provided by cystines and metal ions, respectively. Homologous domains with common functions usually show sequence similarities.
Domain composition	Proteins with the same domain composition have at least one copy of each domain of the query.
Domain organization	Proteins having all the domains as the query in the same order (additional domains are allowed).
Motif	Sequence motifs are short conserved regions of polypeptides. Sets of sequence motifs need not necessarily represent homologs.
Profile	A profile is a table of position-specific scores and gap penalties, representing an homologous family that may be used to search sequence databases (Bork and Gibson, 1996).

Source: Adapted from ► http://smart.embl-heidelberg.de/help/smart_glossary.shtml.
SMART is a tool to allow automatic identification and annotation of domains in user-supplied protein sequences (see Chapter 6).

20 contiguous amino acid residues, although it can be smaller or larger. Some simple and common motifs, such as a stretch of amino acids that form a transmembrane domain or a consensus phosphorylation site, do not imply homology when found in a group of proteins. In other cases a small motif may provide a characteristic signature for a protein family.

To introduce specific examples of domains, Table 10.3 lists the 15 most common domains in the proteins encoded by the human genome. Similar lists are available for the abundant protein domains of other organisms (Chapters 14 to 16). In many cases, two proteins that share a domain also share a common function. For example, the immunoglobulin-like domain (InterPro accession IPR007110, with over 1000

Web document 10.1 at ► http://www.bioinfbook.org/chapter10 lists the 15 most common human protein families and protein repeats, from the InterPro database.

TABLE 10-3   Fifteen Most Common Domains of *Homo sapiens*

InterPro Accession	Proteins Matched	Name
IPR007110	1,176	Immunoglobulin-like
IPR007087	1,055	Zinc finger, C_2H_2-type
IPR003599	977	Immunoglobulin subtype
IPR011009	883	Protein kinase-like
IPR011993	596	Pleckstrin homology-type
IPR011992	436	EF-Hand type
IPR001849	410	Pleckstrin-like
IPR012677	409	Nucleotide-binding, alpha-beta plait
IPR009057	403	Homeodomain-like
IPR001841	389	Zinc finger, RING-type
IPR013151	386	Immunoglobulin
IPR011989	380	Armadillo-like helical
IPR001452	355	Src homology-3
IPR003596	349	Immunoglobulin V-type
IPR011990	335	Tetratricopeptide-like helical

Source: From the European Bioinformatics Institute (EBI) proteome analysis site (► http://www.ebi.ac.uk/integr8) (August 2007), based on the InterPro database (► http://www.ebi.ac.uk/interpro/).

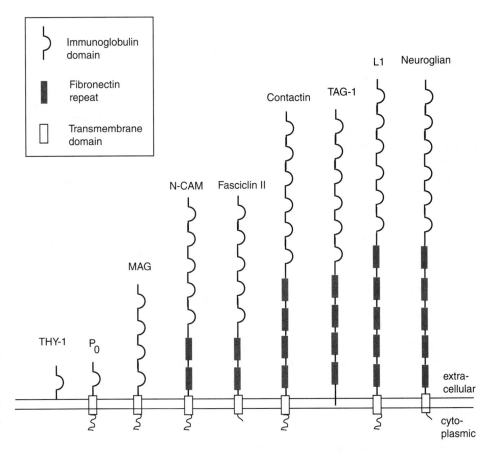

FIGURE 10.7. *Many proteins have multiple copies of distinct domains. The most common domain in humans is the immunoglobulin (Ig) domain, and the fibronectin repeat also commonly occurs. These domains are especially prevalent in the extracellular regions of proteins. Information about domains such as these is summarized in the InterPro database.*

members) is the most common domain encoded by the human genome. Many proteins having this domain have roles in extracellular signaling (Fig. 10.7). As another example, in humans there are hundreds of small guanosine triphosphate- (GTP)-binding proteins (InterPro IPR005225). Many dozens are thought to regulate the intracellular docking and fusion of transport vesicles through a cycle of GTP binding and hydrolysis (Geppert et al., 1997). Other, related low molecular weight GTP-binding proteins function in cell cycle control and cytoskeletal organization (reviewed in Takai et al., 2001). This superfamily is organized into related subfamilies that are usually presumed to share common functions.

Focusing our attention on a single domain, there are many ways in which proteins can share that domain in common. The entire protein may consist of one domain, such as the lipocalin domain or globin domain (Fig. 10.8a). Many other small, globular proteins also consist of a single domain.

It is even more common for a domain to form a subset of a protein. A comparison of two proteins often indicates that the domains occupy different regions of each protein (Fig. 10.8b). A group of six proteins contains a domain that confers the ability of each protein to bind methylated DNA. One of these proteins, methyl-CpG-binding protein 2 (MeCP2), is a transcriptional repressor that binds the regulatory region of a variety of genes. (Mutations in the *MECP2* gene cause Rett syndrome, a neurological disorder that affects girls and is one of the most common causes of mental retardation in females. See Box 20.2.) We can perform a blastp search with the MeCP2 protein sequence to illustrate the concept of protein domains. The BLAST formatting page shows that the methyl-CpG-binding domain (MBD) is present in several databases of protein domains (Fig. 10.9a). The BLAST search result

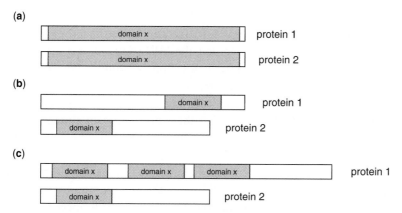

FIGURE 10.8. *Proteins can share a common domain in a number of ways. (a) A domain may extend essentially across the length of a protein. An example of this format is the lipocalin family. (b) Domains may contain highly related stretches of amino acids that form only a subset of each protein's sequence. An example of this situation is found in the family of transcriptional regulators that bind methylated DNA. (c) A domain may be repeated within a single protein (sometimes with many copies). Such a domain may occur in homologous proteins any number of times. An example is the family of proteins containing a fibronectin III–like repeat.*

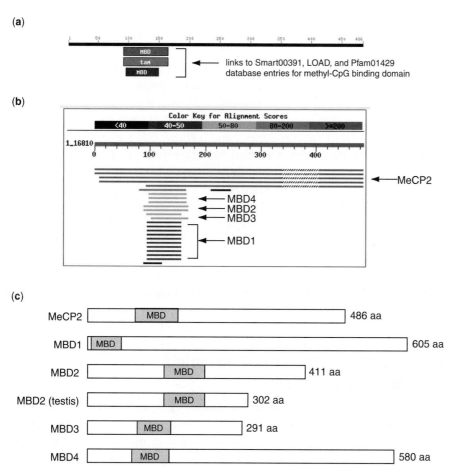

FIGURE 10.9. *A methyl-binding domain is found in several human proteins. To illustrate the concept of domains, methyl-CpG-binding protein 2 (MeCP2; NP_004983) was used as a query in a blastp search restricted to human proteins. (a) The formatting BLAST web page shows that this protein has a domain that is present in several databases. (b) The BLAST search reveals there are five separate MeCP2 entries that match the query (top five alignments). Additionally, there is a region of about 80 amino acids in MeCP2 that matches other methyl-CpG-binding proteins: MBD1 (NP_056671), MBD2 (NP_003918), a testis-specific isoform of MBD2 (NP_056647), MBD3 (NP_003917), and MBD4 (NP_003916). (c) These proteins have different sizes. Also, the methylated DNA-binding domain that these proteins share occurs in different regions of the proteins. Further BLAST searches confirm that together these six proteins share no significant amino acid identity at any region other than the methyl-binding domain.*

shows that a portion of MeCP2 matches five other MBD proteins (Fig. 10.9b). Furthermore, examination of the MeCP2/MBD family shows that the proteins are various different sizes, sharing in common only the MBD domain (Fig. 10.9c).

What is the definition of a family; is a group of proteins homologous if they share only one domain in common? The MBD domains are clearly homologous (descended from a common ancestor), defining this group of proteins as a family. But the regions outside the MBD domain share no significant amino acid identity. A family is a group of evolutionarily related proteins that share one (or more) regions of homology.

A third scenario for proteins containing individual domains is that the domain may be repeated many times (Fig. 10.8c). Two of the most common protein domains in *H. sapiens* are immunoglobulin domains (Table 10.3) and fibronectin repeats. Both of these domains are present in variable numbers in a group of proteins having extracellular domains (Fig. 10.7). Notably, these and other extracellular domains are highly abundant in humans and the multicellular nematode *Caenorhabditis elegans* but nearly absent in the single-celled eukaryote *Saccharomyces cerevisiae* (Copley et al., 1999). Comparison of protein families that are encoded by various genomes sheds light on the biological processes that each organism performs (Chapters 13 to 19).

Added Complexity of Multidomain Proteins

So far we have focused on the subject of single domains. Multidomain proteins provide a common, more complicated scenario. HIV-1 gag-pol is an example of such a protein (Frankel and Young, 1998). The *gag-pol* gene encodes a single large polypeptide that is cleaved into several independent proteins with distinct biochemical activities, including an aspartyl protease, a reverse transcriptase (RNA-dependent DNA polymerase), and an integrase. Note that other multidomain proteins, such as the immunoglobulin domain proteins depicted in Fig. 10.7, maintain separate domains within a mature polypeptide without cleaving them into separate proteins.

You can access the Sequence Retrieval System of ExPASy at ► http://www.expasy.ch/srs/.

To examine the sequence of gag-pol, we will first go to Entrez Gene at NCBI. That entry shows that the protein accession is NP_057849 (corresponding to a protein of 1435 amino acid residues), and it shows that gag-pol encodes at least six mature proteins, each with a RefSeq identifier (Fig. 10.10a). When we use the Sequence Retrieval System (SRS) at ExPASy and enter a search of HIV-1 (organism) and gag-pol, we find 75 entries. Restricting the output to those with a sequence length of 1435 (Fig. 10.10b), we find just eight matches (Fig. 10.10c). By inspection it is difficult to know which of these is prototypical (emphasizing the benefits of a RefSeq-like project that defines reference sequences). Select the first match (SwissProt accession O93215). This SwissProt record includes a variety of links to related databases, including InterPro, Pfam, PROSITE, and ProDom, as well as structure databases (Chapter 11). Follow the ProDom link to a series of proteins sharing at least one domain in common with HIV-1 pol. The ProDom result is a graphical overview of hundreds of proteins that share regions in common with HIV-1 pol (Fig. 10.11a).

Protein Patterns: Motifs or Fingerprints Characteristic of Proteins

Within a domain, there may be a small number of characteristic amino acid residues that occur consistently. These are called motifs (or fingerprints). An example of a motif is the amino acids that are reliably found at the active site of an enzyme. In the aspartyl protease domain of HIV-1 pol, an aspartate residue is crucial for

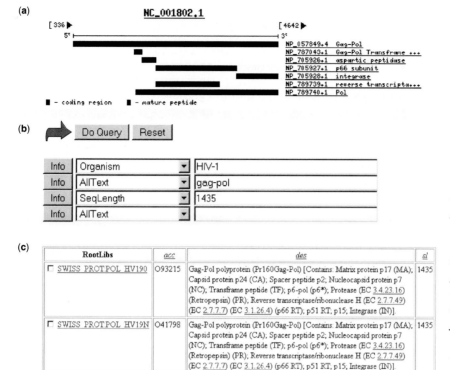

FIGURE 10.10. Searches for a multidomain protein. (a) The Entrez Gene entry for HIV-1 gag-pol provides RefSeq accession numbers for the precursor protein (NP_057849, 1435 amino acids) and for six predicted mature protein products. (b) The Sequence Retrieval System at ExPASy includes a flexible query form with pull-down menus. (c) Two of the eight SRS results are shown for gag-pol proteins of length 1435 residues.

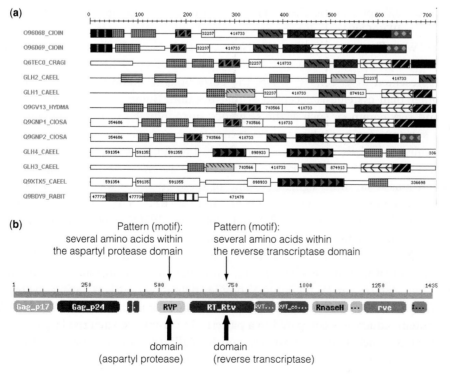

FIGURE 10.11. The UniProtKB/Swiss-Prot entry for HIV-1 gag-pol (O93215) includes links to many other databases, including ProDom. The ProDom entry shows over 1300 proteins that share domains in common with pol, in almost 400 different arrangements, several of which are shown here. This list is obtained by clicking the ProDom link ("List of seq. sharing at least 1 domain") from the UniProtKB/Swiss-Prot entry. It can also be accessed directly from the ProDom database (▶ http://prodom.prabi.fr).

Domains appear as distinct modules within protein sequences. (b) A protein may have domains (modules) which are relatively larger and patterns (motifs) which typically consist of only a few amino acids. Athough a pattern or motif might not adopt a known three-dimensional structural conformation, it may nonetheless contain an amino acid sequence that is characteristic of a protein family.

FIGURE 10.12. PROSITE is a database of patterns. The eukaryotic and viral proteases signature and profile (document PDCO-000128) are shown, including a description of the family, and the consensus pattern.

PROSITE is accessed at ▶ http://www.expasy.org/prosite/. In PROSITE, the term *profile* refers to a quantitative motif description based on a generalized profile syntax. The term *pattern* refers to a qualitative motif description based on a regular expression-like syntax such as those described below. The term *motif* refers to the biological object one attempts to approximate by a pattern or a profile. See web document 10.2 at ▶ http://www.bioinfbook.org/chapter10 for these definitions.

You can use the ScanProsite tool to search a pattern against the PROSITE database, and the PRATT tool to generate a pattern based on an input of unaligned sequences. See computer lab Exercise 10.1.

the proteolytic reaction. PROSITE is a dictionary of protein motifs (Sigrist et al., 2002). Following the link from ExPASy (Fig. 10.3) or searching the site directly, one finds an entry for aspartyl proteases (Fig. 10.12). The motif is defined by a string of 12 amino acid residues: [LIVMFGAC]-[LIVMTADN]-[LIVFSA]-D-[ST]-G-[STAV]-[STAPDENQ]-x-[LIVMFSTNC]-x-[LIVMFGTA]. This format is identical to that used by PHI-BLAST (Chapter 5). A motif may be inside a domain (as illustrated in Fig. 10.11b) or outside.

Motifs are typically subsets of protein domains. A short motif that is found in almost all lipocalins is GXW. The consensus pattern defined in PROSITE (document PDOC00187) incorporates several additional amino acids surrounding GXW. That motif is [DENG]-x-[DENQGSTARK]-x(0,2)-[DENQARK]-[LIVFY]-{CP}-G-{C}-W-[FYWLRH]-x-[LIVMTA]. The GXW sequence is represented as G-{C}-W, where the curly brackets indicate that any amino acid other than cysteine is accepted at that position. Some motifs are extremely short and very common, such as the sequence surrounding a serine or threonine that is a substrate for many kinases. Such motifs are not specific to a particular protein family, and their occurrence in multiple proteins does not reflect homology. A search of PROSITE for "kinase" reveals three dozen entries, including both kinase and kinase substrate signatures. One of these entries is for the protein kinase C (PKC) consensus phosphorylation site, [ST]-x-[RK] (S or T is the phosphorylation site, and x is any residue) (PROSITE document PDOC00005). This simple motif occurs in proteins many thousands of times.

An important aspect of regular expressions (or patterns) in the PROSITE database is that they are qualitative (i.e., either matching or not) and not quantitative. While patterns can accommodate complex definitions, such as having one of several different amino acid residues in a given position, mismatches are not tolerated when a protein sequence is compared to a pattern. In contrast to such rigid patterns, many databases such as Pfam, ProDom, and SMART (described in Chapter 6) use profiles. Profiles, like patterns, are built from multiple sequence alignments, but they employ position-specific scoring matrices. They also span larger stretches of protein sequence than do patterns.

PERSPECTIVE 2. PHYSICAL PROPERTIES OF PROTEINS

Proteins are characterized by a variety of physical properties that derive both from their essential nature as an amino acid polymer and from a variety of posttranslational modifications (Table 10.4). Some of these modifications allow the covalent attachment of a hydrophobic group to a protein to promote insertion into a lipid bilayer. Examples include palmitoylation, farnesylation, myristylation, and inositol glycolipid attachment (Fig. 10.13). The InterPro database also lists categories of posttranslational domains (Table 10.5).

For websites offering protein motif analysis tools, see Table 10.11 under Web Resources.

The COILS server is available at ► http://www.ch.embnet.org/ software/COILS_form.html.

TABLE 10-4 Some Physical Properties of Proteins

Property	Classical Method	Example
Amino acid motifs	—	PDZ domain (e.g., nitric oxide synthase), coiled-coil domain (e.g., hemagglutinin, syntaxin, SNAP-25, myosin)
Isoelectric point (pI)	Derived from isoelectric focusing	—
Molecular weight	Derived from Stokes radius and sedimentation coefficient	—
Posttranslational modifications: phosphorylation	Enzymatic analyses	Synapsin
Posttranslational modifications: glycosylation	Enzymatic analyses	Nerve growth factor, neural cell adhesion molecule
Posttranslational modifications: isoprenylation	Biochemical analyses	Lamin B, G protein γ subunits, *rab3A*
Posttranslational modifications: palmitoylation	Biochemical analyses	β-Adrenergic receptor, *GAP-43*, insulin receptor, rhodopsin, nAChR
Posttranslational modifications: myristoylation	Biochemical analyses	PKA, $G_{i\alpha}$-subunit, MARCKS protein, calcineurin
Posttranslational modifications: GPI-anchored proteins	Enzymatic analyses	Alkaline phosphatase, *thy-1*, prion protein, 5′-nucleotidase, uromodulin
Sedimentation coefficient	Derived from sucrose density gradients	—
Stokes radius	Derived from gel filtration	—
Transmembrane domain	Derived from subcellular fractionation	—

Abbreviations: G protein, guanosine triphosphate-binding protein; GAP-43, growth-associated protein of 43 kDa; MARCKS, myristoylated alanine-rich C-kinase substrate; nAChR, nicotinic acetylcholine receptor; PDZ domain, post-synaptic density protein PSD-95, the *Drosophila* tumor suppressor discs-large, tight-junction protein ZO-1; PKA, protein kinase A; SNAP-25, synaptosomal-associated protein of 25 kDa; Rab3A, rat brain GTP-binding protein 3A; thy-1, thymocyte-1.

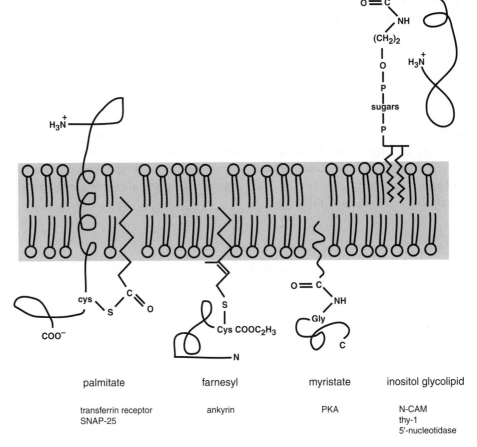

FIGURE 10.13. A variety of post-translational modifications are added to proteins. Examples are palmitoylation (e.g., to the transferrin receptor and SNAP-25), farnesylation (e.g., to ankyrin), myristoylation (e.g., to protein kinase A), and inositol glycolipid anchoring to a membrane (e.g., neural cell adhesion molecule, thy-1, and 5'-nucleotidase). While these covalent modifications can be studied biochemically, a variety of websites offer predictions of possible sites of covalent modification to proteins. Adapted from Austen and Westwood (1991, p. 42). Used with permission.

palmitate	farnesyl	myristate	inositol glycolipid
transferrin receptor SNAP-25	ankyrin	PKA	N-CAM thy-1 5'-nucleotidase

TABLE 10-5	Posttranslational Modifications at InterPro
IPR000042	N-glycosylation site
IPR000134	Amidation site
IPR000152	Aspartic acid and asparagine hydroxylation site
IPR000220	Tyrosine kinase phosphorylation site
IPR000338	N-myristoylation site
IPR000430	Casein kinase II phosphorylation site
IPR000865	Microbodies C-terminal targeting signal
IPR000886	Endoplasmic reticulum targeting sequence
IPR001020	Phosphotransferase system, HPr histidine phosphorylation site
IPR001120	Prokaryotic N-terminal methylation site
IPR001230	Prenyl group, CAAX box, attachment site
IPR001495	Protein kinase C phosphorylation site
IPR001637	Glutamine synthetase class-I, adenylation site
IPR001833	cAMP/cGMP-dependent protein kinase phosphorylation site
IPR002114	Phosphotransferase system, HPr serine phosphorylation site
IPR002332	P-II protein urydylation site
IPR006141	Intein splicing site
IPR006162	Phosphopantetheine attachment site

Source: InterPro (▶ http://www.ebi.ac.uk/interpro/), October 2007.

Compute pI/Mw

HBB_HUMAN (P68871)

DE   Hemoglobin subunit beta (Hemoglobin beta chain) (Beta-globin)
DE   [Contains: LVV-hemorphin-7].

The computation has been carried out on the complete sequence (**147** amino acids).

Molecular weight (average): 15998.41

Theoretical pI: 6.74

A variety of web-based services are available to evaluate the predicted physical properties of proteins. Resources are available to input an individual protein sequence and to predict its physical properties, such as mass and isoelectric point (pI; Fig. 10.14 and Table 10.12 under Web Resources), amino acid composition, glycosylation sites (Table 10.13), phosphorylation sites in which kinases reversibly add a phosphate group to individual serine, threonine, or tyrosine residues (Fig. 10.15), and tyrosine sulfation (Table 10.14 under Web Resources). Many programs predict secondary-structure features of proteins (see Chapter 11). One such feature is coiled-coil regions, which are typically associated with protein–protein interaction domains (Lupas et al., 1991; Lupas, 1997) (Fig. 10.16).

Accuracy of Prediction Programs

For each of these various prediction programs, it is important to assess the accuracy. This is typically done by measuring sensitivity and specificity relative to a "gold

FIGURE 10.14. *The Compute pI/ Mw server at ExPASy calculates the predicted molecular weight and isoelectric point of input proteins. Here, the values for beta globin are calculated. Programs at ExPASy do not accept RefSeq accession numbers as input (e.g., NP_000509 for beta globin), but do accept raw sequence or Uni-Prot accessions (e.g., P68871). Pairwise BLAST (Chapter 3) confirms that these two sequences are identical.*

```
    147 Sequence
MVHLTPEEKSAVTALWGKVNVDEVGGEALGRLLVVYPWTQRFFESFGDLSTPDAVMGNPKVKAHGKKVLGAFSDGLAHLD      80
NLKGTFATLSELHCDKLHVDPENFRLLGNVLVCVLAHHFGKEFTPPVQAAYQKVVAGVANALAHKYH                  160
....T.........................................S....S..........................      80
.......T........................................................................    160

Phosphorylation sites predicted:        Ser: 2   Thr: 2   Tyr: 0


                Serine predictions

Name            Pos     Context     Score   Pred
                         v
Sequence        10      PEEKSAVTA   0.389    .
Sequence        45      RFFESFGDL   0.621   *S*
Sequence        50      FGDLSTPDA   0.987   *S*
Sequence        73      LGAFSDGLA   0.026    .
Sequence        90      FATLSELHC   0.020    .
                         ^

                Threonine predictions

Name            Pos     Context     Score   Pred
                         v
Sequence        5       MVHLTPEEK   0.930   *T*
Sequence        13      KSAVTALWG   0.022    .
Sequence        39      VYPWTQRFF   0.398    .
Sequence        51      GDLSTPDAV   0.489    .
Sequence        85      NLKGTFATL   0.012    .
Sequence        88      GTFATLSEL   0.587   *T*
Sequence        124     GKEFTPPVQ   0.393    .
                         ^
```

FIGURE 10.15. *The ExPASy web server offers a large group of protein analysis tools such as the NetPhos server for prediction of phosphorylation sites (▶ http://www.cbs.dtu. dk/services/NetPhos/). Beta globin protein sequence was input and the output includes two likely sites for phosphorylation on serines, two on threonines, and none on tyrosines based on scores exceeding a threshold value of 0.5. Such information on sulfation, phosphorylation, glycosylation, or other posttranslational modifications may be fundamental in designing experiments to test the function of a protein.*

FIGURE 10.16. The coils program of Lupas et al. (1991) assesses the likelihood that a protein sequence forms a coiled-coil structure. (a) Output of the coils program using human SNAP-25 protein (NP_003072) as input. The result depicts the probability that the protein will form a coiled-coil secondary structure motif (y axis) across the length of the protein (x axis). Coiled-coils often represent protein–protein interaction domains. In this case, the coiled-coils of SNAP-25, a peripherally associated plasma membrane protein, allow it to bind tightly to two other proteins (syntaxin and vesicle-associated membrane protein [VAMP/synaptobrevin]) to coordinate synaptic vesicle docking and neurotransmitter release at the presynaptic nerve terminal. (b) According to the Conserved Domain Database (CDD at NCBI), SNAP-25 has two t-SNARE domains that are known to coordinate binding to syntaxin and VAMP. These domains partially overlap the predicted coiled-coil domains.

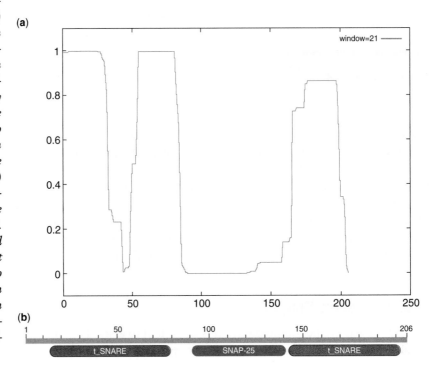

standard" of a set of proteins known to have a particular modification. In recent decades, the physical properties of proteins were assessed at the laboratory bench, one protein at a time (Cooper, 1977). The molecular mass of a protein can be estimated by gel filtration chromatography or by polyacrylamide gel electrophoresis (PAGE). Its shape can be estimated by calculating the frictional coefficient, obtained through a combination of gel filtration and sucrose density gradient centrifugation. Such techniques cannot be applied to large numbers of proteins. Almost all proteins that are studied using the tools of bioinformatics have not been purified, but instead the protein sequence is predicted from genomic DNA or cDNA sequence data.

Prediction programs vary in their accuracy. For proteins with typical amino acid compositions, the prediction of the molecular weight and pI (Fig. 10.14) is likely to be accurate. These protein features can also be confirmed experimentally using techniques such as gel electrophoresis and isoelectric focusing. A prediction algorithm may accurately specify that a protein has a consensus site for phosphorylation or sulfation, but these modifications are not necessarily made in living cells, and their regulation is likely to be dynamic. One can ask whether a protein has a potential site for modification, and a separate question is the conditions under which such modification occurs.

Proteomic Approaches to Phosphorylation

It has been estimated that one third of all proteins are phosphorylated, affording an important mechanism for regulating their function. Also, there are nearly 1000 kinases encoded by the human genome. The accuracy of computational-based ("in silico") prediction programs can be measured. To generate the phosphorylation site predictions used by the NetPhos program (Fig. 10.15), Blom et al. (1999) analyzed a large number of amino acid sequences surrounding known acceptor residues on substrate proteins. They applied an artificial neural network to classify sequence patterns in a training set, and then examined a test set. This allowed them to determine the sensitivity (proportion of positive sites correctly predicted) and specificity (proportion of all positive classifications that are correct). A challenge they addressed is that the sequence databases include sites incorrectly annotated as nonphosphorylated (i.e., the false positive rate of their program was inappropriately high). Some methods Blom et al. (1999) tested surpassed 95% sensitivity and specificity for predictions of phosphorylation on serine, with less accuracy for predictions on threonine or tyrosine.

Some proteins with unusual occurrences of particular amino acids are given in Table 10.15 under Web Resources. We provided other examples in web documents 4.1 to 4.4 at ▶ http://www.bioinfbook.org/chapter4. These proteins may have physical properties (such as pI) that are difficult to predict.

Competitions that are open to the research community allow us to assess the state of proteomics techniques under laboratory conditions. The Association of Biomolecular Resource Facilities (ABRF), discussed above, assessed the ability of 54 laboratories to detect phosphorylation sites (Arnott et al., 2003). They prepared a sample consisting of bovine protein disulfide isomerase (PDI; 5 picomoles), two phosphopeptides corresponding to PDI (length 8 and 17 amino acids; 1 picomole each), and bovine serum albumin (BSA; 200 femtomoles). After proteolytic digestion with trypsin, the samples were distributed blind to the research community and 54 laboratories reported 67 analyses; 96% of the laboratories identified PDI, but only 10% detected BSA. There was a surprisingly low success rate for detecting the phosphopeptides and assigning the phosphorylation site: only 3 of 54 laboratories did so for both phosphopeptides. This study highlights the enormous challenges of experimental protein analyses. Most of the laboratories employed MALDI-TOF or LC-MS.

In addition to considering the phosphorylation of individual proteins, some investigators have examined the total collection of phosphorylated sites in a biological sample (the "phosphoproteome") (Kalume et al., 2003; Ptacek and Snyder, 2006). Advances have occurred in the ability to enrich complex mixtures for phosphoproteins, and in mass spectrometry approaches. For example, Ptacek et al. (2005) determined the substrates recognized by 87 different yeast protein kinases (of the total of 122 annotated kinases from *Saccharomyces cerevisiae*). They used protein chips containing ∼4400 proteins present in duplicate. They catalogued about 4200 phosphorylation events affecting 1325 proteins, and validated some of the findings in vivo.

The data from Ptacek et al. (2005) are available at ▶ http://networks.gersteinlab.org/phosphorylome/.

A variety of databases provide annotation of posttranslational modifications of proteins. The Human Protein Reference Database (HPRD) features expert curation on thousands of proteins, including information on phosphoproteins (Mishra et al., 2006). Phospho3D specifically focuses on three-dimensional structures of phosphorylation sites (Zanzoni et al., 2006).

HPRD is available at ▶ http://www.hprd.org. It includes a PhosphoMotif Finder. Currently (November 2007) HPRD includes over 25,000 protein entries. Phospho3D is at ▶ http://cbm.bio.uniroma2.it/phospho3d/.

Proteomic Approaches to Transmembrane Domains

What is the accuracy of a program that predicts transmembrane topology? It is easy to use a search tool to find a prediction. However, this is fundamentally a cell biological question, and it requires the tools of cell biology to obtain a clear answer. Many proteins have stretches of 10 to 25 hydrophobic amino acid residues that may form transmembrane domains. The most rigorous assessment of the true number of transmembrane domains comes from experimental approaches such as immunocyto-chemistry. Specific antisera can be raised in rabbits, mice, or other species and used to detect an antigen (such as a cell surface receptor) in a sample affixed to a microscope slide. In unpermeabilized cells, the antisera can be used to visualize protein regions that are oriented outside the cell. However, when cells are permeabilized with detergent, the antisera can gain access to the cytosol and thus can visualize intracellular (cytoplasmic) domains. Cell biological analyses such as these have been used to experimentally determine the number of transmembrane domains, and in many cases these results contradict the predictions of hydropathy plots (e.g., Ratnam et al., 1986).

The Membrane Protein Data Bank (Raman et al., 2006) summarizes proteins for which three-dimensional structural data are available. It is online at ▶ http://www.mpdb.ul.ie/.

The TMHMM server is available at ▶ http://www.cbs.dtu.dk/services/TMHMM/. The Phobius web server is at ▶ http://phobius.cgb.ki.se/. SignalP (Emanuelsson et al., 2007) has a server at ▶ http://www.cbs.dtu.dk/services/SignalP/.

One prominent program for transmembrane domain prediction, TMHMM, employs a hidden Markov model whose states include regions spanning the membrane (the core of a transmembrane helix as well as cytoplasmic and noncytoplasmic caps) and globular regions and loops on the cytoplasmic and noncytoplasmic sides of the membrane (Krogh et al., 2001). The accuracy of this program in predicting the topology of 160 proteins was about 78%. A further advance comes from incorporating information about transmembrane domains with signal peptide predictions (Käll et al., 2007). In analyses of various eukaryotic and prokaryotic genomes about 5% to 10% of all proteins had predicted transmembrane segments that overlap predicted signal peptides (as predicted by software such as SignalP). TMHMM improves its accuracy by accounting for this.

Introduction to Perspectives 3 and 4: Gene Ontology Consortium

The Gene Ontology Consortium main web site is ▶ http://www.geneontology.org/.

An ontology is a description of concepts. The GO Consortium is a project that compiles a dynamic, controlled vocabulary of terms related to different aspects of genes and gene products (proteins) (Thomas et al., 2007). A prominent use of this vocabulary is to annotate and interpret the results of microarray experiments that profile RNA transcripts, although many other kinds of high throughput assays are also annotated using GO (Beissbarth, 2006; Whetzel et al., 2006). The consortium was begun by scientists associated with three model organism databases: the *Saccharomyces* Genome Database (SGD), the *Drosophila* genome database (FlyBase), and the Mouse Genome Informatics databases (MGD/GXD) (Ashburner et al., 2000, 2001). Subsequently, databases associated with many other organisms have joined the GO Consortium (Table 10.6). The GO database is not centralized per se but instead relies on external databases (such as a mouse database) in which each gene or gene product is annotated with GO terms. Thus, it represents an ongoing, cooperative effort to unify the way genes and gene products are described. There are several web browsers that serve as principal gateways to search GO terms (Table 10.7). Additionally, Entrez Gene and Entrez Protein entries at NCBI (Chapter 2) contain GO terms.

There are three main organizing principles of GO: (1) molecular function, (2) biological process, and (3) cellular component. Molecular function refers to the tasks performed by individual gene products. For example, a protein can be a transcription factor or a carrier protein. Biological process refers to the broad biological goals that a gene product (protein) is associated with, such as mitosis or purine

TABLE 10-6 Participating Organizations and Databases in Gene Ontology Consortium

Database or Organization	Organism	Common Name	URL
Berkeley *Drosophila* Genome Project	*Drosophila melanogaster*	Fly	▶ http://www.fruitfly.org/
DictyBase	*Dictyostelium discoideum*	Slime mold	▶ http://dictybase.org/
European Bioinformatics Institute (EBI)	Various	—	▶ http://www.ebi.ac.uk/GOA/
FlyBase	*D. melanogaster*	Fly	▶ http://flybase.bio.indiana.edu/
GeneDB (Wellcome Trust Sanger Institute)	Protozoans, fungi	—	▶ http://www.genedb.org/
Gramene	*Oryza sativa;* other grains, monocots	Rice	▶ http://www.gramene.org/
HUGO Gene Nomenclature Committee	Various	—	▶ http://www.genenames.org/
Mouse Genome Database (MGD) and Gene Expression Database (GXD)	*Mus musculus*	Mouse	▶ http://www.informatics.jax.org/
Rat Genome Database (RGD)	*Rattus*	Rat	▶ http://rgd.mcw.edu/
Reactome	—	—	▶ http://www.genomeknowledge.org/
Saccharomyces Genome Database (SGD)	*Saccharomyces cerevisiae*	Baker's yeast	▶ http://www.yeastgenome.org/
The *Arabidopsis* Information Resource (TAIR)	*Arabidopsis thaliana*	Thale cress	▶ http://www.arabidopsis.org/
The J. Craig Venter Institute	Various	—	▶ http://www.jcvi.org/
WormBase	*Caenorhabditis elegans*	Worm	▶ http://www.wormbase.org/
Zebrafish Information Network	*Danio rerio*	Zebrafish	▶ http://zfin.org/

Source: Adapted from ▶ http://www.geneontology.org/. Updated November 2007.

metabolism. Cellular component refers to the subcellular localization of a protein. Examples include nucleus and lysosome. Any protein may participate in more than one molecular function, biological process, and/or cellular component.

Genes and gene products are assigned to GO categories through a process of annotation. The author of each GO annotation supplies an evidence code that indicates the basis for that annotation (Table 10.8). As an example of a GO-annotated protein look at the Entrez Gene entry for human beta globin (HBB; Fig. 10.17). Entrez Gene entries include a section on function that includes information from OMIM (Chapter 20), Enzyme Commission nomenclature (see below), and GO terms. For HBB, the GO terms include heme binding, oxygen binding, and oxygen transporter activity (molecular functions); oxygen transport (a biological process); and hemoglobin complex (a cellular component).

You can also access gene ontology information by entering a query term such as "HBB" or "lipocalin" into a GO web browser. In some cases the output includes a graphical tree view. This displays the relationships between the different levels of

TABLE 10-7   Websites Useful to Access Gene Ontology Data

Browser	Description	URL
AmiGO	GO browser from the Berkeley *Drosophila* Genome Project	▶ http://amigo. geneontology.org/cgi-bin/amigo/go.cgi
Mouse Genome Informatics (MGI) GO Browser	From Jackson Laboratories	▶ http://www. informatics.jax.org/ searches/GO_form. shtml
"QuickGO" at EBI	From the EMBL and European Bioinformatics Institute; integrated with InterPro (Chapter 10)	▶ http://www.ebi.ac.uk/ ego/
Expression Profiler (EP) GO Browser	GO browser and analysis tool that is part of the EP suite at the European Bioinformatics Institute	▶ http://ep.ebi.ac.uk/ EP/GO/
Cancer Gene Anatomy Project (CGAP) GO Browser	From the National Cancer Institute, NIH	▶ http://cgap.nci.nih. gov/Genes/ AllAboutGO

TABLE 10-8   Evidence Codes for Gene Ontology Project

Abbreviation	Evidence Code	Example(s)
IC	Inferred by curator	A protein is annotated as having the function of a "transcription factor." A curator may then infer that the localization is "nucleus"
IDA	Inferred from direct assay	An enzyme assay (for function); immunofluorescence microscopy (for cellular component)
IEA	Inferred from electronic annotation	Annotations based on "hits" in searches such as BLAST (but without confirmation by a curator; compare ISS)
IEP	Inferred from expression pattern	Transcripts levels (e.g., based on Northern blotting or microarrays) or protein levels (e.g., from Western blots)
IGC	Inferred from Genomic Context	Identity of the genes neighboring the gene product in question (i.e., synteny), operon structure, and phylogenetic or other whole genome analysis
IGI	Inferred from genetic interaction	Suppresors; genetic lethals; complementation assays; experiments in which one gene provides information about the function, process, or component of another gene
IMP	Inferred from mutant phenotype	Gene mutation; gene knockout; overexpression; antisense assays
IPI	Inferred from physical interaction	Yeast two-hybrid assays; copurification; co-immunoprecipitation; binding assays
ISS	Inferred from sequence or structural similarity	Sequence similarity; domains; BLAST results that are reviewed for accuracy by a curator
NAS	Nontraceable author statement	Database entries such as a SwissProt record that does not cite a published paper

(Continued)

TABLE 10-8	Continued	
Abbreviation	Evidence Code	Example(s)
ND	No biological data available	Corresponds to "unknown" molecular function, biological process, or cellular compartment
RCA	Inferred from Reviewed Computational Analysis	Predictions based on large-scale experiments (e.g., genome-wide two-hybrid, genome-wide synthetic interactions); predictions based on integration of large-scale datasets of several types; text-based computation (e.g., text mining)
TAS	Traceable author statement	Information in a review article or dictionary
NR	Not recorded	Used for annotations done before curators began tracking evidence types

Source: Adapted from ► http://www.geneontology.org/.

GeneOntology   Provided by GOA

Function	Evidence	
heme binding	IEA	
hemoglobin binding	IDA	PubMed
iron ion binding	IEA	
metal ion binding	IEA	
molecular function	ND	
oxygen binding	IDA	PubMed
oxygen binding	IEA	
oxygen transporter activity	IEA	
oxygen transporter activity	NAS	PubMed
selenium binding	IDA	PubMed

Process	Evidence	
biological process	ND	
nitric oxide transport	NAS	PubMed
oxygen transport	IEA	
oxygen transport	NAS	PubMed
oxygen transport	TAS	PubMed
positive regulation of nitric oxide biosynthetic process	NAS	PubMed
transport	IEA	

Component	Evidence	
hemoglobin complex	IEA	
hemoglobin complex	NAS	PubMed
hemoglobin complex	TAS	PubMed

FIGURE 10.17. *The GO Consortium provides a dynamic, controlled vocabulary that describes genes and gene products from a variety of organisms. Its three organizing principles are molecular function, biological process, and cellular component. GO terms can be accessed through a variety of browsers or through Entrez Gene, as shown for human beta globin. These GO terms at NCBI are obtained from the Gene Ontology Annotation (GOA) Database at the European Bioinformatics Institute (► http://www.ebi.ac.uk/GOA/).*

GO terms. These have the form of a "directed acyclic graph" or network. This differs from a hierarchy in that in a hierarchy each child term can have only one parent, while in a directed acyclic graph it is possible for a child to have more than one parent. A child term may be an instance of its parent term, in which case the graph is labeled "isa," or the child term may be component of the parent term (a "partof" relationship). This complicates the structure of the terms in GO and the evaluation of their biological and statistical significance. Some statistical tests assess the likelihood that each GO category is under- or overrepresented more than is expected by chance. However, a concept such as "mitochondria" occurs in all three categories (biological process, molecular function, cellular compartment) at multiple levels.

We will next consider protein localization and protein function. These topics loosely correspond to the GO categories "cellular component" and "molecular function." In Chapter 12 we will discuss protein pathways, although the GO category "biological process" does not refer specifically to pathways.

PERSPECTIVE 3: PROTEIN LOCALIZATION

In eukaryotic cells, the intracellular organelles account for up to 95% of the cell's membranes.

The cellular localization of a protein is one of its fundamental properties. Proteins are synthesized on ribosomes from mRNA. Some proteins are synthesized in the cytosol. Other proteins, destined for secretion or insertion in the plasma membrane, are inserted into the endoplasmic reticulum (in eukaryotes) or into the plasma membrane (in prokaryotes). This insertion, which occurs either cotranslationally or posttranslationally, is mediated by the signal recognition particle, an RNA–multiprotein complex (Stroud and Walter, 1999). In the endoplasmic reticulum, proteins may be transported through the secretory pathway to the Golgi apparatus and then to further destinations such as intracellular organelles (e.g., endosomes, lysosomes) or to the cell surface.

Proteins may further be secreted into the extracellular milieu. The trafficking of a protein to its appropriate destination is achieved by transport in secretory vesicles. These vesicles are typically 50 to 100 nanometers in diameter, and they transport soluble or membrane-bound cargo to specific compartments.

We may also distinguish two main categories of proteins based on their relationships to phospholipid bilayers: (1) those that are soluble and exist in the cytoplasm, in the lumen of an organelle, or in the extracellular environment, and (2) those that are membrane attached, associated with a lipid bilayer. Those proteins associated with membranes may be integral membrane proteins (having a span of 10 to 25 hydrophobic amino acid residues that traverse the lipid bilayer) or they may be peripherally associated with membranes (attached via a variety of anchors such as those shown in Fig. 10.13).

Many proteins defy categorization into one static location in the cell. For example, the annexins and the low molecular weight GTP-binding proteins are families of proteins that migrate between the cytosol and a membrane compartment. This movement typically depends on the presence of dynamically regulated cellular signals such as calcium or transient phosphorylation.

Proteins are often targeted to their appropriate cellular location because of intrinsic signals embedded in their primary amino acid sequence. For example, the sequence KDEL (lysine–aspartic acid–glutamic acid–leucine) at the carboxy terminus of a soluble protein specifies that it is selectively retained in the endoplasmic reticulum. Other targeting motifs have been identified for import into mitochondria, lysosomes, or peroxisomes and for endocytosis. However, these motifs are typically not as invariant as KDEL.

You can access WoLF PSORT server at ▶ http://wolfpsort.org/.

Several web-based programs predict the intracellular localization of any individual protein sequence (see Web Resources, Tables 10.16 and 10.17). For example, WoLF PSORT accurately predicts the signal sequence at the amino terminus of retinol-binding protein (Fig. 10.18). This signal peptide is characteristic of proteins that enter the secretory pathway in the endoplasmic reticulum. WoLF PSORT analyzes a protein query for localization features based on sorting signals, amino acid composition, and functional motifs (Horton et al., 2007). It then uses a k-nearest neighbor classifier to predict the localization.

Normalized Feature Values

id	site	iPSORT				PSORT Features															Amino Acid Content				Misc.
		-1	25	MxHy1	30	act	alm	dna	gvh	leu	mNt	mip	mit	myr	nuc	rib	rnp	tms	tyr	vac	C	I	K	S	length
queryProtein	extr?		78		75	50	76	44	96	46	49	71	59	49	29	50	50	27	49	48	68	10	51	27	26
CASP_CHICK	extr		78		70	50	82	44	90	46	49	55	80	49	29	50	50	27	49	48	60	26	34	17	36
IL10_HUMAN	extr		64		89	50	93	44	100	46	49	71	61	49	29	50	50	27	49	48	74	35	76	31	23
IL10_MACNE	extr		64		89	50	93	44	100	46	49	84	61	49	29	50	50	27	49	48	74	35	70	41	23
A2HS_HUMAN	extr		64		75	50	57	44	97	46	49	81	56	49	70	50	50	27	49	48	80	12	32	28	53
IL10_CERTO	extr		64		89	50	93	44	100	46	49	84	61	49	29	50	50	27	49	48	74	35	76	41	23
IL10_MACFA	extr		64		89	50	93	44	100	46	49	84	61	49	29	50	50	27	49	48	74	35	76	41	23
IL10_MACMU	extr		64		89	50	93	44	100	46	49	84	61	49	29	50	50	27	49	48	74	35	76	41	23
IBP2_BRARE	extr		46		80	50	50	44	84	46	49	75	67	49	29	50	50	27	49	48	92	15	56	37	36
PPT1_HUMAN	lyso		64		75	50	52	44	96	46	49	76	63	49	29	50	50	27	49	48	61	61	52	37	40
NDDB_CAVPO	extr		46		74	50	47	44	86	46	49	84	77	49	29	50	50	27	49	48	65	7	48	37	32

FIGURE 10.18. *The WoLF PSORT server provides a web-based query form to predict the sub-cellular location of a protein. The program searches for sorting signals and other features that are characteristic of proteins localized to particular compartments. The output of a search using reti-nol-binding protein protein sequence (NP_006735) includes 32 nearest neighbors (of which ten are shown here in rows along with the query). The columns include features analyzed including site (proposing correctly that the query is extracellular), results of the iPSORT program (including calculations of the negative charge and hydrophobicity of the initial 25 to 30 amino acid residues), and results from the PSORT program (including the presence of motifs typical of proteins localized to various subcellular compartments). The output shows that there is strong evidence for a signal peptide with a cleavage site between amino acid residues 16 and 17. Such a signal peptide characterizes proteins that enter the secretory pathway where some (such as RBP) are secreted outside the cell.*

In addition to evaluating the location of individual proteins, it is possible to apply high throughput technologies. Michael Snyder and colleagues attempted the first proteome-scale analysis of protein localization (Kumar et al., 2002). They cloned cDNA encoding several thousand proteins from the budding yeast *S. cerevisiae*, incorporating epitope tags into the carboxy- or amino-terminals. An epitope tag is a short protein fragment, such as the nine-amino-acid hemagglutinin (HA) peptide, that is attached covalently to a protein of interest. An antiserum that detects the epitope tag can then be used to localize the protein of interest by immunofluoresence microscopy or to purify the protein (and its binding partners) by immunoprecipitation with an anti–epitope tag antiserum. The directed cloning approach of Kumar et al. allowed them to systematically evaluate the location of many specific proteins of interest. As a complementary strategy, they used random transposon-mediated tagging of genes.

Kumar et al. 2002 generated over 13,000 yeast strains and determined the sub-cellular localization of 2744 yeast proteins by immunofluorescence microscopy using monoclonal antibodies directed against the epitope tags. Many proteins of unknown function were assigned to intracellular locations. For example, if a protein is localized to the yeast peroxisome, then this suggests it may function in fatty acid metabolism.

Bacterial transposons are mobile DNA elements that can be randomly inserted into genomic DNA. The transposons can be modified to incorporate an epitope tag. This strategy is practically simpler than directed cloning of specific yeast cDNAs, although from an experimental point of view transposon-tagged proteins are difficult to localize in cells.

A database with 2900 fluorescence micrographs of the Kumar et al. 2002 data is available at ▶ http://ygac.med.yale.edu.

PERSPECTIVE 4: PROTEIN FUNCTION

We have described bioinformatics tools to describe protein families, their physical properties, and the cellular localization of proteins. A fourth aspect of proteins is their function (Raes et al., 2007). Function is defined as the role of a protein in a cell (Jacq, 2001). Each protein is a gene product that interacts with the cellular

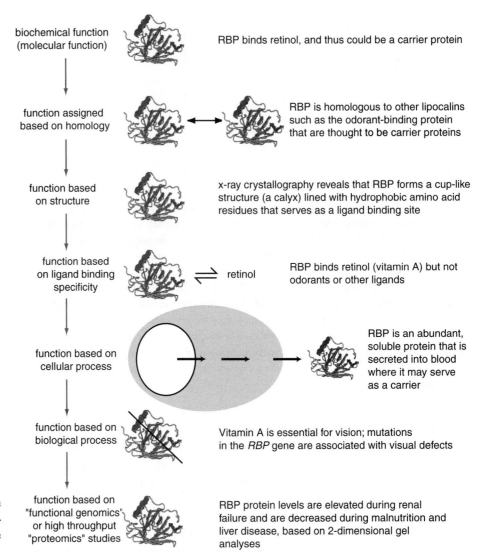

biochemical function
(molecular function)

RBP binds retinol, and thus could be a carrier protein

function assigned
based on homology

RBP is homologous to other lipocalins such as the odorant-binding protein that are thought to be carrier proteins

function based
on structure

x-ray crystallography reveals that RBP forms a cup-like structure (a calyx) lined with hydrophobic amino acid residues that serves as a ligand binding site

function based
on ligand binding
specificity

⇌ retinol

RBP binds retinol (vitamin A) but not odorants or other ligands

function based on
cellular process

RBP is an abundant, soluble protein that is secreted into blood where it may serve as a carrier

function based on
biological process

Vitamin A is essential for vision; mutations in the *RBP* gene are associated with visual defects

function based on
"functional genomics"
or high throughput
"proteomics" studies

RBP protein levels are elevated during renal failure and are decreased during malnutrition and liver disease, based on 2-dimensional gel analyses

FIGURE 10.19. Protein function may be analyzed from several perspectives. Retinol-binding protein (RBP) is used as an example.

environment in some way to promote the cell's growth and function. We can consider the concept of function from seven perspectives (Fig. 10.19):

1. A protein has a biochemical function synonymous with its molecular function. For an enzyme, the biochemical function is to catalyze the conversion of one or more substrates to product(s). For a structural protein such as actin or tubulin, the biochemical function is to influence the shape of a cell. For a transport protein, the biochemical function is to carry a ligand from one location to another. (Such a transport role may even occur in the absence of a requirement for an energy source such as ATP—in such a way, retinol-binding protein transports retinol through serum, and hemoglobin transports oxygen.) For a hypothetical protein that is predicted to be encoded by a gene, the biochemical function is unknown but is presumed to exist. There are thought to be no proteins that exist without a biochemical function.

2. Functional assignment is often made based on homology. Currently when a genome is sequenced the great majority of its predicted proteins can be

functionally assigned based on orthology. If a hypothetical protein is homologous to an enzyme, it is often provisionally assigned that enzymatic function. This is best viewed as a hypothesis that must be tested experimentally. As an example, many globin-like proteins occur in bacteria, protozoa, and fungi having biochemical properties distinct from those of vertebrate globins (Poole and Hughes, 2000).

3. Function may be assigned based on structure (Chapter 11). If a protein has a three-dimensional fold that is related to that of a protein with a known function, this may be the basis for functional assignment. Note, however, that structural similarity does not necessarily imply homology, and homology does not necessarily imply functional equivalence.

4. All proteins function in the context of other proteins and molecules. Thus, a definition of a protein's function may include its ligand (if the protein is a receptor), its substrate (if the protein is an enzyme), its lipid partner (if the protein interacts with membrane), or any other molecule with which it interacts. The odorant-binding protein (OBP) is a lipocalin that binds a variety of odorants in nasal mucus, suggesting that the binding properties of the protein are central to its function (Pevsner et al., 1990). However, the biological function of OBP is not known from its ligand-binding properties alone. The protein could transport odorants toward the olfactory epithelium to promote sensory perception, it could carry odorants from the olfactory epithelium to facilitate odorant clearance, or it could metabolize odorants.

5. Many proteins function as part of a distinct biochemical pathway such as the Krebs cycle, in which discrete steps allow the cell to perform a complex task. Other examples are fatty acid oxidation in peroxisomes or proteolytic degradation that is accomplished by the proteasome.

6. Proteins function as part of some broad cell biological process. Cells divide, grow, and senesce; neurons have axons that display outgrowth, pathfinding, target recognition, and synapse formation; and all cells secrete molecules through discrete pathways. All cellular processes require proteins in order to function, and each individual protein can be defined in the context of

TABLE 10-9 Functional Assignment of 4808 Proteins Based on Their Enzymatic Activity: Partial List of Enzyme Commission Classification System

EC Number	Description of Class	No. of Enzymes	Example of Subclass
1. -. -. -	Oxidoreductases	1,317	
1. 1. -. -	—	—	Acting on the CH–OH group of donors
1. 2. -. -	—	—	Acting on the aldehyde or oxo group of donors
2. -. -. -	Transferases	1,297	
2. 1. -. -	—	—	Transferring one-carbon groups
3. -. -. -	Hydrolases	1,437	
4. -. -. -	Lyases	441	
5. -. -. -	Isomerases	172	
6. -. -. -	Ligases	144	

Source: From ▶ http://www.expasy.org/enzyme/ (November 2007).

the broad cellular function it serves. The Gene Ontology Consortium (Ashburner et al., 2000, p. 27) defines a biological process as "a biological objective to which the gene or gene product contributes. A process is accomplished via one or more ordered assemblies of molecular functions. Processes often involve a chemical or physical transformation, in the sense that something goes into a process and something different comes out of it."

7. Protein function can be considered in the context of all the proteins that are encoded by a genome—that is, in terms of the proteome. The term *functional genomics* refers to the attempt to use experimental approaches and/or

TABLE 10-10 Functional Classification of Proteins in Clusters of Orthologous Groups Database

General Category	Function	Clusters of Orthologous Groups	Domains
Information storage and processing	Translation, ribosomal structure, and biogenesis	245	10,572
	RNA processing and modification	25	137
	Transcription	231	11,271
	Replication, recombination, and repair	238	10,338
	Chromatin structure and dynamics	19	228
Cellular processes and signaling	Cell cycle control, cell division chromosome partitioning	72	1,678
	Defense mechanisms	46	2,380
	Signal transduction mechanisms	152	7,683
	Cell wall/membrane/envelope biogenesis	188	7,858
	Cell motility	96	2,747
	Cytoskeleton	12	128
	Extracellular structures	1	25
	Intracellular trafficking, secretion, vesicular transport	159	3,743
	Posttranslational modification, protein turnover, chaperones	203	6,206
	Energy production and conversion	223	5,584
	Carbohydrate transport and metabolism	170	5,257
Metabolism	Energy production and conversion	258	9,830
	Carbohydrate transport and metabolism	230	10,816
	Amino acid transport and metabolism	270	14,939
	Nucleotide transport and metabolism	95	3,922
	Coenzyme transport and metabolism	179	6,582
	Lipid transport and metabolism	94	5,201
	Inorganic ion transport and metabolism	212	9,232
	Secondary metabolites biosynthesis, transport and catabolism	88	4,055
Poorly characterized	General function prediction only	702	22,721
	Function unknown	1,346	13,883

Source: From ► http://www.ncbi.nlm.nih.gov/COG/ (November 2007).

computational tools to analyze the role of many hundreds or thousands of genes and gene products. Since the ultimate product of transcription is often a protein, the term functional genomics is sometimes applied to large-scale studies of protein function. Chapter 12 addresses the topic of functional genomics.

Thus, protein function can be defined in many ways. Many proteins are enzymes. The Enzyme Commission (EC) system provides a standardized nomenclature for almost 4000 enzymes (Table 10.9). When a genome is sequenced and a potential protein-coding sequence is identified, homology of that protein to an enzyme with a defined EC listing provides a specific, testable hypothesis about the biochemical function of that hypothetical protein.

Another broader approach to the functional assignment of proteins is provided by the Clusters of Orthologous Groups (COGs) database (Tatusov et al., 1997; Chapter 15). The functional groups defined by this system are listed in Table 10.10. While the COGs database has initially focused on prokaryotic genomes, the general categories are relevant to basic cellular processes in all living organisms. Many other functions that are unique to eukaryotes, such as apoptosis and complex developmental processes, are now represented in the eukaryotic portion of the COGs scheme (Tatusov et al., 2003).

Apoptosis is programmed cell death. It occurs in a variety of multicellular organisms, both as a normal process in development and as a homeostatic mechanism in adult tissues. Apoptosis can be triggered by external stimuli (such as infectious agents or toxins) or by internal agents such as those causing oxidative stress. You can visit the COGs database at ▶ http://www.ncbi.nlm.nih.gov/COG/.

PERSPECTIVE

In this chapter we have considered bioinformatics approaches to individual proteins. In Chapter 11 we will next consider protein structure, which provides us with deeper insight into the nature of proteins, including their domains, physical properties, and function. Then in Chapter 12 (functional genomics) we will explore high throughput approaches to studying sets of proteins (e.g., techniques employing gel electrophoresis and mass spectrometry), as well as protein–protein interactions.

In the past decade, our understanding of the properties of proteins has advanced dramatically, from the level of biochemical function to the role of proteins in cellular processes. Advances in instrumentation have propelled mass spectrometry into a leading role for many proteomics applications.

Many web-based tools are available to evaluate the biochemical features of individual proteins. Such programs can predict the existence of glycosylation, phosphorylation, or other sites. These predictions can be extremely valuable in guiding the biologist to experimentally test the possible posttranslational modifications of a protein.

High throughput approaches such as two-dimensional gel electrophoresis and the yeast two-hybrid system have been used in an effort to define the function of all proteins. Large numbers of proteins still have no known function because they lack detectable homology to other characterized proteins. We will continue to obtain a more comprehensive description of protein function as distinct high throughput strategies are applied to model organisms, such as large-scale analyses of protein localization and protein interactions.

PITFALLS

Many of the experimental and computational strategies used to study proteins have limitations. Two-dimensional protein gels are most useful for studying relatively abundant proteins, but thousands of proteins expressed at low levels are harder to characterize. Experimental approaches are extremely challenging in practice, as shown by the ABRF critical assessments.

WEB RESOURCES

TABLE 10-11 Tools to Analyze Protein Motifs

Program	Comment	URL
ExPASy	Source of many tools	► http://www.expasy.org/tools/
InterProScan	At EBI	► http://www.ebi.ac.uk/InterProScan/
ppsearch	At EBI	► http://www2.ebi.ac.uk/ppsearch/
PRATT	At EBI	► http://www2.ebi.ac.uk/pratt/
ProfileScan Server	At ISREC	► http://hits.isb-sib.ch/cgi-bin/PFSCAN
PROSCAN (PROSITE SCAN)	Many tools at PBIL (Pôle Bio-Informatique Lyonnais) (► http://pbil.univ-lyon1.fr/)	► http://npsa-pbil.ibcp.fr/cgi-bin/npsa_automat.pl?page = npsa_prosite.html
ScanProsite tool	At ExPASy	► http://www.expasy.ch/tools/scanprosite/
SMART	At EMBL	► http://smart.embl-heidelberg.de/
TEIRESIAS	At IBM	► http://cbcsrv.watson.ibm.com/Tspd.html

TABLE 10-12 Tools to Analyze Primary and/or Secondary Structure Features of Proteins

Program	Source/Comment	URL
COILS	Prediction of coiled-coil regions in proteins	► http://www.ch.embnet.org/software/COILS_form.htmll
Compute pI/Mw	From ExPASy	► http://www.expasy.org/tools/pi_tool.html
drawhca	Hydrophobic cluster analysis plot	► http://psb00.snv.jussieu.fr/hca/hca-seq.html
Helical Wheel	Draws an helical wheel, i.e., an axial projection of a regular alpha helix	► http://www.site.uottawa.ca/ ~ turcotte/resources/HelixWheel
M.M., pI, composition, titrage	Many tools from the Atelier Bio Informatique de Marseille	► http://www.up.univ-mrs.fr/wabim/english/logligne.html#predi
Paircoil	Prediction of coiled-coil regions in proteins	► http://groups.csail.mit.edu/cb/paircoil/paircoil.html
PeptideMass	From ExPASy	► http://www.expasy.ch/tools/peptide-mass.html
REP	Searches a protein sequence for repeats	► http://www.embl-heidelberg.de/ ~ andrade/papers/rep/search.html
SAPS	Statistical analysis of protein sequences	► http://www.isrec.isb-sib.ch/software/SAPS_form.html

Source: From ExPASy: ► http://www.expasy.org/tools/.

TABLE 10-13 Web Resources for Characterization of Glycosylation Sites on Proteins

Program	Comment/Source	URL
DictyOGlyc 1.1 Prediction Server	Neural network predictions for GlcNAc O-glycosylation sites in *Dictyostelium discoideum* proteins	► http://www.cbs.dtu. dk/services/ DictyOGlyc/
NetGlycate	Prediction of glycation of ε amino groups of lysines in mammalian proteins	► http://www.cbs.dtu. dk/services/ NetGlycate/
NetOGlyc	Prediction of type O-glycosylation sites in mammalian proteins	► http://www.cbs.dtu. dk/services/ NetOGlyc/
YinOYang 1.2	Produces neural network predictions for O-β-GlcNAc attachment sites in eukaryotic protein sequences	► http://www.cbs.dtu. dk/services/ YinOYang/

TABLE 10-14 Tools to Analyze Posttranslational Modifications

Program	Comment	URL
big-PI Predictor	GPI modification site prediction	► http://mendel.imp. ac.at/gpi/gpi_server. html
NetPhos 2.0 Prediction Server	Produces neural network predictions for serine, threonine, and tyrosine phosphorylation sites in eukaryotic proteins	► http://www.cbs.dtu. dk/services/ NetPhos/
Sulfinator	Prediction of tyrosine sulfation sites	► http://www.expasy. org/tools/sulfinator/

Source: From ExPASy: ► http://www.expasy.org/tools/.

TABLE 10-15 Examples of Proteins with Unusually High Occurrences of Specific Amino Acids

Amino Acid(s)	Proteins
C	Disulfide-rich proteins; metallothioneins; zinc finger proteins
D, E	Acidic proteins
G	Collagens (e.g., NP_000079)
H	Hisactophilin; histidine-rich glycoprotein (e.g., XP_629852)
W, L, P, Y, L, V, M, A	Transmembrane domains (e.g., NP_004594, NP_062098)
K, R	Nuclear proteins (nuclear localization signals)
N	*Dictyostelium* proteins
P	Collagens; filaments; SH3/WW/EVHI binding sites
Q	Proteins encoded by genes mutated in triplet repeat disorders (Chapter 20)
S, R	Some RNA-binding motifs
S, T	Mucins; oligosaccharide attachment sites (e.g., XP_855042)
abcdefg	Heptad coiled coils (**a** and **d** are hydrophobic residues), e.g., myosin (NP_005370)

Source: Modified from Ponting (2001). The hydrophobic residues characteristic of transmembrane helices are from Tanford (1980). Used with permission.

TABLE 10-16 Web-Based Programs for Prediction of Protein Localization

Program	Comment	URL
ChloroP	Predicts presence of chloroplast transit peptides (cTP) in protein sequences	► http://www.cbs.dtu.dk/services/ChloroP/
MITOPROT	Calculates the N-terminal protein region that can support a mitochondrial targeting sequence and the cleavage site	► http://ihg.gsf.de/ihg/mitoprot.html
PSORT	Prediction of protein-sorting signals and localization sites; access to PSORT II, WoLF PSORT	► http://psort.nibb.ac.jp/
SignalP	Predicts presence and location of signal peptide cleavage sites in prokaryotes and eukaryotes	► http://www.cbs.dtu.dk/services/SignalP/
TargetP	Predicts subcellular location of eukaryotic protein sequences	► http://www.cbs.dtu.dk/services/TargetP/

TABLE 10-17 Web Servers for Prediction of Transmembrane Domains in Protein Sequences

Program	Comment/Source	URL
DAS server	Prediction of transmembrane regions	► http://www.sbc.su.se/~miklos/DAS/
HMMTOP	Prediction of transmembrane helices and topology of proteins	► http://www.enzim.hu/hmmtop/
Phobius	Combined transmembrane topology and signal peptide predictor	► http://phobius.sbc.su.se/
PredictProtein server	Prediction of transmembrane helix location and topology	► http://www.predictprotein.org/
SOSUI	Classification and secondary structure prediction of membrane proteins	► http://bp.nuap.nagoya-u.ac.jp/sosui/
TMpred	Prediction of membrane-spanning regions and their orientation	► http://www.ch.embnet.org/software/TMPRED_form.html
TopPred2	Topology prediction of membrane proteins	► http://bioweb.pasteur.fr/seqanal/interfaces/toppred.html

Source: ExPASy web server.

DISCUSSION QUESTIONS

[10-1] InterPro is an important resource that coordinates information about protein signatures from a variety of databases. When these databases all describe a particular protein family or a particular signature, what different kinds of information can you obtain? Is the information in InterPro redundant?

[10-2] How do you define the function of a protein? Does the function change over time, or physiological state, or other condition?

PROBLEMS/COMPUTER LAB

[10-1] Select a group of unaligned, divergent globins (web document 6.3 at ▶ http://www.bioinfbook.org/chapter6). Use them as input to the PRATT program at Prosite (▶ http://www.expasy.ch/prosite/) in order to find a representative pattern. Scan this pattern against the PROSITE database using the ScanPROSITE tool. Do you identify globin proteins? Are there non-globin proteins as well?

[10-2] Salmon has a pinkish color, and some lobsters are blue (but turn red when boiled) because a chromophore called astaxanthin binds to a carrier protein called crustacyanin. Examine the protein sequence of crustacyanin from the European lobster *Homarus gammarus*. What are some of its physical properties (e.g., molecular weight, isoelectric point)? Does it have any known domains or motifs that might explain how or why it binds to the chromophore? Use the tools at the ExPASy site. (For more information about this protein, read the article at

ExPASy: ▶ http://www.expasy.org/spotlight/back_issues/sptlt026.shtml.)

[10-3] Evaluate human syntaxin at the ExPASy site. Does it have coiled-coil regions? How many predicted transmembrane domains does it have? What is its function? Use the ExPASy sequence retrieval system first.

[10-4] Olfactory receptors are related to the rhodopsin-like G-protein coupled receptor (GPCR) superfamily. Use the Integr8 proteome tools at EBI (▶ http://www.ebi.ac.uk/integr8) to decide about what percent of the mouse proteome is comprised of these receptors. About what percent of the human proteome is comprised of these receptors?

[10-5] Again use the proteome tools at ▶ http://www.ebi.ac.uk/integr8/. Are any of the 15 most common protein domains in *E. coli* K12 also present in human?

SELF-TEST QUIZ

[10-1] Can a domain be at the amino terminus of one protein and the carboxy terminus of another protein?

(a) Yes

(b) No

[10-2] In general, if you compare the size of a pattern (also called a motif or fingerprint) and a domain:

(a) They are about the same size.

(b) The pattern is larger.

(c) The pattern is smaller.

(d) The comparison always depends on the particular proteins in question.

[10-3] The amino acid sequence [ST]-X-[RK] is the consensus for phosphorylation of a substrate by protein kinase C. This sequence is an example of:

(a) A motif that is characteristic of proteins that are homologous to each other

(b) A motif that is characteristic of proteins that are not necessarily homologous to each other

(c) A domain that is characteristic of proteins that are homologous to each other

(d) A domain that is characteristic of proteins that are not necessarily homologous to each other

[10-4] If you analyze a single, previously uncharacterized protein using programs that predict glycosylation, sulfation, phosphorylation, or other posttranslational modifications:

(a) The predictions of the programs are not likely to be accurate.

(b) The accuracy of the predictions is unknown and difficult to assess.

(c) The predictions of the programs are likely to be accurate concerning the possible presence of particular

modifications, but their biological revelance is unknown until you assess the protein's properties experimentally.

(d) The predictions of the programs are likely to be accurate concerning the possible presence of particular modifications, but it is not feasible to assess the protein's properties experimentally.

[10-5] An underlying assumption of the Gene Ontology Consortium is that the description of a gene or gene product according to three categories (molecular function, biological process, and cellular component):

(a) Is likely to be identical across many species, from plants to worms to human

(b) Is likely to vary greatly across many species, from plants to worms to human

(c) May or may not be identical across many species and thus must be assessed for each gene or gene product individually

(d) May or may not be identical across many species and thus must be assessed for each gene or gene product individually by an expert curator

[10-6] Protein localization is described primarily in which Gene Ontology category?

(a) Molecular function

(b) Cellular component

(c) Cellular localization

(d) Biological process

[10-7] Which of the following is a means of assessing protein function?

(a) Finding structural homologs

(b) Studying bait–prey interactions

(c) Determining the isoelectric point

(d) All of the above

[10-8] A major advantage of two-dimensional protein gels as a high-throughput technology for protein analysis is that:

(a) Sample preparation and the process or running two-dimensional gels is straightforward and can be automated.

(b) The result of two-dimensional gels includes data on both the size and the charge of thousands of proteins.

(c) The technique is well suited to the detection of low-abundance proteins.

(d) The technique is well suited to the detection of hydrophobic proteins.

SUGGESTED READING

Excellent reviews on proteomics are available, including de Hoog and Mann (2004) and Twyman (2004). Bernard Jacq (2001) and Raes et al. (2007) have written superb reviews on protein function. Both articles discuss the complexity of protein function and the use of bioinformatic tools to dissect function. Jacq proposes to consider function from six structural levels, from the structure of a protein to its role in a population of organisms. Raes et al. include a discussion of the impact of genomics projects on assessing protein function.

REFERENCES

Arnott, D. P., Gawinowicz, M., Grant, R. A., Lane, W. S., Packman, L. C., Speicher, K., and Stone, K. Proteomics in mixtures: Study results of ABRF-PRG02. *J. Biomol. Tech.* **13**, 179–186 (2002).

Arnott, D., Gawinowicz, M. A., Grant, R. A., Neubert, T. A., Packman, L. C., Speicher, K. D., Stone, K., and Turck, C. W. ABRF-PRG03: Phosphorylation site determination. *J. Biomol. Tech.* **14**, 205–215 (2003).

Ashburner, M., et al. Gene ontology: Tool for the unification of biology. The Gene Ontology Consortium. *Nat. Genet.* **25**, 25–29 (2000).

Ashburner, M., et al. Creating the gene ontology resource: Design and implementation. *Genome Res.* **11**, 1425–1433 (2001).

Austen, B. M., and Westwood, O. M. *Protein Targeting and Secretion*. IRL Press, Oxford, 1991.

Beissbarth, T. Interpreting experimental results using gene ontologies. *Methods Enzymol.* **411**, 340–352 (2006).

Blom, N., Gammeltoft, S., and Brunak, S. Sequence- and structure-based prediction of eukaryotic protein phosphorylation sites. *J. Mol. Biol.* **294**, 1351–1362 (1999).

Bork, P., and Gibson, T. J. Applying motif and profile searches. *Methods Enzymol.* **266**, 162–184 (1996).

Carrette, O., Burkhard, P. R., Sanchez, J. C., and Hochstrasser, D. F. State-of-the-art two-dimensional gel electrophoresis: A key tool of proteomics research. *Nat. Protoc.* **1**, 812–823 (2006).

Cooper, T. G. *The Tools of Biochemistry*. Wiley, New York, 1977.

Copley, R. R., Doerks, T., Letunic, I., and Bork, P. Protein domain analysis in the era of complete genomes. *FEBS Lett.* **513**, 129–134 (2002).

Copley, R. R., Schultz, J., Ponting, C. P., and Bork, P. Protein families in multicellular organisms. *Curr. Opin. Struct. Biol.* **9**, 408–415 (1999).

Cox, J., and Mann, M. Is proteomics the new genomics? *Cell* **130**, 395–398 (2007).

de Hoog, C. L. and Mann, M. Proteomics. *Annu. Rev. Genomics Hum. Genet.* **5**, 267–293 (2004).

Doolittle, R. F. The multiplicity of domains in proteins. *Annu. Rev. Biochem.* **64**, 287–314 (1995).

Dunn, M. J. (ed.). *From Genome to Proteome: Advances in the Practice and Application of Proteomics*. Wiley-VCH, New York, 2000.

Emanuelsson, O., Brunak, S., von Heijne, G., and Nielsen, H. Locating proteins in the cell using TargetP, SignalP and related tools. *Nat. Protoc.* **2**, 953–971 (2007).

Farmer, T. B., and Caprioli, R. M. Determination of protein–protein interactions by matrix-assisted laser desorption/ionization mass spectrometry. *J. Mass Spectrom.* **33**, 697–704 (1998).

Fountoulakis, M., Schuller, E., Hardmeier, R., Berndt, P., and Lubec, G. Rat brain proteins: Two-dimensional protein database and variations in the expression level. *Electrophoresis* **20**, 3572–3579 (1999).

Frankel, A. D., and Young, J. A. HIV-1: Fifteen proteins and an RNA. *Annu. Rev. Biochem.* **67**, 1–25 (1998).

Geppert, M., Goda, Y., Stevens, C. F., and Sudhof, T. C. The small GTP-binding protein Rab3A regulates a late step in synaptic vesicle fusion. *Nature* **387**, 810–814 (1997).

Görg, A., Weiss, W., and Dunn, M. J. Current two-dimensional electrophoresis technology for proteomics. *Proteomics* **4**, 3665–3685 (2004).

Grünenfelder, B., et al. Proteomic analysis of the bacterial cell cycle. *Proc. Natl. Acad. Sci. USA* **98**, 4681–4686 (2001).

Henikoff, S., et al. Gene families: The taxonomy of protein paralogs and chimeras. *Science* **278**, 609–614 (1997).

Hoogland, C., et al. The SWISS-2DPAGE database: What has changed during the last year. *Nucleic Acids Res.* **27**, 289–291 (1999).

Horton, P., Park, K. J., Obayashi, T., Fujita, N., Harada, H., Adams-Collier, C. J., and Nakai, K. WoLF PSORT: Protein localization predictor. *Nucleic Acids Res.* **35**, W585–W587 (2007).

Jacq, B. Protein function from the perspective of molecular interactions and genetic networks. *Brief. Bioinform.* **2**, 38–50 (2001).

Jones, D. T. Protein structure prediction in genomics. *Brief. Bioinform.* **2**, 111–125 (2001).

Käll, L., Krogh, A., and Sonnhammer, E. L. Advantages of combined transmembrane topology and signal peptide prediction: The Phobius web server. *Nucleic Acids Res.* **35**, W429–W432 (2007).

Kalume, D. E., Molina, H., and Pandey, A. Tackling the phosphoproteome: Tools and strategies. *Curr. Opin. Chem. Biol.* **7**, 64–69 (2003).

Kersey, P. J., Duarte, J., Williams, A., Karavidopoulou, Y., Birney, E., and Apweiler, R. The International Protein Index: An integrated database for proteomics experiments. *Proteomics* **4**, 1985–1988 (2004).

Krogh, A., Larsson, B., von Heijne, G., and Sonnhammer, E. L. Predicting transmembrane protein topology with a hidden Markov model: Application to complete genomes. *J. Mol. Biol.* **305**, 567–580 (2001).

Langen, H., et al. Two-dimensional map of human brain proteins. *Electrophoresis* **20**, 907–916 (1999).

Lupas, A. Predicting coiled-coil regions in proteins. *Curr. Opin. Struct. Biol.* **7**, 388–393 (1997).

Lupas, A., Van Dyke, M., and Stock, J. Predicting coiled coils from protein sequences. *Science* **252**, 1162–1164 (1991).

Mann, G. *The Chemistry of the Proteids.* Macmillan, New York, 1906.

Mann, M., Hendrickson, R. C., and Pandey, A. Analysis of proteins and proteomes by mass spectrometry. *Annu. Rev. Biochem.* **70**, 437–473 (2001).

Marcotte, E. M. How do shotgun proteomics algorithms identify proteins? *Nat. Biotechnol.* **25**, 755–757 (2007).

Martens, L., Orchard, S., Apweiler, R., and Hermjakob, H. Human proteome organization proteomics standards initiative: Data standardization, a view on developments and policy. *Mol. Cell. Proteomics* **6**, 1666–1667 (2007).

Mishra, G. R., et al. Human protein reference database: 2006 update. *Nucleic Acids Res.* **34**(Database issue), D411–D414 (2006).

Mulder, G. J. *The Chemistry of Vegetable and Animal Physiology.* P. F. H. Fromberg, transl. William Blackwood & Sons, London, 1849.

Mulder, N. J., et al. InterPro: An integrated documentation resource for protein families, domains, and functional sites. *Brief. Bioinform.* **3**, 225–235 (2002).

Mulder, N. J., et al. New developments in the InterPro database. *Nucleic Acids Res.* **35**(Database issue), D224–D228 (2007).

O'Farrell, P. H. High resolution two-dimensional electrophoresis of proteins. *J Biol. Chem.* **250**, 4007–4021 (1975).

Østergaard, M., et al. Proteome profiling of bladder squamous cell carcinomas: Identification of markers that define their degree of differentiation. *Cancer Res.* **57**, 4111–4117 (1997).

Perkins, D. N., Pappin, D. J., Creasy, D. M., and Cottrell, J. S. Probability-based protein identification by searching sequence databases using mass spectrometry data. *Electrophoresis* **20**, 3551–3567 (1999).

Pevsner, J., Hou, V., Snowman, A. M., and Snyder, S. H. Odorant-binding protein. Characterization of ligand binding. *J. Biol. Chem.* **265**, 6118–6125 (1990).

Ponting, C. P. Issues in predicting protein function from sequence. *Brief. Bioinform.* **2**, 19–29 (2001).

Poole, R. K., and Hughes, M. N. New functions for the ancient globin family: Bacterial responses to nitric oxide and nitrosative stress. *Mol. Microbiol.* **36**, 775–783 (2000).

Ptacek, J., et al. Global analysis of protein phosphorylation in yeast. *Nature* **438**, 679–684 (2005).

Ptacek, J., and Snyder, M. Charging it up: Global analysis of protein phosphorylation. *Trends Genet.* **22**, 545–554 (2006).

Raes, J., Harrington, E. D., Singh, A. H., and Bork, P. Protein function space: Viewing the limits or limited by our view? *Curr. Opin. Struct. Biol.* **17**, 362–369 (2007).

Raman, P., Cherezov, V., and Caffrey, M. The Membrane Protein Data Bank. *Cell. Mol. Life Sci.* **63**, 36–51 (2006).

Ratnam, M., Nguyen, D. L., Rivier, J., Sargent, P. B., and Lindstrom, J. Transmembrane topography of nicotinic acetylcholine receptor: Immunochemical tests contradict theoretical predictions based on hydrophobicity profiles. *Biochemistry* **25**, 2633–2643 (1986).

Reti, L. Le arti chimiche di Leonardo da Vinci. *La Chimica e l'Industria* **34**, 655–721 (1952).

Sanchez, J. C., et al. The mouse SWISS-2D PAGE database: A tool for proteomics study of diabetes and obesity. *Proteomics* **1**, 136–163 (2001).

Shively, J. E. The chemistry of protein sequence analysis. *EXS* **88**, 99–117 (2000).

Sigrist, C. J., et al. PROSITE: A documented database using patterns and profiles as motif descriptors. *Brief. Bioinform.* **3**, 265–274 (2002).

Sonnhammer, E. L., and Kahn, D. Modular arrangement of proteins as inferred from analysis of homology. *Protein Sci.* **3**, 482–492 (1994).

Stroud, R. M., and Walter, P. Signal sequence recognition and protein targeting. *Curr. Opin. Struct. Biol.* **9**, 754–759 (1999).

Takai, Y., Sasaki, T., and Matozaki, T. Small GTP-binding proteins. *Physiol. Rev.* **81**, 153–208 (2001).

Tanford, C. *The Hydrophobic Effect: Formation of Micelles and Biological Membranes.* Wiley, New York, 1980.

Tatusov, R. L., Koonin, E. V., and Lipman, D. J. A genomic perspective on protein families. *Science* **278**, 631–637 (1997).

Tatusov, R. L., et al. The COG database: An updated version includes eukaryotes. *BMC Bioinformatics* **4**, 41 (2003).

Taylor, C. F., et al. The minimum information about a proteomics experiment (MIAPE). *Nat. Biotechnol.* **25**, 887–893 (2007).

Thomas, P. D., Mi, H., and Lewis, S. Ontology annotation: Mapping genomic regions to biological function. *Curr. Opin. Chem. Biol.* **11**, 4–11 (2007).

Venter, J. C., et al. Environmental genome shotgun sequencing of the Sargasso Sea. *Science* **304**, 66–74 (2004).

Viswanathan, S., Unlü, M., and Minden, J. S. Two-dimensional difference gel electrophoresis. *Nat. Protoc.* **1**, 1351–1358 (2006). PMID: 17406422

Whetzel, P. L., Parkinson, H., and Stoeckert, C. J. Jr. Using ontologies to annotate microarray experiments. *Methods Enzymol.* **411**, 325–339 (2006).

Yooseph, S., et al. The Sorcerer II Global Ocean Sampling expedition: Expanding the universe of protein families. *PLoS Biol.* **5**, e16 (2007).

Zanzoni, A., Ausiello, G., Via, A., Gherardini, P. F., and Helmer-Citterich, M. Phospho3D: A database of three-dimensional structures of protein phosphorylation sites. *Nucleic Acids Res.* **35**, D229–D231 (2006).

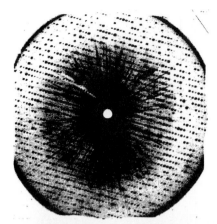

Beginning in the 1940s, Max Perutz and John Kendrew realized the goal of determining the structure of globular proteins by solving the structure of hemoglobin and myoglobin. In recognition of this work, they shared the Nobel Prize in Chemistry in 1962. (top) X-ray precession photograph of a myoglobin crystal (from ▶ http://www.nobel.se/chemistry/laureates/1962/kendrew-lecture.pdf). Kendrew studied myoglobin from the sperm whale (Physeter catodon), and incorporated a heavy metal by the method of isomorphous replacement. He could then bombard the crystals with x-rays in order to obtain an x-ray diffraction pattern (such as that shown here) with which to deduce the electron density throughout the crystal. This required the analysis of 25,000 reflections. (middle) Perutz and Kendrew used the EDSAC I computer (introduced in 1949 and shown here (from ▶ http://www.cl.cam.ac.uk/Relics/jpegs/edsac99. 36.jpg). This computer was essential to interpret the diffraction patterns. For a simulator that shows the capacity of the EDSAC machine, see ▶ http://www.dcs.warwick.ac.uk/~edsac/. (bottom) Photograph by Max Perutz of John Kendrew with his model of myoglobin in 1959. Source: ▶ http://img.cryst.bbk.ac.uk/BCA/obits/jck.html.

11

Protein Structure

A visitor to the Accademia in Florence can see magnificent images that emerged from blocks of marble at the hands of Michelangelo. By analogy, the noncrystallographer can capture the vision that a crystallographer has when admiring a rigorously shaped crystal before exploring the marvelous structure hidden within. So the Protein Data Bank is our museum, with models of molecules reflecting the wonders of nature and complex shapes that may be as old as life itself. With the aid of interactive graphics and networking, the PDB makes these images readily available. What wonders still remain hidden as we build, compare, and extend our database?

—Edgar F. Meyer (1997)

OVERVIEW OF PROTEIN STRUCTURE

Proteins adopt a spectacular range of conformations and interact with their cellular milieu in diverse ways. There are three major classes of proteins: structural proteins (such as tubulin and actin), membrane proteins (such as photoreceptors and ion channels), and globular proteins (such as globins).

The three-dimensional structure of a protein determines its capacity to function. This structure is determined from its primary (linear) amino acid sequence. In the 1950s Christian Anfinsen and others performed a remarkable set of experiments. They purified the enzyme ribonuclease from bovine pancreas, and denatured it with urea. This enzyme includes eight sulfhydryl groups that form four disulfide

Bioinformatics and Functional Genomics, Second Edition. By Jonathan Pevsner
Copyright © 2009 John Wiley & Sons, Inc.

Christian Anfinsen won part of the 1972 Nobel Prize in Chemistry "for his work on ribonuclease, especially concerning the connection between the amino acid sequence and the biologically active conformation" (▶ http://nobelprize.org/nobel_prizes/chemistry/laureates/1972/).

An angstrom (abbreviated Å) is 0.1 nanometers or 10^{-10} meters; a carbon-carbon bond has a distance of about 1.5 Å. John Kendrew and Max Perutz shared the 1962 Nobel Prize in Chemistry "for their studies of the structures of globular proteins." See ▶ http://nobelprize.org/nobel_prizes/chemistry/laureates/1962/.

bonds. After removing the urea, the ribonuclease refolded and adopted a conformation that was indistinguishable from native ribonuclease. Anfinsen stated the thermodynamic hypothesis that the three-dimensional structure of a native protein under physiological conditions is the one in which the Gibbs free energy of the system is lowest (Anfinsen, 1973). Thus we can picture an energy landscape in which many conformations are possible, and proteins tend to adopt the structure(s) that minimize the free energy. Anfinsen's work helped to solidify the concept that the three-dimensional structure of a protein is inherently specified by the linear amino acid sequence.

In the 1950s researchers applying the techniques of x-ray crystallography to proteins focused on the structures of hemoglobin, myoglobin, ribonuclease, and insulin. By 1957 John Kendrew and colleagues reported the three-dimensional structure of myoglobin to 6 Å resolution, sufficient to reconstruct the main outline of the protein. Soon after the resolution was improved to 2 Å. For the first time, all the atoms comprising a protein could be spatially described and the structural basis of the function of a protein—here, myoglobin as an oxygen carrier—was elucidated. Today the central repository of protein structures, the Protein Data Bank, contains 50,000 structures (see below).

In this chapter we will consider the structure of individual proteins from the principles of primary, secondary, tertiary, and quaternary structure. We will also consider structural genomics initiatives in which a very broad range of high resolution tertiary structures are determined for proteins, spanning organisms across the tree of life and also spanning the set of all possible conformations that protein structures can adopt. We will introduce the main repository of protein structures, the Protein Data Bank (PDB), as well as three software tools to visualize structures: WebMol at PDB, Cn3D at NCBI, and DeepView at ExPASy. Many databases provide analyses of structural data and we will describe three prominent ones: CATH, SCOP, and the Dali Domain Dictionary. Finally, we will discuss protein structure prediction which underlies the newly emerging field of structural genomics.

Protein Sequence and Structure

It is difficult to make a pairwise alignment of rat retinol-binding protein (P04916) and rat odorant-binding protein (NP_620258). If you use BLAST 2 Sequences, no significant match is found, even using a large expect value and a scoring matrix appropriate for distantly related proteins (PAM250). If you do a PSI-BLAST search with rat OBP as a query, you will eventually detect retinol-binding protein after many iterations. We will compare the three-dimensional structures of these two proteins in computer lab problem in this chapter using the DaliLite server and see evidence that they are homologous.

As described in Chapter 10, one of the most fundamental questions about a protein is its function. Function is often assigned based on homology to another protein whose function is perhaps already known or inferred (Holm, 1998; Domingues et al., 2000). Two proteins that share a similar structure are usually assumed to also share a similar function. For example, two receptor proteins may share a very similar structure, and even if they differ in their ability to bind ligands or transduce signals, nonetheless they still share the same basic function.

Various types of BLAST searching are employed to identify such relationships of homology (Chapters 4 and 5). However, for many proteins sequence identity is extremely limited. We may take retinol-binding protein and odorant-binding protein as examples: these are both lipocalins of about 20 kDa and are abundant, secreted carrier proteins. They share a GXW motif that is characteristic of lipocalins. However, it is difficult to detect homology based on analysis of the primary amino acid sequences. By pairwise alignment the two proteins share less than 20% identity. Both structure and function are preserved over evolutionary time more than is sequence identity. Thus, the three-dimensional structures of these proteins are extraordinarily similar. We have seen similar relationships for myoglobin relative to alpha globin and beta globin (Fig. 3.1).

Can we generalize about the relationship between amino acid sequence identity and protein structures? It is clear that even a single amino acid substitution can in some instances cause a dramatic change in protein structure, as exemplified by disease-causing mutations (discussed at the end of this chapter). Many other substitutions have no observable effects on protein structure (discussed in Anfinsen, 1973). It is common for amino acid sequence to change more rapidly than three-dimensional structure, as in the case of lipocalins. Wood and Pearson (1999) examined 36 protein families, each having five or more members with known three-dimensional structures. They found a very high correlation between sequence similarity and structural similarity for three-quarters of the protein families. Wood and Pearson concluded that most amino acid sequence changes cause detectable structural changes. Also, the amount of structural change is relatively constant within a protein family.

Biological Questions Addressed by Structural Biology: Globins

We can use the globins to illustrate some of the key questions in structural biology:

- What ligand does each protein transport? For many the answer is unknown. Can structural studies reveal the binding domain to suggest the identity of the ligand? How much structural information is required in order to predict the ligand from sequence information?

- Mutations in globin genes result in a variety of human diseases, including thalassemias and sickle cell anemia (Chapter 20). Can we predict the structural and functional consequences of a specific mutation?

- Globins have been divided into subgroups based on phylogenetic analyses and their localization. To what extent do those groupings reflect structural and functional similarities?

- When a genome is sequenced and a gene encoding a putative novel globin is discovered, can we use information about other globins of known structure in order to predict a new structure?

PRINCIPLES OF PROTEIN STRUCTURE

Protein structure is defined at several levels. Primary structure refers to the linear sequence of amino acid residues in a polypeptide chain, such as human beta globin (Fig. 11.1a). Secondary structure refers to the arrangements of the primary amino acid sequence into motifs such as α helices, β sheets, and coils (or loops) (Fig. 11.1b). The tertiary structure is the three-dimensional arrangement formed by packing secondary structure elements into globular domains (Fig. 11.1c). Finally, quaternary structure involves this arrangement of several polypeptide chains. Figure 11.1d depicts two alpha globin chains and two beta globin chains joined to form mature hemoglobin, with four heme groups attached. Functionally important areas of a protein such as ligand-binding sites or enzymatic active sites are formed at the levels of tertiary and quaternary structure. We will next describe these levels of protein structure using myoglobin and hemoglobin as examples.

(a) Primary structure

```
MVHLTPEEKSAVTALWGKVNVDEVGGEALGRLLVVYPWTQRFFESFGDLSTPDAVMGNPKVKAHGKKVLGAFSD
GLAHLDNLKGTFATLSELHCDKLHVDPENFRLLGNVLVCVLAHHFGKEFTPPVQAAYQKVVAGVANALAHKYH
```

(b) Secondary structure

```
                 10        20        30        40        50        60        70
                  |         |         |         |         |         |         |
UNK_257900 MVHLTPEEKSAVTALWGKVNVDEVGGEALGRLLVVYPWTQRFFESFGDLSTPDAVMGNPKVKAHGKKVLG
DSC        cccchhhhhhhhhhhhhccccchhhhhhhhhhhhhcccchhhhhhhhcccccccccccchhhhhhhhhhh
MLRC       ccccchhhhhhhhhhhhcccccccccchhhhhhhheeecccchhhhhcccccccccccccccccchhhhh
PHD        ccccchhhhhhhhhhhhhcccchhhcchhhhhhheeecccchhhhhhhhhccchhhhhecchhhhhhhhhhh
Sec.Cons.  ccccchhhhhhhhhhhhhcccccchcchhhhhhheeecccchhhhhhhhhcccccccccccchhhhhhhhhh

                 80        90       100       110       120       130       140
                  |         |         |         |         |         |         |
UNK_257900 AFSDGLAHLDNLKGTFATLSELHCDKLHVDPENFRLLGNVLVCVLAHHFGKEFTPPVQAAYQKVVAGVAN
DSC        hhhhhhhhhhhhhhhhhhhhhhhhhhcccchhhhhhhhhhhhhhhhhhhcccccchhhhhhhhhhhhhhhh
MLRC       hhhhhhhhhhhhhhhhhhhhhhhhcccccccccchhhhhhhhhhhhhhhhhhcccccchhhhhhhhhhhhhhh
PHD        hhhhhhhhhhhhhhhhhhhhhhhhhcccchhhhhhhhhhhhhhhhhhhhhhhcccccchhhhhhhhhhhhhhh
Sec.Cons.  hhhhhhhhhhhhhhhhhhhhhhhhhhcccccchhhhhhhhhhhhhhhhhhhhcccccchhhhhhhhhhhhhhhh
```

```
UNK_257900 ALAHKYH
DSC        hhhhccc
MLRC       hhhhccc
PHD        hhhhhcc
Sec.Cons.  hhhhccc
```

FIGURE 11.1. A hierarchy of protein structure. (a) The primary structure of a protein refers to the linear polypeptide chain of amino acids. Here, human beta globin is shown (NP_000539). (b) The secondary structure includes elements such as alpha helices and beta sheets. Here, beta globin protein sequence was input to the POLE server for secondary structure (▶ http://pbil.univ-lyon1.fr/) where three prediction algorithms were run and a consensus was produced. Abbreviations: h, alpha helix; c, random coil; e, extended strand. (c) The tertiary structure is the three-dimensional structure of the protein chain. Alpha helices are represented as thickened cylinders. Arrows labeled N and C point to the amino- and carboxy-terminals, respectively. (d) The quarternary structure includes the interactions of the protein with other subunits and heteroatoms. Here, the four subunits of hemoglobin are shown (with an α2β2 composition and one beta globin chain highlighted) as well as four noncovalently attached heme groups. Panels (c) and (d) were produced using Cn3D software from NCBI.

(c) Tertiary structure

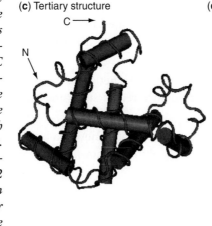

(d) Quaternary structure

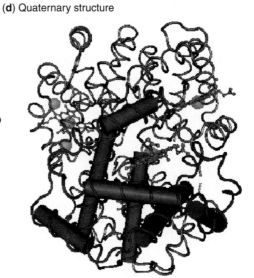

Primary Structure

We will discuss intrinsically disordered proteins later in this chapter; they do not adopt a unique native structure.

In nature, the primary amino acid sequence specifies a three-dimensional structure that forms for each protein. A protein folds to form its native structure(s), sometimes including the participation of chaperones. This process is rapid, typically taking from seconds to minutes; consider for example the bacterium *Escherichia coli* that can double every 20 minutes, requiring all its thousands of proteins to be functionally expressed within that time. Formation of the native structure(s) may depend on some posttranslational modifications, such as the addition of sugars or disulfide bridges. The central issue, called the protein folding problem, is that each cell interprets the information in a primary amino acid sequence to form an appropriate structure. Challenges to structural biologists include (1) how to understand the biological process of protein folding, and (2) how to predict a three-dimensional structure based on primary sequence data alone.

Proteins are synthesized from ribosomes where amino acids are joined by peptide bonds into a polypeptide chain. Each amino acid consists of an amino group, a central carbon atom $C\alpha$ to which a side chain R is attached, and a carboxyl group (Fig. 11.2a). The peptide bond is a carbon-nitrogen amide linkage between the carboxyl group of one amino acid and the amino group of the next amino acid. One water molecule is eliminated during the formation of a peptide bond. The basic repeating unit of a polypeptide chain is thus NH-$C\alpha$H-CO with a different R group extending from various $C\alpha$ of various amino acids. In glycine, the R group is a hydrogen and thus that amino acid is not chiral. For the other amino acids the R group is not a hydrogen and thus there are four different moieties attached to $C\alpha$, allowing chiral (L- and D-) forms of most amino acids.

The amino acid residues of the backbone of the polypeptide chain are constrained to the surface of a plane, and have mobility only around a restricted set of bond angles (Fig. 11.2b) (reviewed by Branden and Tooze, 1991; Shulz and Schirmer, 1979). Phi (ϕ) is the angle around the N-$C\alpha$ bond, and psi (ψ) is the angle around the $C\alpha$-C' bond. Glycine is an exceptional amino acid because it has the flexibility to occur at $\phi\psi$ combinations that are not tolerated for other amino acids. For most amino acids the ϕ and ψ angles are constrained to allowable regions in which there is a high propensity for particular secondary structures to form.

DeepView is a popular software program used to visualize protein structures and to analyze many features of one or more protein structures. It is also used in conjunction with SwissModel, an automated comparative modeling server. DeepView is available for download from the ExPASy website (Chapter 10). When we upload a file in the pdb (Protein Data Bank) format for myoglobin, we can view a control bar with assorted options for manipulating and analyzing the structure (Fig. 11.2c). Using the control panel of DeepView we can select just the first two amino acids of myoglobin (gly-leu) and obtain a description of the bond angles (Fig. 11.2c and d). One reason it is useful to inspect these bond angles is that they provide information about the secondary structure of a protein, which we describe next.

Remarkably, the discovery of the peptide bond was announced at a meeting on the same day (September 22, 1902) by two researchers: Franz Hofmeister and Emil Fisher. Fisher won a 1902 Nobel Prize "in recognition of the extraordinary services he has rendered by his work on sugar and purine syntheses" (▶ http://nobelprize.org/nobel_prizes/chemistry/laureates/1902/). In the area of protein research, he discovered proline and oxyproline, synthesized peptides up to eight amino acids in length, and devised new methods of compositional analysis of proteins such as casein.

We describe how to obtain DeepView in computer lab exercise 11.3 at the end of this chapter. The pdb file for human myoglobin, 2MM1, is available as web document 11.1 at ▶ http://www.bioinfbook.org/chapter11. SwissModel is available at ▶ http://swissmodel.expasy.org/.

Secondary Structure

In general, proteins tend to be arranged with hydrophobic amino acids in the interior and hydrophilic residues exposed to the surface. This hydrophobic core is produced in spite of the highly polar nature of the peptide backbone of a protein. The most common way that a protein solves this problem is to organize the interior amino residues into secondary structures consisting of α helices and pleated β sheets. Linus Pauling and Robert Corey (1951) described these structures from studies of hemoglobin, keratins, and other peptides and proteins. Their models were later confirmed by x-ray crystallography. These secondary structures consist of patterns of interacting amino acid residues in which main chain amino (NH) and carboxy ($C'O$) groups form hydrogen bonds. There are three types of helices: (1) α helices have 3.6 amino acids per turn, and represent $\sim$97% of all helices; (2) 3.10 helices have 3.0 amino acids per turn (and thus are more tightly packed), and account for $\sim$3% of all helices; and (3) π helices, which occur only rarely, have 4.4 amino acids per turn. Myoglobin is an example of a protein with α helices (Fig. 11.3a); these helices typically are formed from contiguous stretches of 4 to 40 amino acid residues in

(a) peptide bond

first amino acid   second amino acid   dipeptide (peptide bond)

(b) phi and psi angles of polypeptide

peptide plane

(c) DeepView control bar

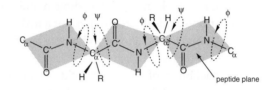

1   2

(d) DeepView viewer

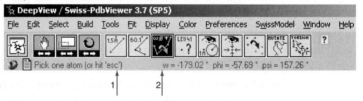

FIGURE 11.2. *Peptide bonds and angles. (a) Each amino acid includes an amino group, an alpha carbon (Cα) to which a side group R is attached, and a carboxyl group (having carbon C′). Two amino acids condense to form a dipeptide with the elimination of water. The peptide bond (highlighted in red) is an amide linkage. (b) Polypeptide chains can be thought of as extending from one Cα atom to the next with the peptide bond constrained to lie along a plain. The N-Cα bond is called phi (φ), and the Cα-C′ bond is called psi (ψ). The angle of rotation around φ and ψ for each peptide defines the entire main chain conformation. (c) The DeepView software from ExPASy includes a control bar with buttons for manipulating a molecule (translation, rotation, and zoom). There are additional tools that measure the following (from left to right): distance between two atoms (arrow 1); angle between three atoms; dihedral angles (arrow 2; here this tool has been selected and the φ, ψ, and ω values are shown); select groups a certain distance from an atom; center the molecule on one atom; fit one molecule onto another; mutation tool; torsion tool. (d) Myoglobin (2MM1) was loaded into DeepView, and using the Control Panel the first three amino acid residues (Gly-Leu-Ser) were selected. The nitrogens, oxygens, Cα carbons (CA) and C′ carbons are indicated. By selecting the dihedral angle tool and clicking the leucine Cα carbon (arrow 3), the bond values in panel (c) were shown.*

(a)

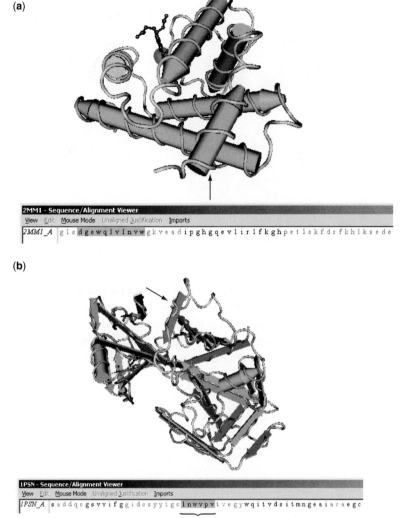

(b)

FIGURE 11.3. Examples of secondary structure. (a) Myoglobin (Protein Data Bank ID 2MM1) is composed of large regions of α helices, shown as strands wrapped around barrel-shaped objects. By entering the accession 2MM1 into NCBI's structure site, one can view this three-dimensional structure using Cn3D software. The accompanying sequence viewer shows the primary amino acid sequence. By clicking on a colored region corresponding to an alpha helix, that structure is highlighted in the structure viewer (arrow). (b) Human pepsin (PDB 1PSN) is an example of a protein primarily composed as β strands, drawn as large arrows. Selecting a region of the primary amino acid sequence (bracket) results in a highlighting of the corresponding β strand.

length. The β sheets are formed from adjacent β strands composed of 2 to 15 residues (typically 5 to 10 residues). They are arranged in either parallel or antiparallel orientations that have distinct hydrogen bonding patterns. Pepsin (1PSN) provides an example of a protein comprised largely of β sheets (Fig. 11.3b). β sheets have higher order properties, including the formation of barrels and sandwiches and "super secondary structure motifs" such as β-α-β loops and α/β barrels. Proteins commonly contain combinations of both α helices and β sheets.

A Ramachandran plot displays the ϕ and ψ angles for essentially all amino acids in a protein (proline and glycine are not displayed). The Ramachandran plot for beta globin shows a preponderance of $\phi\psi$ angle combinations in a region that is typical of proteins with a helical content (Fig. 11.4a). In contrast for pepsin the majority of $\phi\psi$ angles occur in a region that is characteristic of β sheets (Fig. 11.4b). Ramachandran plots can be created using a variety of software packages, including WebMol from the Protein Data Bank and DeepView from ExPASy.

We describe how to use WebMol in computer lab exercise 11.2 near the end of this chapter.

Computational secondary-structure prediction began in the 1970s. Chou and Fasman (1978) developed a method to predict secondary structure based on the frequencies of residues found in α helices, β sheets, and turns. Their algorithm

FIGURE 11.4. A Ramachandran plot displays the φ and ψ angles for each amino acid of a protein (except proline and in some cases glycine). Examples are shown for myoglobin, a protein characterized by alpha helical secondary structure (a) and for pepsin, a protein comprised largely of beta sheets (b). The plots were generated using DeepView software from ExPASy. The arrows in (a) indicate the region of the Ramachandran plot in which φψ angles typical of alpha helices and beta sheets predominate.

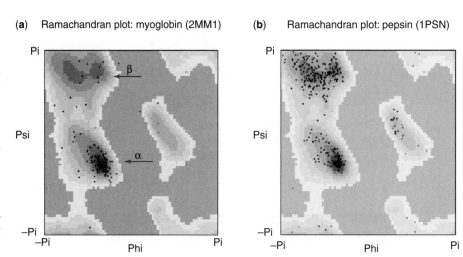

(a) Ramachandran plot: myoglobin (2MM1)

(b) Ramachandran plot: pepsin (1PSN)

calculates the propensity of each residue to form part of a helix, strand, or coil in the context of a sliding window of amino acids. For example, a proline is extremely unlikely to occur in an α helix, and it is often positioned at a turn. The Chou–Fasman algorithm scans through a protein sequence and identifies regions where at least four out of six contiguous residues have a score for α helices above some threshold

TABLE 11-1   Conformational Preferences of the Amino Acids

Amino Acid	Preference			Properties
	Helix	Strand	Turn	
Glu	**1.59**	0.52	1.01	Helical preference; extended flexible
Ala	**1.41**	0.72	0.82	side chains
Leu	**1.34**	1.22	0.57	
Met	**1.30**	1.14	0.52	
Gln	**1.27**	0.98	0.84	
Lys	**1.23**	0.69	1.07	
Arg	**1.21**	0.84	0.90	
His	**1.05**	0.80	0.81	
Val	0.90	**1.87**	0.41	Strand preference; bulky side chains,
Ile	1.09	**1.67**	0.47	beta-branched
Tyr	0.74	**1.45**	0.76	
Cys	0.66	**1.40**	0.54	
Trp	1.02	**1.35**	0.65	
Phe	1.16	**1.33**	0.59	
Thr	0.76	**1.17**	0.90	
Gly	0.43	0.58	**1.77**	Turn preference; restricted
Asn	0.76	0.48	**1.34**	conformations, side chain–main
Pro	0.34	0.31	**1.32**	chain interactions
Ser	0.57	0.96	**1.22**	
Asp	0.99	0.39	**1.24**	

Source: Adapted from Williams et al. (1987).

TABLE 11-2	Secondary Structure Assignment from the DSSP Database
H	Alpha helix
B	Residue in isolated beta-bridge
E	Extended strand, participates in beta ladder
G	3-helix (3/10 helix)
I	5-helix (pi helix)
T	Hydrogen bonded turn
S	Bend
Blank or C	Loop or irregular element, incorrectly called "random coil" or "coil."

Source: DSSP (► http://swift.cmbi.ru.nl/gv/dssp/).

value. The algorithm extends the search in either direction. Similarly, it searches for bends and turns.

Subsequently other approaches have been developed such as the GOR method of Garnier, Osguthorpe, and Robson (1978) (Garnier et al., 1996). In most cases, these algorithms were used to analyze individual sequences (and they are still useful for this purpose). As multiply aligned sequences have become increasingly available, the accuracy of related secondary-structure prediction programs has increased. The PHD program (Rost and Sander, 1993a, 1993b) is an example of an algorithm that uses multiple sequence alignment for this purpose. The accuracy of the various algorithms has been assessed by evaluating their performance using databases of known structures. Typically, the more recently developed algorithms have about 70% to 75% accuracy (Rost, 2001; Przybylski and Rost, 2002). This accuracy far exceeds that of the Chou–Fasman algorithm. Williams et al. (1987) tabulated the conformational preferences of the amino acids (Table 11.1). Their

TABLE 11-3	Secondary-Structure Prediction Programs Available on Internet	
Program	Comment	URL
APSSP	Based on neural networks	► http://imtech.res.in/raghava/apssp/
DSSP		
GOR4	From the Pole Bio-Informatique Lyonnais	► http://npsa-pbil.ibcp.fr/cgi-bin/ npsa_automat.pl?page = npsa_gor4. html
Jpred	From the Barton group (Dundee)	► http://www.compbio.dundee.ac. uk/~www-jpred/
NNPREDICT	An enhanced neural network approach (from UCSF)	► http://alexander.compbio.ucsf. edu/~nomi/nnpredict.html
PredictProtein server	Based on neural networks	► http://www.predictprotein.org/
PSIPRED	From the University College London	► http://bioinf.cs.ucl.ac.uk/psipred/
SAM-T02	Uses hidden Markov models (Chapter 6)	► http://www.soe.ucsc.edu/research/ compbio/HMM-apps/T02-query. html
Sosui	From the Mitaku Group (Tokyo)	► http://bp.nuap.nagoya-u.ac.jp/ sosui/

Note: Additional sites are listed at ExPASy (► http://www.expasy.org/tools/#secondary) and PBIL (► http://npsa-pbil.ibcp.fr).

analysis assessed accuracy relative to known secondary structures as determined by the gold standard of x-ray crystallography. The standard measure for prediction accuracy, called Q3, is the proportion of all amino acids that have correct matches for the three states of helix, strand, and loop. Another measure is the segment overlap (Sov), which is relatively insensitive to small variations in secondary structure assignments, with less emphasis on assigning states to individual residues (Rost et al., 1994).

DSSP software is available from ► http://swift.cmbi.ru.nl/gv/dssp/. A DSSP server is available at ► http://bioweb.pasteur.fr/seqanal/interfaces/dssp-simple.html. The PBIL website is at ► http://npsa-pbil.ibcp.fr.

In 1983 Wolfgang Kabsch and Christian Sander introduced a dictionary of secondary structure, including a standardized code for secondary structure assignment. These are applied in the DSSP database with eight states (Table 11.2). A variety of web servers allow you to input a primary amino acid sequence and delineate the secondary structure, often employing the DSSP codes (Table 11.3). Some of the programs allow you to enter a single sequence, while others allow you to enter a multiple sequence alignment. As an example, the Pôle Bio-Informatique Lyonnais (PBIL) has a web server that offers secondary-structure predictions for a protein query. We used this server to generate the beta globin prediction in Fig. 11.1b. This server also generates predictions using nine different algorithms and calculates a consensus. The various predictions differ somewhat in detail but are generally consistent.

Tertiary Protein Structure: Protein-Folding Problem

How does a protein fold into a three-dimensional structure? As mentioned above, this problem is solved very rapidly in nature. In 1969 Cyrus Levinthal introduced an argument (later called "Levinthal's paradox") that there are far too many possible conformations for a linear sequence of amino acids to adopt its native conformation through random samplings of the energy landscape. To find the most stable thermodynamic structure would require a period of time far greater than the age of the universe. Thus, proteins must adopt their three-dimensional conformations by following specific folding pathways. Dill et al. (2007) have reviewed current progress in understanding protein folding.

In structural biology, there are two main approaches to determining protein structure: x-ray crystallography and nuclear magnetic resonance spectroscopy (NMR). Structures can also be predicted computationally using three approaches described near the end of this chapter (homology modeling, threading, and ab initio prediction).

X-ray crystallography is the most rigorous experimental technique used to determine the structure of a protein (Box 11.1), and about 80% of known structures were

BOX 11.1
X-Ray Crystallography

A protein is obtained in high concentration and crystallized in a solution such as ammonium sulfate. A beam of x-rays is aimed at the protein crystals. The protein is in a highly regular array that causes the x-rays to diffract (scatter) where they are detected on x-ray film. Spot intensities are measured, and an image is generated by Fourier transformation of the intensities. An electron density map is generated corresponding to the arrangements of the atoms that comprise the protein. Individual atoms are distinguishable with 1 to 1.5 Å resolution, and resolution of less than 2 Å is generally required for a detailed structure determination.

(a)

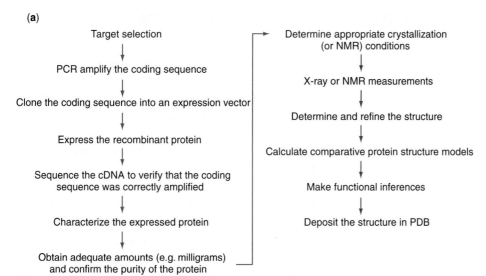

Target selection
↓
PCR amplify the coding sequence
↓
Clone the coding sequence into an expression vector
↓
Express the recombinant protein
↓
Sequence the cDNA to verify that the coding sequence was correctly amplified
↓
Characterize the expressed protein
↓
Obtain adequate amounts (e.g. milligrams) and confirm the purity of the protein

Determine appropriate crystallization (or NMR) conditions
↓
X-ray or NMR measurements
↓
Determine and refine the structure
↓
Calculate comparative protein structure models
↓
Make functional inferences
↓
Deposit the structure in PDB

(b)

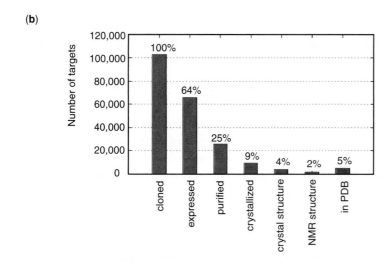

FIGURE 11.5. Obtaining high resolution structures. (a) General procedure for obtaining a three-dimensional protein structure (modified from Burley, 2000). (b) Key stages of structure determination from TargetDB, a target registration database (available at ▶ http://targetdb.pdb.org). While over 100,000 targets have been cloned as part of structural genomics initiatives, only about two-thirds have been successfully expressed, and 25% purified. Five percent of the original targets have been deposited in PDB, either as a crystal structure or from NMR spectroscopy. Source: ▶ http://targetdb.pdb.org/statistics/TargetStatistics.html.

determined using this approach. The basic steps involved in this process are outlined in Fig. 11.5a. A protein must be obtained in high concentration and seeded in conditions that permit crystallization. The crystal scatters x-rays onto a detector, and the structure of the crystal is inferred from the diffraction pattern. The wavelength of x-rays (about 0.5 to 1.5 Å) is useful to measure the distance between atoms, making this technique suitable to trace the amino acid side chains of a protein.

Nuclear magnetic resonance spectroscopy is an important alternative approach to crystallography. A magnetic field is applied to proteins in solution, and characteristic chemical shifts are observed. From these shifts, the structure is deduced. The largest structures that have been determined by NMR are about 350 amino acids ($\approx$40 kDa), considerably smaller than the size of proteins routinely studied by crystallography. Other limitations are that the quality of NMR structures is less than those obtained by crystallography, and NMR yields multiple structure solutions rather than one. However, an advantage of NMR is that it does not require a protein to be crystallized, a notoriously difficult process.

Target Selection and Acquisition of Three-Dimensional Protein Structures

Links to a variety of structural genomics initiatives, including the *Mycobacterium tuberculosis* Structural Genomics Consortium and other bacterial and eukaryotic projects, can be found at the Protein Data Bank website, ▶ http://sg.pdb.org/target_centers.html.

The general procedure for experimentally acquiring protein structural data, outlined in Fig. 11.5a, begins with target selection, the process of choosing which structure to solve (Brenner, 2000). Historically, proteins such as hemoglobin and cytochrome c were selected that were most amenable to experimental study. They are generally small, soluble, abundant, and known to have interesting biological functions. Today, additional criteria are considered in deciding priorities for which protein structures to solve (McGuffin and Jones, 2002):

- All branches of life (eukaryotes, bacteria, archaea, and viruses) are studied.
- Should there be efforts to exhaustively solve all structures within an individual organism? This is being attempted for *Methanococcus jannaschii* and *Mycobacterium tuberculosis*. The Bacterial Structural Genomics Initiative (Matte et al., 2007) includes efforts to determine structures for a large number of *Escherichia coli* proteins.
- Should representatives from previously uncharacterized protein families be selected preferentially? Chandonia and Brenner (2005) proposed that the Pfam5000 set be selected: these are the 5000 largest Pfam families for which no structure has yet been solved.
- Should medically important proteins such as drug discovery targets be chosen first?
- How can structures be solved for more proteins having transmembrane-spanning domains? These are among the most technically challenging proteins to study. Chang and Roth (2001) successfully solved the structure of a multidrug-resistant ABC transporter from *E. coli*. They screened 96,000 crystallization conditions to find several that were adequate for x-ray structure determination.

Structural Genomics and the Protein Structure Initiative

Structural genomics is a newly emerging field of research. Its goal is to determine the three-dimensional structure of all the major protein families throughout the tree of life, spanning fold space (Gerstein and Levitt, 1997; Brenner, 2001; Koonin et al., 2002; Thornton et al., 2000; Burley, 2000). Fold space refers to the total variety of three-dimensional protein structures that occur in nature. This is comprised mostly of proteins having α, β, or $\alpha\beta$ secondary structure composition (Holm and Sander, 1997). This comprehensive approach will permit a deeper understanding of the relatedness of protein domains, and will also enable us to assign function to many proteins. Structure space (or fold space) may be defined in terms of protein sequence families, which generally are defined as containing members having greater than about 30% amino acid identity. Thus, structural genomics ultimately aims to solve at least one high-resolution structure for every sequence family.

The relationship of structural genomics to traditional structural biology is outlined in Fig. 11.6. Traditionally, researchers obtained the structure of individual proteins by starting with information about the known function of the protein. The new approach of structural genomics is based on a reverse strategy: genome sequence

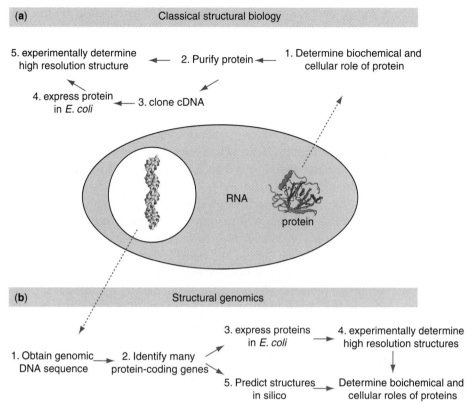

(a) Classical structural biology

5. experimentally determine high resolution structure ← 2. Purify protein ← 1. Determine biochemical and cellular role of protein

4. express protein in *E. coli* ← 3. clone cDNA

RNA

protein

(b) Structural genomics

3. express proteins in *E. coli* → 4. experimentally determine high resolution structures

1. Obtain genomic DNA sequence → 2. Identify many protein-coding genes

5. Predict structures in silico → Determine boichemical and cellular roles of proteins

FIGURE 11.6. *Classical structural biology versus structural genomics. (a) In classical structural biology approaches, a protein is purified based on some known function or activity. After biochemical purification of the protein, if there is sufficient yield, the protein may be crystallized and its structure determined. This in turn allows one to study the biochemical function of the protein and its mechanism of action. Having obtained a protein sequence, the corresponding cDNA may be cloned, allowing recombinant protein to be expressed and purified for structure analyses. (b) The field of structural genomics proceeds from a genomic DNA sequence. Large numbers of protein-coding genes are predicted, often including all those encoded by a genome of interest. Selected proteins are either cloned and expressed for biochemical analysis or the structure is predicted computionally ("in silico") as described later in this chapter. The three-dimensional (3D) structure of a protein may be determined experimentally using techniques such as x-ray crystallography or nuclear magnetic resonance (NMR) spectroscopy. Finally, the biochemical role may be inferred based on the nature of the structure. Additional insight into biochemical function is derived from database searches of the protein sequence (e.g., using PSI-BLAST).*

projects generate predictions of protein-coding sequences (Chapters 13 to 19). One fundamentally important aspect of each predicted protein is its structure. Predicted proteins may be expressed and their structures are solved to high resolution (Fig. 11.6b). The recent identification of literally millions of novel predicted proteins has enabled researchers to choose structures to solve (targets) based on a variety of criteria. Once a target is selected and a cDNA encoding that protein is cloned, there are still many challenges in successfully expressing, purifying, and crystallizing the protein, as well as obtaining its structure by either x-ray crystallography or NMR. Figure 11.5b shows the recent progress in these areas from structural genomics initiatives.

The Protein Structure Initiative (PSI) was established in the United States in 2000, with similar structural genomics projects conducted in other countries

The main PSI website is ► http://
www.nigms.nih.gov/Initiatives/
PSI/. The targets that are currently selected for structural
genomics projects including PSI
are centrally listed at TargetDB
(► http://targetdb.pdb.org/).

(Canada, Israel, Japan, and Europe). The PSI is a coordinated effort by the academic, industry, and federal research communities to develop the technology
needed to determine the three-dimensional structures of most proteins based on
knowledge of the corresponding DNA sequences. The pilot phase of the project
(conducted from 2000 to 2005) involved nine structural genomics centers that
solved more than 1100 structures at high resolution. A key feature of this project is
that solving the structure of proteins that are closely related to those having known
structures is relatively easy, but predicting structures without close structure neighbors can be extremely difficult. Of the 1100 solved structures, over 700 were
unique, that is, the structures shared less than 30% amino acid sequence identity
with other known proteins.

The second phase of the PSI project, the production phase, is currently ongoing.
Chandonia and Brenner (2006) reviewed the progress of the project, including the
novelty of the structures, the cost effectiveness of the project, and the impact of the
project on the community. About half of the novel structures that were reported
recently (in which a novel structure is the first member of a protein family) came
from structural genomics projects. An analysis by Levitt (2007) emphasizes the general decline in the number of novel structures added to the PDB beginning in 1995,
and a reversal of this trend because of contributions from structural genomics initiatives. Others have also assessed the recent progress of these initiatives (Todd et al.,
2005; Xie and Bourne, 2005; Marsden et al., 2007).

In 1992, even before the first genome of a free-living organism had been fully
sequenced, Cyrus Chothia estimated that there may be about 1500 distinct protein
folds. The structural genomics initiatives continue to bring us closer to identifying
all of them.

THE PROTEIN DATA BANK

The PDB was established at
Brookhaven National
Laboratories in Long Island in
1971. Initially, it contained seven
structures. It moved to the
Research Collaboratory for
Structural Bioinformatics (RCSB)
in 1998. PDB is accessed at
► http://www.rcsb.org/pdb/ or
► http://www.pdb.org.

Once a protein sequence is determined, there is one principal repository in which the
structure is deposited: the Protein Data Bank (PDB) (Westbrook et al., 2002;
Berman et al., 2002, 2007). A broad range of primary structural data is collected,
such as atomic coordinates, chemical structures of cofactors, and descriptions of
the crystal structure. The PDB then validates structures by assessing the quality of
the deposited models and by how well they match experimental data.

The main page of the PDB website is shown in Fig. 11.7. This database currently
has about 50,000 structure entries (Table 11.4), with new structures added at a rapid
rate (Fig. 11.8). The database can be accessed directly by entering a PDB identifier
into the query box on the main page, that is, by entering an accession number consisting of one number and three letters (e.g., 4HHB for hemoglobin). The PDB database can also be searched by keyword. The result of a keyword search for hemoglobin
is shown in Fig. 11.9. In this case there are hundreds of results, and the list can be
refined using options on the left sidebar. The result of searching for a specific hemoglobin identifier, 4HHB, links to a typical PDB entry (of which a portion is shown in
Fig. 11.10). By clicking on an icon (arrow 1) the 4HHB.pdb file can be downloaded
locally for further analysis with a variety of tools such as DeepView. Information
provided on the 4HHB page includes the resolution of the experimentally derived
structure, the space group, and the unit cell dimensions of the crystals. There are
links to a series of tools to visualize the three-dimensional structure, including
WebMol (Fig. 11.10, arrow 2). Table 11.5 lists some additional visualization

See computer lab problem 11.2
for instructions on using WebMol.

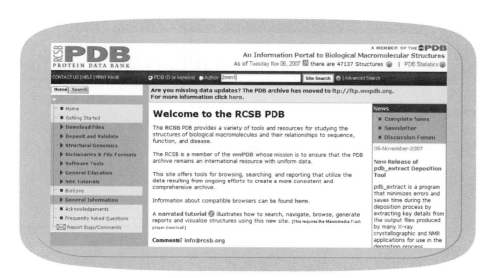

FIGURE 11.7. *The PDB is the main repository for three-dimensional structures of proteins and other macromolecules (Berman et al., 2007) (► http://www.pdb. org/). A PDB accession number (such as 2mm1) can be entered into the search box along the top. The left sidebar links to many useful resources. A variety of sites allow access to PDB data, including NCBI and EMBL. Also, many databases analyze PDB structures to generate classification schemes for all protein folds and for other levels of analysis of protein structures. Examples of these databases are SCOP, CATH, Dali, and FSSP (see below).*

software. Using WebMol does not require the installation of software (other than Java), and WebMol is highly versatile (Fig. 11.11).

It is also possible to search within the PDB website using dozens of advanced search features (accessed via the top of the home page; see Fig. 11.7). This includes the use of BLAST or FASTA programs, allowing convenient access to PDB structures related to a query. Other advanced search features allow you to query based on properties of the molecule (e.g., its molecular weight), PubMed identifier, Medical Subject Heading (MeSH term; Chapter 2), deposit date, or experimental method.

In addition to the PDB, the European Bioinformatics Institute operates the Macromolecular Structure Database. This project is integrated with PDB and represents the European center for the collection of macromolecular structure data.

The PDB database occupies a central position in structural biology. Several dozen other databases and web servers link directly to it or incorporate its data into their local resources. We will next explore NCBI and other sites that allow a single protein structure to be analyzed or several structures to be compared. Then we will explore databases that create comprehensive classification systems or taxonomies for all protein structures.

The Macromolecular Structure Database is at ► http://www.ebi. ac.uk/msd/.

TABLE 11-4   PDB Holdings

	Molecule Type				
Experimental Technique	Proteins	Nucleic Acids	Protein and Nucleic Acid Complexes	Other	Total
X-ray diffraction	37,105	995	1,723	24	39,847
NMR	5,941	789	136	7	6,873
Electron microscopy	109	11	40	0	160
Other	83	4	4	2	93
Total	43,238	1,799	1,903	33	46,973

Source: From ► http://www.pdb.org (November 2007).

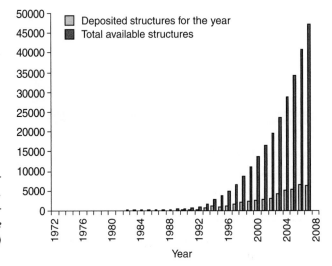

FIGURE 11.8. *Number of search-able structures per year in PDB. The PDB database has grown dramatically in the past decade. The yearly (red) and total (gray) number of structures are shown.*

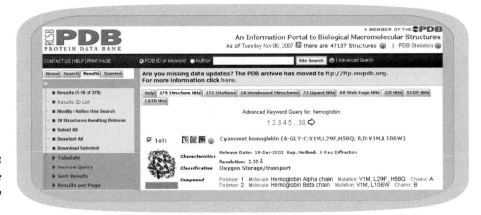

FIGURE 11.9. *Result of a PDB query for hemoglobin. There are several hundred results, of which one is shown.*

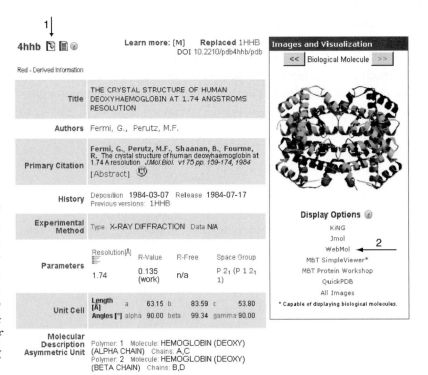

FIGURE 11.10. *Result of a search for a hemoglobin structure, 4HHB. The summary information includes a description of the resolution (1.74 Å), the space group, and the unit cell dimensions. Available links include one to download the 4HHB pdb format file (arrow 1), and a variety of visualization software (including WebMol, arrow 2).*

TABLE 11-5	Interactive Visualization Tools for Protein Structures	
Tool	Comment	URL
Chime	Plug-in for a web browser	Instructions at PDB
Cn3D	From NCBI	► http://www.ncbi.nlm.nih.gov/Structure/CN3D/cn3d.shtml
Mage	Reads Kinemages	► http://kinemage.biochem.duke.edu/ and http://www.ncbi.nlm.nih.gov/Structure/CN3D/mage.html
MICE Java applet		Instructions at PDB
RasMol	A stand-alone package	Instructions at PDB
SwissPDB viewer	At ExPASy	► http://www.expasy.org/spdbv/
VMD	Visual Molecular Dynamics; University of Illinois	► http://www.ks.uiuc.edu/Research/vmd/
VRML	Uses MolScript	Instructions at PDB

Note: The Protein Data Bank maintains a list of molecular graphics software links, accessible from the PDB home page via software tools/Molecular Viewers, at ► http://www.pdb.org/pdb/static.do?p = software/software_links/molecular_graphics.html.

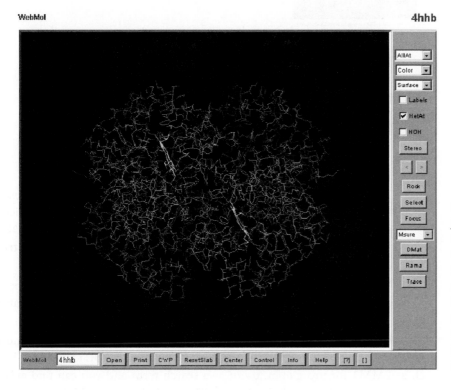

FIGURE 11.11. *WebMol software permits the visualization and analysis of macromolecular structures. Tools on its sidebars allow a molecule to be manipulated (e.g., zoomed or rotated), colored according to criteria such as secondary structure, visualized (e.g., to show van der Waal radii), and analyzed (e.g., by measuring interatomic distances or Ramachandran plots).*

Accessing PDB Entries at the NCBI Website

There are three main ways to find a protein structure in the NCBI databases:

1. Text searches allow access to PDB structures. These searches can be performed on the structure page or through Entrez, and they can consist of keywords or PDB identifiers (Fig. 11.12a).

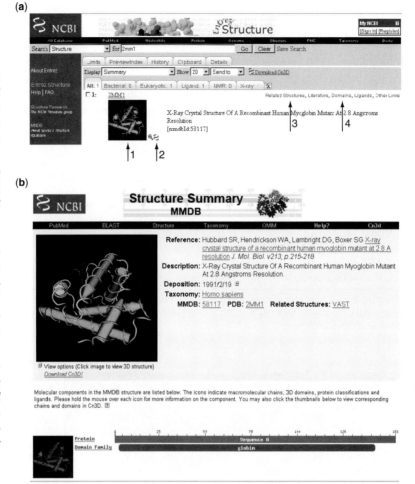

FIGURE 11.12. *The structure site at NCBI offers links to PDB and to tools for structural genomics such as Cn3D (a structure viewer), VAST (a tool to compare structures), and the structure database MMDB (Molecular Modeling DataBase). (a) Structure entries can be retrieved using a keyword (e.g., myoglobin) or a PDB identifier (e.g., 2MM1 as shown here). By clicking on the graphic of the structure in (a), one links to the entry in the Molecular Modeling Database at NCBI (panel b). This includes links (e.g., to the PDB), a graphic of the myoglobin structure which, when clicked on, invokes the Cn3D structure visualization software (or other structure viewers such as RasMol and MAGE). There are also links to related structures and sequences (panel b, bottom). The NCBI structure site is at ▶ http://www. ncbi.nlm.nih.gov/Structure/.*

The NCBI structure page is at ▶ http://www.ncbi.nlm.nih.gov/ Structure/.

2. One can search by protein similarity. To do this, use the Entrez protein database to select a protein of interest and look for a link to "Related Structures." Alternatively, perform a blastp search and restrict the output to the PDB database (Fig. 11.13). All database matches have entries in the Entrez Structure database.

3. One can search using nucleotide queries. It is possible to use a blastx search with a DNA sequence as input, restricting the output to the PDB database.

A keyword search of Entrez structures for hemoglobin yields a list of almost 400 proteins with four-character PDB identifiers. If you know a PDB identifier of interest, such as 2MM1 for myoglobin, use it as a search term and you can find an Entrez Structure entry with useful links (Fig. 11.12a), including to the Molecular Modeling Database (Fig. 11.12b), the Cn3D viewer, the VAST comparison tool (see below), and the conserved domain database. The Molecular Modeling Database is the main NCBI database entry for each protein structure (Wang et al., 2000). It includes literature and taxonomy data, sequence neighbors (as defined by BLAST), structure neighbors (as defined by VAST; see below), and visualization options.

Cn3D is the NCBI software for structure visualization. We describe its use in computer lab problem 11.1, and we used it to generate Fig. 11.3. Upon launching

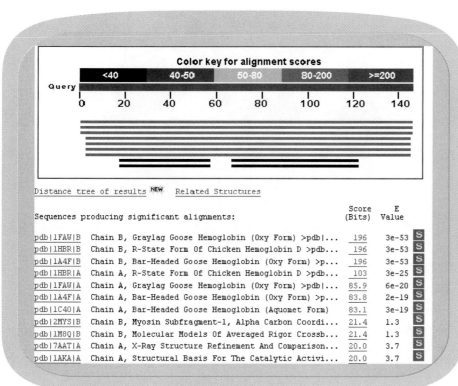

FIGURE 11.13. Structure entries can be retrieved from NCBI by performing a blastp search (with a protein query) or a blastx search (with DNA) restricting the output to the PDB database. Here, a search with human beta globin (NP_000509) restricted to birds (aves) produces matches against a variety of goose and chicken globins of known structure. Note that the boxed S symbol refers to a link to the Entrez Structure database.

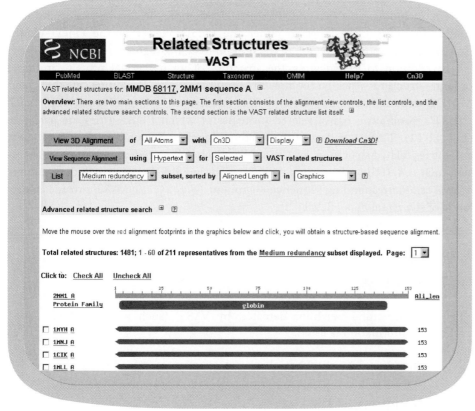

FIGURE 11.14. The Vector Alignment Search Tool (VAST) at NCBI allows the comparison of two or more structures. These may be selected by checking boxes (lower left) or by entering a specific PDB accession number (under the advanced search options). This site also provides links to data on the structures being compared (see Box 11.2) and links to the Conserved Domain Database at NCBI.

FIGURE 11.15. Three structures that are selected in VAST (here myoglobin, 2MM1, and hemoglobin, 4HHB) are compared as overlaid structures in the Cn3D viewer and in the form of a sequence alignment. Despite the relatively low sequence identity between these three proteins, they adopt highly similar three-dimensional folds.

2MM1 neighbors – Sequence/Alignment Viewer

View Edit Mouse Mode Unaligned Justification Imports

2MM1_A ~GLSDGEWQLVLNVWGKVеaDIPGHGQEVLIRLFKGHPETLEKFDRFKHLKSEDEMKASEDLKK
4HHB_B vHLTPEEKSAVTALWGKV~~NVDEVGGEALGRLLVVYPWTQRFFESFGDLSTPDAVMGNPKVKA

Cn3D two windows open: a Cn3D Viewer and a OneD-Viewer (Fig. 11.3). The Cn3D Viewer shows the structure of the protein in seven available formats (such as ball-and-stick or space-filling models), and it can be rotated for exploration of the structure. The corresponding OneD-Viewer shows the amino acid sequence of the protein, including α helices and β sheets. Highlighting any individual amino acid residue or group of residues in either the Cn3D Viewer or the OneD-Viewer causes the corresponding region of the protein to be highlighted in the other viewer.

In addition to investigating the structure of an individual protein, multiple protein structures can be compared simultaneously. Beginning at the main MMDB structure summary for a protein such as myoglobin (Fig. 11.12b), click "VAST" to obtain a list of related proteins for which PDB entries are available (Fig. 11.14). This list is part of the Vector Alignment Search Tool (VAST). Select the entries related structures, or (using the advanced query feature) enter an accession such as 4HHB for hemoglobin. This results in a Cn3D image of both structures as well as a corresponding sequence alignment (Fig. 11.15). VAST provides many kinds of structural data (Box 11.2).

Cn3D is an acronym for "see in 3D."

BOX 11.2
VAST Information

For each structural neighbor detected by VAST (such as Fig. 11.15), the following information is listed:

- Checkbox: The checkbox allows for selection of individual neighbors.
- PDB: The four-character PDB identifier of the structural neighbor.
- The PDB chain name.

- The MMDB domain identifier.
- A VAST structure similarity score based on the number of related secondary structure elements and the quality of the superposition.
- RMSD: The root mean square superposition residual in angstroms. This is a descriptor of overall structural similarity.
- NRES: The number of equivalent pairs of $C\alpha$ atoms superimposed between the two structures. This number gives the alignment length, that is, how many residues have been used to calculate the three-dimensional superposition.
- %Id: Percent identical residues in the aligned sequence region.
- Description: A string parsed from PDB records
- A metric (called Loop Hausdorff Metric) that describes how well two structures match in loop regions
- A gapped score that combines RMSD, the length of the alignment, and the number of gapped regions.

Source: ▶ http://www.ncbi.nlm.nih.gov/Structure/VAST/vasthelp.html#VASTTable.

Integrated Views of the Universe of Protein Folds

The PDB database contains over 50,000 structures. We have examined how to view individual proteins and how to compare small numbers of structures. Chothia (1992) predicted about 1500 folds; how many different protein folds are now thought to exist? How many structural groups are there? Several databases have been established to explore the broad question of the total protein fold space. We will examine several of these databases: SCOP, CATH, and the Dali Domain Dictionary. These databases also permit searches for individual proteins.

Taxonomic System for Protein Structures: The SCOP Database

The Structural Classification of Proteins (SCOP) database provides a comprehensive description of protein structures and evolutionary relationships based on a hierarchical classification scheme (Fig. 11.16) (Andreeva et al., 2008). At the top of the hierarchy are classes that are subsequently subdivided into folds, superfamilies, families, protein domains, and then individual PDB protein structure entries. Myoglobin provides an example in which the class is all alpha proteins, the fold and superfamily are globin-like, the family is the globins, and the protein is myoglobin (Fig. 11.17). The SCOP database can be navigated by browsing the hierarchy, by a keyword query or PDB identifier query, or by a homology search with a protein sequence. A key feature of this database is that it is manually curated by experts, including Alexey Murzin, John-Marc Chandonia, Steven Brenner, Tim Hubbard, and Cyrus Chothia. Because of their expertise, SCOP has a reputation as being one of the most important and trusted databases for classifying protein structures. Automatic classification is now performed in SCOP, in part due to the increase in structures through structural genomics initiatives, with manual annotation for particularly difficult problems.

The main classes are listed in Table 11.6. The folds level of the hierarchy describes proteins sharing a particular secondary structure with the same

The SCOP database is accessed at ▶ http://scop.mrc-lmb.cam.ac.uk/scop/. The seven main classes in release 1.73 contain 92,927 domains organized into 3464 families, 1777 superfamilies, and 1086 folds.

Root: scop

Classes:

1. All alpha proteins [46456] (226)
2. All beta proteins [48724] (149)
3. Alpha and beta proteins (a/b) [51349] (134)
   Mainly parallel beta sheets (beta-alpha-beta units)
4. Alpha and beta proteins (a+b) [53931] (286)
   Mainly antiparallel beta sheets (segregated alpha and beta regions)
5. Multi-domain proteins (alpha and beta) [56572] (48)
   Folds consisting of two or more domains belonging to different classes
6. Membrane and cell surface proteins and peptides [56835] (49)
   Does not include proteins in the immune system
7. Small proteins [56992] (79)
   Usually dominated by metal ligand, heme, and/or disulfide bridges
8. Coiled coil proteins [57942] (7)
   Not a true class
9. Low resolution protein structures [58117] (24)
   Not a true class
10. Peptides [58231] (116)
    Peptides and fragments. Not a true class
11. Designed proteins [58788] (42)
    Experimental structures of proteins with essentially non-natural sequences. Not a true class

FIGURE 11.16. *The Structural Classification of Proteins (SCOP) database includes a hierarchy of terms. From ▶ http://scop.mrc-lmb. cam.ac.uk/scop/ (version 1.71).*

arrangement and topology. However, different proteins with the same fold are not necessarily evolutionarily related.

As we continue down the SCOP hierarchy, we arrive at the level of the superfamily. Here proteins probably do share an evolutionary relationship, even if they share relatively low amino acid sequence identity in pairwise alignments. For example, the lipocalin superfamily in the SCOP database includes both the retinol-binding protein (RBP) family that we have used as an example and another group of carrier

Protein: Myoglobin from Sperm whale (*Physeter catodon*)

Lineage:

1. Root: scop
2. Class: All alpha proteins [46456]
3. Fold: Globin-like [46457]
   core: 6 helices; folded leaf, partly opened
4. Superfamily: Globin-like [46458]
5. Family: Globins [46463]
   Heme-binding protein
6. Protein: Myoglobin [46469]
7. Species: Sperm whale (*Physeter catodon*) [46470]

PDB Entry Domains:

1. 1a6m [15018]
   complexed with hem, oxy, so4
2. 1naz
   complexed with hem, oh, sul; mutant
   1. chain a [85505]
3. 1a6k [15019]
   complexed with hem, so4

FIGURE 11.17. *The SCOP classification of sperm whale myoglobin showing its class, fold, superfamily, family, protein, and species. The entry further lists relevant PDB structures.*

TABLE 11-6	Release Notes from SCOP Database, Release 1.71 (November 2007)			
Class	No. of Folds	No. of Superfamilies	No. of Families	Notes[a]
All alpha proteins	226	392	645	—
All beta proteins	149	300	594	—
Alpha and beta proteins (α/β)	134	221	661	1
Alpha and beta proteins ($\alpha + \beta$)	286	424	753	2
Multidomain proteins	48	48	64	3
Membrane and cell surface proteins	49	90	101	4
Small proteins	79	114	186	5
Total	971	1589	3004	—

For each fold, there are between one and dozens of superfamilies.
[a](1) Mainly parallel beta sheets (beta–alpha–beta units). (2) Mainly antiparallel beta sheets (segregated alpha and beta regions). (3) Folds consisting of two or more domains belonging to different classes. (4) Does not include proteins in the immune system. (5) Usually dominated by metal ligand, heme, and/or disulfide bridges.

proteins exemplified by the fatty acid-binding proteins (FABPs). Like the lipocalins, most FABPs are small (15 kDa), abundant, secreted proteins that bind hydrophobic ligands, and they generally have a glycine–X–tryptophan motif near the amino terminus of each protein. Pairwise sequence alignment fails to reveal significant matches, but the FABP family and the RBP lipocalin family are likely to be homologous. In the SCOP database, the lipocalin superfamily contains three groups: the RBP-like lipocalins, the FABPs, and a thrombin protein. (Some researchers have defined the calycin superfamily as consisting of RBP-like lipocalins, the FABPs, and the avidins [Flower et al., 2000].)

In the SCOP hierarchy members of a family have a clear evolutionary relationship. Usually, the structures of the proteins are related and the pairwise amino acid sequence identity is greater than 30%. In some cases, such as the lipocalins or the globins, some members of each family share as little as 15% identity, but the assignment to the status of a family member is still unambiguous based on structural and evolutionary considerations.

The CATH Database

CATH is a hierarchical classification system that describes all known protein domain structures (Greene et al., 2007). It has been developed by Janet Thornton, Christine Orengo, and colleagues with a particular emphasis on defining domain boundaries. While some parts of the classification system are automated, expert manual curation is also employed for tasks such as classifying remote folds and remote homologs. CATH clusters proteins at four major levels: class (C), architecture (A), topology (T), and homologous superfamily (H) (Fig. 11.18). Five new levels have been added within the homologous superfamily level, abbreviated SOLID, which refer to clustering of domains within the homologous superfamily level having 35%, 60%, 95%, and 100% sequence similarities. A search of CATH with the myoglobin identifier 2MM1 results in an output displaying the various hierarchy levels (Fig. 11.19).

At the highest level (class) the CATH database describes main folds based on secondary-structure prediction: mainly α, mixed α and β, and mainly β, as well as

CATH is accessed at ▶ http://www.biochem.ucl.ac.uk/bsm/cath_new/ or ▶ http://www.cathdb.info. Version 3.1.0 includes about 93,000 domains and 63,000 chains from 30,000 PDB structures (November 2007).

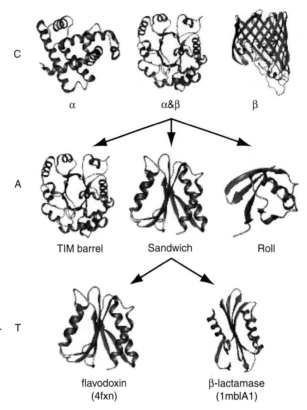

FIGURE 11.18. *The CATH resource organizes protein structures by a hierarchical scheme of class, architecture, topology (fold family), and homologous superfamily. From* ▶ *http://www.cathdb.info.*

The SCOP classification system distinguishes alpha and beta proteins (α/β, consisting of mainly parallel beta sheets with β-α-β units) from $\alpha + \beta$ (mainly antiparallel beta sheets, segregating α and β regions). CATH does not make this distinction.

a category of few secondary structures. Assignment at this level resembles the SCOP database system (Table 11.6). The architecture (A) level of CATH describes the shape of the domain structure as determined by the orientations of the secondary structures. Examples are the TIM barrel (named for triose phosphate isomerase) and jelly roll. These assignments are made by expert judgment rather than by an automated process.

The topology (T) level of CATH describes fold families. Protein domains are clustered into families using several approaches, including the SSAP algorithm of

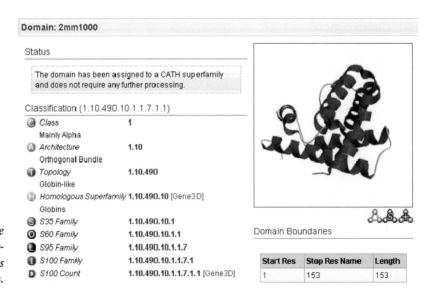

FIGURE 11.19. *A search of the CATH database with the myoglobin PDB accession 2MM1 shows the resulting domain classification.*

Taylor and Orengo (1989a, 1989b). While at the architecture level proteins share structural elements, they may differ in their connectivities; at the topology level structures are assembled into groups sharing both shape and connectivity. Proteins sharing topologies in common are not necessarily homologous. In contrast, the homologous superfamily (H) level clusters proteins that are likely to share homology (i.e., descent from a common ancestor).

The CATH database is a central resource for the hierarchical classification of protein domains. The site includes a large number of software tools and links to structure viewers and databases (e.g., SwissProt and Pfam). The CATH Dictionary of Homologous Superfamilies is also useful. In sum, the CATH database provides a deep and broad set of data on the structure of individual proteins, placing them in the context of a comprehensive taxonomy of protein structure.

The SSAP algorithm compares two protein structures. It can be accessed at ▶ http://www.cathdb.info/cgi-bin/cath/GetSsapRasmol.pl.

The Dali Domain Dictionary

Dali is an acronym for distance matrix alignment. The Dali database provides a classification of all structures in PDB and a description of families of protein

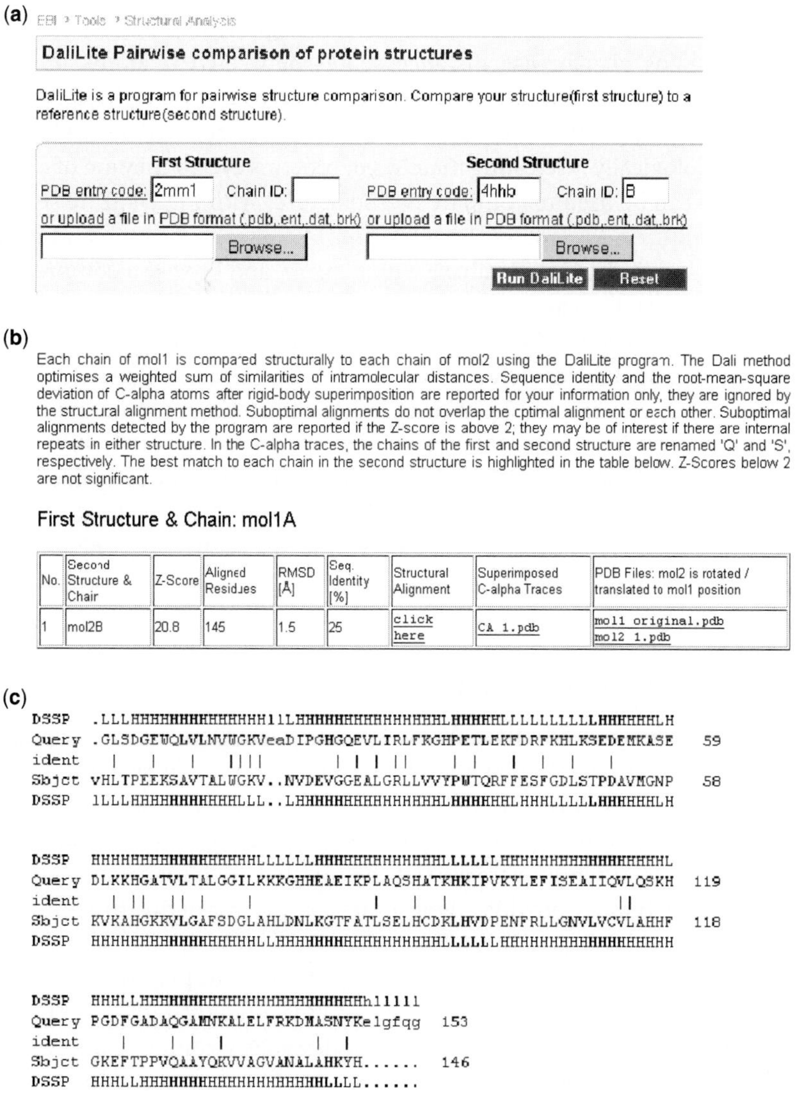

FIGURE 11.20. The DaliLite server at the European Bioinformatics Institute allows a comparison of two three-dimensional structures based on analyses using distance matrices. (a) The PDB identifiers for myoglobin and beta globin are entered in the input form. (b) The output includes a Z score (here a highly significant value of 20.8) based on quality measures such as the resolution and amount of shared secondary structure. (c) The output also links to a pairwise structural alignment. This can be compared to pairwise alignments created with scoring matrices in the absence of structural information (Chapter 3). DaliLite is available at ▶ http://www.ebi.ac.uk/DaliLite/.

FIGURE 11.21. Result of searching myoglobin (2MM1) at the Dali server (▶ http://www.ebi.ac.uk/dali/). (a) The fold index provides a numerical classification scheme based on clustering of similarity scores from a set of comprehensive pairwise comparisons of PDB structures. (b) The browse link shows pairwise structural alignments (such as that shown in Fig. 11.20c) of an entry with its database neighbors. The interact link, shown here, allows a protein structure and/or sequence to be compared and aligned to related structures.

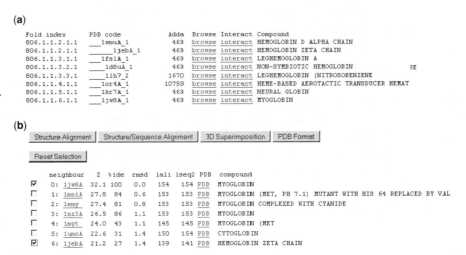

Dali is at ▶ http://www.ebi.ac.uk/dali/ or ▶ http://ekhidna.biocenter.helsinki.fi/dali/start and DaliLite is at ▶ http://www.ebi.ac.uk/DaliLite/. The Families of Structurally Similar Protein (FSSP) database also uses the Dali algorithm (▶ http://www.ebi.ac.uk/dali/fssp/fssp.html) (Holm and Sander, 1996a).

sequences associated with representative proteins of known structure (Holm and Sander, 1996a; Holm, 1998). For pairwise alignments, Dali uses a distance matrix that contains all pairwise distance scores between Cα atoms in two structures. These scores from structural alignments are derived as a weighted sum of similarities of intramolecular distances. The Dali output reports Z scores, which are useful to report biologically interesting matches of proteins even if they are of different lengths.

Dali can be used to compare two structures with the DaliLite server (Holm and Park, 2000). An example is shown in Fig. 11.20 for myoglobin and beta globin. You can also search the Dali database with a query, and browse a comprehensive classification of folds. For example, a search of the Dali fold index at the website in Finland yields a classification of structural domains in PDB90 (a subset of the PDB in which no two chains share more than 90% sequence identity). A portion of the fold index shows a tree constructed by average linkage clustering of structural similarity scores (Fig. 11.21a). For one of the entries (myoglobin, accession 1jw8a), the browse link leads to a set of structural alignments such as that between myoglobin and a hemoglobin chain (Fig. 11.21b) with quality data and Z scores as shown in Fig. 11.20 for DaliLite. The interact link allows you to select structural neighbors in order to analyze their structural relationships (Fig. 11.20c).

Comparison of Resources

You can access the Protein Domain Parser at ▶ http://123d.ncifcrf.gov/pdp.html, and DomainParser at ▶ http://compbio.ornl.gov/structure/domainparser/.

We have described SCOP, CATH, and the Dali Domain Dictionary. Many other databases are available that classify and analyze protein structures. Some of these are listed in Table 11.7. It is notable that for some proteins, such as the four listed in Table 11.8, authoritative resources such as SCOP, CATH, and Dali-based databases provide different estimates of the number of domains in a protein. The field of structural biology provides rigorous measurements of the three-dimensional structure of proteins, and yet classifying domains can be a complex problem requiring expert human judgments. There may be differing interpretations as to whether a particular segment of a protein exists as an independent folding unit, or whether the main principle of domain decomposition involves compactness (as is the case for the Protein Domain Parser) or the density of residue–residue contacts within a

TABLE 11-7 Partial List of Protein Structure Databases

Database	Comment	URL
3dee	Structural domain definitions	▶ http://www.compbio. dundee.ac.uk/3Dee/
CASTp	Computed Atlas of Surface Topography of proteins (CASTp)	▶ http://sts.bioengr.uic. edu/castp/index.php
CE	Complete PDB and representative structure comparison and alignments	▶ http://cl.sdsc.edu/ce. html
Enzyme Structures Databases	Enzyme classifications and nomenclature	▶ http://www.ebi.ac.uk/ thornton-srv/databases/ enzymes/
FATCAT	Flexible structure AlignmenT by Chaining Aligned fragment pairs allowing Twists	▶ http://fatcat.burnham. org/
FSSP	Structurally similar families	▶ http://www.sander.ebi. ac.uk/dali/fssp/
HSSP	Homology-derived secondary structures	▶ http://swift.cmbi.ru.nl/ gv/hssp/
JenaLib	Jena Library of Biological Macromolecules (JenaLib)	▶ http://www.fli-leibniz. de/IMAGE.html
NDB	Database of three-dimensional nucleic acid structures	▶ http://ndbserver. rutgers.edu/
OCA	Browser-database for protein structure/ function	▶ http://oca.ebi.ac.uk/ oca-docs/oca-home. html
PDBSum	Summary information about protein structures	▶ http://www.ebi.ac.uk/ pdbsum/

TABLE 11-8 Proteins Having Different Numbers of Domains Assigned by SCOP, CATH, DALI, and PDP

Name	PDB Accession	SCOP	CATH	DALI	PDP
Glycogen phosphorylase	1gpb	1	2	3	2
Annexin V	1avh_A	1	4	4	2
Submaxillary renin	1smr_A	1	2	1	2
Fructose-1,6-bisphosphatase	5fbp_A	1	2	2	2

Values are the number of domains assigned by each database.
Sources: data from CATH, SCOP, and DALI were from the Protein Data Bank (▶ http://www.pdb.org).
Protein Domain Parser (PDP) data are from ▶ http://123d.ncifcrf.gov/pdp.html (November 2007).

putative domain (as is the case for DomainParser). SCOP is especially oriented towards classifying whole proteins, while CATH is oriented towards classifying domains.

PROTEIN STRUCTURE PREDICTION

Structure prediction is a major goal of proteomics. There are three principal ways to predict the structure of a protein (Fig. 11.22). First, for a protein target that shares substantial similarity to other proteins of known structure, homology modeling (also called comparative modeling) is applied. Second, for proteins that share folds

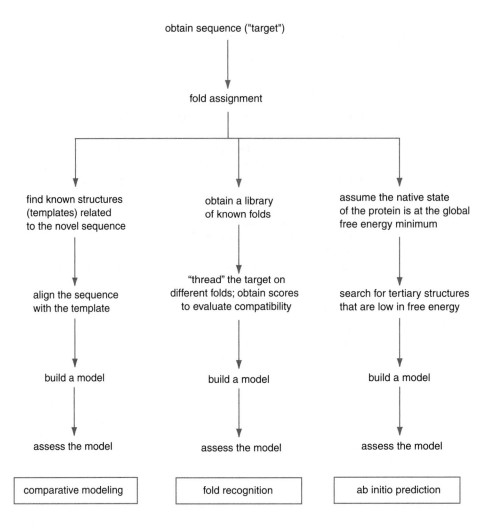

obtain sequence ("target")

fold assignment

| find known structures (templates) related to the novel sequence | obtain a library of known folds | assume the native state of the protein is at the global free energy minimum |

| align the sequence with the template | "thread" the target on different folds; obtain scores to evaluate compatibility | search for tertiary structures that are low in free energy |

build a model | build a model | build a model

assess the model | assess the model | assess the model

| comparative modeling | fold recognition | ab initio prediction |

FIGURE 11.22. *Approaches to predicting protein structures (adapted from Baker and Sali, 2001). Comparative modeling is the most powerful approach when a target sequence has any indications of homology with a known structure. Threading is used to compare segments of a protein to a library of known folds. In the absence of homologous structures, ab initio prediction is used to model protein structure.*

but are not necessarily homologous, threading is a major approach. Proteins that are analogous (related by convergent evolution rather than homology) can be studied this way. Third, for targets lacking identifiable homology (or analogy) to proteins of known structure, ab initio approaches are applied.

Homology Modeling (Comparative Modeling)

While approximately 50,000 protein structures have been deposited in PDB, over five million protein sequences have been deposited in the SwissProt/TrEMBL databases (November 2007). For the vast majority of proteins, the assignment of structural models relies on computational biology approaches rather than experimental determination. As protein structures continue to be solved by x-ray crystallography and NMR spectroscopy, the most reliable method of modeling and evaluating new structures is by comparison to previously known structures (Jones, 2001; Baker and Sali, 2001). This is the method of comparative modeling of protein structure, also called homology modeling. This method is fundamental to the field of structural genomics.

Comparative modeling consists of four sequential steps (Marti-Renom et al., 2000).

1. Template selection and fold assignment are performed. This can be accomplished by searching for homologous protein sequences and/or structures

with tools such as BLAST and PSI-BLAST. The target can be queried against databases described in this chapter, such as PDB, CATH, and SCOP. As part of this analysis, structurally conserved regions and structurally variable regions are identified. It is common for structurally variable regions to correspond to loops and turns, often at the exterior of a protein.

2. The target is aligned with the template. As for any alignment problem, it is especially difficult to determine accurate alignments for distantly related proteins. For 30% sequence identity between a target and a template protein, the two proteins are likely to have a similar structure if the length of the aligned region is sufficient (e.g., more than 60 amino acids). The use of multiple sequence alignments (Chapter 6) can be especially useful.

3. A model is built. A variety of approaches are employed, such as rigid-body assembly and segment matching.

4. The model must be evaluated (see below).

There are several principal types of errors that occur in comparative modeling (see Marti-Renom et al., 2000):

- Errors in side-chain packing
- Distortions within correctly aligned regions
- Errors in regions of a target that lack a match to a template
- Errors in sequence alignment
- Use of incorrect templates

sequence identity	model accuracy	resolution	technique	applications
100%			X-ray crystallography, NMR	Studying catalytic mechanisms
	100%	1.0 Å		Designing and improving ligands
			comparative protein structural modeling	Prediction of protein partners
50%				
	95%	1.5 Å		Defining antibody epitopes
				Supporting site-directed mutagenesis
30%	80%	3.5 Å	threading	Refining NMR structures
				Fitting into low-resolution electron density
<<20%	80 aa	4-8 Å	de novo structure prediction	Identifying regions of conserved surface residues

FIGURE 11.23. *Protein structure prediction and accuracy as a function of the relatedness of a novel structure to a known template. Modified from Baker and Sali (2001). Abbreviation: aa, amino acids. Used with permission.*

TABLE 11-9	Websites for Structure Prediction by Comparative Modeling, and for Quality Assessment	
Website	Comment	URL
3D-JIGSAW	Laboratory of Paul Bates	► http://www.bmm.icnet.uk/servers/ 3djigsaw/
Geno3D	POLE	► http://pbil.ibcp.fr/htm/index.php
MODELLER	From Andrej Sali's group	► http://www.salilab.org/modeller/
PredictProtein	Laboratory of Burkhard Rost	► http://www.predictprotein.org/
SWISS-MODEL	ExPASy	► http://swissmodel.expasy.org/
PROCHECK	Quality assessment	► http://www.biochem.ucl.ac.uk/~roman/ procheck/procheck.html
VERIFY3D	Quality assessment	► http://nihserver.mbi.ucla.edu/Verify_3D/
WHATIF	Quality assessment	► http://swift.cmbi.kun.nl/whatif/

In Chapter 3, we discussed the importance of the length of the alignment in considering percent identity between two proteins.

The accuracy of protein structure prediction is closely related to the percent sequence identity between a target protein and its template (Fig. 11.23). When the two proteins share 50% amino acid identity or more, the quality of the model is usually excellent. For example, the root mean square deviation (RMSD) for the main-chain atoms tends to be 1 Å in such cases. Model accuracy declines when comparative models rely on 30% to 50% identity, and the error rate rises rapidly below 30% identity. De novo models are able to generate low-resolution structure models.

Many web servers offer comparative modeling including quality assessment, such as SWISSMODEL at ExPASy, MODELLER, and the Predict Protein server (Table 11.9). After a model is generated it is necessary to assess its quality. The goal is to assess whether a particular structure is likely, based on a general knowledge of protein structure principles. Criteria for quality assessment may include whether the bond lengths and angles are appropriate; whether peptide bonds are planar; whether the carbon backbone conformations are allowable (e.g., following a Ramachandran plot); whether there are appropriate local environments for hydrophobic and hydrophilic residues; and solvent accessibility. Quality assessment programs include VERIFY3D, PROCHECK, and WHATIF at CMBI (Netherlands) (Table 11.9).

Websites for fold recognition include 3D-PSSM (► http://www.sbg.bio.ic.ac.uk/~3dpssm/index2.html) and its successor PHYRE (► http://www.sbg.bio.ic.ac.uk/~phyre/), FUGUE (► http://www-cryst.bioc.cam.ac.uk/servers.html/), LIBRA I (►), the UCLA-DOE fold server (► http://www.doe-mbi.ucla.edu/Services/FOLD/), 123D (► http://123d.ncifcrf.gov/123D + .html), and the Structure Prediction Meta Server (► http://meta.bioinfo.pl/submit_wizard.pl).

Fold Recognition (Threading)

While there are currently 50,000 entries in the Protein Data Bank, there may be only 1000 to 2000 distinct folds in nature. Fold recognition, also called threading, is useful when a target sequence of interest lacks identifiable sequence matches and yet may have folds in common with proteins of known structure. The target might assume a fold that occurs in a characterized protein because of convergent evolution, or because the two proteins are homologous but extremely distantly related. An input sequence is parsed into subfragments and "threaded" onto a library of known folds. Scoring functions allow an assessment of how compatible the sequence is with known structures. A variety of web servers provide automatic threading.

Ab Initio Prediction (Template-Free Modeling)

In the absence of detectable homologs, protein structure may be assessed by ab initio (or de novo) structure prediction (Fig. 11.22). "Ab initio," meaning "from the beginning," is the most difficult approach to structure prediction (Osguthorpe, 2000; Simons et al., 2001). It is based on two assumptions: (1) All the information about the structure of a protein is contained in its amino acid sequence. (2) A globular protein folds into the structure with the lowest free energy. Finding such a structure requires both a scoring function and a search strategy. While the resolution of ab initio methods is generally low, this approach is useful to provide structural models.

The Rosetta method is one of the most successful ab initio strategies (Simons et al., 2001; Rohl et al., 2004). The target protein is evaluated in fragments of nine amino acids. These fragments are compared to known structures in PDB. From this analysis, structures can be inferred for the entire peptide chain. Typically, models generated with Rosetta have accuracies of 3 to 6 Å RMSD from known structures for aligned segments of 60 or more amino acids (Rohl et al., 2004). Bonneau et al. (2002) used the Rosetta method to model the structure of all Pfam-A sequence families (Chapter 6) for which three-dimensional structures are unknown. By calibrating their method on known structures, they estimated that for 60% of the proteins studied (80 of 131), one of the top five ranked models successfully predicted the structure within 6.0 Å RMSD.

The Robetta server from David Baker's lab is at ▶ http://robetta.bakerlab.org/. It applies the Rosetta method (Kim et al., 2004).

A Competition to Assess Progress in Structure Prediction

How well can the community predict the structures of proteins, particularly those with novel folds? The state of the art of protein prediction is assessed by the structural genomics community at Critical Assessment of Techniques for Protein Structure Prediction (CASP) (Moult, 2005; Kryshtafovych et al., 2007). This is a double-blind structure prediction experiment (or competition) that has occurred every two years since the first competition in 1996. While 35 groups participated in CASP1, over 200 prediction teams from dozens of countries joined CASP8 in 2008. Ninety-five experimentally determined targets (with 123 domains) were evaluated, and tens of thousands of models were deposited with the team of assessors. The structures of the targets were known but withheld from publication so that the community could perform predictions in a blind fashion. Predictors consisted of either scientists who performed modeling of each target, or automatic servers that produced predictions in a short time (48 hours) without human intervention.

The CASP targets include those that require (1) comparative modeling with close evolutionary relationships (e.g., those identifiable by BLAST), (2) comparative modeling to distantly related targets (e.g., those requiring PSI-BLAST or hidden Markov models to detect relationships of a template to proteins having known structure), (3) threading, or (4) template-free modeling. With each successive CASP experiment the ability to accurately model templates in all categories has improved. Major challenges include the need for improved alignments, the need for models of close evolutionary relationships to approach the accuracy obtained by experimental structure determination, the need to better refine models of remote evolutionary relationships, and the need to discriminate among the best template-free models (Moult, 2005; Moult et al., 2007; Tai et al., 2005).

The CASP website provides detailed results of the competition. One measure of the accuracy of a prediction is the GDT_TS measure, which compares the difference in position of the main chain Cα atoms in a model relative to the position in the experimentally determined structure. Figures 11.24a and b show examples of an easy protein target from CASP7 that was solved by most groups and a difficult target

The Protein Structure Prediction Center organizes CASP information (▶ http://predictioncenter.org/) including results from each CASP competition.

FIGURE 11.24. Examples of results from the CASP7 competition. Each plot (called a GDT plot or "Hubbard plot") shows the percent of CA or Cα residues (that is, the percent of the modeled structure; x axis) versus the distance cutoff in Ångstroms (from 0 to 10 Å; y axis). Each line represents a summary of a single prediction of that protein's structure; multiple lines are from the many groups that submitted predictions. (a) Example of a protein target (T0346) whose structure was modeled extremely well by many teams participating in the CASP competition. Note that a very high percentage of the residues in the predictions that could be overlaid on the correct structure (x axis values up to 100%) with only a very small root mean squared deviation (distance cutoff; y axis) as indicated by arrow 1. A small number of predictions were wrong (arrow 2) because they correctly matched the true structure over only a small percent of residues even at large distance cutoffs. (b) Example of a protein target (T0287) whose true structure was not predicted by any group in the CASP competition. A single group's prediction (arrow 3) was better than all others. (c) Example of a target (T0328) that was predicted incorrectly by many teams (arrow 4) but correctly by others (arrow 5). Such a broad discrepancy in prediction accuracy is often attributable to incorrect sequence alignments in homology modeling. Source: CASP7 results at ► http://www.predictioncenter.org/casp/casp7/public/cgi-bin/results.cgi.

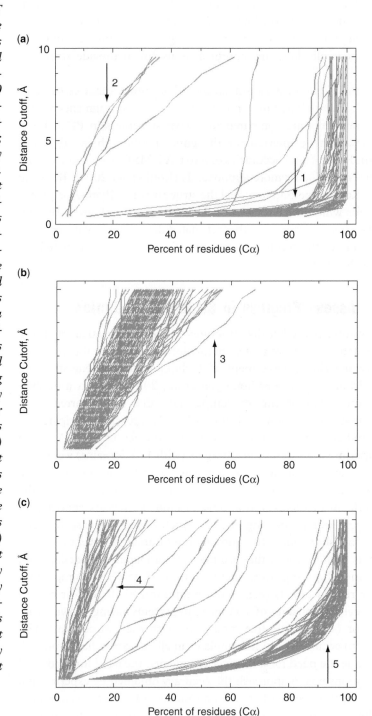

that no group solved. Figure 11.24c shows an example of a target that was aligned either very well or very poorly by many groups; those with poor results misaligned the sequence of the target, highlighting the difficulty of correctly aligning a target sequence onto available template structures for template-based models. The group of David Baker reported that the Rosetta methodology generated highly accurate predictions for some targets under 100 residues for template-free modeling, incorporating an all-atom energy function for refinement of the predicted structures (Das et al., 2007).

INTRINSICALLY DISORDERED PROTEINS

Dyson and Wright (2006) wrote an article entitled "According to Current Textbooks, a Well-Defined Three-Dimensional Structure Is a Prerequisite for the Function of a Protein. Is This Correct?" Many proteins do not adopt stable three-dimensional structures, and this may be an essential aspect of their ability to function properly. Intrinsically disordered proteins are defined as having unstructured regions of significant size such as at least 30 or 50 amino acids (Dyson and Wright, 2005; Le Gall et al., 2007; Radivojac et al., 2007). Such regions do not adopt a fixed three-dimensional structure under physiological conditions, but instead exist as dynamic ensembles in which the backbone amino acid positions vary over time without adopting stable equilibrium values.

Keith Dunker and colleagues have estimated that about 10% of the PDB proteins have disordered regions longer than 30 amino acids (Le Gall et al., 2007). Only ~7% of the protein structures in PDB correspond to the full-length sequence in SwissProt (and only ~25% of the proteins correspond to the structures that match >95% of the length of the protein in SwissProt). The lack of full-length sequences among proteins with solved structures may reflect the common occurrence of intrinsic disorder. Furthermore these authors suggest that >25% of the proteins in SwissProt have disordered regions. DisProt, the Database of Disordered Proteins, centralizes information on this class of proteins (Sickmeier et al., 2007).

Intrinsically disordered regions may have important cellular functions. They may change conformation upon binding to a biological target (a ligand) in a process in which folding and binding are coupled. Many disordered regions of proteins are highly conserved, consistent with their having functionally important roles. Dunker et al. (2005) discuss the role of intrinsic disorder in protein–protein interaction networks, in which it is thought that the average protein has few connections but "hub" proteins serve central roles with many (tens to hundreds) links. Intrinsic disorder in hub proteins could facilitate their ability to bind to structurally diverse protein partners.

The Database of Intrinsic Disorder is available at ▶ http://www.disprot.org/.

PROTEIN STRUCTURE AND DISEASE

The linear sequence of amino acids specifies the three-dimensional structure of a protein. A change in even a single amino acid can cause a profound disruption in structure. For example, cystic fibrosis is caused by mutations in the gene encoding cystic fibrosis transmembrane regulator (CFTR) (Ratjen and Döring, 2003). The most common mutation is *ΔF508*, a deletion of a phenylalanine at position 508. The consequence of removing this residue is to alter the alpha helical content of the protein. This in some way impairs the ability of the CFTR protein to traffic through the secretory pathway to its normal location on the plasma membrane of lung epithelial cells.

A variety of protein structures in PDB have been annotated in terms of diseases they are associated with (Fig. 11.25). Changes in protein sequence that are associated with disease do not necessarily cause large changes in protein structure. An example

TABLE 11-10   Examples of Proteins Associated with Diseases for Which Subtle Change in Protein Sequence Leads to Change in Structure

Disease	Gene/Protein	RefSeq
Cystic fibrosis	CFTR	NP_000483
Sickle cell anemia	Hemoglobin beta	NP_000509
"Mad cow" disease (BSE)	Prion protein	NP_000302
Alzheimer disease	Amyloid precursor protein	NP_000475

Abbreviations: CFTR, cystic fibrosis transmembrane regulator; BSE, bovine spongiform encephalopathy.

You can access a brief definition of the hemoglobin chains at Entrez Gene. You can also find a link there to Online Mendelian Inheritance in Man, which provides a detailed description of the clinical and molecular consequences of globin gene mutations. We discuss sickle cell anemia in Chapter 20.

is provided by sickle cell anemia, the most common inherited blood disorder. It is caused by mutations in the gene encoding beta globin on chromosome 11p15.4. Adult hemoglobin is a tetramer consisting of two alpha chains and two beta chains. The protein carries oxygen in blood from the lungs to various parts of the body. A substitution of a valine for a normally occurring glutamic acid residue forms a hydrophobic patch on the surface of the beta globin, leading to clumping of many hemoglobin molecules. Several examples of proteins associated with human disease are presented in Table 11.10, including CFTR and beta globin.

PERSPECTIVE

The aim of structural genomics is to define structures that span the entire space of protein folds. This project has many parallels to the Human Genome Project. Both are ambitious endeavors that require the international cooperation of many laboratories. Both involve central repositories for the deposit of raw data, and in each the growth of the databases is exponential.

It is realistic to expect that the great majority of protein folds will be defined in the near future. Each year, the proportion of novel folds declines rapidly. A number of lessons are emerging:

- Proteins assume a limited number of folds.
- A single three-dimensional fold may be used by proteins to perform entirely distinct functions.
- The same function may be performed by proteins using entirely different folds.

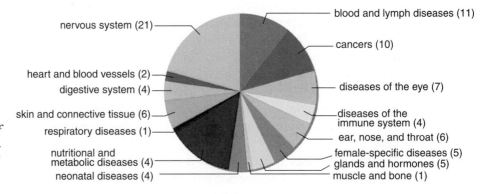

FIGURE 11.25. Distribution of PDB structures annotated according to disease (from ▶ http://function.rcsb.org:8080/pdb/).

PITFALLS

One of the great mysteries of biology is how the linear amino acid sequence of a protein folds quickly into the correct three-dimensional conformation. One set of challenges concerns the experimental solution of three-dimensional structures that span the extent of sequence space. At the present time, no representative structures have been solved for thousands of protein families. Another set of challenges concerns protein structure prediction. While structures can be predicted with high confidence when a closely related template of known structure is available, it is still difficult to predict entirely novel protein structures. Ab initio methods are continually improving, particularly for predicting the structures of small proteins.

DISCUSSION QUESTIONS

[11-1] The Protein Data Bank (PDB) is the central repository of protein structure data. What do databases such as SCOP and CATH offer that PDB lacks?

[11-2] A general rule is that protein structure evolves more slowly than primary amino sequence. Thus, two proteins can have only limited amino acid sequence identity while sharing highly similar structures. (A good example of this is the lipocalins, where retinol-binding protein, odorant-binding protein, and β-lactoglobulin share highly related structures with low sequence identity.) Are there likely to be exceptions to this general rule?

PROBLEMS/COMPUTER LAB

[11-1] View the structure of a protein using Cn3D at NCBI.

(a) Download Cn3D from the NCBI Structure site (▶ http://www.ncbi.nlm.nih.gov/Structure/CN3D/cn3d.shtml).

(b) Go to NCBI Entrez Structures and select a lipocalin. You can access this from the main NCBI page by going to "structure." Alternatively, in Entrez you can type a query, select "limits," and restrict the output to PDB. If you select "odorant-binding protein," there are entries for odorant-binding proteins from several different species. From cow, there are entries deposited independently from different research groups (e.g., PDB identifiers 1OBP, 1PBO).

(c) Select "View 3D Structure" in the MMDB web page. Explore the links on the page. Click "View/Save Structure."

(d) Two windows open: the Cn3D viewer and the 1D-viewer. Click on each of these, and notice how they are interconnected. Change the "style" of the Cn3D viewer. Identify the α helices and β sheets of the protein.

[11-2] View the structure of a protein using WebMol at PDB.

(a) Go to ▶ http://www.pdb.org and enter the term 4HHB (for hemoglobin) in the search box. Note that the title of this page is "the crystal structure of human deoxyhaemoglobin at 1.74 angstroms resolution." An icon at the top includes the option to download the pdb file to your desktop; by doing this you can easily load the 4HHB file into other programs later. Next, under the heading "display options" click WebMol.

(b) The WebMol program opens (running Java) without the need to install software locally. There are pull-down menus and command tabs along the side and bottom of the image of hemoglobin, including a help document. While there are dozens of features, we will select just several. First, explore the mouse options. On a PC, left click to rotate the structure; right click to zoom in and out.

(c) Change the view from all atoms (AllAt) to the main chain (MainCh). Color the hemoglobin molecule by secondary structure. Add (then remove) labels to view the amino acids. Toggle the heteroatoms (HetAt) to see the four heme groups positioned inside each of the four globin chains. Rock to gain a view of the protein.

(d) Click the select tool, and highlight the first three amino acids of the B chain (val-his-leu). Zoom in to view these. Under measure, select the omega angle. Click DMat to invoke a distance matrix, and Rama for a Ramachandran plot. While globins have a high alpha helical content, repeat the Ramachandran plot using a protein with beta sheets.

[11-3] View the structure of a protein using DeepView at ExPASy.

(a) Visit the website for DeepView, the Swiss PDB Viewer, at ▶ http://expasy.org/spdbv/. Select download and install the software locally.

(b) Open the file 2MM1 (a myoglobin pdb file). You can find this by visiting PDB (▶ http://www.pdb.org), querying 2MM1, and downloading the pdb file to your desktop. There is a main toolbar (see Fig. 11.2b); use its File → Open command.

(c) Under the Window pull-down menu, open the control panel. Click the column header "show" to deselect all the amino acid residues, then click the first two to view just them. On the main toolbar, click the ω, ϕ, ψ button (see Fig. 11.2b) to view the bond angles.

[11-4] Compare the structures of two lipocalins using VAST at NCBI:

(a) Go back to the MMDB page for 1PBO and select "Structure neighbors." (This can be accessed by mousing over the protein graphic.) You are now looking at the NCBI VAST (Vector Alignment Search Tool) site. There is a list of proteins related to OBP. Select one or two other proteins, such as β-lactoglobulin or retinol-binding protein, by clicking on the box(es) to the left. Now view/save the alignments.

(b) Notice that two windows open up: Cn3D and DDV (the two-dimensional viewer). Again explore the relationship between these two visualization tools. What are the similarities between the proteins you are comparing? What are their differences? Highlight the regions of conserved amino acids both in the alignment viewer and the graphical viewer. Where are the invariant GXW residues located?

[11-5] Compare the structures of two lipocalins using DaliLite at EBI (▶ http://www.ebi.ac.uk/DaliLite/). Try structures such as 1PBO (for an odorant-binding protein) and 1RBP (for retinol-binding protein). Are the structures significantly related? By what criteria? Are the sequences significantly related, and by what criteria?

[11-6] Mutations in the beta chain of hemoglobin (gene symbol HBB; also called beta globin) can cause sickle cell anemia or other diseases. Try to find the PDB accession numbers for both normal hemoglobin and a mutated form. Try the following:

(a) The NCBI Structure page

(b) The PDB

(c) CATH or SCOP

(d) A blastp search against the PDB at the NCBI website

[11-7] Sickle-cell anemia is caused by a specific mutation in HBB, E6V (i.e., a glutamic acid residue at amino acid position 6 is substituted with a valine). As a consequence of this mutation, hemoglobin tetramers can clump together. This causes the entire red blood cell to deform, adopting a sickled shape. Use PDB identifier 4HHB for wild-type hemoglobin and 2HBS for a mutant form. Compare the structures using the VAST tool at NCBI. Is the glutamate at position 6 on the surface of the protein or is it buried inside? Does the mutation to a valine cause a change in the predicted secondary or tertiary structure of the protein?

SELF-TEST QUIZ

[11-1] In comparing two homologous but distantly related proteins:

(a) They tend to share more three-dimensional structure features in common than percent amino acid identity.

(b) They tend to share more percent amino acid identity in common than three-dimensional structure features.

(c) They tend to share three-dimensional structure features and percent amino acid identity to a comparable extent.

(d) It is not reasonable to generalize about the extent to which they share three-dimensional structure features and percent amino acid identity.

[11-2] Protein secondary structure prediction algorithms typically calculate the likelihood that a protein:

(a) Forms α helices

(b) Forms α helices and β sheets

(c) Forms α helices, β sheets, and coils

(d) Forms α helices, β sheets, coils, and multimers

[11-3] An advantage of x-ray crystallography relative to NMR for structure determination is that using x-ray crystallography:

(a) It is easier to solve the structure of transmembrane domain-containing proteins.

(b) It is easier to grow crystals than to prepare samples for NMR.

(c) It is easier to interpret diffraction data.

(d) It is easier to determine the structures of large proteins.

[11-4] The Protein Data Bank (PDB):

(a) Functions primarily as the major worldwide repository of macromolecular secondary structures.

(b) Contains approximately as many structures as there are protein sequences in SwissProt/TrEMBL.

(c) Includes data on proteins, DNA–protein complexes as well as carbohydrates.

(d) Is operated jointly by the NCBI and EBI.

[11-5] The NCBI VAST algorithm:

(a) Is a web browser tool for the visualization of related protein structures by threading.

(b) Is a visualization tool that allows the simultaneous comparison of as many as two structures.

(c) Allows searches of all the NCBI structure database with queries that have known structures (i.e., having PDB accession numbers), but this tool is not useful for the analysis of uncharacterized structures.

(d) Allows searches of all the NCBI structure database entries against each other and provides a list of "structure neighbors" for a given query.

[11-6] Cn3D is a molecular structure viewer at NCBI. It features

(a) A menu-driven program linked to automated homology modeling

(b) A command line interface useful for a variety of structure analyses

(c) A structure viewer that is accompanied by a sequence viewer

(d) A structure viewer that allows stereoscopic viewing of structure images

[11-7] The CATH database offers a hierarchical classification of protein structures. The first three levels, class (C), architecture (A), and topology (T), all describe:

(a) Protein tertiary structure (e.g., tertiary structure composition, packing, shape, orientation, and connectivity)

(b) Protein secondary structure (e.g., secondary structure composition, packing, shape, orientation, and connectivity)

(c) Protein domain structure

(d) Protein superfamilies grouped according to homologous domains

[11-8] Homology modeling may be distinguished from ab initio prediction because:

(a) Homology modeling requires a model to be built.

(b) Homology modeling requires alignment of a target to a template.

(c) Homology modeling is usefully applied to any protein sequence.

(d) The accuracy of homology modeling is independent of the percent identity between the target and the template.

[11-9] You have a protein sequence, and you want to quickly predict its structure. After performing BLAST and PSI-BLAST searches, you identify the most closely related proteins with known structures as several having 15% amino acid identity to your protein, with a nonsignificant expect value. Which of these options is best?

(a) X-ray crystallography

(b) NMR

(c) Submitting your sequence to a protein structure prediction server that performs homology modeling

(d) Submitting your sequence to a protein structure prediction server that performs ab initio modeling

Suggested Reading

There are many superb overviews of structural genomics and protein structure prediction. For an overview on protein folding, see Dill et al. (2007). The central structure repository PDB is described by Berman et al. (2007). Structural genomics initiatives are reviewed and analyzed by Marsden et al. (2007), Chandonia and Brenner (2006), Levitt (2007), and others cited above. A review by Holm and Sander (1997) includes a graphical representation of multidimensional protein fold space. Holm and Sander (1996b) also wrote an important article, "Mapping the Protein Universe." For the CATH database, see the overview from Christine Orengo and colleagues (Greene et al., 2007). SCOP is described by Murzin and colleagues (Andreeva et al., 2008), and for Dali and FSSP, see Holm and Sander 1993 on distance matrices as well as Holm and Sander (1998).

References

Andreeva, A., Howorth, D., Chandonia, J. M., Brenner, S. E., Hubbard, T. J., Chothia, C., and Murzin, A. G. Data growth and its impact on the SCOP database: New developments. *Nucleic Acids Res.* **36**, D419–D425 (2008).

Anfinsen, C. B. Principles that govern the folding of protein chains. *Science* **181**, 223–230 (1973).

Baker, D., and Sali, A. Protein structure prediction and structural genomics. *Science* **294**, 93–96 (2001).

Berman, H. M., et al. The Protein Data Bank. *Acta Crystallogr. D Biol. Crystallogr.* **58**, 899–907 (2002).

Berman, H., Henrick, K., Nakamura, H., and Markley, J. L. The worldwide Protein Data Bank (wwPDB): Ensuring a single, uniform archive of PDB data. *Nucleic Acids Res.* **35** (Database issue), D301–D303 (2007).

Bonneau, R., Strauss, C. E., Rohl, C. A., Chivian, D., Bradley, P., Malmstrom, L., Robertson, T., and Baker, D. De novo prediction of three-dimensional structures for major protein families. *J. Mol. Biol.* **322**, 65–78 (2002).

Branden, C., and Tooze, J. *Introduction to Protein Structure.* Garland Publishing, New York, 1991.

Brenner, S. E. Target selection for structural genomics. *Nat. Struct. Biol.* **7**(Suppl.), 967–969 (2000).

Brenner, S. E. A tour of structural genomics. *Nat. Rev. Genet.* **2**, 801–809 (2001).

Burley, S. K. An overview of structural genomics. *Nat. Struct. Biol.* **7**(Suppl.), 932–934 (2000).

Chandonia, J. M., and Brenner, S. E. Implications of structural genomics target selection strategies: Pfam5000, whole genome, and random approaches. *Proteins* **58**, 166–179 (2005).

Chandonia, J. M., and Brenner, S. E. The impact of structural genomics: Expectations and outcomes. *Science* **311**, 347–351 (2006).

Chang, G., and Roth, C. B. Structure of MsbA from *E. coli*: A homolog of the multidrug resistance ATP binding cassette (ABC) transporters. *Science* **293**, 1793–1800 (2001).

Chothia, C. Proteins. One thousand families for the molecular biologist. *Nature* **357**, 543–544 (1992).

Chou, P. Y., and Fasman, G. D. Prediction of the secondary structure of proteins from their amino acid sequence. *Adv. Enzymol. Relat. Areas Mol. Biol.* **47**, 45–148 (1978).

Das, R., Qian, B., Raman, S., Vernon, R., Thompson, J., Bradley, P., Khare, S., Tyka, M. D., Bhat, D., Chivian, D., Kim, D. E., Sheffler, W. H., Malmström, L., Wollacott, A. M., Wang, C., Andre, I., and Baker, D. Structure prediction for CASP7 targets using extensive all-atom refinement with Rosetta@home. *Proteins* **69**, 118–128 (2007).

Dill, K. A., Ozkan, S. B., Weikl, T. R., Chodera, J. D., and Voelz, V.A. The protein folding problem: When will it be solved? *Curr. Opin. Struct. Biol.* **17**, 342–346 (2007).

Domingues, F. S., Koppensteiner, W. A., and Sippl, M. J. The role of protein structure in genomics. *FEBS Lett.* **476**, 98–102 (2000).

Dunker, A. K., Cortese, M. S., Romero, P., Iakoucheva, L. M., and Uversky, V. N. Flexible nets. The roles of intrinsic disorder in protein interaction networks. *FEBS J.* **272**, 5129–5148 (2005).

Dyson, H. J., and Wright, P. E. Intrinsically unstructured proteins and their functions. *Nat. Rev. Mol. Cell Biol.* **6**, 197–208 (2005).

Dyson, H. J., and Wright, P. E. According to current textbooks, a well-defined three-dimensional structure is a prerequisite for the function of a protein. Is this correct? *IUBMB Life* **58**, 107–109 (2006).

Flower, D. R., North, A. C., and Sansom, C. E. The lipocalin protein family: Structural and sequence overview. *Biochim. Biophys. Acta* **1482**, 9–24 (2000).

Garnier, J., Gibrat, J. F., and Robson, B. GOR method for predicting protein secondary structure from amino acid sequence. *Methods Enzymol.* **266**, 540–553 (1996).

Garnier, J., Osguthorpe, D. J., and Robson, B. Analysis of the accuracy and implications of simple methods for predicting the secondary structure of globular proteins. *J. Mol. Biol.* **120**, 97–120 (1978).

Gerstein, M., and Levitt, M. A structural census of the current population of protein sequences. *Proc. Natl. Acad. Sci. USA* **94**, 11911–11916 (1997).

Greene, L. H., Lewis, T. E., Addou, S., Cuff, A., Dallman, T., Dibley, M., Redfern, O., Pearl, F., Nambudiry, R., Reid, A., Sillitoe, I., Yeats, C., Thornton, J. M., and Orengo, C. A. The CATH domain structure database: New protocols and classification levels give a more comprehensive resource for exploring evolution. *Nucleic Acids Res.* **35**(Database issue), D291–D297 (2007).

Holm, L. Unification of protein families. *Curr. Opin. Struct. Biol.* **8**, 372–379 (1998).

Holm, L., and Park, J. DaliLite workbench for protein structure comparison. *Bioinformatics* **16**, 566–567 (2000).

Holm, L., and Sander, C. Protein structure comparison by alignment of distance matrices. *J. Mol. Biol.* **233**, 123–138 (1993).

Holm, L., and Sander, C. The FSSP database: Fold classification based on structure–structure alignment of proteins. *Nucleic Acids Res.* **24**, 206–209 (1996a).

Holm, L., and Sander, C. Mapping the protein universe. *Science* **273**, 595–603 (1996b).

Holm, L., and Sander, C. New structure—novel fold? *Structure* **5**, 165–171 (1997).

Jones, D. T. Protein structure prediction in genomics. *Brief. Bioinform.* **2**, 111–125 (2001).

Kim, D. E., Chivian, D., and Baker, D. Protein structure prediction and analysis using the Robetta server. *Nucleic Acids Res.* **32**(Web Server issue), W526–W531 (2004).

Koonin, E. V., Wolf, Y. I., and Karev, G. P. The structure of the protein universe and genome evolution. *Nature* **420**, 218–223 (2002).

Kryshtafovych, A., Fidelis, K., and Moult, J. Progress from CASP6 to CASP7. *Proteins* **69**, 194–207 (2007).

Le Gall, T., Romero, P. R., Cortese, M. S., Uversky, V. N., and Dunker, A. K. Intrinsic disorder in the Protein Data Bank. *J. Biomol. Struct. Dyn.* **24**, 325–342 (2007).

Levitt, M. Growth of novel protein structural data. *Proc. Natl. Acad. Sci. USA* **104**, 3183–3188 (2007).

Marsden, R. L., Lewis, T. A., and Orengo, C. A. Towards a comprehensive structural coverage of completed genomes: A structural genomics viewpoint. *BMC Bioinformatics* **8**, 86 (2007).

Marti-Renom, M. A., et al. Comparative protein structure modeling of genes and genomes. *Annu. Rev. Biophys. Biomol. Struct.* **29**, 291–325 (2000).

Matte, A., Jia, Z., Sunita, S., Sivaraman, J., and Cygler, M. Insights into the biology of *Escherichia coli* through structural proteomics. *J. Struct. Funct. Genomics* **8**, 45–55 (2007).

McGuffin, L. J., and Jones, D. T. Targeting novel folds for structural genomics. *Proteins* **48**, 44–52 (2002).

Meyer, E. F. The first years of the Protein Data Bank. *Protein Sci.* **6**, 1591–1597 (1997).

Moult, J. A decade of CASP: Progress, bottlenecks and prognosis in protein structure prediction. *Curr. Opin. Struct. Biol.* **15**, 285–289 (2005).

Moult, J., Fidelis, K., Kryshtafovych, A., Rost, B., Hubbard, T., and Tramontano, A. Critical assessment of methods of protein structure prediction: Round VII. *Proteins* **69**, 3–9 (2007).

Orengo, C. A., Sillitoe, I., Reeves, G., and Pearl, F. M. Review: What can structural classifications reveal about protein evolution? *J. Struct. Biol.* **134**, 145–165 (2001).

Osguthorpe, D. J. Ab initio protein folding. *Curr. Opin. Struct. Biol.* **10**, 146–152 (2000).

Pauling, L., and Corey, R. B. Configurations of polypeptide chains with favored orientations around single bonds: Two new pleated sheets. *Proc. Natl. Acad. Sci. USA* **37**, 729–740 (1951).

Przybylski, D., and Rost, B. Alignments grow, secondary structure prediction improves. *Proteins* **46**, 197–205 (2002).

Radivojac, P., Iakoucheva, L. M., Oldfield, C. J., Obradovic, Z., Uversky, V. N., and Dunker, A. K. Intrinsic disorder and functional proteomics. *Biophys. J.* **92**, 1439–1456 (2007).

Ratjen, F., and Döring, G. Cystic fibrosis. *Lancet* **361**, 681–689 (2003).

Rohl, C. A., Strauss, C. E., Misura, K. M., and Baker, D. Protein structure prediction using Rosetta. *Methods Enzymol.* **383**, 66–93 (2004).

Rost, B. Review: Protein secondary structure prediction continues to rise. *J. Struct. Biol.* **134**, 204–218 (2001).

Rost, B., and Sander, C. Prediction of protein secondary structure at better than 70% accuracy. *J. Mol. Biol.* **232**, 584–599 (1993a).

Rost, B., and Sander, C. Improved prediction of protein secondary structure by use of sequence profiles and neural networks. *Proc. Natl. Acad. Sci. USA* **90**, 7558–7562 (1993b).

Rost, B., Sander, C., and Schneider, R. Redefining the goals of protein secondary structure prediction. *J. Mol. Biol.* **235**, 13–26 (1994).

Shortle, D. Prediction of protein structure. *Curr. Biol.* **10**, R49–R51 (2000).

Shulz, G. E., and Schirmer, R. H. *Principles of Protein Structure.* Springer-Verlag, New York, 1979.

Sickmeier, M., Hamilton, J. A., LeGall, T., Vacic, V., Cortese, M. S., Tantos, A., Szabo, B., Tompa, P., Chen, J., Uversky, V. N., Obradovic, Z., and Dunker, A. K. DisProt: The Database of Disordered Proteins. *Nucleic Acids Res.* **35**(Database issue), D786–D793 (2007).

Simons, K. T., Strauss, C., and Baker, D. Prospects for ab initio protein structural genomics. *J. Mol. Biol.* **306**, 1191–1199 (2001).

Tai, C. H., Lee, W. J., Vincent, J. J., and Lee, B. Evaluation of domain prediction in CASP6. *Proteins* **61**(Suppl. 7), 183–192 (2005).

Taylor, W. R., and Orengo, C. A. Protein structure alignment. *J. Mol. Biol.* **208**, 1–22 (1989a).

Taylor, W. R., and Orengo, C. A. A holistic approach to protein structure alignment. *Protein Eng.* **2**, 505–519 (1989b).

Thornton, J. M., Todd, A. E., Milburn, D., Borkakoti, N., and Orengo, C. A. From structure to function: Approaches and limitations. *Nat. Struct. Biol.* **7**(Suppl.), 991–994 (2000).

Todd, A. E., Marsden, R. L., Thornton, J. M., and Orengo, C. A. Progress of structural genomics initiatives: An analysis of solved target structures. *J. Mol. Biol.* **348**, 1235–1260 (2005).

Wang, Y., et al. MMDB: 3D structure data in Entrez. *Nucleic Acids Res.* **28**, 243–245 (2000).

Westbrook, J., et al. The Protein Data Bank: Unifying the archive. *Nucleic Acids Res.* **30**, 245–248 (2002).

Williams, R. W., Chang, A., Juretic, D., and Loughran, S. Secondary structure predictions and medium range interactions. *Biochim. Biophys. Acta* **916**, 200–204 (1987).

Wood, T. C., and Pearson, W. R. Evolution of protein sequences and structures. *J. Mol. Biol.* **291**, 977–995 (1999).

Xie, L., and Bourne, P. E. Functional coverage of the human genome by existing structures, structural genomics targets, and homology models. *PLoS Comput. Biol.* **1**, e31 (2005).

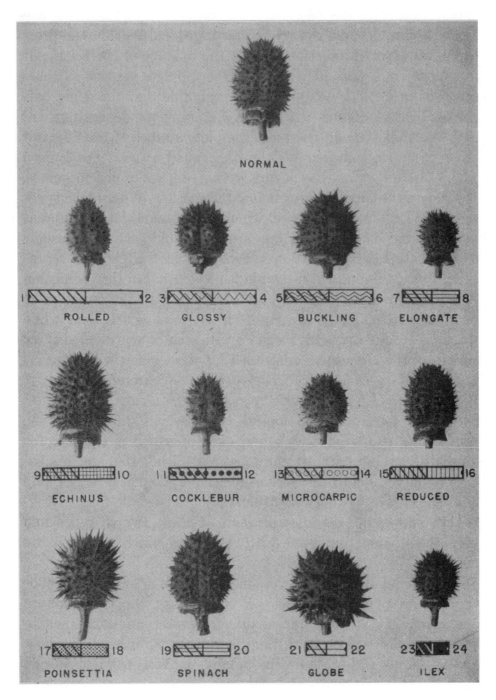

NORMAL

ROLLED GLOSSY BUCKLING ELONGATE

ECHINUS COCKLEBUR MICROCARPIC REDUCED

POINSETTIA SPINACH GLOBE ILEX

It is of great interest to understand the relationship between the genotype (e.g., having an altered chromosome number) and the phenotype (the appearance of the organism, including its fitness). When an organism has an extra copy of a chromosome it is trisomic. By the 1940s, the mechanisms by which trisomy occurs were understood in detail. The jimsonweed (Datura stramonium L.), a flowering plant of the potato family (Solanaceae), normally has 12 pairs of chromosomes. Albert Blakeslee (1874–1954) investigated the seed capsule from wild-type Datura (top) and 12 distinct trisomic types. For each trisomic type, a diagram of the extra chromosome is shown, including a numbering system for the chromosome ends (telomeres). Blakeslee noted that since each chromosome has a distinctive set of genes, each trisomic plant has a distinctive phenotype. From Riley (1948, p. 420). Used with permission.

12

Functional Genomics

Nil adeo quoniam natum'st in corpore, ut uti possemus, sed quod natum'st, id procreat usum. (In fact, nothing in our bodies was born in order that we might be able to use it, but rather, having been born, it begets a use.)
— Lucretius (c.100–c.55 B.C.E.), *De Rerum Natura*, IV, 834–835 (1772, p. 160).

INTRODUCTION TO FUNCTIONAL GENOMICS

A genome is the collection of DNA that comprises an organism. Functional genomics is the genome-wide study of the function of DNA (including genes and nongenic elements), as well as the nucleic acid and protein products encoded by DNA. We may further consider the meaning of the term functional genomics by considering some examples of the ways it has been characterized in recent years.

- Functional genomics may be applied to the complete collection of DNA (the genome), RNA (the transcriptome), or protein (the proteome) of an organism. The assessment of RNA transcripts that are expressed at various times of development or various body regions constitutes an example of functional genomics.
- Functional genomics implies the use of high throughput screens, in contrast to traditional methods of biology in which one gene or protein has been

characterized experimentally in depth. Such traditional methods commonly complement high throughput approaches. For example, after performing a yeast two-hybrid screen to identify thousands of interacting protein partners in some model organism, further validation of selected binding partners is subsequently performed.

- Functional genomics often involves the perturbation of gene function to investigate the consequence on the function of other genes in a genome. For example, in the yeast *Saccharomyces cerevisiae* each gene has been individually knocked out and "bar-coded" as discussed below.

- One of the most challenging and fundamental problems in modern biology is to understand the relationship between genotype and phenotype (discussed below). Connecting the two is a fundamental part of functional genomics.

We provide an overview of functional genomics in Fig. 12.1 with a schematic of a cell. We can consider the three cellular consitutents of genomic DNA (including genes), RNA (including coding and noncoding RNA; Chapter 8); and proteins (Chapters 10 and 11). Other constituents, such as lipids and various metabolites, are also worthy of consideration but are not "informational" in the same sense as the polymers above. The scope of functional genomics includes two levels. (1) Natural variation. How do genes, RNA transcripts, and proteins change across body regions, or across developmental stages? In terms of genomic DNA we will see in Chapter 17 that the genomes of many closely related yeast species have been sequenced, and in Chapter 18 we will describe the recent sequencing of 12 *Drosophila* species and 15 mouse strains. In Chapter 19, we discuss the variation in individual human genome sequences as well. Variation encompasses other aspects such as epigenetics (the study of heritable changes in gene function that occur without a change in DNA sequence, as when DNA is reversibly methylated). In terms of RNA transcripts, techniques such as microarrays and serial analysis of gene expression (Chapter 8) are used to define region- and time-specific features of RNA transcripts. (2) Functional disruptions occur in nature and are studied experimentally. These include deletions, duplications, inversions, and translocations. The scale includes entire genomes (we discuss fish, plant, and *Paramecium* genome duplications in Chapter 18), entire chromosomes (which may become aneuploid, that is, having an abnormal copy number), segments of chromosomes, or single nucleotides. Examples of naturally occurring deletions include the many microdeletion syndromes in which there is a hemizygous loss of chromosomal material, often spanning several million base pairs and including the loss of one copy or dozens of genes. We can find many examples of RNA loss (such as nonsense-mediated decay) and protein loss (for example, in one form of myasthenia gravis, muscle weakness results from an autoimmune reaction that destroys copies of the nicotinic acetylcholine receptor at the neuromuscular junction; reviewed in Drachman, 1994). In this chapter we will describe many experimental approaches to deleting genes as well as to reducing protein levels. Amplifications also commonly occur in nature; Down syndrome is a well-known example in which the presence of three copies of chromosome 21 (instead of the usual two) is associated with increased levels of mRNA and possibly of protein derived from chromosome 21. Experimentally, transgenic or other models can be used to overexpress DNA, RNA, or protein.

We can summarize our focus in this chapter as the consideration of both natural variation and also disrupted cellular function. We will explore how to disrupt gene, gene expression, or protein function, and what the consequences are of such disruptions.

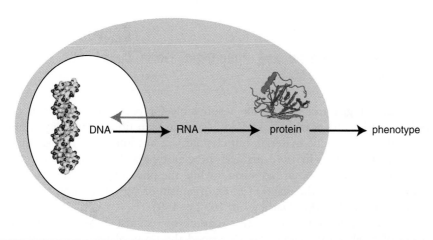

	DNA	RNA	protein
Natural variation --across development --across body regions --across species, strains	SNPs; epigenomics	transcriptome profiling (microarray, SAGE)	protein localization; protein-protein interactions; pathways
Functional disruptions --experimental	knockout collections transgenic animals	RNAi; siRNA	chemical modification
--in nature	Williams syndrome Down syndrome cancers chromosomal changes	nonsense-mediated RNA decay	myasthenia gravis

FIGURE 12.1. *Functional genomics approaches to high throughput protein analysis. From left to right, we can consider several aspects of a cell: the functions associated with DNA, RNA, and protein, as well as higher-order aspects such as protein interactions, biochemical pathways, cell metabolism, and ultimately the phenotype of the cell and of the organism. We can also consider functional genomics approaches in the two broad categories of natural variation and of functional disruptions. Natural variation includes comparisons of the state of DNA, RNA, protein, or other cellular constituents as changes occur over time, under different physiological conditions, or (in the case of multicellular organisms) across different cell types and body regions. Functional disruptions occur in nature (such as chromosomal abnormalities); Williams syndrome is an example of a microdeletion syndrome causing the hemizygous (single-copy) loss of dozens of genes on chromosome 7, and Down syndrome is caused by the gain of an extra copy of chromosome 21. In this chapter we will discuss high throughput experimental approaches to disrupting gene function. Such studies elucidate the normal function of genes.*

The Relationship of Genotype and Phenotype

The genotype of an individual consists of the DNA that comprises the organism. The phenotype is the outward manifestation in terms of properties such as size, shape, movement, and physiology. We can consider the phenotype of a cell (e.g., a precursor cell may develop into a brain cell or liver cell) or the phenotype of an organism (e.g., a person may have a disease phenotype such as sickle-cell anemia). We can trace the history of how genotype and phenotype are defined back to August Weismann in the late nineteenth century (Box 12.1).

A great challenge of biology is to understand the relationship between genotype and phenotype. We can gather information about either one alone. Considering the genotype, we have now sequenced thousands of genomes (including viral and

BOX 12.1
Early Theories of Genotype and Phenotype, Germ Cells and Somatic Cells

The germ cells (eggs and sperm) are responsible for propagating the genetic material. Somatic cells of most eukaryotes (such as skin fibroblasts) have a full complement of the genetic material, but changes are not inherited. Recognizing this, Wilhelm Johannsen in 1908 defined the genotype as "the kind or type of the hereditary properties of an organism," while the phenotype is "the external appearance produced by the reaction of an organism of a given genotype with a given environment" (cited in Darlington, 1932 p. 499).

These ideas were introduced even earlier by August Weismann of the University of Freiburg-in-Baden. He described the units of heredity (what we call genes and DNA) as ids, and he wrote (1893, p. 392): "By acquired characters I mean those which are not preformed in the germ, but which arise only through special influences affecting the body or parts of it. They are due to the reaction of these parts to any external influences apart from the necessary conditions for development. I have called them 'somatogenic' characters, because they are produced by the reaction of the body or soma, and I contrast them with the 'blastogenic' characters of an individual, or those which originate solely in the primary constituents of the germ. It is an inevitable consequence of the theory of the germ-plasm, and of its present elaboration and extension so as to include the doctrine of determinants, that somatogenic variations are not transmissible, and that consequently every permanent variation proceeds from the germ, in which it must be represented by a modification of the primary constituents." Thus, acquired characteristics (such as an injured hand) could not be inherited as Lamarck had hypothesized. Weismann (1893, p. 458) discussed the process of development following fertilization of the egg. "The type of the child is determined by the paternal and maternal ids contained in the corresponding germ-cells meeting together in the process of fertilization, and the blending of the parental and ancestral characters is thus predetermined, and cannot become essentially modified by subsequent influences. The facts relating to identical twins and to plant-hybrids prove that this is so." Weissman described "perfectly homologous ids," one of the earliest references to homology (see also Richard Owen's definition of homology in Chapter 3).

organellar genomes), and defined many of the coding and noncoding genes. Information about the DNA is deposited in GenBank, EMBL, and DDBJ (Chapter 2). It is possible to further describe the transcription of DNA into both coding and noncoding RNA. Protein products are also characterized in depth.

Considering the phenotype, we can describe many categories of phenotype, from variation in the natural state (such as hair color or other quantitative traits) to disease. We discuss the major database for human disease (OMIM) in Chapter 20. As an example of a disease phenotype, Rett syndrome primarily affects girls, leading to hand-wringing, the loss of purposeful hand movements, and autism-like features. The syndrome was recognized (and named) in the 1980s when a group of patients with a similar phenotype were gathered at a meeting in Austria. Eventually, Huda Zhogbi and colleagues identified mutations in the X-linked gene *MECP2* as causing Rett syndrome (Amir et al., 1999). *MECP2* encodes a protein that functions as a transcriptional repressor, regulating gene expression. This case typifies the challenge in understanding the relationship between the genotype (a mutation in a specific gene encoding a transcriptional repressor) and a phenotype (a syndrome having unique features). We have thousands of patients with diagnoses from mental retardation to learning disorders, and beginning with a phenotype how do we find the corresponding genotype for disorders that have a genetic basis? In the case of diseases such as Rett syndrome for which both genotype and phenotype are known, how do we connect them? Through understanding the cellular phenotype we may rationally devise therapeutic strategies aimed at correcting abnormalities that are introduced by a mutant gene product.

The field of functional genomics involves experimental and computational strategies to elucidate the function of DNA and chromosomes in relation to phenotype at the levels of the cell, the tissue, and the organism. There is a large gap in our understanding of how genotype and phenotype are connected. For many diseases, understanding a primary genetic mutation (or insult) has not led to effective treatment or to a cure because of this gap in our understanding. We know that Down syndrome is caused by the occurrence of an extra copy of chromosome 21, but we do not understand why Down syndrome individuals have characteristic symptoms ranging from mental retardation to abnormal facial features to common heart problems, and we do not know why the phenotype ranges from mild to extremely severe (e.g., profound mental retardation and self-injurious behavior).

The remainder of this chapter is organized in three parts. First, we introduce eight model organisms that are prominent in functional genomics studies. We then describe two basic approaches to genetic studies of gene function, reverse and forward genetics. Finally we explore functional genomics as related to proteomics, networks, and pathways as molecular biology intersects with systems biology.

EIGHT MODEL ORGANISMS FOR FUNCTIONAL GENOMICS

The tree of life has three great domains: the bacteria, archaea, and eukaryotes, as well as the separate group of viruses. Thousands of organisms across the tree of life are studied intensively. We can describe eight of them that have particularly important roles in the field of functional genomics. This is not a comprehensive list of model organisms, but helps to define the strengths and limitations of different experimental systems, as well as the types of questions that can be addressed. For the eight species highlighted in this chapter, we will discuss the properties of their genomes in more detail in Chapters 13 (providing an overview of genomes), 15 (*Escherichia coli*), 16 (the eukaryotic chromosome), 18 (various eukaryotic genomes), and 19 and 20 (the human genome).

Leading bioinformatics and genomics organizations have initiated a broad range of functional genomics projects related to model organisms. These include efforts by the Wellcome Trust Sanger Institute, the National Institutes of Health (NIH), and the National Human Genome Resaerch Institute (NHGRI) at NIH. The Encyclopedia of DNA Elements (ENCODE) project, which focused on functionally characterizing 1% of the human genome in great depth, includes efforts to assess function in model organisms as well (ENCODE Project Consortium, 2007). We discuss ENCODE in Chapter 16 and other chapters.

The Model Organism Genetics website at the Wellcome Trust Sanger Institute is available at ► http://www.sanger.ac.uk/ modelorgs/. The NIH offers a website on model organisms for biomedical research (► http:// www.nih.gov/science/models/). The NHGRI Functional Analysis Program is available at ► http:// www.genome.gov/10000612. The website for the ENCODE project at UCSC is ▶ http://genome.ucsc.edu/ENCODE/.

EcoCyc is online at ▶ http://ecocyc.org/, Regulon is at ▶ http://regulondb.ccg.unam.mx/, and EcoGene is available at ▶ http://ecogene.org/.

MetaCyc is available at ▶ http://metacyc.org/index.shtml.

SGD is online at ▶ http://www.yeastgenome.org/. Genome statistics are available at ▶ http://www.yeastgenome.org/cache/genomeSnapshot.html.

The Bacterium *Escherichia coli*

The bacterium *Escherichia coli* serves as the best-characterized prokaryotic organism if not the best-characterized living organism. For decades it served as a leading model organism for bacterial genetics and molecular biology studies. Its 4.6 megabase genome was sequenced by Blattner et al. (1997); we will describe the genome further in Chapter 15. At the time of the initial genome sequencing some function could be assigned to 62% of its genes. The principal website for *E. coli* is EcoCyc, the Encyclopedia of *Escherichia coli* K-12 Genes and Metabolism (Karp et al., 2007). Today EcoCyc assigns a function to 76% of the 4460 annotated genes.

Genome databases are available for all prominent organisms. As an introduction to the use of the EcoCyc database, a query with the term globin links to nitric oxide dioxygenase, a flavohemoglobin. The result includes links to the protein sequence, and functional annotation from the Gene Ontology project (Chapter 10) and Multifun (a classification scheme similar to that of Clusters of Orthologous Groups [COGs] described in Chapter 10). There is extensive annotation of thousands of *E. coli* genes at EcoCyc.

Reed et al. (2006) described four dimensions of genome annotation, encompassing both experimental and computational (in silico) approaches.

1. One-dimensional annotation refers to identifying genes and assigning predicted functions. For *E. coli* this has been achieved to a high degree. For a variety of eukaryotes (Chapters 16 to 19), obtaining a trusted, precise catalog of genes has been extremely challenging because of the difficulty of identifying genes in genomic DNA. The task is becoming easier as more genomes are sequenced and comparative genomics approaches facilitate gene discovery.

2. Two-dimensional annotation refers to specifying the cellular components and their interactions, a topic we will discuss later in this chapter. For *E. coli* this has been achieved to a great extent, for example through the description of transcriptional regulatory networks in the RegulonDB database (Gama-Castro et al. 2008) and protein interactions in Bacteriome.org (Su et al., 2008). The MetaCyc database (Caspi et al., 2008) currently includes over 900 metabolic pathways from over 900 organisms.

3. Three-dimensional annotation is a description of the intracellular arrangement of chromosomes and of cellular components.

4. Four-dimensional annotation refers to characterizing genome changes that occur during evolution. This is a major theme of our study of eukaryotic chromosomes, where comparative genomics approaches have allowed the delineation of evolution from the level of whole genomes and chromosomes to individual DNA segments that are under positive or negative selection (Chapter 7).

The Yeast *Saccharomyces cerevisiae*

The budding yeast *S. cerevisiae* is the best characterized organism among the eukaroytes. This single-celled fungus was the first eukaryote to have its genome sequenced (see Chapters 13 and 17). Its 13 megabase genome encodes about 6000 proteins. The *Saccharomyces* Genome Database (SGD) offers a remarkably deep view into many aspects of the genome, including access to the results of hundreds of functional genomics experiments (Christie et al., 2004). There are currently 6608 annotated genes, including ~4600 that are verified, ~1100 that are uncharacterized (likely to be functional based on conservation across species but not experimentally validated), and ~800 dubious (open reading frames that are neither well conserved nor validated). Approximately 4200 gene products have been annotated to the root gene ontology terms (molecular function, biological process, cellular component; see Chapter 10).

To introduce SGD, we perform a search with a typical query, *SEC1* (Figs. 12.2 and 12.3). *SEC1* is a gene that encodes a protein (Sec1p) involved in vesicle trafficking (Fig. 12.4a). *SEC1* was discovered in a genetic screen (described below) for mutants that fail to secrete the enzyme invertase properly. Later experiments showed that Sec1p is related to *SSO1* (named "suppressor of *SEC1*") and that the Sec1p and Sso1p proteins bind to each other to facilitate vesicle-mediated secretion in yeast. Sso1p, localized to the plasma membrane, is called a SNARE protein (α-soluble NSF attachment protein receptor) that also interacts with the vesicular SNARE protein Snc1p. Thus, Sec1p, Sso1p, and Snc1p are proteins that function in the process of delivering a vesicle and its contents to an appropriate compartment in a eukaryotic cell; in this case, the vesicles deliver proteins to the plasma membrane and they are then secreted outside the cell. All of these yeast trafficking proteins have mammalian counterparts (indicated in Fig. 8.4b). The SGD entry for *SEC1* includes a wealth of information, including a description of its role in vesicle trafficking, and an explanation that the null (or knockout) phenotype is inviable and accumulates secretory vesicles

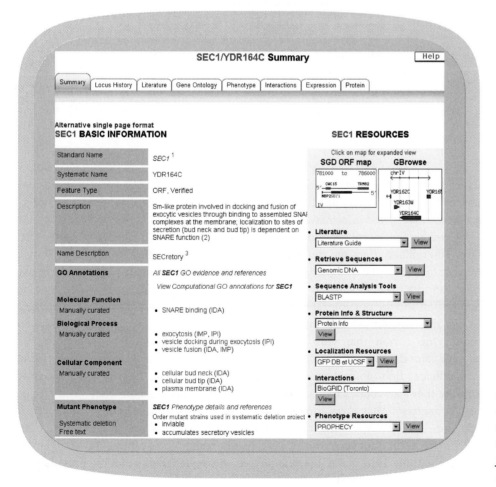

FIGURE 12.2. **The Saccharomyces Genome Database (SGD) offers a wealth of functional genomics information. The top portion of a search for a typical gene, SEC1, is shown. See ▶ http://www.yeastgenome.org.**

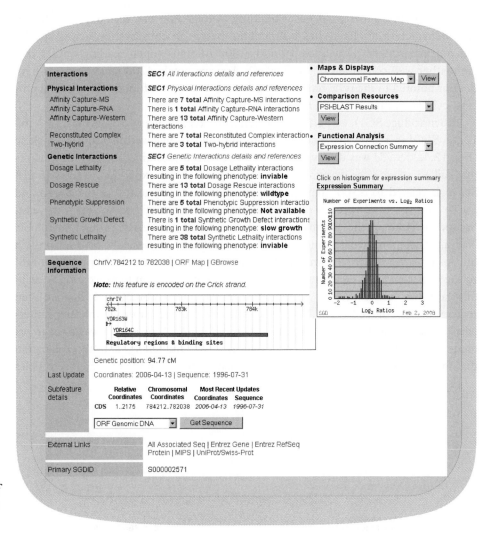

Interactions

SEC1 All interactions details and references

Physical Interactions

SEC1 Physical Interactions details and references

Affinity Capture-MS   There are **7 total** Affinity Capture-MS interactions
Affinity Capture-RNA   There is **1 total** Affinity Capture-RNA interactions
Affinity Capture-Western   There are **13 total** Affinity Capture-Western
                                    interactions

Reconstituted Complex   There are **7 total** Reconstituted Complex interactions
Two-hybrid   There are **3 total** Two-hybrid interactions

Genetic Interactions   *SEC1* Genetic Interactions details and references

Dosage Lethality   There are **5 total** Dosage Lethality interactions
                        resulting in the following phenotype: **inviable**
Dosage Rescue   There are **13 total** Dosage Rescue interactions
                        resulting in the following phenotype: **wildtype**
Phenotypic Suppression   There are **5 total** Phenotypic Suppression interactions
                        resulting in the following phenotype: **Not available**
Synthetic Growth Defect   There is **1 total** Synthetic Growth Defect interactions
                        resulting in the following phenotype: **slow growth**
Synthetic Lethality   There are **38 total** Synthetic Lethality interactions
                        resulting in the following phenotype: **inviable**

• **Maps & Displays**
Chromosomal Features Map ▾ View

• **Comparison Resources**
PSI-BLAST Results ▾
View

• **Functional Analysis**
Expression Connection Summary ▾
View

Click on histogram for expression summary
Expression Summary

Sequence Information   ChrIV:784212 to 782038 | ORF Map | GBrowse

Note: this feature is encoded on the Crick strand.

Genetic position: 94.77 cM

Last Update   Coordinates: 2006-04-13 | Sequence: 1996-07-31

Subfeature details

	Relative Coordinates	**Chromosomal Coordinates**	**Most Recent Updates**	
			Coordinates	**Sequence**
CDS	1..2175	784212..782038	2006-04-13	1996-07-31

ORF Genomic DNA ▾   Get Sequence

External Links   All Associated Seq | Entrez Gene | Entrez RefSeq
                     Protein | MIPS | UniProt/Swiss-Prot

Primary SGDID   S000002571

FIGURE 12.3. **Bottom portion of a search for** SEC1 **in SGD (see Fig. 12.2).**

(consistent with its required role in trafficking) (Fig. 12.2). The SGD page also provides dozens of resources, including links to a genome browser (GBrowse), literature, interaction databases, and information on physical and genetic interactions (Fig. 12.3). As we introduce functional genomics approaches we will return to *SEC1* as an example. Over 100 milllion years ago the entire *S. cerevisiae* genome duplicated, followed by a massive loss of duplicated genes. We will discuss this in Chapter 17, and we will use *SSO1* and its paralog *SSO2* as examples to discuss the evidence for whole genome duplication and the possible fates of duplicated genes.

In this chapter we will introduce a variety of functional genomics assays in yeast. One reason that yeast offers an appealing experimental system is that virtually any desired genomic change can be introduced at the native locus using very efficient homologous recombination based methods. In addition they grow rapidly, they can make colored colonies, and it is easy to construct yeast strains with "reporters" that allow for selection of mutants with interesting traits, even if very rare. A variety of selectable colony color markers are available, such as *MET15* or *ADE2*, in which mutants can be selected for color upon growth in a particular medium. In this way the phenotypic consequence of genetic manipulations can be readily determined.

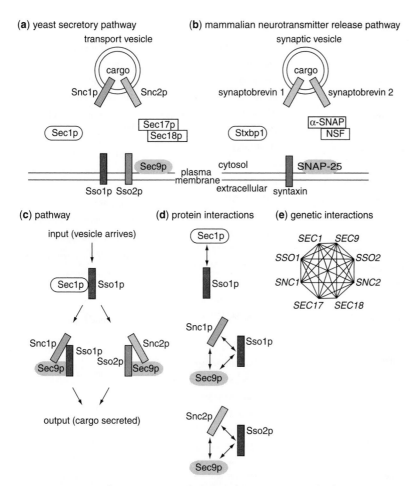

FIGURE 12.4. *Diagram of* S. cerevisiae *and mammalian proteins involved in secretion as an illustration of functional genomics principles and approaches. (a) A constitutive trafficking pathway exists in yeast with a set of proteins, including several of the sec (secretory pathway) mutants. The cytosolic protein Sec1p interacts with Sso1p, a plasma membrane protein and ortholog of mammalian syntaxin. Sso1p also interacts with a protein complex that includes the vesicle-associated proteins Snc1p and Snc2p (mammalian synaptobrevin/VAMP) and the membrane-associated protein Sec9p (mammalian SNAP-25). Sec17p and Sec18p are required for this step and for other intracellular trafficking pathways such as from the Golgi apparatus to the vacuole. In yeast, the paralogous SNC1/SNC2 and SSO1/SSO2 genes arose after an ancient whole genome duplication event (see Chapter 17). The presence of two copies of each molecule could allow functional redundancy, so that if one copy is lost (e.g., through mutation), the organism could be viable. Alternatively, the duplicated genes could acquire distinct functions, such as conferring the specificity of the docking and fusion events of transport vesicles with the appropriate intracellular target membrane. (b) Simplified diagram of proteins in the mammalian nerve terminal. Syntaxin binding protein 1 (Stxbp1, also called Munc18-1/N-sec1) binds tightly to the plasma membrane protein syntaxin. Separately, syntaxin binds to the synaptic vesicle protein synaptobrevin as well as SNAP-25 to form a protein complex, and subsequently the proteins NSF and α-SNAP further bind. Through this pathway synaptic vesicles fuse with the plasma membrane and release their neurotransmitter contents by exocytosis. (c) Hypothetical pathway diagram showing two sets of proteins that could accomplish the task of secretion in yeast using parallel pathways. (d) Biochemical studies can reveal pairwise protein interactions and can also reveal complexes of multiple proteins. However, physical interactions would not reveal the relationship of proteins that do not interact directly but are part of the same pathway (such as Sec1p and Sec9p). (e) Genetic interaction maps reveal functionally related genes, including those involved in parallel pathways and those that do not physically interact. Adapted in part from Ooi et al. (2006).*

Search	Tools	Stocks
Search Overview	Tools Overview	Stocks Overview
DNA/Clones	GBrowse	ABRC Home
Ecotypes	Seqviewer	Browse ABRC Catalog
Genes	Mapviewer	Supplement to ABRC Catalog
GO Annotations	AraCyc Metabolic Pathways	Search ABRC DNA/Clone Stocks
Keywords	BLAST	
Locus History	WU-BLAST	Search ABRC Seed/Germplasm Stocks
Markers	FASTA	ABRC Stock Order History
Microarray Element	Patmatch	ABRC Fee Structure
Microarray Experiment	Motif Analysis	Place ABRC Order
Microarray Expression	VxInsight	Search My ABRC Orders
People/Labs	Java Tree View	Search ABRC Invoices
Polymorphisms/Alleles	Bulk Data Retrieval	How to Make Payments to ABRC
Proteins	Chromosome Map Tool	
Protocols	Gene Hunter	ABRC Stock Donation
Publication	Restriction Analysis	
Seed/Germplasm	Gene Symbol Registry	
Sequences		

Browse	Portals	Download
Browse Overview	Portals Overview	Download Overview
ABRC Catalog	Clones/DNA Resources	Genes
2010 Projects	Education and Outreach	GO and PO Annotations
Monsanto SNP and Ler Collections	Gene Expression Resources	Maps
Gene Families	Genome Annotation	Pathways
Gene Class Symbols	MASC/Functional Genomics	Proteins
Ontologies/Keywords	Mutant and Mapping Resources	Protocols
Archived e-Journals	Nomenclature	Microarray Data
	Proteome Resources	Sequences
		Software
		User Requests

FIGURE 12.5. The Arabidopsis Information Resource (TAIR) is the principal genome database for Arabidopsis (▶ http://www.arabidopsis.org/). The screen capture shows some of the menu options, including search strategies, analysis tools, available stocks, functional classification, and acess to functional genomics initiatives.

The Plant *Arabidopsis thaliana*

The thale cress *Arabidopsis thaliana* was the first plant to have its genome sequenced (and the third finished eukaryotic genome sequence). It has served as a model for eukaryotic functional genomics projects (reviewed in Borevitz and Ecker, 2004). The principal website, the Arabidopsis Information Resource (TAIR), centralizes a vast amount of information about its genome (Swarbreck et al., 2008). Figure 12.5 shows some of the diversity of information that is accessible from the home page pull-down menus. Under the browse menu, a link to "2010 projects" describes dozens of projects designed to reach a National Science Foundation goal to functionally annotate all *Arabidopsis* genes by 2010. As an example of a gene search at TAIR, a query for *Arabidopsis* SEC1A (RefSeq accession NP_563643; locus tag At1g02010) reveals information about its chromosomal location and available mutants.

The TAIR website is ▶ http://www.arabidopsis.org/.

The Nematode *Caenorhabditis elegans*

Among the metazoans (animals), the soil-dwelling nematode *Caenorhabditis elegans* is a key model organism. This was the first multicellular animal to have its genome sequenced. This roundworm, like fruitflies and humans, is capable of complex

WormBase is available at ▶ http://www.wormbase.org. The trans-NIH *C. elegans* initiative website is

behaviors, but its body is simple and all 959 somatic cells in its body have been mapped, including their lineages throughout development. Wormbase is the main on-line information repository (Rogers et al., 2008).

► http://www.nih.gov/science/models/c_elegans/.

The Fruitfly *Drosophila melanogaster*

The fruitfly *Drosophila melanogaster*, another metazoan invertebrate, has long served as a model for genetics. Early studies of *Drosophila* resulted in the descriptions of the nature of the gene, as well as linkage and recombination, producing gene maps a century ago. The recent sequencing of 12 species of the *Drosophila* genus is already providing unprecedented insight into mechanisms of genome evolution (Chapter 18). The central *Drosophila* database, FlyBase, combines molecular and genetic data on the Drosophilidae (Wilson et al., 2008). A strength of *Drosophila* as a model organism is that genomic changes can be induced with extreme precision, from single nucleotide changes to introducing large-scale chromosomal deletions, duplications, inversions, or other modifications. At the same time, it is a multicellular animal that features a complex body plan. Currently, loss of function mutations have been introduced into all of its ~14,000 genes, and over half of these have an identifiable phenotype.

Flybase is at ► http://www.flybase.org.
Two of the giants of genetics research focused their studies on *Drosophila*: Thomas Hunt Morgan and Hermann J. Muller. Morgan was awarded a Nobel Prize in 1933 "for his discoveries concerning the role played by the chromosome in heredity" (► http://nobelprize.org/nobel_prizes/medicine/laureates/1933/). He and his contemporaries A.H. Sturtevant, C.B. Bridges, and H.J. Muller discovered a broad array of properties of genes and chromosomes. They described chromosomal deficiencies, including nondisjunction, balanced lethals, chromosomal duplication (trisomy) and monosomy, and translocations. Muller was awarded a 1946 Nobel Prize "for the discovery of the production of mutations by means of X-ray irradiation." His finding of position effect variegation laid the foundation for modern epigenetics research. The 1995 Nobel Prize in Physiology or Medicine was awarded to Edward B. Lewis, Christiane Nüsslein-Volhard, and Eric F. Wieschaus "for their discoveries concerning the genetic control of early embryonic development." These studies were also performed in *Drosophila* (► http://nobelprize.org/nobel_prizes/medicine/laureates/1995/).

The Zebrafish *Danio rerio*

Although the lineages leading to modern fish and humans diverged approximately 450 million years ago, both are vertebrate species, and orthologs are identifiable for the great majority of their protein-coding genes (with an average of about 80% amino acid identity between orthologs). The first four fish genomes to be sequenced were the pufferfish *Takifugu rubripes* and *Tetraodon nigroviridis*, the medaka *Oryzias latipes*, and the zebrafish *Danio rerio* (Chapter 18). Of these, the zebrafish has emerged as an important model organism for functional genomics (Henken et al., 2004). It is a small tropical freshwater fish having a genome size of 1.8 billion base pairs (Gb) organized into 25 chromsomes. For functional genomics studies, the zebrafish has served as a model for understanding both normal and abnormal development. Mutations in large numbers of human disease gene orthologs have been generated and characterized, and both forward and reverse genetic screens (introduced below) have been applied. Some of the advantages of zebrafish as a model organism include the following:

- Its generation time is short, especially for a vertebrate.
- It produces large numbers of progeny.
- The developing embryo is transparent. Thus, for example, if a transgene is inserted into the genome with a promoter that drives the expression of green fluorescent protein (GFP), it is possible to see this expression from the outside of each animal's body.
- It is a vertebrate and thus a close model for human disease.
- Its genome is well annotated. The vertebrate genome annotation (Vega) database at the Sanger Institute focuses on high quality manual annotation with a particular focus on just three genomes: human, mouse, and zebrafish (Wilming et al., 2008).

The principal zebrafish website is the Zebrafish Information Network (ZFIN) (Sprague et al., 2008).

The Vega database is available at ► http://vega.sanger.ac.uk/.

ZFIN is online at ► http://www.zfin.org. The trans-NIH zebrafish initiative website is ► http://www.nih.gov/science/models/zebrafish/.

The Mouse *Mus musculus*

The rodents diverged from the primate lineage relatively recently (80 million years ago) and share almost all of their genes with humans. The mouse *Mus musculus* is one of the most important model organisms for the study of human gene function because of the close structural and functional relationship between the two genomes, combined with a relatively short generational span, and powerful tools have been developed to manipulate its genome. About 9000 mouse genes have been knocked out. There are currently three major mouse functional genomics initiatives (International Mouse Knockout Consortium, 2007): the Knockout Mouse Project (KOMP), The European Conditional Mouse Mutagenesis Program (EUCOMM), and the North American Conditional Mouse Mutagenesis Project (NorCOMM). We will discuss their strategies for mutating all protein-coding genes in mouse, including gene targeting and gene trapping.

The trans-NIH mouse Initiatives homepage is ▶ http://www.nih.gov/science/models/mouse/.

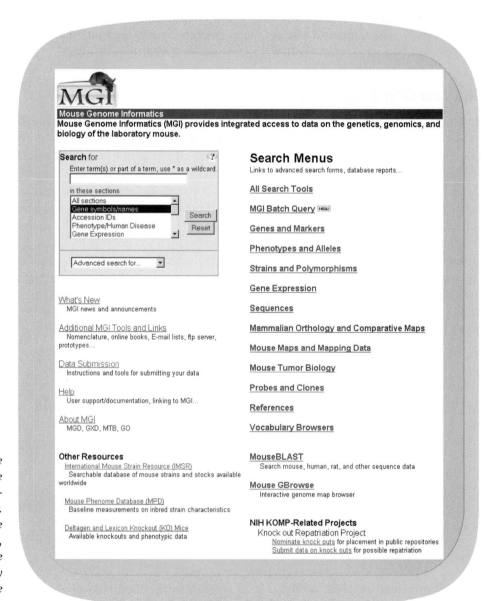

FIGURE 12.6. *The Mouse Genome Informatics (MGI) Database is the principal website for mouse genomics information (▶ http://www.informatics.jax.org/). It includes the Mouse Genome Database (MGD), the Gene Expression Database (GXD), the Mouse Tumor Biology (MTB) Database, and a Gene Ontology project.*

The main mouse genome website is the Mouse Genome Database (MGD) (Bult et al., 2008). In reviewing dozens of organism-specific website resources, those dedicated to three stand out for their breadth and depth: SGD for yeast, the main human genome web browsers (Chapters 16 and 19), and MGD. MGD provides a portal to mouse-specific resources, including sequence data, a web browser, available mutant strains, gene expression studies, and literature (Fig. 12.6).

The MGD website is ▶ http://www.informatics.jax.org/.

In a project called the Collaborative Cross, one thousand recombinant inbred strains of mouse are being bred (Complex Trait Consortium, 2004). This project will produce large numbers of genetically identical mice that have nonlethal phenotypic diversity, and also that can be exposed to manipulations such as phenotypic screens (see below). The 1000 strains are derived from eight inbred founder strains that were systematically crossed. These strains will be fully genotyped and used to model human populations and diseases. If we denote the eight inbred founder strains A to H, then the G_1 generation will consist of AB, CD, EF, and GH genotypes (from mating of AA × BB mice, CC × DD, etc.), the G_2 generation will consist of AB × CD mice yielding ABCD genotypes and EF × GH yielding EFGH. After 23 generations there will be 99% inbreeding with unique recombination events. The 1000 mouse strains are expected to provide an important resource for modeling human populations and diseases.

Homo sapiens: Variation in Humans

To some, humans are not considered to be a model organism, and we do not consider ourselves to be an experimental system per se. And yet we are motivated to understand the range of phenotypic expression to understand how we aquire our characteristic features, how we have evolved, and how we fit into the ecosystem. One of the strongest motivations for studying humans is to understand the causes of disease in order to search for more effective diagnoses, treatments, and ultimately to find cures if possible. And although in most contexts we do not experiment on ourselves invasively, nature does perform functional genomics experiments on us. For example, human fecundity is extraordinarily low, relative to other mammalian and vertebrate species (see Chapter 20). Of all conceptuses that appear normal after one week of development as a zygote, perhaps over 80% are not viable. This is due to massive aneuploidy that commonly occurs, causing trisomy, monosomy, and even tetrasomy (four copies) or nullisomy (zero copies) of many chromosomes. Functional genomics is an experimental science in which gene function is often assessed by perturbing a system. Genes may be selectively deleted or duplicated, and then the functional consequence is measured to infer the function of the gene. Nature produces the equivalent of functional genomics experiments through the many forms of variation that organisms experience.

Aneuploidy refers to a change in chromosomal copy number. A euploid individual has the normal two copies of a set of chromosomes.

FUNCTIONAL GENOMICS USING REVERSE GENETICS AND FORWARD GENETICS

There are many different basic approaches to identifying the function of a gene. Biochemical strategies can be employed. This typically involves studying one gene or gene product at a time. This is often the most rigorous way to study gene function, and it has been the main approach for the past century. For example, in order to understand the function of a globin gene one can purify its protein product to homogeneity and characterize its physical properties (such as molecular mass, isoelectric point, oxygen- and heme-binding properties, and posttranslational modifications; Chapter 10), its interactions with other proteins, its role in cellular pathways, and the consequence of mutating the gene. We described seven different aspects of protein function in Fig 10.19. The analysis of a single gene and its products, while invaluable, is almost always laborious and time consuming and can be myopic. Thus, a variety of complementary high throughput strategies have been introduced. These strategies can produce thousands of mutant alleles that are then available to facilitate the research of scientists who focus on the study of any particular genes.

One high throughput way to assess gene function is to examine messenger RNA levels in various conditions or states using microarrays (described in Chapters 8 and 9) or to measure protein levels (Chapter 10). These studies give only indirect, rather than direct information about gene function. For example, if red blood cells are treated with a drug that inhibits heme biosynthesis in mitochondria, the cell may respond with a complex program of responses that serve to regulate the expression of heme-binding proteins such as the globins. Globin messenger RNA and protein levels might be reduced dramatically, but it would be incorrect to infer that the drug acted directly on the globin gene, messenger RNA, or protein. Similarly, when one measures RNA transcript levels in tissues or cell lines derived from individuals with a disease, significantly regulated transcripts might reflect adaptive changes made in response to a primary insult such as a genetic mutation. Changes might also occur because of downstream effects: a gene defect could disrupt a pathway, leading to degeneration of a brain region, and other cells such as glia could proliferate as a downstream response. Such experiments are not likely to directly reveal the gene-causing mutation although they may reveal information about its secondary consequences and are essentially a molecular phenotype for the mutant.

There are two main kinds of genetic screens that are used to identify gene function in a high throughput fashion: reverse and forward genetics (reviewed in Schulze and McMahon, 2004; Ross-Macdonald, 2005; Alonso and Ecker, 2006; Caspary and Anderson, 2006). These two approaches are illustrated in Fig. 12.7. In reverse genetic screens, a large number of genes (or gene products) is systematically inhibited one by one. This can be accomplished many ways, for example by deleting genes using homologous recombination or by selectively reducing messenger RNA abundance. Then, one or more phenotypes of interest are measured. The main challenge of this approach is that for some organisms it is difficult to disrupt large numbers of genes (such as tens of thousands) in a systematic fashion. It can also be challenging to discern the phenotypic consequences for a gene that is disrupted. As an example of reverse genetics, Thomas Südhoff and colleagues targeted the deletion of mouse *syntaxin binding protein 1* (*Stxb1*; also called *Munc18-1* or *N-sec1*), a gene encoding a nerve terminal protein (Verhage et al., 2000). The phenotype was lethality at the time of birth, with neurons unable to secrete neurotransmitter. Remarkably, brain development appeared normal up to the time of death. This targeted deletion allowed the dissection of the functional role of this gene. We will return to Stxb1 in this chapter to illustrate several principles of functional genomics. Figure 12.4b shows a schematic diagram of its function.

In forward genetic screens, one begins with a defined phenotype of interest, such as the ability of plants to grow in the presence of a drug, or the ability of neurons to extend axons to appropriate targets in the mammalian nervous system, or the ability of a eukaryotic cell to transport cargo. An experimental intervention is made, such as administering a chemical mutagen or radiation to cells (or to an organism). This results in the creation of mutants. The phenotype of interest is observed in rare

As a particularly complex example of reverse genetics, Tumpey et al. (2005) engineered a virus containing all the open reading frames of the deadly 1918 influenza virus that is estimated to have killed 50 million people. They characterized its extraordinary pathogenicity (see Chapter 14).

Reverse genetics (mutate genes then examine phenotypes)

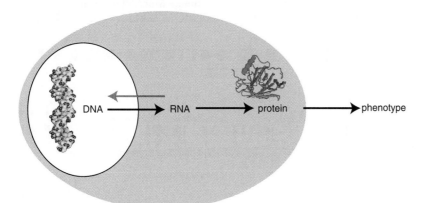

Strategy: systematically inhibit the function of every gene in a genome
approach 1: gene targeting by homologous recombination
approach 2: gene trap mutagenesis
approach 3: inhibit gene expression using RNA interference
measure the effect of gene disruption on a phenotype

DNA → RNA → protein → phenotype

Strategy: identify a phenotype (e.g., growth in the presence of a drug)
mutate genomic DNA (e.g., by chemical mutagenesis)
identify individuals having an altered phenotype
identify the gene(s) that were mutated
confirm those genes have causal roles in influencing the genotype

Forward genetics ("phenotype-driven" screen)

FIGURE 12.7. Reverse and forward genetics. In reverse genetics, genes are targeted for deletion through approaches such as homologous recombination. After a knockout animal is produced, the phenotype is investigated to discern the function of the gene. This is called a "gene-driven" approach because it begins with targeted deletion or disruption of a gene. In forward genetics, one typically begins with a phenotype of interest. The genome is subjected to a process of mutagenesis (typically N-ethyl-N-nitrosurea with a chemical such as or an exogenous DNA transposon). Mutants are collected and screened for those that display an altered phenotype. Next, the genes underlying the altered phenotype are mapped and identified. This is called a "phenotype-driven" approach because one does not begin with particular disrupted genes but instead with an altered phenotype.

representatives among a large collection of mutants. If individuals need to be assayed for the phenotype one at a time (as part of a screen) this can be extremely laborious. If a specific selective condition can be defined in which only the desired mutant grows (a selection) the process is greatly facilitated. A second challenge of forward genetics approaches is to then identify the responsible gene(s) using mapping and sequencing strategies. As an example of this approach, Peter Novick, Randy Schekman, and colleagues characterized temperature-sensitive yeast mutants that accumulate secretory vesicles (Novick and Schekman, 1979; Novick et al., 1980). These secretion (*sec*) mutants occurred in a series of dozens of complementation groups (yeast strains harboring different mutant alleles of the same gene). All the *sec* mutant genes were subsequently identified. For example, the *SEC1* gene encodes that Sec1p protein that functions in vesicle docking at the cell surface. Sec1p is a yeast ortholog of mammalian Stxb1. A schematic showing the role of Sec1p and three other sec proteins in vesicle trafficking is shown in Fig. 12.4a.

The accession number of *S. cerevisiae* Sec1p is NP_010448. The accession of an ortholog, human syntaxin binding protein 1a (Munc18-1; N-sec1), is NP_003156.

Reverse Genetics: Mouse Knockouts and the β-Globin Gene

Knocking out a gene refers to creating an animal model in which a homozygous deletion is created, that is, there are zero copies [denoted $(-/-)$ and referred to as a null allele] instead of the wild-type situation of two copies in a diploid organism $(+/+)$. In a hemizygous deletion one copy is deleted and one copy remains $(+/-)$.

We can illustrate the use of knockouts with the example of the β-globin gene. In normal adult humans, hemoglobin is a tetramer that consists of two α-globin subunits and two β-globin subunits $(\alpha_2\beta_2)$, with a minor amount (~2% to 3%)

consisting of $\alpha_2\delta_2$ tetramers. The β and δ genes are part of a cluster of β-like genes on chromosome 11 (Fig. 12.8a). There is a similar arrangement on mouse chromosome 7 (Fig. 12.8b). The globin genes are expressed at different developmental stages and cell types in a manner that is exquisitely choreographed. Within the β-globin cluster, ε-globin is expressed in the blood island of the yolk sac until 6 to

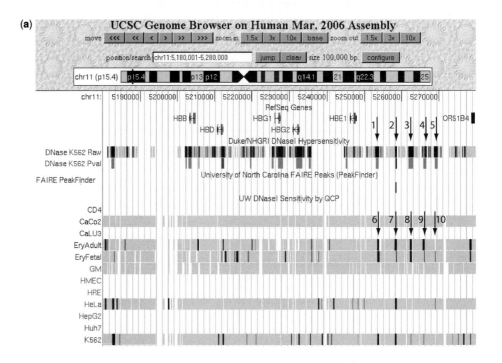

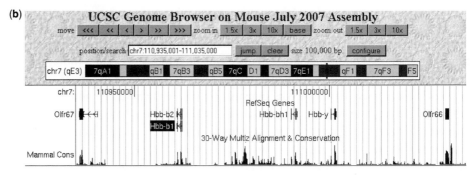

FIGURE 12.8. *The β globin locus (a) on human 11 and (b) on mouse chromosome 7. In (a), a region of 100,000 base pairs is displayed (chr11:5,180,001–5,280,000) on the UCSC Genome Browser (▶ http://genome.ucsc.edu). This region includes five globin RefSeq genes, transcribed from right to left (along the bottom strand towards the 11p telomere). This region is part of the ENCODE project (discussed in Chapter 16) in which 1% of the human genome has been studied in tremendous detail. Annotation tracks from the ENCODE project are shown, including a Duke/NHGRI DNAase I hypersensitivity study, showing genomic loci that are likely to have regulatory functions because they are in a conformation that is susceptible to DNase cleavage. Five of these sites are indicated upstream of the β globin locus (arrows 1 to 5). Other annotation tracks show comparable patterns (e.g., arrows 6 to 10). Note that these studies further show that the properties of gene regulatory regions vary across cell types (e.g., erythrocytes and hematopoietic precursor K562 cells prominently display hypersensitivity sites), as well as at different developmental stages (e.g., fetal versus adult erythrocytes). In (b), mouse globin genes are shown in a 100,000 base pair window. These are flanked by a very large number of olfactory receptor genes (not shown). A conservation track is displayed, showing multispecies conservation corresponding to exons as well as some conservation in noncoding regions, corresponding to cis-regulatory elements.*

8 weeks of gestation when it is silenced and γ-globin genes are activated. At birth, δ-globin and β-globin gene expression increase, while γ-globin expression declines until it is silenced at about age 1. This process is called hemoglobin switching, and it is thought to occur because of interactions between the globin genes and the upstream locus control region (reviewed in Q. Li et al., 2006). Various protein complexes interact with the locus control region (Mahajan and Wissman, 2006). As indicated in Fig. 12.8a, specific regulatory sites have been identified by techniques such as DNase I hypersensitivity assays that reveal regions of exposed chromatin.

Several diseases are associated with perturbations of globin function (discussed in Chapter 20 on human disease). Sickle-cell anemia is caused by mutations in a copy of the β-globin gene. Thalassemias are hereditary anemias that result from an imbalance in the usual one-to-one proportion of α and β chains. In an effort to create an animal model of thalassemias, and to further understand the function of the β globin gene, Oliver Smithies and colleagues used homologous recombination in embryonic stem cells to disrupt the mouse major adult β-globin gene *b1* (Shehee et al., 1993). In homologous recombination, recombinant DNA introduced into the cell recombines with the endogenous, homologous sequence (Capecchi, 1989). The approach, outlined in Fig. 12.9, requires a targeting vector that includes the β-globin gene having a portion modified by insertion of the *neo* gene into exon 2. This targeting vector is introduced into embryonic stem cells by electroporation. When the cells are cultured in the presence of the drug G418, wild-type cells die, whereas cells having the *neo* cassette survive. The successful introduction of an interrupted form of the β-globin gene into stem cells can be confirmed by using the polymerase chain reaction and/or Southern blots (in which a radiolabeled fragment of the insert is hybridized to membranes containing extracts of genomic DNA from wild-type and targeted cells). Targeted embryonic cell lines are injected into mouse blastocysts and implanted into the uterus of a foster mother to generate chimeric offspring. The mice that were heterozygous for the disrupted gene $(+/-)$ appeared normal, while homozygous mutants $(-/-)$ died

The 2007 Nobel Prize in Physiology or Medicine was awarded to Mario Capecchi, Sir Martin Evans, and Oliver Smithies "for their discoveries of principles for introducing specific gene modifications in mice by the use of embryonic stem cells." See ▶ http://nobelprize.org/nobel_prizes/medicine/laureates/2007/.

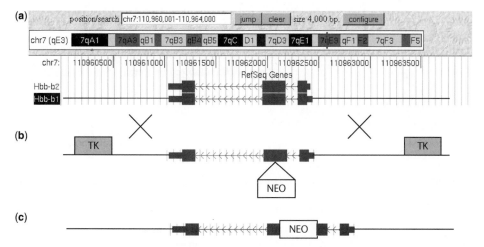

FIGURE 12.9. *Method of gene knockout by homologous recombination. (a) Structure of the β globin gene locus (from the UCSC Genome Browser), showing three exons that are transcribed from right to left. (b) Schematic of the linearized targeting vector used by Shehee et al. (1993). It includes the β globin gene with a neo gene inserted into exon 2 to allow for selection based on conferring resistance to the drug G418. Copies of the thymidine kinase (TK) gene from herpes simplex virus 1 flank the homologous segments and are also used for selection. The large X symbols indicate regions where crossing over can occur between homologous segments. (c) The successfully targeted locus includes a β globin gene that is interrupted by the neo gene.*

in utero or near the time of birth. Thus, the knockout caused a lethal thalassemia, with abnormal red blood cells and lack of protein produced from the deleted *b1* gene.

In nature, the same *b1* gene is sometimes deleted in mice. Surprisingly, this naturally occurring deletion results in only a mild thalassemia, rather than the lethal phenotype that results from the knockout. Shehee et al. (1993) hypothesized that the locus control region normally regulates the *b1* and *b2* genes, but there is a rate-limiting amount of promoter sequence neighboring each gene that the locus control region can regulate. In the naturally occurring deletion associated with nonlethal thalassemia the locus control region interacts with just the *b2* gene (and mediates a compensatory increase in *b2*-derived globin protein). However, in the targeted mutant the locus control region regulates *b2* and also interacts with two more promoters: the inserted *tk* promoter driving the *neo* gene, and the promoter of the deleted *b1* gene. The three promoters compete for factors associated with the locus control region, and so relatively little functional *b2* mRNA is produced and the phenotype is lethal instead of mild.

This example highlights the complexity of creating a null allele with an insertion vector. Many other strategies have been introduced (reviewed in van der Weyden et al., 2002), including a variety of positive and negative selection markers, and the use of replacement vectors instead of insertion vectors that leave behind no selectable markers (and thus are less likely to interfere with endogenous processes). Conditional knockouts permit activation (for "gain-of-function") or inactivation (for "loss-of-function") in vivo, and can be invoked at any time of development or, through the use of tissue-specific promoters, in any region of the body. Conditional knockouts can be used to study the effects of disrupting a gene while avoiding embryonic lethality.

The National Institutes of Health initiated a Knockout Mouse Project (KOMP; Austin et al., 2004). Its ultimate goal is to systematically knock out all mouse genes using several approaches. It is proposed to generate null alleles, including a null-reporter allele for each gene (such has β-galactosidase or green fluorescent protein). The reporter allows the determination of the cell types that normally express that gene.

The NIH Knockout Mouse Project (KOMP) has websites at ► http://www.nih.gov/science/models/mouse/knockout/ and ► http://www.genome.gov/17515708. The data coordination center website is ► http://www.knockoutmouse.org. Currently, about 9800 genes of the ~25,000 mouse genes are on the KOMP target list (February 2008).

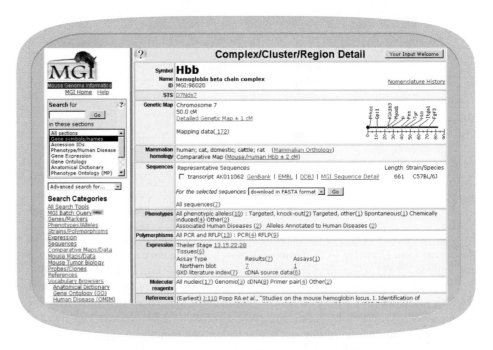

FIGURE 12.10. *The Mouse Genome Informatics (MGI) website entry for the major beta globin gene (Hbb-b1) summarizes molecular data on that gene and includes a phenotype category indicating that three mutant alleles are indexed.*

It is further proposed to make mutated alleles using gene targeting, gene trapping, and RNA interference (discussed below). The KOMP uses mouse strain C57BL/6 because it is widely used and was the first strain to have its genome sequenced. In addition to the effort by KOMP, the European Conditional Mouse Mutagenesis Program (EUCOMM), and the North American Conditional Mouse Mutagenesis Project (NorCOMM) are making targeted conditional mutants. Cumulatively, the three consortia anticipate generating over 18,000 targeted deletion and conditional embryonic stem cell lines (International Mouse Knockout Consortium, 2007).

The Mouse Genome Informatics (MGI) website (Fig. 12.6) provides portals for browsing available knockout resources. This includes Deltagen and Lexicon Knockout Mice, and KOMP genes. As an example of a search for a specific gene, enter "globin" into the main search box at the MGI website and follow the link to

Phenotypic Alleles
Query Results -- Summary

Symbol Hbb
Name hemoglobin beta chain complex
ID MGI:96020

10 matching Alleles (1 Gene/Marker represented)

Allele Symbol Gene; Allele Name	Chr	Synonyms	Category	Observed Phenotypes in Mouse		Allelic Composition (Genetic Background)
				Affected Anatomical Systems	Similar Human Diseases	
Hbb^d hemoglobin beta chain complex; d	7		Not Applicable			
Hbb^{d2} hemoglobin beta chain complex; d2	7		Chemically induced (other)			
Hbb^{d3th} hemoglobin beta chain complex; beta-thalassemia	7	Hbb^{th-1}	Spontaneous	lethality/postnatal, hematopoietic, reproductive	Beta Thalassemia, Dominant Inclusion Body Type 603902; Hemoglobin--Beta Locus; HBB 141900	Hbb^{d3th}/Hbb^{d3th} (involves: C57BL/6 * DBA/2J)
Hbb^{d4} hemoglobin beta chain complex; polycythaemia	7	polycythemia	Chemically induced (ENU)			
Hbb^p hemoglobin beta chain complex; p	7		Not Applicable			
Hbb^p hemoglobin beta chain complex; p	7		Not Applicable			
Hbb^s hemoglobin beta chain complex; s	7		Chemically induced (ENU)			
Hbb^{s2} hemoglobin beta chain complex; s2	7		Chemically induced (ENU)			
Hbb^{tm1Tow} hemoglobin beta chain complex; targeted mutation 1, Timothy Townes	7	Hbb^0, Hbb^{tm1Tmt}	Targeted (knock-out)	lethality/embryonic-perinatal		Hbb^{tm1Tow}/Hbb^{tm1Tow} (involves: 129S2/SvPas * C57BL/6)
				hematopoietic, immune		Hbb^{tm1Tow}/Hbb$^+$ (involves: 129S2/SvPas * C57BL/6)
Hbb^{tm1Unc} hemoglobin beta chain complex; targeted mutation 1, University of North Carolina	7	Hbb^{th-3}	Targeted (knock-out)	lethality/embryonic-perinatal, hematopoietic, skin/coat/nails	Hemoglobin--Beta Locus; HBB 141900	Hbb^{tm1Unc}/Hbb^{tm1Unc} (involves: 129P2/OlaHsd * C57BL/6J)
Hbb^{tm2Unc} hemoglobin beta chain complex; targeted mutation 2, University of North Carolina	7	Hbb^{th-4}, th-4	Targeted (knock-in)			

FIGURE 12.11. *The MGI description of beta globin mutants includes phenotypic data such as the type of mutation (e.g., targeted knockout or conditional knock-in), the observed phenotypes, the human disease relevance, and the allelic composition (genetic background).*

Hbb-b1 (the beta globin adult major chain on chromosome 7) (Fig. 12.10). This page includes information about the gene, as well as a link to phenotypic alleles (Fig. 12.11). Detailed phenotypic data are provided, such as the body weight and the effects on the hematopoietic system.

Reverse Genetics: Knocking Out Genes in Yeast Using Molecular Barcodes

Knockout studies in the yeast *S. cerevisiae* are far more straightforward and also quite more sophisticated than in the mouse for several reasons. The yeast genome is extremely compact, having very short noncoding regions and introns in fewer than 7% of its ~6000 genes. Also, homologous recombination can be performed with high efficiency. A consortium of researchers achieved the remarkable goal of creating yeast strains representing the targeted deletion of virtually every known gene (Winzeler et al., 1999; Giaever et al., 2002). The goals of this project were as follows:

- To create a yeast knockout collection in which all of the ≈6000 ORFs in the *S. cerevisiae* genome are disrupted.

- To provide all nonessential genes (85% of the total) in four useful forms: (1) diploids heterozygous for each yeast knockout (*MAT*a and *MAT*α strains), (2) diploids homozygous for each yeast knockout, (3) **a** mating type (*MAT*a haploid), and (4) α-mating-type (*MAT*α haploid). Knockouts of essential genes are only viable in the heterozygous diploids.

- To provide all essential genes (15% of the total) as diploids heterozygous for each yeast knockout.

Budding yeasts have two mating types: *MAT*a, and *MAT*α. Haploid *MAT*a and *MAT*α cells can mate with each other to form diploid *MAT*aα cells. Both haploid and diploid phases of the life cycle grow mitotically.

Within five years of the creation of the knockout strains, more than 5000 genes were associated with a phenotype based on three dozen publications (reviewed in Scherens and Goffeau, 2004). The strategy employed for this project is gene replacement by polymerase chain reaction (PCR), relying on the high rate of homologous recombination that occurs in yeast (Fig. 12.12a). A short region of DNA (about 50 bp), corresponding to the upstream and downstream portions of each open reading frame, is placed on the end of a selectable marker gene. Additionally, two "molecular barcodes," an UPTAG and a DOWNTAG, unique 20 bp oligonucleotide sequences, are included in each such deletion/substitution strain. This feature allows thousands of deletion strains to be pooled and assayed in parallel in a variety of growth conditions. The molecular barcode approach is extremely powerful. One can grow a collection of thousands of yeast knockouts in routine medium (Fig. 12.12b, unselected population) or in the presence of drug, temperature change, or other experimental condition (selected population). Some of the strains in the selected population might grow slowly (or die), and others might grow favorably. Genomic DNA is isolated, the TAGs (or molecular barcodes) are PCR amplified, labeled with Cy3 or Cy5 dyes (discussed in Chapters 8 and 9), and hybridized to a microarray that contains all 12,000 molecular barcodes (20-mers) on its surface (Fig. 12.12c). Strains that are represented at high or low levels relative to the unselected population are identified based on unequal Cy3/Cy5 ratios on the microarray.

Giaever et al. (2002) used the yeast knockout collection to describe genes that are necessary for optimal growth under six conditions: high salt, sorbitol, galactose,

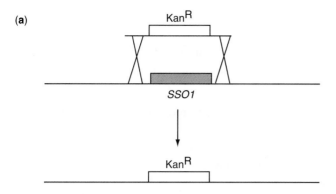

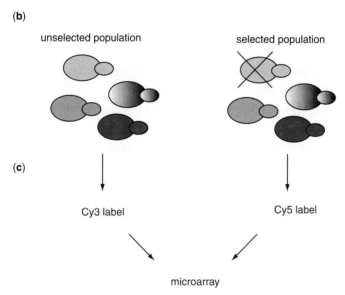

FIGURE 12.12. *Targeted deletion of virtually all* S. cerevisiae *genes. (a) The strategy is to use gene replacement by homologous recombination. Each gene (e.g., SSO1) is deleted and replaced by a Kan^R gene, with unique UPTAG and DOWNTAG primer sequences located on either end. (b) A variety of selection conditions can be used. (c) Genomic DNA is isolated from each condition, labeled with Cy3 or Cy5, and hybridized to a microarray. In this way, genes functionally involved in each growth condition can be identified.*

pH 8, minimal medium, and treatment with the antifungal drug nystatin. Among their findings:

- About 19% of the yeast genes (1105) were essential for growth on rich glucose medium. Only about half of these genes were previously known to be essential. Beyond these 1105 genes, additional genes could be essential in other growth conditions.

- Nonessential ORFs are more likely to encode yeast-specific proteins.

- Essential genes are more likely to have homologs in other organisms.

- Few of the essential genes are duplicated within the yeast genome (8.5% of the nonessential genes have paralogs, while only 1% of the essential genes have paralogs). This supports the hypothesis that duplicated genes have important redundant functions (see Chapter 17).

The systematic deletion method offers a number of important advantages:

- All known genes in the *S. cerevisiae* genome are assayed.

- Each mutation is of a defined, uniform structure.

- Mutations are guaranteed to be null.

- Mutant knockout strains are recovered, banked, and made available to the scientific community.
- Studies of multigene families are facilitated.
- Parallel phenotypic analyses are possible, and many different phenotypes can be assayed.
- Once the strains have been generated, the labor requirement is low when a new phenotype is assessed.

This method also has limitations:

- The labor investment to generate these knockouts was very large.
- For each gene, only null alleles were generated for study. (Additional alleles may be available from other studies.)
- No new genes are discovered with this approach, in contrast to random transposon insertion approaches (described below).
- All nonannotated ORFs are missed. In particular, short ORFs may not be annotated.
- Deletions in overlapping genes may be difficult to interpret.

Since over 80% of the yeast genes are nonessential, this implies that yeast can compensate for their loss through functional redundancy, perhaps by the presence of paralogs (such as *SSO1* and *SSO2*) in which the loss of one is compensated by the presence of the other. A similar scenario explains why deletion of the *b1* beta globin gene in mouse results in a mild disease due to upregulation of the activity of the paralogous *b2* gene. Another possibility is that parallel pathways exist such that if one is compromised the other can compensate; in this scenario, outlined in Fig. 12.4c, the genes encoding members of each pathway need not be homologous. Another idea is that nonessential genes do not have redundancy or compensatory pathways but are functionally required only under highly specific circumstances; thus under some experimental condition they would be found to be essential or at least to confer improved fitness.

How can we determine the functions of nonessential genes in yeast? One approach is to study synthetic lethality, in which a combination of two separate nonlethal mutations causes inviability (reviewed in Ooi et al., 2006). A related concept is synthetic fitness in which two non-lethal mutations combine to confer a growth defect or other disruption that is more severe than that of either single mutation. Tong et al. (2001) devised a high throughput strategy called synthetic genetic array (SGA) analysis to generate haploid double mutants (reviewed in Tong and Boone, 2006). A "query" mutation is crossed to an array of ~4700 "target" deletion mutants, and double mutant meiotic progeny that are inviable indicate that the two mutants are functionally related. Using 132 different query genes, Tong et al. (2004) identified a genetic interaction network having ~1000 genes and ~4000 interactions. The queries included nonessential genes as well as conditional alleles of essential genes. The results were consistent with the behavior of a "small world network" in which immediate neighbors of a gene tend to interact together. In a related TAG array-based approach, Jef Boeke and colleagues defined functionally related networks of genes that are responsible for maintaining DNA integrity, the processes by which cells protect themselves from chromosomal damage (Pan et al., 2006).

They identified ~5000 interactions involving 74 query genes. This illustrates how functional pathways can be inferred using a genetic screen to identify modules of interacting proteins.

Another approach to gene function based on the yeast knockout collection is heterozygous diploid-based synthetic lethality by microarray analysis (dSLAM) (Pan et al., 2007; reviewed in Ooi et al., 2006). In dSLAM, a "query" mutation is introduced into a population of ~6000 heterozygous diploid yeast "target" mutants. The pool of double heterozygotes is then haploidized by sporulation and the haploids are analyzed. A control pool consists of single target mutants, while the experimental pool consists of double (query plus target) mutants. TAGs from these two pools are labeled and analyzed on microarrays to define differential growth properties. Advantages of dSLAM are its use of molecular barcodes to quantify synthetic lethal relationships on microarrays, and its use of heterozygous diploid cells which accumulate fewer suppressor mutations that can confound analysis. A concern for all genetic interaction methods is that the false positive and false negative error rates may vary according to many factors, including the nature of the particular query.

A practical approach to finding genetic relationships between yeast genes is to use the SGD database. As shown for *SEC1* in Fig. 12.2, five different types of genetic interaction were observed using a variety of genetic screens. (1) There were five dosage lethality interactions. These involved *SEC1*, *SEC4*, *SEC8*, and *SEC15* genes, and the identification of additional *SEC* genes suggest that these genes all function in a common pathway. In a dosage lethality experiment, overexpression of one gene causes lethality in a strain that is mutated or deleted for another gene. (2) There were 13 dosage rescue interactions in which overexpression of one gene rescues the deleterious phenotype (lethality or growth defect) caused by deletion of another gene. These interactions included *SEC3*, *SEC5*, *SEC10*, and *SEC15*. (3) There were five phenotypic suppression interactions in which mutation (or overexpression) of one gene suppresses the phenotype (other than a lethality or growth defect) caused by mutation or overexpression of another gene. These interactors included both *SEC* genes (*SEC6*, *SEC14*, *SEC18*) and *SNC1* (Fig. 12.4). (4) There was one synthetic growth defect interaction, in which the expression of two mutant genes in a strain, each of which causes a mild phenotype under some experimental condition, results in the phenotype of slow growth. This occurred between *SEC1* and *SRO7*. (5) There were 38 synthetic lethality interactions that resulted in the phenotype of inviability. These synthetic lethals included a range of genes, both in the *SEC* family and others.

Reverse Genetics: Random Insertional Mutagenesis (Gene Trapping)

We have discussed targeted gene knockouts in mouse and yeast. Many other reverse genetics techniques have been developed (summarized in Table 12.1). Another high throughput approach to disrupting gene function is called gene trapping. When this technique is applied to mouse, insertional mutations are introduced across the genome in embryonic stem cells (reviewed in Stanford et al., 2006; Abuin et al., 2007). Gene trapping is performed using vectors that insert into genomic DNA leaving sequence tags that often include a reporter gene. In this way, mutagenesis of a gene can be accomplished and the gene expression pattern of the mutated gene can be visualized. When the random insertional mutagenesis technique is applied to *Arabidopsis*, DNA is often introduced using the bacterium *Agrobacterium tumefaciens* as a vector (Alonso et al., 2003; reviewed in Alonso and Ecker, 2006).

TABLE 12-1   Reverse Genetics Techniques

Method	Advantages	Disadvantages	Species Studied
Homologous recombination (e.g., gene knockouts)	A targeted gene can be replaced, deleted, or modified precisely Stable mutations are produced Specific (no off-target effects)	Low throughput Low efficiency	Highly useful in yeast; less in mouse; least in plants
Gene silencing (e.g., RNAi)	Can be high throughput Can be used to generate an allelic series Can restrict application to specific tissues or developmental stages	Unpredictable degree of gene silencing Phenotypes not stable Off-target effects are possible	Plants
Insertional mutagenesis	High throughput Used for loss-of-function and gain-of-function studies Results in stable mutations	 Random or transposon-mediated insertions target only a subset of the genome Limited effectiveness on tandemly repeated genes Limited usefulness for essential genes	Plants (>500 publications)
Ectopic expression	Similar to gene silencing	Similar to gene silencing	

Source: Modified from Alonso and Ecker (2006). Used with permission.

Gene trap vectors are typically transfected into mouse embryonic stem cells with subsequent expression of a selectable marker and resistance to antibiotics. Figure 12.13 shows three strategies for using gene traps in mouse. Each gene trap vector lacks an essential transcriptional component. An enhancer trap includes a promoter, neomycin resistance (neo) gene, and polyadenylation signal (Fig. 12.13a). It requires an endogenous enhancer to drive expression of the neo mRNA. A promoter trap lacks a promoter (but includes a splice acceptor and a selectable marker), and its expression is driven by the function of an endogenous promoter (Fig. 12.13b). PolyA traps have their own promoter that drives expression of neo, but they depend on external polyadenylation signals to successfully confer drug resistance (Fig. 12.13c). These traps are useful to trap untranscribed genes since they do not depend on activity of an endogenous promoter.

Gene trapping is a method of random mutagenesis and it is not used to target a specific gene or locus. One strength of the method is that a single vector can be used to both mutate and identify thousands of genes. Also, the technique has the potential to trap genes that were not previously mapped; this contrasts with targeted approaches that require prior knowledge of the gene sequence. A limitation is that one cannot target specific genes of interest. Even a large-scale random mutagenesis

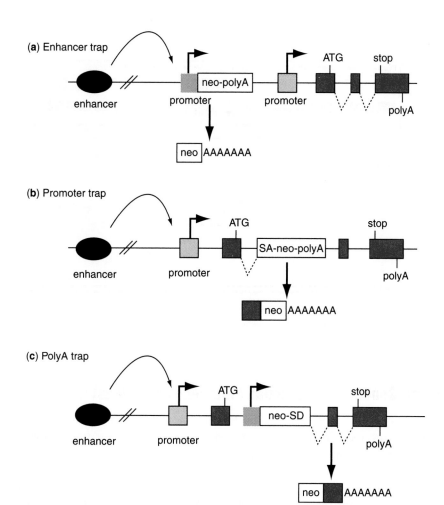

FIGURE 12.13. *Strategies for gene trap mutagenesis. (a) An enhancer traps consist of a vector containing a promoter, a neo gene that confers antibiotic resistance (and thus allows for selection of successfully integrated sequences), and a polyadenylation signal (polyA). This construct is activated by an endogenous enhancer, and disrupts the function of the endogenous gene. The endogenous gene is depicted with its own promoter, start codon (ATG), three exons in this schematic example, a stop codon, and a polyadenylation signal. (b) A promoter trap lacks an exogenous promoter and instead depends on an endogenous enhancer and promoter. It includes a splice acceptor (SA), neo cassette, and polyadenylation site. Integration of this vector disrupts the expression of an endogenous gene. (c) A poly(A) trap vector includes its own promoter and neo cassette but depends on an endogenous polyadenylation signal for successful expression. From Abuin et al. (2007). Used with permission.*

experiment may fail to trap genes because of the nonrandom nature of the insertion sites in the genome (Hansen et al., 2003).

There are several large-scale insertional mutagenesis projects. The International Gene Trap Consortium (IGTC) manages a collection of ~45,000 mouse embryonic stem cell lines that represent ~45% of known mouse genes (Skarnes et al., 2004; Nord et al., 2006). The Mutagenic Insertion and Chromosome Engineering Resource (MICER) includes ~94,000 insertional targeting constructs that can be used to inactivate genes with a high targeting efficiency (28%) (Adams et al., 2004). You can view IGTC gene trap constructs at the UCSC Genome Browser, and both MICER and IGTC resources are available as annotation tracks at the Ensembl mouse genome browser (Fig. 12.14).

The International Gene Trap Consortium website is ► http://www.genetrap.org. The Unitrap resource (Roma et al., 2008) is at ► http://unitrap.cbm.fvg.it.

FIGURE 12.14. Access to information on gene trapped genes at the Ensembl mouse genome browser. From the home page of Ensembl (▶ http://www.ensembl. org) select mouse syntaxin 1a (stx1a) then use the Distributed Annotation System (DAS) pulldown menu (arrow 1) to select GeneTrap (arrow 2) and MICER (arrow 3) data. One gene trap clone is available (arrow 4; sequence tag PST2461-NR) with further links to acquiring this construct. Similarly, several MICER constructs are shown; these are vectors that are useful for generating knockout mice and for chromosome engineering.

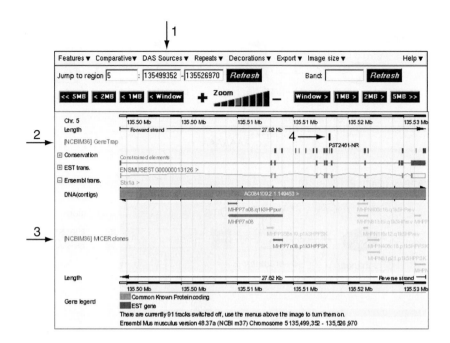

Reverse Genetics: Insertional Mutagenesis in Yeast

We will describe two powerful approaches to gene disruption in yeast, in addition to homologous recombination: (1) genetic footprinting using transposons, and (2) harnessing exogenous transposons.

Transposons are DNA elements that physically move from one physical location to another in the genome (Chapter 16). They accomplish this either with an RNA intermediate (retrotransposons) or without (DNA transposons). The Ty1 element is a yeast retrotransposon that inserts randomly into the genome. Patrick Brown, David Botstein, and colleagues developed a strategy in which populations of yeast are grown under several different conditions (e.g., rich medium versus minimal medium) and subjected to Ty1 transposon-mediated mutagenesis (Smith et al., 1995, 1996) (Fig. 12.15). Following the insertion, PCR is performed using primers that are specific to the gene and to the Ty1 element. This results in a series of DNA products of various molecular weights. The premise of the approach is that an individual gene (e.g., SSO1) might be important for growth under certain conditions. There will be a loss of PCR products (a "genetic footprint") that indicates the importance of that gene for a particular condition.

This approach has several advantages:

- Any gene of interest can be assayed or genes can be selected randomly.
- Multiple mutations can be assayed for any given gene.
- It is possible to perform phenotypic analyses in parallel in a population.
- Many different phenotypes can be selected for analysis.
- The approach can succeed even for overlapping genes.

There are also several disadvantages:

- Mutant strains are not recovered.
- Multiple mutations (alleles) are generated, but they are all insertions (rather than knockouts or other types of mutation).

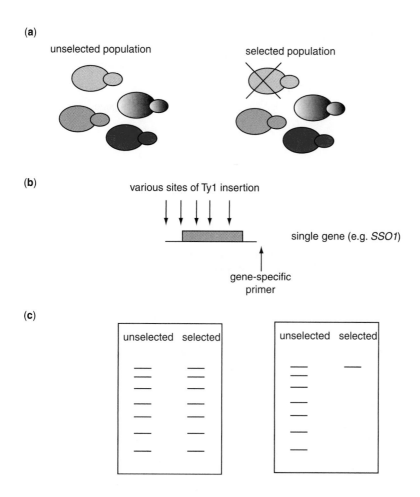

FIGURE 12.15. *Genetic footprinting.* (a) *A population of yeast is selected, e.g., by changing the medium or adding a drug. Some genes will be unaffected by the selection process.* (b) *Random insertion of a transposon allows gene-specific PCR to be performed and* (c) *subsequent visualization of DNA products electrophoresed on a gel. Some genes will be unaffected by the selection process (panel at left). Other genes, tagged by the transposition, will be associated with a reduction in fitness. Less PCR product will be observed (in* [c]*), thus identifying this gene as necessary for survival of yeast in that selection condition.*

- The approach is labor intensive and entails a gene-by-gene analysis.
- The role of duplicated genes with overlapping functions may be missed.

Another mutagenesis approach involves the random insertion of reporter genes and insertional tags into genes using bacterial or yeast transposons (Ross-Macdonald et al., 1999) (Fig. 12.16). A minitransposon derived from a bacterial transposon Tn3 contains a lacZ reporter gene lacking an initiator methionine or upstream promoter sequence. When randomly inserted into a protein-coding gene, it is expected to be translated in-frame in one out of six cases. When this happens, the yeast will produce β-galactosidase, allowing the insertion event to be detected. The construct includes loxP sites that allow a recombination event in which the lacZ is removed and the target gene is tagged with only a short amount of DNA encoding three copies of a hemagglutinin (HA) epitope tag.

An HA-tagged protein can be localized within a cell using an antibody specific to HA.

This minitransposon construct allows a genomewide analysis of disruption phenotypes, gene expression studies, and protein localization. Ross-Macdonald et al. (1999) generated 11,000 yeast strains in which they characterized disruption phenotypes under 20 different growth conditions. These studies resulted in the identification of 300 previously nonannotated ORFs. Data from this study were deposited in the TRIPLES database (Kumar et al., 2002). An example of a search result from this database is shown for *SSO2* (Fig. 12.17).

The TRIPLES database is available at ▶ http://ygac.med.yale.edu/triples/. TRIPLES stands for Transposon-Insertion Phenotypes, Localization and Expression in *Saccharomyces.*

- Surprisingly, 480 expressed insertions were fused to an ORF but in the wrong reading frame. This suggests that frame shifting may be a very common gene expression mechanism.

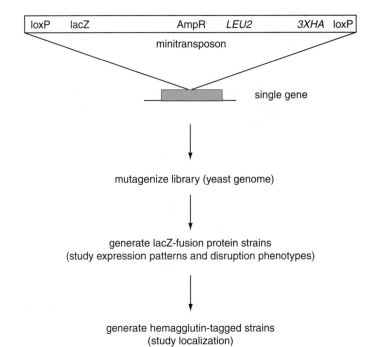

| loxP | lacZ | | AmpR | *LEU2* | *3XHA* | loxP |

minitransposon

single gene

mutagenize library (yeast genome)

generate lacZ-fusion protein strains
(study expression patterns and disruption phenotypes)

generate hemagglutin-tagged strains
(study localization)

FIGURE 12.16. *Transposon tagging and gene disruption to assess gene function in yeast. (Adapted from Ross-Macdonald et al., 1999.)*

(a)

Query Results   Help

The search conditions are: Gene Name: sso2
The output results are sorted by: Clone ID

Clone ID	Gene or ORF (plus synonyms)	mTn insertion	Insertion point	Total protein length
TN7-113D8	SSO2/YM8010.13/YMR183C	in-frame	94	295
TN7-95G9	SSO2/YM8010.13/YMR183C	out-of-frame	157	295
V195A4	SSO2/YM8010.13/YMR183C	in-frame	272	295
V10G6	SSO2/YM8010.13/YMR183C	out-of-frame	-116	0

- Please click on the highlighted clone ID to obtain a composite report of phenotypic, expression, and protein localization data for a given clone.
- Please access our on-line help for a more complete description of the data presented above.

(b)

V195A4   Request reagents

Genbank accession	SGD Link	YPD Link	Genecensus Link (Gerstein's Lab)
none	YMR183C	YMR183C	YMR183C

Potential ORFs Disrupted by mTn Insertion   Help

Gene or ORF (plus synonyms)	mTn insertion/ORF features	Chr.	Chr. Insertion Coord.	Insertion point (codon #)	Total ORF length (amino acids)	Repeated Gene	LacZ Orientation
SSO2/YM8010.13/YMR183C	In frame	XIII	626998	272	295	may be	sense

Gene Expression Data   Help

Growth condition	LacZ expression level	Background strain
vegetative	intense blue	Y800
sporulation	intense blue	Y800

Subcellular Localization   Help

Localization	Cell stage	% of cells staining	Stain intensity	Trials	Comments
cyto. (patches) / Endo. Reticulum	all	35	2	2	patchy staining of the cytoplasm and ER

FIGURE 12.17. *(a) An example of a search result of the TRIPLES database for SSO2, (b) showing a mutant with a transposon insertion and a resultant phenotype.*

- There were 328 in-frame insertions to nonannotated ORFs (from 50 to 247 codons long). Thus, this method is useful to identify novel protein-coding genes.

- Fifty-two percent of these previously nonannotated ORFs are antisense to a known ORF, and 15% overlap a known ORF in a different frame. Thirty-three percent were intergenic. These findings are consistent with the hypothesis that many small genes remain undiscovered.

This approach offers a variety of useful features:

- It includes data on expression levels and protein localization.
- New genes can be discovered, as described above.
- The analysis works for overlapping genes.
- Mutant strains can be recovered and banked. They are made available to the scientific community through the TRIPLES database.

Problems are similar to those involved in genetic footprinting and include the requirement for transposon site specificity, the missing of information on genes with duplicated functions, and the labor-intensive nature of the project.

Reverse Genetics: Gene Silencing by Disrupting RNA

We have discussed reverse genetics approaches in which a gene is deleted by homologous recombination. Another approach to identifying gene function is to disrupt the messenger RNA rather than the genomic DNA. RNA interference (RNAi) is a powerful, versatile, and relatively novel technique that allows genes to be silenced by double-stranded RNA (reviewed in Lehner et al., 2004; Sachidanandam, 2004; Martin and Caplen, 2007). In plants and animals small RNAs (21 to 23 nucleotides) regulate the expression of target genes. The extent of inhibition of gene function may be variable, in contrast to null alleles created by gene knockouts. Mechanistically, RNAi is a form of posttranscriptional gene silencing that is mediated by double-stranded RNA. It may function as a host defense system to protect against viruses, and RNAi also may serve to regulate endogenous gene expression. When double-stranded RNAs are introduced into *Drosophila*, nematode, plant, or human cells they are processed by the endoribonuclease Dicer into small interfering RNAs (siRNAs). These siRNAs cleave target messenger RNAs through the actions of an RNA-induced silencing complex (RISC) composed of proteins (such as Argonaute proteins) and RNA. The endogenous RNAi process seems to involve microRNAs (described in Chapter 8) rather than double-stranded RNAs.

RNAi has been used in genome-wide screens to systematically survey the phenotypic consequence of disrupting almost every gene. In *Drosophila*, Boutros et al. (2004) ascribed functions to 91% of all genes and reported 438 double-stranded RNAs that inhibited the function of essential genes. A further extension of the RNAi approach was provided by creating a transgenic RNAi library in *Drosophila* that permits targeted, conditional gene inactivation in virtually any cell type at any developmental stage. Dietzl et al. (2007) created an RNAi library that targets over 13,000 genes (97% of the predicted protein-coding genes in *Drosophila*). There are many false negative results, based on comparisons to a positive control set consisting of known phenotypes that are expected to occur based on previous classical genetics studies. This may occur because the library was constructed by randomly inserting transgenes into the fly genome, and not all transgenes express at sufficiently high levels. (The false negative rate for the library was ~40% and for the genes was ~35%.) There were also false positive results; some could occur because of off-target effects such as changes in the expression levels of genes flanking the target. As an example of the usefulness of this approach, Dietzl et al. described the use of a neuronal promoter to screen neuronal genes, and reported a lethal phenotype for many, including *n-syb* (a homolog of *SNC1*/synaptobrevin, Fig. 12.4), *Snap* (a homolog of *SEC17*/αSNAP), and Syx5 (a homolog of *SSO1*/syntaxin).

While it is known that false positive results can occur, Ma et al. (2006) emphasized how extensive this problem can be. Off-target effects consist of RNAi constructs that inhibit the expression of endogenous genes other than those that are targeted. It is expected that sequences sharing a high degree of conservation to the small RNA regulator over a span of 19 or more nucleotides will also be targeted. In RNAi studies of *Drosophila* Ma et al. noted off-target effects mediated by short stretches of double-stranded RNA. These false positives often contain tandem trinucleotide repeats (CAN where N represents any of the four nucleotides, with especially strong effects observed with CAA and CAG repeats). Such genes are overrepresented in the results of published RNAi screens. Ma et al. propose that libraries should be designed to avoid even short sequences present in multiple genes, and further that identified phenotypic effects should be independently confirmed using more than one nonoverlapping double-stranded RNA for each candidate.

RNAi screens have been performed in other organisms such as *C. elegans* (e.g., Kamath et al., 2003; Sönnichsen et al., 2005; Kim et al., 2005). Remarkably, *C. elegans* can be fed bacteria that express double-stranded RNA to inhibit gene function (Fraser et al., 2000). Kamath et al. performed a genome-wide RNAi screen and described mutant phenotypes for ~1500 genes, about two-thirds of which did not previously have an assigned phenotype. The most common RNAi phenotype they observed is embryonic lethality, observed in over 900 strains. In human, Berns et al. (2004) targeted ~7900 genes using retroviral vectors that encode over 23,000 short hairpin RNAs, and identified novel modulators of proliferation arrest dependent on p53, a key tumor suppressor and regulator of the cell cycle. Brass et al. (2008) used RNAi to systematically inhibit the function of human genes in a HeLa cell line transfected with short interfering RNAs. They identified 273 messenger RNAs that are required for human immunodeficiency virus (HIV) infection and replication in human cells. These human genes and gene products are potential targets for antiviral drugs. Unlike other antiretroviral drugs, potential drugs targeting these key human host proteins would not be affected by the extraordinary diversity of HIV genotypes (even within a single infected individual there may be one million variant HIV genomes) nor by viral mutations that promote positive selection resulting in drug resistance.

There are several prominent database resources for RNAi data. (1) The GenomeRNAi database integrates sequence data for RNAi reagents with phenotypic data from RNAi screens, primarily in cultured *Drosophila* cells (Horn et al., 2007). A search of the GenomeRNAi database with the query rop (a *Drosophila* homolog of yeast *SEC1*) shows several RNAi probes (including the phenotype, the specificity, the occurrence of off-target effects, and the efficiency), as well as a link to the FlyBase gene entry. (2) FLIGHT also provides data on high throughput RNAi screens (Sims et al., 2006). Its scope and mission are comparable to the GenomeRNAi database. Both include a blast server, and FLIGHT contains additional analysis tools. (3) The RNAi Database is a similar database dedicated to *C. elegans* (Gunsalus et al., 2004). A search for *UNC-18*, the *C. elegans* homolog of *SEC1*/

An HIV interaction database is available at ▶ http://www.ncbi. nlm.nih.gov/RefSeq/ HIVInteractions/. Currently it lists ~1500 human genes whose products interact with HIV. We discuss HIV in Chapter 14.

The GenomeRNAi database is available at ▶ http://www.dkfz. de/signaling2/rnai/ernai.html. Currently it includes over 91,000 double-stranded RNAs and ~6100 phenotype records from 29

rop/syntaxin-binding protein 1, shows a list of phenotypes observed in RNAi screens. For example, RNAi of *UNC-18* leads to resistance to the acetylcholinesterase inhibitor aldicarb, a drug that induces paralysis by preventing the normal breakdown of the neurotransmitter acetylcholine. This result is consistent with a functional role for unc-18 in modulating the release of acetylcholine from vesicles in the presynaptic terminal at the neuromuscular junction.

A second approach to disrupting RNA is to knock down gene expression using morpholinos (Angerer and Angerer, 2004; Pickart et al., 2004). Morpholinos are a form of antisense oligonucleotide consisting of a nucleic acid base with a morpholine ring and a phosphorodiamidate linkage between residues. They specifically bind to messenger RNAs (and microRNAs) and have been used to downregulate transcripts. They have been used extensively in zebrafish, and the ZFIN database describes the results of experiments using morpholinos. The Morpholino DataBase lists morpholinos and their targets, and associated phenotypic data (Knowlton et al., 2008).

large-scale *Drosophila* RNAi studies (February 2008). FLIGHT is online at ▶ http://www.flight.licr.org/ and requires registration for full access. The RNAi Database (RNAiDB) is available at ▶ http://www.rnai.org.

The Morpholino Database (MODB) is available at ▶ http://www.secretomes.umn.edu/MODB/. It currently contains over 700 morpholinos.

Forward Genetics: Chemical Mutagenesis

Forward genetics approaches are sometimes referred to as phenotype-driven screens. They are commonly performed using *N*-ethyl-*N*-nitrosurea (ENU), a powerful chemical mutagen used to alter the male germline (O'Brien and Frankel, 2004; Clark et al., 2004). ENU is more effective than x-irradiation, γ-irradiation, or other chemical mutagens at inducing point mutations in organisms from mice to *Drosophila* to plants (Russell et al., 1979). While the spontaneous mutation rate is about 5 to 10×10^{-6} for the average locus, ENU treatment typically yields a mutation frequency of about 1×10^{-3} per locus. After ENU is administered to mice or other organisms, a phenotype of interest is observed (such as failure of neurons to migrate to an appropriate position in the spinal cord). Recombinant animals are created by inbreeding and the phenotype can then be demonstrated to be heritable. The mutagenized gene is mapped by positional cloning and identified by sequencing the genes in the mapped interval. In mice, ENU is used to mutagenize either spermatogonia or embryonic stem cells. O'Brien and Frankel (2003) reviewed the use of chemical mutagenesis in the mouse and emphasized the need for phenotyping that is both expert and high capacity.

A major limitation of the ENU approach is that the gene(s) whose point mutations are responsible for the observed phenotypic change must be identified without the benefit of tags introduced into the genomic DNA. Although positional cloning used to be a laborious process, the availability of complete genome sequences and dense maps of polymorphic markers has permitted relatively rapid identification of genes of interest.

The use of balancer chromosomes has facilitated the ENU approach (Hentges and Justice, 2004). In a balancer chromosome, a phenotypically marked chromosomal segment is inverted, and this facilitates mapping as well as maintenance of mutations in the heterozygous state. This effect was first described by Hermann Muller (1918). Monica Justice and colleagues used the strategy of a balancer chromosome to characterize dozens of novel recessive lethal mutations on mouse chromosome 11 (Kile et al., 2003). The balancer chromosome consists of mouse chromosome 11 harboring a large inversion (34 megabases). Male mice are treated with ENU, mated to females with the balancer chromosome, and through a strategy of successive intercrosses mice that have a homozygous lethal mutation can be identified and the gene can be easily mapped.

Reverse and forward genetics approaches are both powerful. We can contrast and compare several of their features.

- These approaches ask different questions. Reverse genetics asks "What is the phenotype of this mutant?" Forward genetics asks "What mutants have this particular phenotype?"

- Reverse genetics approaches attempt to generate null alleles as a primary strategy (and conditional alleles in many cases). Forward genetics strategies such as chemical mutagenesis are "blind" in that multiple mutant alleles are generated that affect a phenotype (Guénet, 2002). These alleles include hypomorphs (having reduced function), hypermorphs (having enhanced function), and neomorphs (having novel function), as well as null alleles.

- We introduced techniques such as insertional mutagenesis (see above) as a form of reverse genetics. However, insertional mutagenesis has also been used in the context of forward genetics screens. In each case an attempt is made to infer the function of a set of genes based on the phenotypic consequence of disrupting the expression of a gene.

FUNCTIONAL GENOMICS AND THE CENTRAL DOGMA

See ▶ http://www.genome.gov/10000612 for a description of the NHGRI functional analysis program.

We have discussed reverse and forward genetics approaches to gene function. Another way to describe the scope of the field of functional genomics is to to consider the central dogma that DNA is transcribed to RNA and translated to protein. These levels of analysis are reflected in the organization of functional genomics projects at the National Human Genome Research Institute (NHGRI) and elsewhere.

Functional Genomics and DNA: The ENCODE Project

The ENCODE website at the UCSC Genome Bioinformatics site is ▶ http://genome.ucsc.edu/ENCODE/, and the ENCODE homepage at NHGRI is ▶ http://www.genome.gov/10005107.

Our understanding of the scope of functional genomics has been transformed through large-scale sequencing efforts such as the ENCODE project. The goal of this project is to define all the functional elements of genomic DNA. The initial phase of the ENCODE project centered on studying about 1% of the human genome in depth. Forty-four regions were selected, ranging from 0.5 to 1 megabase in size, and including regions that have been well characterized (such as the globin loci) as well as randomly selected regions having properties such as relatively high or low gene density. Additionally, the ENCODE project involved the analysis of conserved syntenic regions in a variety of other vertebrate organisms. There were four major conclusions from the pilot phase of the ENCODE project (ENCODE Project Consortium, 2007). (1) The genome is pervasively transcribed, a result we discussed in Chapter 8. (2) Features of transcriptional regulation were elucidated, including information about chromatin accessibility and histone modification. (3) Chromatin structure was described in new detail. (4) Mechanistic insights about genome evolution were obtained. We will introduce the ENCODE project in Chapter 16 in our description of the eukaryotic chromosome.

Functional Genomics and RNA

Surveys of RNA transcript levels across different regions (for multicellular organisms) and times of development provide fundamental information about an organism's program of gene expression. (The term gene expression profiling is commonly used, although more precisely it is steady-state mRNA levels that are measured rather than the process of gene expression.) Many studies have surveyed changes in RNA transcripts levels across developmental stages of organisms, or across body regions. We introduced approaches and resources such as serial analysis of gene expression (SAGE) and UniGene in Chapter 8. Microarrays have been used to measure gene expression patterns for thousands of *Drosophila* genes across many developmental stages (Arbeitman et al., 2002). Similar studies have been performed for the mosquito (Koutsos et al., 2007), *C. elegans* (Kim et al., 2001), and other species.

The *Saccharomyces* Genome Database (SGD) offers many resources to describe gene expression in yeast. For each gene, an expression summary plots the $\log_2$ ratio of gene expression (x axis) versus the number of experiments (y axis) (Fig. 12.3, lower right). That plot is clickable, so one can quickly identify experiments in which SEC1 RNA is dramatically up- or downregulated. Another resource is the Function Junction, a resource that provides data on functional analyses for individual *S. cerevisiae* loci from six separate sites, including one for SAGE data and a microarray viewer.

Kim et al. (2001) introduced a gene expression terrain map that resembles a map of the mutation landscape in cancer (see Fig. 20.15).

Functional Genomics and Protein

Classical biochemical approaches to protein function involve an assay for the function of a protein (such as its enzymatic activity or a bioassay for its influence on a cellular process). This assay may be used as the basis of a purification scheme in which the protein is purified to homogeneity. Thousands of proteins have been studied individually with this approach. Each protein has its own personality in terms of biochemical properties and its propensity to interact with a variety of resins that separate proteins on the basis of size, charge, or hydrophobicity. We described several techniques to study proteins in Chapter 10, including two-dimensional gel electrophoresis and mass spectrometry.

In the remainder of this chapter we will introduce proteomics approaches to functional genomics. We first describe protein–protein interactions, and then conclude with a study of protein pathways. The functions of most proteins are unknown. Even for relatively well-studied model organisms such as *Escherichia coli* and *S. cerevisiae*, functions have been assigned to only perhaps two-thirds of all proteins, and the function of the great majority of mouse or human proteins is unknown. The high throughput proteomics projects attempt to assign function on a large scale, identifying the presence of proteins in particular physiological conditions or identifying protein–protein interaction partners.

PROTEOMICS APPROACHES TO FUNCTIONAL GENOMICS

We introduced proteins in Chapter 10 and discussed some of their basic features, including posttranslational modifications, localization, and function. In addition to the study of individual proteins, high throughput analyses of thousands of proteins are possible (Molloy and Witzmann, 2002). We will describe three such approaches: (1) identifying pairwise interactions between protein using the yeast two-hybrid system and other techniques, (2) identifying protein complexes involving two or more proteins using affinity chromatography with mass spectrometry, and (3) analyzing protein pathways. While protein studies have been studied in depth in a variety of model organisms, studies in *S. cerevisieae* are particularly advanced.

We have discussed forward genetics and reverse genetics approaches to gene function. A similar framework can be applied to proteomics (Palcy and Chevet, 2006). Forward proteomics approaches correspond to the classical approach to protein characterization (Fig. 12.18a). A biological system is selected, such as human cells from individuals with or without a disease. Proteins are compared by techniques

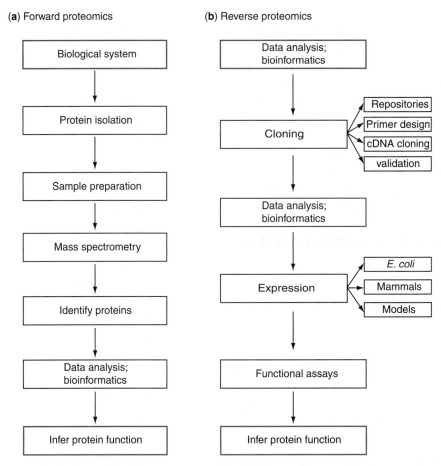

FIGURE 12.18. *Forward and reverse proteomics. (a) Forward proteomics. An experimental system is selected (such as a comparison of two developmental stages or normal versus diseased tissue). Proteins are extracted in a manner depending on the biological question that is addressed (e.g., selecting for membrane proteins or a subcellular organelle). Sample preparation may include steps such as polyacrylamide gel electrophoresis or chromatography columns to separate complex protein mixtures and reduce the complexity of the sample fractions being compared. Proteins may be labeled with fluorescent dyes or a variety of other tags, then they are separated and analyzed by techniques such as mass spectrometry (Chapter 10). Spectra are analyzed and proteins are identified to identify differentially regulated proteins. These regulated proteins may reflect functional differences in the comparison of the original samples. (b) Reverse proteomics. A genome sequence of interest is analyzed, and genes, transcripts, and proteins are predicted based on a combination of computational and experimental evidence (discussed in Chapter 16 for eukaryotes). Complementary DNAs (cDNAs) are cloned based on information about open reading frames available in repositories and based on appropriate primer design. cDNAs are validated by sequence analysis and are then expressed in systems such as* E. coli *(for the production of recombinant proteins), mammalian cells, or other model organism systems. Functional assays are performed in order to assess function; assays include the yeast two-hybrid system or other protein interaction assays. Modified from Palcy and Chevet (2006). Used with permission.*

such as mass spectrometry, differentially regulated proteins are identified, and from this the function of these proteins and their possible roles in the disease state may be inferred and further studied. In reverse proteomics, the starting point is the genomic sequence, from which genes, RNA transcripts, and protein products can be inferred (Fig. 12.18b). Complementary DNA (cDNA) clones can be obtained and expressed in a variety of systems so that their function may be assessed in assays for protein–protein interactions or other behaviors (cellular phenotypes).

Both forward and reverse proteomics approaches may be applied to discover protein function. Both of these may involve high throughput techniques and large numbers of samples and/or proteins are assayed. For example, in the forward proteomics approach of "isobaric tags for relative and absolute quantitation" (iTRAQ; Aggarwal et al., 2006), for four protein samples of interest, the identity and relative quantity of 1000 proteins in each of these samples may be determined with high accuracy. Protein microarrays, analogous to DNA microarrays, consist of affinity reagents (such as specific antibodies) that are attached to a solid support (MacBeath, 2002; Chen and Zhu, 2006). Such technology has not yet reached widespread use because of the inherent difficulty in maintaining the structure (and function) of immobilized proteins. Tissue microarrays represent another high throughput approach that is particularly well suited to molecular pathology studies (Kononen et al., 1998; Kallioniemi et al., 2001). A tissue microarray typically consists of several hundred (or thousand) tissue specimens immobilized on a slide in an orderly array. These samples can be probed in parallel to detect and quantify DNA, RNA, or protein targets.

Protein–Protein Interactions

Proteins are responsible for a dazzling variety of functions, from serving as enzymes to having structural roles. A consistent theme is that most proteins perform their functions in networks associated with other proteins and other biomolecules. As a basic approach to discerning protein function, pairwise interactions between proteins can be characterized. Proteins often interact with partners with high affinity. (The two main parameters of any binding interaction are the affinity, measured by the dissociation constant K_D, and the maximal number of binding sites, measured by the B_{max}). The interactions of two purified proteins can be measured with dozens of techniques, such as the following:

- Co-immunoprecipitation, in which specific antibodies directed against a protein of interest are used to precipate the protein to the bottom of a test tube along with any associated binding proteins.

- Affinity chromatography, in which a cDNA construct is engineered that encodes a protein of interest in frame with glutathione-*S*-transferase (GST) or some other tag, such as polyhistidine. A resin to which glutathione is covalently attached is incubated with a GST fusion protein, and it binds to the resin along with any binding partners. Irrelevant proteins are eluted and then the specific binding complex is eluted and its protein content is identified.

- Cross-linking with chemicals or ultraviolet radiation. A protein is allowed to bind to its partners and then cross-linking is applied and the interactors are identified.

- Surface plasmon resonance (with the BIAcore technology of GE Healthcare) in which a protein is immobilized to a surface and kinetic binding properties of interacting proteins are measured.

- Equilibrium dialysis and filter binding assays in which bound and free ligands (that is, a protein with and without its interacting partner) are separated and quantitated.

- Fluorescent resonance energy transfer (FRET) in which two labeled proteins yield a characteristic change in resonance energy upon sharing a close physical interaction.

We can approach the general issues associated with protein–protein interactions by considering the trafficking proteins shown in Fig. 12.4. Some interactions occur in a pairwise fashion; for example, mammalian syntaxin binds to syntaxin-binding protein 1 in a binary complex (Fig. 12.4d). Syntaxin is also a member of several other complexes to the exclusion of syntaxin-binding protein; for example, syntaxin 1a, synaptobrevin-2/VAMP-2, and SNAP-25 (Fig. 12.4b) bind in a complex so tightly that they are able to migrate together as a trimer even in the harsh condition of polyacrylamide gel electrophoresis that denatures most proteins. If purified syntaxin is immobilized on a column and mixed with an extract of rat brain, it is likely that two or more separate complexes will form, as depicted in Fig. 12.4d for Sso1p and other yeast orthologs. It would be incorrect to infer a direct binding interaction between syntaxin-binding protein and synaptobrevin or SNAP-25. At the same time, it would be reasonable to conclude that all these proteins function as part of a common pathway. Finding genetic interactions can provide even more information about genes whose products function in a pathway or in parallel, related pathways (Figs. 12.4c and e). Genetic interaction data give less information about which particular proteins directly interact or which form protein complexes, but they may provide more information than studies of protein partners and protein complexes in terms of the members of protein pathways.

The Yeast Two-Hybrid System

The yeast two-hybrid system is a high throughput method used to identify protein–protein interactions (Fields and Song, 1989). The assay is extremely versatile and has been used to identify protein-binding partners in many species. It is based on the fact that the yeast *GAL4* transcriptional activator is composed of two independent activation and binding domains (see Box 12.2). The cDNA encoding a protein of interest (the "bait") is fused to the *GAL4* DNA binding domain. A large collection of cDNAs (a library consisting of various "prey") is cloned into a vector containing the *GAL4* activation domain. Alone, the *GAL4* DNA binding domain does not activate transcription. But when the bait binds to another fusion protein expressed from the cDNA library, the proximity of the two proteins enables transcription of a *GAL4* reporter gene. The name "two-hybrid" system refers to the use of two recombinant proteins that must interact.

In addition to the strategy of using a bait protein to screen a library, the yeast two-hybrid system has been used to measure the interaction of a known bait protein with individual, cloned prey proteins. In this way a set of many protein–protein interactions can be assayed. Compared to screening libraries, this approach has the advantage of systematically testing a matrix of possible protein–protein interactions, while it has the disadvantage of not allowing the discovery of novel interacting partners that might be found in a complex cDNA library.

Yeast two-hybrid system technology has been applied to analyses of essentially all possible pairwise protein–protein interactions in the yeast *S. cerevisiae*. Uetz et al. (2000) described 957 interactions involving 1004 yeast proteins, while Ito et al. (2001) identified 4549 interactions among 3278 proteins. These data sets are useful to define possible pathways of interacting proteins. Surprisingly, only about 20% of

BOX 12.2
Yeast Two-Hybrid System

The yeast two-hybrid system allows the identification of the binding partners of a protein. A cDNA encoding a protein of interest (such as huntingtin, the protein that is mutated in Huntington disease) is used as a "bait" to identify interacting proteins in a library of cDNAs encoding human proteins expressed in brain ("prey"). A construct containing huntingtin cDNA, fused to a DNA binding domain (BD), is introduced into yeast cells. The BD interacts with a yeast *GAL1* upstream activating sequence (UAS), but in the absence of an appropriate activator domain (AD) a *lacZ* reporter gene is not activated [see (a) below]. A library of thousands of cDNAs is created, each fused to an activation sequence, but these alone are also unable to activate a reporter gene [see (b)]. When a clone from the library (AD fused to prey 1) binds to the bait/DNA BD construct, the activator domain is able to activate transcription of the *lacZ* reporter gene (c). This reporter allows identification of plasmid DNA from these yeast cells, and the prey 1 cDNA is sequenced. There may be many different binding partners identified from a yeast two-hybrid library. In one application of this technology, X. J. Li et al. (1995) identified huntingtin-associated protein (HAP-1), a protein enriched in brain that may affect the selective neuropathology of expanded polyglutamine repeats in Huntington disease.

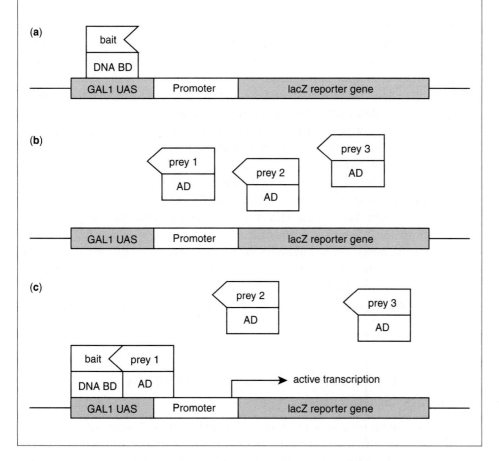

these two data sets overlap. The lack of concordance between these data sets may be due to differences in the physiological conditions in the studies, or to different sources of false positive and false negative errors (discussed below). Other high throughput yeast two-hybrid assays have been applied to *Drosophila* and other organisms (Giot et al., 2003).

This experimental strategy entails a number of assumptions, including reasons for false positive results (biologically nonsignificant interactions) and false negative results (missed biological interactions) (Schächter, 2002). False negative results may occur for the following reasons:

- The bait that is introduced into yeast cells must be localized to the nucleus. If the bait targets its native location, this could explain why some previously known interactions were not observed.
- The fusion protein construct must not interfere with the function of the bait protein.
- Transient protein interactions may be missed.
- Some protein complexes require highly specific physiological conditions in which to form and thus may be missed. Some interactions may fail in the specialized environment of the yeast nucleus.
- There may be a bias against hydrophobic proteins and low molecular weight proteins.

False-positive results may also occur for a variety of reasons. Some proteins may be inherently susceptible to nonspecific binding interactions (i.e., they are "sticky" and activate many bait proteins). Proteins that are denatured may bind nonspecifically. A bait protein may autoactivate a reporter gene. Careful analysis of two-hybrid results allows these sources of false positive and false negative results to be reduced, for example by identifying promiscuous binding proteins.

Information about yeast two-hybrid data is available in several databases. The *Saccharomyces* Genome Database includes a link to physical interaction data, including interactions from two-hybrid screens (Fig. 12.3, upper left). A search for Sec1p reveals several interaction partners, including Sso2p and Mso1p. (When Mso1p was used as a bait in a reciprocal fashion, it was again found to bind to Sec1p.) Another database of two-hybrid data is the MPact data set, a protein interaction and complex database from the Munich Information Center for Protein Sequences (MIPS) (Güldener et al., 2006). MPact includes manually curated data that integrate different sources of high throughput results and includes data on protein–protein interactions taken from the literature by expert curators. MPact follows the HUPO PSI standards for reporting protein–protein interaction data (Hermjakob et al., 2004). A search for sec1 shows a variety of genetic and physical interactions.

MPact is available at ▶ http:// mips.gsf.de/genre/proj/mpact/. The MIPS Comprehensive Yeast Genome Database (CYGD) is at ▶ http://mips.gsf.de/genre/proj/ yeast/ and is comparable to SGD.

Protein Complexes: Affinity Chromatography and Mass Spectrometry

Affinity chromatography is a technique in which a ligand such as a protein is chemically immobilized to a matrix on a column. A major difference between the yeast two-hybrid strategy and the affinity chromatography approach is that the yeast two-hybrid system is only used to detect pairwise interactions between proteins.

In contrast, an affinity chromatography approach allows subunits consisting of many proteins to be isolated and identified.

Several groups employed a strategy of identifying thousands of multiprotein complexes in the yeast *S. cerevisiae* (Gavin et al., 2002; Ho et al., 2002; Gavin et al., 2006; Krogan et al., 2006). Each group selected large numbers of "bait" proteins containing a tag that allowed each bait to be introduced into yeast, where they could form native protein complexes. After complexes were allowed to form under physiologically relevant conditions, the bait was extracted, copurifying associated proteins. These protein complexes were resolved by one-dimensional SDS-PAGE. Thousands of individual protein gel bands (from experiments with many different bait proteins) were excised from the gel with a razor, digested with trypsin to form relatively small protein fragments, and identified by MALDI-TOF mass spectrometry (Chapter 10).

Employing this strategy, Gavin et al. (2002) obtained 1167 yeast strains expressing tagged proteins, from which they purified 589 tagged proteins and identified 232 protein complexes. Ho et al. (2002) selected 725 bait proteins and also detected thousands of protein–protein associations. In each case, a large number of the protein complexes that were identified included proteins of previously unknown function, highlighting the strength of these large-scale approaches. Gavin et al. (2006) performed a more comprehensive screen using tandem affinity purification coupled to mass spectrometry (TAP-MS) to create ~2000 TAP-fusion proteins. Of these, 88% interacted with at least one partner, and the abundance of the identified binding partners ranged from 32 to 500,000 copies per cell. Gavin et al. developed a "socio-affinity" index measuring the log odds of the number of times two proteins are observed interacting divided by the expected occurrence based on the frequency in the data set. Krogan et al. (2006) also used TAP-MS and reported over 7000 protein–protein interactions involving ~2700 proteins. Employing a clustering algorithm they defined ~550 protein complexes averaging 4.9 subunits per complex. There was a large number of complexes with few members (two to four proteins), and few complexes with many members. Each of these various studies reported many complexes that were absent from the MIPS database, and they also revealed new members of previously characterized complexes. Krogan et al. reported enhanced coverage and accuracy because of technical improvements such as (1) avoiding artifacts associated with protein overproduction, (2) systematically tagging and purifying both interacting partners, (3) using two methods of sample preparation and two methods of mass spectrometry, and (4) assigning confidence values to protein interaction predictions.

Data from Gavin et al. (2006) and many other interaction experiments are available in the IntAct database (Kerrien et al., 2007). A search for sec1 shows 16 interactors (although not Sso1p homologs as recorded for yeast two-hybrid screens). The Krogan et al. (2006) data are also available online.

Basic questions about complexes include the stoichiometry (the number of various subunits), the subunit interactions, and the organization. Conventional biochemical techniques can be used to approach all these questions, and in some cases electron microscopy can reveal structural organization. Hernández et al. (2006) applied TAP-MS to several well-characterized complexes: the scavenger decapping and nuclear cap-binding complexes as well as the exosome which contains ten different subunits. They could distinguish dimers from trimers and reveal subunit interactions that were not apparent using the yeast two-hybrid approach.

As for the yeast two-hybrid screens, this approach yields false positive and false negative results for reasons similar to those presented above. While many complexes

IntAct is available at the European Bioinformatics Institute (► http://www.ebi.ac.uk/intact). Currently it contains ~65,000 proteins, ~105,000 interactions and complexes, and ~162,000 binary or higher-order interactions (February 2008). The main species covered in IntAct are *S. cerevisiae*, human, *Drosophila*, *E. coli* strain K12, *C. elegans*, mouse, and *Arabidopsis*. The Krogan et al. (2006) data are available at ► http://tap.med.utoronto.ca.

are identified repeatedly within a given experiment, indicating that saturation has been reached, this does not mean that those complexes are biologically real. Also, when a protein is identified by mass spectrometry it is usually accompanied by a confidence score. Peptides that are identified multiple times are associated with high confidence identifications, while "one hit wonders" that are identified by one peptide observed in a single run are by definition present in low abundance and are more likely to be spurious or misidentified.

The Rosetta Stone Approach

Several groups have adopted a computational approach to protein function prediction. Marcotte and colleagues (1999a, b) as well as Enright et al. (1999) hypothesized that some pairs of interacting proteins are encoded by two distinct genes in one genome that have fused into a single gene in another genome (Fig. 12.19). Marcotte et al. scanned multiple genomes and identified 6809 such cases in *E. coli* and 45,502 in *S. cerevisiae*. This domain function analysis has been called the Rosetta Stone approach. The Rosetta Stone approach makes the prediction that protein pairs generated from gene fusions have related biological functions. For example, they may function in the same protein complex, pathway, or biological

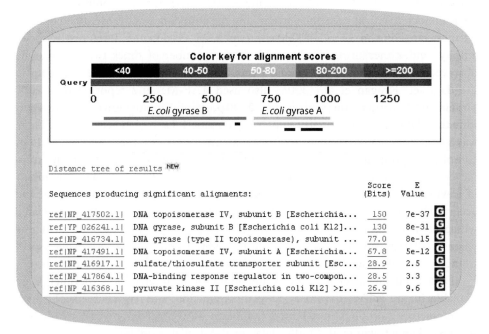

FIGURE 12.19. *The Rosetta Stone method has been used to predict functional interactions between proteins based on analysis of genomic DNA sequences. Genome sequences are scanned for the presence of independent genes from one organism (e.g., E. coli gyrases A and B) that occur in orthologs as part of a single open reading frame. The presence of such a fusion event is interpreted as evidence that the two proteins are part of a functionally related pathway. For the organism that has fused the two genes, an evolutionary advantage could be conferred by the usefulness of co-regulating expression of the two functional units, and/or there may be entropic benefit from the presence of high local concentrations of both proteins. Here, the result of a blastp search is shown. The query was Saccharomyces cerevisiae Top2p (topoisomerase II; NP_014311), and the output was restricted to RefSeq proteins from E. coli strain K12. Note that the query matches to two separate bacterial proteins, gyrase A and gyrase B.*

process. The approach also predicts possible protein–protein interactions. An organism that has fused two genes encoding biologically related proteins may benefit from an entropic contribution afforded by increased effective concentrations of the two proteins in a local environment.

Kuriyan and Eisenberg (2007) discussed the basis of protein–protein interactions in terms of the law of mass action (which states that molecules will tend to bind as their concentrations increase) and the role of colocalization, for which gene fusions represent one possible mechanism. It is also possible that domain fusion does not involve functionally related proteins but occurs for other reasons.

The Protein Link EXplorer (PLEX) is a web resource that includes Rosetta Stone predictions (Date and Marcotte, 2005).

Rosetta Stone refers to the ancient tablet that contains a "fusion" of three distinct scripts (hiero-glyphic, demotic Greek, and classical Greek); the hieroglyphics could be deciphered once all three were discovered together.

PLEX is available at ▶ http://bioinformatics.icmb.utexas.edu/plex.

Protein–Protein Interaction Databases

Many prominent databases store information on protein–protein interactions as well as protein complexes (e.g. Breitkrutz et al., 2008; Ruepp et al., 2008; Sprinzak et al., 2006). Several of these are listed in Table 12.2. Mathivanan et al. (2006) compared the content of eight major databases that include information on human protein–protein interactions. They emphasized the dramatic differences in their content, including the number of reported interactions, the total number of proteins, the curation methodology, and the methods of detecting protein–protein interactions.

TABLE 12-2   Protein–Protein Interaction Databases

Database	Comment	URL
BioGrid	Repository for interaction data sets	▶ http://www.thebiogrid.org/
Biomolecular Object Network Databank (BIND)	Requires log-in; formerly BIND	▶ http://bond.unleashedinformatics.com/
Comprehensive Yeast Genome Database (CYGD)	From the Munich Information Center for Protein Sequences (MIPS)	▶ http://mips.gsf.de/genre/proj/yeast/
Database of Interacting Proteins (DIP)	From UCLA	▶ http://dip.doe-mbi.ucla.edu/
Human Protein Reference Database (HPRD)	From Akhilesh Pandey's group at Johns Hopkins	▶ http://www.hprd.org/
IntAct	At the European Bioinformatics Institute	▶ http://www.ebi.ac.uk/intact/
Molecular Interactions (MINT) Database	Rome	▶ http://mint.bio.uniroma2.it/mint/
PDZBase	Database of PDZ domains	▶ http://icb.med.cornell.edu/services/pdz/start
Reactome	Curated resource of core human pathways and reactions	▶ http://reactome.org/
Search Tool for the Retrieval of Interacting Genes/Proteins (STRING)	Database of known and predicted protein–protein interactions	▶ http://string.embl.de/

Protein Networks

A typical mammalian genome has $\sim$20,000 to 25,000 protein-coding genes, a subset of which (perhaps 10,000 to 15,000) is expressed in any given cell type. These proteins are localized to particular compartments (or are secreted) where many of them interact as part of their function. Some, such as the carrier proteins hemoglobin, myoglobin, retinol-binding protein, and odorant–binding protein, do not rely on protein–protein interactions but instead bind to a ligand (such as oxygen or vitamin A or odorants) and transport it across a compartment by facilitated diffusion. Other proteins function through binary interactions; the majority function via protein complexes. In some cases, these complexes are spatially arranged in what Robinson et al. (2007) call the "molecular sociology of the cell." These authors describe some of the techniques used to determine the structures of complexes, and they further describe the architecture of multisubunit structures such as the nuclear pore complex and the 26S proteasome.

Information about the roles of many proteins in a cell can be integrated in databases and visualized with protein network maps (Schächter, 2002; Bader et al., 2003; Sharom et al., 2004). A pathway is a linked set of biochemical reactions (Karp, 2001). The motivation behind making pathway maps is to visualize complex biological processes, that is, to use high throughput data on protein interactions to generate a model of all functional pathways that is as complete as possible. There are unusual challenges associated with defining protein networks.

(1) One of the basic issues associated with a prediction is the assessment of its accuracy. How likely is it that a false positive or false negative error has occurred? To assess this, benchmark ("gold standard") data sets are required that consist of trustworthy pathways. One can then test whether a particular approach to predicting or reconstructing pathways is specific and sensitive. Unfortunately relatively few interaction networks have been characterized in great detail, and there are no accepted benchmark data sets comparable to those available for fields such as sequence alignment and structural biology. There is little concordance between major benchmark sets such as MIPS, Gene Ontology designations, and KEGG (introduced below) (Bork et al., 2004).

(2) A related issue is that the choice of data is critical. Many researchers integrate data from genomic sequences, expression of RNA transcripts, and protein measurements. It can be challenging to perform this integration since RNA and protein levels are often shown to be poorly correlated. Considering just protein–protein interaction data, for all high throughput techniques the false positive and false negative error rates can be extremely high, as we have seen for example with yeast two-hybrid system data. Nonetheless many projects have proceeded to integrate the largest available datasets, including ones with millions of predicted protein interactions and also interactions as reported in thousands of literature references. For any study, it is essential to carefully evaluate the sources of error and the sensitivity and specificity of the assigned pathways.

(3) The choice of experimental organism is important. Among the eukaryotes, *S. cerevisiae* is the best characterized. Its genome encodes a relatively small number of genes, a tremendous amount of information is known about genes and gene products, and as a unicellular fungus it is simple compared to multicellular metazoans. In considering the use of different organisms to model pathways, a caveat is that even when orthologs of members of a particular pathway are identified, the function of homologs is not necessarily conserved across species. (When a protein has an

established function in one species, an ortholog in a different species is often assigned the same function as a transitive property, and when these orthologs actually do not share the same function this situation has been called "transitive catastrophe.") Mika and Rost (2006) analyzed high throughput data sets from human, *Drosophila*, *C. elegans*, and *S. cerevisiae*. They introduced two metrics: an identity-based overlap measure that describes the overlap between two different data sets in the IntAct database within a single organism, and a homology-based measure that can be used to compare results from data sets in two different organisms. Their unexpected finding was that for all organisms analyzed and at almost all levels of sequence similarity, inference of protein–protein interactions based on homology was dramatically more accurate for pairs of homologs from the same organism than for homologs between different organisms. One significant aspect of this result is that if two proteins are shown to interact in yeast, they do not necessarily interact in animals. Mika and Rost provide examples of protein sequences in *Drosophila* that have different binding partners than in yeast.

(4) In attempting to reconstruct networks on a global scale another consideration is the great variation in the composition and behavior of different pathways. Some, such as the tricarboxylic acid (Krebs) cycle or urea cycle have been examined in depth for many decades; for example, extremely detailed maps of metabolic pathways are available at the ExPASy and KEGG websites. Other pathways are hypothetical or very poorly characterized. Some are constitutive, while others form transiently under particular physiological conditions or developmental stages. Some complexes are highly abundant, while others (such as the exocyst complex) appear to exist in vanishingly small quantities. For others, such as the vault complex (van Zon et al., 2003), the function remains entirely obscure even after extensive studies.

The ExPASy website (Chapter 10) includes detailed maps for metabolic pathways and for cellular and molecular processes (▶ http://www.expasy.org/cgi-bin/search-biochem-index).

(5) There are different categories of network or pathway maps. These include maps based on metabolic pathways, physical and/or genetic interaction data, summaries of the scientific literature, or signalling pathways. For some screens (including the yeast two-hybrid system), information may be accumulated about the interactions of particular domain(s) within a protein that are responsible for interactions in addition to information about the interactions of full-length proteins. Some maps are based on experimental data, while others mix computationally derived results (such as transfers of information from orthologous networks) with experimental data.

We can describe the properties of protein networks. In graphical representations of such complexes, nodes typically represent proteins while edges represent interactions. Most nodes are sparsely connected, while a few nodes are highly connected. Barabási and Albert (1999) suggested that most networks (including biological networks, social networks, and the World Wide Web) follow a scale-free power law distribution:

$$P(k) \sim k^{-\gamma} \tag{12.1}$$

where $P(k)$ is the probability that a node in the network interacts with k other nodes, and $P(k)$ decays following the constant γ. As a consequence, large networks self-organize into a scale-free state. According to this model, this power law distribution is a consequence of the continuous growth of networks, and for the propensity of new nodes to attach preferentially to sites (nodes; here, proteins) that are already well connected. Two basic models that have emerged to describe protein complexes are a "spoke" model, in which a protein bait interacts with multiple partners like the spokes on a wheel, and a "matrix" model in which all proteins are connected (Bader and Hogue, 2002). Either of these models can encompass scale-free properties, although an analysis by Bader and Hogue indicates that a spoke model is more accurate. In reviewing eight databases of human protein interactions, Mathivanan et al. (2006) noted that the Human Protein Reference Database (HPRD) and Reactome databases include a large number of hub proteins that have many binary (direct) protein interactions. (The Reactome database assumes a matrix model with all proteins interconnected within a complex.) A similar finding applies to yeast; as described above, Krogan et al. (2006) described ~550 protein complexes of which about two dozen complexes have ten or more members, while the majority had two to four members. A property of networks having hub proteins is that random disruption of individual nodes (e.g., through mutation) is likely to be well tolerated, although the entire system is vulnerable to some failures at highly connected nodes (Albert et al., 2000).

Many aspects of network properties have been further studied, such as the performance of different ways of creating and assessing confidence scores assigned to particular edges (interactions) (Suthram et al., 2006). Assigning confidence scores requires a benchmark (for example, STRING relies on KEGG, described below) although it is challenging to define adequate benchmarks. Another aspect of protein networks is the nature of hub proteins. Haynes et al. (2006) showed that hub proteins (defined as having ten or more interacting partners) have more intrinsic disorder than do end proteins (those with one interacting partner) in worm, fly, and human. We described intrinsic disorder in Chapter 11. Yet another feature of networks is their modularity (Sharom et al., 2004). One example of modularity is vesicle-mediated exocytosis of neurotransmitter in the mammalian nerve terminal (Fig. 12.4b). The components required for neurotransmitter release function autonomously at a great distance from the cell body, and respond to the arrival of an action potential (an electrical signal) in a local fashion by releasing neurotransmitters. This signaling system has a modular nature. D. Li et al. (2006) estimated the modularity as well as the clustering exponent g for protein interaction networks in yeast, *C. elegans*, and *Drosophila*, reporting that all three have a scale-free nature and varying degrees of modularity.

There is a variety of database resources for global interaction networks. PathGuide is a website that lists 240 biological pathway resources (Bader et al., 2006). These are organized into categories such as protein–protein interactions, metabolic pathways, signaling pathways, pathway diagrams, and genetic interaction networks. For *S. cerevisiae*, the BioGRID database (Reguly et al., 2006) provides manual curation of ~32,000 publications describing physical and genetic interactions. It is available online at its own site and through the SGD (see Fig. 12.2, lower right side). In an effort to standardize the way various database projects present information, the Biological Pathway Exchange (BioPAX) consortium provides a data exchange ontology for biological pathway integration.

Several web servers provide pathway maps. MetaCyc is a database of metabolic pathways (Caspi et al., 2008). It includes experimentally verified enzyme and pathway information, with links from pathways to genes, proteins, reactions, and metabolites. The SGD offers similar metabolic pathway maps for yeast, including data derived from MetaCyc (Fig. 12.20).

A major pathway database is offered by the Kyoto Encyclopedia of Genes and Genomes (KEGG) (Kanehisa et al., 2008) (Fig. 12.21). The KEGG atlas contains a detailed map of metabolism based on 120 metabolic pathways, with links to various organisms. KEGG pathways are a collection of manually drawn maps in six areas (metabolism, genetic information processing, environmental information processing, cellular processes, human diseases, and drug development). An example of a

PathGuide is at ► http://www.pathguide.org/. BioGRID is available at ► http://www.thebiogrid.org. SGD is at ► http://www.yeastgenome.org. BioPAX is online at ► http://www.biopax.org/.

MetaCyc is available at ► http://metacyc.org/index.shtml. There are currently over 1000 pathways, 4000 genes, 1000 organisms, and 15,000 citations in the database (February 2008).

KEGG is available at ► http://www.genome.ad.jp/kegg/. Release 45 (2008) includes about 2.9 million genes from 55 eukaryotes, 584 bacteria, and 49 archaea, and over 71,000 pathways from 356 reference pathways.

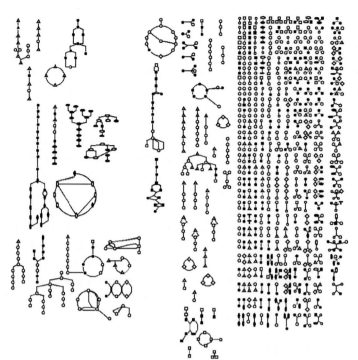

▲ Amino acids
□ Carbohydrates
◇ Proteins
⊓ Purines
⊏ Pyrimidines
▽ Cofactors
⊤ tRNAs
○ Other
● (Filled) Phosphorylated

FIGURE 12.20. *The Saccharomyces Genome Database includes metabolic pathway maps, some of which use data from MetaCyc. See* ▶ *http://pathway. yeastgenome.org/biocyc/.*

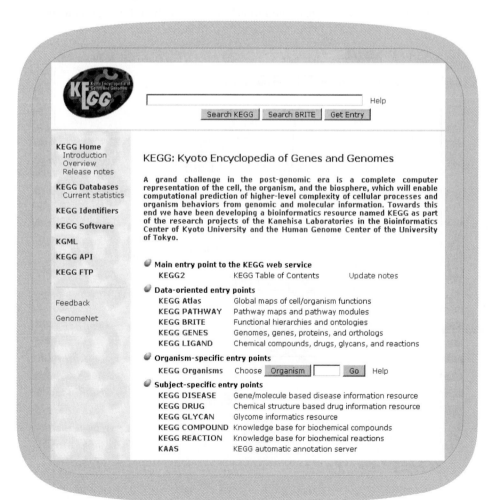

KEGG: Kyoto Encyclopedia of Genes and Genomes

A grand challenge in the post-genomic era is a complete computer representation of the cell, the organism, and the biosphere, which will enable computational prediction of higher-level complexity of cellular processes and organism behaviors from genomic and molecular information. Towards this end we have been developing a bioinformatics resource named KEGG as part of the research projects of the Kanehisa Laboratories in the Bioinformatics Center of Kyoto University and the Human Genome Center of the University of Tokyo.

Main entry point to the KEGG web service
  KEGG2   KEGG Table of Contents   Update notes

Data-oriented entry points
  KEGG Atlas   Global maps of cell/organism functions
  KEGG PATHWAY   Pathway maps and pathway modules
  KEGG BRITE   Functional hierarchies and ontologies
  KEGG GENES   Genomes, genes, proteins, and orthologs
  KEGG LIGAND   Chemical compounds, drugs, glycans, and reactions

Organism-specific entry points
  KEGG Organisms   Choose [Organism] [   ] [Go] Help

Subject-specific entry points
  KEGG DISEASE   Gene/molecule based disease information resource
  KEGG DRUG   Chemical structure based drug information resource
  KEGG GLYCAN   Glycome informatics resource
  KEGG COMPOUND   Knowledge base for biochemical compounds
  KEGG REACTION   Knowledge base for biochemical reactions
  KAAS   KEGG automatic annotation server

KEGG sidebar

KEGG Home
  Introduction
  Overview
  Release notes

KEGG Databases
  Current statistics

KEGG Identifiers

KEGG Software

KGML

KEGG API

KEGG FTP

Feedback

GenomeNet

Search KEGG   Search BRITE   Get Entry   Help

FIGURE 12.21. *The KEGG database includes pathway maps and data for a broad range of organisms (* ▶ *http://www.genome.jp/ kegg/).*

pathway map is shown for vesicular transport (Fig. 12.22); by choosing *S. cerevisiae* from a menu of organisms, clicking on a box such as syntaxin links to an entry on yeast Sso1p. Related maps can be derived from other databases such as GeneGo (Fig. 12.23). For all these pathway maps, the information obtained from biochemical studies is far richer and more accurate in terms of the identities of genes and gene products, their correct subcellular distributions, and the details of their interactions with partner proteins.

As another example of a KEGG pathway, by selecting human neurodegenerative disorders, one can find a pathway description of amyotrophic lateral sclerosis (ALS; Lou Gehrig's disease) (Fig. 12.24). Mutations in the superoxide dismutase gene, *SOD1*, are a common cause of this debilitating disease. *SOD1* is an enzyme that normally converts the toxic oxygen metabolite superoxide (O_2^-) into hydrogen peroxide and water. As shown in the KEGG pathway map, *SOD1* has been shown to interact directly and indirectly with a variety of other proteins, such as those involved in apoptosis (programmed cell death). Clicking on *SOD1*, one finds an entry describing the protein and nucleotide sequence, as well as several external links, such as the Enzyme Commission number and protein structure links, Pfam, Prosite, the Human Protein Reference Database, and Online Mendelian Inheritance in Man (OMIM; Chapter 20).

This example of *SOD1* highlights a strength of KEGG: its coverage of a broad range of proteins and cellular processes is comprehensive. The example also serves to show that some processes described in KEGG are likely to be organism specific.

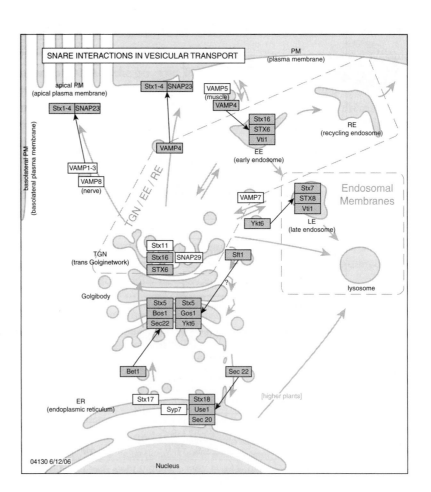

FIGURE 12.22. KEGG pathway map for vesicular transport includes a variety of syntaxin and synaptobrevin homologs (but does not include nerve terminal proteins nor Sec1p homologs).

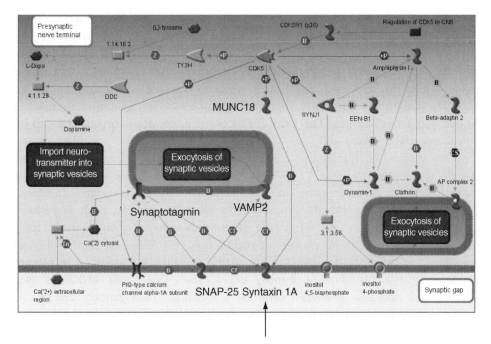

FIGURE 12.23. *Software for pathway maps. The GeneGo commercial product includes a Java module allowing over 500 premade pathway maps to be viewed, or new maps to be created. Mapped objects include genes, proteins, or various compounds. Interactions are depicted such as binding (e.g., between syntaxin, indicated with an arrow, and MUNC18) or complexes (e.g., syntaxin, SNAP-25, synaptotagmin, and VAMP2 form a complex, indicated by Cf). Some labels were redrawn for clarity. From* ► *http://www.genego.com.*

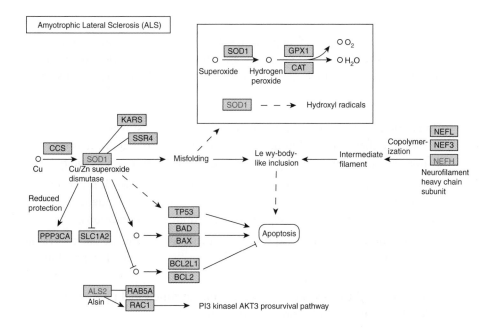

FIGURE 12.24. *KEGG includes pathways for diseases. A pathway for amyotrophic lateral sclerosis (ALS; Lou Gehrig's disease) is shown. Proteins in boxes link to detailed entries.*

KEGG is based primarily on data generated from bacterial genomes, and pathways described in bacteria are not always applicable to eukaryotic organisms.

PERSPECTIVE

Many thousands of genomes have now been sequenced (including viral and organellar genomes). For the genomes of prominent organisms such as human, worms, flies, plants, and yeast, we are acquiring catalogs of the genes and gene products encoded by each genome. Defining the genes and the complete structure of the genome are challenging problems that we will address in the last third of this book. We are already beginning to confront a problem that is perhaps even harder than identifying genes: that is identifying their function. Function has many definitions, as we discussed for proteins in Chapter 10. In this chapter we have described many innovative, high throughput functional genomics approaches to defining gene function. The field of functional genomics is broad, and can be considered using many different categories. (1) What type of organism do we wish to study? We highlighted eight model organisms, although many other models are commonly used. (2) What type of questions do we want to address: natural variation or experimental manipulations used to elucidate gene function? (3) What type of experimental approach do we wish to apply, such as forward versus reverse genetics. (4) What type of molecules do we wish to study, from genomic DNA to RNA to protein or metabolites. (5) What types of biological questions are we trying to address? For many investigators interested in human diseases or the function of human genes, there are yeast orthologs (see Chapters 17 and 20). If a yeast ortholog is identified then genetic screens can suggest many potential interacting partners that may elucidate the function of the human gene.

PITFALLS

We have described a range of approaches to assessing gene function, including analyses at the levels of genes (e.g., creating null alleles or otherwise interfering with gene function), RNA, and proteins. The following caveats should be noted.

- Every method produces false negatives and false positives. It is important to estimate these rates, although it can be difficult to acquire trusted ("gold standard") data sets with which to measure sensitivity and specificity.
- Many methods seem to work well with "knowns" but work much less well with unknown genes. Reasons may include functional redundancy, complex, multiple functions or functions not evident under lab conditions.
- Combinatorial informatic approaches need weighting to help evaluate strength of "links" between genes. Also, any single set of gene "links" is incomplete.
- What is needed to have a better success rate at functional prediction is fewer links of low quality and more links of high quality.

DISCUSSION QUESTIONS

[12-1] Define a functional genomics question. (For example, how can we predict the functions of genes that currently lack functional annotation?) How does the choice of experimental organism affect the approaches you might take to answer the question?

[12-2] Consider a human disease for a gene has been implicated (such as β-globin in sickle cell anemia) and for which an animal model is available. How can forward genetics approaches be used to study this disease? How can reverse genetics approaches be used? What are some of the differences in the kinds of information these two approaches can provide?

PROBLEMS/COMPUTER LABORATORY

[12-1] Suppose you did not know anything about the function of hemoglobin but wanted to use bioinformatics resources to learn about its role in mouse and zebrafish. What information can you find?

[12-2] Select a yeast gene such as *SEC1*. Is it an essential gene? What proteins does it interact with based on physical (biochemical) or genetic assays? Are the interactions observed in yeast also found in mammalian systems?

SELF-TEST QUIZ

[12-1] While there are many definitions of "functional genomics," select the best of these choices:

(a) The assignment of function to genes based primarily on genomewide gene expression data using techniques such as microarrays or SAGE

(b) The assignment of function to genes based primarily on comprehensive surveys of protein–protein interactions and protein networks

(c) The combined use of genetic, biochemical, and cell biological approaches to study the function of a particular gene, its mRNA product, and its corresponding protein product

(d) The assignment of function to genes and proteins using genomewide screens and analyses

[12-2] Reverse genetics approaches involve

(a) Systematically inhibiting the functions of one or many genes (or gene products), and measuring the phenotypic consequences correctly.

(b) Measuring a phenotype of interest (such as cell growth), applying an intervention (such as radiation exposure) to generate a large collection of mutants, and identifying changes to the phenotype of interest.

(c) Treating an organism with a chemical mutagen or other agent to induce mutations, observing a phenotype of interest, and mapping the gene(s) responsible for the phenotype.

(d) All of the above.

[12-3] The "YKO" project is an effort to systematically knock out all yeast ORFs. A potential limitation of this approach is:

(a) Molecular barcodes may sometimes be toxic for yeast genes.

(b) This approach is not suited to finding new genes but instead focuses on already known genes.

(c) Mutant knockout strains cannot be banked for later study by other investigators.

(d) Mutations may not be null.

[12-4] A major advantage of genetic footprinting using transposons is:

(a) The approach is technically easy and can be scaled up to study the function of many genes.

(b) Both insertion alleles and knockout alleles can be studied.

(c) Any known gene of interest can be studied with this approach.

(d) Mutant strains can be banked for later study by other researchers.

[12-5] Forward genetics screens have become increasingly powerful. However, a major limitation is that

(a) Mutations that are introduced through the use of mutagens or radiation do not leave molecular "tags" or barcodes in the genomic DNA, thus adding to the challenge of identifying DNA changes that are responsible for particular phenotypes correctly.

(b) Mutant alleles tend to be null rather than having a broad range of phenotypes.

(c) These screens often involve morpholinos, but these compounds are effective in only a limited number of organisms.

(d) There is no universally preferred method to systematically inhibit the function of each gene in a genome.

[12-6] High throughput screens such as the yeast two-hybrid system and affinity purification experiments can have false positive results because:

(a) Some proteins are inherently sticky.

(b) Some bait proteins that are introduced into cells become mislocalized.

(c) Some protein complexes form only very transiently.

(d) Affinity tags or epitope tags can interfere with protein–protein interactions.

(e) All of the above.

[12-7] Problems in determining protein networks include all of the following EXCEPT which one?

(a) Few benchmark data sets are available with which to assess false positive and false negative results.

(b) False positive and negative error rates tend to be very high.

(c) There is tremendous heterogeneity in the types of protein complexes that form.

(d) Experimental data have been generated for prokaryotes and single-celled eukaryotes such as the yeast *S. cerevisiae*, but it has not yet been possible to obtain high throughput data for organisms such as *Drosophila* and human.

[12-8] Hub proteins are

(a) Proteins that occur at nodes that are highly connected within a protein network.

(b) Proteins that occur at edges that are highly connected within a protein network.

(c) Proteins that occur at nodes that are sparsely connected within a protein network.

(d) Proteins that occur at edges that are sparsely connected within a protein network.

[12-9] Which of the following best describes a major problem in evaluating large-scale cellular pathway diagrams?

(a) The direction of the biochemical pathways is not usually known.

(b) The pathway maps do not employ Gene Ontology nomenclature.

(c) The pathway maps often depend on the correct identification of orthologs, but this can be problematic.

(d) The pathway maps tend to be derived from prokaryotes, but only limited information is available on eukaryotes.

SUGGESTED READING

Excellent reviews of functional genomics approaches are available for the mouse (van der Weyden et al., 2002; Guénet, 2002), plants (Borevitz and Ecker, 2004; Alonso and Ecker, 2006), and yeast. Abuin et al. 2007 thoroughly review gene trap mutagenesis.

REFERENCES

Abuin, A., Hansen, G. M., and Zambrowicz, B. Gene trap mutagenesis. *Handb. Exp. Pharmacol.* **178**,129–147 (2007).

Adams, D. J., et al. Mutagenic insertion and chromosome engineering resource (MICER). *Nat. Genet.* **36**, 867–871 (2004).

Aggarwal, K., Choe, L. H., and Lee, K. H. Shotgun proteomics using the iTRAQ isobaric tags. *Brief. Funct. Genomic. Proteomic.* **5**, 112–120 (2006).

Amir, R. E., Van den Veyver, I. B., Wan, M., Tran, C. Q., Francke, U., Zoghbi, H. Y. Rett syndrome is caused by mutations in X-linked *MECP2*, encoding methyl-CpG-binding protein 2. *Nat. Genet.* **23**, 185–188 (1999).

Albert, R., Jeong, H., and Barabasi, A. L. Error and attack tolerance of complex networks. *Nature* **406**, 378–382 (2000).

Alonso, J. M., and Ecker, J. R. Moving forward in reverse: Genetic technologies to enable genome-wide phenomic screens in *Arabidopsis*. *Nat. Rev. Genet.* **7**, 524–536 (2006).

Alonso, J. M., et al. Genome-wide insertional mutagenesis of *Arabidopsis thaliana*. *Science* **301**, 653–657 (2003).

Angerer, L. M., and Angerer, R. C. Disruption of gene function using antisense morpholinos. *Methods Cell Biol.* **74**, 699–711 (2004).

Arbeitman, M. N., Furlong, E. E., Imam, F., Johnson, E., Null, B. H., Baker, B. S., Krasnow, M. A., Scott, M. P., Davis, R. W., and White, K. P. Gene expression during the life cycle of *Drosophila melanogaster*. *Science* **297**, 2270–2275 (2002).

Austin, C. P., et al. The knockout mouse project. *Nat. Genet.* **36**, 921–924 (2004).

Bader, G. D., Heilbut, A., Andrews, B., Tyers, M., Hughes, T., and Boone, C. Functional genomics and proteomics: Charting a multidimensional map of the yeast cell. *Trends Cell Biol.* **13**, 344–356 (2003).

Bader, G. D., and Hogue, C. W. Analyzing yeast protein-protein interaction data obtained from different sources. *Nat. Biotechnol.* **20**, 991–997 (2002).

Bader, G. D., Cary, M. P., and Sander, C. Pathguide: A pathway resource list. *Nucleic Acids Res.* **34**, D504–D506 (2006).

Barabási, A. L., and Albert, R. Emergence of scaling in random networks. *Science* **286**, 509–512 (1999).

Berns, K., et al. A large-scale RNAi screen in human cells identifies new components of the p53 pathway. *Nature* **428**, 431–437 (2004).

Blattner, F. R., et al. The complete genome sequence *of Escherichia coli* K-12. *Science* **277**, 1453–1474 (1997).

Borevitz, J. O., and Ecker, J. R. Plant genomics: The third wave. *Annu. Rev. Genomics Hum. Genet.* **5**, 443–477 (2004).

Bork, P., Jensen, L. J., von Mering, C., Ramani, A. K., Lee, I., and Marcotte, E. M. Protein interaction networks from yeast to human. *Curr. Opin. Struct. Biol.* **14**, 292–299 (2004).

Boutros, M., Kiger, A. A., Armknecht, S., Kerr, K., Hild, M., Koch, B., Haas, S. A., Paro, R., Perrimon, N., the Heidelberg Fly Array Consortium. Genome-wide RNAi analysis of growth and viability in *Drosophila* cells. *Science* **303**, 832–835 (2004).

Brass, A. L., Dykxhoorn, D. M., Benita, Y., Yan, N., Engelman, A., Xavier, R. J., Lieberman, J., and Elledge, S. J. Identification of host proteins required for HIV infection through a functional genomic screen. *Science* **319**, 921–926 (2008).

Breitkreutz, B. J., Stark, C., Reguly, T., Boucher, L., Breitkreutz, A., Livstone, M., Oughtred, R., Lackner, D. H., Bähler, J., Wood, V., Dolinski, K., and Tyers, M. The BioGRID Interaction Database: 2008 update. *Nucleic Acids Res.* **36**, D637–D640 (2008).

Bult, C. J., Eppig, J. T., Kadin, J. A., Richardson, J. E., Blake, J. A., the Mouse Genome Database Group. The Mouse Genome Database (MGD): Mouse biology and model systems. *Nucleic Acids Res.* **36**, D724–D728 (2008).

Chen, C. S., and Zhu, H. Protein microarrays. *Biotechniques* **40**, 423–427 (2006).

Capecchi, M. R. Altering the genome by homologous recombination. *Science* **244**, 1288–1292 (1989).

Caspary, T., and Anderson, K. V. Uncovering the uncharacterized and unexpected: Unbiased phenotype-driven screens in the mouse. *Dev. Dyn.* **235**, 2412–2423 (2006).

Caspi, R., et al. The MetaCyc Database of metabolic pathways and enzymes and the BioCyc collection of Pathway/Genome Databases. *Nucleic Acids Res.* **36**, D623–D631 (2008).

Christie, K. R., et al. *Saccharomyces* Genome Database (SGD) provides tools to identify and analyze sequences from *Saccharomyces cerevisiae* and related sequences from other organisms. *Nucleic Acids Res.* **32**, D311–D314 (2004).

Clark, A. T., Goldowitz, D., Takahashi, J. S., Vitaterna, M. H., Siepka, S. M., Peters, L. L., Frankel, W. N., Carlson, G. A., Rossant, J., Nadeau, J. H., and Justice, M. J. Implementing large-scale ENU mutagenesis screens in North America. *Genetica* **122**, 51–64 (2004).

Complex Trait Consortium. The Collaborative Cross, a community resource for the genetic analysis of complex traits. *Nat. Genet.* **36**, 1133–1137 (2004).

Darlington, C. D. *Recent Advances in Cytology.* P. Blakiston's Son & Co., Philadelphia, 1932.

Date, S. V., and Marcotte, E. M. Protein function prediction using the Protein Link EXplorer (PLEX). *Bioinformatics* **21**, 2558–2559 (2005).

Dietzl, G., et al. A genome-wide transgenic RNAi library for conditional gene inactivation in *Drosophila*. *Nature* **448**, 151–156 (2007).

Drachman, D. B. Myasthenia gravis. *N. Engl. J. Med.* **330**, 1797–1810 (1994).

ENCODE Project Consortium, et al. Identification and analysis of functional elements in 1% of the human genome by the ENCODE pilot project. *Nature* **447**, 799–816 (2007).

Enright, A. J., Iliopoulos, I., Kyrpides, N. C., and Ouzounis, C. A. Protein interaction maps for complete genomes based on gene fusion events. *Nature* **402**, 86–90 (1999).

Fields, S., and Song, O. A novel genetic system to detect protein-protein interactions. *Nature* **340**, 245–246 (1989).

Fraser, A. G., Kamath, R. S., Zipperlen, P., Martinez-Campos, M., Sohrmann, M., and Ahringer, J. Functional genomic analysis of *C. elegans* chromosome I by systematic RNA interference. *Nature* **408**, 325–330 (2000).

Gama-Castro, S., et al. RegulonDB (version 6.0): Gene regulation model of *Escherichia coli* K-12 beyond transcription, active

(experimental) annotated promoters and Textpresso navigation. *Nucleic Acids Res.* **36**, D120–D124 (2008).

Gavin, A. C., et al. Functional organization of the yeast proteome by systematic analysis of protein complexes. *Nature* **415**, 141–147 (2002).

Gavin, A. C., et al. Proteome survey reveals modularity of the yeast cell machinery. *Nature* **440**, 631–636 (2006).

Giaever, G., et al. Functional profiling of the *Saccharomyces cerevisiae* genome. *Nature* **418**, 387–391 (2002).

Giot, L., et al. A protein interaction map of *Drosophila melanogaster*. *Science* **302**, 1727–1736 (2003).

Guénet, J. L. The mouse genome. *Genome Res.* **15**, 1729–1740 (2005).

Güldener, U., Münsterkötter, M., Oesterheld, M., Pagel, P., Ruepp, A., Mewes, H. W., and Stümpflen, V. MPact: The MIPS protein interaction resource on yeast. *Nucleic Acids Res.* **34**, D436–D441 (2006).

Gunsalus, K. C., Yueh, W. C., MacMenamin, P., and Piano, F. RNAiDB and PhenoBlast: Web tools for genome-wide phenotypic mapping projects. *Nucleic Acids Res.* **32**, D406–D410 (2004).

Hansen, J., Floss, T., Van Sloun, P., Füchtbauer, E. M., Vauti, F., Arnold, H. H., Schnütgen, F., Wurst, W., von Melchner, H., and Ruiz, P. A large-scale, gene-driven mutagenesis approach for the functional analysis of the mouse genome. *Proc. Natl. Acad. Sci. USA* **100**, 9918–9922 (2003).

Haynes, C., Oldfield, C. J., Ji, F., Klitgord, N., Cusick, M. E., Radivojac, P., Uversky, V. N., Vidal, M., and Iakoucheva, L. M. Intrinsic disorder is a common feature of hub proteins from four eukaryotic interactomes. *PLoS Comput Biol.* **2**, e100 (2006).

Henken, D. B., Rasooly, R. S., Javois, L., and Hewitt, A. T., the National Institutes of Health Trans-NIH Zebrafish Coordinating Committee. The National Institutes of Health and the growth of the zebrafish as an experimental model organism. *Zebrafish* **1**, 105–110 (2004).

Hentges, K. E., and Justice, M. J. Checks and balancers: Balancer chromosomes to facilitate genome annotation. *Trends Genet.* **20**, 252–259 (2004).

Hermjakob, H., et al. The HUPO PSI's molecular interaction format: A community standard for the representation of protein interaction data. *Nat. Biotechnol.* **22**, 177–183 (2004).

Hernández, H., Dziembowski, A., Taverner, T., Séraphin, B., and Robinson, C. V. Subunit architecture of multimeric complexes isolated directly from cells. *EMBO Rep.* **7**, 605–610 (2006).

Ho, Y., et al. Systematic identification of protein complexes in *Saccharomyces cerevisiae* by mass spectrometry. *Nature* **415**, 180–183 (2002).

Horn, T., Arziman, Z., Berger, J., and Boutros, M. GenomeRNAi: A database for cell-based RNAi phenotypes. *Nucleic Acids Res.* **35**, D492–D497 (2007).

International Mouse Knockout Consortium, Collins, F. S., Rossant, J., and Wurst, W. A mouse for all reasons. *Cell* **128**, 9–13 (2007).

Ito, T., et al. A comprehensive two-hybrid analysis to explore the yeast protein interactome. *Proc. Natl. Acad. Sci. USA* **98**, 4569–4574 (2001).

Kallioniemi, O. P., Wagner, U., Kononen, J., and Sauter, G. Tissue microarray technology for high-throughput molecular profiling of cancer. *Hum. Mol. Genet.* **10**, 657–662 (2001).

Kamath, R. S., et al. Systematic functional analysis of the *Caenorhabditis elegans* genome using RNAi. *Nature* **421**, 231–237 (2003).

Kanehisa, M., Araki, M., Goto, S., Hattori, M., Hirakawa, M., Itoh, M., Katayama, T., Kawashima, S., Okuda, S., Tokimatsu, T., and Yamanishi, Y. KEGG for linking genomes to life and the environment. *Nucleic Acids Res.* **36**, D480–D484 (2008).

Karp, P. D. Pathway databases: A case study in computational symbolic theories. *Science* **293**, 2040–2044 (2001).

Karp, P. D., et al. Multidimensional annotation of the *Escherichia coli* K-12 genome. *Nucleic Acids Res.* **35**, 7577–7590 (2007).

Kerrien, S., et al. IntAct: Open source resource for molecular interaction data. *Nucleic Acids Res.* **35**, D561–D565 (2007).

Kile, B. T., et al. Functional genetic analysis of mouse chromosome 11. *Nature* **425**, 81–86 (2003).

Kim, J. K., et al. Functional genomic analysis of RNA interference in *C. elegans*. *Science* **308**, 1164–1167 (2005).

Kim, S. K., Lund, J., Kiraly, M., Duke, K., Jiang, M., Stuart, J. M., Eizinger, A., Wylie, B. N., and Davidson, G. S. A gene expression map for *Caenorhabditis elegans*. *Science* **293**, 2087–2092 (2001).

Kononen, J., Bubendorf, L., Kallioniemi, A., Barlund, M., Schraml, P., Leighton, S., Torhorst, J., Mihatsch, M. J., Sauter, G., and Kallioniemi, O. P. Tissue microarrays for high-throughput molecular profiling of tumor specimens. *Nat. Med.* **4**, 844–847 (1998).

Knowlton, M. N., Li, T., Ren, Y., Bill, B. R., Ellis, L. B., and Ekker, S. C. A PATO-compliant zebrafish screening database (MODB): Management of morpholino knockdown screen information. *BMC Bioinformatics* **9**, 7 (2008).

Koutsos, A. C., Blass, C., Meister, S., Schmidt, S., MacCallum, R. M., Soares, M. B., Collins, F. H., Benes, V., Zdobnov, E., Kafatos, F. C., and Christophides, G. K. Life cycle transcriptome of the malaria mosquito *Anopheles gambiae* and comparison with the fruitfly *Drosophila melanogaster*. *Proc. Natl. Acad. Sci. USA* **104**, 11304–11309 (2007).

Krogan, N. J., et al. Global landscape of protein complexes in the yeast *Saccharomyces cerevisiae*. *Nature* **440**, 637–643 (2006).

Kumar, A., et al. The TRIPLES database: A community resource for yeast molecular biology. *Nucleic Acids Res.* **30**, 73–75 (2002).

Kuriyan, J., and Eisenberg, D. The origin of protein interactions and allostery in colocalization. *Nature* **450**, 983–990 (2007).

Lehner, B., Fraser, A. G., and Sanderson, C. M. Technique review: How to use RNA interference. *Brief. Funct. Genomic. Proteomic.* **3**, 68–83 (2004).

Li, D., Li, J., Ouyang, S., Wang, J., Wu, S., Wan, P., Zhu, Y., Xu, X., and He, F. Protein interaction networks of *Saccharomyces cerevisiae, Caenorhabditis elegans* and *Drosophila melanogaster*: Large-scale organization and robustness. *Proteomics* **6**, 456–461 (2006).

Li, Q., Barkess, G., and Qian, H. Chromatin looping and the probability of transcription. *Trends Genet.* **22**, 197–202 (2006).

Li, X. J., Li, S. H., Sharp, A. H., Nucifora, F. C. Jr., Schilling, G., Lanahan, A., Worley, P., Snyder, S. H., and Ross, C. A. A huntingtin-associated protein enriched in brain with implications for pathology. *Nature* **378**, 398–402 (1995).

Lucretius. *De Rerum Natura Libri Sex*. John Baskervile, Birmingham, 1772.

Ma, Y., Creanga, A., Lum, L., and Beachy, P. A. Prevalence of off-target effects in *Drosophila* RNA interference screens. *Nature* **443**, 359–363 (2006).

MacBeath, G. Protein microarrays and proteomics. *Nat. Genet.* Supplement **32**, 526–532 (2002).

Mahajan, M. C., and Weissman, S. M. Multi-protein complexes at the beta-globin locus. *Brief. Funct. Genomic. Proteomic.* **5**, 62–65 (2006).

Marcotte, E. M., et al. Detecting protein function and protein-protein interactions from genome sequences. *Science* **285**, 751–753 (1999a).

Marcotte, E. M, Pellegrini, M., Thompson, M. J., Yeates, T. O., and Eisenberg, D. A combined algorithm for genome-wide prediction of protein function. *Nature* **402**, 83–86 (1999b).

Martin, S. E., and Caplen, N. J. Applications of RNA interference in mammalian systems. *Annu. Rev. Genomics Hum. Genet.* **8**, 81–108 (2007).

Mathivanan, S., Periaswamy, B., Gandhi, T. K., Kandasamy, K., Suresh, S., Mohmood, R., Ramachandra, Y. L., and Pandey, A. An evaluation of human protein-protein interaction data in the public domain. *BMC Bioinformatics* **7**(Suppl. 5), S19 (2006).

Mika, S., and Rost, B. Protein-protein interactions more conserved within species than across species. *PLoS Comput. Biol.* **2**, e79 (2006).

Molloy, M. P., and Witzmann, F. A. Proteomics: Technologies and applications. *Brief. Funct. Genomic. Proteomic.* **1**, 23–39 (2002).

Muller, H. J. Genetic variability, twin hybrids and constant hybrids, in a case of balanced lethal factors. *Genetics* **3**, 422–499 (1918).

Nord, A. S., et al. The International Gene Trap Consortium Website: A portal to all publicly available gene trap cell lines in mouse. *Nucleic Acids Res.* **34**, D642–D648 (2006).

Novick, P., Field, C., and Schekman, R. Identification of 23 complementation groups required for post-translational events in the yeast secretory pathway. *Cell* **21**, 205–215 (1980).

Novick, P., and Schekman, R. Secretion and cell-surface growth are blocked in a temperature-sensitive mutant of *Saccharomyces cerevisiae*. *Proc. Natl. Acad. Sci. USA* **76**, 1858–1862 (1979).

O'Brien, T. P., and Frankel, W. N. Moving forward with chemical mutagenesis in the mouse. *J. Physiol.* **554**, 13–21 (2004).

Ooi, S. L., Pan, X., Peyser, B. D., Ye, P., Meluh, P. B., Yuan, D. S., Irizarry, R. A., Bader, J. S., Spencer, F. A., and Boeke, J. D. Global synthetic-lethality analysis and yeast functional profiling. *Trends Genet.* **22**, 56–63 (2006).

Palcy, S., and Chevet, E. Integrating forward and reverse proteomics to unravel protein function. *Proteomics* **6**, 5467–5480 (2006).

Pan, X., Ye, P., Yuan, D. S., Wang, X., Bader, J. S., and Boeke, J. D. A DNA integrity network in the yeast *Saccharomyces cerevisiae*. *Cell* **124**, 1069–1081 (2006).

Pan, X., Yuan, D. S., Ooi, S. L., Wang, X., Sookhai-Mahadeo, S., Meluh, P., and Boeke, J. D. dSLAM analysis of genome-wide genetic interactions in *Saccharomyces cerevisiae*. *Methods* **41**, 206–221 (2007).

Pickart, M. A., Sivasubbu, S., Nielsen, A. L., Shriram, S., King, R. A., and Ekker, S. C. Functional genomics tools for the analysis of zebrafish pigment. *Pigment Cell Res.* **17**, 461–470 (2004).

Reed, J. L., Famili, I., Thiele, I., and Palsson, B. O. Towards multidimensional genome annotation. *Nat. Rev. Genet.* **7**, 130–141 (2006).

Reguly, T., et al. Comprehensive curation and analysis of global interaction networks in *Saccharomyces cerevisiae*. *J. Biol.* **5**, 11 (2006).

Riley, H. P. Introduction to Genetics and Cytogenetics. John Wiley & Sons, New York, 1948.

Robinson, C. V., Sali, A., and Baumeister, W. The molecular sociology of the cell. *Nature* **450**, 973–982 (2007).

Rogers, A., et al. WormBase 2007. *Nucleic Acids Res.* **36**, D612–D617 (2008).

Roma, G., Sardiello, M., Cobellis, G., Cruz, P., Lago, G., Sanges, R., and Stupka, E. The UniTrap resource: Tools for the biologist enabling optimized use of gene trap clones. *Nucleic Acids Res.* **36**, D741–D746 (2008).

Ross-Macdonald, P. Forward in reverse: How reverse genetics complements chemical genetics. *Pharmacogenomics* **6**, 429–434 (2005).

Ross-Macdonald, P., et al. Large-scale analysis of the yeast genome by transposon tagging and gene disruption. *Nature* **402**, 413–418 (1999).

Ruepp, A., Brauner, B., Dunger-Kaltenbach, I., Frishman, G., Montrone, C., Stransky, M., Waegele, B., Schmidt, T., Doudieu, O. N., Stümpflen, V., and Mewes, H. W. CORUM: The comprehensive resource of mammalian protein complexes. *Nucleic Acids Res.* **36**, D646–D650 (2008).

Russell, W. L., Kelly, E. M., Hunsicker, P. R., Bangham, J. W., Maddux, S. C., and Phipps, E. L. Specific-locus test shows ethylnitrosourea to be the most potent mutagen in the mouse. *Proc. Natl. Acad. Sci. USA* **76**, 5818–5819 (1979).

Sachidanandam, R. RNAi: Design and analysis. *Curr. Prot. Bioinf.* **12**, 12.3.1–12.3.10 (2004).

Schächter, V. Bioinformatics of large-scale protein interaction networks. *Computational Proteomics. A Supplement to BioTechniques*, 16–27 (2002).

Scherens, B., and Goffeau, A. The uses of genome-wide yeast mutant collections. *Genome Biol.* **5**, 229 (2004).

Schulze, T. G., and McMahon, F. J. Defining the phenotype in human genetic studies: Forward genetics and reverse phenotyping. *Hum. Hered.* **58**, 131–138 (2004).

Sharom, J. R., Bellows, D. S., and Tyers, M. From large networks to small molecules. *Curr. Opin. Chem. Biol.* **8**, 81–90 (2004).

Shehee, W. R., Oliver, P., and Smithies, O. Lethal thalassemia after insertional disruption of the mouse major adult β-globin gene. *Proc. Natl. Acad. Sci. USA* **90**, 3177–3181 (1993).

Sims, D., Bursteinas, B., Gao, Q., Zvelebil, M., and Baum, B. FLIGHT: Database and tools for the integration and cross-correlation of large-scale RNAi phenotypic datasets. *Nucleic Acids Res.* **34**, D479–D483 (2006).

Skarnes, W. C., et al. A public gene trap resource for mouse functional genomics. *Nat. Genet.* **36**, 543–544 (2004).

Smith, V., Botstein, D., and Brown, P. O. Genetic footprinting: A genomic strategy for determining a gene's function given its sequence. *Proc. Natl. Acad. Sci. USA* **92**, 6479–6483 (1995).

Smith, V., Chou, K. N., Lashkari, D., Botstein, D., and Brown, P. O. Functional analysis of the genes of yeast chromosome V by genetic footprinting. *Science* **274**, 2069–2074 (1996).

Sönnichsen, B., et al. Full-genome RNAi profiling of early embryogenesis in *Caenorhabditis elegans*. *Nature* **434**, 462–469 (2005).

Sprague, J., et al. The Zebrafish Information Network: The zebrafish model organism database provides expanded support for genotypes and phenotypes. *Nucleic Acids Res.* **36**, D768–D772 (2008).

Sprinzak, E., Altuvia, Y., and Margalit, H. Characterization and prediction of protein-protein interactions within and between complexes. *Proc. Natl. Acad. Sci. USA* **103**, 14718–14723 (2006).

Stanford, W. L., Epp, T., Reid, T., and Rossant, J. Gene trapping in embryonic stem cells. *Methods Enzymol.* **420**, 136–162 (2006).

Su, C., Peregrin-Alvarez, J. M., Butland, G., Phanse, S., Fong, V., Emili, A., and Parkinson, J. Bacteriome.org: An integrated protein interaction database for *E. coli*. *Nucleic Acids Res.* **36**, D632–D636 (2008).

Suthram, S., Shlomi, T., Ruppin, E., Sharan, R., and Ideker, T. A direct comparison of protein interaction confidence assignment schemes. *BMC Bioinformatics* 7, 360 (2006).

Swarbreck, D., et al. The *Arabidopsis* Information Resource (TAIR): Gene structure and function annotation. *Nucleic Acids Res.* **36**, D1009–D1014 (2008).

Tong, A. H., and Boone, C. Synthetic genetic array analysis in *Saccharomyces cerevisiae*. *Methods Mol. Biol.* **313**, 171–192 (2006).

Tong, A. H., et al. Global mapping of the yeast genetic interaction network. *Science* **303**, 808–813 (2004).

Tong, A. H., et al. Systematic genetic analysis with ordered arrays of yeast deletion mutants. *Science* **294**, 2364–2368 (2001).

Tumpey, T. M., Basler, C. F., Aguilar, P. V., Zeng, H., Solórzano, A., Swayne, D. E., Cox, N. J., Katz, J. M., Taubenberger, J. K., Palese, P., and García-Sastre, A. Characterization of the reconstructed 1918 Spanish influenza pandemic virus. *Science* **310**, 77–80 (2005).

Uetz, P., et al. A comprehensive analysis of protein–protein interactions in *Saccharomyces cerevisiae*. *Nature* **403**, 623–627 (2000).

van der Weyden, L., Adams, D. J., and Bradley, A. Tools for targeted manipulation of the mouse genome. *Physiol. Genomics* **11**, 133–164 (2002).

van Zon, A., Mossink, M. H., Scheper, R. J., Sonneveld, P., and Wiemer, E. A. The vault complex. *Cell. Mol. Life Sci.* **60**, 1828–1837 (2003).

Verhage, M., et al. Synaptic assembly of the brain in the absence of neurotransmitter secretion. *Science* **287**, 864–869 (2000).

Weismann, A. *The Germ-Plasm: A Theory of Heredity.* Walter Scott, Ltd., London, 1893.

Wilming, L. G., Gilbert, J. G., Howe, K., Trevanion, S., Hubbard, T., and Harrow, J. L. The vertebrate genome annotation (Vega) database. *Nucleic Acids Res.* **36**, D753–D760 (2008).

Wilson, R. J., Goodman, J. L., Strelets, V. B., the FlyBase Consortium. FlyBase: Integration and improvements to query tools. *Nucleic Acids Res.* **36**, D588–D593 (2008).

Winzeler, E. A., et al. Functional characterization of the *S. cerevisiae* genome by gene deletion and parallel analysis. *Science* **285**, 901–906 (1999).

Part III

Genome Analysis

PEDIGREE OF MAN

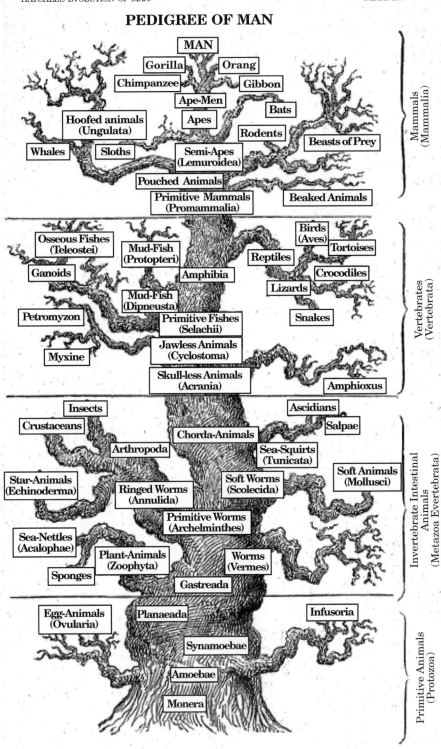

The tree of life from Ernst Haeckel (1879). The figure shows mammals (with humans at the top shown ascending from apes), vertebrates, invertebrates, and primitive animals at the bottom, including Monera (bacteria). (Reproduced with permission of the Institute of the History of Medicine, The Johns Hopkins University.)

13

Completed Genomes

The affinities of all the beings of the same class have sometimes been represented by a great tree. I believe this simile largely speaks the truth. The green and budding twigs may represent existing species; and those produced during each former year may represent the long succession of extinct species. ... The limbs divided into great branches, and these into lesser and lesser branches, were themselves once, when the tree was small, budding twigs; and this connexion of the former and present buds by ramifying branches may well represent the classification of all extinct and living species in groups subordinate to groups. ... From the first growth of the tree, many a limb and branch has decayed and dropped off, and these lost branches of various sizes may represent those whole orders, families, and genera which have now no living representatives, and which are known to us only from having been found in a fossil state. ... As buds give rise by growth to fresh buds, and these, if vigorous, branch out and overtop on all a feebler branch, so by generation I believe it has been with the Tree of Life, which fills with its dead and broken branches the crust of the earth, and covers the surface with its ever branching and beautiful ramifications.

—Charles Darwin, *The Origin of Species* (1859)

INTRODUCTION

A genome is the collection of DNA that comprises an organism. Each individual organism's genome contains the genes and other DNA elements that ultimately define its identity. Genomes range in size from the smallest viruses, which encode fewer than 10 genes, to eukaryotes such as humans that have billions of base pairs of DNA encoding tens of thousands of genes.

The recent sequencing of genomes from all branches of life—including viruses, prokaryotes, fungi, nematodes, plants, and humans—presents us with an extraordinary moment in the history of biology. By analogy, this situation resembles the completion of the periodic table of the elements in the nineteenth century. As it became clear that the periodic table could be arranged in rows and columns, it became possible to predict the properties of individual elements. A logic emerged to explain the properties of the elements. But it still took another century to grasp the significance of the elements and to realize the potential of the organization inherent in the periodic table.

Today we have sequenced the DNA from hundreds of genomes, and we are now searching for a logic to explain their organization and function. This process will take decades. A variety of tools must be applied, including bioinformatics approaches, biochemistry, genetics, and cell biology.

This chapter introduces the tree of life and the sequencing of genomes. We will then proceed to assess the progress in studying the genomes of viruses (Chapter 14); prokaryotes (bacteria and archaea) (Chapter 15); the eukaryotic chromosome (Chapter 16); fungi, including the yeast *Saccharomyces cerevisiae* (Chapter 17); an assortment of eukaryotes from parasites to primates (Chapter 18); and finally the human genome (Chapters 19 and 20). For definitions of several key terms related to the tree of life, see Table 13.1.

TABLE 13-1   Nomenclature for Tree of Life

Name[a]	Synonym(s)	Definition
Archaea (singular: archaeon)	Archaebacteria	One of the three "urkingdoms" or "domains" of life
Bacteria	Eubacteria; Monera (obsolete name)	One of the three "urkingdoms" or "domains" of life; unicellular organisms characterized by lack of a nuclear membrane
Eukaryotes	Eucarya	One of the three "urkingdoms" or "domains" of life; cells characterized by a nuclear membrane
Microbe	—	Microorganisms that cause disease in humans; microbes include bacteria and eukaryotes such protozoa and fungi
Microorganism	—	Unicellular life forms of microscopic size, including bacteria, archaea, and some eukaryotes
Progenote	Last universal common ancestor	The ancient, unicellular life form from which the three domains of life are descended
Prokaryotes	Prokaryotes; formerly synonymous with bacteria	Organism lacking a nuclear membrane; bacteria and archaea

[a]Name refers to the name adopted in this book. See Woese et al. (1990).

Five Perspectives on Genomics

The field of genomics can be surveyed from many perspectives, including the following five:

Perspective 1: Catalog genomic information. What are the basic features of each genome? These include its size; the number of chromosomes; the guanine plus cytosine (GC) content; the presence of isochores (described in Chapter 16); the number of genes, both coding and noncoding; repetitive DNA; and unique features of each genome. The techniques used to answer these questions include topics we introduce in this chapter: genomic DNA sequencing, assembly, and genome annotation including gene prediction. Genome browsers represent a major resource to access catalogs of genomic information, organized into categories such as raw underlying DNA data, as well as models of genes, regulatory elements, and other features of the genomic landscape.

Perspective 2: Catalog comparative genomic information. Our understanding of any genome is dramatically enhanced through comparisons to related genomes (Miller et al., 2004). When did a given species diverge from its relatives? Which genes or other DNA elements are orthologous, or share conserved synteny (Chapter 16)? To what extent did lateral gene transfer occur in each genome? Techniques of comparative genomics used to address these issues include whole genome alignment and analyses with databases such as the UCSC Genome Browser (Karolchik et al., 2007) and Clusters of Orthologous Groups of Proteins (COGs, Chapter 15). This approach also includes phylogenetic reconstruction (Chapter 7).

Perspective 3: Biological principles. For each genome, what are the functions of the organism (e.g., with respect to development, metabolism, and behavior) and how are they served by the genome? What are the mechanisms of evolution of the genome? This includes consideration of how genome size is regulated, whether there is polyploidization (Chapter 16), how the birth and death of genes occurs, and what forces operate on DNA whether they involve positive or negative selection or neutral evolution. What forces shape speciation? What is the role of epigenetics (Chapter 16)? Some of the many techniques used to address these issues include molecular phylogeny (Chapter 7) and BLAST or related tools (Chapters 4 and 5).

Perspective 4: Human disease relevance. What are the mechanisms by which organisms such as viruses or protozoan pathogens cause disease in humans or plants? What are the types of genomic responses and defenses that organisms have to prevent or adapt to avoid becoming subject to disease? What is the genetic basis of autism or arthritis? A variety of techniques are applied to these questions, including the study of single nucleotide polymorphisms (SNPs, Chapter 16) and linkage and association studies (Chapter 20).

Perspective 5: Bioinformatics aspects. What are some of the key databases and websites associated with each genome, and what algorithms have been developed to facilitate the analysis and visualization of data? The functionality of genome browsers has been greatly enhanced in recent years, providing a system with which to store, analyze, and interpret hundreds of categories of genomic data.

For a course on genomics that I teach, we discuss genomes across the tree of life from these five perspectives. Each student selects any genome of interest and writes a report describing the genome according to these approaches. Students may identify an outstanding research problem and describe how genomics approaches are being

Web document 13.1 at ▶ http://www.bioinfbook.org/chapter13 presents a table of these perspectives on genomics, and web document 13.2 outlines the details of a project to analyze a gene in depth from a genomics perspective.

applied to solve it. A related project is to select a single gene of interest and analyze it in depth, again following these five areas.

Brief History of Systematics

Throughout recorded history, philosophers and scientists have grappled with questions regarding the diversity of life on Earth (Mayr, 1982). Aristotle (384–322 B.C.E.) was an active biologist, describing over 500 species in his zoological works. He did not create a general classification scheme for life, but he did describe animals as "blooded" or "bloodless" in his *Historia animalium*. (Eventually, Lamarck [1744–1829] renamed these categories "vertebrates" and "invertebrates.") Aristotle's division of animals into genera and species provides the origin of the taxonomic system we use today.

The greatest advocate of this binomial nomenclature system of genus and species for each organism was the Swedish naturalist Carl Linnaeus (1707–1778). Linnaeus also introduced the notion of the three kingdoms, Animaliae, Plantae, and Mineraliae; in his hierarchical system the four levels were class, order, genus, and species. Ernst Haeckel (1834–1919), who described over 4000 new species, enlarged this system. He described life as a continuum from mere complex molecules to plants and animals, and he described the Moner as formless clumps of life. The monera were later named bacteria, and in 1937 Edouard Chatton made the distinction between prokaryotes (bacteria that lack nuclei) and eukaryotes (organisms with cells that have nuclei). By the end of the 1960s the work of Haeckel (1879), Copeland, Whittaker (1969), and many others led to the standard five-kingdom system of life: animals, plants, single-celled protists, fungi, and monera. Whittaker's 1969 scheme shows monera at the base of the tree representing the prokaryotes, and then eukaryotes (either unicellular or multicellular) represented by the Protista, Plantae, Fungi, and Animalia. An example of the tree of life, from an 1879 book by Haeckel, is shown in the frontis to this chapter.

The tree of life was rewritten in the 1970s and 1980s by Carl Woese and colleagues (Fox et al., 1980; Woese, 1998; Woese et al., 1990). They studied a group of prokaryotes that were presumed to be bacteria because they were single-celled life forms that lack a nucleus. The researchers sequenced small-subunit ribosomal RNAs (SSU rRNA) and performed phylogenetic analyses. This revealed that archaea are as closely related to eukaryotes as they are to bacteria. A phylogenetic analysis of SSU rRNA sequences, which are present in all known life forms, provides one version of the tree of life (Fig. 13.1). There are three main branches. While the exact root of the tree is not known, the deepest branching bacteria and archaea are thermophiles, suggesting that life may have originated in a hot environment.

Many groups have reconstructed the tree of life using large number of taxa and/or concatenations of large numbers of protein (or DNA or RNA) sequences (e.g., Driskell et al., 2004; Ciccarelli et al., 2006). While the tree of life provides an appealing metaphor, there are other global descriptions of life forms such as a bush or reticulated tree (Doolittle, 1999) or a ring of life (Rivera and Lake, 2004). These alternate metaphors attempt to account for the lateral transfer of genetic material between organisms (discussed in Chapter 15) as well as the fusion of ancient genomes.

Viruses do not meet the definition of living organisms, and thus they are excluded from most trees of life. Although they replicate and evolve, viruses only survive by commandeering the cell of a living organism (see Chapter 14).

A species is a group of similar organisms that only breed with one another, under normal conditions. A genus may consist of between one and hundreds of species.

The remarkable tree of life by Ciccarelli et al. (2006), from the group of Peer Bork, is available online at the Interactive Tree of Life webpage at ► http://itol.embl.de/ (see Letunic and Bork, 2007). Another extraordinary tree based on ribosomal RNA from about 3000 species is available from David Hillis and James Bull at ► http://www.zo.utexas.edu/faculty/antisense/DownloadfilesToL.html.

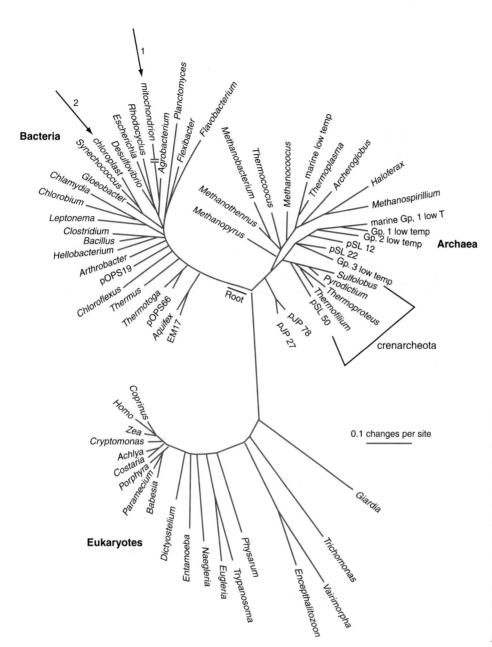

FIGURE 13.1. *A global tree of life, based on phylogenetic analysis of small-subunit rRNA sequences (modified from Barns et al., 1996 and Pace, 1997). Life is thought to have originated about 3.8 BYA in an anaerobic environment. The primordial life form (progenote) displayed the defining features of life (self-replication and evolution). The eukaryotic mitochondrion (arrow 1) and chloroplast (arrow 2) are indicated, showing their bacterial origins. Used with permission.*

History of Life on Earth

Our recent view of the tree of life (Fig. 13.1) is accompanied by new interpretations of the history of life on Earth. All life forms share a common origin and are part of the tree of life. A species has an average half-life of 1 to 10 million years (Graur and Li, 2000), and more than 99% of all species that ever lived are now extinct (Wilson, 1992). In principle, there is one single tree of life that accurately describes the evolution of species. The object of phylogeny is to try to deduce the correct trees both for species and for homologous families of genes and proteins. Another object of phylogeny is to infer the time of divergence between organisms since the time they last shared a common ancestor.

An overview of the history of life is shown in Fig. 13.2. The earliest evidence of life is from about 4 billion years ago (BYA), just 0.5 billion years after the formation of

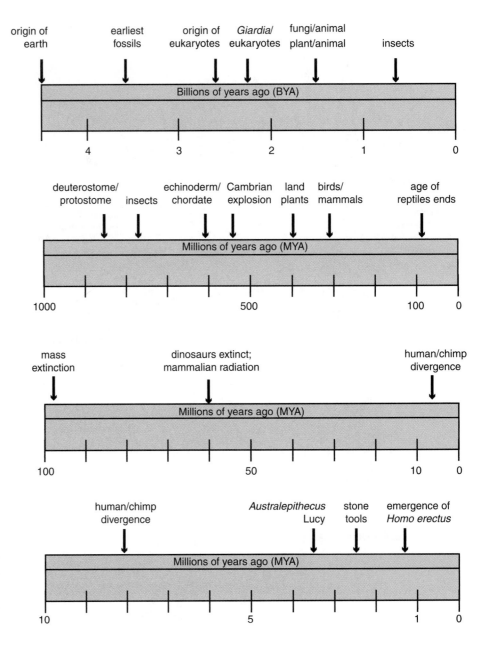

FIGURE 13.2. *History of life on the planet. Sources include Kumar and Hedges (1998), Hedges et al. (2001), and Benton and Ayala (2003).*

Multicellular organisms evolved independently many times. A variety of multicellular bacteria evolved several billion years ago, allowing selective benefits in feeding and in dispersion from predators (Kaiser, 2001).

Earth. This earliest life was centered on RNA (rather than DNA or protein; reviewed in Joyce, 2002). Earth's atmosphere was anaerobic throughout much of early evolution, and this early life form was presumably a unicellular prokaryote. The first fossil evidence of life is dated about 3.5 to 3.8 BYA (e.g., Allwood et al., 2006). The last common ancestor of life, predating the divergence of the lineage that leads to modern bacteria and modern archaea, was probably a hyperthermophile. This is suggested by the deepest branching organisms of trees (see Fig. 13.1), such as the bacterium *Aquifex* and the hyperthermophilic crenarcheota (Chapter 15). Eukaryotes appeared between 2 and 3 BYA and remained unicellular until almost 1 BYA. Approximately 1.5 BYA, plants and animals diverged, as did fungi from the lineage that gave rise to metazoans (animals) (see Fig. 18.11). The most recent billion years of life has seen the evolution of an enormous variety of multicellular organisms. The so-called Cambrian explosion of 550 million years ago (MYA)

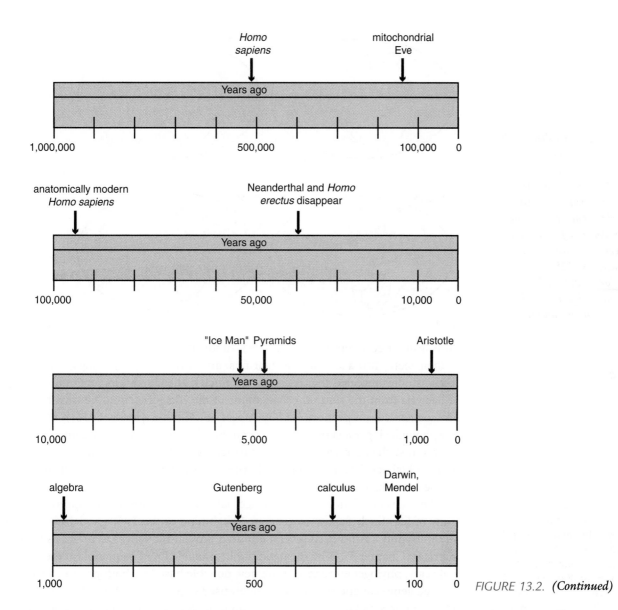

FIGURE 13.2. *(Continued)*

witnessed a tremendous increase in the diversity of animal life forms. In the past 250 million years, the continents coalesced into a giant continent, Pangaea (Fig. 13.3). When Pangaea separated into northern and southern supercontinents (Laurasia and Gondwana), this created natural barriers to reproduction and influenced subsequent evolution of life. By 60 MYA, the dinosaurs were extinct, and the mammalian radiation was well underway.

The lines leading to modern *Homo sapiens*, chimpanzees, and bonobos diverged about 5 MYA (Chapter 18). The earliest human ancestors include "Lucy," the early *Australepithecus*, and early hominids used stone tools over 2 MYA. Further features of recent historical interest are indicated in Fig. 13.2.

Molecular Sequences as the Basis of the Tree of Life

In past decades and centuries, the basis for proposing models of the tree of life was primarily morphology. Thus, Linnaeus divided animals into six classes (mammals,

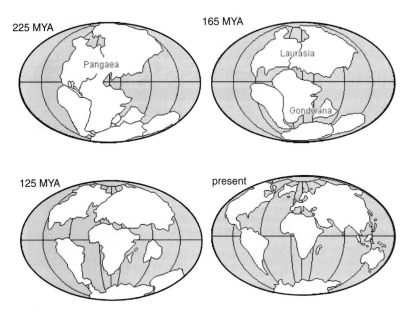

FIGURE 13.3. Geological history of the earth from 225 MYA. At that time, there was one supercontinent, Pangaea. By 165 MYA, Pangaea had separated into Laurasia (modern Asia and North America) and Gondwana (modern Africa and South America). At 125 MYA Laurasia and Gondwana had both begun separations that led to the present divisions among continents.

The European Ribosomal RNA Database (Wuyts et al., 2004) (► http://bioinformatics.psb. ugent.be/webtools/rRNA/), although no longer updated, contains over 20,000 SSU and large subunit rRNA sequences (as of 2007). About 600 are archaeal species, 12,000 bacteria, and 6500 eukaryotes. You can use sequences from this database to generate phylogenetic trees or view trees at the website.

birds, fish, insects, reptiles, and worms), subdividing mammals according to features of their teeth, fish according to their fins, and insects by their wings. Early microscopic studies revealed that bacteria lack nuclei, allowing a fundamental separation of bacteria from the four other kingdoms of life. Bacteria could be classified based on biochemical properties (e.g., by Albert Jan Kluyver [1888–1956]), and from a morphological perspective bacteria can be classified into several major groups. However, such criteria are insufficient to appreciate the dazzling diversity of millions of microbial species. Thus, physical criteria were unavailable by which to discover archaea as a distinct branch of life.

The advent of molecular sequence data has transformed our approach to the study of life. Such data were generated beginning in the 1950s and 1960s, and by 1978 Dayhoff's Atlas used several hundred protein sequences as the basis for PAM matrices (Chapter 3). With the rapid rise in available DNA sequences of the past several years, phylogenetic analyses are now possible based on both phenotypic characters and gene sequences. The most widely used sequences are SSU rRNA molecules, which are present across virtually all extant life forms. The slow rate of evolution of SSU rRNAs and their convenient size makes them appropriate for phylogenetic analyses. Genome-sequencing efforts are now reshaping the field of evolutionary studies, providing thousands of DNA and protein sequences for phylogenetic trees.

Over 1000 prokaryotic genomes have now been sequenced (see below and Chapter 15). We are now beginning to appreciate lateral gene transfer (Chapter 15), a phenomenon in which a species does not acquire a particular gene by descent from an ancestor. Instead, it acquires the gene horizontally (or laterally) from another unrelated species. Thus, genes can be exchanged between species (Eisen, 2000). As a consequence, the use of different individual genes in molecular phylogeny often results in distinctly different tree topologies. Because of the phenomena of lateral gene transfer and gene loss, it might never be possible to construct a single tree of life that reflects the evolution of life on the planet (Wolf et al., 2002).

Role of Bioinformatics in Taxonomy

The field of bioinformatics is concerned with the use of computer algorithms and computer databases to elucidate the principles of biology. The domain of bioinformatics includes the study of genes, proteins, and cells in the context of organisms across the tree of life. Some have advocated a web-based taxonomy intended to catalog an inventory of life (Blackmore, 2002; Pennisi, 2001). Several projects attempt to create a tree of life (see sidebar). Others suggest that although web-based initiatives are useful, the current system is adequate: zoological, botanical, or other specimens are collected, named, and studied according to guidelines established by international conventions (Knapp et al., 2002).

The Convention on Biological Diversity (► http://www.biodiv. org/) and the Global Biodiversity Information Facility (► http:// www.gbif.net/) are examples of organizations that address issues of global biodiversity. The Tree of Life is at ► http://www. panspermia.org/tree.htm and the Tree of Life Web Project (created by David R. Maddison) is at ► http://tolweb.org/tree/ phylogeny.html.

Genome-Sequencing Projects: Overview

The advent of DNA-sequencing technologies in the 1970s, including Frederick Sanger's dideoxynucleotide methodology, enabled large-scale sequencing projects to be performed. This chapter provides a brief history of genome-sequencing projects, including the completion of the genomic sequence of the first free-living organism in 1995, *Haemophilus influenzae*. By 2001, a draft sequence of the human genome was reported by two groups. The most remarkable feature of current efforts to determine the sequence of complete genomes is the dramatic increase in data that are collected each year (Fig. 2.1). The ability to sequence first millions and now billions of nucleotides of genomic DNA presents the scientific community with unprecedented opportunities and challenges.

Several themes have emerged in the past several years:

- The amount of sequence data that are generated continues to accelerate rapidly.

- For many genomes, even unfinished genomic sequence data—that is, versions of genomic sequence that include considerable gaps and sequencing errors— are immediately available and useful to the scientific community. We will also see that a finished sequence (defined below) provides substantially better descriptions of genome features than does an unfinished sequence.

- The value of comparative genome analysis is now appreciated for solving problems such as identifying protein-coding genes in human and mouse or differences in virulent and nonvirulent strains of pathogens (Miller et al., 2004). Comparative analyses will also be useful to define gene regulatory regions and the evolutionary history of species through the analysis of conserved DNA elements.

Four Prominent Web Resources

We will introduce four main web resources for the study of genomes in this and the following chapters. (1) The European Bioinformatics Institute (EBI) offers a variety of genome databases (some of which are listed in Table 13.2). (2) The genomes section of Entrez at the National Center for Biotechnology Information (NCBI) is organized with a search feature on the top bar, links to eukaryotes, bacteria, archaea, and viruses on the left bar, and a variety of specialized genomics resources on the right sidebar (Fig. 13.4). The current NCBI holdings include over 1800 eukaryotic, bacterial, and archaeal genomes, of which 620 have been completely sequenced (Table 13.3). There are an additional 1400 completely sequenced organellar genomes (discussed below). (3) The Comprehensive Microbial Resource (CMR) focuses on prokaryotic projects (see Chapter 15). (4) The genome browser at the

You can access the NCBI Genomes page at NCBI via ► http://www.ncbi.nlm.nih.gov/ Genomes/, or from the home page of NCBI click All Databases then Genomes. The CMR, formerly of the Institute for Genomic Research (TIGR) and now of the J. Craig Venter Institute, is at ► http://cmr.jcvi.org/. The UCSC Genome Browser and Table Browser are at ► http:// genome.ucsc.edu.

TABLE 13-2	EBI Genome Projects	
Ensembl	Genome browser for ~24 mammalian genomes and many nonmammalian metazoan genomes	▶ http://www.ebi.ac.uk/ensembl/ or ▶ http://www.ensembl.org
Genome Reviews	Database of annotated complete genomes (emphasis on archaea and bacteria)	▶ http://www.ebi.ac.uk/ GenomeReviews/
Genomes Server	Gateway to completed genomes	▶ http://www.ebi.ac.uk/genomes/
Integr8	Proteome analysis	▶ http://www.ebi.ac.uk/integr8/ EBI-Integr8-HomePage.do or ▶ http://www.ebi.ac.uk/proteome

University of California, Santa Cruz has a particular emphasis on vertebrate genomes (see Chapters 16 and 18).

Brief Chronology

The progress in completing many hundreds of genome-sequencing projects has been rapid, and we can expect the pace to accelerate in the future. In the following sections we present a chronological overview to provide a framework for these events. When the sequencing of the first bacterial genomes was completed in 1995, there were

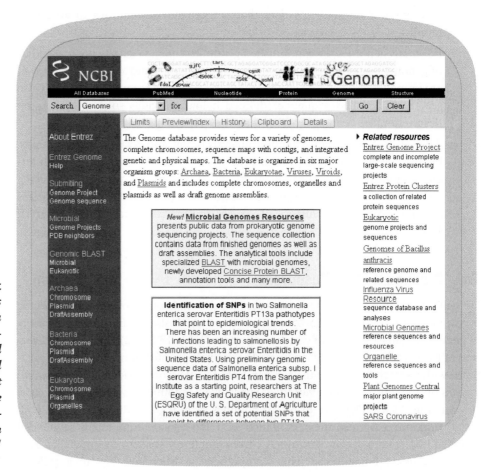

FIGURE 13.4. The Entrez Genome site of NCBI includes links to molecular sequence data from over 200,000 species (left sidebar). Related resources for selected organisms are also provided as well as tools for genome analysis (right sidebar). The nucleotide sequence is available for thousands of genomes. This page is accessible from ▶ http://www.ncbi.nlm.nih.gov/ Entrez/.

TABLE 13-3   Summary of Currently Sequenced Genomes (Excluding Viruses and Organellar Genomes)

	Organism	Complete	Draft Assembly	In Progress	Total
Prokaryotes		**597**	**397**	**492**	**1,486**
	Archaea	47	3	30	80
	Bacteria	550	394	462	1406
Eukaryotes		**23**	**131**	**184**	**338**
	Animals	4	54	89	147
	Mammals	2	22	25	49
	Birds		1	2	3
	Fishes		3	6	9
	Insects	1	19	20	40
	Flatworms		1	3	4
	Roundworms	1	3	13	17
	Amphibians			2	2
	Reptiles			2	2
	Other animals		6	19	25
	Plants	3	3	34	40
	Land plants	2	2	27	31
	Green algae	1	1	7	9
	Fungi	10	53	30	93
	Ascomycetes	8	46	21	75
	Basidiomycetes	1	5	5	11
	Other fungi	1	2	4	7
	Protists	6	19	27	52
	Apicomplexans	1	10	6	17
	Kinetoplasts	1	2	6	9
	Other protists	4	7	14	25
Total		**620**	**528**	**676**	**1,824**

Source: Entrez Genome at NCBI (▶ http://www.ncbi.nlm.nih.gov/genomes/static/gpstat.html), November 2007.

relatively few other genome sequences available for comparison. Now with over 2000 completed genomes available (including organellar genomes) we are better able to annotate and interpret the significance of genome sequences.

First Bacteriophage and Viral Genomes (1976–1978)

Bacteriophage are viruses that infect bacteria. Fiers et al. (1976) reported the first complete bacteriophage genome, MS2. This genome of 3569 base pairs encodes just four genes. The next complete virus genome was Simian virus 40 (SV40) by Fiers et al. (1978). That genome contains 5224 base pairs and contains eight genes (seven of which encode proteins).

The bacteriophage MS2 genome has RefSeq accession NC_001417. The SV40 RefSeq accession is NC_001669.

Frederick Sanger and colleagues also sequenced the genome of bacteriophage φX174 (Sanger et al., 1977a). They developed several DNA-sequencing techniques, including the dideoxynucleotide chain termination procedure discussed below. Bacteriophage φX174 is 5386 base pairs (bp) encoding 11 genes (see GenBank accession J02482). A graphical depiction of a portion of this viral genome is shown in Fig. 13.5. At the time, the most surprising result was the unexpected presence of overlapping genes that are transcribed on different reading frames.

First Eukaryotic Organellar Genome (1981)

The first complete organellar genome to be sequenced was the human mitochondrion (Anderson et al., 1996). The genome is characterized by extremely little

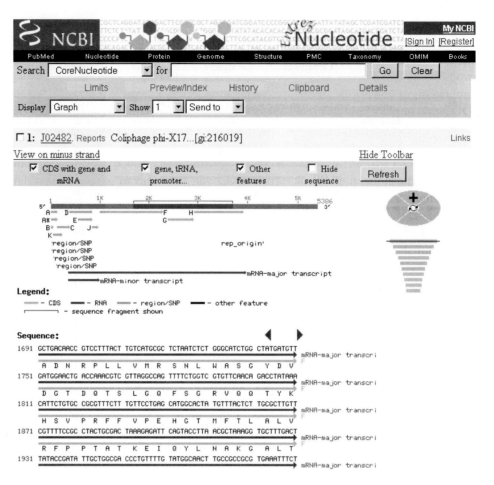

FIGURE 13.5. *Portion of the Entrez nucleotide record for bacteriophage φX174. This format was obtained by viewing the entry for accession J02482 in the graphics display format. This provides an overview of the predicted open reading frames (ORFs). A portion of the nucleotide sequence and corresponding predicted proteins is shown at the bottom.*

Information on organellar genomes is available at ▶ http://www.ncbi.nlm.nih.gov/genomes/ORGANELLES/organelles.html.

noncoding DNA. The great majority of metazoan (i.e., multicellular animal) mitochondrial genomes are about 15 to 20 kb (kilobase) circular genomes. The human mitochrondrial genome is 16,568 bp (base pairs) and encodes 13 proteins, 2 ribosomal RNAs, and 22 transfer RNAs. It can be accessed through the NCBI Entrez genome site (Fig. 13.6). This circular diagram is clickable, allowing you to access the individual genes. Thus, the DNA and corresponding protein sequences of all the mitochondrial genes are easily accessible.

Today, there are over over 1500 completed mitochondrial genome sequences. Several of these are listed in Table 13.4. This table also lists several exceptionally large cases. While the largest sequenced mitochrondrial genome is that of the thale cress *Arabidopsis thaliana* (367 kb), several plants reportedly have even larger mitochondrial genomes. Thus, there is a tremendous diversity of mitochondrial genomes (Lang et al., 1999). Molecular phylogenetic approaches suggest that mitochondria are descendants of an endosymbiotic α-proteobacterium, although it is possible that the origin of mitochondria in eukaryotes was coincident with the evolution of the nuclear genome (Lang et al., 1999).

First Chloroplast Genomes (1986)

The first chloroplast genomes were reported (*Nicotiana tabacum*; Shinozaki et al., 1986), followed by the liverwort *Marchantia polymorpha* (Ohyama et al., 1986).

(a)

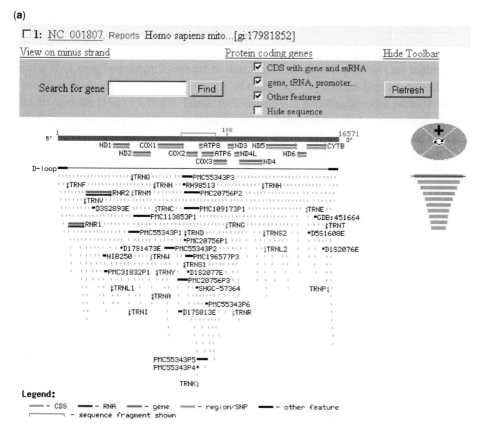

(b)   13 protein(s) shown

Legends:◆ DNA region in flatfile format ◆ DNA region in FASTA format ◆ Protein in FASTA format

Product Name	Start	End	Strand	Length	Gi	GeneID	Locus	Locus_tag	COG(s)	Links
NADH dehydrogenase subunit 1	3308	4264	+	318	17981853	4535	ND1	-	-	◆◆◆
NADH dehydrogenase subunit 2	4471	5512	+	347	17981854	4536	ND2	-	-	◆◆◆
cytochrome c oxidase subunit I	5905	7446	+	513	17981855	4512	COX1	-	-	◆◆◆
cytochrome c oxidase subunit II	7587	8270	+	227	17981856	4513	COX2	-	-	◆◆◆
ATP synthase F0 subunit 8	8367	8573	+	68	17981857	4509	ATP8	-	-	◆◆◆
ATP synthase F0 subunit 6	8528	9208	+	226	17981858	4508	ATP6	-	-	◆◆◆
cytochrome c oxidase subunit III	9208	9988	+	260	17981859	4514	COX3	-	-	◆◆◆
NADH dehydrogenase subunit 3	10060	10405	+	115	17981860	4537	ND3	-	-	◆◆◆
NADH dehydrogenase subunit 4L	10471	10767	+	98	17981861	4539	ND4L	-	-	◆◆◆
NADH dehydrogenase subunit 4	10761	12138	+	459	17981862	4538	ND4	-	-	◆◆◆
NADH dehydrogenase subunit 5	12338	14149	+	603	17981863	4540	ND5	-	-	◆◆◆
NADH dehydrogenase subunit 6	14150	14674	-	174	17981864	4541	ND6	-	-	◆◆◆
cytochrome b	14748	15882	+	378	17981865	4519	CYTB	-	-	◆◆◆

FIGURE 13.6. NCBI entry for the human mitochondrial genome (NC_001807; 16,568 bp). (a) Detailed information is available by clicking on a specific gene (e.g., COX1). (b) A link from the Entrez Genome entry for the mitochondrial genome page provides a list of all mitochondrial protein-coding genes.

Most plant chloroplast genomes are 120,000 to 200,000 bp in size. There are other chloroplast-like organelles in eukaryotic organisms. Unicellular protozoan parasites of the phylum Apicomplexa, such as *Toxoplasma gondii* (Table 13.5), have smaller plastid genomes.

We will discuss chloroplasts and other plastids in the plant section of Chapter 18.

First Eukaryotic Chromosome (1992)

The first eukaryotic chromosome was sequenced in 1992: chromosome III of the budding yeast *S. cerevisiae* (Oliver et al., 1992). There were 182 predicted open

TABLE 13-4 Selected Mitochondrial Genomes Arranged by Size

Kingdom	Species	Accession	Size (bp)
Eukaryote	*Plasmodium falciparum* (malaria parasite)	NC_002375	5,967
Metazoa (Bilateria)	*Caenorhabditis elegans* (worm)	NC_001328	13,794
Plant (Chlorophyta)	*Chlamydomonas reinhardtii* (green alga)	NC_001638	15,758
Metazoa (Bilateria)	*Mus musculus*	NC_001569	16,295
	Pan troglodytes (chimpanzee)	NC_001643	16,554
Metazoa	*Homo sapiens*	NC_001807	16,568
Metazoa (Cnidaria)	*Metridium senile* (sea anenome)	NC_000933	17,443
Metazoa (Bilateria)	*Drosophila melanogaster*	NC_001709	19,517
Fungi (Ascomycota)	*Schizosaccharomyces pombe*	NC_001326	19,431
Fungi	*Candida albicans*	NC_002653	40,420
Eukaryote (stramenopiles)	*Pylaiella littoralis* (brown alga)	NC_003055	58,507
Fungi (Chytridiomycota)	*Rhizophydium* sp. 136	NC_003053	68,834
Eukaryote	*Reclinomonas americana* (protist)	NC_001823	69,034
Fungi (Ascomycota)	*Saccharomyces cerevisiae*	NC_001224	85,779
Plant (Streptophyta)	*Arabidopsis thaliana*	NC_001284	366,923
Plant (Streptophyta)	*Zea mays* (corn)	NC_008332	680,603
Plant (Streptophyta)	*Tripsacum dactyloides*	NC_008362	704,100
	Cucumis melo	—	2,400,000

Source: As of November 2007, 1127 metazoan (multicellular animal) organellar genomes have been sequenced, 50 fungi, and 27 plants (see ▶ http://www.ncbi.nlm.nih.gov/Genomes/). For the size estimate of the muskmelon *C. melo*, see Lilly and Havey (2002)

We describe this bacterial genome as derived from a "free-living organism" to distinguish it from a viral genome or an organellar genome. Viruses (Chapter 14) exist on the borderline of the definition of life, and organellar genomes are usually derived from bacteria that are no longer capable of independent life.

reading frames (for proteins larger than 100 amino acids), and the size of the sequenced DNA was 315 kb. Of the 182 open reading frames that were identified, only 37 corresponded to previously known genes, and 29 showed similarity to known genes. We will explore this genome in Chapter 17.

Complete Genome of Free-Living Organism (1995)

The first genome of a free-living organism to be completed was the bacterium *H. influenzae* Rd (Fleischmann et al., 1995). Its size is 1,830,138 bp (i.e., 1.8 Mb

TABLE 13-5 Selected Chloroplast Genomes

Species	Common Name	Accession	Size (bp)
Arabidopsis thaliana	Thale cress	NC_000932	154,478
Guillardia theta	Red alga	NC_000926	121,524
Marchantia polymorpha	Liverwort; moss	NC_001319	121,024
Nicotiana tabacum	Tobacco	NC_001879	155,939
Oryza sativa	Rice	NC_001320	134,525
Porphyra purpurea	Red alga	NC_000925	191,028
Toxoplasma gondii	Apicomplexan parasite	NC_001799	34,996
Zea mays	corn	NC_001666	140,384

Genome > *cellular organisms* > *Haemophilus influenzae Rd KW20, complete genome*

Lineage: <u>cellular organisms</u>; <u>Bacteria</u>; <u>Proteobacteria</u>; <u>Gammaproteobacteria</u>; <u>Pasteurellales</u>; <u>Pasteurellaceae</u>; <u>Haemophilus</u>; <u>Haemophilus influenzae</u>;
<u>Haemophilus influenzae Rd KW20</u>

Genome Info:	Features:	BLAST homologs:	Links:	Review Info:
Refseq: <u>NC_000907</u>	Genes: <u>1789</u>	<u>COG</u>	<u>Genome Project</u>	Publications: [<u>5</u>]
GenBank: <u>L42023</u>	Protein coding: <u>1657</u>	<u>3D Structure</u>	<u>Refseq FTP</u>	Refseq Status: **Reviewed**
Length: **1,830,138 nt**	Structural RNAs: <u>81</u>	<u>TaxMap</u>	<u>GenBank FTP</u>	Seq.Status: **Completed**
GC Content: **38%**	Pseudo genes: **7**	<u>TaxPlot</u>	<u>BLAST</u>	Sequencing center: <u>TIGR</u>
% Coding: **84%**	Others: **54**	<u>GenePlot</u>	TraceAssembly	Completed: **2001/10/19**
Topology: **circular**	Contigs: <u>1</u>	<u>gMap</u>	<u>CDD</u>	Organism Group
Molecule: **dsDNA**			Other genomes for species	

Gene Classification based on <u>COG functional categories</u>   Search gene, GeneID or locus_tag: [_____] [Find Gene]

FIGURE 13.7. *Entrez Genome record for* H. influenzae *Rd, the first free-living organism for which the complete genomic sequence was determined. This record is obtained from the Entrez Genome resource by clicking "bacteria" on the left sidebar. The top of this entry includes information such as the accession number and the size of the genome. The entire nucleotide sequence is downloadable here. At top right are several resources for studying the 1657 proteins encoded by this genome. An* H. influenzae-*specific BLAST search is available here. At bottom right the entry includes a color-coded circular representation of the genome, along with a functional classification based on COG (Chapter 15). The circular map is clickable, showing detailed information on the genes and proteins in this genome. The record also contains literature references, including the initial report of this genomic sequence by Fleischmann et al. (1995) at the Institute for Genomic Research.*

[megabase pairs]). This organism was sequenced at the Institute for Genomic Research using the whole-genome shotgun sequencing and assembly strategy (see below).

To study this genome in NCBI, go to Entrez, click "genome," and then click "bacteria genome" from the left sidebar. Note that there are over 100 complete bacterial genomes listed; scroll down to *H. influenzae* and click on the accession number. (Note that you can also see the RefSeq accession number, NC_000907; the genome size [1,830,138 bp]; and the date entered, July 25 1995.) The page for this organism contains a wealth of information on the genes and encoded proteins as well as the predicted functional classification of the proteins (Fig. 13.7). This classification scheme, Clusters of Orthologous Groups (COG), will be discussed in Chapter 15. The lineage (Fig. 13.7, top) shows you that this is a bacterium in the gamma division. As with the mitochondrial genome (Fig. 13.6), the circular diagram of the *H. influenzae* genome is clickable to allow a detailed study of its genes and proteins.

By the end of 1995 (Table 13.6), the complete DNA sequence of a second bacterial genome had been obtained, *Mycoplasma genitalium*. Notably, this is one of the smallest known genomes of any free-living organism.

TABLE 13-6 Genome-Sequencing Projects Completed in 1995

Organism	Size (bp)	Accession	Reference
Haemophilus influenzae Rd	1,830,138	NC_000907	Fleischmann et al., 1995
Mycoplasma genitalium	580,074	NC_000908	Fraser et al., 1995

TABLE 13-7   Genome-Sequencing Projects Completed in 1996

Organism	Size	Accession	Reference
Methanococcus jannaschii (A)	1,664,976 bp[a]	NC_000909	Bult et al., 1996
Mycoplasma pneumoniae (B)	816,394 bp	NC_000912	Himmelreich et al., 1996; see Dandekar et al., 2000
Synechocystis PCC6803 (B)	3,573,470 bp	NC_000911	Kaneko et al., 1996
Saccharomyces cerevisiae (E)	12,068 kb	Various, for each chromosome	Goffeau et al., 1996

Abbreviations: A, archaeon; B, bacterium; E, eukaryote.
[a]The size does not include extrachromosomal elements (for *M. jannaschii*).

First Eukaryotic Genome (1996)

The complete genome of the first eukaryote, *S. cerevisiae* (a yeast; Chapter 17) (Goffeau et al., 1996), was sequenced by 1996 (Table 13.7). This was accomplished by a collaboration of over 600 researchers in 100 laboratories spread across Europe, North America, and Japan. To find information about this completed genome, go to Entrez genomes, then use the left sidebar to click "eukaryota."

In 1996, TIGR researchers reported the first complete genome sequence for an archaeon, *Methanococcus jannaschii* (Bult et al., 1996). This offered the first opportunity to compare the three main divisions of life, including the overall metabolic capacity of bacteria, archaea, and eukaryotes.

Escherichia coli (1997)

In 1997, the complete genomic sequences of two archaea were reported (Table 13.8). Of the five bacterial genomes that were reported, the most well known is that of *Escherichia coli* (Blattner et al., 1997; Koonin, 1997), which has served as a model organism in bacteriology for decades. Its 4.6 Mb genome encodes over 4200 proteins, of which 38% had no identified function at the time. We will explore this further in Chapter 15.

First Genome of Multicellular Organism (1998)

The nematode *Caenorhabditis elegans* was the first multicellular organism to have its genome sequenced—although technically, the sequencing is still not complete

TABLE 13-8   Genome-Sequencing Projects Completed in 1997

Organism	Size (bp)	Accession	Reference
Archaeoglobus fulgidus (A)	2,178,400	NC_000917	Klenk et al., 1997
Methanobacterium thermoautotrophicum (A)	1,751,377	NC_000916	Smith et al., 1997
Bacillus subtilis (B)	4,214,814	NC_000964	Kunst et al., 1997
Borrelia burgdorferi (B)	910,724	NC_001318	Fraser et al., 1997; Casjens et al., 2000
Escherichia coli (B)	4,639,221	NC_000913	Blattner et al., 1997
Helicobacter pylori 26695 (B)	1,667,867	NC_000915	Tomb et al., 1997; see Alm et al., 1999

Abbreviations: A, archaeon; B, bacterium.

TABLE 13-9	Genome-Sequencing Projects Completed in 1998		
Organism	Size	Accession	Reference
Pyrococcus horikoshii OT3 (A)	1,738,505 bp	NC_000961	Kawarabayasi et al., 1998
Aquifex aeolicus (B)	1,551,335 bp	NC_000918	Deckert et al., 1998
Chlamydia trachomatis (B)	1,042,519 bp	NC_000117	Stephens et al., 1998
Chlamydophila pneumoniae (B)	1,230,230 bp	NC_000922	
Mycobacterium tuberculosis (B)	4,411,529 bp	NC_000962	Cole et al., 1998
Rickettsia prowazekii (B)	1,111,523 bp	NC_000963	Andersson et al., 1998
Treponema pallidum (B)	1,138,011 bp	NC_000919	Fraser et al., 1998
Caenorhabditis elegans (E)	97 Mb	AE000001	*C. elegans* Sequencing Consortium, 1998
		AE000002	
		AE000003	
		AE000004	
		AE000005	
		AE000006	

Abbreviations: A, archaeon; B, bacterium; E, eukaryote.

(because of the presence of repetitive DNA elements that have been difficult to resolve). The sequence spans 97 Mb and is predicted to encode over 19,000 genes (*C. elegans* Sequencing Consortium, 1998).

Two more archaea brought the total to four sequenced genomes by 1998 (Table 13.9). Six more bacterial genomes were also sequenced. The genome of sequence of *Rickettsia prowazekii*, the α-proteobacterium that causes typhus and was responsible for tens of millions of deaths in the twentieth century, is very closely related to the eukaryotic mitochondrial genome (Andersson et al., 1998).

Human Chromosome (1999)

In 1999, the sequence of the euchromatic portion of human chromosome 22 was published (Table 13.10) (Dunham et al., 1999). This was the first human

TABLE 13-10	Genome-Sequencing Projects Completed in 1999		
Organism	Size (bp)	Accession	Reference
Aeropyrum pernix (A)	1,669,695	NC_000854	Kawarabayasi et al., 1999
Thermoplasma volvanium GSS1 (A)	1,584,804	NC_002689	Kawashima et al., 1999
Chlamydia pneumoniae (B)	1,229,858	NC_002179	Kalman et al., 1999
Deinococcus radiodurans (B)	2,648,638	NC_001263	White et al., 1999
	412,348	NC_001264	White et al., 1999
Helicobacter pylori J99 (B)	1,643,831	NC_000921	Alm et al., 1999; see Tomb et al., 1997
Thermotoga maritima (B)	1,860,725	NC_000853	Nelson et al., 1999
Homo sapiens chromosome 22 (E)	33,792,315	NT_001454, 10 other contigs	Dunham et al., 1999

Abbreviations: A, archaeon; B, bacterium; E, eukaryote.

chromosome to be essentially completely sequenced. We will describe the human genome in Chapter 19.

Fly, Plant, and Human Chromosome 21 (2000)

We describe these and other eukaryotic genomes in Chapter 18.

In this year, the completed genome sequences of the fruit fly *Drosophila melanogaster* and the plant *A. thaliana* were reported, bringing the number of eukaryotic genomes to four (with a yeast and a worm) (Table 13.11). The *Drosophila* sequence was obtained by scientists at Celera Genomics and the Berkeley Drosophila Genome Project (BDGP) (Adams et al., 2000). There are approximately 13,500 annotated

TABLE 13-11 Genome-Sequencing Projects Completed in 2000

Organism	Size	Accession	Reference
Halobacterium sp. NRC-1 (A)	2,014,239 bp	NC_002607	Ng et al., 2000
Thermoplasma acidophilum (A)	1,564,906 bp	NC_002578	Ruepp et al., 2000
Bacillus halodurans (B)	4,202,353 bp	BA000004	Takami et al., 2000
Buchnera sp. APS (B)	640,681 bp	NC_002528	
Campylobacter jejuni (B)	1,641,481 bp	NC_002163	Parkhill et al., 2000a
Chlamydia muridarum (B)	1,069,412 bp	NC_002178	Read et al., 2000
	1,228,267 bp	NC_002491	
Chlamydia pneumoniae AR39 (B)	1,229,853 bp	NC_002179	Read et al., 2000
Chlamydia trachomatis (B)	1,069,412 bp	NC_000117	Read et al., 2000; see Stephens et al., 1998
Neisseria meningitidis MC58 (B)	2,272,351 bp	NC_002183	Tettelin et al., 2000
Neisseria meningitidis Z2491 (B)	2,184,406 bp	NC_002203	Parkhill et al., 2000b
Pseudomonas aeruginosa (B)	6,264,403 bp	NC_002516	Stover et al., 2000
Ureaplasma urealyticum (B)	751,719 bp	NC_002162	Glass et al., 2000
Vibrio cholerae (B)	2,961,149 bp	NC_002505	Heidelberg et al., 2000
	1,072,315 bp	NC_002506	Heidelberg et al., 2000
Xylella fastidiosa (B)	2,679,306 bp	NC_002488	Simpson et al., 2000
Drosophila melanogaster (E)	137 Mb	NC_004354	Adams et al., 2000
		NT_003779	
		NT_003778	
		NT_037436	
		NT_033777	
		NC_004353	
Arabidopsis thaliana (E)	125 Mb	NC_003070	Arabidopsis Genome Initiative, 2000
		NC_003071	
		NC_003074	
		NC_003075	
		NC_003076	
Homo sapiens chromosome 21 (E)	33.8 Mb	NT_002836	Hattori et al., 2000
		NT_002835	
		NT_003545	
		NT_001715	
		NT_001035	

Abbreviations: A, archaeon; B, bacterium; E, eukaryote.

genes. *Arabidopsis* is a thale cress of the mustard family. Its compact genome serves as a model for plant genomics.

Also in the year 2000, human chromosome 21 was the second human chromosome sequence to be reported (Hattori et al., 2000). This is the smallest of the human autosomes. An extra copy of this chromosome causes Down syndrome, the most common inherited form of mental retardation.

Meanwhile, bacterial genomes continued to be sequenced, and many surprising properties emerged. The genome of *Neisseria meningitidis*, which causes bacterial meningitis, contains hundreds of repetitive elements. Such repeats are more typically associated with eukaryotes. The *Pseudomonas aeruginosa* genome is 6.3 Mb, making it the largest of the sequenced bacterial genomes at that time (Stover et al., 2000).

Among the archaea, the genome of *Thermoplasma acidophilum* was sequenced (Ruepp et al., 2000). This organism thrives at 59°C and pH 2. Remarkably, it has undergone extensive lateral gene transfer with *Sulfolobus solfataricus*, an archaeon that is distantly related from a phylogenetic perspective but occupies the same ecological niche as coal heaps.

We discuss lateral gene transfer in Chapter 15.

Draft Sequences of Human Genome (2001)

Two groups published the completion of a draft version of the human genome. This was accomplished by the International Human Genome Sequencing Consortium (2001) and by a consortium led by Celera Genomics (Table 13.12) (Venter et al., 2001). The reports both arrive at the conclusion that there are about 30,000 to 40,000 protein-coding genes in the genome, an unexpectedly small number. Subsequently the number of human genes has been estimated to be 20,000 to 25,000 (International Human Genome Sequencing Consortium, 2004). Analysis of the human genome sequence will have vast implications for all aspects of human biology (see Chapter 19).

The bacterial genomes that are sequenced continue to have interesting features. *Mycoplasma pulmonis* has one of the lowest guanine–cytosine (GC) contents that have been described, 26.6% (Chambaud et al., 2001). The genome of *Mycobacterium leprae*, the bacterium that causes leprosy, has undergone massive gene decay, with only half the genome coding for genes (Cole et al., 2001). Analysis of the *Pasteurella multocida* genome suggests that the radiation of the γ subdivision of proteobacteria, which includes *H. influenzae* and *E. coli* and other pathogenic gram-negative bacteria, occurred about 680 MYA (May et al., 2001). The *Sinorhizobium meliloti* genome consists of a circular chromosome and two additional megaplasmids (Galibert et al., 2001). Together, these three elements total 6.7 Mb, expanding our view of the diversity of bacterial genome organization.

Cryptomonads are a type of algae that contain one distinct eukaryotic cell (a red alga, with a nucleus) nested inside another cell (see Fig. 18.9). This unique arrangement derives from an ancient evolutionary fusion of two organisms. That red algal nucleus, termed a nucleomorph, is the most gene-dense eukaryotic genome known. Its genome was sequenced (Douglas et al., 2001) and found to be dense (1 gene per 977 bp) with ultrashort noncoding regions.

Continuing Rise in Completed Genomes (2002)

In the year 2002, dozens more microbial genomes were sequenced. Of the eukaryotes (Table 13.13), the fission yeast *Schizosaccharomyces pombe* was found to have the smallest number of protein-coding genes (4824) (Wood et al., 2002). The genomes of both

TABLE 13-12   Genome-Sequencing Projects Completed in 2001

Organism	Size	Accession	Reference
Pyrococcus abyssi (A)	1,765,118 bp	NC_000868	R. Heilig, 2001
Sulfolobus solfataricus (A)	2,992,245 bp	NC_002754	She et al., 2001
Sulfolobus tokodaii (A)	2,694,765 bp	NC_003106	Kawarabayasi et al., 2001
Caulobacter crescentus (B)	4,016,942 bp	NC_002696	Nierman et al., 2001
Escherichia coli 0157:H7 (B)	5,498,450 bp	NC_002695	Perna et al., 2001; Hayashi et al., 2001; see Blattner et al., 1997
	5,528,970 bp	AE005174	Perna et al., 2001
Mycobacterium leprae (B)	3,268,203 bp	NC_002677	Cole et al., 2001
Mycoplasma pulmonis (B)	963,879 bp	NC_002771	Chambaud et al., 2001
Pasteurella multocida (B)	2,257,487 bp	AE004439	May et al., 2001
Sinorhizobium meliloti (B)	6.7 Mb	NC_003047	Galibert et al., 2001
Streptococcus pneumoniae (B)	2,160,837	AE005672	Tettelin et al., 2001
Streptococcus pyogenes (B)	1,852,442 bp	AE004092	Ferretti et al., 2001
Encephalitozoon cuniculi (E)	2.5 Mb	AL391737 and AL590442 to AL590451	Katinka et al., 2001
Guillardia theta nucleomorph genome (E)	551,264 bp	NC_002751	Douglas et al., 2001
Homo sapiens (E)	3,300 Mb	Various	International Human Genome Sequencing Consortium, 2001; Venter et al., 2001

Abbreviations: A, archeon; B, bacterium; E, eukaryote.

the malaria parasite *Plasmodium falciparum* and its host, the mosquito *Anopheles gambiae*, were reported (Holt et al., 2002). Additionally, the genome of the rodent malaria parasite *Plasmodium yoelii yoelii* was determined and compared to that of *P. falciparum* (Carlton et al., 2002). These projects are described in Chapter 18.

Expansion of Genome Projects (2003–2009)

In recent years, the number of completed eukaryotic, archaeal, and bacterial genomes has continued to increase, with a particularly large number of genome projects that are currently in the assembly phase (near completion) or otherwise in progress.

TABLE 13-13   Eukaryotic Genome-Sequencing Projects Completed in 2002

Organism	Size	Accession	Reference
Anopheles gambiae (E)	278 Mb	AAAB00000000	Holt et al., 2002
Plasmodium falciparum (E)	22.8 Mb	NC_002375	Gardner et al., 2002
Plasmodium yoelii yoelii (E)	23.1 Mb	AABL00000000	Carlton et al., 2002
Schizosaccharomyces pombe (E)	13.8 Mb	NC_003424 NC_003423 NC_003421	Wood et al., 2002

Abbreviation: E, eukaryote.

TABLE 13-14   Completed Genome Projects (1995–2008)

Year	Eukaryotes	Archaea	Bacteria
1995	0	0	3
1996	1	1	2
1997	0	2	4
1998	2	1	5
1999	1	2	5
2000	2	3	14
2001	1	3	22
2002	2	4	26
2003	1	1	47
2004	7	3	63
2005	3	4	77
2006	2	8	138
2007	2	17	173
2008	0	6	172
Total complete	23	48	551
In progress or assembly (2008)	491	41	1364
Grand total	514	96	2130

Note: The number of complete eukaryotic projects is greater than the cumulative total because not all projects were annotated with a release date. In progress eukaryotic genomes include those in the assembly phase.

Source: Entrez Genome Projects at NCBI, January 2009 (▶ http://www.ncbi.nlm.nih.gov/sites/entrez?db = genome).

Table 13.14 summarizes 2700 such projects, excluding thousands of ongoing and completed viral and organellar genome projects. Several trends contribute to the rapid development of this field. (1) In sequencing a genome of interest, the availability of completed genomes of closely related organisms greatly aids the assembly and annotation process. For example, the assembly of the chimpanzee genome relied heavily on using the very closely related human reference genome as a template (Chapter 18). (2) Sequencing technologies have continuously improved. An entire bacterial genome can be sequenced in just four hours using the 454 Life Sciences technology described below. (3) There has been progress in selecting, obtaining, and preparing genomic DNA from a spectacular range of biological sources. This has led to the creation of the new discipline of the genomics of ancient, extinct organisms (e.g., the Neanderthal genome was deciphered in 2007) to metagenomics projects that define the community of organisms living in sites such as the oceans or the human gut.

GENOME ANALYSIS PROJECTS

We have surveyed completed genome projects from a chronological point of view. There are many questions associated with genome sequencing. Which genomes are sequenced? How is it accomplished? How big are genomes? When genomic DNA is sequenced, what are its main features (e.g., genes, regulatory regions, repetitive elements) and how are they determined? Even the goals of sequence analysis are evolving as we learn what questions to ask and what tools are available to address those questions.

TABLE 13-15 Applications of Genome Sequencing

Purpose	Template	Example
De novo sequencing	Genome sequencing	Sequencing >1000 influenza genomes
	Ancient DNA	Extinct Neanderthal genome
	Metagenomics	Human gut
Resequencing	Whole genomes	Individual humans
	Genomic regions	Assessment of genomic rearrangements or disease-associated regions
	Somatic mutations	Sequencing mutations in cancer
Transcriptome	Full-length transcripts	Defining regulated messenger RNA transcripts
	Serial Analysis of Gene Expression (SAGE)	
	Noncoding RNAs	Identifying and quantifying microRNAs in samples
Epigenetics	Methylation changes	Measuring methylation changes in cancer

Four main types of genome analysis projects are outlined in Table 13.15. (1) De novo sequencing involves determining the DNA sequence of an organism as completely as possible, as described chronologically in the sections above. While many more de novo genome sequencing projects are underway, two recently developed, specialized categories are the sequencing of ancient DNA (from extinct organisms) and metagenomics (sampling the genomes of many organisms from a particular environmental site such as the human gut or an ocean region). (2) Resequencing a genome permits the variation between individuals to be assessed. For example, the genomic sequences of James Watson (a co-discoverer of the double helical nature of DNA) and J. Craig Venter (a pioneer in genome sequencing) have been determined. Applications of resequencing include the assessment of genomic changes in disease-associated regions, the sequencing of all human exons in multiple individuals, or the sequencing of large sets of genes associated with cancer. (3) While we introduced microarray-based gene expression profiling in Chapters 8 and 9, a sequencing-based approach is also possible. Total RNA can be isolated, converted to complementary DNA, packaged into libraries, and then exhaustively sequenced to determine the quantities of RNA transcripts. This strategy can also be applied to serial analysis of gene expression (SAGE; Chapter 8). (4) Epigenetics refers to heritable changes other than those involving the four DNA sequence per se. Such epigenetic changes include the modification of DNA or chromatin through DNA methylation (the addition of methyl groups to cytosine residues in CpG dinucleotides) and/or through the posttranslational modification of histones. High throughput sequencing can be used to assess the methylation status of a genome (Callinan and Feinberg, 2006).

We will discuss these types of projects, and the sequencing technologies that enable them. We will also examine the process of sequencing a genome in a number of distinct phases, from the selection of an appropriate genome to sequencing the DNA to genome annotation (Fig. 13.8).

Criteria for Selection of Genomes for Sequencing

The choice of which genome to sequence depends on several main factors. The selection criteria change over time as technological advances reduce costs and as

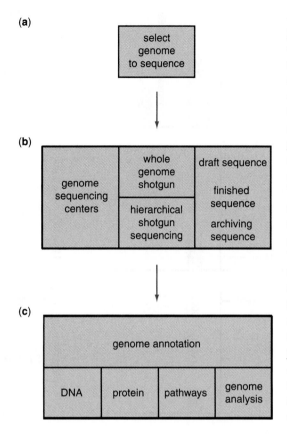

FIGURE 13.8. *Overview of the process of sequencing and analyzing a genome. (a) The selection of which genome to sequence involves decisions about cost, relevance to biological principles, and relevance to disease. (b) A variety of genome-sequencing centers perform genome sequencing by approaches such as whole genome shotgun (WGS) sequencing, hierarchical WGS, or both. This DNA sequencing is performed in stages, including draft and finished sequencing, and the results are archived. (c) The completed sequence is annotated at the level of DNA (e.g., to identify repetitive elements, nucleotide composition, and protein-coding genes), at the level of predicted proteins, and at the level of predicted cellular pathways. Additionally, a variety of genome-wide analyses may be performed, such as comparisons between genomes or phylogenomics (see below).*

genome-sequencing centers gain experience in this new endeavor. One set of criteria is offered by the National Human Genome Research Institute (NHGRI) at the National Institutes of Health. These include projects to study comparative genome evolution, to survey human structural variation, to annotate the human genome, and to perform medical sequencing. The process of selecting sequencing targets includes the submission of proposals ("white papers" that are available on the NHGRI website) as well as working groups that help to set priorities.

The NHGRI large-scale genome sequencing program is described at ► http://www.genome.gov/10001691. A list of white papers and sequencing targets is online at ► http://www.genome.gov/10002154.

Genome Size

For a microbial genome, the size is typically several megabases (millions of base pairs), and a single center often has the resources to complete the entire project. For larger (typically eukaryotic) genomes, international collaborations are often established to share the effort (see Chapter 18). For the Human Genome Project (Chapter 19), about 1000 bp were sequenced per second, 24 hours a day, up to the 2003 announcements of completed draft sequences.

A graphical overview of the sizes of various genomes is presented in Fig. 13.9. Viral genomes range from 1 to 350 kb (Chapter 14). In haploid genomes such as bacteria (Chapter 15), the genome size (or *C* value) is the total amount of DNA in the genome. In diploid or polyploid organisms, the genome size is the amount of DNA in the unreplicated haploid genome (such as the sperm cell nucleus). Bacterial genomes vary in size over about a 22-fold range from about 500,000 bp (*M. genitalium*) to 13 Mb (currently the largest sequenced prokaryotic genome, *Streptomyces coelicolor*, is 8.7 Mb).

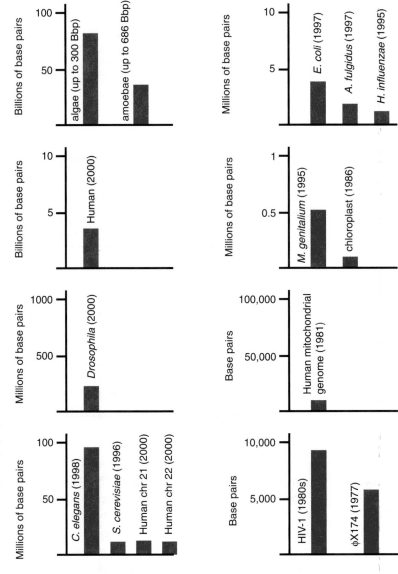

FIGURE 13.9. *Comparison of the sizes of various genomes. Each graph represents a 10-fold change in scale. For bacterial genomes, the genome size ranges from a mere 580,000 bp (M. genitalium, with 470 protein-coding genes is among the smallest sequenced genomes) to cyanobacteria with genome sizes of 13 Mb. This is a 22-fold range. For eukaryotic genomes, the range is from the 8 Mb of some fungi to 686 Gb for some amoebae. This range is over 75,000-fold and has been called the C value paradox (see Chapter 16). The C value is the total amount of DNA in the genome, and the paradox is the relation between complexity of a eukaryote and its amount of genomic DNA.*

As of November 2008, the Wellcome Trust Sanger Institute (▶ http://www.wellcome.ac.uk/) has supported the sequencing of 68 Gb of DNA from several dozen organisms (▶ http://www.sanger.ac.uk/Info/Statistics/).

Among eukaryotes, there is about a 75,000-fold range in genome sizes from 8 Mb for some fungi to 686 Gb (gigabases) for some amoebae. The so-called *C* value paradox is that some organisms with extremely large *C* values are morphologically simple and appear to have a modest number of protein-coding genes. We will explore this paradox in Chapter 16.

Cost

You can read about the NHGRI Genome Technology Program at ▶ http://www.genome.gov/10000368.

The total worldwide cost of producing a draft sequence of the human genome by a public consortium was about $300 million (or $3 billion including development costs). In contrast, completion of a draft sequence of another primate, the rhesus macaque, cost $22 million in 2006. Currently, the cost of sequencing a human genome by Sanger technology is approximately $1 million to $10 million, although Venter's sequence cost ~$70 million. A stated goal of the NHGRI is to promote the development of technology to reduce the cost of sequencing a human genome to $100,000 and then to $1000.

One measure of cost is the number of dollars per Q_{20} bases, that is, high quality base reads (defined below). In fiscal year 2006, the NHGRI budget was $130 million, which enabled the sequencing of 150 billion raw Q_{20} base pairs per year. Funds were distributed to a variety of sequencing centers. Novel high throughput technologies such as pyrosequencing (introduced below) can be more than 10 times less expensive than conventional sequencing, and technological innovations are certain to continue lowering the cost of genome sequencing.

Relevance to Human Disease

All genome projects have yielded information about how an organism causes disease and/or is susceptible to disease. For example, by sequencing the chimpanzee genome, we may learn why these animals are not susceptible to diseases that afflict humans, such as malaria and AIDS. We discuss genomics aspects of human disease in Chapter 20 and we will consider the disease relevance of all parts of the tree of life in Chapters 14 to 18.

Relevance to Basic Biological Questions

Each genome is unique and its analysis enables basic questions about evolution and genome organization to be addressed. As an example, the chicken provides a non-mammalian vertebrate system that is widely used in the study of development. The analysis of protozoan genomes can illustrate the evolutionary history of the eukaryotes.

Relevance to Agriculture

Analyses of the chicken, cow, and honeybee genome sequences are expected to benefit agriculture in a variety of ways, such as leading to strategies to protect these organisms from disease. By 2050, 90% of the world's population will live in developing countries where agriculture is the most important activity. Raven et al. (2006) thus suggest this should guide the choice of genome projects towards those that may benefit resource-poor farmers.

Should an Individual from a Species, Several Individuals, or Many Individuals Be Sequenced?

Ultimately, it will be important to determine the entire genomic sequence from multiple individuals of a species in order to correlate the genotype with the phenotype. In the case of humans (Chapter 19), the public consortium's Human Genome Project initially involved the sequencing and analysis of genomic DNA from individuals from many ethnic backgrounds, both male and female.

For viruses such as human immunodeficiency virus (HIV-1 and HIV-2), the virus rapidly undergoes enormous numbers of DNA changes, making it necessary to sequence many thousands of independent isolates (Chapter 14). This is practical to achieve because the genome is extremely small (<10 kb). In some cases, comparison of different bacterial strains reveals why one is harmless to humans while another is highly pathogenic (see Table 15.10). Such a comparison has been performed for a strain of *E. coli* that normally inhabits the human gut and another strain that causes severe, sometimes fatal disease (Chapter 15).

Resequencing Projects

In studying genomic variation between individual humans, one approach is to resequence the entire human genome (reviewed in Bentley, 2006). This was accomplished recently for two individuals, and plans are underway to sequence thousands more. One goal of such an endeavor is to use genomic information to guide medical decisions. As an alternative strategy it may be cost effective to resequence portions of the genome that are of particular interest, such as globin loci in patients with thalassemia. Another approach is to sequence all human exons, since this focuses on protein coding regions rather than the ~98% of the genome composed of noncoding regions (including introns, intergenic regions, and large expanses of repetitive DNA). Hodges et al. (2007), Albert et al. (2007), and Porreca et al. (2007) independently reported methodology to capture (isolate) from 10,000 to 200,000 protein-coding exons using high-density microarrays. In the Porreca et al., strategy, 55,000 oligonucleotides were synthesized, each of which includes portions that hybridize to genomic loci immediately upstream and downstream of a given small exon of interest. After hybridization of the oligonucleotides to genomic DNA the intervening exon sequence can be amplified and then sequenced. While the cost of sequencing a human genome currently remains in the tens of millions of dollars, the cost of sequencing all exons may be one thousand-fold less expensive.

Ancient DNA Projects

The study of ancient DNA presents a fascinating glimpse into the history of life on earth. It is now possible to isolate genomic and/or mitochondrial DNA from museum specimens, fossils, and other sources of organisms that are now extinct. Svante Pääbo is a pioneer in this field. There are special challenges encountered in these studies (Pääbo et al., 2004; Willerslev and Cooper, 2005):

- Ancient DNA is often degraded by nucleases. Thus, the size fragments of ancient DNA are often small (100 to 500 base pairs), and the nucleotides are often damaged by strand breaks (induced by microorganisms or endogenous nucleases), oxidation (resulting in fragmentation of bases and/or deoxyribose groups), cross-linking of nucleotides, or deamination. There are many strategies available to address these issues, including performing multiple independent polymerase chain reactions (PCR) or sequencing reactions from ancient DNA extracts. C to T and G to A substitutions are particularly prevalent, as shown for example in studies of 11 European cave bears (Hofreiter et al., 2001).

- The majority of DNA isolated from ancient samples derives from unrelated organisms such as bacteria that invaded the specimen after death.

- Much DNA isolated from ancient specimens is contaminated by modern human DNA. Extraordinary measures must be taken to minimize laboratory or other sources of human contamination.

- A large number of criteria must be applied to demonstrate authenticity of ancient DNA samples. These include the use of appropriate control extracts and negative controls; analysis of multiple extracts independently isolated from each specimen; quantitation of the number of amplifiable molecules; inverse correlation between amplification efficiency and the length of amplification which is expected to occur because of the fragmented nature of ancient DNA.

Despite the considerable technical challenges, dramatic progress has been made in the field of ancient DNA analysis. An example is the Neanderthals, who were hominids that thrived from about 400,000 years ago until 30,000 years ago, and who represent the closest known relative of modern humans. Mitochondrial DNA has been extracted and sequenced from approximately one dozen Neanderthal fossils. Green et al. (2006) isolated genomic DNA from the 38,000-year-old fossil of a Neanderthal bone found in modern Croatia. They performed 454 sequencing (described below) and identified over 15,000 sequences of primate origin. Their analysis allowed them to compare patterns of nucleotide substitution between Neanderthals and modern humans and to date the species' divergence to approximately 500,000 years ago.

Other ancient DNA projects include the sequencing of mitochondrial genomes from the moa (flightless birds from New Zealand; Cooper et al., 2001), cave bear (Noonan et al., 2005), and woolly mammoth (Krause et al., 2006), and more recently from hair shafts of the Siberian mammoth *Mammuthus primigenius* (Gilbert et al., 2007). A list of DNA sequences available from extinct organisms is available at the NCBI website. While ancient DNA can be extracted, ancient RNA and proteins have not been extracted. As a notable exception, Schweitzer et al. (2007) found evidence of collagen in the extracellular matrix of bone from a *Tyrannosaurus rex* fossil based on immunohistochemistry (with antisera developed against avian collagen) and mass spectrometry.

To see DNA entries from extinct organisms, visit the Taxonomy home at ▶ http://www.ncbi.nlm.nih.gov/Taxonomy/taxonomyhome.html/ then follow the link to extinct organisms. Currently (January 2009) there are data available from 46 mammals, 32 birds, and assorted plants, lizards, insects, and amphibians.

Metagenomics Projects

The great majority of organisms on the planet are viruses and the prokaryotes (bacteria and archaea). Of these various organisms, most (probably >99%) are not cultivatable, making them extremely difficult to study. Metagenomics is the functional and sequence-based analysis of microorganisms that occur in an environmental sample (Riesenfeld et al., 2004). Genomic sequencing efforts have been directed to a variety of environmental samples. In some cases the goal has been to perform the polymerase chain reaction to amplify ribosomal RNA genes. More recently, high throughput sequencing technologies have been applied to more broadly sample DNA (and genes in particular) in environmental samples.

Metagenomics projects may be grouped into two broad areas: environmental (also called ecological) and organismal (Table 13.16). Environmental projects address the genomic community in an ecological site such as a hot spring, an ocean, sludge, or soil. As an example of an environmental project, Robert Edwards and colleagues (2006) obtained over 70 million base pairs of sequence data (from over 700,000 sequences) sampled from two neighboring sites of an iron-rich mine in Minnesota. The samples were characterized by unexpectedly distinct sets of bacterial microorganisms, based principally on the analysis of 16S ribosomal DNA sequences.

Organismal metagenomics projects include such sites as human or mouse gut, feces, or lung. For example, it is estimated that the human intestinal tract contains on the order of 10^{13} to 10^{14} microorganisms (Gill et al., 2006). Collectively, these bacteria, archaea, and viruses contain 100 times as many genes as the human genome. Gill et al. sequenced 78 million base pairs of DNA from human fecal samples to assess the diversity of microorganisms living in the human gut.

Two primary sources of information on metagenomics projects are the NCBI website and the Genomes Online Database (Liolios et al., 2008).

NCBI summarizes current metagenomics projects, including links to project homepages, GenBank entries and literature, at ▶ http://www.ncbi.nlm.nih.gov/genomes/lenvs.cgi. The Genomes Online (GOLD) database is available at ▶ http://www.genomesonline.org. The homepage for the Human Gut Microbiome Initiative (HGMI) is ▶ http://genome.wustl.edu/hgm/HGM_frontpage.cgi. For the Global Ocean Sampling expedition of J. Craig Venter and colleagues, the key website is the Community Cyberinfrastructure for Advanced Marine Microbial Ecology Research and Analysis (CAMERA; (▶ http://camera.calit2.net).

TABLE 13-16   Metagenomics Projects: Selected Examples

Type	Project	Source	GenBank Accession	Publication
Environmental	Global Ocean Sampling Expedition Metagenome	Marine	AACY000000000	Rusch et al. (2007)
	Soudan Mine Red Sample	Mine drainage	—	Edwards et al. (2006)
	Enhanced biological phosphorus removal (EBPR) sludge community	Sludge	AATO00000000	García Martín et al. (2006)
Organismal	Termite Gut Metagenome	Termite gut	—	Warnecke et al. (2007)
	Uncultured Human Fecal Virus Metagenome	Human gut	AAMG00000000 AAMH00000000 AAMI00000000	Zhang et al. (2006)
	Human Distal Gut Biome	Human gut	AAQK00000000 AAQL00000000	Gill et al. (2006)

Source: ► http://www.ncbi.nlm.nih.gov/genomes/lenvs.cgi (November 2007).

DNA SEQUENCING TECHNOLOGIES

Nucleic acids were discovered by Johann Friedrich Miescher (1844–1895) in 1869. He called them "nucleins" because they were present in all cell nuclei. The first complete nucleic acid sequence of a molecule (an alanine tRNA from yeast) was accomplished by Holley et al. (1965), who purified tRNAs then treated them with a series of ribonucleases. Another milestone was reached in 1970 when Ray Wu developed a primer extension strategy to sequence nucleotides of DNA; this became the basis of Sanger sequencing. Wu determined the sequence of the two cohesive ends of lambda phage DNA in 1971 (Wu, 1970; Wu and Taylor, 1971).

Sanger Sequencing

Sanger and colleagues (1977b) introduced the most commonly used technique for sequencing DNA, now called Sanger sequencing or dideoxy sequencing. The principle is to obtain a template of interest (such as a fragment of genomic DNA or complementary DNA), denature it to yield single-stranded DNA, and add to it an olignonucleotide primer (typically about 20 nucleotides in length and complementary to the strand being sequenced). In the presence of DNA polymerase I (Klenow fragment) and the four $2'$-deoxynucleotides (dNTPs), a second strand is synthesized. This synthesis can be inhibited by the further addition of a dideoxynucleotide such as $2',2'$-dideoxythymidine triphosphate (ddTTP). Separate reactions include ddATP, ddGTP, or ddCTP accompanying the four dNTPs. Each dideoxynucleotide lacks a $3'$ hydroxyl group and so serves as a chain terminator, preventing any further extension. The reaction with ddTTP contains a series of extended fragments, each sharing the same $5'$ end but terminating at various positions having a T residue. Four reactions are performed, each having a trace amount of radioactivity. Upon electrophoresis of the four reactions through an acrylamide gel to separate the DNA based on size, a ladder of nested fragments was visualized using film to detect the radioactivity in a pattern that corresponds to the DNA sequence. In their 1977 paper, Sanger et al. reported that they could read as many as 300 bases in a set of reactions.

The Sanger method has been modified to improve its efficiency (Metzker, 2005). Fluorescently labeled analogs are spiked into the experiment instead of radioactive nucleotides. Samples travel by capillary electrophoresis to a detection area within a DNA sequencing machine where a laser excites the fluorophores producing fluorescence emissions that correspond to the base calls. Recent improvements include better microfluidic separation devices and superior fluorescence detection.

You can read about a standard Sanger sequencing machine, the Applied Biosystems 3730, at ► http://www.3700.com.

In recent decades, Sanger sequencing has been the dominant method for genome sequencing. A typical sequencing facility can produce very high quality reads (having an error rate of less than 1% per base; see below). Most large genome sequencing centers rely on high throughput Sanger sequencing.

Pyrosequencing

Pyrosequencing is one of the powerful new alternative technologies that is gaining prominence. First introduced by Hyman (1988), it forms the core of the 454 Life Sciences Corp. technology that has produced dramatic genome sequencing results (Margulies et al., 2005). That group sequenced and assembled the entire *Mycoplasma gentialium* genome (580,069 bases) with 96% coverage and at 99.96% accuracy with a single run of a sequencing machine.

The 454 Life Sciences Corp. website is ► http://www.454.com/.

A key feature of pyrosequencing is that only one dNTP is added into the reaction at a time. The principle is outlined in Fig. 13.10. DNA is immobilized on beads that capture (on average) one single-stranded template that is amplified using PCR. The template is placed in small (picoliter volume) wells, with 1.6 million wells per plate, and one dNTP is added to the wells per cycle. The reaction mixture contains the template DNA, a sequencing primer, four enzymes (DNA polymerase I, ATP sulfurylase, luciferase, and apyrase) as well as the substrates adenosine 5'-phosphosulfate (APS) and luciferin (Fig. 13.10a). In each cycle a single dNTP is added and is incorporated into the nascent strand until a different dNTP is required (Fig. 13.10b). Upon incorporation of each dNTP, an equimolar amount of pyrophosphate (PPi) is generated. This PPi is converted to ATP by ATP sulfurylase (Fig. 13.10c) and the ATP promotes the luciferase-mediated conversion of luciferin to oxyluciferin with the generation of light (Fig. 13.10d). The emitted light is detected with a charge coupled device (CCD) camera. The amount of light is measured over time (Fig. 13.10e) to indicate at which position a nucleotide was incorporated; because of the quantitative nature of this process, the incorporation of two nucleotides creates twice the light output. Apyrase degrades both unincorporated dNTPs and excess ATP, clearing the system for repeated cycles with low background noise (Fig. 13.10f). In this process dNTPs are systematically added across different cycles, but dATPαS is used in place of the usual dATP because it is efficiently used by DNA polymerase I but is not a substrate for luciferase. A schematic of the output, showing a sequencing read of GACCGTTC, is shown in Fig. 13.10g.

Pyrosequencing offers many advantages. (1) It is very fast and the cost per base is low relative to Sanger sequencing. (2) One experiment can generate up to 40 megabases of raw nucleotide sequence data, a massive amount. (3) DNA molecules are amplified without the need for bacterial cloning; this is especially helpful for metagenomics and ancient genomics projects. (4) The accuracy of the reads is very high.

(a)   sequencing primer hybridized to   enzymes: DNA polymerase
              single stranded DNA template   ATP sulfurylase
                                                             luciferase

```
5' ...GGACATATCG 3' (primer)
3' ...GGACATATCCCTGGCAAG... 5'
```

                                                  apyrase
                                  substrates: adenosine 5' phosphosulfate (APS)
                                                       luciferin

(b)

$$(DNA)_n + dNTP \xrightarrow{\text{DNA polymerase}} (DNA)_{n+1} + PPi$$

(c)

$$PPi + APS \xrightarrow{\text{ATP sulfurylase}} ATP$$

(d) $luciferin + ATP \xrightarrow{\text{luciferase}} oxyluciferin + light$

(e) amount of light | time

(f) $ATP \xrightarrow{\text{apyrase}} ADP + AMP + phosphate$

$dNTP \xrightarrow{\text{apyrase}} dNDP + dNMP + phosphate$

(g)

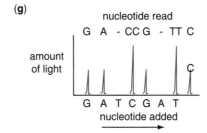

nucleotide read

G   A   -   CC   G   -   TT   C

amount of light

G   A   T   C   G   A   T
nucleotide added →

FIGURE 13.10. *Pyrosequencing. (a) A single-stranded DNA template is immobilized on a bead and amplified by PCR. After transfer to a small well, primer is added as well as additional enzymes and substrates and one of the four deoxynucleotides (dGTP, dCTP, dTTP, or in place of dATP the modified nucleotide dATPαS). (b) DNA polymerase I catalyzes the addition of a single deoxynucleotide, releasing pyrophosphate (PPi). If there is a sequence of n nucleotides in a row in the template DNA, an equimolar amount of PPi will be released. (c) ATP sulfurylase converts a substrate (APS) and PPi to adenosine triphosphate (ATP). (d) Luciferase, in the presence of its substrate luciferin and the ATP, produces a product (oxyluciferin) and light. (e) A charge coupled camera detects the light and provides an intensity measurement over time. The y axis is proportional to the amount of deoxynucleotide that was incorporated, thus specifying whether zero, one, two, or more dNTPs occur in the template DNA in that position. (f) Apyrase cleaves ATP, thus clearing the system for successive cycles. (g) The light patterns emitted from a series of cycles allow the DNA sequence of the template to be read. Typical reads with current technology are near 100 bases. Because of the massively parallel nature of this process, tens of millions of base pairs of high quality sequence can be generated with this technology.*

There are also several major disadvantages of pyrosequencing technology. The sequencing reads are short (several hundred base pairs), making whole genome assembly extremely challenging. It is anticipated that this technology will soon permit reads over 400 base pairs and generation of 1 billion base pairs of sequence per 8 hour day per machine. Another disadvantage is that the machine has difficulty in sequencing homopolymers (e.g., a string of 10 identical nucleotides). Huse et al. (2007) compared about 340,000 sequencing reads to reference templates of known sequence and determined the error rates. Errors involved homopolymer effects, insertions, deletions, and mismatches. While these errors were distributed along the length of each read, they found that 82% of all the reads had no errors, while only a small percent had a disproportionately large number of errors. By identifying and removing such low quality reads they could improve the overall accuracy of the dataset from 99.5% to 99.75% or higher.

Recent applications of 454 technology include some of the ancient DNA and metagenomics sequencing projects described above, such as the sequencing of the Neanderthal genome, and identifying the microbial community in parts of a mine in Minnesota (Edwards et al., 2006).

Cyclic Reversible Termination: Solexa

The Illumina Genome Analyzer (also called the Solexa system) offers another advanced technology that can generate one billion bases of DNA sequence data in a single run. It works on the principle of cycle reversible termination (CRT) which functions as follows. (1) DNA is randomly fragmented and adapters are attached to both ends. (2) Single-stranded DNA fragments are covalently attached to the surface of flow cell channels. (3) The addition of DNA polymerase and unlabeled deoxynucleotides creates solid-phase "bridge amplification" in which the template DNA makes U-shaped loops with both ends attached to the surface of the channel. (4) Double-stranded bridges are formed. The double-stranded molecules are denatured and then amplified to generate dense clusters of template DNA. (5) Four labeled reversible terminators are added (with primer and DNA polymerase). Only a single reversible terminator will be added to each template in a given cycle. As with Sanger sequencing, chain termination will occur at specific bases that cannot elongate. (6) Following laser excitation, the identity of the first base is recorded. (7) For the second cycle, the reversible terminators are removed (by deprotection). All four labeled reversible terminators and the polymerase are again added to the flow cell. The cycles are repeated.

You can learn more about the Solexa system at ▶ http://www. solexa.com/.

The Solexa system is very fast and generates massive amounts of sequence data. Its read lengths are typically only 30 to 50 bases, making it particularly appropriate for resequencing projects (Bentley, 2006). The main advantages of this approach relative to Sanger sequencing are its scalability and the elimination of the need for gel electrophoresis. The main advantanges relative to pyrosequencing are that all four bases are present at each cycle, and the sequential addition of dNTPs allows homopolymer tracts to be accurately read.

THE PROCESS OF GENOME SEQUENCING

Genome-Sequencing Centers

Large-scale sequencing projects are conducted at centers around the world. Twenty sequencing centers contributed to the production of a draft version of the human

TABLE 13-17   Major Genome-Sequencing Centers

Center	URL
Baylor College of Medicine Human Genome Sequencing Center	► http://www.hgsc.bcm.tmc.edu/
Beijing Genomics Institute	► http://www.genomics.org.cn/
The Broad Institute	► http://www.broad.mit.edu/
Genoscope	► http://www.cns.fr/
U.S. Department of Energy Joint Genome Institute	► http://www.jgi.doe.gov/
Washington University Genome Sequencing Center	► http://genome.wustl.edu/
The Wellcome Trust Sanger Institute	► http://www.sanger.ac.uk/

A list of genome-sequencing centers is offered at the NCBI (► http://www.ncbi.nlm.nih.gov/genomes/static/lcenters.html).

genome (2001) (see Table 19.6). These centers were also supported by the NIH and the EBI. All of these centers have also been involved in sequencing the genomes of other organisms as well. Table 13.17 provides links to seven of the major sequencing centers currently in operation.

Sequencing and Assembling Genomes: Strategies

There are two main approaches to sequencing genomes. The first is whole-genome shotgun (WGS) sequencing. Frederick Sanger first applied this approach in the sequencing of bacteriophage φX174: randomly selected fragments of genomic DNA were isolated, sequenced, and then assembled to derive a complete sequence. The application of this approach to an entire organismal genome was pioneered by Hamilton O. Smith of Johns Hopkins and J. Craig Venter of the J. Craig Venter Institute who used this strategy to sequence *H. influenzae* (Fleischmann et al., 1995).

The WGS method has been used successfully for most small genomes (i.e., viruses, bacteria and archaea, and eukaryotic genomes that lack large portions of repetitive DNA). Genomic DNA is isolated from an organism and mechanically sheared (or digested with restriction enzymes). The fragments are subcloned into small-insert libraries (e.g., 2 kb fragments), and large-insert libraries (e.g., 10 to 20 kb). Clones are sequenced from both ends (i.e., both "top" strand and "bottom" strand), and then the sequences are assembled. A typical sequencing reaction generates about 500 to 800 bp of sequence data. These small amounts of sequence are assembled into contiguous transcripts ("contigs") and then into a map of the complete genome. This strategy is employed in the majority of the bacterial and archaeal sequencing projects. Table 13.18 introduces some of the terminology associated with genome sequencing.

A second, related approach is "hierarchical shotgun sequencing" (Fig. 13.11). Genomic DNA is digested and subcloned into bacterial artificial chromosome (BAC) libraries. These libraries contain large inserts (100 to 500 kb). Alternatively, smaller cosmid libraries (with insert sizes of about 50 kb) or plasmid libraries (2 to 10 kb inserts) are generated. Unlike WGS sequencing, this hierarchical strategy employs clones (contigs) that are mapped to known chromosomal locations. Thus, sequence assembly is focused on a small region of the genome. This approach has been taken for many large, eukaryotic genomes, including the public consortium's version of the Human Genome Project (International Human Genome Sequence Consortium, 2001).

The WGS approach requires the computationally difficult task of fitting contigs together, regardless of which chromosomal region they are derived from. It was

TABLE 13-18 Terminology Used in Genome-Sequencing Projects

Term	Definition
BAC end sequence	The ends of a bacterial artificial chromosome (BAC) have been sequenced and submitted to GenBank; the internal BAC sequence may not be available. When both end sequences from the same BAC are available, this information can be used to order contigs into scaffolds.
Contig	A set of overlapping clones or sequences from which a sequence can be obtained. NCBI contig records represent contiguous sequences constructed from many clone sequences. These records may include draft and finished sequences and may contain sequence gaps (within a clone) or gaps between clones when the gap is spanned by another clone that is not sequenced.
Draft sequence	At least three- to fourfold of the estimated clone insert is covered in Phred Q20 bases in the shotgun sequencing stage, as defined for the human genome sequencing project. Note that the exact definition of "draft" may be different for other genome projects. Clone sequence may contain several pieces of the sequence separated by gaps. The true order and orientation of these pieces may not be known.
Finished sequence	The clone insert is contiguously sequenced with a high-quality standard of error rate of 0.01%. There are usually no gaps in the sequence.
Fragment	A contiguous stretch of a sequence within a clone sequence that does not contain a gap, vector, or other contaminating sequence.
Meld	When two or more fragments overlap in the entire alignable region, these sequences are merged together to make a single longer sequence.
Order and orientation	Sequence overlap information is used to order and orient (ONO) fragments within a large clone sequence.
Scaffold	Ordered set of contigs placed on the chromosome.

Source: Adapted from ► http://www.ncbi.nlm.nih.gov/genome/guide/build.html and ► http://www.ncbi.nlm.nih.gov/genome/guide/glossary.htm.

thought by some that this approach could not be practically applied to large eukaryotic genomes. However, it was successfully applied to the 120 Mb *D. melanogaster* genome (Adams et al., 2000) in combination with a hierarchical approach and to the human genome (Weber and Myers, 1997; Venter et al., 2001). The WGS data are processed at GenBank but are not distributed with GenBank releases. Instead, beginning with GenBank release 129 in 2002, WGS entries have been available from GenBank on a per-project basis (and are searchable by BLAST). Release 162 (October 2007) contained over 100 billion base pairs, surpassing the 80 billion base pairs in the corresponding traditional GenBank release.

Regions of heterochromatin contain large segments of highly repetitive DNA (Chapter 16), and in some cases cannot be effectively sequenced using WGS or hierarchical approaches. Skaletsky et al. (2003) applied an alternative technique of iterative mapping and sequencing to determine the extremely repetitive sequence of the human Y chromosome.

Genomic Sequence Data: From Unfinished to Finished

Raw DNA sequence data are deposited in databases such as the Trace Archives at NCBI and EBI (see below). The raw DNA data are typically read by software such as Phred. This program interprets which bases are sequenced by a DNA sequencing machine and further estimates the quality of each read. Phred then writes the sequences in a format such as FASTA (see Fig. 2.10) for further analysis.

At NCBI, raw genomic DNA data are made available through the high throughput genomic (HTG) sequence division. Accession numbers are assigned to each entry. The HTG database contains sequence data in four phases (Table 13.19).

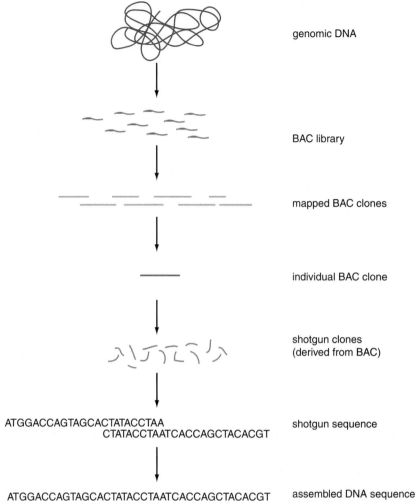

genomic DNA

BAC library

mapped BAC clones

individual BAC clone

shotgun clones
(derived from BAC)

ATGGACCAGTAGCACTATACCTAA
          CTATACCTAATCACCAGCTACACGT   shotgun sequence

ATGGACCAGTAGCACTATACCTAATCACCAGCTACACGT   assembled DNA sequence

FIGURE 13.11. Schematic of the hierarchical shotgun sequencing strategy. Genomic DNA is isolated from an organism of interest, fragmented, and inserted into a BAC library. Each BAC clone is 100 to 500 Kb. BACs are ordered (mapped). Individual BAC clones are fragmented into smaller cDNA clones and sequenced. Individual sequencing reactions are typically 300 to 700 nucleotides. These "shotgun sequences" are assembled. This process is further illustrated in Fig. 19.8. Modified from the International Human Genome Sequencing Consortium (2001, p. 863). Used with permission.

Phase 0 data are typically sequences derived from a single cosmid or BAC. They are likely to have sequencing errors and gaps of indeterminate size. However, the data may still have tremendous usefulness to the scientific community even in this form. For example, if you are performing BLAST searches and are looking for novel homologs to your query, the HTG division may contain useful information. Phase 1 data may consist of sequencing reads from contigs derived from a larger clone (e.g., a BAC clone) in which the order of the contigs is unknown and their orientation (top strand or bottom strand) is also unknown. The sequence is defined as unfinished, and it still contains gaps.

TABLE 13-19   High Throughput Genomic Records at GenBank Defined in Four Phases

Status	Location	Definition
Phase 0	HTG division	Single–few pass reads of a single clone (not contigs)
Phase 1	HTG division	Unfinished, may be unordered, unoriented contigs, with gaps
Phase 2	HTG division	Unfinished, ordered, oriented contigs, with or without gaps
Phase 3	Primary division	Finished, no gaps (with or without annotations)

Source: From ▶ http://www.ncbi.nlm.nih.gov/HTGS/.

In the finished state (phase 2), the contigs are ordered and oriented properly, and the error rate must be 10^{-4} or less.

The assembly process involves the collection of individual sequences (phase 0), the closing of gaps, and the lowering of the error rate. This process can be performed using a variety of software packages, such as Phrap (and its graphical viewer, Consed), Assembler, and Sequencher. For either the whole-genome sequencing or the hierarchical approach, after the shotgun phase is complete, the next step is to assemble contigs. This is accomplished in a process called finishing. The goal of finishing is to identify gaps in the tile path and to close them. Ideally, this process results in a single contiguous DNA sequence that spans all the contigs. The finishing process can be performed manually by experts or in an automated fashion with a program such as Autofinish (Gordon et al., 2001).

Genomic sequencing projects often rely heavily on expressed sequence tags (ESTs) to help define the protein-coding genes. Transcripts that are expressed (i.e., RNA molecules) are converted to cDNA, incorporated into libraries, and sequenced. Such cDNAs are ESTs. While they do not reveal some information about the corresponding genomic DNA, such as the sequence of introns, they are invaluable in identifying expressed genes (see below).

For examples of phases 1, 2, and 3 sequences in GenBank, see ▶ http://www.ncbi.nlm.nih.gov/HTGS/examples.html.

Phred and Phrap (see below) are available at ▶ http://www.phrap.org/. They operate on UNIX-based systems. Many other assembly software programs are available, including Arachne from the Broad Institute (▶ http://www.broad.mit.edu/wga/).

Finishing: When Has a Genome Been Fully Sequenced?

Typically, a genome is sequenced with 5- to 10-fold coverage to maximize the likelihood that it has been completely sequenced. The greatest technical challenge is to resolve the sequences of the long regions of repetitive DNA found in eukaryotic and some prokaryotic genomes (see below). To date, the human genome is one of the few for which large regions of repetitive DNA sequence are being carefully sequenced. For BAC-based assemblies gaps can be caused by low representation of genomic loci in the BAC library.

It is possible to estimate the amount of DNA that is sequenced as a function of fold coverage (Table 13.20). The probability a base is not sequenced was derived by

TABLE 13-20 Probability That a Base Is Sequenced According To Equation 13.1

Fold Coverage	P_0	Percent Not Sequenced	Percent Sequenced
0.25	$e^{-0.25} = 0.78$	78	22
0.5	$e^{-0.5} = 0.61$	61	39
0.75	$e^{-0.75} = 0.47$	47	53
1	$e^{-1} = 0.37$	37	63
2	$e^{-2} = 0.135$	13.5	87.5
3	$e^{-3} = 0.05$	5	95
4	$e^{-4} = 0.018$	1.8	98.2
5	$e^{-5} = 0.0067$	0.6	99.4
6	$e^{-6} = 0.0025$	0.25	99.75
7	$e^{-7} = 0.0009$	0.09	99.91
8	$e^{-8} = 0.0003$	0.03	99.97
9	$e^{-9} = 0.0001$	0.01	99.99
10	$e^{-10} = 0.000045$	0.005	99.995

Source: Adapted from ▶ http://www.genome.ou.edu/poisson_calc.html and Lander and Waterman (1988).

Lander and Waterman (1988) and is given by

$$P_0 = e^{-c} \tag{13.1}$$

where c is the fold coverage and is given by

$$c = \frac{LN}{G} \tag{13.2}$$

and where LN is the number of bases sequenced, L being the **read length** and N the number of reads, G is the target sequence length, and e is the constant ~ 2.718. These results show that to achieve an error rate of 1 in 10,000 (0.01%), it is theoretically necessary to obtain ninefold coverage of the genome. With fivefold coverage, an error rate of 0.6% is expected.

Some regions of genomic DNA yield ambiguous sequencing results. This can occur in regions of extremely high or low GC content or unusual secondary structure. These areas are routinely resolved ("finished") by directed sequencing of the region in question with specific oligonucleotide primers (Tettelin et al., 1999). The sequencing of the malaria parasite *P. falciparum* was especially difficult to achieve because adenine and thymine comprise about 80% of the genome (Gardner et al., 2002). For this reason, a chromosome-based approach was needed to sequence this 23-Mb genome (Chapter 18).

The NHGRI offers standard finishing practices for the human genome. See ► http://www.genome.gov/10001812.

Repository for Genome Sequence Data

Raw sequence data for the genome-sequencing projects of several organisms have been deposited in the Trace Archive located at both NCBI and Ensembl/EBI. All entries in this archive are given a Trace Identifier (Ti) number. The archive can be searched by several criteria (such as query by Ti or sequencing center or by BLAST).

The trace server at Ensembl is at ► http://trace.ensembl.org/, and the NCBI trace archive is at ► http://www.ncbi.nlm.nih.gov/Traces/. A specialized Trace Archive BLAST server is available at ► http://www.ncbi.nlm.nih.gov/blast/Blast.cgi. The Short Read Archive contains next-generation sequencing data.

Search the trace archive with human beta globin and the output contains several Ti matches (Fig. 13.12a). By clicking on the link to a Ti record, the sequence data can be obtained in the FASTA format or as a trace of the dye termination reaction used to sequence the DNA (Fig. 13.12b).

Table 13.21 summarizes several principal organisms for which trace archive data are available. In an innovative approach to using these raw data, Salzberg et al. (2005) studied the genomic DNA records from *Drosophila ananassae*, *D. simulans*, and *D. mojavensis* and searched for matches to bacterial species that might colonize these fruitflies. They identified three new species of the bacterial endosymbiont *Wolbachia pipientis* and were able to assemble sequences that covered substantial portions of the genomes.

Role of Comparative Genomics

Comparative genomics involves the comparison of genome sequences from multiple species, or in some cases from individuals within a species. Miller et al. (2004) have reviewed this discipline and described how genome comparisons have aided the annotation of genomes (discussed below), particularly for the prediction of genes and conserved regulatory elements. They also discuss the impact on evolutionary analysis and function: through comparative analyses we can define DNA segments that are under positive or negative selection (Chapter 7).

The use of whole genome comparisons at various evolutionary distances provides a powerful technique for applying many genomic analyses (Fig. 13.13, adapted from Miller et al., 2004). Phylogenetic footprinting refers to comparisons of genomic

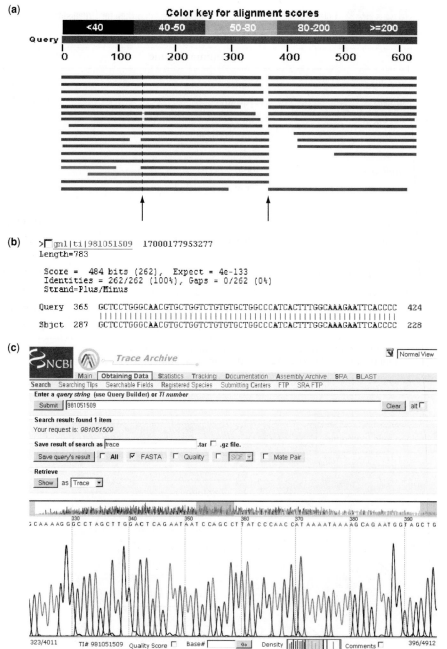

FIGURE 13.12. The trace archive is a repository of raw data from genome-sequencing projects. It is accessed from the front page of NCBI (▶ http://www.ncbi.nlm.nih.gov/Traces/trace.cgi?) or from the Trace Server at EBI (▶ http://trace.ensembl.org/). (a) A blastn search of the human whole genome shotgun (WGS) sequences of the trace archive, using human beta globin (NM_000518) as a query, results in several dozen trace archive matches. The pattern of hits (several to the 3' end of the query and several others to the 5' end) occurs because the query is an expressed transcript, and there are introns corresponding to position ~350 and ~140 (arrows). Clicking on the first database match (ti 328428) provides access to (b) the sequence in FASTA format or (c) the trace from the DNA-sequencing reaction. This trace allows you to evaluate the quality of the raw data that underlie a genomic DNA record. In some cases, the dye termination reaction (or alternate DNA-sequencing technology) yields ambiguous results, and access to the raw data allows the user to make an informed decision about the quality of the sequence call.

sequences from distantly related organisms, such as humans relative to fish, chicken, dog, and rodents. This is especially useful to identify conserved elements (under negative selection), emphasizing the relatively rare coding and noncoding segments of the genome that remain shared even after hundreds of millions of years since species such as human and fish diverged. Phylogenetic shadowing permits comparisons of more closely related species such as humans and chimpanzees that diverged about 5 million years ago. These comparisons between closely related species allow the identification of regions that are different between the two, such as genes under positive selection. Population shadowing refers to sampling multiple genomes from one species (as discussed above for resequencing the human genome from many individuals). We will adopt a comparative genomic approach throughout our exploration of the tree of life in Chapters 14 to 19.

TABLE 13-21   Current Contents of Trace Archives of Sequence Data for Organisms Having the Most Traces

Organism	Common Name	Number of Traces
Mus musculus	Mouse	206,364,140
Homo sapiens	Human	110,941,459
Rattus norvegicus	Rat	51,810,253
Pan troglodytes	Chimpanzee	45,992,278
Danio rerio	Zebrafish	43,454,015
Canis familiaris	Dog	43,096,145
Monodelphis domestica	Gray short-tailed opossum	41,581,643
Bos taurus	Cow	36,203,128
Equus caballus	Horse	31,548,760
Macaca mulatta	Rhesus monkey	31,146,365
Zea mays	Maize	29,978,391
Ornithorhynchus anatinus	Platypus	29,761,138
Callithrix jacchus	White-tufted-ear marmoset	28,287,429
Loxodonta africana	African savanna elephant	27,895,646
Cavia porcellus	Domestic guinea pig	26,942,120

The trace archives contain data from WGS, EST, and BAC-based projects.
Source: From the Ensembl trace server (▶ http://trace.ensembl.org/), November 2007.

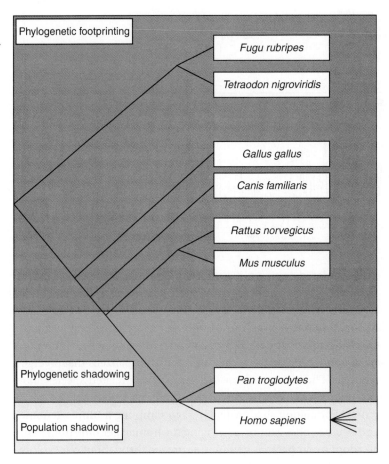

FIGURE 13.13. *Comparative genomics allows the comparison of a genome (such as human) to other genomes of varying evolutionary distance. In phylogenetic footprinting, this includes genomes from organisms that diverged a relatively long time ago, such as fish (*Fugu rubripes, Tetraodon nigoviridis *that diverged from the human lineage >400 million years ago), chicken (*Gallus gallus*), dog (*Canis familiaris*), rat and mouse (*Rattus norvegicus *and* Mus musculus *that diverged from the human lineage ~80 to 100 million years ago). In phylogenetic shadowing, more closely related genomes are compared (e.g., the chimpanzee* Pan troglodytes*). In population shadowing, multiple genomes from one species are compared, permitting analyses of genotype–phenotype correlations. Redrawn from Miller et al. (2004). Used with permission.*

GENOME ANNOTATION: FEATURES OF GENOMIC DNA

When a genome is sequenced, we learn its exact size and we obtain the complete (or nearly complete) nucleotide sequence. Genome annotation is the process by which the landscape of genomic DNA is surveyed, and key features of the DNA are described. An example of an automated pipeline for genome annotation from the Ensembl server is outlined in Fig. 13.14 (see Curwen et al., 2004; Potter et al., 2004).

Three fundamental questions may be asked about the nature of this sequence:

1. What is the overall GC content or other nucleotide composition? Many eukaryotic genomes are characterized by a GC content of about 35% to 45%, while bacteria display a far wider range (Fig. 13.15).

2. What are the repetitive DNA sequences and where are they? Programs such as RepeatMasker can identify and mask repetitive elements such as Alu repeats. We will discuss these repeats in Chapters 16 to 19. Programs such as GLIMMER and GRAIL (see below) incorporate algorithms that identify repetitive elements in genomic DNA.

We will show examples of repetitive DNA, and the software used to identify and mask it, in Chapter 16.

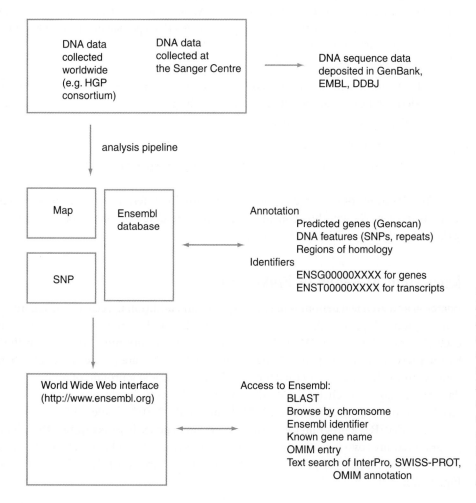

FIGURE 13.14. *Overview of the Ensembl annotation pipeline. Ensembl is a joint EBI-EMBL and Sanger Institute project that automatically tracks and annotates DNA sequence data from the Human Genome Project and other sequencing projects (e.g., mouse, rat zebrafish, fugu, mosquito, fruitfly, and nematodes). (See ▶ http://www.ensembl.org/info/data/docs/genome_annotation.html.)*

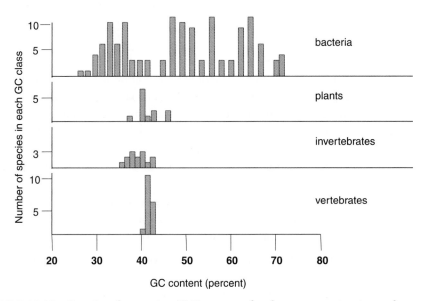

FIGURE 13.15. *Guanine plus cytosine (GC) content of prokaryotes, protists, invertebrates, and vertebrates. Note that most eukaryotic genomes have 40% to 45% G + C content, while bacteria and archaea have a far wider range. This figure is adapted from Bernardi and Bernardi (1990) based on studies in the 1970s and 1980s. Recent eukaryotic genome sequencing projects (described in Chapter 18) reveal that GC content for various organisms includes 19.4% (P. falciparum), 22.2% (the slime mold* Dictyostelium discoideum*), 34.9% (A. thaliana), 36% (C. elegans), 38.3% (S. cerevisiae), 41.1% (human), 42% (M. musculus), and 43.3% (O. sativa). For sequenced prokaryotes, GC content values range from 26% (Ureaplasma urealyticum parvum) to 72% (Streptomyces coelicolor). Used with permission.*

3. How many genes (protein-coding sequences) are present? Genes may be identified by a number of features, including:

- Gene-specific codon bias
- Absence of repetitive DNA sequences
- Presence of signals such as promoter region-specific motifs

These features of genomic DNA are substantially different between prokaryotes and eukaryotes. We will thus consider them in more detail in Chapters 15 (on bacteria and archaea) and 16 to 19 (on eukaryotes).

Annotation of Genes in Prokaryotes

Bacterial and archaeal genomes have both genes and additional, relatively small intergenic regions. Typically, these genomes are circular, and there is about one gene in each kilobase of genomic DNA. For prokaryotes, genes are most simply identified by the presence of long open reading frames (ORFs) that are greater in length than some cutoff value such as 90 nucleotides (30 amino acids; a protein of about 3 kilodaltons). Programs such as GLIMMER and GenMark efficiently locate genes in bacterial genomic sequence (reviewed in Baytaluk et al., 2002) (Table 13.22).

GLIMMER is a program for the identification of genes in prokaryotic DNA. The program requires two inputs: a genomic DNA sequence file (in FASTA format) and a set of Markov models for genes. We will examine a sample GLIMMER output (see Fig. 15.7).

TABLE 13-22 Gene-Finding Programs for Prokaryotes

Resource	Source/Description	URL
GeneMark	Provides several gene prediction programs in prokaryotes, eukaryotes, and viruses	► http://opal.biology.gatech.edu/ GeneMark/
GeneScan	Identification of complete gene structures in vertebrate genomic DNA	► http://genes.mit.edu/GENSCAN. html
GLIMMER	A system for finding genes in microbial DNA	► http://cbcb.umd.edu/software/ glimmer or ► http://www.ncbi.nlm. nih.gov/genomes/MICROBES/ glimmer_3.cgi

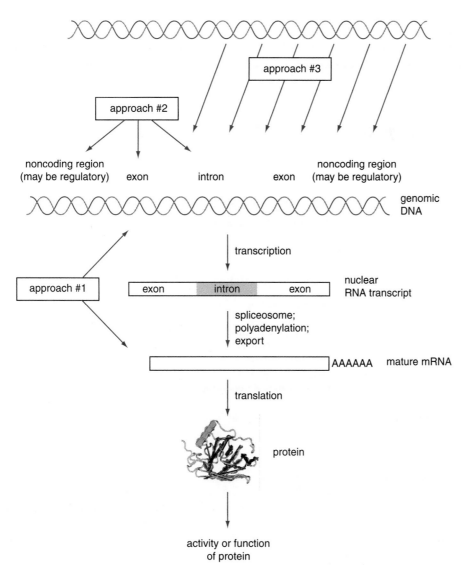

FIGURE 13.16. *There are three principal approaches to finding protein-coding genes in genomic DNA. (1) In homology-based (or extrinsic) gene finding, genomic DNA is compared to ESTs. ESTs are cDNAs that are generated from the RNA of an organism, and thousands to millions of ESTs are available for various organisms. When an EST sequence matches a region of genomic DNA, this provides strong evidence that a protein-coding gene has been identified. (2) In algorithm-based (or intrinsic) gene finding, the nucleotide composition of the genomic DNA is analyzed for features such as the presence of a long open reading frame, that is, a start codon followed by a threshold such as at least 300 nucleotides before a stop codon is encountered. The presence of introns (usually in eukaryotic genomes) complicates this analysis. The base composition of coding regions often differs dramatically from noncoding regions, and this also serves as the basis for gene-finding algorithms using the second approach. (3) In a third approach, comparative genomics is used to guide annotation by aligning conserved segments between two genomes, including genes previously annotated in the reference genome.*

Annotation of Genes in Eukaryotes

In contrast to bacterial genomes, eukaryotic genomes contain both genes and large amounts of noncoding DNA. This noncoding material includes repetitive DNA, genes that have regulatory functions, and introns that interrupt exons and are removed from mature RNA transcripts. A major focus of genome-sequencing projects is to identify all the genes in a genome. However, it is necessary to define the variety of genes and the criteria for identifying them. This includes protein-coding genes, pseudogenes, and a variety of RNA genes. We will discuss these in Chapters 17 (on fungi) and 19 (on human), with a particular emphasis on gene finding in Chapter 16 (on eukaryotic chromosomes). There, we will discuss three principal approaches to gene identification in eukaryotic genomic DNA (Fig. 13.16). The first approach is based on aligning expressed sequences (ESTs or cDNAs) to genomic DNA. Since the ESTs are obtained independently of the genomic DNA sequence, this approach is called "extrinsic." The availability of a full-length cDNA is invaluable in defining the extent of the exons in a gene based on experimental evidence. Second, an "intrinsic" approach is to predict gene structures (exons and introns) solely through analysis of genomic DNA, searching for features such as ORFs, exon/intron boundaries, start and stop codons, and codon usage typical of coding regions. Third, a comparative genomic approach relies on mapping genes from one organism to conserved syntenic regions of a closely related organism whose genome has previously been sequenced. Examples of gene prediction software are presented in Table 13.23 and in Figs. 16.12 to 16.18.

Summary: Questions from Genome-Sequencing Projects

A series of basic questions are associated with virtually all genome sequencing projects:

- Can we identify both protein-coding genes and RNA-coding genes?
- Can we assign a function to these genes?

TABLE 13-23 Gene-Finding Programs for Eukaryotes

Resource	Source/Description	URL
GeneMark	Provides several gene prediction programs in prokaryotes, eukaryotes, and viruses	► http://opal.biology.gatech.edu/GeneMark/
GeneScan	Identification of complete gene structures in vertebrate genomic DNA	► http://genes.mit.edu/GENSCAN.html
GLIMMER	A system for finding genes in microbial DNA	► http://cbcb.umd.edu/software/glimmer or ► http://www.ncbi.nlm.nih.gov/genomes/MICROBES/glimmer_3.cgi
GRAIL	Gene Recognition and Assembly Internet Link	► http://compbio.ornl.gov/Grail-bin/EmptyGrailForm
ORF Finder	Open reading frame finder	► http://www.ncbi.nlm.nih.gov/gorf/gorf.html
Procrustes	Gene recognition via spliced alignment	► http://www-hto.usc.edu/software/procrustes/

- Can we determine (or predict) the structure of all the gene products?
- Can we reconstruct transcriptional networks and metabolic signaling pathways associated with each gene product?
- Can we link genotypes to phenotypes? Thus, for example, can we explain why humans vary greatly in their susceptibility to the same disease-causing organism or environmental toxin? Can we explain why two strains of anthrax or herpesvirus vary in their pathogenicity?
- Can we define the evolutionary history of life? This may be accomplished in part through molecular phylogenetic studies and comparative genomics. This approach has been called phylogenomics (Eisen and Hanawalt, 1999; Eisen and Fraser, 2003).

PERSPECTIVE

Beginning in 1995, we have entered an era in which the completed genome sequence has been determined for many dozens of organisms. Thousands of complete genome sequences are now available. Since the completion of the human genome sequencing in the year 2003, some call the present state of biology the "postgenomic era."

A major consequence of genome-sequencing projects is that molecular phylogeny has been revolutionized. The present version of the tree of life includes three main branches (bacteria, archaea, and eukaryotes). In the coming years, molecular data will clarify some of the key questions about life on Earth:

- How many species exist on the planet?
- How did life evolve, from 4 BYA up to the present time?
- Why are some organisms pathogenic while close relatives are harmless?
- What mutations cause disease in humans and other organisms?

PITFALLS

While the research community is generating massive amounts of DNA sequence data, there are many pitfalls associated with interpretation of those data. There is an error rate associated with genome sequences (typically one nucleotide per 10,000 in finished DNA). Thus, in evaluating possible polymorphisms or mutations in genomic DNA sequences, it is important to assess the quality of the sequence data. Even if the sequence is correct, algorithms do not yet have complete success in problems such as finding protein-coding genes in eukaryotic DNA; in Chapter 16 we will see examples of genome-sequencing projects (such as rice and human) in which the predicted exons and gene models improve dramatically with each subsequent revision of the genome sequence. (For bacterial DNA, which generally lack introns, the success rate is much higher.) Once protein-coding genes or other types of genes are identified, there are very large numbers of errors in genome annotation (Brenner, 1999). It will be important to carefully assess the basis of functional annotation of genes, and ultimately the problem of gene function must be assessed by biological as well as computational criteria.

DISCUSSION QUESTIONS

[13-1] If you could decide which genome-sequencing projects to pursue, how would you prioritize the organisms?

[13-2] If you could sequence the genomes of 50 individual humans, who would they be, what hypotheses would you test, how would you perform data analyses, and what resources would you require in terms of hardware, software, and collaborators? What ethical issues might arise in sequencing human genomes?

PROBLEMS/COMPUTER LAB

[13-1] Figure 13.1 shows a tree of life based on rRNA sequences. Construct a tree of life based on glyceraldehyde-3-phosphate dehydrogenase (GAPDH) protein sequences. One approach is to identify this family in Pfam, export a large number of the sequences in the fasta format, perform a multiple sequence alignment using MUSCLE (Chapter 6), and create and evaluate a neighbor-joining tree using MEGA (Chapter 7). Use at least 30 sequences from each domain of life. How similar is your tree to the one in Fig. 13.1? What might account for their differences?

[13-2] Obtain approximately 1000 bases of DNA sequence in the fasta format from the bacterium *Escherichia coli* K12 (the accession number of the complete genome is NC_000913). Use this as a query in a blastn search of the Trace Archive at NCBI. Can you identify a eukaryotic sequencing project that includes bacterial DNA? For example, search against human whole genome shotgun (WGS) sequences. How would you determine the total amount of bacterial DNA in any given eukaryotic entry in the Trace Archive?

SELF-TEST QUIZ

[13-1] The first complete genome to be sequenced was:

(a) *Saccharomyces cerevisiae* chromosome III

(b) *Haemophilus influenzae*

(c) A bacteriophage

(d) The human mitochondrial genome

[13-2] A typical eukaryotic mitochondrial genome encodes about how many proteins (excluding RNAs)?

(a) From 5 to 20

(b) From 50 to 100

(c) From 500 to 1000

(d) 10,000

[13-3] Thousands of genomes have now been completely sequenced. The majority of these are

(a) Viral

(b) Bacterial

(c) Archaeal

(d) Organellar (mitochondrial and plastid)

(e) Eukaryotic

[13-4] Ancient DNA projects allow the sequencing of historical samples. A special challenge is:

(a) The DNA is often fragmented.

(b) The DNA is often contaminated by modern human DNA.

(c) The DNA is often contaminated by ancient prokaryotic DNA.

(d) All of the above.

[13-5] Pyrosequencing is a powerful next-generation sequencing technology. One of its limitations is that

(a) The length of each read is typically only 33 to 50 nucleotides.

(b) Homopolymers are difficult to sequence accurately.

(c) Bacterial cloning of DNA fragments is required.

(d) The error rate is high.

[13-6] Cycle reversible termination is another next-generation sequencing technology. One of its advantages is that

(a) It does not require gel electrophoresis.

(b) All four nucleotides are added to each reaction.

(c) There are millions of reads, offsetting the very short read lengths.

(d) All of the above.

[13-7] The term "whole-genome shotgun sequencing" refers to:

(a) A strategy to sequence an entire genome by breaking up DNA and sequencing using oligonucleotide primers that span the genomic DNA

(b) A strategy to sequence an entire genome by breaking up DNA, cloning it into libraries, and sequencing using oligonucleotide primers that correspond to known chromosomal locations (contigs)

(c) A strategy to sequence an entire genome by breaking up DNA, cloning it into libraries, hybridizing small fragments, then reassembling the fragments into a complete map

(d) A strategy to sequence an entire genome by breaking up DNA, cloning it into libraries, sequencing small fragments, then reassembling the fragments into a complete map

[13-8] The biggest problem in predicting protein-coding genes from genomic sequences using algorithms is that:

(a) The software is difficult to use.

(b) The false negative rate is high: many exons are missed.

(c) The false positive rate is high: many exons are falsely assigned.

(d) The false positive rate is high: many exons have unknown function.

[13-9] For finished DNA sequence, the error rate must be

(a) .01 or less

(b) .001 or less

(c) .0001 or less

(d) .00001 or less

SUGGESTED READING

Miller et al. (2004) present an exceptional overview of the field of comparative genomics. Metzker (2005) and Bentley (2006) provide reviews of new DNA sequencing technologies.

REFERENCES

Albert, T. J., Molla, M. N., Muzny, D. M., Nazareth, L., Wheeler, D., Song, X., Richmond, T. A., Middle, C. M., Rodesch, M.J, Packard, C. J., Weinstock, G. M., and Gibbs, R. A. Direct selection of human genomic loci by microarray hybridization. *Nat. Methods* **4**, 903–905 (2007).

Adams, M. D., et al. The genome sequence of *Drosophila melanogaster*. *Science* **287**, 2185–2195 (2000).

Allwood, A. C., Walter, M. R., Kamber, B. S., Marshall, C. P., and Burch, I. W. Stromatolite reef from the Early Archaean era of Australia. *Nature* **441**, 714–718 (2006).

Alm, R. A., et al. Genomic-sequence comparison of two unrelated isolates of the human gastric pathogen *Helicobacter pylori*. *Nature* **397**, 176–180 (1999).

Arabidopsis Genome Initiative. Analysis of the genome sequence of the flowering plant *Arabidopsis thaliana*. *Nature* **408**, 796–815 (2000).

Anderson, S., et al. Sequence and organization of the human mitochondrial genome. *Nature* **290**, 457–465 (1981).

Andersson, S. G., et al. The genome sequence of *Rickettsia prowazekii* and the origin of mitochondria. *Nature* **396**, 133–140 (1998).

Barns, S. M., Delwiche, C. F., Palmer, J. D., and Pace, N. R. Perspectives on archaeal diversity, thermophily and monophyly from environmental rRNA sequences. *Proc. Natl. Acad. Sci. USA* **93**, 9188–9193 (1996).

Baytaluk, M. V., Gelfand, M. S., and Mironov, A. A. Exact mapping of prokaryotic gene starts. *Brief Bioinform.* **3**, 181–194 (2002).

Bentley, D. R. Whole-genome re-sequencing. *Curr. Opin. Genet. Dev.* **16**, 545–552 (2006).

Benton, M. J., and Ayala, F. J. Dating the tree of life. *Science* **300**, 1698–1700 (2003).

Bernardi, G., and Bernardi, G. Compositional transitions in the nuclear genomes of cold-blooded vertebrates. *J. Mol. Evol.* **31**, 282–293 (1990).

Blackmore, S. Environment. Biodiversity update—progress in taxonomy. *Science* **298**, 365 (2002).

Blattner, F. R., et al. The complete genome sequence of *Escherichia coli* K-12. *Science* **277**, 1453–1474 (1997).

Brenner, S. E. Errors in genome annotation. *Trends Genet.* **15**, 132–133 (1999).

Bult, C. J., et al. Complete genome sequence of the methanogenic archaeon, *Methanococcus jannaschii*. *Science* **273**, 1058–1073 (1996).

Callinan, P. A., and Feinberg, A. P. The emerging science of epigenomics. *Hum. Mol. Genet.* **15**, R95–R101 (2006).

Carlton, J. M., et al. Genome sequence and comparative analysis of the model rodent malaria parasite *Plasmodium yoelii yoelii*. *Nature* **419**, 512–519 (2002).

Casjens, S., et al. A bacterial genome in flux: The twelve linear and nine circular extrachromosomal DNAs in an infectious isolate of the Lyme disease spirochete *Borrelia burgdorferi*. *Mol. Microbiol.* **35**, 490–516 (2000).

C. elegans Sequencing Consortium. Genome sequence of the nematode *C. elegans*: A platform for investigating biology. *Science* **282**, 2012–2018 (1998).

Chambaud, I., et al. The complete genome sequence of the murine respiratory pathogen *Mycoplasma pulmonis*. *Nucleic Acids Res.* **29**, 2145–2153 (2001).

Ciccarelli, F. D., Doerks, T., von Mering, C., Creevey, C. J., Snel, B., and Bork, P. Toward automatic reconstruction of a highly resolved tree of life. *Science* **311**, 1283–1287 (2006).

Cole, S. T., et al. Deciphering the biology of *Mycobacterium tuberculosis* from the complete genome sequence. *Nature* **393**, 537–544 (1998).

Cole, S. T., et al. Massive gene decay in the leprosy bacillus. *Nature* **409**, 1007–1011 (2001).

Cooper, A., Lalueza-Fox, C., Anderson, S., Rambaut, A., Austin, J., and Ward, R. Complete mitochondrial genome sequences of two extinct moas clarify ratite evolution. *Nature* **409**, 704–707 (2001).

Curwen, V., Eyras, E., Andrews, T. D., Clarke, L., Mongin, E., Searle, S. M., and Clamp, M. The Ensembl automatic gene annotation system. *Genome Res.* **14**, 942–950 (2004).

Dandekar, T., et al. Re-annotating the *Mycoplasma pneumoniae* genome sequence: Adding value, function and reading frames. *Nucleic Acids Res.* **28**, 3278–3288 (2000).

Deckert, G., et al. The complete genome of the hyperthermophilic bacterium *Aquifex aeolicus*. *Nature* **392**, 353–388 (1998).

Doolittle, W. F. Phylogenetic classification and the universal tree. *Science* **284**, 2124–2129 (1999).

Douglas, S., et al. The highly reduced genome of an enslaved algal nucleus. *Nature* **410**, 1091–1096 (2001).

Driskell, A. C., Ané, C., Burleigh, J. G., McMahon, M. M., O'Meara, B. C., and Sanderson, M. J. Prospects for building the tree of life from large sequence databases. *Science* **306**, 1172–1174 (2004).

Dunham, I., et al. The DNA sequence of human chromosome 22. *Nature* **402**, 489–495 (1999).

Edwards, R. A., Rodriguez-Brito, B., Wegley, L., Haynes, M., Breitbart, M., Peterson, D. M., Saar, M. O., Alexander, S., Alexander, E. C. Jr., and Rohwer, F. Using pyrosequencing to shed light on deep mine microbial ecology. *BMC Genomics* **7**, 57 (2006).

Eisen, J. A. Horizontal gene transfer among microbial genomes: New insights from complete genome analysis. *Curr. Opin. Genet. Dev.* **10**, 606–611 (2000).

Eisen, J. A., and Fraser, C. M. Phylogenomics: Intersection of evolution and genomics. *Science* **300**, 1706–1707 (2003).

Eisen, J. A., and Hanawalt, P. C. A phylogenomic study of DNA repair genes, proteins, and processes. *Mutat. Res.* **435**, 171–213 (1999).

Ferretti, J. J., et al. Complete genome sequence of an M1 strain of *Streptococcus pyogenes*. *Proc. Natl. Acad. Sci. USA* **98**, 4658–4663 (2001).

Fiers, W., Contreras, R., Duerinck, F., Haegeman, G., Iserentant, D., Merregaert, J., Min Jou, W., Molemans, F., Raeymaekers, A., Van den Berghe, A., Volckaert, G., and Ysebaert, M. Complete nucleotide sequence of bacteriophage MS2 RNA: Primary and secondary structure of the replicase gene. *Nature* **260**, 500–507 (1976).

Fiers, W., Contreras, R., Haegemann, G., Rogiers, R., Van de Voorde, A., Van Heuverswyn, H., Van Herreweghe, J., Volckaert, G., and Ysebaert, M. Complete nucleotide sequence of SV40 DNA. *Nature* **273**, 113–120 (1978).

Fleischmann, R. D., et al. Whole-genome random sequencing and assembly of *Haemophilus influenzae* Rd. *Science* **269**, 496–512 (1995).

Fox, G. E., et al. The phylogeny of prokaryotes. *Science* **209**, 457–463 (1980).

Fraser, C. M., et al. The minimal gene complement of *Mycoplasma genitalium*. *Science* **270**, 397–403 (1995).

Fraser, C. M., et al. Genomic sequence of a Lyme disease spirochaete, *Borrelia burgdorferi*. *Nature* **390**, 580–586 (1997).

Fraser, C. M., et al. Complete genome sequence of *Treponema pallidum*, the syphilis spirochete. *Science* **281**, 375–388 (1998).

Galibert, F., et al. The composite genome of the legume symbiont *Sinorhizobium meliloti*. *Science* **293**, 668–672 (2001).

García Martín, H., et al. Metagenomic analysis of two enhanced biological phosphorus removal (EBPR) sludge communities. *Nat. Biotechnol.* **24**, 1263–1269 (2006).

Gardner, M. J., et al. Genome sequence of the human malaria parasite *Plasmodium falciparum*. *Nature* **419**, 498–511 (2002).

Gilbert, M. T., et al. Whole-genome shotgun sequencing of mitochondria from ancient hair shafts. *Science* **317**, 1927–1930 (2007).

Gill, S. R., Pop, M., Deboy, R. T., Eckburg, P. B., Turnbaugh, P. J., Samuel, B. S., Gordon, J. I., Relman, D. A., Fraser-Liggett, C. M., and Nelson, K. E. Metagenomic analysis of the human distal gut microbiome. *Science* **312**, 1355–1359 (2006).

Glass, J. I., et al. The complete sequence of the mucosal pathogen *Ureaplasma urealyticum*. *Nature* **407**, 757–762 (2000).

Goffeau, A., et al. Life with 6000 genes. *Science* **274**, 546, 563–577 (1996).

Gordon, D., Desmarais, C., and Green, P. Automated finishing with autofinish. *Genome Res.* **11**, 614–625 (2001).

Graur, D., and Li, W.-H. *Fundamentals of Molecular Evolution*. Sinauer Associates, Sunderland, MA, 2000.

Green, R. E., Krause, J., Ptak, S. E., Briggs, A. W., Ronan, M. T., Simons, J. F., Du, L., Egholm, M., Rothberg, J. M., Paunovic, M., and Pääbo, S. Analysis of one million base pairs of Neanderthal DNA. *Nature* **444**, 330–336 (2006).

Haeckel, E. *The Evolution of Man: A Popular Exposition of the Principal Points of Human Ontogeny and Phylogeny*. D. Appleton and Company, New York, 1879.

Hattori, M., et al. The DNA sequence of human chromosome 21. The chromosome 21 mapping and sequencing consortium. *Nature* **405**, 311–319 (2000).

Hayashi, T., et al. Complete genome sequence of enterohemorrhagic *Escherichia coli* O157:H7 and genomic comparison with a laboratory strain K-12. *DNA Res* **8**, 11–22 (2001).

Hedges, S. B., et al. Genomic timescale for the origin of eukaryotes. *BMC Evol. Biol.* **1**, 1–10 (2001).

Heidelberg, J. F., et al. DNA sequence of both chromosomes of the cholera pathogen *Vibrio cholerae*. *Nature* **406**, 477–483 (2000).

Heilig, R. *Pyrococcus abyssi* genome sequence: Insights into archaeal chromosome structure and evolution. Unpublished GenBank entry NC_000868 (▶ *http://www.ncbi.nlm.nih.gov*), 2001.

Himmelreich, R., et al. Complete sequence analysis of the genome of the bacterium *Mycoplasma pneumoniae*. *Nucleic Acids Res.* **24**, 4420–4449 (1996).

Hodges, E., Xuan, Z., Balija, V., Kramer, M., Molla, M. N., Smith, S. W., Middle, C. M., Rodesch, M. J., Albert, T. J., Hannon, G. J., and McCombie, W. R. Genome-wide in situ exon capture for selective resequencing. *Nat. Genet.* **39**, 1522–1527 (2007).

Hofreiter, M., Jaenicke, V., Serre, D., Haeseler, A. A., and Pääbo, S. DNA sequences from multiple amplifications reveal artifacts induced by cytosine deamination in ancient DNA. *Nucleic Acids Res.* **29**, 4793–4799 (2001).

Holley, R. W., Apgar, J., Everett, G. A., Madison, J. T., Marquisee, M., Merrill, S. H., Penswick, J. R., and Zamir, A. Structure of a ribonucleic acid. *Science* **147**, 1462–1465 (1965).

Holt, R. A., et al. The genome sequence of the malaria mosquito *Anopheles gambiae*. *Science* **298**, 129–149 (2002).

Huse, S. M., Huber, J. A., Morrison, H. G., Sogin, M. L., and Welch, D. M. Accuracy and quality of massively parallel DNA pyrosequencing. *Genome Biol.* **8**, R143 (2007).

Hyman, E. D. A new method of sequencing DNA. *Anal. Biochem.* **174**, 423–436 (1988).

International Human Genome Sequence Consortium. Initial sequencing and analysis of the human genome. *Nature* **409**, 860–921 (2001).

International Human Genome Sequencing Consortium. Finishing the euchromatic sequence of the human genome. *Nature* **431**, 931–945 (2004).

Joyce, G. F. The antiquity of RNA-based evolution. *Nature* **418**, 214–221 (2002).

Kaiser, D. Building a multicellular organism. *Annu. Rev. Genet.* **35**, 103–123 (2001).

Kalman, S., et al. Comparative genomes of *Chlamydia pneumoniae* and *C. trachomatis*. *Nat. Genet.* **21**, 385–389 (1999).

Kaneko, T., et al. Sequence analysis of the genome of the unicellular cyanobacterium *Synechocystis* sp. strain PCC6803. II. Sequence determination of the entire genome and assignment of potential protein-coding regions. *DNA Res.* **3**, 109–136 (1996).

Karolchik, D., Bejerano, G., Hinrichs, A. S., Kuhn, R. M., Miller, W., Rosenbloom, K. R., Zweig, A. S., Haussler, D., and Kent, W. J. Comparative genomic analysis using the UCSC Genome Browser. *Methods Mol. Biol.* **395**, 17–34 (2007).

Katinka, M. D., et al. Genome sequence and gene compaction of the eukaryote parasite *Encephalitozoon cuniculi*. *Nature* **414**, 450–453 (2001).

Kawarabayasi, Y., Hino, Y., Horikawa, H., Yamazaki, S., Haikawa, Y., Jin-no, K., Takahashi, M., Sekine, M., Baba, S., Ankai, et al. Complete genome sequence of an aerobic hyper-thermophilic crenarchaeon, *Aeropyrum pernix* K1. *DNA Res.* **6**, 83–101 (1999).

Kawarabayasi, Y., et al. Complete genome sequence of an aerobic thermoacidophilic crenarchaeon, *Sulfolobus tokodaii* strain 7. *DNA Res.* **8**, 123–140 (2001).

Kawarabayasi, Y., et al. Complete sequence and gene organization of the genome of a hyper-thermophilic archaebacterium, *Pyrococcus horikoshii* OT3. *DNA Res.* **5**, 55–76 (1998).

Kawashima, T., et al. Determination of the complete genomic DNA sequence of *Thermoplasma volvanium* GSS1. *Proc. Jpn. Acad.* **75**, 213–218 (1999).

Klenk, H. P., et al. The complete genome sequence of the hyperthermophilic, sulphate-reducing archaeon *Archaeoglobus fulgidus*. *Nature* **390**, 364–370 (1997).

Knapp, S., et al. Taxonomy needs evolution, not revolution. *Nature* **419**, 559 (2002).

Krause, J., Dear, P. H., Pollack, J. L., Slatkin, M., Spriggs, H., Barnes, I., Lister, A. M., Ebersberger, I., Pääbo, S., and Hofreiter, M. Multiplex amplification of the mammoth mitochondrial genome and the evolution of Elephantidae. *Nature* **439**, 724–727 (2006).

Kumar, S. and Hedges, S. B. A molecular timescale for vertebrate evolution. *Nature* **392**, 917–920 (1998).

Kunst, F., et al. The complete genome sequence of the gram-positive bacterium *Bacillus subtilis*. *Nature* **390**, 249–256 (1997).

Lander, E. S., and Waterman, M. S. Genomic mapping by fingerprinting random clones: A mathematical analysis. *Genomics* **2**, 231–239 (1988).

Lang, B. F., Gray, M. W., and Burger, G. Mitochondrial genome evolution and the origin of eukaryotes. *Annu. Rev. Genet.* **33**, 351–397 (1999).

Letunic, I., and Bork, P. Interactive Tree Of Life (iTOL): An online tool for phylogenetic tree display and annotation. *Bioinformatics* **23**, 127–128 (2007).

Lilly, J. W., and Havey, M. J. Small, repetitive DNAs contribute significantly to the expanded mitochondrial genome of cucumber. *Genetics* **159**, 317–328 (2001).

Liolios, K., Mavromatis, K., Tavernarakis, N., and Kyrpides, N. C. The Genomes On Line Database (GOLD) in 2007: Status of genomic and metagenomic projects and their associated metadata. *Nucleic Acids Res.* **36**, D475–D479 (2008).

Margulies, M., et al. Genome sequencing in microfabricated high-density picolitre reactors. *Nature* **437**, 376–380 (2005).

May, B. J., et al. Complete genomic sequence of *Pasteurella multocida*, Pm70. *Proc. Natl. Acad. Sci. USA* **98**, 3460–3465 (2001).

Mayr, E. *The Growth of Biological Thought: Diversity, Evolution, and Inheritance*. Belknap Harvard, Cambridge, MA, 1982.

Metzker, M. L. Emerging technologies in DNA sequencing. *Genome Res.* **15**, 1767–1776 (2005).

Miller, W., Makova, K. D., Nekrutenko, A., and Hardison, R. C. Comparative genomics. *Annu. Rev. Genomics Hum. Genet.* **5**, 15–56 (2004).

Nelson, K. E., et al. Evidence for lateral gene transfer between *Archaea* and bacteria from genome sequence of *Thermotoga maritima*. *Nature* **399**, 323–329 (1999).

Ng, W. V., et al. Genome sequence of *Halobacterium* species NRC-1. *Proc. Natl. Acad. Sci. USA* **97**, 12176–12181 (2000).

Nierman, W. C., et al. Complete genome sequence of *Caulobacter crescentus*. *Proc. Natl. Acad. Sci. USA* **98**, 4136–4141 (2001).

Noonan, J. P., Hofreiter, M., Smith, D., Priest, J. R., Rohland, N., Rabeder, G., Krause, J., Detter, J. C., Pääbo, S., and Rubin, E.

M. Genomic sequencing of Pleistocene cave bears. *Science* **309**, 597–599 (2005).

Ohyama, K., et al. Chloroplast gene organization deduced from complete sequence of liverwort *Marchantia polymorpha* chloroplast DNA. *Nature* **322**, 572–574 (1986).

Oliver, S. G., et al. The complete DNA sequence of yeast chromosome III. *Nature* **357**, 38–46 (1992).

Pääbo, S., Poinar, H., Serre, D., Jaenicke-Despres, V., Hebler, J., Rohland, N., Kuch, M., Krause, J., Vigilant, L., and Hofreiter, M. Genetic analyses from ancient DNA. *Annu. Rev. Genet.* **38**, 645–679 (2004).

Pace, N. R. A molecular view of microbial diversity and the biosphere. *Science* **276**, 734–740 (1997).

Parkhill, J., et al. The genome sequence of the food-borne pathogen *Campylobacter jejuni* reveals hypervariable sequences. *Nature* **403**, 665–668 (2000a).

Parkhill, J., et al. Complete DNA sequence of a serogroup A strain of *Neisseria meningitidis* Z2491. *Nature* **404**, 502–506 (2000b).

Pennisi, E. Taxonomy. Linnaeus's last stand? *Science* **291**, 2304–2307 (2001).

Perna, N. T., et al. Genome sequence of enterohaemorrhagic *Escherichia coli* O157:H7. *Nature* **409**, 529–533 (2001).

Porreca, G. J., Zhang, K., Li, J. B., Xie, B., Austin, D., Vassallo, S. L., Leproust, E. M., Peck, B. J., Emig, C. J., Dahl, F., Gao, Y., Church, G. M., and Shendure, J. Multiplex amplification of large sets of human exons. *Nat. Methods* **4**, 931–936 (2007).

Potter, S. C., Clarke, L., Curwen, V., Keenan, S., Mongin, E., Searle, S. M., Stabenau, A., Storey, R., and Clamp, M. The Ensembl analysis pipeline. *Genome Res.* **14**, 934–941 (2004).

Raven, P., Fauquet, C., Swaminathan, M. S., Borlaug, N., and Samper, C. Where next for genome sequencing? *Science* **311**, 468 (2006).

Read, T. D., et al. Genome sequences of *Chlamydia trachomatis* MoPn and *Chlamydia pneumoniae* AR39. *Nucleic Acids Res.* **28**, 1397–1406 (2000).

Riesenfeld, C. S., Schloss, P. D., and Handelsman, J. Metagenomics: Genomic analysis of microbial communities. *Annu. Rev. Genet.* **38**, 525–552 (2004).

Rivera, M. C., and Lake, J. A. The ring of life provides evidence for a genome fusion origin of eukaryotes. *Nature* **431**, 152–155 (2004).

Ruepp, A., et al. The genome sequence of the thermoacidophilic scavenger *Thermoplasma acidophilum*. *Nature* **407**, 508–513 (2000).

Rusch, D. B., et al. The Sorcerer II Global Ocean Sampling expedition: Northwest Atlantic through eastern tropical Pacific. *PloS Biol.* **5**, e77 (2007).

Salzberg, S. L., Hotopp, J. C., Delcher, A. L., Pop, M., Smith, D. R., Eisen, M. B., and Nelson, W. C. Serendipitous discovery of *Wolbachia* genomes in multiple *Drosophila* species. *Genome Biol.* **6**, R23 (2005).

Sanger, F., et al. Nucleotide sequence of bacteriophage phi X174 DNA. *Nature* **265**, 687–895 (1977a).

Sanger, F., Nicklen, S., and Coulson, A. R. DNA sequencing with chain-terminating inhibitors. *Proc. Natl. Acad. Sci. USA* **74**, 5463–5467 (1977b).

Schweitzer, M. H., Suo, Z., Avci, R., Asara, J. M., Allen, M. A., Arce, F. T., and Horner, J. R. Analyses of soft tissue from *Tyrannosaurus rex* suggest the presence of protein. *Science* **316**, 277–280 (2007).

She, Q., et al. The complete genome of the crenarchaeon *Sulfolobus solfataricus* P2. *Proc. Natl. Acad. Sci. USA* **98**, 7835–7840 (2001).

Shinozaki, K. M., et al. The complete nucleotide sequence of the tobacco chloroplast genome: Its gene organization and expression. *EMBO J.* **5**, 2043–2049 (1986).

Simpson, A. J., et al. The genome sequence of the plant pathogen *Xylella fastidiosa*. *Nature* **406**, 151–157 (2000).

Skaletsky, H., et al. The male-specific region of the human Y chromosome is a mosaic of discrete sequence classes. *Nature* **423**, 825–837 (2003).

Smith, D. R., et al. Complete genome sequence of *Methanobacterium thermoautotrophicum* deltaH: Functional analysis and comparative genomics. *J. Bacteriol.* **179**, 7135–7155 (1997).

Stephens, R. S., et al. Genome sequence of an obligate intracellular pathogen of humans: *Chlamydia trachomatis*. *Science* **282**, 754–759 (1998).

Stover, C. K., et al. Complete genome sequence of *Pseudomonas aeruginosa* PA01, an opportunistic pathogen. *Nature* **406**, 959–964 (2000).

Takami, H., et al. Complete genome sequence of the alkaliphilic bacterium *Bacillus halodurans* and genomic sequence comparison with *Bacillus subtilis*. *Nucleic Acids Res.* **28**, 4317–4331 (2000).

Tettelin, H., Radune, D., Kasif, S., Khouri, H., and Salzberg, S. L. Optimized multiplex PCR: Efficiently closing a whole-genome shotgun sequencing project. *Genomics* **62**, 500–507 (1999).

Tettelin, H., et al. Complete genome sequence of *Neisseria meningitidis* serogroup B strain MC58. *Science* **287**, 1809–1815 (2000).

Tettelin, H., et al. Complete genome sequence of a virulent isolate of *Streptococcus pneumoniae*. *Science* **293**, 498–506 (2001).

Tomb, J. F., et al. The complete genome sequence of the gastric pathogen *Helicobacter pylori*. *Nature* **388**, 539–547 (1997).

Venter, J. C., et al. The sequence of the human genome. *Science* **291**, 1304–1351 (2001).

Warnecke, F., Luginbühl, P., Ivanova, N., Ghassemian, M., Richardson, T. H., Stege, J. T., Cayouette, M., McHardy, A. C., Djordjevic, G., Aboushadi, N., et al. Metagenomic and functional analysis of hindgut microbiota of a wood-feeding higher termite. *Nature* **450**, 560–565 (2007).

Weber, J. L., and Myers, E. W. Human whole-genome shotgun sequencing. *Genome Res.* **7**, 401–409 (1997).

White, O., et al. Genome sequence of the radioresistant bacterium *Deinococcus radiodurans* R1. *Science* **286**, 1571–1577 (1999).

Whittaker, R. H. New concepts of kingdoms or organisms. Evolutionary relations are better represented by new classifications than by the traditional two kingdoms. *Science* **163**, 150–160 (1969).

Willerslev, E., and Cooper, A. Ancient DNA. *Proc. Biol. Sci.* **272**, 3–16 (2005).

Wilson, E. O. *The Diversity of Life.* W.W. Norton, New York, 1992.

Woese, C. R. Default taxonomy: Ernst Mayr's view of the microbial world. *Proc. Natl. Acad. Sci. USA* **95**, 11043–11046 (1998).

Woese, C. R., Kandler, O., and Wheelis, M. L. Towards a natural system of organisms: Proposal for the domains Archaea, Bacteria, and Eucarya. *Proc. Natl. Acad. Sci. USA* **87**, 4576–4579 (1990).

Wolf, Y. I., Rogozin, I. B., Grishin, N. V., and Koonin, E. V. Genome trees and the tree of life. *Trends Genet.* **18**, 472–479 (2002).

Wood, V., et al. The genome sequence of *Schizosaccharomyces pombe. Nature* **415**, 871–880 (2002).

Wu, R. Nucleotide sequence analysis of DNA. I. Partial sequence of the cohesive ends of bacteriophage lambda and 186 DNA. *J. Mol. Biol.* **51**, 501–521 (1970).

Wu, R., and Taylor, E. Nucleotide sequence analysis of DNA. II. Complete nucleotide sequence of the cohesive ends of bacteriophage lambda DNA. *J. Mol. Biol.* **57**, 491–511 (1971).

Wuyts, J., Perriere, G., and Van de Peer, Y. The European ribosomal RNA database. *Nucleic Acids Res.* **32**, D101–D103 (2004).

Zhang, T., Breitbart, M., Lee, W. H., Run, J. Q., Wei, C. L., Soh, S. W., Hibberd, M. L., Liu, E. T., Rohwer, F., and Ruan, Y. RNA viral community in human feces: Prevalence of plant pathogenic viruses. *PloS Biol.* **4**, e3 (2006).

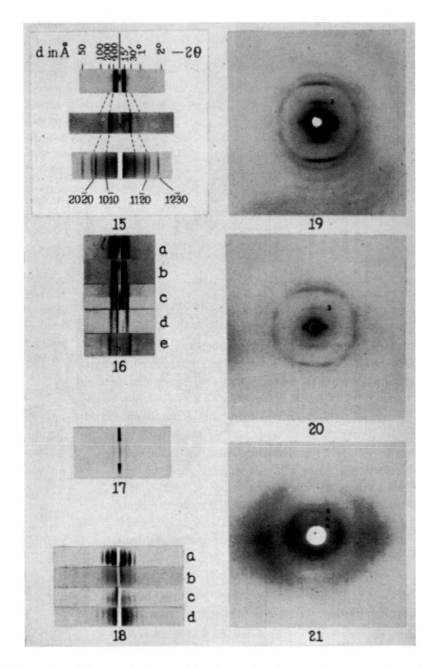

As early as 1885 Adolf Mayer showed that mosaic disease of the tobacco plant is contagious; we now know that it is caused by tobacco mosaic virus. Martinus Beijerinck (1851–1931) further isolated a "contagium vivum fluidum" (virus) from tobacco leaves, distinguishing the causative agent from bacteria. Due to their small size, almost all viruses cannot be visualized by conventional microscopy. Beginning in the 1930s Helmut Ruska pioneered the use of the electron microscope to visualize viruses (Kruger et al., 2000). Early studies of the structure of viruses based on x-ray crystallography were performed by John D. Bernal (1901–1971). He also trained Maurice Wilkins and Rosalind Franklin (who confirmed the structure of the double helix of DNA) and Nobel laureate Dorothy Crowfoot Hodgkin (who solved the structure of vitamin B_{12}). Together with Rosalind Franklin, Bernel studied tobacco mosaic virus in the 1950s. Bernal and Fankuchen 1941 obtained a variety of purified viruses and performed x-ray analyses. This set of images shows figures 15 (demonstrating shifts of intermolecular reflections), 16 (showing varying concentrations of viruses), 17 (enation mosaic virus), 18 (dry gels of various virus proteins), 19 (tobacco mosaic virus), 20 (cucumber mosaic virus), and 21 (potato virus X).

14

Completed Genomes: Viruses

INTRODUCTION

In this chapter we will consider bioinformatic approaches to viruses. Viruses are small, infectious, obligate intracellular parasites. They depend on host cells for their ability to replicate. The virion (virus particle) consists of a nucleic acid genome surrounded by coat proteins (capsid) that may be enveloped in a lipid bilayer (derived from the host cell) studded with viral glycoproteins. Unlike other genomes, viral genomes can consist of either DNA or RNA. Furthermore, they can be single, double, or partially double stranded, and can be circular, linear, or segmented (having different genes on distinct nucleic acid segments).

Viruses lack the biochemical machinery that is necessary for independent existence. This is the fundamental distinction between viruses and free-living organisms. Thus, while they replicate and evolve, viruses exist on the borderline of the definition of life. The largest virus has a genome size of over 1 megabase (Mimivirus; see below), and other large viruses (such as pox viruses) have genome sizes of several hundred kilobases. These are nearly the same size as the smallest archaeal and bacterial genomes (such as *Nanoarchaeum equitans* and *Mycoplasma genitalium*) (Chapter 15). It is not a coincidence that those smallest prokaryotic genomes are from organisms that (like viruses) are small, infectious, obligate intracellular agents (see Chapter 15).

While there may be tens or hundreds of millions of species of bacteria and archaea, only a few thousand species of virus are known. This disparity probably reflects their specialized requirement for invading a host. Also, recent metagenomics projects (described below) suggest that we have an extremely limited understanding of both the number of virus species and the diversity of viral genes and genomes. Viruses infect all forms of life, including bacteria, archaea (Prangishvili et al., 2006), and eukaryotes from plants to humans to fungi. Although we have catalogued relatively few viral species, viruses are nonetheless the most abundant biological entities on the earth (Edwards and Rohwer, 2005).

Although viruses are relatively simple agents, they are more complex than two other pathogenic agents: viroids and prions. Viroids are small, circular RNA molecules of 200 to 400 nucleotides that cause diseases in plants (Flores, 2001; Daròs et al., 2006). This minuscule genome does not encode any proteins, and the RNA itself has enzymatic activity. Prions are infectious protein molecules (Prusiner, 1998; DeArmond and Prusiner, 2003). Cruetzfeld–Jakob disease is the most common human prion disease (Johnson and Gibbs, 1998). It has a worldwide incidence of one in one million individuals and usually presents as dementia. Scrapie in sheep and bovine spongiform encephalopathy (BSE; "mad cow" disease) are the most common prion diseases in animals.

Classification of Viruses

Before the sequencing era, morphology was the most important criterion for the classification of viruses. Since 1959, electron microscopy has been employed to describe the structure of over 5500 bacteriophages (viruses that invade bacteria; Ackermann, 2007), as well as additional viruses that invade plants and animals. Ninety-six percent of bacteriophages are tailed viruses, with the remainder having filamentous, icosahedral, or pleiomorphic shapes. Many electron microscopic images of viruses are available at ICTVdb, the database of the International Committee on Taxonomy of Viruses (ICTV) (Büchen-Osmond, 1997, 2003). Several of these images are presented in Fig. 14.1. ICTV is responsible for classifying viruses and its website includes resources such as an index of viruses, viral characters, and a variety of software tools (Fauquet et al., 2005).

In addition to morphology, another fundamental basis for classifying viruses is to define the type of nucleic acid genome that is packaged into the virion. Virions contain DNA or RNA; the nucleic acid may be single or double stranded, and translation may occur from the sense strand, the antisense strand, or both. Double-stranded viral genomes replicate by using the individual strands of the DNA or RNA duplex as a template to synthesize daughter strands. Single-stranded DNA or RNA viruses use their strand of nucleic acid as a template for a polymerase to copy a complementary strand. Replication may involve the stable or transient formation of double-stranded intermediates. Some viruses with single-stranded RNA genomes convert the RNA strand to DNA using reverse transcriptase (RNA-dependent DNA polymerase). In the case of HIV-1, the *pol* gene encodes reverse transcriptase.

The ICTV regularly meets to refine an accepted standard for virus classification. The ICTV recognizes the taxa of order, family, genus, and species. The ICTV database (eighth report) subdivides viruses into some 73 families, 9 subfamilies, 287 genera and 1938 species (summarized by Mayo and Pringle, 1998). An example of an online description of viruses on the ICTV website is provided in Fig. 14.2. For a discussion of issues in virus taxonomy, including the concept of a viral species, see van Regenmortel and Mahy 2004.

Stanley Prusiner won the Nobel Prize in Physiology or Medicine in 1997 "for his discovery of Prions — a new biological principle of infection." See ► http://nobelprize.org/nobel_prizes/medicine/laureates/1997/.

The ICTV website is at ► http://www.ncbi.nlm.nih.gov/ICTVdb/. The ICTVdb was constructed by Cornelia Büchen-Osmond (Columbia University). The Büchen-Osmond 2003 reference is online at ► http://www.ncbi.nlm.nih.gov/ICTVdb/c3buch.lo.pdf.

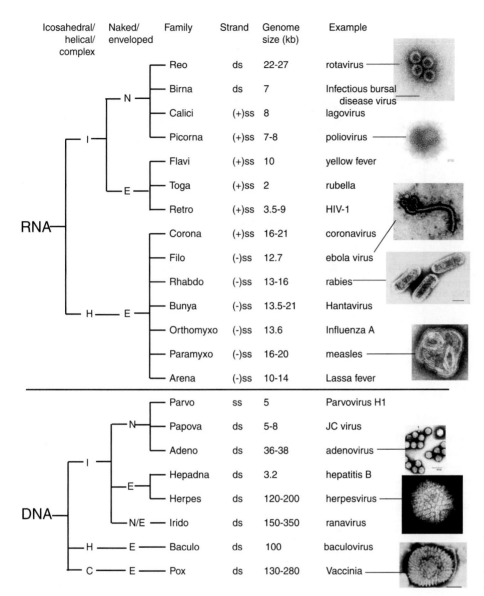

Icosahedral/ helical/ complex	Naked/ enveloped	Family	Strand	Genome size (kb)	Example
		Reo	ds	22-27	rotavirus
	N	Birna	ds	7	Infectious bursal disease virus
		Calici	(+)ss	8	lagovirus
I		Picorna	(+)ss	7-8	poliovirus
		Flavi	(+)ss	10	yellow fever
	E	Toga	(+)ss	2	rubella
RNA		Retro	(+)ss	3.5-9	HIV-1
		Corona	(+)ss	16-21	coronavirus
		Filo	(-)ss	12.7	ebola virus
		Rhabdo	(-)ss	13-16	rabies
H — E		Bunya	(-)ss	13.5-21	Hantavirus
		Orthomyxo	(-)ss	13.6	Influenza A
		Paramyxo	(-)ss	16-20	measles
		Arena	(-)ss	10-14	Lassa fever
		Parvo	ss	5	Parvovirus H1
	N	Papova	ds	5-8	JC virus
		Adeno	ds	36-38	adenovirus
I		Hepadna	ds	3.2	hepatitis B
	E	Herpes	ds	120-200	herpesvirus
DNA	N/E	Irido	ds	150-350	ranavirus
H — E		Baculo	ds	100	baculovirus
C — E		Pox	ds	130-280	Vaccinia

FIGURE 14.1. *Classification of viruses. Adapted from ICTVdb and Flint et al. (2000, pp. 16–17). Electron micrographs are from the ICTV website.*

Some of the major groups of viruses are shown in Fig. 14.1 and Table 14.1. They range in genome size from very small viruses such as rubella ($\approx$2 kb) to several viruses over 350 kb in size. A giant virus (called Mimivirus for *Mimicking microbe*) has been described, having a double-stranded circular genome of 1,181,404 base pairs (1.2 megabases) (La Scola et al., 2003; Raoult et al., 2004). Its mature particles are 400 nanometers in diameter. It is thus larger than many bacteria (the *Mycoplasma genitalium* genome is 580 kilobases) and archaea (the *Nanoarchaeum equitans* genome is 490 kb) and it is almost half the size of the smallest eukaryotic genome (that of *Encephalitozoon cuniculi*, 2.5 million base pairs). Of its 1262 open reading frames of length $\geq$100 amino acids, just 194 have similarity to proteins of known function.

An entirely different approach to classifying viruses is to identify those that cause human disease. Many viral diseases can be prevented by vaccination (Table 14.2). Others, such as smallpox, are of concern because of their potential use by bioterrorists (Cieslak et al., 2002). Smallpox, caused by the variola virus, was eradicated in 1977, and routine vaccination was discontinued in 1972 in the United States.

The Mimivirus GenBank accession number is NC_006450.

The National Institute of Allergy and Infectious Diseases (NIAID) at the National Institutes of Health offers information on viral and other diseases at ▶ http://www.niaid.nih.gov/publications/.

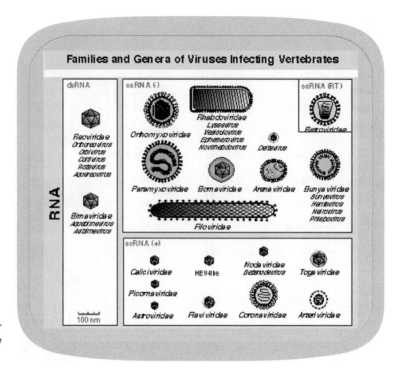

FIGURE 14.2. *Example of a diagram illustrating virus morphology from the ICTV database.*

TABLE 14-1   Classification of Viruses Based on Nucleic Acid Composition

Nucleic Acid	Strands	Family	Example	Accession	Base Pairs
RNA	Single	Picornaviridae	Human poliovirus 1	NC_002058	7,440
		Togaviridae	Rubella virus	NC_001545	9,755
		Flaviviridae	Yellow fever virus	NC_002031	10,862
		Coronaviridae	Coronavirus	NC_002645	27,317
		Rhabdoviridae	Rabies virus	NC_001542	11,932
		Paramyxoviridae	Measles virus	NC_001498	15,894
		Orthomyxoviridae	Influenza A virus (segment 1)	NC_002023	13,585
		Bunyaviridae	Hantavirus	—	—
		Arenaviridae	Lassa fever virus	J04324	3,402
		Retroviridae	HIV	NC_001802	9,181
	Double	Reoviridae	Rotavirus		
DNA	Single	Parvoviridae	Parvovirus H1	NC_001358	5,176
	Mixed	Hepadnaviridae	Hepatitis B	NC_001707	3,215
	Double	Papovaviridae	JC virus	NC_001699	5,130
		Adenoviridae	Human adenovirus, type 17	NC_002067	35,100
		Herpesviruses	Human herpesvirus 1	NC_001806	152,261
		Poxviridae	Vaccinia	NC_001559	191,737

Source: Adapted in part from Schaechter et al. (1999, p. 292). Used with permission.

TABLE 14-2 Vaccine-Preventable Viral Diseases

Disease	Virus	Comment
Hepatitis A	Hepatitis A virus	Causes liver disease
Hepatitis B	Hepatitis B virus	Causes liver disease
Influenza	Influenza type A or B	Causes 20,000 deaths per year (U.S.)
Measles	Measles virus	See below
Mumps	Rubulavirus	A disease of the lymph nodes
Poliomyelitis	Poliovirus (three serotypes)	Inflammation of the gray matter of the spinal cord; kills neurons
Rotavirus	Rotavirus	Most common cause of diarrhea in children; kills 600,000 children annually worldwide
Rubella	Genus Rubivirus	Also called German measles.
Smallpox	Variola virus	Eradicated in 1977
Varicella	Varicella-zoster virus	About 75% of all children contract varicella by age 15

Source: Adapted from ► http://www.cdc.gov/nip/diseases/disease-chart-hcp.htm.

Diversity and Evolution of Viruses

A practical way to access the diversity of known viruses is through the National Center for Biotechnology Information (NCBI) website. We introduced the NCBI Entrez Genome resources in Chapter 13 (Fig. 13.4). This site includes dedicated resources for viruses (Fig. 14.3), as well as specialized sites for influenza virus, retroviruses, SARS, and links to the ICTV database.

A premise of taxonomy is that it should represent phylogeny. In the case of viruses, their unique, elusive, and sometimes fragile nature makes it difficult to trace their evolution in as comprehensive a fashion as can be accomplished with archaea, bacteria, and eukaryotes. Like living organisms, viruses are subject to mutation (genetic variability) and selection. But viral genomes evolve far faster than cellular genomes and present special difficulties for evolutionary studies:

- Viruses tend not to survive in archeological or historical samples. There is considerable evidence for the existence of viruses over 10,000 years ago, based on human skeletal remains, historical accounts, and other historical artifacts. However, ancient viral DNA or RNA has not been recovered. As discussed below, influenza virus from the deadly 1918 pandemic has been isolated, sequenced, and functionally analyzed.

- Viral polymerases of RNA genomes typically lack proofreading activity. This leads to a mutation rate that may be 1 million to 10 million times greater than that of DNA genomes (McClure, 2000). For viruses having DNA genomes, the mutation rates are typically 20- to 100-fold higher than that of the host cell. As an example, the mutation rate of hepatitis C virus is 10^{-3} per nucleotide per generation (Chisari, 2005).

- In addition to a high mutation rate, many viruses also have an extremely high rate of replication. A single cell can produce 10,000 poliovirus particles, and an HIV-infected individual can produce 10^9 virus particles per day. For

Currently (January 2009) there are over 2700 viral genomes and 500 phage genomes listed at the NCBI Genome site. The Entrez Genome homepage for viruses is ► http://www.ncbi.nlm.nih.gov/genomes/VIRUSES/viruses.html.

According to George Gaylord Simpson (1963, p. 7), "Species are groups of actually or potentially inbreeding populations, which are reproductively isolated from other such groups. An evolutionary species is a lineage (an ancestral-descendant sequence of populations) evolving separately from others and with its own unitary evolutionary role and tendencies."

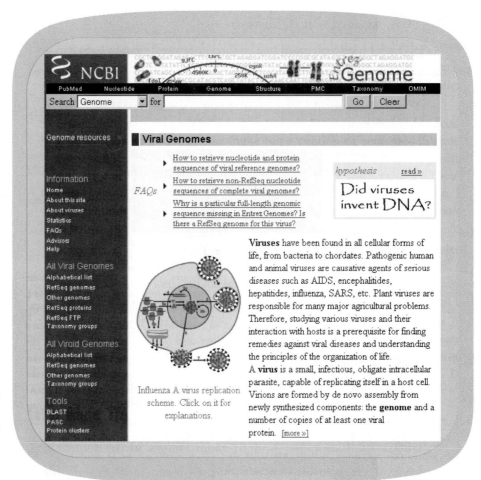

FIGURE 14.3. *The viral genomes page at NCBI provides information and resources for the study of viruses. There are links to tools (such as PASC for comparisons of viral genomes) and to specialized NCBI sites on retroviruses, SARS, and influenza viruses.*

hepatitis C, 10^{12} virions per day can be produced (Neumann et al., 1998). This can lead to the formation of quasi-species (a population of related but nonidentical viruses).

- Many viral genomes are segmented. This allows segments to be shuffled among progeny, producing a great diversity of viral subtypes (see influenza virus and HIV sections below).

- Viruses are often subjected to intense selective pressures such as host immune responses or antiviral drug therapies. The rapid mutation rate of HIV-1 ensures that some versions of the virus are likely to contain mutations conferring resistance to retroviral drugs, and these HIV-1 molecules will be selected for.

- Viruses have evolved to invade diverse species across the entire tree of life: archaea, bacteria, and eukaryotes. Viruses that infect plants (e.g., tomato bushy stunt virus), animals (e.g., SV40, rhinovirus, and poliovirus), as well as bacteria (e.g., bacteriophage ϕX174) all share a "viral β-barrel" or "viral jelly roll" fold in the capsid protein structure (Hendrix, 1999). Unless a remarkable case of convergent evolution occurred, this suggests that these viruses are homologous. A group of reoviruses that infect both plants and animals has a characteristic double-stranded RNA genome packaged in an unusual capsid. They share these features in common with a family of bacteriophages (such as $\phi6$), again suggesting homology between viruses that infect different branches of life (Hendrix, 1999). Notably sequence identity has not been detected in analysis of genes and/or proteins from these viral genomes, highlighting the rapid rate of viral genome evolution.

The great diversity of viral genomes precludes us from making comprehensive phylogenetic trees based on molecular sequence data that span the entire universe of viruses. This reflects the complex molecular evolutionary events that form viral genomes (McClure, 2000).

For a variety of viral families, phylogenetic trees have been generated. These are indispensable in establishing the evolution, host specificity, virulence, and other biological properties of viral species. We will examine phylogenetic reconstructions of the herpesviruses. Phylogenetic trees have been generated for HIV (see below) and for other viruses from measles to hepatitis.

Metagenomics and Virus Diversity

Historically we have classified viruses based on observation of their effects (e.g. by studying plant or human diseases caused by viruses), based on morphology, or based on the nature of the nucleic acid in purified virus particles. Metagenomics projects survey large amounts of genomic sequence from environmental samples or from host organisms (Chapter 13). Several metagenomics studies have resulted in the identification of large numbers of virus genomes (reviewed in Edwards and Rohwer, 2005).

A major metagenomics approach is to characterize DNA sequences in environmental samples. Recently Craig Venter and colleagues surveyed marine planktonic microbiota in a Global Ocean Sampling expedition (Rusch et al., 2007). Forty-one samples were collected over a range of 8000 kilometers, and 7.7 million sequencing reads were obtained. Combining their results with the previous Sargasso Sea survey (Venter et al., 2004) they reported the identification of 6.1 million proteins. There was a disproportionately large number of novel protein sequences assigned to viral genomes, consistent with the view that we have not yet achieved a broad sampling of viral diversity. Culley et al. 2006 also reported a diverse set of previously unknown RNA viruses in seawater.

Another metagenomic approach is to sample genomic DNA from individual organisms. For example, the human gut is colonized by hundreds or thousands of microbial species, including bacteria and archaea. Many of these prokaryotes are infected by viruses. Breitbart et al. 2003 analyzed 532 clones from a library created from viral samples (excluding prokaryotic cells) in the feces of a healthy adult. The majority of the sequences (59%) were not significantly similar to other known sequences based on tblastx. Zhang et al. (2006) further described plant pathogenic RNA viruses in human fecal samples. In a separate metagenomic study, Cox-Foster et al. (2007) determined DNA sequences associated with colony collapse disorder, an apparently recent phenomenon in which honey bee colonies collapse. This now affects about a quarter of bee-keeping operations in the United States. RNA samples were collected from hives that are either affected or not, and pyrosequencing was performed. In addition to bacterial and fungal sequences, a group of RNA viruses were identified, including one (Israeli acute paralysis virus) associated with risk for colony collapse disorder.

The accession for Israeli acute paralysis virus is NC_009025. It is a picorna-like virus with a genome that encodes two large proteins.

BIOINFORMATICS APPROACHES TO PROBLEMS IN VIROLOGY

The tools of bioinformatics are well suited to address some of the outstanding problems in virology:

- Why does a virus such as HIV-1 infect one species selectively (human) while a closely related virus (simian immunodeficiency virus) infects monkeys but not humans? Analysis of the sequence of the viruses as well as the host cell receptors can address this question.

- Why do some viruses change their natural host? In 1997 a chicken influenza virus infected 18 humans, killing 6. Are there changes in the genome of the virus, or of the host, or both that facilitate cross-species changes in specificity?

- Why are some viral strains deadlier than others? We will explore the properties of the 1918 influenza virus that killed an estimated 50 million people.

- What are the mechanisms of viral evasion of host immune systems? We will see below how some herpesviruses acquire viral homologs of human immune system molecules and thus interfere with human antiviral mechanisms.

- Where did viruses originate? There are three main theories:
  1. The regressive theory suggests that viruses are derived from more complex intracellular parasites that eliminated many nonessential features.
  2. Viruses could be derived from normal cellular components that now replicate autonomously.
  3. Viruses could have coevolved with their host cells, possibly originating from self-replicating RNA molecules.

  Phylogenetic analyses could help resolve these theories.

- Which vaccines are most likely to be effective? There are two main approaches to developing vaccines for viruses that display a great amount of molecular sequence diversity. One approach is to select isolates of a particular subtype based on regional prevalence. A second approach is to deduce an ancestral sequence or a consensus sequence for use as an antigen in vaccine development (Gaschen et al., 2002). These approaches depend on molecular phylogeny.

INFLUENZA VIRUS

The World Health Organization (WHO) maintains a listing of confirmed human cases of avian influenza A (H5N1) (▶ http://www.who.int/csr/disease/avian_influenza/country/en/). As of December 2007 there were 336 laboratory-confirmed cases and 207 deaths.

The "Spanish" influenza pandemic of 1918–1919 infected hundreds of millions of people, and is estimated to have killed 50 million people. The death rate among otherwise healthy young adults was especially high. Why was it so deadly? Influenza virus pandemics returned in the 1957 "Asian" flu, and the 1968 "Hong Kong" flu. More recently, the avian influenza subtype H5N1 infected over 300 humans and killed over 200 of them, and also led to the slaughter of millions of birds. Many wild birds such as ducks, geese, swans, and gulls are infected with influenza A (Olsen et al., 2006). Will an avian influenza virus like H5N1 infect humans globally? What are the properties of the influenza genome, and how can genome analyses help us to predict the next epidemic and devise strategies to prevent and/or treat its effects? In addition to the deadly avian flu strain, other subtypes of influenza virus are estimated to cause 250,000 to 500,000 deaths annually (36,000 deaths annually in the United States).

The influenza genome exists in several strains, each consisting of about 12,500 to 14,500 bases of single-stranded negative sense RNA and encoding 9 to 11 genes (Table 14.3). The genome of influenza A consists of eight segments (ranging in length from 890 to 2341 nucleotides) named PB1, PB2, PA, HA, NP, NA, M, and NS (Fig. 14.4 and Table 14.4). The hemagglutinin (HA) and neuraminidase (NA) segments encode two key surface glycoproteins that together define viral subtypes. The HA and NA segments occur in particular combinations that account for the antigenic variation of the virus. These combinations include H1N1, H2N2, and H3N2. The 1918 pandemic was of the H1N1 subtype, while subsequent pandemics in 1957 and 1968 were dominated by the H2N2 and H3N2 subtypes, respectively

TABLE 14-3	Influenza Viruses: Family *Orthomyxoviridae* Complete Genomes			
Virus	Source Information	Segments	Length (nt)	Proteins
Influenzavirus A				
Influenza A virus (A/Goose/ Guangdong/1/ 96(H5N1))	Strain: A/Goose/ Guangdong/1/96(H5N1)	8	13,590	11
Influenza A virus (A/Hong Kong/1073/ 99(H9N2))	Serotype: H9N2			
	Strain: A/Hong Kong/ 1073/99	8	13,498	11
Influenza A virus (A/Korea/426/ 68(H2N2))	Serotype: H2N2			
	Strain: A/Korea/426/68	8	13,460	11
Influenza A virus (A/New York/ 392/ 2004(H3N2))	Serotype: H3N2			
	Strain: A/New York/392/ 2004	8	13,627	11
Influenza A virus (A/Puerto Rico/8/ 34(H1N1))	Serotype: H1N1			
	Strain: A/Puerto Rico/8/34	8	13,588	11
Influenzavirus B				
Influenza B virus	Strain: B/Lee/40	8	14,452	11
Influenzavirus C				
Influenza C virus	Strain: C/Ann Arbor/1/50	7	12,501	9
Isavirus				
Infectious salmon anemia virus	Isolate: CCBB	8	12,716	10
Thogotovirus				
Thogoto virus	Strain: SiAr 126	6	10,461	7

Source: NCBI (▶ http://www.ncbi.nlm.nih.gov/genomes/VIRUSES/11308.html), December 2007. Genus ranks are in bold.

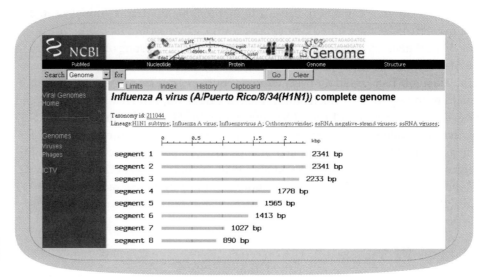

FIGURE 14.4. Schematic of the eight segments from a typical Influenza A virus (from NCBI).

As of January 2009 there have been 5000 human and avian influenza genomes sequenced or in progress. These genome sequences have been deposited in GenBank and can be accessed through the NCBI influenza virus resource (▶ http://www.ncbi.nlm.nih.gov/genomes/FLU/FLU.html).

(Fig. 14.5). In the 1957 and 1968 pandemics the viruses resembled human strains into which avian HA, NA, and PB1 molecules became incorporated, while the recent Asian outbreaks are caused by avian strains that infected humans.

An Influenza Genome Sequencing Project has achieved the remarkable accomplishment of sequencing 2800 full influenza genomes. All the sequence data are available through GenBank (Bao et al., 2008). This project provides an opportunity to address a range of fundamental questions about influenza viruses. One approach that has been taken is to characterize the genomes of avian influenza isolates. Obenauer et al. 2006 analyzed 169 complete avian influenza genomes and reported strong positive selection for an alternatively spliced transcript of the PB1 gene (the nonsynonymous to synonymous substitution rate ratio dN/dS

TABLE 14-4 Genes in a Representative Influenza A Virus Complete Genome (A/Puerto Rico/8/34(H1N1)), Taxonomy Identifier 211044

Gene	Segment	Protein Accession	Length (Amino Acids)	Name
PB2	1	NP_040987	759	RNA-dependent RNA polymerase subunit PB2
PB1	2	NP_040985	757	RNA-dependent RNA polymerase subunit PB1
PA	3	NP_040986	716	RNA-dependent RNA polymerase subunit PA
HA	4	NP_040980	566	Hemagglutinin
NP	5	NP_040982	498	Nucleocapsid protein
NA	6	NP_040981	454	Neuraminidase
M2	7	NP_040979	97	Matrix protein 2
M1	7	NP_040978	252	Matrix protein 1
NS1	8	NP_040984	230	Nonstructural protein NS1
NS2	8	NP_040983	121	Nonstructural protein NS2

Source: NCBI (December 2007).

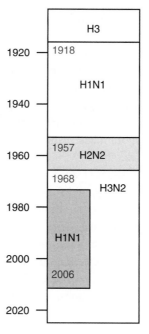

FIGURE 14.5. *Summary of influenza A strains. Analysis of archived tissue samples indicates that prior to 1918, the H3 strain predominated, while the great pandemic of 1918 was of the H1N1 subtype. Subsequent pandemics were associated with the H2N2 and H3N2 subtypes, while the H1N1 subtype has gained in recent decades. Adapted from Enserink 2006. Used with permission.*

[see Chapter 7] was over 9). In addition to performing phylogenetic analyses to distinguish emerging viral clades, Obenauer et al. described "proteotyping" in which unique amino acid signatures of viral proteins are identified and used to define molecular subtypes.

The analysis of human influenza virus strains has provided information about influenza evolution and diversity. In one approach Ghedin et al. (2005) sequenced 209 human influenza A genomes taken from one geographic location (New York State) over a period of several years (1998–2004). They plotted the amino acid positions from 207 viruses as a function of year and presented evidence for segment exchange between viruses. Reassortment among recent H3N2 strains was also reported by Holmes et al. (2005). Large-scale surveillance through genome sequencing permits the frequency of mutations and segment exchanges to be estimated, both within human influenza strains and between avian and human subtypes.

In a dramatic effort to understand the nature of the 1918 influenza virus, Jeffery Taubenberger, Terrence Tumpey, and colleagues isolated it and determined its full genome sequence. Viral nucleic acid was purified from historic samples, including an Alaskan woman and several soldiers who died of the 1918 flu. Taubenberger et al. 2005 proposed that the 1918 virus was entirely of avian origin (in contrast to the 1957 and 1968 strains that were reassortment viruses). Tumpey et al. (2005) created a viral strain having the complete coding sequences of the eight viral segments of the 1918 virus. They introduced the 1918 virus into mice, where it caused a titer from 125 to 39,000 higher than in mice exposed to a contemporary, less virulent strain. Lethality was 100-fold greater, with all mice dying within six days of infection (but none dying from the less virulent strain). This work carries considerable risk, but allows analysis of mutations that confer virulence. For example, a mutation found in the polymerase gene PB2 was also found in the virus isolated from a recent fatal case of bird flu involving the H7N7 subtype (von Bubnoff, 2005). Such analyses may aid surveillance efforts as we prepare for the next influenza pandemic (Taubenberger et al., 2007).

HERPESVIRUS: FROM PHYLOGENY TO GENE EXPRESSION

Herpesviruses are a diverse group of double-stranded DNA viruses that include herpes simplex, cytomegalovirus, and Epstein-Barr virus (McGeoch et al., 2006). As an example of viral phylogeny, McGeoch et al. analyzed well-conserved genes to deduce a phylogeny of the herpesviruses. Their phylogenetic reconstruction agrees with a proposed major taxonomy of the herpesviruses (family Herpesviridae) in three subfamilies: α-herpesviruses (formally called Alphaherpesvirinae), β-herpesviruses (Betaherpesvirinae), and γ-herpesviruses (Gammaherpesvirinae). This and similar analyses (Davison, 2002; McGeoch et al., 1995) provide great insight into the origin, diversity, and function of herpesviruses. Each herpesvirus is associated with a single host species (although some hosts, including humans, are infected by a variety of herpesviruses). This specificity suggests that herpesviruses have coevolved with their hosts over millions of years. Within each of the three subfamilies, the branching order showing the emergence of various herpesvirus subtypes corresponds to the emergence of the corresponding host organisms (Fig. 14.6). This suggests coevolution of the virus and host lineages. Figure 14.6a shows the timescale for the emergence of major Eutherian (placental mammal) lineages. Figure 14.6b to d shows the three herpesvirus subfamilies with molecular clocks. Note for example that in Fig. 14.6b there is a clade (thick red lines) of herpesviruses of the genus *Varicellovirus* (containing artiodactyls, perissocdactyl, and carnivore viruses). Note the correspondence of this clade structure to the evolution of those host organisms in Fig. 14.6a. McGeoch et al. 2006 estimate that the herpesiviruses shown in Fig. 14.6 arose about 400 million years ago.

The Alloherpesviridae comprise a distinct family of piscine and amphibian herpesviruses, and a single known virus that infects invertebrates is assigned to a third family, the Malacoherpesviridae.

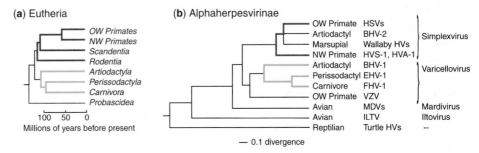

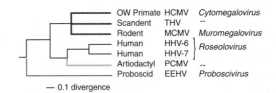

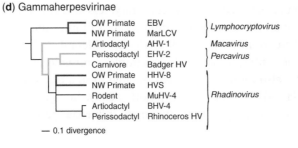

FIGURE 14.6. *Phylogeny of the herpesviruses and comparison to the evolution of host genomes. (a) Phylogenetic tree for eight orders of the Eutheria (placental mammals), all of which are hosts to herpesviruses. Three deep clades are indicated in thick red, thin red, and gray. (b) Alpha-, (c) Beta-, and (d) Gammaherpesvirinae are indicated with the hosts and examples of viruses. The divergence scales (in units of substitutions per site) are indicated. Abbreviations: NW, New World; OW, Old World. For virus abbreviations see the source of this figure, McGeoch et al. 2006. Used with permission.*

Consider human herpesvirus 8 (HHV-8), a γ-herpesvirus (Fig. 14.6d). HHV-8 is also called Kaposi's sarcoma-associated herpesvirus, and it was initially identified by representational difference analysis in Kaposi's sarcoma lesions of AIDS patients (Chang et al., 1994). HHV-8 causes AIDS-associated Kaposi's sarcoma and other disorders, such as primary effusion lymphoma and multicentric Castleman's disease. HHV-8 is closely related to rhesus rhadinovirus (RRV). The divergence of the HHV-8 and RRV may have coincided with speciation of humans and rhesus monkeys (Davison, 2002). The presence of both HHV-8 and an additional HHV-8-related virus in chimpanzees suggests that an additional virus may be identified that infects humans.

Kaposi's sarcoma is the most common tumor related to AIDS. It is a vascular malignancy that is typically first apparent in the skin.

What is the molecular basis for the cycle of latent and lytic infection by HHV-8? The genome is about 140,000 bp (NC_003409) and encodes over 80 proteins (Russo et al., 1996). We can explore the genome at the NCBI website using the Entrez genomes tool (Chapter 13). There are additional NCBI resources for the study of viruses. From a viral genomes home page (Fig. 14.3) you can link to double-stranded DNA viruses (such as the herpesviruses) or you can select the protein clusters tool (Fig. 14.7). From this site, you can browse the double-stranded DNA viruses (Fig. 14.8) and obtain a list of several dozen herpesvirus genomes. The protein clusters tool also allows you to search by functional categories (Fig. 14.9), analogous to the COGs tool (Chapter 15). Select HHV-8 and you can view the open reading frames encoded by its genome in a graphic form or as a table (Fig. 14.10).

The HHV-8 proteins include virion structural and metabolic proteins. Interestingly, it also contains a variety of viral homologs of human host proteins, such as complement-binding proteins, the apoptosis inhibitor Bcl-2, dihydrofolate reductase, interferon regulatory factors, an interleukin 8 (IL-8) receptor, a neural cell adhesion molecule–like adhesin, and a D-type cyclin.

How can viral genomes acquire a motif or an entire gene from a host organism? This can occur by a variety of mechanisms, including recombination, transposition, splicing, translocation, and inversion (McClure, 2000). Consider the IL-8 receptor, a eukaryotic protein that functions in cell growth and survival. This receptor is a member of the large family of G-protein-coupled receptors, including rhodopsin

Viral COG (VOG) - clusters of related viral proteins

Resource	dsDNA	ssDNA	sspRNA	ssnRNA	Phage
Clusters list/description	●	●	●	●	●
Functional categories	●	●	●	●	●
CDD search results on clusters	●	●	●	●	●
BLAST against viral proteins	●	●	●	●	●

FIGURE 14.7. *The NCBI clusters of related viral proteins (VCOG) website. See* ▶ *http://www.ncbi.nlm.nih.gov/genomes/VIRUSES/vog.html.*

Clusters of Related *Viral* Proteins (VOG): dsDNA virus proteins

dsDNA virus families involved:

Family	# of genomes	# of proteins	# of proteins in clusters	# of clusters
☐ Adenoviridae	19	549	249	29
☐ Asfarviridae	1	151	38	35
☐ Baculoviridae	15	2162	910	132
☐ Herpesviridae	27	2679	1437	147
■ Iridoviridae	4	726	25	23
☐ Nimaviridae	1	531	0	0
■ Papillomaviridae	89	645	539	14
■ Phycodnaviridae	2	938	103	40
☐ Polyomaviridae	7	42	40	5
☐ Poxviridae	15	3016	1287	163
All dsDNA viruses	180	11439	4628	433

FIGURE 14.8. From the NCBI VCOG page, you can obtain a list of double-stranded DNA virus protein families. This includes the Herpesviridae.

(that responds to light), the beta-adrenergic receptor (that binds adrenalin), and a variety of neurotransmitter receptors. Several viruses in addition to HHV-8 contain genes that encode a viral IL-8 receptor. A blastp search using this protein as a query reveals that several viruses have proteins that are distinct from but closely related to mammalian IL-8 receptors (Fig. 14.11). Presumably, when the virus infects a mammalian cell, this viral IL-8 protein is expressed and confers growth and survival that is advantageous to the virus (Wakeling et al., 2001).

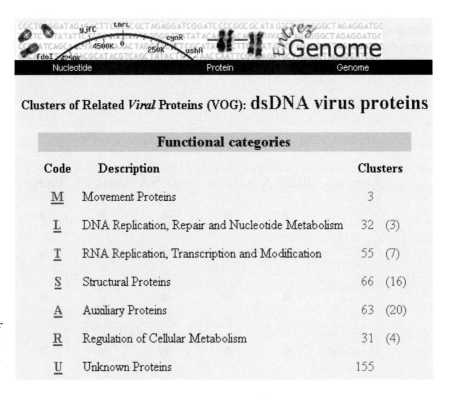

Clusters of Related *Viral* Proteins (VOG): dsDNA virus proteins

Functional categories

Code	Description	Clusters	
M	Movement Proteins	3	
L	DNA Replication, Repair and Nucleotide Metabolism	32	(3)
T	RNA Replication, Transcription and Modification	55	(7)
S	Structural Proteins	66	(16)
A	Auxiliary Proteins	63	(20)
R	Regulation of Cellular Metabolism	31	(4)
U	Unknown Proteins	155	

FIGURE 14.9. The Clusters of Related Viral Proteins page allows searches based on assigned functional categories.

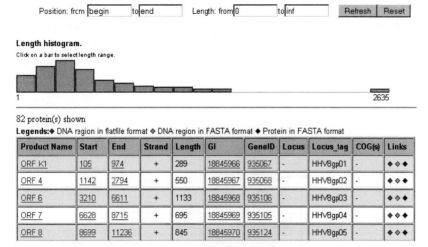

Human herpesvirus 8 type M, complete genome

FIGURE 14.10. The HHV-8 genome at NCBI. Five of the 82 proteins are shown. By clicking the length histograms, particular proteins can be selected based on size.

Two complementary approaches have been taken to further study the function of viral genes (such as v-*IL-8* receptor) as well as mechanisms of HHV-8 infection. Paulose-Murphy et al. (2001) synthesized a microarray that represents 88 HHV-8 open reading frames and measured the transcriptional response of viral genes that

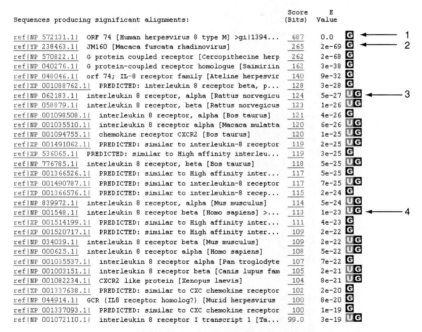

FIGURE 14.11. A viral protein is a G-protein-coupled receptor that is homologous to a superfamily of mammalian G-protein-coupled receptors, including a high affinity interleukin 8 (IL-8) receptor. Closely related homologs of this viral protein (open reading frame 74 or ORF74; RefSeq accession NP_572131) exist in several viruses, including Kaposi's sarcoma-associated herpesvirus (also called HSV-8) and a murine γ-herpesvirus. Database matches include HHV8 ORF74 (arrow 1), other viral proteins such as a Macaca fuscata rhadinovirus (arrow 2), interleukin 8 receptor from a variety of vertebrates including rat (arrow 3) and human (arrow 4). The gene encoding this receptor was presumably of mammalian origin and integrated into the genomes of several viruses. Upon viral infection, this receptor may promote growth and survival of infected cells.

Apoptosis is a type of programmed cell death in which the cell actively commits suicide. It serves as a mechanism by which a host cell can destroy infected cells, preventing a pathogen from spreading throughout the body. However, viruses have adapted to manipulate the cellular death pathway.

are activated during the lytic replication cycle of HHV8 in human cells. They measured gene expression across a time series after inducing lytic infection and described clusters of genes that are coexpressed. Such genes may be functionally related. Clusters of genes coexpressed at early time points include several implicated in activation of the lytic viral cycle; another group of genes encode proteins that function in virion assembly (Fig. 14.12). The viral homologs of human proteins were expressed throughout the induced lytic cycle.

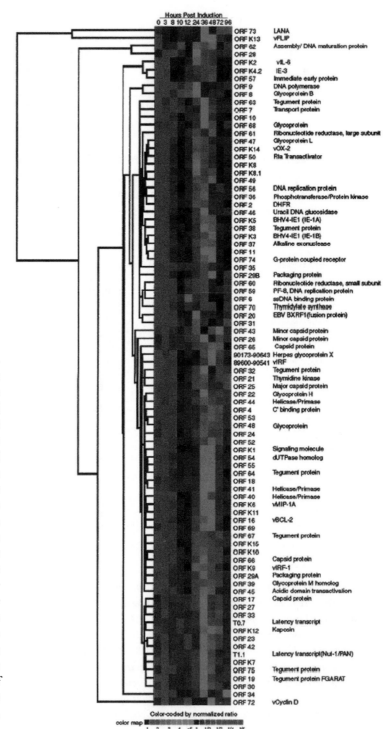

FIGURE 14.12. *Two-way hierarchical clustering of microarray data using an HHV-8 array to measure HHV-8 gene expression in infected human cells. The temporal expression ratios of genes were compared pairwise and grouped according to their similarity. The columns indicate separate time points (hours postinduction of the viral lytic cycle). The row displays the expression profile of each single open reading frame. The normalized expression ratios across all time points are color coded to denote the level of up- or downregulation. The dendrogram at left groups genes based on similar gene expression patterns across time. This approach is useful to discern the function of viral genes during infection. From Paulose-Murphy et al. (2001). Used with permission.*

In an independent study, Poole et al. (2002) infected human dermal microvascular endothelial cells with HHV-8 and measured the transcriptional response of host cells to both latent and lytic virus infection. HHV-8 transforms the endothelial cells from a cobblestone shape to a characteristic spindle shape. Kaposi's sarcoma is associated with many additional pathological features, including angiogenesis and immune dysregulation. The endothelial genes regulated by HHV-8 infection included those such as interferon-responsive genes involved in immune function and genes encoding proteins with roles in cytoskeletal function, apoptosis, and angiogenesis. Such studies may be useful in defining the cellular response to viral infection.

Angiogenesis is the development of blood vessels. Infectious viruses (and cancerous tumors) require the presence of an adequate blood supply and sometimes promote angiogenesis.

HUMAN IMMUNODEFICIENCY VIRUS

Human immunodeficiency virus is the cause of AIDS (reviewed in Meissner and Coffin, 1999). Until recently, HIV has been uniformly fatal. Most of the symptoms of AIDS are not caused directly by the virus but instead are a consequence of the ability of the virus to compromise the host immune system. Thus, HIV infection leads to disease caused by opportunistic organisms.

Information about AIDS is available at ▶ http://www.niaid.nih.gov/factsheets/aidsstat.htm, an NIH website. Information on prevalence is from the Centers for Disease Control and Prevention at ▶ http://www.cdc.gov/hiv/resources/factsheets/ and UNAIDS and the World Health Organization at ▶ http://www.unaids.org/.

At the end of the year 2006, close to 33 million people were infected with AIDS worldwide, and an additional 16 million people have died from AIDS. The prevalence of AIDS is increasing by about 3% per year. There have been many multinational efforts to combat HIV/AIDS across disciplines from treatment to prevention (Piot et al., 2004).

HIV-1 and HIV-2 are retroviruses of the group lentivirus. The viruses probably originated in sub-Saharan Africa, where the diversity of viral strains is greatest and the infection rates are highest (Sharp et al., 2001). The primate lentiviruses occur in five major lineages, as shown by a phylogenetic tree based on full-length pol protein sequences (Fig. 14.13a; see arrows 1 to 5) (Hahn et al., 2000; see also Rambaut et al., 2004; Heeney et al., 2006). These five lineages are:

Prevalence of a disease (or infection) is the proportion of individuals in a population who have a disease at a particular time. Prevalence does not describe when individuals contracted a disease. Incidence is the frequency of new cases of a disease that occur over a particular time. For example, the incidence of a disease might be described as 10 new cases per 1000 people in the general population in a given year.

1. Simian immunodeficiency virus (SIV) from the chimpanzee *Pan troglodytes* (SIVcpz), together with HIV-1

2. SIV from the sooty mangabeys *Cerecocebus atys* (SIVsm), together with HIV-2 and SIV from the macaques (genus *Macaca;* SIVmac)

3. SIV from African green monkeys (genus *Chlorocebus;* SIVagm)

4. SIV from Sykes' monkeys, *Cercopithecus albogularis* (SIVsyk)

5. SIV from l'Hoest monkeys, *Cercopithecus lhoesti* (SIVlhoest); SIV from suntailed monkeys (*Cercopithecus solatus;* SIVsun); and SIV from a mandrill (*Mandrillus sphinx;* SIVmnd)

As of November 2007, GenBank contains over 200,000 nucleotide records for HIV-1. To see this, go to the Taxonomy browser page and enter HIV-1. If you limit the output in a search of the Entrez nucleotide database to RefSeq entries, there is only one entry: the complete HIV-1 genome (NC_001802).

A prominent feature of phylogenetic analyses such as those in Fig. 14.13a is that viruses appear to have evolved in a host-dependent manner (Hahn et al., 2000). Viruses infecting any particular nonhuman primate species are more closely related to one another than they are related to viruses from other species. For HIV-2, transmission from the sooty mangabeys was indicated by five lines of evidence (Hahn et al., 2000):

1. Similarities in the genome structures of HIV-2 and SIVsm

2. Phylogenetic relatedness of HIV-2 and SIVsm (see Fig. 14.12, arrow 4)

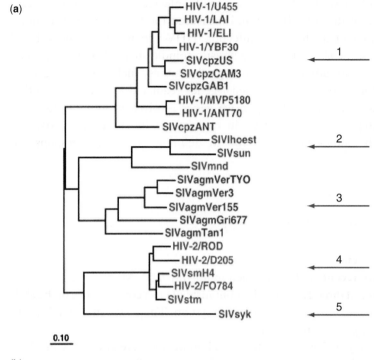

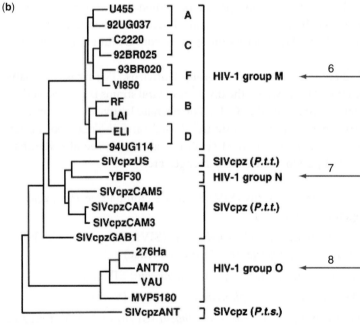

FIGURE 14.13. Evolutionary relationships of primate lentiviruses. (a) Full-length Pol protein sequences were aligned and a tree was created using the maximum-likelihood method. There are five major lineages (arrows 1 to 5). The scale bar indicates 0.1 amino acid replacements per site after correction for multiple hits. (b) The HIV-1/SIVcpz lineage is displayed based on a maximum-likelihood tree using Env protein sequences. Note that the three major HIV-1 groups (M, N, O; arrows 6 to 8) are distinguished. The scale bar is the same as in (a). From Hahn et al. (2000). Used with permission.

The Entrez Genome section (► http://www.ncbi.nlm.nih.gov/ entrez/query.fcgi?db = Genome) includes a listing of thousands of viruses. As of November 2007, there are almost 2500 completed virus genome sequences available

3. Prevalence of SIVsm in the natural host

4. Geographic coincidence of those affected and the natural host

5. Plausible routes of transmission, such as exposure of humans to chimpanzee blood in markets

Similar arguments have been applied to HIV-1, which probably appeared in Africa in 1930 to 1940 as a cross-species contamination by SIVcpz. HIV-1 occurs in three major subtypes, called M, N, and O. This is consistent with the occurrence of three separate SIVcpz transmissions to humans: M is the main group of HIV-1 viruses; O is an outlier group; and N is also distinct from M and O. The three main HIV-1 subtypes are apparent in a phylogenetic tree generated from full-length Env protein sequences (Fig. 14.13b, arrows 6 to 8) (Hahn et al., 2000).

We saw an NCBI entry for HIV-1 in Fig. 2.12; the genome is 9181 bases and encodes nine proteins. While the HIV-1 genome is small and there are few gene products, GenBank currently has about 200,000 nucleotide sequence records and an equal number of protein records. The reason for this enormous quantity of data is that HIV-1 mutates extremely rapidly, producing many subtypes of the M, N, and O variants. Thus, researchers sequence HIV variants very often. A major challenge for virologists is to learn how to manipulate such large amounts of data and how to use those data to find meaningful approaches to treating or curing AIDS. We will next describe two bioinformatics resources for the study of HIV molecular sequence data: NCBI and LANL.

Bioinformatic Approaches to HIV-1

The NCBI website offers several ways to study retroviruses, including HIV. You can access information on HIV-1 via the Entrez Genome site at NCBI, as we have described for HHV-8 above.

NCBI also offers a dedicated resource for the study of retroviruses (Fig. 14.14). This site includes the following:

- A genotyping tool based on BLAST searching
- A multiple sequence alignment tool specific for retroviral sequences
- A reference set of retroviral genomes

at this site. Additionally there are 500 phage sequences and three dozen viroids (infectious agents with RNA genomes that cause diseases in plants). Under the virus category of "Entrez genomes" a link is provided to this virus (listed alphabetically) and to the HIV-1 accession number (NC_001802). By clicking on the name of the virus, one is linked to the NCBI taxonomy browser, which includes information on the lineage of HIV-1 (Viruses; Retroid viruses; Retroviridae; Lentivirus; Primate lentivirus group) as well as links to dozens of HIV-1 variants. From the Entrez Genome page, by clicking on the accession number NC_001802, one links to the Entrez Nucleotide (GenBank). entry for HIV-1. As indicated in the entry, this is a 9181-base single-stranded RNA molecule. On the left sidebar, by clicking "Coding Regions," one links to a table listing the coding regions in the virus. This is a convenient way to obtain the DNA (or amino acid) sequence corresponding to a specific gene (or protein) of interest.

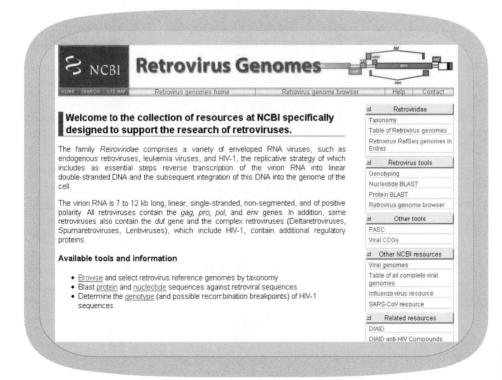

FIGURE 14.14. Retroviruses resource from NCBI (▶ http://www.ncbi.nlm.nih.gov/retroviruses/).

From the NCBI home page (▶http://www.ncbi.nlm.nih.gov/), select "All Databases" then "Genomes" to find viruses resources, or go directly to "Retrovirus Resources" (▶http://www.ncbi.nlm.nih.gov/retroviruses/). NCBI also offers a database of interactions between HIV and human proteins (▶http://www.ncbi.nlm.nih.gov/RefSeq/HIVInteractions/).

- Specific pages with tools to study HIV-1, HIV-2, SIV, human T-cell lymphotropic virus type 1 (HTLV), and STLV
- A listing of the previous week's publications on retroviruses
- A listing of the previous week's GenBank releases (many hundreds of new HIV-1 sequences are deposited weekly)
- Links to external retroviral website resources

A fundamental resource for the study of several virus types including HIV is the Los Alamos National Laboratory (LANL) which operates a group of four HIV

(a)

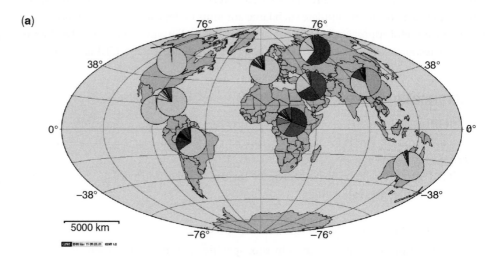

(b)

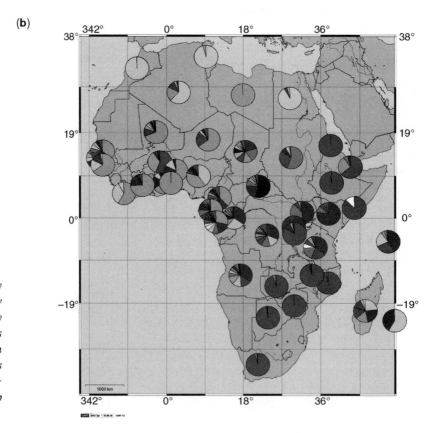

FIGURE 14.15. The geography tool at LANL allows you to view HIV infection subtypes (a) globally or (b) by continent (Africa is shown). The subtype distribution is displayed using pie charts. This geography tool is available at ▶ http://www.hiv.lanl.gov/ through the tools section.

databases. The HIV Sequence Database is an important, comprehensive repository of HIV sequence data. It allows searches for sequences by common names, accession number, PubMed identifier, country in which each case was sampled, and likely country in which infection occurred. Sequences may be retrieved as part of a multiple sequence alignment or unaligned, and groups of sequences derived from an individual patient may be retrieved. The site includes a variety of specialized tools, including:

- An HIV BLAST server

- SNAP (Synonymous/Non-synonymous Analysis Program), a program that calculates synonymous and nonsynonymous substitution rates

- Recombinant Identification Program (RIP), a program that identifies mosaic viral sequences that may have arisen through recombination

- A multiple alignment program called MPAlign (Gaschen et al., 2001) that uses HMMER software (Chapter 6)

- PCoord (Principal Coordinate Analysis), a program that performs a procedure similar to principal components analysis (Chapter 9) on sequence data based on distance scores

- A geography tool that shows both total HIV infection levels (either worldwide or by continent) as well as the subtype distribution of HIV (Figs. 14.15a and b)

The LANL HIV databases are available at ▶ http://hiv-web.lanl.gov/. This site offers four databases: sequence, resistance, immunology, and vaccine trials. In the HIV Sequence Database you can find the geography tool by selecting Tools and then Geography.

The LANL website includes other databases that provide important tools for the bioinformatic analysis of HIV-1 and related viruses. The HIV Drug Resistance Database allows you to browse HIV-1 genes to identify specific drugs that are affected by amino acid substitutions in that gene product (Fig. 14.16). This information is

Viewing Records: 311 through 320 of 607 records in database.

Gene (Click for details)	Drug Class	Compound	AA Mutation	Codon Mutation	Cite
Protease	Protease Inhibitor	MK-639	90 L M	TTG -> ATG	Condra96
Protease	Protease Inhibitor	Ro 31-8959	90 L M	TTG -> ATG	Jacobsen94
Protease	Protease Inhibitor	ABT-378	90 L M	TTG -> ATG	Kempf01
Protease	Protease Inhibitor	ABT-378	91 T S	ACT -> TCT	Carrillo98
Protease	Protease Inhibitor	DMP-323	97 L V	TTA -> GTA	King95
RT	Nucleoside RT Inhibitor	AZT	41 M L	ATG -> TTG/CTG	Larder89 Larder91 Kellam92
RT	HIV-1 Specific RT Inhibitor (NNRTI)	AZT + 3TC	44 E D	GAA -> GAC	Hertogs00
RT	Nucleoside RT Inhibitor	3TC	44 E A	GAA -> GCA	Montes02
RT	Multiple Nucleoside		62 A V	GCC -> GTC	Iversen96 Shirasaka95
RT	HIV-1 Specific RT Inhibitor (NNRTI)	BHAP U-90152	63 I M	ATA -> ATG	Pelemans01

[◀ Go First] [◀ Back] [▶ Forward] [▶ Go Last] [🔍 Find]

Resort by: [FoldResist ▾]   Sort order [Ascending ▾]   [Sort Again]

AAPosition
Gene
WildtypeAA
AAPosition
MutantAA
WildtypeCodon
MutantCodon
CrugClass
Compound
IrVitro
IrVivo

FIGURE 14.16. *The HIV Drug Resistance Database compiles amino acid substitutions in HIV genes that confer resistance to anti-HIV drugs. The browse tool (available via ▶ http://resdb.lanl.gov/Resist_DB/default.htm) allows you to search any HIV-1 gene for substitutions. The data can be sorted by many criteria as shown in the pull-down menu.*

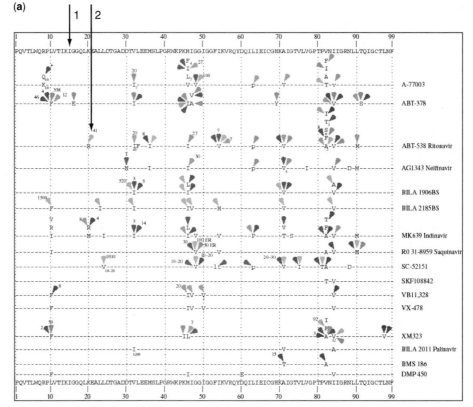

FIGURE 14.17. (a) The LANL website offers a map of HIV-1 protease mutations versus drugs. Each row represents a drug (labeled at right). The wild-type (strain HXB2) HIV-1 protease sequence is listed at top and bottom (arrow 1). Dashes indicate wild-type amino acid positions, while mutations that confer resistance to the drug are indicated. An example of a K-to-R (lysine-to-arginine) mutation is indicated (arrow 2). The small number (41) indicates the "fold resistance" of that particular mutation. Mutations that have a colored shape pointing to them are also part of a synergistic combination of mutations. (b) By clicking on the position of a mutation (arrow 2), the map links to a detailed report of the effects of that mutation.

A. Molla, M. Korneyeva, Q. Gao, S. Vasavanonda, P. J. Schipper, H. M. Mo, M. Markowitz, T. Chernyavskiy, P. Niu, N. Lyons, A. Hsu, G. R. Granneman, D. D. Ho, C. A. Boucher, J. M. Leonard, D. W. Norbeck, D. J. Kempf
Ordered accumulation of mutations in HIV protease confers resistance to ritonavir.
Nat Med 2: 760-6 (1996) Medline link: 96266327

The LANL Protease Mutations-by-Drug Map is available at ▶ http://resdb.lanl.gov/Resist_DB/protease_mutation_map.htm.

also displayed graphically with the mapping tool in the HIV Drug Resistance Database. This clickable plot shows the complete amino acid sequence of an HIV protein (e.g., the protease; Fig. 14.17a, arrow 1). Each row contains a different drug, and amino acid substitutions associated with the resistance of HIV to each of the drugs are also displayed. Clicking on a substituted amino acid (Fig. 14.17, arrow 2) leads to a report describing the mutation in detail (Fig. 14.17b).

Before the measles vaccination was introduced in the United States, there were 450,000 cases annually (and about 450 deaths). See ▶ http://www.cdc.gov/nchs/fastats/measles.htm.

MEASLES VIRUS

Measles virus is one of the deadliest viruses in human history. Today, it is the leading cause of death in children in many countries, killing over one million infants each year (Johnson et al., 2000). Vaccines have helped to reduce the mortality and morbidity rates, but the presence of an immature immune system and maternal antibodies prevent successful immunization in newborns before nine months of age. The virus spreads by respiratory droplets, infecting epithelial cells in the respiratory tract. In 2004 measles caused over 450,000 deaths in sub-Saharan Africa, and thus this disease is a leading vaccine-preventable cause of child mortality (Moss and Griffin, 2006).

The measles virus is a Morbillivirus of the Paramyxoviridae family, which includes mumps and respiratory syncytial virus. Rota and Bellini (2003) reviewed the worldwide distribution of 14 different measles virus genotypes. You can access a reference genome through the NCBI Entrez genomes resource (accession NC_001498). Measles virus consists of a nonsegmented, negative sense RNA genome protected by nucleocapsids and an envelope. The genome has 15,894 bases and encodes seven proteins. These sequences can be accessed by clicking on the "coding regions" option on the left sidebar of the Entrez record (Fig. 14.18a). Six genes are designated N (nucleocapsid), P (phosphoprotein), M (matrix),

Another member of the Paramyxoviridae family is the cause of rinderpest, an ancient plague of cattle (Barrett and Rossiter, 1999). These viruses have had a devastating impact on both humans and ruminants.

(a)

(b)

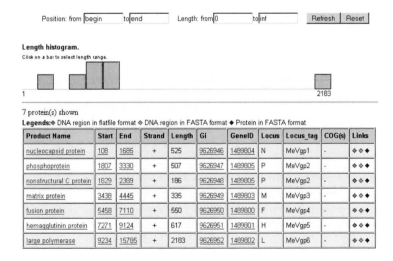

FIGURE 14.18. Analysis of the measles virus genome. (a) The measles virus entry is accessed from the NCBI Entrez website. By clicking on the "Coding Regions" option on the left sidebar, (b) a list of the protein-coding genes is obtained. These genes are designated N (nucleocapsid), P (phosphoprotein), M (matrix), F (fusion), H (hemagglutinin), and L (large polymerase). Note that the P gene is predicted to encode another protein (nonstructural C protein) using an alternative start site on a different reading frame.

F (fusion), H (hemagglutinin), and L (large polymerase) (Fig. 14.18b). The P gene is predicted to encode another protein (nonstructural C protein) using an alternative start site on a different reading frame. It is easy to visualize this by clicking on the Entrez nucleotide record, then choosing the "Graphics" display option (Fig. 14.19). This shows where the measles virus genome encodes the nonstructural C protein.

The functions of the six measles virus proteins have been assigned: N binds to genomic RNA and surrounds it, P and L form a complex involved in RNA synthesis, M links the ribonucleoprotein to the envelope glycoproteins H and F which are inserted in the virus membrane on the surface of the virion, H binds the cell surface receptor through which the virus enters its host, and F is a fusion protein that promotes insertion of the virus into the host cell membrane. The functions of each of these proteins can be assessed by performing BLAST searches. For the nonstructural C protein, a blastp nonredundant (nr) search reveals homology to proteins encoded by the genomes of rinderpest virus, canine and phocine (seal) distemper virus, and dolphin morbillivirus. A blastp nr search with the viral hemagglutinin reveals membership in a Pfam family (pfam00423, Hemagglutinin-neuraminidase), and there are several hundred matches to measles virus hemagglutinin. Repeat the search with the Entrez limit "hemagglutinin NOT measles virus[Organism]" and the results are reduced to several dozen hemagglutinins from the homologous morbilliviruses other than measles. A PSI-BLAST search identifies hundreds of additional

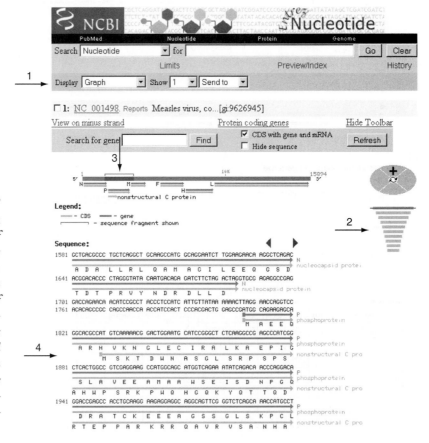

FIGURE 14.19. The "graphics" display (arrow 1) of any Entrez nucleotide entry shows a map of the DNA sequence along with the corresponding protein sequences. It is easy to zoom in or out for a more detailed or global view of the sequence (arrow 2). The portion of the measles genome that is displayed is indicated in a graphical overview (arrow 3). Usage of an alternative start site allows the measles genome to encode two distinct proteins using nonoverlapping reading frames (arrow 4).

hemagglutinins from viruses such as human parainfluenza, mumps, and a turkey rhinotracheitis virus.

PERSPECTIVES

Several thousand species of viruses are known. In contrast, there may be tens or hundreds of millions of species of bacteria and archaea (Chapter 15) and perhaps tens of millions of eukaryotic species (Chapters 16 to 18). There are probably relatively few species of viruses because of their specialized requirements for replication in host cells.

Essentially all the bioinformatic tools that are applied to eukaryotic or prokaryotic protein and nucleic acid sequences are applicable to the study of viruses as well (Kellam, 2001).

- BLAST, PSI-BLAST, and other database searches may be applied to define the homology of viral sequences to other molecules.
- Microarrays have been used to represent viral genes, allowing an assessment of viral gene transcription during different phases of the viral life cycle.
- In independent approaches, the transcriptional response of host cells to viral infection has begun to be characterized.
- Structural genomics approaches to viruses result in the identification of three-dimensional structures of viral proteins. Some structures are solved in the presence of pharmacological inhibitors. The Entrez protein division of NCBI currently includes over 3200 virus structural records.

PITFALLS

Viruses evolve extremely rapidly, in large part because some viral polymerases tend to operate with low fidelity. It is for this reason that a person infected with HIV may harbor millions of distinct forms of the virus, each with its own unique RNA sequence. Thus, it may be difficult to define a single canonical sequence for some viruses. This complicates attempts to study the evolution of viruses and the functions of their genes.

While the tree of life has been described using rRNA or other sequences (Chapter 13), viruses are almost entirely absent from this tree. This is because there are no genes or proteins that all viruses share in common with other life forms or with each other.

WEB RESOURCES

We have focused on ICTVdb, NCBI, and LANL tools. Many specialized databases have been established for the study of viruses, including those listed in Table 14.5. Project VirgO offers software tools, including the Viral Genome Organizer, for the graphical display of viral sequences (Upton et al., 2000). This site also contains a Viral Genome DataBase (VGDB) with analyses of the properties of viral genomes such as GC content. The Stanford HIV RT and Protease Sequence Database offers an algorithm that can be queried with an input viral DNA sequence (Rhee et al., 2003). The output describes possible mutations in the viral gene and an interpretation of likely susceptibility of that protein to drug resistance.

TABLE 14-5   Virus Resources Available on the Web

Resource	Description	URL
ICTVdb	Universal virus database	▶ http://www.ncbi.nlm.nih.gov/ICTVdb/
All the Virology on the WWW	Provides many virology links and resources	▶ http://www.virology.net/
The Big Picture Book of Viruses	General virus resource	▶ http://www.virology.net/Big_Virology/BVHomePage.html
VIrus Particle ExploreER (VIPER)	High-resolution virus structures in the Protein Data Bank (PDB)	▶ http://viperdb.scripps.edu/
Viral Genome Organizer	Analyses of large poxviruses and other viruses	▶ http://athena.bioc.uvic.ca/tools/VGO
Institute for Molecular Virology	A research institute at the University of Wisconsin-Madison	▶ http://virology.wisc.edu/virusworld/
Stanford HIV Drug Resistance Database	A curated database with information on drug targets	▶ http://hivdb.stanford.edu/

DISCUSSION QUESTIONS

[14-1] There is no comprehensive molecular phylogenetic tree of all viruses. Why not?

[14-2] If you wanted to generate phylogenetic trees that are as comprehensive as possible, using DNA or RNA or protein sequences available in GenBank, what molecule(s) would you select? What database(s) would you search?

PROBLEMS/COMPUTER LAB

[14-1] How many HIV-1 proteins are in Entrez at NCBI? Given the tremendous heterogeneity of HIV-1, you might expect there to be thousands of variant forms of each protein. How many are actually assigned RefSeq accession numbers? How many measles virus RefSeq proteins are there?

[14-2] Find an HIV-1 protein with a RefSeq identifier in Entrez Protein (such as the Vif protein, NP_057851; you should select your own example). Perform a blastp search with it, and inspect the results using the taxonomy report. Next, repeat the search, excluding HIV from the output. As an example of how to do this, enter "vif NOT txid11676[Organism]" or "vif NOT Hiv[Organism]" into the advanced search option "Limit by Entrez query." (Note that you can find the taxonomy identifier txid11676[Organism] by using the NCBI taxonomy browser.) How broadly is the gene or protein you selected represented among viruses? Do you expect some genes to be HIV specific while other genes are shared broadly by viruses?

[14-3] Analyze a set of influenza viruses using the NCBI Influenza Virus Resource (▶ http://www.ncbi.nlm.nih.gov/genomes/FLU/FLU.html).

(a) Click tree to begin choosing sequences. Select the virus species (Influenza A), host (human), country/region (e.g., Europe), and segment (HA). Include the options of full-length sequences only, and remove identical sequences. Click Get sequences.

(b) Construct a multiple sequence alignment and phylogenetic tree. Use neighbor-joining. In the case of HA, does the tree form clades corresponding to H1N1, H3N2, and H7N7 subtypes? Optionally, export the sequences in the fasta format, perform your own multiple sequence alignments using MAFFT or MUSCLE (Chapter 6), then import the alignment into MEGA (or other software) to perform phylogenetic analyses yourself.

[14-4] Analyze HIV sequences at the HIV Sequence Database (http://www.hiv.lanl.gov/). Select the search interface, then choose genomic regions with the Vif coding sequence (Vif CDS). Restrict the output to ten sequences. Select these, and click "Make tree." Include the reference sequences HXB2. Choose a distance model (the default is Felsenstein 1984) and either equal site rates or a gamma distribution. How many clades do you observe? What do these clades represent? Note that you can download the multiple sequence alignment used to generate the tree to perform further phylogenetic analyses.

SELF-TEST QUIZ

[14-1] There are several thousand known viruses, while there are many millions of prokaryotes and eukaryotes. The most likely explanation for the small number of viruses is that

(a) we have not yet learned how to detect most viruses

(b) we have not yet learned how to sequence most viruses

(c) there are few viruses because their needs for survival are highly specialized

(d) viruses use an alternative genetic code

[14-2] The HIV genome contains nine protein-coding genes. The number of GenBank accession numbers for these nine genes is approximately

(a) 9

(b) 900

(c) 9000

(d) 90,000

[14-3] For functional genomics analyses of viruses, it is possible to measure gene expression

(a) of viral genes upon viral infection of human tissues

(b) of human genes upon viral infection of human tissues

(c) of viral genes and human genes, simultaneously measured upon viral infection of human tissue

(d) of viral genes or human genes, separately measured upon viral infection of human tissue

[14-4] Herpesviruses probably first appeared about

(a) 200 million years ago

(b) 2 million years ago

(c) 20,000 years ago

(d) 200 years ago

[14-5] HIV probably first appeared about

(a) 70 million years ago

(b) 7 million years ago

(c) 7,000 years ago

(d) 70 years ago

[14-6] Phylogeny of HIV virus subtypes

(a) establishes that HIV emerged from a cattle virus

(b) can be used to develop vaccines directed against ancestral protein sequences

(c) establishes which human tissues are most susceptible to infection

[14-7] Specialized virus databases such as that at Oak Ridge National Laboratory offer resources for the study of HIV that are not available at NCBI or EBI. An example is:

(a) a listing of thousands of variant forms for each HIV gene

(b) a listing of literature and citations from the previous week

(c) graphical displays of the genome

(d) a description of where HIV variants have been identified across the world

SUGGESTED READING

Kellam (2001) has reviewed bioinformatics and functional genomics approaches to virology. He describes a broad range of subjects from viral genome structure to protein structure and gene expression studies of both viral pathogens and hosts. For a review of viral metagenomics see Edwards and Rohwer (2005).

REFERENCES

Ackermann, H. W. 5500 Phages examined in the electron microscope. *Arch. Virol.* **152**, 227–243 (2007).

Bao, Y., Bolotov, P., Dernovoy, D., Kiryutin, B., Zaslavsky, L., Tatusova, T., Ostell, J., and Lipman, D. The Influenza Virus Resource at the National Center for Biotechnology Information. *J. Virol.* **82**, 596–601 (2008).

Barrett, T., and Rossiter, P. B. Rinderpest: The disease and its impact on humans and animals. *Adv. Virus Res.* **53**, 89–110 (1999).

Bernal, J. D., and Fankuchen, I. X-ray and crystallographic studies of plant virus preparations. *J. Gen. Physiol.* **25**, 111–146 (1941).

Breitbart, M., Hewson, I., Felts, B., Mahaffy, J. M., Nulton, J., Salamon, P., and Rohwer, F. Metagenomic analyses of an uncultured viral community from human feces. *J. Bacteriol.* **185**, 6220–6223 (2003).

Büchen-Osmond, C. Further progress in ICTVdB, a universal virus database. *Arch. Virol.* **142**, 1734–1739 (1997).

Büchen-Osmond C. The Universal Virus Database ICTVdB. *Comput. Sci. Eng.* **5**, 16–25 (2003).

Chang, Y., et al. Identification of herpesvirus-like DNA sequences in AIDS-associated Kaposi's sarcoma. *Science* **266**, 1865–1869 (1994).

Chisari, F. V. Unscrambling hepatitis C virus-host interactions. *Nature* **436**, 930–932 (2005).

Cieslak, T. J., Christopher, G. W., and Ottolini, M. G. Biological warfare and the skin. II. Viruses. *Clin. Dermatol.* **20**, 355–364 (2002).

Cox-Foster, D. L., et al. A metagenomic survey of microbes in honey bee colony collapse disorder. *Science* **318**, 283–287 (2007).

Culley, A. I., Lang, A. S., and Suttle, C. A. Metagenomic analysis of coastal RNA virus communities. *Science* **312**, 1795–1798 (2006).

Daròs, J. A., Elena, S. F., and Flores, R. Viroids: An Ariadne's thread into the RNA labyrinth. *EMBO Rep.* 7, 593–598 (2006).

Davison, A. J. Evolution of the herpesviruses. *Vet. Microbiol.* **86**, 69–88 (2002).

DeArmond, S. J., and Prusiner, S. B. Perspectives on prion biology, prion disease pathogenesis, and pharmacologic approaches to treatment. *Clin. Lab. Med.* **23**, 1–41 (2003).

Edwards, R. A., and Rohwer, F. Viral metagenomics. *Nat. Rev. Microbiol.* **3**, 504–510 (2005).

Enserink, M. Influenza. What came before 1918? Archaeovirologist offers a first glimpse. *Science* **312**, 1725 (2006).

Fauquet, C. M., Mayo, M. A., Maniloff, J., Desselberger, U., and Ball, L. A. (eds.).*Virus Taxonomy: VIIIth Report of the International Committee on Taxonomy of Viruses.* Elsevier Academic Press, New York, 2005.

Flint, S. J., Enquist, L. W., Krug, R. M., Racaniello, V. R., and Skalka, A. M. *Principles of Virology Molecular Biology, Pathogenesis, and Control.* American Society for Microbiology Press, Washington, DC, 2000.

Flores, R. A naked plant-specific RNA ten-fold smaller than the smallest known viral RNA: The viroid. *C. R. Acad. Sci. III* **324**, 943–952 (2001).

Gaschen, B., Kuiken, C., Korber, B., and Foley, B. Retrieval and on-the-fly alignment of sequence fragments from the HIV database. *Bioinformatics* **17**, 415–418 (2001).

Gaschen, B., et al. Diversity considerations in HIV-1 vaccine selection. *Science* **296**, 2354–2360 (2002).

Ghedin, E., et al. Large-scale sequencing of human influenza reveals the dynamic nature of viral genome evolution. *Nature* **437**, 1162–1166 (2005).

Hahn, B. H., Shaw, G. M., De Cock, K. M., and Sharp, P. M. AIDS as a zoonosis: Scientific and public health implications. *Science* **287**, 607–614 (2000).

Heeney, J. L., Dalgleish, A. G., and Weiss, R. A. Origins of HIV and the evolution of resistance to AIDS. *Science* **313**, 462–466 (2006).

Hendrix, R. W. Evolution: The long evolutionary reach of viruses. *Curr. Biol.* **9**, R914–R917 (1999).

Holmes, E. C., Ghedin, E., Miller, N., Taylor, J., Bao, Y., St. George, K., Grenfell, B. T., Salzberg, S. L., Fraser, C. M., Lipman, D. J., and Taubenberger, J. K. Whole-genome analysis of human influenza A virus reveals multiple persistent lineages and reassortment among recent H3N2 viruses. *PLoS Biol.* **3**, e300 (2005).

International Human Genome Sequencing Consortium. Initial sequencing and analysis of the human genome. *Nature* **409**, 860–921 (2001).

Johnson, R. T., and Gibbs, C. J., Jr. Creutzfeldt–Jakob disease and related transmissible spongiform encephalopathies. *N. Engl. J. Med.* **339**, 1994–2004 (1998).

Johnson, C. E., et al. Measles vaccine immunogenicity and antibody persistence in 12 vs 15-month old infants. *Vaccine* **18**, 2411–2415 (2000).

Kellam, P. Post-genomic virology: The impact of bioinformatics, microarrays and proteomics on investigating host and pathogen interactions. *Rev. Med. Virol.* **11**, 313–329 (2001).

Kruger, D. H., Schneck, P., and Gelderblom, H. R. Helmut Ruska and the visualisation of viruses. *Lancet* **355**, 1713–1717 (2000).

La Scola, B., et al. A giant virus in Amoebae. *Science* **299**, 2033 (2003).

Lamichhane, G., Zignol, M., Blades, N. J., Geiman, D. E., Dougherty, A., Grosset, J., Broman, K. W., and Bishai, W. R. A postgenomic method for predicting essential genes at subsaturation levels of mutagenesis: Application to *Mycobacterium tuberculosis*. *Proc. Natl. Acad. Sci. USA* **100**, 7213–7218 (2003).

Mayo, M. A., and Pringle, C. R. Virus taxonomy—1997. *J. Gen. Virol.* **79**, 649–657 (1998).

McClure, M. A. The complexities of genome analysis, the Retroid agent perspective. *Bioinformatics* **16**, 79–95 (2000).

McGeoch, D. J., Cook, S., Dolan, A., Jamieson, F. E., and Telford, E. A. Molecular phylogeny and evolutionary timescale for the family of mammalian herpesviruses. *J. Mol. Biol.* **247**, 443–458 (1995).

McGeoch, D. J., Rixon, F. J., and Davison, A. J. Topics in herpesvirus genomics and evolution. *Virus Res.* **117**, 90–104 (2006).

Meissner, C., and Coffin, J. M. The human retroviruses: AIDS and other diseases. In M. Schaechter, N. C. Engleberg, B. I. Eisenstein, and G. Medoff (eds.), *Mechanisms of Microbial Disease*, Chapter 38 Lippincott Williams & Wilkins, Baltimore, MD, 1999.

Moss, W. J., and Griffin, D. E. Global measles elimination. *Nat. Rev. Microbiol.* **4**, 900–908 (2006).

Neumann, A. U., Lam, N. P., Dahari, H., Gretch, D. R., Wiley, T. E., Layden, T. J., and Perelson, A. S. Hepatitis C viral dynamics in vivo and the antiviral efficacy of interferon-alpha therapy. *Science* **282**, 103–107 (1998).

Obenauer, J. C., Denson, J., Mehta, P. K., Su, X., Mukatira, S., Finkelstein, D. B., Xu, X., Wang, J., Ma, J., Fan, Y., Rakestraw, K. M., Webster, R. G., Hoffmann, E., Krauss, S., Zheng, J., Zhang, Z., and Naeve, C. W. Large-scale sequence analysis of avian influenza isolates. *Science* **311**, 1576–1580 (2006).

Olsen, B., Munster, V. J., Wallensten, A., Waldenström, J., Osterhaus, A. D., and Fouchier, R. A. Global patterns of influenza a virus in wild birds. *Science* **312**, 384–388 (2006).

Paulose-Murphy, M., et al. Transcription program of human herpesvirus 8 (Kaposi's sarcoma-associated herpesvirus). *J. Virol.* **75**, 4843–4853 (2001).

Piot, P., Feachem, R. G., Lee, J. W., and Wolfensohn, J. D. Public health. A global response to AIDS: Lessons learned, next steps. *Science* **304**, 1909–1910 (2004).

Poole, L. J., et al. Altered patterns of cellular gene expression in dermal microvascular endothelial cells infected with Kaposi's sarcoma-associated herpesvirus. *J. Virol.* **76**, 3395–3420 (2002).

Prangishvili, D., Garrett, R. A., and Koonin, E. V. Evolutionary genomics of archaeal viruses: Unique viral genomes in the third domain of life. *Virus Res.* **117**, 52–67 (2006).

Prusiner, S. B. Prions. *Proc. Natl. Acad. Sci. USA* **95**, 13363–13383 (1998).

Rambaut, A., Posada, D., Crandall, K. A., and Holmes, E. C. The causes and consequences of HIV evolution. *Nat. Rev. Genet.* **5**, 52–61 (2004).

Raoult, D., Audic, S., Robert, C., Abergel, C., Renesto, P., Ogata, H., La Scola, B., Suzan, M., and Claverie, J. M. The 1.2-megabase genome sequence of Mimivirus. *Science* **306**, 1344–1350 (2004).

Rhee, S. Y., Gonzales, M. J., Kantor, R., Betts, B. J., Ravela, J., and Shafer, R. W. Human immunodeficiency virus reverse transcriptase and protease sequence database. *Nucleic Acids Res.* **31**, 298–303 (2003).

Rota, P. A., and Bellini, W. J. Update on the global distribution of genotypes of wild type measles viruses. *J. Infect. Dis.* **187**, S270–S276 (2003).

Rusch, D. B., et al. The Sorcerer II Global Ocean Sampling expedition: Northwest Atlantic through eastern tropical Pacific. *PLoS Biol.* **5**, e77 (2007).

Russo, J. J., et al. Nucleotide sequence of the Kaposi sarcoma-associated herpesvirus (HHV8). *Proc. Natl. Acad. Sci. USA* **93**, 14862–14867 (1996).

Schaechter, M., Engleberg, N. C., Eisenstein, B. I., and Medoff, G. *Mechanisms of Microbial Disease.* Lippincott Williams & Wilkins, Baltimore, MD, 1999.

Sharp, P. M., et al. The origins of acquired immune deficiency syndrome viruses: Where and when? *Philos. Trans. R. Soc. Lond. B Biol. Sci.* **356**, 867–876 (2001).

Simpson, G. G. The meaning of taxonomic statements. In S. L. Washburn (ed.). *Classification and Human Evolution*, pp. 1–31. Aldine Publishing Co., Chicago, 1963.

Taubenberger, J. K., Morens, D. M., and Fauci, A. S. The next influenza pandemic: Can it be predicted? *JAMA* **297**, 2025–2027 (2007).

Taubenberger, J. K., Reid, A. H., Lourens, R. M., Wang, R., Jin, G., and Fanning, T. G. Characterization of the 1918 influenza virus polymerase genes. *Nature* **437**, 889–893 (2005).

Tumpey, T. M., Basler, C. F., Aguilar, P. V., Zeng, H., Solórzano, A., Swayne, D. E., Cox, N. J., Katz, J. M., Taubenberger, J. K., Palese, P., and García-Sastre, A. Characterization of the reconstructed 1918 Spanish influenza pandemic virus. *Science* **310**, 77–80 (2005).

Upton, C., Hogg, D., Perrin, D., Boone, M., and Harris, N. L. Viral genome organizer: A system for analyzing complete viral genomes. *Virus. Res.* **70**, 55–64 (2000).

van Regenmortel, M. H., and Mahy, B. W. Emerging issues in virus taxonomy. *Emerg. Infect. Dis.* **10**, 8–13 (2004).

Venter, J. C., et al. Environmental genome shotgun sequencing of the Sargasso Sea. *Science* **304**, 66–74 (2004).

von Bubnoff, A. The 1918 flu virus is resurrected. *Nature* **437**, 794–795 (2005).

Wakeling, M. N., Roy, D. J., Nash, A. A., and Stewart, J. P. Characterization of the murine gammaherpesvirus 68 ORF74 product: A novel oncogenic G protein-coupled receptor. *J. Gen. Virol.* **82**, 1187–1197 (2001).

Zhang, T., Breitbart, M., Lee, W. H., Run, J. Q., Wei, C. L., Soh, S. W., Hibberd, M. L., Liu, E. T., Rohwer, F., and Ruan, Y. RNA viral community in human feces: Prevalence of plant pathogenic viruses. *PLoS Biol.* **4**, e3 (2006).

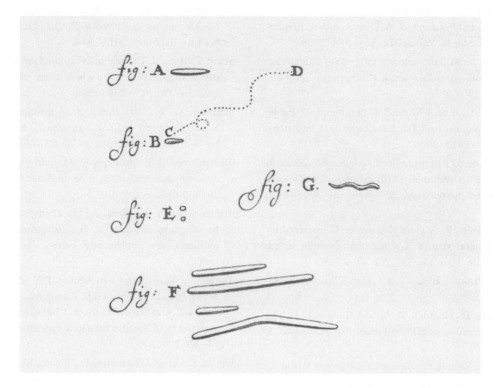

Antony van Leeuwenhoek (1622–1723) has been called the father of protozoology and bacteriology. This figure shows bacteria he observed taken from his own mouth. Figure A indicates a motile Bacillus. Figure B shows Selenomonas sputigena, *while C and D show the path of its motion. Figure E shows two micrococci; F shows* Leptotrichia buccalis, *and G shows a spirochete. He describes these "animalcules," found in his and others' mouths, in a letter written 17 September 1683. "While I was talking to an old man (who leads a sober life, and never drinks brandy or [smokes] tobacco, and very seldom any wine), my eye fell upon his teeth, which were all coated over; so I asked him when he had last cleaned his mouth? And I got for answer that he'd never washed his mouth in all his life. So I took some spittle out of his mouth and examined it; but I could find in it nought but what I had found in my own and other people's. I also took some of the matter that was lodged between and against his teeth, and mixing it with his own spit, and also with fair water (in which there were no animalcules), I found an unbelievably great company of living animalcules, a-swimming more nimbly than any I had ever seen up to this time. The biggest sort (where of there were a great plenty) bent their body into curves in going forwards, as in Fig. G. Moreover, the other animalcules were in such enormous numbers, that all the water (notwithstanding only a very little of the matter taken from between the teeth was mingled with it) seemed to be alive" (translated from the Dutch by Dobell, 1932, pp. 242–243). Used with permission.*

15

Completed Genomes: Bacteria and Archaea

"And now you may be disposed to ask: To what end is this discourse on the anatomy of beings too minute for ordinary vision, and of whose very existence we should be ignorant unless it were revealed to us by a powerful microscope? What part in nature can such apparently insignificant animalcules play, that can in any way interest us in their organization, or repay us for the pains of acquiring a knowledge of it? I shall endeavour briefly to answer these questions. The Polygastric Infusoria, notwithstanding their extreme minuteness, take a great share in important offices of the economy of nature, on which our own well-being more or less immediately depends.

Consider their incredible numbers, their universal distribution, their insatiable voracity; and that it is the particles of decaying vegetable and animal bodies which they are appointed to devour and assimilate.

Surely we must in some degree be indebted to those ever active invisible scavengers for the salubrity of our atmosphere. Nor is this all: they perform a still more important office, in preventing the gradual diminution of the present amount of organized matter upon the earth. For when this matter is dissolved or suspended in water, in that state of comminution and decay which immediately precedes its final decomposition into the elementary gases, and its consequent return from the organic to the inorganic world, these wakeful members of nature's invisible police are every where ready to arrest the fugitive organized particles, and turn them back into the ascending stream of animal life."

—Richard Owen (1843, p. 27)

Bioinformatics and Functional Genomics, Second Edition. By Jonathan Pevsner
Copyright © 2009 John Wiley & Sons, Inc.

INTRODUCTION

William Martin and Eugene Koonin (2006) briefly discuss the definition of the term prokaryote. We will contrast prokaryotes and eukaryotes in Chapter 16.

It has been estimated that there are 10^{30} bacteria, comprising the majority of the biomass on the planet (Sherratt, 2001).

In this chapter we will consider bioinformatic approaches to two of the three main branches of life: bacteria and archaea. Bacteria and archaea are grouped together because they are prokaryotes, that is, single-celled organisms that lack nuclei. Bacteria and archaea are sometimes also termed microorganisms. The term microbe refers to those microorganisms that cause disease in humans; microbes include many eukaryotes such as fungi and protozoa (Chapters 17 and 18) as well as some prokaryotes.

It has been estimated that bacteria account for 60% of Earth's biomass. Bacteria occupy every conceivable ecological niche in the planet, and there may be from 10^7 to 10^9 distinct bacterial species (Fraser et al., 2000), although some suggest there may be fewer species (Schloss and Handelsman, 2004). The great majority of bacteria and archaea (>99%) have never been cultured or characterized (DeLong and Pace, 2001). A compelling reason to study bacteria is that many cause disease in humans and other animals.

This chapter provides an overview of bioinformatic approaches to the study of bacteria and archaea. We review aspects of prokaryotic biology such as genome size and complexity, and tools for the analysis and comparison of prokaryotic genomes. The analysis of whole genome sequencing has had profound effects on our understanding of bacteria and archaea (reviewed in Bentley and Parkhill, 2004; Fraser-Liggett, 2005; Ward and Fraser, 2005; Binnewies et al., 2006). Some of the main issues are (1) an improved sampling of the diversity of the prokaryotes through genomic sequence analyses, along with improved phylogeny and classification; (2) a better understanding of the forces that shape microbial genomes. These forces include the following:

- Loss of genes and reductions in genome size, especially in species that are dependent on their hosts for survival, such as obligate intracellular parasites; gains in genome size, especially in free-living organisms that may require many genomic resources to cope with variable environmental conditions;
- Lateral gene transfer, in which genetic material is transferred horizontally between organisms that share an environmental niche and not vertically through descent from ancestors;
- Chromosomal rearragements such as inversions often occur in related species.

In this chapter we will discuss these topics as well as bioinformatics tools that are available to investigate them.

CLASSIFICATION OF BACTERIA AND ARCHAEA

In Chapter 13, we described many of the genome-sequencing projects for bacteria and archaea in chronological order, beginning with the sequencing of *Haemophilus influenzae* in 1995. We will now consider the classification of bacteria and archaea by six different criteria: (1) morphology, (2) genome size, (3) lifestyle, (4) relevance to human disease, (5) molecular phylogeny using rRNA, and (6) molecular phylogeny using other molecules. There are many other ways to classify bacteria and archaea (Box 15.1).

BOX 15.1
Classification of Prokaryotes

While we will choose six basic ways to classify the prokaryotes, there are many other approaches. These include the energy source (respiration, fermentation, photosynthesis), their formation of characteristic products (e.g., acids), the presence of immunological markers such as proteins or lipopolysaccharides, their ecological niche (also related to the lifestyle), and their nutritional growth requirements. The types based on growth requirements include obligate and/or facultative aerobes (requiring oxygen) or anaerobes (growing in environments without oxygen), chemotrophs (deriving energy from the breakdown of organic molecules such as proteins, lipids, and carbohydrates), and autotrophs (synthesizing organic molecules through the use of an external energy source and inorganic compounds such as carbon dioxide and nitrates). Autotrophs (from the Greek for "self feeder") are either photoautotrophs (obtaining energy through photosynthesis; requiring carbon dioxide and expiring oxygen) or chemautotrophs (obtaining energy from inorganic compounds and carbon from carbon dioxide). Heterotrophs, unlike autotrophs, must feed on other organisms to obtain energy.

In describing prokaryotes and their genomes, we will examine bioinformatics tools to analyze individual microbial genomes and tools for the comparison of two or more genomes. It is through comparative genomics that we are beginning to appreciate some of the important principles of microbial biology, such as the adaptation of microbes to highly specific ecological niches, the lateral transfer of genes between microbes, genome expansion and reduction, and the molecular basis of pathogenicity (Bentley and Parkhill, 2004; Binnewies et al., 2006).

The Comprehensive Microbial Resource (CMR) and the National Center for Biotechnology Information (NCBI) describe major divisions of bacteria (Table 15.1), as well as the two major divisions of archaea: crenarchaeota and euryarchaeota (Table 15.2). These tables provide an overview of the prokaryotes as we begin to classify them by various criteria.

Classification of Bacteria by Morphological Criteria

Most bacteria are classified into four main types: Gram-positive and Gram-negative cocci or rods (reviewed in Schaechter, 1999). Examples of these different bacteria are presented in Table 15.3. The Gram stain is absorbed by about half of all bacteria and reflects the protein and peptidoglycan composition of the cell wall. Many other bacteria do not fit the categories of Gram-positive or Gram-negative cocci or rods because they have atypical shapes or staining patterns. As an example, spirochetes such as the Lyme disease agent *Borrelia burgdorferi* have a characteristic outer membrane sheath, protoplasmic cell cylinder, and periplasmic flagella (Charon and Goldstein, 2002).

The classification of microbes based on molecular phylogeny is far more comprehensive, as described below. Molecular differences can reveal the extent of microbial diversity both between species (showing the breadth of the prokaryotic tree of life) and within species (e.g., showing molecular differences in pathogenic isolates and in closely related, nonvirulent strains). However, beyond molecular criteria there are many additional ways to differentiate bacteria based on microscopy and

Pathogenicity is the ability of an organism to cause disease. Virulence is the degree of pathogenicity.

The CMR was developed at the Institute for Genomic Research (TIGR) which is now part of the J. Craig Venter Institute (see databases at ▶ http://cmr.jcvi.org). We also introduced major genomics resources from the Europan Bioinformatics Institute in Chapter 13, such as Genome Reviews (▶ http://www.ebi.ac.uk/GenomeReviews/). Another major resource for prokaryotic and eukaryotic genomes is PEDANT at the Munich Information Center for Protein Sequences (MIPS; ▶ http://pedant.gsf.de/).

TABLE 15-1 Classification of Bacteria

Intermediate Rank 1	Intermediate Rank 2	Genus, Species, and Strain (Examples)	Genome Size (Mb)	GenBank Accession
Actinobacteria	Actinobacteridae	*Mycobacterium tuberculosis* CDC1551	4.4	NC_002755
Aquificae	Aquificales	*Aquifex aeolicus* VF5	1.5	NC_000918
Bacteroidetes	Bacteroides	*Porphyromonas gingivalis* W83	2.3	NC_002950
Chlamydiae	Chlamydiales	*Chlamydia trachomatis* serovar D	1.0	NC_000117
Chlorobi	Chlorobia	*Chlorobium tepidum* TLS	2.1	NC_002932
Cyanobacteria	Chroococcales	*Synechocystis* sp. PCC6803	3.5	NC_000911
	Nostocales	*Nostoc* sp. PCC 7120	6.4	NC_003272
Deinococcus-Thermus	Deinococci	*Deinococcus radiodurans* R1	2.6	NC_001263
Firmicutes	Bacillales	*Bacillus subtilis* 168	4.2	NC_000964
	Clostridia	*Clostridium perfringens* 13	3.0	NC_003366
	Lactobacillales	*Streptococcus pneumoniae* R6	2.0	NC_003098
	Mollicutes	*Mycoplasma genitalium* G-37	0.580	NC_000908
Fusobacteria	Fusobacteria	*Fusobacterium nucleatum* ATCC 25586	2.1	NC_003454
Proteobacteria	Alphaproteobacteria	*Rickettsia prowazekii* Madrid E	1.1	NC_000963
	Betaproteobacteria	*Neisseria meningitidis* MC58	2.2	NC_003112
	Epsilon subdivision	*Helicobacter pylori* J99	1.6	NC_000921
	Gamma subdivision	*Escherichia coli* K12-MG1655	4.6	NC_000913
	Magnetotactic cocci	*Magnetococcus* sp. MC-1	4.7	NC_008576
Spirochaetales	Spirochaetaceae	*Borrelia burgdorferi* B31	0.91	NC_001318
Thermotogales	Thermotoga	*Thermotoga maritima* MSB8	1.8	NC_000853

Bacteria are described as a kingdom, followed by "intermediate ranks."
Sources: TIGR Comprehensive Microbial Resource (▶ http://cmr.jcvi.org/) and NCBI (▶ http://www.ncbi.nlm.nih.gov).

TABLE 15-2	Classification of Archaea			
Intermediate Rank 1	Intermediate Rank 2	Genus, Species, and Strain (Examples)	Genome Size (Mb)	GenBank Accession
Crenarchaeota	Thermoprotei	*Aeropyrum pernix* K1	1.6	NC_000854
Euryarchaeota	Archaeoglobi	*Archaeoglobus fulgidus* DSM4304	2.2	NC_000917
	Halobacteria	*Halobacterium* sp. NRC-1	2.0	NC_002607
	Methanobacteria	*Methanobacterium thermoautotrophicum* delta H	1.7	NC_000916
	Methanococci	*Methanococcus jannaschii* DSM2661	1.6	NC_000909
	Methanopyri	*Methanopyrus kandleri* AV19	1.6	NC_003551
	Thermococci	*Pyrococcus abyssi* GE5	1.7	NC_000868
	Thermoplasmata	*Thermoplasma volcanium* GSS1	1.5	NC_002689

Archaea are described as a kingdom, followed by "intermediate ranks."
Sources: TIGR Comprehensive Microbial Resource (► http://cmr.jcvi.org/) and NCBI (► http://www.ncbi.nlm.nih.gov).

TABLE 15-3	Major Categories of Bacteria Based on Morphological Criteria
Type	Examples[a]
Gram-positive cocci	*Streptococcus pyogenes*, *Staphylococcus aureus*
Gram-positive rods	*Corynebacterium diphtheriae*, *Bacillus anthracis* (anthrax), *Clostriduium botulinum*
Gram-negative cocci	*Neisseria*, *Gonococcus*
Gram-negative rods	*Escherichia coli*, *Vibrio cholerae*, *Helicobacter pylori*
Other	*Mycobacterium leprae* (leprosy), *Borrelia burgdorferi* (Lyme disease), *Chlamydia trachomatis* (sexually transmitted disease), *Mycoplasma pneumoniae*

[a]The disease is indicated in parentheses.

studies of physiology—for example, distinguishing those microbes that are capable of oxygenic photosynthesis (cyanobacteria) or those that produce methane.

The diversity of morphologies in prokaryotic life forms is spectacular. We can provide examples of two predatory bacteria that prey on other bacteria. Each of these examples is intended to highlight both the diversity of morphologies that may occur, and the role that genome sequence analysis may have in elucidating mechanisms of structural change.

(1) The myxobacteria are single-celled δ-proteobacteria organisms that are highly successful, with millions of cells per gram of cultivated soil. Upon encountering low nutrient conditions up to 100,000 individuals of *Myxococcus xanthus* join to form a fruiting body, which is essentially a multicellular organism having a spherical

The *M. xanthus* DK 1622 complete, circular genome (length 9,139,763 nucleotides) has accession NC_008095. Note that by entering that accession number into the Entrez search engine from the home page of NCBI you can link to the Genome Project page that provides an overview of the organism. The slime mold *Dictyostelium discoideum*, a eukaryote, also includes a lifestyle that can alternate between single-celled and multicellular (Chapter 18).

The *B. bacteriovorus* accession is NC_005363. Its life cycle is described at the NCBI Entrez Genome Project page for this organism.

In diploid or polyploid organisms, the genome size is the amount of DNA in the unreplicated haploid genome (such as the sperm cell nucleus). We discuss eukaryotic genome sizes in Chapters 16 to 19.

shape and that is resistant to different kinds of stress. In favorable nutrient conditions individual spores within the fruiting body germinate and thousands of *M. xanthus* spores swarm. This swarm can surround, lyse, and consume prey bacteria. Goldman et al. (2006) reported the complete genome sequence of *M. xanthus* and provided insight into genes that encode motor proteins and allow the organism to glide, use retractable pili, and secrete mucus. Also, the large genome size (9.1 megabases [Mb]) contrasts with the much smaller size of other, related δ subgroup proteobacteria (3.7 to 5.0 Mb). Goldman et al. characterized the nature of the *M. xanthus* genome expansion and its possible relation to this organism's extraordinary behavior and morphology.

(2) *Bdellovibrio bacteriovorus* provides a second example of a prokaryote with an extraordinary morphology. This is also a predatory δ-proteobacterium that eats Gram-negative bacteria. Its genome of about 3.8 Mb is predicted to encode over 3500 proteins (Rendulic et al., 2004). The bacterium attacks its prey (by swimming to them at high speed), adheres irreversibly, opens a pore in the prey's outer membrane and peptidoglycan layer, then enters the periplasm and replicates. *B. bacteriovorus* then forms a structure called a bdelloplast in which the rod-shaped prey becomes rounded and the predator grows to several multiples of its normal size as it consumes the prey nutrients. Later, the predator exits the bdelloplast. The analysis of this genome allowed Rendulic et al. to identify genes encoding catabolic enzymes (e.g., proteases, nucleases, glycanases, and lipases) implicated in its lifestyle, as well as a host interaction locus containing genes implicated in pilus and adherence genes.

Classification of Bacteria and Archaea Based on Genome Size and Geometry

In haploid organisms such as bacteria and archaea, the genome size (or *C* value) is the total amount of DNA in the genome. Bacterial and archaeal genomes vary in size from under 500,000 bp [0.5 megabases (Mb)] to over 10 Mb (Table 15.4) (Casjens, 1998). The genome sizes of 23 named major bacterial phyla and some of their subgroups are shown in Fig. 15.1. As indicated in the figure, most bacterial genomes are circular, although some are linear; some bacterial genomes consist of multiple circular chromosomes. Plasmids (small circular extrachromosomal elements) have been found in most bacterial phyla, although linear extrachromosomal elements are more rare.

Some bacterial genomes are comparable in size to or even larger than eukaryotic genomes. The genome of the fungus *Encephalitozoon cuniculi* is just 2.5 Mb and encodes about 2000 proteins (see Chapter 17), and at least a dozen eukaryotic genomes that are currently being sequenced are under 10 Mb. The *Sorangium cellulosum* genome, the largest bacterial genome that has been sequenced to date, is

TABLE 15-4	Range of Genome Sizes in Bacteria and Archaea	
Taxon	Genome Size Range (Mb)	Ratio (Highest/Lowest)
Bacteria	0.16–13.2	83
Mollicutes	0.58–2.2	4
Gram-negative	0.16–9.5	59
Gram-positive	1.6–11.6	7
Cyanobacteria	3.1–13.2	4
Archaea	0.49–5.75	12

Source: Modified from Graur and Li (2000, p. 36). Used with permission.

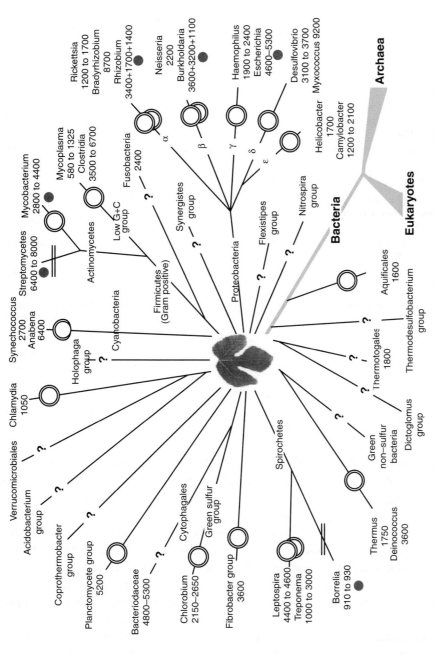

FIGURE 15.1. *Bacterial chromosome size and geometry. The 23 named major bacterial phyla are represented, as well as some of their subgroups. The tree is based on rRNA sequences and is unrooted. The branch lengths do not depict phylogenetic distances, and the fig leaf at the center indicates uncertain branching patterns. The chromosome geometry (circular or linear, in some cases with multiple chromosomes) is indicated at the end of each branch. The chromosome sizes of representative genera are given (in kilobases). Linear extrachromosomal elements, common in borrelias and actinomycetes, are indicated. This figure was modified from Casjens (1998). Used with permission.*

TABLE 15-5   Genome Size of Selected Bacteria and Archaea Having Relatively Large or Small Genomes

Species	Genome Size (Mb)	Coding Regions	GC Content	Reference
Sorangium cellulosum	13	9703	71	Unpublished; accession NC_010162
Solibacter usitatus [B]	10	7,888	61.9	Unpublished; accession NC_008536
Myxococcus xanthus DK 1622 [B]	9.1	7,388	68.9	Goldman et al., 2006
Streptomyces coelicolor [B]	8.67	7,825	72	Bentley et al., 2002
Methanosarcina acetivorans C2A [A]	5.75	4,524	42.7	Galagan et al., 2002
Ureaplasma urealyticum parvum biovar serovar 3 [B]	0.752	613	26	Glass et al., 2000
Mycoplasma pneumoniae M129 [B]	0.816	677	40	Himmelreich et al., 1996
Mycoplasma genitalium G-37 [B]	0.58	470	32	Fraser et al., 1995
Nanarchaeum equitans [A]	0.49	552	31.6	Huber et al., 2002; Waters et al., 2003
Buchnera aphidicola [B]	0.42	362	20	Pérez-Brocal et al., 2006
Carsonella ruddii [B]	0.16	182	16.5	Nakabachi et al., 2006

Abbreviations: [A], archaeal; [B], eubacterial.
Source: Adapted from ► http://www.sanger.ac.uk/Projects/Microbes/ and the NCBI website (PubMed, Entrez Genome).

13 Mb and includes over 9700 genes (Table 15.5). In general, those prokaryotes having notably large genome sizes exhibit great behavioral or phenotypic complexity, participating in complex social behavior (such as multicellular interactions) or processes such as differentiation.

Overall, the number of genes encoded in a bacterial genome ranges from the exceptionally small number of 182 to 8000. This range is comparable to the range in *C* values. For a large number of bacteria with completely sequenced genomes, protein-coding genes constitute about 85% to 95% of the genome. Thus, intergenic and nongenic fractions are small. (An exception is the pathogen that causes leprosy, *Mycobacterium leprae*. Its genome underwent massive gene decay, and protein-coding genes constitute only 49.5% of the genome [Cole et al., 2001].) The density of genes in microbial genomes is consistently about one gene per kilobase. As an example, the genome of *Escherichia coli* K12 (accession NC_000913) is 4.6 Mb and it encodes 4243 proteins (one gene per 1093 base pairs). Even in very small genomes such as *Mycoplasma genitalium*, reduced genome sizes are not associated with changes either in gene density or in the average size of genes (Fraser et al., 1995). The genome sizes of selected large or small bacteria and archaea are shown in Table 15.5.

Examination of the sizes of several hundred prokaryotic genomes in relation to the number of genes shows a linear relationship (Fig. 15.2). This figure (adapted from Giovannoni et al., 2005) further distinguishes free-living, host-associated, and obligate symbiont organisms. The smallest bacterial and archaeal genomes are

Another exception is the parasite *Rickettsia prowazekii*, described below, that has 24% noncoding DNA.

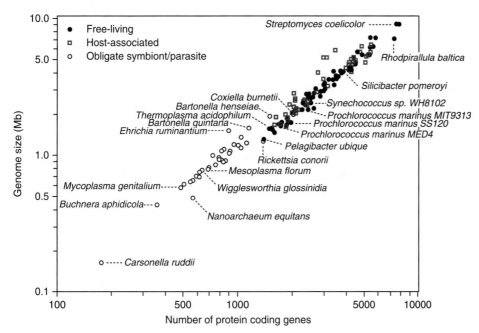

FIGURE 15.2. Number of predicted protein-encoding genes versus genome size for 246 complete published genomes from bacteria and archaea. This figure is adapted from Giovannoni et al. (2005) who reported that P. ubique *has the smallest number of genes (1354 open reading frames) for any free-living organism that has been studied in the laboratory. Recent data from the smallest prokaryotic genomes are indicated (*B. aphidicola, C. ruddii*). Used with permission.*

from intracellular bacteria that are parasites or symbionts having an obligate relationship with a host that provides nutrition. In general, prokaryotes having very small genome sizes live in extremely stable environments in which the host provides reliable resources (e.g., nutrients) and homeostatic benefits (e.g., a constant pH). Organisms with small genomes evolved from ancestors with larger genomes. One of the smallest sequenced genomes of a free-living organism (and one of the first genomes to have been sequenced) is that of *Mycoplasma genitalium*, a urogenital pathogen. The *M. genitalium* has 580,070 bp encoding 470 protein-coding genes, 3 rRNA genes, and 33 tRNA genes (Fraser et al., 1995). Mycoplasmas are bacteria of the class Mollicutes. They lack a cell wall and have a low GC content (32%) characteristic of this class.

Of the very smallest bacterial genomes, *Buchnera aphidicola* has a genome of just 422,434 base pairs with 362 protein-coding genes (Pérez-Brocal et al., 2006). The genome is organized in a circular chromosome and an additional 6 kilobase plasmid for leucine biosynthesis. There is an obligate endosymbiotic relationship between *B. aphidicola* and the cedar aphid *Cinara cedri*. The bacterium has lost most of its metabolic functions, depending on those provided by its host, while in turn it provides metabolites since the aphid diet is restricted to plant sap and so it needs essential amino acids and other nutrients. The relationship between host and bacterium is thought to have been established over 200 million years ago, with a continual reduction in the size of the bacterial genome such that it no longer possesses the capability to synthesize its own cell wall.

The smallest prokaryotic genome that has been found to date is that of another endosymbiont, *Carsonella ruddii* (indicated in Fig. 15.2). Its genome consists of a single circular chromosome of 159,662 base pairs with only 182 open reading frames (Nakabachi et al., 2006). Both the small genome size and the low guanine plus cytosine content (GC content 16.5%) are exceptional. Half of the open reading frames encode proteins implicated in translation and amino acid metabolism. Like *B. aphidicola*, *C. ruddii* is an obligate endosymbiont of a sap-feeding insect, the psyllid *Pachypsylla venusta*.

Aphids are metazoans (animals) within the class Insecta. The *B. aphidicola* accession is NC_008513.

Among the archaea the smallest genome is that of a hyperthermophilic organism that was cultured from a submarine hot vent, *Nanoarchaeum equitans* (Huber et al., 2002). This archaeon appears to grow attached to another archeon, *Ignicoccus*. Because of its small cell size (400 nm) and small genome size, Huber et al. 2002 suggested that *N. equitans* resembles an intermediate between the smallest living organisms (such as *M. genitalium*) and large viruses (such as the pox virus). Nonetheless, even parasitic intracellular bacteria and archaea are classified as organisms distinct from viruses.

By comparing small prokaryotic genomes, it is possible to estimate the minimal number of genes required for life (Box 15.2). The *B. aphidicola* and *C. ruddii* genomes do not encode many proteins that serve transport functions, suggesting that their metabolites may freely diffuse to their hosts. Many required gene products could have been transferred to their hosts' nuclear genomes. Such a process has occurred in mitochondria, which depends on many proteins encoded by a eukaryotic nuclear genome.

See Andersson (2006) for a review of the *B. aphidicola* and *C. ruddii* genomes.

BOX 15.2
Small Genome sizes, Minimal Genome Sizes, and Essential Genes

How many genes are required in the genome of the smallest living organism—that is, the smallest autonomous self-replicating organism? One approach is to identify the smallest genomes in nature. The *B. aphidicola* and *C. ruddii* genomes encode only 362 and 182 proteins, respectively, although they are constrained to living within particular insect cells (Andersson, 2006). Prokaryotes of the genus *Mycoplasma* tend to have both small sizes and small genomes, and thus have been studied in terms of minimal gene sets. At present, 34 species from this genus are at least partially sequenced, including *M. pneumoniae* and *M. genitalium* (Fadiel et al., 2007). The forces driving the evolution of small genome size include genome reduction from larger ancestral genomes in a process that may promote fitness of the organism. In thinking about a minimal genome size we must always consider the ecological niche occupied by the organism, which will have an enormous influence on the particular genes of the endosymbiont as well as the mechanisms of reductive evolution.

A second approach involves comparative genomics by identifying the orthologs in common between several microbes. In the earliest days of complete genome sequencing, Mushegian and Koonin 1996 identified 239 genes in common between *Escherichia coli*, *H. influenzae*, and *M. genitalium*. This is considered one estimate of the minimal genome size. The functions of these 239 genes include several basic categories: translation, DNA replication, recombination and DNA repair, transcription, anaerobic metabolism, lipid and cofactor biosynthesis, and transmembrane transporters.

A third approach to determining the minimal number of genes required for life is experimental. Itaya 1995 randomly knocked out protein-coding genes in the bacterium *Bacillus subtilis*. Mutations in only 6 of 79 loci prevented growth of the bacteria and were indispensible. Extrapolating to the size of the complete *B. subtilis* genome, about 250 genes were estimated to be essential for life. Attempts are underway to create life forms from a specific set of genes. Pósfai et al. (2006) from the group of Frederick Blattner have experimentally reduced the genome size to *Escherichia coli* K-12 (by 20% to about 4 Mb), targeting the removal of insertion sequence elements and other mobile DNA elements as well as repeats that mediate structural changes (such as inversions, duplications, and deletions). For *M. tuberculosis*, random transposon mutagenesis has been employed to identify essential genes (Lamichhane et al., 2003). This and related approaches can provide information on which genes and gene products are likely to be most useful as drug targets (Lamichhane and Bishai, 2007).

Several groups have reviewed progress toward identifying core sets of genes required for life, including Koonin (2003) and Gil et al. (2004). Koonin lists 63 genes that are present across all of ~100 genomes sequenced at the time. These include genes having functions in translation (e.g., ribosomal proteins and aminoacyl-transfer RNA synthetases, and translation factors), transcription (RNA polymerase subunits), and replication and repair (DNA polymerase subunits, exonuclease, topoisomerase). The COGs database (see below) provides access to many of these highly conserved proteins.

Classification of Bacteria and Archaea Based on Lifestyle

In addition to the criteria of morphology and genome size and geometry, a third approach to classifying bacteria (and archaea) is based on their lifestyle. One main advantage of this approach is that it conveniently highlights the principle of extreme reduction in genome size that is associated with three lifestyles: extremophiles and intracellular and epicellular prokaryotes:

- Extremophiles are microbes that live in extreme environments. Archaea have been identified in hypersaline conditions (halophilic archaea), geothermal areas such as hot vents (hyperthermophilic archaea), and anoxic habitats (methanogens) (DeLong and Pace, 2001). One of the most extraordinary extremophiles is *Deinoccoccus radiodurans*, which can survive dessication as well as massive doses of ionizing radiation (it thrives in nuclear waste). It achieves this feat by reassembling shattered chromosomes through a novel repair mechanism (Zahradka et al., 2006).

- Intracellular bacteria invade eukaryotic cells; a well-known example is the α-proteobacterium that is thought to have invaded eukaryotic cells and evolved into the present-day mitochondrion.

- Epicellular bacteria (and archaea) are parasites that live in close proximity to their hosts but not inside host cells.

We may distinguish six basic lifestyles of bacteria and archaea (Table 15.6):

1. Extracellular: For example, *E. coli* commonly inhabits the human intestine without entering cells. Many free-living prokaryotes have relatively large genomes (as indicated in Fig. 15.2), such as the δ-proteobacterium *Myxococcus xanthus* described above. Having a larger genome may provide a reservoir of genes that can be utilized to meet the needs of changing environments. As another example the Gram-positive bacterium *Propionibacterium acnes* inhabits human skin and can cause acne. Its 2.5 Mb genome allows *P. acnes* the flexibility to grow under aerobic or aneaerobic conditions and to utilize a variety of substrates available from skin cells (Brüggemann et al., 2004).

| TABLE 15-6 | Classification of Bacteria and Archaea Based on Ecological Niche |

Lifestyle	Bacterium	Genome Size (Mb)	Reference
Extracellular	*Escherichia coli*	4.6	Blattner et al., 1997
	Vibrio cholerae	4.0	Heidelberg et al., 2000
	Pseudomonas aeruginosa	6.3	Stover et al., 2000
	Bacillus subtilis	4.2	Kunst et al., 1997
	Clostridium acetobutylicum	4.0	Nolling et al., 2001
	Deinococcus radiodurans	3.3	White et al., 1999
Facultatively intracellular	*Salmonella enterica*	4.8	Parkhill et al., 2001a
	Yersinia pestis	4.7	Parkhill et al., 2001b
	Legionella pneumophila	3.9	Bender et al., 1990
	Mycobacterium tuberculosis	4.4	Cole et al., 1998
	Listeria monocytogenes	2.9	Glaser et al., 2001
Extremophile	*Aeropyrum pernix*	1.7	Kawarabayasi et al., 1999
	Methanococcus janneschi	1.7	Bult et al., 1996
	Archeoglobus fulgidus	2.2	Klenk et al., 1997
	Thermotoga maritima	1.9	Nelson et al., 1999
	Aquifex aeolius	1.6	Deckert et al., 1998
Epicellular	*Neisseria meningitidis*	2.2	Tettelin et al., 2000
	Haemophilus influenzae	1.8	Fleischmann et al., 1995
	Mycoplasma genitalium	0.6	Fraser et al., 1995
	Mycoplasma pneumoniae	0.8	Himmelreich et al., 1996
	Ureaplasma urealyticum	0.8	Glass et al., 2000
	Mycoplasma pulmonis	1.0	Chambaud et al., 2001
	Borrelia burgdorferi	0.9	Fraser et al., 1997; Casjens et al., 2000
	Treponema pallidum	1.1	Fraser et al., 1998
	Helicobacter pylori	1.7	Tomb et al., 1997; Alm et al., 1999
	Pasteurella multocida	2.3	May et al., 2001
Obligate intracellular, symbiotic	*Buchnera* sp.	0.6	Shigenobu et al., 2000
	Wolbachia spp.	1.1	Sun et al., 2001
	Wigglesworthia glossinidia	0.7	Akman et al., 2002
	Sodalis glossinidius	2.0	Akman et al., 2001

(Continued)

TABLE 15-6   Continued

Lifestyle	Bacterium	Genome Size (Mb)	Reference
Obligate intracellular, parasitic	*Rickettsia prowazekii*	1.1	Andersson et al., 1998
	Rickettsia conorii	1.3	Ogata et al., 2001
	Ehrlichia chaffeensis	1.2	Hotopp et al., 2006
	Cowdria ruminantium	1.6	de Villiers et al., 2000
	Chlamydia trachomatis	1.1	Stephens et al., 1998; Read et al., 2000
	Chlamydophila pneumoniae	1.3	Kalman et al., 1999; Read et al., 2000; Shirai et al., 2000

Source: Adapted from ▶ http://www.chlamydiae.com.

2. Facultatively intracellular bacteria can enter host cells, but this behavior depends on environmental conditions. *Mycobacterium tuberculosis*, the cause of tuberculosis, can remain dormant within infected macrophages, only to activate and cause disease many decades later.

3. Extremophilic microbes: Initially, archaea were all identified in extreme environmental conditions. Some archaea have been found to grow at temperatures as high as 113°C, at pH 0, and in salt concentrations as high as 5 *M* sodium chloride. *Methanococcus janneschi*, the first archeal organism to have its genome completely sequenced (Bult et al., 1996), grows at pressures over 200 atm and at an optimum temperature near 85°C. Archaea have subsequently been identified in less extreme habitats, including forest soil and ocean seawater (DeLong, 1998; Robertson et al., 2005).

4. Epicellular prokaryotes grow outside of their hosts, but in association with them. *Mycoplasma pneumoniae*, a bacterium with a genome size of ≈816,000 bp, is a major cause of respiratory infections. The bacterium is a surface parasite that attaches to the respiratory epithelium of its host. The genome was sequenced (Himmelreich et al., 1996) and subsequently reannotated by Peer Bork and colleagues (Dandekar et al., 2000).

5. Obligately intracellular and symbiotic: Tamas et al. 2002 compared the complete genome sequences of two bacteria, *Buchnera aphidicola* (Sg) and *Buchnera aphidicola* (Ap), that are endosymbionts of the aphids *Schizaphis graminum* (Sg) and *Acyrthosiphon pisum* (Ap). Each of these bacteria has a small genome size of about 640,000 bp. They have 564 and 545 genes, respectively, of which they share almost all (526). Remarkably, these bacteria diverged about 50 MYA, yet they share complete conservation of genome architecture. There have been no inversions, translocations, duplications, or gene acquisitions in either bacterial genome since their divergence (Tamas et al., 2002). This provides a dramatic example of genomic stasis. Although it is extremely rare for obligate intracellular bacteria to share such genome conservation, it is common for endosymbionts to have relatively small genome sizes. This may reflect the dependence of these bacteria on nutrients derived from the host.

6. Obligately intracellular and parasitic: *Rickettsia prowazekii* is the bacterium that causes epidemic typhus. Its genome is relatively small, consisting of

Each year, 1.9 million people die of tuberculosis and 1.9 billion people are infected worldwide (▶ http://www.cdc.gov/ncidod/eid/vol8no11/02-0468.htm). The *M. tuberculosis* genome was sequenced by Cole et al. (1998).

1.1 Mb (Andersson et al., 1998). Like other *Rickettsia*, it is an α-proteobacterium that infects eukaryotic cells selectively. It is also of interest because it is closely related to the mitochondrial genome. A closely related species, *Rickettsia conorii*, is an obligate intracellular parasite that causes Mediterranean spotted fever in humans. Its genome was sequenced by Ogata et al. (2001). Similar to the *Buchnera aphidicola* subspecies, the genome organization of the two *Rickettsia* parasites is well conserved.

Why are some bacterial genome sizes severely reduced? Intracellular parasites are subject to deleterious mutations and substitutions that cause gene loss, tending toward genome reduction (Andersson and Kurland, 1998). A similar process occurred as a primordial α-proteobacterium evolved into the modern mitochondrion, maintaining only a minuscule mitochondrial genome size (Chapter 13).

While we are interested in surveying different lifestyles of prokaryotes and the specific niches they inhabit, the vast majority are not cultivatable in vitro and thus have been difficult to sample. Metagenomics projects have enabled the large-scale, culture-independent characterization of the prokaryotes inhabiting a particular environment (reviewed in Riesenfeld et al., 2004; Tringe and Rubin, 2005). Some metagenomics studies have employed high throughput sequencing technologies (as discussed in Chapter 13). Other studies have focused on sampling ribosomal RNA (discussed below).

- Walker and Pace (2007) review studies of prokaryotes in endolithic ecosystems, that is, living in the pore space in rocks.

- Tringe et al. (2005) sampled a range of nutrient-rich environments from agriculture soil and from deep-sea "whale fall" carcasses.

- Trillions of bacteria live in the human gut; Gill et al. (2006) estimate that 10^{13} to 10^{14} microorganisms inhabit the intestine, collectively containing 100 times as many genes as the human genome. Several studies have investigated the diversity of microbial life in feces. Palmer et al. (2007) used a microarray-based approach to identify ribosomal RNAs in the stool samples of 14 healthy human infants across the first year of life. Gill et al. (2006) sequenced ~78 million base pairs of DNA from the fecal samples of two adults. They identified commonly occurring bacteria such as *Bifidobacterium longum* and archaea such as *Methanobrevibacter smithii*. Sixty of 72 main bacterial phylotypes were from uncultivatable organisms. Both studies identified bacteria from several clades, including the Gram-positive bacteria (*Firmicutes* and *Actinobacteria*), Proteobacteria, and Bacteroidetes.

Classification of Bacteria Based on Human Disease Relevance

Bacteria and eukaryotes have engaged in an ongoing war for millions of years. Bacteria occupy the nutritive environment of the human body in an effort to reproduce. Typical sites of bacterial colonization include the skin, respiratory tract, digestive tract (mouth, large intestine), urinary tract, and genital system (Eisenstein and Schaechter, 1999). It has been estimated that each human has ten times more bacterial cells than human cells in the body. In the majority of cases, these bacteria are harmless to humans. However, many bacteria cause infections, often with devastating consequences.

In recent years, the widespread use of antibiotics has led to an increased prevalence of drug resistance among bacteria. It is thus imperative to identify bacterial

TABLE 15-7   Vaccine-Preventable Bacterial Diseases

Disease	Species
Anthrax	*Bacillus anthracis*
Diarrheal disease (cholera)	*Vibrio cholerae*
Diphtheria	*Corynebacterium diphtheriae*
Community acquired pneumonia	*Haemophilus influenzae* type B, *Streptococcus pneumoniae*
Lyme disease	*Borrelia burgdorferi*
Meningitis	*Haemophilus influenzae* type B (HIB), *Streptococcus pneumoniae*, *Neisseria meningitidis*
Pertussis	*Bordetella pertussis*
Tetanus	*Clostridium tetani*
Tuberculosis	*Mycobacterium tuberculosis*
Typhoid	*Salmonella typhi*

Source: Adapted from ► http://www.cdc.gov/vaccines/vpd-vac/vpd-list.htm and ► http://www.cdc.gov/ncidod/dbmd/diseaseinfo/default.htm.

virulence factors and to develop strategies for vaccination. One approach to this problem is to compare pathogenic and nonpathogenic strains of bacteria (see below). Table 15.7 lists some of the bacterial diseases for which vaccinations are routinely administered. The worldwide disease burden caused by bacteria is enormous.

As of late 2008, there have been 770 completed prokaryotic genome-sequencing projects, and another 1300 are in progress. At least one representative strain of the major bacteria known to cause human disease is being sequenced. The National Microbial Pathogen Data Resource centralizes information about 50 strains of pathogenic bacteria, including expert curation and analyses of metabolic pathways in those organisms (McNeil et al., 2007).

The NMPDR is available online at ► http://www.nmpdr.org/.

There has been a strong bias toward sequencing prokaryotic genomes of known medical relevance. For example, there are 690,000 new cases of leprosy reported annually worldwide; the causative agent is *Mycobacterium leprae*. There are millions of cases of salmonellosis each year, caused by *Salmonella enterica*. A pathogenic strain of *E. coli* (O157:H7) causes hemorrhagic colitis and infects 75,000 individuals in the United States each year. As mentioned above, *M. tuberculosis* infects billions of people and kills millions.

You can read about a variety of bacterial diseases at the Centers for Disease Control and Prevention website (► http://www.cdc.gov/DataStatistics/).

An emerging theme in the biology of prokaryotes is that in addition to mutation bacterial populations undergo recombination, causing genetic diversification (Fraser et al., 2007). Species can be defined as clusters of genetically related strains, and the exchange of DNA by homologous recombination or other processes can complicate species definitions. Joyce et al. (2002) have reviewed recombination in the context of pathogenic bacteria such as *Helicobacter pylori* (a leading cause of gastric ulcers), *Streptococcus pneumoniae*, and *Salmonella enterica*. While eukaryotes achieve genetic diversity through sexual reproduction, prokaryotes also achieve tremendous genetic diversity through both recombination and lateral gene transfer (discussed below).

Classification of Bacteria and Archaea Based on Ribosomal RNA Sequences

The main way we know to analyze the diversity of microbial life is by molecular phylogeny. Trees have been generated based on multiple sequence alignments of

Brochier and Philippe (2002) have contested the view that hyperthermophilic bacteria (such as Aquificales and Thermotaogales) are the most deeply branching. Instead, they suggest that Planctomycetales are positioned at the base of the tree.

Reysenbach and Shock (2002) described a phylogenetic tree of extremophilic microbes based on 16S rRNA sequences. They used a software package designed for rRNA studies, called ARB (described in Chapter 8). You can obtain this software at ▶ http://www.arb-home.de/.

16S rRNA and other small rRNAs from various species. Ribosomal RNA has excellent characteristics as a molecule of choice for phylogeny: it is distributed globally, it is highly conserved yet still exhibits enough variability to reveal informative differences, and it is only rarely transferred between species. An example of a rRNA based tree is shown in Fig. 15.1, and we saw a similar tree reconstruction in Fig. 13.1.

A major conclusion of early rRNA studies, by Carl Woese and colleagues (Woese and Fox, 1977; Fox et al., 1980), is that bacteria and archaea are distinct groups. The deepest branching phyla are hyperthermophilic microbes, consistent with the hypothesis that the universal ancestor of life existed at hot temperatures (Achenbach-Richter et al., 1987).

A great advance in our appreciation of microbial diversity has come from the realization that the vast majority of bacteria and archaea are noncultivatable (Hugenholtz et al., 1998). It is straightforward to obtain microbes from natural sources and grow some of them in the presence of different kinds of culture medium. But for the great majority of microbes, perhaps >99%, culture conditions are not known. It is still possible to sample uncultivated (or uncultivatable) microbes by extracting nucleic acids directly from naturally occurring habitats (Delong and Pace, 2001). Norman Pace and colleagues pioneered the analysis of rRNA to characterize uncultivated species.

Because of sampling bias, four bacterial phyla have been characterized most fully: Proteobacteria, Firmicutes, Actinobacteria, and Bacteroidetes (Hugenholtz, 2002). These major groups account for over 90% of all known bacteria (discussed in Gupta and Griffiths, 2002). However, 35 bacterial and 18 archaeal phylum-level lineages are currently known (Hugenholtz, 2002). Analyses of uncultivated microbes will expand our view of bacterial and archaeal diversity. Among the archaea, early studies described them as extremophiles, but more recent samplings indicate that they are present in nonextreme habitats such as surface waters of the oceans and agricultural soils (DeLong, 1998). Twenty percent of all microbes in the oceans may be archaea (DeLong and Pace, 2001).

Classification of Bacteria and Archaea Based on Other Molecular Sequences

RibAlign is available at ▶ http://www.megx.net/ribalign. Its multiple sequence alignments of ribosomal proteins use MAFFT (Chapter 6). HOGENOM: Homologous Sequences in Complete Genomes Database (HOGENOM) is available at ▶ http://pbil.univ-lyon1.fr/databases/hogenom.php.

In addition to rRNA, many other DNA, RNA, or protein sequences can be used for molecular phylogeny studies. One motivation to do this is that the analysis of 16S ribosomal RNA sequence occasionally yields conflicting results. For example, the α-proteobacterium *Hyphomonas neptunium* is classified as a member of the order *Rhodobacterales* based on 16S rRNA but *Caulobacterales* based on 23S rRNA as well as according to ribosomal proteins, HSP70 and EF-Tu (Badger et al., 2005). This is potentially due to lateral gene transfer (discussed below). In other instances, 16S rRNA of unusual composition has been identified (Baker et al., 2006). Because of concerns about the properties of 16S rRNA for phylogenetic analysis, Teeling and Gloeckner (2006) introduced RibAlign, a database of ribosomal protein sequences. The HOGENOM database is another resource that is useful for phylogenetic studies. It includes large numbers of protein families across the tree of life.

We will study eukaryotes from the perspective of a tree that uses a combined protein data set (Fig. 16.1).

The use of individual proteins (or genes) for such studies sometimes yields tree topologies that conflict with each other and with topologies obtained using rRNA sequences. These discrepancies are usually attributed either to lateral gene transfer (see below), which can confound phylogenetic reconstruction, or to the loss of phylogenetic signals due to saturating levels of substitutions in the gene or protein

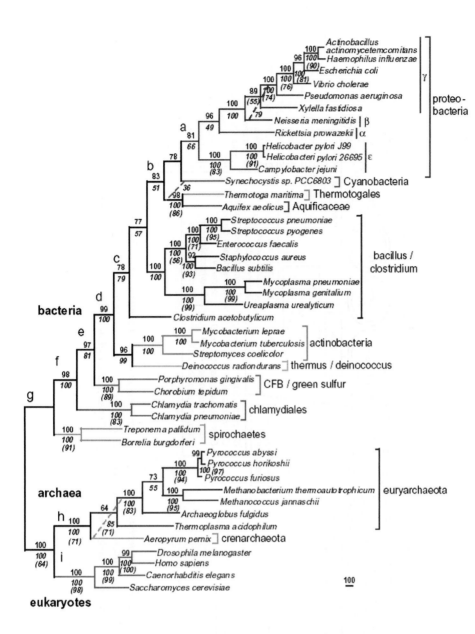

FIGURE 15.3. *An unrooted tree of life adapted from Brown et al. (2001) is based on an alignment of 23 proteins (spanning 6591 amino acid residues). These proteins are conserved across 45 species and include tRNA synthetases, elongation factors, and DNA polymerase III subunit. By combining these proteins, there are many phylogenetically informative sites. The tree consists of three major, monophyletic branches of life as described in Chapter 13. The tree was generated in PAUP by maximum parsimony. Used with permission.*

sequences. A strategy to circumvent this problem is to use combined gene or protein sets. Brown et al. (2001) aligned 23 orthologous proteins conserved across 45 species. Their trees supported thermophiles as the earliest evolved bacteria lineages (Fig. 15.3).

There are many other approaches to bacterial phylogeny. One is to identify conserved insertions and deletions in a large group of proteins. Such "signature sequences" can distinguish bacterial groups and form the basis of a tree (Fig. 15.4) (Gupta and Griffiths, 2002). This tree shows the relative branching order of bacterial species from completed genomes. Eugene Koonin and colleagues (Wolf et al., 2001) used five independent approaches to construct trees for 30 completely sequenced bacterial genomes and 10 sequenced archaeal genomes:

1. They assessed genes that are present or absent in each of these genomes using the COG database (see below). Seventeen invariant genes were identified (all of which encode ribosomal proteins and RNA polymerase subunits).

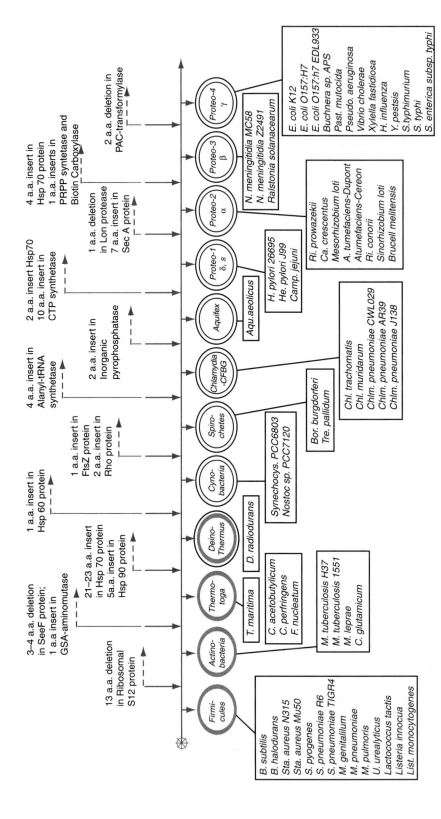

FIGURE 15.4. Phylogenetic relationships of bacterial species based on signature sequences in different proteins. The solid arrows above the line indicate the evolutionary point at which each insertion or deletion is proposed to have occurred. Thus, all bacterial groups to the right of each arrow are predicted to share that signature insertion/deletion while all groups to the left lack those changes. From Gupta and Griffiths (2002). Used with permission.

2. They assessed the conservation of local gene order (i.e., pairs of adjacent genes) among the genomes.

3. They measured the distribution of percent identity between likely orthologs.

4. They aligned 32 ribosomal proteins into a multiple sequence alignment consisting of 4821 columns (characters) and then generated a tree using the maximum-likelihood approach.

5. They compared multiple trees generated from a series of protein alignments. Wolf et al. (2001) concluded that traditional alignment-based methods were as effective as newer approaches based on genomic data such as local gene order. However, these approaches can yield different kinds of information (e.g., analysis of orthologs can identify genes that have been lost or horizontally transferred between lineages).

ANALYSIS OF PROKARYOTIC GENOMES

Some of the main attributes of a prokaryotic genome are its genome size, nucleotide composition, gene content, extent of lateral gene transfer, and functional annotation. We can approach this subject by considering *Escherichia coli*, arguably the best characterized bacterium. Following its intial genome sequence analysis by Blattner et al. (1997), it continues to be annotated and used as a reference genome (Riley et al., 2006), and related K12 strains have been sequenced. The annotation process includes an effort by the community to correct sequence errors, to update the boundaries for genes and transcripts (based for example on models for gene structures in related bacteria), and to assign functional descriptions for all genes (as described in Chapter 12). There are online resources that centralize information about *E. coli*, such as EcoCyc (Karp et al., 2007), RegulonDB (Salgado et al., 2006), and EcoGene (Rudd, 2000).

A search of the Comprehensive Microbial Resource (CMR) shows that there are currently seven *E. coli* genomes sequenced (Fig. 15.5a). Following the link to *E. coli* K12 (strain MG1655) provides a wealth of information on its genome, including summaries of the DNA molecule (4,639,221 base pairs; 50.78% GC content) and of the primary annotation (e.g., 4289 protein coding genes, of which almost half have been assigned a functional role category) (Fig. 15.5b).

EcoCyc is online at ► http://ecocyc.org/, Regulon is at ► http://regulondb.ccg.unam.mx/, and EcoGene is available at ► http://ecogene.org/. For each database try entering a query for the gene BLC and you will see a variety of data, including its genomic context, links to structural genomics projects, and blast links. Julio Collado-Vides and colleagues have expertly curated the transcription initiation sites and operon organization of *E. coli* with an emphasis on elucidating the regulatory networks.

Visit the CMR at ► http://cmr.jcvi.org.

Nucleotide Composition

In the analysis of a completed genome, the nucleotide composition has characteristic properties. The GC content is the mean percentage of guanine and cytosine, and as first reported by Noboru Sueoka (1961) it typically varies from 25% to 75% in prokaryotes (Fig. 15.6). Eukaryotes almost always have a larger and more variable genome size than bacteria, but their GC content is very uniform (around 40% to 45%). Within each species, nucleotide composition tends to be uniform. We showed the range of GC content in Fig. 13.15.

GC content varies within an individual genome. The CMR website includes a tool to plot GC content as shown for *E. coli* (Fig. 15.7). Regions having atypical GC content sometimes reflect invasions of foreign DNA (such as phage DNA incorporating into bacterial genomes). The GC content is highest (AT content lowest) in intergenic regions, possibly because of the requirements of transcription factor

You can determine GC content with the Emboss program GEECEE (► http://bioweb.pasteur.fr/seqanal/interfaces/geecee.html) or with other programs such as GLIMMER (see below).

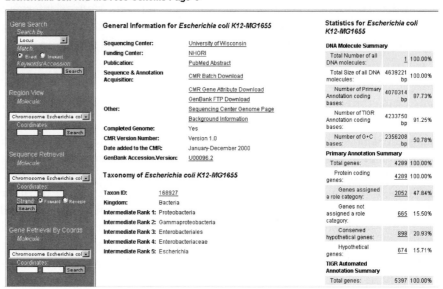

(a)

Organism Name (sort)	Kingdom (sort)	Taxon ID (sort)	Size (sort)	Complete Genome (sort)	Sequencing Center (sort)	Links (sort)
Escherichia coli 536	Bacteria	362663	4.93 Mb	Yes	Univ of Wuerzburg|Goettingen Genomics Laboratory	[G] [T] [S] [N] [P]
Escherichia coli CFT073	Bacteria	199310	5.23 Mb	Yes	Univ of Wisconsin	[G] [T] [S] [N] [P]
Escherichia coli K12-MG1655	Bacteria	168927	4.63 Mb	Yes	University of Wisconsin	[G] [T] [S] [N] [P]
Escherichia coli O157:H7 EDL933	Bacteria	155864	5.52 Mb	Yes	Univ. of Wisconsin	[G] [T] [S] [N] [P]
Escherichia coli O157:H7 VT2-Sakai	Bacteria	83334	5.49 Mb	Yes	Japanese Consortium	[G] [T] [S] [N] [P]
Escherichia coli UTI89	Bacteria	364106	5.17 Mb	Yes	Washington Univ	[G] [T] [S] [N] [P]
Escherichia coli W3110	Bacteria	316407	4.64 Mb	Yes	NARA	[G] [T] [S] [N] [P]

(b) *Escherichia coli K12-MG1655* **Genome Page** ⊕

FIGURE 15.5. *The Comprehensive Microbial Resource at the Institute for Genomic Research provides one of the most important websites for the study of microbes (▶ http://cmr.tigr.org/). The site includes a wide range of tools. (a) The results are shown for a search for* Escherichia coli. *(b) The genome page for* E. coli K12 *(strain MG1655) is shown.*

binding sites (Mitchison, 2005). GC content is also related to the frequency of codon utilization; we will explore this in a computer lab exercise at the end of this chapter.

Two-dimensional bacterial genomic display (2DBGD) is an electrophoresis technique analogous to two-dimensional protein gels (Chapter 10) in which bacterial genomic DNA is fragmented and separated first by size and then by sequence composition. Malloff et al. (2002) used 2DBGD to compare the genomes of three respiratory pathogens with very high GC content: *Bordetella pertussis*, the cause of whooping cough (68% GC); *M. tuberculosis* (66% GC); and *Mycobacterium avium* (66% GC). This technique can be used to detect insertions and deletions in prokaryotic strains.

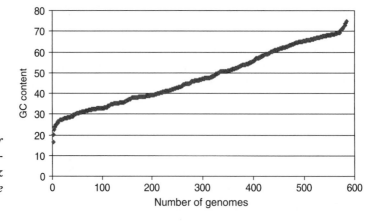

FIGURE 15.6. *GC content for 584 sequenced prokaryotic genomes (data from NCBI Entrez Genome, December 2007 were plotted).*

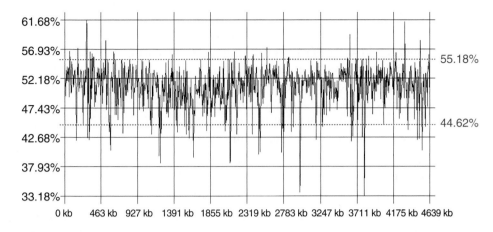

FIGURE 15.7. GC content across the E. coli *K12 genome (in 5000 base pair windows). This plot is available at the CMR website. The window size is adjustable, and there are links to the genes in each region.*

Finding Genes

Bacteria and archaea are characterized by a high gene density (about one gene per kilobase), absence of introns, and very little repetitive DNA. Thus the problem of finding genes is relatively simple in comparison to searching eukaryotic DNA (Chapter 16). Several programs are available for microbial gene identification (Table 15.8).

There are four main features of genomic DNA that are useful for gene recognition (Baytaluk et al., 2002). These features apply to both bacterial and eukaryotic gene finding:

1. *Open reading frame (ORF) length.* An ORF is not necessarily a gene; for example, many short ORFs are not part of authentic genes (discussed further below). An ORF is defined by a start codon (i.e., ATG encoding a methionine) and a stop codon (TAA, TAG, TGA). However, in bacteria, alternative start codons may be employed, such as GTG or TTG, and there are rarely used alternative stop codons.

2. *Presence of a consensus sequence for ribosome binding in the immediate vicinity of the start codon.* In some cases, it is possible to identify two in-frame ATG codons, either of which could represent the start codon. Identifying a ribosome binding site can be an important indicator of which is the likely start site. In bacteria, the ribosome binding site is called a Shine–Dalgarno sequence. It is a purine-rich stretch of nucleotides that is complementary to the 3′ end of

TABLE 15-8	Programs for Gene Finding in Prokaryotic Genomes	
Program	Description	URL
EasyGene	A web server from Anders Krogh and colleaguges	▶ http://www.cbs.dtu.dk/services/ EasyGene/
FrameD	Locates genes and frameshifts; optimized for GC-rich genomes	▶ http://bioinfo.genopole-toulouse.prd.fr/apps/FrameD/ FrameD.html
GeneMarkP, GeneMarkS	Uses hidden Markov models	▶ http://exon.gatech.edu/ GeneMark/
GLIMMER	At the University of Maryland	▶ http://www.cbcb.umd.edu/ software/glimmer/

16S rRNA, extending from the -20 position (i.e., $5'$ to the initiation codon) to the $+13$ position (i.e., 13 nucleotides downstream in the $3'$ direction). Samuel Karlin and colleagues (Ma et al., 2002) studied 30 prokaryotic genomes and correlated the features of the Shine–Dalgarno sequence with expression levels of genes based on codon usage bias (see below), type of codon, functional gene class, and type of start codon. They have shown a positive correlation between the presence of a strong Shine–Dalgarno sequence and high levels of gene expression.

3. *Presence of a pattern of codon usage that is consistent with genes.* Hidden Markov models (Chapter 6 and see below) have been particularly useful in defining the coding potential of putative protein-coding DNA sequences.

4. *Homology of the putative gene to other, known genes.* Genomic DNA sequences, including putative genes, can be searched against protein databases using blastx (see Chapter 5). This approach is especially helpful in finding genes in eukaryotic organisms. For example, exons can be matched to expressed sequence tags (Chapter 16).

> Intrinsic approaches are also sometimes called ab initio approaches.

The first three of these features are studied using intrinsic approaches to gene finding. They are called intrinsic because the features do not necessarily depend on comparisons to gene sequences from other organisms. The fourth feature, relationship to other genes, is called an extrinsic approach. Prokaryotic gene-finding programs sometimes combine both intrinsic and extrinsic approaches.

The GLIMMER system is one of the premier gene-finding algorithms. It identifies over 99% of all genes in a bacterial genome (Delcher et al., 1999a, 2007). The latest version has excellent sensitivity (determined based on comparisons to well-annotated bacterial genomes) and specificity (there are relatively few false positive results, i.e., gene predictions that do not correspond to authentic genes). The algorithm uses interpolated Markov models (IMMs). A Markov chain can describe the probability distribution for each nucleotide in a genomic DNA sequence. This probability can depend on the preceding k variables (nucleotides) in the sequence. A fixed-order Markov chain would describe the k-base context for each nucleotide position; for example, a fixed fifth-order Markov chain model describes $4^5 = 1024$ probability distributions, one for each possible 5-mer. GLIMMER uses a fifth-order Markov chain because that corresponds to a model of two consecutive codons (six nucleotide positions). The k-mers are used as a training set to teach the algorithm the rules for which probability distributions are most likely to be relevant to this particular genomic sequence. Larger values for k are more informative, but since they occur more rarely, it is more difficult to sample enough data for a training set in order to model the probability of the next base in the sequence. IMMs are a specialization of Markov models in which rare k-mers tend to be ignored, and more common k-mers are weighted more heavily.

> GLIMMER was written by Owen White, Steven Salzberg and colleagues while at the Institute for Genomic Research. GLIMMER is an acronym for Gene Locator and Interpolated Markov Modeler.

GLIMMER builds an IMM from a training set, then scans a genomic DNA sequence to predict genes. Criteria for gene finding include the presence of an initiation codon and some particular minimal length for an open reading frame. GLIMMER further assigns functions to predicted genes through BLAST searches and HMM searches, and also searches for noncoding RNAs (e.g., using tRNAscan; Chapter 8), paralogs, and PROSITE motifs (Chapter 10).

A simplified form of GLIMMER is available online at the NCBI website. To use the full GLIMMER program, it is necessary to run the software on a UNIX operating system. First, enter a data set of genomic DNA from the organism of interest (e.g.,

(a)

```
$ /usr/local/glimmer2.02/build-icm < ecoli_first100.txt > trainecoli

$ /usr/local/glimmer2.02/glimmer2 ecoli76k.fasta colitrain
```

(b)

```
GC Proportion = 51.5%
Minimum gene length = 90
Minimum overlap length = 30
Minimum overlap percent = 10.0%
Use independent scores = True
Ignore independent score on orfs longer than 765
Use first start codon = True
```

ID#	Fr	Orf Start	Gene Start	End	Lengths Orf	Lengths Gene	Gene Score	F1	F2	F3	R1	R2	R3	Indep Score	
	F2	35	44	178	144	135	0	0	0	0	_	0	_	99	0
	F1	226	247	402	177	156	0	0	_	_	_	0	_	99	0
	F3	273	420	563	291	144	0	_	_	0	_	0	_	99	0
	R2	740	713	609	132	105	0	_	6	_	0	2		89	0
	F2	515	548	916	402	369	0	_	0	_	_	_	0	99	6
	R1	1149	1143	1036	114	108	0	_	_	0	0	_	0	99	0
	F3	888	936	1265	378	330	0	_	_	0	_	_	0	99	0
	F1	1162	1210	1347	186	138	0	0	_	_	_	0	_	99	0
	F3	1365	1377	1592	228	216	0	_	_	0	_	0	_	99	0
	F3	1707	1710	1823	117	114	0	_	_	0	_	_	0	99	0
1	R3	1951	1909	380	1572	1530	99	_	_	_	_	_	99	0	101
	F3	1857	1872	1994	138	123	0	_	_	0	_	_	_	99	0
	R1	2124	2121	2029	96	93	0	_	_	_	0	_	_	99	0
	F3	2043	2043	2273	231	231	0	_	_	0	_	_	0	99	0
	F1	2098	2140	2319	222	180	0	0	0	_	_	_	0	99	0
	R1	2604	2589	2182	423	408	0	_	0	_	0	_	0	99	0
	F3	2313	2349	2645	333	297	0	_	0	0	_	_	0	99	0
	F2	2057	2183	2692	636	510	0	_	0	_	_	_	0	99	0
2	R1	2844	2835	2692	153	144	90	_	_	_	90	_	1	7	90
	F1	2809	2815	2946	138	132	7	7	_	0	_	_	4	87	7
	F3	2769	2835	2987	219	153	0	_	_	0	_	5	94	0	
3	R3	3034	2971	2039	996	933	99	_	_	_	_	_	99	0	109

(c)

```
48    66485    65460    [-3 L=1026]
49    69967    68642    [-2 L=1326]
50    71402    70614    [-3 L= 789]
51    73146    71341    [-1 L=1806]
52    73426    74445    [+1 L=1020]
53    74625    76166    [+3 L=1542]
54    80198    76878    [-3 L=3321]
55    80118    81140    [+3 L=1023]    [LowScoreBy #56 L=918 S=2]
56    81188    80223    [-3 L= 966]    [OlapWith #55 L=918 S=97]
57    82370    81288    [-3 L=1083]
58    84895    83759    [-2 L=1137]    [DelayedBy #59 L=108]
59    84873    86438    [+3 L=1566]
```

FIGURE 15.8. The GLIMMER *program is useful to find genes in bacterial DNA. The program is run on a UNIX operating system. (a) A data set of genomic DNA must first be trained to generate Markov models, and then the program is run. (b) The output includes a list of identifiers with ORF data on the forward (F1, F2, F3) and reverse (R1, R2, R3) strands and scores for the likelihood that a gene has been identified. (c) The output also includes a list of several genes. The DNA used was from* E. coli *K12 (accession number U14003).*

E. coli) in order to train the algorithm. The command to do this is shown in Fig. 15.8a, and the output for the analysis of 76,000 nucleotides of *E. coli* genomic DNA is shown in Fig. 15.8b. This shows the GC content of the DNA fragment (51.5% in this case), the parameters (e.g., the minimum gene length is set to 90 nucleotides), and a list of the predicted genes, including orientation (on the forward or reverse strand), length, and score. The GLIMMER output also has a summary of the predicted genes, including notations on possible overlaps (Fig. 15.8c).

There are several pitfalls associated with prokaryotic gene prediction:

- There may be multiple genes that are encoded by one genomic DNA segment, in an alternate reading frame on the same strand or opposite strand. GLIMMER includes features to address this situation.

- It is difficult to assess whether a short ORF is genuinely transcribed. According to Skovgaard et al. 2001, there are far too many short genes annotated in many genomes. For *E. coli*, they suggest that there are 3800 true

The NCBI GLIMMER program is available in the tools menu of the Microbial Genomes page found at ► http://www.ncbi.nlm.nih.gov/genomes/, or visit ► http://www.ncbi.nlm.nih.gov/genomes/MICROBES/glimmer_3.html.

An operon is a cluster of contiguous genes, transcribed from one promoter, that gives rise to a polycistronic mRNA. The predicted gene pairs from this study, encompassing 73 bacterial and archaeal genomes, are available on the web at OperonDB (▶ http://www.cbcb.umd.edu/cgi-bin/operons/operons.cgi).

protein-coding genes rather than the 4300 genes that have been annotated. Since stop codons (TAA, TAG, TGA) are AT rich, genomes that are GC rich tend to have fewer stop codons and more predicted long ORFs. For all predicted proteins in a genome, the proportion of hypothetical proteins (defined as predicted proteins for which there is no experimental evidence that they are expressed) rises greatly as sequence length is smaller.

- Frame shifts can occur, in which the genomic DNA is predicted to encode a gene with a stop codon in one frame but a continuing sequence in another frame on the same strand. A frame shift could be present because of a sequencing error or because of a mutation that leads to the formation of a pseudogene (a nonfunctional gene). GLIMMER extends gene prediction loci several hundred base pairs upstream and downstream to search for homology to known proteins and thus is designed to detect possible frameshifts.

- Some genes are part of operons that often have related functional roles in prokaryotes. Operons have promoter and terminator sequence motifs, but these are not well characterized. Steven Salzberg and colleagues (Ermolaeva et al., 2001) analyzed 7600 pairs of genes in 34 bacterial and archaeal genomes that are likely to belong to the same operon.

- Lateral gene transfer, also called horizontal gene transfer, commonly occurs in bacteria and archaea. We will discuss this next.

Lateral Gene Transfer

Lateral, or horizontal, gene transfer (LGT) is the phenomenon in which a genome acquires a gene from another organism directly, rather than by descent (Eisen, 2000; Koonin et al., 2001; Boucher et al., 2003). There are many situations in which examination of a genome shows that a particular gene is very closely related to orthologs in distantly related organisms. The simplest explanation for how a species acquired such a gene is through lateral gene transfer. This mechanism represents a major force in genome evolution. The gene transfer is unidirectional, rather than involving a reciprocal exchange of DNA, and it does not involve the usual pattern of inheritance from a parental lineage. Over 50% of archaeal and a smaller percentage of bacterial species have one or more protein domains acquired by lateral gene transfer, in contrast to <10% of eukaryotic species (Choi and Kim, 2007).

Lateral gene transfer is a significant phenomenon for several reasons:

1. This mechanism vastly differs from the normal mode of inheritance in which genes are transmitted from parent to offspring. Thus, lateral gene transfer represents a major shift in our conception of evolution.

2. This mechanism is very common in prokaryotes, and many examples have been described in eukaryotes as well. It has been observed within and between each of the three main branches of life but is particularly prevalent in prokaryotes relative to eukaryotes (Choi and Kim, 2007).

3. Lateral gene transfer can greatly confound phylogenetic studies. If a DNA, RNA, or protein is selected for phylogenetic analysis that has undergone lateral gene transfer, then the tree will not accurately represent the natural history of the species under consideration. An extreme interpretation of lateral gene transfer is that if it is common enough, then it is impossible in principle to derive a single true tree of life. Daubin et al. (2003) and Choi and Kim

(2007) have suggested that although lateral gene transfer is common, it is not so prevalent that it greatly interferes with phylogenetic studies of organisms. Huang and Gogarten (2006) offer the perspective that lateral gene transfer can be useful in phylogenetic studies to infer monophyletic groups and to elucidate the evolutionary history of both donor and recipient species.

4. Lateral gene transfer can profoundly affect the properties of basic biological processes, as reviewed extensively by Boucher et al. (2003). They describe its importance in a variety of processes such as photosynthesis, aerobic respiration, nitrogen fixation, sulfate reduction, and isoprenoid biosynthesis.

Lateral gene transfer occurs as a multistep process (Fig. 15.9) (Eisen, 2000). A gene that evolves in one lineage (by the traditional Darwinian process of vertical descent) may transfer to the lineage of a second species. This DNA transfer could be mediated by a viral vector or by a mechanism such as homologous recombination. Once a new gene is incorporated into the genome of individuals with a population (e.g., species 3 in Fig. 15.9), positive selection maintains its presence within those individuals. A transferred gene presumably must confer benefits to the new species in order to be maintained, propagated, and spread throughout the population of the new species. Finally, the new gene adapts to its new lineage, a process called "amelioration" (Eisen, 2000) (Fig. 15.9, arrow 6).

Carl Woese (2002) has suggested that in early evolution lateral gene transfer predominated to such an extent that primitive cellular evolution was a communal process, followed only later by vertical (Darwinian) evolution.

How is lateral gene transfer identified? The main criterion is that a gene has an unusual nucleotide composition, codon usage, phylogenetic position, or other feature that distinguishes it from most other genes in a genome. There are three principal methods by which lateral gene transfer may be inferred:

1. Phylogenetic trees of different genes may be compared. This is the favored approach (Eisen, 2000). If a tree based on a gene (or protein) has a topology

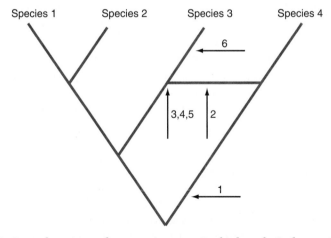

FIGURE 15.9. *Lateral gene transfer occurs in stages. In this hypothetical scenario, four species evolved from a common ancestor. Genes in each species descend in a horizontal fashion over time (arrow 1). At some point in time, a gene transfers horizontally from the lineage of species 4 to the lineage of species 3 (arrow 2). Transferred genes must then be fixed in some individual genomes (arrow 3), maintained under strong positive selection (arrow 4), and spread through the population of species 3 (arrow 5). The laterally transferred gene then evolves as an integral part of the new genome (arrow 6). This gene may be distinguished from other genes in species 3 by having a nucleotide composition or codon usage profile that is characteristic of species 4. This figure is adapted from Eisen (2000). Used with permission.*

different than that observed using ribosomal RNA, this discrepancy could be caused by lateral gene transfer.

2. Patterns of best matches for each gene in a genome may be used. A gene may have a highly unusual nucleotide composition or frequency of codon utilization, consistent with its origin in a distantly related genome.

3. The distribution pattern of genes across species can be assessed to search for genes that have undergone lateral gene transfer. If a gene is present in crenarcheota and a group of plants but not in other archaea, bacteria, or eukaryotes, this may be taken as evidence favoring a lateral gene transfer mechanism from crenarcheota to plants.

There are several reasons for caution in assigning a mechanism of lateral gene transfer. Consider the case of a gene widely distributed in bacteria that is observed in humans.

- If orthologs of the bacterial gene were present in an insect such as *Drosophila* or a plant, then the argument in favor of lateral gene transfer to humans would be considerably weakened. A concern in positing lateral gene transfer has been that the candidate gene might be present throughout the tree of life, but we might have insufficient sequence data to find it in other species; the recent flood of sequence data makes the possible lack of data less likely. Over time it will be progressively easier to assess evolutionary relationships.

- It is also possible that the gene in question has undergone rapid mutation, such that the phylogenetic signal is lost. This mechanism may lead to artifactual results (false positives) if gene loss or rapid mutation has occurred but not lateral gene transfer.

As a specific example of a gene that has undergone lateral gene transfer, consider proteorhodopsin. This protein is found in proteobacteria such as *Candidatus Pelagibacter ubique* (accession ZP_01264205). A blastp search using this protein as a query, restricted to the RefSeq database, shows many matches to other proteobacteria but also a match to Archaea of the order Thermoplasmatales. Frigaard et al. (2006) reported that these Archaea inhabit the upper water column of the oceans where planktonic bacteria laterally transferred proteorhodopsin genes to them. Frigaard et al. speculate that there is strong selective pressure for the use of proteorhodopsins in the presence of light, leading to a broad spatial distribution of the gene across species.

See the computer laboratory exercises at the end of this chapter for another example of lateral gene transfer.

Functional Annotation: COGs

As prokaryotic genomes are sequenced, they are annotated (see Chapter 13). This process is far more straightforward in bacteria and archaea than in eukaryotes. A large collection of tools is available at CMR, at NCBI, and at EBI. An example of the functional groups assigned to *E. coli* genes by the EcoCyc database is shown in Fig. 15.10.

At NCBI, The Clusters of Orthologous Groups of Proteins (COG) database organizes information collected from dozens of prokaryotic genomes as well as the yeast *Saccharomyces cerevisiae* (Chapter 17) and other eukaryotes (Tatusov et al., 1997, 2003; Koonin et al., 2004). The goal of the COG project is to provide a phylogenetic classification of prokaryotic proteins. The approach is to classify the relationships of proteins in groups based on "best-hit" BLAST search results.

The COG URL is ▶ http://www. ncbi.nlm.nih.gov/COG/. About 200,000 proteins are organized into about 5000 clusters of orthologous groups.

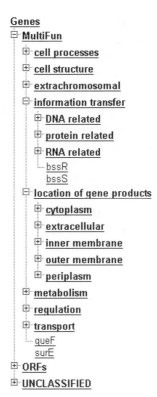

Genes
- **MultiFun**
  - ⊞ **cell processes**
  - ⊞ **cell structure**
  - ⊞ **extrachromosomal**
  - ⊟ **information transfer**
    - ⊞ **DNA related**
    - ⊞ **protein related**
    - ⊟ **RNA related**
      - bssR
      - bssS
  - ⊟ **location of gene products**
    - ⊞ **cytoplasm**
    - ⊞ **extracellular**
    - ⊞ **inner membrane**
    - ⊞ **outer membrane**
    - ⊞ **periplasm**
  - ⊞ **metabolism**
  - ⊞ **regulation**
  - ⊞ **transport**
    - queF
    - surE
- ⊞ **ORFs**
- ⊞ **UNCLASSIFIED**

FIGURE 15.10. The EcoCyc database includes a Pathways link. This site organizes E. coli *proteins according to function. From* ▶ *http://ecocyc.org/.*

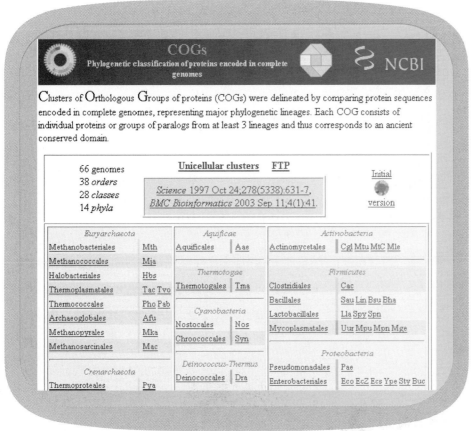

FIGURE 15.11. The COG page at NCBI provides analyses of functionally related genes and proteins from completely sequenced genomes (▶ http://www.ncbi.nlm.nih.gov/COG/).

The COG main page is shown in Fig. 15.11. Information is organized by taxonomy (there are currently 14 prokaryotic phyla, 28 classes, 38 orders, and 66 genomes [species] represented). Following the link to unicellular clusters, one can perform a text-based query such as for globins (Fig. 15.12a). This result shows how many phyla, classes, orders, and genomes have globin-related domains, and also assigns a functional category to that cluster. We presented a list of the COGs functional categories in Table 10.10, including 25 categories organized into the general areas of information storage and processing, cellular processes and signaling, metabolism, and poorly characterized. There is a link to the two COGs that are related to globins. In each case, the COG entries show the relationship of a group of bacterial globins in various species, as well as a multiple sequence alignment of these proteins and phylogenetic tree.

The COGs website summarizes the distribution of clusters of orthologous genes as a function of the number of species (Fig. 15.12b). This is a useful way to identify groups

FIGURE 15.12. (a) A text search for the term "globin" resulted in matches for two Clusters of Orthologous Groups of proteins (COGs): hemoglobin-like flavoproteins and truncated hemoglobins. The output includes a list of the total numbers of phyla, classes, orders, and genomes (i.e., species) having these clusters and a graphical view of the orders having these globin clusters. The functional categories are indicated, following the COGs schema. There is also a link to the specific COGs; that page includes accession numbers of the individual prokaryotic globins and a phylogenetic tree showing the relations of the proteins in the clusters. (b) The COGs site summarizes the distribution of clusters, from those occurring very rarely (arrow 1) to those that occur in all genomes covered in the database (arrow 2). The bars in this plot are clickable, leading to the results shown in (c) and (d). (c) Examples of COGs that occur only rarely. Many of these are annotated as uncharacterized conserved proteins. (d) COGs that occur across all prokaryotic orders include ATPases, GTPases, amino acid tRNA synthetases, and ribosomal proteins. All these proteins are highly conserved and thus may be useful for large-scale phylogenetic studies.

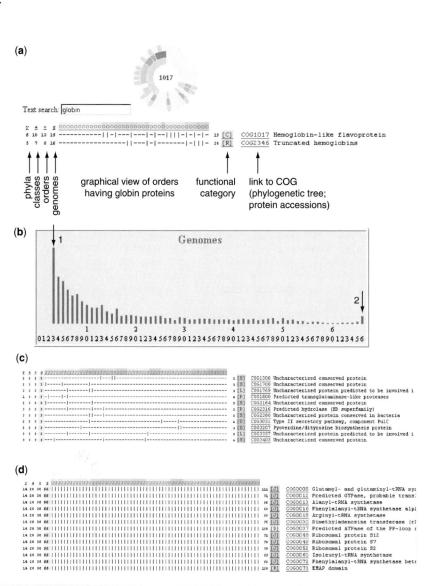

of related proteins that occur very rarely or very frequently. Those COGs that occur infrequently tend to be annotated as uncharacterized conserved proteins (Fig. 15.12c). If you are studying a particular prokaryote, it could be of interest to study these proteins because they may be relatively unique to that organism. Other proteins are highly conserved. For example, there are 63 clusters of orthologous proteins that are found in 26 different species (Fig. 15.12d). These well-conserved protein families include tRNA synthetases, ribosomal proteins, and other enzymes such as signal recognition particle GTPase and S-adenosyl methionine-dependent methyltransferases. Each of these is thus a good candidate for phylogenetic studies across the bacteria and/or archaea.

COMPARISON OF PROKARYOTIC GENOMES

One of the most important lessons of whole-genome sequencing is that comparative analyses greatly enhance our understanding of genomes. It can be useful to compare genomes whether they are closely or distantly related organisms. Some of the species that have had the genomes of closely related strains completely sequenced are indicated in Table 15.9. It will be significant to compare such genomes for several reasons:

- We may be able to discover why some strains are pathogenic.

- Eventually, we may be able to predict clinical outcome of infections based on the genotype of the pathogen.

- We may develop strategies for vaccine development.

TABLE 15-9   Prokaryotic Species for Which Genome of at Least Two Closely Related Strains Have Been Determined

Organism	Accession	Genome Size (bp)
Chlamydophila pneumoniae AR39	NC_002179	1,229,858
C. pneumoniae CWL029	NC_000922	1,230,230
C. pneumoniae J138	NC_002491	1,226,565
Escherichia coli K12	NC_000913	4,639,221
E. coli O157:H7	NC_002695	5,498,450
E. coli O157:H7 EDL933	NC_002655	5,528,445
Helicobacter pylori 26695	NC_000915	1,667,867
H. pylori J99	NC_000921	1,643,831
Mycobacterium tuberculosis CDC1551	NC_002755	4,403,836
M. tuberculosis H37Rv	NC_000962	4,411,529
Neisseria meningitidis MC58	NC_003112	2,272,351
N. meningitidis Z2491	NC_003116	2,184,406
Staphylococcus aureus aureus MW2	NC_003923	2,820,462
S. aureus aureus Mu50	NC_002758	2,878,040
S. aureus aureus N315	NC_002745	2,813,641
Streptococcus agalactiae 2603V/R	NC_004116	2,160,267
S. agalactiae NEM316	NC_004368	2,211,485
S. pneumoniae R6	NC_003098	2,038,615
S. pneumoniae TIGR4	NC_003028	2,160,837
S. pyogenes M1 GAS	NC_002737	1,852,441
S. pyogenes MGAS315	NC_004070	1,900,521
S. pyogenes MGAS8232	NC_003485	1,895,017

In the United States, 10% of all pneumonia cases and 5% of bronchitis cases are attributed to *C. pneumoniae*.

For an example of comparisons of prokaryotic genomes (and proteomes) we can consider Chlamydiae which are obligate intracellular bacteria that are phylogenetically distinct from other bacterial divisions. *Chlamydia pneumoniae* infects humans, causing pneumonia and bronchitis. *Chlamydia trachomatis* causes trachoma (an ocular disease that leads to blindness) and sexually transmitted diseases. Why do these closely related bacteria affect different body regions and cause such distinct pathologies? Their genomes have been sequenced and compared (Stephens et al., 1998; Kalman et al., 1999; Read et al., 2000). There are hundreds of genes present uniquely in each bacterium, including a family of outer membrane proteins that could be important in tissue tropism (Kalman et al., 1999).

TaxPlot

The NCBI offers a powerful tool for genome comparison that is easy to use. From the Entrez Genome page, select *C. trachomatis* to obtain a page such as that shown in Fig. 13.7. Select TaxPlot, and you will be able to compare two genomes (such as *C. trachomatis* and *C. pneumoniae* AR39) against a reference genome (the anthrax bacterium *B. anthracis* in the example of Fig. 15.13). In this plot, each point represents a protein in the reference genome. The *x* and *y* coordinates show the BLAST score for the closest match of each protein to the two *Chlamydia* proteomes

FIGURE 15.13. The TaxPlot tool at NCBI (Entrez) allows the comparison of two bacteria (C. trachomatis A/HAR-13 and C. pneumoniae AR39) to a reference genome (B. anthracis strain Ames in this case). The plot shows the distribution of blastp scores of each bacterium against the reference genome. Thirty matches are identical, while 459 hits are at least marginally closer to C. trachomatis and 633 hits are closer to C. pneumoniae. The two queries are selected using a pull-down menu (arrows 1 and 2). A match of interest that has a higher pairwise blastp score in one proteome relative to the other query can be clicked (arrow 3) leading to a zoom feature (arrow 4). The highlighted protein is identified in all three species (arrow 5) and there are links to the pairwise alignments from blast 2 sequences (Chapter 3). Optionally, the user can select from a set of functional categories to further focus the analysis (arrow 6).

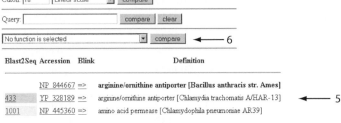

being compared. Most proteins are found along a diagonal line, indicating that they have equal (or nearly equal) scores between the reference protein and either of the *Chlamydia* proteins. However, there are notable outliers, which could represent genes important in the distinctive behavior of these two organisms. These points are clickable (see circled data point in Fig. 15.13, arrow 3), and the selected data point is highlighted (Fig. 15.13, arrow 4). This protein is identified as an arginine/ornithine antiporter in *B. anthracis* and *C. trachomatis*, and as an amino acid permease in *C. pneumoniae*. There are further links to the pairwise BLAST comparisons (arrow 5). The displays in TaxPlot can further be color coded according to the COG classification scheme (arrow 6).

Another powerful application of TaxPlot is to select a genome for both reference and for one of the queries, then select a second genome for the second query. This is illustrated in Fig. 15.14 for a *C. trachomatis* strain versus *C. pneumoniae*. All the data points fall on the diagonal (indicating that they share identity between the two species) or in the upper left section. No data points are in the lower right section because no *C. trachomatis* query protein can possibly be more related to *C. pneumoniae* than to its own protein sequence. The outliers, such as those indicated with arrows, are of particular interest because they are particularly highly divergent

There are several ways to access TaxPlot, including from the Tools link on the left sidebar of the main NCBI home page, as well as from ► http://www.ncbi.nlm.nih.gov/Genomes.

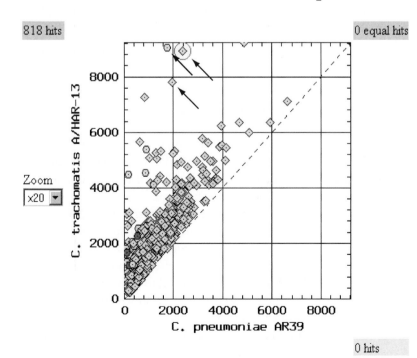

FIGURE 15.14. *TaxPlot can be used with one proteome serving as both the reference and the first query (in this case,* C. trachomatis *A/HAR-13) while another proteome forms the second query (in this case,* C. pneumoniae *AR39). Points that fall off the diagonal line (e.g., see arrows) have a high blastp score in one proteome but a relatively low score in the other, indicating that they are relatively poorly conserved. Such proteins may be of great interest in explaining the particular physiology or behavior of a strain or species.*

between the two species, having high blastp scores in one but low scores in the other. All three of the arrows point to polymorphic outer membrane proteins. Several additional outlying data points correspond to proteins that are annotated as hypothetical and thus for which function has not been assigned. These are potentially important in distinguishing the functional differences between these two species.

TaxPlot is thus an easy way to identify proteins that are different in two microbial genomes of interest. The tool has been extended to eukaryotes as well (Chapter 16).

MUMmer

MUMmer was written by Steven Salzberg and colleagues at TIGR. You can download the software from ► http://mummer. sourceforge.net/, and there is an interactive web browser at the Comprehensive Microbial Resource. From the CMR homepage (► http://cmr.jcvi.org/) follow the links to comparative tools then alignment tools.

A major challenge in aligning whole microbial genomes is the excessive amount of time required to perform an alignment of millions of base pairs using dynamic programming (Chapter 3). We introduced several fast algorithms such as BLAT in Chapter 5. Still, additional tools to accomplish large-scale genome alignment are needed (Miller, 2001). MUMmer is a software package that offers rapid, accurate alignments of microbial genomes (Delcher et al., 1999b). It has been adapted to aligning eukaryotic sequences (Delcher et al., 2002; Kurtz et al., 2004).

MUMmer accepts two sequences as input. The algorithm finds all subsequences that are longer than a specified minimum length k and that are perfectly matched. By definition, these matches are maximal because extending them further in either direction causes a mismatch. The algorithm uses a suffix tree, which is a search structure that identifies all the maximal unique matches ("MUM"s) in the pairwise alignment. The MUMs are ordered, and the algorithm closes gaps by identifying large inserts, repeats, small mutated regions, and single-nucleotide polymorphisms (SNPs).

MUMmer output consists of a dot matrix plot (Fig. 15.15) showing the alignment of the two genomic sequences with some minimum alignment length (e.g., 15 or 100 bp). The kinds of results that can be obtained include:

1. SNPs
2. Regions where sequences diverge by more than an SNP
3. Large insertions (e.g., by transposition, sequence reversal, or lateral gene transfer)
4. Repeats (e.g., a duplication in one genome)
5. Tandem repeats (in different copy number)

In the example of Fig. 15.15, two strains of *E. coli* are compared: a harmless *E. coli* K12 strain and the *E. coli* O157:H7 strain that appears in contaminated food, causing disease such as hemorrhagic colitis. These strains diverged about 4.5 MYA (Reid et al., 2000). Both genomes were sequenced and compared (Blattner et al., 1997; Perna et al., 2001; Hayashi et al., 2001; reviewed in Eisen, 2001). *Escherichia coli* O157:H7 is about 859,000 bp larger than *E. coli* K12. The two bacteria share a common backbone of about 4.1 Mb, while *E. coli* O157:H7 has an additional 1.4-Mb sequence comprised largely of genes acquired by lateral gene transfer. The MUMmer output is useful to identify regions of the two genomes that are shared in common as well as regions in which the orientation is inverted. Eisen et al. 2000 used such analyses to describe symmetrical chromosomal inversions around the origin of replication in comparisons of closely related species including *C. pneumoniae* versus *C. trachomatis*.

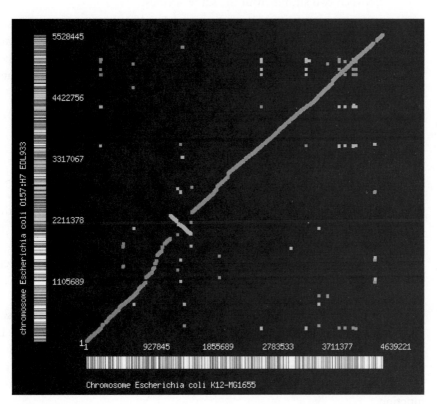

FIGURE 15.15. The MUMmer program allows you to select two microbial genomes of interest for comparison on a dot plot. The minimal alignment length can be adjusted. The MUMmer output consists of a dot plot that displays maximally unique matching subsequences (MUMs) between two genomes. This tool rapidly describes the relationship between two genomes, including information on the relative orientation of the genomic DNA and the presence of insertions or deletions. Here E. coli K12-MG1655 is represented on the x axis, and the pathogenic strain E. coli 0157:H7 EDL933 is on the y axis. There is a major 45° line where the two closely related genomes align. A line segment near the center is oriented at a 90° angle. This represents an inversion in which the orientation of a genomic segment in one of the two strains is reversed relative to the other.

There are two further extensions of MUMmer. NUCmer (NUCleotide MUMmer) allows multiple reference and query sequences to be aligned. One application is to align a group of contigs. PROtein MUMmer (PROmer) is similar to NUCmer but uses six-frame translations of each nucleotide sequence, thus offering superior sensitivity in aligning distantly related sequences.

PERSPECTIVE

The recent sequencing of several thousand bacterial and archaeal genomes has had a profound effect on virtually all aspects of microbiology. We can summarize the benefits of whole-genome sequencing of microbes as follows:

- Upon identifying the entire DNA sequence of a bacterial or archaeal genome, we obtain a comprehensive survey of all the genes and regulatory elements. This is similar to obtaining a parts list of a machine, although we do not also have the instruction manual.
- Through comparative genomics, we may learn the principles by which the "machine" is assembled and by which it functions.
- We can understand the diversity of microbial species through comparative genomics. Thus, we can begin to uncover the principles of genome

organization, and we can compare pathogenic versus nonpathogenic strains. We can also appreciate the dramatic differences in genome properties between two strains from the same species.

- We are gaining insights into the evolution of both genes and species. We can now appreciate lateral gene transfer as one of the driving forces of microbial evolution. We can study gene duplication and gene loss. Having the complete genome available is important both to learn what genes comprise an organism and to learn what genes are absent.

- Complete genome sequences offer a starting point for biological investigations.

PITFALLS

As complete bacterial and archaeal genomes are sequenced, two of the most important tasks are gene identification and genome annotation. Gene identification has become routine, but can be difficult for several reasons: It is difficult to assess whether short ORFs correspond to transcripts that are actively transcribed, and (in contrast to eukaryotes) prokaryotes do not always use AUG as a start codon.

Genome annotation is the critical process by which functions are assigned to predicted proteins. When genome sequences were first identified in the 1990s it was common for half of all predicted proteins to have no known homologs, and their function was entirely obscure. Perhaps surprisingly this situation has persisted to a large extent, with many genes annotated as "hypothetical" or having unknown function.

Gene annotation performed computationally should always be viewed as generating a hypothesis that needs to be experimentally tested. There are several kinds of common errors (Brenner, 1999; Mural, 1999; Peri et al., 2001):

- Transitive catastrophes: inappropriately assigning a function to a gene based on homology to another gene with a known function.

- Identification of small ORFs as authentic genes when they are not transcribed. Devos and Valencia 2001 estimate that about 5% of the genes annotated for general functions are incorrect, while about 33% of the gene annotations for specific functions are erroneous.

WEB RESOURCES

The Comprehensive Microbial Resource (▶ http://cmr.jcvi.org) provides an important starting point for any study of microbial genomes. A useful link on microbes is ▶ http://www.microbes.info.

This site includes a broad variety of resources, including introductory articles on microbiology.

DISCUSSION QUESTIONS

[15-1] Anthrax strains vary in their pathogenicity. What bioinformatics approaches could you take to understand the basis of this difference? What specific proteins are involved in its pathogenicity?

[15-2] How can you assess whether bacterial genes have incorporated into the human genome through lateral gene transfer? What alternative explanations could there be for the presence of a human protein that is most closely related to a

group of bacterial proteins, without having other eukaryotic orthologs?

[15-3] Consider the differences between *E. coli* K12 and *E. coli* O157:H7 and other closely related pairs of bacteria. They undergo lateral gene transfer to different degrees, they have distinct patterns of pathogenicity, and these two strains even differ in genome size by over a million base pairs. What is the definition of a species? Is *E. coli* a species?

PROBLEMS/COMPUTER LAB

[15-1] Analyze the genome of *E. coli*. Begin at Entrez Genomes. Find a gene that is known to have a homolog in eukaryotes. Use the TaxPlot tool of Entrez genomes. Now use the Clusters of Orthologous Genes (COG) site to find a gene that is known to have a homolog in eukaryotes. In addition to NCBI, there are two excellent resources for completed genomes:

- Explore this same bacterial genome at the TIGR website. Go to ► http://cmr.jcvi.org. Then (from the left sidebar) click TIGR databases. Click the link for "projects completed" and find your genome.

- Explore this same bacterial genome at the Wellcome Trust Sanger Institute website. Go to ► http://www.sanger.ac.uk/Projects/.

[15-2] Explore the GC content and codon utilization of a bacterium. From the CMR (► http://cmr.jcvi.org) search the genomes for *Yersinia pestis*, and select *Y. pestis* CO92. How many chromosomes and plasmids does it contain? Under genome tools/analysis tools select GC content display tool, choose *Y. pestis*, and compare the range of GC percent across the main chromosome and across plasmid pPCP1. Which has a higher GC content? Under genome tools/analysis tools select the codon usage tool, and choose *Y. pestis*. Look at the codon utilization for both the main chromosome and for one of its plasmids (you may choose to print them out to study them). Are there differences in the codon usages?

[15-3] The bacterium *Wolbachia pipientis* is an endosymbiont that lives inside insect and nematode hosts. A large fraction of its genome has transferred to the nuclear genome of some hosts (Hotopp et al., 2007). Select a *Wolbachia* protein (e.g., NP_965857) and provide evidence that an ortholog has been laterally transferred to a *Drosophila* species. As one strategy, first perform blastp with the protein as a query restricting the output to bacteria, and then restricting the output to eukaryotes. Try performing a tblastn search against the trace archives (a link is provided on the main NCBI blast web page). Try a tblastn search against the whole-genome shotgun read database restricted to the insects.

[15-4] Compare two completed genomes. Begin at Entrez Genomes. Choose bacteria, then choose an organism such as *Rickettsia prowazekii*. Use TaxPlot to perform a three-way genome comparison. Try clicking on a point on the graph. Restrict your analysis to a functional group of genes ("transcription"). Repeat your search with the group "function unknown." Are the profiles different?

SELF-TEST QUIZ

[15-1] A typical bacterial genome is composed of approximately how many base pairs of DNA?

(a) 20,000 bp

(b) 200,000 bp

(c) 2,000,000 bp (2 Mb)

(d) 20,000,000 bp (20 Mb)

[15-2] *Myxococcus xanthus* has a relatively large genome size, even compared to other proteobacteria. One reason for this size may be the following:

(a) *M. xanthus* acquired repetitive DNA sequences

(b) *M. xanthus* lives is a bacterium with a relatively large diameter size

(c) *M. xanthus* has a complex social lifestyle requiring large numbers of genes

(d) *M. xanthus* acquired a large number of plasmids

[15-3] The *E. coli* genome encodes about 4300 protein-coding genes. The total number of *E. coli* introns is approximately

(a) 10

(b) 430

(c) 4,300

(d) 43,000

[15-4] The smallest prokaryotic genomes tend to be those of

(a) Extremophiles

(b) Viruses

(c) Intracellular bacteria

(d) Bacilli

[15-5] Which of the following constitutes strongest evidence that an *E. coli* gene became incorporated into the *E. coli* genome by lateral gene transfer?

(a) The GC content of the gene varies greatly relative to other *E. coli* genes.

(b) The frequency of codon utilization of the gene varies greatly relative to other *E. coli* genes/.

(c) Phylogenetic analysis shows that proteobacteria closely related to *E. coli* lack this gene.

(d) Any of the above.

[15-6] The main idea of the Clusters of Orthologous Groups of Proteins (COGs) database is

    (a) To classify proteins from completely sequenced prokaryotic genomes based on orthologous relationships

    (b) To provide multiple sequence alignments of completed prokaryotic genomes

    (c) To provide a functional classification system for proteins

    (d) To predict the functions of individual eukaryotic proteins based on the conserved families in prokaryotes

[15-7] We noted that the *Candidatus Carsonella ruddii* genome is extremely small (see accession NC_008512). First note how many genes are annotated based on NCBI's Entrez database. Next, obtain the sequence (159,662 nucleotides) in FASTA format, input it to the GLIMMER program for gene prediction (obtained via the NCBI prokaryotic genomes site). How many genes does the GLIMMER program annotate relative to the NCBI annotation?

    (a) More genes

    (b) The same number of genes

    (c) Fewer genes

[15-8] The pathogenic strain *E. coli* O157:H7 EDL933 is substantially larger than *E. coli* K12 substr. DH10B (discussed above). Use TaxPlot, MUMmer, or NCBI Genomes to determine approximately how many more genes it has.

    (a) 1000

    (b) 2000

    (c) 3000

    (d) 8000

SUGGESTED READING

There is a large literature on bacterial genomics. Important reviews are by Claire Fraser-Liggett (2005), Ward and Fraser (2005), and Bentley and Parkhill (2004). Casjens (1998) provided an excellent introduction, and this was updated and expanded by Bentley and Parkhill on comparative genomics. Susannah Tringe and Edward Rubin (2005) as well as Riesenfeld et al. (2004) provide key introductions to metagenomics. Boucher et al. (2003) from the group of W. Ford Doolittle have reviewed lateral gene transfer. For an overview of the meaning of bacterial populations from a genomics perspective, see Joyce et al. (2002).

REFERENCES

Achenbach-Richter, L., Gupta, R., Stetter, K. O., and Woese, C. R. Were the original eubacteria thermophiles? *Syst. Appl. Microbiol.* **9**, 34–39 (1987).

Akman, L., Rio, R. V., Beard, C. B., and Aksoy, S. Genome size determination and coding capacity of *Sodalis glossinidius*, an enteric symbiont of tsetse flies, as revealed by hybridization to *Escherichia coli* gene arrays. *J. Bacteriol.* **183**, 4517–4525 (2001).

Akman, L., Yamashita, A., Watanabe, H., Oshima, K., Shiba, T., Hattori, M., and Aksoy, S. Genome sequence of the endocellular obligate symbiont of tsetse flies, *Wigglesworthia glossinidia*. *Nat Genet.* **32**, 402–407 (2002).

Alm, R. A., et al. Genomic-sequence comparison of two unrelated isolates of the human gastric pathogen *Helicobacter pylori*. *Nature* **397**, 176–180 (1999).

Andersson, S. G. The bacterial world gets smaller. *Science* **314**, 259–260 (2006).

Andersson, S. G., and Kurland, C. G. Reductive evolution of resident genomes. *Trends Microbiol.* **6**, 263–268 (1998).

Andersson, S. G., et al. The genome sequence of *Rickettsia prowazekii* and the origin of mitochondria. *Nature* **396**, 133–140 (1998).

Badger, J. H., Eisen, J. A., and Ward, N. L. Genomic analysis of *Hyphomonas neptunium* contradicts 16S rRNA gene-based phylogenetic analysis: Implications for the taxonomy of the orders "*Rhodobacterales*" and *Caulobacterales*. *Int. J. Syst. Evol. Microbiol.* **55**, 1021–1026 (2005).

Baker, B. J., Tyson, G. W., Webb, R. I., Flanagan, J., Hugenholtz, P., Allen, E. E., and Banfield, J. F. Lineages of acidophilic archaea revealed by community genomic analysis. *Science* **314**, 1933–1935 (2006).

Baytaluk, M. V., Gelfand, M. S., and Mironov, A. A. Exact mapping of prokaryotic gene starts. *Brief. Bioinform.* **3**, 181–194 (2002).

Bender, L., Ott, M., Marre, R., and Hacker, J. Genome analysis of *Legionella* ssp. by orthogonal field alternation gel electrophoresis (OFAGE). *FEMS Microbiol. Lett.* **60**, 253–257 (1990).

Bentley, S. D., et al. Complete genome sequence of the model actinomycete *Streptomyces coelicolor* A3(2). *Nature* **417**, 141–147 (2002).

Bentley, S. D., and Parkhill, J. Comparative genomic structure of prokaryotes. *Annu. Rev. Genet.* **38**, 771–792 (2004).

Binnewies, T. T., Motro, Y., Hallin, P. F., Lund, O., Dunn, D., La, T., Hampson, D. J., Bellgard, M., Wassenaar, T. M., and Ussery, D. W. Ten years of bacterial genome sequencing: Comparative-genomics-based discoveries. *Funct. Integr. Genomics* **6**, 165–185 (2006).

Blattner, F. R., et al. The complete genome sequence of *Escherichia coli* K-12. *Science* 277, 1453–1474 (1997).

Boucher, Y., Douady, C. J., Papke, R. T., Walsh, D. A., Boudreau, M. E., Nesbø, C. L., Case, R. J., and Doolittle, W. F. Lateral gene transfer and the origins of prokaryotic groups. *Annu. Rev. Genet.* 37, 283–328 (2003).

Brenner, S. E. Errors in genome annotation. *Trends Genet.* 15, 132–133 (1999).

Brochier, C., and Philippe, H. Phylogeny: A non-hyperthermophilic ancestor for bacteria. *Nature* 417, 244 (2002).

Brown, J. R., Douady, C. J., Italia, M. J., Marshall, W. E., and Stanhope, M. J. Universal trees based on large combined protein sequence data sets. *Nat. Genet.* 28, 281–285 (2001).

Brüggemann, H., Henne, A., Hoster, F., Liesegang, H., Wiezer, A., Strittmatter, A., Hujer, S., Dürre, P., and Gottschalk, G. The complete genome sequence of *Propionibacterium acnes*, a commensal of human skin. *Science* 305, 671–673 (2004).

Bult, C. J., et al. Complete genome sequence of the methanogenic archaeon, *Methanococcus jannaschii*. *Science* 273, 1058–1073 (1996).

Casjens, S. The diverse and dynamic structure of bacterial genomes. *Annu. Rev. Genet.* 32, 339–377 (1998).

Casjens, S., et al. A bacterial genome in flux: The twelve linear and nine circular extrachromosomal DNAs in an infectious isolate of the Lyme disease spirochete *Borrelia burgdorferi*. *Mol. Microbiol.* 35, 490–516 (2000).

Chambaud, I., et al. The complete genome sequence of the murine respiratory pathogen *Mycoplasma pulmonis*. *Nucleic Acids Res.* 29, 2145–2153 (2001).

Charon, N. W., and Goldstein, S. F. Genetics of motility and chemotaxis of a fascinating group of bacteria: The Spirochetes. *Annu. Rev. Genet.* 36, 47–73 (2002).

Choi, I. G., and Kim, S. H. Global extent of horizontal gene transfer. *Proc. Natl. Acad. Sci. USA* 104, 4489–4494 (2007).

Cole, S. T., et al. Deciphering the biology of *Mycobacterium tuberculosis* from the complete genome sequence. *Nature* 393, 537–544 (1998).

Cole, S. T., et al. Massive gene decay in the leprosy bacillus. *Nature* 409, 1007–1011 (2001).

Dandekar, T., et al. Re-annotating the *Mycoplasma pneumoniae* genome sequence: Adding value, function and reading frames. *Nucleic Acids Res.* 28, 3278–3288 (2000).

Daubin, V., Moran, N. A., and Ochman, H. Phylogenetics and the cohesion of bacterial genomes. *Science* 301, 829–832 (2003).

Deckert, G., et al. The complete genome of the hyperthermophilic bacterium *Aquifex aeolicus*. *Nature* 392, 353–358 (1998).

Delcher, A. L., Harmon, D., Kasif, S., White, O., and Salzberg, S. L. Improved microbial gene identification with GLIMMER. *Nucleic Acids Res.* 27, 4636–4641 (1999a).

Delcher, A. L., et al. Alignment of whole genomes. *Nucleic Acids Res.* 27, 2369–2376 (1999b).

Delcher, A. L., Phillippy, A., Carlton, J., and Salzberg, S. L. Fast algorithms for large-scale genome alignment and comparison. *Nucleic Acids Res.* 30, 2478–2483 (2002).

Delcher, A. L., Bratke, K. A., Powers, E. C., and Salzberg, S. L. Identifying bacterial genes and endosymbiont DNA with GLIMMER. *Bioinformatics* 23, 673–679 (2007).

DeLong, E. F. Everything in moderation: Archaea as "non-extremophiles." *Curr. Opin. Genet. Dev.* 8, 649–654 (1998).

DeLong, E. F., and Pace, N. R. Environmental diversity of bacteria and archaea. *Syst. Biol.* 50, 470–478 (2001).

de Villiers, E. P., Brayton, K. A., Zweygarth, E., and Allsopp, B. A. Genome size and genetic map of *Cowdria ruminantium*. *Microbiology* 146, 2627–2634 (2000).

Devos, D., and Valencia, A. Intrinsic errors in genome annotation. *Trends Genet.* 17, 429–431 (2001).

Dobell, C. *Antony van Leeuwenhoek and His "Little Animals."* Harcourt, Brace and Company, New York, 1932.

Eisen, J. A. Horizontal gene transfer among microbial genomes: New insights from complete genome analysis. *Curr. Opin. Genet. Dev.* 10, 606–611 (2000).

Eisen, J. A. Gastrogenomics. *Nature* 409, 463, 465–466 (2001).

Eisen, J. A., Heidelberg, J. F., White, O., and Salzberg, S. L. Evidence for symmetric chromosomal inversions around the replication origin in bacteria. *Genome Biol.* 1, RESEARCH0011 (2000).

Eisenstein, B. I., and Schaechter, M. Normal microbial flora. In M. Schaechter, N. C. Engleberg, B. I. Eisenstein, and G. Medoff (eds.), *Mechanisms of Microbial Disease*, Chapter 20. Lippincott Williams and Wilkins, Baltimore, MD, 1999.

Ermolaeva, M. D., White, O., and Salzberg, S. L. Prediction of operons in microbial genomes. *Nucleic Acids Res.* 29, 1216–1221 (2001).

Fadiel, A., Eichenbaum, K. D., El Semary, N., and Epperson, B. Mycoplasma genomics: Tailoring the genome for minimal life requirements through reductive evolution. *Front Biosci.* 12, 2020–2028 (2007).

Fleischmann, R. D., et al. Whole-genome random sequencing and assembly of *Haemophilus influenzae* Rd. *Science* 269, 496–512 (1995).

Fox, G. E., et al. The phylogeny of prokaryotes. *Science* 209, 457–463 (1980).

Fraser, C. M., et al. The minimal gene complement of *Mycoplasma genitalium*. *Science* 270, 397–403 (1995).

Fraser, C. M., et al. Genomic sequence of a Lyme disease spirochaete, *Borrelia burgdorferi*. *Nature* 390, 580–586 (1997).

Fraser, C. M., et al. Complete genome sequence of *Treponema pallidum*, the syphilis spirochete. *Science* 281, 375–388 (1998).

Fraser, C. M., Eisen, J. A., and Salzberg, S. L. Microbial genome sequencing. *Nature* 406, 799–803 (2000).

Fraser, C., Hanage, W. P., and Spratt, B. G. Recombination and the nature of bacterial speciation. *Science* **315**, 476–480 (2007).

Fraser-Liggett, C. M. Insights on biology and evolution from microbial genome sequencing. *Genome Res.* **15**, 1603–1610 (2005).

Frigaard, N. U., Martinez, A., Mincer, T. J., and DeLong, E. F. Proteorhodopsin lateral gene transfer between marine planktonic Bacteria and Archaea. *Nature* **439**, 847–850 (2006).

Galagan, J. E., et al. The genome of *M. acetivorans* reveals extensive metabolic and physiological diversity. *Genome Res.* **12**, 532–542 (2002).

Gil, R., Silva, F. J., Pereto, J., and Moya, A. Determination of the core of a minimal bacterial gene set. *Microbiol. Mol. Biol. Rev.* **68**, 518–537 (2004).

Gill, S. R., Pop, M., Deboy, R. T., Eckburg, P. B., Turnbaugh, P. J., Samuel, B. S., Gordon, J. I., Relman, D. A., Fraser-Liggett, C. M., and Nelson, K. E. Metagenomic analysis of the human distal gut microbiome. *Science* **312**, 1355–1359 (2006).

Giovannoni, S. J., Tripp, H. J., Givan, S., Podar, M., Vergin, K. L., Baptista, D., Bibbs, L., Eads, J., Richardson, T. H., Noordewier, M., Rappé, M. S., Short, J. M., Carrington, J. C., and Mathur, E. J. Genome streamlining in a cosmopolitan oceanic bacterium. *Science* **309**, 1242–1245 (2005).

Glaser, P., et al. Comparative genomics of *Listeria* species. *Science* **294**, 849–852 (2001).

Glass, J. I., et al. The complete sequence of the mucosal pathogen *Ureaplasma urealyticum*. *Nature* **407**, 757–762 (2000).

Goldman, B. S., et al. Evolution of sensory complexity recorded in a myxobacterial genome. *Proc. Natl. Acad. Sci. USA* **103**, 15200–15205 (2006).

Graur, D., and Li, W.-H. *Fundamentals of Molecular Evolution.* Sinauer Associates, Sunderland, MA, 2000.

Gupta, R. S., and Griffiths, E. Critical issues in bacterial phylogeny. *Theor. Popul. Biol.* **61**, 423–434 (2002).

Hayashi, T., et al. Complete genome sequence of enterohemorrhagic *Escherichia coli* O157:H7 and genomic comparison with a laboratory strain K-12. *DNA Res.* **8**, 11–22 (2001).

Heidelberg, J. F., et al. DNA sequence of both chromosomes of the cholera pathogen *Vibrio cholerae*. *Nature* **406**, 477–483 (2000).

Himmelreich, R., et al. Complete sequence analysis of the genome of the bacterium *Mycoplasma pneumoniae*. *Nucleic Acids Res.* **24**, 4420–4449 (1996).

Hotopp, J. C., Clark, M. E., Oliveira, D. C., Foster, J. M., Fischer, P., Torres, M. C., Giebel, J. D., Kumar, N., Ishmael, N., Wang, S., Ingram, J., Nene, R. V., Shepard, J., Tomkins, J., Richards, S., Spiro, D. J., Ghedin, E., Slatko, B. E., Tettelin, H., and Werren, J. H. Widespread lateral gene transfer from intracellular bacteria to multicellular eukaryotes. *Science* **317**, 1753–1756 (2007).

Hotopp, J. C., et al. Comparative genomics of emerging human ehrlichiosis agents. *PLoS Genet.* **2**, e21 (2006). PMID: 16482227

Huang, J., and Gogarten, J. P. Ancient horizontal gene transfer can benefit phylogenetic reconstruction. *Trends Genet.* **22**, 361–366 (2006).

Huber, H., et al. A new phylum of *Archaea* represented by a nanosized hyperthermophilic symbiont. *Nature* **417**, 63–67 (2002).

Hugenholtz, P. Exploring prokaryotic diversity in the genomic era. *Genome Biol.* **3**, 0003.1–0003.8 (2002).

Hugenholtz, P., Goebel, B. M., and Pace, N. R. Impact of culture-independent studies on the emerging phylogenetic view of bacterial diversity. *J. Bacteriol.* **180**, 4765–4774 (1998).

Itaya, M. An estimation of minimal genome size required for life. *FEBS Lett.* **362**, 257–260 (1995).

Joyce, E. A., Chan, K., Salama, N. R., and Falkow, S. Redefining bacterial populations: A post-genomic reformation. *Nat. Rev. Genet.* **3**, 462–473 (2002).

Kalman, S., et al. Comparative genomes of *Chlamydia pneumoniae* and *C. trachomatis*. *Nat. Genet.* **21**, 385–389 (1999).

Karp, P. D., et al. Multidimensional annotation of the *Escherichia coli* K-12 genome. *Nucleic Acids Res.* (2007).

Kawarabayasi, Y., et al. Complete genome sequence of an aerobic hyperthermophilic crenarchaeon, *Aeropyrum pernix* K1. *DNA Res.* **6**, 83–101, 145–152 (1999).

Klenk, H. P., et al. The complete genome sequence of the hyperthermophilic, sulphate-reducing archaeon *Archaeoglobus fulgidus*. *Nature* **390**, 364–370 (1997).

Koonin, E. V. Comparative genomics, minimal gene-sets and the last universal common ancestor. *Nat. Rev. Microbiol.* **1**, 127–136 (2003).

Koonin, E. V., Makarova, K. S., and Aravind, L. Horizontal gene transfer in prokaryotes: Quantification and classification. *Annu. Rev. Microbiol.* **55**, 709–742 (2001).

Koonin, E. V., et al. A comprehensive evolutionary classification of proteins encoded in complete eukaryotic genomes. *Genome Biol.* **5**, R7 (2004).

Kunst, F., et al. The complete genome sequence of the gram-positive bacterium *Bacillus subtilis*. *Nature* **390**, 249–256 (1997).

Kurtz, S., Phillippy, A., Delcher, A. L., Smoot, M., Shumway, M., Antonescu, C., and Salzberg, S. L. Versatile and open software for comparing large genomes. *Genome Biol.* **5**, R12 (2004).

Lamichhane, G., and Bishai, W. Defining the "survivasome" of *Mycobacterium tuberculosis*. *Nat. Med.* **13**, 280–282 (2007).

Ma, J., Campbell, A., and Karlin, S. Correlations between Shine-Dalgarno sequences and gene features such as predicted expression levels and operon structures. *J. Bacteriol.* **184**, 5733–5745 (2002).

Malloff, C. A., Fernandez, R. C., Dullaghan, E. M., Stokes, R. W., and Lam, W. L. Two-dimensional display and whole genome comparison of bacterial pathogen genomes of high G + C DNA content. *Gene* **293**, 205–211 (2002).

Martin, W., and Koonin, E. V. A positive definition of prokaryotes. *Nature* **442**, 868 (2006).

May, B. J., et al. Complete genomic sequence of *Pasteurella multocida*, Pm70. *Proc. Natl. Acad. Sci. USA* **98**, 3460–3465 (2001).

McNeil, L. K., et al. The National Microbial Pathogen Database Resource (NMPDR): A genomics platform based on subsystem annotation. *Nucleic Acids Res.* **35**, D347–D353 (2007).

Miller, W. Comparison of genomic DNA sequences: Solved and unsolved problems. *Bioinformatics* **17**, 391–397 (2001).

Mitchison, G. The regional rule for bacterial base composition. *Trends Genet.* **21**, 440–443 (2005).

Mural, R. J. Current status of computational gene finding: A perspective. *Methods Enzymol.* **303**, 77–83 (1999).

Mushegian, A. R., and Koonin, E. V. A minimal gene set for cellular life derived by comparison of complete bacterial genomes. *Proc. Natl. Acad. Sci. USA* **93**, 10268–10273 (1996).

Nakabachi, A., Yamashita, A., Toh, H., Ishikawa, H., Dunbar, H. E., Moran, N. A., and Hattori, M. The 160-kilobase genome of the bacterial endosymbiont *Carsonella*. *Science* **314**, 267 (2006).

Nelson, K. E., et al. Evidence for lateral gene transfer between *Archaea* and bacteria from genome sequence of *Thermotoga maritima*. *Nature* **399**, 323–329 (1999).

Nolling, J., et al. Genome sequence and comparative analysis of the solvent-producing bacterium *Clostridium acetobutylicum*. *J. Bacteriol.* **183**, 4823–4838 (2001).

Ogata, H., et al. Mechanisms of evolution in *Rickettsia conorii* and *R. prowazekii*. *Science* **293**, 2093–2098 (2001).

Owen, R. *Lectures on the Comparative Anatomy and Physiology of the Invertebrate Animals*. Longman, Brown, Green, and Longmans, London, 1843.

Palmer, C., Bik, E. M., Digiulio, D. B., Relman, D. A., and Brown, P. O. Development of the human infant intestinal microbiota. *PLoS Biol.* **5**, e177 (2007).

Parkhill, J., et al. Complete genome sequence of a multiple drug resistant *Salmonella enterica* serovar *typhi* CT18. *Nature* **413**, 848–852 (2001a).

Parkhill, J., et al. Genome sequence of *Yersinia pestis*, the causative agent of plague. *Nature* **413**, 523–527 (2001b).

Pérez-Brocal, V., Gil, R., Ramos, S., Lamelas, A., Postigo, M., Michelena, J. M., Silva, F. J., Moya, A., and Latorre, A. A small microbial genome: The end of a long symbiotic relationship? *Science* **314**, 312–313 (2006).

Peri, S., Ibarrola, N., Blagoev, B., Mann, M., and Pandey, A. Common pitfalls in bioinformatics-based analyses: Look before you leap. *Trends Genet.* **17**, 541–545 (2001).

Perna, N. T., et al. Genome sequence of enterohaemorrhagic *Escherichia coli* O157:H7. *Nature* **409**, 529–533 (2001).

Pósfai, G., Plunkett, G. III, Feh ér, T., Frisch, D., Keil, G. M., Umenhoffer, K., Kolisnychenko, V., Stahl, B., Sharma, S. S., de Arruda, M., Burland, V., Harcum, S. W., and Blattner, F. R. Emergent properties of reduced-genome *Escherichia coli*. *Science* **312**, 1044–1046 (2006).

Read, T. D., et al. Genome sequences of *Chlamydia trachomatis* MoPn and *Chlamydia pneumoniae* AR39. *Nucleic Acids Res.* **28**, 1397–1406 (2000).

Read, T. D., et al. Comparative genome sequencing for discovery of novel polymorphisms in *Bacillus anthracis*. *Science* **296**, 2028–2033 (2002).

Reid, S. D., Herbelin, C. J., Bumbaugh, A. C., Selander, R. K., and Whittam, T. S. Parallel evolution of virulence in pathogenic *Escherichia coli*. *Nature* **406**, 64–67 (2000).

Rendulic, S., Jagtap, P., Rosinus, A., Eppinger, M., Baar, C., Lanz, C., Keller, H., Lambert, C., Evans, K. J., Goesmann, A., Meyer, F., Sockett, R. E., and Schuster, S. C. A predator unmasked: Life cycle of *Bdellovibrio bacteriovorus* from a genomic perspective. *Science* **303**, 689–692 (2004).

Reysenbach, A. L., and Shock, E. Merging genomes with geochemistry in hydrothermal ecosystems. *Science* **296**, 1077–1082 (2002).

Riesenfeld, C. S., Schloss, P. D., and Handelsman, J. Metagenomics: Genomic analysis of microbial communities. *Annu. Rev. Genet.* **38**, 525–552 (2004).

Riley, M., et al. *Escherichia coli* K-12: A cooperatively developed annotation snapshot–2005. *Nucleic Acids Res.* **34**, 1–9 (2006).

Robertson, C. E., Harris, J. K., Spear, J. R., and Pace, N. R. Phylogenetic diversity and ecology of environmental Archaea. *Curr. Opin. Microbiol.* **8**, 638–642 (2005).

Rudd, K. E. EcoGene: A genome sequence database for *Escherichia coli* K-12. *Nucleic Acids Res.* **28**, 60–64 (2000).

Salgado, H., Santos-Zavaleta, A., Gama-Castro, S., Peralta-Gil, M., Penaloza-Spinola, M. I., Martinez-Antonio, A., Karp, P. D., and Collado-Vides, J. The comprehensive updated regulatory network of *Escherichia coli* K-12. *BMC Bioinformatics* **7**, 5 (2006).

Schaechter, M. Introduction to the pathogenic bacteria. In M. Schaechter, N. C. Engleberg, B. I. Eisenstein, and G. Medoff (eds.), *Mechanisms of Microbial Disease*, Chapter 10. Lippincott Williams and Wilkins, Baltimore, MD, 1999.

Schloss, P. D., and Handelsman, J. Status of the microbial census. *Microbiol. Mol. Biol. Rev.* **68**, 686–691 (2004).

Sherratt, D. Divide and rule: The bacterial chromosome. *Trends Genet.* **17**, 312–313 (2001).

Shigenobu, S., Watanabe, H., Hattori, M., Sakaki, Y., and Ishikawa, H. Genome sequence of the endocellular bacterial symbiont of aphids *Buchnera* sp. APS. *Nature* **407**, 81–86 (2000).

Shirai, M., et al. Comparison of whole genome sequences of *Chlamydia pneumoniae* J138 from Japan and CWL029 from USA. *Nucleic Acids Res.* **28**, 2311–2314 (2000).

Skovgaard, M., Jensen, L. J., Brunak, S., Ussery, D., and Krogh, A. On the total number of genes and their length distribution in complete microbial genomes. *Trends Genet.* **17**, 425–428 (2001).

Stephens, R. S., et al. Genome sequence of an obligate intracellular pathogen of humans: *Chlamydia trachomatis*. *Science* **282**, 754–759 (1998).

Stover, C. K., et al. Complete genome sequence of *Pseudomonas aeruginosa* PA01, an opportunistic pathogen. *Nature* **406**, 959–964 (2000).

Sueoka, N. Correlation between base composition of deoxyribonucleic acid and amino acid composition of protein. *Proc. Natl. Acad. Sci. USA* **47**, 1141–1149 (1961).

Sun, L. V., et al. Determination of *Wolbachia* genome size by pulsed-field gel electrophoresis. *J. Bacteriol.* **183**, 2219–2225 (2001).

Tamas, I., et al. 50 million years of genomic stasis in endosymbiotic bacteria. *Science* **296**, 2376–2379 (2002).

Tatusov, R. L., Koonin, E. V., and Lipman, D. J. A genomic perspective on protein families. *Science* **278**, 631–637 (1997).

Tatusov, R. L., et al. The COG database: An updated version includes eukaryotes. *BMC Bioinformatics* **4**, 41 (2003).

Teeling, H., and Gloeckner, F. O. RibAlign: A software tool and database for eubacterial phylogeny based on concatenated ribosomal protein subunits. *BMC Bioinformatics* **7**, 66 (2006).

Tettelin, H., et al. Complete genome sequence of *Neisseria meningitidis* serogroup B strain MC58. *Science* **287**, 1809–1815 (2000).

Tomb, J. F., et al. The complete genome sequence of the gastric pathogen *Helicobacter pylori*. *Nature* **388**, 539–547 (1997).

Tringe, S. G., and Rubin, E. M. Metagenomics: DNA sequencing of environmental samples. *Nat. Rev. Genet.* **6**, 805–814 (2005).

Tringe, S. G., von Mering, C., Kobayashi, A., Salamov, A. A., Chen, K., Chang, H. W., Podar, M., Short, J. M., Mathur, E. J., Detter, J. C., Bork, P., Hugenholtz, P., and Rubin, E. M. Comparative metagenomics of microbial communities. *Science* **308**, 554–557 (2005).

Walker, J. J., and Pace, N. R. Endolithic microbial ecosystems. *Annu. Rev. Microbiol.* **61**, 331–347 (2007).

Ward, N., and Fraser, C. M. How genomics has affected the concept of microbiology. *Curr. Opin. Microbiol.* **8**, 564–571 (2005).

Waters, E., et al. The genome of *Nanoarchaeum equitans*: Insights into early archaeal evolution and derived parasitism. *Proc. Natl. Acad. Sci. USA* **100**, 12984–12988 (2003).

White, O., et al. Genome sequence of the radioresistant bacterium *Deinococcus radiodurans* R1. *Science* **286**, 1571–1577 (1999).

Woese, C. R., and Fox, G. E. The concept of cellular evolution. *J. Mol. Evol.* **10**, 1–6 (1977).

Woese, C. R. On the evolution of cells. *Proc. Natl. Acad. Sci. USA* **99**, 8742–8747 (2002).

Wolf, Y. I., Rogozin, I. B., Grishin, N. V., Tatusov, R. L., and Koonin, E. V. Genome trees constructed using five different approaches suggest new major bacterial clades. *BMC Evol. Biol.* **1**, 8 (2001).

Zahradka, K., Slade, D., Bailone, A., Sommer, S., Averbeck, D., Petranovic, M., Lindner, A. B., and Radman, M. Reassembly of shattered chromosomes in *Deinococcus radiodurans*. *Nature* **443**, 569–573 (2006).

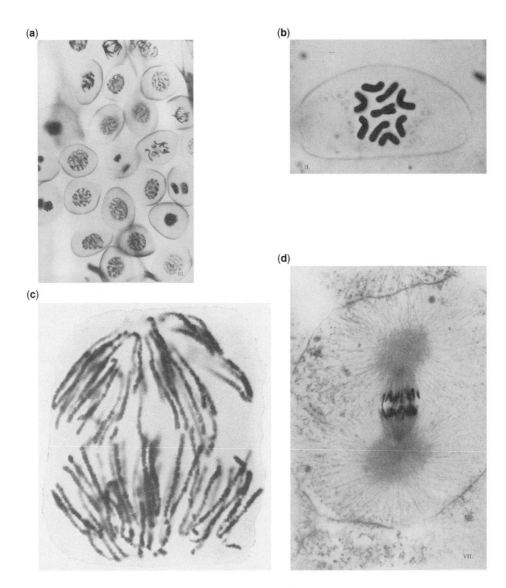

Through the first half of the twentieth century, Charles Darlington performed brilliant studies of the chromosomes. (a) First P.G. mitosis in Paris quadrifolia, Liliaceae, showing all stages from prophase to telophase. n = 10. ×800 magnification. (b) First P.G. mitosis in polar view. Tradescantia virginiana, Commelinaceae, *n = 9 (from aberrrant plant with 22 chromosomes). ×1200. (c) Root tip squashes showing anaphase separation in* Fritillaria pudica, *3x = 39. Note the spiral structure of chromatids (daughter chromosomes). ×3000. (d) Cleavage mitosis in the morula of the teleostean fish,* Coregonus clupeoides, *in the middle of anaphase. Spindle structure revealed by slow fixation. ×4000.*

16

The Eukaryotic Chromosome

Science is about building causal relations between natural phenomena (for instance, between a mutation in a gene and a disease). The development of instruments to increase our capacity to observe natural phenomena has, therefore, played a crucial role in the development of science—the microscope being the paradigmatic example in biology. With the human genome, the natural world takes an unprecedented turn: it is better described as a sequence of symbols. Besides high-throughput machines such as sequencers and DNA chip readers, the computer and the associated software becomes the instrument to observe it, and the discipline of bioinformatics flourishes. However, as the separation between us (the observers) and the phenomena observed increases (from organism to cell to genome, for instance), instruments may capture phenomena only indirectly, through the footprints they leave. Instruments therefore need to be calibrated: the distance between the reality and the observation (through the instrument) needs to be accounted for. [We are] calibrating instruments to observe gene sequences; more specifically, computer programs to identify human genes in the sequence of the human genome.

—Martin Reese and Roderic Guigó (2006, p. S1.1),
introducing EGASP, the Encyclopedia of DNA Elements (ENCODE)
Genome Annotation Assessment Project

INTRODUCTION

Synonyms of eukaryotes include eucaryotae, eucarya, eukarya, and eukaryotae. The word derives from the Greek *eu-* ("true") and *karutos* ("having nuts"; this refers to the nucleus).

The eukaryotes are single-celled or multicellular organisms that are characterized by the presence of a membrane-bound nucleus and a cytoskeleton. We will begin our examination of specific eukaryotes with the fungi (Chapter 17), including *Saccharomyces cerevisiae*. We then broadly survey the eukaryotes (Chapter 18), from the simplest primitive single-celled organisms to plants and metazoans (animals).

At the start of Chapter 13 we addressed five basic perpectives on the field of genomics. With respect to the topic of eukaryotic chromosomes, we may briefly reiterate the five perspectives as follows:

Perspective 1: Catalog genomic information. We will examine genome sizes, noncoding DNA (e.g., repetitive DNA), and coding DNA (genes). For a given segment of genomic DNA, we will address the problem of annotation: how much repetitive DNA is present and of what type? How many protein-coding genes or RNA genes are present?

Perspective 2: Catalog comparative genomic information. How can comparative genomics help us to understand chromosomal rearrangements that have occurred over time?

Perspective 3: Biological principles. What are the mechanisms underlying chromosomal functions and chromosomal variations such as duplications, inversions, and translocations? More broadly, as we examine genomic DNA, we want to address the molecular basis of how organisms and species evolve.

Perspective 4: Human disease relevance. In what ways are chromosomal variants associated with disease?

Perspective 5: Bioinformatics aspects. What tools are available to understand chromosomes, from genome browsers to gene-finding algorithms?

A focus of this chapter is on the analysis of completely sequenced eukaryotic genomes. The *C. elegans* sequencing consortium 1998 described why we want to obtain the complete genomic sequences:

- The complete genome sequence provides the basis for discovering all the genes that are encoded in a genome. Other approaches, such as characterizing expressed sequence tags, can never be as comprehensive.

- The comparative genomic sequence shows the structural and regulatory elements associated with genes.

- It provides the basis to assess the molecular evolution of a species as well as the extent of its variation between individuals, populations, and other species.

- It provides a set of tools for future experimentation.

At the same time that eukaryotic genomes are completely sequenced, a parallel molecular approach is the characterization of individual genes from many hundreds or thousands of species. The use of model genes complements the use of model organisms. For example, a search of GenBank for the gene ribulose-1,5-bisphosphate carboxylase (rubisco; *rbcL*) currently reveals about 50,000 entries. Rubisco is a major plant gene (discussed in Chapter 18), and the availability of molecular sequence data from many species for this and selected other genes is crucial for phylogenetic reconstructions and structure–activity studies. Other commonly studied genes include highly conserved molecules such as those described in Figs. 17.2 and 18.1.

Major Differences between Eukaryotes and Prokaryotes

Eukaryotes share a common ancestry with prokaryotes, but when we compare them, we find several outstanding differences (Cavalier-Smith, 2002; Vellai and Vida, 1999; Watt and Dean, 2000). Some of these genomic features are highlighted in Table 16.1.

- There is a tremendous diversity of both prokaryotic and eukaryotic life forms. However, very few bacterial or archaeal life forms are visible to the human eye. Many eukaryotes are single-celled, microscopic organisms. Nonetheless, most life forms that we can see are multicellular eukaryotes (e.g., plants and metazoans).

- Eukaryotic cells have three cellular features that are lacking in prokaryotes: (1) a membrane-bound nucleus, (2) an extensive system of organelles bound by intracellular membranes, and (3) a cytoskeleton, including elements such as actin and tubulin, and molecular motors. Notably, prokaryotes lack energy-producing organelles and are incapable of endocytosis, the process by which extracellular cargo is internalized (Vellai and Vida, 1999).

- Most eukaryotes undergo sexual reproduction, although some are asexual. Bacteria lack gamete fusion and do not exchange DNA by sex.

- The genome size of eukaryotes varies widely, spanning five orders of magnitude (Table 16.2). In contrast, most archaeal and bacterial genomes are between about 0.2 and 13 Mb in size (see Chapters 13 and 15).

- Prokaryotic genomes tend to have a relatively high density of protein-coding genes and little repetitive or other noncoding DNA. For example, 0.7% of the *Escherichia coli* genome consists of noncoding repeats (Blattner et al., 1997). In contrast, many eukaryotic genomes include large tracts of noncoding DNA. Several examples are provided in Table 16.1.

Sexual reproduction is called syngamy, the process by which the haploid chromosomes of the male and female gametes combine to form the zygote (i.e., the fertilized ovum).

TABLE 16-1 Features of Several Sequenced Bacterial and Eukaryotic Genomes

Feature	*E. coli* K-12	Parasite[a]	Yeast[b]	Slime Mold[c]	Plant[d]	Human[e]
Genome size, Mb	4.64	22.8	12.5	8.1	115	3289
GC content, %	50.8	19.4	38.3	22.2	34.9	41
Number of genes	4288	5268	5770	2799	25,498	20,000–25,000
Gene density, kb per gene	0.95	4.34	2.09	2.60	4.53	27
Percent coding	87.8	52.6	70.5	56.3	28.8	1.3
Number of introns	0	7406	272	3578	107,784	53,295
Repeat %	<1	<1	2.4	<1	14	46

[a]*Plasmodium falciparum.*
[b]*Saccharomyces cerevisiae.*
[c]*Dictyostelium discoideum.*
[d]*Arabidopsis thaliana.*
[e]*Homo sapiens.*
Abbreviations: bp, base pairs; Mb, millions of base pairs (megabases).
Source: Adapted from Gardner et al. (2002); Blattner et al. (1997); International Human Genome Sequencing Consortium (2001, 2004).

TABLE 16-2	Genome Size of Selected Phyla or Classes of Eukaryotes		
Taxon	Phylum, Class, or Division	Genome Size Range (Gb)	Ratio of genome sizes (Highest/Lowest)
All eukaryotes	—	0.003–686	228,667
Alveolata	—	—	22,333
	Apicomplexians	0.009–201	22,333
	Ciliates	0.024–8.62	359
	Dinoflagellates	1.37–98	72
Diatoms		0.035–24.5	700
Amoebae		0.035–686	19,600
Euglenozoa		0.098–2.35	24
Fungi/microsporidia		0.003–1.47	490
Animals	—	—	3,325
	Sponges	0.059–1.78	30
	Cnidarians	0.227–1.83	8
	Insects	0.089–9.47	106
	Elasmobranchs	1.47–15.8	11
	Bony fishes	0.345–133	386
	Amphibians	0.93–84.3	91
	Reptiles	1.23–5.34	4
	Birds	1.67–2.25	1
	Mammals	1.7–6.7	4
	Placozoa	0.04	—
Plants	—	—	6,140
	Algae	0.080–30	375
	Pteridophytes	0.098–307	3,133
	Gymnosperms	4.12–76.9	19
	Angiosperms	0.050–125	2,500

Note: 0.001 Gb (gigabases) equals 1 Mb. Values in picograms were multiplied times 0.9869×10^9 to obtain gigabases.

Sources: Adapted from Graur and Li 2000, Animal Genome Size Database of T. R. Gregory (▶ http://www.genomesize.com), and the National Center for Biotechnology Information (▶ http://www.ncbi.nlm.nih.gov).

- Prokaryotes are haploid, that is, the organism has one set of chromosomes. Eukaryotes may be haploid or diploid ($2x$; having two sets of chromosomes) or have other ploidy states (such as triploid [$3x$]). This higher level of ploidy offers eukaryotes a variety of evolutionary mechanisms such as heterozygous advantage (Watt and Dean, 2000).

- The genomes are organized differently. The majority of bacterial and archaeal genomes are organized in circular chromosomes, often with small accompanying plasmids (see Fig. 15.1). Eukaryotic nuclear genomes are organized primarily into linear chromosomes. These eukaryotic chromosomes are typically numerous (ranging from a few to over 100) and each has a centromere (defined below) as well as telomeres at either end. These features are absent from prokaryotic chromosomes, although centromere-like elements have been described (Ben-Yehuda et al., 2002; Moller-Jensen et al., 2002). The mechanisms by which bacteria segregate DNA are relatively obscure.

GENERAL FEATURES OF EUKARYOTIC GENOMES AND CHROMOSOMES

C Value Paradox: Why Eukaryotic Genome Sizes Vary So Greatly

In eukaryotic genomes, the haploid genome size (*C* value) varies enormously. This is shown in Table 16.2 for various taxa of eukaryotes and in Table 16.3 for specific eukaryotic species. Some genomes are relatively quite small, such as the microsporidian *Encephalitozoon cuniculi* (2.9 Mb; Chapter 17). Others have genome sizes in the range of hundreds of billions of base pairs. Tremendous variation in *C* values occurs among the unicellular protists such as amoebae, with a 20,000-fold range. Within the animal kingdom, the range is about 3000-fold.

The *C* value is measured in base pairs or in picograms (pg) of DNA. One picogram of DNA corresponds to approximately 1 Gb.

An online database of plant *C* values is available at ► http://data.kew.org/cvalues/homepage.html. Currently (December 2007) it lists data for over 5000 species. The Animal Genome Size Database (from T. Ryan Gregory) is online at ► http://www.genomesize.com/. Another resource is the Database of Genome Size (DOGS) (► http://www.cbs.dtu.dk/databases/DOGS/).

TABLE 16-3 Genome Size (*C* Value) for Various Eukaryotic pecies

Species	Common Name	*C* Value (Gb)
Saccharomyces cerevisiae	Yeast	0.012
Neurospora crassa	Fungus	0.043
Dysidea crawshagi	Sponge	0.054
Caenorhabditis elegans	Nematode	0.097
Drosophila melanogaster	Fruitfly	0.12
Paramecium aurelia	Ciliate	0.19
Oryza sativa	Rice	0.47
Strongylocentrotus purpuratus	Sea urchin	0.80
Gallus domesticus	Chicken	1.23
Erysiphe cichoracearum	Powdery mildew	1.5
Boa constrictor	Snake	2.1
Parascaris equorum	Roundworm	2.5
Carcharias obscurus	Sand-tiger shark	2.7
Canis familiaris	Dog	2.9
Rattus norvegicus	Rat	2.9
Xenopus laevis	African clawed frog	3.1
Homo sapiens	**Human**	**3.3**
Nicotania tabacum	Tobacco plant	3.8
Locusta migratoria	Migratory locust	6.6
Paramecium caudatum	Ciliate	8.6
Allium cepa	Onion	15
Truturus cristatus	Warty newt	19
Thuja occidentalis	Western giant cedar	19
Coscinodisucus asteromphalus	Centric diatom	25
Lilium formosanum	Lily	36
Amphiuma means	Two-toed salamander	84
Pinus resinosa	Canadian red pine	68
Protopterus aethiopicus	Marbled lungfish	140
Amoeba proteus	Amoeba	290
Amoeba dubia	Amoeba	690

Sources: Adapted from Graur and Li (2000), NCBI (► http://www.ncbi.nlm.nih.gov), Cameron et al. (2000), and the Database of Genome Sizes (► http://www.cbs.dtu.dk/databases/DOGS/index.php).

Alternative solutions to the *C* value paradox do not fit. The number of protein-coding genes in eukaryotes varies over a ≈10-fold range, but this variation is far smaller than the range of genome sizes. Also, interspecies variation in the lengths of mRNA molecules does not explain the *C* value paradox because no correlation exists between mean gene length and genome size.

Remarkably, the range in *C* values does not correlate well with the complexity of organisms. Some organisms such as *A. thaliana* (a plant) and *Fugu rubripes* (a pufferfish) have extremely compact genomes while closely related organisms of similar biological complexity have genomes that are orders of magnitude larger. This lack of correlation is called the *C* value paradox (Hartl, 2000; Knight, 2002; Hancock, 2002; Kidwell, 2002). The genomes of many eukaryotes have now been sequenced, including *Caenorhabditis elegans* (1998), *Drosophila melanogaster* (2000), *Homo sapiens* (2001), and *Mus musculus* (2002) (see below and Chapters 13 and 18). These whole-genome studies provide one clear answer to the *C* value paradox: genomes are filled with large tracts of noncoding DNA sequences in varying amounts. This accounts for the variation in genome size. We will explore this noncoding DNA below.

Organization of Eukaryotic Genomes into Chromosomes

Genomic DNA is organized in chromosomes. Originally, chromosomes were defined morphologically as the bodies into which the nucleus resolves itself at the beginning of mitosis and from which it is derived at the end of mitosis (Waldeyer, 1888; Darlington, 1932). It was clear by the 1880s that the nucleus is the cellular organelle that directs the cell division process, and that mitosis occurs in both plants and animals (Lima-de-Faria, 2003). Visualizing chromosomes cytogenetically was challenging, and reports from the 1920s that there are 48 human chromosomes were not corrected until Joe Hin Tjio and Albert Levan (1956) reported that the diploid number of chromosomes is 46, that is, there are 23 pairs of human chromosomes.

Chromosomes are often studied at metaphase, when they are thickest and most condensed. For human studies, a sample is typically collected from blood cells or amniotic fluid. Chromosomes are most often visualized using dyes or using specific DNA probes by fluorescence in situ hybridization (FISH).

As we explore a variety of eukaryotic genomes that have been completely sequenced, it is helpful to describe the structure and content of chromosomes. We will refer to a karyotype of human metaphase chromosomes visualized with Wright's stain (Fig. 16.1a). A variety of stains produce banding patterns on chromosomes. These include Q bands (based on stains using quinacrine mustard or derivatives) and G bands (based on the Giemsa dye; Wright's stain is an example of such a dye). These dyes stain the entire length of each chromosome and produce a characteristic banding pattern. A band is defined as a portion of a chromosome that is distinguishable from adjacent segments by appearing lighter or darker.

There are several major features of eukaryotic chromosomes. The most apparent landmarks are the two telomeres (the chromosome ends) and the centromere. Telomeres are structures characterized by tandem arrays of repetitive sequences found at the chromosome ends. They provide stability to chromosomes by preventing the degradation of the chromosome end and by blocking the fusion of chromosome ends (Blackburn et al., 1989). The centromere, a region that remains unstained with many dyes, appears as a constriction. Centromeres may be metacentric (located near the middle of the chromosome) or acrocentric (located close to a telomere). In humans, the five acrocentric chromosomes are 13, 14, 15, 21, and 22. In some species such as the mouse, *Mus musculus*, all chromosomes are acrocentric.

Deletion 11q syndrome results in trigonencephaly (a triangle-shaped head), a carp-shaped mouth, and cardiac defects (Jones, 1997).

The autosomes consist of chromosomes 1 to 22, while the X and Y are the sex chromosomes. In the particular karyotype shown in Fig. 16.1a there is a hemizygous deletion of the terminus of chromosome 11q. In a euploid (apparently normal) individual there are two copies of each autosome; in a hemizygous deletion there is only one copy, and in a homozygous deletion there are zero copies. Using conventional karyotyping, deletions or duplications as small as several million base pairs can be observed by inspection of the banding patterns. Figure 16.1b shows a trisomy of

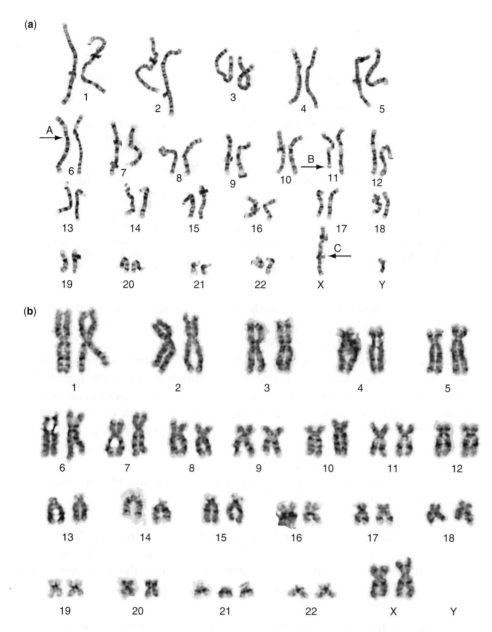

FIGURE 16.1. *Example of human karyotypes. (a) The chromosomes are visualized with Wright's stain. Centromeres are visible as an indentation in the chromosome (e.g., see arrows A and C). This karyotype is of a person with a hemizygous deletion of a telomeric portion of chromosome 11q, resulting in a loss of several million base pairs of DNA (arrow B). (b) Karyotype of a female with trisomy 21 (Down syndrome). Note that there are three copies of chromosome 21.*

chromosome 21 in which the entire chromosome (~46 million base pairs) is present in three copies.

Inside a nucleus, chromosomes tend to be unraveled structures that occupy restricted spaces called chromosome territories. Meaburn and Misteli (2007) provide an overview of the spatial organization of chromosomes and genomes, including visualization with chromosome-specific fluorescent probes. Trask 2002 has written an overview of the field of human cytogenetics.

Analysis of Chromosomes Using Genome Browsers

The diploid number of chromosomes is constant in each species, although there may be individual variation. We will explore the 16 *S. cerevisiae* chromosomes (Chapter 17), including a variety of databases such as NCBI, MIPS, and SGD that provide graphic displays. In humans, the diploid number is 46 (i.e., there are 23 pairs of

TABLE 16-4   Web-Based Databases of Chromosomes

Resource	Comment	URL
Ensembl genome browser	Ideograms for human (Chapter 19), mouse, rat, zebrafish, fugu, mosquito, other	▶ http://www.ensembl.org/
Ideogram Album	Human, mouse, and horse ideograms from the University of Washington	▶ http://www.pathology. washington.edu/research/ cytopages/
Human Chromosome Launchpad	From the Oak Ridge National Laboratory	▶ http://www.ornl.gov/sci/ techresources/Human_ Genome/launchpad/
KaryotypeDB	From Mario Nenno	▶ http://www.nenno.it/ karyotypedb/
SKY/M-FISH and CGH Database	From the National Cancer Institute and NCBI	▶ http://www.ncbi.nlm.nih. gov/sky/skyweb.cgi

An ideogram is a diagram of a karyotype. A karyotype is an image (often a photograph) of the chromosomes from a cell during metaphase, when each chromosome is a pair of sister chromatids. Karyotypes display the chromosomes in numerical order, with the short arm (p arm) oriented upward. For humans, the short arm is called "p" for *petit* (French for "small"), while the q arm (long arm) is named as the letter following p.

chromosomes in almost all somatic cells). We will explore databases that display ideograms of chromosomes in Chapters 18 and 19. Ideograms of karyotypes for some other organisms are available online (Table 16.4).

Databases such as GenBank, EMBL, and DDBJ (Chapter 2) store hundreds of billions of base pairs of DNA from various organisms. For a particular organism of interest, whether a fungus, plant, or animal, genome browsers represent an essential tool to store, centralize, process, and display both raw sequence data and analyses based on annotation of the data. Annotation consists of adding information about features such as the experimentally determined or computationally predicted repetitive elements or genes or sites of variation.

There are three major genome browsers that provide broad and deep coverage of a variety of eukaryotic genomes, as follows.

1. NCBI offers a map viewer for dozens of species. An example is shown for human chromosome 21 (Fig. 16.2). There is an ideogram (arrow 1). Vertical tracks can be added or removed, such as the UniGene entries (arrow 2). The symbols column (arrow 3) provides a link for each gene to Entrez Gene. A variety of other links are provided (arrow 4).

Ensembl (▶ http://www.ensembl. org) is a joint project between EMBL-EBI and the Sanger Institute.

2. The Ensembl project offers a map viewer filled with annotation data (Hubbard et al., 2007). A view of human chromosome 21 includes summaries of the genomic features such as GC content, single nucleotide polymorphisms (SNPs), and coding and noncoding gene content (Fig. 16.3a). A link to the cytoviewer (Fig. 16.3b) provides access to dozens of additional tracks of features that can be viewed or downloaded for detailed chromosomal analyses.

The UCSC Genome Bioinformatics site is ▶ http:// genome.ucsc.edu/.

3. We focus on the UCSC Genome Browser in this chapter. It includes a gateway to select a genome and chromosomal region of interest (Fig. 16.4a). The main genome browser page depicts the chromosome of interest (human chromosome 21 in Fig. 16.4b) along with a series of user-selected annotation tracks. In this example, tracks are displayed showing the chromosome band, gaps in the genome assembly, GC percent, and RefSeq genes. Recent literature on the UCSC Genome Browser includes an overview of its function (Kuhn et al., 2007), its resources for analyzing variation (Thomas et al., 2007b), its Table Browser (Karolchik et al., 2004), and BLAT (Kent, 2002).

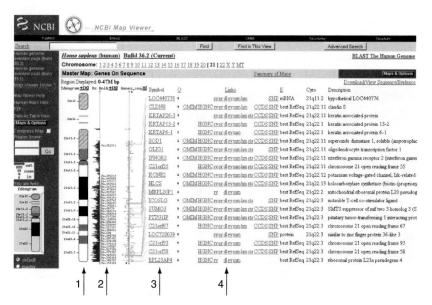

FIGURE 16.2. *Chromosomes can be explored using the NCBI Map Viewer, accessible from the main page of NCBI (▶ http://www.ncbi.nlm.nih.gov). Here, chromosome 21 is depicted. There is an ideogram of the chromosome (arrow 1); upon clicking one can zoom or display from 10,000 to 10 million base pairs of DNA. Optional tracks such as UniGene entries (arrow 2) can be added using the Maps and Options button (top right of web page). Gene symbols (arrow 3) link to Entrez Gene entries. Many other links are provided (arrow 4), such as "pr" to access the protein sequence, "dl" to download genomic DNA in the vicinity of a gene, or "ev" to link to the Evidence Viewer presenting models of exon structure for each gene.*

Analysis of Chromosomes by the ENCODE Project

An initial version of the human genome sequence was reported by a public consortium (International Human Genome Sequencing Consortium, 2001) and by Venter et al. 2001. It was immediately clear that the annotation of the functional elements embedded in the genomic DNA is extraordinarily complex. The Encyclopedia of DNA Elements (ENCODE) project was initated to investigate the properties of the human and other genomes (ENCODE Project Consortium, 2004). Forty-four regions of the human genome were selected, spanning 30 megabases or about 1% of the human genome. These ENCODE regions include a mixture of about half randomly selected loci as well as half containing well-known genes (such as alpha and beta globins, and cystic fibrosis transmembrane regulator).

The ENCODE Project Consortium (2007) released its findings on 1% of the genome in a paper with over 250 coauthors. This represented the generation of over 200 data sets by 35 groups. Their 11 main conclusions were as follows:

The National Human Genome Research Institute offers information on the ENCODE project at ▶ http://www.genome.gov/10005107.

1. The human genome is pervasively transcribed. We discussed this in Chapter 8.

2. Many novel noncoding transcripts were identified, sometimes overlapping protein-coding genes.

3. Novel transcriptional start sites were identified and characterized in detail.

4. Regulatory sequences surrounding transcription start sites are symmetrically distributed. Previously, it had been thought that there is a bias towards the location of regulatory sequences upstream of genes.

5. Histone modification and chromatin accessibility predict the presence and activity of transcription start sites.

6. Some genomic DNA sites are hypersensitive to digestion with the endonuclease DNaseI. Such sites have histone modification patterns that distinguish them from promoters.

7. DNA replication timing correlates with chromatin structure.

(a)

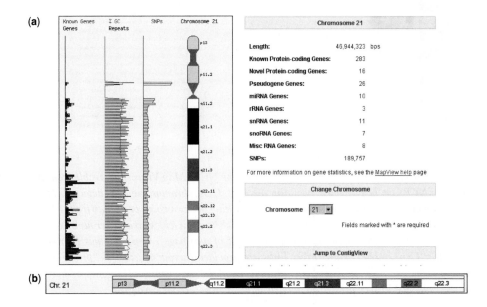

(b)

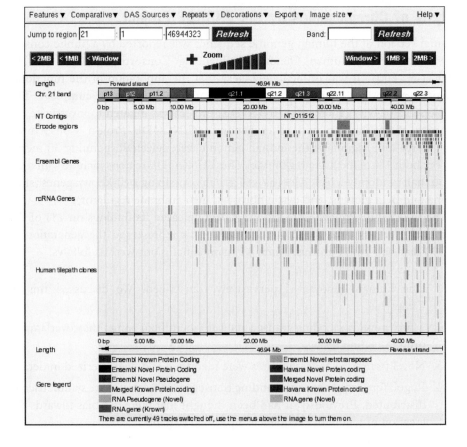

FIGURE 16.3. *View of human chromosome 21 using Ensembl. This is one of the three important genome browsers (with UCSC and NCBI) and it offers an exceptionally wide range of viewing and analysis options. (a) Overview of human chromosome 21. (b) Ideogram and view of selected chromosome features.*

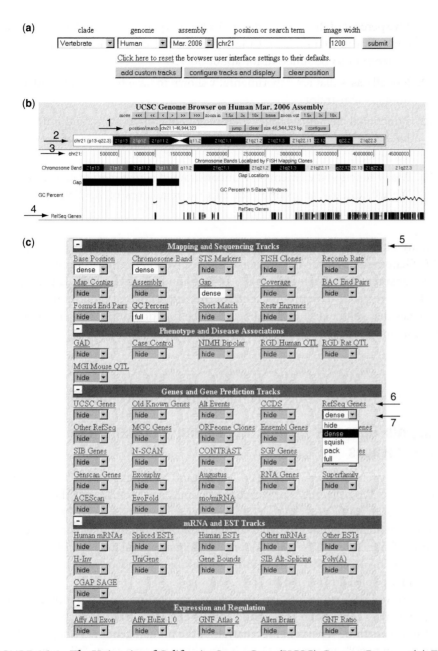

FIGURE 16.4. *The University of California, Santa Cruz (UCSC) Genome Browser. (a) From the genome browser portal you can select the clade (e.g., vertebrate, deuterostome such as sea squirts, insect, nematode), the assembly (which may vary as genome sequencing and annotation progress), and the position or search term (e.g., the name myoglobin or an accession number). (b) The genome browser includes a position/search box that allows you to specify a name, accession number, or physical map position to view. Here, all of chromosome 21 is selected (about 47 million base pairs). The browser includes an ideogram (arrow 2) and a box showing user-selected annotation tracks. By clicking the top row (arrow 3), the view is zoomed in threefold. An example of an optional annotation track is the GC percent or the RefSeq genes (arrow 4). (c) There are many dozens of available annotation tracks, arranged into categories such as Mapping and Sequencing Tracks (arrow 5), Phenotype and Disease Associations, Genes and Gene Prediction Tracks, mRNA and EST Tracks, Expression and Regulation, Comparative Genomics, and Variation and Repeats. Additional sections correspond to the ENCODE project (discussed below). For any annotation track you can click its title (e.g., RefSeq, arrow 6) to link to detailed information on options for displaying that track, the method by which the genome was annotated, and literature references. A pull-down menu (arrow 7) allows you to hide each track or to view in relatively condensed or extended forms.*

8. Five percent of the nucleotides in the human genome are under evolutionary constraint in mammals. Of these constrained bases, there is experimental evidence of some function for about 60%.

9. Not all bases that are experimentally shown to have function are under evolutionary constraint.

10. Functional elements vary in their degree of conservation, and in their likelihood of being located in a structurally variable region of the human genome.

11. Many functional elements are unconstrained across mammalian evolution. We will discuss an example of this for regulators of *RET* gene function between human and fish.

To view the ENCODE regions, visit ▶ http://genome.ucsc.edu, click ENCODE, then select from the list.

Many of the findings of the ENCODE consortium are available through the UCSC Genome Browser (Thomas et al., 2007a). We will provide examples throughout this chapter. With its focus on 1% of the human, the ENCODE project has helped the research community define the experimental and computational approaches that will be useful to characterize the function of the remaining portion of the human genome as well as other eukaryotic genomes.

REPETITIVE DNA CONTENT OF EUKARYOTIC CHROMOSOMES

Eukaryotic Genomes Include Noncoding and Repetitive DNA Sequences

Bacterial and archaeal genomes have both genes and additional, relatively small intergenic regions. Typically, these prokaryotic genomes are circular, and there is about one gene in each kilobase of genomic DNA (Chapter 14 and Table 16.1). In contrast, eukaryotic genomes contain a smaller proportion of protein-coding genes and large amounts of noncoding DNA. This noncoding material includes repetitive DNA, genes encoding RNAs that have regulatory functions, and introns that interrupt exons and are spliced from mature RNA transcripts.

Repetitive DNA sequences can occupy vast proportions of eukaryotic genomes. These sequences consist of repeated nucleotides of various lengths (Jurka, 1998). We will also discuss these repeats in our analysis of the human genome (Chapter 19). In mammals, up to 60% of genomic DNA is repetitive; in some yeasts 20% is repetitive. Identifying repetitive DNA elements in eukaryotic DNA is essential in genome analysis. Such repeats can powerfully influence the structure of the genome, including the capacity of chromosomes to rearrange and to regulate transcription. They are often important in disease, serving as substrates for recombination events that delete or duplicate chromosomal segments. Repeats are also useful as "molecular fossils" in evolutionary studies based on comparative analysis of genomes from different species (Chapter 19).

Britten and Kohne (1968) performed some of the earliest experiments that defined the repetitive nature of eukaryotic DNA. They purified genomic DNA from a wide variety of species, sheared it, and dissociated the DNA strands. Under appropriate conditions of salt, temperature, and time, the DNA strands reanneal. They measured the rate at which the DNA reassociates and found that for dozens of eukaryotes—but not for several viruses or bacteria—DNA reassociates in several distinct fractions.

Large amounts of eukaryotic DNA reassociate extremely rapidly. For the mouse genome, about 10% of genomic DNA reassociates rapidly and consists of about one million copies (Fig. 16.5, arrow A). This highly repetitive DNA is localized to the highly condensed portion of chromosomes referred to as heterochromatin (Redi et al., 2001; Avramova, 2002). A further 20% of the DNA reassociates in a fraction containing from 1000 to 100,000 distinct DNA species (arrow B). Finally, about 70% of the DNA is unique, consisting of only a single copy (arrow C). This DNA forms the euchromatin, a portion of the chromosome that is not condensed and thus is accessible for the transcription of genes. The banding pattern of chromosomes (Fig. 16.1) corresponds to regions of heterochromatin and euchromatin. Heterochromatic regions tend to lack (or actively inhibit) gene expression, although some expressed genes have been identified in the heterochromatin of a variety of species from *Drosophila* to human (Yasuhara and Wakimoto, 2006).

The origin of these repeats and their function present fascinating questions. What different kinds of repeats occur? From where did they originate and when? Is there a logic to their promiscuous growth or do they multiply without purpose? We are beginning to understand the extent and nature of the repeat content of eukaryotic genomes, including the human genome. Repetitive DNA has sometimes been called "junk DNA" or "selfish DNA," reflecting its propensity to expand throughout

Britten and Kohne (1968) used several techniques to distinguish single-stranded from double-stranded DNA, such as hydroxyapatite chromatography (a calcium phosphate column), binding of radiolabeled DNA fragments to immobilized DNA on filters, and spectrophotometry. The rate of DNA reassociation is a function of the incubation time (t) and the DNA concentration (C_0). The C_0t plot displays the fraction of DNA that remains single stranded versus the C_0t value, and it is the basis for the data shown in Fig. 16.5.

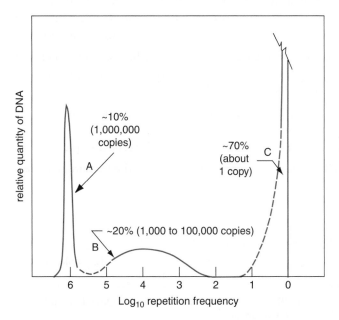

FIGURE 16.5. The complexity of genomic DNA can be estimated by denaturing then renaturing DNA. This figure (redrawn from Britten and Kohne, 1968) depicts the relative quantity of mouse genomic DNA (y axis) versus the logarithm of the frequency with which the DNA is repeated. The data are derived from a $C_0t_{1/2}$ curve, which describes the percent of genomic DNA that reassociates at particular times and DNA concentrations. A large $C_0t_{1/2}$ value implies a slower reassociation reaction. Three classes are apparent. The fast component accounts for 10% of mouse genomic DNA (arrow A), and it represents highly repetitive satellite DNA. An intermediate component accounts for about 20% of mouse genomic DNA and contains repeats having from 1000 to 100,000 copies. The slowly reassociating component, comprising 70% of the mouse genome, corresponds to unique, single-copy DNA. Britten and Kohne (1968) obtained similar profiles from other eukaryotes, although distinct differences were evident between species. Used with permission.

A retrotransposon (also called a retroposon or retroelement) is a transposable element that copies itself to genomic locations through a process of reverse transcription with an RNA intermediate. This process is similar to that of a retrovirus.

Barbara McClintock was awarded a Nobel Prize in 1983 for her discovery of mobile genetic elements in maize (*Zea mays*). You can read more about this pioneering work at ► http://www.nobel.se/medicine/laureates/1983/.

A search of Entrez nucleotide with the term "retropseudogene" yields 65 hits (December 2007), while "retrotransposed" yields 53 hits. But a search with the term "retrotransposon" yields >20,000 core nucleotide matches, >14,000 expressed sequence tags, and almost 40,000 genome survey sequences.

The mouse genome contains one functional gene encoding glyceraldehyde-3-phosphate dehydrogenase (*Gapdh;* NM_008084) and at least 400 pseudogenes distributed across 19 chromosomes (Mouse Genome Sequencing Consortium, 2002). The functional *Gapdh* gene was listed as assigned to mouse chromosomes 7 (Mouse Genome Sequencing Consortium, 2002), but currently (December 2007) it is assigned to chromosome 6 by Entrez Gene at NCBI and by the Ensembl mouse genome Contig Viewer. The presence of many pseudogenes contributes to the difficulty of assigning correct chromosomal loci.

genomes. However, it is likely that repetitive DNA has important roles in chromosome structure, recombination events, and the function of some genes (Makalowski, 2000; and see below).

There are five main classes of repetitive DNA in eukaryotes (International Human Genome Sequencing Consortium, 2001; Jurka, 1998; Kidwell, 2002; Makalowski, 2000).

1. Interspersed Repeats (Transposon-Derived Repeats).

Together, interspersed repeats constitute about 45% of the human genome (see Chapter 19). These repeats can be generated by elements that copy RNA intermediates (retroelements) or DNA intermediates (DNA transposons) (Table 16.5). Genes may be copied by retrotransposition when an mRNA is reverse transcribed and then integrated into the genome. Such genes can be identified because they usually lack introns, while they do have short direct flanking repeats. Examples of some mammalian retrotransposed genes are presented in Table 16.6.

Interspersed repeats can be divided into four categories (Ostertag and Kazazian, 2001; Kidwell, 2002) (see also Fig. 19.18):

- Long-terminal-repeat (LTR) transposons, which are RNA-mediated elements. These are also called retrovirus-like elements. LTR transposons have LTRs of several hundred base pairs at either end of the element.
- Long interspersed elements (LINEs), which encode an enzyme with reverse transcriptase activity (and possibly additional proteins). In mammals, LINE1 and LINE2 families are most prevalent.
- Short interspersed elements (SINEs), which are also RNA-mediated elements. *Alu* repeats, found in primates, are well-known examples of SINEs. We will see an example of an *Alu* repeat sequence below.
- DNA transposons comprise about 3% of the human genome.

TABLE 16-5 Examples of Classes and Transposable Elements

Class	Subclass	Superfamily	Examples of Family	Approximate Size Range (bp)
Retroelements (RNA-mediated elements)	LTR retrotransposons	Ty1-copia	Opie-1 (maize)	3,000–12,000
	Non-LTR retrotransposons	LINEs	*LINE-1* (human)	1,000–7,000
		SINEs	*Alu* (human)	100–500
DNA transposons	Cut-and-paste transposition	Mariner-Tc1	*Tc1* in *C. elegans*	1,000–2,000
		P	*P* in *Drosophila*	500–4,600
	Rolling circle transposition	*Helitrons*	*Helitrons* in *A. thaliana*, *O. sativa*, and *C. elegans*	5,500–17,500

Source: Adapted from Kidwell (2002). Used with permission.

TABLE 16-6   Examples of Mammalian Genes Generated by Retrotransposition

| Retrotransposed Gene | | | Original Gene | | | | Age |
Name	RefSeq	Chr	Name	RefSeq	Chr	Distribution	(MYA)
ADAM20	NM_003814	14q	*ADAM9*	NM_003816	8p	Human, not macaque	<20
Cetn1	NM_004066	18p	*Cetn2*	NM_004344	Xq28	Mammals	>75
Glud2	NM_012084	Xq	*Glud1*	NM_005271	10q	Human, not mouse	<70
Pdha2	NM_005390	4q	*Pdha1*	NM_000284	Xp	Placentals	~70
SRP46	NM_032102	11q	*PR264/ SC35*	NM_003016	17q	Human, simians	<70
Supt4h2	NM_011509	10	*Supt4 h*	NM_009296	11	Mouse	<70

Retrotransposed genes lack introns, and they often have flanking direct repeats and a polyadenine tail.

Abbreviations: Chr, chromosome; MYA, millions of years ago; *ADAM*, a disintegrin and metalloproteinase; *Cetn*, centrin, EF-hand protein; *Glud*, glutamate dehydrogenase; *Pdha2*, pyruvate dehydrogenase (lipoamide) alpha 2; *Supt4h*, suppressor of Ty 4 homolog (*S. cerevisiae*).

Sources: Adapted from Betrán and Long 2002 (see that article for additional genes and literature references) and from a search of Entrez (NCBI) with the term *retropseudogene*.

We can illustrate interspersed repeats using the UCSC Genome Browser. A region of 15,000 base pairs including the beta globin (HBB) and delta globin (HBD) genes is shown (Fig. 16.6a). Information on repeats that is precomputed using the RepeatMasker software package shows SINE, LINE, LTR, and DNA transposon elements as well as several other categories of repetitive DNA (simple repeats, low complexity DNA, satellite DNA). By clicking the Table link on the top sidebar of the UCSC Genome Browser, you can access the Table Browser (Fig. 16.6b). By clicking "get output" you can obtain a tab-delimited file listing all the elements detected by RepeatMasker as well as their genomic coordinates.

RepeatMasker searches a DNA query of interest against RepBase, a database of known repeats and low-complexity regions in eukaryotic DNA. Several programs, including RepeatMasker and the Censor Server at GIRI, effectively allow searches of DNA query sequences against this database (Smit, 1999; Jurka, 2000).

To identify and mask repetitive DNA sequences, you can use a RepeatMasker web server. Several servers are listed in Table 16.7, making it unnecessary to install either the program or the database locally. We will explore this with 50,000 bp of genomic DNA from human chromosome 11 in the beta globin locus. Paste your sequence into a box that is provided and select the output options. The RepeatMasker output includes a list of scores using the Smith–Waterman algorithm, the position of the repeat, and information on the type of repeat (e.g., SINE/*Alu*, LTR, or simple repeat) (Fig. 16.7). In this example about 17,500 bases were masked (~35%), the majority of which were interspersed repeats. Pairwise alignments are provided between canonical (reference) repeat sequences and the query; examples are shown for an Alu repeat (Fig. 16.8) and a DNA transposon (Fig. 16.9). RepeatMasker also returns the input sequence in the FASTA format, with the repetitive residues masked with the letter N or X. This version of the sequence is especially useful for subsequent database searches.

RepBase Update has been developed since 1990 by Jerzy Jurka and colleagues. RepeatMasker was written by Arian Smit and Phil Green. It is available at ▶ http://www.repeatmasker.org/. The Censor Server at the Genetic Information Research Institute (GIRI) is available at ▶ http://www.girinst.org/censor/index.php.

The 50,000 bases of genomic DNA we are using is available as web document 16.1 at ▶ http://www.bioinfbook.org/chapter16.

Recall that a BLAST search uses the SEG and/or DUST programs to define and mask repetitive DNA sequences and also to detect and mask low-complexity protein sequences (Chapter 4).

2. Processed Pseudogenes.

These are genes that are not actively transcribed or translated (Harrison and Gerstein, 2002; Echols et al., 2002). They

(a)

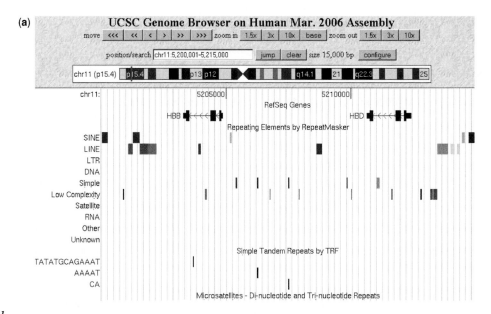

(b)

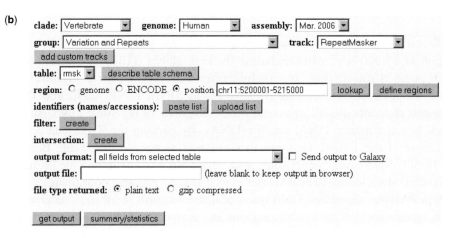

FIGURE 16.6. *Interspersed and other repetitive DNA elements are visualized using the UCSC Genome Browser. (a) A region of 15,000 bases in the beta globin region of chromosome 11 is shown. The RepeatMasker track is set to "full," displaying the location of several repetitive DNA elements such as SINE, LINE, LTR, and DNA transposons. (b) A link from the Genome Browser to the Table Browser allows you to access this (or other) information as a tabular output.*

	TABLE 16-7	Web Servers That Provide Access to Software for Identifying Repetitive Elements in Genomic DNA

Program	Description	URL
RepeatFinder	A computational system for analysis of repetitive structure of genomic sequences	► http://www.cbcb.umd.edu/software/RepeatFinder/
RepeatMasker	University of Washington Genome Center	► http://www.repeatmasker.org/
RepeatMasker	NCKU Bioinformatics Center (Taiwan)	► http://www.binfo.ncku.edu.tw/RM/RepeatMasker.php
RepeatMasker	For zebrafish; at the Wellcome Trust Sanger Instiutte	► http://www.sanger.ac.uk/Projects/D_rerio/fishmask.shtml
Censor Server	Genetic Information Research Institute	► http://www.girinst.org/

```
total length:        50000 bp  (50000 bp excl N/X-runs)
GC level:            38.87 %
bases masked:        17595 bp ( 35.19 %)
===================================================
                 number of    length    percentage
                 elements*    occupied   of sequence
---------------------------------------------------
SINEs:               9         2169 bp     4.34 %
      ALUs           7         1993 bp     3.99 %
      MIRs           2          176 bp     0.35 %

LINEs:              16        12381 bp    24.76 %
      LINE1         11        11160 bp    22.32 %
      LINE2          5         1221 bp     2.44 %
      L3/CR1         0            0 bp     0.00 %

LTR elements:        5         1591 bp     3.18 %
      MaLRs          2          669 bp     1.34 %
      ERVL           1          548 bp     1.10 %
      ERV_classI     2          374 bp     0.75 %
      ERV_classII    0            0 bp     0.00 %

DNA elements:        1          248 bp     0.50 %
      MER1_type      1          248 bp     0.50 %
      MER2_type      0            0 bp     0.00 %

Unclassified:        0            0 bp     0.00 %

Total interspersed repeats:   16389 bp    32.78 %


Small RNA:           0            0 bp     0.00 %

Satellites:          0            0 bp     0.00 %
Simple repeats:     11          519 bp     1.04 %
Low complexity:     12          700 bp     1.40 %
```

FIGURE 16.7. RepeatMasker output. A summary of the identified repetitive elements is provided, consisting primarily of interspersed repeats.

```
ANNOTATION EVIDENCE:
 2002    7.72 9.19 0.00  UnnamedSequence   1241  1512   48324 + AluSp    SINE/Alu

UnnamedSequen  1241 GGCCAGGCCCGGTGGCTCACGCCTGTAATCCCAGCCCTTTGGGTGGCCGA 1290
                      ?  i   v                         v        v
AluSp#SINE/Al     1 RGCCGGGCGCGGTGGCTCACGCCTGTAATCCCAGCACTTTGGGAGGCCGA 50

UnnamedSequen  1291 GGCAGGCGGATCA-----------------------GCCTGACCAACAT 1316
                      i        -----------------------
AluSp#SINE/Al    51 GGCGGGCGGATCACCTGAGGTCGGGAGTTCGAGACCAGCCTGACCAACAT 100

UnnamedSequen  1317 GGAGAAACCCCGTCTCTACTAAAAATACAAAA-TTAGCTGGGCATGGTGT 1365
                                                     -      i  i    v
AluSp#SINE/Al   101 GGAGAAACCCCGTCTCTACTAAAAATACAAAAATTAGCCGGGCGTGGTGG 150

UnnamedSequen  1366 CGCATGCCTGTAATCCCAGCTACTCGGGAGGCTGAGGCAGGAGATTTGCT 1415
                       ii                                       v i
AluSp#SINE/Al   151 CGCGCGCCTGTAATCCCAGCTACTCGGGAGGCTGAGGCAGGAGAATCGCT 200

UnnamedSequen  1416 TGAACCCAGGAGGCGGAGGTTGCAGTGAGCCAAGATCTTGCCATTGCACT 1465
                          i          i         i    vi
AluSp#SINE/Al   201 TGAACCCGGGAGGCGGAGGTTGCGGTGAGCCGAGATCGCGCCATTGCACT 250

UnnamedSequen  1466 CCAGCCTGGACAACAAGAGCAAAACTCCATCTAAAAAAAAAAAAAAA 1512
                        i          i      i  v
AluSp#SINE/Al   251 CCAGCCTGGGCAACAAGAGCGAAACTCCGTCTCAAAAAAAAAAAAAA 297

Matrix = 14p37g.matrix
Transitions / transversions = 2.00 (14 / 7)
Gap_init rate = 0.01 (2 / 271), avg. gap size = 12.50 (25 / 2)
```

FIGURE 16.8. The genomic DNA in the beta globin locus includes an Alu SINE element that is identified by RepeatMasker and aligned to a canonical member of that family.

```
ANNOTATION EVIDENCE:
    630  21.49 11.69 2.42  UnnamedSequence 49753 50000         0 C Charlie4a DNA/MER1_type      238   508    0  65

 C UnnamedSequen    50000 TTTATCTGCATCACTACTTGGATCTTAAGGTAGCTGT-AGACCCAATCCT 49952
                          viv            v vv  v     i i-       iviv
   Charlie4a#DNA      238 TTTATGCTCATCACTACTTCGAAATTACGGTAGTTATTAGACCCGCCGCT 287

 C UnnamedSequen    49951 AGATCT--------AATGCTTTCATAAAGAAGCAAATATAATAAATACTA 49910
                          --------     v v        v     - ---
   Charlie4a#DNA      288 AGATCTTGTTATTTAATGCATTAATAAAGAAGCACATATA-T---TACTA 333

 C UnnamedSequen    49909 TACCACAAATGTAATGTTT-GATGTCT-GATAATGATATTTCAGTGTAAT 49862
                          i      v ii v  -i i i -    ivi       i i
   Charlie4a#DNA      334 TATCACAAATTTGGTTTTTTAATATTTTGATAACTGTATTTCAATATAAT 383

 C UnnamedSequen    49861 TAA---ACTTAGCACTCCTATGTATATTATTTGATGCAAT-AAAAACATA 49816
                          ii---v  v i v         v     v    v-         -
   Charlie4a#DNA      384 TGGTTTCCTTTGTAATCCTATGTATTTTATTTTATGCATTTAAAAACAT- 432

 C UnnamedSequen    49815 TTTTTTTAG--------CA----CTTA--CAGTCTGCCAAACTGGCCTGT 49780
                          v i v  ---------  ----  v-- v      vv  ii
   Charlie4a#DNA      433 TATTCTGAGAAGGGGTCCATAGGCTTCACCAGACTGCCAAAGGGGCCCAT 482

 C UnnamedSequen    49779 GACACAAAAAAAGTTTAGGAATTCCTG 49753
                          i      i  - i  ii
   Charlie4a#DNA      483 GGCACAAAAAAGGTT-AAGAACCCCTG 508
Matrix = 20p37g.matrix
Transitions / transversions = 1.00 (26 / 26)
Gap_init rate = 0.05 (13 / 247), avg. gap size = 2.69 (35 / 13)
```

FIGURE 16.9. *Example of a DNA transposon (transposable element) in the beta globin locus. As identified by RepeatMasker, this region matches the MER80 element called CHARLIE4A, a 508 base pair sequence having 16 base pair terminal inverted repeats.*

Mark Gerstein's laboratory offers a website on pseudogenes (▶ http://www.pseudogene.org/). This includes a browser and descriptions of pseudogenes in human, worm, fly, yeast, and plant.

Web document 16.2 shows a pairwise alignment between HBB and its pseudogene HBBP1 (▶ http://www.bioinfbook.org/chapter16).

represent genes that were once functional, but they are defined by their lack of protein product. They can be recognized because of the presence of a stop codon or frameshift that interrupts an open reading frame. There are two main classes of pseudogenes. Processed pseudogenes arise through retrotransposition events (i.e., random insertion events mediated by LINEs having reverse transcriptase activity) via an RNA intermediate. Nonprocessed pseudogenes are remnants of duplicated genes. We describe mechanisms for the origin of pseudogenes later in this chapter, and in Chapter 17 we discuss the duplication of entire yeast genomes followed by rapid, subsequent gene loss to generate pseudogenes.

The number of pseudogenes in the human genome is remarkably close to the number of predicted protein-coding genes. For example, chromosome 1 has 3141 protein-coding genes and 991 pseudogenes (Gregory et al., 2006); chromosome 2 has 1346 genes and 1239 pseudogenes (Hillier et al., 2005); while chromosome 7 has 1150 genes and 941 pseudogenes (Hillier et al., 2003); and the smallest autosome, chromosome 21, has 225 known and predicted genes and 59 pseudogenes (Hattori et al., 2000).

While pseudogenes are defined as nonfunctional, many recent studies have emphasized their possible functional roles (Balakirev and Ayala, 2003; Castillo-Davis, 2005; Pavlicek et al., 2006). These include gene expression, the regulation of gene function, and roles in recombination. Evolutionary studies suggest that some pseudogenes do not evolve at the neutral rate (compared for example to extinct repeat elements), consistent with some functional role. The ENCODE Project Consortium (2007) and Zheng et al. 2007 reported that there are 201 pseudogenes in the ENCODE regions (124 processed and 77 nonprocessed). Of these, at least 19% are transcribed.

You can activate the pseudogene track at the UCSC Genome Browser. This is shown for a segment of 100,000 base pairs within the ENCODE region for beta globin (Fig. 16.10a). The RefSeq track shows six genes (five globin genes and one other), while the UCSC genes track shows several additional gene models. (The UCSC Genes track includes predictions from RefSeq, GenBank, and UniProt, and is somewhat less conservative than the RefSeq annotations.) The Pseudogenes track is also displayed, showing three pseudogenes as a consensus from several independent prediction methods. As for any UCSC Genome Browser

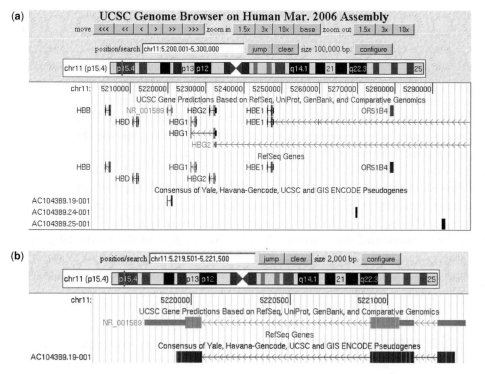

FIGURE 16.10. Viewing pseudogenes at the UCSC Genome Browser. (a) 100,000 base pair view of the ENCODE beta globin region (chr11:5,200,001-5,300,000). Note that a consensus annotation track for pseudogenes is activated as well as a RefSeq gene track. Three pseudogenes are evident, one of which matches transcript NR_001589. This is beta globin pseudogene 1 (HBBP1). (b) View of HBBP1 in a 2000 base pair window. The structure of the pseudogene is indicated by blocks, and its orientation (with transcription from right to left).

track, you can click on the title "pseudogenes" above the pull-down menu to access more details on the methodology as well as literature citations. The first of the three pseudogenes in this region, between HBD (delta globin) and HBG1, corresponds to accession NR_001589 and is annotated in Entrez Gene as beta hemoglobin pseudogene 1 (official symbol HBBP1). Viewing HBBP1 at higher magnification (a 2000 base pair view; Fig. 16.10b) shows more details of the structure of the pseudogene, and by clicking it one accesses details of a model of the gene and information on its expression and RNA folding properties.

3. Simple Sequence Repeats.
These microsatellites (typically from 1 to 6 bp in length) and minisatellites (typically from a dozen to 500 bp repeats) include short sequences such as $(A)_n$, $(CA)_n$, or $(CGG)_n$. An example of a CA repeat from our RepeatMasker analysis of human genomic DNA is shown in Fig. 16.11. Replication slippage is a mechanism by which simple sequence repeats may occur. Many functions have been ascribed to simple sequence repeats, from

Some authors define microsatellites as having a length of 1 to 6 bp, while others suggest 1 to 12 bp.

```
ANNOTATION EVIDENCE:
  330 12.28 0.00 0.00  UnnamedSequence 31470 31526   18474 + (CA)n    Simple_repeat    2    58   122 48

  UnnamedSequen   31470 ACACACGCTCTCACACACACACAAACACACGCGCGCACACACACACAC 31519
                             i v v             v    i i i
  (CA)n#Simple_       2 ACACACACACACACACACACACACACACACACACACACACACACACAC 51

  UnnamedSequen   31520 ACACACA 31526
  (CA)n#Simple_      52 ACACACA 58

Matrix = simple1.matrix
Transitions / transversions = 1.33 (4 / 3)
Gap_init rate = 0.00 (0 / 56), avg. gap size = 0.0 (0 / 0)
```

FIGURE 16.11. Simple sequence repeats. A region of 50,000 base pairs of genomic DNA from the beta globin locus includes simple sequence repeats such as the CA motif identified by RepeatMasker by comparison of the query to its database of known repeats.

To see specific examples of simple sequence repeats, go to Entrez Nucleotide and enter "microsatellite." There are over 150,000 entries from which to choose. For more information on SCA10 and its repeats, enter the query SCA10 at NCBI and see the Online Mendelian Inheritance in Man (OMIM) entry #603516.

The Tandem Repeats Finder is an online tool that allows you to search a sequence for tandem repeats of up to 2000 bp (▶ http://tandem.bu.edu/trf/trf.html) (Benson, 1999).

A duplication browser from Evan Eichler's group allows you to identify segmental duplications in the human genome. It is available at ▶ http://humanparalogy.gs.washington.edu/SDD/, and it uses a locally installed version of the UCSC Genome Browser. This site offers a database of over 8500 segmental duplications in the human genome.

Lipocalins localized to human chromsome 9q32-34 include α-1-microglobulin/bikunin (NM_001633), complement component 8, gamma polypeptide (NM_000606), lipocalin 1 (protein migrating faster than albumin, tear prealbumin; *LCN1*) (NM_002297), lipocalin 2 (oncogene 24p3)(NM_005564), odorant-binding protein (OBP) 2A (NM_014582) and 2B (NM_014581), orosomucoid 1 (NM_000607) and 2 (NM_000608), progestagen-associated endometrial protein (NM_002571), and prostaglandin D2 synthase (NM_000954). See Chan et al. (1994) and Dewald et al. (1996).

The function of *LCN1* is not known; it was identified by cloning a cDNA from a tear gland library. Rat and bovine OBPs selectively

influencing transcription factor binding to influencing morphological traits in dogs and yeast (reviewed in Kashi and King, 2006).

Simple sequence repeats of particular length and composition occur preferentially in different species. For example, $(AT)_n$ is especially common in *A. thaliana*, and $(CT/GA)_n$ occurs preferentially in *C. elegans* (Schlötterer and Harr, 2000). In *Drosophila virilis*, the density and length of microsatellites are considerably greater than in *D. melanogaster* or *H. sapiens* (Schlötterer and Harr, 2000). In humans, simple sequence repeats are of particular interest because they are highly polymorphic between individuals and thus serve as useful genetic markers. Also, the expansion of triplet repeats such as CAG is associated with over a dozen diseases, including Huntington disease (Cummings and Zoghbi, 2000). We will discuss these issues in Chapter 20 (on human disease). A disease characterized by cerebellar ataxia and seizures (spinocerebellar ataxia type 10; SCA10) is caused by the expansion of the sequence ATTCT in intron 9 of the ataxin 10 gene on chromosome 22q13.31 (Matsuura et al., 2000). While there are 10 to 29 repeats in apparently normal individuals, those with SCA10 have from hundreds to as many as 4500 repeats.

4. Segmental Duplications.

Segmental duplications are often defined as two genomic regions sharing at least 90% nucleotide identity over a span of one kilobase, although they sometimes consist of blocks of 200 or 300 kilobases (kb) in length (Bailey et al., 2001). These duplications occur both within and between chromosomes (intra- and interchromosomally). The euchromatic portion of the human genome consists of about 5.3% duplicated regions (She et al., 2004). This includes about 150 megabases. Later in this chapter we will discuss mechanisms by which segmental duplications (also called low copy repeats) may cause genes to become deleted, duplicated, or inverted. A practical consideration is that after whole genome shotgun sequencing, the assembly of segmentally duplicated regions (especially those >15 kilobases in length and sharing >97% sequence identity) is problematic (She et al., 2004). As a consequence, assemblies based on whole genome shotgun assembly may underestimate the extent of duplications (including duplicated genes), underestimate the length of euchromatin, and underrepresent duplication-rich regions including pericentromeric and subtelomeric areas.

As an example of a segmental duplication, we will consider a cluster of lipocalin genes on human chromosome 9. The lipocalins of all species have been divided into 14 monophyletic clades (Gutierrez et al., 2000). In humans, the lipocalins include at least 10 genes localized to chromosome 9q32-34. Figure 16.12 presents a schematic view of the genomic DNA, including the tear lipocalin (*LCN1*) and odorant-binding protein genes (adapted from Lacazette et al., 2000). Based on their analysis of this genomic region, Lacazette et al. (2000) proposed a model to account for the lipocalin genes and pseudogenes observed today in a portion of chromosome 9q34 (Fig. 16.12, bottom). A hypothetical ancestral lipocalin gene had seven exons and six introns (Fig. 16.12, top), a gene structure typical of mammalian lipocalins (Salier, 2000). This gene duplicated by tandem duplication (Fig. 16.12, step 1), after which the two ancestral genes differentiated to assume distinct functions (step 2). This locus then duplicated twice, generating *LCN1* and *OBPII* paralogs (step 3). However, only two *OBPII* genes are present in this locus today (Fig. 16.12, bottom), and the *LCN1* gene is accompanied by two pseudogenes (*LCN1b* and *LCN1c*). Thus, partial duplications may have occurred (Fig. 16.12,

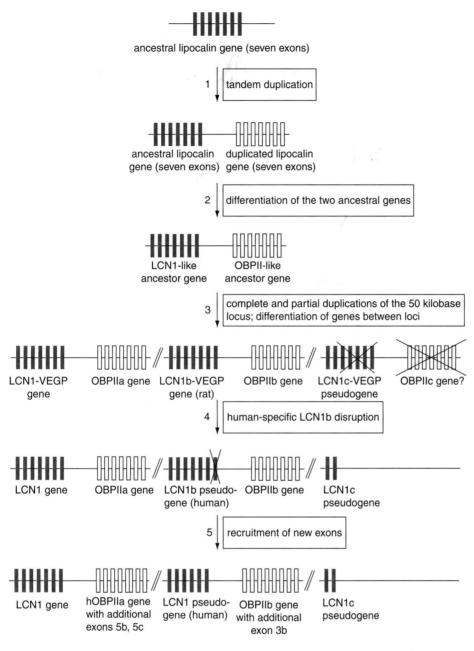

FIGURE 16.12. *Genes evolve by successive tandem duplications. Lacazette et al. (2000) proposed this model to explain how a hypothetical lipocalin gene (top of figure) could have evolved to account for the extant pattern of genes and pseudogenes determined by sequence analysis of this locus. First, an ancestral lipocalin having seven exons duplicated (step 1) and functionally diverged (step 2). This region, containing two genes, duplicated twice (step 3) after which one gene was deleted (the hypothetical OBPIIc gene) and portions of an LCN1 gene were deleted (step 4). Finally, several new exons were recruited (step 5). Used with permission.*

step 3) followed by disruption of the *LCN1b* gene in human (but not mouse) (step 4). Finally, the presence of new exons in human *OBPIIa* and *OBPIIb* suggests a selective duplication of individual exons (step 5).

We can view segmental duplications using the UCSC Genome Browser for both the alpha globin locus (Fig. 16.13a) and the beta globin locus (Fig. 16.13b). For the alpha globin locus on chromosome 16, the *HBZ* gene (zeta globin) is tandemly duplicated to generate a pseudogene less than 10,000 base pairs apart. At the beta globin locus, the immediately adjacent *HBG1* and *HBG2* genes represent a segmental duplication. By clicking on the segmental duplication block on the Genome Browser output, you can access the exact genomic coordinates of the duplicated blocks as well as a global pairwise alignment of the two.

bind odorants of many diverse chemical classes (e.g., terpenes, aldehydes, esters, and musks) (Pevsner et al., 1990; Pelosi, 1996). Thus, it is assumed that human OBP gene products also transport hydrophobic ligands.

The Mouse Genome Sequencing Consortium (2002) described a group of eight lipocalin genes on the mouse X chromosome that are absent from primates. These may have been generated by local gene duplication.

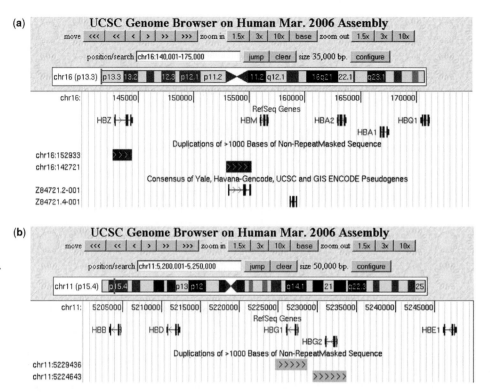

FIGURE 16.13. Segmental duplications visualized at the UCSC Genome Browser. (a) The alpha globin locus. In the Variation and Repeats category of annotation tracks, the Segmental Duplication track was set to full for genomic coordinates 140,001–175,000 on chromosome 16. Note that there are two segmental duplications of size >1000 bases for non-RepeatMasked sequence. One corresponds to the HBZ gene and the other matches a pseudogene. (b) The beta globin locus. The adjacent HBG1 and HBG2 genes are segmentally duplicated.

Web document 16.3 shows a global pairwise alignment between the two segmentally duplicated blocks at the beta globin locus. See ▶ http://www.bioinfbook.org/chapter16.

Telomeric repeats are synthesized by telomerase, a ribonucleoprotein that has specialized reverse transcriptase activity.

5. Blocks of Tandemly Repeated Sequences Such as Are Found at Telomeres, Centromeres, and Ribosomal Gene Clusters.

Several telomere repeat sequences are listed in Table 16.8. In human telomeres, the short sequence TTAGGG is repeated thousands of times. Try a blastn search using TTAGGG TTAGGG TTAGGG as a query, restricting the output to human, and remove the filter for low complexity. The result is several thousand BLAST hits, most from telomeric sequences such as that shown in Fig. 16.14.

The centromere is a constricted site of a chromosome that serves as an attachment point for spindle microtubules, allowing chromosomal segregation during mitotic and meiotic cell divisions (Choo, 2001). All eukaryotic chromosomes have a functional centromere, although the primary nucleotide sequence is not well conserved between species. In humans, this DNA consists largely of a 171 bp repeat of α-satellite DNA extending for 1 to 4 Mb. Almost all eukaryotic centromeres are able to bind a histone H3-related protein (called CENP-A in vertebrates). This protein–DNA complex forms a building block of centromeric chromatin that is essential for the function of the kinetochore, the site of attachment of the spindle fiber.

The GenBank accession number for a human α-satellite consensus sequence is X07685. An alignment of this sequence (171 base pairs) with a typical bacterial artificial chromosome (BAC) clone from a pericentromeric region dramatically shows how often the satellite sequence is repeated (Fig. 16.15). A blastn search of the nonredundant database, using this as a query and turning off filtering, results in over 30,000 database hits (December 2007). If you exclude human entries from the output of your search (with the Entrez command "satellite NOT human[organism]"), you will find that the human α-satellite sequence matches other primates.

TABLE 16-8 Telomeric Repeat Sequences from Several Eukaryotic Organisms

Organism	Telomeric Repeat	Reference
Arabidopsis thaliana, other plants	TTTAGGG	McKnight et al., 1997
Ascaris suum (nematode)	TTAGGC	Jentsch et al., 2002
Euplotes aediculatus, Euplotes crassus, Oxytricha nova (ciliates)	TTTTGGGG	Jarstfer and Cech, 2002; Shippen-Lentz and Blackburn, 1989; Melek et al., 1994
Giardia duodenalis, Giardia lamblia	TAGGG	Upcroft et al., 1997; Hou et al., 1995
Guillardia theta (cryptomonad nucleomorph)	$[AG]_7AAG_6A$	Douglas et al., 2001
Homo sapiens, other vertebrates	TTAGGG	Nanda et al., 2002
Hymenoptera, Formicidae (ants)	TTAGG	Lorite et al., 2002
Paramecium, Tetrahymena	TTGGGG, TTTGGG	McCormick-Graham and Romero, 1996
Plasmodium falciparum	AACCCTA	Gardner et al., 2002
Plasmodium yoelii yoelii	AACCCTG	Carlton et al., 2002

However, the human sequence has only very little conservation to nonprimate sequences, with nonsignificant expect values.

We described expect values in Chapter 4.

Satellite DNA is a feature of every known eukaryotic centromere, with only two documented exceptions. In the yeast *S. cerevisiae*, the entire centromere sequence extends only several hundred base pairs. A second exception is the neocentromere, an ectopic centromere that assembles a functional kinetochore, is stable in mitosis, but lacks α-satellite DNA (Amor and Choo, 2002). About 60 human neocentromeres have been described, many involving trisomy or tetrasomy (extra chromosomal copies). As part of the analysis of the genome of the rhesus macaque *Macaca mulatta*, Ventura et al. (2007) described evolutionarily new centromeres that appeared while the conventional centromere was inactivated. They reported that in the 25 million years since macaque and human lineages diverged, 14 evolutionarily new centromeres have emerged and become fixed in one or the other species.

We discuss a possible mechanism for neocentromere formation in Fig. 16.25 below.

```
>gnl|ti|1745943411 name:1094791574190 mate:1745266475
AGGGTGGCGAATACGCGACTACCTACCTACCCTAACCCTAACCCTAACCCTAACCCTAACCCTA
ACCCTAACCCTAACCCTAACCCTAACCCTAACCCTAACCCTAACCCTAACCCTAACCCTAACCC
TAACCCTAACCCTAACCCTAACCCTAACCCTAACCCTAACCCTAACCCTAACCCTAACCCTAAC
CCTAACCCTAACCCTAACCCTAACCCTAACCCTAACCCTAACCCTAACCCTAACCCTAACCCTA
ACCCTAACCCTAACCCTAACCCTAACCCTAACCCTAACCCTAACCCTAACCCTAACCCTAACCC
TAACCCTAACCCTAACCCTAACCCTAACCCTAACCCTAACCCGAACCCTAACCCTAAC
CCTAACCCTAACCCTAACCCTAACCCTAACCCGAACGCTAACCCTGACCCTAGCCCTAACCCTA
ACCCTAACCCTAACCCTAACCCTAAGCCTAACCCTAACCCTAACCGTAAGCCTAACCCTAAGCCTAACCC
TAACCCTAACCCTAACCCTAGCCGTAACCCTAACCGGAACCCTAACCCTAGCGCTAGCCGTAGCCCTAGC
CGTAACCCTAACGCTGACCCTAGCGGTAGCCCGAACCCTAACCCTAACCCTAACCCTAACCCTGACCCTA
ACCCTGACCCTAAGCGTAACCCTAACCCTGACCCTAAGCGTAACCCTAACGCTAACCCTAACGGTAACCC
TAACCCTAAGCCTAACCGTAACGCTAACCCTGACCCTAACTCTGAGCCTGAGCCGGCCCGTAATCCTAAC
GGGACGGTACGCTAACGCAGAGTGTGCGGTAGCCCTGCGGCTGACCGTAACCCGAGGCCTAACCGAGCCC
GACCGTAGGCCGAGGCCTAGCCTGAGCGGTGACCTGAGGCATAGCCCTAGGGTATCGCGTAACGTGAGCC
TAACC
```

FIGURE 16.14. A blastn search of the Trace Archives was performed using TTAGGGTTAGGGTTAGGG as query, setting the database to Homo sapiens *WGS, with a word size of 15 and removing the filtering. There were over 4000 matches, including the one shown here. Note the many TTAGGG repeats; this clone matches the query on the reverse strand orientation so the observed repeat pattern is CCCTAA. Thousands of such repeats occupy the telomeres.*

FIGURE 16.15. *The repetitive nature of* α*-satellite DNA. A consensus sequence for human* α*-satellite DNA (X07685) was compared to a BAC clone (AC125634) assigned to a pericentromeric region of chromosome 9q. Pairwise BLAST at NCBI was used, and the dotplot is shown. Note that a consecutive 60 kilobases of the BAC clone (y axis) matches the satellite consensus sequence repeatedly.*

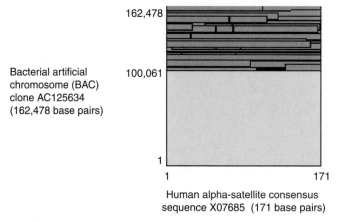

Bacterial artificial chromosome (BAC) clone AC125634 (162,478 base pairs)

Human alpha-satellite consensus sequence X07685 (171 base pairs)

GENE CONTENT OF EUKARYOTIC CHROMOSOMES

Definition of Gene

We have begun our analysis of eukaryotic genomes by considering noncoding and repetitive DNA. The coding portions of a genome are of particular interest, as they largely determine the phenotype of all organisms. Two of the biggest challenges in understanding any eukaryotic genome are defining what a gene is and identifying genes within genomic DNA. We will first define the variety of genes and then give the criteria for identifying them:

- Protein-coding genes form a major category of genes. Several criteria are applied to the assignment of a DNA sequence as a protein-coding gene. The principal requirement is that there must be an open reading frame (ORF) of at least some minimum length such as 90 bp (corresponding to 30 codons encoding amino acids, or a 3 kDa protein). Frith et al. (2006) identified large numbers of short proteins (less than 100 amino acids). Of 3701 proteins they identified, only 232 matched a mouse International Protein Index or Swiss-Prot database.

- Pseudogenes do not encode functional gene products, although as discussed above some important exceptions have been reported.

- Many kinds of noncoding genes do not encode protein, but instead encode functional RNA molecules (Eddy, 2001, 2002). These include transfer RNA (tRNA) genes. These translate information from the triplet codons in mRNA to amino acids. tRNAscan-SE identifies 99% to 100% of RNA genes in genomic DNA sequence with an error rate of one false positive per 15 Gb (Lowe and Eddy, 1997). We showed an example of the tRNAscan-SE server in Fig. 8.5.

- We discussed a variety of other noncoding genes in Chapter 8. These include ribosomal RNA (rRNA) genes that function in translation; small nucleolar RNAs (snoRNAs) that function in the nucleolus; small nuclear RNAs that function in spliceosomes to remove introns from primary RNA transcripts; and microRNAs (miRNAs) of about 21 to 25 nucleotides in length that are widely conserved among species and may serve as antisense regulators of other RNAs (Ambros, 2001; Ruvkun, 2001).

In annotating genomic DNA, an emphasis is often placed on describing the protein-coding genes. However, it is now clear that noncoding genes encoding various types of RNA products have diverse and important functions. Furthermore, it is not as straightforward to identify noncoding RNAs (Eddy, 2002). Their full size might be extremely small, as in the case of miRNAs. There is no ORF to help define the boundaries of noncoding genes. Database searches may be less sensitive than is possible for protein-coding genes, because the scoring matrices for amino acids are more sensitive and specific. We discussed databases of noncoding RNAs such as Rfam (Griffiths-Jones et al., 2003) in Chapter 8.

Given the insights of the ENCODE project (ENCODE Project Consortium, 2007) as well as the analysis of completed genome sequences, Gerstein et al. (2007) reviewed the historical definitions of a gene. Classically, a gene has been defined as a unit of hereditary information localized to a particular chromosome position and encoding one protein. More recently, we have become aware of alternative splicing to produce multiple transcripts from one gene locus, we have identified large numbers of noncoding RNAs, and we have observed pervasive transcription throughout the genome (including transcriptionally active regions or transfrags that have not been annotated as genes). The ENCODE project revealed many previously unknown transcription start sites. Gerstein et al. (2007, p. 677) thus proposed a new definition. They noted that a gene is a nucleotide sequence that encodes functional products, whether RNA or protein (or both), and they considered that one gene structure may have multiple functional products. They then proposed that "the gene is a union of genomic sequences encoding a coherent set of potentially overlapping functional products." In the simplest scenario a gene is a DNA sequence that codes for an RNA and/or protein product. The genomic sequence constitutes the genotype that is related to the phenotype of a cell or ultimately of an organism. The ENCODE project has helped to redefine the complexity of the genotype.

Finding Genes in Eukaryotic Genomes

Finding protein-coding genes in eukaryotic genomes is a far more complex problem than for prokaryotes (Burge and Karlin, 1998; Mural, 1999; Claverie, 1997). While bacterial genes typically correspond to long ORFs, most eukaryotic genes have exons and introns. The structure of a typical eukaryotic gene that is transcribed by RNA polymerase II is summarized in Fig. 16.16. Distal upstream and/or downstream enhancers and silencers as well as proximal (more neighboring) promoter elements regulate transcription. A CCAAT box and a TATA box are promoter elements, with the TATA box typically located 20 to 30 base pairs upstream of the transcription start site and the CCAAT box further to the 5′ side. There are several kinds of exons (Fig. 16.16):

RNA polymerase I synthesizes most ribosomal RNAs; RNA polymerase II synthesizes messenger RNAs and small nuclear RNAs (snRNAs); and RNA polymerase III synthesizes 5S rRNA and transfer RNAs (Chapter 8).

1. Noncoding exons correspond to the untranslated 5′ or 3′ region of DNA.
2. Initial coding exons include the start methionine and continue to the first 5′ splice junction.
3. Internal exons begin with a 3′ splice site and continue to a 5′ splice site.
4. Terminal exons proceed from a 3′ splice site to a termination codon.
5. Single-exon genes are intronless, beginning with a start codon and ending with a stop codon (Table 16.6).

FIGURE 16.16. (a) Eukaryotic gene prediction algorithms differentiate several kinds of exons, including those in noncoding regions; initial coding exons that include a start codon; internal exons; and terminal exons that include a stop codon. These exons are built into a model for a predicted gene. (b) In some cases, genes have a single exon and are intronless. The border of exons and introns typically has a GT/AG boundary, but the structure of genes is still difficult to predict ab initio.

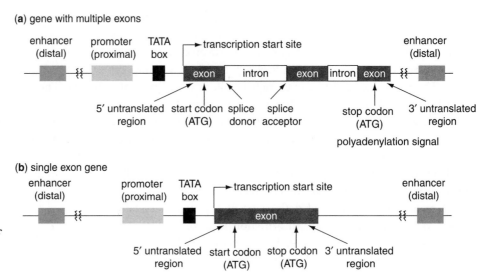

(a) gene with multiple exons

(b) single exon gene

About one-third of all human genes are alternatively spliced. If ESTs are available corresponding to alternatively spliced isoforms, these sequences can be mapped to exons.

Introns have been categorized into four groups based on their splicing mechanism: (1) autocatalytic group I, found in protists, bacteria, and bacteriophages; (2) group II, found in fungal and land plant mitochondria and in prokaryotes; (3) spliceosomal introns, found in nuclear pre-mRNA genes; and (4) tRNA introns, found in eukaryotic nuclei and in archaea (Haugen et al., 2005; Roy and Gilbert, 2006). Eukaryotic spliceosomal introns vary by two orders of magnitude in their density, from <0.1 to 5.5 introns per gene in fungi to 2.6–9.3 introns per gene in the metazoans (Roy, 2006). Fascinating questions include the mechanisms by which introns are gained and lost, the selective pressures on intron size, and their evolutionary history (Jeffares et al., 2006; Pozzoli et al., 2007). While introns were thought to have arisen late in eukaryotic evolution, a single intron was discovered in the genome of the primitive protozoan *Giardia lamblia* (see Chapter 18), as well as several introns in its close relative *Carpediemonas membranifera* (Nixon et al., 2002; Simpson et al., 2002).

In addition to the issue of introns, eukaryotic genes also occupy a far smaller proportion of the genome than do prokaryotic genes. Eukaryotic protein-coding genes occupy just 25% of the nematode and insect genomes and less than 3% of the human and mouse genomes. In the human genome, exons span 1.5% of the genome. Chromosome 13 has the lowest gene density (6.5 genes per megabase, with a region of 38 megabases having just 3.1 genes/Mb) (Dunham et al., 2004). Chromosome 19 has the highest gene density, with 26 loci per megabase (Grimwood et al., 2004).

Algorithms for finding protein-coding genes in eukaryotes can be divided into three categories: homology based (also called extrinsic), algorithm based (also called intrinsic), and comparative (Stein, 2001). These approaches are outlined in Fig. 13.17. Homology-based approaches typically involve the alignment of expressed genes (ESTs from cDNA libraries; see Chapters 2 and 8) with genomic DNA. In these cases, the ESTs can help to define the exon/intron structure in genomic DNA. Thus, homology-based approaches are generally very successful. An additional form of homology-based gene identification is to compare genomic DNA of two related organisms (Morgenstern et al., 2002; Novichkov et al., 2001). By comparing human DNA to pufferfish (*F. rubripes*) DNA, it was possible to discover nearly 1000 putative human genes (Hedges and Kumar, 2002; Aparicio et al., 2002).

While the use of EST data is extremely helpful in annotating eukaryotic genes, there are notable limitations to this approach.

- The quality of EST sequence is sometimes low, as clones are often sequenced on only one strand and sequencing errors are common.
- Highly expressed genes are often disproportionately represented, although some cDNA libraries are normalized (Chapter 8).
- ESTs provide no information regarding the genomic location.

Intrinsic programs are also widely used to annotate genomic DNA. A large fraction of predicted genes do not have identifiable orthologs, nor are EST sequences available. It is thus essential to identify protein-coding genes using ab initio (intrinsic) approaches. We discussed the GLIMMER program for prokaryotes in Chapter 15.

Many web-based eukaryotic gene prediction programs are available (Table 16.9). These include GENSCAN (Burge and Karlin, 1997) and GRAILEXP, and several studies have compared their accuracy (Rogic et al., 2001; Makarov et al., 2002). These programs typically produce models of gene structures (exons, introns, alternative splicing) and identify other features such as CpG islands (regions of a higher than expected occurrence of CpG dinucleotides over a particular distance such as 300 base pairs). Often these programs include searches with RepeatMasker to identify classes of repetitive DNA as well as BLAST or BLAST-like searches to identify known genes, proteins, and expressed sequence tags that help to model the gene structure.

The difficulty of finding protein-coding genes in genomic DNA is illustrated by the efforts to annotate a typical eukaryotic genome: the *indica* and *japonica* subspecies of the rice genome. Yu et al. (2002) obtained 75,659 gene predictions when they submitted their assembled draft version of the rice genome (*indica*) to an FGeneSH web server (see Table 16.9). Only 53,398 of these predictions were complete (having both initial and terminal exons): about 7500 had only an initial exon, 11,000 had only a terminal exon, and 3400 predicted genes had neither. Additionally, they reported that exon–intron boundaries were often not precisely defined. However, when the finished sequence was obtained rather than the draft sequence, the estimate of gene content improved dramatically. Sasaki et al. (2002) obtained the finished sequence of rice chromosome 1 (subspecies *japonica*) and predicted 6756 genes on this chromosome. In contrast, the draft version of this genome predicted just 4467 genes. Sasaki et al. (2002) suggest that the presence of several thousand gaps in the draft sequence precluded the ability to accurately predict complete genes.

As another example of an approach to annotating genes, the *Drosophila* 12 Genomes Consortium (2007) reported the sequencing of ten *Drosophila* species yielding a total of 12 *Drosophila*-related genomes. The genomes were sequenced to varying depths, from over 10X coverage to just 2.9X coverage. They used four different de novo gene prediction algorithms, three homology-based predictors that relied on the well-annotated *Drosophila melanogaster* genome sequence, one predictor (called Gnomon) that combined de novo and homology-based evidence, and a gene model combiner (called GLEAN) that reconciled all the predicted genes into

The Oak Ridge National Laboratory (ORNL) Genome Analysis Pipeline is a web-based tool for the annotation of genomic DNA from several species. It includes the GrailEXP program. This pipeline is available at ▶ http://compbio.ornl.gov/tools/pipeline/.

TABLE 16-9	Algorithms for Finding Genes in Eukaryotic DNA	
Program	Description	URL
AAT	Analysis and Automation Tool; web-based server	▶ http://www.tigr.org/software/alignment.shtml
AUGUSTUS	University of Göttingen	▶ http://augustus.gobics.de/
FgeneSH	Ab initio gene finder	▶ http://www.softberry.com/berry.phtml
Gene Finder	For human, mouse, *Arabidopsis*, and fission yeast	▶ http://argon.cshl.org/genefinder/
Geneid	Roderic Guigó and colleagues	▶ http://www1.imim.es/geneid.html
GeneMark	Georgia Institute of Technology	▶ http://exon.gatech.edu/GeneMark/
Genie	Based on HMMs	▶ http://www.cse.ucsc.edu/~dkulp/cgi-bin/genie
GenLang	Syntactic pattern recognition system; uses computational linguistics to find genes	▶ http://www.cbil.upenn.edu/genlang/genlang_home.html
Genscan	Based on HMMs; rule based rather than homology based	▶ http://genes.mit.edu/GENSCAN.html
GlimmerM	From TIGR and the University of Maryland	▶ http://www.cbcb.umd.edu/software/glimmerm/index.shtml
GlimmerM web server	Trained for *Arabidopsis thaliana*, *Oryza sativa* (rice), *Plasmodium falciparum* (malaria parasite)	▶ http://www.tigr.org/software/glimmerm/
GRAILEXP	One of the most widely used algorithms	▶ http://compbio.ornl.gov/
MORGAN	A decision tree system for finding genes in vertebrate DNA	▶ http://www.tigr.org/~salzberg/morgan.html
WebGene	Consiglio Nazionale delle Ricerche, Milano	▶ http://www.itba.mi.cnr.it/webgene/
Xpound	A probabilistic model for detecting coding regions	▶ http://bioweb.pasteur.fr/seqanal/interfaces/xpound-simple.html

Abbreviation: HMM, hidden Markov model.

We have discussed other competitions for proteomics (Chapter 10) and protein structure (CASP, Chapter 11). The GENCODE Project website is ▶ http://genome.imim.es/gencode/, including a genome browser. The GENCODE team worked in collaboration with the Human And Vertebrate Analysis aNd Annotation (HAVANA) team at the Sanger Institute (▶ http://www.sanger.ac.uk/HGP/havana/).

a set of consensus models. Quality was assessed in part by measuring RNA transcript levels with microarrays (Chapter 9).

EGASP Competition and JIGSAW

The ENCODE Genome Annotation Assessment Project (EGASP) was a competition designed to objectively test the performance of a set of gene-finding software. The GENCODE consortium created a "gold standard" by rigorously mapping all the protein-coding genes with the ENCODE regions (Harrow et al., 2006). This was achieved by carefully applying a range of experimental techniques such as 5′ rapid amplification of complementary DNA ends (RACE) and the polymerase chain reaction with reverse transcription (RT-PCR); 434 coding loci were annotated as part of the GENCODE reference set. Only 40% of the GENCODE annotations were within the RefSeq and Ensembl annotation sets, reflecting the discovery of a large number of alternatively spliced isoforms with unique exons.

Given this deep level of annotation of ENCODE regions based on experimental evidence, the EGASP competition consisted of groups that predicted gene structures with the raw sequence data but without prior access to the annotation results (Harrow et al., 2006; Guigo et al., 2006). This allowed false positive and false negative error rates to be assessed. Sensitivity was defined as the proportion of annotated features (nucleotides, exons, or genes) that are predicted correctly, while specificity was defined as the proportion of predicted features that are annotated. The most successful gene prediction methods achieved a maximum sensitivity of 70% at the gene level (for finding at least one correct exon/intron structure), 45% at the transcript level (for correctly predicting all alternatively spliced variants), and 90% at the coding nucleotide level. Only about 3% of the many computationally predicted exons could be experimentally validated, suggesting that overprediction remains a fundamental problem.

We can view the results of the EGASP competition at the UCSC Genome Browser website (Fig. 16.17). There is generally good agreement on the identification of exons, although there is considerable variation in the prediction of complete gene models.

One of the best-performing programs in the GENCODE competition was JIGSAW from Jonathan Allen, Steven Salzberg and colleagues (Allen and Salzberg, 2005; Allen et al., 2004, 2006). JIGSAW is an integrative program that combines different sources of evidence into a model of a gene structure. It incorporates models from other gene prediction programs (typically three or more) as well as sequence alignment data and intron splice site prediction programs. It allows separate signal types, including start codons, stop codons, and splice junctions (acceptor and donor sites at the 5′ and 3′ ends of introns). In one mode JIGSAW uses a linear combiner to assign a weight to each evidence source, and it maximizes the sum of the evidence (Fig. 16.18) (Allen et al., 2004). This can be accomplished without

JIGSAW can be downloaded from
► http://cbcb.umd.edu/software/jigsaw/.

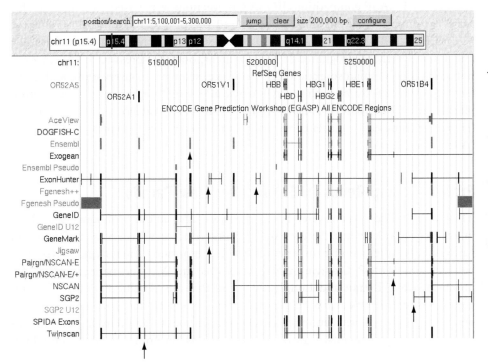

FIGURE 16.17. *In the EGASP competition, protein coding genes were experimentally validated in ENCODE regions. Various gene-finding software tools were used to independently predict gene structures. The beta globin ENCODE region consists of one million base pairs on human chromosome 11p. A portion of 200,00 base pairs is shown (x axis) with tracks for RefSeq genes and EGASP predictions from 19 software programs (y axis tracks). Many of the programs predict exons and/or entire gene structures that are not experimentally confirmed; examples are shown (arrows). Thus overfitting remains a problem for prediction software. An even greater problem is that a complete, correct gene model is generated for fewer than half of all genes.*

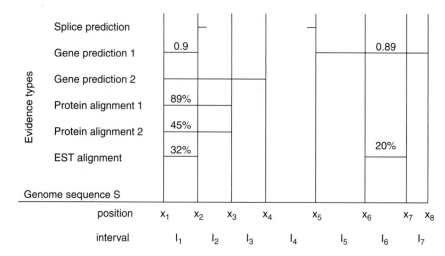

FIGURE 16.18. *Combining sources of information in gene-finding software. The JIGSAW program accepts different sources of evidence to generate a gene model. These sources may include output from a splice prediction program, varying gene prediction programs (two are indicated, the first of which includes an exon prediction confidence score and the second of which does not provide a confidence score), alignments to two different proteins based on BLAST, and alignments to expressed sequence tags sharing varying percent identity between the genomic DNA sequence S and the human (or nonhuman) EST. The Combiner algorithm used by JIGSAW divides the genomic sequence into intervals (here I_1 to I_7) with boundaries $x_1 \ldots x_8$. These are used to compile a model of a gene structure including start and stop sites, exons and introns. From Allen et al. (2006). Used with permission.*

using a training set. In another mode JIGSAW uses a statistical combiner which requires a training set (with examples of known genes) that are used to evaluate the accuracy of various combinations of evidence. Once a model is trained it is applied to a data set.

For the EGASP competition, JIGSAW predictions were based on training with a variety of inputs including gene finders used by the UCSC annotation database (GENEID, SGP, TWINSCAN, and GENSCAN) as well as the GeneZilla and GlimmerHMM programs. It further incorporated expression evidence from human and nonhuman sources, GC percentage, sequence conservation, and a variety of genomic features such as TATA box and signal peptide sequences, intron phase, and CpG islands. Surprisingly, adding some categories of information (such as training on untranslated regions) diminished rather than improved accuracy (Allen et al., 2006).

Protein-Coding Genes in Eukaryotes: New Paradox

The *C* value paradox is answered based on the variable amounts of noncoding DNA in a variety of eukaryotes. A new paradox is introduced: Why are the proteomes of various eukaryotes similar in size, given the enormous phenotypic differences between eukaryotes? Claverie (2001) calls this the *N* value paradox (*N* is for number), while Betrán and Long (2002) call this the *G* value paradox (*G* is for genes). As we survey eukaryotic genomes in Chapters 18 and 19, we will see that organisms such as worms and flies appear to have about 13,000 to 20,000 protein-coding genes, while plants, fish, mice, and humans have only slightly more (about

20,000 to 40,000 genes) (Harrison et al., 2002). Why do organisms such as humans, having so much greater biological complexity than insects and nematodes, have not even twice as many genes? The genes of higher eukaryotes employ more complex forms of gene regulation, such as alternative splicing. Also the architecture of individual genes tends to be more complex, for example with more domains present in an average human protein relative to insect.

REGULATORY REGIONS OF EUKARYOTIC CHROMOSOMES

Transcription Factor Databases and Other Genomic DNA Databases

In addition to predicting the presence of genes, it is also important to predict the presence of genomic DNA features such as promoters, enhancers, silencers, insulators, and locus control regions (Maston et al., 2006). Such regulatory elements are sometimes called cis-regulatory modules (CRMs). Identifying them is difficult compared to finding protein-coding genes because the DNA sequences of interest may be very short (e.g., fewer than a dozen base pairs for transcription factor binding sites), and conserved between species to variable extents. Algorithms are available for identifying regulatory elements, as well as databases storing compilations of genomic features (Table 16.10).

CpG islands represent an example of a regulatory element. The dinucleotide cytosine followed by guanosine (CpG) is approximately fivefold underrepresented in many genomes, in part because the cytosine residue can be exchanged for thymidine by spontaneous deamination. Cytosine residues on CpG dinucleotides are often methylated. This in turn leads to the recruitment of protein complexes that include histone deacetylases capable of removing acetyl groups of histones and thus inhibiting active transcription. CpG islands are regions of high density of unmethylated CpG dinucleotides and are commonly found in upstream (5′) regulatory regions near the transcription start sites of constitutively active "housekeeping" genes. By one criterion, a CpG island is defined as having a GC content $\geq 50\%$, a length ≥ 200 base pairs, and a ratio of observed to expected number of CpG dinucleotides >0.6. Figure 16.19a shows five CpG islands in the human alpha globin locus, visualized using the UCSC Genome Browser, each in the vicinity of an alpha globin gene. The extraordinarily dense number of CpG dinucleotides is evident in one of these islands (Fig. 16.19b).

The UCSC Genome Browser offers access to dozens of additional resources related to transcriptional regulation in the "Expression and Regulation" category of annotation tracks (Fig. 16.20a). Some of these elements are shown for a small region (15,000 base pairs) of the beta globin locus (Fig. 16.20b). For example, the Open REGulatory ANNOtation database (ORegAnno) compiles regulatory elements from the literature and includes a validation process by expert curators (Griffith et al., 2008). Information in ORegAnno includes promoters, enhancers, transcription factor bindings sites, and regulatory polymorphisms. Eight ORegAnno features are included in the UCSC Genome Browser output in Fig. 16.20b. As another example, that figure illustrates the $7\times$ regulatory potential track based on regulatory potential scores computed from alignments of seven organisms (human, chimpanzee, rhesus macaque, mouse, rat, dog, and cow)

ORegAnno is available online at ▶ http://www.oreganno.org. Web document 16.4 lists definitions of several categories of regulatory elements within ORegAnno. Evolutionary and sequence pattern extraction through reduced representation (ESPERR) software and data sets are available at ▶ http://www.bx.psu.edu/projects/esperr.

TABLE 16-10 Software for Identifying Features of Promoter Regions in Genomic DNA

Program	Description	URL
Ancient conserved untranslated DNA sequences (ACUTS)	Analyzes genes from metazoan species	▶ http://pbil.univ-lyon1.fr/acuts/ACUTS.html
AliBaba2	Predicts binding sites of transcription factor binding sites in an unknown DNA sequence	▶ http://www.gene-regulation.com/pub/programs.html
Eukaryotic Promoter Database (EPD)	Annotated nonredundant collection of eukaryotic POL II promoters, for which the transcription start site has been determined experimentally	▶ http://www.epd.isb-sib.ch/
Open REGulatory ANNOtation database (ORegAnno)	Comprehensive, open access, community-based resource	▶ http://www.oreganno.org
PlantProm	Plant promoter database	▶ http://mendel.cs.rhul.ac.uk
Promoter 2.0 Prediction Server	Technical University of Denmark	▶ http://www.cbs.dtu.dk/services/promoter/
Regulatory Sequence Analysis Tools (RSAT)	Université Libre de Bruxelles	▶ http://rsat.ulb.ac.be/rsat/
TESS	Transcription Element Search System, University of Pennsylvania	▶ http://www.cbil.upenn.edu/cgi-bin/tess/tess
Transcriptional Regulatory Element Database (TRED)	Cold Spring Harbor Laboratory	▶ http://rulai.cshl.edu/cgi-bin/TRED/tred.cgi?process=home
TRANSFAC	Database of transcription factors, their genomic binding sites, and DNA-binding profiles	▶ http://www.gene-regulation.de/

Note: Additional resources are summarized at ▶ http://www.oreganno.org/oregano/OtherResources.jsp.

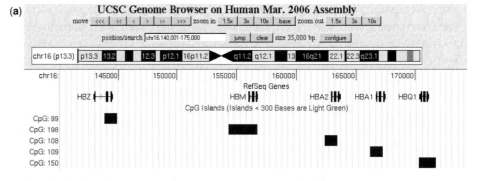

FIGURE 16.19. CpG islands are associated with the regulation of expression of many eukaryotic genes. (a) The alpha globin gene cluster on human chromosome 16 is shown (in a window of 35,000 base pairs from chr16:140,001-175,000 on the UCSC Genome Browser). Each of the five genes has an associated CpG island, defined as having a GC content of 50% or greater, a length greater than 200 base pairs, and a ratio >0.6 of observed to expected CpG dinucleotides. (b) By clicking on the HBA2 CpG island, one accesses its DNA sequence. CpG dinucleotides are highlighted in red.

(b) >hg18_cpgIslandExt_CpG: 108 range=chr16:162370-163447
CGTCCGGGTGCGCGCATTCCTCTCCGCCCCAGGATTGGGCGAAGCCCTCCGGCTCGCACTCGCTCGCCCGTG
TGTTCCCCGATCCCGCTGGAGTCGATGCGCGTCCAGCGCGTGCCAGGCCGGGGCGGGGGTGCGGGCTGACTT
TCTCCCTCGCTAGGGACGCTCCGGCGCCCGAAAGGAAAGGGTGGCGCTGCGCTCCGGGGTGCACGAGCCGAC
AGCGCCCGACCCCAACGGGCCGGCCCCGCCAGCGCCGCTACCGCCCTGCCCCCGGGCGAGCGGGATGGGCGG
GAGTGGAGTGGCGGGTGGAGGGTGGAGACGTCCTggccccgccccgcgtgcacccccaggggaggccgagc
ccgccgccCggcccCGcgcaggcccCGcccgggACTCCCCTGCGGTCCAGGCCGCGCCCCGGGCTCCGCGCC
AGCCAATGAgcgccgcccggccgggcgtgccccCGcgccccAAGCATAAACCCTGGCGCGCTCGCGGGCCGG
CACTCTTCTGGTCCCCACAGACTCAGAGAGAACCCACCATGGTGCTGTCTCCTGCCGACAAGACCAACGTCA
AGGCCGCCTGGGGTAAGGTCGGCGCGCACGCTGGCGAGTATGGTGCGGAGGCCCTGGAGAGGTGAGGCTCCC
TCCCCTGCTCCGACCCGGGCTCCTCGCCCGCCCGGACCCACAGGCCACCCTCAACCGTCCTGGCCCCGGACC
CAAACCCCACCCCTCACTCTGCTTCTCCCCGCAGGATGTTCCTGTCCTTCCCCACCACCAAGACCTACTTCC
CGCACTTCGACCTGAGCCACGGCTCTGCCCAGGTTAAGGGCCACGGCAAGAAGGTGGCCGACGCGCTGACCA
ACGCCGTGGCGCACGTGGACGACATGCCCAACGCGCTGTCCGCCCTGAGCGACCTGCACGCGCACAAGCTTC
GGGTGGACCCGGTCAACTTCAAGGTGAGCGGCGGGCCGGGAGCGATCTGGGTCGAGGGGCGAGATGGCGCCT
TCCTCTCAGGGCAGAGGATCACGCGGGTTGCGGGAGGTGTAGCGCAGGCGGCGGCTGCGGGCCTGGGCCG

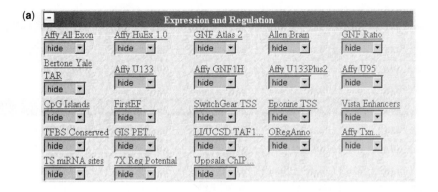

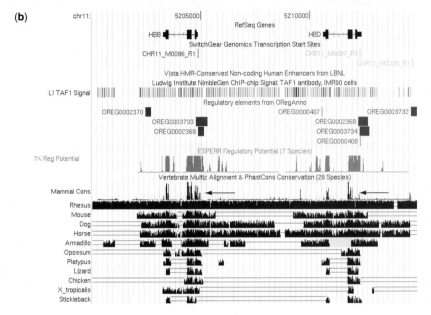

FIGURE 16.20. Regulatory elements in genomic DNA. (a) The UCSC Genome Browser (March 2006 assembly) includes two dozen annotation tracks in the "expression and regulation" category, many of which include analyses of transcriptional regulation. (b) The beta globin and delta globin gene loci are shown (15,000 bases at the location chr11:5,200,001-5,215,000) with some of these annotation tracks opened. This highlights regulatory elements surrounding these genes.

(King et al., 2005; Taylor et al., 2006). Scores are based on log ratios of transition probabilities from variable order Markov models, based on the use of a training set. Constrained (conserved) residues in a multiple sequence alignment may have regulatory potential if they are more similar to known regulatory elements than to ancestral repeats (which serve as a model for neutrally evolving DNA). King et al. evaluated regulatory regions of the beta globin locus, which includes 23 experimentally determined CRMs, all but three or four of which are conserved in rat and mouse, and of which just four are conserved in chicken. The regulatory potential method performed better (based on estimates of sensitivity and specificity) than other methods that rely exclusively on conservation of loci among species. Figure 16.20b also shows a conservation track that has partially overlapping results with the 7× regulatory potential track.

In addition to the standard UCSC Genome Browser options for the "Expression and Regulation" category, there are many additional tracks in the ENCODE regions (Fig. 16.21). Clicking on any of these headers provides access to track display features as well as the methodology and literature citations. These options include chromatin immunoprecipitation (ChIP) experiments, in which antisera directed against specific proteins (such as DNA-binding transcription factors) are used to immunoprecipitate those proteins with their target DNA. This DNA can be amplified by

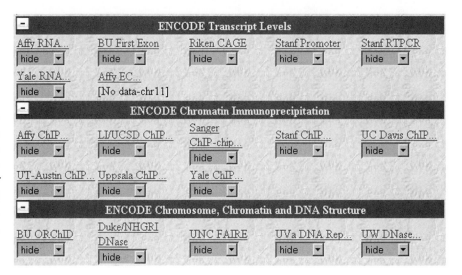

FIGURE 16.21. *Annotation tracks available at the UCSC Genome Browser site for functional studies of regulatory elements in the ENCODE project. A variety of tracks can be viewed and analyzed in the ENCODE categories of transcript levels, chromatin immunoprecipitation, and chromosome, chromatin and DNA structure.*

the polymerase chain reaction (PCR) to be identified. In another scenario, the immuno-precipitated DNA is amplified, labeled with fluorescence, and hybridized to tiling oligonucleotide arrays. Mikkelsen et al. (2007) mapped precipitated DNA by Solexa sequencing (Chapter 13), and thus described distinct categories of promoters in different cell populations (embryonic stem cells, neural progenitor cells, and embryonic fibroblasts) based on their chromatin state including histone methylation profile. ChiP data using a variety of approaches can be displayed on the UCSC Genome Browser.

Another set of data are from DNase I sensitivity experiments. DNase I hypersensitive sites reveal accessible genomic regions that are characteristic of active *cis*-regulatory sequences and transcription start sites in particular. Sabo et al. (2006) developed a tiling microarray to detect such sites on a genome-wide basis with high sensitivity (>91%) and specificity (>99%).

Ultraconserved Elements

Comparisons of eukaryotic genome sequences have revealed some highly conserved coding and noncoding DNA sequences. The UCSC Genome Browser offers a set of comparative genomics annotation tracks including one for conservation. That track shows the extent of conservation in 17 vertebrate species (including mammals, amphibians, birds, and fish) based on phastCons, a phylogenetic hidden Markov model (Siepel et al., 2005). We showed an example of the conservation track in Chapter 5.

Comparison of the human and *Fugu rubripes* genomes that last shared a common ancestor about 450 million years ago revealed many ultraconserved sequences (also called highly conserved elements). Ultraconserved elements are sometimes defined as having a length ≥200 base pairs that match identically with corresponding regions of the human, mouse, and rat genomes. Bejerano et al. (2004) identified 481 such segments, most of which were also highly conserved with the dog and chicken genomes. Many of these elements are distant from any protein-coding gene. These regions are highly constrained evolutionarily (Katzman et al., 2007). Dermitzakis et al. (2002) also described ultraconserved sequences on human chromosome 21. In a computer laboratory exercise at the end of this chapter, we will identify a series of DNA sequences that share 100% nucleotide identity between human and chicken (species that last shared a common ancestor over 300 million years ago).

Nonconserved Elements

In analyzing regulatory regions of genomic DNA, a focus has been on identifying conserved noncoding regions as candidates for functionally important loci. Fisher et al. (2006) studied regulatory regions near the *RET* gene in zebrafish, and used a transgenic assay to identify a series of teleost sequences that direct ret-specific reporter gene expression. Surprisingly, a series of human noncoding sequences were also able to drive zebrafish gene expression, even though there was no detectable conservation between the human and zebrafish sequences. This highlights how little we understand about transcription factor binding, and suggests that vast amounts of functionally important regulatory sequences are not detectable based on sequence conservation (Elgar, 2006).

We gain another perspective on the importance of conserved elements by asking the consequence of deleting them. Nóbrega et al. (2004) deleted two large noncoding regions from the mouse genome (consisting of 1511 and 845 kilobases) and created viable homozygous deletion mice. They detected no altered phenotype (and only very minor differences in the expression of neighboring genes). These deletion regions harbored over 1200 noncoding sequences conserved between humans and rodents. It is possible that under some physiological conditions the deletions would have large phenotypic consequences, but nonetheless this study suggests that large portions of chromosomal DNA are potentially dispensable.

COMPARISON OF EUKARYOTIC DNA

Comparative genomics is a powerful approach to annotating and interpreting the meaning of genomic DNA from multiple organisms. When we analyze the genomes of organisms that diverged recently (e.g., humans and chimpanzees diverged 5 MYA) or in the distant past (e.g., mosquitoes and fruit flies diverged 250 MYA; Zdobnov et al., 2002), it is helpful to align the genomic sequences in order to define conserved regions. Such analyses can provide a wealth of information about the existence and evolution of protein-coding genes and other DNA features as well as information about chromosomal evolution.

Genes from different organisms that are derived from a common ancestor and that share a common function are orthologs (Chapter 3). In comparing genomic sequences from two (or more) organisms, we may wish to analyze regions in each species having orthologous genes. Such regions are said to have conserved synteny. Synteny denotes the occurrence of two or more gene loci on the same chromosome, regardless of whether or not they are genetically linked. This definition refers to an arrangement of genes along a chromosome within a single species. "Conserved synteny" refers to the occurrence of orthologous genes (i.e., in two species) that are syntenic. As an example, the occurrence of the neighboring genes *RBP4* and *CYP26A1* on human chromosome 10 and mouse chromosome 19 represents conserved synteny.

In order to analyze regions of conserved synteny—or even larger regions of genomic DNA that do not necessarily contain protein-coding genes—it is necessary to perform pairwise alignment and multiple sequence alignment of genomic DNA.

Synteny derives from Greek roots meaning "same thread" or "same ribbon." A common error is to refer to orthologous genes as being syntenic when instead they share conserved synteny (Passarge et al., 1999). The National Institutes of Health Intramural Sequencing Center (NISC) is currently sequencing BAC contig genomic DNA from dozens of genomic regions of conserved synteny in a variety of species, including human, baboon, chimpanzee, cow, mouse, rat, dog, cat, chicken, zebrafish, and *Fugu*. See ▶ http://www.nisc.nih.gov.

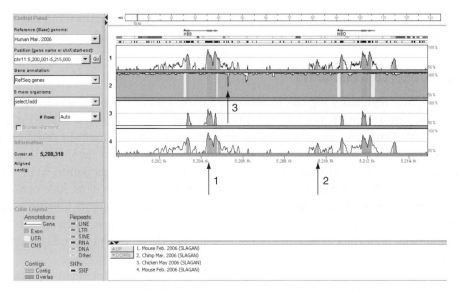

FIGURE 16.22. The VISTA program for aligning genomic DNA sequences is available through a web browser that can be queried with text or DNA sequence (up to 300,000 bases). The output for a query of the human beta and delta globin gene region is shown here. The x axis shows the nucleotide position along human chromosome 11, and the y axis shows the percent nucleotide identity between human and chimpanzee, mouse, and chicken. A variety of exons (e.g., arrow 1) and conserved noncoding sequences (e.g., arrow 2) are shown. Human and chimpanzee have nearly identical sequences, but divergent regions are easily seen (e.g., arrow 3). By clicking a link (not shown), VISTA data can be output on a version of the UCSC Genome Browser.

PipMaker and MultiPipMaker are available at ▶ http://bio.cse.psu.edu/pipmaker/. ("Pip" stands for "percent identity plot.") VISTA (Visualization Tools for Alignments) is at ▶ http://genome.lbl.gov/vista/index.shtml. mVISTA (main VISTA) is a program for visualizing genomic alignments, while rVISTA (regulatory VISTA) is used to align transcription factor binding sites. AVID is an alignment algorithm used by the VISTA tools (Bray et al., 2003). The Berekeley Genome Pipeline includes a VISTA browser (▶ http://pipeline.lbl.gov/). This allows human–mouse, mouse–rat, and human–rat genomic DNA comparisons. VISTA also offers a browser for enhancer elements (▶ http://enhancer.lbl.gov/).

We discussed approaches to this for prokaryotes (Chapter 15), and in Chapter 5 we discussed algorithms that are useful for the comparison of large DNA queries to databases containing genomic DNA, including PatternHunter, BLASTZ, MegaBLAST, BLAT, and LAGAN.

There are other powerful tools for the comparison of genomic DNA in eukaryotes, including PipMaker (Schwartz et al., 2000), VISTA (Mayor et al., 2000; reviewed in Frazer et al., 2003) and MUMmer (Kurtz et al., 2004). The goal of each program is to align long sequences (e.g., thousands to millions of base pairs) while visualizing conserved segments (exons and presumed regulatory regions) as well as large-scale genomic changes (inversions, rearrangements, duplications). It is important to learn both the order and orientation of conserved sequence features. The VISTA browser output for human chromosome 11, including the beta globin and delta globin genes, is shown in Fig. 16.22. This includes an alignment to the chimpanzee, mouse, and chicken genomes, highlighting conserved exons and conserved noncoding regions.

VARIATION IN CHROMOSOMAL DNA

We might think of chromosomes as unchanging entities that define the genome of each species. However, they are dynamic in many ways across large time scales (millions of years), between generations, between individuals in a population, and even within individual lifetimes. A broad variety of cytogenetic changes occur in eukaryotes, allowing an assessment of different types, mechanisms, and consequences of rearrangement (Coghlan, 2005).

Dynamic Nature of Chromosomes: Whole Genome Duplication

When we compare the genomes of related species, we can observe many types of chromosomal changes. One level is ploidy. In eukaryotes, normal germ cells are haploid while somatic cells are usually diploid. Thus, different cells within an individual can have different ploidy. Ploidy is the number of chromosome sets in a cell. It can vary in many ways. Some single-celled eukaryotes such as *S. cerevisiae* can grow in either the haploid or diploid state. Triploid *Drosophila* are viable (but with reduced fertility). Although we distinguish the ploidy state in germ cells and somatic cells, ploidy can also vary in somatic cells within an individual. For example, in humans a small fraction of liver cells is typically triploid. In general an extra germline copy of even one chromosome is usually lethal in mammals.

One of the dramatic ways that ploidy can change for an entire species is through whole genome duplication. Mechanistically, a mitotic or meiotic error may cause diploid gametes to form, having two sets of chromosomes. These may fuse with haploid gametes to form triploid zygotes, which are unstable but may lead to the formation of stable tetraploid zygotes. When whole genome duplication occurs within a species, the result is termed autopolyploidy. Such a massive event may have happened in yeast; in Chapter 17 we will review evidence for whole genome duplication and computational tools to analyze and visualize it. A variety of fish and plant genomes also underwent whole genome duplication. In the case of the ciliate *Paramecium tetraurelia*, analysis of the genome sequence suggests that there have been at least three whole genome duplication events (Aury et al., 2006; see Chapter 18).

Paramecium tetraurelia is exceptional because most of its duplicated genes have not been deleted (Aury et al., 2006 and see Chapter 18).

The genomes of two distinct species may merge to generate a novel species (allopolyploidy) (Hall et al., 2002). This phenomenon has been described in many plants (Comai, 2000), animals, and fungi. For example, the plant *Arabidopsis suecica* derives from the *A. thaliana* and *Cardaminopsis aerenosa* genomes (Lee and Chen, 2001; Lewis and Pikaard, 2001). Another example of allopolyploidy is the mule, which is the result of a cross between a male donkey (*Equus asinus*, $2n = 62$) and a female horse (*Equus caballus*, $2n = 64$). Mules cannot propagate because they are sterile (they cannot produce functional halploid gametes) (see Ohno, 1970).

Ohno 1970 hypothesized that the increased complexity of vertebrates is due to two rounds of whole genome duplication in early vertebrate evolution. This has been called the 2R hypothesis (reviewed in Panopoulou and Poustka, 2005; Dehal and Boore, 2005). Ohno argued that duplication provided the genetic material that can be shaped by mutation and selection to introduce novel functions to organisms (Prince and Pickett, 2002; Taylor and Raes, 2004). There are three advantages of becoming polyploid (Comai, 2005). (1) Hybrids sometimes exhibit an increase in performance relative to their inbred parents, a phenomenon termed heterosis. (2) Gene redundancy occurs, offering the opportunity to mask recessive deleterious alleles by dominant wild-type alleles. Also, one member of a duplicated gene pair may be silenced, up- or downregulated in its expression level, or regulated in a tissue-specific manner (Adams and Wendel, 2005; Li et al., 2005). The most common fate of duplicated genes is that they become deleted (discussed in Chapter 17), as has been shown in the plants *Arabidopsis thaliana* and *Oryza sativa* (Thomas et al., 2006), and fish (Brunet et al., 2006; Paterson et al., 2006). (3) Self-fertilization may become possible (asexual reproduction).

Another type of chromosomal change that can be fixed in a species is the fusion of two chromosomes. For example, acrocentric chromosomes may be subject to Robertsonian translocation, in which two centromeres fuse (Slijepcevic, 1998).

In human trisomy 21 (Down syndrome), it is not uncommon for a copy of chromosome 21 to fuse with another acrocentric chromosome.

Human chromosome 2, the second largest human chromosome, is derived from two ancestral great ape acrocentric chromosomes (chimpanzee chromosomes 2a and 2b [formerly named 12 and 13]) (Fig. 19.19) (Ijdo et al., 1991a; Martin et al., 2002; Fan et al., 2002). The human 2q13 band, near the centromere, contains telomeric repeats in a head-to-head orientation. Over 50 interstitial telomeres have been described (Azzalin et al., 2001).

In addition to fusion, chromosomes can split (fission). As an example, human chromosomes 3 and 21 derive from a larger ancestral chromosome (Muzny et al., 2006). Chromosomal inversions represent another change that can lead to speciation. There are five distinct subtypes of the mosquito *Anopheles gambiae* having varying kinds of paracentric inversions on chromosome 2 (Holt et al., 2002), and these inversions may lead to speciation by preventing successful chromosomal pairing among members of different subtypes.

The recent availability of draft sequences from dozens of eukaryotic genomes has led to the reconstruction of many ancestral genomes. For example, Kohn et al. (2006) decribed the eutherian karyotype from 100 million years ago, prior to the radiation of mammalian species. Murphy et al. (2005) compared the chromosomal architecture of eight species (human, horse, cat, dog, pig, cattle, rat, and mouse) and inferred the structure of their ancestral chromosomes. They characterized the sites of evolutionary breakages, which included subtelomeric and pericentromeric regions in particular.

Large-scale chromosomal changes may lead to the establishment of a new species (speciation). Ohno (1970) provided an example. The karyotypes of the tobacco mouse *Mus poschiavinus* ($2n = 26$) and the house mouse *Mus musculus* ($2n = 40$) are shown (Fig. 16.23a and b). The ancestral *M. poschiavinus* may have become physically isolated from *M. musculus* and thus was not able to interbreed. At this time its chromosomes underwent Robertsonian translocations, thus forming a new genome with a reduced number of chromosomes. The F1 progeny form a series of seven trivalents (each from one *poschiavinus* metacentric and two *musculus* acrocentrics; Fig. 16.23c) which are not compatible with survival.

Chromosomal Variation in Individual Genomes

Comparison of closely related species has revealed many chromosomal changes involving single chromosomes. At the level of the individual organism, many changes to chromosomes occur, sometimes causing disease:

- An individual may acquire an extra copy of an entire chromosome. For example, Down syndrome is caused by a trisomy (triplicated copy) of chromosome 21 (Fig. 16.1b). We discuss this type of disorder in Chapter 20. Aneuploidy (the presence of an abnormal number of chromosomal copies) occurs commonly and is often caused by nondisjunction (Hassold and Hunt, 2001).

- Uniparental disomy may occur, in which both homologous chromosomes are inherited from one parent. We discuss this in more detail below. Uniparental disomy is often associated with disease in humans (Kotzot, 2001).

- A portion of a chromosome may be deleted. Deletions may be terminal or interstitial; an example of a terminal deletion of chromosome 11q is shown in Fig. 16.1 (arrow B).

- Segmental duplications commonly occur (see Chapter 19).

(a)   ordinary male house mouse (*Mus musculus*, 2n = 40)

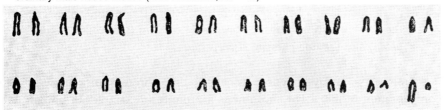

(b)   male tobacco mouse (*Mus poschiavinus*, 2n = 26)

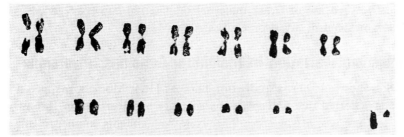

(c)   Male first meiotic metaphase from an interspecific F1-hybrid.

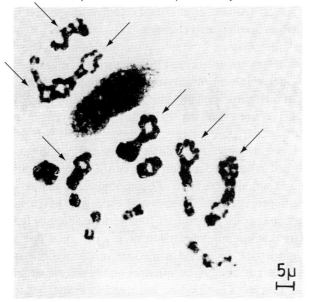

5μ

FIGURE 16.23. *Robertsonian fusion creates one metacentric chromosome by the fusion of two acrocentrics. (a) Karyotype of the normal mouse, Mus musculus. (b) Karytoype of the male tobacco mouse (Mus poschiavinus, 2n = 26). Its smaller chromosome number derives from Robertsonian fusion events. (c) Male first meiotic metaphase from an interspecific F1-hybrid. Note seven trivalents (indicated with arrows). Each represents one poschiavinus metacentric and two musculus acrocentrics. From Ohno (1970). Used with permission.*

- Normal chromosomes from any eukaryotic species can vary between individuals in length, number, and position of heterochromatic segments. For example, the ribosomal DNA repeat segments on the short arms of the five human acrocentric chromosomes vary greatly in length between individuals. A variety of human chromosomes show tremendous polymorphisms in the population, such as portions of chromosome 7 (Chapter 19).

- Fragile sites often occur, sometimes causing chromosomal breaks. These fragile sites can be inherited in a dominant Mendelian fashion.

- At least some eukaryotes display chromatin diminution, a form of developmentally programmed DNA rearrangement. Remarkably, chromosomes in somatic cells can fragment, then lose some chromosomal material. Thus somatic chromosomes can have a different structural organization and a smaller gene number than germline cells. Chromatin diminution could represent an unusual gene-silencing mechanism (Müller and Tobler, 2000). This phenomenon has been observed in at least 10 nematode species, including the horse intestinal parasite *Parascaris univalens* (also called *Ascaris megalocephala*) and the hog parasite *Ascaris suum*.

Among the many functional changes that chromosomes undergo, dosage compensation of the X chromosome is a prominent example. In human females, one copy of each X chromosome is functionally inactivated through the action of an X-chromosome inactivation center (XCI)(Latham, 2005). Genomic imprinting, the selective silencing of either maternal or paternal copies of genes, is another regulatory mechanism (Morison et al., 2005).

Chromosomal Variation in Individual Genomes: Inversions

A. H. Sturtevant (1921), a student of Thomas Hunt Morgan, mapped a series of genes and reported that *Drosophila simulans* has an inversion on chromosome III relative to *Drosophila melanogaster*. This example highlights another feature of chromosomal plasticity: while 500 unique inversions are known in *D. melanogaster* (a highly polymorphic species), only 14 unique inversions are known in *D. simulans* (a monomorphic species) (Aulard et al., 2004). Different species present varying propensities to undergo chromosomal changes.

In humans and other species, inversions commonly occur. They can be extraordinarily difficult to detect, because even DNA sequencing may not reveal changes, and they may be undetectable using conventional cytogenetics. Stefansson et al. (2005) described an inversion polymorphism of 900 kilobases that occurs on chromosome 17q21.31 (from 44.1 to 45.0 Mb). This inversion is common in Europeans where it is under positive selective pressure. Surprisingly, the inverted segment occurs in chromosomes having different orientations in two lineages (H1 and H2) which diverged as long as 3 million years ago. As another example, an inversion of a single gene causes a severe form of hemophilia (Antonarakis et al., 1995).

You can read about this hemophilia at the Online Mendelian Inheritance in Man (OMIM) site at NCBI (entry 306700). We describe OMIM in Chapter 20.

In an innovative approach, Pavel Pevzner and colleagues have used small inversions as evolutionary characters to perform phylogenetic analyses (Chaisson et al., 2006). They estimate that one microinversion occurs per megabase per 66 million years of evolution, and they developed a method to distinguish microinversions (local alignments between orthologous sequences on the reverse strand) from palindromes and inverted repeats. This method is limited to analysis of sequences with sufficient conservation to permit clear assignment of orthology, but its phylogenetic reconstruction matches traditional approaches.

Models for Creating Gene Families

One prominent aspect of genomes is the occurrence of multigene families. Multigene families (also called superfamilies) consist of a group of paralogs such as the globins. Nei and Rooney (2005) reviewed this topic and described three separate models for their evolution.

1. According to a divergent evolution model, members of a gene family gradually diverge as duplicate genes assume new functions (Fig. 16.24a). For example, the alpha and beta globin groups each have multiple members, as shown in the phylogenetic tree of Fig. 3.2. Some of these globins are expressed at specific developmental stages.

2. According to the concerted evolution model, all the members of a gene family evolve in a concerted manner, rather than independently (Fig. 16.24b). An example of this scenario is the tandemly repeated ribosomal DNA genes. We described the structure of human rDNA repeats in Chapter 8 (Fig. 8.7). Work by Donald Brown and others showed that intergenic regions of ribosomal DNA clusters were more similar within a species than between two related *Xenopus* (frog) species. When one member of such a gene cluster acquires a mutation, that change spreads to other members. One mechanism by which this can occur is unequal crossing over (discussed below). Another proposed mechanism is gene conversion. In gene conversion, one gene (or other DNA element) serves as a donor, and through a form of nonreciprocal recombination, it mediates the conversion of a second gene to form a copy of the first gene. Examples of gene families that have evolved by concerted evolution include the primate U2 snRNA genes, 5S RNA genes in *Xenopus* (which has 9000 to 24,000 members) or humans (which has ~500 members), and heatshock protein genes in *Drosophila*. The *hsp70Aa* and *hsp70Ab* genes are a pair of inverted tandem repeats that are virtually identical in *D. melanogaster* as well as in *D. simulans*. Their within-species identity could provide an example of gene conversion.

The DNA and protein RefSeq accession numbers for *hsp70Aa* are NM_169441 and NP_731651, while for *hslp70Ab* they are NM_080059 and NP_524798.

3. The birth-and-death evolution model was proposed by Masatoshi Nei and others (reviewed in Nei and Rooney, 2005) (Fig. 16.24c). According to this model, new genes are created by gene duplication. Some duplicates remain in the genome, while others are inactivated (becoming pseudogenes) or deleted. This model was proposed to explain the evolution of the major histocompatibility complex (MHC) genes. MHC proteins bind foreign or self peptides and present them to T-lymphocytes as part of the immune response. MHC class I genes in particular are highly polymorphic due to

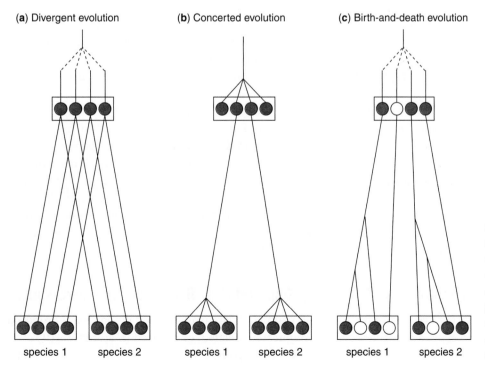

(a) Divergent evolution   **(b)** Concerted evolution   **(c)** Birth-and-death evolution

species 1   species 2   species 1   species 2   species 1   species 2

FIGURE 16.24. Three models for the creation of duplicate genes in multigene families. (a) Divergent evolution; (b) concerted evolution; (c) birth-and-death evolution. Red-shaded circles refer to functional genes; unfilled circles correspond to pseudogenes. Redrawn from Nei and Rooney (2005). Used with permission.

TABLE 16-11 Examples of Multigene Families that Undergo Birth-and-Death Evolution

Category	Family	Organism
Immune system	MHC	Vertebrates
	Immunoglobulins	Vertebrates
	T-cell receptors	Vertebrates
	Natural killer cell receptors	Mammals
	Eosinophilic RNases	Rodents
	Disease resistance (R) loci	Plants
	α-Defensins	Mammals
Sensory system	Chemoreceptors	Nematodes
	Taste receptors	Mammals
	Sex pheromone desaturases	Insects
	Olfactory receptors	Mammals
Development	Homeobox genes	Mammals
	MADS-box	Plants
	WAK-like kinase	*Arabidopsis*
Highly conserved	Histones	Eukaryotes
	Amylases	*Drosophila*
	Peroxidases	All kingdoms
	Ubiquitins	Eukaryotes
	Nuclear ribosomal RNA	Protists, Fungi
Miscellaneous	DUP240 genes	Yeast
	Polygalacturonases	Fungi
	3-Finger venom toxins	Snakes
	Replication proteins	Nanoviruses
	ABC transporters	Eukaryotes

Source: Nei and Rooney (2005). Used with permission.

positive selection on the peptide-binding region (Hughes and Nei, 1989). The birth-and-death model presents a mechanism for the generation of gene diversity that is distinct from concerted evolution or divergent evolution, and explains how new functions can be acquired by duplicate genes.

According to Nei and Rooney, most gene families are subject to birth-and-death evolution. Some examples are listed in Table 16.11. In some cases such as histone genes and the ubiquitins, the birth-and-death process is accompanied by very strong purifying selection that conserves the protein seqences. This selective pressure, rather than the homogenizing properties of gene conversion or unequal crossover, accounts for the tremendous conservation of these proteins. In other cases a mixed process of concerted evolution and birth-and-death evolution occurs, such as in the alpha globin genes in which *HBA1* and *HBA2* genes encode identical proteins, possibly because of gene conversion.

Mechanisms of Creating Duplications, Deletions, and Inversions

In the first half of the twentieth century, a variety of detailed models were proposed to explain how genes become duplicated, deleted, or inverted (Darlington, 1932). A major current model is nonallelic homologous recombination mediated by low-copy

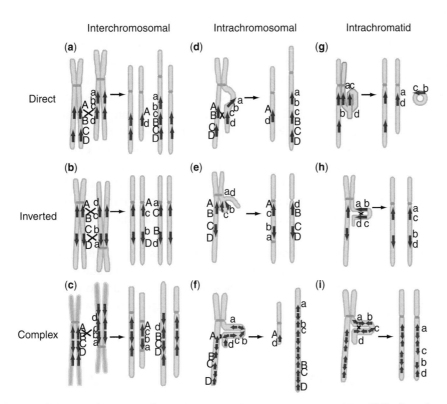

FIGURE 16.25. *Mechanisms of creating genomic rearrangements. Nonallelic homologous recombination (NAHR) based on low-copy repeats (LCRs) or segmental duplications cause these changes. The orientation of the LCRs may be head-to-head (top row), head-to-toe (middle row), or complex (bottom row) involving DNA exchanges that are interchromosomal (left column), intrachromosomal (middle column), or intrachromatid (right column). For each of the nine scenarios the chromosomal configuration is shown as well as the products of unequal crossing over. (a) Unequal crossovers between directly ordered repeats lead to a duplication and a deletion. (b) Mechanism of forming an inversion. (c) Interchromosomal exchange between inverted repeats causes inversions and can result in duplications and deletions. (d) Mispairing of direct repeats leads to an intrachromosomal deletion/duplication. (e) An inversion results from intrachromosomal unequal exchange between inverted repeats. (f) Complex repeats lead to an intrachromosomal deletion/duplication. (g) A deletion and an acentric fragment result from intrachromatid mispairing due to direct low copy repeats.(h) An intrachromatid loop of inverted repeats results in an inversion. (i) Complex repeats lead to intrachromatid mispairing and an inversion. Adapted from Stankiewicz and Lupski (2002). Used with permission.*

repeats (that is, by segmental duplications) (Stankiewicz and Lupski, 2002; Bailey and Eichler, 2006). Repetitive DNA of about 10 kilobases to 500 kilobases that occurs in two (or more) distinct chromosomal loci can lead to unequal crossing over (Fig. 16.25). These crossovers can occur interchromosomally, intrachromosomally, or between sister chromatids (Fig. 16.25, columns). The orientation of the low-copy repeats influences the nature of the rearrangement that occurs; these repeats may occur in a direct orientation, they may be inverted repeats, or they may have a complex structure (Fig. 16.25, rows).

We can examine the case of direct repeats in Fig. 16.25a. The phrase "nonallelic homologous recombination" refers to meiotic recombination between chromosomes. One chromosome has two repeated segments labeled AB and CD, while the other has ab cd. The repeats can combine even when they are nonallelic (e.g., example, AB and ab are allelic but AB and CD are nonallelic). Nonetheless they are homologous and thus able to pair. Following the crossover event indicated by the X in Fig. 16.25a, one copy contains ab cB CD and thus has a duplication, while the other copy has Ad from the crossover event and thus has a deletion.

As indicated in Fig. 16.25, many other products can result from unequal exchanges. In this way, segmental duplications (low-copy repeats) have been a major force in shaping genome evolution, including the emergence of gene families. In Chapter 20 we will present six models by which deletions (or duplications or inversions) may cause disease (Lupski and Stankiewicz, 2005). In other cases, the genomic rearrangements, such as altering the dosage of a gene or fusing two genes together, may present an organism with an innovation that is advantageous and selected for.

The boundaries of segmentally duplicated regions often contain *Alu* repetitive sequences (Bailey and Eichler, 2006). Pericentromeric and subtelomeric regions are also enriched for segmental duplications, with interchromosomal segmental duplications present in 30 out of 42 subtelomeric regions (reviewed in Bailey and Eichler, 2006).

TECHNIQUES TO MEASURE CHROMOSOMAL CHANGE

For several decades, karyotyping has been the preeminent technique to visualize chromosomes. Today, clinical genetics laboratories routinely use karyotyping to assess the occurrence of aneuploidy as well as smaller changes such as microdeletions and microduplications. Typically, deletions that are smaller than about 3 million base pairs are too small to detect. Chromosomal inversions can only be detected if they are large enough to disrupt the banding pattern. Translocations may be balanced (if two chromosomal regions exchange) or unbalanced (if material is gained or lost).

Fluorescence in situ hybridization (FISH) offers greatly increased resolution. A bacterial artificial chromosome (BAC) clone, typically consisting of about 200,000 base pairs of genomic DNA inserted into a cloning vector of about 10,000 base pairs, can be labeled with a fluorescent dye then used to probe a spread of metaphase chromosomes on a microscope slide. FISH has been used to refine information about chromosomal anomalies such as microdeletions and translocations.

In 1992 Kallioniemi and colleagues performed comparative genome hybriziation (CGH) in which genomic DNA from two samples (such as one diseased and one apparently normal) is isolated, labeled with a green or red fluorescent dye, and hybridized to a normal chromosomal spread. This technique showed regions of gain or loss of DNA sequences, including amplifications seen in tumor cell lines.

Array Comparative Genomic Hybridization

Array CGH (aCGH) is a high throughput extension of the CGH technique to microarrays that is useful to detect copy number changes at defined chromosomal loci. It combines the high resolution of FISH with the broad chromosome-wide perspective of karyotyping. An aCGH platform may consist of thousands of BAC clones or oligonucleotides immobilized on the surface of a glass microarray. Genomic DNA is purified from a test sample (e.g., the DNA is isolated from a cell line or blood sample) and a reference sample. If the test sample DNA is labeled with a red dye and the reference is labeled with a green dye, then upon hybridization the signal intensities are comparable. If an amplification or deletion occurs, the log signal intensities deviate from a value of zero. The region of copy number gain or loss may be as small as one single probe (e.g., one base pair for an SNP array, or about 200,000 base pairs for a BAC array). The change may also extend across an entire chromosome arm or entire chromosome. Figure 16.26 shows an example of a microdeletion on chromosome 2. This resulted in the hemizygous loss of many genes, and mental retardation in the patient.

A simple approach has been to apply a ratio threshold to define a region of amplification or deletion. For a gain of one copy, the amount of signal is expected to increase 1.5-fold (from two copies in the euploid state to three copies), while a hemizygous deletion reduces the copy number twofold (from two copies to one). On a $\log_2$ scale, unchanged copy number corresponds to a value of 0 (i.e., a 1:1 ratio), while a gain and loss correspond to $+1$ and -1 $\log_2$ intensity values, respectively.

Many statistical approaches have been developed to analyze aCGH data. Two estimation problems must be addressed: inferring the number of chromosomal alterations and their statistical significance, and locating the boundaries of such events. Lai et al. 2005 tested the accuracy of a group of 11 algorithms. Their comparative study included receiver operating characteristic (ROC curves) plotting the false positive rate versus the true positive rate. For many test data sets, the 11 algorithms produced dramatically different estimates of copy number changes. The algorithms were all better at detecting large-scale aberrations with a good signal to noise ratio, but faltered with smaller aberrations and noisy data. Some algorithms did not detect particular amplifications or deletions; others either merged a group of alterations or splintered them inappropriately. Overall, one of the best-performing algorithms in this (Lai et al., 2005) and another comparative study (Willenbrock and Fridlyand, 2005) was the circular binary segmentation method (CBS)(Olshen et al., 2004; Venkatraman and Olshen, 2007). This method divides the genome into regions of equal copy number, assuming that chromosomal gains or losses occur in discrete, contiguous regions. The goal is to identify copy number changepoints which partition the chromosome into segments. A likelihood ratio statistic tests the null hypothesis that there is no change against the alternative hypothesis that there is one change at a given location. The null hypothesis is rejected if the test statistic exceeds some threshold; the variance can be estimated from the data by Monte Carlo simulations using a permuted reference distribution.

aCGH is one of the techniques that has been used to discover copy number variants (CNVs) in the human genome. There is an astonishing amount of variation between even apparently normal individuals, with large numbers of megabase-sized deletions and duplications. We address this topic in Chapter 19.

Single Nucleotide Polymorphism (SNP) Microarrays

SNPs represent one of the most commonly occurring forms of variation in all genomes. Figure 16.27 shows an example of two SNPs from the beta globin gene at the Entrez database of SNPs (dbSNP) at NCBI. By convention, each of the variants (C or G in these two cases) is represented as A or B for the major and minor alleles in the population. Most SNPs are biallelic (i.e., there are two rather than three or four variants at a given position) with a range of population frequencies. Thus possible genotype calls for a diploid sample (such as human) are AA or BB (homozygous) or AB (heterozygous) (Fig. 16.28). In regions of hemizygous deletion

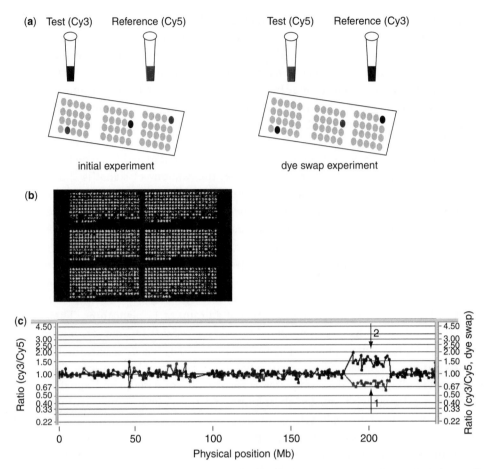

FIGURE 16.26. *Array comparative genome hybridization (aCGH) allows detections of chromosomal gains and losses. (a) Experimental design. Genomic DNA is isolated from a test sample (e.g., from a patient) and a reference (e.g., from a pool of apparently normal controls). The DNA is fragmented then labeled with differently colored fluorescent dyes such as Cy3 and Cy5. In a parallel dye swap experiment, the test and reference samples are labeled with opposite dyes. The samples are coincubated with a microscope slide containing up to tens of thousands of bacterial artificial chromosome (BAC) clones, each of which typically spans 200,000 base pairs and has known chromosomal position. Following hybridization, washing, and image analysis, most BACs on the array have a comparable amount of Cy3 and Cy5 dye (indicated as gray spots on the two slides). A deletion in the test sample is associated with relatively more Cy5 dye in the reference; see the two red spots in the slide at the left. In the dye swap, these two spots appear black, providing an independent validation. An amplification in the test sample results in relatively more Cy3 dye (see the black spot in the slide at left, which appears red in the dye swap experiment to the right). (b) Example of an aCGH image from a scanner. The output includes a spreadsheet that includes quantities of the signal intensities in the Cy3 and Cy5 channels for each BAC clone. (c) Example of the output for chromosome 2. The x axis corresponds to chromosome 2 (from the p terminus to the q terminus). The y axis corresponds to the Cy3/Cy5 ratio from the initial experiment and from the dye swap. Thus there are two sets of data points that are superimposed. The test sample is from a patient who has a deletion of about 23 megabases (from 190.5 to 213.8 Mb in chromosome 2q32.2-q34). This deletion is evident as a reduced signal intensity ratio across a group of adjacent BACs (arrow 1). As expected the dye swap experiment shows a mirror image deviation (arrow 2).*

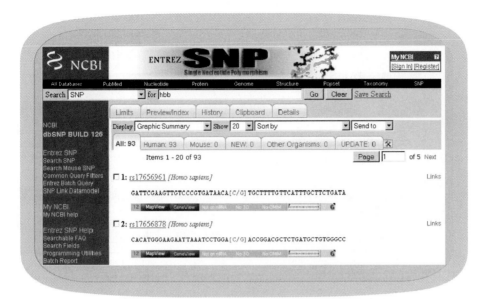

FIGURE 16.27. Example of two single nucleotide polymorphisms (SNPs) from the database of SNPs at NCBI (dbSNP). A portion of the Entrez SNP results for human beta globin is shown. For the two SNPs, the nucleotide C or G occurs in the population (indicated [C/G]). The flanking sequence is also shown, as well as the RefSNP accession number (e.g., rs17656961) and a variety of links.

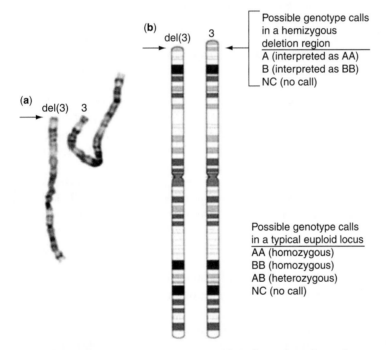

FIGURE 16.28. SNP microarray experiments provide information about chromosomal copy number (based on the intensity of hybridization) and genotype (based on alleles detected at each SNP position). (a) Karyotype of chromosome 3 from a patient with a hemizygous deletion (i.e., loss of a portion of one of the two chromosomal copies). The deletion region is indicated with an arrow. (b) Ideogram of chromosome 3. Throughout most of the chromosome there are four possible genotype calls: AA or BB (homozygous calls), AB (heterozygous), or NC (no call). In the deletion region there are three possible calls: an underlying state of A (interpreted by current software packages as a biallelic call, AA), B (interpreted as BB) or no call. There can be no AB calls (unless there is a technical failure). Some software packages detect stretches of homozygous SNPs, which in the presence of a reduced copy number corresponds to a hemizygous deletion. Note that the human male X chromosome is by its nature hemizygous, and no AB calls are expected other than those that represent genotyping errors (or, in some instances, pseudoautosomal regions).

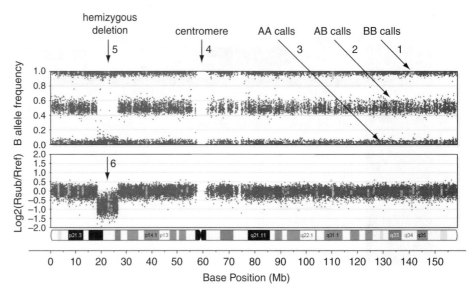

FIGURE 16.29. SNP profile from chromosome 7 in a patient with a hemizygous deletion. The upper panel shows the B allele frequency from thousands of SNPs across the chromosome, including BB calls (B allele frequency near 1.0; arrow 1), heterozygous AB calls (arrow 2), or homozygous AA calls (arrow 3). In some heterochromatic regions such as the centromere (arrow 4), there are no SNPs and thus the plot lacks data points (arrow 4). In the region of a hemizygous deletion on 7p (arrow 5), there are essentially no AB calls. The lower panel shows the intensity values corresponding to chromosomal copy number. The y axis is $Log_2(Rsub/Rref)$, corresponding to the log_2 ratio of the intensity value for the subject (i.e., this patient sample) to the intensity values for a reference set, such as mean intensity values for a large set of apparently normal individuals. $Log_2(Rsub/Rref)$ tends to have a value near 0.0 (thus the subject and reference data have a one-to-one correspondence), but in the deletion region the log_2 value is -1.0 (see arrow 6). In regions of homozygous deletion (i.e., two copies deleted; not shown), the log_2 value tends to be close to -5.0. In cases of trisomy (not shown), the extra copy causes the B allele frequency to split into four tracks (corresponding to AAA, AAB, ABB, and BBB genotypes) and the intensity values are elevated. Data are from an Illumina microarray with 550,000 SNPs.

(where one of two chromosomal copies are deleted), or on the male X chromosome which is by its nature hemizygous, the genotypes are A or B but should never be heterozygous (Fig. 16.28).

The HapMap project was created to identify SNPs in the human genome. It resulted in the determination of over three million SNPs (International HapMap Consortium, 2005, 2007). This resource, available through a HapMap database, initially centered on genotyping of four diverse populations (from northern Europe, Africa, Japan, and China). The SNP data are useful to describe variation between and within populations, including the structure of shared alleles (haplotypes), to characterize recombination rates, and to characterize the evolution of both nonsynonymous and synonymous SNPs in coding regions.

There are many applications of SNPs, including mapping polymorphisms in genes and genomes, selecting markers to identify individuals having alleles of interest in large segregating populations, and finding associations between genomic regions and segregating traits (Chapter 20). A basic application is to measure chromosomal changes in genomic DNA samples. Several technologies exist to measure vast numbers of SNPs on microarrays, such as a single-base extension strategy from Illumina

The HapMap website is ▶ http://www.hapmap.org.

and an oligonucleotide-based hybridization strategy from Affymetrix. An example of an SNP data set using the Illumina platform is shown in Fig. 16.29. The experiment provides information on chromosomal copy number (based on hybridization intensity measurements) and genotype (based on AA, AB, or BB calls). There is a characteristic profile for hemizygous deletions (as shown in Fig. 16.29).

SNP arrays can provide information on a variety of chromosomal changes beyond those detectable by aCGH or conventional cytogenetics. An example is uniparental disomy in which both homologous chromosomes are inherited from one parent. The term disomy refers to two copies, as opposed to zero (nullsomy), one (monosomy), three (trisomy), or four (tetrasomy). There are two copies of each chromosome, as usual, but the two copies of a single chromosome are derived from just one parent (uniparental disomy). Since each parent has two copies of a given autosome, the result may be uniparental heterodisomy (in which the two copies derived from the mother or the father are different) or uniparental isodisomy (in which the two copies are identical). This is also associated with disease in humans (Kotzot, 2001). SNP arrays can show regions of homozygosity without copy number change. In the absence of copy number change, the cause can be uniparental disomy (Ting et al., 2007).

Jason Ting in my laboratory developed SNPtrio, a program that identifies uniparental disomy using SNP data.

PERSPECTIVE

One of the broadest goals of biology is to understand the nature of each species of life: What are the mechanisms of development, metabolism, homeostasis, reproduction, and behavior? Sequencing of a genome does not answer these questions directly. Instead, we must first try to annotate the genome sequence in order to estimate its contents, and then we try to interpret the function of these parts in a variety of physiological processes.

The genome sequences of representative species from all major eukaryotic divisions are now becoming available. This will have dramatic implications for all aspects of eukaryotic biology. For studies of evolution, we will further understand mutation and selection, the forces that shape genome evolution.

As complete genomes are sequenced, we are becoming aware of the nature of noncoding and coding DNA. Major portions of the eukaryotic genomic landscape are occupied by repetitive DNA, including transposable elements. The number of protein-coding genes varies from about 6000 in fungi to tens of thousands in plants and mammals. Many of these protein-coding genes are paralogous within each species, such that the "core proteome" size is likely to be on the order of 10,000 genes for many eukaryotes. New proteins are invented in evolution through expansions of gene families or through the use of novel combinations of DNA encoding protein domains.

PITFALLS

A tremendous need in genomics research is the continued development of algorithms to find protein-coding genes, noncoding RNAs, repetitive sequences, duplicated blocks of sequence within genomes, and conserved syntenic regions shared between genomes. We may then characterize gene function in different developmental stages, body regions, and physiological states. Through these approaches we may generate and test hypotheses about the function, evolution, and biological adaptations of eukaryotes. Thus, we may extract meaning from the genomic data.

We are now in the earliest years of the field of genomics. Many new lessons are emerging:

- Draft versions of genome sequences are extremely useful resources, but gene annotation often improves dramatically as a sequence becomes finished.

- It is extraordinarily difficult to predict the presence of protein-coding genes in genomic DNA. This is especially true in the absence of complementary experimental data on gene expression, such as expressed sequence tag information.

- We know little about the nature of noncoding RNA molecules.

- Large portions of eukaryotic genomes consist of repetitive DNA elements. Segmental duplications offer a creative evolutionary opportunity to shuffle DNA within and between chromosomes.

- Comparative genomics is extraordinarily useful in defining the features of each eukaryotic genome.

Most publications describing genomes (both eukaryotic and prokaryotic) define orthologs as descended by speciation from a single gene in a common ancestor. Typically, the predicted proteins from an organism are searched by BLAST against the complete proteome of other species using an E value cutoff such as 10^{-4}. However, two orthologous proteins could have species-specific functions.

WEB RESOURCES

We have presented key resources for many eukaryotic organisms and their genome-sequencing websites. An excellent starting point is the Ensembl website (▶ http://www.ensembl.org/), which currently includes gateways for the mouse, rat, zebrafish, fugu, mosquito, and other genomes.

DISCUSSION QUESTIONS

[16-1] If there were no repetitive DNA of any kind, how would the genomes of various eukaryotes (human, mouse, a plant, a parasite) compare in terms of size, gene content, gene order, nucleotide composition, or other features?

[16-2] If someone gave you 1 Mb of genomic DNA sequence from a eukaryote, how could you identify the species? (Assume you cannot use BLAST to directly identify the species.) What features distinguish the genomic DNA sequence of a protozoan parasite from an insect or a fish?

PROBLEMS/COMPUTER LAB

[16-1]

(a) Retrieve a typical *Arabadopsis thaliana* bacterial artificial chromosome (BAC) from Entrez (e.g., choose BAC T18A20, GenBank accession AC009324).

- Note the approximate size (in kilobases). Is this a large or a small BAC?

- Note the approximate number of protein products in it. Bacteria have about one gene per kilobase. How many genes are there per kilobase in this eukaryotic DNA?

(b) Go to the ORF Finder at NCBI:

- From the main page, look at the left sidebar. Choose "Tools for data mining"; then you will see the ORF Finder.

- Alternatively, from the main page, look at the left sidebar at the top. Choose "Site map" and you will also find a link to the ORF Finder.

- Paste in the accession number for your BAC. Click OrfFind.

(c) At the ORF Finder at NCBI, Click on the largest ORF.

- How many amino acids long is it?

- What is its molecular weight (in kilodaltons)?

- Is this protein small, average, or large?

- From which strand of the BAC is this putative gene transcribed? Overall, are there more ORFs on the top or bottom strand or is it about the same?

(d) Using the ORF Finder at NCBI, BLAST search the ORF of (c) using the default parameters that are given to you.

- Note that the results page is NOT updated automatically so you may need to reload your page.

- This BLAST result reveals many matches to *Arabidopsis* proteins. However, note that if you do a standard blastp search using this ORF as a query, you will find matches to many dozens of species. Also you will see a match to the Conserved Domain Database. Thus, the BLAST tool within OrfFinder is not as thorough as a regular BLAST search.

[16-2] Human centromeres typically contain several thousand base pairs of a 171 bp repeat called α-satellite (accession X07685).

First perform a blastn search against the nonredundant database. What kinds of database matches do you observe? Second, restrict your BLAST search to nonhumans. (In the options section that allows you to limit by Entrez query, try typing "satellite NOT human[organism].") Are there matches in primates, rodents, or plants? Why might centromeric repeats have this phylogenetic distribution; would you expect each species to have its own, unique centromeric signature?

[16-3] Identify ultraconserved elements that share 100% identity between the chicken and human genomes. While there are several approaches, try the following. (1) Go the the UCSC Genome Bioinformatics site (▶ http://genome.ucsc.edu). Select the Table Browser. Set the clade to vertebrate clade, the genome to Chicken, the group to "Comparative Genomics," and the track to "Most Conserved." Under "region" select whole genome. (2) If you get the summary statistics at this point, there are over 950,000 items, which include a range of conservation levels. The output format is "all fields from selected table." Click the filter button, and select scores that are ≥900 (on a scale from 1 to 1000). There are now only six items (on chicken chromosomes 1, 2, 5, and 7). These are listed in web document 16.4 at ▶ http://www. bioinfbook.org/chapter16. (3) Change the output format to "hyperlinks to Genome Browser." You can now access the Genome Browser showing these ultraconserved elements, and by clicking the annotation tracks you can view multiple sequence alignments of the highly conserved DNA.

SELF-TEST QUIZ

[16-1] The *C* value paradox is that

(a) The nucleotide C is underrepresented in some genomes.

(b) The genome size of various eukaryotes correlates poorly with the number of protein-coding genes of the organism.

(c) The genome size of various eukaryotes correlates poorly with the biological complexity of the organism.

(d) The genome size of various eukaryotes correlates poorly with the evolutionary age of the organism.

[16-2] Hundreds or thousands of sequence repeats, each consisting of a unit of about four to eight nucleotides, are commonly found where?

(a) In interspersed repeats

(b) In processed pseudogenes

(c) In telomeres

(d) In segmentally duplicated regions

[16-3] You are sequencing the genome of a newly described organism (a slime mold). What is likely to happen if you use RepeatMasker to assess its repetitive DNA content? You set the default setting of RepeatMasker to the settings for human DNA.

(a) RepeatMasker should successfully identify essentially all of the repetitive DNA. Various repetitive DNA elements are similar enough between organisms to allow this software to work on your slime mold DNA.

(b) RepeatMasker should identify most of the repetitive DNA. However, because some types of repeats are species specific, it is likely that there will be many false positive and false negative results.

(c) RepeatMasker would fail to identify most of the repetitive DNA. Most types of repeats are highly species specific. It is necessary for you to train the RepeatMasker algorithm on your slime mold DNA in order for the program to work.

(d) It is not possible to predict, because repetitive DNA may or may not be variable between organisms.

[16-4] What is the definition of a gene? Use a recent definition introduced as part of the ENCODE project.

(a) A gene is a unit of hereditary information localized to a particular chromosome position and encoding one protein.

(b) A gene is a unit of hereditary information localized to a particular chromosome position and encoding one or more protein products.

(c) A gene is a union of genomic sequences encoding a coherent set of potentially overlapping functional products.

(d) A gene is a unit of hereditary information encoding one or more functional products.

[16-5] It is extremely difficult for intrinsic (ab initio) gene-finding algorithms to predict protein-coding genes in eukaryotic genomic DNA . What is the main problem?

  (a) Exon/intron borders are hard to predict.

  (b) Introns may be many kilobases in length.

  (c) The GC content of coding regions is not always differentiated from the GC content of noncoding regions.

  (d) All of the above.

[16-6] What are some of the properties of ultraconserved elements?

  (a) They have variable lengths (from 50 to >1000 base pairs) and are nearly perfectly conserved.

  (b) They have variable lengths (from 50 to >1000 base pairs), are nearly perfectly conserved, and typically correspond to protein-coding regions.

  (c) They have lengths ≥200 base pairs and are perfectly or nearly perfectly conserved between relatively closely related species such as rats and mice.

  (d) They have lengths ≥200 base pairs and are perfectly or nearly perfectly conserved between relatively distantly related species such as humans and rodents.

[16-7] The genomes of two distinct eukaryotic species can sometimes merge to create an entirely new species.

  (a) True

  (b) False

[16-8] According to Ohno's 2R hypothesis, whole genome duplication (polyploidy) offers several advantages. Which of the following is NOT an advantage?

  (a) Hybrids may propagate more successfully than their parents.

  (b) Genes may become redundant, allowing novel functions to emerge.

  (c) Self-fertilization may become possible.

  (d) Self-fertilizing organisms may become able to interbreed.

[16-9] Several mechanisms have been proposed by which new gene families are formed. According to the birth-and-death evolution model,

  (a) New genes arise by gene duplication followed by either functional diversification or inactivation.

  (b) Genes acquire novel functions as a gradual process that follows gene duplication.

  (c) Members of a gene family evolve in a concerted manner.

  (d) New genes arise and acquire new functions in a coordinated manner dependent on the death of other duplicated genes.

[16-10] Single nucleotide polymorphism (SNP) arrays can reliably detect all of the following phenomena except which one?

  (a) Deletions

  (b) Duplications

  (c) Inversions

  (d) Uniparental isodisomy

Suggested Reading

Our understanding of eukaryotic chromosomes has been transformed by the sequencing and analysis of genomes. The ENCODE Project Consortium papers of 2004 (introducing the project) and 2007 (describing an overview of the results of analyzing 1% of the human genome) are essential resources.

Evan Eichler and David Sankoff (2003) and Coghlan (2005) provide recommended reviews of eukaryotic chromosome evolution from a genomics perspective. The classic studies of Britten and Kohne 1968 are highly recommended for explanations of repetitive DNA. For eukaryotic gene prediction, a review of software programs by Makarov (2002) is recommended.

References

Adams, K. L., and Wendel, J. F. Novel patterns of gene expression in polyploid plants. *Trends Genet.* **21**, 539–543 (2005).

Allen, J. E., Majoros, W. H., Pertea, M., and Salzberg, S. L. JIGSAW, GeneZilla, and GlimmerHMM: Puzzling out the features of human genes in the ENCODE regions. *Genome Biol.* **7** (Suppl 1), S9.1–S9.13 (2006).

Allen, J. E., Pertea, M., and Salzberg, S. L. Computational gene prediction using multiple sources of evidence. *Genome Res.* **14**, 142–148 (2004).

Allen, J. E., and Salzberg, S. L. JIGSAW: Integration of multiple sources of evidence for gene prediction. *Bioinformatics* **21**, 3596–3603 (2005).

Allen, K. D. Assaying gene content in *Arabidopsis*. *Proc. Natl. Acad. Sci. USA* **99**, 9568–9572 (2002).

Ambros, V. microRNAs: Tiny regulators with great potential. *Cell* **107**, 823–826 (2001).

Amor, D. J., and Choo, K. H. Neocentromeres: Role in human disease, evolution, and centromere study. *Am. J. Hum. Genet.* **71**, 695–714 (2002).

Antonarakis, S. E., et al. Factor VIII gene inversions in severe hemophilia A: Results of an international consortium study. *Blood* **86**, 2206–2212 (1995).

Aparicio, S., et al. Whole-genome shotgun assembly and analysis of the genome of *Fugu rubripes*. *Science* **297**, 1301–1310 (2002).

Arabidopsis Genome Initiative. Analysis of the genome sequence of the flowering plant *Arabidopsis thaliana*. *Nature* **408**, 796–815 (2000).

Aulard, S., Monti, L., Chaminade, N., and Lemeunier, F. Mitotic and polytene chromosomes: Comparisons between *Drosophila melanogaster* and *Drosophila simulans*. *Genetica* **120**, 137–150 (2004).

Aury, J. M., et al. Global trends of whole-genome duplications revealed by the ciliate *Paramecium tetraurelia*. *Nature* **444**, 171–178 (2006).

Avramova, Z. V. Heterochromatin in animals and plants. Similarities and differences. *Plant Physiol.* **129**, 40–49 (2002).

Azzalin, C. M., Nergadze, S. G., and Giulotto, E. Human intra-chromosomal telomeric-like repeats: Sequence organization and mechanisms of origin. *Chromosoma* **110**, 75–82 (2001).

Bailey, J. A., and Eichler, E. E. Primate segmental duplications: Crucibles of evolution, diversity and disease. *Nat. Rev. Genet.* **7**, 552–564 (2006).

Bailey, J. A., Yavor, A. M., Massa, H. F., Trask, B. J., and Eichler, E. E. Segmental duplications: Organization and impact within the current human genome project assembly. *Genome Res.* **11**, 1005–1017 (2001).

Balakirev, E. S., and Ayala, F. J. Pseudogenes: Are they "junk" or functional DNA? *Annu. Rev. Genet.* **37**, 123–151 (2003).

Bejerano, G., Pheasant, M., Makunin, I., Stephen, S., Kent, W. J., Mattick, J. S., and Haussler, D. Ultraconserved elements in the human genome. *Science* **304**, 1321–1325 (2004).

Benson, G. Tandem repeats finder: A program to analyze DNA sequences. *Nucleic Acids Res.* **27**, 573–580 (1999).

Ben-Yehuda, S., Rudner, D. Z., and Losick, R. RacA, a bacterial protein that anchors chromosomes to the cell poles. *Science* **19**, 19 (2002).

Betrán, E., and Long, M. Expansion of genome coding regions by acquisition of new genes. *Genetica* **115**, 65–80 (2002).

Blackburn, E. H., et al. Recognition and elongation of telomeres by telomerase. *Genome* **31**, 553–560 (1989).

Blattner, F. R., et al. The complete genome sequence of *Escherichia coli* K-12. *Science* **277**, 1453–1474 (1997).

Bray, N., Dubchak, I., and Pachter, L. AVID: A Global Alignment Program. *Genome Res.* **13**, 97–102 (2003).

Britten, R. J., and Kohne, D. E. Repeated sequences in DNA. *Science* **161**, 529–540 (1968).

Brooks, D. R., and Isaac, R. E. Functional genomics of parasitic worms: The dawn of a new era. *Parasitol. Int.* **51**, 319–325 (2002).

Brunet, F. G., et al. Gene loss and evolutionary rates following whole-genome duplication in teleost fishes. *Mol. Biol. Evol.* **23**, 1808–1816 (2006).

Burge, C., and Karlin, S. Prediction of complete gene structures in human genomic DNA. *J. Mol. Biol.* **268**, 78–94 (1997).

Burge, C. B., and Karlin, S. Finding the genes in genomic DNA. *Curr. Opin. Struct. Biol.* **8**, 346–354 (1998).

Cameron, R. A., et al. A sea urchin genome project: Sequence scan, virtual map, and additional resources. *Proc. Natl. Acad. Sci. USA* **97**, 9514–9518 (2000).

Carlton, J. M., et al. Genome sequence and comparative analysis of the model rodent malaria parasite *Plasmodium yoelii yoelii*. *Nature* **419**, 512–519 (2002).

Castillo-Davis, C. I. The evolution of noncoding DNA: How much junk, how much func? *Trends Genet.* **21**, 533–536 (2005).

Cavalier-Smith, T. Origins of the machinery of recombination and sex. *Heredity* **88**, 125–141 (2002).

C. elegans Sequencing Consortium. Genome sequence of the nematode *C. elegans*: A platform for investigating biology. *Science* **282**, 2012–2018 (1998).

Chaisson, M. J., Raphael, B. J., and Pevzner, P. A. Microinversions in mammalian evolution. *Proc. Natl. Acad. Sci. USA* **103**, 19824–19829 (2006).

Chan, P., Simon-Chazottes, D., Mattei, M. G., Guenet, J. L., and Salier, J. P. Comparative mapping of lipocalin genes in human and mouse: The four genes for complement C8 gamma chain, prostaglandin-D-synthase, oncogene-24p3, and progestagen-associated endometrial protein map to HSA9 and MMU2. *Genomics* **23**, 145–150 (1994).

Choo, K. H. Domain organization at the centromere and neocentromere. *Dev. Cell.* **1**, 165–177 (2001).

Claverie, J. M. Computational methods for the identification of genes in vertebrate genomic sequences. *Hum. Mol. Genet.* **6**, 1735–1744 (1997).

Claverie, J. M. Gene number. What if there are only 30,000 human genes? *Science* **291**, 1255–1257 (2001).

Coghlan, A. Chromosome evolution in eukaryotes: A multi-kingdom perspective. *Trends Geneti* **21**, 673–682 (2005).

Comai, L. Genetic and epigenetic interactions in allopolyploid plants. *Plant Mol. Biol.* **43**, 387–399 (2000).

Comai, L. The advantages and disadvantages of being polyploid. *Nat. Rev. Genet.* **6**, 836–846 (2005).

Cummings, C. J., and Zoghbi, H. Y. Trinucleotide repeats: Mechanisms and pathophysiology. *Annu. Rev. Genomics Hum. Genet.* **1**, 281–328 (2000).

Darlington, C. D. *Recent Advances in Cytology*. P. Blakiston's Son & Co., Philadelphia, 1932.

Dehal, P., and Boore, J. L. Two rounds of whole genome duplication in the ancestral vertebrate. *PLoS Biol.* **3**, e314 (2005).

Dermitzakis, E. T., et al. Numerous potentially functional but non-genic conserved sequences on human chromosome 21. *Nature* **420**, 578–582 (2002).

Dewald, G., et al. The human complement C8G gene, a member of the lipocalin gene family: Polymorphisms and mapping to chromosome 9q34.3. *Ann. Hum. Genet.* **60**, 281–291 (1996).

Douglas, S., et al. The highly reduced genome of an enslaved algal nucleus. *Nature* **410**, 1091–1096 (2001).

Drosophila 12 Genomes Consortium. Evolution of genes and genomes on the *Drosophila* phylogeny. *Nature* **450**, 203–218 (2007).

Dunham, A., et al. The DNA sequence and analysis of human chromosome 13. *Nature* **428**, 522–528 (2004).

Echols, N., et al. Comprehensive analysis of amino acid and nucleotide composition in eukaryotic genomes, comparing genes and pseudogenes. *Nucleic Acids Res.* **30**, 2515–2523 (2002).

Eddy, S. R. Non-coding RNA genes and the modern RNA world. *Nat. Rev. Genet.* **2**, 919–929 (2001).

Eddy, S. R. Computational genomics of noncoding RNA genes. *Cell* **109**, 137–140 (2002).

Eichler, E. E., and Sankoff, D. Structural dynamics of eukaryotic chromosome evolution. *Science* **301**, 793–797 (2003).

Elgar, G. Different words, same meaning: understanding the languages of the genome. *Trends Genet.* **22**, 639–641 (2006).

ENCODE Project Consortium. The ENCODE (ENCyclopedia Of DNA Elements) Project. *Science* **306**, 636–640 (2004).

ENCODE Project Consortium, et al. Identification and analysis of functional elements in 1% of the human genome by the ENCODE pilot project. *Nature* **447**, 799–816 (2007).

Fan, Y., Linardopoulou, E., Friedman, C., Williams, E., and Trask, B. J. Genomic structure and evolution of the ancestral chromosome fusion site in 2q13–2q14.1 and paralogous regions on other human chromosomes. *Genome Res.* **12**, 1651–1662 (2002).

Fisher, S., Grice, E. A., Vinton, R. M., Bessling, S. L., and McCallion, A. S. Conservation of RET regulatory function from human to zebrafish without sequence similarity. *Science* **312**, 276–279 (2006).

Frazer, K. A., Elnitski, L., Church, D. M., Dubchak, I., and Hardison, R. C. Cross-species sequence comparisons: A review of methods and available resources. *Genome Res.* **13**, 1–12 (2003).

Frith, M. C., Forrest, A. R., Nourbakhsh, E., Pang, K. C., Kai, C., Kawai, J., Carninci, P., Hayashizaki, Y., Bailey, T. L., and Grimmond, S. M. The abundance of short proteins in the mammalian proteome. *PLoS Genet.* **2**, e52 (2006).

Gardner, M. J., et al. Genome sequence of the human malaria parasite *Plasmodium falciparum*. *Nature* **419**, 498–511 (2002).

Gerstein, M. B., Bruce, C., Rozowsky, J. S., Zheng, D., Du, J., Korbel, J. O., Emanuelsson, O., Zhang, Z. D., Weissman, S., and Snyder, M. What is a gene, post-ENCODE? History and updated definition. *Genome Res.* **17**, 669–681 (2007).

Graur, D., and Li, W.-H. *Fundamentals of Molecular Evolution*. Sinauer Associates, Sunderland, MA, 2000.

Griffith, O. L., et al. ORegAnno: An open-access community-driven resource for regulatory annotation. *Nucleic Acids Res.* **36**, D107–D113 (2008).

Griffiths-Jones, S., Bateman, A., Marshall, M., Khanna, A., and Eddy, S. R. Rfam: An RNA family database. *Nucleic Acids Res.* **31**, 439–441 (2003).

Grimwood, J., et al. The DNA sequence and biology of human chromosome 19. *Nature* **428**, 529–535 (2004).

Guigo, R., et al. EGASP: The human ENCODE Genome Annotation Assessment Project. *Genome Biol.* **7**, S2.1–31 (2006).

Gutierrez, G., Ganfornina, M. D., and Sanchez, D. Evolution of the lipocalin family as inferred from a protein sequence phylogeny. *Biochim. Biophys. Acta* **1482**, 35–45 (2000).

Hall, A. E., Fiebig, A., and Preuss, D. Beyond the *Arabidopsis* genome: Opportunities for comparative genomics. *Plant Physiol.* **129**, 1439–1447 (2002).

Hancock, J. M. Genome size and the accumulation of simple sequence repeats: Implications of new data from genome sequencing projects. *Genetica* **115**, 93–103 (2002).

Harrison, P. M., and Gerstein, M. Studying genomes through the aeons: Protein families, pseudogenes and proteome evolution. *J. Mol. Biol.* **318**, 1155–1174 (2002).

Harrison, P. M., Kumar, A., Lang, N., Snyder, M., and Gerstein, M. A question of size: The eukaryotic proteome and the problems in defining it. *Nucleic Acids Res.* **30**, 1083–1090 (2002).

Hartl, D. L. Molecular melodies in high and low C. *Nat. Rev. Genet.* **1**, 145–149 (2000).

Hassold, T., and Hunt, P. To err (meiotically) is human: The genesis of human aneuploidy. *Nat. Rev. Genet.* **2**, 280–291 (2001).

Haugen, P., Simon, D. M., and Bhattacharya, D. The natural history of group I introns. *Trends Genet.* **21**, 111–119 (2005).

Hedges, S. B., and Kumar, S. Genomics. Vertebrate genomes compared. *Science* **297**, 1283–1285 (2002).

Holt, R. A., et al. The genome sequence of the malaria mosquito *Anopheles gambiae*. *Science* **298**, 129–149 (2002).

Hou, G., Le Blancq, S. M., Yaping, E., Zhu, H., and Lee, M. G. Structure of a frequently rearranged rRNA-encoding chromosome in *Giardia lamblia*. *Nucleic Acids Res.* **23**, 3310–3317 (1995).

Hubbard, T. J., et al. Ensembl 2007. *Nucleic Acids Res.* **35**(Database issue), D610–D617 (2007).

Hughes, A. L., and Nei, M. Nucleotide substitution at major histocompatibility complex class II loci: Evidence for overdominant selection. *Proc. Natl. Acad. Sci. USA* **86**, 958–962 (1989).

Ijdo, J. W., Baldini, A., Ward, D. C., Reeders, S. T., and Wells, R. A. Origin of human chromosome 2: An ancestral telomere–telomere fusion. *Proc. Natl. Acad. Sci. USA* **88**, 9051–9055 (1991a).

Ijdo, J. W., Wells, R. A., Baldini, A., and Reeders, S. T. Improved telomere detection using a telomere repeat probe (TTAGGG)n generated by PCR. *Nucleic Acids Res.* **19**, 4780 (1991b).

International HapMap Consortium. A haplotype map of the human genome. *Nature* **437**, 1299–1320 (2005).

International HapMap Consortium, et al. A second generation human haplotype map of over 3.1 million SNPs. *Nature* **449**, 851–861 (2007).

International Human Genome Sequencing Consortium. Initial sequencing and analysis of the human genome. *Nature* **409**, 860–921 (2001).

International Human Genome Sequencing Consortium. Finishing the euchromatic sequence of the human genome. *Nature* **431**, 931–945 (2004).

Jarstfer, M. B., and Cech, T. R. Effects of nucleotide analogues on *Euplotes aediculatus* telomerase processivity: Evidence for product-assisted translocation. *Biochemistry* **41**, 151–161 (2002).

Jeffares, D. C., Mourier, T., and Penny, D. The biology of intron gain and loss. *Trends Genet.* **22**, 16–22 (2006).

Jentsch, S., Tobler, H., and Muller, F. New telomere formation during the process of chromatin diminution in *Ascaris suum*. *Int. J. Dev. Biol.* **46**, 143–148 (2002).

Jones, K. L. *Smith's Recognizable Patterns of Human Malformation*, W. B. Saunders, New York, 1997.

Jurka, J. Repeats in genomic DNA: Mining and meaning. *Curr. Opin. Struct. Biol.* **8**, 333–337 (1998).

Jurka, J. Repbase update: A database and an electronic journal of repetitive elements. *Trends Genet.* **16**, 418–420 (2000).

Kallioniemi, A., Kallioniemi, O. P., Sudar, D., Rutovitz, D., Gray, J. W., Waldman, F., and Pinkel, D. Comparative genomic hybridization for molecular cytogenetic analysis of solid tumors. *Science* **258**, 818–821 (1992).

Karolchik, D., Hinrichs, A. S., Furey, T. S., Roskin, K. M., Sugnet, C. W., Haussler, D., and Kent, W. J. The UCSC Table Browser data retrieval tool. *Nucleic Acids Res.* **32**(Database issue), D493–D496 (2004).

Kashi, Y., and King, D. G. Simple sequence repeats as advantageous mutators in evolution. *Trends Genet.* **22**, 253–259 (2006).

Katzman, S., Kern, A. D., Bejerano, G., Fewell, G., Fulton, L., Wilson, R. K., Salama, S. R., and Haussler, D. Human genome ultraconserved elements are ultraselected. *Science* **317**, 915 (2007).

Kent, W. J. BLAT: The BLAST-like alignment tool. *Genome Res.* **12**, 656–664 (2002).

Kidwell, M. G. Transposable elements and the evolution of genome size in eukaryotes. *Genetica* **115**, 49–63 (2002).

King, D. C., Taylor, J., Elnitski, L., Chiaromonte, F., Miller, W., and Hardison, R. C. Evaluation of regulatory potential and conservation scores for detecting cis-regulatory modules in aligned mammalian genome sequences. *Genome Res.* **15**, 1051–1060 (2005).

Kohn, M., Högel, J., Vogel, W., Minich, P., Kehrer-Sawatzki, H., Graves, J. A., and Hameister, H. Reconstruction of a 450-MY-old ancestral vertebrate protokaryotype. *Trends Genet.* **22**, 203–210 (2006).

Knight, J. All genomes great and small. *Nature* **417**, 374–376 (2002).

Kotzot, D. Complex and segmental uniparental disomy (UPD): Review and lessons from rare chromosomal complements. *J. Med. Genet.* **38**, 497–507 (2001).

Kuhn, R. M., et al. The UCSC genome browser database: Update 2007. *Nucleic Acids Res.* **35**(Database issue), D668–D673 (2007).

Kurtz, S., Phillippy, A., Delcher, A. L., Smoot, M., Shumway, M., Antonescu, C., and Salzberg, S. L. Versatile and open software for comparing large genomes. *Genome Biol.* **5**, R12 (2004).

Lacazette, E., Gachon, A. M., and Pitiot, G. A novel human odorant-binding protein gene family resulting from genomic duplicons at 9q34: Differential expression in the oral and genital spheres. *Hum. Mol. Genet.* **9**, 289–301 (2000).

Lai, W. R., Johnson, M. D., Kucherlapati, R., and Park, P. J. Comparative analysis of algorithms for identifying amplifications and deletions in array CGH data. *Bioinformatics* **21**, 3763–3770 (2005).

Latham, K. E. X chromosome imprinting and inactivation in pre-implantation mammalian embryos. *Trends Genet.* **21**, 120–127 (2005).

Lee, H. S., and Chen, Z. J. Protein-coding genes are epigenetically regulated in *Arabidopsis* polyploids. *Proc. Natl. Acad. Sci. USA* **98**, 6753–6758 (2001).

Lewis, M. S., and Pikaard, C. S. Restricted chromosomal silencing in nucleolar dominance. *Proc. Natl. Acad. Sci. USA* **98**, 14536–14540 (2001).

Li, W. H., Yang, J., and Gu, X. Expression divergence between duplicate genes. *Trends Genet.* **21**, 602–607 (2005).

Lima-de-Faria, A. *One Hundred Years of Chromosome Research and What Remains to Be Learned*. Kluwer Academic Publishers, Boston, 2003.

Lorite, P., Carrillo, J. A., and Palomeque, T. Conservation of (TTAGG)(n) telomeric sequences among ants (Hymenoptera, Formicidae). *J. Hered.* **93**, 282–285 (2002).

Lowe, T. M., and Eddy, S. R. tRNAscan-SE: A program for improved detection of transfer RNA genes in genomic sequence. *Nucleic Acids Res.* **25**, 955–964 (1997).

Makalowski, W. Genomic scrap yard: How genomes utilize all that junk. *Gene* **259**, 61–67 (2000).

Makarov, V. Computer programs for eukaryotic gene prediction. *Brief. Bioinform.* **3**, 195–199 (2002).

Martin, C. L., et al. The evolutionary origin of human subtelomeric homologies—or where the ends begin. *Am. J. Hum. Genet.* **70**, 972–984 (2002).

Maston, G. A., Evans, S. K., and Green, M. R. Transcriptional regulatory elements in the human genome. *Annu. Rev. Genomics Hum. Genet.* **7**, 29–59 (2006).

Mayor, C., et al. VISTA: Visualizing global DNA sequence alignments of arbitrary length. *Bioinformatics* **16**, 1046–1047 (2000).

McCormick-Graham, M., and Romero, D. P. A single telomerase RNA is sufficient for the synthesis of variable telomeric DNA repeats in ciliates of the genus *Paramecium. Mol. Cell. Biol.* **16**, 1871–1879 (1996).

McKnight, T. D., Fitzgerald, M. S., and Shippen, D. E. Plant telomeres and telomerases. A review. *Biochemistry (Mosc.)* **62**, 1224–1231 (1997).

Meaburn, K. J., and Misteli, T. Cell biology: Chromosome territories. *Nature* **445**, 379–381 (2007).

Melek, M., Davis, B. T., and Shippen, D. E. Oligonucleotides complementary to the *Oxytricha nova* telomerase RNA delineate the template domain and uncover a novel mode of primer utilization. *Mol. Cell. Biol.* **14**, 7827–7838 (1994).

Mikkelsen, T. S., et al. Genome-wide maps of chromatin state in pluripotent and lineage-committed cells. *Nature* **448**, 553–560 (2007).

Moller-Jensen, J., Jensen, R. B., Lowe, J., and Gerdes, K. Prokaryotic DNA segregation by an actin-like filament. *EMBO J.* **21**, 3119–3127 (2002).

Morgenstern, B., et al. Exon discovery by genomic sequence alignment. *Bioinformatics* **18**, 777–787 (2002).

Morison, I. M., Ramsay, J. P., and Spencer, H. G. A census of mammalian imprinting. *Trends Genet.* **21**, 457–465 (2005).

Müller, F., and Tobler, H. Chromatin diminution in the parasitic nematodes *Ascaris suum* and *Parascaris univalens. Int. J. Parasitol.* **30**, 391–399 (2000).

Mural, R. J. Current status of computational gene finding: A perspective. *Methods Enzymol.* **303**, 77–83 (1999).

Murphy, W. J., et al. Dynamics of mammalian chromosome evolution inferred from multispecies comparative maps. *Science* **309**, 613–617 (2005).

Muzny, D. M., et al. The DNA sequence, annotation and analysis of human chromosome 3. *Nature* **440**, 1194–1198 (2006).

Nanda, I., et al. Distribution of telomeric (TTAGGG)(n) sequences in avian chromosomes. *Chromosoma* **111**, 215–227 (2002).

Nei, M., and Rooney, A. P. Concerted and birth-and-death evolution of multigene families. *Annu. Rev. Genet.* **39**, 121–152 (2005).

Nixon, J. E., et al. A spliceosomal intron in *Giardia lamblia. Proc. Natl. Acad. Sci. USA* **99**, 3701–3705 (2002).

Nóbrega, M. A., Zhu, Y., Plajzer-Frick, I., Afzal, V., and Rubin, E. M. Megabase deletions of gene deserts result in viable mice. *Nature* **431**, 988–993 (2004).

Novichkov, P. S., Gelfand, M. S., and Mironov, A. A. Gene recognition in eukaryotic DNA by comparison of genomic sequences. *Bioinformatics* **17**, 1011–1018 (2001).

Ohno, S. *Evolution by Gene Duplication.* Springer Verlag, Berlin, 1970.

Olshen, A. B., Venkatraman, E. S., Lucito, R., and Wigler, M. Circular binary segmentation for the analysis of array-based DNA copy number data. *Biostatistics* **5**, 557–572 (2004).

Ostertag, E. M., and Kazazian, H. H., Jr. Twin priming: A proposed mechanism for the creation of inversions in L1 retrotransposition. *Genome Res.* **11**, 2059–2065 (2001).

Panopoulou, G., and Poustka, A. J. Timing and mechanism of ancient vertebrate genome duplications: The adventure of a hypothesis. *Trends Genet.* **21**, 559–567 (2005).

Passarge, E., Horsthemke, B., and Farber, R. A. Incorrect use of the term synteny. *Nat. Genet.* **23**, 387 (1999).

Paterson, A. H., Chapman, B. A., Kissinger, J. C., Bowers, J. E., Feltus, F. A., and Estill, J. C. Many gene and domain families have convergent fates following independent whole-genome duplication events in *Arabidopsis, Oryza, Saccharomyces* and *Tetraodon. Trends Genet.* **22**, 597–602 (2006).

Pavlicek, A., Gentles, A. J., Paces, J., Paces, V., and Jurka, J. Retroposition of processed pseudogenes: The impact of RNA stability and translational control. *Trends Genet.* **22**, 69–73 (2006).

Pelosi, P. Perireceptor events in olfaction. *J. Neurobiol.* **30**, 3–19 (1996).

Pevsner, J., Hou, V., Snowman, A. M., and Snyder, S. H. Odorant-binding protein. Characterization of ligand binding. *J. Biol. Chem.* **265**, 6118–6125 (1990).

Prince, V. E., and Pickett, F. B. Splitting pairs: The diverging fates of duplicated genes. *Nature Rev. Genet.* **3**, 827–837 (2002).

Redi, C. A., Garagna, S., Zacharias, H., Zuccotti, M., and Capanna, E. The other chromatin. *Chromosoma* **110**, 136–147 (2001).

Reese, M. G., and Guigó, R. EGASP: Introduction. *Genome Biol.* **7** (Suppl 1), S1.1–1.3 (2006).

Rogic, S., Mackworth, A. K., and Ouellette, F. B. Evaluation of gene-finding programs on mammalian sequences. *Genome Res.* **11**, 817–832 (2001).

Roy, S. W. Intron-rich ancestors. *Trends Genet.* **22**, 468–471 (2006).

Roy, S. W., and Gilbert, W. The evolution of spliceosomal introns: Patterns, puzzles and progress. *Nat. Rev. Genet.* **7**, 211–221 (2006).

Ruvkun, G. Molecular biology. Glimpses of a tiny RNA world. *Science* **294**, 797–799 (2001).

Sabo, P. J., et al. Genome-scale mapping of DNase I sensitivity in vivo using tiling DNA microarrays. *Nat. Methods* **3**, 511–518 (2006).

Salier, J. P. Chromosomal location, exon/intron organization and evolution of lipocalin genes. *Biochim. Biophys. Acta* **1482**, 25–34 (2000).

Sasaki, T., et al. The genome sequence and structure of rice chromosome 1. *Nature* **420**, 312–316 (2002).

Schlötterer, C., and Harr, B. *Drosophila virilis* has long and highly polymorphic microsatellites. *Mol. Biol. Evol.* **17**, 1641–1646 (2000).

Schwartz, S., et al. PipMaker: A web server for aligning two genomic DNA sequences. *Genome Res.* **10**, 577–586 (2000).

She, X., Jiang, Z., Clark, R. A., Liu, G., Cheng, Z., Tuzun, E., Church, D. M., Sutton, G., Halpern, A. L., and Eichler, E. E. Shotgun sequence assembly and recent segmental duplications within the human genome. *Nature* **431**, 927–930 (2004).

Shippen-Lentz, D., and Blackburn, E. H. Telomere terminal transferase activity from *Euplotes crassus* adds large numbers of TTTTGGGG repeats onto telomeric primers. *Mol. Cell. Biol.* **9**, 2761–2764 (1989).

Siepel, A., et al. Evolutionarily conserved elements in vertebrate, insect, worm, and yeast genomes. *Genome Res.* **15**, 1034–1050 (2005).

Simpson, A. G., MacQuarrie, E. K., and Roger, A. J. Eukaryotic evolution: Early origin of canonical introns. *Nature* **419**, 270 (2002).

Slijepcevic, P. Telomeres and mechanisms of Robertsonian fusion. *Chromosoma* **107**, 136–140 (1998).

Smit, A. F. Interspersed repeats and other mementos of transposable elements in mammalian genomes. *Curr. Opin. Genet. Dev.* **9**, 657–663 (1999).

Stein, L. Genome annotation: From sequence to biology. *Nat. Rev. Genet.* **2**, 493–503 (2001).

Stefansson, H. et al. A common inversion under selection in Europeans. *Nature Genet.* **37**, 129–137 (2005).

Sturtevant, A. H. A case of rearrangement of genes in *Drosophila*. *Proc. Natl. Acad. Sci. USA* 7, 235–237 (1921).

Taylor, J., Tyekucheva, S., King, D. C., Hardison, R. C., Miller, W., and Chiaromonte, F. ESPERR: Learning strong and weak signals in genomic sequence alignments to identify functional elements. *Genome Res.* **16**, 1596–1604 (2006).

Taylor, J. S., and Raes, J. Duplication and divergence: The evolution of new genes and old ideas. *Annu. Rev. Genet.* **38**, 615–643 (2004).

Thomas, B. C., Pedersen, B., and Freeling, M. Following tetraploidy in an *Arabidopsis* ancestor, genes were removed preferentially from one homeolog leaving clusters enriched in dose-sensitive genes. *Genome Res.* **16**, 934–946 (2006).

Thomas, D. J., et al. The ENCODE Project at UC Santa Cruz. *Nucleic Acids Res.* **35**(Database issue), D663–D667 (2007a).

Thomas, D. J., Trumbower, H., Kern, A. D., Rhead, B. L., Kuhn, R. M., Haussler, D., and Kent, W. J. Variation resources at UC Santa Cruz. *Nucleic Acids Res.* **35**(Database issue), D716–D720 (2007b).

Ting, J. C., Roberson, E. D., Miller, N. D., Lysholm-Bernacchi, A., Stephan, D. A., Capone, G. T., Ruczinski, I., Thomas, G. H., and Pevsner, J. Visualization of uniparental inheritance, Mendelian inconsistencies, deletions, and parent of origin effects in single nucleotide polymorphism trio data with SNPtrio. *Hum. Mutat.* **28**, 1225–1235 (2007).

Trask, B. J. Human cytogenetics: 46 chromosomes, 46 years and counting. *Nature Rev. Genet.* **3**, 769–778 (2002).

Upcroft, P., Chen, N., and Upcroft, J. A. Telomeric organization of a variable and inducible toxin gene family in the ancient eukaryote *Giardia duodenalis*. *Genome Res.* **7**, 37–46 (1997).

Vellai, T., and Vida, G. The origin of eukaryotes: The difference between prokaryotic and eukaryotic cells. *Proc. R. Soc. Lond. B Biol. Sci.* **266**, 1571–1577 (1999).

Venkatraman, E. S., and Olshen, A. B. A faster circular binary segmentation algorithm for the analysis of array CGH data. *Bioinformatics* **23**, 657–663 (2007).

Venter, J. C., et al. The sequence of the human genome. *Science* **291**, 1304–1351 (2001).

Ventura, M. et al. Evolutionary formation of new centromeres in macaque. *Science* **316**, 243–246 (2007).

Waldeyer, H., Über Karyokinese und ihre Beziehungen zu den Befruchtungsvorgängen. *Archiv für mikroskopische Anatomie und Entwicklungsmechanik* **32**, 1–122 (1888).

Watt, W. B., and Dean, A. M. Molecular-functional studies of adaptive genetic variation in prokaryotes and eukaryotes. *Annu. Rev. Genet.* **34**, 593–622 (2000).

Willenbrock, H. and Fridlyand, J. A comparison study: Applying segmentation to array CGH data for downstream analyses. *Bioinformatics* **21**, 4084–4091 (2005).

Yasuhara, J. C., and Wakimoto, B. T. Oxymoron no more: The expanding world of heterochromatic genes. *Trends Genet.* **22**, 330–338 (2006).

Yu, J., et al. A draft sequence of the rice genome (*Oryza sativa* L. ssp. *indica*). *Science* **296**, 79–92 (2002).

Zdobnov, E. M., et al. Comparative genome and proteome analysis of *Anopheles gambiae* and *Drosophila melanogaster*. *Science* **298**, 149–159 (2002).

Zheng, D., Frankish, A., Baertsch, R., Kapranov, P., Reymond, A., Choo, S. W., Lu, Y., Denoeud, F., Antonarakis, S. E., Snyder, M., Ruan, Y., Wei, C. L., Gingeras, T. R., Guigó, R., Harrow, J., and Gerstein, M. B. Pseudogenes in the ENCODE regions: Consensus annotation, analysis of transcription, and evolution. *Genome Res.* **17**, 839–851 (2007).

Some 200 fungal species are known to be pathogenic for humans, distressing millions of people. From the times of Greek and Roman antiquity to the middle of the nineteenth century only two fungal diseases were known: ringworm (tinea) and thrush (oral candidiasis) (Ainsworth, 1993). Ringworm is caused by fungi of the genera Microsporum, Trichophyton, *and* Epidermophyton. *Candidiasis (including thrush) is caused by* Candida albicans *and other* Candida *species. This image from Kuchenmeister (1857, plate IV) shows the thrush fungus, at that time called* Oidium albicans *(Figs. 3 to 8).*

17

Eukaryotic Genomes: Fungi

INTRODUCTION

According to the classification system of Whittaker (1969), there are five kingdoms of life: monera (prokaryotes), protoctists, animals, fungi, and plants. We have examined the prokaryotes in Chapter 15, and introduced the eukaryotic chromosome in Chapter 16. In this chapter we begin our exploration of eukaryotes by studying one of the kingdoms, Fungi. This diverse and interesting group of organisms last shared a common ancestor with plants and animals 1.5 billion years ago (BYA) (Wang et al., 1999, discussed in Chapter 18). We may think of fungi as organisms such as mushrooms that might be studied by botanists. Surprisingly, fungi are far more closely related to animals than to plants. In Chapter 18 we will extend our study to the entire kingdom of eukaryotes, including animals, plants, and a variety of protozoa. We will then discuss humans (Chapter 19).

The first eukaryotic genome to be fully sequenced was the 13 million base pair (Mb) genome of a fungus, the budding yeast *Saccharomyces cerevisiae*. Its genome is very small compared with that of humans (3 billion base pairs, or gigabase pairs [Gb]), and its size is only severalfold larger than a typical bacterial genome. This yeast has served as a model eukaryotic organism for genetics studies because it grows rapidly, it can be genetically modified easily, and many of its cellular functions are conserved with metazoans and other eukaryotes. More recently, it has become a model organism for functional genomics studies (Chapter 12). Every one of its

Mycology (from the Greek word *mukes*, "fungus") is the study of fungi. Mycosis is a disease or ailment caused by fungi. The suffix −*mycota* refers to fungi: the kingdom Fungi is also called the kingdom Eumycota ("true fungi").

approximately 6000 genes has been characterized, deleted, overexpressed, and characterized functionally using a variety of assays.

Now, as whole genome sequencing has become routine, the sequencing of yeasts and other fungi has progressed at an accelerated pace. While the fungi are eukaryotes, and share many properties in common with the metazoans (animals), most have relatively small genome sizes. Through comparative analysis we are gaining new insights into many basic properties of genome structure and evolution, including whole genome duplications and the fate of duplicated genes (Dujon, 2006).

This chapter begins with an overview of the fungi. We will then describe bioinformatic approaches to analyzing the *S. cerevisiae* genome. Finally, we will describe the sequencing of other fungal genomes and the early lessons of comparative genomics in fungi.

Description and Classification of Fungi

Morphologically, fungi are characterized by hyphae (filaments) that grow and may branch. The Museum of Paleontology at the University of California, Berkeley, offers an introduction to fungi, including photographs of many species (► http://www. ucmp.berkeley.edu/fungi/fungi. html). The American Museum of Natural History (New York) also provides an overview of fungi (► http://ology.amnh.org/ biodiversity/treeoflife/pages/ fungi.html).

Fungi are eukaryotic organisms that can be filamentous (as in the case of molds) or unicellular (as in the case of yeasts such as *S. cerevisiae*). The main criteria for classifying fungi are based on morphology (e.g., ultrastructure), biochemistry (e.g., growth properties or cell wall composition), and molecular sequence data (DNA, RNA, and protein sequences). Most fungi are aerobic, and all are heterotrophs that absorb their food. Fungi are typically very hardy, forming spores composed of chitin that are immobile throughout their lifespan. They have a major role in the ecosystem in degrading organic waste material. Fungi are important causative agents of disease in humans, other animals, and plants. Fungi also have key roles in fermentation; the fungal mold *Rhizopus nigricans* is used in the manufacture of steroids such as cortisone, and *Penicillium chrysogenum* produces the antibiotic penicillin.

The relationships of many species throughout the tree of life have been described in phylogenetic analyses based on small-subunit ribosomal RNA (Fig. 13.1). In a complementary approach, W. F. Doolittle and colleagues defined a phylogeny of the eukaryotes based on the concatenated amino acid sequences from four proteins: elongation factor-1α, actin, α-tubulin, and β-tubulin (Baldauf et al., 2000). A portion of the tree shows that fungi form a monophyletic clade that is a sister group to

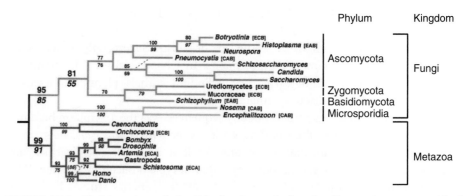

FIGURE 17.1. *Phylogenetic analysis of the fungi reveals that they form a sister group with the metazoa (animals). This tree is a detailed view of a broad analysis of the eukaryotes (see Fig. 18.1) by Baldauf et al. (2000). The tree was generated using a multiple sequence alignment of four concatenated protein sequences: elongation factor 1α (EF-1α) (abbreviated E in tree), actin (C), α-tubulin (A), and β-tubulin (B). Microsporidia were formerly classified as deep-branching eukaryotes but are now grouped with fungi. The fungal phylum Chytridiomycota is not shown in this tree.*

animals (metazoa) (Fig. 17.1). This close relationship between fungi and animals has been considered somewhat surprising, given the apparently simple, unicellular nature of many fungi. However, fungi and animals share many similarities. Chitin is the main component of the fungal cell wall, and it is also a constituent of the arthropod exoskeleton. (Plant cell walls use cellulose.) Many of the fundamental processes of yeast, such as cell cycle control, DNA repair, and intracellular vesicle trafficking, are closely conserved with mammalian cells.

According to the phylogenetic classification of Hibbett et al. (2007), the kingdom *Fungi* has seven phyla (Box 17.1). Of these the subkingdom Dikarya includes the *Ascomycota* (including *Saccharomyces cerevisieae*) and *Basidiomycota*. The Hibbett et al. classification was consistent with a sampling of nearly 200 fungal species of every major clade of Fungi by James et al. (2006). Phylogenetic analysis relied on a set of six genes (Box 17.1). Figure 17.2 presents a phylogenetic tree based on James et al.

See Box 17.1 for a discussion of fungal taxonomy.

Fungi are grown on food products such as Camembert and Brie cheeses to provide flavor. Fungi are used to produce soy sauce and many other foods and medicines.

Box 17.1
Fungal Taxonomy

Approximately 70,000 fungal species have been described, although the total number of species is estimated to be at least 1.5 million. These fungi were classified in four phyla: Ascomycota, Basidiomycota, Chytridiomycota, and Zygomycota (Guarro et al., 1999). (1) Ascomycota includes yeasts, blue-green molds, truffles, and lichens; about 30,000 species are known, including the genera *Aspergillus*, *Candida*, *Cryptosporium*, *Histoplasma*, *Neurospora*, and *Saccharomyces*. (2) Basidiomycota includes rusts, smuts, and mushrooms; they are distinguished by club-shaped reproductive structures called basidia. (3) The phylum Chytridiomycota, sometimes classified in the kingdom Protoctista (Margulis and Schwartz, 1998), includes the genera *Allomyces* and *Polyphagus*. (4) Finally, fungi of the phylum Zygomycota lack septa (cross walls), typically feed on decaying vegetation, and include the genera *Glomus*, *Mucor*, and *Rhizopus*.

The phylum Ascomycota is of particular interest because it includes the yeasts. The phylum is further divided into four classes: Hemiascomycetae (e.g., *S. cerevisiae*), Euascomycetae (e.g., *Neurospora crassa*), Loculoascomycetae (e.g., *Elsinoe proteae*), and Laboulbeniomycetae (parasites of insects).

Recently Hibbett et al. (2007), in a paper with 67 authors, proposed a reclassification of the Fungi into one kingdom (Fungi), one subkingdom (Dikarya, encompassing the clade containing Ascomycota and Basidiomycota), seven phyla, 35 classes, and 129 orders. The seven phyla are the Chytridiomycota, Neocallimastigomycota, Blastocladiiomycota, Microsporidia, Glomeromycota, Ascomycota, and Basidiomycota.

The Dikarya encompass about 98% of all known fungal species. The Hibbett et al. classification is consistent with the phylogeny of James et al. (2006), who analyzed sequence data from six genes in 199 taxa: 18S rRNA, 28S rRNA, 5.8S rRNA, elongation factor 1-α, and the two RNA polymerase II subunits *RPB1* and *RPB2*.

For web resources that address fungal taxonomy, visit the Index Fungorum (▶ http://www.indexfungorum.org/), MycoBank (▶ http://www.mycobank.org/), and the Global Biodiversity Information Facility (▶ http://www.gbif.org).

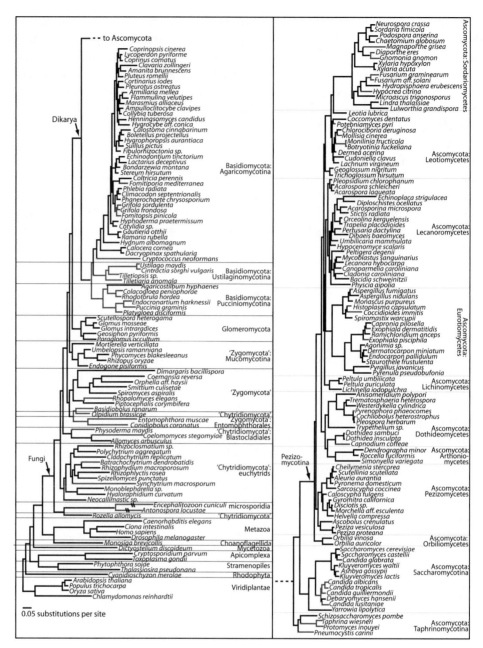

FIGURE 17.2. Fungal phylogeny. Nearly 200 fungal species were sampled, and six molecules were analyzed (see Box 17.1). The majority of known fungal species are from the phyla Ascomycota and Basidiomycota of the subkingdom Dikarya. Adapted from James et al. (2006). Used with permission.

INTRODUCTION TO BUDDING YEAST
SACCHAROMYCES CEREVISIAE

From the time of Anton van Leeuwenhoek (1632–1723), yeast was thought to be a chemical substance that was not living. Theodor Schwann (1810–1882) and Baron Charles

The budding yeast *S. cerevisiae* was the first species domesticated by humans at least 10,000 years ago. It is commonly called brewer's yeast or baker's yeast, and it ferments glucose to ethanol and carbon dioxide. For more than 100 years, researchers have exploited this organism for biochemical, genetic, molecular, and cell biological studies. Because many of its characteristics are conserved also in human cells, yeast has emerged as a powerful instrument for basic research.

Sequencing the Yeast Genome

Currently, relatively small genomes are sequenced using the whole-genome shotgun method and Sanger sequencing or relatively newer technologies such as pyrosequencing (Chapter 13). In contrast, the yeast genome was sequenced in the early to mid-1990s by chromosome. This was accomplished by a worldwide consortium of over 600 researchers (Mewes et al., 1997). The work proceeded in several phases. First, a crude physical map of its 16 chromosomes was constructed using rare-cutter restriction enzymes. Second, a library of ~10 kilobase genomic DNA inserts was constructed in phage lambda, and the inserts were fingerprinted using restriction enzymes. Computer analysis identified clones with overlapping inserts, which were then assembled into 16 large contigs. A set of clones covering the genome with minimal overlap was selected and parsed out to individual laboratories for sequencing followed by assembly and annotation using a standardized nomenclature. (The final error rate was less than 3 per 10,000 bases, or 0.03% [Mewes et al., 1997].) Today, this approach would be considered arduous, inefficient, and expensive. However, the collaboration worked extremely well.

Features of the Budding Yeast Genome

The *S. cerevisiae* genome consists of about 13 Mb of DNA in 16 chromosomes. With the complete sequencing of the genome, the physical map (determined directly from DNA sequencing) was unified with the genetic map (determined by tetrad analysis to derive genetic distances between genes) (Cherry et al., 1997). The final sequence was assembled from 300,000 independent sequence reads (Mewes et al., 1997). Some of the features of the *S. cerevisiae* sequence are listed in Table 17.1, based on the initial annotation of the genome (Goffeau et al., 1996) as well as recent updates. In the decade since the initial sequence analysis, the annotation was regularly updated as models of genes were corrected and additional information (e.g., based on comparative analyses with other fungal genomes) allowed a more accurate assessment of genome features.

A notable feature of the yeast genome is its high gene density (about one gene every 2 kb). While bacteria have a density of about one gene per kilobase, most higher eukaryotes have a much sparser density of genes. Also, only 4% of the genes are interrupted by introns. In contrast, in the fission yeast *S. pombe*, 40% of the genes have introns (see below). The lack of introns makes *S. cerevisiae* an attractive model organism for the identification of genes from genomic DNA. The most common protein families and protein domains in *S. cerevisiae* are listed in Tables 17.2 and 17.3. The EBI offers a variety of proteomics analyses of this and dozens of other organisms, such as an analysis of protein composition and lengths (Fig. 17.3).

At the time the genomic sequence was initially annotated, there were 6275 predicted open reading frames (ORFs). An ORF was defined as ≥100 codons (300 nucleotides) in length, thus specifying a protein of at least ≈11,500 daltons. Of these, 390 were listed as questionable (Table 17.1) because they were short and unlikely to encode proteins (Dujon et al., 1994). Questionable ORFs display an unlikely preference for codon usage based on a "codon adaptation index" of <0.11.

How many protein-coding genes are present in *S. cerevisiae*? Is it possible that short ORFs encode authentic proteins? These questions are fundamental to our understanding of any eukaryotic genome. In annotating the yeast genome, there are false positives (identified ORFs that do not encode an authentic gene) and false negatives (true genes

Cagniard-Latour (1777–1859) independently discovered in 1836 to 1837 that yeast is composed of living cells. Schwann studied fermenting yeast and called it *Zuckerpilz* (sugar fungus), from which the term *Saccharomyces* is derived (Bulloch, 1938).

Saccharomyces cerevisiae is often called a "budding yeast" to distinguish it from a "fission yeast," *Schizosaccharomyces pombe*, the second fungal genome to be sequenced (see below). *Saccharomyces cerevisiae* is a single-celled organism that "buds" off in the process of replication. The sequence of *S. cerevisiae* was first released April 24, 1996.

By definition, all ORFs begin with a start codon (typically AUG encoding methionine) and end with a stop codon (usually UAG, UAA, or UGA).

TABLE 17-1	Features of the *S. cerevisiae* Genome
Feature	Amount
Sequenced length*	12,156,679 base pairs
Length of repeats	1321 kb
Total length	13,389 kb
Total ORFs*	6,608
Verified ORFs*	4,667
Uncharacterized ORFs*	1,127
Dubious ORFs*	814
Introns in ORFs	220
Introns in UTRs	15
Pseudogenes*	21
ARS*	274
Intact Ty elements*	50
tRNA genes*	299
snRNA genes *	6
snoRNA genes*	77
noncoding RNA*	9

Abbreviations: ARS, autonomously replicating sequences (chromosomal replication origins); ORF, open reading frame; snoRNA, small nucleolar RNA; tRNA, transfer RNA; Ty, retrotransposons; UTR, untranslated region.

Source: Adapted from Goffeau et al. (1996). Entries that were updated from the *Saccharomyces* Genome Database (► http://www.yeastgenome.org, December 2007) are indicated with an asterisk and include the nuclear and mitochondrial genomes.

TABLE 17-2	Fifteen Most Common Protein Families for *S. cerevisiae*

InterPro ID	No. of Proteins Matched	Name
IPR007114	66	Major facilitator superfamily
IPR011701	45	Major facilitator superfamily MFS-1
IPR005829	39	Sugar transporter superfamily
IPR001993	35	Mitochondrial substrate carrier
IPR005828	33	General substrate transporter
IPR001806	33	Ras GTPase
IPR003663	30	Sugar transporter
IPR000992	27	Stress-induced protein SRP1/TIP1
IPR013753	24	Ras
IPR002293	24	Amino acid/polyamine transporter I
IPR002067	24	Mitochondrial carrier protein
IPR001142	22	Yeast membrane protein DUP
IPR004840	21	Amino acid permease
IPR002085	21	Alcohol dehydrogenase superfamily, zinc-containing
IPR015609	19	Molecular chaperone, heat shock protein, Hsp40, DnaJ

Source: From Integr8 at the European Bioinformatics Institute (► http://www.ebi.ac.uk/proteome/), December 2007.

TABLE 17-3 Fifteen Most Common Domains of the Completed Genome of Yeast *S. cerevisiae*

InterPro	Proteins Matched	Name
IPR011009	133	Protein kinase-like
IPR000719	124	Protein kinase, core
IPR011046	113	WD40 repeat-like
IPR015943	105	WD40/YVTN repeat-like
IPR016040	103	NAD(P)-binding
IPR016024	93	Armadillo-type fold
IPR003593	85	AAA+ ATPase, core
IPR002290	76	Serine/threonine protein kinase
IPR014001	74	DEAD-like helicase, N-terminal
IPR014021	72	Helicase, superfamily 1 and 2, ATP-binding
IPR001650	72	DNA/RNA helicase, C-terminal
IPR015820	68	Retrotransposon Ty1 A, N-terminal
IPR001042	68	Peptidase A11B, Ty1 B
IPR012337	58	Polynucleotidyl transferase, Ribonuclease H fold
IPR000504	56	RNA recognition motif, RNP-1

Source: From the European Bioinformatics Institute (EBI) Integr8 proteome analysis site (▶ http://www. ebi.ac.uk/proteome/, November 2007).

with short ORFs that are not annotated). The *Saccharomyces* Genome Database (introduced below) lists categories of verified ORFs, uncharacterized ORFs, and dubious ORFs. There are 40,000 ORFs longer than 20 codons (Mackiewicz et al., 2002). Below the arbitrary cutoff of 100 codons, there are many ORFs that meet the criteria of having a codon adaptation index of >0.11 and which do not overlap a

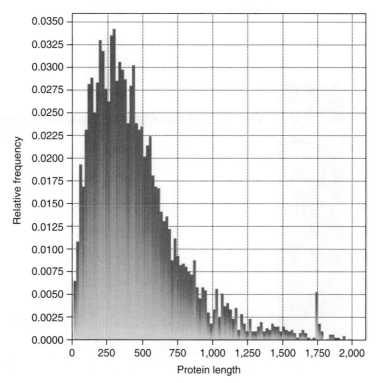

FIGURE 17.3. The European Bioinformatics Institute offers proteome analysis tools for S. cerevisiae and over 1000 other organisms through the Integr8 project (▶ http://www.ebi.ac.uk/proteome/). This plot shows the number of S. cerevisiae proteins as a function of the length of the protein. The average protein length is 496 amino acid residues, with a range of 25 to 4910 residues. In the case of small predicted proteins (e.g., <100 codons) it is important to confirm that the gene is transcribed and translated in vivo and does not represent a fortuitous open reading frame that is not biologically meaningful.

longer ORF (Harrison et al., 2002). The main criteria for deciding whether they are protein-coding genes are (1) evidence of conservation in other organisms and/or (2) experimental evidence of gene expression (Chapter 16). For *S. cerevisiae*, Winzeler and colleagues used a combination of gene expression profiling with oligonucleotide arrays and mass spectrometry to verify the transcription of 138 and the translation of 50 previously nonannotated genes (Oshiro et al., 2002). Michael Snyder and colleagues combined expression profiling, transposon-mediated gene trapping (Chapter 12), and homology searching to identify 137 genes (Kumar et al., 2002).

In addition to protein-coding genes, there are many transcribed genes that encode functional RNA molecules but are not subsequently translated into protein. In addition to the 299 tRNA genes shown in Table 17.1, there are 140 tandemly repeated copies of rRNA genes as well as small nucleolar (snoRNA) (Lowe and Eddy, 1999) and other RNA species.

> Over half the human genome is composed of transposable elements; we will explore them in more detail in Chapter 19.

The *S. cerevisiae* genome encodes 50 intact retrotransposons (called Ty1, Ty2, Ty3, Ty4, and Ty5). These are endogenous retrovirus-like elements that mediate transposition (i.e., insertion into a new genomic location) (Roth, 2000). They are flanked by long terminal repeats (LTRs) that function in integration of the retrotransposon into a new genomic site. Retrotransposons have shaped the genomic landscape of all eukaryotic genomes.

Exploring a Typical Yeast Chromosome

> Chromosome XII (accession number NC_001144) has 1,078,173 bp. To view a map of chromosome XII at the MIPS site, go to ▶ http://mips.gsf.de/genre/proj/yeast/. For SGD, visit ▶ http://www.yeastgenome.org/. NCBI also offers specialized resources on fungi (▶ http://www.ncbi.nlm.nih.gov/projects/genome/guide/saccharomyces/).

You can access the DNA sequence of any *S. cerevisiae* chromosome through several websites. We begin with NCBI; click "Map Viewer" on the home page to link to *S. cerevisiae* and then select any of the 16 chromosomes (Fig. 17.4). The result is an entry similar to the one we saw in Fig. 16.2. Consider the specific features of chromosome XII, a typical chromosome consisting of just over 1 Mb (Johnston et al., 1997). The NCBI page for this chromosome offers features such as TaxPlot and COG that are similar to the features we have seen for prokaryotes (Chapter 15). You can also

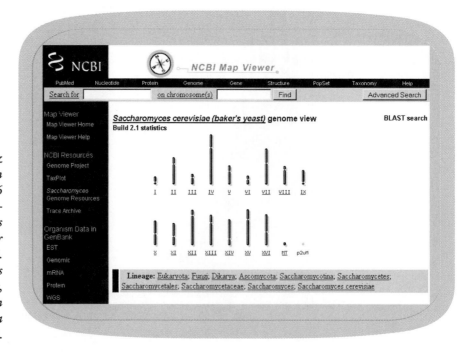

FIGURE 17.4. The NCBI Entrez Genomes site includes this page on S. cerevisiae. Each of the 16 chromosomes can be explored separately. The left sidebar includes links to major web resources for S. cerevisiae and other fungi. Additional links are to the COGs database, protein structure links, and TaxPlot that we have seen for bacteria and archaea (Chapter 15 and Fig. 17.11 below).

view chromosome XII at the major yeast-specific databases such as the MIPS Comprehensive Yeast Genome Database (Mewes et al., 2002) (Fig. 17.5), the *Saccharomyces* Genome Database (SGD) (Dwight et al., 2002) (Fig. 17.6), as well as the UCSC Genome Browser (Fig. 17.7). Information in these databases is often cross-referenced; for example, the UCSC tracks include SGD data. Computer lab problem 17.1 at the end of this chapter describes how to download a General Feature Format (.gff) file from the MIPS database and upload it as a custom track

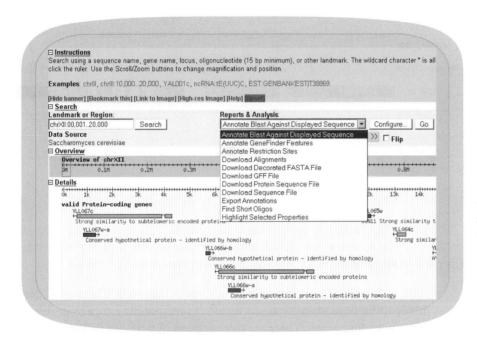

FIGURE 17.5. *MIPS offers the Comprehensive Yeast Genome Database (▶ http://mips.gsf.de/ projects/fungi). Each chromosome can be viewed on a map. The features are clickable for detailed information on each element, and the genes are annotated. The reports and analysis menu includes options to download information including a GFF file (see text).*

S. cerevisiae Genomic View

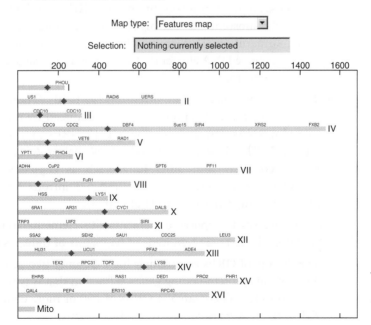

FIGURE 17.6. *The* Saccharomyces *Genome Database (SGD) is a central resource for yeast genomics. It includes a genome viewer allowing you to browse by chromosome using a GMOD browser (as shown for the MIPS database, Fig. 17.5).*

FIGURE 17.7. The UCSC Genome Browser includes a database for S. cerevisiae. Here 100,000 base pairs from chromosome XII are shown, highlighting SGD gene annotations, a track of human orthologs, several tracks identifying transcriptional regulation in yeast, a conservation track highlighting similarities among seven yeast species (discussed later in this chapter), and a "most conserved" track showing highly conserved DNA regions. By linking from the Genome Browser to the Table Browser, any of these categories of information can be summarized and downloaded as a tabular output.

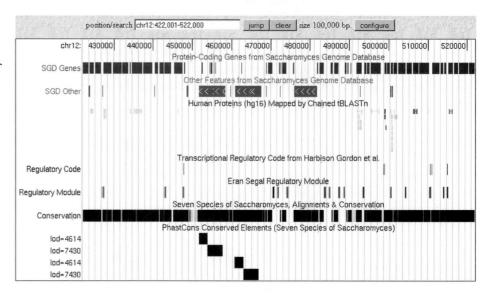

The centromere is the site at which chromosomes attach to the mitotic or meiotic spindle. In yeast, the centromere divides each chromosome into the left and right arm; in humans, it divides each chromosome into a short (or p) arm and a long (or q) arm.

The telomere is the terminal region of each chromosome arm. These arms are important in the maintenance of chromosome structure. They have been implicated in processes ranging from aging to mental retardation (Chapters 16 and 20).

on the UCSC genome browser. Each of these various browsers offers unique features (Box 17.2). Chromosome XII has the following properties:

- The overall G + C content of chromosome XII is 38%. The G + C content tends to be highest in localized regions corresponding to a high density of protein-coding genes. There are three regions of particularly low G + C content (below 37%); one of these corresponds to the centromere. This feature is typical of all eukaryotic centromeres.

- Overall, there is very little repetitive DNA throughout the *S. cerevisiae* genome. The rDNA repeats are all on chromosome XII (encoding rRNAs). This region of the chromosome has the highest G + C content as well (approximately 42%). In addition, *S. cerevisiae* chromosomes have telomeric and subtelomeric repetitive DNA elements. This feature is typical of essentially all eukaryotic chromosomes.

- There are few spliceosomal introns (~235 total). These are probably due to homologous recombination of cDNAs produced by reverse transcription of spliced mRNAs. On chromosome XII, 17 ORFs (3.2% of the total) contain introns; half of these genes encode ribosomal proteins. The extreme lack of introns contrasts with other fungi such as *Cryptococcus neoformans* (see below) which averages 6.3 exons and 5.3 introns for its 6572 predicted protein-coding genes (Loftus et al., 2005).

- There are six transposable elements (Ty elements) on chromosome XII. Additionally there are hundreds of fragments of transposable elements.

- The density of ORFs is extremely high. Seventy-two percent of chromosome XII contains protein-coding genes, a fraction that is typical of the other yeast chromosomes. There are 534 ORFs of 100 or more codons on chromosome XII, with an average codon size of 485 codons.

Box 17.2
Multiple Yeast Genome Browsers

We have illustrated four yeast genome browsers: those at NCBI, MIPS, SGD, and UCSC. Each offers different advantages, and there is no single best resource. The SGD is arguably the central web resource for the yeast genomics community. The strength of NCBI is its critical role in the bioinformatics community. The UCSC Genome Browser is an increasingly essential resource for the visualization, annotation, and analysis of vertebrate genomes (Chapters 18 and 19), but its application to fungi is currently limited. MIPS offers expert curation. Notably, its web browser is based on the Generic Model Organism Database project (GMOD; ► http://www.gmod.org/). GMOD is a set of interconnected applications and databases, including the Generic Genome Browser (GBrowse). The research communities involved in a variety of organisms have contributed to GMOD (including the SGD and model organism projects described in Chapter 18 such as FlyBase, WormBase, and TAIR).

For the nomenclature system used for *S. cerevisiae* genes and proteins, see Box 17.3.

Box 17.3
Gene Nomenclature in *Saccharomyces cerevisiae*

All ORFs that are ≥100 codons were assigned unique names consisting of three letters followed by a numeral and a subscript to describe its genomic position. For example, the gene name *YKL159c* refers to the ORF number 159 (from the centromere) on the left arm (*L*) of chromosome XI (*K*) of yeast (*Y*). The designations *c* or *w* ("Crick" or "Watson") reflect the orientation of the gene on the chromosome. Once a gene has been characterized and assigned some kind of function, the investigators may assign a new name that reflects the function, in this case *RCN1* for "regulator of calcineurin." Dominant alleles (typically the wild-type allele) are listed with three uppercase letters while recessive alleles (typically knockout mutations or loss of function alleles) are listed with three lowercase letters. The protein product of the gene is designated without italics and with only the first letter in uppercase and with "p" appended to designate protein. Many genes have multiple names (synonyms) because investigators have identified them in independent functional screens. Some examples of nomenclature are given in Table 17.4.

Chromosome XII includes the largest gene in the *S. cerevisiae* genome, *YLR106c*. This gene encodes a protein with 4910 amino acids (MDN1p; accession NP_013207) (Garbarino and Gibbons, 2002). Midasin, a human ortholog, is 5596 amino acids long (over 600 kD; RefSeq accession NP_055426).

TABLE 17-4	Examples of Nomenclature for Yeast Genes	
Wild-Type Allele	Protein Product	Mutant Alleles
CNA1	Cna1p	*cna1Δ*
RCN1	Rcn1p	*rcn1, rcn1::URA3*
YKL159c	Ykl159cp	*ykl159c*

GENE DUPLICATION AND GENOME DUPLICATION OF S. *CEREVISIAE*

As the genome sequence of *S. cerevisiae* was analyzed, it became apparent that there are many duplications of DNA sequence, involving both ORFs and larger genomic regions. In many cases, the gene order and orientation (top or bottom strand) is preserved between the duplicated regions. The duplications are both intrachromosomal (occurring within a chromosome) and interchromosomal (occurring between chromosomes).

These changes in genetic material are fundamental in explaining the evolution of species in yeast or in any branch of life. We will see that in the human genome and a variety of other eukaryotic genomes, as many as 25% of the genes are duplicated (Chapters 18 and 19). From where can new, duplicate genes arise? Several mechanisms are outlined in Fig. 17.8:

- Genes can arise by tandem repeat slippage during replication.

- New genes can arise by gene conversion. In this process, genes are transferred nonreciprocally from one genomic region to another. This occurs between repetitive regions of the human Y chromosome (Rozen et al., 2003).

- Genes can be introduced into a genome by lateral (horizontal) gene transfer, as discussed in Chapter 15.

- Segments of a genome can duplicate. We discuss segmental duplication of the human genome in Chapter 19; it is sometimes defined as consisting of two loci sharing 90% or more identity over a length of 1000 base pairs or more.

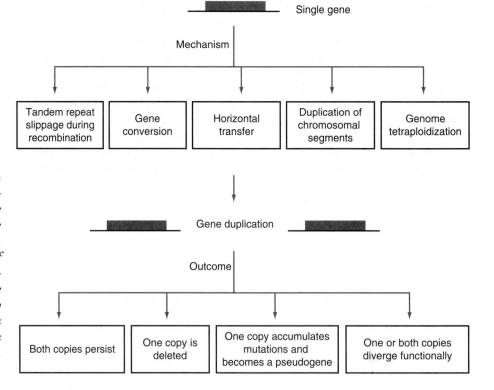

FIGURE 17.8. Gene duplication can occur by a variety of mechanisms. The duplicated copy may be localized on the same chromosome or on a different chromosome. There are several possible fates of a gene pair that arises by duplication: copies may persist, one copy may be deleted, one copy may become nonfunctional (a pseudogene), or the two genes may acquire distinct functions. See Sankoff (2001).

- An entire genome can duplicate, a process called polyploidy. In the case of *S. cerevisiae*, this is a tetraploidization. If this resulted from the combining of two genomes from one species it is called autopolyploidy; if two distinct species fused it is allopolyploidy.

Tetraploidy is the presence of four haploid sets of chromosomes in the nucleus.

In 1970, Susumu Ohno published the book *Evolution by Gene Duplication*. He proposed that vertebrate genomes evolved by two rounds of whole-genome duplication. These duplication events, according to this hypothesis, occurred early in vertebrate evolution and allowed the development of a variety of cellular functions. Ohno (1970) wrote: "Had evolution been entirely dependent upon natural selection, from a bacterium only numerous forms of bacteria would have emerged. The creation of metazoans, vertebrates, and finally mammals from unicellular organisms would have been quite impossible, for such big leaps in evolution required the creation of new gene loci with previously nonexistent function. Only the cistron that became redundant was able to escape from the relentless pressure of natural selection. By escaping, it accumulated formerly forbidden mutations to emerge as a new gene locus."

Which mechanism of gene duplication might have occurred in *S. cerevisiae*? Wolfe and Shields (1997) provided support for Ohno's whole-genome duplication paradigm. They assessed the duplicated regions of the yeast genome by performing systematic blastp searches of all yeast proteins against each other and plotting the matches on dot matrices. Duplicate regions were observed as diagonal lines, such as the three duplicated regions seen in a comparison of proteins derived from chromosomes X and XI (Fig. 17.9). In the whole genome, they identified 55 duplicated regions and 376 pairs of homologous genes. In subsequent studies, they employed the more sensitive Smith–Waterman algorithm and identified a few additional regions of duplication (Seoighe and Wolfe, 1999). Based on these results, they proposed a single, ancient duplication of the *S. cerevisiae* genome, approximately 100 million years ago (Wolfe and Shields, 1997). Subsequent to this duplication event, many duplicated genes were deleted. Other genes were rearranged by reciprocal translocation.

Wolfe and Shields (1997) used Blastp rather than Blastn to study duplicated regions of chromosomes. This is because protein sequence data are more informative than DNA for the detection of distantly related sequences. See Chapter 3.

You can view chromosome comparisons at the SGD website by selecting Analysis and Tools and then Pairwise Chromosome Similarity View. This generates plots such as that of Fig. 17.9.

There are two main explanations for the presence of so many duplicated regions. There could have been whole-genome duplication (tetraploidy) followed by translocations as well as gene loss, or alternatively there could have been a series of independent duplications. Wolfe and Shields (1997) favored the tetraploidy model for two reasons:

1. For 50 of the 55 duplicate regions, the orientation of the entire block was preserved with respect to the centromere. If each block were generated independently, a random orientation would be expected.

2. Fifty-five successive, independent duplications of blocks would be expected to result in about seven triplicated regions, but only zero (or possibly one) such triplicated region was observed.

What is the fate of genes after duplication? The presence of extra copies of genes is usually deleterious to an organism. In the model of Wolfe and colleagues, the genome of an ancestral yeast doubled (from the diploid number of about 5000 to the tetraploid number of 10,000 genes) then lost the majority of its duplicated genes, yielding the present-day number of about 6200 ORFs. Overall, between 50% and 92% of duplicated genes are eventually lost (Wagner, 2001).

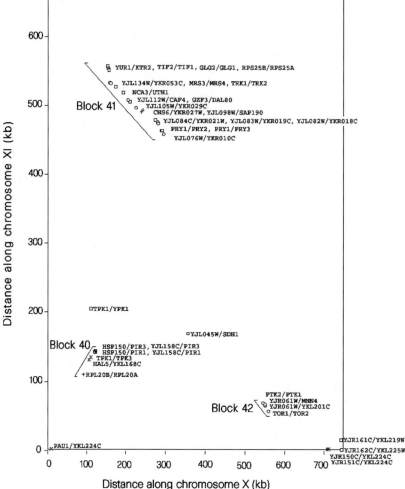

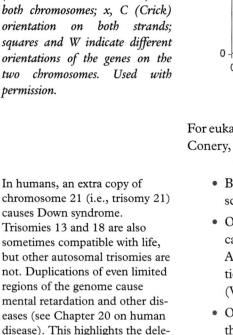

FIGURE 17.9. *Wolfe and Shields (1997) performed blastp searches of proteins from* S. cerevisiae *and found 55 blocks of duplicate regions. This provides strong evidence that the entire genome underwent an ancient duplication. This figure depicts the result of BLAST searches of proteins encoded by genes on chromosomes X and XI. Matches with scores >200 are shown, arranged in several blocks of genes. Symbols indicate the gene orientations: +, W (Watson strand orientation) on both chromosomes; x, C (Crick) orientation on both strands; squares and W indicate different orientations of the genes on the two chromosomes. Used with permission.*

For eukaryotes, the half-life of duplicated genes is only a few million years (Lynch and Conery, 2000) (see Chapter 16). There are four main possibilities (Fig. 17.8):

- Both copies can persist, maintaining the function of the original gene. In this scenario, there is a gene dosage effect because of the extra copy of the gene.

- One copy could be completely deleted. This is the most common fate of duplicated genes, as confirmed by recent whole genome studies (described below). A rationale for this fate is that since the duplicated genes share identical functions initially, either one of them may be subject to loss-of-function mutations (Wagner, 2001).

- One copy can accumulate mutations and evolve into a pseuodogene (a gene that does not encode a functional gene product). This represents a loss of gene function, although it occurs without the complete deletion of the duplicate copy. Over time, the pseudogene may be lost entirely.

- One or both copies of the gene could diverge functionally. According to this hypothesis, gene duplications (regardless of mechanism) provide an organism with the raw material needed to expand its repertoire of functions. Furthermore, loss of either gene having overlapping functions might not be tolerated. Thus, the functionally diverged genes would both be positively selected.

In humans, an extra copy of chromosome 21 (i.e., trisomy 21) causes Down syndrome. Trisomies 13 and 18 are also sometimes compatible with life, but other autosomal trisomies are not. Duplications of even limited regions of the genome cause mental retardation and other diseases (see Chapter 20 on human disease). This highlights the deleterious nature of duplications at the level of individual organisms.

After a gene duplicates, why does one of the members of the newly formed gene pair often become inactivated? At first glance, it might seem highly advantageous to have two copies, because one may functionally diverge (driving the process of evolution to allow a cell to perform new functions), or one may be present in an extra copy in case the other undergoes mutation. However, gene duplication instead appears to be generally deleterious, leading to the loss of duplicated genes. The logic is that some mutations in a gene are *forbidden* rather than *tolerable* (these terms were used by Ohno [1970] in describing gene duplication). Forbidden mutations severely affect the function of a gene product, for instance by altering the properties of the active site of an enzyme. (A tolerable mutation causes a change that remains compatible with the function of the gene product.) Natural selection can eliminate forbidden mutations, because the individual is less fit to reproduce. After a gene duplicates, a deleterious mutation in one copy of a gene might now be tolerated because the second gene can assume its function. A second reason that duplicated genes may be deleterious is that in their presence the crossing over of homologous chromosomes during meiosis may be mismatched, causing unequal crossing over.

We can consider the possible fates of duplicated genes with the specific example of genes encoding proteins that are essential for vesicle trafficking. We introduced *SSO1* in Chapter 12 (Fig. 12.4). In yeast and all other eukaryotes, spherical intracellular vesicles transport various cargo to destinations within the cell. These vesicles traffic cargo to the appropriate target membrane through the binding of vesicle proteins (e.g., Snc1p in yeast or VAMP/synaptobrevin in mammals) to target membrane proteins (e.g., Sso1p in yeast or syntaxin in mammals) (Protopopov et al., 1993; Aalto et al., 1993). In *S. cerevisiae*, genome duplications presumably caused the appearance of two paralogous genes in each case: *SNC1* and *SNC2* as well as *SSO1* and *SSO2*. The *SNC1* and *SNC2* genes are on corresponding regions of chromosomes I and XV, while the *SSO1* and *SSO2* genes are on chromosomes XVI and XIII, respectively.

What could the consequences of genome duplication have been? The two pairs of syntaxin-like and VAMP/synaptobrevin-like yeast proteins might have maintained the same function of the original proteins (before genome duplication). A search for *SSO1* at the SGD website shows that the gene is nonessential (the null mutant is viable), but the double knockout is lethal (see Fig. 12.4). Thus, it is likely that these paralogs offer functional redundancy for the organism; in the event a gene is lost (e.g., through mutation), the organism can survive because of the presence of the other gene. Similarly, the *SNC1* null mutant is viable, but the double knockout of *SNC1* and *SNC2* is deficient in secretion.

As an explanation of the mechanism by which these genes duplicated, it is possible that whole-genome duplication provided the new genetic materials with which the intracellular secretion machinery could be diversified. Syntaxin and VAMP/synaptobrevin proteins function at a variety of intracellular trafficking steps, and these gene families diversified throughout eukaryotic evolution (Dacks and Doolittle, 2002).

Andreas Wagner (2000) addressed the question of how *S. cerevisiae* protects itself against mutations by one of two mechanisms: (1) having genes with overlapping functions (such as paralogs that maintain related functions) or (2) through the interactions of nonhomologous genes in regulatory networks. He found that genes whose loss of function caused mild rather than severe effects on fitness did not tend to have closely related paralogs. This is consistent with a model in which gene duplication does not provide robustness against mutations.

COMPARATIVE ANALYSES OF HEMIASCOMYCETES

Analysis of *S. cerevisiae* has elucidated many fundamental principles concerning genome structure, function, and evolution. Comparison of phylogenetically related genomes has opened an entirely new dimension on genome analysis. Some of the first genomes selected were hemiascomycetes phylogenetically close to *S. cerevisiae*, such as *Candida glabrata*, *Kluyveromyces lactis*, *Debaryomyces hansenii*, and *Yarrowia lipolytica* (Dujon et al., 2004). In all, hundreds of fungal genomes are currently being sequenced.

Analysis of Whole Genome Duplication

The hypothesis that yeast underwent a whole-genome duplication event has been tested by analyzing whole genome sequences. By becoming polyploid, an organism doubles its complement of chromosomes (and thus genes). This might appear to be an appealing mechanism to increase the repertoire of genes available for adaptation to new environments. However, polyploidy leads to genome instability, in part because of difficulties for the cell to perform proper chromosome segregation.

To understand the whole genome duplication of *S. cerevisiae*, Kellis et al. (2004) sequenced the genome of *Kluyverocmyces waltii*, a related yeast that diverged before the whole genome duplication event (Fig. 17.10). They sequenced the eight chromosomes of *K. waltii*, and annotated 5230 putative protein-coding genes. They identified blocks of conserved synteny (loci containing orthologous genes in the same order between the two species). Most regions of *K. waltii* mapped to two separate regions of *S. cerevisiae*. However, these regions of *S. cerevisiae* show evidence of massive gene loss (with 12% of the paralogous gene pairs retained, and 88% of paralogous genes deleted to leave one copy remaining).

Kellis et al. considered the rate of evolution of 457 gene pairs in *S. cerevisiae* that arose by whole genome duplication; 76 of these gene pairs displayed accelerated evolution (based on amino acid substitution rates in the *S. cerevisiae* lineage relative to *K. waltii*). Remarkably, in 95% of these cases, the accelerated evolution was restricted to just one of the two paralogs. This supports Ohno's suggestion that after duplication one copy of a gene can preserve the original function while the other may diverge to acquire a novel function.

With the continuing production of new genome sequencing data, Scannell et al. (2006) considered six yeast species: three that descended from a common ancestor that is thought to have undergone a whole genome duplication (*S. cerevisiae*,

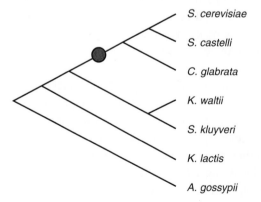

FIGURE 17.10. Phylogeny of several yeasts, after Kurtzman and Robnett (2003) (adapted from ▶ http://wolfe.gen.tcd.ie/ygob/). A red circle indicates the likely place in which a whole genome duplication (WGD) occurred.

Saccharomyces castellii, and *Candida glabrata*), as well as three additional yeasts that diverged before the whole genome duplication event (*Kluyveromyces waltii*, *Kluyveromyces lactis*, and *Ashbya gossypii*). They used the Yeast Gene Order Browser to compare the six species. This browser is available online (Byrne and Wolfe, 2005, 2006). An example is shown in Fig. 17.11 for the query SSO1 as well as six adjacent upstream and downstream genes. There are seven horizontal tracks in this example. Three in the center show the genes in the reference species that diverged before the whole genome duplication event (*A. gossypii*, *K. waltii*, and *K. lactis*). For *S. cerevisiae* and *C. glabrata* there are pairs of tracks, both above and below the reference species. For genes such as *SSO1*, *YPL230W*, and *WPL228W* (Fig. 17.11, arrows 2 to 4) there are two copies in both *S. cerevisiae* and *C. glabrata* but only one copy in the reference genomes. These two copies occur in adjacent positions along separate chromosomes.

A variety of patterns of loss can occur. Scannell et al. (2006) described 14 patterns by which gene loss can occur (outlined in Fig. 17.12). Out of 2723 ancestral loci that aligned appropriately, in only 210 cases was there no gene loss among the

The Yeast Gene Order Browser is online at the website of Kenneth Wolfe (► http://wolfe.gen.tcd.ie/ygob/). *Saccharomyces cerevisiae* can live under anaerobic conditions, while *K. lactis* cannot. It is possible that the *S. cerevisiae* genome duplication resulted in physiological changes that allowed this organism to acquire the new growth phenotype (Piskur, 2001).

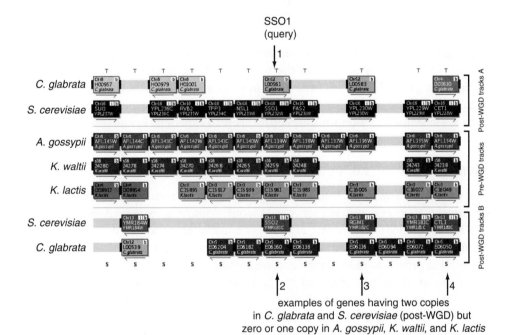

FIGURE 17.11. *The Yeast Gene Order Browser of Kevin Byrne in the group of Kenneth Wolfe provides evidence supporting whole genome duplication events. Upon entering the query (SSO1; top arrow) and selecting the species to display, the query and varying numbers of adjacent genes are displayed. Each box represents a gene, and boxes are color-coded to correspond to particular chromosomes. Solid bars connect genes that are immediately adjacent. Here, the first and seventh rows correspond to C. glabrata, and the second and sixth rows correspond to S. cerevisiae (chromosome 16, including SSO1 gene, on row 2; chromosome 13, including the paralog SSO2, on row six). In this view there are three genes that have two copies in C. glabrata and S. cerevisiae that may have resulted from whole genome duplication (see arrows 2–4). For yeast lineages that are hypothesized to have not undergone whole genome duplication (A. gossypii, K. waltii, K. lactis) there tends to be only one copy of these genes. For all species, occasional gene losses are evident (e.g. K. waltii, the gene indicated by arrow 3). Yeast Gene Order Browser includes additional features such as links to the raw sequences and to phylogenetic reconstructions of each gene family. See ► http://wolfe.gen.tcd.ie/ygob/.*

FIGURE 17.12. Patterns of gene loss after whole genome duplication in three species. For three species that underwent whole genome duplication (C. glabrata, S. cerevisiae, and S. castellii) there are 14 possible fates, including loss of no genes (class 0), loss of one gene from any one of the three lineages (class 1A, 1B, 1C), loss of two genes (class 2), loss of three genes from different loci (class 3), or loss of three genes in a convergent manner (class 4; loss of duplicated orthologs). Class 4 represents the most common fate of duplicated genes. Redrawn from Scannell et al. (2006). Used with permission.

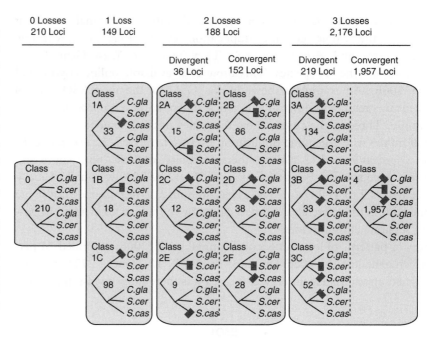

three genomes that underwent whole genome duplication. In the great majority of cases (1957 instances or 72% of the total), all three species lost one of the two copies of a given duplicated gene, and most commonly all three species lost the same copy of the gene. Genes involved in highly conserved biological processes such as ribosome function were especially likely to experience gene loss.

Identification of Functional Elements

It is extraordinarily difficult to identify genes and gene regulatory regions (such as promoters) from genomic sequence data alone. Matching expressed sequence tags (ESTs; Chapter 8) to genomic DNA is one useful approach to defining protein-coding genes. Comparative analyses between genomic sequences also provide a powerful approach to identifying functionally important elements.

Kellis et al. (2003) obtained the draft sequences of *Saccharomyces paradoxus*, *S. mikatae*, and *S. bayanus* which diverged from *S. cerevisiae* some five to 20 million years ago. Almost all of the 6235 ORFs in the SGD annotation of *S. cerevisiae* had clear orthologous matches in each of the other three species. A noticeable exception is at all 32 telomeres (i.e., both ends of the 16 chromosomes) where matches are often ambiguous. Genes assigned to subtelomeric regions are often present in different number, order, or orientation, and these regions have undergone multiple reciprocal translocations. Kellis et al. refer to changes in the telomeric regions as "genomic churning." For all ORFs in the four *Saccharomyces* genomes, Kellis et al. introduced a reading frame conservation test to classify each ORF as authentic (if conserved) or spurious (if not well conserved). As a result of their analysis, Kellis et al. proposed revising the entire *S. cerevisiae* gene catalog to 5538 ORFs of ≥100 amino acids. Their analyses further revised the count of introns (predicting 58 new ones beyond the 240 previously predicted).

Another aspect of the comparison of four *Saccharomyces* genome sequences is the opportunity to identify regulatory elements. Gal4 is one of the best characterized transcription factors. It regulates genes involved in galactose metabolism including

S. cerevisiae and *S. bayanus* share 62% nucleotide identity in conserved regions, while for comparison human and mouse share 66% nucleotide identity in conserved regions.

Dramatic genomic changes that occur in subtelomeric regions have also been observed in the malaria parasite *Plasmodium falciparum* (see Chapter 18), and in humans subtelomeric deletions are a major cause of mental retardation.

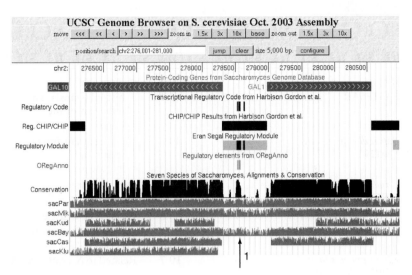

FIGURE 17.13. *View of the transcription factor Gal4 binding site region between the GAL1 and GAL10 genes of S. cerevisiae. A 5000 base pair view of yeast chromosome 2 is shown (from the UCSC Genome Browser). The short intergenic region (arrow 1) includes regions defined as having regulatory properties by several databases (see annotation tracks). The conservation track shows that some of the intergenic region is highly conserved among four Saccharomyces species. That conserved region contains binding sites for Gal4.*

the *GAL1* and *GAL10* genes. These two genes can be viewed at the UCSC Genome Browser (Fig. 17.13). They are transcribed from a short intergenic region that includes the Gal4 binding motif $CGGn_{(11)}CCG$ where n(11) refers to any 11 nucleotides. By clicking the conservation track, you can see several copies of this motif in a multiple sequence alignment of DNA from the four *Saccharomyces* species. Kellis et al. studied both previously known and predicted motifs, and predicted 52 new motifs. Other groups such as Cliften et al. (2003) and Harbison et al. (2004) have also identified yeast functional elements through comparative genomics.

ANALYSIS OF FUNGAL GENOMES

In addition to *S. cerevisiae*, the genomes of many other fungi are now being sequenced, including *Ascomycetes* (Table 17.5), *Basidiomycetes* (Table 17.6), and others (Table 17.7). We will discuss some of these fascinating projects: *Aspergillus*, *Candida albicans*, *Cryptococcus neoformans*, the microsporidial parasite *Encephalitozoon cuniculi*, *Neurospora crassa*, the Basidiomycete *Phanerochaete chrysosporium*, and the fission yeast *Schizosaccharomyces pombe* (the second fungal genome to be completely sequenced). All these projects highlight the remarkable diversity of fungal life. In Chapter 18 we will describe comparative genomics projects on more familiar organisms such as humans and fish (diverged ~450 million years ago), the fruit fly and mosquito (estimated to have diverged ~250 million years ago), as well as closely related species that diverged more recently. The fungi offer an opportunity to analyze highly divergent species (e.g., *S. cerevisiae* and *S. pombe* diverged ~400 million years ago), as well as closely related species.

Aspergillus

The genus *Aspergillus* consists of filamentous Ascomycetes. Of the 185 known species of *Aspergillus*, 20 are human pathogens. Three have now been sequenced, and dozens of other genomes are being sequenced. (1) *Aspergillus nidulans* has had a long-standing role as a model organism in genetics. Its genome was sequenced by

TABLE 17-5 Fungal Genome Projects: Representative Examples of the Ascomycetes

Organism	Chromosomes	Genome Size (Mb)	Comment	ID[a]	Total Projects
Ajellomyces capsulatus G186AR	7	~23 to 25	Causes histoplasmosis, an infection of the lungs	12635	3
Aspergillus fumigatus Af293	8	30	Most frequent fungal infection worldwide	131	10
Candida albicans SC5314	8	16	Diploid fungal pathogen	10701	6
Coccidioides immitis RS	4	28.8	Causes the disease coccidioidomycosis (valley fever)	12883	15
Kluyveromyces lactis NRRL Y-1140	6	10.7	Related to *S. cerevisiae*	13835	4
Magnaporthe grisea 70-15	7	40	Rice blast fungus	13840	2
Pichia angusta CBS 4732	6		Related to *S. cerevisiae*	12500	5
Pneumocystis carinii	15	7.7	Opportunistic pathogen; causes pneumonia in rats	125	3
Saccharomyces cerevisiae S288c	16	12.1	Baker's yeast	13838	14
Schizosaccharomyces pombe 972 h-	3	12.5	Fission yeast	13836	3
Yarrowia lipolytica CLIB122	6	20.5	Nonpathogenic yeast, distantly related to other yeasts	13837	1

[a]ID refers to the NCBI Genome Project identifier; by entering this into the search box at the home page of NCBI you can link to information on this genome project. Total projects refers to the number of genome sequencing projects for the same genus.

Galagan et al. (2005). (2) *Aspergillus fumigatus* is the most common mold that causes infection worldwide. It is an opportunistic pathogen to which immunocompromised individuals are particularly susceptible. Nierman et al. (2005) sequenced its genome and identified candidate pathogenicity genes as well as genes that may facilitate its unusual lifestyle (e.g., thriving at temperatures up to 70°C). One of the many unique features of this genome is the presence of *A. fumigatus*-specific proteins that are closely related to a class of arsenate reductases previously seen only in bacteria. (3) *Aspergillus oryzae* is a fungus from which sake, miso, and soy sauce are prepared. Like *A. nidulans* and *A. fumigatus* its genome is organized into eight chromosomes, but the total genome size is 7 to 9 megabases larger (29% to 34% larger)(Machida et al., 2005). This is due to blocks of sequence that are dispersed throughout the *A. oryzae* genome.

Comparative analyses revealed the presence of conserved noncoding DNA elements (Galagan et al., 2005), analogous to the studies of *Saccharomyces* described above. Of the three *Aspergilli*, *A. fumigatus* and *A. oryzae* reproduce through asexual mitotic spores, while *A. nidulans* has a sexual cycle. Comparative analysis of the three

TABLE 17-6   Fungal Genome Projects: Representative Examples of the Basidiomycetes

Organism	Chromosomes	Genome Size (Mb)	Comment	ID[a]	Total Projects
Coprinopsis cinerea okayama7#130	13	37.5	Multicellular basidiomycete, undergoes complete sexual cycle	1447	1
Cryptococcus neoformans var. neoformans JEC21	14	19.1	Pathogenic fungus, causes cryptococcosis	13856	5
Lentinula edodes L-54	8	33	Edible shiitake mushroom	17581	1
Phanerochaete chrysosporium RP-78	10	30	Wood-decaying white rot fungus	135	1
Puccinia graminis f. sp. tritici CRL 75-36-700-3	18	81.5	Pathogenic fungus causes stem rust in cereal crops	18535	1
Ustilago maydis 521	23	20	Causes corn smut disease	1446	1

[a]ID refers to the NCBI Genome Project identifier; by entering this into the search box at the home page of NCBI you can link to information on this genome project. Total projects refers to the number of genome sequencing projects for the same genus.

Source: Entrez Genome at NCBI (December, 2007).

TABLE 17-7   Fungal Genome Projects: Representative Examples of Fungi Other than Ascomycetes and Basidiomycetes

Organism	Chromosomes	Genome Size (Mb)	Comment	ID[a]	Total Projects
Allomyces macrogynus	nd	30	Filamentous chytrid fungus	20563	1
Antonospora locustae	nd	2.9	Intracellular microsporidian parasite	12903	1
Batrachochytrium dendrobatidis JEL423	20	23.7	Aquatic chytrid fungus kills amphibians	13653	1
Encephalitozoon cuniculi GB-M1	11	2.5	Intracellular parasite, infects mammals	13833	1
Rhizopus oryzae RA 99-880	nd	40	Opportunistic pathogen causes mucormycosis	13066	1

[a]ID refers to the NCBI Genome Project identifier; by entering this into the search box at the home page of NCBI you can link to information on this genome project. Total projects refers to the number of genome sequencing projects for the same genus.

Abbreviation: nd, not determined.

Source: Entrez Genome at NCBI (December, 2007).

Distribution of *Saccharomyces cerevisiae (baker's yeast)* homologs

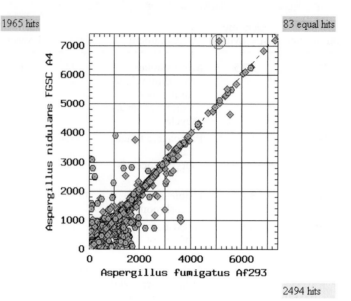

FIGURE 17.14. The TaxPlot tool at NCBI shows proteins from A. nidulans and A. fumigatus in relation to a reference proteome of S. cerevisiae. TaxPlot can help to identify organism-specific innovations that may underlie the distinct physiologies of these Aspergilli. A midasin homolog that is more closely related to S. cerevisiae in A. nidulans is circled.

The *Aspergillus* website (▶ http://www.aspergillus.org.uk/) provides detailed information.

genomes suggested that, surprisingly, *A. fumigatus* and *A. oryzae* have the necessary genes for a sexual cycle (reviewed in Scazzochio, 2006). Another surprising aspect of the comparative analyses is that peroxisomes in *Aspergilli* (organelles responsible for fatty acid β-oxidation) resemble those of mammalian cells more than yeasts because (1) β-oxidation occurs in both peroxisomes and mitochondria, and both *Aspergilli* and mammals have two sets of the necessary genes, and (2) both *Aspergilli* and mammalian genomes encode peroxisomal acyl-CoA dehydrogenases. The yeasts have served as important model systems for the study of human peroxisomal disorders such as adrenoleukodystrophy, but genomic analyses highlight the importance of the *Aspergilli*.

A comparison of *A. nidulans* and *A. fumigatus* using TaxPlot at NCBI (Chapter 15), with *S. cerevisiae* as a reference, shows that many proteins are conserved between those three species (Fig. 17.14). Of those that differ, a notable example is midasin (circled), the giant protein from *S. cerevisiae* chromosome XII.

Candida albicans

The *C. albicans* genome was sequenced by Ron Davis and colleagues at Stanford University.

Candida albicans is a diploid sexual fungus that frequently causes opportunistic infections in humans. The skin, nails, and mucosal surfaces are typical targets, but deep tissues can also be infected. The genome size is approximately 14.8 Mb, which is typical for many fungi, but the chromosomal arrangement is unusual: the genome has eight chromosome pairs, seven of which are constant and one of which is variable (ranging from about 3 to 4 Mb). Another unusual feature is that it has no known haploid state. Thus, the diploid genome was sequenced (Jones et al., 2004; reviewed by Odds et al., 2004). This was challenging because heterozygosity commonly occurs at many alleles, making it difficult to assign a sequence to one heterozygous locus rather than two independent loci. On average there is one polymorphism every 237 bases, a considerably higher frequency than occurs in human or the mosquito *Anopheles gambiae* (Chapter 18).

Information on the *Candida* genome is centralized at the CandidaDB database (Rossignol et al., 2008). The reference haploid genome initially contained 7677 ORFs (of size 100 amino acids or greater), although as is routine for any genome project the annotation process is ongoing. About half the predicted proteins match human, *S. cerevisiae*, and *Schizosaccharomyces pombe*, and only 22% of the ORFs did not match any of those three genomes. A specialized feature of *C. albicans* (shared by *Debaryomyces hansenii*; Dujon et al., 2004) is that the codon CUG is translated as serine rather than the usual product, leucine.

CandidaDB is available online at ► http://www.candidagenome. org/.

Cryptococcus neoformans: Model Fungal Pathogen

C. neoformans is a soil-dwelling fungus that causes cryptococcosis, one of the most life-threatening infections in AIDS patients. Its genome of 20 megabases is organized into 14 chromosomes as well as a mitochondrial genome. Loftus et al. (2005) sequenced two separate strains. Transposons constitute about 5% of the genome and are dispersed among all 14 chromosomes. In contrast to *S. cerevisiae* there is no evidence for a whole genome duplication. Another difference between the two fungi is that *C. neoformans* gene organization is more complex. Its 5672 predicted protein-coding genes are characterized by introns (an average of 5.3 per gene of 67 base pairs), alternatively spliced transcripts, and endogenous antisense transcripts.

Atypical Fungus: Microsporidial Parasite *Encephalitozoon cuniculi*

Microsporidia are single-celled eukaryotes that lack mitochondria and peroxisomes. These organisms infect animals (including humans) as obligate intracellular parasites. The complete genome of the microsporidium *E. cuniculi* was determined by several research groups in France (Katinka et al., 2001). The genome is highly compacted, having about 2000 protein-coding genes in 2.9 Mb. Thus, analogous to parasitic bacteria (Chapter 15), these pathogens have undergone a reduction in genome size. Phylogenetic analyses using several *E. cuniculi* proteins suggest that these parasites are atypical fungi that once possessed but subsequently lost their mitochondria (Fig. 17.15) (Katinka et al., 2001).

Neurospora crassa

The orange bread mold *Neurospora* has served as a beautiful and simple model organism for genetic and biochemical studies since Beadle and Tatum used it to establish the one gene–one enzyme model in the 1940s. *Neurospora* is the best characterized of the filamentous fungi, a group of organisms critically important to agriculture, medicine, and the environment (Perkins and Davis, 2000). The developmental complexity of *Neurospora* contrasts with other unicellular yeasts (Casselton and Zolan, 2002). *Neurospora* is widespread in nature and thus, like the fly *Drosophila*, it is exceptionally suited as a subject for population studies.

Like *S. cerevisiae*, *Neurospora* is an ascomycete and thus shares the advantage of this group of organisms in yielding complete tetrads for genetic analyses. However, it is more similar to animals than yeasts in many important ways. For example, unlike yeast but like mammals, it contains complex I in its respiratory chain, it has a clearly discernable circadian rhythm, and it methylates DNA to control gene expression.

Neurospora crassa genome database websites are available at the University of New Mexico (► http://www.unm.edu/ ~ ngp/), at the Whitehead Institute (► http://www-genome.wi.mit. edu/annotation/fungi/ neurospora/), and at MIPS (► http://www.mips.biochem. mpg.de/proj/neurospora/).

George Beadle and Edward Tatum shared a Nobel Prize in 1958 (with Joshua Lederberg) "for their discovery that genes act by regulating definite chemical events" (► http://www.nobel.se/ medicine/laureates/1958/). They irradiated *N. crassa* with x-rays to study gene function.

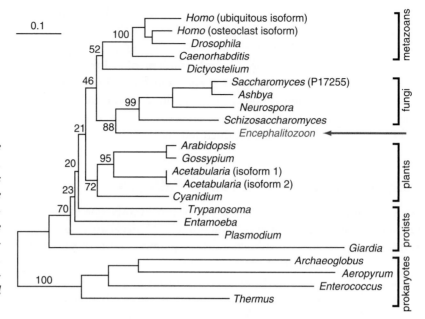

FIGURE 17.15. *Phylogenetic analysis of vacuolar ATPase subunit A from animals, plants, fungi, protists, and prokaryotes supports a fungal origin for the microsporidial parasite Encephalitozoon cuniculi (arrow). This tree was generated using a neighbor-joining method by Katinka et al. (2001). Values are bootstrap percentages (see Chapter 7). Used with permission.*

The six decades of intensive studies on the genetics, biochemistry, and cell biology of *Neurospora* establish this organism as an important source of biological knowledge.

Galagan et al. (2003) reported the complete genome sequence of *Neurospora*. They sequenced about 39 Mb of DNA on seven chromosomes, and identified 10,082 protein-coding genes (9200 longer than 100 amino acids). Of these proteins, 41% have no similarity to known sequences, and 57% do not have identifiable orthologs in *S. cerevisiae* or *S. pombe*.

The *Neurospora* genome has only 10% repetitive DNA, including ∼185 copies of rDNA genes (Krumlauf and Marzluf, 1980). Other repeated DNA is dispersed and tends to be short and/or diverged, presumably because of the phenomenon of "RIP" (repeat-induced point mutation). RIP is a mechanism by which the genome is scanned for duplicated (repeated) sequences in haploid nuclei of special premeiotic cells. The RIP machinery efficiently finds them, and then litters them with numerous GC-to-AT mutations (Selker, 1990). Apparently RIP serves as a genome defense system for *Neurospora*, inactivating transposons and resisting genome expansion (Kinsey et al., 1994). Galagan et al. (2003) found relatively few *Neurospora* genes that are in multigene families, and a mere eight pairs of duplicated genes that encode proteins >100 amino acids. Also, 81% of the repetitive DNA sequences were mutated by RIP. Thus, RIP has suppressed the creation of new genes through duplication in *Neurospora* (Galagan et al., 2003; Perkins et al., 2001).

First Basidiomycete: *Phanerochaete chrysosporium*

The *P. chrysosporium* genome-sequencing project was undertaken by the U.S. Department of Energy (▶ http://genome.jgi-psf.org/Phchr1/Phchr1.home.html).

Phanerochaete chrysosporium is the first fungus of the phylum Basidiomycota to have its genome completely sequenced. This is a white rot fungus that degrades many biomaterials, including pollutants. Since fungi appeared about 1 to 1.5 billion years ago, and the Basidiomycota diverged from the better characterized Ascomycota over 500 million years ago, there were relatively little sequence data available from closely related organisms, and annotation of this genome was particularly difficult. The genome consists of about 30 Mb of DNA arranged in 10 chromosomes. Martinez

TABLE 17-8 Features of the *S. pombe* Genome

Chromosome Number	Length (Mb)	Number of Genes	Mean Gene Length (bp)	Coding (%)
1	5.599	2255	1446	58.6
2	4.398	1790	1411	57.5
3	2.466	884	1407	54.5
Whole genome	12.462	4929	1426	57.5

Source: From Wood et al. (2002).

et al. (2004) predicted 11,777 genes, of which three quarters had significant matches to previously known proteins. White rot fungi are able to degrade the major components of plant cell walls, including cellulose and lignins, using a series of oxidases and peroxidases. The genome encodes hundreds of enzymes that are able to cleave carbohydrates.

Fission Yeast *Schizosaccharomyces pombe*

The fission yeast *S. pombe* has a genome size of 13.8 Mb. The complete sequencing of this genome was reported by a large European consortium (Wood et al., 2002). The genome is divided into three chromosomes (Table 17.8).

Notably, there are 4940 predicted protein-coding genes (including 11 mitochondrial genes) and 33 pseudogenes. This is substantially fewer genes than is found in *S. cerevisiae* and is among the smallest number of protein-coding genes observed for any eukaryote. Some bacterial genomes encode more proteins, such as *Mesorhizobium loti* (6752 predicted genes) and *Streptomyces coelicolor* (7825 predicted genes).

The gene density in *S. pombe* is about one gene per 2400 bp, which is slightly less dense than is seen for *S. cerevisiae*. The intergenic regions are longer, and about 4730 introns were predicted. In *S. cerevisiae*, only 4% of the genes have introns.

Schizosaccharomyces pombe and *S. cerevisiae* diverged between 330 and 420 MYA. Some gene and protein sequences are equally divergent between these two fungi as they are between fungi and their vertebrate (e.g., human) orthologs. To identify such genes, you can use the TaxPlot tool on the NCBI Entrez genomes website. Comparative analyses are likely to elucidate the genetic basis for differences in the biology of these fungi, such as the propensity of *S. pombe* to divide by binary fission and the relatively fewer number of transposable elements in *S. pombe*.

For extensive information on *S. pombe* genome sequence analysis, see the Wellcome Trust Sanger Institute website (► http://www.sanger.ac.uk/Projects/S_pombe/).

Leland Hartwell, Timothy Hunt, and Sir Paul Nurse won the Nobel Prize in Physiology or Medicine in 2001 for their work on cell cycle control. Nurse's studies employed *S. pombe*, while Hartwell studied *S. cerevisiae* and Hunt studied sea urchins and other organisms. See ► http://www.nobel.se/medicine/laureates/2001/.

PERSPECTIVE

The budding yeast *S. cerevisiae* is one of the most significant organisms in biology for several reasons:

- It represents the first eukaryotic genome to have been sequenced. It was selected because of its compact genome size and structure.

- As a single-celled eukaryotic organism, its biology is simple relative to humans and other metazoans.

- The biology community has acquired a deep knowledge of yeast genetics and has collected a variety of molecular tools that are useful to elucidate the function of yeast genes. Functional genomics approaches based on genomewide analysis of gene function have been implemented as described in Chapter 12. For example, each of its >6000 genes has been knocked out and tagged with molecular barcodes, allowing massive, parallel studies of gene function.

Many additional fungal genomes are now being sequenced. In each branch of biology, we are learning that comparative genomic analyses are essential in helping to identify protein-coding genes (by homology searching), in evolutionary studies such as analyses of genome duplications, and in helping us to uncover biochemical pathways that allow cells to survive.

PITFALLS

At the same time that *S. cerevisiae* serves as an important model organism, it is important to realize the scope of our ignorance. How does the genotype of a single gene knock-out lead to a particular phenotype? We urgently need to answer this question for gene mutations in humans that cause disease, but even in a so-called simple model organism such as yeast we do not understand the full repertoire of protein–protein interactions that underlie cell function. If we think of the genome as a blueprint of a machine, we now have a "parts list" in the form of a list of the gene products. We must next figure out how the parts fit together to allow the machine to function in a variety of contexts. Gene annotation in yeast databases such as SGD, including the results of broad functional genomics screens, provides an excellent starting point for functional analyses.

WEB RESOURCES

The SGD (▶ http://www.yeastgenome.org/) lists a series of yeast resources. Another useful gateway is the Virtual Library—Yeast (▶ http://www.yeastgenome.org/VL-yeast.html).

DISCUSSION QUESTIONS

[17-1] The budding yeast *Saccharomyces cerevisiae* is sometimes described as a simple organism because it is unicellular, its genome encodes a relatively small number of genes (about 6000), and it has served as a model organism for genetics studies. Still, we understand the function of only about half its genes. Many functional genomics tools are now available, such as a complete collection of yeast knock-out strains (i.e., null alleles of each gene). How would you use such functional genomics tools to further our knowledge of gene function in yeast?

[17-2] The fungi are a sister group to the metazoans (animals) (Fig. 17.1). Do you expect the principles of genome evolution, gene function, and comparative genomics that are elucidated by studies of fungi to be closely applicable to metazoans such as humans, worms, and flies? For example, we discussed the whole-genome duplication of some fungi; how would you test the hypothesis that the human genome also underwent a similar duplication? In comparative genomics, do you expect fungi to be far more similar to each other in their biological properties than metazoans are to each other?

PROBLEMS/COMPUTER LAB

[17-1] Use the MIPS and UCSC Genome Browsers. (1) Visit MIPS (► http://mips.gsf.de/genre/proj/yeast/) and select *S. cerevisiae* chromosome XII. By default, the window size is 20,000 base pairs (as indicated in Fig. 17.5). Click the box labeled "3-frame translation (forward)" or add any other tracks of interest, and then update the image. You should see an extra track(s) appear on the browser. (2) Use the pull-down menu to select "Download GFF file" (also shown in Fig. 17.5). This is a General Feature Format (.gff) file that you can save to the desktop as a text file. Note that you may need to reformat the file by finding all instances of the chromosome designation "XII" and replacing them with "12". While you should create your own .gff file, a working version is available as web document 17.1 at ► http://www.bioinfbook.org/chapter17. (3) Visit the UCSC Genome Bioinformatics site (► http://genome.ucsc.edu/), click Genome Browser, select "other" for clade, Saccharomyces cerevisiae as organism, and enter "chr12:1-20000" for the query. Next, click the box labeled "add custom tracks" and upload your .gff file, and click submit. You can then view your custom track on the genome browser.

[17-2] ABC transporters constitute a large family of transmembrane-spanning proteins that hydrolyze ATP and drive the transport of ligands such as chloride across a membrane. How many ABC transporters are there in yeast?

[17-3] Use the *Saccharomyces* Genome Database:

- Go to the SGD site (► http://www.yeastgenome.org/).
- Pick any uncharacterized ORF. To find one, use the Gene/Seq Resources (one of the analysis tools), pick a chromosome (e.g, XII), then select Chromosomal Features Table. The first hypothetical ORF listed is YLL067C.
- Explore what its function might be. For some uncharacterized ORFs there will be relatively little information available; for others you may find a lot. From the Chromosomal Features Table click "Info" to view a page similar to that shown in Chapter 12.
   - What are the physical properties of the protein (e.g., molecular weight, isoelectric point)?
   - Does the protein have known domains?
   - Have interactions been characterized between this and other proteins?
   - Is the gene either induced or repressed in various physiological states, such as stress response or during sporulation?
- Try using Function Junction (at the bottom of the information page for your ORF). This will simultaneously search six databases:
   - Yeast Path Calling (two-hybrid analysis)
   - SGD SAGE query
   - Worm-Yeast protein comparison
   - Yeast Microarray Global Viewer
   - Yeast Protein Function Assignment
   - Triples database
- In what other organisms is this gene present? Compare the usefulness of exploring SGD versus performing your own BLAST searches to answer this question. Which is better?

[17-4] Create a phylogenetic tree of the fungi. Web document 17.2 at ► http://www.bioinfbook.org/chapter17 includes a set of 18S ribosomal RNA sequences; try using these or other sequences. Align them, and create a tree using MEGA or related software (Chapter 7). Does the tree agree with those shown in this chapter? If not, why not?

SELF-TEST QUIZ

[17-1] The *Saccharomyces cerevisiae* is characterized by the following properties except which one?

(a) Very high gene density (2000 base pairs per gene)

(b) Very low number of introns

(c) High degree of polymorphism

(d) 16 chromosomes

[17-2] The yeast *Saccharomyces cerevisiae* is an attractive model organism for many reasons. Which one of the following is NOT a useful feature of yeast?

(a) The genome size is relatively small.

(b) Gene knock-outs by homologous recombination are possible.

(c) Large repetitive DNA sequences serve as a good model for higher eukaryotes.

(d) There is high open reading frame (ORF) density.

[17-3] The *Saccharomyces cerevisiae* genome is small (it encodes about 6000 genes). It is thought that, about 100 MYA:

(a) The entire genome duplicated, followed by tetraploidization.

(b) The genome underwent many segmental duplications, followed by gene loss.

(c) The entire genome duplicated, followed by gene loss.

(d) The genome duplicated, followed by gene conversion.

[17-4] After gene duplication, the most common outcome is the loss of the duplicated gene. A reasonable explanation of why this might occur is that

(a) This second copy is superfluous.

(b) This second gene may acquire forbidden mutations that are deleterious to the fitness of the organism.

(c) This second copy is under intense negative selection.

(d) This second copy is a substrate for nonallelic homologous recombination.

[17-5] Comparative analyses of *S. cerevisiae* and two closely related species (*S. castelli* and *C. glabrata*) allow a description of the patterns of gene retention and gene loss in multiple organisms following whole-genome duplication. Across thousands of gene loci in three genomes that underwent genome duplication, which of the following occurred?

(a) For about three quarters of the loci, all three species lost one of the two copies of a duplicated gene.

(b) For about half of the loci, no gene loss occurred.

(c) For about half of the loci, there was partial loss of both copies of a duplicated gene.

(d) For about three quarters of all loci, all three loci lost both copies of the duplicated gene.

[17-6] Features of the *Candida albicans* genome include:

(a) An accessory plasmid

(b) One of its chromosomes has a highly variable length

(c) The DNA is characterized by an extraordinarily high amount of polymorphism

(d) The CTG codon that encodes leucine in most organisms encodes serine in *C. albicans*

[17-7] The filamentous fungus *Neurospora crassa* has an extremely low amount of repetitive DNA (spanning only 10% of its 39 megabase genome). Why?

(a) It uses chromatin diminution

(b) It uses repetitive DNA inversion

(c) It uses repeat-induced point mutations, a phenomenon in which repeats are inactivated

(d) It uses repeat-induced synchronization to inactivate repeats

[17-8] One of the most remarkable features of the *Schizosaccharomyces pombe* genome is that:

(a) It is predicted to encode fewer than 5000 proteins, making its genome (and proteome) smaller than even some bacterial genomes.

(b) The number of predicted introns is about the same as the number of predicted ORFs.

(c) It has as many genes that are homologous to bacterial genes as it has genes that are homologous to *S. cerevisiae* genes.

(d) Its genome size is approximately the same as that of *S. cerevisiae*, even though these species diverged hundreds of millions of years ago.

[17-9] Yeast is the only major research organism approved by the U.S. Food and Drug Administration (FDA) for human consumption.

(a) True

(b) False

SUGGESTED READING

A superb overview of fungal taxonomy is provided by Guarro et al. (1999), while important, recent papers are by Hibbett et al. (2007) and James et al. (2006).

An excellent overview of the *S. cerevisiae* genome is provided by Johnston (2000). This article discusses strategies to assign functions to yeast genes. Bernard Dujon (2006) reviews yeast genomics in relation to eukaryotic genome evolution.

REFERENCES

Aalto, M. K., Ronne, H., and Keranen, S. Yeast syntaxins Sso1p and Sso2p belong to a family of related membrane proteins that function in vesicular transport. *EMBO J.* **12**, 4095–4104 (1993).

Ainsworth, G. C. Fungus infections (mycoses). In Kiple, K. F., ed. *The Cambridge World History of Human Disease*, pp. 730–736. Cambridge University Press, New York, 1993.

Baldauf, S. L., Roger, A. J., Wenk-Siefert, I., and Doolittle, W. F. A kingdom-level phylogeny of eukaryotes based on combined protein data. *Science* **290**, 972–977 (2000).

Bulloch, W. *The History of Bacteriology*. Oxford University Press, New York, 1938.

Byrne, K. P., and Wolfe, K. H. The Yeast Gene Order Browser: Combining curated homology and syntenic context reveals gene fate in polyploid species. *Genome Res.* **15**, 1456–1461 (2005).

Byrne, K. P., and Wolfe, K. H. Visualizing syntenic relationships among the hemiascomycetes with the Yeast Gene Order Browser. *Nucleic Acids Res.* **34**, D452–D455 (2006).

Casselton, L., and Zolan, M. The art and design of genetic screens: Filamentous fungi. *Nat. Rev. Genet.* **3**, 683–697 (2002).

Cherry, J. M., et al. Genetic and physical maps of *Saccharomyces cerevisiae*. *Nature* **387**, 67–73 (1997).

Cliften, P., Sudarsanam, P., Desikan, A., Fulton, L., Fulton, B., Majors, J., Waterston, R., Cohen, B. A., and Johnston, M. Finding functional features in *Saccharomyces* genomes by phylogenetic footprinting. *Science* **301**, 71–76 (2003).

Dacks, J. B., and Doolittle, W. F. Novel syntaxin gene sequences from *Giardia, Trypanosoma* and algae: Implications for the ancient evolution of the eukaryotic endomembrane system. *J. Cell Sci.* **115**, 1635–1642 (2002).

Dujon, B. Yeasts illustrate the molecular mechanisms of eukaryotic genome evolution. *Trends Genet.* **22**, 375–387 (2006).

Dujon, B., et al. Complete DNA sequence of yeast chromosome XI. *Nature* **369**, 371–378 (1994).

Dujon, B., et al. Genome evolution in yeasts. *Nature* **430**, 35–44 (2004).

Dwight, S. S., et al. Saccharomyces Genome Database (SGD) provides secondary gene annotation using the Gene Ontology (GO). *Nucleic Acids Res.* **30**, 69–72 (2002).

Galagan, J. E., et al. Sequencing of *Aspergillus nidulans* and comparative analysis with *A. fumigatus* and *A. oryzae*. *Nature* **438**, 1105–1115 (2005).

Garbarino, J. E., and Gibbons, I. R. Expression and genomic analysis of midasin, a novel and highly conserved AAA protein distantly related to dynein. *BMC Genomics* **3**, 18 (2002).

Goffeau, A., et al. Life with 6000 genes. *Science* **274**, 546, 563–567 (1996).

Guarro, J., Gene J., and Stchigel, A. M. Developments in fungal taxonomy. *Clin. Microbiol. Rev.* **12**, 454–500 (1999).

Harbison, C. T., et al. Transcriptional regulatory code of a eukaryotic genome. *Nature* **431**, 99–104 (2004).

Harrison, P. M., Kumar, A., Lang, N., Snyder, M., and Gerstein, M. A question of size: The eukaryotic proteome and the problems in defining it. *Nucleic Acids Res.* **30**, 1083–1090 (2002).

Hibbett, D. S., et al. A higher-level phylogenetic classification of the Fungi. *Mycol. Res.* **111**, 509–547 (2007).

James, T. Y., et al. Reconstructing the early evolution of Fungi using a six-gene phylogeny. *Nature* **443**, 818–822 (2006).

Johnston, M. The yeast genome: On the road to the Golden Age. *Curr. Opin. Genet. Dev.* **10**, 617–623 (2000).

Johnston, M., et al. The nucleotide sequence of *Saccharomyces cerevisiae* chromosome XII. *Nature* **387**, 87–90 (1997).

Jones, T., Federspiel, N. A., Chibana, H., Dungan, J., Kalman, S., Magee, B. B., Newport, G., Thorstenson, Y. R., Agabian, N., Magee, P. T., Davis, R. W., and Scherer, S. The diploid genome sequence of *Candida albicans*. *Proc. Natl. Acad. Sci. USA* **101**, 7329–7334 (2004).

Katinka, M. D., et al. Genome sequence and gene compaction of the eukaryote parasite *Encephalitozoon cuniculi*. *Nature* **414**, 450–453 (2001).

Kellis, M., Birren, B. W., and Lander, E. S. Proof and evolutionary analysis of ancient genome duplication in the yeast *Saccharomyces cerevisiae*. *Nature* **428**, 617–624 (2004).

Kellis, M., Patterson, N., Endrizzi, M., Birren, B., and Lander, E. S. Sequencing and comparison of yeast species to identify genes and regulatory elements. *Nature* **423**, 241–254 (2003).

Kinsey, J. A., Garrett-Engele, P. W., Cambareri, E. B., and Selker, E. U. The *Neurospora* transposon Tad is sensitive to repeat-induced point mutation (RIP). *Genetics* **138**, 657–664 (1994).

Krumlauf, R., and Marzluf, G. A. Genome organization and characterization of the repetitive and inverted repeat DNA sequences in *Neurospora crassa*. *J. Biol. Chem.* **255**, 1138–1145 (1980).

Kuchenmeister, F. *On Animal and Vegetable Parasites of the Human Body, a Manual of Their Natural History, Diagnosis, and Treatment*. Sydenham Society, London, 1857.

Kumar, A., et al. An integrated approach for finding overlooked genes in yeast. *Nat. Biotechnol.* **20**, 58–63 (2002).

Kurtzman, C. P., and Robnett, C. J. Phylogenetic relationships among yeasts of the "Saccharomyces complex" determined from multigene sequence analyses. *FEMS Yeast Res.* **3**, 417–432 (2003).

Loftus, B. J., et al. The genome of the basidiomycetous yeast and human pathogen *Cryptococcus neoformans*. *Science* **307**, 1321–1324 (2005).

Lowe, T. M., and Eddy, S. R. A computational screen for methylation guide snoRNAs in yeast. *Science* **283**, 1168–1171 (1999).

Lynch, M., and Conery, J. S. The evolutionary fate and consequences of duplicate genes. *Science* **290**, 1151–1155 (2000).

Machida, M., et al. Genome sequencing and analysis of *Aspergillus oryzae*. *Nature* **438**, 1157–1161 (2005).

Mackiewicz, P., et al. How many protein-coding genes are there in the *Saccharomyces cerevisiae* genome? *Yeast* **19**, 619–629 (2002).

Margulis, L., and Schwartz, K. V. *Five Kingdoms. An Illustrated Guide to the Phyla of Life on Earth*. W.H. Freeman and Company, New York, 1998.

Martinez, D., et al. Genome sequence of the lignocellulose degrading fungus *Phanerochaete chrysosporium* strain RP78. *Nat. Biotechnol.* **22**, 695–700 (2004).

Mewes, H. W., et al. Overview of the yeast genome. *Nature* **387**, 7–65 (1997).

Mewes, H. W., et al. MIPS: A database for genomes and protein sequences. *Nucleic Acids Res.* **30**, 31–34 (2002).

Nierman, W. C., et al. Genomic sequence of the pathogenic and allergenic filamentous fungus *Aspergillus fumigatus*. *Nature* **438**, 1151–1156 (2005).

Odds, F. C., Brown, A. J., and Gow, N. A. *Candida albicans* genome sequence: A platform for genomics in the absence of genetics. *Genome Biol.* **5**, 230 (2004).

Ohno S., *Evolution by Gene Duplication*. Springer Verlag, Berlin, 1970.

Oshiro, G., et al. Parallel identification of new genes in *Saccharomyces cerevisiae*. *Genome Res.* **12**, 1210–1220 (2002).

Perkins, D. D., Radford, A., and Sachs, M. S. *The Neurospora Compendium: Chromosomal Loci*. Academic Press, San Diego, CA, 2001.

Piskur, J. Origin of the duplicated regions in the yeast genomes. *Trends Genet.* **17**, 302–303 (2001).

Protopopov, V., Govindan, B., Novick, P., and Gerst, J. E. Homologs of the synaptobrevin/VAMP family of synaptic vesicle proteins function on the late secretory pathway in *S. cerevisiae*. *Cell* **74**, 855–861 (1993).

Rossignol, T., Lechat, P., Cuomo, C., Zeng, Q., Moszer, I., d'Enfert, C. CandidaDB: A multi-genome database for *Candida* species and related *Saccharomycotina*. *Nucleic Acids Res.* **36**, D557–D561 (2008).

Roth, J. F. The yeast Ty virus-like particles. *Yeast* **16**, 785–795 (2000).

Rozen, S., et al. Abundant gene conversion between arms of palindromes in human and ape Y chromosomes. *Nature* **423**, 873–876 (2003).

Sankoff, D. Gene and genome duplication. *Curr. Opin. Genet. Dev.* **11**, 681–684 (2001).

Scannell, D. R., Byrne, K. P., Gordon, J. L., Wong, S., and Wolfe, K. H. Multiple rounds of speciation associated with reciprocal gene loss in polyploid yeasts. *Nature* **440**, 341–345 (2006).

Scazzocchio, C. *Aspergillus* genomes: Secret sex and the secrets of sex. *Trends Genet.* **22**, 521–525 (2006).

Selker, E. U. Premeiotic instability of repeated sequences in *Neurospora crassa*. *Annu. Rev. Genet.* **24**, 579–613 (1990).

Seoighe, C., and Wolfe, K. H. Updated map of duplicated regions in the yeast genome. *Gene* **238**, 253–261 (1999).

Wagner, A. Robustness against mutations in genetic networks of yeast. *Nat. Genet.* **24**, 355–361 (2000).

Wagner, A. Birth and death of duplicated genes in completely sequenced eukaryotes. *Trends Genet.* **17**, 237–239 (2001).

Wang, D. Y., Kumar, S., and Hedges, S. B. Divergence time estimates for the early history of animal phyla and the origin of plants, animals and fungi. *Proc. R. Soc. Lond. B Biol. Sci.* **266**, 163–171 (1999).

Whittaker, R. H. New concepts of kingdoms or organisms. Evolutionary relations are better represented by new classifications than by the traditional two kingdoms. *Science* **163**, 150–160 (1969).

Wolfe, K. H., and Shields, D. C. Molecular evidence for an ancient duplication of the entire yeast genome. *Nature* **387**, 708–713 (1997).

Wood, V., et al. The genome sequence of *Schizosaccharomyces pombe*. *Nature* **415**, 871–880 (2002).

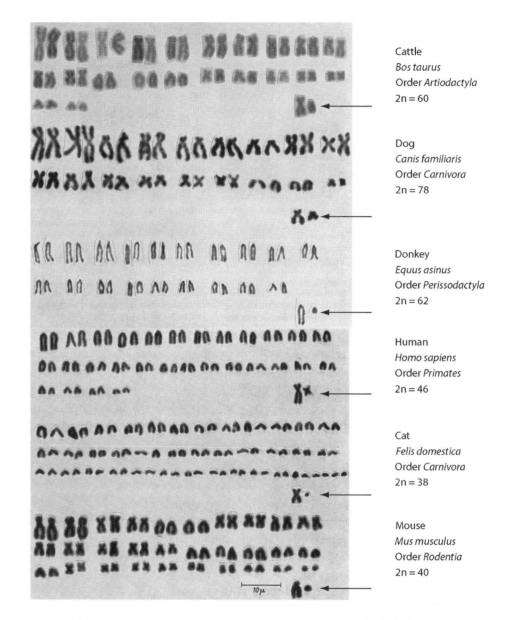

Cattle
Bos taurus
Order *Artiodactyla*
2n = 60

Dog
Canis familiaris
Order *Carnivora*
2n = 78

Donkey
Equus asinus
Order *Perissodactyla*
2n = 62

Human
Homo sapiens
Order *Primates*
2n = 46

Cat
Felis domestica
Order *Carnivora*
2n = 38

Mouse
Mus musculus
Order *Rodentia*
2n = 40

Plants, fungi, Paramecia, and many other organisms have undergone one or more rounds of whole genome duplication. For placental mammals this is no longer an option because the sex differentiation system (with females having XX sex chromosomes and males XY) would be disrupted by whole genome duplication, leading to sterility. Susumu Ohno (1928–2000) discussed this in his 1967 book Sex Chromosomes and Sex-Linked Genes. *He reviewed evidence that the X and Y chromosomes (or Z and W in avian and ophidian [snake] species) were derived from an ancient pair of homologous chromosomes. Of these the X chromosome has maintained a remarkably similar size, while the Y chromosome has undergone rapid reduction in size. This figure (Ohno, 1967, fig. 11) shows the male diploid karyotypes of six species of placental mammals (cattle, dog, donkey, human, cat, and mouse). While there is great variability in the number of chromosomes among mammals, the X chromosomes (indicated by arrows next to the Y chromosomes) have maintained a relatively constant size. Used with permission.*

18

Eukaryotic Genomes: From Parasites to Primates

INTRODUCTION

In this chapter we will explore individual eukaryotic genomes, from parasites to primates. We will refer to a phylogenetic tree of the eukaryotes that was produced by Baldauf et al. (2000) (Fig. 18.1). This tree was created by parsimony analysis using four concatenated protein sequences: elongation factor 1a (EF-1α), actin, α-tubulin, and β-tubulin. We already discussed fungal genomes in Chapter 17; they are represented in a group that is adjacent to the metazoa (animals). We will examine representative organisms in this tree, moving from the bottom up. This includes the diplomonad *Giardia lamblia* and other protozoans, such as the malaria parasite, *Plasmodium falciparum*; the plants, including the first sequenced plant genome (that of the thale cress, *Arabidopsis thaliana*) and rice (*Oryza sativa*); and the metazoans, from worms and insects to fish and mammals. We will address the human genome in Chapter 19.

The phylogenetic tree of Fig. 18.1 has also been reproduced by Tyler et al. (2006) in their description of *Phytophthora* genomes of the kingdom Stramenopila, described below.

Bioinformatics and Functional Genomics, Second Edition. By Jonathan Pevsner
Copyright © 2009 John Wiley & Sons, Inc.

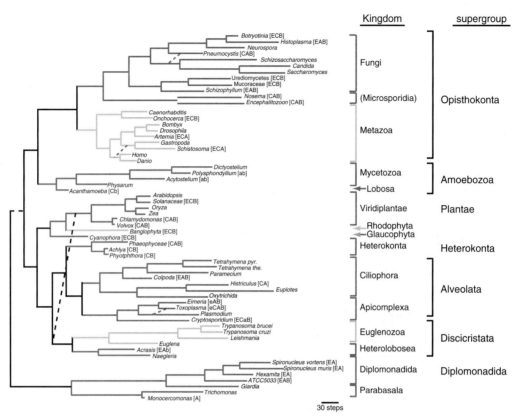

FIGURE 18.1. *A phylogeny of eukaryotes based on parsimony analysis of concatenated protein sequences. The proteins analyzed were EF-1α (abbreviated E in tree), actin (C), α-tubulin (A), and β-tubulin (B). This tree may be compared to the eukaryotic portion of the global tree of life based on small subunit ribosomal RNA sequences (Fig. 13.1). In this tree, 14 kingdoms are indicated as well as seven supergroups. One of the supergroups, Opisthokonta, includes fungi and microsporidia (Chapter 17) and metazoa (vertebrate and invertebrate animals). The tree was constructed by maximum parsimony and by maximum-likelihood analysis of second-codon-position nucleotides. For taxa with missing data, the sequences used are indicated in brackets (e.g., [EAB]). Modified from Baldauf et al. (2000). Used with permission.*

Following the outline introduced in Chapter 13, we will consider five aspects of various genomes.

1. Cataloguing information includes describing the complete sequence of each chromosome, annotating the DNA to identify and characterize noncoding DNA, and identifying protein-coding genes and other noncoding genes. We will survey chromosome number and structure (such as regions of duplication or deletion). This chapter provides a large amount of information about genome sizes. In many cases the exact size of a genome in megabases or the exact number of genes are unknown; in some cases, even the number of chromosomes is unknown. A goal of this chapter is to provide a survey of currently available information about eukaryotic genomes that orients you to the scales of genome sizes.

2. Comparative genomics is an essential part of any genome analysis. The availability of closely related species (such as 12 Drosopholids) permits

a series of questions to be addressed about recent evolutionary changes, such as lineage-specific expansions or contractions of gene families. The availability of distantly related species (such as fish genome sequences that last shared a common ancestor with humans over 400 million years ago) permits different kinds of questions to be addressed, such as the presence of conserved gene structures and regulatory elements.

3. Biological principles can be explored through genome sequences. For example, the genome of an underwater sea urchin unexpectedly encodes receptors that in other animals facilitate hearing and chemoreception, suggesting unsuspected sensory abilities of these animals. In general, genome sequence analysis can be used in an attempt to relate the genomic sequence to the phenotype of the organism. This phenotype includes an organism's strategies for adaptation to its environment, evolution, metabolism, growth, development, maintenance of homeostasis, and reproduction.

4. Bioinformatics approaches are constantly evolving, such as techniques for whole genome sequencing and assembly as well as analytic tools. Analysis of genomes involves the use of many of the tools we introduced in Chapters 2 to 7, including BLAST and molecular phylogeny. In the first third of the book we discussed many of the complexities of multiple sequence alignment and phylogeny, and showed that the same raw data can be used to generate many alternative results. As you read about various genomes in this chapter, accession numbers (for genome projects and/or genes and proteins) are provided that will allow you to independently analyze many sequence analysis problems.

5. Analysis of genomic sequences offers a unique perspective on human disease (and diseases afflicting other organisms). In the case of many eukaryotes— from the protozoans such as *Plasmodium* to pathogenic fungi and parasitic worms—we also want to understand the genetic basis of how the organism causes disease and how we can counterattack. At present, there are no vaccines available to prevent diseases caused by any eukaryotic parasites that infect humans, including protozoans (such as trypanosomes) and helminths (parasitic nematodes). The availability of whole genome sequences may provide clues as to which antigens are promising targets for vaccine development and pharmacological intervention. For example, predicted secreted surface proteins can be expressed in bacteria and used to immunize mice in order to develop potential vaccines (Fraser et al., 2000).

The word protozoan derives from the Greek *proto* ("early") and *zoion* ("animal"). This contrasts with the word metazoan (animal) from the Greek *meta* ("after"; at a later stage of development) and *zoion*.

A phylogenetic description of the eukaryotes is essential for our understanding of both evolutionary processes that shaped the development of species and the diversity of life today. Evolutionary reconstructions that are based on molecular sequence data typically use small subunit ribosomal RNA because it has many sites that are phylogenetically informative across all life forms (Van de Peer et al., 2000). We saw an example of such a tree in Fig. 13.2. However, there is no uniform consensus on the optimal approach to making a tree (Box 18.1 and Chapter 7). For other phylogenetic trees of the eukaryotes, differing in some details from Fig. 18.1, see Keeling (2007) in an introduction to the *Giardia lamblia* genome project, and a detailed review by Embley and Martin (2006).

BOX 18.1
Inconsistent Phylogenies

It is important to note that many phylogenetic reconstructions are inconsistent with each other. There are three main sources of conflicting results (Philippe and Laurent, 1998):

1. Gene duplication followed by random gene loss can cause artifacts in tree reconstruction. This occurred at the whole-genome level in yeast (Chapter 17) and other eukaryotes such as plants and fish (see below).

2. Lateral gene transfer can confuse phylogenetic interpretation (Chapter 15).

3. The technical artifact of long branch chain attraction can confuse phylogenetic analyses. This is a phenomenon where the longest branches of a tree are grouped together, regardless of the true tree topology (See Fig. 7.29). It is essential to account for differences in substitution rates among sites within a molecule. Reyes et al. (2000) consider this problem in their phylogeny of the order Rodentia.

Researchers often overcome these potential problems by concatenating multiple protein (or nucleic acid) sequences. For example, the tree in Fig. 18.1 is based on four concatenated proteins. Wang et al. (1999) used 75 genes in their comprehensive phylogeny of animals, plants, and fungi (described below); Kumar and Hedges (1998) studied 658 genes in 207 vertebrate species. In another strategy, several groups have made trees based on gene content or gene fusion events (Snel et al., 1999; Stechmann and Cavalier-Smith, 2002).

PROTOZOANS AT THE BASE OF THE TREE LACKING MITOCHONDRIA

The eukaryotes include deep-branching protozoan species from the parabasala (e.g., *Trichomonas*), diplomonadida (such as *Giardia*), discicristata (e.g., *Euglena*, *Leishmania*, and *Trypanosoma*), alveolata (e.g., *Toxoplasma* and *Plasmodium*), and heterokonta (Fig. 18.1). We begin at the bottom of the tree of Fig. 18.1 by describing *Trichomonas* and *Giardia*.

The microsporidia such as *Encephalitozoon* used to be classified as deep-branching eukaryotes. Subsequent analysis of the complete *E. cuniculi* genome revealed that this microsporidial parasite is closely related to the fungi (Chapter 17 and Fig. 18.1).

There is strong evidence that mitochondrial genes, present in most eukaryotes, are derived from an α-proteobacterium (see Chapter 15). Previously, it was hypothesized that deep-branching organisms such as *Giardia* and *Trichomonas* lack mitochondria. They were thought to have evolved from other eukaryotes prior to the symbiotic invasion of an α-proteobacterium. However, analyses of *Giardia*, *Trichomonas*, and microsporidia such as *Trachipleistophora hominis* suggest the presence of mitochondrial genes (Embley and Hirt, 1998; Williams et al., 2002; Lloyd and Harris, 2002). Some protists (including trichomonads and ciliates) lack typical mitochondria but have a derived organelle called the hydrogenosome. This membrane-bound structure produces adenosine triphosphate (ATP) and molecular hydrogen via fermentation.

Trichomonas

Trichomonas vaginalis, a flagellated protist and member of the parabasilids, is a sexually transmitted pathogen (Fig. 18.2). The World Health Organization estimates that

Genus, species: *Trichomonas vaginalis*

Lineage: Eukaryota; Parabasalidea; Trichomonada;
Trichomonadida; Trichomonadidae; Trichomonadinae;
Trichomonas; *Trichomonas vaginalis* G3

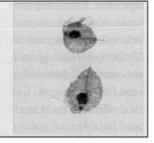

Haploid genome size: ~160 Mb
GC content: 32.7%
Number of chromosomes: 6
Number of protein-coding genes: ~60,000

Disease association: *Trichomonas* causes the sexually transmitted infection trichomoniasis
　　　　　(~170 million cases worldwide annually).
Key genomic features: The *Trichomonas* genome encodes an extraordinarily large number of genes
　　　　　(~60,000). A total of 65 genes have introns. 65% of the genome consists of repetitive DNA.
Entrez Genome Project: 16084
RefSeq accession number: NZ_AAHC00000000

FIGURE 18.2. The parabasala (see Fig. 18.1) are protozoans including Trichomonas vaginalis. Photograph from the Centers for Disease Control (CDC) Parasite Image Library (▸ http://www.dpd.cdc.gov/dpdx/HTML/Image_Library.htm) shows two trophozoites obtained from in vitro culture.

there are ~170 million cases annually worldwide. *Trichomonas* is a single-celled organism that resides in the genitourinary tract, where it phagocytoses vaginal epithelial cells, erythrocytes, and bacteria. Its genome of ~160 Mb has several remarkable features (Carlton et al., 2007). Sixty-two percent of the genome consists of repetitive DNA, confounding efforts to characterize the genome architecture. Many of these repeats are of viral, transposon, or retrotransposon origin. There are 60,000 predicted protein-coding genes, one of the highest numbers among all life forms. Several gene families have undergone massive expansion, such as protein kinases ($n = 927$), the *BspA-like* gene family ($n = 658$), and small GTPases ($n = 328$). The BspA-like proteins are surface antigens that participate in host cell adherence and aggregation. *T. vaginalis* has apparently acquired 152 genes by lateral gene transfer from bacteria that thrive in the intestinal flora; most of these genes encode metabolic enzymes.

Information on trichomoniasis is available at ▸ http://www.trichomoniasis.org/.

Analysis of the draft genome sequence by Carlton et al. suggests mechanisms by which *T. vaginalis* obtains its energy, functions as a parasite adhering to and invading host cells, and degrades proteins (via a complex degradome).

Giardia lamblia: A Human Intestinal Parasite

Giardia lamblia (also called *Giardia intestinalis*) is a protozoan, water-borne parasite that lives in the intestines of mammals and birds (Adam, 2001). It is the cause of giardiasis, the most frequent source of nonbacterial diarrhea in North America. Like some other unicellular protozoans, *Giardia* lack not only mitochondria but also peroxisomes (responsible for fatty acid oxidation) and nucleoli. Thus, the genome of *Giardia* could reflect the adaptations that led to the early emergence of eukaryotic cells.

The *Giardia* genome is ~11.7 Mb and was sequenced by the whole-genome shotgun method (Morrison et al., 2007) (Fig. 18.3). Each cell has two morphologically identical nuclei, each nucleus having five chromosomes ranging from 0.7 to over

Giardia was the first parasitic protozoan of humans observed with a microscope by Antony van Leeuwenhoek (in 1681). The diplomonadida are also called diplomonads. This group includes the family Hexamitidae, which further includes the genus *Giardia*. Information on *Giardia* is available at the U.S. Food and Drug Administration (▸ http://vm.cfsan.fda.gov/ ~ mow/chap22.html) and the CDC (▸ http://www.cdc.gov/healthyswimming/giardiafacts.htm).

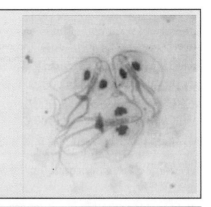

Genus, species: *Giardia lamblia*

Lineage: Eukaryota; Diplomonadida; Hexamitidae;
Giardiinae; Giardia; *Giardia lamblia* ATCC 50803

Haploid genome size: 12 Mb
GC content: 49%
Number of chromosomes: 5
Number of protein-coding genes: 6,470
Number of genes per kilobase: 0.58

Disease association: *Giardia* causes ~100 million infections annually, and is the most prevalent
      parasitic protist in the United States.
Key genomic features: *Giardia* lacks mitochondria, hydrogenosomes, and peroxisomes. The
      organism has two similar, active, diploid nuclei. The genome encodes simplified machinery
      for DNA replication, transcription and RNA processing. There are no Krebs cycle proteins
      and few genes encoding proteins involved in amino acid metabolism.
Entrez Genome Project: 1439
Project accession number: AACB02000000
Key website: GiardiaDB (http://www.giardiadb.org)

FIGURE 18.3. *The Diplomonadida (see Fig. 18.1) are protozoans including Giardia. Image shows three trophozoites stained with Giemsa (from the CDC Parasite Image Library, ▶ http://www.dpd.cdc.gov/dpdx/HTML/Image_Library.htm). Each protist has two prominent nuclei.*

Organisms that lack peroxisomes could provide us insight into fatty acid metabolism or other metabolic processes. This in turn could prove helpful to our understanding of human diseases that affect such organelles. The most common human genetic disorder affecting peroxisomes is adrenoleukodystrophy, caused by mutations in the *ABCD1* gene (RefSeq accession NM_000033). Does *Giardia* have an ortholog of this gene?

The *Giardia* genome project website is at ▶ http://www.giardiadb.org/giardiadb/.

You can study the *Giardia* ferredoxin gene at GenBank (DNA accession AF393829). To find the intron, try using BLAST (Chapter 4) to compare the protein (or the DNA encoding the protein) to the genomic DNA. Note that the project accession number for this organism (given in Fig. 18.3; AACB00000000) points to a set of whole genome shotgun sequence reads (accessions AACB02000001 to AACB02000306). To perform the blast search, go to blastn, use the

3 Mb. 6470 open reading frames were identified, spanning 77% of the genome, with 1800 overlapping genes and an additional 1500 open reading frames spaced within 100 nucleotides of an adjacent open reading frame.

As we consider the genomes of various eukaryotes, a consistent theme is that transposable elements are extremely abundant, occupying half the entire human genome (Chapter 19) and causing massive genomic rearrangements. Thus, in order to understand their origins and their function, it is of interest to find eukaryotes that lack these elements. *Giardia* provides such an example. Arkhipova and Meselson (2000) examined 24 eukaryotic species for the presence of two major classes of transposable elements (retrotransposon reverse transcriptases and DNA transposons). They found them present in all species except bdelloid rotifers, an asexual animal. Deleterious transposable elements thrive in sexual species, but they are unlikely to propagate in asexual species because of strong selective pressure against having active elements. Further inspection of the asexual *Giardia* by Arkhipova and Morrison (2001) revealed just three retrotransposon families. One of these is inactive, and the other two are telomeric. This location could provide a buffer between protein-coding genes and the telomeres, and these elements could contribute to the ability of *Giardia* to vary the length of its chromosomes in response to environmental pressures—for example, chromosome 1 can expand from 1.1 to 1.9 Mb (Pardue et al., 2001).

Another basic question about eukaryotic genomes is the origin of introns. Spliceosomal introns occur commonly in the "crown group" of eukaryotes (the kingdoms Animalia, Plantae, and Fungi). However, their presence in the earliest branching protozoa has been disputed (Johnson, 2002), and introns have not been detected in parabasalids such as *Trichomonas*. Nixon et al. (2002) identified a 35 bp intron in a gene encoding a putative [2Fe-2S] ferredoxin, and analysis of the draft genome sequence by Morrison et al. (2007) identified three more. Simpson et al. (2002) also identified several introns in *Carpediemonas membranifera*, a eukaryote thought to be a close relative of *Giardia*. These findings suggest that if introns were a eukaryotic adaptation, they arrived early in evolution and possibly in the last common eukaryotic ancestor.

Genomes of Unicellular Pathogens: Trypanosomes and *Leishmania*

Trypanosomes

There are about 20 species in the protozoan genus *Trypanosoma* (reviewed in Donelson, 1996). Two of these are pathogenic in humans (Cox, 2002). *Trypanosoma brucei* subspecies cause several forms of sleeping sickness, a fatal disease that infects hundreds of thousands of people in Africa (Fig. 18.4). *Trypanosoma cruzi* causes Chagas disease, prevalent in South and Central America. The adverse impact of these trypanosomes is even greater because they also afflict livestock. Tsetse flies or other insects transmit the trypanosomes to humans.

Several *Trypanosome* genome project websites offer information on the biology of these parasites as well as sequencing information (Table 18.1). Berriman et al. (2005) reported the genome sequence of *T. brucei*. The genome is 26 Mb, although its size varies by up to 25% in different isolates (reviewed in El-Sayed et al., 2000). There are at least 11 pairs of large, diploid, nuclear chromosomes (ranging in size from about 1 Mb to >6 Mb). Additionally, there are variable numbers of intermediate chromosomes (200 to 900 kb), and there are about 100 linear minichromosomal DNA molecules (50 to 150 kb). Some of these minichromosomes contain a 177 bp repeat that comprises more than 90% of the total sequence (El-Sayed et al., 2000). The genome includes 9068 predicted genes, of which about 900 are pseudogenes and ~1700 are specific to *T. brucei*.

query AF393829, set the database to WGS, and include the Entrez Query AACB02000001: AACB02000306[PACC].

Tsetse flies are insects that feed on vertebrate blood. To obtain additional nutrients beyond what is available in blood, tsetse flies harbor two obligate intracellular bacteria: *Wigglesworthia glossinidia* and *Sodalis glossinidius*. The *W. glossinidia* genome was sequenced (see RefSeq accession NC_004344). Similar to other intracellular bacteria (Chapter 15), it has a reduced genome size of only 700,000 bp (Akman et al., 2002). For a description of the *T. brucei* lifecycle, see ▶ http://www. dpd.cdc.gov/dpdx/HTML/ TrypanosomiasisAfrican.htm.

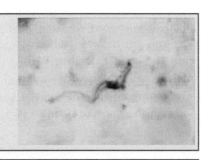

Genus, species: *Trypanosoma brucei*
      Trypanosoma cruzi
      Leishmania major (Friedlin strain)

Lineage: Eukaryota; Euglenozoa; Kinetoplastida (order); Trypanosomatidae (family); Trypanosoma

	T. brucei	T. cruzi	L. major
Haploid genome size:	35 Mb	60 Mb	32.8 Mb
GC content	46.4%	51%	59.7%
Number of chromosomes:	11*	~28 (variable)	36
Number of genes (incl. pseudogenes)	9,068	~12,000	8,311
	* includes ~100 mini- and intermediate size chromosomes		

Disease association: *T. brucei* causes trypanosomiasis (sleeping sickness). The incidence is 300,000 to 500,000 cases per year. *T. cruzi* causes Chagas disease in humans; 16–18 million people are infected, with 21,000 deaths per year. Leishmaniasis is an infectious disease with 2 million new cases annually and 350 million people at risk; 20 *Leishmania* species infect humans. No vaccines and few drugs are available.
Key genomic features: These three species share a conserved core proteome of ~6,200 proteins.
Entrez Genome project identifiers: 11756 (*T. brucei*), 11755 (T. *cruzi*), 10724 (*L. major*).

FIGURE 18.4. The Euglenozoa (see Fig. 18.1) include the kinetoplastid parasitic protozoa Trypanosoma brucei, T. cruzi, *and* Leishmania major. *The image (from the CDC Parasite Image Library) shows a* T. brucei *from a blood smear in the trypomastigotes stage. There is a centrally located nucleus, a small kinetoplast at the posterior end (upper right), and an undulating membrane with a flagellum exiting the body at the anterior end. The length ranges from 14 to 33 μm.*

TABLE 18-1   Web Resources for *Trypanosome Genomics*

Resource	Comment	URL
The *Trypanosoma brucei* Genome Project	Sponsored by the Wellcome Trust Sanger Institute	▶ http://www.sanger.ac.uk/Projects/T_brucei/
Trypanosoma brucei Genome Project	TIGR Database	▶ http://www.tigr.org/tdb/e2k1/tba1/intro.shtml
Trypanosoma cruzi Genome Initiative Information Server	From the Oswaldu Cruz Institute, Brazil	▶ http://www.dbbm.fiocruz.br/TcruziDB/

For information on sleeping sickness, see the World Health Organization website at ▶ http://www.who.int/tdr/diseases/tryp/default.htm.

The accession number of a typical VSG protein from *T. brucei* is XP_822273. Try a PSI-BLAST searching using it as a query, and contrast the results of the first and second or further iterations. The *Trypanosoma brucei* GeneDB is available at ▶ http://www.genedb.org/genedb/tryp.

See problem 18.3 for an exercise on a trypanosome universal minicircle binding protein. For an example of a maxicircle sequence and the genes it encodes, see GenBank accession M94286.

Another remarkable feature of trypanosomes is the presence of a massive network of circular rings of mitochondrial DNA, termed kinetoplast DNA. Thousands of rings of kinetoplast DNA interlock in a shape resembling medieval armor (Shapiro and Englund, 1995). Kinetoplast DNA occurs as maxicircles (present in several dozen copies) and minicircles (present in thousands of copies). These include a universal minicircle sequence of 12 nucleotides that serves as a replication origin (Morris et al., 2001).

For a major portion of their life cycles, trypanosomes thrive in the bloodstreams of their hosts. They evade assault from the immune system by densely coating their exteriors with variant surface glygoprotein (VSG) homodimers. There are over 1000 VSG genes and pseudogenes encoded in the *T. brucei* genome, of which only one is expressed at a time (Berriman et al., 2005; reviewed by Taylor and Rudenko, 2006). Remarkably, fewer than 7% of these encode functional proteins, while 66% encode full-length pseudogenes and the remainder are gene fragments or otherwise atypical. Most of the VSG genes are located in subtelomeric arrays of from three to 250 copies. Taylor and Rudenko suggest that the pseudogenes could be advantageous in the generation of antigenic diversity during chronic infections of the bloodstream. The limited number of intact *VSG* genes could be used, but also segmental gene conversion of pseudogenes could create novel, intact, mosaic, *VSG* genes.

T. cruzi infects 16 to 18 million people and is the cause of 21,000 deaths per year from Chagas disease. El-Sayed et al. (2005a) reported the diploid genome sequence of two different haplotypes that averaged 5.4% sequence divergence. The diploid genome size is ~106 to 111 Mb and is predicted to contain 22,570 genes, while the haploid genome contains ~12,000 genes. There is a notable large family of 1377 copies of mucin-associated surface protein (*masp*) genes. They may be involved in immune system evasion.

Leishmania

Leishmania major is another deadly protozoan parasite in the Euglenozoa (Fig. 18.1). Twenty different species of *Leishmania* cause the disease leishmaniasis, for which there is no effective vaccine and limited pharmacological intervention available. The annual incidence is two million cases. These various *Leishmania* species have from 34 to 36 chromosomes (Myler et al., 2000). While the Old World groups *L. major* and *L. donovani* have 36 chromosome pairs (ranging from 0.28 to 2.8 Mb), New World groups *L. mexicana* and *L. braziliensis* have undergone chromosomal fusions (Chapter 16) and have 34 or 35 chromosomal pairs.

The *Leishmania major* genome is about 34 Mb with 36 chromosomes (from 0.3 to 2.5 Mb). Several *Leishmania* genome web resources are listed in Table 18.2. The

The World Health Organization offers information on leishmaniasis at ▶ http://www.who.int/tdr/diseases/leish/default.htm.

TABLE 18-2   Web Resources for *Leishmania* Genomics

Resource	Comment	URL
The *Leishmania major* Friedlin Genome Project	At the Wellcome Trust Sanger Institute	► http://www.sanger.ac.uk/ Projects/L_major/
Seattle Biomedical Research Institute (SBRI)	Information on various infectious diseases	► http://www.genome.sbri. org/
The European *Leishmania major* Friedlin Genome Sequencing Consortium	A listing of participating laboratories	► http://www.sanger.ac.uk/ Projects/L_major/ EUseqlabs.shtml

nucleotide sequence was determined for chromosome 1 (the smallest chromosome) and was found to have a remarkable genomic organization (Myler et al., 1999). The first 29 genes (from the left telomere) are all transcribed from the same DNA strand, while the remaining 50 genes are all transcribed from the opposite strand. This polarity is unprecedented in eukaryotes and resembles prokaryote-like operons. It has a 257 kb region that is filled with 79 protein-coding genes ($\approx$1 gene each 3200 bp). Ivens et al. (2005) reported the *L. major* genome sequence (Fig. 18.4). There are 8272 predicted protein-coding genes, including $\sim$3000 that cluster into 662 different families of paralogs. These families arose principally by tandem gene duplication. The *L. major* genome encodes relatively few proteins involved in transcriptional control, and gene duplication may be a mechanism for increasing expression levels.

In addition to *L. major* (32.8 Mb), Peacock et al. (2007) sequenced the genomes of *Leishmania infantum* (32.0 Mb) and *L. braziliensis* (32.0 Mb). The three genomes share a comparable GC percentage and number of predicted genes. *L. major* and *L. braziliensis* diverged between 20 and 100 million years ago; this broad range of estimates reflects uncertainty as to whether the *Leishmania* genus speciated due to migration events or to the breakup of the supercontinent Gondwanda (see Fig. 13.3). *L. braziliensis* has 35 chromosomes rather than 36 because of the fusion of chromosomes 20 and 34. There is conserved synteny for more than 99% of the genes across the three genomes, and the average nucleotide and amino acid identities are high (e.g., 92% amino acid identity between *L. major* and *L. infantum*). Although many pathogenic protozoans have large gene families involved in immune evasion localized to subtelomeric regions, such families are not evident in the *Leishmania* species. Remarkably, Peacock et al. identified only 5 genes that are specific to *L. major*, 26 *L. infantum*-specific genes, and $\sim$47 *L. braziliensis*-specific genes.

Comparisons of the three trypanosomatid genomes of *L. major*, *T. brucei*, and *T. cruzi* have revealed a shared core of 6200 genes (El-Sayed et al., 2005b). Some protein domains are specific to just one group, such as the variant surface glycoprotein (VSG) expression site-associated domains (Pfam families PF03238 and PF05446) in *T. brucei*. Some domains appear to have expanded or contracted selectively, and insertions, deletions, and substitutions occurred. However, there is a notable high overall gene conservation between the three species. El-Sayed et al. (2005b) measured the number of nonsynonymous substitutions per nonsynonymous site (d_N; see Chapter 7) for every COG (Chapter 15) for which there was a 1:1:1 orthologous relationship between the genomes. This provided a measure of which genes are evolving rapidly and may be under positive selection. Most of these had no functional annotation, while other genes encoded transport proteins that may be exposed to the host immune system and thus evolve rapidly.

THE CHROMALVEOLATES

The Chromalveolates are a supergroup of unicellular eukaryotes, distinct from the Excavates (such as *Giardia*). Many of them have cryptic mitochondria (e.g., hydrogenosomes rather than traditional mitochondria). They include six groups or phyla (Keeling, 2007): (1) the *Apicomplexa* consist of protozoan pathogens that invade host cells using a specialized apical complex. They are typically transmitted by an invertebrate vector such as mosquitoes or flies. This phylum includes parasites such as *P. falciparum* and *Toxoplasma gondii*. (2) Dinoflagellates; these include a cause of paralytic shellfish poisoning, *Alexandrium*. (3) Ciliates include *Paramecium* and *Tetrahymena thermophila*; (4) Heterokonts; (5) Haptophytes; and (6) Cryptomonads. In the tree of Fig. 18.1, these groups are organized as the Apicomplexa, Ciliophora, and Heterokonta. In the following sections of this chapter we will turn to the Virdiplantae (plants), the Mycetozoa, and the Metazoa (animals).

Malaria Parasite *Plasmodium falciparum* and Other Apicomplexans

Malaria kills about 2.7 million people each year, mostly children in Africa, and almost 500 million people are newly infected each year. It is caused by the apicomplexan parasite *Plasmodium falciparum*. While there are 120 species of *Plasmodium*, only four infect humans: *P. falciparum*, *P. vivax*, *P. ovale*, and *P. malariae*. The main vector for malaria in Africa is the mosquito, *A. gambiae*.

Plasmodium falciparum has a complex lifestyle, contributing to the challenge of developing a successful vaccine (Cowman and Crabb, 2002; Wirth, 2002; Long and Hoffman, 2002). *Plasmodium* resides in the salivary glands and gut of the mosquito *A. gambiae*. When a mosquito bites a human, it introduces the parasite in the sporozoite form that infects the liver. *Plasmodium* then matures to the merozoite form, which attaches to and invades human erythrocytes through host cell receptors. Within erythrocytes, trophozoites form. Some merozoites transform into gametocytes, which are captured when mosquitoes feed on infected individuals. A goal of sequencing the *P. falciparum* genome is to find gene products that function at selective stages of the parasite life cycle, offering targets for drug therapy or vaccine development.

The complete genome sequence of *P. falciparum* was reported by an international consortium (Gardner et al., 2002) (see Fig. 18.5). The sequencing was extraordinarily challenging because the AT (adenine and thymine) content of the genome is 80.6% overall, which is the highest for any eukaryotic genome. In intergenic regions and introns, the AT content reached 90% in some cases. A whole-chromosome (rather than a whole-genome) shotgun sequencing strategy was employed. With this approach, chromosomes were separated on pulsed-field gels, DNA was extracted, and shotgun libraries containing 1 to 3 kb of DNA were constructed and sequenced. The genome is 22.8 Mb, with 14 chromosomes from 0.6 to 3.3 Mb.

Gardner et al. (2002) identified 5268 protein-coding genes in *P. falciparum*. This is the same number as is predicted for *Schizosaccharomyces pombe* (Chapter 17), although the genome size is twice as large. There is one gene approximately every 4300 bp overall. Gene Ontology Consortium terms (Chapter 10) were assigned to about 40% of the gene products ($\approx$2100). However, about 60% of the predicted proteins have no detectable homology to proteins in other eukaryotes. These proteins are potential targets for drug therapies. For example, some are essential for the function

The name Apicomplexa derives from a characteristic apical complex of microtubules. You can read more about apicomplexans online at ► http://www.ucmp.berkeley.edu/protista/apicomplexa.html or ► http://www.tulane.edu/~wiser/protozoology/notes/api.html. For online facts on malaria, see ► http://malaria.wellcome.ac.uk/ and ► http://www.who.int/tdr/diseases/malaria/default.htm. Charles Louis Alphonse Laveran won a Nobel Prize in 1907 for his work on malaria-causing parasites (► http://nobelprize.org/nobel_prizes/medicine/laureates/1907/). Earlier, Ronald Ross was awarded a Nobel Prize for his studies of malaria (► http://nobelprize.org/nobel_prizes/medicine/laureates/1902/).

The *P. falciparum* genome was sequenced by a consortium including the Wellcome Trust Sanger Institute, the Institute for Genomic Research, the U.S. Naval Medical Research Center (NMRC, Maryland), and Stanford University. The genome

| TABLE 18-3 | Genomics Resources for *Plasmodium falciparum* and Malaria |

Resource	Comment	URL
PlasmoDB	Main web resource for *P. falciparum*	► http://www.plasmodb.org/
Links page	At NCBI	► http://www.ncbi.nlm.nih.gov/projects/Malaria/related_links.html
P. falciparum Genome Project	At the Sanger Institute	► http://www.sanger.ac.uk/Projects/P_falciparum/

of the apicoplast. This is a plastid, unique to Apicomplexa and homologous to the chloroplast, that functions in fatty acid and isoprenoid biosynthesis.

There are several main resources on the web to study *P. falciparum* (Table 18.3). PlasmoDB (Stoeckert et al., 2006) is the centralized resource for genomic data. There are many complementary resources such as ProtozoaDB (Dávila et al., 2008).

In addition to the initial *P. falciparum* genome project, the genomes of 20 *Plasmodium* species or strains have been partially or completely sequenced as of 2008. A consortium sequenced the genome of the rodent malaria parasite, *Plasmodium yoelii yoelii* (Carlton et al., 2002); and Hall et al. (2005) sequenced the genomes of the rodent malaria parasites *Plasmodium berghei* and *P. chabaudi*.

of the slime mold *Dictyostelium discoideum* also has a high AT content (see below).

A plastid is any photosynthetic organelle. The most well-known plastid is the chloroplast, found in green algae and land plants (Gilson and McFadden, 2001). See the section on plants below.

Genus, species: *Plasmodium falciparum*

Selected lineages: Eukaryota; Alveolata; Apicomplexa; Aconoidasida; Haemosporida; Plasmodium; Plasmodium (Laverania); *Plasmodium falciparum*

Eukaryota; Alveolata; Apicomplexa; Aconoidasida; Piroplasmida; Theileriidae; Theileria; *Theileria annulata*

Eukaryota; Alveolata; Apicomplexa; Coccidia; Eucoccidiorida; Eimeriorina; Sarcocystidae; Toxoplasma; *Toxoplasma gondii* RH

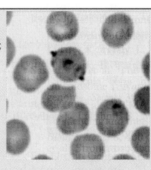

	Haploid genome size	GC content	Number of chromos.	Number of genes	Entrez Genome ID
P. falciparum 3D7	22.8 Mb	19.4%	14	5,268	13173
P. yoelii yoelii	23.1 Mb	22.6%	14	5,878	1436
Babesia bovis	8.2 Mb	41.8%	4	3,671	18731
Cryptosporidium hominis	9.2 Mb	31.7%	8	3,956	13200
Cryptosporidium parvum	9.1 Mb	30.3%	8	3,886	144
Theileria annulata	8.4 Mb	32.5%	4	3,792	153
Theileria parva	8.3 Mb	34.1%	4	4,035	16138
Toxoplasma gondii	65 Mb	n/a	9	8,032	16727

Selected divergence dates: the Apicomplexa lineage originated less than 1,000 million years ago.
Disease association: Each of these apicomplexans is a parasite that causes disease in mammals. *B. babesi* causes babesiosis, a tick-borne disease that threatens half the cattle in the world. *P. falciparum* causes malaria. *T. gondii* causes toxoplasmosis.
Key genomic features: *Theileria* parasites are the only eukaryotes that transform lymphocytes (and thus induce lymphoma).
Organism-specific web resources: ApiDB for apicomplexans (http://www.apidb.org/apidb/); PlasmodDB for Plasmodium (http://plasmodb.org).

FIGURE 18.5. The Apicomplexa (see Fig. 18.1) include the malaria parasite Plasmodium falciparum. *This image shows multiply infected red blood cells in thin blood smears (from the CDC Parasite Image Library).*

Jeffares et al. (2007) performed shallow (e.g., one- to fivefold rather than the traditional sevenfold to 12-fold) coverage of the *P. falciparum* genome from a clinical isolate in Ghana, a laboratory isolate, and the chimpanzee parasite *P. reichenowi*.

What is the significance of sequencing additional *Plasmodium* genomes? In the case of *P. yoelli yoelli*, *P. berghei*, and *P. chabaudi* this is an extremely important accomplishment because the complete life cycle of *P. falciparum* cannot be maintained in vitro, while the rodent parasites can. The *P. yoelli yoelli* genome is 23.1 Mb and has 14 chromosomes, as does *P. falciparum*. The AT content is comparably high (77.4%). The genomes are also predicted to encode a comparable number of genes. When the full set of predicted *P. falciparum* proteins (5268) were searched against the predicted *P. yoelii yoelii* proteins (5878 proteins) by BLAST searching (with an *E* value cutoff of 10^{-15}), 3310 orthologs were identified. These include vaccine antigen candidates known to elicit immune responses in exposed humans (Carlton et al., 2002).

Having the genome sequences of *P. falciparum* and several rodent parasites available, how can bioinformatics and genomics approaches be used to understand the basic biology of these organisms? Data are now available on thousands of previously unknown genes, offering many new potential strategies to combat malaria (Hoffman et al., 2002).

For an NCBI website on malaria genetics and genomics, visit ▶ http://www.ncbi.nih.gov/ projects/Malaria/.

- The apicoplast is a potential drug target. Zuegge et al. (2001) analyzed the amino-terminal sequences of 84 proteins targeted to apicoplasts and 102 non-apicoplast (e.g., cytoplasmic, secretory, or mitochondrial) sequences. They used principal components analysis, neural networks, and self-organizing maps (Chapter 9) to build a predictive model for apicoplast targeting signals.

- Comparative genomics approaches yield important insight into the genome structure, gene content, and other genomic features of closely related species. Carlton et al. (2001) compared ESTs and genome survey sequences (see Chapter 2) from *P. falciparum*, *P. vivax*, and *P. berghei*. As part of this analysis, they identified the most highly expressed genes, such as the *rif* gene family of *P. falciparum* that is implicated in antigenic variation.

- Hall et al. (2005) measured synonymous versus nonsynonymous substitution rates in genes from three rodent *Plasmodium* species in comparison to *P. falciparum*. They measured gene expression, categorizing transcripts according to the four categories of housekeeping; host-related; invasion, replication, and development-related; or stage-specific.

The Prediction of Apicoplast Targeted Sequences (PATS) database is available at ▶ http://gecco. org.chemie.uni-frankfurt.de/ pats/pats-index.php.

- A map of conserved syntenic regions between *P. yoelii yoelii* and *P. falciparum*, covering over 16 Mb overall, provides insight into the evolution of these parasites. Carlton et al. (2002) used the MUMmer program (Chapter 15) to align protein-coding regions. The conserved synteny map reveals regions of conserved gene order, allows analysis of chromosomal break points, and confirms the absence of some genes (such as *var* and *rif* in *P. yoelii yoelii*).

We encountered *vir* in Chapter 5 (problem 5.2) where we used both BLAST and PSI-BLAST to evaluate the family.

- Genes that function in antigenic variation and immune system evasion can be investigated. In *P. vivax*, there are as many as 1000 copies of *vir*, a gene family localized to subtelomeric regions. *Plasmodium yoelii yoelii* has 838 copies of a related gene, *yir* (Carlton et al., 2002).

- Several groups applied proteomics approaches to analyze the proteins of *P. falciparum* at four stages of the life cycle (sporozoites, merozoites, trophozoites, and gametocytes). Florens et al. (2002) identified 2415 expressed proteins, about half of which are annotated as hypothetical. An unexpected finding was that the *var* and *rif* genes—thought to be involved in immune system invasion—were abundantly present in the sporozoite stage. Together, these studies define stage-specific expression of proteins, suggesting possible protein functions. Proteomics approaches also validate the gene-finding approaches from genomic DNA. Lasonder et al. (2002) identified some protein sequences by mass spectrometry that were not initially predicted using gene-finding algorithms to analyze genomic DNA.

- It is possible to identify *Plasmodium* metabolic pathways as therapeutic targets (Gardner et al., 2002; Hoffman et al., 2002). All organisms studied to date synthesize isoprenoids using isopentyl diphosphate as a building block. An atypical pathway employed by some plants and bacteria involves 1-deoxy-D-xylulose 5-phosphate (DOXP). This DOXP pathway is absent in mammals. Jomaa et al. (1999) used tblastn (with a bacterial DOXP reductoisomerase protein as a query against a *Plasmodium* genomic DNA database) and found an orthologous *Plasmodium* gene. They showed that this protein is likely localized to the apicoplast and that *P. falciparum* survival is sensitive to low levels of two inhibitors of the enzyme. They further showed that these drugs have antimalarial activity in mice infected with *Plasmodium vinckei*. This type of bioinformatics-based approach holds great promise in the search for additional antimalarial drugs.

Isoprenes are five-carbon chemical molecules that combine to form many thousands of natural compounds, including steroids, retinol, and odorants. (RBP and OBP are lipocalins that transport isoprenoids.)

There are 5000 species in the phylum Apicomplexa, causing a wide range of diseases by mechanisms that are now being elucidated through genome sequence analysis (reviewed in Roos, 2005). Other apicomplexan genomes that have been sequenced include the following (summarized in Fig. 18.5):

- *Babesia bovis*, the cause of tick fever in cattle, threatens livestock globally. Brayton et al. (2007) reported its genome sequence. It has extremely limited metabolic potential, lacking genes encoding proteins that are required for gluconeogenesis, the urea cycle, fatty acid oxidation, and heme, nucleotide, and amino acid biosynthesis. Thus it relies on its host for many nutrients, and the *B. bovis* genome encodes many transporters. Analogous to *Plasmodium falciparum*, its genome encodes about 150 copies of a polymorphic variant erythrocyte surface antigen protein (ves1 gene) family.

- *Theileria annulata* and *T. parva* are tick-borne parasites that cause tropical theilorisosis and East Coast fever, respectively, in cattle. Pain et al. (2005) and Gardner et al. (2005) reported their genome sequences. *T. parva* reversibly, malignantly transforms its host cell, the bovine lymphocyte, causing lymphoma; *T. annulata* transforms macrophages. The *T. parva* genome encodes about 20% fewer genes than *P. falciparum*, but it has a higher density of genes.

- *Cryptosporidium hominis* causes diarrhea and acute gastroenteritis. Unlike other Apicomplexans that are transmitted via an invertebrate host, *C. hominis* is transmitted by ingestion of oocytes in water. Xu et al. (2004) sequenced the *C. hominis* genome, while Abrahamsen et al. (2004) sequenced the related parasite *C. parvum* that infects humans and other mammals. Like *B. bovis* and many other parasites, these genomes have very limited metabolic capabilities and rely on host cells for nutrients.

- *Toxoplasma gondii* causes toxoplasmosis. The Centers for Disease Control estimates that 60 million people in the United States are infected, although most

The *T. gondii* database ToxoDB is available at ► http://toxodb.org/toxo/ (Gajria et al., 2008).

are asymptomatic. After infection, oocysts and tissue cysts transform into tachyzoites and localize in neural and muscle tissue. The *T. gondii* genome is currently being sequenced.

Astonishing Ciliophora: *Paramecium* and *Tetrahymena*

Ciliates are unicellular eukaryotes that are part of the monophyletic alveolate clade that includes the Apicomplexans (see Fig. 18.1). Ciliates share two properties: they use vibrating cilia for locomotion and food capture, and they have two nuclei with separate germline and somatic functions (nuclear dimorphism). One nucleus is a diploid germinal micronucleus that undergoes meiosis and thus is responsible for transmitting genetic information to the progeny (but is otherwise silent). The other is a polyploid somatic macronucleus that is responsible for gene expression. This macronucleus is lost with each generation and is replenished following meiosis and development of the micronuclear lineage.

We discussed chromatin diminution in nematodes in Chapter 16.

Paramecium tetraurelia is a ciliate that lives in freshwater environments. It has long served as a model organism for many aspects of eukaryotic biology. *Paramecium* has an unknown number of micronuclear chromosomes (>50) (Fig. 18.6). As the macronuclear chromosome develops, it is amplified to ~800 copies and it is rearranged extensively through a process of DNA elimination. Tens of thousands of unique copy elements are removed, and in a separate process transposable elements and other repeats are deleted. This leads to a fragmented set of about 200 acentric chromosomes, ranging in size from ~50 kilobases to 981 kilobases. Aury et al. (2006) sequenced the *Paramecium* macronuclear genome which, although fragmented, is genetically homogeneous because of the sexual process of autogamy

Genus, species: *Paramecium tetraurelia*
  Tetrahymena thermophila
  Sterkiella histriomuscorum (also called *Oxytricha trifallax*)

Lineage: Eukaryota; Alveolata; Ciliophora; Intramacronucleata; Oligohymenophorea; Peniculida; Parameciidae; Paramecium; *Paramecium tetraurelia*

Lineage: Eukaryota; Alveolata; Ciliophora; Intramacronucleata; Oligohymenophorea; Hymenostomatida; Tetrahymenina; Tetrahymenidae; Tetrahymena; *Tetrahymena thermophila*

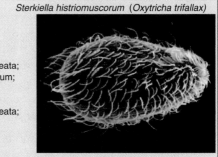
Sterkiella histriomuscorum (*Oxytricha trifallax*)

	Haploid genome size	GC content	Number of chromos.	Number of genes	Entrez Genome ID
Paramecium tetraurelia (macronuclear genome)	~72 Mb	28%	~200	39,642	18363
Tetrahymena thermophila (macronuclear genome)	~104 Mb	22%	~225	27,424	12564
Sterkiella histriomuscorum (macronuclear genome)	~50 Mb	not avail.	~24,500	~26,800	12857

Selected divergence dates: the ciliates diverged from other eukaryotes ~one billion years ago.
Key genomic features: *Paramecium* has a macronuclear nucleus (with somatic functions) and a diploid micronuclear nucleus (with germline functions). The gene content is extraordinarily high, and the genome underwent at least three whole genome duplications.
Organism-specific web resources: http//www.ciliate.org; http://paramecium.cgm.cnrs-gif.fr/.

FIGURE 18.6. The Ciliophora (see Fig. 18.1) include Paramecium *and* Tetrahymena. *In some classifications, the Apicomplexa and Ciliophora are grouped together to form the Alveolata. Image from the National Human Genome Research Institute (► http://www.genome.gov/17516871).*

by which it arose. The total coverage was 72 Mb, and most of the 188 largest scaffolds likely represent macronuclear chromosomes because they contain telomeric repeats.

While the presence of two nuclei and the process of DNA rearrangement and elimination are extraordinary, another startling finding is that *Paramecium* encodes about 40,000 protein-coding genes (a far greater number than is found in animals or fungi). The genome sequencing process resulted in the creation of several hundred scaffolds. As we view the scaffold 1, corresponding to the longest chromosome observed by pulsed-field gel electrophoresis, we can see the compact nature of the coding portion of the genome (Fig. 18.7). Across the genome, 78% of the nucleotides occur in genes, and the intergenic regions average 352 bases.

Yet another surprising finding is the series of three whole genome duplications that Aury et al. (2006) inferred (Fig. 18.8). All proteins were searched against each other using the Smith–Waterman algorithm (Chapter 3). Two thirds of the predicted proteins occur in paralog pairs, maintaining conserved synteny across large portions of the chromosomes. The other third of the proteins presumably lost their duplicates after the whole genome duplication event(s). The situation contrasts with the fungi (Chapter 17) and plant and fish genomes (see below) in which whole genome duplication events are followed by rapid gene loss and large-scale chromosomal rearrangements. By inferring ancestral blocks and then iteratively repeating the within-proteome alignments to search for conserved blocks sharing progressively less conservation, Aury et al. inferred the occurrence of three whole-genome duplications (Fig. 18.8). For a discussion of the software used to make the figure, see Box 18.2.

Tetrahymena thermophila is another ciliate that has long served as a model organism for biological research (Collins and Gorovsky, 2005). Discoveries made using *Tetrahymena* include catalytic RNA, telomeric repeats, telomerase, and the function of histone acetylation. Eisen et al. (2006) reported the sequence of its macronuclear genome which is 104 Mb and composed of about 225 chromosomes with a ploidy of ~45. In marked contrast to *Paramecium*, they did not find evidence for either segmental or whole genome duplications. The relatively high gene count is explained by extensive tandem duplication of genes. The availability of the macronuclear genome

The *Paramecium* genome project website, including the ParameciumDB genome browser, is at ▶ http://paramecium.cgm.cnrs-gif.fr/.

A primary *Tetrahymena* genome database is at ▶ http://www.ciliate.org/, while a *Tetrahymena* genome sequencing website is at ▶ http://lifesci.ucsb.edu/~genome/Tetrahymena/.

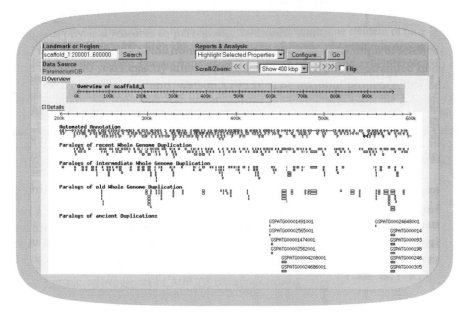

FIGURE 18.7. *The* Paramecium *genome is proposed to have undergone at least three whole genome duplications. The longest chromosome (scaffold 1 of the genome assembly) is viewed in the genome browser of ParameciumDB. A region of 400,000 base pairs is displayed, and the annotation tracks show the conservation to many paralogs reflecting recent, intermediate, and old whole genome duplications. The browser is available at ▶ http://paramecium.cgm.cnrs-gif.fr.*

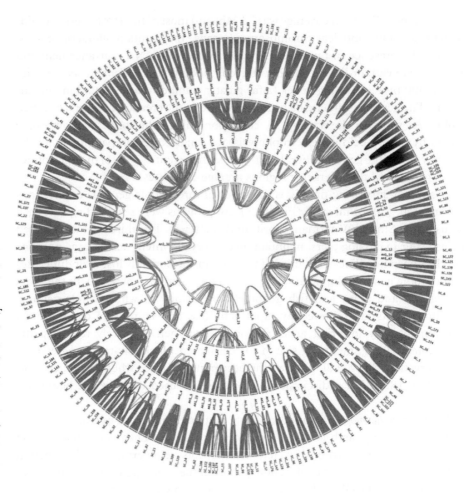

FIGURE 18.8. *Whole genome duplication in the ciliate Paramecium tetraurelia is inferred by analysis of protein paralogs. The outer circle displays all chromosome-sized scaffolds from the genome sequencing project. Lines link pairs of genes with a "best reciprocal hit" match. The three interior circles show the reconstructed ancestral sequences obtained by combining the paired sequences from each previous step. The inner circles are progressively smaller and reflect fewer conserved genes with a smaller average similarity. From Aury et al. (2006). Used with permission.*

BOX 18.2
Graphically Representing Whole Genome Duplications

We introduced the ideogram as a representation of a karyotype in Chapter 16. Traditionally, linear eukaryotic chromosomes are depicted as straight bars. However, when depicting the relationships between genes (or proteins or other elements) on multiple chromosomes, the patterns of relationships can be so complex that the visual presentation is confusing. Circular plots offer a concise way to view relationships between chromosomal elements. Figure 18.8 shows an example of *Paramecium* chromosomes made by Aury et al. (2006) using Circos software developed by Martin Krzywinski (available as free software at ▶ http://mkweb.bcgsc.ca/circos/?home). This website also offers a tutorial and a gallery of visually stunning samples.

Chromowheel is a related tool, developed by Ekdahl and Sonnhammer (2004) and available at Karolinska Institutet as a web service (▶ http://chromowheel.cgb.ki.se). The user can submit a generic data definition format file which is then converted into an image in the Scalable Vector Graphics (SVG) format. Other software (such as the Circular Genome Viewer CGView, ▶ http://wishart.biology.ualberta.ca/cgview/; Stothard and Wishart, 2005) allow representation of circular genomes such as those of prokaryotes.

will facilitate future sequencing of the micronuclear genome, which contains substantially more repetitive DNA. Such studies may elucidate the fascinating relationship between macro- and micronuclear chromosomes in the ciliates. This in turn may reveal fundamental mechanisms by which genome-wide rearrangement occurs.

A third ciliate genome is that of *Sterkiella histriomuscorum*, formerly called *Oxytricha trifallax* and indicated in the *Oxytrichida* group in Fig. 18.1. *Sterkiella histriomuscorum* is of the class Spirotrichea. This macronuclear genome fragments into an astonishing number of about 24,500 minichromosomes (called nanochromosomes). Doak et al. (2003) described its ongoing genome project, including evidence of a ploidy of ∼1000 per macronuclear genome.

Nucleomorphs

The chloroplast is a plastid (photosynthetic organelle) in plants that contains the green pigment chlorophyll. Chloroplasts convert light to energy. A major hypothesis about their origin is that a eukaryotic cell acquired a cyanobacterium soon after the divergence of plants from animals and fungi (see below). But a radically different mechanism is also common: a eukaryote can ingest an alga (i.e., another eukaryote) that already has a chloroplast (Gilson and McFadden, 2002). This process, called endosymbiosis, may have occurred independently in at least seven separate eukaryotic groups: apicomplexa (discussed above), chlorarachniophytes, cryptomonads, dinoflagellates, euglenophytes, heterokonts, and haptophytes (reviewed in Gilson and McFadden, 2002).

Most chloroplast-containing plants and some algae have three genomes in each cell: a nuclear genome, a mitochondrial genome, and a chloroplast genome. In cryptomonads (such as *Guillardia theta*) and chlorarachniophytes (such as *Bigelowiella natans*), there is an additional, fourth distinct genome: the vestigial nuclear genome of the engulfed alga. This second nucleus is called a nucleomorph. The process of sequential endosymbioses is outlined in Fig. 18.9.

The lineage of *G. theta* is Eukaryota; Cryptophyta; Cryptomonadaceae; Guillardia. the lineage of *B. natans* is Eukaryota; Cercozoa; Chlorarachniophyceae; Bigelowiella.

Just as the genome of intracellular bacteria is highly reduced, the nucleomorph genome is extremely small. Douglas et al. (2001) sequenced the nucleomorph genome of *G. theta*. It is only 551,264 bp. The gene density is extraordinarily high, with one gene per 977 bp. The noncoding regions are extremely short, and there is only one pseudogene. Some genes, such as those encoding DNA polymerases, are absent and the gene product must be imported to the plastid across four separate membranes.

The circular plastid DNA of *G. theta* is also very compacted. Douglas and Penny (1999) sequenced this genome of 121,524 bp and found that 90% of the DNA is coding, with no pseudogenes or introns. (In contrast, only 68% of the rice plastid genome is coding.) You can explore the *G. theta* plastid genome at NCBI (accession NC_000926) and compare it with the plastid genome of the red alga *Pophyra purea*, a rhodophyte (accession NC_000925). These two genomes show a high degree of conserved synteny. You can also compare the *G. theta* plastid genome to that of the diatom *Odontella sinensis* (accession NC_001713). This is a related alga that also acquired its plastid by secondary endosymbiosis but lacks a nucleomorph.

The smallest known eukaryotic genome belongs to the nucleomorph genome of the chlorarachniophyte *Bigelowiella natans*. Its size is 373,000 base pairs, containing 331 genes on three chromosomes (Gilson et al., 2006). Its nature is clearly eukaryotic, including the presence of 852 introns—although these "pygmy introns" are the smallest known, having lengths of 18 to 21 nucleotides. This genome offers a model for

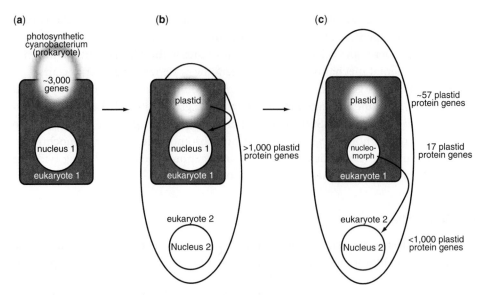

FIGURE 18.9. *Sequential endosymbioses result in a eukaryote with three genomes. (a) In a primary endosymbiotic event, a eukaryotic host (eukaryote 1) acquires a photosynthetic bacterium such as a cyanobacterium. (b) Over time, the nuclear genome of eukaryote 1 acquires over 1000 plastid protein-coding genes. The plastid is the engulfed prokaryotic genome, that is, the chloroplast. Secondary endosymbiosis occurs when another nonphotosynthetic organism (eukaryote 2) engulfs and retains eukaryote 1 and so acquires photosynthetic capability. (c) Over time plastid protein genes are transferred to the nuclear genome of organism 2, resulting in the emergence of a severely reduced nucleomorph genome. The numbers of genes in the figure are for the chlorarachniophyte* Bigelowiella natans, *whose nucleomorph genome is the smallest known nucleus. Adapted from Gilson et al. (2006). Used with permission.*

extreme reduction (Fig. 18.9). Although *G. theta* and *B. natans* are phylogenetically distinct, Patron et al. (2006) compared their highly reduced nucleomorph genomes relative to their corresponding nuclear and plastid genomes. They concluded that *B. natans* nucleomorph genes are evolving at a rapid rate, while *G. theta* has stabilized.

Kingdom Stramenopila

The kingdom Stramenopila includes a wide range of fascinating organisms such as the oömycetes (e.g., the *Phytophthora* plant pathogens) and photosynthetic algae (e.g., diatoms, brown algae such as kelp, and the golden-brown algae). The Stramenopila group is represented in Fig. 18.1 as part of the Heterokonta, and we summarize several genomes in Fig. 18.10.

Diatoms are single-celled algae that occupy vast expanses of the oceans and are responsible for ∼20% of global carbon fixation (see Armbrust et al., 2004). They have an intricately patterned silicified (glass) cell wall called the frustule that displays beautiful, species-specific patterns as seen for example in Fig. 18.10. Armbrust et al. (2004) determined the sequences of the three genomes of the diatom *Thalassiosira pseudonana*: a diploid nuclear genome of 34.5 megabases organized in 24 pairs, a plastid genome acquired by secondary endosymbiosis perhaps 1300 million years ago, and a mitochondrial genome. The plastid was acquired when a nonphotosynthetic, eukaryotic diatom ancestor engulfed a photosynthetic eukaryote (probably a red algal endosymbiont), a remarkable process described above (Fig. 18.9). Half

A principal website for *Thalassiosira pseudonana* is at ▶ http://genome.jgi-psf.org/ Thaps3/Thaps3.home.html (from the Joint Genome Institute).

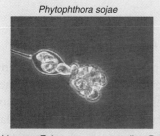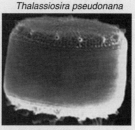

Lineage: Eukaryota; stramenopiles; Bacillariophyta; Coscinodiscophyceae; Thalassiosirophycidae; Thalassiosirales; Thalassiosiraceae; Thalassiosira; *Thalassiosira pseudonana CCMP1335* (diatom)

Lineage: Eukaryota; stramenopiles; Oomycetes; Peronosporales; Phytophthora; *Phytophthora sojae*

Lineage: Eukaryota; stramenopiles; Oomycetes; Peronosporales; Phytophthora; *Phytophthora ramorum*

	Haploid genome size	GC content	Number of chromos.	Number of genes	Entrez Genome ID
Thalassiosira pseudonana	34.5 Mb	47%	24	11,242	191
plastid genome	128,813 bp	31%	1	144	
mitochondrial genome	43,827 bp	30.5%	1	40	
Phytophthora sojae	95 Mb	22%	~225	19,027	17989
Phytophthora ramorum	65 Mb		~24,500	15,743	12571

Genome features
--*P. ramorum* is heterothallic (outbreeding); ~13,600 SNPs were identified
--*P. sojae* is homothallic (inbreeding); only 499 SNPs were identified
--These are the only eukaryotic genomes for which no gene encoding phospholipase C has been identified, nor have *Phytophthora* expressed sequence tags corresponding to PLC been found.
Disease relevance: *P. sojae* (potato pink rot agent) is a soybean pathogen.
     P. ramorum causes sudden oak death.

FIGURE 18.10. *The Heterokonta (see Fig. 18.1) include the Phytophthora and the diatom. Photographs are from the NCBI Entrez Genomes website (Phytophthora sojae by Edward Braun, Iowa State University; Phytophthora ramorum by Edwin R. Florance, Lewis Clark College&; Thalassiosira pseudonana by DOE-Genomes to Life).*

of the *T. pseudonana* genes cannot be assigned function based on homology to other organisms. This may reflect their unique capability of metabolizing silicone to form the frustules.

Another two members of the kingdom Stramenopila are the soybean pathogen *Phytophthora sojae* and the sudden oak death pathogen *Phytophthora ramorum*. There are 59 known species of the genus *Phytophthora*, and together these cost tens of billions of dollars per year because of their destruction of plant species, including crops. Tyler et al. (2006) reported draft genome sequences for both these plant pathogens (summarized in Fig. 18.10). *P. sojae* and *P. ramorum* are oömycetes (also called water molds) which, in contrast to the diatoms, are nonphotosynthetic stramenopiles. The two genomes encode comparable numbers of genes, including about 9700 pairs of orthologs (with extensive colinearity of orthologs spanning up to several megabases per block). While neither organism is photosynthetic, both contain many hundreds of genes that are derived from a red alga or cyanobacterium, suggesting that there was a photosynthetic ancestor. Since both are cellular pathogens having different host ranges, Tyler et al. searched for genes encoding secreted proteins. Of the more than 1000 predicted secreted proteins in each organism, many show evidence of rapid diversification in terms of sequence conservation and the evolution of multigene families. These include secreted proteases that could relate to nectrotrophic growth, that is, feeding on dead plants after infection of living plant tissue. Of particular note is the *Avh* (avirulence homolog) family of genes that has 350 members in each genome whose products suppress plant defense responses.

The Department of Energy Joint Genome Institute (DOE JGI) website for *P. ramorum* is ▶ http://genome.jgi-psf.org/Phyra1_1/Phyra1_1.home.html. The Phytophthora Functional Genomics Database is online at ▶ http://www.pfgd.org/. For an example of an Avh protein from *P. sojae*, see AAR05402.

PLANT GENOMES

Overview

The *Epifagus virginiana* chloroplast genome has been sequenced (NC_001568) (Wolfe et al., 1992). *Epifagus* is parasitic on the roots of beech trees. The original major function of its chloroplast genome, photosynthesis, has become obsolete. It lacks six ribosomal protein and 13 tRNA genes that are present in the chloroplast genomes of photosynthetic plants (Wolfe et al., 1992).

Plants and animals differ greatly in their content of selected genes. For example, plants lack intermediate filaments and the genes that encode intermediate filament proteins such as cytokeratin and vimentin.
The use of 18S RNA has suggested an animal–fungi clade (Fig. 18.1), consistent with Fig. 18.11.

The earliest known plant fossils date from the Silurian period (430–408 MYA) (Margulis and Schwartz, 1998).

Hundreds of thousands of plant species occupy the planet. Molecular phylogeny shows us that plants form a distinct clade within the eukaryotes (see Viridiplantae, Fig. 18.1). These include algae and the familiar green plants. All plants (but not algae) are multicellular because they develop from embryos, which are multicellular structures enclosed in maternal tissue (Margulis and Schwartz, 1998). Most plants have the capacity to perform photosynthesis, although some (such as the beech drop, *Epifagus*) do not.

The analysis of plant genomes allows us to address the molecular genetic basis of characteristics that distinguish plants from animals, such as the presence of specialized cell walls, vacuoles, plastids, and cytoskeleton. Plants are sessile and depend on photosynthesis. The sequencing of plant genomes is likely to lead to explanations for many of these basic features.

When did the lineages leading to today's plants diverge from animals, fungi, and other organisms? The earliest evidence of life is from about 3.8 billion years ago (BYA), while eukaryotic fossils have been dated to 2.7 BYA. These events are depicted in the schematic tree of Fig. 18.11, based on separate studies by Meyerowitz (2002) and Wang et al. (1999). There are no very early plant fossils extant, and it is thus difficult to assess the dates that species diverged from each other. Various researchers have used molecular clocks based on protein, DNA (nuclear or mitochondrial), or RNA data. A study by Wang et al. (1999) used a combined analysis of 75 nuclear genes to estimate the divergence times of plants, fungi, and several animal phyla. Their estimates of divergence time were calibrated based on evidence from the fossil record that birds and mammals diverged 310 MYA. They found that animals and plants diverged 1547 MYA, at almost exactly the same time that animals and fungi diverged (1538 MYA) (Fig. 18.11).

The early appearance of plants, animals, and fungi may have occurred with the divergence of a unicellular progenitor. Thus, a comparison of plants and animals allows us to see how plants and animals independently evolved into multicellular forms (Meyerowitz, 2002). The mitochondrial genes of plants and animals are homologous, indicating that their common ancestor was invaded by an α-proteobacterium (Fig. 18.11). After their divergence, in another endosymbiotic event, a cyanobacterium occupied plant cells to ultimately form the chloroplast. This occurred independently several times. Still, it has proven difficult to date these events (Meyerowitz, 2002). The first appearance of most animal phyla in the fossil record occurs in many samples dated 530 MYA—the time of "Cambrian explosion."

We begin our bioinformatics and genomics approaches to plants by exploring their position among the eukaryotes (Fig. 18.1) and from a phylogenetic tree based on sequences of a key plant enzyme, rubisco (Fig. 18.12). The two main groups of Viridiplantae are *Chlorophyta* (green algae such as the generum *Chlamydomonas*) and *Streptophyta*. *Streptophyta* is further subdivided into additional groups such as mosses, liverworts, and the angiosperms (flowering plants), including the familiar monocots and eudicots. We begin with the green algae, then proceed to the flowering plants.

Green Algae (*Chlorophyta*)

Chlamydomonas reinhardtii is a unicellular alga that lives in soil and water. Among the unicellular green algae, *Chlamydomonas* has served as a model organism for studying

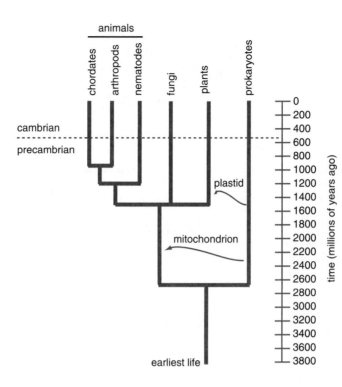

FIGURE 18.11. The evolution of plants, animals, and fungi. The estimated time of divergence of plants, fungi, and animals is 1.5 BYA according to a phylogenetic study (adapted from Wang et al., 1999). Prior to this divergence event, a single-celled eukaryotic organism acquired an α-proteobacterium (the modern mitochondrion, present today in animals, fungi, and plants). After the divergence of plants from animals and fungi about 1.5 BYA, the plant lineage acquired a plastid (the chloroplast). According to this model, metazoans diverged about 400 million years earlier than predicted by the fossil record. Also, nematodes (e.g., C. elegans) diverged earlier than chordates (e.g., vertebrates) and arthropods (e.g., insects). Adapted from separate studies by Meyerowitz (2002) and Wang et al. (1999). Used with permission.

photosynthesis and chloroplast biogenesis (unlike the flowering plants it grows in the dark). The genome is 121 megabases with a very high GC content (64%) and contains about 15,000 protein-coding genes (Merchant et al., 2007) (Fig. 18.13). We can perform comparative genomic analyses of the *Chlamydomonas* genome to infer the properties of the ancestor of the green plants (Viridiplantae) and the opithokonts

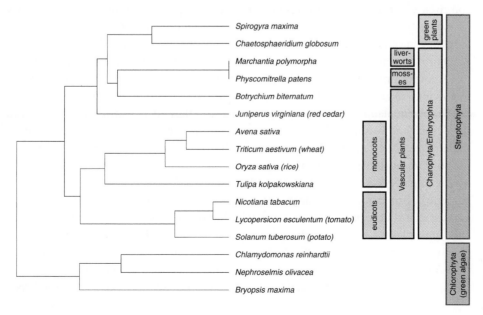

FIGURE 18.12. Phylogenetic tree of the plants. A neighbor-joining tree of the plants using rubisco protein.

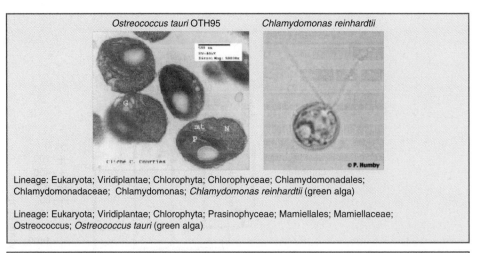

Ostreococcus tauri OTH95   Chlamydomonas reinhardtii

Lineage: Eukaryota; Viridiplantae; Chlorophyta; Chlorophyceae; Chlamydomonadales; Chlamydomonadaceae; Chlamydomonas; *Chlamydomonas reinhardtii* (green alga)

Lineage: Eukaryota; Viridiplantae; Chlorophyta; Prasinophyceae; Mamiellales; Mamiellaceae; Ostreococcus; *Ostreococcus tauri* (green alga)

	Haploid genome size	GC content	Number of chromos.	Number of genes	Entrez Genome ID
Chlamydomonas reinhardtii	121 Mb	64%	17	15,143	12260
Ostreococcus tauri OTH95	12.6 Mb	58%	20	8,166	12912

Genome features: *Chlamydomonas* has 0.125 genes per kilobase, comparable to *Arabidopsis*.
In contrast, *O. tauri* has 0.648 genes per kilobase. (Humans have ~0.0008 genes per kilobase.)
O. tauri is the smallest free-living eukaryote.
Websites: http://www.chlamy.org/; http://genome.jgi-psf.org/Chlre3/Chlre3.home.html

FIGURE 18.13. *One major division of the plants (Viridiplantae) is the green algae including* Chlamydomonas *(see Fig. 18.1). Photographs are from the NCBI Entrez Genomes website (of* Ostreococcus tauri *by O. O.Banyuls- CNRS Courties; of* Chlamydomonas reinhardtii *by Dr. Durnford, University of New Brunswick).*

(animals, fungi [Chapter 17], and Choanozoa). Many genes are shared by *Chlamydomonas* and animals but have been lost in angiosperms, such as those encoding the flagellum (or cilium) and the basal body (or centriole). For example, the *Chlamydomonas* genome encodes 486 membrane transporters, including many shared in common with animals (e.g., voltage-gated ion channels involved in flagellar function). We will explore further examples in computer lab exercise 18.3 at the end of this chapter. There are several possible explanations for the proteins that occur in *Chlamydomonas* and plants but not animals: (1) they may have been present in the common plant-animal ancestor and lost or diverged in the animal lineage, (2) they may have been horizontally transferred between plants and *Chlamydomonas*, or (3) they may have arisen in the plant lineage before the divergence of *Chlamydomonas*. Such proteins include many involved in chloroplast function (Merchant et al., 2007).

Another unicellular green alga, *Ostreococcus tauri*, is thought to be the smallest free-living eukaryote (Fig. 18.13). *O. tauri* presents a simple, naked, nonflagellated cell with a nucleous, mitochondrion, and chloroplast. It is distributed throughout the oceans and was first identified in 1994 as a common component of marine phytoplankton. Derelle et al. (2006) sequenced its 12.6 Mb genome which is distributed on 20 chromosomes. There are 8166 protein-coding genes with a density of 1.3 kilobases per gene, greater than any other eukaryote sequenced to date. Thus, the genome has an extraordinary degree of compaction, with very short intergenic regions, many gene fusion events, and a reduction in the size of gene families. Another remarkable, unexplained feature of the genome is that two of the chromosomes (a large portion of 2 and all of 19) differ from all others in GC content (52% to 54% rather than 59% on the other chromosomes), and these two loci also contain most of the transposable elements in the genome (321 of 417). Chromosome 2 also employs a different frequency of codon utilization, and has much smaller introns (40 to 65 base pairs in contrast to an average of 187 base pairs elsewhere). The origin of these various differences is unknown but these data suggest horizontal transfer from another organism.

Arabidopsis thaliana Genome

Angiosperms are flowering plants in which the seeds are enclosed in an ovary that ripens into a fruit. Monocots are characterized by an embryo with a single cotyledon (seed leaf); examples are rice, wheat, and oats. Eudicots (also called dicotyledons) have an embryo with two seed leafs; examples are tomato and potato. Eudicots include the majority of flowers and trees (but not conifers).

Arabidopsis thaliana is a thale cress and eudicot that is prominent as having the first plant genome to be sequenced (Fig. 18.14). *Arabidopsis* has been adopted by the plant research community as a model organism to study because it is small (about 12 inches tall), has a short generation time (about 5 weeks), has many offspring, and is convenient for genetic manipulations. It is a member of the

The Angiosperm Phylogeny website is at ▶ http://www.mobot.org/MOBOT/Research/APweb/welcome.html. It includes dozens of phylogenetic trees, with access to text, photographs of plants, and extensive references. In contrast to angiosperms, gymnosperms develop their seeds in cones. The eudicots (such as *Arabidopsis*) diverged from the monocots (such as *O. sativa*) about 200 MYA. Among the eudicots, the rosids and the asterids diverged about 100 to 150 MYA (Allen, 2002). The rosids include *Arabidopsis*, *Glycine max* (soybean), and *M. trunculata*. The asterids include *Lycopersicon esculentum* (tomato).

Populus trichocarpa (black cottonwood)   **Vitis vinifera (wine grape)**   **Arabidopsis thaliana (mouse-ear cress)**

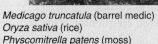

Medicago truncatula (barrel medic)
Oryza sativa (rice)
Physcomitrella patens (moss)

Selected lineages: Eukaryota; Viridiplantae; Streptophyta; Embryophyta; Tracheophyta; Spermatophyta; Magnoliophyta; eudicotyledons; core eudicotyledons; rosids; eurosids II; Brassicales; Brassicaceae; Arabidopsis; *Arabidopsis thaliana*

Eukaryota; Viridiplantae; Streptophyta; Embryophyta; Tracheophyta; Spermatophyta; Magnoliophyta; eudicotyledons; core eudicotyledons; Vitales; Vitaceae; Vitis; *Vitis vinifera*

	Haploid genome size	GC content	Number of chromos.	Number of genes	Entrez Genome ID
Arabidopsis thaliana	125 Mb	34.9%	5	~25,498	13190
M. truncatula	470-580 Mb	not avail.	8	~19,000	10791
O. sativa	389 Mb	43.3%	12	37,544	13139, 13174, 13141, 361
Physcomitrella patens	480 Mb	not avail.	27	35,938	13064
Populus trichocarpa	485 Mb	not avail.	19	45,555	10772
Vitis vinifera	487 Mb	~35%	19	30,434	18357, 18785

Key dates: Emergence of flowering plants 200 million years ago (MYA). *Arabidopsis* and the moss *P. patens* diverged ~450 MYA; *Arabidopsis* and *Populus* diverged ~120 MYA.
Disease relevance: Worldwide, up to 30% of crop yield is lost to pathogens. Plant genome sequencing projects can reveal disease resistance mechanisms.
Genome features: While the *Arabidopsis* genome is ~93% euchromatin, *Populus* is ~70% euchromatin. *Populus* has far more genes than *Vitis vinifera* although the two genomes have a similar size.
Web resources: http://www.medicago.org (*Medicago*).

FIGURE 18.14. Overview of plant genomes. Photographs are from the NCBI Entrez Genomes website (P. trichocarpa by J. S. Peterson, USDA-NRCS PLANTS Database; V. vinifera by Dr. Kurt Stueber, Max Planck Institute for Plant Breeding Research, Cologne; A. thaliana by Luca Comai, University of Washington, Seattle, WA).

Brassicaceae (mustard) family of vegetables, which includes horseradish, broccoli, cauliflower, and turnips. It is one of about 250,000 species of flowering plants, a group that emerged 200 MYA (Walbot, 2000). Comparative genomics analyses will allow the comparison of the *Arabidopsis* genome to the genomes of other flowering plants in order to learn more about plant genomics (Hall et al., 2002).

The *Arabidopsis* genome is about 125 Mb. Its genome size is thus very small compared to agriculturally important plants such as wheat and barley (18.5 and 5 Gb, respectively) (Table 18.4). This made it an attractive choice as the first plant genome to be sequenced. The *Arabidopsis* Genome Initiative (2000) reported the sequence of most (115 Mb) of the genome. There are five chromosomes, initially predicted to contain 25,498 protein-coding genes. The *Arabidopsis* genome has an average density of one gene per 4.5 kb.

The estimated number of predicted genes in *Arabidopsis* has increased slightly to ~26,800, following reannotation of the genome (Crowe et al., 2003; TAIR database, described below). *Arabidopsis* has considerably more genes than *Drosophila* (about 13,000 genes) and *C. elegans* (about 19,000 genes; see below). The larger number of plant genes can be accounted for by a far greater extent of tandem gene duplications and segmental duplications. There is a core of about 11,600 distinct proteins, while the remaining genes are paralogs (*Arabidopsis* Genome Initiative, 2000).

A further feature of *Arabidopsis* is that the whole genome may have duplicated twice. (For an overview of ploidy in plants see Box 18.3.) Within the genome there are 24 large, duplicated segments of 100 kb or more, spanning 58% of the genome (*Arabidopsis* Genome Initiative, 2000). A comparison of tomato genomic DNA with *Arabidopsis* revealed conserved gene content and gene order with four different *Arabidopsis* chromosomes (Ku et al., 2000). The presence of duplicated and triplicated genomic regions suggests that two (or more) large-scale genome duplication events occurred. One event was ancient, while another occurred about 112 MYA. Following whole-genome duplication, gene loss occurred frequently. This reduces the amount of gene colinearity observed today and hinders our ability to decipher the nature and timing of past polyploidization events (Simillion et al., 2002). The pattern of gene loss following genome duplication is typical of fungi (Chapter 17) and fish (see below) but not *Paramecium*, described above.

Several *Arabidopsis* genomics resources are listed in Table 18.5. The most comprehensive site is TAIR, with a wide range of services (Swarbreck et al., 2008).

Rubisco is ribulose-1,5-diphosphate carboxylase. It is an enzyme localized to chloroplasts that catalyzes the first step of carbon fixation in photosynthesizing plants. The enzyme irreversibly converts ribulose diphosphate and carbon dioxide (CO_2) to two 3-phosphoglycerate molecules. The gene name for rubisco is *rbcL*, and for a typical example see the rice protein (RefSeq accession NP_039391).

Online databases are available for model plant genome projects, such as MtDB for *Medicago trunculata* (Lamblin et al., 2003) (▶ http://www.medicago.org/) and MaizeGDB for maize (▶ http://www.maizegdb.org/). More comprehensive plant genomics databases include Unité de Recherche Génomique Info (URGI) (▶ http://urgi.versailles.inra.fr/) and Sputnik (Rudd et al., 2003) in Turku (Åbo; ▶ http://sputnik.btk.fi/). GrainGenes, a database for wheat, barley, rye, and oat, is available at ▶ http://wheat.pw.usda.gov/GG2/index.shtml (Matthews et al., 2003).

TABLE 18-4	Major Plant Genome-Sequencing Projects		
Plant	Common Name	Genome Size (Gb)	Size Relative to Human
Arabidopsis thaliana	Thale cress	0.125	25.6-fold smaller
Avena sativa	Oat	16	5-fold larger
Glycine max	Soybean	1–2	About 2-fold smaller
Hordeum vulgare	Barley	5	1.7-fold larger
Lycopersicon esculentum	Tomato	1	3.2-fold smaller
Medicago truncatula	Barrel medic	0.5	6.4-fold smaller
Oryza sativa	Rice	0.466	6.9-fold smaller
Triticum aestivum	Bread wheat	18.5	5-fold larger
Zea mays	Corn	2.365	1.4-fold smaller

See the NCBI plant resources at ▶ http://www.ncbi.nlm.nih.gov/genomes/PLANTS/PlantList.html. Many plant genomes are about the same size as the human genome (3.2 Gb).

BOX 18.3
Ploidy in Plants

Many plants are polyploid, that is, the nuclear genome is more than diploid. This includes autopolyploids such as *Saccharum* spp. (sugarcane) and *Medicago sativa* (alfalfa). Such species are often intolerant of inbreeding (see Paterson, 2006). Allopolyploids include wheat and cotton. In many naturally occurring allotetraploids (such as the tetraploid *Arabidopsis suecica*), the flowers are distinctly different than those of the diploid parents (*Cardaminopsis* and *Arabidopsis*). Polyploid plants are usually bigger and more vigorous than diploid plants. Examples of polyploid species include banana and apple (triploid), potato, cotton, tobacco, and peanut (all tetraploid), wheat and oat (hexaploid), and sugar cane and strawberry (octaploid).

Plant genome sequencing projects have allowed paralogs to be identified. Whole genome duplication events have been inferred, including two or three events in both *Arabidopsis* and the poplar *Populus*, and one or two in the rice genome.

For an introduction to polyploidy in plants, see ▶ http://polyploid.agronomy. wisc.edu/.

This site includes a genome browser that provides access to genomic DNA sequence from the broadest chromosome-level view to descriptions of single-nucleotide polymorphisms. The format of this site, GBrowse, is shared by a variety of genome projects (Box 18.4). Other databases include SeedGenes, which describes essential genes of *Arabidopsis* that give a seed phenotype when disrupted by mutation (Tzafrir et al., 2003).

The Complete *Arabidopsis* Transcriptome Micro Array (CATMA) database is online at ▶ http://www.catma.org/ (Sclep et al., 2007).

Large segmental duplications in *Arabidopsis* were identified using MUMmer (see Chapter 15) and tblastx searches (see Chapter 4).

The Second Plant Genome: Rice

By some estimates, rice (*O. sativa*) is the staple food for half the human population. The rice genome was the second plant genome to be sequenced (Fig. 18.14). At approximately 389 Mb, this genome size is about one-eighth that of the human genome. Still it is one of the smallest genomes among the grasses, and rice is studied as a model monocot species.

Four groups generated draft versions of the rice genome (Buell, 2002), including two subspecies. A consortium led by the Beijing Genomics Institute reported a draft

Grasses include rice, wheat, maize, sorghum, barley, sugarcane, millet, oat, and rye. There are over 10,000 species of grasses (Bennetzen and Freeling, 1997). Cereals are seeds of flowering plants of the grass family (Gramineae, also called Poaceae) that are cultivated for the food value of their grains. Grasses are monocotyledonous plants that range from small, twisted, erect, or creeping annuals to perennials.

TABLE 18-5	Genomics Resources for *Arabidopsis thaliana*	
Resource	Comment	URL
TAIR	The Arabidopsis Information Resource	▶ http://www.arabidopsis.org/
Arabidopsis thaliana Database	Includes a map of segmental duplications	▶ http://www.tigr.org/tdb/e2k1/ ath1/
Arabidopsis thaliana Project	At MIPS	▶ http://mips.gsf.de/proj/plant/ jsf/index.jsp
SeedGenes	Essential genes in *Arabidopsis* development	▶ http://www.seedgenes.org

The International Rice Genome Sequencing Project (IRGSP) produced a draft version of the *O. sativa* ssp. *japonica* genome (▶ http://genome.sinica.edu.tw/). The Beijing Genomics Institute led a consortium that generated a draft version of the subspecies *indica* genome (see ▶ http://rise.genomics.org.cn/rice/index2.jsp). The principal international consortium for rice genomics from Japan has a website (▶ http://rgp.dna.affrc.go.jp/). The *indica* and *japonica* subspecies are thought to have been domesticated separately from an ancestral species, *Oryza rufipogon*.

sequence of the rice genome (*O. sativa* L. ssp. *indica*) (Yu et al., 2002). Another consortium reported a draft genome sequence of a different rice subspecies, *O. sativa* L. ssp. *japonica*) (Goff et al., 2002), and Monsanto generated another genome sequence. As discussed in Chapter 16, the annotation of genes and other features is far superior in finished sequence of chromosomes 1 (Sasaki et al., 2002) and 4 (Feng et al., 2002) relative to draft sequence. A finished quality sequence was

BOX 18.4
Databases for Eukaryotic Genomes

The main *Arabidopsis* database, TAIR, uses a database template shared by other major sequencing projects (Table 18.6). We already explored EcoCyc in Chapter 12 and the yeast database SGD in Chapters 12 and 17. These databases offer both detailed and extremely broad views of the genomic landscape. The Genomics Unified Schema (GUS) is another commonly used platform (Table 18.7). Many databases use a distributed annotation system (DAS) that allows a computer server to integrate genomic data from a variety of external computer systems. DAS, written by Lincoln Stein and Robin Dowell, is described at biodas.org (▶ http://www.biodas.org/). It is employed at WormBase, FlyBase, Ensembl, and TIGR sites, among others.

TABLE 18-6 Variety of Databases Employing Template from Generic Model Organism Project (GMOD) (▶ http://www.gmod.org/)

Database	Comment	URL
EcoCyc	Encyclopedia of *Escherichia coli* Genes and Metabolism	▶ http://EcoCyc.org/
FlyBase	*Drosophila* site	▶ http://www.flybase.org/
Mouse Genome Informatics	Main mouse resource	▶ http://www.informatics.jax.org/
Rat Genome Database (RGD)	Rat resource	▶ http://rgd.mcw.edu/
SGD	See Chapter 17	▶ http://genome-www.stanford.edu/Saccharomyces/
TAIR	The *Arabidopsis* Information Resource	▶ http://www.arabidopsis.org/
Wormbase	*C. Elegans* Site	▶ http://www.wormbase.org/

TABLE 18-7 Genomics Unified Schema Platform (▶ http://www.gusdb.org/)

Database	Comment	URL
AllGenes	Human and mouse gene index	▶ http://www.allgenes.org/
EPConDB	Endocrine Pancreas Consortium	▶ http://www.cbil.upenn.edu/EPConDB/
GeneDB	Curated database for *S. pombe*, *Leishmania major*, and *T. brucei*	▶ http://www.genedb.org/
PlasmoDB	Genomic database for *P. falciparum*	▶ http://plasmodb.org/
RAD	RNA abundance database	▶ http://www.cbil.upenn.edu/RAD2/

This platform is used for some organism databases.

reported by Yu et al. (2005) and separately by the International Rice Genome Sequencing Project (2005) for a single inbred cultivar, *O. sativa* L. ssp. *japonica* cv. Nipponbare. Yu et al. (2005) reported that relative to their 2002 initial publication they achieved a lower error rate and a 1000-fold improvement in long-range contiguity. The N50 sequence (the length above which half the total length of the sequence data set is found) improved to 8.3 megabases, about a 1000-fold improvement, as the coverage increased from 4.2x to 6.3x.

Several databases provide comprehensive collections of genomic and other data on rice, such as TIGR and MOsDB. These sites provide extensive genome annotation, including updates that reflect information provided through comparative genomics projects.

The rice genome (subspecies *indica*) displays an unusual feature of a gradient in GC content. The mean GC content is 43.3%, higher than in *Arabidopsis* (34.9%) or human (41.1%) (Yu et al., 2002). A plot of the number of 500 bp sequences (y axis) versus the percent GC content (x axis) revealed a tail of many GC-rich sequences. These GC-rich regions occurred selectively in rice exons (rather than introns), and at least one exon of extremely high GC content was found in almost every rice gene (Yu et al., 2002). The GC content of the 5′ end of each gene was typically 25% more GC rich than the 3′ end. These unique features of the rice genome present another major challenge for the use of ab initio gene-finding software.

The TIGR Rice Genome Project database is online at ▶ http://www.tigr.org/tdb/e2k1/osa1/ (Yuan et al., 2003). The MIPS (*O. sativa*) database (MOsDB) is available at ▶ http://mips.gsf.de/proj/plant/jsf/index.jsp (Karlowski et al., 2003).

The Third Plant Genome: Poplar

The black cottonwood tree *Populus trichocarpa* was the third plant genome to be sequenced (Fig. 18.14). *Populus* was selected for sequencing because its haploid nuclear genome is relatively small (480 megabases), it grows quickly relative to other trees (~5 years), and it is economically important as a source of wood and paper products.

Analysis of the genome indicates that *Populus* underwent a relatively recent whole-genome duplication about 65 million years ago, as well as experiencing tandem duplications and chromosomal rearrangements (Tuskan et al., 2006). In contrast to *Arabidopsis*, *Populus* is predominantly dioecious (having male and female reproductive structures on separate plants) such that it must outcross and achieves high levels of heterozygosity. Tuskan et al. (2006) identified 1.2 million single nucleotide polymorphisms (SNPs; Chapter 16), and with insertion/deletion events estimated 2.6 polymorphisms per kilobase. This will enable further genetics and population biology studies.

The Fourth Plant Genome: Grapevine

The gravevines are highly heterozygous, with as much as 13% sequence divergence between alleles. The French-Italian Public Consortium for Grapevine Genome Characterization (Jaillon et al., 2007) bred a grapevine variety derived from Pinot Noir to a high level of homozygosity and then determined its genome sequence (Fig. 18.14). Using an inbred variety was necessary to facilitate the assembly process. There are ~30,000 protein-coding genes predicted, which is fewer than in *Populus* even though the two organisms have similar sized genomes. In *Arabidopsis* and rice genes are evenly distributed across the genome, while in *V. vinifera* as in *Populus* there are gene-rich and gene-poor regions, with transposable elements (such as SINEs) occupying complementary positions.

One notable feature of the *V. vinifera* genome is that it encodes more than twice as many proteins related to terpene synthesis as other sequenced plant genomes. There are tens of thousands of terpenes in nature, typically containing two to four isoprene units, and many of these are highly odorous.

Analysis of the haploid grapevine genome showed that most gene regions have two different paralogous regions, thus forming homologous triplets and suggesting that the present genome derives from three ancestral genomes (Jaillon et al., 2007). There may have been three successive whole genome duplications, or a single hexaploidization event. To address this question they compared the *Vitis* gene order to poplar (its closest relative), *Arabidopsis* (a more distantly related dicotyledon), and rice (as a monocotyledon its most distant relative). Grapevine aligned with two poplar segments, consistent with a recent whole-genome duplication in poplar (described above). Also, the grapevine homologous triplets aligned with different pairs of poplar segments, suggesting that a hexaploidy of ancient origin was already present in the common ancestor of grapevine and poplar.

Moss

The Moss Genome website is ▶ http://www.mossgenome.org/. A Joint Genomes Initiative website on *P. patens* is at ▶ http://genome.jgi-psf.org/Phypa1_1/Phypa1_1.home.html.

The bryophytes, encompassing mosses, hornworts, and liverworts, diverged from the embryophytes (land plants) about 450 million years ago (near the time of divergence of the fish and human lineages). Rensing et al. (2008) sequenced the genome of the bryophyte moss *Physcomitrella patens*. Through comparisons to the genomes of water-dwelling plants, they propose that the movement of plants from aquatic to land environments involved the following components: (1) loss of genes that are associated with aquatic environments, such as those involved in flagellar function; (2) loss of dynein-mediated transport (as discussed above, *Chlamydomonas* and animals share dyneins); (3) gain of genes involved in signaling capabilities such as auxin, many of which are absent in *Chlamydomonas* and *O. tauri* genomes; (4) capability of adapting to conditions of drought, radiation, and temperature extremes; (5) gain of transport capabilities; and (6) gain in gene family complexity, reflected in the large numbers of genes in the moss and other plant genomes.

Many additional plant genome sequencing projects are in progress. Figure 18.14 provides an overview of the barrel medic project, *Medicago trunatula*.

SLIME AND FRUITING BODIES AT THE FEET OF METAZOANS

As we examine the upper part of the tree of the eukaryotes in Fig. 18.1, we see three great clades: the Mycetozoa, the Metazoa (animals), and the Fungi (Chapter 17). The metazoans are familiar to us as animals, including worms, insects, fish, and mammals. The Mycetozoa form a sister clade. The slime mold *Dictyostelium discoideum* is a social amoeba that is of great interest as a eukaryote that is an outgroup of the metazoa.

Social Slime Mold *Dictyostelium discoideum*

The principal website for information on *Dictyostelium* is ▶ http://www.dictybase.org/. The social, multicellular lifestyle of this eukaryote is reminiscent of the similar behavior of the proteobacterium *Myxococcus xanthus* (Chapter 15).

Biologists have studied *Dictyostelium* because of its remarkable life cycle. In normal conditions it is a single-celled organism that occupies a niche in soil. Upon conditions of starvation, it emits pulses of cyclic AMP (cAMP), promoting the aggregation of large numbers of amoebae. This results in the formation of an organism having

Genus, species: *Dictyostelium discoideum*
Lineage: Eukaryota; Mycetozoa; Dictyosteliida;
Dictyostelium; *Dictyostelium discoideum* AX4
(social amoeba; slime mold)

	Haploid genome size	GC content	Number of chromos.	Number of genes	Entrez Genome ID
Dictyostelium discoideum	34 Mb	22.4%	6	~12,500	201

Disease relevance: *Dictyostelium* has many hundreds of orthologs of human disease genes, and can reveal the principles of the evolution of these genes.
Genome features: The GC content is extraordinarily low and impacts many features of the genome.
Web resources: http://www.dictybase.org

FIGURE 18.15. *The slime mold D. discoideum is closely related to the metazoans, as shown in Fig. 18.1. This summary includes a photograph from the NHGRI (▶ http://www.genome.gov/17516871).*

the properties of other multicellular eukaryotes: It differentiates into several cell types, responds to heat and light, and undergoes a developmental profile.

The *Dictyostelium* genome is 34 Mb, localized on six chromosomes (Fig. 18.15). There are also about 100 copies per nucleus of an 88 kilobase palindromic extrachromosomal element containing the rRNA genes, and a mitochrondrial genome (55 kb). The largest of the main chromosomes, chromosome 2, consists of 8 Mb and was sequenced by an international consortium (Glockner et al., 2002) while the complete genome was reported by Eichinger et al. (2005) (reviewed in Williams et al., 2005).

Because the genome consists of about 78% AT content—similar to *P. falciparum*—as well as many repetitive DNA sequences, large-insert bacterial clones are unstable, and a whole-chromosome shotgun strategy was adopted. The genome is compact: the gene density is high (there are ~12,500 genes, with one gene per 2.6 kb, spanning 62% of the genome), there are relatively few introns (1.2 per gene), and both introns and intergenic regions are short. The introns have an AT content of 87%, while in exons the AT content is 72%. This discrepant compositional bias may represent a mechanism by which introns are spliced out (Glockner et al., 2002). Reflecting the AT richness of the genome, the codons NNT or NNA are used preferentially relative to the synonymous codons NNG or NNC. Amino acids encoded by codons having A or T in the first two positions and any nucleotide in the third position (asn, lys, ile, tyr, and phe) are far more common in *Dictyostelium* proteins than human ones.

An unusual feature of the genome is that 11% is comprised of simple sequence repeats (Chapter 16), more than for any other sequenced genome. There is a bias toward repeat units of 3 to 6 base pairs. Noncoding simple sequence repeats and homopolymer tracts have 99.2% AT content.

Perhaps the most unusual feature of *Dictyostelium* is that it achieves multicellularity upon conditions of starvation. This is reminiscent of the bacterium *Myxococcus xanthus*, described in Chapter 15. Eichinger et al. (2005) and Insall (2005) discuss genes that *Dictyostelium* has retained or acquired that facilitate this lifestyle. These

Chromosome 2 is characterized by an inverted 1.51 megabase duplication that is present in only some wild-type isolates.

include genes encoding proteins that are sometimes associated with metazoans or plants selectively: cell adhesion and signaling molecules (such as G-protein coupled receptors for signal transduction), ATP-binding cassette (ABC) transporters, and some genes encoding enzymes that are used in cellulose metabolism.

METAZOANS

Introduction to Metazoans

The diagram in Fig. 18.16 is consistent with that of Fig. 18.11, although it differs from Fig. 18.1 where nematodes form an outgroup. For discussions of bilaterian phylogeny see Lartillot and Philippe (2008) and Peterson et al. (2000). For alternative classification systems, see Cavalier-Smith (1998) and Margulis and Schwartz (1998). For an animal phylogeny based on cytochrome *c* oxidase I see Hebert et al. (2003). Karl Leuckart (1822–1898) first divided the metazoa into six phyla. For a table describing the metazoan (animal) kingdom superphyla and phyla, see web document 18.1 at ▶ http://www.bioinfbook.org/chapter18, while for a table describing the phylum bilateria including the Coelomata (animals with a body cavity), Acoelomata (animals lacking a body cavity such as flatworms) and Pseudocoelomata (such as the roundworm *C. elegans*) see web document 18.2.

The metazoans include most of the animals that are familiar to us, and in particular the main group of animals are bilaterians, that is, bilaterally symmetric animals (Fig. 18.16). The bilaterian animals are further divided into two major groups. (1) The protostomes include the Ecdysozoa (arthropods and nematodes), as well as the Lophotrochozoa (annelids and mollusks). We will survey the first protostome genomes that have been sequenced, such as the insects *D. melanogaster* and *A. gambiae*, and the nematode *C. elegans*. (2) The deuterostomes form a superclade consisting of the phylum of echinoderms (such as the sea urchin *Strongylocentrotus purpuratus*), the phylum of the hemichordates (such as acorn worms), and the chordates (vertebrates as well as the invertebrate cephalochordates and urochordates). These three deuterostome phyla descended from a common ancestor about 550 MYA, the time of the Cambrian explosion. We will discuss a basal member of the deuterostomes (the sea urchin *S. purpuratus*) and a basal member of the chordates (the urochordate sea squirt *Ciona intestinalis*), and we will then examine the vertebrate genomes, such as the fish, mouse, and chimpanzee.

As we seek to understand the human genome and what makes us unique as a species from a genomic perspective, one approach has been to determine whether our complexity and advanced features can be accounted for by a relatively large collection of genes. It is now clear that this is not the case; our gene numbers are comparable to those of other species across the eukaryotic domain. Another notion has been that humans, and vertebrates in general, have a large collection of unique genes that are not present in invertebrates. This notion is correct to a limited

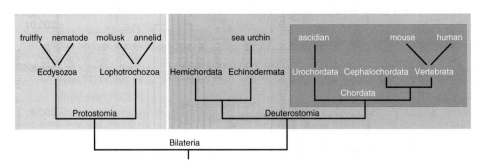

FIGURE 18.16. Phylogenetic relationships of the bilaterians, which have a bilateral body organization. The Protostomia include the arthropods or insects such as the fruitfly Drosophila melanogaster, *and the nematode worms such as* Caenorhabditis elegans, *as well as the mollusks and annelids. The Deuterostomia include the sister phyla Hemichordata and Echinodermata (including the sea urchin* Strongylocentrotus purpuratus) *as well as the Chordata. The chordates are further divided into the three groups including the vertebrates. This figure was redrawn from the Sea Urchin Genome Sequencing Consortium (2006). Used with permission.*

extent, but it too is being challenged. As metazoan genomes become sequenced we find many vertebrate genetic features shared with simpler animals (from insects to the invertebrate sea urchin to the sea squirt, a simple chordate).

Analysis of a Simple Animal: The Nematode *Caenorhabditis elegans*

Caenorhabditis elegans is a free-living soil nematode. It has served as a model organism because it is small (about 1 mm in length), easy to propagate (its life cycle is three days), has an invariant cell lineage that is fully described, and is suitable for many genetic manipulations. Furthermore, it has a variety of complex physiological traits characteristic of higher metazoans such as vertebrates, including an advanced central nervous system. Many nematodes are parasitic, and an understanding of *C. elegans* biology may lead to treatments for a variety of human diseases.

The soma of an adult hermaphrodite worm consists of 959 cells, including 302 cells in the central nervous system.

Another advantage of studying *C. elegans* is that its genome size of ~100 Mb is relatively small (Fig. 18.17). This genome was the first of an animal and the first of a multicellular organism to be sequenced (*C. elegans* Sequencing Consortium, 1998). The genome sequencing was based on physical maps of the five autosomes and single X chromosome. The GC content is an unremarkable 36%. It was predicted that there are 19,099 protein-coding genes, with 27% of the genome consisting of exons. About 42% of *C. elegans* proteins have predicted orthologs outside Nematoda, while 34% match only other nematode proteins.

The 2002 Nobel Prize in Physiology or Medicine was awarded to three researchers who pioneered the use of *C. elegans* as a model organism: Sydney Brenner, H. Robert Horvitz, and John E. Sulston. See ▶ http://www.nobel.se/medicine/laureates/2002/.

The *C. elegans* proteome contains a large number of predicted seven-transmembrane-domain (7TM) receptors of both the chemoreceptor family and rhodopsin family. This illustrates the principle that new protein functions can emerge following gene duplication (Sonnhammer and Durbin, 1997). It is also notable that many nematode proteins are absent from nonmetazoan species (plants and fungi).

About 300 species of parasitic worms infect human (Cox, 2002). While 20,000 nematode species have been described, it is thought that there may be one million species (Blaxter, 1998, 2003).

Genus, species	*Brugia malayi*
	Caenorhabditis briggsae
	Caenorhabditis elegans

Selected lineages: Eukaryota; Metazoa; Nematoda; Chromadorea; Spirurida; Filarioidea; Onchocercidae; Brugia; *Brugia malayi*

Lineage: Eukaryota; Metazoa; Nematoda; Chromadorea; Rhabditida; Rhabditoidea; Rhabditidae; Peloderinae; Caenorhabditis; *Caenorhabditis briggsae*

Eukaryota; Metazoa; Nematoda; Chromadorea; Rhabditida; Rhabditoidea; Rhabditidae; Peloderinae; Caenorhabditis; *Caenorhabditis elegans*

	Haploid genome size	GC content	Number of chromos.	Number of genes	Entrez Genome ID
Brugia malayi	90–95 Mb	30.5%	6	11,508	10729
Caenorhabditis briggsae	~104 Mb	37.4%	6	19,500	10731
Caenorhabditis elegans	100 Mb	35.4%	6	18,808	13758

Divergence dates: Nematodes diverged from arthropods (insects) 800–1000 million years ago (MYA).
    C. elegans diverged from *C. briggsae* ~80–110 MYA.
Disease relevance: *Brugia malayi* is the agent of lymphatic filariasis which infects 120 million people.
Key website: http://www.wormbase.org

FIGURE 18.17. Overview of roundworm genomes. Image of the anterior end of a Brugia malayi *microfilaria in a thick blood smear using Giemsa stain is from the CDC (▶ http://phil.cdc.gov/phil/details.asp; content provider Dr. Mae Melvin).*

The principal web resource for *C. elegans* is WormBase, a comprehensive database (Rogers et al., 2008). WormBase features a variety of data, including genomic sequence data; the developmental lineage; the connectivity of the nervous system; mutant phenotypes, genetic markers, and genetic map data; gene expression data; and bibliographic resources.

WormBase is available at ▶ http://www.wormbase.org.

Caenorhabditis elegans has been the subject of many functional genomics projects (Chapter 12) (Fields et al., 1999; Kim, 2001; Brooks and Isaac, 2002). Gene expression has been measured using microarrays at six different developmental stages (Hill et al., 2000). About 10,700 open reading frames (56%) were detected in at least one hybridization. This number is comparable to the complement of expressed sequence tags, and the remaining thousands of genes may be expressed in specialized body regions, developmental stages, or physiological conditions. In another approach to defining gene function, Kamath et al. (2003) inhibited the function of 86% of the >19,000 predicted *C. elegans* genes using RNA interference (RNAi). They identified mutant phenotypes for 1722 genes.

After *C. elegans*, the genome of the related soil nematode *Caenorhabditis briggsae* was sequenced (Stein et al., 2003; reviewed in Gupta and Sternberg, 2003). Remarkably, these organisms speciated about 100 million years ago, but they are indistinguishable by eye. Each genome is about 100 megabases and encodes a comparable number of genes. The availability of *C. briggsae* sequence facilitated an improved annotation of the *C. elegans* genome and the discovery of about 1300 novel *C. elegans* genes. The genomes share extensive colinearity. An example is shown for a 100,000 base pair region including a globin gene using the WormBase synteny viewer (Fig. 18.18).

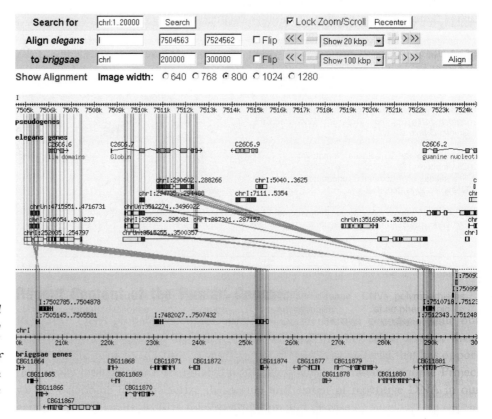

FIGURE 18.18. Alignment of C. elegans and C. briggsae conserved syntenic regions using the synteny viewer at WormBase (▶ http://www.wormbase.org). Regions of chromosome I are aligned from C. elegans (above) and C. briggsae (below).

Brugia malayi was the first parasitic nematode to have its genome sequenced (Ghedin et al., 2007). This parasite causes lymphatic filariasis, a chronic disease that is debilitating although associated with low mortality. The *B. malayi* genome contains fewer genes than *C. elegans* (~11,500 versus ~18,500) primarily because of lineage-specific expansions in *C. elegans*. There is a need for drugs to treat filariasis. Ghedin et al. identified a number of gene products that are potential targets for therapeutic intervention. For example, *B. malayi* lacks most enzymes required for de novo purine biosynthesis, heme biosynthesis, and de novo riboflavin synthesis, probably obtaining these compounds from its host or its endosymbiont *Wolbachia*. Drugs that interfere with these synthetic pathways are potential targets.

The genomes of several dozen nematodes are currently being sequenced, and the UCSC Genome Browser currently includes assemblies for five nematodes (*C. elegans, C. briggsaei, C. brenneri, C. remanei*, and *Pristionchus pacificus*). Annotation tracks are available for conservation among these worms and to human proteins.

The First Insect Genome: *Drosophila melanogaster*

The arthropods may be the most successful set of eukaryotes on the planet in terms of the number of species. They include the Chelicerates—such as the scorpions, spiders, and mites—and the Mandibulata, animals with modified appendages (mandibles) such as the insects (Table 18.8). While insects first appear in the fossil record from about 350 million years ago, their lineage is thought to have emerged 600 million years ago.

The fruitfly *D. melanogaster* has been an important model organism in biology for a century (Rubin and Lewis, 2000). The fly is ideal for studies of genetics because of its short life cycle (two weeks), varied phenotypes (from changes in eye color to changes in behavior, development, or morphology), and large polytene chromosomes that are easily observed under a microscope.

The *Drosophila* genome was sequenced based in large part on the whole-genome shotgun sequencing strategy (Adams et al., 2000) (Fig. 18.19). Prior to this effort, the whole-genome shotgun strategy had only been applied to far smaller genomes, and thus this success represented a significant breakthrough. The 180 Mb genome is organized into an X chromosome (numbered 1), two principal autosomes (numbered 2 and 3), a very small third autosome (numbered 4; about 1 Mb in length), and a Y chromosome. Approximately one-third of the genome contains heterochromatin (mostly simple sequence repeats as well as transposable elements and tandem arrays of rRNA genes). This heterochromatin is distributed around the centromeres

Thomas Hunt Morgan was awarded a Nobel Prize in 1933 "for his discoveries concerning the role played by the chromosome in heredity." See ► http://www. nobel.se/medicine/laureates/ 1933/. In 1995, Edward B. Lewis, Christiane Nüsslein-Volhard, and Eric F. Wieschaus shared a Nobel Prize "for their discoveries concerning the genetic control of early embryonic development." These studies concerned *Drosophila* development (► http://www. nobel.se/medicine/laureates/ 1995/).

About 1 million arthropod species have been described, but there are an estimated 3 to 30 million species (Blaxter, 2003).

The *Drosophila* genome was sequenced through a collaborative effort that included Celera Genomics, the Berkeley *Drosophila* Genome Project (BDGP; ► http://www.fruitfly.org), and the European *Drosophila* Genome Project (EDGP) (Adams et al., 2000). As part of the effort, three million random genomic fragments of ≈500 bp were sequenced.

TABLE 18-8 Arthropods (Phylum Arthropoda) as Classified at NCBI (► http://www.ncbi.nlm.nih.gov/Taxonomy/)

Subphylum	Class
Chelicerata	Arachnida (mites, ticks, spiders)
	Merostomata (horseshoe crabs)
	Pycnogonida (sea spiders)
Mandibulata	Myriapoda (centipedes)
	Pancrustacea (crustaceans, insects)

Arthropods are invertebrate protostomes (see Fig. 18.16). Pancrustacea (boldface) is further divided into the superclasses Crustacea (crustaceans) and Hexapoda (insects). Insecta includes *D. melanogaster* and *A. gambiae*.

| Tribolium castaneum | Anopheles gambiae | Aedes aegypti |

Selected lineages: Eukaryota; Metazoa; Arthropoda; Hexapoda; Insecta; Pterygota; Neoptera; Endopterygota; Diptera; Nematocera; Culicoidea; Culicidae; Anophelinae; Anopheles; *Anopheles gambiae* str. PEST (African malaria mosquito)

Eukaryota; Metazoa; Arthropoda; Hexapoda; Insecta; Pterygota; Neoptera; Endopterygota; Diptera; Brachycera; Muscomorpha; Ephydroidea; Drosophilidae; Drosophila; *Drosophila melanogaster* (fruit fly)

Eukaryota; Metazoa; Arthropoda; Hexapoda; Insecta; Pterygota; Neoptera; Endopterygota; Coleoptera; Polyphaga; Cucujiformia; Tenebrionidae; Tribolium; *Tribolium castaneum* (red flour beetle)

	Haploid genome size	GC content	Number of chromos.	Number of genes	Entrez Genome ID
Anopheles gambiae	278 Mb	44%	3	13,683	1438
Apis mellifera DH4	262 Mb	33%	16	10,157	10625
Bombyx mori	429 Mb		28	18,510	12259, 13125
Drosophila ananassae	217 Mb		4	22,551	12651
Drosophila erecta	135 Mb		4	16,880	12661, 12662
Drosophila grimshawi	231 Mb		4	16,901	12678 , 12679
Drosophila melanogaster	200 Mb	42%	4	13,733	13812
Drosophila mojavensis	130 Mb		4	17,738	12682, 12685
Drosophila persimilis	193 Mb		4	23,029	12705, 12708
Drosophila pseudoobscura	193 Mb		5	17,328	10626
Drosophila sechellia	171 Mb		4	21,332	12711, 12712
Drosophila simulans	162 Mb		4	17,049	12464
Drosophila virilis	364 Mb		4	17,679	12688
Drosophila willistoni	222 Mb		4	20,211	12664
Drosophila yakuba	190 Mb		4	18,816	12366
Tribolium castaneum	200 Mb		10	not avail.	12540

Selected divergence dates: The insect lineage diverged from the human lineage ~600 million years ago (MYA). Hymenoptera (such as the honeybee *A. mellifera*) diverged from Lepidopterans (such as the silkworm *B. mori*) and dipterans (such as fruitfly and mosquito) 300 MYA; silkworm and fruitfly lineages split 280–350 MYA.
Disease association: mosquitos are vectors for many diseases including dengue and yellow fever.
Organism-specific web resources: http://www.flybase.org; http://www.anobase.org.

FIGURE 18.19. Overview of insect genomes. Photo of a mosquito (Aedes) and scanning electron micrograph of Anopheles gambiae from the CDC image library (http://phil.cdc.gov/phil/details.asp) by CDC/Paul I. Howell, MPH and Prof. Frank Hadley Collins. Tribolium photo from the NHGRI (▶ http://www.genome.gov/17516871).

The principal database for *D. melanogaster* (and for other species of the family Drosophilidae) is FlyBase (▶ http://www.flybase.org/) (Wilson et al., 2008). Many gene prediction algorithms, including GenScan, overpredict the correct number of genes (see Chapter 16). Currently, the Ensembl database lists ~23,000 mouse genes and ~49,000 GenScan predictions (▶ http://www.ensembl.org/Mus_musculus/).

and across the length of the Y chromosome. The transition zones at the boundary of heterochromatin and euchromatin contained many protein-coding genes that were previously unknown.

The initial annotation of the *Drosophila* genome described 17,464 genes predicted with Genscan and 13,189 genes predicted with the Genie algorithm (Reese et al., 2000; Adams et al., 2000). The authors believed that Genscan overestimated the true number of genes. Subsequently, the genome sequence was finished to close gaps and to improve sequence quality. This resulted in a series of releases of updated annotation efforts. Release 5.1 of the *D. melanogaster* heterochromatin covered 24 megabases, and indicated a set of 230 to 254 protein-coding genes as well as several dozen pseudogenes and noncoding genes (Smith et al., 2007; Hoskins et al., 2007). During the reannotation process there have been changes to the models for 85% of the transcripts and about half of the predicted proteins. The improved annotation can be attributed to the availability of more expressed sequence tags, complete cDNAs that can be aligned to genomic DNA, and continued genomic DNA sequencing. These studies have clarified the extent of untranslated regions, which are difficult to assign using ab initio gene-finding algorithms.

While one task of genome annotation is the description of protein-coding genes, another task is to assign function to those genes. This can be approached by assessing the extent to which predicted proteins have identifiable orthologs. Rubin et al. (2000) systematically blastp searched the proteomes of the fly, *C. elegans*, and *S. cerevisiae*. They drew several conclusions which appear valid today although the exact numbers of annotated genes improves over time:

- The "core proteomes" of these organisms represent the set of unique proteins, excluding paralogs. These sizes are 4383 proteins (yeast), 8065 proteins (fly), and 9453 proteins (worm). Thus, despite the fact that *S. cerevisiae* is unicellular while fly and worm are multicellular, the core proteomes are only twofold different.

- About 30% of the fly genes have orthologs in worm; 20% have an ortholog in both worm and yeast. Such proteins may be in common to all eukaryotic cells.

- Half of the fly proteins have a mammalian homolog (at an *E* value cutoff below 10^{-10}), consistent with a model in which the fly is more closely related to humans than is the worm (see Fig. 18.11).

- A substantial number of *Drosophila* proteins are not significantly related to proteins from yeast, worm, or mammals.

- The fly and worm each have about 2200 multidomain proteins. However, yeast has only 672. Proteomic analyses such as these may elucidate the molecular basis of phenotypic differences between organisms. For example, in contrast to yeast, the fly and worm have many proteins with extracellular domains having roles in cell–cell contact and cell–substrate contact.

Rubin et al. (2000) defined orthologs as having significant similarity (based on *E* values) over at least 80% of the query protein length. This leads to an underestimate of the number of orthologs because some matched regions are small.

Twelve *Drosophila*-related genomes have currently been sequenced (Figs. 18.19 and 18.20). Following the sequencing of *D. melanogaster* and *D. pseudoobscura* (Richards et al., 2005), a consortium of 250 researchers sequenced ten more genomes (*Drosophila* 12 Genomes Consortium, 2007). Seven genomes were sequenced to deep coverage (8.4x to 11.0x) and others to intermediate or low coverage to provide population variation data. These include several species that are closely related (e.g., *D. yakuba* and *D. erecta*, or *D. pseudoobscura* and *D. persimilis*) as well as some distantly related (e.g., *D. grimshawi* is a species restricted to Hawaii). Total genome size varies less than threefold among the 12 species, and the gene content ranges from ~14,000 to ~17,000. Based on comparative annotation of protein-coding genes, Stark et al. (2007) identified almost 1200 new protein-coding exons and resulted in a modification of 10% of the annotated protein-coding genes in *D. melanogaster*.

The availability of this many related genome sequences permits a deeper understanding of many areas of evolution, including genomic rearrangements, the acquisition of transposable elements, and protein evolution. Most genes evolve under evolutionary constraint at most of their sites, so that the ratio ω of nonsynonymous to synonymous mutaions (d_N/d_S) tends to be low. Of all *D. melanogaster* proteins, the majority (77%) are conserved across all 12 species. The number of noncoding RNA genes is also conserved, ranging from ~600 to ~900.

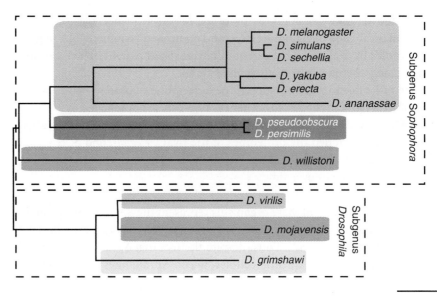

FIGURE 18.20. Phylogeny of 12 sequenced Drosophila species. The tree was created using the neighbor-joining method with additional strong support from Bayesian and maximum parsimony analyses (see Chapter 7). The branch lengths indicate the number of mutations per site at four-fold degenerate sites. Redrawn from Drosophila 12 Genomes Consortium (2007). Used with permission.

0.1 mutations per site

The sequencing and analysis of 12 *Drosophila* genomes as well as multiple fungal genomes (Chapter 17) represent important, pioneering effort in eukaryotic comparative genomics. Such approaches will result in improved catalogs of coding and noncoding genes, regulatory features, and functional regions of genomic DNA, as well as a clearer understanding of evolutionary events, including when species diverged, and how and when genomes have been sculpted by forces from chromosomal alterations to lateral transfer of transposable elements.

The Second Insect Genome: *Anopheles gambiae*

A haplotype is a combination of alleles of closely linked loci that are found in a single chromosome and tend to be inherited together.

AnoBase is a major resource for anopheline species (Topalis et al., 2005) (► http://www.anobase. org). The Ensembl genome browser for the mosquito is available at ► http://www.ensembl. org/Anopheles_gambiae/.

The mosquito *A. gambiae* is most well known as the malaria vector that carries the protozoan parasite *P. falciparum* (as well as *P. vivax*, *P. malariae*, and *P. ovale*). Mosquitoes are responsible for a variety of human diseases, although most of these (except West Nile) are generally restricted to the tropics (Table 18.9).

Holt et al. (2002) reported the genomic sequence of a strain of *A. gambiae* using the whole-genome sequencing strategy. The genome is 278 Mb arranged in an X chromosome (numbered 1) and two autosomes (numbered 2 and 3). A particular challenge in sequencing this genome is the high degree of genetic variation, as manifested in "single-nucleotide discrepancies." Thus there is a mosaic genome

TABLE 18-9 Human Diseases Borne by Mosquitoes

Disease	Mosquito Species	Number of Cases
Malaria	*Anopheles gambiae*	500 million
Dengue	*Aedes aegypti*	50 million per year
Lymphatic filariasis	*Culex quinquefasciatus, Anopheles gambiae*	120 million
Yellow fever	*Aedes aegypti*	200,000 per year
West Nile virus disease	*Culex tarsalis, Culex pipiens,* other	≈4200 per year

West Nile virus disease data are for the year 2006 in the United States (Centers for Disease Control and Prevention, http://www.cdc.gov).
Source: Adapted from Budiansky (2002) and Holt et al. (2002).

structure caused by two haplotypes of approximately equal abundance. In contrast, the *D. melanogaster* and *M. musculus* genomes are relatively homozygous.

You can view and explore the *A. gambiae* genome at the Ensembl genome browser. Annotation by the Ensembl pipeline (Chapter 13) and Celera suggests the existence of 13,683 genes. As with all eukaryotic genome projects, the *A. gambiae* annotation is known to contain many incomplete or incorrect gene assignments (Holt et al., 2002).

The *A. gambiae* genome is more than twice the size of that of *Drosophila*. This difference is largely accounted for by intergenic DNA, and *Drosophila* appears to have undergone a genome size reduction relative to *Anopheles* species (Holt et al., 2002). *Anopheles gambiae* and *D. melanogaster* diverged about 250 MYA (Zdobnov et al., 2002). Almost half the genes in these genomes are orthologs, with an average amino acid sequence identity of 56%. By comparison, the lineage leading to modern humans and pufferfish (see below) diverged 450 MYA, but proteins from those two species share even slightly higher sequence identity (61%). Thus, insect proteins diverge at a faster rate than vertebrate proteins. An outstanding problem is to understand the ability of *Anopheles* to feed on human blood selectively and to identify therapeutic targets. For this effort, it is important to identify arthropod-specific and *Anopheles*-specific genes (Zdobnov et al., 2002).

We described the *Drosophila* Down syndrome cell adhesion molecule (DSCAM) in Chapter 8, a gene that potentially encodes up to 38,000 distinct proteins through alternative splicing (NP_523649). The *A. gambiae* ortholog appears to share the same potential for massive alternative splicing (Zdobnov et al., 2002). See GenBank protein accession XP_309810.

Silkworm

The cocoon of the domesticated silkworm *Bombyx mori* is the source of silk fibers. Xia et al. (2004) determined the sequence of its genome (see Fig. 18.19). At 429 Mb it is 3.6 times larger than that of fruitfly, and 1.5 times larger than mosquito; some of this size can be attributed to the presence of more genes (18,510 relative to ~13,700 in *D. melanogaster*) and also larger genes. Transposable elements have also shaped the genome, comprising 21%. Of that fraction, half arrived just 5 million years ago as a single *gypsy-Ty3*-like retrotransposon insertion. Analysis of the *B. mori* genome has helped to elucidate the function of the silk gland (a modified salivary gland), and although silkworms do not fly nor do they have colorful wing patterns, there are homologs of genes implicated in wing development and pattern formation.

Honeybee

The western honeybee *Apis mellifera* is of special interest because of its highly social behavior. Bee hives are organized around a queen and her workers who transition from roles in the hive (such as nurses and hive maintainers) to the outside (such as foragers and defenders). The queens typically live ten times longer than the workers and lay up to 2000 eggs per day. The workers have brains with only a million neurons but display highly intricate behaviors. Somehow all these differentiated phenotypes are directed by a single underlying genome. The Honeybee Genome Sequencing Consortium (2006) sequenced the *A. mellifera* genome. There are 15 acrocentric chromosomes, and a large metacentric chromosome 1; as is the case for human chromosome 2 (Chapter 19; Fan et al., 2002), this is thought to represent a fusion of two acrocentrics. Relative to other insect genomes it has a lower GC content, and fewer predicted protein-coding genes (summarized in Fig. 18.19). The relatively low gene count could be influenced both by the limited number of expressed sequence tags and by the divergence between *Apis* and other sequenced insect genomes; in general, a typical pattern is that additional genes are discovered as the genomes of more closely related species are sequenced. Although CpG dinucleotides are underrepresented fivefold in the human genome (Chapter 19) and other genomes (Chapter 16), they are surprisingly overrepresented 1.67-fold in honeybee. The Honeybee Genome Sequencing Consortium (2006) provided a detailed analysis of the proteins in honeybee relative to other insects and to other animals; for example, the *A. mellifera* gene catalog illuminates those highly conserved proteins that are absent in a particular species such as *Drosophila*.

The Road to Chordates: The Sea Urchin

For a brief and useful overview of how to interpret the relatedness of different species by inspection of a phylogenetic tree, see Baum et al. (2005).

As we survey the metazoan animals and move from the Protostomia (including the insects, nematodes, molluscs and annelids) to the Deuterostomia (Fig. 18.16), we first come to the sister phyla of the hemichordates and echinoderms. The purple sea urchin *Strongylocentrotus purpuratus* is an echinoderm that has served as a model organism for studies of cell biology (including embryology and gene regulation) and evolution. The sea urchin serves as an outgroup for the chordates. This creature is a marine invertebrate, has a radial adult body plan (as shown in the photograph in Fig. 18.21), and has no apparent brain although there are neurons and brain functions. An individual can have a lifespan over a century. And so it may be surprising to consider that it is more closely related to humans than nematodes or fruitflies with their well-defined brains and complex behaviors.

The assembled *S. purpuratus* genome is 814 megabases (Sea Urchin Genome Sequencing Consortium, 2006). The assembly includes hundreds of scaffolds (also called supercontigs), and although linkage and cytogenetic maps are unavailable the number of chromosomes has been estimated to be ~40. There were several outstanding technical issues in sequencing the genome (reviewed in Sodergren et al., 2006). One is that the sea urchin exhibits tremendous heterozygosity, with 4% to 5%

Ciona intestinalis   *Strongylocentrotus purpuratus*

Genus, species:
Ciona intestinalis (sea squirt)
Ciona savignyi
Oikopleura dioica (tunicate)
Strongylocentrotus purpuratus
  (purple sea urchin)

Selected lineages: Eukaryota; Metazoa; Chordata; Urochordata; Ascidiacea; Enterogona; Phlebobranchia; Cionidae; Ciona; *Ciona intestinalis*

Eukaryota; Metazoa; Echinodermata; Eleutherozoa; Echinozoa; Echinoidea; Euechinoidea; Echinacea; Echinoida; Strongylocentrotidae; Strongylocentrotus; *Strongylocentrotus purpuratus*

	Haploid genome size	GC	Number of chromos.	Number of genes	Entrez Genome ID
Ciona intestinalis	~160 Mb	35%	14	15,852	166
Ciona savignyi	190 Mb				1435
Oikopleura dioica	72 Mb			~15,000	12901
S. purpuratus	814 Mb	36.9%	~40	23,300	10736

Key dates: divergence from human lineage: ~550 million years ago
Disease relevance: These organisms have many orthologs of human disease genes.
Genome features: The average gene density is one gene per 7.5 kb in *Ciona*, 1 gene per 9 kb in fruitfly, and 1 gene per 100 kb in human. Some *C. intestinalis* and sea urchin genome features are intermediate between protostomes and vertebrates (e.g. 5 exons/gene in *Drosophila*, 6.8/gene in *Ciona*, and 8.8/gene in human).
Websites: Sea Urchin Genome Project (http://sugp.caltech.edu/; http://www.spbase.org)

FIGURE 18.21. Overview of simple (nonvertebrate) deuterostome genomes. Photograph of purple sea urchin from NCBI Entrez Genomes website (by Andy Cameron).

nucleotide differences between single copy DNA of different individuals (this includes SNPs and insertions/deletions, and contrasts with ~0.5% heterozygosity in humans). The single male sea urchin that was sequenced displayed tremendous heterozygosity between its two haplotypes, making it challenging to distinguish sequencing errors from haplotype variants or from segmentally duplicated regions. One way this problem was overcome was to sequence bacterial artificial chromosome (BAC) clones of ~150,000 base pairs each, in which each BAC corresponds to a single haplotype. A minimal tiling path of BAC clones spanned the genome and was sequenced at low (2x) coverage. This complemented a deep whole-genome shotgun assembly. This combined approach was introduced in the sequencing of the rat genome and has become an increasingly common strategy for genome sequencing.

The Sea Urchin Genome Sequencing Consortium (2006) predicted about 23,300 genes for *S. purpuratus*. Some InterPro and Pfam domains (Chapter 10) are especially overrepresented in sea urchin relative to mouse, *Drosophila*, *C. elegans*, and sea squirt. Most dramatic are three families of receptor proteins that function in the innate immune response (Toll-like receptors; NACHT and leucine-rich repeat-containing proteins; and scavenger receptor cysteine-rich domain proteins). Each of these genes is present in over 200 copies, while other animals from humans to fruit-fly and nematode typically have about 0 to 20 copies. Another surprising finding is the presence of over 600 genes encoding G-protein coupled chemoreceptors, as well as genes involved in photoreception, expressed on the tube feet.

750 Million Years Ago: *Ciona intestinalis* and the Road to Vertebrates

The vertebrates include fish, amphibians, reptiles, birds, and mammals. All these creatures have in common a segmented spinal column. From where did the vertebrates originate? Vertebrates are members of the chordates, animals having a notochord (Fig. 18.16). The sea squirt *C. intestinalis* is a urochordate (also called tunicate), one of the subphyla of chordates but not a vertebrate. *Ciona* is a hermaphroditic invertebrate that offers us a window on the transition to vertebrates (Holland, 2002). The title of this section begins "750 million years ago" referring to the approximate date of a last common ancestor with the human lineage; in the remaining sections of this chapter we will continue to track the relatedness of each group to humans.

Dehal et al. (2002) produced a draft sequence of the *C. intestinalis* genome by the whole-genome shotgun strategy. At 160 Mb it is about 12 times larger than typical fungal genomes and 20 times smaller than the human genome. There are 15,852 predicted genes organized on 14 chromosomes. Most of these predicted genes are supported by evidence from expressed sequence tags.

The availability of the *Ciona* genome sequence allows a comparison with protostomes and other deuterostomes and supports its position as related to an ancestral chordate (Dehal et al., 2002). Almost 60% of *Ciona* genes have protostome orthologs; these presumably represent ancient bilaterian genes. Several hundred genes have invertebrate but not vertebrate homologs, such as the oxygen carrier hemocyanin. These comparative studies will be augmented by the genome sequencing of the related urochordates *Ciona savignyi* and *Oikopleura dioica*. *O. dioca* has one of the smallest chordate genomes (about 72 Mb; Seo et al., 2001), and it is an attractive experimental organism because its lifespan is two to four days, it can be maintained

The phylum Cnidaria is an outgroup to the bilateria, having diverged about 600 to 750 MYA. Its members include sea anemones, hydras, corals, and jellyfishes. CnidBase organizes genomic and other information on diverse cnidarians (▶ http://cnidbase.bu.edu) (Ryan and Finnerty, 2003).

The Department of Energy Joint Genome Institute operates the *C. intestinalis* genome home page (▶ http://genome.jgi-psf.org/Cioin2/Cioin2.home.html). The GenBank accession number for the genome is AABS00000000, and you can find a *Ciona* BLAST server through the NCBI Genomes page of eukaryotic projects. The Ghost database, a *Ciona* EST project that includes a BLAST server and gene expression data, is available at ▶ http://ghost.zool.kyoto-u.ac.jp/indexr1.html.

The Broad Institute offers a *Ciona savignyi* database (▶ http://www.broad.mit.edu/annotation/ciona/).

in culture, and its females are fecund. *C. savignyi*, a sea squirt, exhibits considerable heterozygosity, with variable degrees of heterozygosity across the genome. Eric Lander and colleagues (Vinson et al., 2005) introduced an algorithmic approach to assembling genome sequences from diploid genomes. This method assembles the two haplotypes separately, and thus requires twice the sequencing depth of other whole genome sequencing projects. The result is substantial improvement in sequence quality and contiguity. Such approaches will be increasingly useful as more outbred genomes are sequenced.

There are 2570 *Ciona intestinalis* genes (one-sixth) that have orthologs in vertebrates but none in protostomes; these genes arose in the deuterostome lineage before the last common ancestor diverged into vertebrates, cephalochordates, and urochordates (e.g., *Ciona*). There are 3399 *Ciona* genes (one-fifth) that have no identifiable homolog in vertebrates or invertebrates and thus may be tunicate-specific genes that evolved after the divergence of the urochordate lineage.

Ciona has genes involved in processes such as apoptosis (programmed cell death), thyroid function, neural function, and muscle action. This provides an opportunity for comparative analyses of fundamentally important genes within the chordate lineage. For example, nerves communicate with muscles by releasing the neurotransmitter acetylcholine from synaptic vesicles in presynaptic nerve terminals. This transmitter diffuses across the synapse (a gap between cells) to bind and activate postsynaptic receptors. *Ciona* has genes encoding proteins that function in neurotransmission, including a transferase enzyme that synthesizes acetycholine, an acetycholine transporter that pumps the neurotransmitter into vesicles, synaptic vesicle proteins, and neurotransmitter receptors. Similar genes are present in sea urchin as well, such as the agrin gene encoding a protein that clusters acetylcholine receptors postsynaptically.

A *Ciona* protein (NP_001027621) has 46% identity to human choline acetyltransferase (NP_065574) and 51% identity to a sea urchin ortholog (XP_001185550). A *Ciona* gene (accession AB071998) encodes a protein with 56% identity to a human vesicular acetylcholine transporter (NP_003046). Many such genes also function in neurotransmission in invertebrates.

450 Million Years Ago: Vertebrate Genomes of Fish

The teleosts (or ray-finned fishes, *Actinopterygii*) are the largest group of vertebrates, with ~24,000 known species (more than half the total number of vertebrate species). The ray-finned fishes diverged from the lobe-finned fishes (*Sarcopyerygii*) about 450 million years ago. These relationships are depicted in the phylogenetic tree of Fig. 18.22a. The teleosts are further shown in Fig. 18.22b, including the first four sequenced fish genomes: those of the pufferfishes *Takifugu rubripes* and *Tetraodon nigroviridis*, the medaka *Oryzias latipes*, and the zebrafish *Danio rerio* (see Fig. 18.23). Shallow sequencing (1.4x coverage) of the elephant shark *Callorhinchus milii* has also been reported (Venkatesh et al., 2007), representing a cartilaginous fish that is an outgroup to the teleosts (Fig. 18.22a).

The second vertebrate genome sequencing project (after human) was that of the Japanese pufferfish *T. rubripes*, in part because it has a remarkably compact genome. This teleost fish has a genome size of 365 Mb, about one-eighth the size of the human genome (Aparicio et al., 2002). However, *Takifugu* and humans have comparable numbers of predicted protein-coding genes.

There are several reasons that the *Takifugu* genome is relatively compact (Aparicio et al., 2002):

Fugu rubripes is also called *Takifugu rubripes*. The International *Fugu* Genome Consortium was responsible for the sequencing of its genome. The *Takifugu* Browser is at ▶ http://www.ensembl.org/Takifugu_rubripes/. Produced by the Wellcome Trust Sanger Institute and the European Bioinformatics Institute, it is a major portal to this genome and others. Other major gateways to *Fugu* resources are at the U.S. Department of Energy Joint Genome Institute site (▶ http://genome.jgi-psf.org/Takru4/Takru4.home.html), ▶ http://www.genoscope.cns.fr/externe/English/Projets/Projet_C/C.html, and ▶ http://www.fugu-sg.org/.

- Only 2.7% of the *Takifugu* genome consists of interspersed repeats, based on analyses with RepeatMasker. This contrasts with 45% interspersed repeats in the human genome (Chapter 19). Still, every known class of eukaryotic

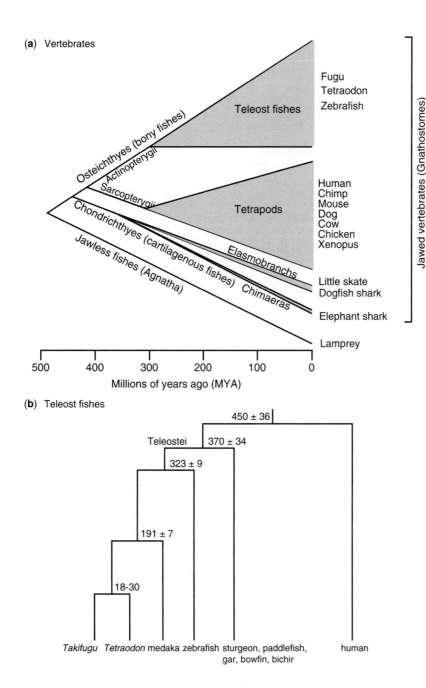

(a) Vertebrates

FIGURE 18.22. (a) *Phylogenetic tree of the vertebrates. The vertical axis corresponds to the abundance of extant species in each group, with representative names given. The Sarcopterygii (lobe finned fishes) include coelacanths, lungfish, and tetrapods (amphibians, birds, reptiles, mammals); a more detailed tree of the tetrapods is presented in Fig. 18.25 below. The x axis shows the divergence times based on fossil records, which differ somewhat from estimates made by molecular sequence analyses. Redrawn from Venkatesh et al. (2007). Used with permission. (b) Phylogenetic tree of the teleosts showing the relationships of the first four sequenced fish genomes. Redrawn and modified from Kasahara et al. (2007). Used with permission.*

transposable elements is represented in *Takifugu*. The most common *Takifugu* repeat is the LINE-like element *Maui* (6400 copies), while in humans there are over one million copies of the most common repeat, *Alu*.

- Introns are relatively short. Seventy-five percent of *Takifugu* introns are <425 bp in length, while in humans 75% of introns are <2609 bp. In *Takifugu*, about 500 introns have a length greater than 10 kb, while in humans more than 12,000 introns are greater than 10 kb.

- Gene loci occupy about 108 Mb of the total euchromatic DNA (320 Mb). This represents about one-third of the genome, a far higher fraction than in mouse or human.

Genus, species (common name):   *Danio rerio*
Callorhinchus milii (elephantfish)
Danio rerio (zebrafish)
Gasterosteus aculeatus (three spined stickleback)
Oryzias latipes (Japanese medaka)
Takifugu rubripes (pufferfish)
Tetraodon nigroviridis (freshwater pufferfish)

Selected lineages: *Eukaryota; Metazoa; Chordata; Craniata; Vertebrata; Euteleostomi; Actinopterygii; Neopterygii; Teleostei; Ostariophysi; Cypriniformes; Cyprinidae; Danio; Danio rerio*

Eukaryota; Metazoa; Chordata; Craniata; Vertebrata; Euteleostomi; Actinopterygii; Neopterygii; Teleostei; Euteleostei; Neoteleostei; Acanthomorpha; Acanthopterygii; Percomorpha; Tetraodontiformes; Tetradontoidea; Tetraodontidae; Takifugu; Takifugu rubripes (pufferfish)

	Haploid genome size	GC content	Number of chromosomes	Number of genes	Entrez Genome ID
Callorhinchus milii	910 Mb			~60,000	18361
Danio rerio Tuebingen	1700 Mb		25		11776
Gasterosteus aculeatus	~500 Mb		22		13579
Oryzias latipes HNI	700 Mb		24	20,141	19569, 16702
Takifugu rubripes	390 Mb	~44%		20,796	1434
Tetraodon nigroviridis	342 Mb		21	27,918	12350

Key dates: divergence from human lineage: ~450 million years ago. *T. nigroviridis* and *T. rubripes* diverged 18–30 MYA.
Disease relevance: *D. rerio* and other fish provide models for human genetic diseases.
Genome features: Some fish species are characterized by relatively small genome sizes; *T. rubripes* is only one-eighth the size of the human genome (while having a comparable number of genes), and the medaka *O. latipes* (medaka) is half the size of zebrafish. The stickleback (*G. aculeatus*) has undergone extreme adaptive radiation in recent millennia by occupying different environments and acquiring diverse phenotypes, making it an excellent model organism in which to study adaptive evolution. Many fish, such as the rainbow trout *O. mykiss*, are recent tetraploids.
Websites: http://www.ensembl.org/ (for multiple fish genomes); http://zfin.org/ (zebrafish).

FIGURE 18.23. Overview of fish genomes. The D. rerio *image is from the NHGRI (▶ http://www. genome.gov/17516871).*

After the *Takifugu* genome was completed, Jaillon et al. (2004) reported the sequence of another pufferfish, *Tetraodon nigroviridis*. This permitted comparative analyses between *Tetraodon* and human (resulting in the prediction of ~900 novel human genes). A main focus of the genome analysis was on the evidence that telosts are descendants of an organism that underwent ancient whole genome duplication. This was followed by massive gene loss, as we described for separate whole genome duplication events in the fungi (Chapter 17). Jaillon et al. further inferred that the ancestral vertebrate genome had 12 chromosomes.

Yuji Kohara and colleagues (Kasahara et al., 2007) generated a draft sequence of the medaka. Upon comparing the four available fish genomes with the human genome they proposed a model of genome evolution in which a fish/human ancestor had 13 chromosomes. There are other models of the ancestral karyotype. However, there is a consensus that several whole genome duplications occurred in the teleost lineage (e.g., Van de Peer, 2004; Christoffels et al., 2004; Postlethwait, 2007). Once duplicate genes are identified within and between genomes (such as fish and human), the date of the duplication events can be estimated by using phylogenetic trees (e.g., neighbor-joining trees assuming a constant molecular clock). About one-third of the duplicated genes in *Takifugu* seem to derive from a whole genome duplication event that occurred ~320 MYA, as suggested by Ohno (1970). Approximately 1000 pairs of duplicated genes (paralogs) were identified in both *Tetraodon* and *Takifugu*, and based on K_s frequencies, 75% represent ancient duplications that occurred prior to the divergence of the *Takifugu* and *Tetraodon* lineages. Two other whole genome duplication events occurred earlier (at the time of divergence of jawless and jawed vertebrates, ~500 MYA) and more recently in the salmonid lineage ~50 MYA (reviewed in Postlethwait, 2007).

According to Kasahara et al. (2007), about 336 to 404 MYA the teleost lineage underwent a whole genome duplication, followed quickly (within 50 million years) by major chromosomsal rearrangements (fissions, fusions, translocations) and reduction to 24 chromosomes. Today, half of all teleosts have 24 or 25 chromosomes. Medaka is unusual because after its divergence from zebrafish its genome did not undergo rearrangements for ~300 million years.

310 Million Years Ago: Dinosaurs and the Chicken Genome

Our approach in this chapter has been to proceed through the tree of the eukaryotes in Fig. 18.1 moving towards the primates. When the chicken genome was sequenced by the International Chicken Genome Sequencing Consortium (2004) it provided a unique perspective on the human genome because those lineages diverged about 310 million years ago (Fig. 18.24). The chicken is far closer to humans than fish, but farther than the rodents (diverged ~80 MYA). Thus it provided an excellent distance for identifying highly conserved functional elements (Chapter 16). The genome is 1200 megabases and is organized in 38 autosomes and a pair of sex chromosomes (ZW is female and ZZ is male, that is, the female is heterogametic; chromosome W is extremely small). Thus the karyotype is $2n = 78$. The autosomes include many minichromosomes, typically having a high GC content, a high gene content, and very high recombination rates (a median value of 6.4 centmorgans [cM] per megabase; by comparison the human genome ranges from 1 to 2 cM/Mb and the mouse genome ranges from 0.5 to 1.0 cM/Mb).

A reason that the chicken genome is smaller than the human genome by a factor of three is that it has relatively few repetitive elements. Interspersed repeats occur as transposable elements in decay. There is no evidence for active short interspersed line

The red jungle fowl, for which the genome was sequenced, is the precursor to the domesticated chicken. NCBI offers a guide to chicken genome resources at ► http://www.ncbi.nlm.nih.gov/ projects/genome/guide/chicken/.

Genus, species:
Gallus gallus

Lineage: Eukaryota; Metazoa; Chordata; Craniata; Vertebrata; Euteleostomi; Archosauria; Dinosauria; Saurischia; Theropoda; Coelurosauria; Aves; Neognathae; Galliformes; Phasianidae; Phasianinae; Gallus; *Gallus gallus* (red jungle fowl)

	Haploid genome size	GC content	Number of chromos.	Number of genes	Entrez Genome ID
Gallus gallus	1200 Mb		39	20–23,000	13342

Key dates: divergence from human lineage: ~310 million years ago
Disease relevance: chicken is an important nonmammalian vertebrate model organism for studies of embryonic development, virus infection (the first tumor virus, Rous sarcoma virus, and the first oncogene, src were identified in the chicken).
Genome features: the genome is ~threefold smaller than other mammalian genomes, and has a relatively small proportion of interspersed repeat content. About 70 Mb of the sequence is alignable with human.

FIGURE 18.24. Overview of the chicken genome. Photograph from the NHGRI (► http://www.genome.gov/17516871).

elements (SINEs) in the past 50 million years, in contrast to their active roles in the human genome. Expansions and reductions of protein-coding gene families occur; for example, an avian-specific family of keratins are used to create claws, scales, and feathers. One surprising expansion is a family of 218 genes that are predicted to encode olfactory receptors and are orthologous to two human genes (*OR5U1* and *OR5BF1*).

180 Million Years Ago: The Opposum Genome

There are three main groups of mammals: (1) eutherians (placental mammals such as humans), (2) metatherians (marsupials) such as the opossum, koala, and kangaroo (3) prototherians such as the platypus (Fig. 18.25). The genome sequence of the gray, short-tailed opossum *Monodelphis domestica* is the first from the metatherians (Mikkelsen et al., 2007) (Fig. 18.26). Its genome size is comparable to that of humans, organized into eight autosomes (257 megabases to 748 megabases). These autosomes are extremely large (the shortest one is longer than the longest human one, chromosome 1). In contrast, the opossum X chromosome is extremely short (~76 megabases), smaller than that of any known eutherian.

The GC content of the *M. domestica* genome is 37.7%, lower than that of other amniote genomes (40.9% to 41.8%). Mikkelsen et al. note that the average recombination rate for the autosomes (~0.2 to 0.3 cM/Mb) is lower than in other amniotes, consistent with a model in which the genome has undergone limited recombination.

In eutherian mammals, females achieve dosage compensation of the X chromosome by the random inactivation of either the maternal or paternal X in female embryos. This is accomplished by an X inactivation center (XIC) that includes the *XIST* gene. Its RNA product coats and silences one X chromosome copy. In contrast, metatherian mammals such as the opossum inactivate the paternal X. Mikkelsen et al. found no evidence for *XIST* in the opossum genome. While the human

FIGURE 18.25. *Phylogenetic tree depicting mammalian genomes. The genomes of some of these organisms have been sequenced, and the others are beginning to be sequenced. Note that the branch lengths of the rat and mouse lineages are long relative to other members of the clade containing humans (the Euarchontoglires), reflecting a faster evolutionary rate. The data are available at* ▶ *http://www.nisc.nih.gov/data/ and in Margulies et al. (2005).*

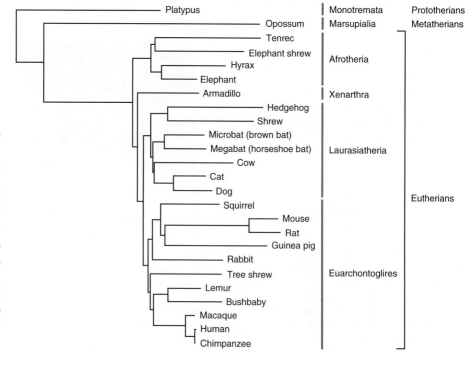

Lineage: Eukaryota; Metazoa; Chordata; Craniata; Vertebrata; Euteleostomi; Mammalia; Metatheria; Didelphimorphia; Didelphidae; Monodelphis; *Monodelphis domestica* (gray short-tailed opossum)

	Haploid genome size	GC content	Numberof chromos.	Numberof genes	Entrez Genome ID
Monodelphis domestica	3600 Mb	37.7%	9	18–20,000	12561

Selected divergence dates: The marsupial lineage diverged from the human lineage ~180 million years ago (MYA).
Genome features: the autosomes are extremely large (the smallest, at 257 Mb, is larger than human chromosome 1).
Disease association: *M. domestica* is a model for radiation-induced malignant melanoma. Newborn opposums are unique in their ability to recover from complete transections of the spinal cord.
Organism-specific web resources: http://www.broad.mit.edu/mammals/opossum

FIGURE 18.26. *Overview of the genome of the short-tailed opossum* Monodelphis domestica, *a marsupial. Photograph from the NHGRI (► http://www.genome.gov/17516871).*

X chromosome has undergone remarkably little change since the eutherian radiation ~100 million years ago, the opossum X chromosome has undergone large-scale rearrangments (affecting the XIC and X-linked pseudoautosomal region).

The predicted gene content of *M. domestica* (18,000 to 20,000) is comparable to that of humans, with relatively small numbers of organism-specific genes. Conserved noncoding elements (CNEs; Chapter 16), rather than genes, comprise the majority of the well-conserved sequence elements.

100 Million Years Ago: Mammalian Radiation from Dog to Cow

A spectacular radiation of mammalian species approximately 100 million years ago (Fig. 18.25). In creating that tree, Margulies et al. (2005) considered the question of how many separate genomes need to be sequenced, even at low coverage, in order to identify highly conserved regions. Eddy (2005) estimated that for a given element of length 50 bases, it is sufficient to compare just human and mouse sequences (having a total branch length D of 0.45). For an element having a length of 8 bases it is necessary to have $D \approx 4$ (e.g. having 40 species each with branch length 0.1 from a root). Margulies et al. estimated the divergence of each species from human, calibrating the mouse/human distance of 0.45 (Table 18.10). The values in this table are branch lengths. Column A of the table shows the divergence from human; some organisms such as the rhesus macaque are closely related (branch length 0.05) while others are very divergent (e.g. opossum, ~0.95). Column B shows the increase in total divergence D that is added, if the genome sequence of each successive species (i.e. each row) is added to the total set of sequenced genomes. Column C shows the total divergence D which would approach a value of 4.6 once all these genomes are sequenced. For example, by sequencing the rat genome (row 3) in addition to human and mouse the total divergence rises from 0.45 to 0.53. The purpose of this analysis is to devise a strategy for choosing genomes to sequence to systematically identify conserved elements (such as gene regulatory sequences) that are likely to be functional. This depends on the length of the element, the extent of its conservation, and its rate of evolution (Kellis et al., 2003; Eddy, 2005).

TABLE 18-10   Divergence of Mammalian Species in Fig. 18.25 from Human

Species	(A) Divergence from Human	(B) Divergence Added	(C) Total Divergence
Human	0	0	0
Mouse	0.45	0.45	0.45
Rat	0.456016	0.0800223	0.530022
Chimpanzee	0.00870256	0.00385479	0.533877
Dog	0.308722	0.187998	0.721875
Rhesus macaque	0.0506105	0.0235448	0.74542
Opossum	0.945002	0.810721	1.55614
Cow	0.362861	0.202799	1.75894
Elephant	0.322285	0.161317	1.92026
Armadillo	0.306693	0.156254	2.07651
Free-tailed bat	0.289354	0.131334	2.20785
Cat	0.292095	0.0815974	2.28944
Ring-tailed lemur	0.225151	0.118897	2.40834
Shrew	0.413912	0.26018	2.66852
Tenrec	0.483987	0.277295	2.94581
Caviomorph	0.423043	0.26171	3.20752
Bushbaby	0.277728	0.136447	3.34397
Hedgehog	0.437285	0.241742	3.58571
Mousette fruit bat	0.293293	0.106385	3.6921
Sciurid	0.299576	0.147832	3.83993
Tree shrew tupaia	0.300747	0.183026	4.02296
Hyrax	0.395407	0.162658	4.18562
Rabbit	0.309612	0.178759	4.36437
Short-eared elephant shrew	0.441402	0.225011	4.58939

Divergence corresponds to branch lengths in units of substitutions per site.
Data are from Margulies et al. (2005) and ▶ http://www.nisc.nih.gov/data/.

The genomes of many of these organisms are now being sequenced. Following a 1.5x coverage of the dog genome by Craig Venter and colleagues (Kirkness et al., 2003), Lindblad-Toh et al. (2005) reported a high quality draft genome sequence. There are ~400 modern dog breeds and many have a high prevalence of particular diseases due to breeding. A boxer was selected because that breed has relatively high homozygosity. There are fewer predicted genes than in human (see Fig. 18.27), and almost all dog genes have human counterparts. Lindblad-Toh et al. noted that many functionally related genes that undergo rapid acceleration in the human genome relative to mouse have been proposed to represent human-specific innovations, but often these genes also show rapid evolution in the canine lineage. Thus, it is important to be cautious about the significance of changes in comparative studies with limited numbers of species.

80 Million Years Ago: The Mouse and Rat

The sequencing and analysis of the mouse genome represents a landmark in the history of biology. Following the human, the mouse is the second mammal to have its

Lineage: Eukaryota; Metazoa; Chordata; Craniata; Vertebrata;
Euteleostomi; Mammalia; Eutheria; Laurasiatheria; Carnivora;
Caniformia; Canidae; Canis; *Canis lupus familiaris* (dog)

Haploid genome size	GC content	Number of chromos.	Number of genes	Entrez Genome ID
2500 Mb	41%	39	19,300	13179

Selected divergence dates: The Canidae (dogs) include 34 closely related species that diverged in the past ~10 million years.
Genome features: About 5.3% of the human and dog lineages contain functional elements that have been under purifying constraint. These have almost all been retained in the mouse as well.

FIGURE 18.27. Overview of the dog genome. The photograph is of a boxer (Tasha) whose genome was sequenced. Photograph from the NHGRI website (▶ http:// www.genome.gov/17516871).

genome sequenced. The mouse is an excellent model for understanding human biology (Fig. 18.28):

- Remarkably, although these two organisms diverged about 80 MYA, only about 300 of the annotated genes in the mouse genome have no counterpart in the human genome.

- In addition to sharing thousands of orthologous protein-coding genes, the mouse and human genomes have large tracts of homologous non-protein-coding DNA. These conserved sequences provide insight into regulatory regions of the genome or noncoding genes (Hardison et al., 1997; Dermitzakis et al., 2002).

Lineage: Eukaryota; Metazoa; Chordata; Craniata; Vertebrata;
Euteleostomi; Mammalia; Eutheria; Euarchontoglires; Glires;
Rodentia; Sciurognathi; Muroidea; Muridae; Murinae; Mus;
Mus musculus

Eukaryota; Metazoa; Chordata; Craniata; Vertebrata;
Euteleostomi; Mammalia; Eutheria; Euarchontoglires;
Glires; Rodentia; Sciurognathi; Muroidea; Muridae; Murinae;
Rattus; *Rattus norvegicus* (see photo)

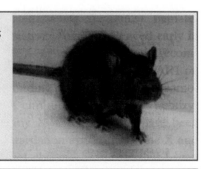

	Haploid genome size	GC content	Number of chromos.	Number of genes	Entrez Genome ID
Mus musculus	2600 Mb	42%	20	23,049	13183
Rattus norvegicus	2750 Mb	~42%	21	20,973	10629

Selected divergence dates: Mouse and rat last shared a common ancestor 12–24 MYA. The rodent lineage diverged from the human lineage ~80 MYA.
Genome features: About 5% of the human and rodent lineages contain functional elements that have been under purifying constraint.
Disease relevance. There are over 450 inbred mouse strains, and many of these serve as disease models. Knockouts and other manipulations of mouse genes allow studies of human diseases. Rats (like mice) are host to many pathogens, and are carriers for over 70 human diseases.
Web resources: Mouse Genome Informatics (http://www.informatics.jax.org/);
Rat Genome Database (http://rgd.mcw.edu/wg/)

FIGURE 18.28. Overview of rodent genomes. The photograph of a rat is from the NHGRI website (▶ http://www.genome. gov/17516871).

- The mouse and human share many physiological features. Thus, mice make an important model for hundreds (or thousands) of human diseases, from infectious diseases to complex disorders.

- There are over 1000 mouse strains having spontaneous mutations. Mutations can be introduced into the mouse through random mutagenesis approaches such as chemical mutagenesis or radiation treatment (Chapter 12). Mutations and other genetic modifications can also be introduced through directed approaches such as transgenic, knock-out, and knock-in technologies.

Two groups independently sequenced the mouse genome: the Mouse Genome Sequencing Consortium (Waterston et al., 2002) and Celera Genomics. These versions of the genome were directly compared by Xuan et al. (2003). They selected over 8300 mouse entries having RefSeq accession numbers as queries and used BLAT (Chapter 5) to compare the coverage and accuracy of the two assemblies. Most mRNAs were matched with both assemblies.

Waterston et al. (2002) sequenced the genome of a female mouse of the B6 strain. The sequencing strategy entailed a combined whole-genome shotgun approach (with sevenfold coverage) and a hierarchical shotgun approach (with sequencing of BAC clones that were physically mapped to chromosomes). The assembly covers most of the mouse genome. Of the RefSeq cDNAs, 99.3% could be aligned to the genomic sequence. Also, the Waterston et al. (2002) assembly closely matches an independent draft sequence of mouse chromosome 16 (Mural et al., 2002).

Waterston et al. (2002) described 11 main conclusions of the mouse genome-sequencing project:

> Sequencing a female assured equal coverage of the X chromosome and each autosome. The Y chromosome in both human and mouse is small and contains highly repetitive DNA elements (see Chapter 19).
> The GenBank accession number of the mouse genome is CAAA01000000. It is accessible through the three main genome browser sites: ▶ http://www.ensembl.org/Mus_musculus/, ▶ http://genome.ucsc.edu/, and ▶ http://www.ncbi.nlm.nih.gov.

1. The total length of the euchromatic mouse genome is 2.5 Gb in size, about 14% smaller than the human genome (2.9 Gb). In contrast to other, more compact genomes we have discussed, the mouse genome (like the human genome) averages about one gene every 100,000 bp of genomic DNA. The GC content is comparable, with mean values of 42% (mouse) versus 41% (human). There are 15,500 CpG islands, about half the number observed in humans (see Chapter 19).

2. Over 90% of the mouse and human genomes can be aligned into conserved synteny regions. After the divergence of mouse and human about 75 to 80 MYA, chromosomal DNA was shuffled in each species. However, large regions of DNA obviously correspond. As an example of how to visualize this, the EBI offers a human/mouse conserved synteny viewer that we will explore in Chapter 19.

3. About 40% of the human genome can be aligned to the mouse genome at the nucleotide level. This represents most of the orthologous sequence shared by these genomes. For 12,845 orthologous gene pairs, 70.1% of the corresponding amino acid residues were identical.

4. The neutral substitution rate in each genome can be estimated by comparing thousands of repetitive DNA elements to the inferred ancestral consensus sequence. The average substitution rate is 0.17 per site in humans and 0.34 per site in mouse. The mouse genome also shows a twofold higher rate of acquisition of small (less than 50 bp) insertions and deletions.

> The mouse sequencing consortium (Waterston et al., 2002, p. 526) defined a syntenic segment as "a maximal region in which a series of landmarks occur in the same order on a single chromosome in both species." They identified 558,000 orthologous and highly conserved landmarks in the mouse assembly, comprising 7.5% of the mouse assembly.

5. The proportion of small (50 to 100 bp) segments in the mammalian genome that is under purifying selection is about 5%. This is estimated by

comparing the neutral rate to the extent of sequence conservation in the genome. Since this 5% value is greater than the proportion of protein-coding genes in the genome, genomic regions that do not code for genes must be selected for, such as regulatory elements. Regulatory regions such as those that control liver-specific and muscle-specific expression were conserved between mouse and human to an extent greater than regions of neutral DNA, although less than regions that are protein coding.

6. The mammalian genome is evolving in a nonuniform manner, with variation in the rates of sequence divergence across the genome. The neutral substitution rate varied across all chromosomes (and was lowest on the X chromosome), with a higher substitution rate associated with extremes of GC content.

7. The mouse and human genomes were each proposed to contain about 30,000 protein-coding genes. (Note that these 2002 estimates have been revised with ongoing annotation and comparative genomics efforts, as summarized in Fig. 18.28.) About 80% of mouse genes have a single identifiable human ortholog. Less than 1% of human genes have no identifiable ortholog in the mouse, and vice versa. The sequencing effort revealed the existence of 9000 previously unknown mouse genes as well as 1200 new human genes.

8. Dozens of local gene family expansions have occurred in the mouse genome, such as the olfactory receptor gene family. About 20% of this family are pseudogenes in mouse, suggesting a dynamic interplay between gene expansion and gene deletion. The lipocalins also underwent a mouse lineage-specific expansion. For example, the mouse X chromosome contains a cluster of genes related to odorant-binding protein that are absent in humans. Such expansions may account in part for the physiological differences between primates and rodents in terms of reproductive processes, feeding, or other behaviors.

9. Particular proteins evolve at a rapid rate in mammals. For example, genes involved in the immune response appear to be under positive selection, which drives their evolution.

10. Similar types of repetitive DNA sequences are found in both human and mouse. We will discuss human repetitive sequences in Chapter 19.

11. The public consortium described 80,000 single-nucleotide polymorphisms (SNPs). We introduced SNPs in Chapter 16 and will discuss them further in Chapter 19.

Nadeau and Taylor (1984) estimated that there are about 180 conserved synteny regions of the mouse and human genomes. Waterston et al. (2002) provided evidence for 342 such regions, each greater than 300 kb in size.

A fundamental problem is to understand the genetic variation that underlies the phenotype differences of different mouse strains. Frazer et al. (2007) resequenced the genomes of 15 mouse subspecies or strains. These included four wild-derived strains (*M. m. musculus*, *M. m. castaneus*, *M. m. domesticus*, and *M. m. molossinus*). They also sequenced 11 wild-derived strains which were genetically more pure because they have been bred to homozygosity. Frazer et al. resequenced almost 1.5 billion bases (58%) of these genomes and, by comparing them to the reference strain C57BL/6J, they identified 8.3 million SNPs. (The false positive rate of discovery was 2%, the accuracy of genotype calls was >99%, and the false negative rate was assessed as roughly half.) They generated a haplotype map across the mouse genome, defining ancestry breakpoints at which pairwise comparisons indicated a transition to (or from) high SNP densities. The genomewide SNP map included over 40,000 segments with

The SNP resequencing data are available at NCBI and at ▶ http://mouse.perlegen.com/mouse/. The Perlegen website includes a genome browser that displays haplotype blocks. As an example, enter a query hbb-b1 to see the beta globin locus on mouse chromosome 7 and the associated haplotype blocks across all 15 mouse strains.

an average length of 58 kilobases and a range of 1 kilobase to 3 megabases. The significance of this project is that it describes the genetic basis of variation in these 15 strains, all of which have unique properties such as behaviors or disease susceptibilities.

The most comprehensive mouse resource on the World Wide Web is the Mouse Genome Informatics (MGI) database and its associated sites (see Chapter 12) (Blake et al., 2003).

Rats and mice last shared a common ancestor about 12 to 24 MYA. The Rat Genome Sequencing Project Consortium (2004) described a high-quality draft genome sequence of the Norway rat, allowing comparisons of the rat, mouse, and human genomes. All have comparable sizes (2.6 to 2.9 billion bases) and encode similar numbers of genes (see Fig. 18.28). Some properties differ: segmental duplications span over 5% of the human genome (Chapters 16 and 19) but just 3% of the rat genome and 1% to 2% of the mouse genome. About 40% of the euchromatic rat genome (or ~1 billion bases) aligns to orthologous regions of both mouse and human, containing most exons and known regulatory elements. A portion of this alignable sequence, spanning about 5% of each genome, is under selective constraint (negative selection) while the remainder evolves at the neutral rate. Another 30% of the rat genome aligns only with the mouse but not human, and is largely comprised of rodent-specific repeats.

The rodent lineage is evolving at a faster rate than the human lineage, as indicated by the longer rodent branch lengths in Fig. 18.25. This includes a threefold higher rate of nucleotide substitution in neutrally evolving DNA, based on analyses of repetitive elements shared since the last common ancestor of humans and rodents.

MGI is available at ▶ http://www.informatics.jax.org and is operated by the Jackson Laboratory (▶ http://www.jax.org). MGI has multiple components, including the Mouse Genome Database (MGD), the Gene Expression Database (GDX), the Mouse Genome Sequencing (MGS) project, and the Mouse Tumor Biology (MTB) database (▶ http://www.informatics.jax.org/mtb). Web document 18.3 at ▶ http://www.bioinfbook.org/chapter18 lists additional web resources for the study of the mouse and rat.

5 to 50 Million Years Ago: Primate Genomes

How did humans evolve from other primates? What features of the human genome account for our distinct traits, such as language and higher cognitive skills? A comparison of several primate genomes may elucidate the molecular basis of our unique traits—or, depending on one's perspective, such a comparison may highlight how closely similar we are to the great apes at a genetic level.

For an overview of primates, we can begin by making a phylogenetic tree with lysozyme protein sequences (Fig. 18.29). The chimpanzee (*Pan troglodytes*) and the bonobo (pygmy chimpanzee, *Pan paniscus*) are the two species most closely related to humans. These three species diverged from a common ancestor 5.4 ± 1.1 MYA, based on analyses of 36 nuclear genes (Stauffer et al., 2001). Our next closest species is the gorilla, which diverged an estimated 6.4 ± 1.5 MYA. Next in the branching order are the orangutan *Pongo pygmaeus* (11.3 ± 1.3 MYA) and the gibbon (14.9 ± 2.0 MYA) (Stauffer et al., 2001). The hominoids diverged from the Old World monkeys (e.g., the macaque and baboon) 23 MYA, close to the age of the earliest extant hominoid fossils. New World monkeys (such as the tamarin) are even more distantly related.

Following humans, the next two genomes to be sequenced were the chimpanzee and the rhesus macaque (Fig. 18.30). The Chimpanzee Sequencing and Analysis Consortium (2005) described the genome sequence of Clint, a captive-born male. By comparing a human reference to an individual chimpanzee, the analysis focused on those relatively few differences that could be found. (In contrast, comparisons of the human genome to the fish or chicken focused on the relatively few similarities that could be detected, such as ultraconserved regions or coding sequences.) The

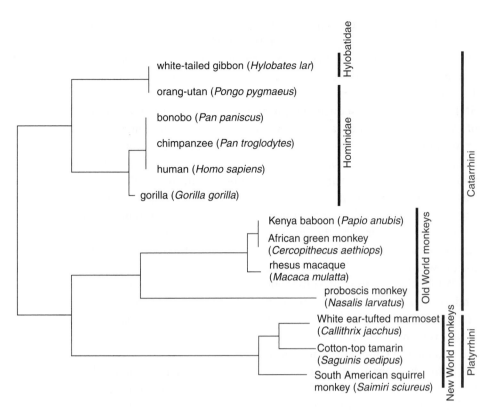

FIGURE 18.29. Phylogeny of the primates. A neighbor-joining tree representing primate phylogeny based on lysozyme protein sequences. These sequences were aligned using ClustalW and displayed as a neighbor-joining tree. The accession numbers are as follows: gibbon (P79180), orangutan (P79180), bonobo (AAB41214), chimpanzee (AAB41209), human (P00695), gorilla (P79179), Kenya baboon (P00696), African green monkey (P00696), rhesus macaque (P30201), proboscis monkey (P79811), marmoset (P79158), tamarin (P79268), and South American squirrel monkey (P79294). The sequences are available in web document 18.4 at ▶ http://www.bioinfbook.org/chapter18.

Lineage: Eukaryota; Metazoa; Chordata; Craniata; Vertebrata; Euteleostomi; Mammalia; Eutheria; Euarchontoglires; Primates; Haplorrhini; Catarrhini; Hominidae; Gorilla; *Gorilla gorilla*

... Catarrhini; Cercopithecidae; Cercopithecinae; Macaca; *Macaca mulatta* (rhesus macaque)

...Catarrhini; Hominidae; Pan; *Pan troglodytes* (chimpanzee)

...Catarrhini; Hominidae; Pongo; *Pongo pygmaeus* (orangutan)

Pongo pygmaeus

	Haploid genome size	GC content	Number of chromos.	Number of genes	Entrez Genome ID
Gorilla gorilla					18247
Homo sapiens	3038 Mb	40.7%	23	~20–25,000	13178
Macaca mulatta	2870 Mb	40.7%	21	~20,000	12537
troglodytes	3350 Mb	~41%	24	~20–25,000	13184
Pongo pygmaeus					18245

Selected divergence dates: The rhesus macaque and human lineages diverged ~25 MYA; chimpanzee and human lineages diverged ~6 MYA, also at the time of divergence from the bonobo.

Genome features: In aligned regions, DNA shares ~98% identity from chimp to human, and 93.5% identity from macaque to human. High confidence macaque–human orthologs share an average of 97.5% identity.

Disease relevance. Macaques are a widely used model for human disease because of their recent divergence (25 MYA rather than 80 MYA for rodents), similar anatomy, physiology, susceptibility to infectious agents related to human pathogens.

Web resources: see the Ensembl database at http://www.ensembl.org.

FIGURE 18.30. Overview of primate genomes. The photograph of an orangutan is from the NHGRI website (▶ http://www.genome.gov/17516871).

assembly represents a consensus of two haplotypes from the diploid individual (with one allele from heterozygous sites arbitrarily selected for the assembled sequence); the situation is similar to that of the first sequence of a diploid human individual genome (Chapter 19). Nucleotide divergence was found to occur at a mean rate of 1.23%, with 35 million SNPs catalogued (including ~1.7 million high quality SNPs determined by sequencing portions of seven additional chimpanzees). Most of these changes reflect random genetic drift rather than being shaped by positive or negative selection pressures. The 1.23% nucleotide divergence rate includes both fixed divergence between humans and chimpanzees (~1.06%) and polymorphic sites within each species. Variation in the nucleotide substitution rates was especially prominent in subtelomeric regions. Of all the observed substitutions, substitutions at CpG dinucleotide sites were most common. Considering the chromosomes separately, the human/chimpanzee divergence is greatest for the Y chromosome (1.9%, perhaps reflecting the greater mutation rate in male) and least for the well-conserved X chromosome (0.94% divergence).

While the number of substitutions is large (35 million), insertion/deletion (indel) events are notable for being fewer (~5 million events) but spanning more of the genomes (there are 40 to 45 megabases of species-specific euchromatic DNA, totaling ~90 megabases and corresponding to a ~3% difference between the human and chimpanzee genomes).

While humans have a haploid set of 23 chromosomes, chimpanzees have one more, reflecting the fusion of two chromosomes corresponding to chimpanzee 2a and 2b. Additionally there have been nine pericentric inversions (Chapter 16). Many other features have been characterized. Among the repetitive elements, SINEs have been threefold more active in humans, while several new retroviral elements (PtERV1, PtERV2) have invaded the chimpanzee genome selectively. Most of the protein-coding genes are highly conserved, with ~29% being identical. However, 585 out of 13,454 chimpanzee-human ortholog pairs have a K_N/K_S ratio greater than 1, suggestive of positive selection. These include glycophorin C, which mediates a *P. falciparum* invasion pathway in human erythrocytes, and granulysin which is involved in defense against pathogens such as *Mycobacterium tuberculosis* (Chapter 15).

Accessions for glycophorin C are NM_002101, NP_002092 (human) and XM_001135559, XP_001135559 (chimpanzee).

A comparison of sequences in humans and chimpanzees does not reveal which genes or other elements evolved rapidly. A phylogenetic reconstruction is necessary in order to infer lineage-specific changes that occurred, leading to the present-day sequences that we can observe. This is one reason that the sequencing of the second nonhuman primate, the rhesus macaque *Macaca mulatta*, was so significant. The rhesus macaque is an Old World monkey (superfamily Cercopithecoidea, family Cercopithecidae; Fig. 18.30) that diverged from the human/chimpanzee lineage ~25 million years ago. Its DNA has an average nucleotide identity of ~93% compared to human, in contast to the ~99% identity between human and chimpanzee. The Rhesus Macaque Genome Sequencing and Analysis Consortium (2007) sequenced the genome using whole genome shotgun sequences. They predicted ~20,000 genes, of which high-confidence orthologs share 97.5% identity to human sequences at the DNA and protein levels. Using the macaque as an outgroup, it was possible to analyze many features of the human and chimpanzee genomes. For example, of the nine pericentric inversions that occurred, seven could now be assigned to the chimpanzee lineage and two to the humans (on chromosomes 1 and 18).

Primate resources are listed in web document 18.5 at ► http://www.bioinfbook.org/chapter16. The genome browsers at Ensembl (► http://www.ensembl.org), UCSC (► http://genome.ucsc.edu), and NCBI are important starting points.

The sequencing consortium detailed many features of the rhesus macaque genome, including that 66.7 megabases (2.3%) consist of segmental duplications, and there are many lineage-specific expansions and contractions of gene families. Ultimately, as for other genome sequencing projects outlined in this chapter, this may permit the analysis of the cellular processes that ultimately underlie the unique biology of this primate.

Perspective

One of the broadest goals of biology is to understand the nature of each species of life: What are the mechanisms of development, metabolism, homeostasis, reproduction, and behavior? Sequencing of a genome does not answer these questions directly. Instead, we must first try to annotate the genome sequence in order to estimate its contents, and then we try to interpret the function of these parts in a variety of physiological processes.

The genomes of representative species from all major eukaryotic divisions are now becoming available. This will have dramatic implications for all aspects of eukaryotic biology. For pathogenic organisms, it is hoped that the genome sequence will lead to an understanding of their cellular mechanisms of toxicity, their mechanisms of host immune system evasion, and their pharmacological response to drug treatments. From studies of evolution, we will further understand mutation and selection, the forces that shape genome evolution. The reconstruction of ancestral karyotypes is a newly emerging discipline.

As complete genomes are sequenced, we are becoming aware of the nature of noncoding and coding DNA. Major portions of the eukaryotic genomic landscape are occupied by repetitive DNA, including transposable elements. The number of protein-coding genes varies from several thousand in fungi to tens of thousands in plants and mammals. Many of these protein-coding genes are paralogous within each species, such that the "core proteome" size is likely to be on the order of 10,000 genes for many eukaryotes. New proteins are invented in evolution through expansions of gene families or through the use of novel combinations of DNA encoding protein domains.

Pitfalls

An urgent need in genomics research is the continued development of algorithms to find protein-coding genes, noncoding RNAs, repetitive sequences, duplicated blocks of sequence within genomes, and conserved syntenic regions shared between genomes. We may then characterize gene function in different developmental stages, body regions, and physiological states. Through these approaches we may generate and test hypotheses about the function, evolution, and biological adaptations of eukaryotes. Thus we may extract meaning from the genomic data.

We are now in the earliest years of the field of genomics. Many new lessons are emerging:

- Draft versions of genome sequences are extremely useful resources, but gene annotation often improves dramatically as a sequence becomes finished.

- It is extraordinarily difficult to predict the presence of protein-coding genes in genomic DNA ab initio. It is important to use complementary experimental data on gene expression, such as expressed sequence tag information, and comparative genomics to align orthologous sequences has become the norm.

- We still know relatively little about the nature of noncoding RNA molecules, but comparative genomic studies have demonstrated their conservation across hundreds of millions of years of evolution (e.g., between opossum and human).

- Large portions of eukaryotic genomes consist of repetitive DNA elements.

- Comparative genomics is extraordinarily useful in defining the features of each eukaryotic genome.

Most publications describing genomes (both eukaryotic and prokaryotic) define orthologs as descended by speciation from a single gene in a common ancestor. Typically, the predicted proteins from an organism are searched by BLAST against the complete proteome of other species using an E value cutoff such as 10^{-4}. However, two orthologous proteins could have species-specific functions. Thus it is appropriate to remain cautious in assigning functions to genes.

WEB RESOURCES

We have presented key resources for many eukaryotic organisms and their genome-sequencing websites. An excellent starting point is the Ensembl website (► http://www.ensembl.org/), which currently includes gateways for the mouse, rat, zebrafish, fugu, mosquito, and other genomes. The Department of Energy Joint Genome Institute (DOE JGI) includes web resources for many of the organisms discussed in this chapter (► http://genome.jgi-psf.org/euk_home.html).

DISCUSSION QUESTIONS

[18-1] If there were no repetitive DNA of any kind, how would the genomes of various eukaryotes (human, mouse, a plant, a parasite) compare in terms of size, gene content, gene order, nucleotide composition, or other features?

[18-2] Web document 18.6 at ► http://www.bioinfbook.org/chapter18 consists of a word document with 256,157 bases of DNA from a eukaryotic genome in the fasta format. How could you identify the species? Assume you cannot use BLAST to directly identify the species. Also, the accession number is given so that you can later look up the species, but assume you cannot use that information at first. What features distinguish the genomic DNA sequence of a protozoan parasite from an insect, or a plant from a human, or one fish from another?

PROBLEMS/COMPUTER LAB

[18-1] A universal minicircle binding protein (GenBank accession A54598) has been purified from a trypanosome that infects insects, *Crithidia fasciculata*. A blastp search reveals that there are homologous proteins in plants, fungi, and metazoans (such as the worm *Caenorhabditis elegans*). How is this protein named in various organisms? What is its presumed function? What is its domain called in the Conserved Domain Database?

[18-2] *Leishmania major* has repetitive DNA elements (e.g., accession AF421497). How can you decide how common this element is and where it is localized (e.g., to a particular chromosome or to a chromosomal region).

[18-3] The green algae (such as *Chlamydomonas* and *Ostreococcus*) are Viridiplantae that share some genes in common with the animals but not the angiosperms. Use TaxPlot at NCBI (from the home page, select Tools on the left sidebar). Set the query genome to *Ostreococcus lucimarinus*, then set the comparison genomes to *Homo sapiens* and *Arabidopsis thaliana* (as examples of an animal and a plant). Several proteins are dramatically absent from either human or *Arabidopsis*. What are they? What is their function?

SELF-TEST QUIZ

[18-1] The *Giardia lamblia* genome is unusual because

    (a) It contains hardly any transposable elements or introns

    (b) It is circular

    (c) It contains extremely little nonrepetitive DNA

    (d) Its AT content is nearly 80%

[18-2] The genome of the trypanosome *T. brucei*

    (a) Has an intricate network of circular rings of genomic DNA

    (b) Almost completely lacks introns

    (c) Almost completetly lacks pseudogenes

    (d) Varies in size by up to 25% in different isolates

[18-3] The genome of the malaria parasite *Plasmodium falciparum* is notable for having a AT content of 80.6%. Which amino acids are overrepresented in its encoded proteins?

    (a) F, L, I, Y, N, K

    (b) F, L, I, Y, V, M

    (c) A, P, C, G, T, R

    (d) N, S, Y, I, M, H

[18-4] The *Paramecium tetraurelia* genome has the following properties except which one?

    (a) It has about 800 macronuclear chromosomes.

    (b) It has two nuclei, each with distinct functions.

    (c) Its genome encodes about twice as many proteins as the human genome.

    (d) It has undergone whole genome duplication with massive gene loss.

[18-5] Plant genomes from species such as *Arabidopsis* (125 Mb) and the black cottonwood tree *Populus trichocarpa* (480 Mb) were selected because they are relatively small. Nonetheless, each of these genomes is characterized by large amounts of repetitive DNA, and each whole genome duplicated one or more times.

    (a) True

    (b) False

[18-6] Which of these pairs of organisms diverged the longest time ago?

    (a) *Caenorhabditis elegans* and *Caenorhabditis briggsae*

    (b) *Drosophila melanogaster* (fruitfly) and *Anopheles gambiae* (mosquito)

    (c) *Homo sapiens* and *Canis familiaris* (dog)

    (d) *Arabidopsis thaliana* and *Oryza sativa* (rice)

[18-7] What do the *Takifugu rubripes* (pufferfish) and *Gallus gallus* (chicken) genomes have in common that distinguishes them from the human genome?

    (a) They have genome sizes 3- to 10-fold smaller than that of human, but a comparable number of genes.

    (b) They have a smaller total genome size but dozens more chromosomes.

    (c) They have smaller genome sizes and approximately half as many protein-coding genes.

    (d) They have a series of minichromosomes of variable size.

[18-8] How are the mouse and human genomes different?

    (a) The mouse genome has a lower GC content.

    (b) The mouse genome has more protein-coding genes.

    (c) The mouse genome has undergone specific expansions of genes encoding particular protein families such as olfactory receptors.

    (d) The mouse genome has fewer telomeric repeats per chromosome, on average.

[18-9] Many features distinguish the chimpanzee and human genomes, including all of the following except which one?

    (a) Chimpanzees have more chromosomes.

    (b) About 35 million nucleotide substitutions have been described.

    (c) There have been hundreds of pericentric inversions.

    (d) Over 500 chimpanzee-human ortholog pairs may be under positive selection.

SUGGESTED READING

We presented a phylogentic tree from Bauldauf et al. (2000). For a more recent evolutionary analysis of eukaryotic evolution, including a discussion of models of eukaryotic origins and the role of mitochondria, see Embley and Martin (2006). For a brief review of the significance of Apicomplexan genome projects, see Roos (2005). Paterson (2006) provides an excellent overview of plant genomics.

For a review of the methods applied to sequencing a particular eukaryotic genome, including BAC and whole genome shotgun sequencing and assembly issues, see the review of the sea urchin project by Sodergren et al. (2006).

REFERENCES

Abrahamsen, M. S., et al. Complete genome sequence of the apicomplexan, *Cryptosporidium parvum*. *Science* **304**, 441–445 (2004).

Adam, R. D. Biology of *Giardia lamblia*. *Clin. Microbiol. Rev.* **14**, 447–475 (2001).

Adams, M. D., et al. The genome sequence of *Drosophila melanogaster*. *Science* **287**, 2185–2195 (2000).

Akman, L., et al. Genome sequence of the endocellular obligate symbiont of tsetse flies, *Wigglesworthia glossinidia*. *Nat. Genet.* **32**, 402–407 (2002).

Allen, K. D. Assaying gene content in *Arabidopsis*. *Proc. Natl. Acad. Sci. USA* **99**, 9568–9572 (2002).

Aparicio, S., et al. Whole-genome shotgun assembly and analysis of the genome of *Fugu rubripes*. *Science* **297**, 1301–1310 (2002).

Arabidopsis Genome Initiative. Analysis of the genome sequence of the flowering plant *Arabidopsis thaliana*. *Nature* **408**, 796–815 (2000).

Arkhipova, I., and Meselson, M. Transposable elements in sexual and ancient asexual taxa. *Proc. Natl. Acad. Sci. USA* **97**, 14473–14477 (2000).

Arkhipova, I. R., and Morrison, H. G. Three retrotransposon families in the genome of *Giardia lamblia:* Two telomeric, one dead. *Proc. Natl. Acad. Sci. USA* **98**, 14497–14502 (2001).

Armbrust, E. V., et al. The genome of the diatom *Thalassiosira pseudonana*: Ecology, evolution, and metabolism. *Science* **306**, 79–86 (2004).

Aury, J. M., et al. Global trends of whole-genome duplications revealed by the ciliate *Paramecium tetraurelia*. *Nature* **444**, 171–178 (2006).

Baldauf, S. L., Roger, A. J., Wenk-Siefert, I., and Doolittle, W. F. A kingdom-level phylogeny of eukaryotes based on combined protein data. *Science* **290**, 972–977 (2000).

Baum, D. A., Smith, S. D., and Donovan, S. S. Evolution. The tree-thinking challenge. *Science* **310**, 979–980 (2005).

Bennetzen, J. L., and Freeling, M. The unified grass genome: Synergy in synteny. *Genome Res.* **7**, 301–306 (1997).

Berriman, M., et al. The genome of the African trypanosome *Trypanosoma brucei*. *Science* **309**, 416–422 (2005).

Blake, J. A., Richardson, J. E., Bult, C. J., Kadin, J. A., and Eppig, J. T. MGD: The Mouse Genome Database. *Nucleic Acids Res.* **31**, 193–195 (2003).

Blaxter, M. *Caenorhabditis elegans* is a nematode. *Science* **282**, 2041–2046 (1998).

Blaxter, M. Molecular systematics: Counting angels with DNA. *Nature* **421**, 122–124 (2003).

Brayton, K. A., et al. Genome sequence of *Babesia bovis* and comparative analysis of apicomplexan hemoprotozoa. *PLoS Pathog.* **3**, 1401–1413 (2007).

Brooks, D. R., and Isaac, R. E. Functional genomics of parasitic worms: The dawn of a new era. *Parasitol. Int.* **51**, 319–325 (2002).

Budiansky, S. Creatures of our own making. *Science* **298**, 80–86 (2002).

Buell, C. R. Current status of the sequence of the rice genome and prospects for finishing the first monocot genome. *Plant Physiol.* **130**, 1585–1586 (2002).

C. elegans Sequencing Consortium. Genome sequence of the nematode *C. elegans:* A platform for investigating biology. *Science* **282**, 2012–2018 (1998).

Carlton, J. M., et al. Profiling the malaria genome: A gene survey of three species of malaria parasite with comparison to other apicomplexan species. *Mol. Biochem. Parasitol.* **118**, 201–210 (2001).

Carlton, J. M., et al. Genome sequence and comparative analysis of the model rodent malaria parasite *Plasmodium yoelii yoelii*. *Nature* **419**, 512–519 (2002).

Carlton, J. M., et al. Draft genome sequence of the sexually transmitted pathogen *Trichomonas vaginalis*. *Science* **315**, 207–212 (2007).

Cavalier-Smith, T. A revised six-kingdom system of life. *Biol. Rev. Camb. Philos. Soc.* **73**, 203–266 (1998).

Chimpanzee Sequencing and Analysis Consortium. Initial sequence of the chimpanzee genome and comparison with the human genome. *Nature* **437**, 69–87 (2005).

Christoffels, A., Koh, E. G., Chia, J. M., Brenner, S., Aparicio, S., and Venkatesh, B. *Fugu* genome analysis provides evidence for a whole-genome duplication early during the evolution of ray-finned fishes. *Mol. Biol. Evol.* **21**, 1146–11451 (2004).

Collins, K., and Gorovsky, M. A. *Tetrahymena thermophila*. *Curr. Biol.* **15**, R317–R318 (2005).

Comai, L. Genetic and epigenetic interactions in allopolyploid plants. *Plant Mol. Biol.* **43**, 387–399 (2000).

Cowman, A. F., and Crabb, B. S. The *Plasmodium falciparum* genome: A blueprint for erythrocyte invasion. *Science* **298**, 126–128 (2002).

Cox, F. E. History of human parasitology. *Clin. Microbiol. Rev.* **15**, 595–612 (2002).

Crowe, M. L., Serizet, C., Thareau, V., Aubourg, S., Rouze, P., Hilson, P., Beynon, J., Weisbeek, P., van Hummelen, P., Reymond, P., Paz-Ares, J., Nietfeld, W., and Trick, M. CATMA: A complete *Arabidopsis* GST database. *Nucleic Acids Res.* **31**, 156–158 (2003).

Dávila, A. M., et al. ProtozoaDB: Dynamic visualization and exploration of protozoan genomes. *Nucleic Acids Res.* **36**, D547–D552 (2008).

Dehal, P., et al. Human chromosome 19 and related regions in mouse: Conservative and lineage-specific evolution. *Science* **293**, 104–111 (2001).

Dehal, P., et al. The draft genome of *Ciona intestinalis:* Insights into chordate and vertebrate origins. *Science* **298,** 2157–2167 (2002).

Derelle, E., et al. Genome analysis of the smallest free-living eukaryote *Ostreococcus tauri* unveils many unique features. *Proc. Natl. Acad. Sci. USA* **103,** 11647–11652 (2006).

Dermitzakis, E. T., et al. Numerous potentially functional but non-genic conserved sequences on human chromosome 21. *Nature* **420,** 578–582 (2002).

Doak, T. G., Cavalcanti, A. R., Stover, N. A., Dunn, D. M., Weiss, R., Herrick, G., and Landweber, L. F. Sequencing the *Oxytricha trifallax* macronuclear genome: A pilot project. *Trends Genet.* **19,** 603–607 (2003).

Donelson, J. E. Genome research and evolution in trypanosomes. *Curr. Opin. Genet. Dev.* **6,** 699–703 (1996).

Douglas, S. E., and Penny, S. L. The plastid genome of the cryptophyte alga, *Guillardia theta:* Complete sequence and conserved synteny groups confirm its common ancestry with red algae. *J. Mol. Evol.* **48,** 236–244 (1999).

Douglas, S., et al. The highly reduced genome of an enslaved algal nucleus. *Nature* **410,** 1091–1096 (2001).

Drosophila 12 Genomes Consortium. Evolution of genes and genomes on the *Drosophila* phylogeny. *Nature* **450,** 203–218 (2007).

Eddy, S. R. A model of the statistical power of comparative genome sequence analysis. *PLoS Biol.* **3,** e10 (2005).

Eichinger, L., et al. The genome of the social amoeba *Dictyostelium discoideum*. *Nature* **435,** 43–57 (2005).

Eisen, J. A., et al. Macronuclear genome sequence of the ciliate *Tetrahymena thermophila*, a model eukaryote. *PLoS Biol.* **4,** e286 (2006).

Ekdahl, S., and Sonnhammer, E. L. ChromoWheel: A new spin on eukaryotic chromosome visualization. *Bioinformatics* **20,** 576–577 (2004).

El-Sayed, N. M., Hegde, P., Quackenbush, J., Melville, S. E., and Donelson, J. E. The African trypanosome genome. *Int. J. Parasitol.* **30,** 329–345 (2000).

El-Sayed, N. M., et al. The genome sequence of *Trypanosoma cruzi*, etiologic agent of Chagas disease. *Science* **309,** 409–415 (2005a).

El-Sayed, N. M., et al. Comparative genomics of trypanosomatid parasitic protozoa. *Science* **309,** 404–409 (2005b).

Embley, T. M., and Hirt, R. P. Early branching eukaryotes? *Curr. Opin. Genet. Dev.* **8,** 624–629 (1998).

Embley, T. M., and Martin, W. Eukaryotic evolution, changes and challenges. *Nature* **440,** 623–630 (2006).

Fan, Y., Linardopoulou, E., Friedman, C., Williams, E., and Trask, B. J. Genomic structure and evolution of the ancestral chromosome fusion site in 2q13–2q14.1 and paralogous regions on other human chromosomes. *Genome Res.* **12,** 1651–1662 (2002).

Feng, Q., et al. Sequence and analysis of rice chromosome 4. *Nature* **420,** 316–320 (2002).

Fields, S., Kohara, Y., and Lockhart, D. J. Functional genomics. *Proc. Natl. Acad. Sci. USA* **96,** 8825–8256 (1999).

Florens, L., et al. A proteomic view of the *Plasmodium falciparum* life cycle. *Nature* **419,** 520–526 (2002).

Fraser, C. M., Eisen, J. A., and Salzberg, S. L. Microbial genome sequencing. *Nature* **406,** 799–803 (2000).

Frazer, K. A., et al. A sequence-based variation map of 8.27 million SNPs in inbred mouse strains. *Nature* **448,** 1050–1053 (2007).

Gajria, B., et al. ToxoDB: An integrated *Toxoplasma gondii* database resource. *Nucleic Acids Res.* **36,** D553–D556 (2008).

Gardner, M. J., et al. Genome sequence of the human malaria parasite *Plasmodium falciparum*. *Nature* **419,** 498–511 (2002).

Gardner, M. J., et al. Genome sequence of *Theileria parva*, a bovine pathogen that transforms lymphocytes. *Science* **309,** 134–137 (2005).

Ghedin, E., et al. Draft genome of the filarial nematode parasite *Brugia malayi*. *Science* **317,** 1756–1760 (2007).

Gilson, P. R., and McFadden, G. I. A grin without a cat. *Nature* **410,** 1040–1041 (2001).

Gilson, P. R., and McFadden, G. I. Jam packed genomes: A preliminary, comparative analysis of nucleomorphs. *Genetica* **115,** 13–28 (2002).

Gilson, P. R., Su, V., Slamovits, C. H., Reith, M. E., Keeling, P. J., and McFadden, G. I. Complete nucleotide sequence of the chlorarachniophyte nucleomorph: Nature's smallest nucleus. *Proc. Natl. Acad. Sci. USA.* **103,** 9566–9571 (2006).

Glockner, G., et al. Sequence and analysis of chromosome 2 of *Dictyostelium discoideum*. *Nature* **418,** 79–85 (2002).

Goff, S. A., et al. A draft sequence of the rice genome (*Oryza sativa* L. ssp. *japonica*). *Science* **296,** 92–100 (2002).

Gupta, B. P., and Sternberg, P. W. The draft genome sequence of the nematode *Caenorhabditis briggsae*, a companion to *C. elegans*. *Genome Biol.* **4,** 238 (2003).

Hall, A. E., Fiebig, A., and Preuss, D. Beyond the *Arabidopsis* genome: Opportunities for comparative genomics. *Plant Physiol.* **129,** 1439–1447 (2002).

Hall, N., et al. A comprehensive survey of the *Plasmodium* life cycle by genomic, transcriptomic, and proteomic analyses. *Science* **307,** 82–86 (2005).

Hardison, R. C., Oeltjen, J., and Miller, W. Long human–mouse sequence alignments reveal novel regulatory elements: A reason to sequence the mouse genome. *Genome Res.* **7,** 959–966 (1997).

Hebert, P. D. N., Cywinska, A., Ball, S. L., and deWaard, J. R. Biological identification through DNA barcodes. *Proc. R. Soc. Lond.* **270,** 313–321 (2003).

Hill, A. A., Hunter, C. P., Tsung, B. T., Tucker-Kellogg, G., and Brown, E. L. Genomic analysis of gene expression in *C. elegans*. *Science* **290,** 809–812 (2000).

Hoffman, S. L., Subramanian, G. M., Collins, F. H., and Venter, J. C. *Plasmodium*, human and *Anopheles* genomics and malaria. *Nature* **415**, 702–709 (2002).

Holland, P. W. *Ciona. Curr. Biol.* **12**, R609 (2002).

Holt, R. A., et al. The genome sequence of the malaria mosquito *Anopheles gambiae. Science* **298**, 129–149 (2002).

Honeybee Genome Sequencing Consortium. Insights into social insects from the genome of the honeybee *Apis mellifera. Nature* **443**, 931–949 (2006).

Hoskins, R. A., et al. Sequence finishing and mapping of *Drosophila melanogaster* heterochromatin. *Science* **316**, 1625–1628 (2007).

Insall, R. The *Dictyostelium* genome: The private life of a social model revealed? *Genome Biol.* **6**, 222 (2005).

International Chicken Genome Sequencing Consortium. Sequence and comparative analysis of the chicken genome provide unique perspectives on vertebrate evolution. *Nature* **432**, 695–716 (2004).

International Rice Genome Sequencing Project. The map-based sequence of the rice genome. *Nature* **436**, 793–800 (2005).

Ivens, A. C., et al. The genome of the kinetoplastid parasite, *Leishmania major. Science* **309**, 436–442 (2005).

Jaillon, O., et al. Genome duplication in the teleost fish *Tetraodon nigroviridis* reveals the early vertebrate proto-karyotype. *Nature* **431**, 946–957 (2004).

Jaillon, O., et al. The grapevine genome sequence suggests ancestral hexaploidization in major angiosperm phyla. *Nature* **449**, 463–467 (2007).

Jeffares, D. C., et al. Genome variation and evolution of the malaria parasite *Plasmodium falciparum. Nat. Genet.* **39**, 120–125 (2007).

Johnson, P. J. Spliceosomal introns in a deep-branching eukaryote: The splice of life. *Proc. Natl. Acad. Sci. USA* **99**, 3359–3361 (2002).

Jomaa, H., et al. Inhibitors of the nonmevalonate pathway of isoprenoid biosynthesis as antimalarial drugs. *Science* **285**, 1573–1576 (1999).

Kamath, R. S., et al. Systematic functional analysis of the *Caenorhabditis elegans* genome using RNAi. *Nature* **421**, 231–237 (2003).

Karlowski, W. M., Schoof, H., Janakiraman, V., Stuempflen, V., and Mayer, K. F. MOsDB: An integrated information resource for rice genomics. *Nucleic Acids Res.* **31**, 190–192 (2003).

Kasahara, M., et al. The medaka draft genome and insights into vertebrate genome evolution. *Nature* **447**, 714–719 (2007).

Keeling, P. J. Genomics. Deep questions in the tree of life. *Science* **317**, 1875–1876 (2007).

Kellis, M., Patterson, N., Endrizzi, M., Birren, B., and Lander, E. S. Sequencing and comparison of yeast species to identify genes and regulatory elements. *Nature* **423**, 241–254 (2003).

Kim, S. K. Functional genomics: The worm scores a knockout. *Curr. Biol.* **11**, R85–R87 (2001).

Kirkness, E. F., Bafna, V., Halpern, A. L., Levy, S., Remington, K., Rusch, D. B., Delcher, A. L., Pop, M., Wang, W., Fraser, C. M., and Venter, J. C. The dog genome: Survey sequencing and comparative analysis. *Science* **301**, 1898–1903 (2003).

Ku, H. M., Vision, T., Liu, J., and Tanksley, S. D. Comparing sequenced segments of the tomato and *Arabidopsis* genomes: Large-scale duplication followed by selective gene loss creates a network of synteny. *Proc. Natl. Acad. Sci. USA* **97**, 9121–9126 (2000).

Kumar, S., and Hedges, S. B. A molecular timescale for vertebrate evolution. *Nature* **392**, 917–920 (1998).

Lamblin, A. F., et al. MtDB: A database for personalized data mining of the model legume *Medicago truncatula* transcriptome. *Nucleic Acids Res.* **31**, 196–201 (2003).

Lartillot, N., and Philippe, H. Improvement of molecular phylogenetic inference and the phylogeny of Bilateria. *Philos. Trans. R. Soc. Lond. B Biol. Sci.* **363**, 1463–1472 (2008).

Lasonder, E., et al. Analysis of the *Plasmodium falciparum* proteome by high-accuracy mass spectrometry. *Nature* **419**, 537–542 (2002).

Lindblad-Toh, K., et al. Genome sequence, comparative analysis and haplotype structure of the domestic dog. *Nature* **438**, 803–819 (2005).

Lloyd, D., and Harris, J. C. *Giardia:* Highly evolved parasite or early branching eukaryote? *Trends Microbiol.* **10**, 122–127 (2002).

Long, C. A., and Hoffman, S. L. Malaria: From infants to genomics to vaccines. *Science* **297**, 345–347 (2002).

Margulis, L., and Schwartz, K. V. *Five Kingdoms. An Illustrated Guide to the Phyla of Life on Earth.* W. H. Freeman, New York, 1998.

Margulies, E. H., Vinson, J. P., NISC Comparative Sequencing Program, Miller, W., Jaffe, D. B., Lindblad-Toh, K., Chang, J. L., Green, E. D., Lander, E. S., Mullikin, J. C., and Clamp, M. An initial strategy for the systematic identification of functional elements in the human genome by low-redundancy comparative sequencing. *Proc. Natl. Acad. Sci. USA* **102**, 4795–4800 (2005).

Matthews, D. E., Carollo, V. L., Lazo, G. R., and Anderson, O. D. GrainGenes, the genome database for small-grain crops. *Nucleic Acids Res.* **31**, 183–186 (2003).

Merchant, S. S., et al. The *Chlamydomonas* genome reveals the evolution of key animal and plant functions. *Science* **318**, 245–250 (2007).

Meyerowitz, E. M. Plants compared to animals: The broadest comparative study of development. *Science* **295**, 1482–1485 (2002).

Mikkelsen, T. S., et al. Genome of the marsupial *Monodelphis domestica* reveals innovation in non-coding sequences. *Nature* **447**, 167–177 (2007).

Morris, J. C., et al. Replication of kinetoplast DNA: An update for the new millennium. *Int. J. Parasitol.* **31**, 453–458 (2001).

Morrison, H. G., et al. Genomic minimalism in the early diverging intestinal parasite *Giardia lamblia*. *Science* **317**, 1921–1926 (2007).

Mural, R. J., et al. A comparison of whole-genome shotgun-derived mouse chromosome 16 and the human genome. *Science* **296**, 1661–1671 (2002).

Myler, P. J., et al. *Leishmania major* Friedlin chromosome 1 has an unusual distribution of protein-coding genes. *Proc. Natl. Acad. Sci. USA* **96**, 2902–2906 (1999).

Myler, P. J., et al. Genomic organization and gene function in *Leishmania*. *Biochem. Soc. Trans.* **28**, 527–531 (2000).

Nadeau, J. H., and Taylor, B. A. Lengths of chromosomal segments conserved since divergence of man and mouse. *Proc. Natl. Acad. Sci. USA* **81**, 814–818 (1984).

Nixon, J. E., et al. A spliceosomal intron in *Giardia lamblia*. *Proc. Natl. Acad. Sci. USA* **99**, 3701–3705 (2002).

Ohno, S. *Sex Chromosmes and Sex-Linked Genes*. Springer-Verlag, New York, 1967.

Ohno, S. *Evolution by Gene Duplication*. Springer Verlag, Berlin, 1970.

Pain, A., et al. Genome of the host-cell transforming parasite *Theileria annulata* compared with *T. parva*. *Science* **309**, 131–133 (2005).

Pardue, M. L., DeBaryshe, P. G., and Lowenhaupt, K. Another protozoan contributes to understanding telomeres and transposable elements. *Proc. Natl. Acad. Sci. USA* **98**, 14195–14197 (2001).

Paterson, A. H. Leafing through the genomes of our major crop plants: Strategies for capturing unique information. *Nature Rev. Genet.* **7**, 174–184 (2006).

Patron, N. J., Rogers, M. B., and Keeling, P. J. Comparative rates of evolution in endosymbiotic nuclear genomes. *BMC Evol. Biol.* **6**, 46 (2006).

Peacock, C. S., et al. Comparative genomic analysis of three *Leishmania* species that cause diverse human disease. *Nat. Genet.* **39**, 839–847 (2007).

Peterson, K. J., Cameron, R. A., and Davidson, E. H. Bilaterian origins: Significance of new experimental observations. *Dev. Biol.* **219**, 1–17 (2000).

Philippe, H., and Laurent, J. How good are deep phylogenetic trees? *Curr. Opin. Genet. Dev.* **8**, 616–623 (1998).

Postlethwait, J. H. The zebrafish genome in context: Ohnologs gone missing. *J. Exp. Zool. B Mol. Dev. Evol.* **308**, 563–577 (2007).

Rat Genome Sequencing Project Consortium. Genome sequence of the Brown Norway rat yields insights into mammalian evolution. *Nature* **428**, 493–521 (2004).

Reese, M. G., Kulp, D., Tammana, H., and Haussler, D. Genie: Gene finding in *Drosophila melanogaster*. *Genome Res.* **10**, 529–538 (2000).

Rensing, S. A., et al. The *Physcomitrella* genome reveals evolutionary insights into the conquest of land by plants. *Science* **319**, 64–69 (2008).

Reyes, A., Pesole, G., and Saccone, C. Long-branch attraction phenomenon and the impact of among-site rate variation on rodent phylogeny. *Gene* **259**, 177–187 (2000).

Rhesus Macaque Genome Sequencing and Analysis Consortium, et al. Evolutionary and biomedical insights from the rhesus macaque genome. *Science* **316**, 222–234 (2007).

Richards, S., et al. Comparative genome sequencing of *Drosophila pseudoobscura*: Chromosomal, gene, and cis-element evolution. *Genome Res.* **15**, 1–18 (2005).

Rogers, A., et al. WormBase 2007. *Nucleic Acids Res.* D612–D617 (2008).

Roos, D. S. Themes and variations in apicomplexan parasite biology. *Science* **309**, 72–73 (2005).

Rubin, G. M., and Lewis, E. B. A brief history of *Drosophila*'s contributions to genome research. *Science* **287**, 2216–2218 (2000).

Rubin, G. M., et al. Comparative genomics of the eukaryotes. *Science* **287**, 2204–2215 (2000).

Rudd, S., Mewes, H. W., and Mayer, K. F. Sputnik: A database platform for comparative plant genomics. *Nucleic Acids Res.* **31**, 128–132 (2003).

Ryan, J. F., and Finnerty, J. R. CnidBase: The Cnidarian Evolutionary Genomics Database. *Nucleic Acids Res.* **31**, 159–163 (2003).

Sasaki, T., et al. The genome sequence and structure of rice chromosome 1. *Nature* **420**, 312–316 (2002).

Sclep, G., et al. CATMA, a comprehensive genome-scale resource for silencing and transcript profiling of *Arabidopsis* genes. *BMC Bioinformatics* **8**, 400 (2007).

Sea Urchin Genome Sequencing Consortium, et al. The genome of the sea urchin *Strongylocentrotus purpuratus*. *Science* **314**, 941–952 (2006).

Seo, H. C., et al. Miniature genome in the marine chordate *Oikopleura dioica*. *Science* **294**, 2506 (2001).

Shapiro, T. A., and Englund, P. T. The structure and replication of kinetoplast DNA. *Annu. Rev. Microbiol.* **49**, 117–143 (1995).

Simillion, C., Vandepoele, K., Van Montagu, M. C., Zabeau, M., and Van de Peer, Y. The hidden duplication past of *Arabidopsis thaliana*. *Proc. Natl. Acad. Sci. USA* **99**, 13627–13632 (2002).

Simpson, A. G., MacQuarrie, E. K., and Roger, A. J. Eukaryotic evolution: Early origin of canonical introns. *Nature* **419**, 270 (2002).

Smith, C. D., Shu, S., Mungall, C. J., and Karpen, G. H. The Release 5.1 annotation of *Drosophila melanogaster* heterochromatin. *Science* **316**, 1586–1591 (2007).

Snel, B., Bork, P., and Huynen, M. A. Genome phylogeny based on gene content. *Nat. Genet.* **21**, 108–110 (1999).

Sodergren, E., Shen, Y., Song, X., Zhang, L., Gibbs, R. A., and Weinstock, G. M. Shedding genomic light on Aristotle's lantern. *Dev. Biol.* **300**, 2–8 (2006).

Sonnhammer, E. L., and Durbin, R. Analysis of protein domain families in *Caenorhabditis elegans*. *Genomics* **46**, 200–216 (1997).

Stark, A., et al. Discovery of functional elements in 12 *Drosophila* genomes using evolutionary signatures. *Nature* **450**, 219–232 (2007).

Stauffer, R. L., Walker, A., Ryder, O. A., Lyons-Weiler, M., and Hedges, S. B. Human and ape molecular clocks and constraints on paleontological hypotheses. *J. Hered.* **92**, 469–474 (2001).

Stechmann, A., and Cavalier-Smith, T. Rooting the eukaryote tree by using a derived gene fusion. *Science* **297**, 89–91 (2002).

Stein, L. D., et al. The genome sequence of *Caenorhabditis briggsae*: A platform for comparative genomics. *PLoS Biol.* **1**, E45 (2003).

Stoeckert, C. J., Jr., Fischer, S., Kissinger, J. C., Heiges, M., Aurrecoechea, C., Gajria, B., and Roos, D. S. PlasmoDB v5: New looks, new genomes. *Trends Parasitol.* **22**, 543–546 (2006).

Stothard, P., and Wishart, D. S. Circular genome visualization and exploration using CGView. *Bioinformatics* **21**, 537–539 (2005).

Swarbreck, D., et al. The *Arabidopsis* Information Resource (TAIR): Gene structure and function annotation. *Nucleic Acids Res.* **36**, D1009–D1014 (2008).

Taylor, J. E., and Rudenko, G. Switching trypanosome coats: What's in the wardrobe? *Trends Genet.* **22**, 614–620 (2006).

Topalis, P., Koutsos, A., Dialynas, E., Kiamos, C., Hope, L. K., Strode, C., Hemingway, J., and Louis, C. AnoBase: A genetic and biological database of anophelines. *Insect Mol. Biol.* **14**, 591–597 (2005).

Tuskan G. A., et al. The genome of black cottonwood, *Populus trichocarpa* (Torr. Gray). *Science* **313**, 1596–1604 (2006).

Tyler, B. M., et al. *Phytophthora* genome sequences uncover evolutionary origins and mechanisms of pathogenesis. *Science* **313**, 1261–1266 (2006).

Tzafrir, I., et al. The *Arabidopsis* SeedGenes Project. *Nucleic Acids Res.* **31**, 90–93 (2003).

Van de Peer, Y. *Tetraodon* genome confirms *Takifugu* findings: Most fish are ancient polyploids. *Genome Biol.* **5**, 250 (2004).

Van de Peer, Y., Baldauf, S. L., Doolittle, W. F., and Meyer, A. An updated and comprehensive rRNA phylogeny of (crown) eukaryotes based on rate-calibrated evolutionary distances. *J. Mol. Evol.* **51**, 565–576 (2000).

Venkatesh, B., et al. Survey sequencing and comparative analysis of the elephant shark (*Callorhinchus milii*) genome. *PLoS Biol.* **5**, e101 (2007).

Vinson, J. P., et al. Assembly of polymorphic genomes: Algorithms and application to *Ciona savignyi*. *Genome Res.* **15**, 1127–1135 (2005).

Walbot, V. *Arabidopsis thaliana* genome. A green chapter in the book of life. *Nature* **408**, 794–795 (2000).

Wang, D. Y., Kumar, S., and Hedges, S. B. Divergence time estimates for the early history of animal phyla and the origin of plants, animals and fungi. *Proc. R. Soc. Lond. B Biol. Sci.* **266**, 163–171 (1999).

Waterston, R. H., et al. Initial sequencing and comparative analysis of the mouse genome. *Nature* **420**, 520–562 (2002).

Williams, B. A., Hirt, R. P., Lucocq, J. M., and Embley, T. M. A mitochondrial remnant in the microsporidian *Trachipleistophora hominis*. *Nature* **418**, 865–869 (2002).

Williams, J. G., Noegel, A. A., and Eichinger, L. Manifestations of multicellularity: *Dictyostelium* reports in. *Trends Genet.* **21**, 392–398 (2005).

Wilson, R. J., Goodman, J. L., and Strelets, V. B. The FlyBase Consortium. FlyBase: Integration and improvements to query tools. *Nucleic Acids Res.* **36**, D588–D593 (2008).

Wirth, D. F. Biological revelations. *Nature* **419**, 495–496 (2002).

Wolfe, K. H., Morden, C. W., and Palmer, J. D. Function and evolution of a minimal plastid genome from a nonphotosynthetic parasitic plant. *Proc. Natl. Acad. Sci. USA* **89**, 10648–10652 (1992).

Xia, Q., et al. A draft sequence for the genome of the domesticated silkworm (*Bombyx mori*). *Science* **306**, 1937–1940 (2004).

Xu, P., et al. The genome of *Cryptosporidium hominis*. *Nature* **431**, 1107–1112 (2004).

Xuan, Z., Wang, J., and Zhang, M. Q. Computational comparison of two mouse draft genomes and the human golden path. *Genome Biol.* **4**, R1.1–R1.10 (2003).

Yu, J., et al. A draft sequence of the rice genome (*Oryza sativa* L. ssp. *indica*). *Science* **296**, 79–92 (2002).

Yu, J., et al. The genomes of *Oryza sativa*: A history of duplications. *PLoS Biol.* **3**, e38 (2005).

Yuan, Q., et al. The TIGR rice genome annotation resource: Annotating the rice genome and creating resources for plant biologists. *Nucleic Acids Res.* **31**, 229–233 (2003).

Zdobnov, E. M., et al. Comparative genome and proteome analysis of *Anopheles gambiae* and *Drosophila melanogaster*. *Science* **298**, 149–159 (2002).

Zuegge, J., Ralph, S., Schmuker, M., McFadden, G. I., and Schneider, G. Deciphering apicoplast targeting signals: Feature extraction from nuclear-encoded precursors of *Plasmodium falciparum* apicoplast proteins. *Gene* **280**, 19–26 (2001).

Our current efforts to understand the human genome include a focus on the genetic similarities and differences among various ethnic groups. The International HapMap Project has initially generated detailed genotype information on 270 individuals from four groups with diverse geographic ancestry: the Yoruba from Ibadan, Nigeria; Utah residents of Northern and Western European ancestry; Han Chinese in Beijing; and Japanese in Tokyo. In past centuries there have been many attempts to understand the bases of phenotypic differences among humans. Baron Georges Cuvier (1769–1832) attempted a systematic classification of animals, describing four great divisions of the animal kingdom: vertebrate animals, molluscous animals, articulate animals, and radiate animals (also called Zoophytes). His work included a classification of humans based on anatomical differences. These images are from The Animal Kingdom, Arranged According to Its Organization *by Cuvier (1849, plates I–IV) and depict varieties of human races.*

19

Human Genome

INTRODUCTION

The human genome is the complete set of DNA in *Homo sapiens*. This DNA encodes the proteins and other products that define our cells and ultimately define who we are as biological entities. Through the genomic DNA, protein-coding genes are expressed that form the architecture of the trillions of cells that comprise each of our bodies. It is variations in the genome that account for the differences between people, from physical features to personality to disease states.

The initial sequencing of the human genome in 2003 was a triumph of science. It followed almost 50 years exactly after the publication of the double-stranded helical structure of DNA by Crick and Watson (1953). The genome sequence has been achieved through an international collaboration involving hundreds of scientists. (In the case of the publicly funded version, this was the International Human Genome Sequencing Consortium [IHGSC], described below.) This project could not have been possible without fundamental advances in the emerging fields of bioinformatics and genomics.

In this chapter we will first summarize some of the major findings of the human genome project. Second, we will review web-based resources for the study of the human genome at three sites: the National Center for Biotechnology Information (NCBI), the Ensembl project, and the genome center at the University of California, Santa Cruz.

Bioinformatics and Functional Genomics, Second Edition. By Jonathan Pevsner
Copyright © 2009 John Wiley & Sons, Inc.

In 2001 the sequencing and analysis of a draft version of the human genome was reported by a public consortium (International Human Genome Sequencing Consortium [IHGSC], 2001) and Celera Genomics (Venter et al., 2001). In the third part of this chapter, we will follow the outline of the public consortium's 62-page article to describe the human genome from a bioinformatics perspective. We will also describe subsequent findings on finishing the euchromatic sequence (IHGSC, 2004) and characterizing each of the 22 autosomes and two sex chromosomes (as well as the mitochondrial genome). Finally, we will describe variation in the human genome, including the analysis of individual genomes.

MAIN CONCLUSIONS OF HUMAN GENOME PROJECT

As an introduction to the Human Genome Project, we begin with a summary of its main findings. These are from the IHGSC (2001) paper, supplemented with more recent observations:

These findings are summarized from several sources, including IHGSC (2001), Venter et al. (2001), and the Wellcome Trust Sanger Institute (▶ http://www.sanger.ac.uk/HGP/publication2001/facts.shtml).

The Ensembl website currently lists 22,730 human protein coding genes (▶ http://www.ensembl.org/Homo_sapiens/, database version 48.36j, January 2008).

1. There were reported to be about 30,000 to 40,000 predicted protein-coding genes in the human genome. However, the initial sequencing and annotation were incomplete, and in subsequent years a variety of new tools were developed (Chapter 16) as well as comparative approaches as more vertebrate genomes were sequenced. A revised estimate suggests that there are 20,000 to 25,000 protein-coding genes (IHGSC, 2004). These estimates are surprising because we have about the same number of genes as much simpler organisms such as *Arabidopsis thaliana* (26,000 genes) and pufferfish (21,000 genes), and marginally more genes than are found in many nematode and insect genomes.

2. The human proteome is far more complex than the set of proteins encoded by invertebrate genomes. Vertebrates have a more complex mixture of protein domain architectures. Additionally, the human genome displays greater complexity in its processing of mRNA transcripts by alternative splicing.

3. Hundreds of human genes were acquired from bacteria by lateral gene transfer, according to the initial report (IHGSC, 2001; Ponting, 2001). Subsequently Salzberg et al. (2001) suggested a revised estimate of 40 genes that underwent horizontal transfer. These genes are homologous to bacterial sequences but appear to lack orthologous genes in other vertebrate and invertebrate species. In recent years the emphasis has changed from laterally acquired genes (discussed in Chapter 15) to the vast number of prokaryotic and viral genes from organisms living inside the human body, called the human microbiome. It has been estimated that there are at least 10 times more microbial cells than human cells in a human body, and perhaps 10 times more prokaryotic and viral genes than human genes. In 2007 the Human Microbiome Project was launched to characterize these organisms and their roles in human health.

Information about the Human Microbiome Project is available at ▶ http://nihroadmap.nih.gov/hmp/.

We introduced these various types of repetitive elements in Chapter 16, and will further define them below.

4. More than 98% of the human genome does not code for genes. Much of this genomic landscape is occupied by repetitive DNA elements such as long interspersed elements (LINEs) (20%), short interspersed elements (SINEs) (13%), long terminal repeat (LTR) retrotransposons (8%), and DNA transposons (3%). Thus half the human genome is derived from transposable elements. However, there has been a decline in the activity of these elements in the hominid lineage. At the same time, the mouse genome displays a continued vigorous activity of transposable elements.

5. Segmental duplication is a frequent occurrence in the human genome, particularly in pericentromeric and subtelomeric regions. This phenomenon is more common than in yeast, fruitfly, or worm genomes. There are three principal ways that gene duplications arise in the human genome (Green and Chakravarti, 2001). First, tandem duplications (created from sequence repeats in a localized region) occur rarely. Second, processed mRNAs are duplicated by retrotransposition. This produces intronless paralogs that are present at one or many sites. Third, segmental duplications occur in which large sections of a chromosome transfer to a new site. We introduced these concepts in Chapter 16.

6. There are several hundred thousand *Alu* repeats in the human genome. These have been thought to represent elements that replicate promiscuously. However, their distribution is nonrandom: they are retained in GC-rich regions and thus may confer some benefit on the human genome.

7. The mutation rate is about twice as high in male meiosis than in female meiosis. This suggests that most mutation occurs in males.

8. More than 1.4 million single nucleotide polymorphisms (SNPs) were identified. SNPs are single nucleotide variations that occur once every 100 to 300 base pairs (bp). The International HapMap Consortium (2007) reported a haplotype map of 3.1 million SNPs, and today the genotype and copy number of one million SNPs are routinely measured on a single sample using a microarray. This is already having a profound impact on studies of variation in the human genome.

The NCBI database of SNPs currently lists over 11.8 million RefSNPs, of which over 6.2 million have been validated (build 128, January 2008; see below and ► http://www.ncbi.nlm.nih.gov/SNP/).

The ENCODE Project

In the years since the initial sequencing of the human genome, an important movement has been the ENCyclopedia Of DNA Elements (ENCODE) Project (ENCODE Project Consortium, 2004, 2007). We described this project and its principal conclusions in Chapter 16. That project initially focused on 1% of the human genome. The results so far have shown that the genome is pervasively transcribed (see Chapter 8). We now have a deeper understanding of transcriptional regulation (including transcription start sites) and chromatin structure. In the coming years the project will continue to expand. Its major goals include the following:

- High throughput sequencing of chromatin regulatory elements including transcription factor binding sites, using chromatin immunoprecipitation followed by high throughput DNA sequencing.
- Comprehensively identifying active functional elements in human chromatin, in part using DNase I hypersensitivity assays.
- Characterizing the human transcriptome.
- Developing a reference gene set for protein-coding genes, noncoding genes, and pseudogenes.

The results of the project will continue to be centralized at the UCSC ENCODE data coordination center. The experimental and computational approaches spawned by ENCODE and related projects will help to reveal the architecture of the genome and its functional properties, from its gene content to its regulatory roles.

You can read about the ENCODE project at ► http://www.genome.gov/26023194. The UCSC ENCODE project website is ► http://genome.ucsc.edu/ENCODE/.

GATEWAYS TO ACCESS THE HUMAN GENOME

Although there are many ways to access information about the human genome, we will briefly describe several, including the three principal browsers for the human genome: NCBI, Ensembl, and UCSC.

NCBI

The NCBI offers two main ways to access data on the human genome. From the main page of NCBI, you can select "human genome resources," which provides links to each chromosome and a variety of web resources. Alternatively, you can select the Map Viewer (Fig. 19.1). This page allows searches by clicking on a chromosome or by entering a text query.

The human Map Viewer integrates human sequence and data from cytogenetic maps, genetic linkage maps, radiation hybrid maps, and YAC chromosomes. A query with "hbb" (Fig. 19.1) links to the Map Viewer (Fig. 19.2). This map can display dozens of kinds of information and links to Entrez Gene, Entrez Nucleotide, and Entrez Protein. Other features include links to human and mouse UniGene and a human–mouse homology map. There is also an "evidence viewer" (Fig. 19.3) that displays evidence supporting the proposed structure of a gene and highlights possible discrepancies in the nucleotide sequence, exon–intron boundaries, or other aspects of an annotated gene. As an example, the evidence viewer shows the density of ESTs that have been identified corresponding to each predicted exon of beta globin (Fig. 19.3).

Ensembl

Ensembl is a comprehensive resource for information about the human genome as well as many other genomes (Flicek et al., 2008). This resource effectively interconnects a wide range of genomics tools with a focus on annotation of known and newly

A Human Genome Resources page is available at ▶ http://www. ncbi.nlm.nih.gov/genome/guide/ human/. The Map Viewer (for human and other organisms) is accessed via ▶ http://www.ncbi. nlm.nih.gov/mapview/. You can read about the genome resources at NCBI at ▶ http://www.ncbi. nih.gov/About/Doc/hs_ genomeintro.html.

The Map Viewer provides a link to Entrez Gene. If instead you begin with any Entrez Gene entry (Fig. 2.8), you can click "map" to move directly to the Map Viewer.

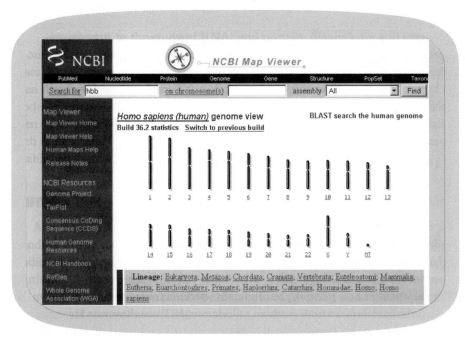

*FIGURE 19.1. **The Human Map Viewer** is accessible from the main page of NCBI. This resource displays cytogenetic, genetic, physical, and radiation hybrid maps of human genome sequence. Here, entering the query "hbb" results in links to beta globin on chromosome 11 (see Fig. 19.2).*

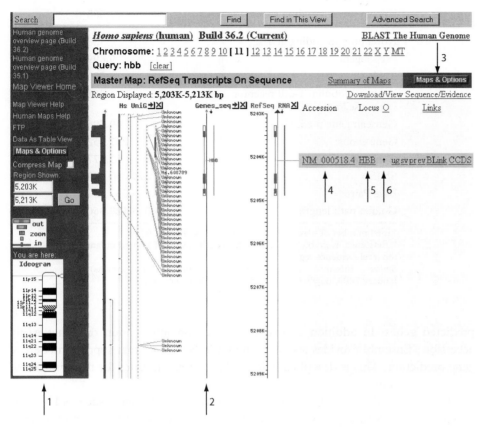

FIGURE 19.2. (a) The Human Map Viewer from NCBI shows the region of chromosome 11 containing the beta globin gene. The location is shown on an ideogram on the left sidebar (arrow 1), above which is a tool to zoom in or out or to specify genomic coordinates. The main information is displayed in vertical tracks (columns from left to right across the page) such as a track of gene structures (arrow 2) and RefSeq genes. The Maps & Options link (arrow 3) allows annotation tracks to be added or removed. A series of links includes Entrez Nucleotide (arrow 4), Entrez Gene (arrow 5), an indication of the orientation of the gene (arrow 6), and the following links: ug (UniGene), sv (Entrez Nucleotide sequence view), pr (Entrez Protein), ev (the Evidence Viewer, shown in Fig. 19.3), BLink (precomputed BLAST results), and CCDS (consensus coding sequence database).

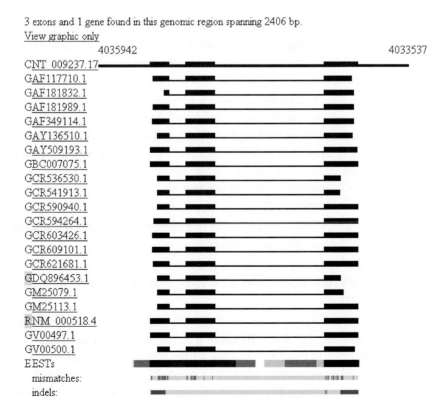

FIGURE 19.3. The NCBI evidence viewer provides data concerning the structure of human genes. The viewer provides links to the genomic contig (C), GenBank mRNAs (G), a RefSeq mRNA (R), and expressed sequence tags (ESTs) (E) with a color-coding scheme showing the density of ESTs at different positions of the gene. The evidence viewer also describes experimental evidence for each exon, including mismatches and insertions/deletions (indels). The evidence viewer is useful to conveniently assess the basis for a particular gene model.

TABLE 19-1   Human Genome Statistics from Ensembl

Known protein-coding genes:	21,388
Novel protein-coding genes:	28
Pseudogenes:	9,899
RNA genes:	5,732
Immunoglobulin/T-cell receptor gene segments:	388
Genscan gene predictions:	49,796
Gene exons:	297,252
Gene transcripts:	62,877
SNPs:	15,040,632
Base Pairs:*	3,253,037,807
Golden path length:**	3,093,120,360

*Total number of base pairs = sum of lengths of DNA table.
**Reference assembly (Golden path) length = sum of nonredundant top level sequence regions.
Source: Ensembl (► http://www.ensembl.org/Homo_sapiens), January 2009, database version 52 (NCBI 36 assembly).

Ensembl, a joint project between the European Molecular Biology Laboratory-European Bioinformatics Institute (EMBL-EBI) and the Sanger Institute is available at ► http://www.ensembl.org. The human database is at ► http://www.ensembl.org/Homo_sapiens/. We described Ensembl projects for mouse, rat, zebrafish, fugu, mosquito, and other organisms in Chapter 18.

We saw an example of the Ensembl BLAST server in Figs. 5.1 and 5.2.

predicted genes. In addition to making annotation information on genes easily accessible, Ensembl provides access to the underlying data that support models of gene prediction. This is described below. The current statistics for the contents of the Ensembl human build are shown in Table 19.1.

From the main page of Ensembl, you can type a text query (such as HBB for human beta globin), perform a BLAST search, or browse by chromosome (Fig. 19.4). There are six main entry points to access the Ensembl database:

1. Contig view allows you to search across an entire chromosome (Fig. 19.5), while also viewing a smaller region in detail (Fig. 19.6). The Contig view integrates features from a variety of external data sources such as UniGene and RefSeq. The tile path shows BAC clones used in the current assembly.

2. Gene view allows a text query for hbb that leads to a typical Ensembl gene report or to the contig view, "transview" evidence report, protein report, database links (such as RefSeq, SwissProt, and InterPro) and Ensembl homology matches. This view includes the transcript DNA sequence and information on exon–intron boundaries (splice sites). The links to evidence reports are especially important in helping you to evaluate the experimental support for a given gene structure.

3. Anchor view allows you to select two features from a chromosome as "anchor points" and to display the intervening region.

4. Disease view links to disease entries in OMIM (Chapter 20). In the case of *RBP4*, this view option leads to the relevant OMIM entry for deficiency of RBP4.

5. Map view shows an ideogram of each chromosome, including the known genes, GC content, and SNPs. By clicking on the synteny link (also accessed from the Contig view of Fig. 19.5), you can see the corresponding region of chromosomes from other organisms where a gene such as *HBB* is localized (Fig. 19.7). This figure shows the correspondence of four mouse chromosomal regions

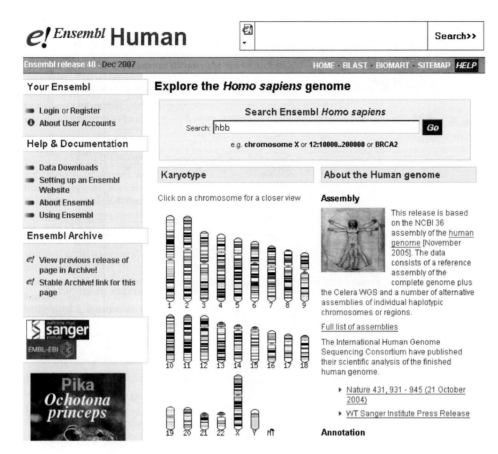

*FIGURE 19.4. **Front page of the Ensembl human genome browser** (▶ http://www.ensembl.org/Homo_sapiens/). A direct way to begin searching the site is to enter a search term such as HBB (top) for beta globin. Other ways to begin searching include the chromosome ideograms or a BLAST server.*

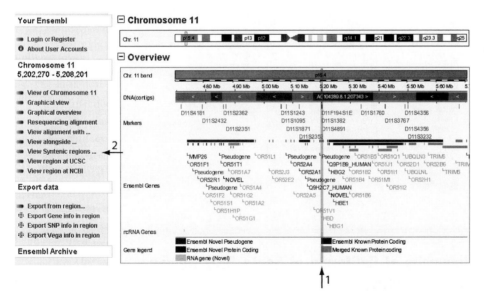

*FIGURE 19.5. **The Ensembl ContigView for human beta globin** (linked from the HBB gene report) shows an ideogram of chromosome 11 at the top, including the position of the HBB gene near the 11p telomere in band 11p15.4 (boxed). The overview section includes an overview of contigs and markers in the 11p15.4 region and annotated genes including HBB (arrow 1). The left sidebar offers a variety of links such as views of conserved syntenic regions (arrow 2), shown in Fig. 19.7 below, as well as links to the UCSC and NCBI genome browsers.*

to human chromosome 11. For contrast, it shows the most-conserved chromosome (the X chromosome) and the least-conserved chromosome (the Y chromosome).

6. Cyto view displays genes, BAC end clones, repetitive elements, and the tiling path across genomic DNA regions.

(a) ⊟ **Detailed view**

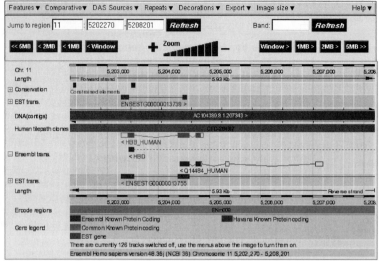

(b)

FIGURE 19.6. (a) The Ensembl ContigView for human beta globin includes a detailed view. This contains a series of horizontal annotation tracks (analogous to the UCSC Genome Browser). The top row includes a series of pull-down menus, shown in (b).

University of California at Santa Cruz Human Genome Browser

The UCSC Genome Bioinformatics site is accessible at ► http://genome.ucsc.edu/. It was developed by David Haussler's group (Kent et al., 2002).

The "Golden Path" is the human genome sequence annotated at UCSC. Along with the Ensembl and NCBI sites, the human genome browser at UCSC is one of the three main web-based sources of information for both the human genome and other genomes. It has become a basic resource in the fields of bioinformatics and genomics, and we have relied on it throughout this book (particularly in Chapter 16).

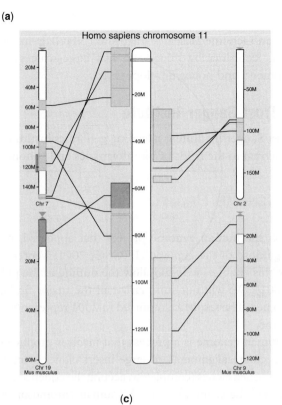

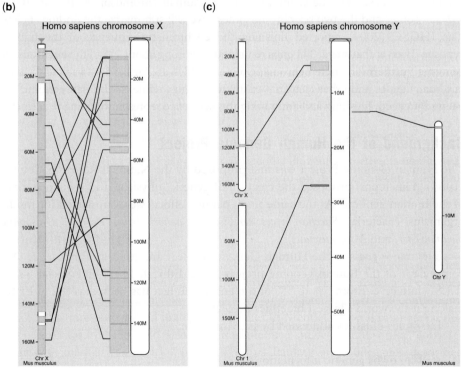

FIGURE 19.7. *The MapView and ContigView at Ensembl link to conserved syntenic maps including those for human/mouse. (a) Human chromosome 11 (including the HBB gene, boxed) is shown in the center as an ideogram. It corresponds to mouse chromosomes 7, 2, 19, and 9. Although the lineages leading to modern humans and mice diverged about 80 million years ago, it is still straightforward to identify regions of conserved synteny. (b) The human X chromosome is extremely closely conserved with the mouse X chromosome. (c) The Y chromosomes of human and mouse are extraordinarily poorly conserved.*

An NHGRI introduction to the human genome project is available at ▶ http://www.genome.gov/10001772. A document describing an NHGRI vision for the future of genomics research is at ▶ http://www.genome.gov/11007524 or Collins et al. (2003).

The website for human genetics at the Wellcome Trust Sanger Institute (WTSI) is ▶ http://www.sanger.ac.uk/humgen/. The Human Genome Project gateway is at ▶ http://www.sanger.ac.uk/HGP/.

The euchromatin is the primary gene-containing part of the genome, although there are also genes in heterochromatin.

The National Human Genome Research Institute describes the finishing process at ▶ http://www.genome.gov/10000923.

The National Academy Press (▶ http://www.nap.edu) offers this 1988 book free online at ▶ http://www.nap.edu/openbook.php?isbn=0309038405/.

You can read about ELSI at ▶ http://www.genome.gov/10001618 or ▶ http://www.ornl.gov/hgmis/elsi/elsi.html.

NHGRI

The National Human Genome Research Institute (NHGRI) has a leading role in genome sequencing, coordinating pilot-scale and large-scale sequencing efforts, technology development, and policy development.

The Wellcome Trust Sanger Institute

The Wellcome Trust Sanger Institute is a leading genomics institute that, like NCBI and NHGRI, is essential to the fields of bioinformatics and genomics.

THE HUMAN GENOME PROJECT

The two articles on the human genome project that appeared in February 2001 provide an initial glimpse of the genome (IHGSC, 2001; Venter et al., 2001). In the next portion of this chapter, we will follow the outline of the public consortium paper (IHGSC, 2001). We will not summarize all the major findings, but we will focus on selected topics. The sequence reported in 2001 represents 90% completion of the human genome.

Finishing the human genome is a process that involves producing finished maps (with continuous, accurate alignments of large-insert clones spanning euchromatic loci) and producing finished clones (completely, accurately sequenced). Additional publications have described the sequence of all 24 human chromosomes in more detail (22 autosomes and the two sex chromosomes); we will summarize the findings below. The IHGSC (2004) reported finishing the euchromatic sequence of the human genome. Even at that stage, 341 gaps remained, spanning about 1% of the euchromatic genome. Furthermore, heterochromatic regions which are far harder to sequence contain many genes and other elements of interest. Thus, while the human genome has been sequenced, finishing and annotating this sequence represent ongoing processes.

Background of the Human Genome Project

The Human Genome Project was first proposed by the National Research Council (1988). This report proposed the creation of genetic, physical, and sequence maps of the human genome. At the same time, parallel efforts were supported for model organisms (bacteria, *Saccharomyces cerevisiae*, *Caenorhabditis elegans*, *Drosophila melanogaster*, and *Mus musculus*).

The major goals of the Human Genome Project are listed in Table 19.2. One component of the human genome project is the Ethical, Legal and Social Issues (ELSI) initiative. From 3% to 5% of the annual budget has been devoted to ELSI, making it the world's largest bioethics project.

Examples of issues addressed by ELSI include:

- Who owns genetic information?
- Who should have access to genetic information?
- How does genomic information affect members of minority communities?
- What societal issues are raised by new reproductive technologies?
- How should genetic tests be regulated for reliability and validity?
- To what extent do genes determine behavior?
- Are there health risks associated with genetically modified foods?

TABLE 19-2   Eight Goals of the Human Genome Project (1998 – 2003)

1. Human DNA sequence

- Finish the complete human genome sequence by the end of 2003.
- Achieve coverage of at least 90% of the genome in a working draft based on mapped clones by the end of 2001.
- Make the sequence totally and freely accessible.

2. Sequencing technology

- Continue to increase the throughput and reduce the cost of current sequencing technology.
- Support research on novel technologies that can lead to significant improvements in sequencing technology.
- Develop effective methods for the development and introduction of new sequencing technologies.

3. Human genome sequence variation

- Develop technologies for rapid, large-scale identification and/or scoring of single-nucleotide polymorphisms and other DNA sequence variants.
- Identify common variants in the coding regions of the majority of identified genes during this five-year period.
- Create an SNP map of at least 100,000 markers.
- Create public resources of DNA samples and cell lines.

4. Functional genomics technology

- Generate sets of full-length cDNA clones and sequences that represent human genes and model organisms.
- Support research on methods for studying functions of nonprotein-coding sequences.
- Develop technology for comprehensive analysis of gene expression.
- Improve methods for genomewide mutagenesis.
- Develop technology for large-scale protein analyses.

5. Comparative genomics

- Complete the sequence of the roundworm *C. elegans* genome and the fruitfly *Drosophila* genome.
- Develop an integrated physical and genetic map for the mouse, generate additional mouse cDNA resources, and complete the sequence of the mouse genome by 2008.

6. Ethical, legal, and social issues

- Examine issues surrounding completion of the human DNA sequence and the study of genetic variation.
- Examine issues raised by the integration of genetic technologies and information into health care and public health activities.
- Examine issues raised by the integration of knowledge about genomics and gene–environment interactions in nonclinical settings.
- Explore how new genetic knowledge may interact with a variety of philosophical, theological, and ethical perspectives.
- Explore how racial, ethnic, and socioeconomic factors affect the use, understanding, and interpretation of

(Continued)

TABLE 19-2   Continued

genetic information, the use of genetic services, and the development of policy.

7. Bioinformatics and computational biology	• Improve content and utility of databases. • Develop better tools for data generation, capture, and annotation. • Develop and improve tools and databases for comprehensive functional studies. • Develop and improve tools for representing and analyzing sequence similarity and variation. • Create mechanisms to support effective approaches for producing robust, exportable software that can be widely shared.
8. Training and manpower	• Nurture the training of scientists skilled in genomics research. • Encourage the establishment of academic career paths for genomic scientists. • Increase the number of scholars who are knowledgeable in both genomic and genetic sciences and in ethics, law, or the social sciences.

Source: Adapted from ► http://www.ornl.gov/sci/techresources/Human_Genome/hg5yp/goal.shtml.

All these issues are becoming increasingly important, particularly as we begin to obtain the nearly complete genomic DNA sequence of individuals (described below).

Strategic Issues: Hierarchical Shotgun Sequencing to Generate Draft Sequence

The public consortium approach to sequencing the human genome was to employ the hierarchical shotgun sequencing strategy. The rationale for taking this approach was as follows:

- Shotgun sequencing can be applied to DNA molecules of many sizes, including plasmids (typically several kilobases), cosmid clones (40 kilobases [kb]), yeast, and BACs (up to 1 or 2 megabases [Mb]).

- The human genome has large amounts of repetitive DNA (about 50% of the genome; see below). Whole-genome shotgun sequencing, the main approach taken by Celera Genomics, was not adopted by the public consortium because of the difficulties associated with assembling repetitive DNA fragments. In the public consortium approach, large-insert clones (typically 100 to 200 kb) from defined chromosomes were sequenced.

- The reduction of the sequencing project to specific chromosomes allowed the international team to reduce and distribute the sequencing project to a set of sequencing centers. These centers are listed in Table 19.3.

The 2001 draft version of the human genome was based on the sequence and assembly of over 29,000 BAC clones with a total length of 4.26 billion base pairs (Gb). There were 23 Gb of raw shotgun sequence data.

TABLE 19-3 Twenty Institutions That Form the Human Genome Sequencing Consortium

Genome Sequencing Center	Location/ Description	URL
Baylor College of Medicine	Houston, Texas	▶ http://www.hgsc.bcm.tmc.edu/
Beijing Human Genome Center, Institute of Genetics, Chinese Academy of Sciences	Beijing, China	▶ http://www.chgb.org.cn/en_index.htm
Cold Spring Harbor Laboratory, Lita Annenberg Hazen Genome Center	Cold Spring Harbor, New York	▶ http://www.cshl.edu/public/genome.html
Gesellschaft fur Biotechnologische Forschung mbH	Braunschweig, Germany	▶ http://genome.gbf.de/
Genoscope	Evry, France	▶ http://www.genoscope.cns.fr/externe/English/Projets/projets.html
Genome Therapeutics Corporation	Waltham, Massachusetts	▶ http://www.genomecorp.com/
Institute for Molecular Biotechnology	Jena, Germany	▶ http://genome.imb-jena.de/
Joint Genome Institute, U.S. Department of Energy	Walnut Creek, California	▶ http://www.jgi.doe.gov/
Keio University	Tokyo, Japan	▶ http://www-alis.tokyo.jst.go.jp/HGS/top.pl
Max Planck Institute for Molecular Genetics	Berlin, Germany	▶ http://seq.mpimg-berlin-dahlem.mpg.de/
Multimegabase Sequencing Center, Institute for Systems Biology	Seattle, Washington	▶ http://www.systemsbiology.org/
RIKEN Genomic Sciences Center	Saitama, Japan	▶ http://hgp.gsc.riken.go.jp/index.php/Main_Page
The Sanger Centre	Hinxton, United Kingdom	▶ http://www.sanger.ac.uk/HGP/
Stanford Genome Technology Center	Palo Alto, California	▶ http://www-sequence.stanford.edu/
Stanford Human Genome Center	Palo Alto, California	▶ http://shgc.stanford.edu/
University of Oklahoma's Advanced Center for Genome Technology	Norman, Oklahoma	▶ http://www.genome.ou.edu/
University of Texas Southwestern Medical Center	Dallas, Texas	▶ http://www8.utsouthwestern.edu/
University of Washington Genome Center	Seattle, Washington	▶ http://www.genome.washington.edu/UWGC/
Washington University Genome Sequencing Center	St. Louis, Missouri	▶ http://genome.wustl.edu/
Whitehead Institute for Biomedical Research, MIT (now the Broad Institute)	Cambridge, Massachusetts	▶ http://www.broad.mit.edu/

Early in the evolution of the Human Genome Project, it was thought that breakthroughs in DNA sequencing technology would be necessary to allow the completion of such a large-scale project. This did not occur until the recent arrival of next-generation sequencing. Instead, the basic principles of dideoxynucleotide sequencing by the method of Sanger (see Chapter 13) were improved upon. Some

TABLE 19-4 Contigs Categorized by Size			
Range (kb)	Number	Length (base pairs)	Percent of Total
<300	74	12,100,100	0.42
300–1000	48	29,373,700	1.02
1000–5000	59	139,894,000	4.89
>5000	99	2,676,790,000	93.65

See Build 36 version 2 statistics (linked from ▶ http://www.ncbi.nlm.nih.gov/genome/guide/human/release_notes.html, January 2008).

innovations to Sanger sequencing (see Chapter 13 and Green, 2001) include capillary electrophoresis-based sequencing machines for the automated detection of DNA molecules, improved thermostable polymerases, and fluorescent dye-labeled dideoxynucleotide terminators.

The public consortium draft genome sequence was generated by selecting, sequencing, and assembling BAC clones. Most libraries contained BAC clones or P1-derived artificial clones (PACs). These libraries were prepared from DNA obtained from anonymous donors. Selected clones were subjected to shotgun sequencing. In conjunction with sequencing of BAC and other large-insert clones, the sequence data were assembled into an integrated draft sequence (Table 19.4). An example of the procedure is shown in Fig. 19.8.

The whole genome shotgun assembly approach that was championed by Celera was proven successful by the sequencing of the *Drosophila melanogaster* genome in 2000 as well as by the initial sequence of the human genome (Venter et al., 2001).

The NCBI Contig Assembly and Annotation Process is described at ▶ http://www.ncbi.nlm.nih.gov/genome/guide/build.html.

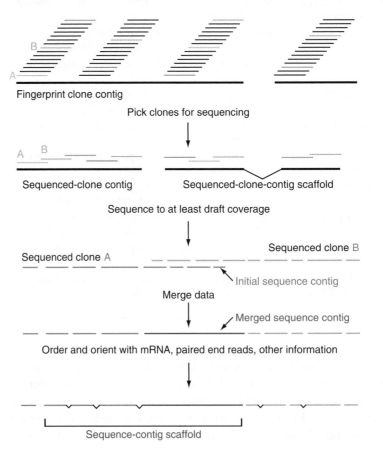

FIGURE 19.8. Clone and sequence coverage of the human genome. A fingerprint clone contig is assembled based on restriction endonuclease digestion patterns in order to select clones for sequencing that are inferred to overlap. These sequenced-clone contigs are merged to generate scaffolds in which the order and orientation of each clone is established. (From IHGSC, 2001.) Used with permission.

Since then it has been widely adopted for hundreds of prokaryotic and eukaryotic genome sequencing projects. A caveat noted by Evan Eichler and colleagues is that whole genome shotgun sequencing and assembly performs poorly at correctly assembling repetitive DNA elements such as the segmental duplications that occupy over 5% of the human genome (She et al., 2004). They compared a whole-genome shotgun sequence assembly to the assembly based on ordered clones and found that 38.2 megabases of pericentromeric DNA (about 80% of the size of a small autosome) was either not assembled, not assigned, or misassigned. Additionally, She et al. suggested that 40% of the duplicated sequence might be misassembled. Correctly resolving these structures will require a targeted approach to supplement whole genome shotgun sequencing and assembly.

Features of the Genome Sequence

A draft genome sequence contains a mixture of finished, draft, and predraft data. A key aspect of the sequence is the extent to which the sequenced fragments are contiguous. The average length of a clone or a contig is not a consistently useful measure of the extent to which a genome has been sequenced and assembled. Instead the N50 length describes the largest length L such that 50% of all nucleotides are contained in contigs or scaffolds of at least size L. For the draft version of the human genome sequence, half of all nucleotides were present in a fingerprint clone contig of at least 8.4 megabases (Table 19.5). The N50 length rose to 38.5 megabases with the most recent freeze of the genome assembly (Table 19.5).

The quality of the genome sequence is assessed by counting the number of gaps and by measuring the nucleotide accuracy. About 91% of the unfinished draft sequence had an error rate of less than 1 per 10,000 bases (PHRAP score >40). A PHRAP score of 40 corresponds to an error probability of $10^{-40/10}$, or 99.99% accuracy (see Chapter 13).

Another fundamental category of error is misassembly. This can be especially problematic for regions of highly repetitive DNA or for duplicated regions of the genome (see below; Eichler, 2001). Comparisons of the genome sequences produced by the IHGSC and Celera Genomics indicate substantial differences that reflect differences in assembly (Li et al., 2002).

N50 statistics are reported at the UCSC genome browser (▶ http://genome.ucsc.edu/goldenPath/stats.html) as well as at NCBI.

TABLE 19-5 Continuity of Draft Genome Sequence[a]	
Clone	Length (L), kilobases
Initial sequencing contig	21.7
Sequence contig	82
Sequence-contig scaffold	274
Sequenced-clone contig	826
Fingerprint clone contig	8,400
Reference contig length (Build 36 version 2)	38,510

[a]This is described by N50 statistics, which report the length of various clone types for which 50% of the nucleotides reside.

Source: Data are adapted from IHGSC (2001) except for Reference contig length from January 2008 (linked from ▶ http://www.ncbi.nlm.nih.gov/genome/guide/human/release_notes.html).

The Broad Genomic Landscape

We will discuss the 24 human chromosomes in more detail below, based on projects focused on finishing the sequence of each one. The autosomes are numbered approximately in order of size. The largest chromosome, chromosome 1, is 223 megabases in length; the smallest, chromosome 21, is about 47 megabases.

Having a nearly complete view of the nucleotide sequence of the human genome, we can explore its broad features. These include:

- The distribution of GC content
- CpG islands and recombination rates
- The repeat content
- The gene content

We will next examine each of these four features of the genome. Using the genome browsers at UCSC as well as Ensembl and NCBI, we can explore the genomic landscape from the level of single nucleotides to entire chromosomes.

Long-Range Variation in GC Content

The average GC content of the human genome is 41%. However, there are regions that are relatively GC rich and GC poor. A histogram of the overall GC content (in 20 kb windows) shows a broad profile with skewing to the right (Fig. 19.9). Fifty-eight percent of the GC content bins are below the average, while 42% are above the average, including a long tail of highly GC-rich regions.

Giorgio Bernardi and colleagues have proposed that mammalian genomes are organized into a mosaic of large DNA segments (e.g., >300 kb) called isochores. These isochores are fairly homogeneous compositionally and can be divided into GC-poor families (L1 and L2) or GC-rich families (H1, H2, and H3). The

You can view GC content across any chromosome in the NCBI, Ensembl, or UCSC genome browsers. For example, in the Ensembl browser (Fig. 19.6) click "decorations" to add a GC content layer.

The L (light) and H (heavy) designations for isochores refer to the sedimentation behavior of genomic DNA in cesium chloride gradients. Genomic DNA fragments migrate to different positions based on their percent GC content.

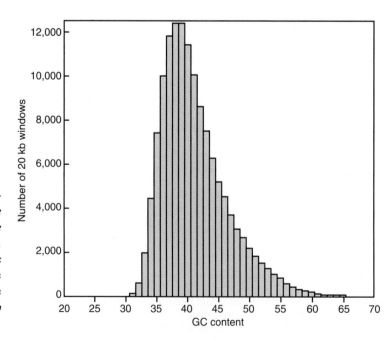

FIGURE 19.9. *Histogram of percent GC content versus the number of 20 kb windows in the draft human genome sequence. Note that the distribution is skewed to the right, with a mean GC content of 41% (from IHGSC, 2001). Used with permission.*

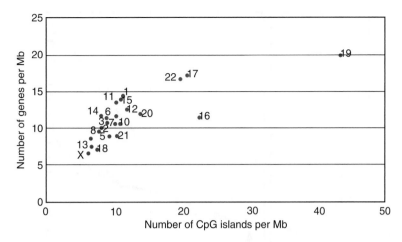

FIGURE 19.10. The number of CpG islands per megabase is plotted versus the number of genes per megabase as a function of chromosome. Note that chromosome 19, the most gene-rich chromosome, has the greatest number of CpG islands per megabase (from IHGSC, 2001). Used with permission.

IHGSC (2001) report did not identify clearly defined isochores, and Haring and Kypr (2001) did not detect isochores in human chromosomes 21 and 22. Subsequent analyses by Bernardi and colleagues (Bernardi, 2001; Pavlicek et al., 2002) do support the mosaic organization of the human genome by GC content. The discrepancies depend in part on the size of the window of genomic DNA that is analyzed.

CpG Islands

The dinucleotide CpG is greatly underrepresented in genomic DNA, occurring at about one-fifth its expected frequency. We introduced this topic in Chapter 16. Most CpG dinucleotides are methylated on the cytosine and subsequently are deaminated to thymine bases. However, the genome contains many "CpG islands" which are typically associated with the promoter and exonic regions of housekeeping genes (Gardiner-Garden and Frommer, 1987). CpG islands have roles in processes such as gene silencing, genomic imprinting (Tycko and Morison, 2002), and X chromosome inactivation (Avner and Heard, 2001).

You can display predicted CpG islands in genomic DNA at the NCBI, Ensembl, and UCSC genome browser websites (see Fig. 16.19). According to the IHGSC (2001), there are 50,267 predicted CpG islands in the human genome. After blocking repetitive DNA sequences with RepeatMasker, there were 28,890 CpG islands. (This lower number reflects the high GC content of *Alu* repeats as seen in Fig. 16.8.) There are 5 to 15 CpG islands per megabase of DNA on most chromosomes, although chromosome 19 (the most gene-dense chromosome) contains 43 CpG islands per megabase (Fig. 19.10).

Comparison of Genetic and Physical Distance

It is possible to compare the genetic maps and physical maps of the chromosomes to estimate the rate of recombination per nucleotide (Yu et al., 2001). Genetic maps, also known as linkage maps, are chromosome maps based on meiotic recombination. During meiosis the two copies of each chromosome present in each cell are reduced to one. The homologous parental chromosomes recombine (exchange DNA) during this process. Genetic maps describe the distances between DNA sequences (genes) based on their frequency of recombination. Thus, genetic maps

Gene silencing refers to transcriptional repression. We briefly described MeCP2, a protein that binds to methylated CpG islands (Fig. 10.9). MeCP2 further recruits proteins such as a histone deactylase that alters chromatin structure and represses transcription. Mutations in *MECP2*, the X-linked gene encoding MeCP2, cause Rett sydrome (Amir et al., 1999). This disease causes distinctive neurological symptoms in girls, including loss of purposeful hand movements, seizures, and autistic-like behavior (Chapter 20). X chromosome inactivation is a dosage compensation mechanism in which cells in a female body selectively silence the expression of genes from either the maternally or paternally derived X chromosome (Avner and Heard, 2001).

The UCSC Table Browser lists 28,226 CpG islands in the human genome. To see this, visit the table browser (via ▶ http://genome. ucsc.edu). Set the clade to vertebrate, the genome to human, the assembly to March 2006 (or another assembly), the group to Expression and Regulation, the track to CpG islands, and click summary statistics.

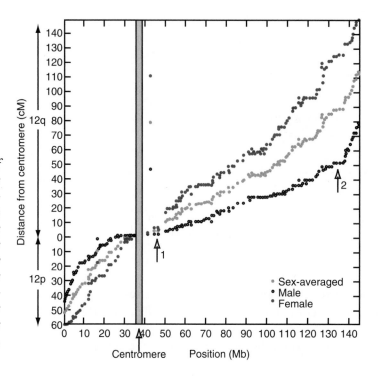

FIGURE 19.11. Comparison of physical distance (in megabases, x axis) with genetic distance (in centimorgans, y axis) for human chromosome 12. Note that the recombination rate tends to be lower near the centromere (arrow 1) and higher near the telomeres (distal portion of each chromosome). The recombination is especially high in the male meiotic map (arrow 2). (From IHGSC, 2001.) Used with permission.

Genomic imprinting is the differential expression of genes from maternal and paternal alleles. Tycko and Morison (2002) offer a database of imprinted genes (► http://igc.otago.ac.nz/home.html).

The NCBI, Ensembl, and UCSC genome browsers allow you to view both physical maps and different kinds of genetic maps.

report DNA sequences in units of centimorgans (cM), which describe relative distance. One centimorgan corresponds to 1% recombination.

In contrast to genetic maps, physical maps describe the physical position of nucleotide sequences along each chromosome. With the completion of draft versions of the human genome, it became possible to compare genetic and physical maps.

Figure 19.11 shows a plot of genetic distance (y axis; in centimorgans) versus physical distance for human chromosome 12 (x axis; in megabases) (IHGSC, 2001). There are two main conclusions. First, the recombination rate tends to be suppressed near the centromeres (note the flat slope in Fig. 19.11, arrow 1), while the recombination rate is far higher near the telomeres. This effect is especially pronounced in males. Second, long chromosome arms tend to have an average recombination rate of 1 cM/Mb, while the shortest arms have a much higher average recombination rate (above 2 cM/Mb). The range of the recombination rate throughout the genome varies from 0 to 9 cM/Mb (Yu et al., 2001). These researchers identified 19 recombination "deserts" (up to 5 Mb in length with sex-average recombination rates below 0.3 cM/Mb) and 12 recombination "jungles" (up to 6 Mb in length with sex-average recombination rates above 3.0 cM/Mb). In Computer Lab exercise 19.3 at the end of this chapter, we will identify regions of high (or low) recombination on the UCSC Genome Browser.

Repeat Content of the Human Genome

Repetitive DNA probably occupies over 50% of the human genome. While our genome has the most fundamental role in defining our biological nature, these repeats might seem to constitute a vast amount of irrelevant material that has opportunistically infiltrated and assimilated itself. Through the Human Genome Project we are beginning to characterize the nature and extent of repetitive DNA in our genome, including its decisive roles in evolution and disease.

There are five main classes of repetitive DNA in humans (IHGSC, 2001; Jurka, 1998), as discussed in Chapter 16:

1. Interspersed repeats (transposon-derived repeats)

2. Processed pseudogenes: inactive, partially retroposed copies of protein-coding genes

3. Simple sequence repeats: microsatellites and minisatellites, including short sequences such as $(A)_n$, $(CA)_n$, or $(CGG)_n$

4. Segmental duplications, consisting of blocks of 10 to 300 kb that are copied from one genomic region to another

5. Blocks of tandemly repeated sequences such as are found at centromeres, telomeres, and ribosomal gene clusters

We will briefly explore each of these repeats.

Transposon-Derived Repeats

Incredibly, 45% of the human genome or more consists of repeats derived from transposons. These are often called interspersed repeats. Many transposon-derived repeats replicated in the human genome in the distant past (hundreds of millions of years ago), and thus because of sequence divergence it is possible that the 45% value is an underestimate. Transposon-derived repeats can be classified in four categories (Jurka, 1998; Ostertag and Kazazian, 2001):

- LINEs occupy 21% of the human genome.

- SINEs occupy 13% of the human genome.

- LTR transposons account for 8% of the human genome.

- DNA transposons comprise about 3% of the human genome.

The structure of these repeats is shown in Fig. 19.12, as well as their abundance in the human genome. LINEs, SINEs, and LTR transposons are all retrotransposons that encode a reverse transcriptase activity. They integrate into the genome through an RNA intermediate. In contrast, DNA transposons have inverted terminal repeats and encode a bacterial transposon-like transposase activity.

Retrotransposons can further be classified into those that are autonomous (encoding activities necessary for their mobility) and those that are nonautonomous

The number of interspersed repeats was estimated using RepeatMasker to search RepBase (see Chapter 16).

Alu elements are so named because the restriction enzyme *Alu* I digests them in the middle of the sequence. In mouse, these are called B1 elements.

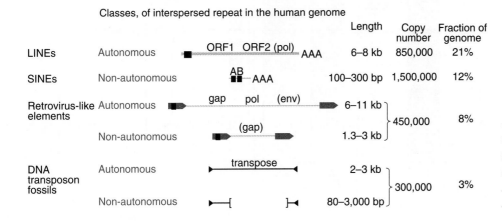

FIGURE 19.12. *There are four types of transposable elements in the human genome: LINEs, SINEs, LTR transposons, and DNA transposons (from IHGSC, 2001). Used with permission.*

TABLE 19-6 Interspersed Repeats in Four Eukaryotic Genomes

	Human		Drosophila		C. elegans		A. thaliana	
	Bases	Families	Bases	Families	Bases	Families	Bases	Families
LINE/SINE	33.4%	6	0.7%	20	0.4%	10	0.5%	10
LTR	8.1%	100	1.5%	50	0%	4	4.8%	70
DNA	2.8%	60	0.7%	20	5.3%	80	5.1%	80
Total	44.4%	166	3.1%	90	6.5%	94	10.5%	160

"Bases" refers to percentage of bases in the genome, "families" to approximate number of families in the genome.
Source: Adapted from IHGSC (2001). Used with permission.

(depending on exogenous activities such as DNA repair enzymes). The most common nonautonomous retrotransposons are *Alu* elements.

Interspersed repeats occupy a far greater proportion of the human genome than in other eukaryotic genomes (Table 19.6). The total number of interspersed repeats is estimated to be 3 million. These repeats offer an important opportunity to study molecular evolution. Each repeat element, even if functionally inactive, represents a "fossil record" which can be used to study genome changes within and between species. Transposons accumulate mutations randomly and independently. It is possible to perform a multiple sequence alignment of transposons and to calculate the percent sequence divergence. Transposon evolution is assumed to behave like a molecular clock, which can be calibrated based on the known age of divergence of species such as humans and Old World monkeys (23 million years ago [MYA]). Based on such phylogenetic analyses, several conclusions can be made (IHGSC, 2001) (Fig. 19.13):

- Most interspersed repeats in the human genome are ancient, predating the mammalian eutherian radiation 100 MYA. These elements are removed from the genome only slowly.

- SINEs and LINEs have long lineages, some dating back 150 MYA.

- There is no evidence for DNA transposon activity in the human genome in the past 50 million years; thus, they are extinct fossils.

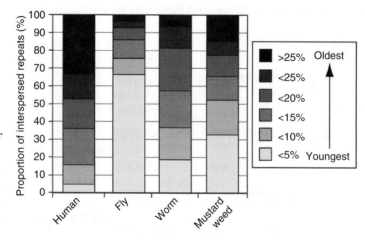

FIGURE 19.13. *Comparison of the age of interspersed repeats in four eukaryotic genomes. Humans have a small proportion of recent interspersed repeats (from IHGSC, 2001). Used with permission.*

TABLE 19-7 Simple Sequence Repeats (Microsatellites) in the Human Genome

Length of Repeat	Average Bases per Megabase	Average Number of SSR Elements per Megabase
1	1660	33.7
2	5046	43.1
3	1013	11.8
4	3383	32.5
5	2686	17.6
6	1376	15.2
7	906	8.4
8	1139	11.1
9	900	8.6
10	1576	8.6
11	770	8.7

Abbreviation: SSR, simple sequence repeat.
Source: IHGSC, 2001.

Simple Sequence Repeats

Simple sequence repeats are repetitive DNA elements that consist of perfect (or slightly imperfect) tandem repeats of k-mers. When the repeat unit is short (k is about 1 to 12 bases), the simple sequence repeat is called a microsatellite. When the repeat unit is longer (from about 12 to 500 bases), it is called a minisatellite (Toth et al., 2000).

Micro- and minisatellites comprise about 3% of the human genome (IHGSC, 2001). The most common repeat lengths are shown in Table 19.7. The most common repeat units are the dinucleotides AC, AT, and AG. We saw examples of these with the RepeatMasker program (Chapter 16).

Segmental Duplications

Segmental duplications occur when the human genome contains duplicated blocks of from 1 to 200 kb of genomic sequence. About 5.5% of the finished human genome sequence consists of segmental duplications, typically of 10 to 50 kb (Bailey et al., 2001). Many of these duplication events are recent, because both introns and coding regions are highly conserved. (For ancient duplication events, less conservation is expected between duplicated intronic regions.) Segmental duplications may be interchromosomal or intrachromosomal. The centromeres contain large amounts of interchromosomal duplicated segments, with almost 90% of a 1.5 Mb region containing these repeats (Fig. 19.14). Smaller regions of these repeats also occur near the telomeres.

Gene Content of the Human Genome

It is of great interest to characterize the gene content of the human genome because of the critical role of genes in human biology. However, the genes are the hardest features of genomic DNA to identify (see Chapter 16). This is a challenging task for many reasons:

- The average exon is only 50 codons (150 nucleotides). Such small elements are hard to identify as exons unambiguously.

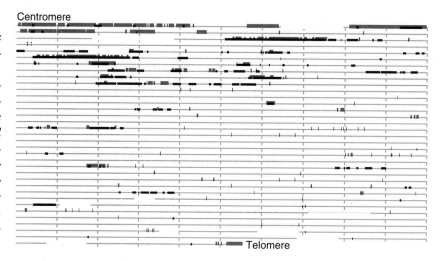

FIGURE 19.14. The centromeres consist of large amounts of interchromosomal duplicated segments. The size and location of intrachromosomal (black) and interchromosomal (red) segmental duplications are indicated. Each horizontal line represents 1 Mb of chromosome 22q; the tick marks indicate 100 kb intervals. The centromere is at top left, and the telomere is at the lower right (adapted from IHGSC, 2001). Used with permission.

We described cDNA projects in Chapter 8.

- Exons are interrupted by introns, some many kilobases in length. In the extreme case, the human dystrophin gene extends over 2.4 Mb, the size of an entire genome of a typical prokaryote. Thus, the use of complementary DNAs continues to provide an essential approach to gene identification.

- There are many pseudogenes that may be difficult to distinguish from functional protein-coding genes.

- The nature of noncoding genes is poorly understood (see Chapter 16 and below).

Noncoding RNAs

There are many classes of human genes that do not encode proteins. Noncoding RNAs can be difficult to identify in genomic DNA because they lack open reading frames, they may be small, and they are not polyadenylated. Thus, they are difficult to detect by gene-finding algorithms, and they are often not present in cDNA libraries. These noncoding RNAs include the following:

- Transfer RNAs, required as adapters to translate mRNA into the amino acid sequence of proteins

- Ribosomal RNAs, required for mRNA translation

- Small nucleolar RNAs (snoRNAs), required for RNA processing in the nucleolus

- Small nuclear RNAs (snRNAs), required for spliceosome function

Hundreds of noncoding RNAs were identified in the draft version of the human genome (Table 19.8). The tRNA genes were most predominant, with 497 such genes and an additional 324 tRNA-derived pseudogenes. The tRNA genes associated with the human genetic code can now be described. This version of the genetic code includes the frequency of codon utilization for each amino acid and the number of tRNA genes that are associated with each codon. The total number of tRNA genes is comparable to that observed in other eukaryotes (Table 8.2).

Protein-Coding Genes

Protein-coding genes are characterized by exons, introns, and regulatory elements. These basic features are summarized in Table 19.9. The average coding sequence

TABLE 19-8	Noncoding Genes in the Human Genome		

RNA Gene	No. of Noncoding Genes	No. of Related Genes	Function
tRNA	497	324	Protein synthesis
SSU (18S) RNA	0	40	Protein synthesis
5.8S rRNA	1	11	Protein synthesis
LSU (28S) rRNA	0	181	Protein synthesis
5S RNA	4	520	Protein synthesis
U1	16	134	Spliceosome component
U2	6	94	Spliceosome component
U4	4	87	Spliceosome component
U4atac	1	20	Minor (U11/U12) spliceosome component
U5	1	31	Spliceosome component
U6	44	1135	Spliceosome component
U6atac	4	32	Minor (U11/U12) spliceosome component
U7	1	3	Histone mRNA 3′ processing
U11	0	6	Minor (U11/U12) spliceosome component
U12	1	0	Minor (U11/U12) spliceosome component
SRP (7SL) RNA	3	773	Component of signal recognition particle
RNAse P	1	2	tRNA 5′ end processing
RNAse MRP	1	6	rRNA processing
Telomerase RNA	1	4	Template for addition of telomeres
hY1	1	353	Component of Ro RNP, function unknown
hY3	25	414	Component of Ro RNP, function unknown
hY4	3	115	Component of Ro RNP, function unknown
hY5 (4.5S RNA)	1	9	Component of Ro RNP, function unknown
Vault RNAs	3	1	Component of 13 Mda vault RNP
7SK	1	330	Unknown
H19	1	2	Unknown
Xist	1	0	Initiation of X chromosome inactivation
Known C/D snoRNAs	69	558	Pre-rRNA processing or site-specific ribose methylation of rRNA
Known H/ACA snoRNAs	15	87	Pre-rRNA processing or site-specific pseudouridylation of rRNA

Source: Adapted from IHGSC (2001). Used with permission.

TABLE 19-9 Characteristics of Human Genes

Feature	Size (median)	Size (mean)
Internal exon	122 bp	145 bp
Exon number	7	8.8
Introns	1,023 bp	3,365 bp
3′-untranslated region	400 bp	770 bp
5′-untranslated region	240 bp	300 bp
Coding sequence	1,100 bp	1,340 bp
Coding sequence	367 aa	447 aa
Genomic extent	14 kb	27 kb

Abbreviations: aa, amino acids; bp, base pairs; kb, kilobases.
Source: Adapted from IHGSC (2001).

The longest coding sequence is titin (80,780 bp; NM_003319). The gene for titin, on chromosome 2q24.3, has 178 exons and encodes a muscle protein of 26,926 amino acids (about 3 million Da). By contrast, a typical protein encoded by an mRNA of 1340 bp is about 50,000 Da.

Chromosome 19, the most GC-rich chromosome, also houses the greatest density of genes (26.8 per megabase). The average density of gene predictions across the genome is 11.1 per megabase. The Y chromosome is least dense, having 6.4 predicted genes per megabase.

EBI proteome analysis via Integr8 is available at ► http://www.ebi.ac.uk/integr8.

We discussed the GO Consortium and InterPro in Chapter 10.

for human genes is 1340 bp (IHGSC, 2001). This is comparable to the size of an average coding sequence in nematode (1311 bp) and *Drosophila* (1497 bp). Most internal exons are about 50 to 200 bp in length in all three species (Fig. 19.15a), although worm and fly have a greater proportion of longer exons (note the flatter tail in Fig. 19.15a). However, the size of human introns is far more variable (Figs. 19.15b and c). This results in a more variable overall gene size in humans than in worm and fly.

Protein-coding genes are associated with a high GC content (Fig. 19.16). While the overall GC content of the human genome is about 41%, the GC content of known genes (having RefSeq identifiers) is higher (Fig. 19.16a). Gene density increases 10-fold as GC content rises from 30% to 50% (Fig. 19.16b).

In an effort to catalog all protein-coding genes and their protein products, the IHGSC (2001) created an integrated gene index (IGI) and a corresponding integrated protein index (IPI). According to the European Bioinformatics Institute Integr8 database, there are 40,014 proteins in the human proteome. We listed the 15 most common protein domains in Table 10.3. Ensembl predictions are shown for the 15 most common protein families (Table 19.10) and most common repeats (Table 19.11).

Comparative Proteome Analysis

The importance of comparative analyses has emerged as one of the fundamental tenets of genomics. A comparison of human proteins to proteins from the completed genomes of *S. cerevisiae*, *A. thaliana*, *C. elegans*, and *D. melanogaster* is shown in Table 19.12. The IHGSC (2001) analyzed functional groups of these proteins based on InterPro and Gene Ontology (GO) Consortium classifications. Humans have relatively more genes that encode proteins predicted to function in cytoskeleton, transcription/translation, and defense and immunity (Fig. 19.17).

The human proteome was further studied by blastp searching every predicted protein against the nonredundant database. The distribution of homologs is shown in Fig. 19.18. Overall, 74% of the proteins were significantly related to other known proteins. As more sequences are accumulated in databases over time, the matches between human proteins and other eukaryotes (and prokaryotes) will continue to increase.

Complexity of Human Proteome

The number of protein-coding genes in humans is comparable to the number of genes in other metazoans and plants and only fivefold greater than the number in

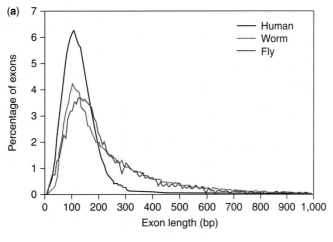

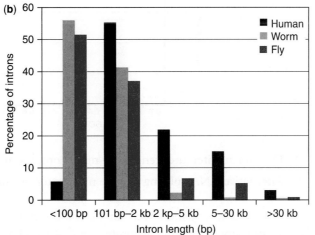

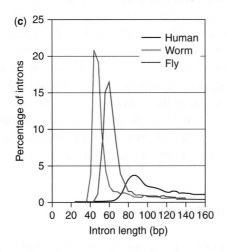

FIGURE 19.15. Size distribution of (a) exons, (b) introns, and (c) short introns (enlarged from [b]) in human, worm, and fly (from IHGSC, 2001). Used with permission.

unicellular fungi. Nonetheless the human proteome may be far more complex for several reasons (IHGSC, 2001):

1. There are relatively more domains and protein families in humans than in other organisms.

2. The human genome encodes relatively more paralogs, potentially yielding more functional diversity.

3. There are relatively more multidomain proteins having multiple functions.

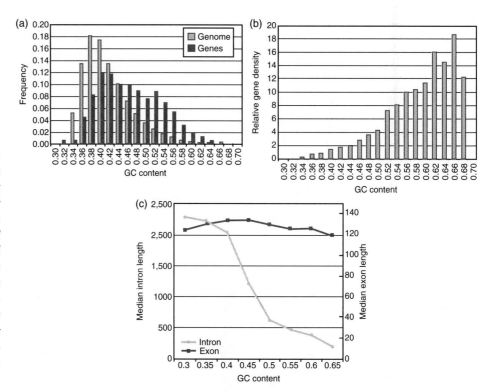

FIGURE 19.16. (a) Distribution of GC content in genes and in the genome shows that protein-coding genes are associated with a higher GC content. (b) The gene density is plotted as a function of the GC content. (The density is obtained by taking the ratio of the values in [a].) As GC content rises, the relative gene density increases dramatically. (c) Mean exon length is unaffected by GC content, but introns are far shorter as GC content rises (from IHGSC, 2001). Used with permission.

4. Domain architectures tend to be more complex in the human proteome.

5. Alternative RNA splicing may be more extensive in humans.

There may be a synergistic effect among these factors, leading to a substantially greater complexity of the human proteome that could account for the phenotypic complexity of vertebrates, including humans.

In its reannotation of the human genome, the IHGSC (2004) identified the largest clusters of human paralogous genes that involve recent gene duplications (Table 19.13). These genes are neighboring (indicating local gene duplication). The selected sites displayed near neutrality (estimated substitution rate per synonymous site $K_s < 0.30$, such that each homolog differs from a common ancestral gene by an average $K_s < 0.15$). These represent genes that were recently born in the human lineage (after the divergence from rodents), and many have functions in olfaction, immune function, and the reproductive system.

24 HUMAN CHROMOSOMES

Each human chromosome has been finished (or nearly finished) by a dedicated research team. For each chromosome, this has resulted in a publication in the journal *Nature* (or *Science*). There are seven traditional cytogenetic groups A to G, which categorize the chromosomes (other than the mitochondrial genome) according to morphological properties (Table 19.14). We next briefly summarize key aspects of each chromosome, following this organization (Tables 19.15 to 19.21).

The exact number of genes is not yet known (for typical projects comparative studies suggest that over 95% of the genes have been annotated). The EGASP

TABLE 19-10	Fifteen Most Common Families for *Homo sapiens*	
InterPro	Proteins Matched	Name
IPR000276	872	Rhodopsin-like GPCR superfamily
IPR000725	547	Olfactory receptor
IPR001909	350	KRAB box
IPR001806	173	Ras GTPase
IPR013753	134	Ras
IPR007114	98	Major facilitator superfamily
IPR001664	88	Intermediate filament protein
IPR011701	82	Major facilitator superfamily MFS-1
IPR001128	74	Cytochrome P450
IPR000832	68	GPCR, family 2, secretin-like
IPR002198	66	Short-chain dehydrogenase/reductase SDR
IPR003579	65	Ras small GTPase, Rab type
IPR001993	61	Mitochondrial substrate carrier
IPR004000	57	Actin/actin-like
IPR002494	55	Keratin, high sulfur B2 protein

Source: From ► http://www.ebi.ac.uk/proteome/, January 2008.

competition, described in Chapter 16, highlights the computational challenges in correctly identifying genes with good sensitivity and specificity. The values of gap lengths in Tables 19.15 to 19.21 are changeable over time. In almost every case they represent regions that are refractory to cloning and sequencing because of the highly repetitive nature of the underlying DNA sequence, even when up to 100-fold coverage of the chromosome is obtained. Overall, the finishing of

TABLE 19-11	Fifteen Most Common Repeats for *Homo sapiens*	
InterPro	Proteins Matched	Name
IPR001680	304	WD40 repeat
IPR002110	280	Ankyrin
IPR001611	270	Leucine-rich repeat
IPR003591	131	Leucine-rich repeat, typical subtype
IPR001440	108	Tetratricopeptide TPR-1
IPR008160	81	Collagen triple helix repeat
IPR000357	80	HEAT
IPR006652	72	Kelch repeat type 1
IPR013105	70	Tetratricopeptide TPR2
IPR000884	65	Thrombospondin, type I
IPR008161	52	Collagen helix repeat
IPR002172	50	Low density lipoprotein-receptor, class A
IPR003659	46	Plexin/semaphorin/integrin
IPR000225	44	Armadillo
IPR002165	33	Plexin

Source: From ► http://www.ebi.ac.uk/proteome/, January 2008.

TABLE 19-12   Proteome Comparisons between Human and *Arabidopsis, C. elegans, Drosophila,* and *S. cerevisiae*

Organism	No. of Proteins in Proteome	Proteins with InterPro Matches	Percent of all Proteins	No. of Signatures	No. of InterPro Entries
H. sapiens	26,146	18,946	72.5	56,344	6,202
M. musculus	23,429	18,203	77.7	53,657	6,125
A. thaliana	31,892	24,315	76.2	56,148	4,494
C. elegans	20,265	13,919	68.7	31,377	4,232
D. melanogaster	13,967	10,482	75.0	27,712	4,472
S. cerevisiae	5,862	4,542	77.5	11,103	3,379

Source: From ▶ http://www.ebi.ac.uk/proteome/, January 2008.

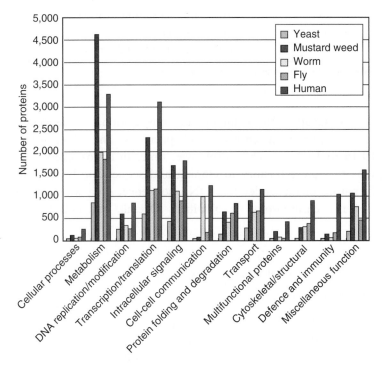

FIGURE 19.17. Functional categories in eukaryotic proteomes of yeast (S. cerevisiae), mustard weed (A. thaliana), worm (C. elegans), fly (D. melanogaster), and human. The classification categories were derived from InterPro (Chapter 10) (from IHGSC, 2001). Used with permission.

the euchromatic portion of the human genome included 250 gaps spanning 25 megabases, while the heterochromatic portion had far fewer gaps (just 33) spanning a vast size (200 megabases) (IHGSC, 2004).

Group A (Chromosomes 1, 2, 3)

Chromosome 1, the largest chromosome, has 3141 genes and 991 pseudogenes (Gregory et al., 2006) (Table 19.15). Its gene density (14.2 genes per megabase) is nearly twice the genome-wide average (7.8 genes per megabase). Typical for essentially all the chromosome finishing projects, sequence integrity and completeness were assessed three ways: (1) by determining whether all RefSeq genes assigned to

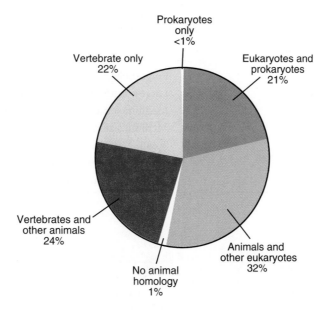

FIGURE 19.18. *Taxonomic distribution of the protein homologs of predicted human proteins. Each protein was searched by blastp, and proteins with an E value less than 0.001 were called homologs. Additional PSI-BLAST searches were performed (with three iterations) (from IHGSC, 2001). Used with permission.*

| **TABLE 19-13** | Human Paralogous Genes with Largest Cluster Sizes Involved in Recent Gene Duplications ($K_s \leq 0.3$) |

Cluster Size	Minimum Size in Ancestral Genome	Genes Involved in Recent Duplications	Chromosome	Gene Family
64	50	23	11	Olfactory receptor
59	54	10	11	Olfactory receptor
34	25	13	1	Olfactory receptor
30	8	26	2	Immunoglobulin K chain V
23	5	19	19	KRAB zinc-finger protein
23	19	6	11	Olfactory receptor
21	9	15	14	Immunoglobulin heavy chain
20	11	12	22	Immunoglobulin λ chain V-region
18	9	13	19	Leukocyte and NK cell immunoglobulin-like receptors
18	14	6	19	Gonadotropin-inducible transcription repressor-2-like
16	4	13	9	Interferon α
16	10	7	19	FDZF2-like KRAB zinc-finger protein
14	8	7	12	Taste receptor, type 2
13	3	11	1	PRAME/MAPE family (cancer/germ line antigen)
13	9	8	17	Olfactory receptor
11	2	11	16	Immunoglobulin heavy chain
10	1	10	19	Pregnancy-specific β-1-glycoprotein

Source: Modified from IHGSC (2004). Used with permission.

TABLE 19-14 Human Chromosome Groups

Group	Chromosomes	Description
A	1–3	Largest chromosomes; 1,3 are metacentric; 2 is submetacentric
B	4, 5	Large chromosomes; submetacentric
C	6–12, X	Medium size chromosomes; submetacentric
D	13–15	Medium size chromosomes; acrocentric with satellites
E	16–18	Small; 16 is metacentric; 17,18 are submetacentric
F	19, 20	Small, metacentric chromosomes
G	21, 22, Y	Smallest chromosomes; acrocentric; satellites on 21 and 22

TABLE 19-15 Group A Chromosomes

Chromosome	Length (Mb)	No. of Genes	No. of Pseudogenes	Gap Size (kb)	Accession
1	247	3,141	991	21,115	NC_000001
2	243	1,346	1,239	5,412	NC_000002
3	200	1,463	122	3,435	NC_000003

Sources: Gregory et al. (2006); Hillier et al. (2005); Muzny et al. (2006a). Length is from NCBI build 36 version 2. Gap sizes are from IHGSC (2004).

TABLE 19-16 Group B Chromosomes

Chromosome	Length (Mb)	No. of Genes	No. of Pseudogenes	Gap Size (kb)	Accession
4	191	796	778	4,250	NC_000004
5	181	923	577	432	NC_000005

Sources: Hillier et al. (2005); Schmutz et al. (2004). Length is from NCBI build 36 version 2. Gap sizes are from IHGSC (2004).

TABLE 19-17 Group C Chromosomes

Chromosome	Length (Mb)	No. of Genes	No. of Pseudogenes	Gap Size (kb)	Accession
6	171	1,557	633	2,958	NC_000006
7	159	1,150	941	5,499	NC_000007
8	146	793	301	2,852	NC_000008
9	140	1,149	426	19,955	NC_000009
10	135	816	430	3,535	NC_000010
11	134	1,524	765	5,082	NC_000011
12	132	1,342	93	5,095	NC_000012
X	155	1,098	700	3,750	NC_000023

Sources: Mungall et al. (2003); Hillier et al. (2003); Nusbaum et al. (2006); Humphray et al. (2004); Deloukas et al. (2004); Taylor et al. (2006); Scherer et al. (2003, 2006); Ross et al. (2005). Length is from NCBI build 36 version 2. Gap sizes are from IHGSC (2004).

TABLE 19-18 Group D Chromosomes

Chromosome	Length (Mb)	No. of Genes	No. of Pseudogenes	Gap Size (kb)	Accession
13	114	633	296	17,915	NC_000013
14	106	1,050	393	17,228	NC_000014
15	100	695	250	18,997	NC_000015

Sources: Dunham et al. (2004); Heilig et al. (2003); Zody et al. (2006). Length is from NCBI build 36 version 2. Gap sizes are from IHGSC (2004).

TABLE 19-19 Group E Chromosomes

Chromosome	Length (Mb)	No. of Genes	No. of Pseudogenes	Gap Size (kb)	Accession
16	89	796	778	10,143	NC_000016
17	79	1,266	274	8,375	NC_000017
18	76	337	171	1,465	NC_000018

Sources: Martin et al. (2004); Zody et al. (2006); Nusbaum et al. (2006). Length is from NCBI build 36 version 2. Gap sizes are from IHGSC (2004).

TABLE 19-20 Group F Chromosomes

Chromosome	Length (Mb)	No. of Genes	No. of Pseudogenes	Gap Size (kb)	Accession
19	64	1,461	321	5,355	NC_000019
20	62	727	168	2,923	NC_000020

Sources: Grimwood et al. (2003); Deloukas et al. (2001). Length is from NCBI build 36 version 2. Gap sizes are from IHGSC (2004).

TABLE 19-21 Group G Chromosomes

Chromosome	Length (Mb)	No. of Genes	No. of Pseudogenes	Gap Size (kb)	Accession
21	47	796	778	11,673	NC_000021
22	50	545	134	14,790	NC_000022
Y	58	78	n/a	33,098	NC_000024

Sources: Hattori et al. (2000); Dunham et al. (1999); Skaletsky et al. (2003). Length is from NCBI build 36 version 2. Gap sizes are from IHGSC (2004).

the chromosome were accounted for, (2) by comparing the order of hundreds of chromosome markers to the Decode genetic map to search for discrepancies, and (3) by aligning over 32,000 pairs of fosmid end sequences to unique positions in the sequence. This resulted in the identification of several misassemblies caused by low-copy repeats. In some cases, naturally occurring polymorphisms confound the analysis; for example, 50% of individuals lack the *GSTM1* gene.

Chromosome 2, the second largest chromosome, is remarkable because it corresponds to two intermediate-sized ancestral, acrocentric chromosomes that fused head-to-head. In other primates these chromosomes remained separate, as in the case of chimpanzee chromosomes 2A and 2B (Fig. 19.19). In its finished sequence,

Homo sapiens chromosome 2

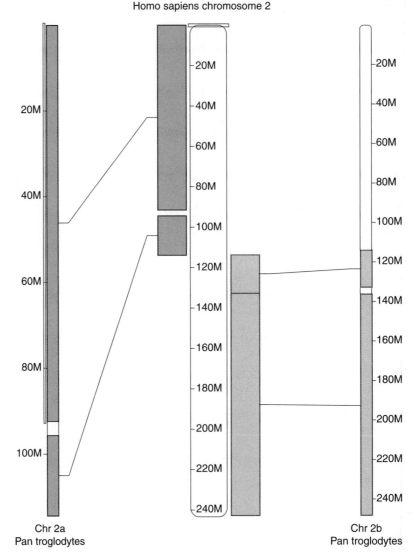

FIGURE 19.19. Conserved synteny between human chromosome 2 and two smaller chimpanzee chromosomes provides evidence that two ancestral human acrocentric chromosomes fused. This image is from the ensembl synteny viewer (linked from ▶ http://www.ensembl.org/Homo_sapiens/syntenyview).

the fusion site in 2q13-2q14.1 has been localized to hg16:114,455,823-114,455,838 (Hillier et al., 2005). One of the two centromeres (at 2q21) became inactivated, and contains α-satellite remnants.

Although chromosome 3 is large, it contains the lowest rate of segmental duplications in the genome (1.7% compared to a genome-wide average of 5.3% of nucleotides segmentally duplicated) (Muzny et al., 2006a). Chromosomes 3 and 21 derive from a larger ancestral chromosome that split. It also includes a large pericentric inversion (also present in chimpanzee and gorilla, but not orangutan or Old World monkeys).

Group B (Chromosomes 4, 5)

Chromosome 4 has an unusually low GC content of 38.2%, compared to the genome-wide average of 41% (Hillier et al., 2005) (Table 19.16). Over 19% of the chromosome has a GC content less than 35%. However, portions of the chromosome have a GC content >70%. You can view these using the UCSC Genome Browser's GC content annotation track, or the Table Browser.

Chromosome 5 has both a very low gene density and a very high rate of intrachromosomal duplications (Schmutz et al., 2004). It includes 923 gene loci, and 577 pseudogenes. There are many gene-poor loci that are highly conserved and thus are thought to be functionally constrained.

Group C (Chromosomes 6 to 12, X)

The largest transfer RNA gene cluster is localized to chromosome 6p, with 157 tRNA genes out of 616 across the entire genome (Mungall et al., 2003). Chromosome 6 (Table 19.17) also contains HLA-B, the most polymorphic gene in the human genome. We will explore this polymorphism further in Computer Lab exercise 19.5 at the end of this chapter.

Chromosome 7 was sequenced by the public consortium (Hillier et al., 2003) and by Scherer et al. (2003) using a mixture of Celera whole-genome scaffolds and International Human Genome Sequencing Consortium data. The centromere is polymorphic, ranging from 1.5 to 3.8 megabases at one locus (marker D7Z1) and from 100 to 500 kilobases at another site (D7Z2). There is an unusually large amount of segmentally duplicated sequence (8.2%). As an example of the consequence of this, Williams–Beuren syndrome results from the hemizygous deletion of 1.5 million base pairs on chromosome 7q11.23, a region containing about 17 genes. There are flanking repeats that mediate unequal meiotic recombination (Fig. 16.25) or, in some cases, hemizygous inversions (Osborne et al., 2001).

Other group C chromosomes are 8 (Nusbaum et al., 2006), 9 (Humphray et al., 2004), 10 (Deloukas et al., 2004), 11 (Taylor et al., 2006), 12 (Scherer et al., 2006), and X (Ross et al., 2005). Chromosome 9 contains the largest autosomal block of heterochromatin. Chromosome 11 is notable for having the beta globin gene cluster as well as the insulin gene.

The X chromosome joins the group C chromosomes because of its comparable size. It is unique in many ways. Mammals are classified into three groups, in all of which males have X and Y chromosomes: the eutherians (placental mammals), the metatheria (marsupials), and the prototheria (egg-laying mammals). Females undergo X chromosome inactivation (XCI) in which one copy is silenced early in development. In contrast to the autosomes, the male X chromosome does not recombine during meiosis, except for short pseudoautosomal regions at the tips (PAR1 on Xp and PAR2 on Xq) that recombine with corresponding portions of the Y chromosome. Since males have only a single copy of the X chromosome (thus it is hemizygous), recessive phenotypes are exposed and many X-linked diseases have been described, from hemophilia to X-linked mental retardation syndromes. The X and Y chromosomes derive from an ancient autosomal chromosome pair that began transforming into sex chromosomes over 300 million years ago, and sequencing of the X (and Y) chromosomes has revealed traces of evolutionary conservation between the two (Ross et al., 2005 and see below).

Group D (Chromosomes 13 to 15)

The five human acrocentric chromosomes are 13, 14, and 15 (Table 19.18), as well as 21 and 22. For each, the p arm is almost entirely heterochromatic. These regions have a highly repetitive structure, and all five include arrays of ribosomal DNA genes as shown in Fig. 8.7. Sequencing and accurately assembling these regions is so challenging that they were not targeted by the Human Genome Project and are still not part of the standard human genome assemblies.

Group E (Chromosomes 16 to 18)

Of this group of chromosomes, 16 and 17 are notable for above-average levels of segmental duplication (Table 19.19). Chromosome 18 has the lowest gene density of any autosome (4.4 genes per megabase) and encodes only 337 genes (about a quarter of the number of the similar-sized chromosome 17). One region of chromosome 18 has only 3 genes across 4.5 megabases. The sparse number of genes may explain why some individuals with trisomy 18 (Edwards syndrome) survive to birth, while all other autosomal trisomies (except trisomy 13 and trisomy 21) are embryonic lethal.

Group F (Chromosomes 19, 20)

Chromosome 19 has the highest gene density with 26 protein-coding genes per megabase (Table 19.20). It also has an unusually high density of repeats (55% of the chromosome, in contrast to a genome-wide average of about 45%). Almost 26% of the chromosome is composed of *Alu* repeats, consistent with the high gene density.

Group G (Chromosomes 21, 22, Y)

Group G chromosomes are the smallest (Table 19.21). While the short arms of the five acrocentric chromosomes are nearly entirely heterochromatic, an exception is 21p11.2, which includes a very small euchromatic region. A view of 12 megabases extending across the p arm of chromosome 21 and its centromere highlights how little annotated information is currently available (Fig. 19.20). There are only two genes there: *BAGE* (B melanoma antigen) and *TPTE* (transmembrane phosphatase with tensin homology). The other acrocentric arms have no genes annotated.

The Y chromosome was the most technically difficult to sequence because of its extraordinarily repetitive nature (Skaletsky et al., 2003; Jobling and Tyler-Smith, 2003). It has short pseudoautosomal regions at the ends that recombine with the X chromosome. A large central region, spanning 95% of its length, is termed the male-specific region (MSY). There are 23 megabases of euchromatin, including 8 Mb on Yp and 14.5 Mb on Yq. There are three notable heterochromatic regions:

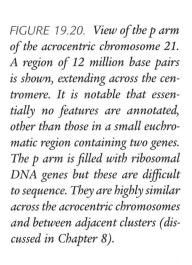

FIGURE 19.20. *View of the p arm of the acrocentric chromosome 21. A region of 12 million base pairs is shown, extending across the centromere. It is notable that essentially no features are annotated, other than those in a small euchromatic region containing two genes. The p arm is filled with ribosomal DNA genes but these are difficult to sequence. They are highly similar across the acrocentric chromosomes and between adjacent clusters (discussed in Chapter 8).*

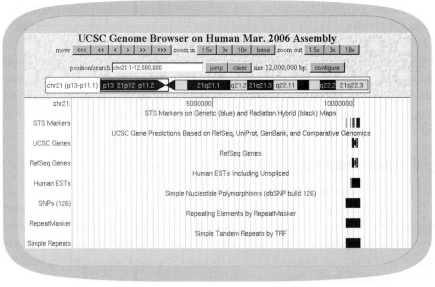

(1) a centromeric region of about 1 Mb, (2) a block of ~40 Mb on the long arm, and (3) an island of 400 kilobases comprised of over 3000 tandem repeats of 125 base pairs. Of 156 transcription units, about half encode proteins. Skaletsky et al. defined three classes of euchromatic sequences: (1) X-transposed sequences total 3.4 megabases and share 99% identity to Xq21 DNA sequences. Just 3 to 4 million years ago, after the human-chimpanzee divergence, there was a massive transposition of X chromosome sequences to the Y chromosome, followed by an inversion that dispersed these sequences on the Y. (2) X-degenerate sequences share 60% to 90% identity to 27 different X chromosome genes, and represent relics of the ancient autosomes from which X and Y evolved. (3) Ampliconic sequences span over 10 megabases and consist of blocks of sequences sharing as much as 99.9% nucleotide identity over spans of tens or hundreds of kilobases. The amplicons are the most gene-dense regions of the Y chromosome, and have a low content of interspersed repeats. The ampliconic regions contain eight giant palindromes, collectively spanning 5.7 megabases, each with two long arms interrupted by a unique, central spacer. The extraordinary conservation of the palindromic arms is due to gene conversion, the nonreciprocal transfer of sequences from one DNA duplex to another (Skaletsky et al., 2003; Rozen et al., 2003).

The Mitochondrial Genome

In addition to 22 autosomes and two sex chromosomes, humans have a mitochondrial genome. Mitochondrial genomes have a number of fascinating properties that also make them useful for phylogentic studies (reviewed in Pakendorf and Stoneking, 2005). They are present in high copy number, typically with hundreds or even thousands of genomes per cell. They are maternally inherited; all (or almost all) sperm-derived mitochondria are targeted for destruction in the fertilized oocyte. One consequence is that molecular phylogenetic studies of mitochondria follow the history of the maternal lineage, and thus have been traced to a "mitochondrial Eve" or proposed earliest human female ancestor. Another consequence of maternal inheritance is that mitochrondrial DNA does not undergo recombination. The mutation rate is higher than in nuclear DNA, providing a useful signal for molecular phylogenetic studies. Excluding the D-loop (which has not evolved at a constant rate across human lineages), Ingman et al. (2000) estimated the mitochondrial mutation rate to be 1.70×10^{-8} substitutions per site per year.

The reference genome (at NCBI) is 16,571 base pairs in a circular genome, obtained from a Yoruba individual (from Nigeria). The GC content is 44.5%, higher than for the other human chromosomes. The genome includes 37 annotated genes, spanning 68% of the genome. These include 13 protein-coding genes (encoding proteins involved in oxidative phosphorylation) and 24 structural RNAs (two ribosomal RNAs and 22 transfer RNAs; see Chapter 8). A region of about 1100 base pairs called the control region has regulatory functions.

The RefSeq human mitochondrial accession number is NC_001807. The Entrez Genome Project identifier is 168.

VARIATION: SEQUENCING INDIVIDUAL GENOMES

While sequencing the human genome was a massive project that has been compared in magnitude to landing a human on the moon, resequencing an individual human genome is relatively easier. The National Human Genome Research Institute (NHGRI) of the National Institutes of Health has launched programs to reduce

The NHGRI genome technology program website is ▶ http://www.genome.gov/10000368.

the cost of sequencing an individual genome from the recent value (tens of millions of dollars) to $100,000 and eventually to $1000 each.

The significance of individual genome sequencing is that it has the potential to facilitate the start of an era of individualized medicine in which DNA changes that are associated with a disease condition are identified. As discussed in Chapter 20, most diseases involve an interplay between genetic and environmental factors. Even for diseases that are seemingly caused by environmental factors, from lead poisoning to malnutrition to infectious disease, an individual's genetic constitution is likely to have a large effect on the disease process. Another significant aspect of individual genome sequencing is that it will help to elucidate the genetic diversity and history of the species.

In 2007 the first two individual human genome sequences were announced: that of J. Craig Venter (Levy et al., 2007) and that of James Watson, Nobel laureate and co-discoverer of the structure of DNA. The Venter genome was reported (Levy et al. 2007) as the diploid sequence of an individual. In contrast, the Celera human genome sequence (Venter et al., 2001) was based on a consensus of DNA sequences from five individuals, and the public consortium sequence (IHGSC, 2001) was also based on genomes from multiple individuals. Thus, these were composite efforts that represented sequence data that were essentially averaged to yield information on 23 pairs of chromosomes. They did not assess the variation that occurs in an individual having each autosome derived from maternal and paternal alleles. The surprising finding of Levy et al. was that there were four million variants between the parental chromosomes, about fivefold more than had been anticipated. Also it was not until the years 2004 to 2006 that the great diversity of copy number variants as well as smaller indels and SNPs became more fully appreciated.

In Computer Lab exercise 19.1 (below) we will perform BLAST searches against the Venter genome.

The strategy employed by Levy et al. (2007) to sequence, assemble, and analyze the genome included seven steps: (1) obtaining informed consent to collect the DNA sample; (2) genome sequencing; (3) genome assembly; (4) comparative mapping of the individual genome to an NCBI reference genome; (5) DNA variation detection and filtering; (6) haplotype assembly; and (7) data annotation and interpretation.

The assembly of Venter's genome was based on 32 million sequence reads generating ~20 billion base pairs of DNA sequence with a 7.5-fold depth of coverage. Sanger dideoxynucleotide sequencing technology was used because each read is longer than currently available 454 technology (used to sequence Watson's genome) or Solexa technology (see Chapter 13). The assembly included 2,782,357,138 bases of DNA. Comparison to the NCBI reference genome revealed 4.1 million variants. These included 3.2 million SNPs (slightly more than one per 1000 base pairs), over 50,000 block substitutions (of length 2 to 206 base pairs), almost 300,000 insertions/deletions (indels) of 1 to 571 base pairs, ~560,000 homozygous indels (ranging up to ~80,000 base pairs), 90 inversions, and many copy number variants. The majority of variants relative to the reference human genome were SNPs. Insertions and deletions accounted for a smaller proportion of the variable events (22%) but because they tend to involve larger genomic regions they accounted for 74% of the variant nucleotides relative to the reference NCBI genome.

VARIATION: SNPs TO COPY NUMBER VARIANTS

Karyotyping is a traditional approach to identifying chromosomal changes, typically using metaphase spreads of lymphoblastoid (or other) cell lines (see Chapter 16). G-banding (Giemsa staining) can detect events such as balanced translocations (which are relatively infrequent) and inversions (which often have severe phenotypic effects), as well as deletions and duplications. The resolution of standard G-banding approaches a limit of detection of approximately 3 megabases for deletions or duplications.

SNPs represent a fundamental form of variation in the human population. The International HapMap Project was begun in 2002 and reported the genotypes of 1.3 million SNPs in four geographically diverse populations (International HapMap Consortium, 2002, 2005): (1) 30 trios (consisting of mother, father, and an adult child) from the Yoruba tribe in Nigeria, abbreviated YRI; (2) 30 trios of northern and western European ancestry living in Utah and obtained from the Centre d'Etude du Polymorphisme Humain (CEPH) collection (abbreviated CEU); (3) 45 unrelated Han Chinese individuals in Beijing, China (CHB); and (4) 45 unrelated Japanese individuals in Tokyo, Japan (abbreviated JPT). In some studies, data from the Chinese and Japanese populations are pooled to yield three groups of 90 (YRI, CEU, CHB + JPT). A second generation haplotype map increased the number of characterized SNPs to 3.1 million (International HapMap Consortium, 2007). On average, the SNPs are spaced apart 875 base pairs across the genome. Each SNP has characteristic properties, including the sequence of its two (or more) alleles (although most SNPs are biallelic), its major and minor allele frequencies, and its relationship to neighboring SNPs. SNPs have varying extents of linkage disequilibrium (LD). In regions of high LD, SNPs are tightly linked to each other and form blocks in which the behavior of one SNP can serve as a proxy for the genotypes of neighboring SNPs. Commonly used measures of LD include D', r^2, and LOD. The HapMap Project includes a website from which all SNP data can be downloaded or viewed in a browser (Thorisson et al., 2005). An example is shown in Fig. 19.21 for the beta globin locus. The recombination rate (in cM/Mb) is plotted, and two recombination hotspots are evident (rectangles). In a triangle plot, LD measures for every pair of SNPs are plotted along lines at 45° to the horizontal track. Here, darker colors correspond to higher LD.

There are many uses of the dense map of SNPs provided by the HapMap project (see International HapMap Consortium, 2007; McVean et al., 2005). (1) SNP microarray analyses are used for genome-wide studies of disease association (Chapter 20). (2) SNPs reveal patterns of variation, such as shared ancestry, in human populations. Extended regions of homozygosity can be caused by inbreeding (in which the phenomenon is termed autozygosity) or it can reflect a genetic change often associated with disease, uniparental isodisomy (Chapter 20). Svante Pääbo (2003) has reviewed human genetic diversity, stressing the nature of each individual genome as a mosaic of haplotype blocks. (3) SNP analyses can reveal regions of the genome under strong positive selection. Wang et al. (2006) sorted high-frequency SNP alleles by homozygosity then searched for patterns of LD in neighboring alleles in order to identify many regions under selection. Also, the HapMap Consortium (2007) genotyped as many nonsynonymous SNPs as possible (over 17,000 passed quality control criteria). Nonsynonymous SNPs display an increased frequency of rare variants and a slight decrease of common variants relative to synonymous

The HapMap website is ▶ http://www.hapmap.org. The HapMap samples are available from the Coriell Cell Repositories (▶ http://ccr.coriell.org/). Accession numbers beginning with NA refer to genomic DNA samples, while GM accession numbers refer to cell lines. SNP data are available at dbSNP at NCBI (▶ http://www.ncbi.nlm.nih.gov/SNP/).

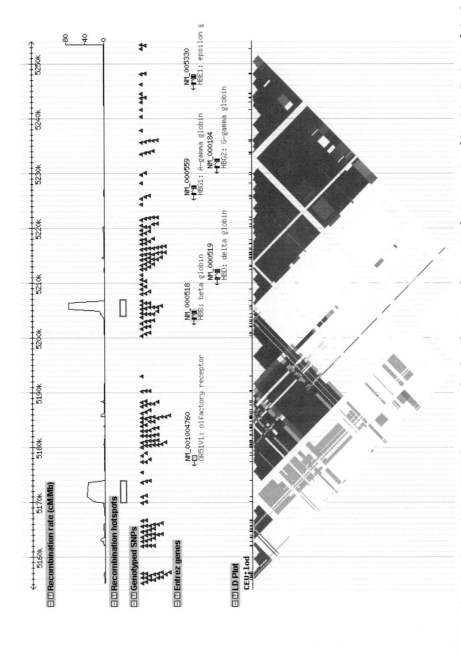

FIGURE 19.21. *Single nucleotide polymorphism (SNP) data are visualized on the genome browser at the International HapMap Project website (▶ http://www.hapmap. org). The displayed annotation tracks are the recombination rate (in cM/Mb; note two peaks), the recombination hotspots (rectangles), the genotyped SNPs, the NCBI Entrez genes, and a linkage disequilibrium (LD) triangle plot for the CEU population. Annotation tracks can be added or removed. For example, by displaying the YRI, CHB and JPT LD plots one can see differences in LD patterns across populations. Note that these data, including LD plots, can also be displayed using the UCSC Genome Browser.*

SNPs. Sabeti et al. (2007), including members of the HapMap Consortium, used three criteria to identify SNPs under strong positive selection: they were newly arisen (derived) alleles, based on comparisons to primate outgroups; they were highly differentiated between human populations, since recent positive selection is likely to reflect a local environmental adaptation; and they focused on nonsynonymous coding SNPs and SNPs in evolutionarily conserved sequences (Chapter 16) since those are most likely to have biological effects. Sabeti et al. described 300 candidate regions. In some cases they identified pairs of genes that have related functions and have undergone positive selection in the same populations (e.g., *LARGE* and *DMD* in the YRI population; both encode proteins implicated in Lass fever virus binding and infection).

Another basic use of SNPs is to identify chromosomal deletions, duplications, inversions and other abnormalities. We introduced SNP microarray and array comparative hybridization (aCGH) technologies in Chapter 16. They have been used to characterize single nucleotide changes (SNPs) or other very small changes as well as copy number variants (CNVs, defined by Scherer et al. [2007] as copy number differences extending over 1000 base pairs). Such changes are studied in apparently normal individuals, as well as in a variety of diseases such as cancers and idiopathic mental retardation. Using these platforms, fundamental questions can be addressed concerning human DNA variation, including the following.

1. What is the extent of variation (including SNPs and copy number variants) within the apparently normal population? That is, how many variants exist across the genome, what are their frequencies in the population, and what are their sizes (Pinto et al., 2007)?

2. Are the copy number variants inherited or do they occur de novo? This is assessed by investigating the corresponding genomic regions of the parents. If one (or both) of the parents have the deletion then it is inherited, and if it is established that the parents are phenotypically unaffected then that variant is presumed to be benign. If the parents do not have that variant then it occurs de novo and if it occurs in a patient then it is potentially a causal abnormality.

3. Are there significantly more copy number variants in one group relative to another (e.g., autism relative to controls)? This hypothesis was tested by Sebat et al. (2007), who reported that there were significantly more copy number variants in individuals with autism than controls, and Weiss et al. (2008), who reported significantly more 16p11 deletions and duplications in individuals with autism than controls.

Experimentally, SNP microarrays as well as aCGH have been used in a large number of studies to assess structural variation in the human genome (Iafrate et al., 2004; Sebat et al., 2004; McCarroll et al., 2006; Redon et al., 2006; reviewed in Feuk et al., 2006). Other techniques have been applied as well, such as the sequencing of fosmid paired-end sequences and their comparisons to a human genome reference sequence (Tuzun et al., 2005). This approach is useful to identify deletions, insertions, and (in contrast to SNP arrays and aCGH) inversions based on discrepancies that are observed between the sequenced fosmid clones and the reference genome. Korbel et al. (2007) introduced a related paired-end mapping strategy that uses 454 sequencing technology (Chapter 16) to sequence 3 kilobase fragments.

The phrase "apparently normal" refers to the idea that normalcy is difficult to define and establish. An individual called "normal" has no apparent disease such as a childhood-onset disorder or cancer, but could get a disease in the future and could have an undiagnosed condition in the present. The phrase "apparently normal" acknowledges this difficulty of demonstrating normalcy. Idiopathic means of unknown origin.

They reported 1300 structural variants in two HapMap females (one YRI, one CEU); 45% of these were shared.

The Database of Genomic Variants is available at ▶ http://projects.tcag.ca/variation/.

Information derived from a variety of approaches is centralized in several databases, including the Database of Genomic Variants (Zhang et al., 2006) and at the UCSC Genome Browser. Figure 19.22 shows an example of copy number variants in the beta globin region. Approximately 12 studies have reported copy number variants (e.g., Iafrate et al., 2004; Sebat et al., 2004; Sharp et al., 2005; McCarroll et al., 2006; reviewed in Scherer et al., 2007). For example, Redon et al. (2006) used a combination of SNP and aCGH arrays and reported almost 1500 copy number variable regions spanning 360 megabases (12% of the human genome) in 270 HapMap samples, including three that are shown in Fig. 19.22.

The various approaches for the detection of structural variation have not used standardized approaches to data collection, data analysis, or quality assessment (discussed by Scherer et al., 2007). There are discrepancies in the results of various studies (as is evident from inspection of the nine structural variation tracks in Fig. 19.22). Analysis tools differ in their sensitivity and specificity. Experimental approaches vary widely in their resolution (in terms of the size of a deleted or duplicated region); aCGH, based on bacterial artificial chromosome arrays, tends to be useful to detect relatively large events (greater than several hundred kilobases). SNP arrays can potentially be used to detect much smaller deletions/duplications (e.g., just tens of kilobases), but require estimation of both genotype and copy number for each SNP, with variable genotyping error rates. (For example, heterozygous calls on the male X chromosome represent presumed genotyping errors.) Some analysis packages estimate discrete numbers of chromosomal copies (1 for a hemizygous deletion, 2 for euploid, and 3 for a duplication) but biologically, mosaicism commonly occurs. For example, some number such as 60% of the cells in a sample could have a deletion.

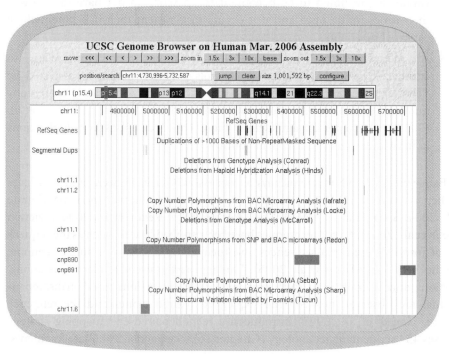

FIGURE 19.22. Copy number variants in the human genome. The ENCODE beta globin region on chromosome 11 is shown (~one megabase on 11p). The annotation tracks include RefSeq genes, segmental duplications (of which several are evident), and results for copy number variants from nine different publications, some of which show regions where deletions or duplications have been observed in apparently normal individuals.

Despite the many concerns about the quality and consistency of data, it is clear that the human genome has a vast number of copy number variants, with a typical individual having from a dozen to several hundred. This highlights the extent to which there is not a single human genome, but each of us has a unique pattern of both sequence (particularly in terms of SNPs as well as larger polymorphisms) and copy number variants such as deletions, duplications, and inversions. The phenotypic consequences of most of these alterations are unknown.

Perspective

The sequencing of the human genome represents one of the great accomplishments in the history of science. This effort is the culmination of decades of work in an international effort. Two major technological advances enabled the human genome to be sequenced: (1) the invention of automated DNA sequencing machines in the 1980s allowed nucleotide data to be collected on a large scale and (2) the computational biology tools necessary to analyze those sequencing data were created by biologists and computer scientists. In the coming years, we can expect the pace of DNA sequencing to continue to increase. It will soon be possible to compare the complete genome sequence of many individuals in an effort to relate genotype to phenotype. The genomic sequence permits analyses of sequence variation such as SNPs and copy number variants; disease-causing mutations; evolutionary forces; and genomic properties such as recombination, replication, and the regulation of gene function.

Pitfalls

As each chromosome has been finished, there have been many technical problems to solve regarding sequencing depth, assembly (particularly in regions with highly repetitive DNA), and annotation. There are discrepancies between the results of gene finding algorithms (as revealed by the ENCODE project, Chapter 16) and there are often discrepancies between different databases. Copy number variants can be particularly difficult to identify and assemble because they are often associated with repetitive DNA, and segmental duplications are difficult to resolve using whole genome shotgun assembly.

There are a number of outstanding problems that have yet to be solved:

- How can we accurately determine the number of protein-coding genes?
- How can we determine the number of noncoding genes?
- How can we determine the function of genes and proteins?
- What is the evolutionary history of our species?
- What is the degree of heterogeneity between individuals at the nucleotide level?

Thus, as we take our first look at the human genome, it is appropriate to see this moment as a beginning rather than an end. Having the sequence in hand, and having the opportunity to compare the human genome sequence to that of many other genomes, we are now in a position to pose a new generation of questions.

DISCUSSION QUESTIONS

[19-1] If you had the resources and facilities to sequence the entire genome of five individuals, which would you select? Why? Describe how you would approach the data analysis.

[19-2] The *Saccharomyces cerevisiae* genome duplicated about 100 MYA, as indicated by BLAST searching (Chapter 17), and we discussed whole-genome duplication in fish, *Paramecium*, and plants (Chapters 16 and 18). Why is it not equally straightforward to identify large duplications of the human genome? Is it because they did not occur, or because the evolutionary history of humans obscures such events, or because we lack the tools to detect such large-scale genomic changes? For a thoughtful discussion of duplications in the human genome, see an article by Evan Eichler (2001).

PROBLEMS/COMPUTER LAB

[19-1] Determine the sequence of beta globin in Craig Venter's genome. First, identify the accession number for beta globin (NM_000518). Next, identify the accession number for the genome; Levy et al. (2007) list it as ABBA00000000. By viewing that record, note that ABBA00000000 itself does not directly refer to DNA sequences, but it lists the accessions ABBA01000001 to ABBA01255300 that do contain whole-genome shotgun sequence data. Perform a blastn search at NCBI, using the beta globin query NM_000518 and setting the database to whole-genome shotgun reads (WGS). In the Entrez query box enter ABBA01000001:ABBA01255300[PACC] in order to limit the search to just Venter's genome sequences. (You can visit the Entrez help link to learn the appropriate formats for limits.) Extra problem: the *ABCC11* gene (ATP-binding cassette, subfamily C, member 11; NM_032583) encodes a protein that Venter has in a variant form that predisposes one to wet rather than dry earwax. Identify the variant nucleotides and/or amino acids.

[19-2] Go to Entrez Gene, and select a human gene of interest, such as alpha-2 globin. Examine the features of this gene at the Ensembl, NCBI, and UCSC websites. Make a table of various properties (e.g., exon/intron structure, number of ESTs corresponding to the expressed gene, polymorphisms identified in the gene, neighboring genes). Are there discrepancies between the data reported in the three databases? Next, obtain a portion of genomic DNA (about 100,000 base pairs in the FASTA format) from the region including this gene. Use the Oak Ridge National Laboratory pipeline to characterize the genomic DNA and potential protein-coding regions (▶ http:// compbio.ornl.gov/tools/pipeline).

[19-3] The recombination rate is higher near the telomeres (see Fig. 19.11). Use the UCSC Table Browser to identify regions having very high recombination rates. (1) Go to ▶ http:// genome.ucsc.edu and select Table Browser. Select the human genome, mapping and sequencing group, Recomb Rate track. Clicking on the summary statistics button shows that there are 2822 entries (one per megabase). (2) Select filter and set the decodeAvg (DeCode genetic map average value) to greater than 5. You can try setting the filter using other genetic maps. (3) When you submit this, the summary statistics show that there are now just 12 entries (on chromosomes 4, 9, 10, 12, 14, 17, 19, 20, and X). A list of these results is also provided as web document 19.1 at ▶ http://www.bioinfbook.org/ chapter19. You can also set the output to hyperlinks to the Genome Browser, showing that most of these regions are indeed subtelomeric. (4) Identify sites with the lowest recombination rate in the genome using a similar strategy. (5) Identify RefSeq genes that are close to the highest (or lowest) recombination rates. Use the intersection tool at the UCSC Table Browser site.

[19-4] Compare the extent of conserved synteny between human and the rhesus macaque (*Macaca mulatta*) on chromosomes 1 (the largest chromosome in humans), 21 (the smallest autosome), X, and Y. Which shows the most conservation? What specific genes are conserved between human and rhesus macaque on the Y chromosome? Why is the extent of conservation on that chromosome so low? One way to accomplish this exercise is to visit the Ensembl human genome browser (▶ http://www. ensembl.org/Homo_sapiens), click on a chromosome (e.g., Y), then use the pull-down menu "View Chr Y Synteny" on the left sidebar.

[19-5] HLA-B is the most polymorphic gene in the human genome. Explore its properties. (1) Set the UCSC Genome Browser (March 2006 assembly) to coordinates chr6:31,429,000-31,433,000 and view the SNPs. You can see the spectacular amount of polymorphism. (2) Obtain a broader perspective by viewing the SNPs across a one million base pair region, chr6:31,000,001-32,000,000. (3) Use the Table Browser and its intersection feature to find the five most polymorphic genes across the entire genome.

[19-6] Human mitochondrial DNA (RefSeq identifier NC_001807) has a bacterial origin. (1) Perform a blastn search of the nonredundant (nr) database, restricting the output to bacteria. To which group of bacteria is the human sequence most related? (You may view the Taxonomy Report for a convenient summary.) (2) To which genes is the human sequence most related? You may inspect your blastn results. (3) There is just one bacterial protein that is related to the proteins encoded by the human mitochondrial genome. What is it? You may inspect your blastn results, or to search specifically for proteins encoded by human mitochondrial DNA use NC_001807 as a query in a blastx search restricted to bacteria.

SELF-TEST QUIZ

[19-1] Approximately how large is the human genome?

(a) 3 Mb

(b) 300 Mb

(c) 3000 Mb

(d) 30,000 Mb

[19-2] Approximately what percentage of the human genome consists of repetitive elements of various kinds?

(a) 5%

(b) 25%

(c) 50%

(d) 85%

[19-3] What percentage of the human genome is devoted to the protein-coding regions?

(a) 1%–5%

(b) 5%–10%

(c) 10%–20%

(d) 20%–40%

[19-4] The UCSC human genome browser differs from the NCBI human genome Map Viewer site because

(a) It offers a large number of annotation tracks, about half of which are supplied by external users of the site

(b) It offers a large number of chromosome maps, including maps of conserved syntenic regions

(c) It offers a genome assembly based on BLAST

(d) It offers a genome assembly incorporating both clone-based and whole-genome shotgun assembly data

[19-5] The human genome contains many transposon-derived repeats. These are described as

(a) Dead fossils

(b) Young, active elements

(c) Human-specific elements

(d) Inverted repeats

[19-6] Approximately how much of the human genome do segmental duplications occupy?

(a) <1%

(b) 3%–5%

(c) 20%–30%

(d) 50%

[19-7] In areas of high GC content of the human genome,

(a) Gene density tends to be low

(b) Gene density tends to be high

(c) Gene density is variable

(d) Genes tend to have fewer introns

[19-8] In comparison to other metazoan genomes (such as nematodes, insects and mouse),

(a) The human genome contains considerably more protein-coding genes

(b) The human genome has considerably more unique genes that lack identifiable orthologs

(c) The human genome has a higher GC content

(d) The human genome has somewhat more multi-domain proteins, paralogous genes, and alternative splicing.

[19-9] When the human genome project was completed by 2001 to 2004, how much of the genome remained impossible to sequence due to repetitive content and other technical challenges?

(a) Essentially none

(b) About 2 megabases (Mb)

(c) About 25 Mb

(d) About 225 Mb

[19-10] Single nucleotide polymorphisms (SNPs) are useful to characterize all these aspects of the human genome except which one?

(a) Disease association

(b) Microduplications

(c) Inverse selection

(d) Population migration

SUGGESTED READING

In this chapter, we have focused on the public consortium description of the human genome (IHGSC, 2001) and the finishing of the euchromatic portion of the genome (IHGSC, 2004). The companion Celera article (Venter et al., 2001) is also of great interest, as are the many accompanying articles in those issues of *Science* and *Nature*. We also discussed the Levy et al. (2007) article on the genome of an individual, with an emphasis on variants of assorted sizes.

For each of the 22 autosomes and two sex chromosomes, there has been a paper published in *Nature* that describes the chromosome in detail. We provide links to these papers at
▶ http://www.bioinfbook.org/chapter19. These important papers describe the in-depth analyses of finished (or nearly finished) chromosomal sequences. They highlight the need for complete sequencing in order to perform more accurate annotation and comparative analyses.

REFERENCES

Amir, R. E., et al. Rett syndrome is caused by mutations in X-linked *MECP2*, encoding methyl-CpG-binding protein 2. *Nat. Genet.* **23**, 185–188 (1999).

Avner, P., and Heard, E. X-chromosome inactivation: Counting, choice and initiation. *Nat. Rev. Genet.* **2**, 59–67 (2001).

Bailey, J. A., Yavor, A. M., Massa, H. F., Trask, B. J., and Eichler, E. E. Segmental duplications: Organization and impact within the current human genome project assembly. *Genome Res.* **11**, 1005–1017 (2001).

Bernardi, G. Misunderstandings about isochores. Part 1. *Gene* **276**, 3–13 (2001).

Collins, F. S., Green, E. D., Guttmacher, A. E., Guyer, M. S.; US National Human Genome Research Institute. A vision for the future of genomics research. *Nature* **422**, 835–847 (2003).

Crick, F. H., and Watson, J. D. Molecular structure of nucleic acids. A structure for deoxyribose nucleic acid. *Nature* **171**, 737–738 (1953).

Cuvier, G. *The Animal Kingdom, Arranged According to Its Organization.* Williams S. Orr & Co., London, 1849.

Deloukas, P., et al. The DNA sequence and comparative analysis of human chromosome 20. *Nature* **414**, 865–871 (2001).

Deloukas, P., et al. The DNA sequence and comparative analysis of human chromosome 10. *Nature* **429**, 375–382 (2004).

Dunham, A., et al. The DNA sequence and analysis of human chromosome 13. *Nature* **428**, 522–528 (2004).

Dunham, I., et al. The DNA sequence of human chromosome 22. *Nature* **402**, 489–495 (1999).

Eichler, E. E. Segmental duplications: What's missing, misassigned, and misassembled—and should we care? *Genome Res.* **11**, 653–656 (2001).

ENCODE Project Consortium. The ENCODE (ENCyclopedia Of DNA Elements) Project. *Science* **306**, 636–640 (2004).

ENCODE Project Consortium, et al. Identification and analysis of functional elements in 1% of the human genome by the ENCODE pilot project. *Nature* **447**, 799–816 (2007).

Feuk, L., Carson, A. R., and Scherer, S. W. Structural variation in the human genome. *Nat. Rev. Genet.* **7**, 85–97 (2006).

Flicek, P., et al. Ensembl 2008. *Nucleic Acids Res.* **36**, D707–D714 (2008).

Gardiner-Garden, M., and Frommer, M. CpG islands in vertebrate genomes. *J. Mol. Biol.* **196**, 261–282 (1987).

Green, E. D. Strategies for the systematic sequencing of complex genomes. *Natl. Rev. Genet.* **2**, 573–583 (2001).

Green, E. D., and Chakravarti, A. The human genome sequence expedition: Views from the "base camp." *Genome Res.* **11**, 645–651 (2001).

Gregory, S. G., et al. The DNA sequence and biological annotation of human chromosome 1. *Nature* **441**, 315–321 (2006).

Grimwood, J., et al. The DNA sequence and biology of human chromosome 19. *Nature* **428**, 529–625 (2003).

Haring, D., and Kypr, J. Mosaic structure of the DNA molecules of the human chromosomes 21 and 22. *Mol. Biol. Rep.* **28**, 9–17 (2001).

Hattori, M., et al. The DNA sequence of human chromosome 21. *Nature* **405**, 311–319 (2000).

Heilig, R., et al. The DNA sequence and analysis of human chromosome 14. *Nature* **421**, 601–607 (2003).

Hillier, L. W., et al. The DNA sequence of human chromosome 7. *Nature* **424**, 157–164 (2003).

Hillier, L. W., et al. Generation and annotation of the DNA sequences of human chromosomes 2 and 4. *Nature* **434**, 724–731 (2005).

Humphray, S. J., et al. DNA sequence and analysis of human chromosome 9. *Nature* **429**, 369–375 (2004).

Iafrate, A. J., Feuk, L., Rivera, M. N., Listewnik, M. L., Donahoe, P. K., Qi, Y., Scherer, S. W., and Lee, C. Detection of large-scale variation in the human genome. *Nat. Genet.* **36**, 949–951 (2004).

Ingman, M., Kaessmann, H., Pääbo, S., and Gyllensten, U. Mitochondrial genome variation and the origin of modern humans. *Nature* **408**, 708–713 (2000).

International HapMap Consortium. A haplotype map of the human genome. *Nature* **437**, 1299–1320 (2005).

International HapMap Consortium, et al. A second generation human haplotype map of over 3.1 million SNPs. *Nature* **449**, 851–861 (2007).

International Human Genome Sequencing Consortium (1HGSC). Initial sequencing and analysis of the human genome. *Nature* **409**, 860–921 (2001).

International Human Genome Sequencing Consortium (1HGSC). Finishing the euchromatic sequence of the human genome. *Nature* **431**, 931–945 (2004).

Jobling, M. A., and Tyler-Smith, C. The human Y chromosome: An evolutionary marker comes of age. *Nature Rev. Genet.* **4**, 598–612 (2003).

Jurka, J. Repeats in genomic DNA: Mining and meaning. *Curr. Opin. Struct. Biol.* **8**, 333–337 (1998).

Kent, W. J. BLAT: The BLAST-like alignment tool. *Genome Res.* **12**, 656–664 (2002).

Kent, W. J., et al. The human genome browser at UCSC. *Genome Res.* **12**, 996–1006 (2002).

Korbel, J. O., Urban, A. E., Affourtit, J. P., Godwin, B., Grubert, F., Simons, J. F., Kim, P. M., Palejev, D., Carriero, N. J., Du, L., Taillon, B. E., Chen, Z., Tanzer, A., Saunders, A. C., Chi, J., Yang, F., Carter, N. P., Hurles, M. E., Weissman, S.

M., Harkins, T. T., Gerstein, M. B., Egholm, M., and Snyder, M. Paired-end mapping reveals extensive structural variation in the human genome. *Science* **318**, 420–426 (2007).

Levy, S., et al. The diploid genome sequence of an individual human. *PLoS Biol.* **5**, e254 (2007). PMID: 17803354.

Li, S., et al. Comparative analysis of human genome assemblies reveals genome-level differences. *Genomics* **80**, 138–139 (2002).

Martin, J., et al. The sequence and analysis of duplication-rich human chromosome 16. *Nature* **432**, 988–994 (2004).

McCarroll, S. A., Hadnott, T. N., Perry, G. H., Sabeti, P. C., Zody, M. C., Barrett, J. C., Dallaire, S., Gabriel, S. B., Lee, C., Daly, M. J., and Altshuler, D. M. International HapMap Consortium. Common deletion polymorphisms in the human genome. *Nat. Genet.* **38**, 86–92 (2006).

McVean, G., Spencer, C. C., and Chaix, R. Perspectives on human genetic variation from the HapMap Project. *PLoS Genet.* **1**, e54 (2005).

Mungall, A. J., et al. The DNA sequence and analysis of human chromosome 6. *Nature* **425**, 805–811 (2003).

Muzny, D. M., et al. The DNA sequence, annotation and analysis of human chromosome 3. *Nature* **440**, 1194–1198 (2006a).

Muzny, D. M., et al. DNA sequence of human chromosome 17 and analysis of rearrangement in the human lineage. *Nature* **440**, 1045–1049 (2006b).

National Research Council. *Mapping and Sequencing the Human Genome*. National Academy Press, Washington, DC, 1988.

Nusbaum, C., et al. DNA sequence and analysis of human chromosome 8. *Nature* **439**, 331–335 (2006).

Osborne, L. R., Li, M., Pober, B., Chitayat, D., Bodurtha, J., Mandel, A., Costa, T., Grebe, T., Cox, S., Tsui, L. C., and Scherer, S. W. A 1.5 million-base pair inversion polymorphism in families with Williams–Beuren syndrome. *Nat. Genet.* **29**, 321–325 (2001).

Ostertag, E. M., and Kazazian, H. H., Jr. Twin priming: A proposed mechanism for the creation of inversions in L1 retrotransposition. *Genome Res.* **11**, 2059–2065 (2001).

Pääbo, S. The mosaic that is our genome. *Nature* **421**, 409–412 (2003).

Pakendorf, B., and Stoneking, M. Mitochondrial DNA and human evolution. *Annu. Rev. Genomics Hum. Genet.* **6**, 165–183 (2005).

Pavlicek, A., Paces, J., Clay, O., and Bernardi, G. A compact view of isochores in the draft human genome sequence. *FEBS Lett.* **511**, 165–169 (2002).

Pinto, D., Marshall, C., Feuk, L., and Scherer, S. W. Copy-number variation in control population cohorts. *Hum. Mol. Genet.* **16**, R168–R173 (2007).

Ponting, C. P. Plagiarized bacterial genes in the human book of life. *Trends Genet.* **17**, 235–237 (2001).

Redon, R., et al. Global variation in copy number in the human genome. *Nature* **444**, 444–454 (2006).

Ross, M. T., et al. The DNA sequence of the human X chromosome. *Nature* **434**, 325–337 (2005).

Rozen, S., et al. Abundant gene conversion between arms of palindromes in human and ape Y chromosomes. *Nature* **423**, 873–876 (2003).

Sabeti, P. C., et al. Genome-wide detection and characterization of positive selection in human populations. *Nature* **449**, 913–918 (2007).

Salzberg, S. L., White, O., Peterson, J., and Eisen, J. A. Microbial genes in the human genome: Lateral transfer or gene loss? *Science* **292**, 1903–1906 (2001).

Scherer, S. W., Cheung, J., MacDonald, J. R., Osborne, L. R., Nakabayashi, K., Herbrick, J. A., Carson, A. R., Parker-Katiraee, L., Skaug, J., Khaja, R., et al. Human chromosome 7: DNA sequence and biology. *Science* **300**, 767–772 (2003).

Scherer, S. E., et al. The finished DNA sequence of human chromosome 12. *Nature* **440**, 346–351 (2006).

Scherer, S. W., Lee, C., Birney, E., Altshuler, D. M., Eichler, E. E., Carter, N. P., Hurles, M. E., and Feuk, L. Challenges and standards in integrating surveys of structural variation. *Nat. Genet.* **39**, S7–S15 (2007).

Schmutz, J., et al. The DNA sequence and comparative analysis of human chromosome 5. *Nature* **431**, 268–274 (2004).

Sebat, J., Lakshmi, B., Troge, J., Alexander, J., Young, J., Lundin, P., Månér, S., Massa, H., Walker, M., Chi, M., Navin, N., Lucito, R., Healy, J., Hicks, J., Ye, K., Reiner, A., Gilliam, T. C., Trask, B., Patterson, N., Zetterberg, A., and Wigler, M. Large-scale copy number polymorphism in the human genome. *Science* **305**, 525–528 (2004).

Sebat, J., Lakshmi, B., Malhotra, D., Troge, J., Lese-Martin, C., Walsh, T., Yamrom, B., Yoon, S., Krasnitz, A., Kendall, J., et al. Strong association of de novo copy number mutations with autism. *Science* **316**, 445–449 (2007).

Sharp, A. J., Locke, D. P., McGrath, S. D., Cheng, Z., Bailey, J. A., Vallente, R. U., Pertz, L. M., Clark, R. A., Schwartz, S., Segraves, R., Oseroff, V. V., Albertson, D. G., Pinkel, D., and Eichler, E. E. Segmental duplications and copy-number variation in the human genome. *Am. J. Hum. Genet.* **77**, 78–88 (2005).

She, X., Horvath, J. E., Jiang, Z., Liu, G., Furey, T. S., Christ, L., Clark, R., Graves, T., Gulden, C. L., Alkan, C., Bailey, J. A., Sahinalp, C., Rocchi, M., Haussler, D., Wilson, R. K., Miller, W., Schwartz, S., and Eichler, E. E. The structure and evolution of centromeric transition regions within the human genome. *Nature* **430**, 857–864 (2004).

Skaletsky, H., et al. The male-specific region of the human Y chromosome is a mosaic of discrete sequence classes. *Nature* **423**, 825–837 (2003).

Taylor, T. D., et al. Human chromosome 11 DNA sequence and analysis including novel gene identification. *Nature* **440**, 497–500 (2006).

Thorisson, G. A., Smith, A. V., Krishnan, L., and Stein, L. D. The International HapMap Project Web site. *Genome Res.* **15**, 1592–1593 (2005).

Toth, G., Gaspari, Z., and Jurka, J. Microsatellites in different eukaryotic genomes: Survey and analysis. *Genome Res.* **10**, 967–981 (2000).

Tuzun, E., Sharp, A. J., Bailey, J. A., Kaul, R., Morrison, V. A., Pertz, L. M., Haugen, E., Hayden, H., Albertson, D., Pinkel, D., Olson, M. V., and Eichler, E. E. Fine-scale structural variation of the human genome. *Nat. Genet.* **37**, 727–732 (2005).

Tycko, B., and Morison, I. M. Physiological functions of imprinted genes. *J. Cell. Physiol.* **192**, 245–258 (2002).

Venter, J. C., et al. The sequence of the human genome. *Science* **291**, 1304–1351 (2001).

Wang, E. T., Kodama, G., Baldi, P., and Moyzis, R. K. Global landscape of recent inferred Darwinian selection for *Homo sapiens*. *Proc. Natl. Acad. Sci. USA* **103**, 135–140 (2006).

Weiss, L. A., Shen, Y., Korn, J. M., Arking, D. E., Miller, D. T., Fossdal, R., Saemundsen, E., Stefansson, H., Ferreira, M. A., Green, T., et al. Association between microdeletion and microduplication at 16p11.2 and autism. *N. Engl. J. Med.* **358**, 667–675 (2008).

Yu, A., et al. Comparison of human genetic and sequence-based physical maps. *Nature* **409**, 951–953 (2001).

Zhang, J., Feuk, L., Duggan, G. E., Khaja, R., and Scherer, S. W. Development of bioinformatics resources for display and analysis of copy number and other structural variants in the human genome. *Cytogenet. Genome Res.* **115**, 205–214 (2006).

Zody, M. C., et al. Analysis of the DNA sequence and duplication history of human chromosome 15. *Nature* **440**, 671–675 (2006).

CHAPITRE IX.

Des modifications chimiques qu'éprouvent les ma-
tières albumineuses des solides et des fluides
organiques chez l'homme malade.

Nous venons de terminer l'étude des matières albumineuses,
telles qu'on les trouve dans l'organisation saine. Nous avons re-
marqué qu'elles y existent en grande abondance, tant dans les
tissus que dans les humeurs. Elles y ont donc une grande importance. Aussi aucun dérangement fonctionnel ne doit pas probable-
ment s'effectuer, sans les léser, soit dans leurs états isomériques,
soit dans leur quantité habituelle, soit dans leurs combinaisons,
soit même dans leur composition élémentaire. En effet, les
désordres qui viennent si souvent tourmenter l'économie, se mani-
festent fréquemment dans ces matières; c'est ce que l'observation
expérimentale nous apprend depuis quelques années. Nous avons déjà
pu signaler chez l'homme malade, tantôt que l'albumine solide et
non dissoute y passe vicieusement, en certains cas, à l'état de disso-
lution, ou bien qu'elle s'y convertit en albumin, mais que cepen-
dant, jamais ou que rarement du moins, l'albumin n'y devient
albumine; tantôt que l'albumine combinée, soit l'albumine à
laquelle on donne l'épithète de soluble, soit celle qu'on nomme
caséine, y éprouve quelquefois, vicieusement encore, un change-
ment qui la rend neutre ou acide, ou qui finit par lui ôter sa
solubilité, et la laisse à l'état solide, désormais indissoute; tantôt
même, que cette albumine ou cet albumin, libre ou combiné,
y augmente ou y diminue irrégulièrement, et par fois y disparaît
dans quelques fluides, dans certains organes; tantôt, enfin, que les
sels et l'alcali ou l'acide unis à l'albumine soluble, y varient de
proportion au-delà du terme normal. De telles connaissances déjà
acquises, prouvent que nous parviendrons à découvrir, un jour,
toutes les modifications chimiques que les maladies apportent dans
les matières albumineuses. C'est par de semblables connaissances,

SERUM SAIN réuni à sa fibrine.		SERUM DU SANG COUENNEUX réuni également à sa fibrine.			
Eau... 900	Eau..............	900	900	Eau.... 900.	
Matière albumineuse 79 {	Albumine........	77	67	} Matière albu-	
	Fibrine..........	2	13	mineuse.. 80.	
Alcali.. 1	Soude...........	1	2	Alcali... 2.	
Sels.... 6 {	Sulfates et phospha-tes à bases alcalines	2	2	} Sels.... 4.	
	Chlorure de sodium	4	2		

As soon as proteins were discovered, investigators studied their role in disease. Prosper Sylvain Denis (1799–1863) wrote Études Chimiques, Physiologiques, et Médicales, Faites de 1835 à 1840, sur les Matières Albumineuses *(1842) (Chemical, Physiological and Medical Studies, done from 1835 to 1840, on the Albuminous Materials). Chapter 9 (p. 141) is entitled "On the chemical modifications in albuminous materials of organic solids and fluids in the sick person." He wrote (see arrow 1): "In effect, the disorders that so often torment the economy, frequently manifest themselves in these materials." This passage includes a reference to caseine ("caséine") and concludes (arrow 2) "Such knowledge already acquired proves that we will come to discover, one day, all the chemical modifications that illnesses carry in the albuminous materials." The lower panel (from p. 144) shows a table comparing the water, proteins, alkali, and salts from healthy and diseased serum.*

20

Human Disease

Life is a relationship between molecules, not a property of any one molecule. So is therefore disease, which endangers life. While there are molecular diseases, there are no diseased molecules. At the level of the molecules we find only variations in structure and physicochemical properties. Likewise, at that level we rarely detect any criterion by virtue of which to place a given molecule "higher" or "lower" on the evolutionary scale. Human hemoglobin, although different to some extent from that of the horse (Braunitzer and Matsuda, 1961), appears in no way more highly organized. Molecular disease and evolution are realities belonging to superior levels of biological integration. There they are found to be closely linked, with no sharp borderline between them. The mechanism of molecular disease represents one element of the mechanism of evolution. Even subjectively the two phenomena of disease and evolution may at times lead to identical experiences. The appearance of the concept of good and evil, interpreted by man as his painful expulsion from Paradise, was probably a molecular disease that turned out to be evolution. Subjectively, to evolve must most often have amounted to suffering from a disease. And these diseases were of course molecular.
　　　　　　　　—Emile Zuckerkandl and Linus Pauling (1962, pp. 189–190)

HUMAN GENETIC DISEASE: A CONSEQUENCE OF DNA VARIATION

Variation in DNA sequence is a defining feature of life on Earth. For each species, genetic variation is responsible for the adaptive changes that underlie evolution.

Mutation is the alteration of DNA sequence. The cause may be

Bioinformatics and Functional Genomics, Second Edition. By Jonathan Pevsner
Copyright © 2009 John Wiley & Sons, Inc.

errors in DNA replication or repair, the effects of chemical mutagens, or radiation. While there may be negative connotations associated with the concept of mutations, mutation and fixation are the essential driving forces behind evolution. From a medical perspective, disease is "a pathological condition of the body that presents a group of clinical signs, symptoms, and laboratory findings peculiar to it and setting the condition apart as an abnormal entity differing from other normal or pathological condition" (Thomas, 1997, p. 552). Disorder is a "pathological condition of the mind or body" (Thomas, 1997, p. 559). A syndrome is "a group of symptoms and signs of disordered function related to one another by means of some anatomical, physiological, or biochemical peculiarity. This definition does not include a precise cause of an illness but does provide a framework of reference for investigating it" (Thomas, 1997, p. 1185).

Evolution is a process by which species adapt to their environment. When changes in DNA improve the fitness of a species, its population reproduces more successfully. When changes are relatively maladaptive, the species may become extinct. At the level of the individual within a species, some mutations improve fitness, most mutations have no effect on fitness, and some are maladaptive (relative to some norm). Disease may be defined as maladaptive changes that afflict individuals within a population. Disease is also defined as an abnormal condition in which physiological function is impaired. Our focus is on the molecular basis of physiological defects at the levels of DNA, RNA, and protein.

There is a tremendous diversity to the nature of human diseases. This is for several reasons:

- Mutations affect all parts of the human genome. There are limitless opportunities for maladaptive mutations to occur. These may be point mutations, affecting just a single nucleotide, or large mutations, affecting as much as an entire chromosome or multiple chromosomes.

- There are many mechanisms by which mutations can cause disease (summarized in Table 20.1). These include disruptions of gene function by point mutations that change the identity of amino acid residues; by deletions or insertions of DNA, ranging in size from one nucleotide to an entire chromosome that is over 100 million base pairs (Mb); or inversions of the orientation of a DNA fragment. In many cases, different kinds of mutations affecting the same gene cause distinct phenotypes.

TABLE 20-1 Mechanisms of Genetic Mutation

Mechanism[a]	Usual Effect	Example
Large Mutation		
Deletion	Null	Duchenne dystrophy
Insertion	Null	Hemophila A/LINE
Duplication	Null, gene disrupted	Duchenne dystrophy
	Dosage, gene intact	Charcot–Marie–Tooth
Inversion	Null	Hemophila A
Expanding triplet	Null	Fragile X
	Gain of function?	Huntington
Point Mutation		
Silent	None	Cystic fibrosis
Missense or in-frame deletion	Null, hypomorphic, altered function, benign	Globin
Nonsense	Null	Cystic fibrosis
Frame shift	Null	Cystic fibrosis
Splicing (AG/GT)	Null	Globin
Splicing (outside AG/GT)	Hypomorphic	Globin
Regulatory (TATA, other)	Hypomorphic	Globin
Regulatory (poly A site)	Hypomorphic	Globin

[a]AG/GT indicates mutations in the canonical first two and last two base pairs of an intron. Outside AG/GT indicates mutations in less canonical sequences.
Source: Adapted from Beaudet et al. (2001, p. 9). Used with permission.

- Most genes function by producing a protein as a gene product. A disease-causing mutation in a gene results in the failure to produce the gene product with normal function. This has profound consequences on the ability of the cells in which the gene product is normally expressed to function.

- The interaction of an individual with his or her environment has profound effects on disease phenotype. Genetically identical twins may have entirely different phenotypes. Such differences are attributable to environmental influences or to epigenetic effects. The concordance rate between monozygotic twins is an indication of the relative extent to which genetic and environmental effects influence disease. Even for highly genetic disorders, such as autism (see below) and schizophrenia, the concordance rate is never 100%.

A Bioinformatics Perspective on Human Disease

In Chapter 1, we defined bioinformatics as a discipline that uses computer databases and computer algorithms to analyze proteins, genes, and genomes. Our approach to human disease is reductionist, in that we seek to describe genes and gene products that cause disease. However, an appreciation of the molecular basis of disease may be integrated with a holistic approach to uncover the logic of disease in the entire human population (Childs and Valle, 2000). As we explore bioinformatics approaches to human disease, we are constantly faced with the complexity of all biological systems. Even when we uncover the gene that when mutated causes a disease, our challenge is to attempt to connect the genotype to the phenotype. We can only accomplish this by synthesizing information about the biological context in which each gene functions and in which each gene product contributes to cellular function (Childs and Valle, 2000; Dipple et al., 2001).

The field of bioinformatics offers approaches to human disease that may help us to understand basic questions about the influence of genes and the environment on all aspects of the disease process. Some examples of ways in which this field can have an impact on our knowledge of disease will be highlighted throughout the chapter, and include the following.

- To the extent that the genetic basis of disease is a function of variation in DNA sequences, DNA databases offer us the basic material necessary to compare DNA sequences. These databases include major, general repositories of DNA sequence such as GenBank/EMBL/DDBJ (Chapter 2), general resources such as Online Mendelian Inheritance in Man (OMIM), and locus-specific databases that provide data on sequence variations at individual loci.

- Geneticists who search for disease-causing genes through linkage studies, association studies, or other tests (described below) depend on physical and genetic maps in their efforts to identify mutant genes.

- When a protein-coding gene is mutated, there is a consequence on the three-dimensional structure of the protein product. Bioinformatics tools described in Chapter 11 allow us to predict the structure of protein variants, and from such analyses we may infer changes in function.

- Once a mutant gene is identified, we want to understand the consequence of that mutation on cellular function. We have described a variety of approaches to understanding protein function in Chapters 10 to 12. And in our discussion

level	bioinformatics resources
molecular level	
DNA	general resources: OMIM locus-specific mutation databases
RNA	databases of gene expression (Chapter 8)
protein	UniProt; databases of mutant proteins
systems level	
organelles	databases of peroxisomal, mitochondrial, lysosomal disease
organs/systems	disease databases focused on blood, neuromuscular, retinal, cardiovascular, gastrointestinal, other
organismal level	
clinical phenotype	databases with information on age of onset; frequency; severity; malformations; tissue involvement; other features
animal model	human disease orthologs in various deuterostomes (mouse, sea urchin), protostomes (fly, worm), plants, other species
organizations and foundations	general organizations (NORD) disease-specific organizations

FIGURE 20.1. *Bioinformatics resources for the study of human disease are organized at a variety of levels.*

of *Saccharomyces cerevisiae*, we discussed additional high throughput approaches to understanding eukaryotic protein function (Chapter 12). Gene expression studies (Chapters 8 and 9) have been employed to study the transcriptional response to disease states.

- We may obtain great insight into the role of a particular human gene by identifying orthologs in simpler organisms. We will discuss orthologs of human disease genes found in a variety of model systems.

In this chapter, we will first provide an overview of human disease, including approaches to disease classification. Next, we will consider the subject of human disease at several levels (outlined in Fig. 20.1). First, at the molecular level, we will focus on the role of genes in disease. In discussing monogenic (single-gene) disorders, we will introduce Online Mendelian Inheritance in Man (OMIM), which is the principal disease database. There are also several hundred locus-specific mutation databases, and we will discuss these. We will also examine both bioinformatics approaches and databases relevant to the study of RNA, and protein. Second, we will examine web resources for diseases at the cellular and systems levels, such as organellar disease databases. Third, we will consider the level of the organism: what bioinformatic tools have been developed to characterize the clinical phenotype of disease (e.g., age of onset, mode of inheritance, frequency, and severity)? What animal models of disease have been developed? We will explore orthologs of human disease genes in model organisms such as fungi and lower metazoans. Finally, we will consider databases that have been established to provide general information on human disease.

Garrod's View of Disease

The OMIM entry for alkaptonuria is #20355; the # sign is defined in Table 20.10 below. The RefSeq accession of HGD is NP_000178. The gene is localized to chromosome 3q21–q23. You can read Garrod's (1902) paper on alkaptonuria online as web document 20.1 at ► http://www.bioinfbook.org/chapter20.

Sir Archibald Garrod (1857–1936) made important contributions to our understanding of the nature of human disease. In a 1902 paper, Garrod described his studies of alkaptonuria, a rare inherited disorder. In alkaptonuria, the enzyme homogentisate 1,2-dioxygenase (HGD) is defective or missing. As a result, the amino acids phenylalanine and tyrosine cannot be metabolized properly, and a metabolite (homogentisic acid) accumulates. This metabolite oxidizes in urine and turns dark. Garrod considered this phenotype from the perspective of evolution, noting the influence of natural selection on chemical processes. Variations in metabolic processes between individuals might include those changes that cause disease.

Garrod had the insight that for each of the rare disorders he studied, the disease phenotype reflects the chemical individuality of the individual. He further realized that this trait was inherited—he proposed that alkaptonuria is transmitted by recessive Mendelian inheritance.

At the time, it was thought that most diseases were caused by external forces such as bacterial infection. In studying this and related recessive disorders (such as cystinuria and albinism) he instead proposed that the manifestation of the disease is caused by an inherited enzyme deficiency or biochemical error (Scriver and Childs, 1989). He described this point of view in his first book, *Inborn Errors of Metabolism* (1909). Garrod wrote in 1923 (cited in Scriver and Childs, 1989, p. 7):

> A trait is a characteristic or property of an individual that is the outcome of the action of a gene or genes.

"If it may be granted that the individual members of a species vary from the normal of the species in chemical structure and chemical behaviour, it is obvious that such variations or mutations are capable of being perpetuated by natural selection; and not a few biologists of the present day assign to chemical structure and function a most important share in the evolution of species ... Very few individuals exhibit such striking deviations from normal metabolism as porphyrinurics and cystinurics show, but I suspect strongly that minimal deviations which escape notice are almost universal. How else can be explained the part played by heredity in disease? There are some diseases which are handed down from generation to generation ... which tend to develop in later childhood and early adult life ... It is difficult to escape the conclusion that although these maladies are not congenital, their underlying causes are inborn peculiarities."

Garrod thus presented a new view of how inborn factors cause disease. He worked at a time before Beadle and Tatum offered the hypothesis that one gene encodes one protein, and Garrod never used the word "gene." But we now understand that the "inborn peculiarities" he described are mutated genes (or other chromosomal loci). A main conclusion of his work is that chemical individuality, achieved through genetic differences, is a major determinant of human health and disease. Although the phrase "chemical individuality" is not used often today, the concept is of tremendous interest in the field of pharmacogenomics. Not everyone who is exposed to an infectious agent gets sick, and it is imperative to understand why. Not everyone who takes a drug responds in a similar way.

Garrod further developed these ideas in a second book, *Inborn Factors in Disease* (1931). Here he addressed the question of why certain individuals are susceptible to diseases—whether the disease is clearly inherited or whether it derives from another cause such as an environmental agent. He argued that chemical individuality predisposes us to disease. Every disease process is affected by both internal and external forces: our genetic complement and the environmental factors we face. In some cases, such as inborn errors of metabolism, genetic factors have a more prominent role. In other cases, such as multifactorial disease, mutations in many genes are responsible for the disease. And in infectious disease, genes also have an important role in defining the individual's susceptibility and bodily response to the infectious agent. We will next proceed to discuss these various kinds of disease.

Classification of Disease

We will describe several general categories of disease below, such as single gene disorders, complex disorders, chromosomal disorders, and environmental disease.

From the perspective of bioinformatics, we are interested in understanding the mechanism of disease in relation to genomic DNA, genes, and their gene products. We are further interested in the consequences of mutations on cell function and on the comparative genomics of disease-causing genes throughout evolution. This perspective is complementary to and yet quite different from that of the clinician or epidemiologist.

For any study of disease a classification system is useful, and many approaches are available. One is to describe mortality statistics. These data (based on death certificates in the United States from the year 2005) include rankings of the cause of death (Table 20.2). This information is helpful in identifying the most common diseases, and projections of the most common causes of death in the future have been made (Fig. 20.2). According to the World Health Organization the four leading causes of death globally in 2030 are projected to be ischemic heart disease, stroke, HIV/AIDS, and chronic obstructive pulmonary disease (Mathers and Loncar, 2006). Tobacco is projected to kill 50% more people than HIV/AIDS in 2015 and it will be responsible for 10% of all deaths.

Another approach to describing the scope of human disease is to measure the global burden of disease in terms of the percentage of affected individuals or in terms of disability-adjusted life years (DALYs) (Lopez et al., 2006). Worldwide, noncommunicable diseases such as depression and heart disease are rapidly replacing infectious diseases and malnutrition as the leading causes of disability and premature death (Murray and Lopez, 1996). The World Health Organization ranks the leading worldwide causes of disease burden (in DALYs) for males and females age 15 years and older. For the year 2002, unipolar depressive disorders ranked first in females (and fourth in males) while HIV/AIDS ranked first in males (and second in females).

The data in Table 20.2 are available from the National Center for Health Statistics. Their website is at ► http://www.cdc.gov/nchs/.

A summary of the Global Burden of Disease report is available at ► http://www.who.int/whr/2003/en/whr03_en.pdf. DALYs are calculated by adding the years of life lost through all deaths in a year plus the years of life expected to be lived with a disability for all cases beginning in that year. The DALYs metric was introduced in the 1990 Global Burden of Disease study (Murray and Lopez, 1996).

TABLE 20-2 Leading Causes of Death in the United States (Year 2005)

Rank	Cause of Death	Number	Percent of All Deaths
–	All causes	2,448,017	100.0
1	Diseases of the heart	652,091	26.6
2	Malignant neoplasms	559,312	22.8
3	Cerebrovascular diseases	143,579	5.9
4	Chronic lower respiratory diseases	130,933	5.3
5	Accidents (unintentional injuries)	117,809	4.8
6	Diabetes mellitus	75,119	3.1
7	Alzheimer's disease	71,599	2.9
8	Influenza and pneumonia	63,001	2.6
9	Nephritis, nephrotic syndrome, and nephrosis	43,901	1.8
10	Septicemia	34,136	1.4
11	Intentional self-harm (suicide)	32,637	1.3
12	Chronic liver disease and cirrhosis	27,530	1.1
13	Essential (primary) hypertension and hypertensive renal disease	24,902	1.0
14	Parkinson disease	19,544	0.8
15	Assault (homicide)	18,124	0.7
–	All other causes (residual)	433,800	17.7

Cause of death is based on the International Classification of Diseases, Tenth Revision, 1992.
Source: National Vital Statistics Reports, 56(10), January 2008 (► http://www.cdc.gov/nchs/data/nvsr/nvsr56/nvsr56_10.pdf).

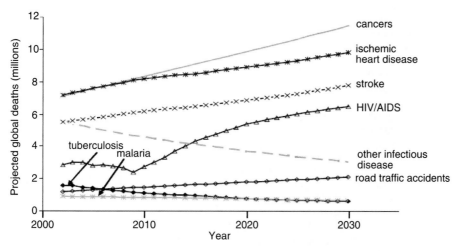

FIGURE 20.2. Projected global deaths for selected causes of death, 2002–2030. From the World Health Organization (World Health Statistics 2007, online at ▶ http://www.who.int/whosis/whostat2007.pdf). Used with permission.

A far more extensive listing of morbidity data is provided by the International Statistical Classification of Diseases and Related Health Problems (abbreviated ICD). This resource, published by the World Health Organization (WHO), is used to classify diseases (Table 20.3). It provides a standard for coding patients at most hospitals.

Mortality statistics list the most common diseases. We are interested in the full spectrum of disease, including rare diseases. These are defined as diseases affecting fewer than 200,000 people. In the United States, an estimated 25 million individuals (almost 10% of the population) suffer from one or more of 6000 rare diseases.

NIH Disease Classification: MeSH Terms

The National Library of Medicine (NLM) has developed Medical Subjects Heading (MeSH) terms as a unified language for biomedical literature database searches.

The WHO ICD website is at ▶ http://www.who.int/classifications/icd/en/. This resource was begun in 1893 as the International List of Causes of Death.

The Office of Rare Diseases at the National Institutes of Health (NIH) has a website that serves as a portal to information on rare diseases (▶ http://rarediseases.info.nih.gov/).

TABLE 20-3   ICD Classification System

1. Infectious and parasitic diseases
2. Neoplasms
3. Endocrine, nutritional, and metabolic diseases and immunity disorders
4. Diseases of the blood and blood-forming organs
5. Mental disorders
6. Diseases of the nervous system and sense organs
7. Diseases of the circulatory system
8. Diseases of the respiratory system
9. Diseases of the digestive system
10. Diseases of the genitourinary system
11. Complications of pregnancy, childbirth, and the puerperium
12. Diseases of the skin and subcutaneous tissue
13. Diseases of the musculoskeletal system and connective tissue
14. Congenital anomalies
15. Certain conditions originating in the perinatal period
16. Symptoms, signs, and ill-defined conditions
17. Injury and poisoning

Source: From ICD-9 as described in the KEGG database, ▶ http://www.genome.ad.jp and http://icd9 cm.chrisendres.com/.

1. ⊞ Anatomy [A]
2. ⊞ Organisms [B]
3. ⊟ Diseases [C]
   ○ Bacterial Infections and Mycoses [C01] +
   ○ Virus Diseases [C02] +
   ○ Parasitic Diseases [C03] +
   ○ Neoplasms [C04] +
   ○ Musculoskeletal Diseases [C05] +
   ○ Digestive System Diseases [C06] +
   ○ Stomatognathic Diseases [C07] +
   ○ Respiratory Tract Diseases [C08] +
   ○ Otorhinolaryngologic Diseases [C09] +
   ○ Nervous System Diseases [C10] +
   ○ Eye Diseases [C11] +
   ○ Male Urogenital Diseases [C12] +
   ○ Female Urogenital Diseases and Pregnancy Complications [C13] +
   ○ Cardiovascular Diseases [C14] +
   ○ Hemic and Lymphatic Diseases [C15] +
   ○ Congenital, Hereditary, and Neonatal Diseases and Abnormalities [C16] +
   ○ Skin and Connective Tissue Diseases [C17] +
   ○ Nutritional and Metabolic Diseases [C18] +
   ○ Endocrine System Diseases [C19] +
   ○ Immune System Diseases [C20] +
   ○ Disorders of Environmental Origin [C21] +
   ○ Animal Diseases [C22] +
   ○ Pathological Conditions, Signs and Symptoms [C23] +
4. ⊞ Chemicals and Drugs [D]
5. ⊞ Analytical, Diagnostic and Therapeutic Techniques and Equipment [E]
6. ⊞ Psychiatry and Psychology [F]
7. ⊞ Biological Sciences [G]
8. ⊞ Natural Sciences [H]
9. ⊞ Anthropology, Education, Sociology and Social Phenomena [I]
10. ⊞ Technology, Industry, Agriculture [J]
11. ⊞ Humanities [K]
12. ⊞ Information Science [L]
13. ⊞ Named Groups [M]
14. ⊞ Health Care [N]
15. ⊞ Publication Characteristics [V]
16. ⊞ Geographicals [Z]

FIGURE 20.3. *The Medical Subject Heading (MeSH) term system at the National Library of Medicine includes 16 major categories (2008 version). The disease category further includes the 23 headings shown here. See* ▶ *http://www.nlm.nih.gov/mesh/.*

We will discuss Rett syndrome below. You can access the MeSH system at NLM (▶ http://www.nlm.nih.gov/mesh/mesh) or at NCBI (from PubMed, select MeSH terms on the left sidebar, then enter a query such as "disease."

Pathology is the study of the nature and cause of disease. Pathophysiology is the study of how disease alters normal physiological processes.

The 2003 MeSH term system includes 23 disease categories (Fig. 20.3). PubMed at NCBI also uses this classification system.

MeSH terms are a controlled vocabulary used to index MEDLINE (and PubMed, which is based on MEDLINE). A search for the term "Rett Syndrome" at the NLM MeSH site shows the hierarchical tree structure of the MeSH terms: Rett syndrome is listed separately under categories such as mental retardation, neurodegenerative disorders, and inborn genetic diseases.

FOUR CATEGORIES OF DISEASE

What kinds of diseases afflict humans? We can describe four main categories: single gene (monogenic) disease, complex disease, genomic disease, and environmental disease (Fig. 20.4). These categories are interconnected in many ways, as we will discuss next. Consistent with Garrod's perspective, the pathophysiology of any disease may be considered multigenic. Two individuals who are exposed to the same disease-causing stimulus—whether it is a virus or lead paint or a mutated gene—may have

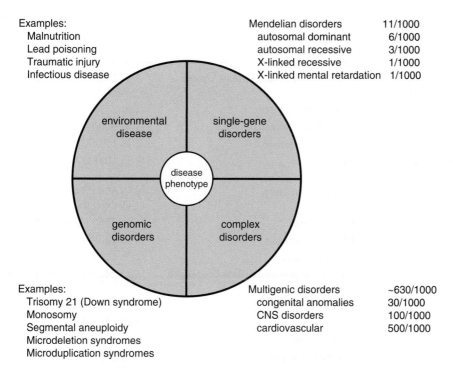

Examples:
- Malnutrition
- Lead poisoning
- Traumatic injury
- Infectious disease

Mendelian disorders	11/1000
autosomal dominant	6/1000
autosomal recessive	3/1000
X-linked recessive	1/1000
X-linked mental retardation	1/1000

Examples:
- Trisomy 21 (Down syndrome)
- Monosomy
- Segmental aneuploidy
- Microdeletion syndromes
- Microduplication syndromes

Multigenic disorders	~630/1000
congenital anomalies	30/1000
CNS disorders	100/1000
cardiovascular	500/1000

FIGURE 20.4. *Human disease can be categorized based on the cause. These include single-gene disorders (mutations in a single gene; examples include phenylketonuria and sickle cell anemia); complex disorders (having mutations in two or more genes, such as cancer or schizophrenia); genomic disorders (such as Down syndrome, involving chromosomal abnormalities); and environmental disease (including infectious disease). The values for the incidence of these disorders are only approximate estimates. The four quadrants of the circle are not intended to reflect incidence. Overall, complex disorders are far more common than single-gene disorders. However, it is far easier to discover the genetic defect that underlies single-gene disorders. For all categories of disease, the pathophysiology (i.e., the disease-altered physiological processes) depends on the influence of many genetic and environmental factors.*

entirely different reactions. One person may become ill, while the other is unaffected. There is a large genetic component to the responses to any disease-causing condition.

Monogenic Disorders

Our perspectives on the molecular nature of disease have evolved in recent decades. Previously, geneticists recognized a dichotomy between simple traits and complex traits. More recently, all traits have come to be appreciated as part of a continuum. Simple traits are transmitted following the rules of Mendel. Several monogenic disorders are listed in Table 20.4. As an example of a single-gene disorder, consider sickle cell anemia (Box 20.1). In 1949 Linus Pauling and colleagues described the abnormal electrophoretic behavior of sickle cell hemoglobin (Pauling et al., 1949). It was subsequently shown that a single amino acid substitution accounts for the abnormal behavior of the sickle cell and is the basis of sickle cell anemia. This is a single-gene disorder that is inherited in an autosomal recessive fashion. Single-gene disorders tend to be rare in the general population. Note that sickle cell disease is the outcome of having a particular mutant hemoglobin protein. While there are common features of sickle cell disease, such as sickling of the red

We examined the structure of normal beta globin (HBB) as well as the most common mutated form (HBS) in Chapter 11. The E6V substitution (valine in place of glutamate as the sixth amino acid) adds a hydrophobic patch to the protein, promoting the aggregation of globin molecules and the formation of sickle-shaped red blood cells. Sickle cell anemia is unusually common for a single-gene disorder. This is presumably because of the protection it confers to heterozygotes exposed to malaria (Box 20.1). You can read

TABLE 20-4   Examples of Monogenic Disorders

Mechanism	Disorder	Frequency
Autosomal dominant	*BRCA1* and *BRCA2* breast cancer	1 in 1000 (1 in 100 for Ashkenazim)
	Huntington chorea	1 in 2500
	Neurofibromatosis I	1 in 3000
	Tuberous sclerosis	1 in 15,000
Autosomal recessive	Albinism	1 in 10,000
	Sickle cell anemia	1 in 655 (U.S. African Americans)
	Cystic fibrosis	1 in 2500 (Europeans)
	Phenylketonuria	1 in 12,000
X linked	Hemophilia A	1 in 10,000 (males)
	Glucose 6-phosphate dehydrogenase deficiency	Variable; up to 1 in 10 males
	Fragile X syndrome	1 in 1250 males
	Color blindness	1 in 12 males
	Rett syndrome	1 in 20,000 females
	Adrenoleukodystrophy	1 in 17,000

Source: Adapted from Beaudet et al. (2001). Used with permission.

BOX 20.1
Sickle Cell Anemia and Thalassemias

Our cells depend on oxygen to live, and blood transports oxygen throughout the body. However, oxygen is a hydrophobic molecule that requires the carrier protein hemoglobin to transport it through blood. (The homologous protein myoglobin transports oxygen in muscle cells.) Adult hemoglobin is composed of two α chains and two β chains. Other α- and β-type chains are used at different developmental stages, such as α2/γ2 in fetal hemoglobin and α2/ε2 in embryonic hemoglobin. Mutation in the β chain (NM_000518 and NP_000509) on chromosome 11p15.5 causes sickle cell anemia (OMIM 603903). Red blood cells in patients can assume a curved, "sickled" appearance that reflects hemoglobin aggregation in the presence of low oxygen levels.

Sickle cell anemia is the most common inherited blood disorder in the United States, affecting 1 in 500 African Americans. It is inherited as an autosomal recessive disease. Heterozygotes (individuals with one normal copy of hemoglobin beta and one mutant copy; the HBS mutation) are somewhat protected against the malaria parasite, *Plasmodium falciparum*. This may be because normal red blood cells infected by the parasite are destroyed. Thus there is a selective evolutionary pressure to preserve the HBS mutation in the population that is at risk for malaria.

Red blood cells closely regulate the proportions of α and β globin that are produced, as well as the heme moiety that is inserted into the globin tetramer to form hemoglobin. The absence of the β chain causes beta-zero-thalassemia, while the production of reduced amounts of β globin causes beta-plus-thalassemia. Reduced levels of α globin cause alpha thalassemias. Thalassemia can cause severe anemia in which hemoglobin levels are low.

Following are web resources for sickle cell disease:

Resource	URL
NIH fact sheet	► http://www.nhlbi.nih.gov/health/dci/ Diseases/Sca/SCA_WhatIs.html
Genes and Disease (NCBI)	► http://www.ncbi.nlm.nih.gov/disease/sickle. html
Sickle Cell Disease Association of America	► http://www.sicklecelldisease.org/

blood cells, there is not a single disease phenotype. The pleiotrophic phenotype is caused by the influence of other genes.

Rett syndrome is another example of a single-gene disorder (Box 20.2). This disease affects girls almost exclusively. While they are apparently born healthy, Rett syndrome girls acquire a constellation of symptoms beginning at 6 to 18 months of age. They lose the ability to make purposeful hand movements, and they typically exhibit hand-wringing behavior. Whatever language skills they have acquired are lost, and they may display autistic-like behaviors. Rett syndrome is caused by

the Pauling et al. (1949) article online at ► http://profiles.nlm. nih.gov/MM/B/B/R/L/. The National Library of Medicine (NLM) offers online access to all the publications of several prominent biologists through its Profiles in Science site (► http://profiles. nlm.nih.gov/). The scientists include Linus Pauling and other Nobel Prize laureates such as Barbara McClintock, Julius Axelrod, and Oswald Avery.

BOX 20.2
Rett Syndrome

Rett syndrome (RTT; OMIM #312750) is a progressive developmental neurological syndrome that occurs almost exclusively in females (Hagberg et al., 1983; Johnston et al., 2005; Percy, 2008). Affected females are apparently normal through pre- and perinatal development, following which there is a developmental arrest. This is accompanied by decelerated head and brain growth, loss of speech and social skills, severe mental retardation, truncal ataxia, and characteristic hand-wringing motions. Prominent neuropathological features include reductions in cortical thickness in multiple cerebral cortical regions, reduced neuronal soma size, and dramatically decreased dendritic arborization (Jellinger et al., 1988; Bauman et al., 1995).

Mutations in the methyl-CpG-binding protein 2 (*MECP2*) gene located in Xq28 have been found in many cases of RS (Amir et al., 1999; Bird, 2008). MeCP2 binds to methylated CpG dinucleotides throughout the genome and is involved in methylation-dependent repression of gene expression via the recruitment of the corepressor mSin3A, and the chromatin remodeling histone deacetylases HDAC1 and HDAC2. The expression of MeCP2 mRNA in many tissues and its interaction with regulatory DNA elements in multiple chromosomes suggest that MeCP2 is a global repressor of gene expression (Nan et al., 1997). DNA methylation-dependent repression of gene expression has been associated with genetic imprinting, X-chromosome inactivation, carcinogenesis, and tissue-specific gene expression (Razin, 1998; Ng and Bird, 1999).

Several groups generated mouse models of RTT. Guy et al. (2007) in Adrian Bird's laboratory showed that knockout of RTT causes a neurological phenotype and that, remarkably, this phenotype can be reversed by subsequent conditional activation of *MECP2* expression. Even functional changes, such as a reduction in hippocampal long-term potentiation (a strengthening of synapses in a brain region implicated in memory consolidation), were improved upon activation. This suggests that it is at least conceivable for the human disease to one day be reversed.

mutations in the gene encoding MeCP2, a transcriptional repressor that binds methylated CpG islands (Amir et al., 1999) (see Chapter 19). It is not yet known why mutations affecting a transcriptional repressor that functions throughout the body cause a primarily neurological disorder.

Although Rett syndrome is a disease caused by a mutation in a single gene, it exemplifies the extraordinary complexity of human disease and even monogenic disorders:

- The disease occurs primarily in females. It was thought that this could be explained by the location of the *MECP2* gene on the X chromosome: a mutation in this gene might be lethal for males in utero (having only a single X chromosome), while females might have the disease phenotype because they have one normal and one mutant copy of the gene. Instead, the more likely explanation is that most mutations occur in fathers. The father is healthy, but a new germline mutation arises and is passed to daughters. Thus all sons (XY) receive a normal Y chromosome from the father, while daughters receive a mutant copy of the X chromosome from the father.

- After the discovery that mutations in *MECP2* cause Rett syndrome, it was discovered that some males with mental retardation also have mutations in this gene (Hammer et al., 2002; Geerdink et al., 2002; Zeev et al., 2002). However, the phenotype of mutations in the male is distinctly different than in females, often involving severe neonatal encephalopathy. In males, having a single X chromosome, the mutant gene is expected to adversely affect virtually every cell in the body. In contrast, females undergo random X chromosome inactivation: Having two copies of the X chromosome, every cell expresses only one chromosome (either the maternal or paternal chromosome, randomly selected early in development). Thus, females are a mosaic in terms of X chromosome allelic expression, and a Rett syndrome female typically has on average 50% normal cells throughout her body.

- While Rett syndrome is caused by mutations in a gene encoding a transcriptional repressor, it is almost certain that the consequence of this mutation involves subsequent effects on the expression of many other genes. Thus, like any other monogenic disorder, many other genes are involved and may influence the phenotype of the disease. Huda Zoghbi and colleagues even showed that MECP2 can function as both a repressor and an activator of transcription in a mouse model (Chahrour et al., 2008).

- Two females having the identical mutation in *MECP2* may have entirely different phenotypes (in terms of severity of the disease). There are two main explanations for this observation, which is seen for many other single-gene disorders as well. (1) There may be modifier genes that influence the disease process (Dipple and McCabe, 2000). Modifier genes have been identified for patients with sickle cell anemia, adrenoleukodystrophy, cystic fibrosis, and Hirschsprung disease. Most (if not all) apparently monogenic disorders are complex. (2) A variety of epigenetic influences may drastically affect the clinical phenotype. For example, the methylation status of genomic DNA could determine the molecular consequences of mutations in *MECP2*. X chromosome inactivation is sometimes skewed, such that the phenotype is more severe (if the X chromosome copy with mutant *MECP2* is preferentially expressed) or less severe (if the healthy X chromosome is selectively expressed).

Complex Disorders

Complex disorders such as Alzheimer's disease and cardiovascular disease are caused by defects in multiple genes. These disorders are also called multifactorial, reflecting that they are expressed as a function of both genetic and environmental factors. In the United States, chronic diseases such as heart disease, senile dementia, cancer, and diabetes are among the leading causes of death and disability. Other examples are asthma, autism (Box 20.3), depression, diabetes, high blood pressure, obesity, and osteoporosis. These all have some degree of genetic basis.

In contrast to single-gene disorders, complex disorders are very common in the population (Todd, 2001). These traits do not segregate in a simple, discrete, Mendelian manner. It is likely that the vast majority of human diseases involve multiple genes. Complex disorders are characterized by the following features:

- Multiple genes are thought to be involved. It is the combination of mutations in multiple genes that defines the disease. In single-gene disorders, even if there are modifying loci, one gene has a dramatic influence on the disease phenotype.

- Complex diseases involve the combined effect of multiple genes, but they also are caused by both environmental factors and behaviors that elevate the risk of disease.

A quantitative trait locus (QTL) is an allele that contributes to a multifactorial disease.

BOX 20.3
Autism: Complex Disorder of Unknown Etiology

Autism (OMIM %209850) is a lifelong neurological disorder with onset before three years of age (Kanner, 1943; reviewed in Bailey et al., 1996; Ciaranello and Ciaranello, 1995; Rapin and Tuchman, 2008; Lintas and Persico, 2009). It is characterized by a triad of deficits: (1) an individual's failure to have normal reciprocal social interaction, (2) impaired language or communication skills, and (3) restricted, stereotyped patterns of interests and activities. Autistic children's play is abnormal beginning in infancy, and there is a notable lack of imaginative play. Approximately 30% of autistic children appear to develop normally but then undergo a period of regression in language skills between 18 and 24 months of age. In addition, cognitive function may be impaired. Seventy-five percent of autistic individuals have mental retardation. Approximately 10% of autistic individuals have savant-like superior abilities in areas such as mathematical calculation, rote memory, or musical performance. Autism is accompanied by seizures; by adulthood about one-third of autistic individuals will have had at least two unprovoked seizures (Olsson et al., 1988; Volkmar and Nelson, 1990).

In the 1990s, the prevalence of autism was estimated to be between 0.2 and 2 per 1000 individuals (Fombonne, 1999; Gillberg and Wing, 1999). More recently the prevalence has been estimated to be about 1:150. However, the definition of autism has broadened considerably in recent years, with a large number of patients formerly defined as having mental retardation now diagnosed as having autism or autism spectrum disorder. About three to four times more males are affected than females (Fombonne, 1999).

The cause of autism is unknown, but there is strong evidence that the disorder is genetic (Szatmari et al., 1998; Turner et al., 2000). The concordance between monozygotic twins is approximately 60%, and >90% if coaffected twins are defined as having classically defined autism or more generalized impairments in social skills, language, and cognition (Bailey et al., 1995). Autism has a far stronger genetic basis than most other common neuropsychiatric disorders such as schizophrenia or depression. Many genetic linkage studies have implicated genes that are significantly associated with autism, and studies of individuals having translocations or deletions have also led to the identification of genes harboring mutations. These genes include neuroligins 3 and 4, SHANK3, and neurexin 1. Other studies have suggested that individuals with autism have increased numbers of copy number variants. Ramocki and Zoghbi (2008) discuss a variety of chromosomal loci which when either duplicated or deleted cause symptoms involving mental retardation and autism. Known medical conditions affecting the central nervous system, such as fragile X syndrome and seizure disorder, may account for 10–30% of autistic cases (Barton and Volkmar, 1998). Indeed, autism may be considered a label for a large collection of distinct disorders, each involving a related phenotype.

- Complex diseases are non-Mendelian: they show familial aggregation but not segregation. For example, autism is a highly heritable condition (if one identical twin is affected, there is a very high probability that the other is also affected).

- Susceptibility alleles have a high population frequency; that is, complex diseases are generally more frequent than single-gene disorders. Sickle cell anemia is unusually frequent in the African American population, for a single-gene disorder, but the heterozygous condition confers a selective advantage (see Box 20.1 above).

- Susceptibility alleles have low penetrance. Penetrance is the frequency with which a dominant or homozygous recessive gene produces its characteristic phenotype in a population. At the extremes, it is an all-or-none phenomenon: a genotype is either expressed or it is not. In complex disorders, partial penetrance is common.

Penetrance is the frequency of manifestation of a hereditary condition in individuals. Having the genotype for a disease does not imply that the phenotype will occur, especially if multiple genes have modifying effects on the presentation of the phenotype.

An aneuploidy is the condition of having an abnormal number of chromosomes. Segmental aneuploidy affects a portion of a chromosome.

Genomic Disorders

Large-scale chromosomal abnormalities are extremely common causes of disease in humans. Lupski (1998) has defined genomic disorders as those changes in the structure of the genome that cause disease. Some genomic disorders involve large-scale changes such as aneuploidies in which a chromosomal copy is gained (trisomy) or lost (monosomy). More rarely, two copies are gained (tetrasomy) or lost (nullsomy). Trisomies 13 (Patau syndrome), 18 (Edwards syndrome), and 21 (Down syndrome)

TABLE 20-5 Frequency of Chromosomal Aneuploidies among Liveborn Infants

Abnormalities	Disorder	Frequency
Autosomal	Trisomy 13 (Patau syndrome)	1 in 15,000
	Trisomy 18 (Edwards syndrome)	1 in 5000
	Trisomy 21 (Down syndrome)	1 in 600
Sex chromosome	Klinefelter syndrome (47,XXY)	1 in 700 males
	XYY syndrome (47,XYY)	1 in 800 males
	Triple X syndrome (47,XXX)	1 in 1000 females
	Turner syndrome (45,X or 45X/46XX or 45X/ 46,XY or isochromosome Xq)	1 in 1500 females

Source: From Beaudet et al. (2001). Used with permission.

are the only autosomal trisomies that are compatible with life (Table 20.5). Of these, trisomies 13 and 18 are typically fatal in the first years of life. A variety of X chromosome aneuploidies are compatible with life.

Many developmental abnormalities involve a portion of a chromosome. Some involve cytogenetically detectable changes and span millions of base pairs. If they are too small to be cytogenetically visible (e.g., smaller than three megabases) they are usually referred to as cryptic changes. Examples of microdeletion syndromes include Cri-du-chat syndrome, Angelman syndrome, Prader-Willi syndrome, Smith-Magenis syndrome, and various forms of mental retardation that result from the gain (microduplication) or loss (microdeletion) of chromosomal regions. Table 20.6 lists examples of genomic disorders that are inherited in a Mendelian fashion and involve only one or several genes (Stankiewicz and Lupski, 2002). Table 20.7 provides a similar list of common structural variations that are associated with disease, reported by the Human Genome Structural Variation Working Group (2007). That group reported an initiative to characterize structural variation in phenotypically normal individuals using fosmid libraries.

We considered several mechanisms by which nonallelic homologous recombination causes deletions or duplications of chromosomal segments in Chapter 16 (see Fig. 16.25). Figure 20.5 shows six possible consequences of such events, such as loss of normal gene function, the fusion of two genes, or the exposure of a recessive allele.

Chromosomal alterations may be considered to occur along a spectrum from having little or no adverse effects to causing disease (Fig. 20.6). Copy number variants (described in Chapters 16 and 19) may have no phenotypic consequences and may be thought of as chromosomal alterations (in contrast to chromosomal abnormalities). Some copy number variants may increase disease susceptibility, perhaps contributing to common complex (multigenic) disorders. Some common and relatively benign traits such as color blindness can be attributed to copy number variants. At the extreme end of the spectrum, chromosomal changes may cause or contribute to a variety of genomic disorders, including aneuploidies, microdeletion syndromes, and microduplication syndromes. Genomic disorders are also notably common in cancers, with occurrence of amplifications and deletions of loci. We will discuss cancer in more detail below.

Chromosomal disorders are an extremely common feature of normal human development. Humans have a very low fecundity even relative to other mammals,

The DatabasE of Chromosomal Imbalance and Phenotype in Humans using Ensembl Resources (DECIPHER) is a major database resource for genomic disease. It is available at ▶ http://www.sanger.ac.uk/ PostGenomics/decipher/.

TABLE 20-6 Examples of Mendelian Genomic Disorders

Disorders	OMIM	Inheritance Pattern	Chrom. Location	Gene(s)	Rearrangement Type	Rearrangement Size (kb)	Recombination Substrates Repeat Size (kb)	% Identity	Orientation	Type
Bartter syndrome type III	601678	AD	1p36	CLCNKA/B	del	11		91	D	G/ψ
Gaucher disease	230800	AR	1q21	GBA	del	16	14		D	G/ψ
Spinal muscular atrophy	253300	AR	5q13.2	SMN	inv/dup	500			I	
β-thalassemia	141900	AR	11p15.5	β-globin	del	4, (7?)			D	G
α-thalassemia	141800		16p13.3	α-globin	del	3.7 or 4.2	4		D	S
Polycystic kidney disease 1	601313	AD	16p13.3	PKD1			50	95		
Charcot–Marie–Tooth (CMT1A)	118220	AD	17p12	PMP22	dup	1400	24	98.7	D	S
Neurofibromatosis type 1	162200	AD	17q11.2	NF1	del	1500			D	G
Hunter syndrome (mucopolysaccharidosis type II)	309900	XL	Xq28	IDS	inv/del	20	3	>88		G/ψ
Hemophilia A	306700	XL	Xq28	FB	inv	300–500	9.5	99.9	I	

Abbreviations: AD, autosomal dominant; AR, autosomal recessive; del, deletion; inv/dup, inversion/duplication; OMIM, Online Mendelian Inheritance in Man; Orientation D, direct; Orientation I, inverted; XL, X chromosome linked; G, gene; ψ, pseudogene; S, segment of genome.
Source: Stankiewicz and Lupski (2002). Used with permission.

TABLE 20-7   Common Structural Polymorhisms and Disease

Gene	Type	Locus	Size (kb)	Phenotype	Copy Number Variation
UGT2B17	Deletion	4q13	150	Variable testosterone levels, risk of prostate cancer	0–2
DEFB4	VNTR	8p23.1	20	Colonic Crohn's disease	2–10
FCGR3	Deletion	1q23.3	>5	Glomerulonephritis, systemic lupus erythematosus	0–14
OPN1LW/ OPN1MW	VNTR	Xq28	13–15	Red/green color blindness	0–4/0–7
LPA	VNTR	6q25.3	5.5	Altered coronary heart disease risk	2–38
CCL3L1/ CCL4L1	VNTR	17q12	Not known	Reduced HIV infection; reduced AIDS susceptibility	0–14
RHD	Deletion	1p36.11	60	Rhesus blood group sensitivity	0–2
CYP2A6	Deletion	19q13.2	7	Altered nicotine metabolism	2–3

Abbreviation: VNTR, variable number tandem repeats.
Source: Human Genome Structural Variation Working Group (2007). Used with permission.

with perhaps 50% to 80% of all human conceptions resulting in miscarriage. This low fecundity is primarily due to the common occurrence of chromosomal abnormalities (Wells and Delhanty, 2000; Voullaire et al., 2000):

- A woman who has already had one child (and thus is of established fertility) has only a 25% chance of achieving a viable pregnancy in any given menstrual cycle.

- 52% of all women who conceive have an early miscarriage.

- Following in vitro fertilization, pregnancies that are confirmed positive in the first two weeks result in miscarriage 30% of the time.

- Over 60% of spontaneous abortions that occur at 12 weeks gestation or earlier are aneuploid, suggesting that early pregnancy failures are likely due to lethal chromosome abnormalities.

Environmentally Caused Disease

Environmental diseases are extremely common. We may consider two types. (1) Infectious diseases are caused by a pathogen (such as a virus, bacterium, protozoan, fungus, or nematode). From birth to old age, infectious disease is the leading cause of death worldwide. We described the most common diseases caused by viruses (e.g., Table 14.2) and by bacteria (Table 15.7), and we discussed fungal pathogens (Chapter 17) and a variety of protozoan pathogens (Chapter 18). (2) Many diseases or other conditions are not caused by an infectious agent. These include malnutrition

About 8% of all children in the United States have blood levels that are defined as "alarming," according to the Centers for Disease Control and Prevention. See ▶ http://www.cdc.gov/nceh/lead/.

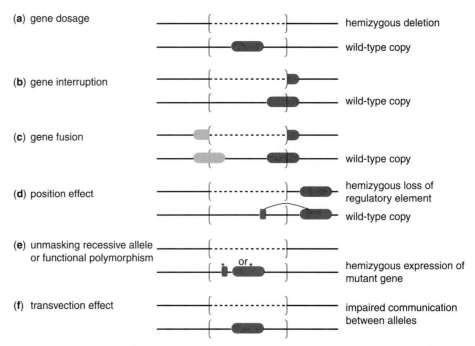

(a) gene dosage — hemizygous deletion / wild-type copy

(b) gene interruption — wild-type copy

(c) gene fusion — wild-type copy

(d) position effect — hemizygous loss of regulatory element / wild-type copy

(e) unmasking recessive allele or functional polymorphism — *or* * — hemizygous expression of mutant gene

(f) transvection effect — impaired communication between alleles

FIGURE 20.5. *Models for the molecular mechanisms of genomic disorders. For each, a hemizygous deletion is depicted (i.e., loss of one of the normal two copies of an allele) in brackets, and the two chromosomal homologs are indicated by horizontal lines. Note that duplications also potentially cause disease, and also, homozygous deletions (resulting in zero copies of a gene) typically have more severe consequences than hemizygous deletions. (a) Gene dosage effect in which one (of two) copies is deleted. Genes vary in their dosage sensitivity. (b) Gene interruption. A rearrangement breakpoint interrupts a gene. (c) Gene fusion in which two genes (and/or regulatory elements such as enhancers or promoters) are fused following a deletion. (d) Position effect: the expression or function of a gene near a breakpoint is disrupted by loss of a regulatory element. (e) Unmasking a recessive allele. The deletion results in hemizygous expression of a recessive mutation (asterisk) in a gene or a regulatory sequence. (f) Transvection, in which a deletion impairs communication between two alleles. Genes are indicated as red (or gray) filled ovals, while regulatory sequences are smaller ovals. Adapted from Lupski and Stankiewicz (2005). Used with permission.*

FIGURE 20.6. *Spectrum of effects of copy number variants. At one extreme, copy number variants cause genomic diseases such as microdeletion and microduplication syndromes. At the other extreme, copy number variants have no known phenotypic effects and occur in the apparently normal population. For example, many of the 270 HapMap individuals (who are defined as normal although everyone is susceptible to some diseases) have hemizygous and homozygous deletions as well as extended tracts of homozygosity. Adapted from Lupski and Stankiewicz (2005). Used with permission.*

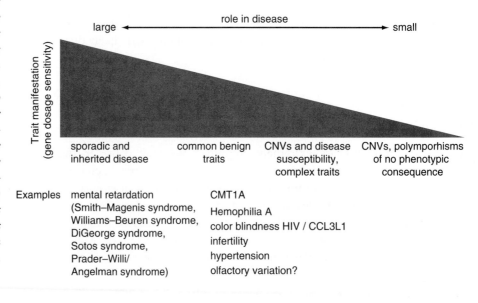

(whether maternal, fetal, or in an independent individual), poisoning by toxicants such as lead or mercury, or injury.

Other Categories of Disease

While we have presented four disease categories (monogenic, complex, genomic, and environmental), these are interrelated categories. If one examines four children who have the same highly elevated blood lead levels due to lead poisoning, they may display entirely different responses. One might be aggressive, another mentally retarded, another hyperactive, and another might appear unaffected. Four individuals exposed to the same pathogen might have different responses. It is likely that the genetic background has a key role in responses to environmental insults. Similarly, four children who have the identical single base pair mutation in the *ABCD1* gene might have entirely different severities of adrenoleukodystrophy, or the identical mutation in *MECP2* may cause very different forms of Rett syndrome. Modifier genes are likely to be involved (highlighting the concept that monogenic disorders may be caused primarily by the abnormal function of a single gene yet they always involve multiple genes), and environmental factors are certain to have large roles in genetic diseases.

There are other ways of classifying basic disease types. For example, particular ethnic groups or other discrete groups have high susceptibility to some genetic diseases. Examples include the following:

- Tay-Sachs disease is prevalent among Ashkenazi Jews.
- About 8% of the African American population are carriers of a mutant *HBB* gene.
- Males rather than females are susceptible to Alport disease, male pattern baldness, and prostate cancer.
- Cystic fibrosis affects ∼30,000 people in the United States with ∼12 million carriers, and is the most common fatal genetic disease in that country. While it affects all groups, Caucasians of northern European ancestry are particularly susceptible.

Another basis for classifying disease is according to tissue type, organ system, or subcellular organelle. Eukaryotic cells are organized into organelles, such as the nucleus, endoplasmic reticulum, Golgi complex, peroxisome, and mitochondrion. Each organelle serves a specialized function, gathering particular protein products to form enzymatic reactions necessary for cell survival, separating metabolic processes, and segregating harmful products. We have considered human disease from the perspective of genes and gene products. We can also examine disease in the context of the higher organizational level of organelles and pathways.

Let us consider the mitochondrion. This organelle was described as the site of respiration in the 1940s, and mitochondrial DNA was first reported by Nass and Nass (1963). But it was not until 1988 that the first disease-causing mutations in mitochondria were described (Wallace et al., 1988a, 1988b; Holt et al., 1988). Today, over 100 disease-causing point mutations have been described (reviewed in DiMauro and Schon, 2001, 2008; Schon, 2000). The mitochondrial genome contains 37 genes, and encodes 13 proteins (see Fig. 13.6). Any of these can be associated with disease. Figure 20.7 shows a morbidity map of the human mitochondrial genome.

GIDEON (Global Infectious Disease and Epidemiology Network) is a commercial database of infectious diseases available at ▶ http://www.gideononline.com.

Most (∼1500) mitochondrial proteins are the product of nuclear genes, and most mitochondrial diseases are caused by mutations in nuclear genes. Normally all mitochondrial genomes are the same, a condition called homoplasy. Pathogenic mutations may be heteroplasmic (having a mixture of normal and mutated genomes).

FIGURE 20.7. *Morbidity map of the human mitochondrial genome. Abbreviations are for the genes encoding seven subunits of complex I (ND), three subunits of cytochrome c oxidase (COX), cytochrome b (Cyt b), and the two subunits of ATP synthase (ATPase 6 and 8). 12S and 16S refer to ribosomal RNAs; 22 transfer RNAs are identified by the one-letter codes for the corresponding amino acids. FBSN, familial bilateral striatal necrosis; KSS, Kearns–Sayre syndrome; LHON, Leber hereditary optic neuropathy; MELAS, mitochondrial encephalomyopathy, lactic acidosis, and stroke-like episodes; MERRF, myoclonic epilepsy with ragged-red fibers; MILS, maternally inherited Leigh syndrome; NARP, neuropathy, ataxia, retinitis pigmentosa; PEO, progressive external ophthalmoplegia. From DiMauro and Schon (2001). Used with permission.*

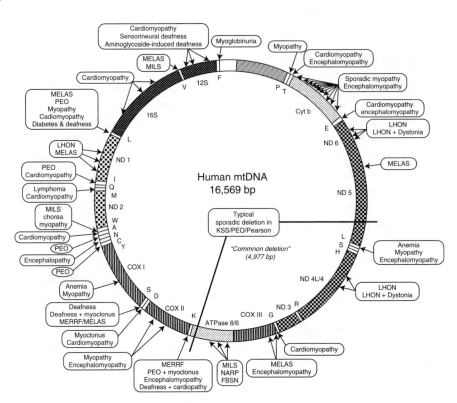

Mitochondrial genetics differs from Mendelian genetics in three main ways (DiMauro and Schon, 2001):

1. Mitochondrial DNA is maternally inherited. Mitochondria in the embryo are derived primarily from the ovum, while sperm mitochondria fail to enter the egg and are actively degraded. Thus, a woman having a mitochondrial DNA mutation may transmit it to her children, but only her daughters will further transmit the mutation to their children.

2. While nuclear genes exist with two alleles (one maternal and one paternal), mitochondrial genes exist in hundreds or thousands of copies per cell. (A typical mitochondrion contains about ten copies of the mitochondrial genome.) An individual may harbor varying ratios of normal and mutated mitochondrial genomes. Some critical threshold of mutated mitochondrial genomes is required before a disease is manifested.

3. As cells divide, the proportion of mitochondria having mutated genomes can change, thus affecting the phenotypic expression of mitochondrial disorders. Clinically, mitochondrial disorders are expressed at different times and in different regions of the body. An extremely broad variety of diseases are associated with mutations in mitochondrial DNA.

MITOMAP is online at ► http://www.mitomap.org/).

MITOMAP is a useful mitochondrial genome database (Ruiz-Pesini et al., 2007). The site lists a broad variety of information on mutations and polymorphisms in mitochondrial genomes involving all known genetic mechanisms (inversions, insertions, deletions, etc.).

DISEASE DATABASES

We next describe two major types of human disease database: (1) central databases such as OMIM provide great breadth in surveying thousands of diseases, and (2) locus-specific mutation databases provide great depth in reporting mutations associated with genes, with a focus on either one specific gene and/or one disease. Patrinos and Brookes (2005) reviewed these two types of databases, emphasizing the great challenges associated with relating genotype to phenotype (that is, relating data on DNA mutations to clinical phenotypes).

OMIM: Central Bioinformatics Resource for Human Disease

OMIM is a comprehensive database for human genes and genetic disorders, particularly monogenic disorders (Hamosh et al., 2005; McKusick, 2007). The OMIM database contains bibliographic entries for over 18,000 human diseases and relevant genes. The focus of OMIM is inherited genetic diseases. As indicated by its name, the OMIM database is concerned with Mendelian genetics. These are inherited traits that are transmitted between generations. There is relatively little information in the database about genetic mutations in complex disorders, or chromosomal disorders. Thus, its focus is a comprehensive survey of single-gene disorders, with richly detailed descriptions as well as links to many database resources.

We can examine OMIM using sickle cell anemia and *HBB* as examples of a disease and a gene implicated in a disease. OMIM can be searched from the NCBI Entrez site, and it is linked from Entrez Gene. Within the OMIM site, there is a search page that allows you to query a variety of fields, including chromosome,

Mendelian Inheritance in Man (MIM) was started in 1966 by Victor A. McKusick. The online version OMIM became integrated with NCBI in 1995. It is available at ▶ http://www.ncbi.nlm.nih.gov/omim/ or through Entrez at NCBI. The director of OMIM is Ada Hamosh of the Johns Hopkins Medical Institutions.

FIGURE 20.8. *Online Mendelian Inheritance in Man (OMIM), accessible via the NCBI website (▶ http://www.ncbi.nlm.nih.gov/omim), allows text searches by criteria such as author, gene identifier, or chromosome. A search of OMIM for "beta globin" produces over 100 results, including entries on that gene, related globin genes, and diseases such as thalassemias and sickle cell anemia.*

FIGURE 20.9. **The OMIM entry for beta globin includes the OMIM identifier (+141900) and a variety of information, indexed on the sidebar, such as clinical features, a description of available animal models, and allelic variants.**

map position, or clinical information. The result of a search for "beta globin" includes both the relevant gene (Fig. 20.8) and relevant diseases (e.g., sickle cell anemia and thalassemias).

We can next view the entry for beta globin (Fig. 20.9), with its OMIM identifier +141900. Each entry in OMIM is associated with a numbering system. There is a six-digit code in which the first digit indicates the mode of inheritance of the gene involved (Table 20.8). The beta globin entry is preceded by a plus sign to indicate that the the entry contains the description of a gene of known sequence and a phenotype. The first number (1) indicates that this gene has an autosomal locus (and the entry was created by 1994). The entry includes bibliographic data such as available information on an animal model for globinopathies. OMIM entries link to a gene map, which provides a tabular listing of the cytogenetic position of disease loci. This gene map further links to the NCBI Map Viewer and to resources for the orthologous mouse gene. The OMIM morbid map also provides cytogenetic loci but is organized alphabetically. The current holdings of OMIM, arranged by chromosome, are given in Table 20.9.

An important feature of OMIM entries is that many contain a list of allelic variants. Most of these represent disease-causing mutations. An example of several allelic variant entries is shown for *HBB* (Fig. 20.10). These allelic variants provide a glimpse of all the human genes that are known to contain disease-causing mutations. Allelic variants are selected based on criteria such as being the first mutation to be discovered, having a high population frequency, or having an unusual pathogenetic mechanism. Some allelic variants in OMIM represent polymorphisms. These may be of particular interest if they show a positive correlation with common disorders (see below). In the particular case of *HBB*, hundreds of allelic variants are included.

The current holdings of OMIM based on disease mechanism are summarized in Table 20.10. OMIM continues to be a crucial and comprehensive resource for

| TABLE 20-8 | OMIM Numbering System |

OMIM Number	Phenotype	OMIM Identifier	Disorder (Example)	Chromosome
1___	Autosomal dominant	+143100	Huntington disease	4p16.3
2___	Autosomal recessive	%209850	Autism, susceptibility to, (AUTS1)	7q
3___	X-linked loci or phenotypes	#312750	Rett syndrome	Xq28
4___	Y-linked loci or phenotypes	*480000	Sex-determining region Y	Yp11.3
5___	Mitochondrial loci or phenotypes	#556500	Parkinson disease	—
6___	Autosomal loci or phenotypes	#603903	Sickle cell anemia	—

Note: The entries beginning 1 and 2 entered the database before May 1994; those beginning with 6 were created after May 1994. An asterisk (*) preceding an entry indicates a gene of known sequence. A number symbol (#) indicates a descriptive entry, usually of a phenotype. For the AUTS1 entry, the number 1 indicates that this is the first listing of several autism susceptibility loci (e.g., AUTS2). A plus sign (+) indicates an entry with a gene of known sequence and a phenotype. A percent sign (%) indicates an entry describing a confirmed mendelian phenotype (or phenotypic locus) for which the underlying molecular basis is not known.

Source: Adapted from ► http://www.ncbi.nlm.nih.gov/Omim/omimfaq.html#mim_number_symbols (January 2008).

information on the human genome. The database is maintained and updated by expert curators (Hamosh et al., 2005). Many other disease databases incorporate OMIM identifiers to provide a common reference to disease-related genes.

The Human Gene Mutation Database (HGMD) is another major source of information on disease-associated mutations (Stenson et al., 2008). The database is partly commercial (requiring payment for full access). George et al. (2008) compared OMIM and HGMD, noting differences in their approaches (such as the OMIM emphasis on detailed descriptions of genes and disorders, and the HGMD emphasis on more comprehensive cataloguing of mutations). George et al. further reviewed projects such as GeneCards (Safran et al., 2003). This is a human gene compendium that includes a wealth of information on human disease genes. GeneCards differs from OMIM in that it collects and integrates data from several

HGMD is a project of David Cooper and colleagues at Cardiff University. It is available at ► http://www.hgmd.cf.ac.uk/ac. There are ~57,000 mutation entries for public release and ~76,000 entries for commercial release (January 2008). GeneCards, a project of Doron Lancet and colleagues at the Weizmann Institute, is available at ► http://www.genecards.org/.

| TABLE 20-9 | Synopsis of OMIM Human Genes per Chromosome (December 2007) |

Chromosome	Loci	Chromosome	Loci	Chromosome	Loci
1	1025	9	390	17	622
2	655	10	374	18	151
3	572	11	659	19	684
4	400	12	555	20	260
5	498	13	191	21	134
6	630	14	328	22	265
7	466	15	307	X	579
8	368	16	418	Y	44

Note: Total number of loci: 10,575.
See ► http://www.ncbi.nlm.nih.gov/Omim/mimstats.html.

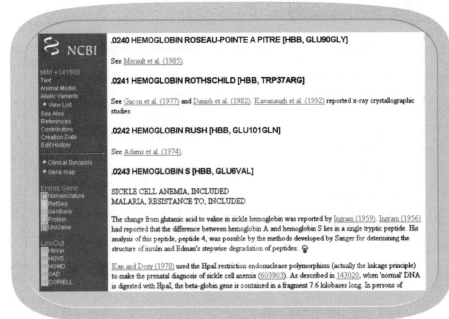

.0240 HEMOGLOBIN ROSEAU-POINTE A PITRE [HBB, GLU90GLY]

See Merault et al. (1985).

.0241 HEMOGLOBIN ROTHSCHILD [HBB, TRP37ARG]

See Gacon et al. (1977) and Danish et al. (1982). Kavanaugh et al. (1992) reported x-ray crystallographic studies.

.0242 HEMOGLOBIN RUSH [HBB, GLU101GLN]

See Adams et al. (1974).

.0243 HEMOGLOBIN S [HBB, GLU6VAL]

SICKLE CELL ANEMIA, INCLUDED
MALARIA, RESISTANCE TO, INCLUDED

The change from glutamic acid to valine in sickle hemoglobin was reported by Ingram (1959). Ingram (1956) had reported that the difference between hemoglobin A and hemoglobin S lies in a single tryptic peptide. His analysis of this peptide, peptide 4, was possible by the methods developed by Sanger for determining the structure of insulin and Edman's stepwise degradation of peptides.

Kan and Dozy (1978) used the HpaI restriction endonuclease polymorphism (actually the linkage principle) to make the prenatal diagnosis of sickle cell anemia (603903). As described in 143020, when 'normal' DNA is digested with HpaI, the beta-globin gene is contained in a fragment 7.6 kilobases long. In persons of

FIGURE 20.10. The OMIM entry for beta globin includes hundreds of allelic variants, most of which are disease-causing mutations. Some allelic variants reflect polymorphisms that are not associated with disease.

dozen independent databases, including OMIM, GenBank, UniGene, Ensembl, the University of California at Santa Cruz (UCSC), and the Munich Information Center for Protein Sequences (MIPS). Thus, relative to OMIM, GeneCards uses relatively less descriptive text of human diseases, and it provides relatively more functional genomics data.

Locus-Specific Mutation Databases

In the context of mutation databases, a mutation is defined as an allelic variant (Scriver et al., 1999). The allele (or the unique sequence change) may be disease causing; such an allele tends to occur at low frequency. The allele may also be neutral, not having any apparent effect on phenotype.

Central databases such as OMIM and HGMD attempt to comprehensively describe all disease-related genes. In contrast, locus-specific mutation databases describe variations in a single gene (or sometimes in several genes) in depth. Curators of these databases provide particular expertise on the genetic aspects of one specific gene,

TABLE 20-10	Current Holdings of OMIM				
	Autosomal	X-Linked	Y-Linked	Mitochondrial	Total
* Gene with known sequence	11,407	525	48	37	12,017
+ Gene with known sequence and phenotype	356	30	0	0	386
# Phenotype description, molecular basis known	2,014	187	2	26	2,229
% Mendelian phenotype or locus, molecular basis unknown	1,470	129	4	0	1,603
Other, mainly phenotypes with suspected mendelian basis	1,965	142	2	0	2,109
Total	17,212	1,013	56	63	18,344

Source: http://www.ncbi.nlm.nih.gov/Omim/mimstats.html (December 2007).

locus, or disease. Also, the coverage of known mutations tends to be far deeper in locus-specific databases as a group than in central databases (Scriver et al., 1999). Thus, these two types of databases serve complementary purposes.

A locus-specific mutation database is a repository for allelic variations. There are hundreds of such databases. The essential components of a locus-specific database include the following (Scriver et al., 1999, 2000; Claustres et al., 2002; Cotton et al., 2008):

- A unique identifier for each allele
- Information on the source of the data
- The context of the allele
- Information on the allele (e.g., its name, type, and nucleotide variation)

Mutation databases have an important role in gathering information about mutations, but there have not been uniform standards for their creation until recently. Claustres et al. (2002) surveyed 94 websites that encompassed 262 locus-specific databases; Cotton et al. (2008) noted over 700 such databases. Both studies noted great variability in the way data are collected, presented, linked, named, and updated. Scriver et al. (1999, 2000) and Cotton et al. (2008) described guidelines for the content, structure, and deployment of mutation databases.

- There is now increased uniformity in naming alleles (Antonarakis, 1998; den Dunnen and Antonarakis, 2000). For example, the A of the ATG of the initiator Met codon is denoted nucleotide +1. Many such rules have been explicitly stated to allow uniform descriptions of mutations.
- Ethical guidelines have been described, such as the obligation of preserving the confidentiality of information (Knoppers and Laberge, 2000). Lowrance and Collins (2007) have reviewed issues of identifiability in genomic research.
- Generic software to build and analyze locus-specific databases has been provided, such as the Universal Mutation Database template (Beroud et al., 2000; Brown and McKie, 2000).

To see the Universal Mutation Database template of Beroud et al. (2000), visit ▶ http://www.umd.be/.

Several websites provide gateways to access locus-specific databases (Table 20.11). A main point of entry is the Human Genome Variation Society (HGVS). This provides access to about 700 locus-specific mutation databases. Its major categories include (1) locus-specific mutation databases, organized by HUGO approved gene symbols; (2) disease-centered central mutation databases, such as the Asthma Gene Database; (3) central mutation and SNP databases, such as OMIM, dbSNP, HGMD, and PharmGKB; (4) national and ethnic mutation databases, such as databases for diseases affecting Finns or Turks; (5) mitochondrial mutation databases, such as MITOMAP; (6) chromosomal variation databases, such as the Mitelman Database of Chromosome aberrations in Cancer; (7) nonhuman mutation databases, such as OMIA (Online Mendelian Inheritance in Animals); and (8) clinical databases such as the National Organization for Rare Disorders (NORD).

HGVS is accessible at ▶ http://www.hgvs.org/. The Mitelman database is available at ▶ http://cgap.nci.nih.gov/Chromosomes/Mitelman. The OMIA website is ▶ http://omia.angis.org.au/.

As an example of a locus-specific database, we can examine HbVar (Giardine et al., 2007a). The database is a useful resource for sequence variation associated

TABLE 20-11   Gateways to Locus-Specific Databases

Site	Description	URL
GeneDis	From Tel Aviv University; performs pairwise alignments against a disease database	▶ http://life2.tau.ac.il/GeneDis/
HUGO Mutation Database Initiative	Comprehensive list of locus-specific mutation databases	▶ http://www.hgvs.org/
Human Gene Mutation Database	From the Institute of Medical Genetics in Cardiff	▶ http://www.hgmd.cf.ac.uk/ac/index.php
Universal Mutation Database	Software and databases for mutations in human genes, from INSERM	▶ http://www.umd.necker.fr/
The Mammalian Gene Mutation Database (MGMD)	Database of published mutagen-induced gene mutations in mammalian tissues	▶ http://lisntweb.swan.ac.uk/cmgt/

HbVar is available at ▶ http://globin.cse.psu.edu/hbvar/menu.html. It is a collaboration between investigators at Penn State University, INSERM Creteil (France), and Boston University Medical Center.

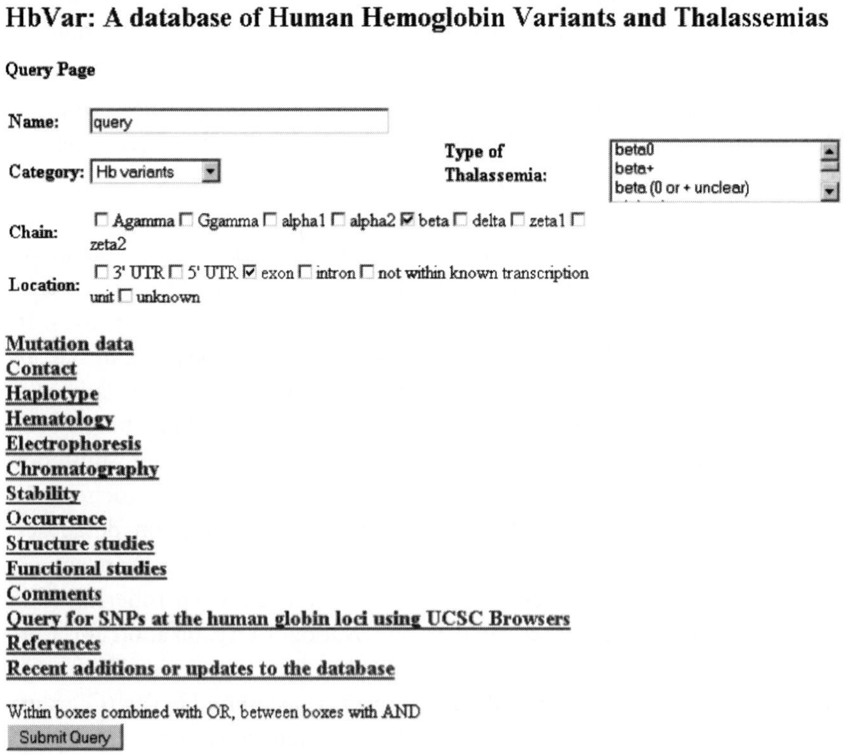

with hemoglobinopathies, and it is designed for both research purposes and clinical utility. The search page (Fig. 20.11) includes over a dozen fields that can be expanded to focus the search on a particular aspect of the globins, such as those with particular physical properties (stability, chromatographic behavior, structural alterations) or functional properties (e.g., sickling of red blood cells, affinity of oxygen binding) or epidemiological aspects (ethnic background, frequency). There are currently over 1300 entries, including categories such as entries involving hemoglobin variants ($\sim$980); thalassemia ($\sim$400 entries); the α1, α2, β, δ, Aγ, and Gγ genes; and mutations involving insertions, deletions, substitutions, gene fusions, or altered stability or oxygen binding properties.

HbVar: A database of Human Hemoglobin Variants and Thalassemias

Query Page

Name: [query]

Category: [Hb variants ▾]

Type of Thalassemia: [beta0 / beta+ / beta (0 or + unclear)]

Chain: ☐ Agamma ☐ Ggamma ☐ alpha1 ☐ alpha2 ☑ beta ☐ delta ☐ zeta1 ☐ zeta2

Location: ☐ 3' UTR ☐ 5' UTR ☑ exon ☐ intron ☐ not within known transcription unit ☐ unknown

Mutation data
Contact
Haplotype
Hematology
Electrophoresis
Chromatography
Stability
Occurrence
Structure studies
Functional studies
Comments
Query for SNPs at the human globin loci using UCSC Browsers
References
Recent additions or updates to the database

Within boxes combined with OR, between boxes with AND
[Submit Query]

FIGURE 20.11. *The HbVar database is a resource for information on globin mutations and thalassemias. It is created and maintained by experts who annotate many structural and functional properties of the globins and globinopathies. See* ▶ *http://globin.bx.psu.edu/hbvar/menu.html.*

BROWSER CHOICES

Destination: | UCSC Genome Browser ▾ | PhenCode at Ensembl (preliminary)

Human genome assembly: | hg18: March 2006 ▾ |

Browser position: | mecp2 | [e.g. "chr12:101,761,637-101,761,687" or "HBB"]

DATA CHOICES

Data sources:
- ☑ Locus-Specific Databases
- ☐ UniProt (Swiss-Prot/TrEMBL)

Mutation types:
- ☑ substitutions
- ☑ deletions
- ☑ insertions
- ☑ duplications
- ☑ complex

Mutation locations:
- ☑ exon
- ☑ intron
- ☑ 5' UTR
- ☑ 3' UTR
- ☑ not within known transcription unit

Estimated coordinates:
- ☑ mutations whose coordinates are estimated

FIGURE 20.12. The PhenCode (Phenotypes for ENCODE) project connects human phenotype and clinical data from locus-specific databases to the vast resources of the UCSC Genome Browser. Users can thus make connections between clinical data, mutation data, and genome properties.

The PhenCode Project

Locus-specific mutation databases provide tremendous depth and breadth of information about one gene and/or disease. However, the information in these databases is usually separate from the wealth of information contained in major genome

The PhenCode website is
▶ http://www.bx.psu.edu/
phencode.

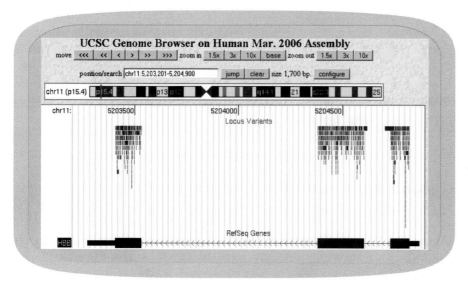

FIGURE 20.13. A sample image from the UCSC Genome Browser showing a connection from the HbVar locus-specific mutation database using PhenCode. Here, many beta globin mutations affecting the exons are displayed on a custom track. The RefSeq gene track is also displayed.

browsers. The PhenCode project connects data in locus-specific databases with genomic data from the UCSC Genome Browser (Giardine et al., 2007b), including the ENCODE project (described in Chapter 16). For a variety of locus-specific mutation databases, properties of interest can be selected, such as the type and location of the mutation (Fig. 20.12). This information is then displayed as a custom track on the UCSC Genome Browser (Fig. 20.13). The significance of PhenCode is that it facilitates the exploration and discovery of genomic features associated with disease-causing mutations. For example, the genomic landscape could include ultraconserved elements in noncoding regions (Chapter 16) that are associated with disease, or repetitive elements that serve as substrates for recombination in deleted or duplicated regions.

FOUR APPROACHES TO IDENTIFYING DISEASE-ASSOCIATED GENES

How can we determine the causes of diseases? There are many approaches to finding genes that confer risk for the disease. By identifying such genes we may rationally develop treatments (or, ultimately, find cures). For example, phenylketonuria (PKU; OMIM +261600) is an inborn error of metabolism that results in mental retardation and other symptoms. It is caused by a deficiency in phenylalanine hydroxylase activity. Knowing this it is possible to screen newborns and if PKU is found then to provide a diet lacking phenylalanine. PKU provides another example of the complexity of any disease. The enzyme phenylalanine hydroxylase is localized to the liver, and yet the symptoms of mental retardation are neurological; if one were searching for the cause by studying brain tissue it would be challenging to discover any biochemical defects. Also, while phenylalanine hydroxylase is overwhelmingly the major cause, it is not the only cause of PKU.

We will next discuss several approaches that are used to identity disease-associated genes (or other genetic elements). Once a gene has been associated with a disease, it is further necessary to determine how susceptibility genes confer risk.

Linkage Analysis

A genetic linkage map displays genetic information in reference to linkage groups (chromosomes) in a genome. The mapping units are centimorgans, based on recombination frequency between polymorphic markers such as SNPs or microsatellites. (One cM equals one recombination event in 100 meioses; for the human genome, the recombination rate is typically 1 to 2 cM/megabase.) In linkage studies, genetic markers are used to search for coinheritance of chromosomal regions within families; that is, polymorphic markers that flank a disease gene segregate with the disease in families. Two genes that are in proximity on a chromosome will usually cosegregate during meiosis. By following the pattern of transmission of a large set of markers in a large pedigree, linkage analysis can be used to localize a disease gene based on its linkage to a genetic marker locus. Huntington disease (OMIM), a progressive degenerative disorder, was the first autosomal disorder for which linkage analysis was used to identify the disease locus (reviewed in Gusella, 1989).

Linkage is usually performed for single-gene disease models rather than for complex traits. It also typically involves studies of large pedigrees. For Mendelian diseases the LOD score approach is used, providing a maximum likelihood estimate of the position of the disease locus (Ott, 2001). A LOD score of three implies that there is a 1 in 1000 chance that a given unlinked locus could have given rise to the observed cosegregation data. Hundreds of software packages are available for linkage analysis. Among the most widely used is Merlin (Multipoint Engine for Rapid Likelihood INference) (Abecasis et al., 2002).

Genome-Wide Association Studies

While the genetic basis of over a thousand single gene disorders has been found, it is far more difficult to identify the genetic causes of common human diseases that involve multiple genes. Part of the challenge is that a large number of genes may each make only a small contribution to the disease risk. Association studies provide an important approach (reviewed in Hirschhorn and Daly, 2005; Altshuler et al., 2008; McCarthy et al., 2008). Genomewide association studies provide a powerful new approach that can rely on SNP microarrays (Chapter 16) having 500,000 to 1 million SNPs represented on a single array. There are two main experimental designs used in association studies (Laird and Lange, 2006). In family-based designs, markers are measured in affected individuals (probands) and unaffected individuals to identify differences in the frequency of variants. In population-based designs, a large number of unrelated cases and controls are studied (typically hundreds in each group). Larger samples sizes offer increased statistical power (Chapter 9).

The Laboratory of Statistical Genetics at Rockefeller University, directed by Jürg Ott, offers a website listing dozens of software packages useful for linkage analysis. The Rockefeller website is ► http://linkage.rockefeller.edu/. Merlin was developed by Gonçalo Abecasis and colleagues and is available at ► http://www.sph. umich.edu/csg/abecasis/Merlin. Another popular software package, Plink, was developed by Shaun Purcell and colleagues (2007) and is at ► http://pngu. mgh.harvard.edu/ ∼ purcell/ plink/.

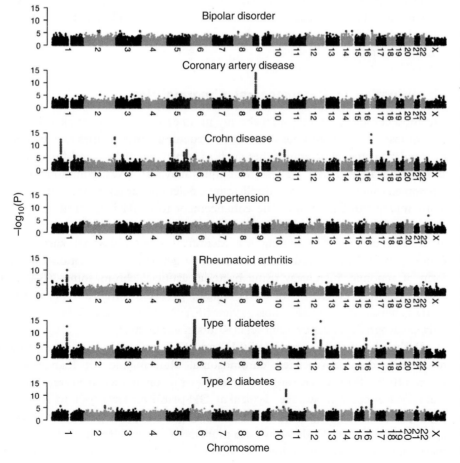

FIGURE 20.14. Results of a genome-wide association study using 16,179 individuals to search for genes contributing to seven common familial disorders. For each of seven diseases, the y axis shows the $-\log_{10}$P *value for SNPs that were positive for quality control criteria. The x axis shows the chromosomes.* P *values* $<1 \times 10^{-5}$ *are highlighted in red. Panels are truncated at* $-\log_{10}$(P value) = 15. *As noted by McCarthy et al. (2008), for the type 2 diabetes study, strong associations were observed on chromosomes 10 (transcription factor 7-like 2; TCF7L2), 16 (fat mass and obesity associated; FTO) and 6 (CDK5 regulatory subunit associated protein 1-like 1; CDKAL1). However, signals on chromosomes 1, 2, and 12 were not replicated. Redrawn from Fig. 4 of the Wellome Trust Case Control Consortium (2007). Used with permission.*

We can illustrate the genomewide association approach with an extremely large-scale study by the Wellcome Trust Case Control Consortium (2007) involving 50 research groups from the United Kingdom and 16,179 individuals (reviewed by Bowcock, 2007). Approximately 2000 affected individuals were studied having one of seven common familial diseases: bipolar disorder, coronary artery disease, Crohn's disease, hypertension, rheumatoid arthritis, type 1 diabetes, and type 2 diabetes. There were ~3000 control individuals. About 500,000 SNPs were measured for each individual, and the relationship between each SNP and the phenotypic trait (disease status) was measured. Twenty-four strong association signals were found for six of the seven diseases (Fig. 20.14). Many of these signals corresponded to previously characterized susceptibility loci, and many novel loci were also identified.

A key aspect of genomewide association studies is that replication studies are required to confirm that positive signals are authentic. The NCI-NHGRI Working Group on Replication in Association Studies (2007) has addressed many of the issues relevant to replication studies, emphasizing the need to eliminate false positive results that often occur. Proper experimental design is especially important, with efforts to assess phenotypes in a standard way, and a need to account for biases such as population stratification.

dbGaP is available at ▶ http://www.ncbi.nlm.nih.gov/dbgap.

The National Library of Medicine (NLM) offers the database of Genotype and Phenotype (dbGaP), a database of archived genomewide association studies (Mailman et al., 2007). dbGaP contains four types of data: (1) study documentation (e.g., protocols and data collection instruments), (2) phenotypic data (of individuals and as a summary), (3) genetic data (genotypes, pedigrees, mapping results), and (4) statistical results (e.g., linkage and association results). Permission from a committee is required to access information such as pedigrees or phenotypic data associated with genotype data.

Identification of Chromosomal Abnormalities

NCBI's SKY/M-FISH & CGH Database is available at ▶ http://www.ncbi.nlm.nih.gov/sky/.

The most common chromosomal aberrations in early development likely involve the gain or loss of whole chromosomes. Such structural abnormalities may be detected by standard cytogenetic approaches such as karyotype analysis and fluorescence in situ hybridization (FISH). These techniques may also reveal commonly observed phenomena such as large-scale duplications, deletions, or rearrangements involving many millions of base pairs. One enhancement to FISH is spectral karyotyping/multiplex-FISH (SKY/M-FISH). This permits each chromosome to be depicted in a different color, facilitating the identification of abnormal karyotypes. In Chapter 16 we introduced array comparative genomic hybridization (aCGH), a form of genomic microarray using bacterial artificial chromosomes (BACs) that also represents an extension of FISH technology. NCBI offers a SKY/M-FISH & CGH Database that includes tools to view SKY/M-FISH and aCGH data, particularly as ideograms of cancer data sets (Knutsen et al., 2005).

Both genomic microarrays (aCGH) and SNP microarrays are used routinely to identify disease-associated chromosomal abnormalities at high resolution. (Currently, SNP arrays have approximately one million markers per array spaced several kilobases apart on average. Typical aCGH platforms have 3000 to 30,000 BAC clones, each with a length of ~200 kilobases, spaced approximately one megabase apart or less. Other aCGH arrays use millions of oligonucleotides.) In addition to measuring copy number based on fluorescence intensity measurements, SNP technologies also permit estimates of genotypes, which provides information about inheritance patterns and homozygosity. Both aCGH and SNP microarrays have been used to measure chromosomal variations in cancer, idiopathic mental retardation, and a variety of other diseases.

Genomic DNA Sequencing

We introduced high throughput DNA sequencing in Chapter 13, and described genome sequencing projects in subsequent chapters. Another use of high throughput sequencing is to resequence genomic regions in patients in order to define nucleotide differences that may be associated with disease.

We can explore sequencing initiatives in the realm of cancer. Cancer occurs when DNA mutations confer selective advantage to cells that proliferate, often uncontrollably (Varmus, 2006). Knudson (1971) introduced a two-hit hypothesis of cancer, suggesting that for dominantly inherited retinoblastoma one mutation is inherited through the germ cells while a second somatic mutation occurs; for a nonhereditary form of cancer two mutations occur in somatic cells. There are many types of cancer and many disease mechanisms, and a growing number of key tumor suppressor genes and other oncogenic genes has been identified, summarized in a Cancer Gene Census. Given the completion of the human genome project and the availability of improved sequencing capabilities, a human cancer genome project has been launched to determine the DNA sequence of a variety of cancer genomes.

In studies of cancer, high throughput sequencing has revealed vast numbers of somatic mutations (in contrast to germline mutations) that might sometimes arise due to environmental exposures to carcinogens or ultraviolet radiation. Greenman et al. (2007) described two types of somatic mutations. "Driver" mutations confer growth advantage, are implicated as causing the neoplastic process, and are positively selected for during tumorigenesis. "Passenger" mutations are retained by chance but confer no selective advantange and do not contribute to oncogenesis. A challenge is to identify driver mutations throughout the genome of a cancer cell and to distinguish them from passenger mutations.

The Cancer Genome Project at the Wellcome Trust Sanger Institute is at ▶ http://www.sanger.ac.uk/genetics/CGP/ with links to many cancer resources, and the Cancer Gene Census is available at ▶ http://www.sanger.ac.uk/genetics/CGP/Census/. The National Cancer Institute (NCI) at the National Institutes of Health website is ▶ http://www.cancer.gov/.

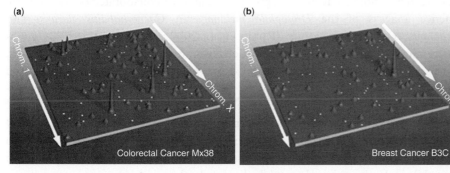

(a)   (b)

Colorectal Cancer Mx38   Breast Cancer B3C

FIGURE 20.15. *The landscape of mutations in cancer. Nonsilent somatic mutations are plotted in two dimensions representing the chromosomal positions of RefSeq genes for (a) a colorectal cancer sample and (b) a breast cancer sample. The telomere of chromosome 1p is at the upper left, continuing to the telomere of Xq at the bottom right. Peaks correspond to the 60 highest-ranking candidate cancer genes that are proposed to be "drivers" (rather than "passengers"). Some genes are shared such as PIK3CA (chromosome 3). Adapted from Wood et al. (2007). Used with permission.*

As an example of this approach, Bert Vogelstein, Victor Velculescu, Ken Kinzler, and colleagues sequenced 20,857 transcripts from 18,191 genes in 11 breast and 11 colorectal cancer samples as well as matched normal tissue (Wood et al., 2007). These corresponded to Consensus Coding Sequence (CCDS) as well as RefSeq genes, and followed an earlier large-scale sequencing effort (Sjöblom et al., 2006). They found 1718 genes (9.4% of those analyzed) that had at least one nonsilent mutation, mostly consisting of single base substitutions. Wood et al. measured synonymous and nonsynonymous mutation rates and predicted 280 genes in which mutations are likely to be drivers rather than passengers. They proposed a cancer landscape for breast and colorectal tumors in which there are several mountains (corresponding to genes that were found to be mutated in many cancer samples) interspersed with many hills (corresponding to driver mutations that occur with lower frequency) (Fig. 20.15). The large number of infrequently mutated genes represented in the hills may be even more important than the mountains, and may represent the relevant mutational signature of each cancer. A goal is to relate such a molecular profile of a cancer to an appropriate therapy to eradicate the cancer.

HUMAN DISEASE GENES IN MODEL ORGANISMS

The study of human disease genes and gene products in other organisms is of fundamental importance in our efforts to understand the pathophysiology of human disease. While mutations in genes cause many diseases, it is the aberrant protein product that has the proximal functional consequence on the cell and ultimately on the organism. Once a human disease gene is identified in a model organism, it can often be knocked out or otherwise manipulated. This allows the phenotypic consequences of specific mutations to be assessed.

Human Disease Orthologs in Nonvertebrate Species

At the time that *C. elegans* was sequenced, about 65% of human disease genes had identifiable *C. elegans* orthologs (Ahringer, 1997).

The Homophila website is at ▶ http://superfly.ucsd.edu/homophila/. It lists ∼2100 *Drosophila* orthologs of human genes having OMIM allelic variants (January 2008).

A basic question then is to identify which known human disease genes have orthologs in model organisms. This approach is of interest even though the consequence of mutating that ortholog may differ. A group of 55 authors collaborated on a systematic sequence analysis of the *Drosophila melanogaster*, *Caenorhabditis elegans*, and *Saccharomyces cerevisiae* genomes (Rubin et al., 2000). They identified 289 genes that are mutated, altered, amplified, or deleted in human disease. Of these genes, 177 (61%) were found to have an ortholog in *Drosophila*. These data are displayed in Fig. 20.16, showing the presence of fly, worm, and yeast orthologs to human disease genes that are functionally categorized in cancer, neurological, cardiovascular, endocrine, and other disease types. Reiter et al. (2001) extended this study to 929 human disease genes in OMIM, 714 of which (77%) matched 548 *Drosophila* protein sequences (Table 20.12). The Reiter et al. (2001) data have been deposited in Homophila, a *Homo sapiens*/*Drosophila* disease database.

The cataloguing of human disease genes in model organisms is important in our efforts to establish functional assays for these genes. In addition to the results in *S. cerevisiae*, *D. melanogaster*, and *C. elegans*, similar descriptions have been made in other eukaryotes, such as *Schizosaccharomyces pombe* (Wood et al., 2002), *Arabidopsis* (*Arabidopsis* Genome Initiative, 2000), and the amoeba

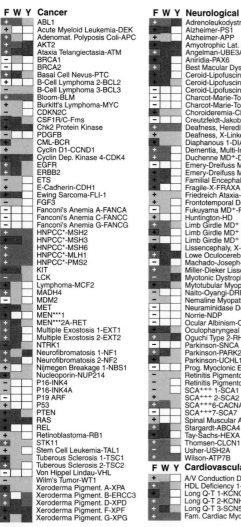

Figure 20.16 — Human disease genes in model organisms. The figure presents lists of human disease genes grouped by category, each with F (fly), W (worm), and Y (yeast) database-match columns.

F W Y Cancer
- ABL1
- Acute Myeloid Leukemia-DEK
- Adenomat. Polyposis Coli-APC
- AKT2
- Ataxia Telangiectasia-ATM
- BRCA1
- BRCA2
- Basal Cell Nevus-PTC
- B-Cell Lymphoma 2-BCL2
- B-Cell Lymphoma 3-BCL3
- Bloom-BLM
- Burkitt's Lymphoma-MYC
- CDKN2C
- CSF1R/C-Fms
- Chk2 Protein Kinase
- PDGFB
- CML-BCR
- Cyclin D1-CCND1
- Cyclin Dep. Kinase 4-CDK4
- EGFR
- ERBB2
- ETS
- E-Cadherin-CDH1
- Ewing Sarcoma-FLI-1
- FGF3
- Fanconi's Anemia A-FANCA
- Fanconi's Anemia C-FANCC
- Fanconi's Anemia G-FANCG
- HNPCC*-MSH2
- HNPCC*-MSH3
- HNPCC*-MSH6
- HNPCC*-MLH1
- HNPCC*-PMS2
- KIT
- LCK
- Lymphoma-MCF2
- MADH4
- MDM2
- MET
- MEN***1
- MEN***2A-RET
- Multiple Exostosis 1-EXT1
- Multiple Exostosis 2-EXT2
- NTRK1
- Neurofibromatosis 1-NF1
- Neurofibromatosis 2-NF2
- Nijmegen Breakage 1-NBS1
- Nucleoporin-NUP214
- P16-INK4
- P16-INK4A
- P19 ARF
- P53
- PTEN
- RAS
- REL
- Retinoblastoma-RB1
- STK11
- Stem Cell Leukemia-TAL1
- Tuberous Sclerosis 1-TSC1
- Tuberous Sclerosis 2-TSC2
- Von Hippel Landau-VHL
- Wilm's Tumor-WT1
- Xeroderma Pigment. A-XPA
- Xeroderma Pigment. B-ERCC3
- Xeroderma Pigment. D-XPD
- Xeroderma Pigment. F-XPF
- Xeroderma Pigment. G-XPG

F W Y Renal
- Alport-COL4A5
- Bartter-SLC12A1
- Congenital Nephrotic-NPHS1
- Dent-CLCN5
- Diabetes Insipidus 2-AQP2
- Gitelman-SLC12A3
- 1° Hyperoxaluria 1-AGXT
- 1° Hypomagnesemia-CLDN16
- Hypophosphatasia-ALPL
- Nephronophthisis 1-NPHP1
- Polycystic Kidney 1-PKD1
- Polycystic Kidney 2-PKD2
- Pseudohypoaldoster-NR3C2
- Renal Tubul. Acidosis-ATP6B1
- Vitamin D Resis. Rickets-PHEX
- Williams-Beuren-ELN

F W Y Hematological
- Chediak-Higashi-CHS1
- Diamond-Blackfan Anem.-RPS19
- Essen. Thrombocythemia-THPO
- G6PD Deficiency-G6PD
- HPLH2-PRF1
- Hemophilia A-F8C
- Hemophilia B-F9
- Hered. Spherocytosis-ANK1
- Megaloblas. Anemia-SLC19A2
- Myeloperoxidase Defic.-MPO
- Osler-Rendu-Weber-ENG
- α-Thalassemia-HBA1
- β-Thalassemia-HBB
- δ-Thalassemia-HBD
- ε-Thalassemia-HBE
- Thrombophilia-PLG
- Von Willebrand-VWF
- Wiskott-Aldrich-WAS

F W Y Neurological
- Adrenoleukodystrophy-ABCD1
- Alzheimer-PS1
- Alzheimer-APP
- Amyotrophic Lat. Sclero.-SOD1
- Angelman-UBE3A
- Aniridia-PAX6
- Best Macular Dystrophy-VMD2
- Ceroid-Lipofuscinosis-PPT
- Ceroid-Lipofuscinosis-CLN3
- Ceroid-Lipofuscinosis-CLN2
- Charcot-Marie-Tooth 1A-PMP22
- Charcot-Marie-Tooth 1B-MPZ
- Choroideremia-CHM
- Creutzfeldt-Jakob-PRNP
- Deafness, Hereditary-MYO15
- Deafness, X-Linked-TIMM8
- Diaphanous 1-DIAPH1
- Dementia, Multi-Infarct-NOTCH3
- Duchenne MD+-DMD
- Emery-Dreifuss MD+-EMD
- Emery-Dreifuss MD+-LMNA
- Familial Encephalopathy-PI12
- Fragile-X-FRAXA
- Friedreich Ataxia-FRDA
- Frontotemporal Dement.-TAU
- Fukuyama MD+-FCMD
- Huntington-HD
- Limb Girdle MD+ 2A-CAPN3
- Limb Girdle MD+ 2B-YSF
- Limb Girdle MD+ 2E-BSG
- Lissencephaly, X-Linked-DCX
- Lowe Oculocerebroren.-OCRL
- Machado-Joseph-MJD1
- Miller-Dieker Lissen.-PAF
- Myotonic Dystrophy-DM1
- Mytotubular Myopathy 1-MTM1
- Naito-Oyangi-DRPLA
- Nemaline Myopathy 2-NEB
- Neuraminidase Defic.-NEU1
- Norrie-NDP
- Ocular Albinism-OA1
- Oculopharyngeal MD+-PABPN1
- Oguchi Type 2-RH KIN
- Parkinson-SNCA
- Parkinson-PARK2
- Parkinson-UCHL1
- Prog. Myoclonic Epilepsy-CSTB
- Retinitis Pigmentosa-RPGR
- Retinitis Pigmentosa 2-RP2
- SCA+++ 1-SCA1
- SCA+++ 2-SCA2
- SCA+++6-CACNA1A
- SCA+++7-SCA7
- Spinal Muscular Atrophy-SMN1
- Stargardt-ABCA4
- Tay-Sachs-HEXA
- Thomsen-CLCN1
- Usher-USH2A
- Wilson-ATP7B

F W Y Cardiovascular
- A/V Conduction Defects-CSX
- HDL Deficiency 1-ABCA1
- Long Q-T 1-KCNQ1
- Long Q-T 2-KCNH2
- Long Q-T 3-SCN5A
- Fam. Cardiac Myopathy-MYH7

F W Y Immune
- Bare Lymphocyte-ABCB3
- Bare Lymphocyte-RFX5
- Bare Lymphocyte-RFX5AP
- Bare Lymphocyte-MHC2TA
- Bruton Agammaglobulin.-BTK
- Chronic Granulom.-NCF1
- Chronic Granulom.-CYBB
- Immunodeficiency-DNA Ligase 1
- Immunodeficiency-CD3G
- SCID**-IL2RG
- SCID**-IL7R
- SCID**-JAK3
- SCID**-RAG1
- SCID**-RAG2
- SCID**-ZAP70
- T-Cell Immunodefic.-CD3E
- X-Linked Lymphoprol.-SH2D1A

F W Y Metabolic
- CPT2 Deficiency-CPT2
- 1° Carnitine Defic.-SLC22A5
- Citrullinemia, Type I-ASS
- Cystinuria, Type 1-SLC3A1
- Hypercalcemia-CASR
- Galactokinase-GALK1
- Gaucher-GBA
- Hemochromatosis-HFE
- Lesch-Nyhan-HPRT1
- Liddle-SCNN1G
- Liddle-SCNN1B
- Menkes-ATP7A
- Niemann-Pick C-NPC1
- SCID**-ADA
- Trimetylaminuria-FMO3
- Variegate Porphyria-PPOX
- Wernicke-Korsakoff-TKT

F W Y Malformation Syndromes
- Aarskog-Scott-FGD1
- Achondroplasia-FGFR3
- Alagille-JAG1
- Barth-TAZ
- Beckwith-Wiedemann-CDKN1C
- Cerebral Cavern. Malf.-CCM1
- Chondrodyspl. Punct. 1-ARSE
- Cleidocranial Dysplasia-OFC1
- Cockayne I-CKN1
- Coffin-Lowry-RPS6KA3
- Diastrophic Dyspl.-SLC26A2
- EEC 3-Ket. P63
- Greig Cephalopolysynd.-GLI3
- Hand-Foot-Genital-HOXA13
- Holoprosencephaly 3-SHH
- Holoprosencephaly-SIX3
- Holt-Oram-TBX5
- ICF-DNMT3B
- Kallman-KAL1
- Laterality, X-Linked-ZIC3
- Melnick-Fraser-EYA1
- Nail Patella-LMX1B
- Opitz-MID1
- Renal Coloboma-PAX2
- Rieger, Type 1-PITX2
- Rubinstein-Taybi-CREBBP
- Saethre-Chotzen-TWIST
- Septooptic Dysplasia-HESX1
- Simpson-Golabi-Behmel-GPC3
- Townes-Brockes-SALL1
- Treacher-Collins-TCOF1
- WMCM-TEK
- Wardenburg-PAX3
- Zellweger-PEX1

F W Y Endocrine
- Adrenal Hypoplasia-NR0B1
- Androgen Receptor-AR
- Adrenal Hyperplas. III-CYP21A2
- Diabetes-INS
- Diabetes-INSR
- Diabet. Ins. Neurohypop.-AVP
- Diabet. w/ Hypertens.-PPARG
- Dwarfism-GH1
- Dwarfism-GHR
- Gonadal Dysgenesis-SRY
- Hyperinsulinism-ABCC8
- Hyperinsulinism-KCNJ11
- Hypothyroidism-TRH
- Hypothyroidism-TSHR
- Leydig Cell Hypoplasia-LHCGR
- MODY++ 1-HNF-4A
- MODY++ 2-GCK
- MODY++ 3-TCF1
- MODY++ 4-IPF1
- MODY++ 5-TCF2
- McCune-Albright-GNAS1
- Non-Insulin Dep. Diabet.-PCSK1
- Obesity-LEP
- Obesity-LEPR
- Obesity-MC4R
- Obesity-POMC
- Pendred-PDS
- Thyr. Resistance-THRA
- Thyr. Resistance-THRB
- Thyrotropin Deficiency-TSHB
- Vitamin-D Resis. Rickets-VDR

F W Y Other
- α-1-Antitrypsin Deficiency-PI
- Alveolar Proteinosis-SFTPB
- Corneal Dystrophy-TGFBI
- Cystic Fibrosis-ABCC7
- Cystinosis-CTNS
- Darier-White-SERCA
- Downreg. in Adenoma-DRA
- Ehlers-Danlos IV-COL3A1
- Fam. Mediterr. Fever-MEFV
- Finnish Amyloidosis-GSN
- Glycerol Kinase Defic.-GK
- Hereditary Pancreatitis-PRSS1
- Hermansky-Pudlak-HPS
- Hyperexplexia-GLRA2
- Juvenile Glaucoma-GLC1A
- Keratoderma-KRT9
- Marfan-FBN1
- Mcleod-XK
- Monilethrix-KRTHB
- Monilethrix-KRTHB6
- Osteogenesis Imperf.-COL1A1
- Spondyloepip. Dysp.-COL2A1
- Vohwinkel-LOR
- Wolfram-WFS1

FIGURE 20.16. A set of 289 proteins encoded by human disease genes were used as blastp queries against a set of 38,860 proteins from the complete genomes of a fly (F), a yeast (Y), and a worm (W). Database matches are presented according to their level of statistical significance. White boxes represent E values greater than 1×10^{-6} (no or weak similarity). Light gray boxes represent E values from 1×10^{-6} to 1×10^{-40}. Red boxes represent E values from 1×10^{-40} to 1×10^{-100}. Dark gray boxes represent E values below 1×10^{-100}. A plus sign indicates that the Drosophila protein is the functional equivalent of the human protein (based on criteria including sequence similarity, InterPro domain composition, and supporting biological evidence). A minus sign indicates that evidence was not obtained for functional equivalence to the human protein. Adapted from Rubin et al. (2000). Used with permission.

TABLE 20-12 Classification of 714 *Drosophila* Genes According to Human Disease Phenotypes

Disorder	No. of Genes
Neurological	
Neuromuscular	20
Neuropsychiatric	9
CNS/developmental	8
CNS/ataxia	9
Mental retardation	6
Other	22
Total	74
Endocrine	
Diabetes	10
Other	40
Total	50
Deafness	
Syndromic	7
Nonsyndromic	6
Total	13
Cardiovascular	
Cardiomyopathy	10
Conduction defects	4
Hypertension	7
Atherosclerosis	3
Vascular malformations	2
Total	26
Ophthalmological	
Anterior segment	
Aniridia	1
Rieger syndrome	1
Mesenchymal dysgenesis	2
Iridogoniodysgenesis	2
Corneal dystrophy	2
Cataract	3
Glaucoma	2
Subtotal	13
Retina	
Retinal dystrophy	1
Choroideremia	1
Color vision defects	4
Cone dystrophy	2
Cone rod dystrophy	1
Night blindness	8
Leber congenital amaurosis	2
Macular dystrophy	4
Retinitis pigmentosa	7
Subtotal	30
Total	43
Pulmonary	4
Gastrointestinal	13

(Continued)

TABLE 20-12 Continued

Disorder	No. of Genes
Renal	13
Immunological	
Complement mediated	11
Other	22
Total	33
Hematological	
Erythrocyte, general	29
Porphyrias	7
Platelets	6
Total	42
Coagulation abnormalities	28
Malignancies	
Brain	3
Breast	4
Colon	11
Other gastrointestinal	3
Genitourinary	5
Gynecological	3
Endocrine	3
Dermatological	3
Xeroderma pigmentosa	6
Other/sarcomas	9
Hematological malignancies	29
Total	79
Skeletal development	
Craniosynostosis	5
Skeletal dysplasia	13
Other	8
Total	26
Soft tissue	2
Connective tissue	18
Dermatological	25
Metabolic/mitochondrial	123
Pharmacological	12
Peroxisomal	9
Storage	
Glycogen storage	11
Lipid storage	13
Mucopolysaccharidosis	10
Other	3
Total	37
Pleiotropic developmental	
Growth, immune, cancer	7
Apoptosis	1
Other	27
Total	35
Complex other	9
Total	714

Source: Adapted from Reiter et al. (2001). Used with permission.

TABLE 20-13   *Schizosaccharomyces pombe* Genes Related to Human Cancer Genes

Human Cancer Gene	Score[a]	*S. pombe* Gene/ Product	Systematic Name
Xeroderma pigmentosum D; *XPD*	$<1 \times 10^{-100}$	rad15, rhp3	SPAC1D4.12
Xeroderma pigmentosum B; *ERCC3*	$<1 \times 10^{-100}$	rad25	SPAC17A5.06
Hereditary nonpolyposis colorectal cancer (HNPCC); *MSH2*	$<1 \times 10^{-100}$	rad16, rad10, rad20, swi9	SPBC24C6.12C
Xeroderma pigmentosum F; *XPF*	$<1 \times 10^{-100}$	cdc17	SPCC970.01
HNPCC; *PMS2*	$<1 \times 10^{-100}$	pms1	SPAC57A10.13C
HNPCC; *MSH6*	$<1 \times 10^{-100}$	msh6	SPAC19G12.02C
HNPCC; *MSH3*	$<1 \times 10^{-100}$	swi4	SPCC285.16C
HNPCC; *MLH1*	$<1 \times 10^{-100}$	mlh1	SPAC8F11.03
Hematological Chediak–Higashi syndrome; *CHS1*	$<1 \times 10^{-100}$	—	SPBC1703.4
Darier–White disease; *SERCA*	$<1 \times 10^{-100}$	Pgak	SPBC28E12.06C
Bloom syndrome; *BLM*	$<1 \times 10^{-100}$	Hus2, rqh1, rad12	SPBC31E1.02C
Ataxia telangiectasia; *ATM*	$<1 \times 10^{-100}$	Tel1	SPAC2G11.12
Xeroderma pigmentosum G; *XPG*	$<1 \times 10^{-40}$	rad13	SPBC3E7.08C
Tuberous sclerosis 2; *TSC2*	$<1 \times 10^{-40}$	—	SPAC630.13C
Immune bare lymphocyte; *ABCB3*	$<1 \times 10^{-40}$	—	SPBC9B6.09C
Downregulated in adenoma; *DRA*	$<1 \times 10^{-40}$	—	SPAC869.05C
Diamond–Blackfan anemia; *RPS19*	$<1 \times 10^{-40}$	rps19	SPBC649.02
Cockayne syndrome 1; *CKN1*	$<1 \times 10^{-40}$	—	SPBC577.09
RAS	$<1 \times 10^{-40}$	Ste5, ras1	SPAC17H9.09C
Cyclin-dependent kinase 4; *CDK4*	$<1 \times 10^{-40}$	Cdc2	SPBC11B10.09
CHK2 protein kinase	$<1 \times 10^{-40}$	Cds1	SPCC18B5.11C
AKT2	$<1 \times 10^{-40}$	Pck2, sts6, pkc1	SPBC12D12.04C

[a]Score is the expect value from a BLAST search; a score of $<1 \times 10^{-40}$ refers to a score between $<1 \times 10^{-40}$ and 1×10^{-100}
Source: Adapted from Wood et al. (2002). Used with permission.

Dictyostelium discoideum (Eichinger et al., 2005). For *S. pombe*, orthologs were identified both for human cancer genes (Table 20.13) and a variety of neurological, metabolic, and other disorders (Table 20.14). In *Dictyostelium*, which is intermediate in complexity between fungi and multicellular animals, many human disease orthologs were identified, including nine that were absent in *S. pombe* and/or *S. cerevisiae*.

It is perhaps expected that human genes involved in cancer are also present in fungi; examples include genes encoding proteins involved in DNA damage and repair and the cell cycle. It might seem surprising that genes implicated in neurological disorders are present in single-celled fungi. However, the explanation may be that neurons are a particularly susceptible cell type with unique metabolic requirements. For example, most lysosomal disorders are caused by the loss of an enzyme that normally contributes to lysosomal function or to intracellular trafficking to lysosomes. Multiple organ systems are typically compromised, but neurological

TABLE 20-14   *Schizosaccharomyces pombe* Genes Related to Human Disease Genes

Human Cancer Gene	Disease	Score[a]	*S. pombe* Gene/Product
Wilson disease; *ATP7B*	Metabolic	$<1 \times 10^{-100}$	P-type copper ATPase
Non-insulin-dependent diabetes; *PCSK1*	Metabolic	$<1 \times 10^{-100}$	Krp1, kinesin related
Hyperinsulinism; *ABCC8*	Metabolic	$<1 \times 10^{-100}$	ABC transporter
G6PD deficiency; *G6PD*	Metabolic	$<1 \times 10^{-100}$	Zwf1 GP6 dehydrogenase
Citrullinemia type I; *ASS*	Metabolic	$<1 \times 10^{-100}$	Arginosuccinate synthase
Wernicke–Korsakoff syndrome; *TKT*	Metabolic	$<1 \times 10^{-40}$	Transketolase
Variegate pophyria; *PPOX*	Metabolic	$<1 \times 10^{-40}$	Protoporphyrinogen oxidase
Maturity-onset diabetes of the young (MODY2); *GCK*	Metabolic	$<1 \times 10^{-40}$	Hxk1, hexokinase
Gitelman syndrome; *SLC12A3*	Metabolic	$<1 \times 10^{-40}$	CCC Na-K-Cl transporter
Cystinuria type 1; *SLC3A1*	Metabolic	$<1 \times 10^{-40}$	α-Glucosidase
Cystic fibrosis; *ABCC7*	Metabolic	$<1 \times 10^{-40}$	ABC transporter
Bartter syndrome; *SLC12A1*	Metabolic	$<1 \times 10^{-40}$	CCC Na-K-Cl transporter
Menkes syndrome; *ATP7A*	Neurological	$<1 \times 10^{-100}$	P-type copper ATPase
Deafness, hereditary; *MYO15*	Neurological	$<1 \times 10^{-100}$	Myo51 class V myosin
Zellweger syndrome; *PEX1*	Neurological	$<1 \times 10^{-40}$	AAA-family ATPase
Thomsen disease; *CLCN1*	Neurological	$<1 \times 10^{-40}$	ClC chloride channel
Spinocerebellar ataxia type 6 (SCA6); *CACNA1A*	Neurological	$<1 \times 10^{-40}$	VIC sodium channel
Myotonic dystrophy; *DM1*	Neurological	$<1 \times 10^{-40}$	Orb6 Ser/Thr protein kinase
McCune–Albright syndrome; *GNAS1*	Neurological	$<1 \times 10^{-40}$	Gpa1 GNP
Lowe's oculocerebrorenal syndrome; *OCRL*	Neurological	$<1 \times 10^{-40}$	PIP phosphatase
Dents; *CLCN5*	Neurological	$<1 \times 10^{-40}$	ClC chloride channel
Coffin–Lowry; *RPS6KA3*	Neurological	$<1 \times 10^{-40}$	Ser/Thr protein kinase
Angelman; *UBE3A*	Neurological	$<1 \times 10^{-40}$	Ubiquitin–protein lgase
Amyotrophic lateral sclerosis; *SOD1*	Neurological	$<1 \times 10^{-40}$	Sod1, superoxide dismutase
Oguschi type 2; *RHKIN*	Neurological	$<1 \times 10^{-40}$	Ser/Thr protein kinase
Familial cardiac myopathy; *MYH7*	Cardiac	$<1 \times 10^{-100}$	Myo2, myosin II
Renal tubular acidosis; *ATP6B1*	Renal	$<1 \times 10^{-100}$	V-type ATPase

Abbreviation: GNP, guanine nucleotide binding.
[a]Score is the expected value from a BLAST search.
Source: Adapted from Wood et al. (2002). Used with permission.

features such as mental retardation are a common consequence of these disorders. The lysosome is a primary site for catabolism in the cell. In fungi, the vacuole performs similar functions, and many human homologs of fungal vacuolar proteins have been identified.

Human Disease Orthologs in Rodents

The Jackson Laboratory website "Mouse Models for Human Disease: Mouse/Human Gene Homologs" is available online (► http://jaxmice.jax.org/jaxmicedb/html/model_975.shtml).

You can access this mouse information at ► http://www.rodentia.com/wmc/domain_genome.html#transgenics and ► http://www.rodentia.com/wmc/domain_mouse.html.

The mouse genome, reported by the Mouse Genome Sequencing Consortium (Waterston et al., 2002), presents us with the most important animal model of human disease. A number of important resources are available:

- The FANTOM database, part of the RIKEN Mouse Gene Encyclopedia Project, contains information on full-length mouse cDNA clones (Bono et al., 2002).
- The Jackson Laboratory website offers a list of mouse/human gene homologs, including mouse models for human disease.
- High-efficiency mutagens such as N-ethyl-N-nitrosourea (ENU) or radiation have been applied to mice to generate models of human disease (see Chapter 12) (Hrabe de Angelis and Strivens, 2001). Nolan et al. (2002) discuss this approach in detail, including web resources for high throughput screening centers and strategies for finding gene mutations that correspond to novel phenotypes.
- The Whole Mouse Catalog describes mouse models of human disease.

The sequencing of the mouse genome was achieved by both Celera Genomics and by a public consortium (Chapter 18). Celera sequenced the genomic DNA of several mouse strains and noted their differences in susceptibility to infectious disease (Table 20.15) and complex inherited disease (Table 20.16). Comparative genomic data will likely help explain why some mouse strains vary in their disease

TABLE 20-15 Infectious Disease Susceptibility of Mouse Strains

	Inbred Mouse Strain	
Infectious Disease	A/J	C57BL/6J
Legionnaire's pneumonia	Susceptible	Resistant
Malaria	Susceptible	Resistant
Viral (MHV3) hepatitis	Resistant	Susceptible
Murine AIDS	Resistant	Susceptible

TABLE 20-16 Common Complex Disease Susceptibility of Mouse Strains

	Inbred Mouse Strain	
Complex Disease	A/J	C57BL/6J
Arthritis	Susceptible	Resistant
Colon cancer	Susceptible	Resistant
Lung cancer	Susceptible	Resistant
Asthma	Susceptible	Resistant
Atherosclerosis	Resistant	Susceptible
Hypertension	Resistant	Susceptible
Type II diabetes	Resistant	Susceptible
Osteoporosis	Susceptible	Resistant
Obesity	Resistant	Susceptible

susceptibility. Hill (2001) has reviewed the genomics and genetics of infectious disease susceptiblility in humans.

The public consortium that sequenced the mouse genome reported that 687 human disease genes have clear orthologs in mouse (Waterston et al., 2002). Surprisingly, for several dozen genes, the wild-type mouse gene sequence was identical to the sequence that is associated with disease in humans. These genes are listed in Table 20.17. This suggests that, assuming the mouse does not have these diseases, any mouse model for human disease must be used with caution. Conceivably, mice have modifying genes (or paralogous genes) not present in humans. Also, inbred strains of laboratory mice are exposed to different environmental stressors than mice in the wild, and their disease susceptibility could vary.

Sequencing of the genome of the Norway rat (Chapter 18; Rat Genome Sequencing Project Consortium, 2004) allowed the detailed comparison of human, mouse, and rat disease genes. Of 1112 well-characterized human disease genes from HGMD (described above), 76% have orthologs in rat. This is a higher percentage than for all rat versus all human genes (of which 46% have 1:1 orthologous matches). Only six human disease genes lack rat orthologs. In general, the Consortium concluded that human disease genes tend to be well conserved in mouse as also indicated by measurement of K_N/K_S ratios.

TABLE 20-17 Human Disease-Associated Sequence Variants for Which Wild-Type Mouse Sequence Matches Diseased Human Sequence

Disease	OMIM	Mutation
Hirschsprung disease	142623	E251K
Leukencephaly with vanishing white matter	603896	R113H
Mucopolysaccharidosis type IVA	253000	R376Q
Breast cancer	113705	L892S
	600185	V211A, Q2421H
Parkinson disease	601508	A53T
Tuberous sclerosis	605284	Q654E
Bardet–Biedl syndrome, type 6	209900	T57A
Mesothelioma	156240	N93S
Long QT syndrome 5	176261	V109I
Cystic fibrosis	602421	F87L, V754M
Porphyria variegata	176200	Q127H
Non-Hodgkin's lymphoma	605027	A25T, P183L
Severe combined immunodeficiency disease	102700	R142Q
Limb-girdle muscular dystrophy type 2D	254110	P30L
Long-chain acyl-CoA dehydrogenase deficiency	201460	Q333K
Usher syndrome type 1B	276902	G955S
Chronic nonspherocytic hemolytic anemia	206400	A295V
Mantle cell lymphoma	208900	N750K
Becker muscular dystrophy	300377	H2921R
Complete androgen insensitivity syndrome	300068	G491S
Prostate cancer	176807	P269S, S647 N
Crohn disease	266600	W157R

Source: Adapted from Waterston et al. (2002). Used with permission.

TABLE 20-18	Human Disease Variants Matching the Wild-Type Chimpanzee Allele			
Gene	Variant	Disease Association	Ancestral	Frequency
AIRE	P252L	Autoimmune syndrome	Unresolved	0
MKKS	R518H	Bardet–Biedl syndrome	Wild type	0
MLH1	A441T	Colorectal cancer	Wild type	0
MYOC	Q48H	Glaucoma	Wild type	0
OTC	T125M	Hyperammonemia	Wild type	0
PRSS1	N29T	Pancreatitis	Disease	0
ABCA1	I883M	Coronary artery disease	Unresolved	0.136
APOE	C130R	Coronary artery disease and Alzheimer's disease	Disease	0.15
DIO2	T92A	Insulin resistance	Disease	0.35
ENPP1	K121Q	Insulin resistance	Disease	0.17
GSTP1	I105V	Oral cancer	Disease	0.348
PON1	I102V	Prostate cancer	Wild type	0.016
PON1	Q192R	Coronary artery disease	Disease	0.3
PPARG	A12P	Type 2 diabetes	Disease	0.85
SLC2A2	T110I	Type 2 diabetes	Disease	0.12
UCP1	A64T	Waist-to-hip ratio	Disease	0.12

Notes: Variants are listed as benign variant, codon number, disease/chimpanzee variant. Ancestral variants are inferred using primate outgroups. Frequency is of the disease allele in humans. *PON1* (Q192R) is polymorphic in chimpanzee.
Source: Chimpanzee Sequencing and Analysis Consortium (2005). Used with permission.

Human Disease Orthologs in Primates

While the chimpanzee and human genomes are extremely closely related (Chimpanzee Sequencing and Analysis Consortium, 2005; Chapter 18), it is surprising that many common human disease variants correspond to the wild-type allele in the chimpanzee. Sixteen examples are presented in Table 20.18. It is possible that not all of these mutations are true positive disease-associated alleles in humans. When a particular sequence occurs in both chimpanzee and macaque this indicates that it is an ancestral allele. Conceivably, specific changes in the human environment in the past several million years have made such ancestral sequences deleterious, such that an altered sequence in humans is adaptive. Other compensatory mutations may be important as well in interpreting the findings. Similar results were reported by the Rhesus Macaque Genome Sequencing and Analysis Consortium (2007), including 229 amino acid substitutions for which the amino acid identified as mutant in human corresponds to the wild-type allele in macaque, chimpanzee, and/or a reconstructed ancestral genome.

Human Disease Genes and Substitution Rates

What is the significance of having identified human disease gene homologs? Beyond cataloguing the presence of orthologs, a next step is to relate the information on mutations in human disease to the conservation of amino acid residues in orthologs. Miller and Kumar (2001) selected seven genes that when mutated cause disease in humans: the cystic fibrosis transmembrane regulator (*CFTR*), glucose-6-phosphate dehydrogenase (*G6PD*), neural cell adhesion molecule L1 (*L1CAM*), phenylalanine hydroxylase (*PAH*), paired box 6 (*PAX6*), the X-linked retinoschisis gene (RS1), and a tuberous sclerosis gene (*TSC2*). For each of these genes, two resources are available: locus-specific databases of mutations that occur in patients and the sequences of a variety of metazoan homologs (e.g., primates, rodents, fish, insects, and nematodes). They generated multiple sequence alignments for each of these seven genes to test the null hypothesis that point mutations occur randomly throughout each gene. Following statistical tets (χ^2 analysis), they determined that most amino acids that can produce human disease mutations are conserved (at least among mammals). Variable sites correspond to positions where amino acid changes are tolerated due to relaxed selection constraints.

As we discussed in Chapter 3, PAM and BLOSUM matrices (Figs. 3.14 and 3.17) reveal that the relative rates of evolutionary substitution vary for different pairs of amino acids. Glutamic acid commonly changes to aspartic acid (in the PAM250 matrix, the score is +2); these two residues are both acidic and thus share common physiochemical properties. However, glutamic acid rarely changes to lysine (the PAM250 score is 0). In human disease, a glutamic acid-to-lysine mutation commonly causes disease. Miller and Kumar (2001) displayed these findings in a table showing the relative frequencies of amino acid changes observed in a variety of eukaryotes (Fig. 20.17, circles) versus amino acid changes that have been detected in patients (Fig. 20.17, squares).

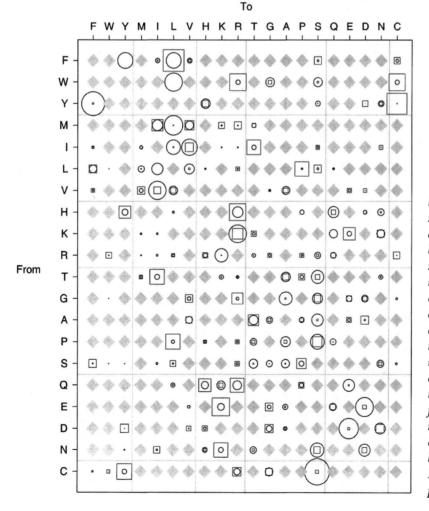

FIGURE 20.17. Amino acid substitutions that occur in human disease are generally not allowed by natural selection. The figure shows a table of the amino acids, indicating the relative frequencies of amino acid changes observed in comparisons between various eukaryotic species (circles) and those changes detected in patients with diseases (squares). The size of the symbols is proportional to the relative frequency of change for a given amino acid. Diamonds indicate changes that cannot be observed as a result of a single base mutation. From Miller and Kumar (2001). Used with permission.

These analyses suggest that disease-associated changes tend to occur at conserved residues. Furthermore, the amino acid changes found in human disease do not commonly occur in comparisons between species. Sudhir Kumar and colleagues further extended these analyses to characterize over 8000 disease-associated mutations in 541 genes (Subramanian and Kumar, 2006). They also showed a nonrandom distribution of disease-causing nucleotide mutations within functional domains (Miller et al., 2003).

FUNCTIONAL CLASSIFICATION OF DISEASE GENES

You can see a list of positionally cloned genes at ▶ http://genome.nhgri.nih.gov/clone/.

We conclude our study of human disease by considering the principles of human disease. The variety of human diseases is extraordinarily broad, yet the field of bioinformatics may provide insight into a logic of disease. One such attempt was by Jimenez-Sanchez et al. (2001), who analyzed 923 human genes that are associated with human disease. These genes primarily cause monogenic disorders, as expected, since at present we know of relatively few genes that are mutated in complex disorders. They classified each disease gene according to the function of its protein product (Fig. 20.18a). Enzymes represent the largest functional category and account for 31% of the total gene products associated with disease. In contrast, only 15% of positionally cloned disease genes encode enzymes. Thus, there may

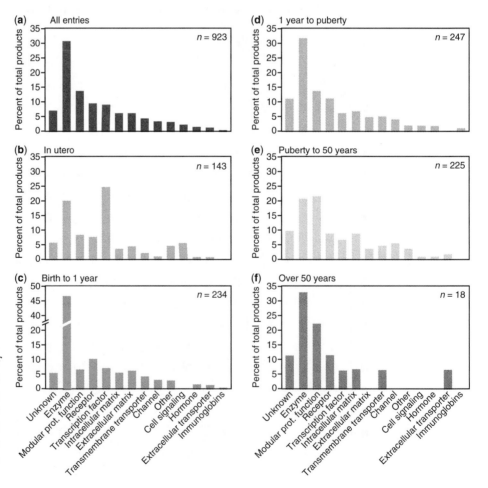

FIGURE 20.18. *The functions of the protein products of disease genes (from Jimenez-Sanchez et al., 2001): (a) all genes (n = 923); (b–f) disease genes listed according to the typical age of onset of the disease phenotype. Used with permission.*

be some historical bias toward our knowledge of disease-causing mutations that are based on enzymatic defects.

Jimenez-Sanchez et al. (2001) further analyzed the correlation between the function of a gene product and the age of disease onset (Figs. 20.18b to f). Genes encoding enzymes and transcription factors are especially likely to be involved in disease in utero, reflecting the importance of transcription factors in early development. Enzymes are particularly involved in disease up to puberty (Figs. 20.18b to d). The developing fetus has access to its mother's metabolic systems and thus may be viable even if it has a gene defect. After birth, such diseases are manifested. Disease genes encoding enzymes are less prevalent in diseases having a later onset in life (Fig. 20.18e).

All of the common diseases in this sample occur with only a rare frequency when analyzed for any of four functional categories of disease—frequency, mode of inheritance, age of onset, and reduction of life expectancy (Fig. 20.19, leftmost column). This rare frequency reflects the population of disease genes that are currently available to study, that is, genes implicated in single-gene disorders. The mode of inheritance tends to be autosomal recessive, particularly for genes encoding enzymes. As described in Fig. 20.18 as well, the age of onset tends to be in utero for transcription factors, from birth to one year for genes encoding enzymes, between one year and puberty and into adulthood for receptors, and early adulthood for modifiers of

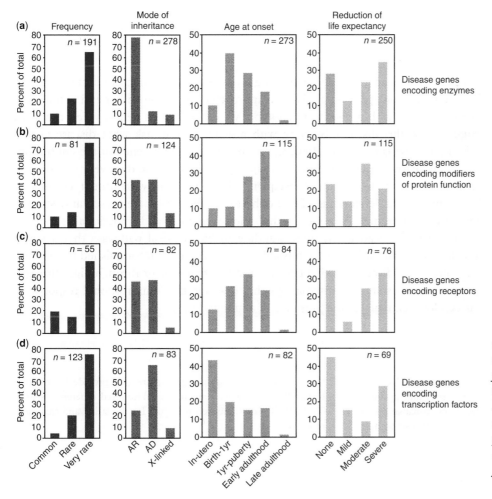

FIGURE 20.19. The characteristics of diseases, organized by the function of the protein encoded by the disease gene. Abbreviations: AR, autosomal recessive; AD, autosomal dominant; early adulthood, puberty to <50 years old; late adulthood, >50 years old. From Jimenez-Sanchez et al. (2001). Used with permission.

protein function (such as proteins that stabilize, activate, or fold other proteins). The severity of the disease, reflected in reduction of life expectancy, varies for diseases without a strong pattern based on functional categories.

These studies represent an early attempt to define a logic of disease. Such genomic-scale efforts will be enhanced when we have more information available on the genetic basis of complex disorders. Functional analyses may be combined using all the tools of bioinformatics and genomics to help elucidate the relationship between genotype and disease phenotype.

PERSPECTIVE

There are several kinds of bioinformatics approaches to human disease:

- Human disease is a consequence of variation in DNA sequence. These variations are catalogued in databases of molecular sequences (such as GenBank).
- Human disease databases have a major role in organizing information about disease genes. There are centralized databases, most notably OMIM and HGMD, as well as locus-specific mutation databases.
- Functional genomics screens provide insight into the mechanisms of disease genes and disease processes.

PITFALLS

A fundamental gap in our understanding is how a genotype such as a mutated gene is related to a disease phenotype. We can approach disease from either end of the spectrum. Starting with a disease phenotype, we can ask what genes, when mutated, might cause this disease? Starting with a gene, we can ask what disease occurs when this gene is mutated? However, connecting these two ends of the continuum has been nearly impossible. For the majority of diseases, the discovery of a disease gene has not yet led to the subsequent discovery of new treatment options or cures, or to an understanding of pathophysiology. Examples are muscular dystrophy and Rett syndrome. A hope is that bioinformatics and functional genomics approaches may lead to an understanding of biochemical pathways that account for the molecular basis of pathophysiology. This could be accomplished by learning the function of disease-causing genes in model organisms or through high through-put technologies such as microarrays that describe the transcriptional response of susceptible cell types to the presence of a mutated gene.

WEB RESOURCES

We have defined bioinformatics as the use of computer algorithms and computer databases to study genes, proteins, and genomes. For human disease, a number of databases are available on the World Wide Web (Guttmacher, 2001). Table 20.19 lists some of these resources, including organizations that provide information to families of those with any of several hundred different diseases. Table 20.20 lists some web resources for the study and treatment of cancer.

TABLE 20-19 General Web Resources for Study of Human Diseases

Site	Description	URL
Diseases, Disorders and Related Topics	Karolinska Institute (Stockholm)	► http://www.mic.ki.se/Diseases/
The Frequency of Inherited Disorders Database (FIDD)	From the Institute of Medical Genetics University of Wales College of Medicine	► http://archive.uwcm.ac.uk/uwcm/mg/fidd/
GeneCards	A database of human genes, their products, and their involvement in diseases	► http://bioinfo.weizmann.ac.il/cards/
Genes and Disease (NCBI)	Organized by chromosome, provides descriptions of 60 diseases	► http://www.ncbi.nlm.nih.gov/disease/
GeneClinics	A clinical information resource from the University of Washington, Seattle	► http://www.geneclinics.org/
Genetic Alliance	International coalition of individuals, professionals, and genetic support organizations	► http://www.geneticalliance.org/; search form ► http://www.geneticalliance.org/diseaseinfo/search.html
Inherited Disease Genes Identified by Positional Cloning	From the National Human Genome Research Institute (NHGRI) at the NIH	► http://genome.nhgri.nih.gov/clone/
The National Information Center for Children and Youth with Disabilities	An information and referral center in the United States	► http://www.nichcy.org/
National Organization for Rare Disorders (NORD)	Federation of voluntary health organizations dedicated to helping people with rare "orphan" diseases and assisting the organizations that serve them	► http://www.rarediseases.org/
Online Mendelian Inheritance in Man (OMIM)	Over 12,000 entries	► http://www.ncbi.nlm.nih.gov/entrez/query.fcgi?db = OMIM

TABLE 20-20 Web Resources for Study of Cancer

Resource	Description	URL
ACOR	Association of Cancer Online Resources	► http://www.acor.org/
Atlas of Genetics and Cytogenetics in Oncology and Haematology	A peer-reviewed online journal and cancer database	► http://www.infobiogen.fr/services/chromcancer/
Cancer Chromosome Aberration Project	Tools to define and characterize chromosomal alterations in cancer	► http://cgap.nci.nih.gov/Chromosomes/CCAP
The Cancer Gene Anatomy Project (CGAP)	At NCBI	► http://www.ncbi.nlm.nih.gov/ncicgap/
The Cancer Genome Project	The Wellcome Trust Sanger Institute	► http://www.sanger.ac.uk/CGP/

(Continued)

TABLE 20-20   Continued

Resource	Description	URL
CancerNet	At the National Cancer Institute (of the NIH)	► http://cancernet.nci. nih.gov/
CancerWEB	Cancer resource site	► http://cancerweb.ncl.ac. uk/
Children's Cancer Web	Directory of childhood cancer resources	► http://www. CancerIndex.org/ccw/
Mitelman Database of Chromosome Aberrations in Cancer	Relates chromosomal aberrations to tumor characteristics	► http://cgap.nci.nih.gov/ Chromosomes/ Mitelman
OncoLink	Cancer information for the general public	► http://oncolink.upenn. edu/

Discussion Questions

[20-1] Many neurological diseases such as Rett syndrome, vanishing white matter syndrome, and Huntington disease have devastating consequences on brain function. For some of these diseases, the responsible genes have homologs in single-celled organisms such as fungi. Why do you think this is so?

[20-2] How have microarrays been used to study human disease? What are some specific examples of progress that has been made?

Problems

[20-1] How many inherited diseases have a known sequence associated with them? Visit OMIM and search for the number of genes having allelic variants.

[20-2] Mutations in *MECP2* cause Rett syndrome.

- Explore this gene and this disease in OMIM. What is the phenotype of the disease? What chromosome is *MECP2* localized to? How many allelic variants are reported? Are mouse models available?

- Explore *MECP2* at a locus-specific mutation database, RettBase. Compare the types of information you obtain from this resource versus OMIM.

- Explore *MECP2* at dbSNP. Are there any SNPs that correspond to disease-associated substitutions? Do any SNPs alter the amino acid sequence?

- Explore MECP2 at the UCSC Genome Browser. Again compare the types of information you obtain from this resource versus OMIM.

[20-3] Some human disease alleles correspond to the wild-type sequence of closely related chimpanzee and/or rhesus macaque proteins. Align the human, macaque, chimpanzee, rat, and mouse sequences for the proteins encoded by the following genes. For each, the description N > A:CHMT refers to the consensus human amino acid N (normal), the disease associated form A (altered), C (chimpanzee), H (inferred human/chimpanzee ancestor), M (rhesus macaque), T (inferred human/rhesus ancestor using mouse and dog as outgroup species).

- *ABCA4* (Stargardt disease; chromosome 1), R > Q:RRQR; H > R:RRRR

- *CFTR* (cystic fibrosis; chromosome 7), F > L:FFLL; K > R:KKRK

- *PAH* (phenylketonuria; chromosome 12), Y > H:YYHY; I > T:IITI

- *OTC* (ornithine hyperammonemia; chromosome X), R > H:RRHH; T > M:MTTT

Resources to find these alignments include the HomoloGene project at NCBI and the comparative genomics tracks at the UCSC Genome Browser.

Self-Test Quiz

[20-1] In humans, disorders that are inherited by simple Mendelian inheritance account for about what percentage of all human disease?

(a) 1%

(b) 10%

(c) 50%

(d) It is impossible to accurately measure the percentage.

[20-2] To a significant extent, susceptibility to an environmentally caused disorder such as poisoning from lead paint is determined by an individual's genes.

(a) True

(b) False

[20-3] Which of the following best describes single gene disorders? Each single gene disorder:

(a) Is caused by a mutation in a single gene. They represent a basic category of disease that is in contrast to complex disorders.

(b) Is caused primarily by a mutation in a single gene, but the disease process always involves the contribution of many genes. Thus, they represent a category of disease along a continuum with complex disorders.

(c) Is primarily caused by a mutation in a single gene in which the mutation almost always introduces a synonymous substitution.

(d) Is primarily caused by a mutation in a single gene in which the mutation almost always introduces a nonsynonymous substitution.

[20-4] In total, rare diseases in the United States affect about how many people?

(a) 200,000

(b) 2 million

(c) 25 million

(d) 100 million

[20-5] Single gene disorders tend to be

(a) Rare in the general population, with an early onset in life

(b) Common in the general population, with an early onset in life

(c) Rare in the general population, with a late onset in life

(d) Common in the general population, with a late onset in life

[20-6] Online Mendelian Inheritance in Man (OMIM) includes entries that focus on:

(a) Particular diseases

(b) Particular genes

(c) Either genes or diseases

(d) Complex chromosomal disorders

[20-7] There are several hundred locus-specific databases. What information do they offer that is not available in central databases such as OMIM and GeneCards?

(a) Comprehensive descriptions of the gene implicated in a disease

(b) Comprehensive lists of mutations associated with disease

(c) Links to foundations and other organizations

(d) Links to chromosome maps displaying the disease-causing gene

[20-8] Genome-wide association studies

(a) Often use array CGH technology

(b) Sometimes include family-based designs

(c) Sometimes include over 10,000 subjects

(d) Are often applied to single-gene disorders

[20-9] Human disease genes have orthologs in a variety of organisms, including worms, insects, and fungi. For a number of human proteins that are implicated in disease, multiple sequence alignments with orthologous proteins have been made. These show that amino acid positions associated with disease-causing mutations in human proteins tend to be residues that are

(a) Strongly conserved in other organisms

(b) Sometimes conserved in other organisms

(c) Poorly conserved in other organisms

(d) Only sometimes aligned with orthologous sequences

Suggested Reading

An essential resource for the study of human disease is *The Metabolic and Molecular Basis of Inherited Disease* (Scriver et al., 2001). This four-volume tome has hundreds of chapters, including introductions to disease from a variety of perspectives (e.g., Mendelian disorders, complex disorders, a logic of disease, mutation mechanisms, and animal models). A recommended introduction to disease is an essay by Barton Childs and David Valle (2000) in the inaugural volume of *Annual Review of Genomics and Human Genetics*. The February 2001 issues of *Science* and *Nature* on the human genome included brief articles on human disease by Leena Peltonen and Victor McKusick (2001) and Jimenez-Sanchez et al. (2001). For a review of linkage and association approaches, see Hirschhorn and Daly (2005), Altshuler et al. (2008), and McCarthy et al. (2008).

REFERENCES

Abecasis, G. R., Cherny, S. S., Cookson, W. O., and Cardon, L. R. Merlin: Rapid analysis of dense genetic maps using sparse gene flow trees. *Nat Genet.* **30**, 97–101 (2002).

Ahringer, J. Turn to the worm! *Curr. Opin. Genet. Dev.* **7**, 410–415 (1997).

Altshuler, D., Daly, M. J., and Lander, E. S. Genetic mapping in human disease. *Science* **322**, 881–888 (2008).

Amir, R. E., et al. Rett syndrome is caused by mutations in X-linked *MECP2*, encoding methyl-CpG-binding protein 2. *Nat. Genet.* **23**, 185–188 (1999).

Amir, R. E., and Zoghbi, H. Y. Rett syndrome: Methyl-CpG-binding protein 2 mutations and phenotype–genotype correlations. *Am. J. Med. Genet.* **97**, 147–152 (2000).

Antonarakis, S. E. Recommendations for a nomenclature system for human gene mutations. Nomenclature Working Group. *Hum. Mutat.* **11**, 1–3 (1998).

Arabidopsis Genome Initiative. Analysis of the genome sequence of the flowering plant *Arabidopsis thaliana*. *Nature* **408**, 796–815 (2000).

Bailey, A., et al. Autism as a strongly genetic disorder: Evidence from a British twin study. *Psychol. Med.* **25**, 63–77 (1995).

Bailey, A., Phillips, W., and Rutter, M. Autism: Towards an integration of clinical, genetic, neuropsychological, and neurobiological perspectives. *J. Child Psychol. Psychiatry* **37**, 89–126 (1996).

Baker, P., Piven, J., Schwartz, S., and Patil, S. Brief report: Duplication of chromosome 15q11-13 in two individuals with autistic disorder. *J. Autism Dev. Disord.* **24**, 529–535 (1994).

Barton, M., and Volkmar, F. How commonly are known medical conditions associated with autism? *J. Autism Dev. Disord.* **28**, 273–278 (1998).

Bauman, M. L., Kemper, T. L., and Arin, D. M. Microscopic observations of the brain in Rett syndrome. *Neuropediatrics* **26**, 105–108 (1995).

Beaudet, A. L., Scriver, C. R., Sly, W. S., and Valle, D. Genetics, biochemistry, and molecular bases of variant human phenotypes. In Scriver, C. et al. (eds.), *The Metabolic and Molecular Bases of Inherited Disease*, Vol. 1, pp. 3–45. McGraw-Hill, New York, 2001.

Beroud, C., Collod-Beroud, G., Boileau, C., Soussi, T., and Junien, C. UMD (Universal mutation database): A generic software to build and analyze locus-specific databases. *Hum. Mutat.* **15**, 86–94 (2000).

Bird, A. The methyl-CpG-binding protein MeCP2 and neurological disease. *Biochem. Soc. Trans.* **36**, 575–583 (2008).

Bono, H., Kasukawa, T., Furuno, M., Hayashizaki, Y., and Okazaki, Y. FANTOM DB: Database of Functional Annotation of RIKEN Mouse cDNA Clones. *Nucleic Acids Res.* **30**, 116–118 (2002).

Bowcock, A. M. Guilt by association. *Nature* **447**, 645–646 (2007).

Brown, A. F., and McKie, M. A. MuStaR and other software for locus-specific mutation databases. *Hum. Mutat.* **15**, 76–85 (2000).

Bundey, S., Hardy, C., Vickers, S., Kilpatrick, M. W., and Corbett, J. A. Duplication of the 15q11–13 region in a patient with autism, epilepsy and ataxia. *Dev. Med. Child. Neurol.* **36**, 736–742 (1994).

Chahrour, M., Jung, S. Y., Shaw, C., Zhou, X., Wong, S. T., Qin, J., and Zoghbi, H. Y. MeCP2, a key contributor to neurological disease, activates and represses transcription. *Science* **320**, 1224–1229 (2008).

Childs, B., and Valle, D. *Genetics, Biology and Disease.* Annual Reviews, Palo Alto, CA, 2000, pp. 1–19.

Chimpanzee Sequencing and Analysis Consortium. Initial sequence of the chimpanzee genome and comparison with the human genome. *Nature* **437**, 69–87 (2005).

Ciaranello, A. L., and Ciaranello, R. D. The neurobiology of infantile autism. *Annu. Rev. Neurosci.* **18**, 101–128 (1995).

Claustres, M., Horaitis, O., Vanevski, M., and Cotton, R. G. Time for a unified system of mutation description and reporting: A review of locus-specific mutation databases. *Genome Res.* **12**, 680–688 (2002).

Cook, E. H., Jr. et al. Autism or atypical autism in maternally but not paternally derived proximal 15q duplication. *Am. J. Hum. Genet.* **60**, 928–934 (1997).

Cotton, R. G., et al. Recommendations for locus-specific databases and their curation. *Hum. Mutat.* **29**, 2–5 (2008).

den Dunnen, J. T., and Antonarakis, S. E. Mutation nomenclature extensions and suggestions to describe complex mutations: A discussion. *Hum. Mutat.* **15**, 7–12 (2000).

Denis, P.-S. *Études Chimiques, Physiologiques, et Médicals, Faites de 1835 à 1840 sur les Matières Albumineuses.* Imprimerie C.-F. Denis, Commercy, 1842.

DiMauro, S., and Schon, E. A. Mitochondrial DNA mutations in human disease. *Am. J. Med. Genet.* **106**, 18–26 (2001).

DiMauro, S., and Schon, E. A. Mitochondrial disorders in the nervous system. *Annu. Rev. Neurosci.* **31**, 91–123 (2008).

Dipple, K. M., and McCabe, E. R. Modifier genes convert "simple" Mendelian disorders to complex traits. *Mol. Genet. Metab.* **71**, 43–50, (2000).

Dipple, K. M., Phelan, J. K., and McCabe, E. R. Consequences of complexity within biological networks: Robustness and health, or vulnerability and disease. *Mol. Genet. Metab.* **74**, 45–50 (2001).

Eichinger, L., et al. The genome of the social amoeba *Dictyostelium discoideum*. *Nature* **435**, 43–57 (2005).

Flejter, W. L., et al. Cytogenetic and molecular analysis of inv dup(15) chromosomes observed in two patients with autistic disorder and mental retardation. *Am. J. Med. Genet.* **61**, 182–187 (1996).

Fombonne, E. The epidemiology of autism: A review. *Psychol. Med.* **29**, 769–786 (1999).

Garrod, A. E. The incidence of alkaptonuria: A study in chemical individuality. *Lancet* **ii**, 1616–1620 (1902).

Garrod, A. E. Inborn errors of metabolism: The Croonian Lectures delivered before the Royal College of Physicians of London, in June, 1908. Frowde, Hodder and Stoughton, London, 1909.

Garrod, A. E. *Inborn Factors in Disease: An Essay.* Clarendon Press, Oxford, 1931.

Geerdink, N., et al. *MECP2* mutation in a boy with severe neonatal encephalopathy: Clinical, neuropathological and molecular findings. *Neuropediatrics* **33**, 33–36 (2002).

George, R. A., Smith, T. D., Callaghan, S., Hardman, L., Pierides, C., Horaitis, O., Wouters, M. A., and Cotton, R. G. General mutation databases: Analysis and review. *J. Med. Genet.* **45**, 65–70 (2008).

Giardine, B., et al. HbVar database of human hemoglobin variants and thalassemia mutations: 2007 update. *Hum. Mutat.* **28**, 206 (2007a).

Giardine, B., et al. PhenCode: Connecting ENCODE data with mutations and phenotype. *Hum. Mutat.* **28**, 554–562 (2007b).

Gillberg, C., and Wing, L. Autism: Not an extremely rare disorder. *Acta Psychiatr. Scand.* **99**, 399–406 (1999).

Greenman, C., et al. Patterns of somatic mutation in human cancer genomes. *Nature* **446**, 153–158 (2007).

Gusella, J. F. Location cloning strategy for characterizing genetic defects in Huntington's disease and Alzheimer's disease. *FASEB J.* **3**, 2036–2041 (1989).

Guttmacher, A. E. Human genetics on the web. *Annu. Rev. Genomics Hum. Genet.* **2**, 213–233 (2001).

Guy, J., Gan, J., Selfridge, J., Cobb, S., and Bird, A. Reversal of neurological defects in a mouse model of Rett syndrome. *Science* **315**, 1143–1147 (2007).

Hagberg, B., Aicardi, J., Dias, K., and Ramos, O. A progressive syndrome of autism, dementia, ataxia, and loss of purposeful hand use in girls: Rett's syndrome: report of 35 cases. *Ann. Neurol.* **14**, 471–479 (1983).

Hammer, S., Dorrani, N., Dragich, J., Kudo, S., and Schanen, C. The phenotypic consequences of *MECP2* mutations extend beyond Rett syndrome. *Ment. Retard. Dev. Disabil. Res. Rev.* **8**, 94–98 (2002).

Hamosh, A., Scott, A. F., Amberger, J. S., Bocchini, C. A., and McKusick, V. A. Online Mendelian Inheritance in Man (OMIM), a knowledgebase of human genes and genetic disorders. *Nucleic Acids Res.* **33**, D514–D517 (2005).

Hill, A. V., The genomics and genetics of human infectious disease susceptibility. *Annu. Rev. Genomics Hum. Genet.* **2**, 373–400 (2001).

Hirschhorn, J. N., and Daly, M. J. Genome-wide association studies for common diseases and complex traits. *Nat. Rev. Genet.* **6**, 95–108 (2005).

Holt, I. J., Harding, A. E., and Morgan-Hughes, J. A. Deletions of muscle mitochondrial DNA in patients with mitochondrial myopathies. *Nature* **331**, 717–719, 1988.

Hrabe de Angelis, M., and Strivens, M. Large-scale production of mouse phenotypes: The search for animal models for inherited diseases in humans. *Brief. Bioinform.* **2**, 170–180 (2001).

Human Genome Structural Variation Working Group, et al. Completing the map of human genetic variation. *Nature* **447**, 161–165 (2007).

International Molecular Genetic Study of Autism Consortium. A full genome screen for autism with evidence for linkage to a region on chromosome 7q. *Hum. Mol. Genet.* **7**, 571–578 (1998).

Jellinger, K., Armstrong, D., Zoghbi, H. Y., and Percy, A. K. Neuropathology of Rett syndrome. *Acta Neuropathol.* **76**, 142–158 (1988).

Jimenez-Sanchez, G., Childs, B., and Valle, D. Human disease genes. *Nature* **409**, 853–855 (2001).

Johnston, M. V., Blue, M. E., and Naidu, S. Rett syndrome and neuronal development. *J. Child Neurol.* **20**, 759–763 (2005).

Knoppers, B. M., and Laberge, C. M. Ethical guideposts for allelic variation databases. *Hum. Mutat.* **15**, 30–35 (2000).

Knudson, A. G., Jr. Mutation and cancer: Statistical study of retinoblastoma. *Proc. Natl. Acad. Sci. USA* **68**, 820–823 (1971).

Knutsen, T., Gobu, V., Knaus, R., Padilla-Nash, H., Augustus, M., Strausberg, R. L., Kirsch, I. R., Sirotkin, K., and Ried, T. The interactive online SKY/M-FISH & CGH database and the Entrez cancer chromosomes search database: Linkage of chromosomal aberrations with the genome sequence. *Genes Chromosomes Cancer* **44**, 52–64 (2005).

Laird, N. M., and Lange, C. Family-based designs in the age of large-scale gene-association studies. *Nat. Rev. Genet.* **7**, 385–394 (2006).

Lamb, J. A., Moore, J., Bailey, A., and Monaco, A. P. Autism: Recent molecular genetic advances. *Hum. Mol. Genet.* **9**, 861–868 (2000).

Lintas, C., and Persico, A. M. Autistic phenotypes and genetic testing: state-of-the-art for the clinical geneticist. *J. Med. Genet.* **46**, 1–8 (2009).

Lopez, A. D., Mathers, C. D., Ezzati, M., Jamison, D. T., and Murray, C. J. Global and regional burden of disease and risk factors, 2001: Systematic analysis of population health data. *Lancet* **367**, 1747–1757 (2006).

Lowrance, W. W., and Collins, F. S. Identifiability in genomic research. *Science* **317**, 600–602 (2007).

Lupski, J. R. Genomic disorders: Structural features of the genome can lead to DNA rearrangements and human disease traits. *Trends Genet.* **14**, 417–422 (1998).

Lupski, J. R., and Stankiewicz, P. Genomic disorders: Molecular mechanisms for rearrangements and conveyed phenotypes. *PLoS Genet.* **1**, e49 (2005).

Mailman, M. D., et al. The NCBI dbGaP database of genotypes and phenotypes. *Nat. Genet.* **39**, 1181–1186 (2007).

Mathers, C. D., and Loncar, D. Projections of global mortality and burden of disease from 2002 to 2030. *PLoS Med.* **3**, e442 (2006).

McCarthy, M. I., Abecasis, G. R., Cardon, L. R., Goldstein, D. B., Little, J., Ioannidis, J. P., and Hirschhorn, J. N. Genome-wide association studies for complex traits: consensus, uncertainty and challenges. *Nat. Rev. Genet.* **9**, 356–369 (2008).

McKusick, V. A. Mendelian Inheritance in Man and its online version, OMIM. *Am. J. Hum. Genet.* **80**, 588–604 (2007).

Miller, M. P., and Kumar, S. Understanding human disease mutations through the use of interspecific genetic variation. *Hum. Mol. Genet.* **10**, 2319–2328 (2001).

Miller, M. P., Parker, J. D., Rissing, S. W., and Kumar, S. Quantifying the intragenic distribution of human disease mutations. *Ann. Hum. Genet.* **67**, 567–579 (2003).

Murray, C. J. L., and Lopez, A. D. (eds.) *The Global Burden of Disease.* Harvard University Press, Cambridge, 1996.

Nan, X., Campoy, F. J., and Bird, A. MeCP2 is a transcriptional repressor with abundant binding sites in genomic chromatin. *Cell* **88**, 471–481 (1997).

Nass, S., and Nass, M. M. K. Intramitochondrial fibers with DNA characteristics. *J. Cell Biol.* **19**, 613–629 (1963).

NCI-NHGRI Working Group on Replication in Association Studies, et al. Replicating genotype-phenotype associations. *Nature* **447**, 655–660 (2007).

Ng, H. H., and Bird, A. DNA methylation and chromatin modification. *Curr. Opin. Genet. Dev.* **9**, 158–163 (1999).

Nolan, P. M., Hugill, A., and Cox, R. D. ENU mutagenesis in the mouse: Application to human genetic disease. *Briefings Functional Genomics Proteomics* **1**, 278–289 (2002).

Olsson, I., Steffenburg, S., and Gillberg, C. Epilepsy in autism and autisticlike conditions. A population-based study. *Arch. Neurol.* **45**, 666–668 (1988).

Ott, J. Major strengths and weaknesses of the LOD score method. *Adv. Genet.* **42**, 125–132 (2001).

Patrinos, G. P., and Brookes, A. J. DNA, diseases and databases: Disastrously deficient. *Trends Genet.* **21**, 333–338 (2005).

Pauling, L., Itano, H. A., Singer, S. J., and Wells, I. C. Sickle cell anemia, a molecular disease. *Science* **110**, 543–548 (1949).

Peltonen, L., and McKusick, V. A. Genomics and medicine. Dissecting human disease in the postgenomic era. *Science* **291**, 1224–1229 (2001).

Percy, A. K. Rett syndrome: recent research progress. *J. Child Neurol.* **23**, 543–549 (2008).

Philippe, A., et al. Genome-wide scan for autism susceptibility genes. Paris Autism Research International Sibpair Study. *Hum. Mol. Genet.* **8**, 805–812 (1999).

Purcell, S., et al. PLINK: A tool set for whole-genome association and population-based linkage analyses. *Am. J. Hum. Genet.* **81**, 559–575 (2007).

Ramocki, M. B., and Zoghbi, H. Y. Failure of neuronal homeostasis results in common neuropsychiatric phenotypes. *Nature* **455**, 912–918 (2008).

Rapin, I., and Tuchman, R. F. What is new in autism? *Curr. Opin. Neurol.* **21**, 143–149 (2008).

Rat Genome Sequencing Project Consortium. Genome sequence of the Brown Norway rat yields insights into mammalian evolution. *Nature* **428**, 493–521 (2004).

Razin, A. CpG methylation, chromatin structure and gene silencing: A three-way connection. *EMBO J* **17**, 4905–4908 (1998).

Reiter, L. T., Potocki, L., Chien, S., Gribskov, M., and Bier, E. A systematic analysis of human disease-associated gene sequences in *Drosophila melanogaster. Genome Res.* **11**, 1114–1125 (2001).

Rhesus Macaque Genome Sequencing and Analysis Consortium, et al. Evolutionary and biomedical insights from the rhesus macaque genome. *Science* **316**, 222–234 (2007).

Rubin, G. M., et al. Comparative genomics of the eukaryotes. *Science* **287**, 2204–2215 (2000).

Ruiz-Pesini, E., Lott, M. T., Procaccio, V., Poole, J. C., Brandon, M. C., Mishmar, D., Yi, C., Kreuziger, J., Baldi, P., and Wallace, D. C. An enhanced MITOMAP with a global mtDNA mutational phylogeny. *Nucleic Acids Res.* **35**, D823–D828 (2007).

Safran, M., et al. Human Gene-Centric Databases at the Weizmann Institute of Science: GeneCards, UDB, CroW 21 and HORDE. *Nucleic Acids Res.* **31**, 142–146 (2003).

Schon, E. A. Mitochondrial genetics and disease. *Trends Biochem. Sci.* **25**, 555–560 (2000).

Scriver, C. R., and Childs, B. *Garrod's Inborn Factors in Disease.* Oxford University Press, New York, 1989.

Scriver, C. R., Nowacki, P. M., and Lehvaslaiho, H. Guidelines and recommendations for content, structure, and deployment of mutation databases. *Hum. Mutat.* **13**, 344–350 (1999).

Scriver, C. R., Nowacki, P. M., and Lehvaslaiho, H. Guidelines and recommendations for content, structure, and deployment of mutation databases: II. Journey in progress. *Hum. Mutat.* **15**, 13–15 (2000).

Scriver, C., Beaudet, A., Sly, W., and Valle, D. (eds.). *The Metabolic and Molecular Basis of Inherited Disease.* McGraw-Hill, New York, 2001.

Sjöblom, T., et al. The consensus coding sequences of human breast and colorectal cancers. *Science* **314**, 268–274 (2006).

Stankiewicz, P., and Lupski, J. R. Genome architecture, rearrangements and genomic disorders. *Trends Genet.* **18**, 74–82 (2002).

Subramanian, S., and Kumar, S. Evolutionary anatomies of positions and types of disease-associated and neutral amino acid mutations in the human genome. *BMC Genomics* **7**, 306 (2006).

Szatmari, P., Jones, M. B., Zwaigenbaum, L., and MacLean, J. E. Genetics of autism: Overview and new directions. *J. Autism Dev. Disord.* **28**, 351–368 (1998).

Thomas, C. L. (ed.). *Taber's Cyclopedic Medical Dictionary.* F. A. Davis Company, Philadelphia, 1997.

Todd, J. A. Multifactorial diseases: Ancient gene polymorphism at quantitative trait loci and a legacy of survival during our evolution. In Scrivner, C. et al. (eds.), *The Metabolic and Molecular Bases of Inherited Disease*, Vol. 1, pp. 193–201. McGraw-Hill, New York, 2001.

Turner, M., Barnby, G., and Bailey, A. Genetic clues to the biological basis of autism. *Mol. Med. Today* **6**, 238–244 (2000).

Varmus, H. The new era in cancer research. *Science* **312**, 1162–1165 (2006).

Volkmar, F. R., and Nelson, D. S. Seizure disorders in autism. *J. Am. Acad. Child. Adolesc. Psychiatry* **29**, 127–129 (1990).

Voullaire, L., Slater, H., Williamson, R., and Wilton, L. Chromosome analysis of blastomeres from human embryos by using comparative genomic hybridization. *Hum. Genet.* **106**, 210–217 (2000).

Wallace, D. C., Singh, G., Lott, M. T., Hodge, J. A., Schurr, T. G., Lezza, A. M., Elsas, L. J. II, and Nikoskelainen, E. K. Mitochondrial DNA mutation associated with Leber's hereditary optic neuropathy. *Science* **242**, 1427–1430, 1988a.

Wallace, D. C., Zheng, X. X., Lott, M. T., Shoffner, J. M., Hodge, J. A., Kelley, R. I., Epstein, C. M., and Hopkins, L. C. Familial mitochondrial encephalomyopathy (MERRF): Genetic, pathophysiological, and biochemical characterization of a mitochondrial DNA disease: *Cell* **55**, 601–610, 1988b.

Waterston, R. H., et al. Initial sequencing and comparative analysis of the mouse genome. *Nature* **420**, 520–562 (2002).

Wellcome Trust Case Control Consortium. Genome-wide association study of 14,000 cases of seven common diseases and 3,000 shared controls. *Nature* **447**, 661–678 (2007).

Wells, D., and Delhanty, J. D. Comprehensive chromosomal analysis of human preimplantation embryos using whole genome amplification and single cell comparative genomic hybridization. *Mol. Hum. Reprod.* **6**, 1055–1062 (2000).

Wood, L. D., et al. The genomic landscapes of human breast and colorectal cancers. *Science* **318**, 1108–1113 (2007).

Wood, V., et al. The genome sequence of *Schizosaccharomyces pombe*. *Nature* **415**, 871–880 (2002).

Zeev, B. B., et al. Rett syndrome: Clinical manifestations in males with *MECP2* mutations. *J. Child Neurol.* **17**, 20–24 (2002).

Zuckerlandl, E., and Pauling, L. Molecular disease, evolution, and genic heterogeneity. In Kasha, M. and Pullman, B. (eds.), *Horizons in Biochemistry*, Albert Szent-Gyorgyi Dedicatory Volume. Academic Press, New York, 1962.

Glossary

This glossary is combined from five web-based glossaries and each entry is marked accordingly: (1) the National Center for Biotechnology Information (NCBI), (2) the Oak Ridge National Laboratory (ORNL), (3) the talking glossary at the National Human Genome Research Institute (NHGRI), (4) the SMART database, and (5) the protein folds glossary from the Structural Classification of Proteins website (SCOP) (these entries are modified). Additional web-based glossaries are listed in a table at the end of this glossary.

The glossaries are online at:

- ▶ http://www.ncbi.nlm.nih.gov/ Education/BLASTinfo/glossary2. html
- ▶ http://www.ornl.gov/ TechResources/Human_Genome/ glossary/
- ▶ http://www.genome.gov/glossary. cfm
- ▶ http://smart.embl-heidelberg.de/ help/smart_glossary.shtml
- ▶ http://scop.mrc-lmb.cam.ac.uk/ scop/gloss.html

A

Additive genetic effects
When the combined effects of alleles at different loci are equal to the sum of their individual effects. (ORNL)

Adenine (A)
A nitrogenous base, one member of the base pair AT (adenine–thymine). *See also:* base pair. (ORNL)

Algorithm
A fixed procedure embodied in a computer program. (NCBI)

Alignment
(a) The process of lining up two or more sequences to achieve maximal levels of identity (and conservation, in the case of amino acid sequences) for the purpose of assessing the degree of similarity and the possibility of homology. (NCBI) (b) Representation of a prediction of the amino acids in tertiary structures of homologs that overlay in three dimensions. Alignments held by SMART are mostly based on published observations (see domain annotations for details) but are updated and edited manually. (SMART)

All alpha
A class that has the number of secondary structures in the domain or common core described as 3-, 4-, 5-, 6-, or multihelical. (SCOP)

All beta
A class that includes two major fold groups: sandwiches and barrels. The sandwich folds are made of two β sheets which are usually twisted and packed so their strands are aligned. The barrel folds are made of a single β sheet that twists and coils upon itself so, in most cases, the first strand in the β sheet hydrogen bonds to the last strand. The strand directions in the two opposite sides of a barrel fold are roughly orthogonal. Orthogonal packing of sheets is also seen in a few special cases of sandwich folds. (SCOP)

Allele
(a) Alternative form of a genetic locus; a single allele for each locus is inherited from each parent (e.g., at a locus for eye color the allele might result in blue or brown eyes). (ORNL) (b) One of the variant forms of a gene at a particular locus, or location, on a chromosome. Different alleles produce variation in inherited characteristics such as hair color or blood type. In an individual, one form of the allele (the dominant one) may be expressed more than another form (the recessive one). (NHGRI)

Allogeneic
Variation in alleles among members of the same species. (ORNL)

Alternative splicing
Different ways of combining a gene's exons to make variants of the complete protein. (ORNL)

Amino acid
Any of a class of 20 molecules that are combined to form proteins in living things. The sequence of amino acids in a protein and hence protein function are determined by the genetic code. (ORNL)

Amplification
An increase in the number of copies of a specific DNA fragment; can be in vivo or in vitro. *See also:* cloning. (ORNL)

Animal model
See: model organisms. (ORNL)

Annotation
Adding pertinent information such as gene coded for, amino acid sequence, or other commentary to the database entry of raw sequence of DNA bases. *See also:* bioinformatics. (ORNL)

Anticipation
Each generation of offspring has increased severity of a genetic disorder; e.g., a grandchild may have earlier onset and more severe symptoms than the parent, who had earlier onset than the grandparent. *See also:* additive genetic effects, complex trait. (ORNL)

Antisense
Nucleic acid that has a sequence exactly opposite to an mRNA molecule made by the body; binds to the mRNA molecule to prevent a protein from being made. *See also:* transcription. (ORNL)

Apoptosis
Programmed cell death, the body's normal method of disposing of damaged, unwanted, or unneeded cells. (ORNL)

Array (of hairpins)
An assemble of α helices that cannot be described as a bundle or a folded leaf. (SCOP)

Arrayed library
Individual primary recombinant clones (hosted in phage, cosmid, YAC, or other vector) that are placed in two-dimensional arrays in microtiter dishes. Each primary clone can be identified by the identity of the plate and the clone location (row and column) on that plate. Arrayed libraries of clones can be used for many applications, including screening for a specific gene or genomic region of interest. *See also:* library, genomic library, gene chip technology. (ORNL)

Assembly
Putting sequenced fragments of DNA into their correct chromosomal positions. (ORNL)

Autoradiography
A technique that uses x-ray film to visualize radioactively labeled molecules or fragments of molecules; used in analyzing length and number of DNA fragments after they are separated by gel electrophoresis. (ORNL)

Autosomal dominant
A gene on one of the non-sex chromosomes that is always expressed, even if only one copy is present. The chance of passing the gene to offspring is 50% for each pregnancy. *See also:* autosome, dominant, gene (ORNL)

Autosome
A chromosome not involved in sex determination. The diploid human genome consists of a total of 46 chromosomes: 22 pairs of autosomes and 1 pair of sex chromosomes (the X and Y chromosomes). *See also:* sex chromosome. (ORNL)

B

Backcross
A cross between an animal that is heterozygous for alleles obtained from two parental strains and a second animal from one of those parental strains. Also used to describe the breeding protocol of an outcross followed by a backcross. *See also:* model organisms. (ORNL)

Bacterial artificial chromosome (BAC)
(a) A vector used to clone DNA fragments (100 to 300 kb insert size; average, 150 kb) in *Escherichia coli* cells. Based on naturally occurring F-factor plasmid found in the bacterium *E. coli. See also:* cloning vector. (ORNL) (b) Large segments of DNA, 100,000 to 200,000 bases, from another species cloned into bacteria. Once the foreign DNA has been cloned into the host bacteria, many copies of it can be made. (NHGRI)

Bacteriophage
See: phage. (ORNL)

Barrel
Structures are usually closed by main-chain hydrogen bonds between the first and last strands of the β sheet; in this case it is defined by the two integer numbers: the number of strands in the β sheet, n, and a measure of the extent to which the strands in the sheet are staggered, the shear number S. (SCOP)

Base
One of the molecules that form DNA and RNA molecules. *See also:* nucleotide, base pair, base sequence. (ORNL)

Base pair (bp)
Two nitrogenous bases (adenine and thymine or guanine and cytosine) held together by weak bonds. Two strands of DNA are held together in the shape of a double helix by the bonds between base pairs. (ORNL)

Base sequence
The order of nucleotide bases in a DNA molecule; determines the structure of proteins encoded by that DNA. (ORNL)

Base sequence analysis
A method, sometimes automated, for determining the base sequence. (ORNL)

Behavioral genetics
The study of genes that may influence behavior. (ORNL)

Beta (β) sheet
Can be antiparallel (i.e., the strand direction in any two adjacent strands are antiparallel), parallel (all strands are parallel to each other), and mixed (there is one strand at least that is parallel to one of its two neighbors and antiparallel to the other). (SCOP)

Bioinformatics
(a) The merger of biotechnology and information technology with the goal of revealing new insights and principles in biology. (NCBI) (b) The science of managing and analyzing biological data using advanced computing techniques. Especially important in analyzing genomic research data. (ORNL)

Bioremediation
The use of biological organisms such as plants or microbes to aid in removing hazardous substances from an area. (ORNL)

Biotechnology
A set of biological techniques developed through basic research and now applied to research and product development. In particular, biotechnology refers to the use by industry of recombinant DNA, cell fusion, and new bioprocessing techniques. (ORNL)

Birth defect
Any harmful trait, physical or biochemical, present at birth, whether a result of a genetic mutation or some other nongenetic factor. *See also:* congenital, gene, mutation, syndrome. (ORNL)

Bit score
(a) The value S' is derived from the raw alignment score S in which the statistical properties of the scoring system used have been taken into account. Because bit scores have been normalized with respect to the scoring system, they can be used to compare alignment scores from different searches. (NCBI) (b) Alignment scores are reported by HMMer and BLAST as bit scores. The likelihood that the query sequence is a bona fide homolog of the database sequence is compared to the likelihood that the sequence was instead generated by a "random" model. Taking the logarithm (to base 2) of this likelihood ratio gives the bit score. (SMART)

BLAST
(a) Basic Local Alignment Search Tool. A sequence comparison algorithm optimized for speed used to search sequence databases for optimal local alignments to a query. The initial search is done for a word

of length *W* that scores at least *T* when compared to the query using a substitution matrix. Word hits are then extended in either direction in an attempt to generate an alignment with a score exceeding the threshold of *S*. The *T* parameter dictates the speed and sensitivity of the search. For additional details, see one of the BLAST tutorials (Query or BLAST) or the narrative guide to BLAST. (NCBI) (b) A computer program that identifies homologous (similar) genes in different organisms, such as human, fruit fly, or nematode. (ORNL)

BLOSUM
Blocks Substitution Matrix. A substitution matrix in which scores for each position are derived from observations of the frequencies of substitutions in blocks of local alignments in related proteins. Each matrix is tailored to a particular evolutionary distance. In the BLOSUM62 matrix, for example, the alignment from which scores were derived was created using sequences sharing no more than 62% identity. Sequences more identical than 62% are represented by a single sequence in the alignment so as to avoid overweighting closely related family members. (NCBI)

Bundle
An array of *α* helices each oriented roughly along the same (bundle) axis. It may have twist, left handed if each helix makes a positive angle to the bundle axis or right handed if each helix makes a negative angle to the bundle axis. (SCOP)

C

Cancer
Diseases in which abnormal cells divide and grow unchecked. Cancer can spread from its original site to other parts of the body and can be fatal. *See also:* hereditary cancer, sporadic cancer. (ORNL)

Candidate gene
A gene located in a chromosome region suspected of being involved in a disease. *See also:* positional cloning, protein. (ORNL)

Capillary array
Gel-filled silica capillaries used to separate fragments for DNA sequencing. The small diameter of the capillaries permit the application of higher electric fields, providing high speed, high throughput separations that are significantly faster than traditional slab gels. (ORNL)

Carcinogen
Something that causes cancer to occur by causing changes in a cell's DNA. *See also:* mutagen. (ORNL)

Carrier
An individual who possesses an unexpressed, recessive trait. (ORNL)

cDNA library
A collection of DNA sequences that code for genes. The sequences are generated in the laboratory from mRNA sequences. *See also:* messenger RNA. (ORNL)

Cell
The basic unit of any living organism that carries on the biochemical processes of life. *See also:* genome, nucleus. (ORNL)

Centimorgan (cM)
A unit of measure of recombination frequency. One centimorgan is equal to a 1% chance that a marker at one genetic locus will be separated from a marker at a second locus due to crossing over in a single generation. In human beings, one centimorgan is equivalent, on average, to one million base pairs. *See also:* megabase. (ORNL)

Centromere
A specialized chromosome region to which spindle fibers attach during cell division. (ORNL)

Chimera (plural chimaera)
An organism that contains cells or tissues with a different genotype. These can be mutated cells of the host organism or cells from a different organism or species. (ORNL)

Chloroplast chromosome
Circular DNA found in the photosynthesizing organelle (chloroplast) of plants instead of the cell nucleus, where most genetic material is located. (ORNL)

Chromosomal deletion
The loss of part of a chromosome's DNA. (ORNL)

Chromosomal inversion
Chromosome segments that have been turned 180°. The gene sequence for the segment is reversed with respect to the rest of the chromosome. (ORNL)

Chromosome
The self-replicating genetic structure of cells containing the cellular DNA that bears in its nucleotide sequence the linear array of genes. In prokaryotes, chromosomal DNA is circular and the entire genome is carried on one chromosome. Eukaryotic genomes consist of a number of chromosomes whose DNA is associated with different kinds of proteins. (ORNL)

Chromosome painting
Attachment of certain fluorescent dyes to targeted parts of the chromosome. Used as a diagnostic for particular diseases, e.g., types of leukemia. (ORNL)

Chromosome region p
A designation for the short arm of a chromosome. (ORNL)

Chromosome region q
A designation for the long arm of a chromosome. (ORNL)

Clone
An exact copy made of biological material such as a DNA segment (e.g., a gene or other region), a whole cell, or a complete organism. (ORNL)

Clone bank
See: genomic library. (ORNL)

Cloning
Using specialized DNA technology to produce multiple, exact copies of a single gene or other segment of DNA to obtain enough material for further study. This process, used by researchers in the Human Genome Project, is referred to as cloning DNA. The resulting cloned (copied) collections of DNA molecules are called clone libraries. A second type of cloning exploits the natural process of cell division to make many copies of an entire cell. The genetic makeup of these cloned cells, called a cell line, is identical to the original cell. A third type of cloning produces complete, genetically identical animals such as the famous Scottish sheep, Dolly. *See also:* cloning vector. (ORNL)

Cloning vector

DNA molecule originating from a virus, a plasmid, or the cell of a higher organism into which another DNA fragment of appropriate size can be integrated without loss of the vector's capacity for self-replication; vectors introduce foreign DNA into host cells, where the DNA can be reproduced in large quantities. Examples are plasmids, cosmids, and yeast artificial chromosomes; vectors are often recombinant molecules containing DNA sequences from several sources. (ORNL)

Closed, Partly Opened, and Opened

For all-alpha structures, the extent to which the hydrophobic core is screened by the comprising α helices. *Opened* means that there is space for at least one more helix to be easily attached to the core. (SCOP)

Code

See: genetic code. (ORNL)

Codominance

Situation in which two different alleles for a genetic trait are both expressed. *See also:* autosomal dominant, recessive gene. (ORNL)

Codon

See: genetic code. (ORNL)

Coisogenic or congenic

Nearly identical strains of an organism; they vary at only a single locus. (ORNL)

Comparative genomics

The study of human genetics by comparisons with model organisms such as mice, the fruitfly, and the bacterium *Escherichia coli*. (ORNL)

Complementary DNA (cDNA)

DNA that is synthesized in the laboratory from a messenger RNA template. (ORNL)

Complementary sequence

Nucleic acid–base sequence that can form a double-stranded structure with another DNA fragment by following base-pairing rules (A pairs with T and C with G). The complementary sequence to GTAC, for example, is CATG. (ORNL)

Complex trait

Trait that has a genetic component that does not follow strict Mendelian inheritance. May involve the interaction of two or more genes or gene–environment interactions. *See also:* Mendelian inheritance, additive genetic effects. (ORNL)

Computational biology

See: bioinformatics. (ORNL)

Confidentiality

In genetics, the expectation that genetic material and the information gained from testing that material will not be available without the donor's consent. (ORNL)

Congenital

Any trait present at birth, whether the result of a genetic or nongenetic factor. *See also:* birth defect. (ORNL)

Conservation

Changes at a specific position of an amino acid or (less commonly, DNA) sequence that preserve the physicochemical properties of the original residue. (NCBI)

Conserved sequence

A base sequence in a DNA molecule (or an amino acid sequence in a protein) that has remained essentially unchanged throughout evolution. (ORNL)

Contig

Group of cloned (copied) pieces of DNA representing overlapping regions of a particular chromosome. (ORNL)

Contig map

A map depicting the relative order of a linked library of overlapping clones representing a complete chromosomal segment. (ORNL)

Cosmid

Artificially constructed cloning vector containing the *cos* gene of phage lambda. Cosmids can be packaged in lambda phage particles for infection into *Escherichia coli*; this permits cloning of larger DNA fragments (up to 45 kb) that can be introduced into bacterial hosts in plasmid vectors. (ORNL)

Crossing over

The breaking during meiosis of one maternal and one paternal chromosome, the exchange of corresponding sections of DNA, and the rejoining of the chromosomes. This process can result in an exchange of alleles between chromosomes. *See also:* recombination. (ORNL)

Crossover

Connection that links secondary structures at the opposite ends of the structural core and goes across the surface of the domain. (SCOP)

Cytogenetics

The study of the physical appearance of chromosomes. *See also:* karyotype. (ORNL)

Cytological band

An area of the chromosome that stains differently from areas around it. *See also:* cytological map. (ORNL)

Cytological map

A type of chromosome map whereby genes are located on the basis of cytological findings obtained with the aid of chromosome mutations. (ORNL)

Cytoplasmic trait

A genetic characteristic in which the genes are found outside the nucleus, in chloroplasts or mitochondria. Results in offspring inheriting genetic material from only one parent. (ORNL)

Cytoplasmic (uniparental) inheritance

See: cytoplasmic trait. (ORNL)

Cytosine (C)

A nitrogenous base, one member of the base pair GC (guanine and cytosine) in DNA. *See also:* base pair, nucleotide. (ORNL)

D

Data warehouse
A collection of databases, data tables, and mechanisms to access the data on a single subject. (ORNL)

Deletion
A loss of part of the DNA from a chromosome; can lead to a disease or abnormality. *See also:* chromosome, mutation. (ORNL)

Deletion map
A description of a specific chromosome that uses defined mutations—specific deleted areas in the genome—as "biochemical signposts," or markers for specific areas. (ORNL)

Deoxyribonucleotide
See: nucleotide. (ORNL)

Deoxyribose
A type of sugar that is one component of DNA (deoxyribonucleic acid). (ORNL)

Diploid
A full set of genetic material consisting of paired chromosomes, one from each parental set. Most animal cells except the gametes have a diploid set of chromosomes. The diploid human genome has 46 chromosomes. *See also:* haploid. (ORNL)

Directed evolution
A laboratory process used on isolated molecules or microbes to cause mutations and identify subsequent adaptations to novel environments. (ORNL)

Directed mutagenesis
Alteration of DNA at a specific site and its reinsertion into an organism to study any effects of the change. (ORNL)

Directed sequencing
Successively sequencing DNA from adjacent stretches of chromosome. (ORNL)

Disease-associated genes
Alleles carrying particular DNA sequences associated with the presence of disease. (ORNL)

DNA (deoxyribonucleic acid)
The molecule that encodes genetic information. DNA is a double-stranded molecule held together by weak bonds between base pairs of nucleotides. The four nucleotides in DNA contain the bases adenine (A), guanine (G), cytosine (C), and thymine (T). In nature, base pairs form only between A and T and between G and C; thus the base sequence of each single strand can be deduced from that of its partner. (ORNL)

DNA bank
A service that stores DNA extracted from blood samples or other human tissue. (ORNL)

DNA probe
See: probe. (ORNL)

DNA repair genes
Genes encoding proteins that correct errors in DNA sequencing. (ORNL)

DNA replication
The use of existing DNA as a template for the synthesis of new DNA strands. In humans and other eukaryotes, replication occurs in the cell nucleus. (ORNL)

DNA sequence
The relative order of base pairs, whether in a DNA fragment, gene, chromosome, or an entire genome. *See also:* base sequence analysis. (ORNL)

Domain
(a) A discrete portion of a protein assumed to fold independently of the rest of the protein and possessing its own function. (NCBI) (b) A discrete portion of a protein with its own function. The combination of domains in a single protein determines its overall function. (ORNL) (c) Conserved structural entities with distinctive secondary structure content and an hydrophobic core. In small disulfide-rich and Zn^{2+}-binding or Ca^{2+}-binding domains, the hydrophobic core may be provided by cystines and metal ions, respectively. Homologous domains with common functions usually show sequence similarities. (SMART)

Domain composition
Proteins with the same domain composition have at least one copy of each of the domains of the query. (SMART)

Domain organization
Proteins having all the domains as the query in the same order. (Additional domains are allowed.) (SMART)

Dominant
An allele that is almost always expressed, even if only one copy is present. *See also:* gene, genome. (ORNL)

Double helix
The twisted-ladder shape that two linear strands of DNA assume when complementary nucleotides on opposing strands bond together. (ORNL)

Draft sequence
The sequence generated by the Human Genome Project that, while incomplete, offers a virtual road map to an estimated 95% of all human genes. Draft sequence data are mostly in the form of 10,000 bp-sized fragments whose approximate chromosomal locations are known. *See also:* sequencing, finished DNA sequence, working draft DNA sequence. (ORNL)

DUST
A program for filtering low-complexity regions from nucleic acid sequences. (NCBI)

E

E **value**
(a) Expectation value. The number of different alignments with scores equivalent to or better than S that are expected to occur in a database search by chance. The lower the E value, the more significant the score. (NCBI) (b) This represents the number of sequences with a score greater than or equal to X expected absolutely by chance. The E value connects the score (X) of an alignment between a user-supplied sequence and a database sequence, generated by any algorithm, with how many alignments with similar or greater scores that would be expected from a search of a random-sequence database of equivalent size. Since version 2.0, E values are calculated using hidden Markov models, leading to more accurate estimates than before. (SMART)

Electrophoresis
A method of separating large molecules (such as DNA fragments or proteins) from a mixture of similar molecules. An electric current is passed through a medium containing the mixture, and each kind of molecule travels through the medium at a different rate, depending on its electrical charge and size. Agarose and acrylamide gels are the media commonly used for electrophoresis of proteins and nucleic acids. (ORNL)

Electroporation
A process using high-voltage current to make cell membranes permeable to allow the introduction of new DNA; commonly used in recombinant DNA technology. *See also:* transfection. (ORNL)

Embryonic stem (ES) cells
An embryonic cell that can replicate indefinitely, transform into other types of cells, and serve as a continuous source of new cells. (ORNL)

Endonuclease
See: restriction enzyme. (ORNL)

Enzyme
A protein that acts as a catalyst, speeding the rate at which a biochemical reaction proceeds but not altering the direction or nature of the reaction. (ORNL)

Epistasis
One gene interferes with or prevents the expression of another gene located at a different locus. (ORNL)

Escherichia coli
Common bacterium that has been studied intensively by geneticists because of its small genome size, normal lack of pathogenicity, and ease of growth in the laboratory. (ORNL)

Eugenics
The study of improving a species by artificial selection; usually refers to the selective breeding of humans. (ORNL)

Eukaryote
Cell or organism with membrane-bound, structurally discrete nucleus and other well-developed subcellular compartments. Eukaryotes include all organisms except viruses, bacteria, and blue-green algae. *See also:* prokaryote, chromosome. (ORNL)

Evolutionarily conserved
See: conserved sequence. (ORNL)

Exogenous DNA
DNA originating outside an organism that has been introduced into the organism. (ORNL)

Exon
The protein-coding DNA sequence of a gene. *See also:* intron. (ORNL)

Exonuclease
An enzyme that cleaves nucleotides sequentially from free ends of a linear nucleic acid substrate. (ORNL)

Expressed gene
See: gene expression. (ORNL)

Expressed sequence tag (EST)
A short strand of DNA that is a part of a cDNA molecule and can act as identifier of a gene. Used in locating and mapping genes. *See also:* cDNA, sequence-tagged site. (ORNL)

F

FASTA
(a) The first widely used algorithm for database similarity searching. The program looks for optimal local alignments by scanning the sequence for small matches called "words." Initially, the scores of segments in which there are multiple word hits are calculated ("init1"). Later the scores of several segments may be summed to generate an "initn" score. An optimized alignment that includes gaps is shown in the output as "opt." The sensitivity and speed of the search are inversely related and controlled by the "k-tup" variable, which specifies the size of a word. (NCBI) (b) An output format for nucleic acid or protein sequences.

Filial generation (F1, F2)
Each generation of offspring in a breeding program, designated F1, F2, etc. (ORNL)

Filtering
Also known as masking. The process of hiding regions of (nucleic acid or amino acid) sequence having characteristics that frequently lead to spurious high scores. *See also:* SEG and DUST. (NCBI)

Fingerprinting
In genetics, the identification of multiple specific alleles on a person's DNA to produce a unique identifier for that person. *See also:* forensics. (ORNL)

Finished DNA sequence
High-quality, low-error, gap-free DNA sequence of the human genome. Achieving this ultimate 2003 Human Genome Project (HGP) goal requires additional sequencing to close gaps, reduce ambiguities, and allow for only a single error every 10,000 bases, the agreed-upon standard for HGP finished sequence. *See also:* sequencing, draft sequence. (ORNL)

Flow cytometry
Analysis of biological material by detection of the light-absorbing or fluorescing properties of cells or subcellular fractions (i.e., chromosomes) passing in a narrow stream through a laser beam. An absorbance or fluorescence profile of the sample is produced. Automated sorting devices, used to fractionate samples, sort successive droplets of the analyzed stream into different fractions depending on the fluorescence emitted by each droplet. (ORNL)

Flow karyotyping
Use of flow cytometry to analyze and separate chromosomes according to their DNA content. (ORNL)

Fluorescence in situ hybridization (FISH)
A physical mapping approach that uses fluorescein tags to detect hybridization of probes with metaphase chromosomes and with the less-condensed somatic interphase chromatin. (ORNL)

Folded leaf
A layer of α helices wrapped around a single hydrophobic core but not with the simple geometry of a bundle. (SCOP)

Forensics

The use of DNA for identification. Some examples of DNA use are to establish paternity in child support cases, establish the presence of a suspect at a crime scene, and identify accident victims. (ORNL)

Fraternal twin

Siblings born at the same time as the result of fertilization of two ova by two sperm. They share the same genetic relationship to each other as any other siblings. *See also:* identical twin. (ORNL)

Full gene sequence

The complete order of bases in a gene. This order determines which protein a gene will produce. (ORNL)

Functional genomics

The study of genes, their resulting proteins, and the role played by the proteins in the body's biochemical processes. (ORNL)

G

Gamete

Mature male or female reproductive cell (sperm or ovum) with a haploid set of chromosomes (23 for humans). (ORNL)

Gap

(a) A space introduced into an alignment to compensate for insertions and deletions in one sequence relative to another. To prevent the accumulation of too many gaps in an alignment, introduction of a gap causes the deduction of a fixed amount (the gap score) from the alignment score. Extension of the gap to encompass additional nucleotides or amino acid is also penalized in the scoring of an alignment. (NCBI) (b) A position in an alignment that represents a deletion within one sequence relative to another. Gap penalties are requirements for alignment algorithms in order to reduce excessively gapped regions. Gaps in alignments represent insertions that usually occur in protruding loops or beta-bulges within protein structures. (SMART)

GC-rich area

Many DNA sequences carry long stretches of repeated G and C, which often indicates a gene-rich region. (ORNL)

Gel electrophoresis

See: electrophoresis. (ORNL)

Gene

The fundamental physical and functional unit of heredity. A gene is an ordered sequence of nucleotides located in a particular position on a particular chromosome that encodes a specific functional product (i.e., a protein or RNA molecule). *See also:* gene expression. (ORNL)

Gene amplification

Repeated copying of a piece of DNA; a characteristic of tumor cells. *See also:* gene, oncogene. (ORNL)

Gene chip technology

Development of cDNA microarrays from a large number of genes. Used to monitor and measure changes in gene expression for each gene represented on the chip. (ORNL)

Gene expression

The process by which a gene's coded information is converted into the structures present and operating in the cell. Expressed genes include those that are transcribed into mRNA and then translated into protein and those that are transcribed into RNA but not translated into protein (e.g., transfer and ribosomal RNAs). (ORNL)

Gene family

Group of closely related genes that make similar products. (ORNL)

Gene library

See: genomic library (ORNL)

Gene mapping

Determination of the relative positions of genes on a DNA molecule (chromosome or plasmid) and of the distance, in linkage units or physical units, between them. (ORNL)

Gene pool

All the variations of genes in a species. *See also:* allele, gene, polymorphism. (ORNL)

Gene prediction

Predictions of possible genes made by a computer program based on how well a stretch of DNA sequence matches known gene sequences. (ORNL)

Gene product

The biochemical material, either RNA or protein, resulting from expression of a gene. The amount of gene product is used to measure how active a gene is; abnormal amounts can be correlated with disease-causing alleles. (ORNL)

Gene testing

See: genetic testing, genetic screening. (ORNL)

Gene therapy

An experimental procedure aimed at replacing, manipulating, or supplementing nonfunctional or misfunctioning genes with healthy genes. *See also:* gene, inherit, somatic cell gene therapy, germ line gene therapy. (ORNL)

Genetic code

The sequence of nucleotides, coded in triplets (codons) along the mRNA, that determines the sequence of amino acids in protein synthesis. A gene's DNA sequence can be used to predict the mRNA sequence, and the genetic code can in turn be used to predict the amino acid sequence. (ORNL)

Genetic counseling

Provides patients and their families with education and information about genetic-related conditions and helps them make informed decisions. (ORNL)

Genetic discrimination

Prejudice against those who have or are likely to develop an inherited disorder. (ORNL)

Genetic engineering

Altering the genetic material of cells or organisms to enable them to make new substances or perform new functions. (ORNL)

Genetic engineering technology

See: recombinant DNA technology. (ORNL)

Genetic illness

Sickness, physical disability, or other disorder resulting from the inheritance of one or more deleterious alleles. (ORNL)

Genetic informatics
See: bioinformatics. (ORNL)

Genetic map
See: linkage map. (ORNL)

Genetic marker
A gene or other identifiable portion of DNA whose inheritance can be followed. *See also:* chromosome, DNA, gene, inherit. (ORNL)

Genetic material
See: genome. (ORNL)

Genetic mosaic
An organism in which different cells contain different genetic sequence. This can be the result of a mutation during development or fusion of embryos at an early developmental stage. (ORNL)

Genetic polymorphism
Difference in DNA sequence among individuals, groups, or populations (e.g., genes for blue eyes versus brown eyes). (ORNL)

Genetic predisposition
Susceptibility to a genetic disease. May or may not result in actual development of the disease. (ORNL)

Genetic screening
Testing a group of people to identify individuals at high risk of having or passing on a specific genetic disorder. (ORNL)

Genetic testing
Analyzing an individual's genetic material to determine predisposition to a particular health condition or to confirm a diagnosis of genetic disease. (ORNL)

Genetics
The study of inheritance patterns of specific traits. (ORNL)

Gene transfer
Incorporation of new DNA into an organism's cells, usually by a vector such as a modified virus. Used in gene therapy. *See also:* mutation, gene therapy, vector. (ORNL)

Genome
All the genetic material in the chromosomes of a particular organism; its size is generally given as its total number of base pairs. (ORNL)

Genome project
Research and technology development effort aimed at mapping and sequencing the genome of human beings and certain model organisms. *See also:* Human Genome Initiative. (ORNL)

Genomic library
A collection of clones made from a set of randomly generated overlapping DNA fragments that represent the entire genome of an organism. *See also:* library, arrayed library. (ORNL)

Genomics
The study of genes and their function. (ORNL)

Genomic sequence
See: DNA. (ORNL)

Genotype
The genetic constitution of an organism, as distinguished from its physical appearance (its phenotype). (ORNL)

Germ cell
Sperm and egg cells and their precursors. Germ cells are haploid and have only one set of chromosomes (23 in all), while all other cells have two copies (46 in all). (ORNL)

Germ line
The continuation of a set of genetic information from one generation to the next. *See also:* inherit. (ORNL)

Germ line gene therapy
An experimental process of inserting genes into germ cells or fertilized eggs to cause a genetic change that can be passed on to offspring. May be used to alleviate effects associated with a genetic disease. *See also:* genomics, somatic cell gene therapy. (ORNL)

Germ line genetic mutation
See: mutation. (ORNL)

Global alignment
The alignment of two nucleic acid or protein sequences over their entire length. (NCBI)

Greek key
A topology for a small number of β-sheet strands in which some interstrand connections go across the end of a barrel or, in a sandwich fold, between β sheets. (SCOP)

Guanine (G)
A nitrogenous base, one member of the base pair GC (guanine and cytosine) in DNA. *See also:* base pair, nucleotide. (ORNL)

H

H
The relative entropy of the target and background residue frequencies, *H* can be thought of as a measure of the average information (in bits) available per position that distinguishes an alignment from chance. At high values of *H*, short alignments can be distinguished by chance, whereas at lower *H* values, a longer alignment may be necessary. (NCBI)

Haploid
A single set of chromosomes (half the full set of genetic material) present in the egg and sperm cells of animals and in the egg and pollen cells of plants. Human beings have 23 chromosomes in their reproductive cells. *See also:* diploid. (ORNL)

Haplotype
A way of denoting the collective genotype of a number of closely linked loci on a chromosome. (ORNL)

Hemizygous
Having only one copy of a particular gene. For example, in humans, males are hemizygous for genes found on the Y chromosome. (ORNL)

Hereditary cancer
Cancer that occurs due to the inheritance of an altered gene within a family. *See also:* sporadic cancer. (ORNL)

Heterozygosity
The presence of different alleles at one or more loci on homologous chromosomes. (ORNL)

Heterozygote
See: heterozygosity. (ORNL)

Highly conserved sequence
DNA sequence that is very similar across several different types of organisms. *See also:* gene, mutation. (ORNL)

High throughput sequencing
A fast method of determining the order of bases in DNA. *See also:* sequencing. (ORNL)

Homeobox
A short stretch of nucleotides whose base sequence is virtually identical in all the genes that contain it. Homeoboxes have been found in many organisms from fruitflies to human beings. In the fruitfly, a homeobox appears to determine when particular groups of genes are expressed during development. (ORNL)

Homolog
A member of a chromosome pair in diploid organisms or a gene that has the same origin and functions in two or more species. (ORNL)

Homologous chromosome
A chromosome containing the same linear gene sequences as another, each derived from one parent. (ORNL)

Homologous recombination
Swapping of DNA fragments between paired chromosomes. (ORNL)

Homology
(a) Similarity attributed to descent from a common ancestor. (NCBI) (b) Similarity in DNA or protein sequences between individuals of the same species or among different species. (ORNL) (c) Evolutionary descent from a common ancestor due to gene duplication. (SMART)

Homozygote
An organism that has two identical alleles of a gene. *See also:* heterozygote. (ORNL)

Homozygous
See: homozygote. (ORNL)

HSP
High-scoring segment pair. Local alignments with no gaps that achieve one of the top alignment scores in a given search. (NCBI)

Human gene therapy
See: gene therapy. (ORNL)

Human Genome Initiative
Collective name for several projects begun in 1986 by the U.S. Department of Energy (DOE) to create an ordered set of DNA segments from known chromosomal locations, develop new computational methods for analyzing genetic map and DNA sequence data, and develop new techniques and instruments for detecting and analyzing DNA. This DOE initiative is now known as the Human Genome Program. The joint national effort, led by the DOE and National Institutes of Health, is known as the Human Genome Project. (ORNL)

Human Genome Project (HGP)
Formerly titled Human Genome Initiative. *See also:* Human Genome Initiative. (ORNL)

Hybrid
The offspring of genetically different parents. *See also:* heterozygote. (ORNL)

Hybridization
The process of joining two complementary strands of DNA or one each of DNA and RNA to form a double-stranded molecule. (ORNL)

I

Identical twin
Twins produced by the division of a single zygote; both have identical genotypes. *See also:* fraternal twin. (ORNL)

Identity
The extent to which two (nucleotide or amino acid) sequences are invariant. (NCBI)

Immunotherapy
Using the immune system to treat disease, for example, in the development of vaccines. May also refer to the therapy of diseases caused by the immune system. *See also:* cancer. (ORNL)

Imprinting
A phenomenon in which the disease phenotype depends on which parent passed on the disease gene. For instance, both Prader–Willi and Angelman syndromes are inherited when the same part of chromosome 15 is missing. When the father's complement of 15 is missing, the child has Prader–Willi, but when the mother's complement of 15 is missing, the child has Angelman syndrome. (ORNL)

Independent assortment
During meiosis each of the two copies of a gene is distributed to the germ cells independently of the distribution of other genes. *See also:* linkage. (ORNL)

Informatics
See: bioinformatics. (ORNL)

Informed consent
An individual willingly agrees to participate in an activity after first being advised of the risks and benefits. *See also:* privacy. (ORNL)

Inherit
In genetics, to receive genetic material from parents through biological processes. (ORNL)

Inherited
See: inherit. (ORNL)

Insertion
A chromosome abnormality in which a piece of DNA is incorporated into a gene and thereby disrupts the gene's normal function. *See also:* chromosome, DNA, gene, mutation. (ORNL)

Insertional mutation
See: insertion. (ORNL)

In situ hybridization
Use of a DNA or RNA probe to detect the presence of the complementary DNA sequence in cloned bacterial or cultured eukaryotic cells. (ORNL)

Intellectual property rights
Patents, copyrights, and trademarks. *See also:* patent. (ORNL)

Interference
One crossover event inhibits the chances of another crossover event. Also known as positive interference. Negative interference increases the chance of a second crossover. *See also:* crossing over. (ORNL)

Interphase
The period in the cell cycle when DNA is replicated in the nucleus; followed by mitosis. (ORNL)

Intracellular domains
Domain families that are most prevalent in proteins within the cytoplasm. (SMART)

Intron
DNA sequence that interrupts the protein-coding sequence of a gene; an intron is transcribed into RNA but is cut out of the message before it is translated into protein. *See also:* exon. (ORNL)

In vitro
Studies performed outside a living organism, such as in a laboratory. (ORNL)

In vivo
Studies carried out in living organisms. (ORNL)

Isoenzyme
An enzyme performing the same function as another enzyme but having a different set of amino acids. The two enzymes may function at different speeds. (ORNL)

J

Jelly roll
A variant of Greek-key topology with both ends of a sandwich or a barrel fold being crossed by two interstrand connections. *See also:* Greek key. (SCOP)

Junk DNA
Stretches of DNA that do not code for genes; most of the genome consists of so-called junk DNA which may have regulatory and other functions. Also called noncoding DNA. (ORNL)

K

K
A statistical parameter used in calculating BLAST scores that can be thought of as a natural scale for search space size. The value K is used in converting a raw score (S) to a bit score (S'). (NCBI)

Karyotype
A photomicrograph of an individual's chromosomes arranged in a standard format showing the number, size, and shape of each chromosome type; used in low-resolution physical mapping to correlate gross chromosomal abnormalities with the characteristics of specific diseases. (ORNL)

Kilobase (kb)
Unit of length for DNA fragments equal to 1000 nucleotides. (ORNL)

Knock-out
Deactivation of specific genes; used in laboratory organisms to study gene function. *See also:* gene, locus, model organisms. (ORNL)

L

Lambda
A statistical parameter used in calculating BLAST scores that can be thought of as a natural scale for a scoring system. The value lambda is used in converting a raw score (S) to a bit score (S'). (NCBI)

Library
An unordered collection of clones (i.e., cloned DNA from a particular organism) whose relationship to each other can be established by physical mapping. *See also:* genomic library, arrayed library. (ORNL)

Linkage
The proximity of two or more markers (e.g., genes, restriction fragment length polymorphism markers) on a chromosome; the closer the markers, the lower the probability that they will be separated during DNA repair or replication processes (binary fission in prokaryotes, mitosis or meiosis in eukaryotes), and hence the greater the probability that they will be inherited together. (ORNL)

Linkage disequilibrium
Where alleles occur together more often than can be accounted for by chance. Indicates that the two alleles are physically close on the DNA strand. *See also:* Mendelian inheritance. (ORNL)

Linkage map
A map of the relative positions of genetic loci on a chromosome, determined on the basis of how often the loci are inherited together. Distance is measured in centimorgans (cM). (ORNL)

Local alignment
The alignment of some portion of two nucleic acid or protein sequences. (NCBI)

Localization
Numbers of domains that are thought from SwissProt annotations to be present in different cellular compartments (cytoplasm, extracellular space, nucleus, and membrane associated) are shown in annotation pages. (SMART)

Localize
Determination of the original position (locus) of a gene or other marker on a chromosome. (ORNL)

Locus (plural loci)
The position on a chromosome of a gene or other chromosome marker; also, the DNA at that position. The use of locus is sometimes restricted to mean expressed DNA regions. *See also:* gene expression. (ORNL)

Long-range restriction mapping
Restriction enzymes are proteins that cut DNA at precise locations. Restriction maps depict the chromosomal positions of restriction enzyme cutting sites. These are used as biochemical "signposts," or markers of specific areas along the chromosomes. The map will detail the positions where the DNA molecule is cut by particular restriction enzymes. (ORNL)

Low-complexity region (LCR)
Regions of biased composition including homopolymeric runs, short-period repeats, and more subtle overrepresentation of one or a few residues. The SEG program is used to mask or filter LCRs in amino acid queries. The DUST program is used to mask or filter LCRs in nucleic acid queries. (NCBI)

M

Macrorestriction map
Map depicting the order of and distance between sites at which restriction enzymes cleave chromosomes. (ORNL)

Mapping
See: gene mapping, linkage map, physical map. (ORNL)

Mapping population
The group of related organisms used in constructing a genetic map. (ORNL)

Marker
See: genetic marker. (ORNL)

Masking
Also known as filtering. The removal of repeated or low-complexity regions from a sequence in order to improve the sensitivity of sequence similarity searches performed with that sequence. (NCBI)

Mass spectrometer
An instrument used to identify chemicals in a substance by their mass and charge. (ORNL)

Meander
A simple topology of a β sheet where any two consecutive strands are adjacent and antiparallel. (SCOP)

Megabase (Mb)
Unit of length for DNA fragments equal to 1 million nucleotides and roughly equal to 1 cM. *See also:* centimorgan. (ORNL)

Meiosis
The process of two consecutive cell divisions in the diploid progenitors of sex cells. Meiosis results in four rather than two daughter cells, each with a haploid set of chromosomes. *See also:* mitosis. (ORNL)

Mendelian inheritance
One method in which genetic traits are passed from parents to offspring. Named for Gregor Mendel, who first studied and recognized the existence of genes and this method of inheritance. *See also:* autosomal dominant, recessive gene, sex linked. (ORNL)

Messenger RNA (mRNA)
RNA that serves as a template for protein synthesis. *See also:* genetic code. (ORNL)

Metaphase
A stage in mitosis or meiosis during which the chromosomes are aligned along the equatorial plane of the cell. (ORNL)

Microarray
Sets of miniaturized chemical reaction areas that may also be used to test DNA fragments, antibodies, or proteins. (ORNL)

Microbial genetics
The study of genes and gene function in bacteria, archaea, and other microorganisms. Often used in research in the fields of bioremediation, alternative energy, and disease prevention. *See also:* model organisms, biotechnology, bioremediation. (ORNL)

Microinjection
A technique for introducing a solution of DNA into a cell using a fine microcapillary pipet. (ORNL)

Mitochondrial DNA
The genetic material found in mitochondria, the organelles that generate energy for the cell. Not inherited in the same fashion as nucleic DNA. *See also:* cell, DNA, genome, nucleus. (ORNL)

Mitosis
The process of nuclear division in cells that produces daughter cells that are genetically identical to each other and to the parent cell. *See also:* meiosis. (ORNL)

Modeling
The use of statistical analysis, computer analysis, or model organisms to predict outcomes of research. (ORNL)

Model organisms
A laboratory animal or other organism useful for research. (ORNL)

Molecular biology
The study of the structure, function, and makeup of biologically important molecules. (ORNL)

Molecular farming
The development of transgenic animals to produce human proteins for medical use. (ORNL)

Molecular genetics
The study of macromolecules important in biological inheritance. (ORNL)

Molecular medicine
The treatment of injury or disease at the molecular level. Examples include the use of DNA-based diagnostic tests or medicine derived from DNA sequence information. (ORNL)

Monogenic disorder
A disorder caused by mutation of a single gene. *See also:* mutation, polygenic disorder. (ORNL)

Monogenic inheritance
See: monogenic disorder. (ORNL)

Monosomy
Possessing only one copy of a particular chromosome instead of the normal two copies. *See also:* cell, chromosome, gene expression, trisomy. (ORNL)

Morbid map
A diagram showing the chromosomal location of genes associated with disease. (ORNL)

Motif
(a) A short conserved region in a protein sequence. Motifs are frequently highly conserved parts of domains. (NCBI) (b) Sequence

motifs are short conserved regions of polypeptides. Sets of sequence motifs need not necessarily represent homologs. (SMART)

Mouse model
See: model organisms. (ORNL)

Multifactorial or multigenic disorder
See: polygenic disorder. (ORNL)

Multiple sequence alignment
An alignment of three or more sequences with gaps inserted in the sequences such that residues with common structural positions and/or ancestral residues are aligned in the same column. ClustalW is one of the most widely used multiple sequence alignment programs. (NCBI)

Multiplexing
A laboratory approach that performs multiple sets of reactions in parallel (simultaneously); greatly increasing speed and throughput. (ORNL)

Murine
Organism in the genus *Mus*. A rat or mouse. (ORNL)

Mutagen
An agent that causes a permanent genetic change in a cell. Does not include changes occurring during normal genetic recombination. (ORNL)

Mutagenicity
The capacity of a chemical or physical agent to cause permanent genetic alterations. *See also:* somatic cell genetic mutation. (ORNL)

Mutation
Any heritable change in DNA sequence. *See also:* polymorphism. (ORNL)

N

Nitrogenous base
A nitrogen-containing molecule having the chemical properties of a base. DNA contains the nitrogenous bases adenine (A), guanine (G), cytosine (C), and thymine (T). *See also:* DNA. (ORNL)

Northern blot
A gel-based laboratory procedure that locates mRNA sequences on a gel that are complementary to a piece of DNA used as a probe. *See also:* DNA, library. (ORNL)

Nuclear transfer
A laboratory procedure in which a cell's nucleus is removed and placed into an oocyte with its own nucleus removed so the genetic information from the donor nucleus controls the resulting cell. Such cells can be induced to form embryos. This process was used to create the cloned sheep Dolly. *See also:* cloning. (ORNL)

Nucleic acid
A large molecule composed of nucleotide subunits. *See also:* DNA. (ORNL)

Nucleolar organizing region
A part of the chromosome containing rRNA genes. (ORNL)

Nucleotide
A subunit of DNA or RNA consisting of a nitrogenous base (adenine, guanine, thymine, or cytosine in DNA; adenine, guanine, uracil, or cytosine in RNA), a phosphate molecule, and a sugar molecule (deoxyribose in DNA and ribose in RNA). Thousands of nucleotides are linked to form a DNA or RNA molecule. *See also:* DNA, base pair, RNA. (ORNL)

Nucleus
The cellular organelle in eukaryotes that contains most of the genetic material. (ORNL)

O

Oligo
See: oligonucleotide. (ORNL)

Oligogenic
A phenotypic trait produced by two or more genes working together. *See also:* polygenic disorder. (ORNL)

Oligonucleotide
A molecule usually composed of 25 or fewer nucleotides; used as a DNA synthesis primer. *See also:* nucleotide. (ORNL)

Oncogene
A gene, one or more forms of which are associated with cancer. Many oncogenes are involved, directly or indirectly, in controlling the rate of cell growth. (ORNL)

Open reading frame (ORF)
The sequence of DNA or RNA located between the start-code sequence (initiation codon) and the stop-code sequence (termination codon). (ORNL)

Operon
A set of genes transcribed under the control of an operator gene. (ORNL)

Optimal alignment
An alignment of two sequences with the highest possible score. (NCBI)

ORF
See: open reading frame. (SMART)

Orthologous
Homologous sequences in different species that arose from a common ancestral gene during speciation; may or may not be responsible for a similar function. (NCBI)

Overlapping clones
See: genomic library. (ORNL)

P

P value
The probability of an alignment occurring with the score in question or better. The *P* value is calculated by relating the observed alignment score, *S*, to the expected distribution of high scoring segment pair scores from comparisons of random sequences of the same length and composition as the query to the database. The most highly significant *P* values will be those close to zero. The *P* and *E* values are different ways of representing the significance of the alignment. (NCBI)

P1-derived artificial chromosome (PAC)
One type of vector used to clone DNA fragments (insert size 100 to 300 kb; average 150 kb) in *Escherichia coli* cells. Based on bacteriophage (a virus) P1 genome. *See also:* cloning vector. (ORNL)

PAM
Point accepted mutation. A unit used to quantify the amount of evolutionary change in a protein sequence. The amount of evolution which will change, on average, 1% of amino acids in a protein sequence is 1.0 PAM units. A PAM(x) substitution matrix is a look-up table in which scores for each amino acid substitution have been calculated based on the frequency of that substitution in closely related proteins that have experienced a certain amount (x) of evolutionary divergence. (NCBI)

Paralogous
Homologous sequences within a single species that arose by gene duplication. (NCBI)

Partly open barrel
Has the edge strands not properly hydrogen bonded because one of the strands is in two parts connected with a linker of more than that one residue. These edge strands can be treated as a single but interrupted strand, allowing classification with the effective strand and shear numbers, n^* and S^*. In the few open barrels the β sheets are connected by only a few *side-chain* hydrogen bonds between the edge strands. (SCOP)

Patent
In genetics, conferring the right or title to genes, gene variations, or identifiable portions of sequenced genetic material to an individual or organization. *See also:* gene. (ORNL)

Pedigree
A family tree diagram that shows how a particular genetic trait or disease has been inherited. *See also:* inherit. (ORNL)

Penetrance
The probability of a gene or genetic trait being expressed. "Complete" penetrance means the gene or genes for a trait are expressed in the whole population that has the genes. "Incomplete" penetrance means the genetic trait is expressed in only part of the population. The percent penetrance also may change with the age range of the population. (ORNL)

Peptide
Two or more amino acids joined by a bond called a "peptide bond." *See also:* polypeptide. (ORNL)

Phage
A virus for which the natural host is a bacterial cell. (ORNL)

Pharmacogenomics
The study of the interaction of an individual's genetic makeup and response to a drug. (ORNL)

Phenocopy
A trait not caused by inheritance of a gene but that appears to be identical to a genetic trait. (ORNL)

Phenotype
The physical characteristics of an organism or the presence of a disease that may or may not be genetic. *See also:* genotype. (ORNL)

Physical map
A map of the locations of identifiable landmarks on DNA (e.g., restriction enzyme cutting sites, genes), regardless of inheritance. Distance is measured in base pairs. For the human genome, the lowest resolution physical map is the banding patterns on the 24 different chromosomes; the highest resolution map is the complete nucleotide sequence of the chromosomes. (ORNL)

Plasmid
Autonomously replicating extrachromosomal circular DNA molecules, distinct from the normal bacterial genome and nonessential for cell survival under nonselective conditions. Some plasmids are capable of integrating into the host genome. A number of artificially constructed plasmids are used as cloning vectors. (ORNL)

Pleiotropy
One gene that causes many different physical traits such as multiple disease symptoms. (ORNL)

Pluripotency
The potential of a cell to develop into more than one type of mature cell, depending on environment. (ORNL)

Polygenic disorder
Genetic disorder resulting from the combined action of alleles of more than one gene (e.g., heart disease, diabetes, and some cancers). Although such disorders are inherited, they depend on the simultaneous presence of several alleles; thus the hereditary patterns usually are more complex than those of single-gene disorders. *See also:* single-gene disorder. (ORNL)

Polymerase chain reaction (PCR)
A method for amplifying a DNA base sequence using a heat-stable polymerase and two 20-base primers, one complementary to the (+) strand at one end of the sequence to be amplified and one complementary to the (−) strand at the other end. Because the newly synthesized DNA strands can subsequently serve as additional templates for the same primer sequences, successive rounds of primer annealing, strand elongation, and dissociation produce rapid and highly specific amplification of the desired sequence. PCR also can be used to detect the existence of the defined sequence in a DNA sample. (ORNL)

Polymerase, DNA or RNA
Enzyme that catalyzes the synthesis of nucleic acids on preexisting nucleic acid templates, assembling RNA from ribonucleotides or DNA from deoxyribonucleotides. (ORNL)

Polymorphism
Difference in DNA sequence among individuals that may underlie differences in health. Genetic variations occurring in more than 1% of a population would be considered useful polymorphisms for genetic linkage analysis. *See also:* mutation. (ORNL)

Polypeptide
A protein or part of a protein made of a chain of amino acids joined by a peptide bond. (ORNL)

Population genetics
The study of variation in genes among a group of individuals. (ORNL)

Positional cloning
A technique used to identify genes, usually those that are associated with diseases, based on their location on a chromosome. (ORNL)

Primer

Short preexisting polynucleotide chain to which new deoxyribonucleotides can be added by DNA polymerase. (ORNL)

Privacy

In genetics, the right of people to restrict access to their genetic information. (ORNL)

Probe

Single-stranded DNA or RNA molecules of specific base sequence, labeled either radioactively or immunologically, that are used to detect the complementary base sequence by hybridization. (ORNL)

Profile

(a) A table that lists the frequencies of each amino acid in each position of protein sequence. Frequencies are calculated from multiple alignments of sequences containing a domain of interest. *See also:* PSSM. (NCBI) (b) A table of position-specific scores and gap penalties, representing an homologous family, that may be used to search sequence databases. In ClustalW-derived profiles those sequences that are more distantly related are assigned higher weights. (SMART)

Prokaryote

Cell or organism lacking a membrane-bound, structurally discrete nucleus and other subcellular compartments. Bacteria are examples of prokaryotes. *See also:* chromosome, eukaryote. (ORNL)

Promoter

A DNA site to which RNA polymerase will bind and initiate transcription. (ORNL)

Pronucleus

The nucleus of a sperm or egg prior to fertilization. *See also:* nucleus, transgenic. (ORNL)

Protein

A large molecule composed of one or more chains of amino acids in a specific order; the order is determined by the base sequence of nucleotides in the gene that codes for the protein. Proteins are required for the structure, function, and regulation of the body's cells, tissues, and organs; each protein has unique functions. Examples are hormones, enzymes, and antibodies. (ORNL)

Proteome

Proteins expressed by a cell or organ at a particular time and under specific conditions. (ORNL)

Proteomics

Systematic analysis of protein expression of normal and diseased tissues that involves the separation, identification, and characterization of all of the proteins in an organism. (NCBI)

Pseudogene

A sequence of DNA similar to a gene but nonfunctional; probably the remnant of a once functional gene that accumulated mutations. (ORNL)

PSI-BLAST

Position-Specific Iterative BLAST. An iterative search using the BLAST algorithm. A profile is built after the initial search, which is then used in subsequent searches. The process may be repeated, if desired, with new sequences found in each cycle used to refine the profile. (NCBI)

PSSM

Position-specific scoring matrix. The PSSM gives the log-odds score for finding a particular matching amino acid in a target sequence. *See also:* profile. (NCBI)

Purine

A nitrogen-containing, double-ring, basic compound that occurs in nucleic acids. The purines in DNA and RNA are adenine and guanine. *See also:* base pair. (ORNL)

Pyrimidine

A nitrogen-containing, single-ring, basic compound that occurs in nucleic acids. The pyrimidines in DNA are cytosine and thymine; in RNA, cytosine and uracil. *See also:* base pair. (ORNL)

Q

Query

The input sequence (or other type of search term) with which all of the entries in a database are to be compared. (NCBI)

R

Radiation hybrid

A hybrid cell containing small fragments of irradiated human chromosomes. Maps of irradiation sites on chromosomes for the human, rat, mouse, and other genomes provide important markers, allowing the construction of very precise sequence-tagged site maps indispensable to studying multifactorial diseases. *See also:* sequence-tagged site. (ORNL)

Rare-cutter enzyme

See: restriction enzyme cutting site. (ORNL)

Raw score

The score of an alignment, S, calculated as the sum of substitution and gap scores. Substitution scores are given by a look-up table. Gap scores are typically calculated as the sum of G, the gap opening penalty, and L, the gap extension penalty. For a gap of length n, the gap cost would be $G + Ln$. The choice of gap costs G and L is empirical, but it is customary to choose a high value for G (10 to 15) and a low value for L (1 to 2). *See also:* PAM, BLOSUM. (NCBI)

Recessive gene

A gene that will be expressed only if there are two identical copies or, for a male, if one copy is present on the X chromosome. (ORNL)

Reciprocal translocation

When a pair of chromosomes exchange exactly the same length and area of DNA. Results in a shuffling of genes. (ORNL)

Recombinant clone

Clone containing recombinant DNA molecules. *See also:* recombinant DNA technology. (ORNL)

Recombinant DNA molecules

A combination of DNA molecules of different origin that are joined using recombinant DNA technologies. (ORNL)

Recombinant DNA technology

Procedure used to join together DNA segments in a cell-free system (an environment outside a cell or organism). Under appropriate conditions, a recombinant DNA molecule can enter a cell and replicate there, either autonomously or after it has become integrated into a cellular chromosome. (ORNL)

Recombination

The process by which progeny derives a combination of genes different from that of either parent. In higher organisms, this can occur by crossing over. *See also:* crossing over, mutation. (ORNL)

Regulatory region or sequence

A DNA base sequence that controls gene expression. (ORNL)

Repetitive DNA

Sequences of varying lengths that occur in multiple copies in the genome; it represents much of the human genome. (ORNL)

Reporter gene

See: marker. (ORNL)

Resolution

Degree of molecular detail on a physical map of DNA, ranging from low to high. (ORNL)

Restriction enzyme cutting site

A specific nucleotide sequence of DNA at which a particular restriction enzyme cuts the DNA. Some sites occur frequently in DNA (e.g., every several hundred base pairs); others much less frequently (rare cutter; e.g., every 10,000 bp). (ORNL)

Restriction enzyme, endonuclease

A protein that recognizes specific, short nucleotide sequences and cuts DNA at those sites. Bacteria contain over 400 such enzymes that recognize and cut more than 100 different DNA sequences. *See also:* restriction enzyme cutting site. (ORNL)

Restriction fragment length polymorphism (RFLP)

Variation between individuals in DNA fragment sizes cut by specific restriction enzymes; polymorphic sequences that result in RFLPs are used as markers on both physical maps and genetic linkage maps. RFLPs usually are caused by mutation at a cutting site. *See also:* marker, polymorphism. (ORNL)

Retroviral infection

The presence of retroviral vectors, such as some viruses, which use their recombinant DNA to insert their genetic material into the chromosomes of the host's cells. The virus is then propagated by the host cell. (ORNL)

Reverse transcriptase

An enzyme used by retroviruses to form a complementary DNA sequence (cDNA) from their RNA. The resulting DNA is then inserted into the chromosome of the host cell. (ORNL)

Ribonucleotide

See: nucleotide. (ORNL)

Ribose

The five-carbon sugar that serves as a component of RNA. *See also:* ribonucleic acid, deoxyribose. (ORNL)

Ribosomal RNA (rRNA)

A class of RNA found in the ribosomes of cells. (ORNL)

Ribosomes

Small cellular components composed of specialized ribosomal RNA and protein; site of protein synthesis. *See also:* RNA. (ORNL)

Risk communication

In genetics, a process in which a genetic counselor or other medical professional interprets genetic test results and advises patients of the consequences for them and their offspring. (ORNL)

RNA (ribonucleic acid)

A chemical found in the nucleus and cytoplasm of cells; it plays an important role in protein synthesis and other chemical activities of the cell. The structure of RNA is similar to that of DNA. There are several classes of RNA molecules, including messenger RNA, transfer RNA, ribosomal RNA, and other small RNAs, each serving a different purpose. (ORNL)

S

Sanger sequencing

A widely used method of determining the order of bases in DNA. *See also:* sequencing, shotgun sequencing. (ORNL)

Satellite

A chromosomal segment that branches off from the rest of the chromosome but is still connected by a thin filament or stalk. (ORNL)

Scaffold

In genomic mapping, a series of contigs that are in the right order but not necessarily connected in one continuous stretch of sequence. (ORNL)

Seed alignment

Alignment that contains only one of each pair of homologs that are represented in a ClustalW-derived phylogenetic tree linked by a branch of length less than a distance of 0.2. (SMART)

SEG

A program for filtering low-complexity regions in amino acid sequences. Residues that have been masked are represented as "X" in an alignment. SEG filtering is performed by default in the blastp subroutine of BLAST 2.0. (NCBI)

Segregation

The normal biological process whereby the two pieces of a chromosome pair are separated during meiosis and randomly distributed to the germ cells. (ORNL)

Sequence

See: base sequence. (ORNL)

Sequence assembly

A process whereby the order of multiple sequenced DNA fragments is determined. (ORNL)

Sequence-tagged site (STS)

Short (200 to 500 bp) DNA sequence that has a single occurrence in the human genome and whose location and base sequence are known. Detectable by polymerase chain reaction, STSs are useful for localizing and orienting the mapping and sequence data reported from many different laboratories and serve as landmarks on the developing physical map of the human genome. Expressed sequence tags (ESTs) are STSs derived from cDNAs. (ORNL)

Sequencing
Determination of the order of nucleotides (base sequences) in a DNA or RNA molecule or the order of amino acids in a protein. (ORNL)

Sequencing technology
The instrumentation and procedures used to determine the order of nucleotides in DNA. (ORNL)

Sex chromosome
The X or Y chromosome in human beings that determines the sex of an individual. Females have two X chromosomes in diploid cells; males have an X and a Y chromosome. The sex chromosomes comprise the 23rd chromosome pair in a karyotype. *See also:* autosome. (ORNL)

Sex linked
Traits or diseases associated with the X or Y chromosome; generally seen in males. *See also:* gene, mutation, sex chromosome. (ORNL)

Shotgun method
Sequencing method that involves randomly sequenced cloned pieces of the genome, with no foreknowledge of where the piece originally came from. This can be contrasted with "directed" strategies, in which pieces of DNA from known chromosomal locations are sequenced. Because there are advantages to both strategies, researchers use both random (or shotgun) and directed strategies in combination to sequence the human genome. *See also:* library, genomic library. (ORNL)

Similarity
The extent to which nucleotide or protein sequences are related. The extent of similarity between two sequences can be based on percent sequence identity and/or conservation. In BLAST similarity refers to a positive matrix score. (NCBI)

Single-gene disorder
Hereditary disorder caused by a mutant allele of a single gene (e.g., Duchenne muscular dystrophy, retinoblastoma, sickle cell disease). *See also:* polygenic disorders. (ORNL)

Single-nucleotide polymorphism (SNP)
DNA sequence variations that occur when a single nucleotide (A, T, C, or G) in the genome sequence is altered. *See also:* mutation, polymorphism, single-gene disorder. (ORNL)

Somatic cell
Any cell in the body except gametes and their precursors. *See also:* gamete. (ORNL)

Somatic cell gene therapy
Incorporating new genetic material into cells for therapeutic purposes. The new genetic material cannot be passed to offspring. *See also:* gene therapy. (ORNL)

Somatic cell genetic mutation
A change in the genetic structure that is neither inherited nor passed to offspring. Also called acquired mutations. *See also:* germ line genetic mutation. (ORNL)

Southern blotting
Transfer by absorption of DNA fragments separated in electrophoretic gels to membrane filters for detection of specific base sequences by radiolabeled complementary probes. (ORNL)

Spectral karyotype (SKY)
A graphic of all an organism's chromosomes, each labeled with a different color. Useful for identifying chromosomal abnormalities. *See also:* chromosome. (ORNL)

Splice site
Location in the DNA sequence where RNA removes the noncoding areas to form a continuous gene transcript for translation into a protein. (ORNL)

Sporadic cancer
Cancer that occurs randomly and is not inherited from parents. Caused by DNA changes in one cell that grows and divides, spreading throughout the body. *See also:* hereditary cancer. (ORNL)

Stem cell
Undifferentiated, primitive cells in the bone marrow that have the ability both to multiply and to differentiate into specific blood cells. (ORNL)

Structural genomics
The effort to determine the three-dimensional structures of large numbers of proteins using both experimental techniques and computer simulation. (ORNL)

Substitution
(a) The presence of a nonidentical amino acid at a given position in an alignment. If the aligned residues have similar physicochemical properties, the substitution is said to be "conservative." (NCBI) (b) In genetics, a type of mutation due to replacement of one nucleotide in a DNA sequence by another nucleotide or replacement of one amino acid in a protein by another amino acid. *See also:* mutation. (ORNL)

Substitution matrix
A substitution matrix containing values proportional to the probability that amino acid i mutates into amino acid j for all pairs of amino acids. Such matrices are constructed by assembling a large and diverse sample of verified pairwise alignments of amino acids. If the sample is large enough to be statistically significant, the resulting matrices should reflect the true probabilities of mutations occurring through a period of evolution. (NCBI)

Suppressor gene
A gene that can suppress the action of another gene. (ORNL)

Syndrome
The group or recognizable pattern of symptoms or abnormalities that indicate a particular trait or disease. (ORNL)

Syngeneic
Genetically identical members of the same species. (ORNL)

Synteny
Genes occurring in the same order on chromosomes of different species. *See also:* linkage, conserved sequence. (ORNL)

T

Tandem repeat sequences
Multiple copies of the same base sequence on a chromosome; used as markers in physical mapping. *See also:* physical map. (ORNL)

Targeted mutagenesis
Deliberate change in the genetic structure directed at a specific site on the chromosome. Used in research to determine the targeted region's function. *See also:* mutation, polymorphism. (ORNL)

Technology transfer
The process of transferring scientific findings from research laboratories to the commercial sector. (ORNL)

Telomerase
The enzyme that directs the replication of telomeres. (ORNL)

Telomere
The end of a chromosome. This specialized structure is involved in the replication and stability of linear DNA molecules. *See also:* DNA replication. (ORNL)

Teratogenic
Substances such as chemicals or radiation that cause abnormal development of an embryo. *See also:* mutagen. (ORNL)

Thymine (T)
A nitrogenous base, one member of the base pair AT (adenine–thymine). *See also:* base pair, nucleotide. (ORNL)

Toxicogenomics
The study of how genomes respond to environmental stressors or toxicants. Combines genomewide mRNA expression profiling with protein expression patterns using bioinformatics to understand the role of gene–environment interactions in disease and dysfunction. (ORNL)

Transcription
The synthesis of an RNA copy from a sequence of DNA (a gene); the first step in gene expression. *See also:* translation. (ORNL)

Transcription factor
A protein that binds to regulatory regions and helps control gene expression. (ORNL)

Transcriptome
The full complement of activated genes, mRNAs, or transcripts in a particular tissue at a particular time. (ORNL)

Transfection
The introduction of foreign DNA into a host cell. *See also:* cloning vector, gene therapy. (ORNL)

Transfer RNA (tRNA)
A class of RNA having structures with triplet nucleotide sequences that are complementary to the triplet nucleotide coding sequences of mRNA. The role of tRNAs in protein synthesis is to bond with amino acids and transfer them to the ribosomes, where proteins are assembled according to the genetic code carried by mRNA. (ORNL)

Transformation
A process by which the genetic material carried by an individual cell is altered by incorporation of exogenous DNA into its genome. (ORNL)

Transgenic
An experimentally produced organism in which DNA has been artificially introduced and incorporated into the organism's germ line. *See also:* cell, DNA, gene, nucleus, germ line. (ORNL)

Translation
The process in which the genetic code carried by mRNA directs the synthesis of proteins from amino acids. *See also:* transcription. (ORNL)

Translocation
A mutation in which a large segment of one chromosome breaks off and attaches to another chromosome. *See also:* mutation. (ORNL)

Transposable element
A class of DNA sequences that can move from one chromosomal site to another. (ORNL)

Trisomy
Possessing three copies of a particular chromosome instead of the normal two copies. *See also:* cell, gene, gene expression, chromosome. (ORNL)

U

Unitary matrix
Also known as identity matrix. A scoring system in which only identical characters receive a positive score. (NCBI)

Up and down
The simplest topology for a helical bundle or folded leaf, in which consecutive helices are adjacent and antiparallel; it is approximately equivalent to the meander topology of a β sheet. (SCOP)

Uracil
A nitrogenous base normally found in RNA but not DNA; it is capable of forming a base pair with adenine. *See also:* base pair, nucleotide. (ORNL)

V

Vector
See: cloning vector. (ORNL)

Virus
A noncellular biological entity that can reproduce only within a host cell. Viruses consist of nucleic acid covered by protein; some animal viruses are also surrounded by membrane. Inside the infected cell, the virus uses the synthetic capability of the host to produce progeny virus. *See also:* cloning vector. (ORNL)

W

Western blot
A technique used to identify and locate proteins based on their ability to bind to specific antibodies. *See also:* DNA, Northern blot, protein, RNA, Southern blotting. (ORNL)

Wild type
The form of an organism that occurs most frequently in nature. (ORNL)

Working draft DNA sequence
See: Draft DNA sequence. (ORNL)

X

X chromosome
One of the two sex chromosomes, X and Y. *See also:* Y chromosome, sex chromosome. (ORNL)

Xenograft
Tissue or organs from an individual of one species transplanted into or grafted onto an organism of another species, genus, or family. A common example is the use of pig heart valves in humans. (ORNL)

Y

Y chromosome
One of the two sex chromosomes, X and Y. *See also:* X chromosome, sex chromosome. (ORNL)

Yeast artificial chromosome (YAC)
Constructed from yeast DNA, it is a vector used to clone large DNA fragments. *See also:* cloning vector, cosmid. (ORNL)

Z

Zinc-finger protein
A secondary feature of some proteins containing a zinc atom; a DNA-binding protein. (ORNL)

TABLE 1 Glossaries Available on Internet

Source	URL
Genomics Glossary from Cambridge Healthtech Institute (requires sign-in)	► http://www.genomicglossaries.com/
Glossary from CancerPage.com	► http://www.cancerpage.com/glossary/
Bioremediation Glossary from U.S. Department of Energy	► http://www.lbl.gov/ERSP/generalinfo/glossary.html
Joint Genome Institute	► http://www.jgi.doe.gov/education/links.html
The Dictionary of Cell and Molecular Biology	► http://www.mblab.gla.ac.uk/dictionary/

Answers to Self-Test Quizzes

[2-1] b

[2-2] c

[2-3] a

[2-4] a

[2-5] a

[2-6] c

[2-7] d

[2-8] c

[2-9] c

[3-1] asparagine   N

      glutamine   Q

      tryptophan   W

      tyrosine   Y

      phenylalanine   F

[3-2] a

[3-3] d

[3-4] c

[3-5] d

[3-6] a

[3-7] c

[3-8] false

[3-9] c

[3-10] d

[4-1] d

[4-2] c

[4-3] a

[4-4] blastp   d

      blastn   a

      blastx   c

      tblastn b

      tblastx e

[4-5] c

[4-6] a

[4-7] a

[4-8] b

[4-9] b

[4-10] c

[5-1] b

[5-2] b

[5-3] b

[5-4] a

[5-5] a

[5-6] b

[5-7] a

[5-8] d

[5-9] b

[6-1] b

[6-2] b

[6-3] c

[6-4] d

[6-5] d

[6-6] a

[6-7] a

[6-8] a

[6-9] a

[6-10] c

[7-1] d

[7-2] b

[7-3] c

[7-4] a

[7-5] b

[7-6] a

[7-7] b

[7-8] a

[7-9] a

[7-10] c

[8-1] a

[8-2] d

[8-3] c

[8-4] c

[8-5] d

[8-6] c

[8-7] c

[8-8] a

[8-9] b

[8-10] c

[9-1] c

[9-2] c

[9-3] a

[9-4] b

[9-5] d

[9-6] d

[9-7] a

[9-8] d

[9-9] a

[10-1] a

[10-2] c

[10-3] b

[10-4] c

[10-5] c

[10-6] b

[10-7] d

[10-8] b

[11-1] a

[11-2] c

[11-3] d

[11-4] c

[11-5] d

[11-6] c

[11-7] a

[11-8] b

[11-9] d

[12-1]	d	[15-1]	c	[18-1]	a
[12-2]	a	[15-2]	c	[18-2]	d
[12-3]	b	[15-3]	a	[18-3]	a
[12-4]	c	[15-4]	c	[18-4]	d
[12-5]	a	[15-5]	d	[18-5]	a
[12-6]	e	[15-6]	a	[18-6]	b
[12-7]	d	[15-7]	c	[18-7]	a
[12-8]	a	[15-8]	a	[18-8]	c
[12-9]	c			[18-9]	c
		[16-1]	c		
[13-1]	c	[16-2]	c	[19-1]	c
[13-2]	a	[16-3]	b	[19-2]	c
[13-3]	d	[16-4]	c	[19-3]	a
[13-4]	d	[16-5]	d	[19-4]	a
[13-5]	b	[16-6]	d	[19-5]	a
[13-6]	d	[16-7]	a	[19-6]	b
[13-7]	d	[16-8]	d	[19-7]	b
[13-8]	c	[16-9]	a	[19-8]	d
[13-9]	c	[16-10]	c	[19-9]	d
				[19-10]	c
		[17-1]	c		
[14-1]	c	[17-2]	c	[20-1]	a
[14-2]	a	[17-3]	c	[20-2]	a
[14-3]	d	[17-4]	b	[20-3]	b
[14-4]	a	[17-5]	a	[20-4]	c
[14-5]	d	[17-6]	a	[20-5]	a
[14-6]	b	[17-7]	c	[20-6]	c
[14-7]	d	[17-8]	b	[20-7]	b
[14-8]	d	[17-9]	a	[20-8]	c
				[20-9]	a

Author Index

Subject Index

Bioinformatics and Functional Genomics, Second Edition. By Jonathan Pevsner
Copyright © 2009 John Wiley & Sons, Inc.